TABLE IV
Values of t_α

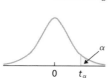

NOTE: *See the version of Table IV in Appendix A for additional values of t_α.*

df	$t_{0.10}$	$t_{0.05}$	$t_{0.025}$	$t_{0.01}$	$t_{0.005}$	df
1	3.078	6.314	12.706	31.821	63.657	1
2	1.886	2.920	4.303	6.965	9.925	2
3	1.638	2.353	3.182	4.541	5.841	3
4	1.533	2.132	2.776	3.747	4.604	4
5	1.476	2.015	2.571	3.365	4.032	5
6	1.440	1.943	2.447	3.143	3.707	6
7	1.415	1.895	2.365	2.998	3.499	7
8	1.397	1.860	2.306	2.896	3.355	8
9	1.383	1.833	2.262	2.821	3.250	9
10	1.372	1.812	2.228	2.764	3.169	10
11	1.363	1.796	2.201	2.718	3.106	11
12	1.356	1.782	2.179	2.681	3.055	12
13	1.350	1.771	2.160	2.650	3.012	13
14	1.345	1.761	2.145	2.624	2.977	14
15	1.341	1.753	2.131	2.602	2.947	15
16	1.337	1.746	2.120	2.583	2.921	16
17	1.333	1.740	2.110	2.567	2.898	17
18	1.330	1.734	2.101	2.552	2.878	18
19	1.328	1.729	2.093	2.539	2.861	19
20	1.325	1.725	2.086	2.528	2.845	20
21	1.323	1.721	2.080	2.518	2.831	21
22	1.321	1.717	2.074	2.508	2.819	22
23	1.319	1.714	2.069	2.500	2.807	23
24	1.318	1.711	2.064	2.492	2.797	24
25	1.316	1.708	2.060	2.485	2.787	25
26	1.315	1.706	2.056	2.479	2.779	26
27	1.314	1.703	2.052	2.473	2.771	27
28	1.313	1.701	2.048	2.467	2.763	28
29	1.311	1.699	2.045	2.462	2.756	29
30	1.310	1.697	2.042	2.457	2.750	30
35	1.306	1.690	2.030	2.438	2.724	35
40	1.303	1.684	2.021	2.423	2.704	40
50	1.299	1.676	2.009	2.403	2.678	50
60	1.296	1.671	2.000	2.390	2.660	60
70	1.294	1.667	1.994	2.381	2.648	70
80	1.292	1.664	1.990	2.374	2.639	80
90	1.291	1.662	1.987	2.369	2.632	90
100	1.290	1.660	1.984	2.364	2.626	100
1000	1.282	1.646	1.962	2.330	2.581	1000
2000	1.282	1.646	1.961	2.328	2.578	2000

1.282	1.645	1.960	2.326	2.576
$z_{0.10}$	$z_{0.05}$	$z_{0.025}$	$z_{0.01}$	$z_{0.005}$

Introductory Statistics

Seventh Edition

Introductory Statistics

Seventh Edition

Neil A. Weiss

Department of Mathematics and Statistics
Arizona State University

Biographies by Carol A. Weiss

PEARSON

Addison
Wesley

Boston   San Francisco   New York
London   Toronto   Sydney   Tokyo   Singapore   Madrid
Mexico City   Munich   Paris   Cape Town   Hong Kong   Montreal

Publisher: Greg Tobin
Senior Acquistions Editor: Deirdre Lynch
Managing Editor: Ron Hampton
Executive Project Manager: Christine O'Brien
Associate Project Editor: Joanne Ha
Editorial Assistant: Keren Blankfeld
Production Supervisor: Julie LaChance
Marketing Manager: Yolanda Cossio
Marketing Assistant: Heather Peck
Senior Manufacturing Buyer: Evelyn Beaton
Media Buyer: Ginny Michaud
Senior Prepress Supervisor: Caroline Fell
Associate Producer: Sara Anderson
Photo Research: Beth Anderson
Software Editors: David Malone, Jan Wann
Senior Author Support/Technology Specialist: Joe Vetere
Supplements Production Supervisor: Sheila Spinney
Illustration: Techsetters, Inc.
Compositor: Techsetters, Inc.
Packager: WestWords, Inc.
Senior Designer: Barbara T. Atkinson
Interior Design: Andrea Menza
Cover Design: Night & Day Design
Cover Photograph: Oyster pearl in shell against turquoise
  textured background © Stone/John Lund

For permission to use copyrighted material, grateful acknowledgment is made to the copyright holders on page C-1, which is hereby made part of this copyright page.

Many of the designations used by manufacturers and sellers to distinguish their products are claimed as trademarks. Where those designations appear in this book, and Addison-Wesley was aware of a trademark claim, the designations have been printed in initial caps or all caps.

The Library of Congress has already catalogued the student edition as follows:

Library of Congress Cataloging-in-Publication Data

Weiss, N. A. (Neil A.)
    Introductory statistics.—7th ed. / Neil A. Weiss.
        p. cm.
    Includes index.
      ISBN 0-201-77131-4
    1. Statistics. I. Title.

QA276.12.W445 2004
519.5–dc22

2004044596

Printed in the United States of America

4 5 6 7 8 9 10-VH-07 06

To Aaron and Greg

About the Author

Neil A. Weiss received his Ph.D. from UCLA in 1970 and subsequently accepted an assistant-professor position at Arizona State University (ASU), where he was ultimately promoted to the rank of full professor. Dr. Weiss has taught statistics, probability, and mathematics—from the freshman level to the advanced graduate level—for more than 30 years. In recognition of his excellence in teaching, he received the *Dean's Quality Teaching Award* from the ASU College of Liberal Arts and Sciences. Dr. Weiss' comprehensive knowledge and experience ensures that his texts are mathematically and statistically accurate, as well as pedagogically sound.

In addition to his numerous research publications, Dr. Weiss has authored or coauthored books in finite mathematics, statistics, and real analysis, and is currently working on two new books: one on probability theory and the other on applied regression analysis and the analysis of variance. His texts—well known for their precision, readability, and pedagogical excellence—are used worldwide.

Dr. Weiss is a pioneer of the integration of statistical software into textbooks and the classroom, first providing such integration over 20 years ago in the book *Introductory Statistics* (Addison-Wesley, 1982). Weiss and Addison-Wesley continue that pioneering spirit to this day with the inclusion of some of the most comprehensive Web sites in the field.

In his spare time, Dr. Weiss enjoys walking, studying and practicing meditation, and playing hold'em poker. He is married and has two sons.

Using and understanding statistics and statistical procedures have become required skills in virtually every profession and academic discipline. The purpose of this book is to help students grasp basic statistical concepts and techniques, and to present real-life opportunities for applying them.

About This Book

The text is intended for a one- or two-semester course and for quarter-system courses as well. Instructors can easily fit the text to the pace and depth they prefer. Introductory high school algebra is a sufficient prerequisite. Although mathematically and statistically sound, the approach doesn't require students to examine complex concepts such as probability theory and random variables. Students need only understand basic ideas such as percentages and histograms.

Advances in technology and new insights into the practice of teaching statistics have inspired many of the pedagogical strategies used in the seventh edition of *Introductory Statistics*, leading to more emphasis on conceptual understanding and less emphasis on computation.

Highlights of the Approach

ASA/MAA-Guidelines Compliant. We follow ASA/MAA guidelines to stress the interpretation of statistical results, the contemporary applications of statistics, and the importance of critical thinking.

Unique Variable-Centered Approach. By consistent and proper use of the terms *variable* and *population*, we unified and clarified the various statistical concepts.

Data Analysis and Exploration. We incorporate an extensive amount of data analysis and exploration in the text and exercises. Recognizing that not all readers have access to technology, we provide ample opportunity to analyze and explore data without the use of a computer or statistical calculator.

Detailed and Careful Explanations. We include every step of explanation that a typical reader might need. Our guiding principle is to avoid cognitive jumps, making the learning process smooth and enjoyable. We believe that detailed and careful explanations result in better understanding.

Emphasis on Application. We concentrate on the application of statistical techniques to the analysis of data. Although statistical theory has been kept to a minimum, we provide a thorough explanation of the rationale for the use of each statistical procedure.

Parallel Critical-Value/P-Value Approaches. Through a parallel presentation, the book offers complete flexibility in the coverage of the critical-value and *P*-value approaches to hypothesis testing—either one or both approaches can be explored and compared.

Parallel Presentations of Technology. The book offers complete flexibility in the coverage of technology, including options for the use of Minitab, Excel, and the TI-83/84 Plus. One or more technologies can be explored and compared.

New and Hallmark Features

UPDATED! **Chapter-Opening Features.** Included at the beginning of each chapter is a general description of the chapter, an explanation of how the chapter relates to the text as a whole, and an outline that lists the sections in the chapter. Each chapter opens with a classic or contemporary case study that highlights the real-world relevance of the material. (Case studies are reviewed and discussed at the end of the chapter.) More than one-third of the case studies are new or updated.

Real-World Examples. Every concept discussed in the text is illustrated by at least one detailed example. The examples are based on real-life situations and were chosen for their interest level as well as for their illustrative value.

NEW! **Interpretation Boxes.** This feature presents the meaning and significance of statistical results in everyday language. Instead of just obtaining the answers or results, students are shown the importance of interpretation.

What Does It Mean? This feature, found in the margin at appropriate places, states in "plain English" the meanings of definitions, formulas, and key facts. It is also used to summarize various expository discussions.

NEW! **Data Sets.** In most examples and many exercises, we present both raw data and summary statistics. This practice gives a more realistic view of statistics and provides an opportunity for students to solve problems by computer or statistical calculator if so desired. Hundreds of data sets are included, many of which are new or updated. All data sets, including large ones, are available in multiple formats on the WeissStats CD.

Procedure Boxes: Why, When, and How. To help students learn statistical procedures, we developed easy-to-follow, step-by-step methods for carrying them out. Each step is highlighted and presented again within the illustrating example. This approach shows how the procedure is applied and helps students master its steps.

   The procedure boxes have been reformatted to include the "why, when, and how" of the methods. Usually, a procedure has a brief identifying title followed by a statement of its purpose (why it's used), the assumptions for its use (when it's used), and the steps for applying the procedure (how it's used). The dual procedures, which provide both the critical-value and *P*-value approaches to a hypothesis-testing method, have been combined into a new, single split format for ease of use and comparison.

UPDATED! **The Technology Center.** The in-text coverage of statistical technology includes three of the most popular applications: Minitab, Excel, and the TI-83/84 Plus graphing calculators. We provide instructions and output for the most recent versions of these applications, including Release 14 of Minitab. The Technology Centers are integrated as optional material.

Computer Simulations. Computer simulations appear in both the text and the exercises. The simulations serve as pedagogical aids for understanding complex concepts such as sampling distributions.

NEW! **Exercises.** Over 1700 exercises that provide current, real-world applications were constructed from an extensive variety of articles in newspapers, magazines, statistical abstracts, journals, and Web sites; sources are explicitly cited. The exercises help students learn the material and, moreover, show that statistics is a lively and relevant discipline. We updated exercises wherever appropriate and have provided many new ones. Exercises related to optional materials are marked with asterisks, unless the entire section is optional. Most section exercise sets are divided into three categories:

- *Statistical Concepts and Skills* exercises help students master the skills and concepts explicitly discussed in the section.
- *Extending the Concepts and Skills* exercises invite students to extend their skills by examining material not necessarily covered in the text. Exercises that introduce new concepts are highlighted in blue.
- *Using Technology* exercises provide students with an opportunity to apply and interpret the computing and statistical capabilities of Minitab, Excel, the TI-83/84 Plus, SPSS, JMP, or any other statistical technology.

Procedure Index. Because of the numerous statistical procedures available, finding a specific one is sometimes difficult. Consequently, we include a *Procedure Index* (located near the end of the book), which provides a quick and easy way to find the right procedure for performing any particular statistical analysis.

WeissStats CD. This PC- and Macintosh-compatible CD, included with every new copy of the book, contains a wealth of resources. Among them are the following:

- All appropriate data sets in the book in relevant formats, including Minitab, Excel, SPSS, ASCII, and JMP, and as a TI-83/84 Plus application.
- Three modular chapters that provide optional extended coverage of regression analysis, experimental design, and ANOVA.
- Data Desk/XL (DDXL) software, an Excel add-in from Data Description, Inc., that enhances Excel's standard statistics and graphics capabilities.
- Adobe Acrobat Reader software for reading the optional modular chapters.

Formula/Table Card. A detachable formula/table card (FTC) is provided with the book. This card contains all the formulas and many of the tables that appear in the text. The FTC is helpful for quick-reference purposes; many instructors also find it convenient for use with examinations.

End-of-Chapter Features

Chapter Reviews.   Each chapter review includes **chapter objectives,** a list of **key terms** with page references, and a **review test** to help students review and study the chapter. Items related to optional materials are marked with asterisks, unless the entire chapter is optional.

Award-Winning Internet Projects.   Each chapter includes an Internet Project to engage students in active and collaborative learning through simulations, demonstrations, and other activities, and to guide them through applications by using Internet links to access data and other information provided by the vast resources of the World Wide Web. The Internet Projects are featured on the Weiss Web site at www.aw-bc.com/weiss.

NEW! **Focusing on Data Analysis.**   A brand new database has been constructed for the *Focusing on Data Analysis* feature, which appears in every chapter. Students can conduct various statistical analyses on these data sets, using the technology of their choice. This feature gives students an opportunity to work with large data sets, to practice using technology, and to discover the many methods of exploring and analyzing data.

The *Focus Database* contains information on 13 variables for the undergraduate students attending the University of Wisconsin - Eau Claire (UWEC).

- Statistical analyses can be performed for the entire population of UWEC students, on topics that require population data.
- The database can be sampled to perform statistical analyses that require sample data (e.g., inference).
- The database lends itself to interesting and informative class projects, as illustrated in the *Focusing on Data Analysis* sections of Chapters 8 and 9.
- The *Focus Database* is included on the WeissStats CD.

Case Study Discussion.   At the end of each chapter, the chapter-opening case study is reviewed and discussed in light of the chapter's major points, and then problems are presented for students to solve.

NEW! **StatExplore in MyMathLab.**   StatExplore is online statistical software available through the Weiss MyMathLab Course. At the end of each chapter of the book, we illustrate the use of StatExplore to perform a statistical analysis discussed in the chapter. For best results, students should implement the steps we present and thereby obtain the StatExplore output for themselves. Exercises encourage students to further apply StatExplore to other statistical analyses examined in the chapter.

Biographical Sketches.   Each chapter ends with a brief biography of a famous statistician. Besides being of general interest, these biographies help students obtain a perspective on the development of the science of statistics.

Flexible Syllabus

The text offers a great deal of flexibility in choosing material to cover. The flowchart on page xv indicates chapter-coverage flexibility. Here are some additional noteworthy items.

NEW!   **Option for Brief Sampling Coverage.**   New to this edition, the only sampling design required for study is simple random sampling, which is presented in Section 1.3. Further sampling designs (systematic random sampling, stratified sampling, cluster sampling, and multistage sampling) are available in Section 1.4 for coverage at the instructor's discretion.

NEW!   **Option for Brief Experimental Design Coverage.**   New to this edition, coverage of experimental designs (which appears in Section 1.5) is optional. An introduction to the principles of experimental design, the terminology of experimental design, and basic statistical designs can be covered at the instructor's discretion.

Option for Brief Probability Coverage.   Only a rudimentary coverage of probability is required, mostly for the frequentist interpretation of probability and for standard statistical terminology such as Type I and Type II error probabilities and P-values. More probability, including probability theory and random variables, can be examined at the discretion of the instructor. The probability concepts required for statistical inference can now be covered in two or three class periods. The option for brief probability coverage is effected by omitting sections marked as optional (with an asterisk) in Chapters 4 and 5, as identified in the table of contents and in the chapter outlines for those two chapters.

Option for Early Regression Coverage.   The chapter discussing descriptive methods in regression and correlation (Chapter 14) is written so that it can be covered at any time after Chapter 3.

Option for Further Regression and ANOVA.   Three chapters that contain additional material on regression analysis and the analysis of variance are available for customizing your course. These chapters, written by Professor Dennis Young of Arizona State University, are provided on the WeissStats CD, and are as follows:

- Module A: *Multiple Regression Analysis*
- Module B: *Model Building in Regression*
- Module C: *Design of Experiments and Analysis of Variance*

Organization

As mentioned, the text offers a great deal of flexibility in choosing material to cover. Following is a brief chapter-by-chapter summary:

- Chapter 1 presents the nature of statistics, sampling designs, and an introduction to experimental designs. The material in Section 1.4 ("Other Sampling Designs") is now optional, as is the material on experimental designs in Section 1.5. The optional chapter "Design of Experiments and Analysis of Variance" (Module C), on the WeissStats CD, provides a more comprehensive treatment of experimental designs.

- Chapters 2 and 3 present the fundamentals of descriptive statistics. Quartiles are defined intuitively and are consistent with the definition of hinges used in boxplots.
- Chapters 4 and 5 examine probability and discrete random variables. Only the first three sections of Chapter 4 are prerequisite to coverage of inferential statistics; the remaining five sections of Chapter 4 and all four sections of Chapter 5 are optional.
- Chapter 6 provides a concise discussion of the normal distribution, including an optional section on the normal approximation to the binomial distribution.
- Chapter 7 introduces the concept of sampling distributions and presents an improved and simplified introduction to the sampling distribution of the sample mean.
- Chapters 8 and 9 give an easily accessible introduction to confidence intervals and hypothesis tests for one population mean by using the terminology of variables and avoiding formal probability. Both chapters employ the σ-known versus σ-unknown criterion for deciding which parametric procedure to use; this approach makes confidence intervals and hypothesis tests easier to understand and apply, and provides a method consistent with most statistical software, including Minitab, Excel (DDXL), and the TI-83/84 Plus. Chapter 9 also presents an optional section on the one-sample Wilcoxon signed-rank test. We consider Chapters 1–9 the core of an introductory statistics course.
- Chapter 10 examines inferences for two population means. It contains a detailed discussion of the meaning of independent samples, including graphics for quick assimilation. The two-sample z-procedures are covered in the exercises so that the presentation can focus on the more practical two-sample t-procedures. Included in the chapter are separate optional sections devoted to the Mann–Whitney test and the Wilcoxon signed-rank test for paired samples.
- Chapter 11, which is optional, presents material on inferences for one and two population standard deviations (or variances).
- Chapter 12 examines inferences for one and two population proportions, and contains discussions of margin of error and sample-size determination.
- Chapter 13 introduces the chi-square goodness-of-fit test and the chi-square independence test. Included are a section on grouping bivariate data into contingency tables and an easy-to-understand presentation of the concept of association.
- Chapter 14 gives an informal, but precise, treatment of descriptive methods in regression and correlation, relying on intuitive and graphical presentations of important concepts. The placement is flexible—this chapter can be covered any time after Chapter 3.
- Chapter 15 examines inferential methods in regression and correlation. Multiple regression and model building are covered in optional chapters on both the WeissStats CD and Weiss Web site: "Multiple Regression Analysis" (Module A) and "Model Building in Regression" (Module B). These optional

chapters include topics such as transformations, polynomial models, qualitative predictors, and model selection.

- Chapter 16 introduces the analysis of variance with sections on one-way ANOVA, multiple comparisons, and the Kruskal–Wallis test. Other types of ANOVA, including two-way ANOVA and randomized block design, are discussed in the optional chapter "Design of Experiments and Analysis of Variance" (Module C) on the WeissStats CD.

The following flowchart summarizes the preceding discussion and shows the interdependence among chapters. In the flowchart, the prerequisites for a given chapter consist of all chapters that have a path that leads to that chapter. Optional sections and chapters can be identified by consulting the table of contents.

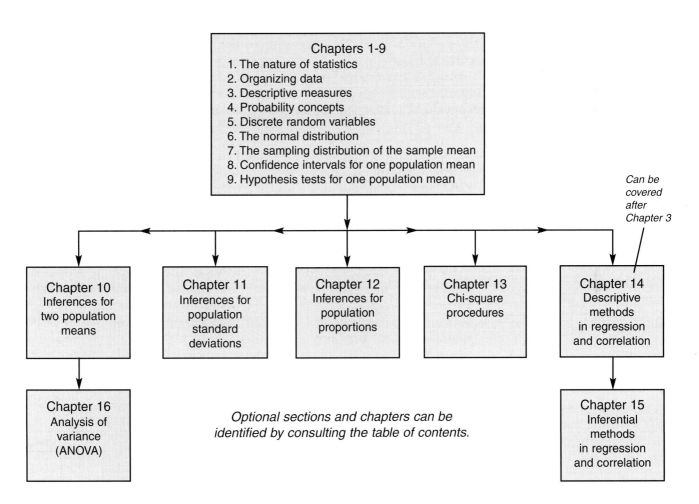

Supplements and Support

For the Instructor

Instructor's Edition (0-321-24138-X). This version of the text includes answers to all basic exercises presented in the book.

Instructor's Solutions Manual (0-321-24139-8).   This supplement contains detailed, worked-out solutions to all section exercises, review-test problems, *Focusing on Data Analysis* exercises, and case studies in the text.

Printed Test Bank (0-321-24140-1).   This supplement provides three printed examinations for each chapter of the text.

TestGen-EQ with QuizMaster-EQ (0-321-24618-7).   TestGen enables instructors to build, edit, print, and administer tests using a computerized bank of questions developed to cover all the objectives of the text. TestGen is algorithmically based, allowing instructors to create multiple, but equivalent, versions of the same question or test with the click of a button. Instructors can also modify test bank questions or add new questions by using the built-in question editor, which allows users to create graphs, import graphics, and insert math notation, variable numbers, or text. Tests can be printed or administered online via the Internet or another network. TestGen comes packaged with QuizMaster, which allows students to take tests on a local area network. The software is available on a dual-platform Windows/Macintosh CD-ROM.

MyMathLab for Statistics.   MyMathLab® is a series of text-specific, easily customizable online courses for Addison-Wesley textbooks in mathematics and statistics, and for this book in particular. MyMathLab is powered by CourseCompass™—Pearson Education's online teaching and learning environment—and by MathXL®—our online homework, tutorial, and assessment system. MyMathLab gives you the tools you need to deliver all or a portion of your statistics course online, whether your students are in a lab setting or working from home. MyMathLab provides a rich and flexible set of course materials, featuring free-response exercises that are algorithmically generated for unlimited practice and mastery. Students can also use online tools, such as animations and a multimedia textbook, to independently improve their understanding and performance. Instructors can use MyMathLab's homework and test managers to select and assign online exercises correlated directly with the textbook, and can import TestGen tests into MyMathLab for added flexibility. MyMathLab's online gradebook—designed specifically for mathematics and statistics—automatically tracks students' homework and test results and gives the instructor control over how to calculate final grades.

MyMathLab is available to qualified adopters. For more information, visit our Web site at www.mymathlab.com, or contact your Addison-Wesley sales representative for a product demonstration.

StatExplore.   StatExplore is online statistical software that offers an easy-to-use interface and is an alternative to commercial statistical software. It is customized for this textbook. StatExplore requires a Java-capable Web browser, such as Internet Explorer or Netscape Navigator, and can be accessed in the student content area of your Weiss MyMathLab course.

For the Student

The following technology manuals include instructions, examples from the main text, and interpretations to complement those given in the text.

Minitab Manual (0-321-24153-3).

Excel Manual (0-321-24142-8).

TI-83/84 Plus Manual (0-321-24152-5).

SPSS Manual (0-321-24133-9).

JMP Manual (0-321-24137-1).

Student's Solutions Manual (0-321-24141-X). This manual includes detailed solutions to all odd-numbered section exercises and all review-test problems in the text.

MathXL for Statistics. MathXL® for Statistics is a powerful online homework, tutorial, and assessment system that accompanies this textbook. With MathXL for Statistics, instructors can create, edit, and assign online homework and tests using algorithmically generated exercises correlated with your textbook at the objective level. All student work is tracked in MathXL's online gradebook. Students can take chapter tests in MathXL and receive personalized study plans based on their test results. The study plan diagnoses weaknesses and links students directly to tutorial exercises for the objectives they need to study and retest. Students can also access supplemental activities directly from selected exercises. MathXL for Statistics is available to qualified adopters. For more information, visit our Web site at www.mathxl.com, or contact your Addison-Wesley sales representative for a product demonstration.

Weiss Web Site. The Weiss Web site includes data sets, the formula/table card, and access to the Internet projects and Case Study extensions. The URL is www.aw-bc.com/weiss. For more information about the Weiss Web site, contact your Pearson/Addison-Wesley representative.

ActivStats (0-201-77139-X). Developed by Paul Velleman and Data Description, Inc., *ActivStats* presents a complete introductory statistics course on CD-ROM, using a full range of multimedia. Integrating video, simulation, animation, narration, text, interactive experiments, World Wide Web access, and Data Desk (a fully functioning statistics package), this product gives students a rich learning environment. Also included are exercises for the reinforcement of key concepts, an index, and a glossary. This program is a strong complement to the text for use in both lecture and Web-based courses. *ActivStats* is PC and Macintosh compatible. Also available are

- *ActivStats for Minitab* (PC) (0-321-26900-4),
- *ActivStats for Excel* (Mac and PC) (0-201-77141-1),
- *ActivStats for SPSS* (PC) (0-201-77571-9),
- *ActivStats for JMP* (Mac and PC) (0-201-77572-7).

The Student Edition of MINITAB (0-321-11313-6).   MINITAB is an easy-to-use general-purpose statistical computing package for analyzing data. It is a flexible and powerful tool that was designed from the beginning to be used by students and researchers new to statistics. It is now one of the most widely used statistics packages in the world. MINITAB performs tedious computations and produces accurate and professional-quality graphs almost instantly. This power frees the user to focus on the exploration of the structure of the data and the interpretation of the output.

AW Tutor Center.   The Addison-Wesley Math and Statistics Tutor Center is staffed by qualified mathematics and statistics instructors who provide students with tutoring on examples and odd-numbered exercises from the textbook. Tutoring is available via toll-free telephone, toll-free fax, e-mail, and the Internet. Interactive, Web-based technology allows tutors and students to view and work through problems together in real time over the Internet. For more information, visit our website at www.aw-bc.com/tutorcenter, or call us at 1-888-777-0463.

Acknowledgments

First, we want to express our sincere appreciation to all reviewers of previous editions for their many contributions to the evolution of the book. For this and the previous few editions of the book, it is our pleasure to thank the following reviewers, whose comments and suggestions resulted in significant improvements:

James Albert
Bowling Green State University

Yvonne Brown
Pima Community College

Beth Chance
California Polytechnic State
   University

Brant Deppa
Winona State University

Carol DeVille
Louisiana Tech University

Jacqueline Fesq
Raritan Valley Community College

Richard Gilman
Holy Cross College

Joel Haack
University of Northern Iowa

Susan Herring
Sonoma State University

David Holmes
The College of New Jersey

Satish Iyengar
University of Pittsburgh

Christopher Lacke
Rowan University

Tze-San Lee
Western Illinois University

Ennis Donice McCune
Stephen F. Austin State University

Jacqueline B. Miller
Drury University

Bernard J. Morzuch
University of Massachusetts,
   Amherst

Dennis M. O'Brien
University of Wisconsin, La Crosse

Dwight M. Olson
John Carroll University

JoAnn Paderi
Lourdes College

Melissa Pedone
Valencia Community College

Alan Polansky
Northern Illinois University

Cathy D. Poliak
Northern Illinois University

Kimberley A. Polly
Parkland College

Geetha Ramachandran
California State University

B. Madhu Rao
Bowling Green State University

Gina F. Reed
Gainesville College

Steven E. Rigdon
Southern Illinois University,
    Edwardsville

Sharon Ross
Georgia Perimeter College

Edward Rothman
University of Michigan

George W. Schultz
St. Petersburg Jr. College

Arvind Shah
University of South Alabama

Cid Srinivasan
University of Kentucky, Lexington

W. Ed Stephens
McNeese State University

Kathy Taylor
Clackamas Community College

Bill Vaughters
Valencia Community College

Brani Vidakovic
Georgia Institute of Technology

Dawn White
California State University,
    Bakersfield

Marlene Will
Spalding University

Matthew Wood
University of Missouri, Columbia

Our thanks as well are extended to Professor Michael Driscoll for his help in selecting the statisticians for the biographical sketches; and Professors Fuchun Huang, Charles Kaufman, Sharon Lohr, Kathy Prewitt, Walter Reid, and Bill Steed with whom we have had several illuminating discussions. Thanks also go to Professors Matthew Hassett and Ronald Jacobowitz for their many helpful comments and suggestions.

Several other people provided useful input and resources. They include Professor Thomas A. Ryan, Jr., Professor Webster West, Dr. William Feldman and Mr. Frank Crosswhite, Dr. Lawrence W. Harding, Jr., Dr. George McManus and Mr. Gregory Weiss, Professor Jeanne Sholl, Professor R. B. Campbell, Mr. Howard Blaut, Mr. Rick Hanna, Ms. Alison Stern-Dunyak, Mr. Dale Phibrick, Ms. Christine Sarris, and Ms. Maureen Quinn. Our sincere thanks to all of them for their help in making a better book.

To Professor Larry Griffey, we convey our appreciation for his formula/table card. We are grateful to the following people for preparing the technology manuals to accompany the book: Professors Jacqueline Fesq (*Minitab Manual*), Ellen Fischer (*TI-83/84 Plus Manual*), Roger Peck (*SPSS Manual*), and Ian Walters (*Excel Manual* and *JMP Manual*). Our gratitude goes as well to Professor David Lund for writing the *Instructor's Solutions Manual* and the *Student's Solutions Manual*.

issues. Thanks also go to Professor Kyle Siegrist for his Internet projects and case-study extensions. For checking the accuracy of the solutions to the examples and exercises, we extend our gratitude to Professors Jackie Miller and Kimberley Polly.

We are also grateful to Professor David Lund and Ms. Patricia Lee for obtaining the new database for the Focusing on Data Analysis sections. Our thanks are extended to the following people for their research in finding myriad new and interesting statistical studies and data for the examples, exercises, and case studies: Ms. Toni Coombs, Ms. Traci Gust, Professor David Lund, Ms. Betty Weiss, and Dr. Morris B. Weiss.

Many thanks to Sara Anderson for directing the development and construction of the WeissStats CD and the Weiss Web site, and to Mr. Brian Dean, Ms. Ellen Keohane, and Professors Roger Peck, Ian Walters, and Jackie Miller for their help in these matters. Our appreciation also goes to our software editors David Malone and Jan Wann.

We are grateful to Pat McCutcheon of WestWords, Inc., who, along with Deirdre Lynch, Julie LaChance, and Christine O'Brien of Addison-Wesley, coordinated the development and production of the book. Thanks as well to our copyeditor Gordon LaTourette of Write With, Inc., and to our proofreaders Cindy Bowles, Carol Sawyer of The Perfect Proof, and Carol Weiss.

To Barbara Atkinson of Addison-Wesley, Andrea Menza (interior design), and Night and Day Design (cover design), we express our thanks for an awesome design and cover. And our sincere thanks go as well to all the people at Techsetters, Inc., for a terrific job of composition and illustration, in particular, John Rogosich (LaTeX design creation), Rena Lam (typesetting and page makeup), Mike Lafferty (illustrations), Amy De Santo (proofreading), and Carol Sawyer (quality control and project manager). We also thank Beth Anderson for her photo research.

Without the help of many people at Addison-Wesley, this book and its numerous ancillaries would not have been possible; to all of them go our heartfelt thanks. We would, however, like to give special thanks to Greg Tobin, Deirdre Lynch, and to the following other people at Addison-Wesley: Christine O'Brien, Keren Blankfeld, Joanne Ha, Sarah Santoro, Sara Anderson, Julie LaChance, Barbara Atkinson, Ron Hampton, Joe Vetere, Sheila Spinney, Yolanda Cossio, Phyllis Hubbard, Heather Peck, Caroline Fell, and Evelyn Beaton.

Finally, we convey our appreciation to Carol A. Weiss. Apart from writing the text, she was involved in every aspect of development and production. Moreover, Carol did a superb job of researching and writing the biographies.

Prescott, Arizona   N.A.W.

Contents

---

*indicates an optional section

---

*indicates an optional section

*indicates an optional section

---

*indicates an optional section

---

*indicates an optional section

*indicates an optional section

---

*indicates an optional section

A.C. Nielsen Company
AAA Daily Fuel Gauge Report
AAA Foundation for Traffic Safety
ABCNews.com
Academy of Medicine Cleveland Survey
Accounttemps
Acta Opthalmologica
Administration for Children and Families
Advances in Cancer Research
African Entomology
Agency for International Development
Agricultural Research Service
Alan Guttmacher Institute
American Association for the Advancement
    of Science
American Association of University
    Professors
American Banker
American Bar Foundation
American College Testing Program
American Council of Life Insurance
American Demographics
American Elasmobranch Society
American Film Institute
American Hospital Association
American Industrial Hygiene Association
    Journal
American Journal of Clinical Nutrition
American Journal of Obstetrics and Gynecology
American Medical Association
Amusement Business
An Aging World: 2001
Analytical Chemistry
Animal Behaviour
Annals of Epidemiology
Annals of the Association of American
    Geographers
Appetite
Arizona Chapter of the American Lung
    Association
Arizona Republic
Arizona State University Focus Student
    Database
Arizona State University Main Facts Book
Arizona State University Statistical Summary
Arthritis Today
Associated Press
Association of Community Organizations
    for Reform Now
ASU Insight
Austin360.com
Auto Trader
Avis Rent-A-Car
Barron's National Business and Financial
    Weekly
Behavior Research Center
Bell Systems Technical Journal
Biological Conservation
Biometrika
BioScience
Board of Governors of the Federal Reserve
    System
Boston Globe
Boston.com

Bottom Line
Boyce Thompson Southwestern Arboretum
Bride's Magazine
British Journal of Educational Psychology
British Medical Journal
Bruskin-Goldring for Bank One
Bureau of Crime Statistics and Research of
    Australia
Bureau of Economic Analysis
Bureau of Justice Statistics
Bureau of Justice Statistics Special Report
Bureau of Labor Statistics
Bureau of Prisons
Bureau of Transportation Statistics
Business Failure Record
Business Times
Buyers of New Cars
California Agriculture
CBS Sportsline
Cellular Telecommunications Industry
    Association
Census Bureau
Center for Disease Control
Center for Housing Policy
Chance
Cheetah Conservation of Southern Africa
Chesapeake and Ohio Railroad Company
Chicago Title Insurance Company
CNN/Sports Illustrated Online
Coleman & Associates, Inc.
College Entrance Examination Board
Communications Industry Forecast
Computer Industry Almanac Inc., 1999
Congressional Directory
Congressional Quarterly Inc.
Consumer Reports
Contribution to Boyce Thompson Institute
Crain's Cleveland Business
Daily Racing Form
Demography
Department of Agriculture
Department of Commerce
Department of Health and Human Services
Department of Housing and Urban
    Development
Department of Justice
Department of Transportation
Desert Samaritan Hospital
Dictionary of Scientific Biography
Diet for a New America
Dietary Guidelines for Americans
Directory of Governors of the American
    States, Commonwealths & Territories
Dr. Thomas Stanley (GSU)
Edinburgh Medical and Surgical Journal
Educational Research
Effects of Environmental Pollutants Upon
    Animals Other Than Man
Employment and Training Administration
Energy Information Administration
Entertainment Weekly
Environmental Pollution
Environmental Pollution (Series A)
Environmental Protection Agency

Environmental Science and Technology
ESPN
Euromonitor Publications Limited, London
Excite Sports
Excite.com
Experimental Agriculture
Federal Bureau of Investigation
Federal Deposit Insurance Company
Federal Highway Administration
Federation Internationale de Football
    Association (web site)
Financial Planning
Florida Museum of Natural History
Food Consumption, Prices, and Expenditures
Food and Nutrition Board of the National
    Academy of Sciences
Forbes Magazine
Forest Mensuration
Forrester Research
Fortune Magazine
Gallup Organization
Giving and Volunteering in the U.S.
Global Financial Data
Global Source Marketing
Greg D. Adams and Chris Fastnow
Handbook of Small Data Sets
Hanna Properties
Harris Poll
Health Letter
Higher Education Research Institute
Hilton Hotels Corporation
HIV/AIDS Surveillance Report
Human Biology
Human Nutrition and Metabolism
Hydrobiologia
ICR
Industry Research
Information and Communications, University
    of California
Information Please Almanac
Inside MS
Interep Research
Internal Revenue Service
International Civil Aviation Organization
International Classification of Diseases
International Communications Research
International Data Base
International Journal for Quality in Health Care
International Waterpower and Dam
    Construction Handbook
Journal of Abnormal Psychology
Journal of Advertising Research
Journal of Agricultural Sciences
Journal of American College Health
Journal of Applied Ecology
Journal of Arachnology
Journal of Athletic Training
Journal of Business
Journal of Chemical Ecology
Journal of Chronic Diseases
Journal of Clinical Oncology
Journal of Counseling Psychology
Journal of Environmental Psychology
Journal of Environmental Science and Health

Journal of Herpetology
Journal of Nutrition
Journal of Organizational Behavior
Journal of Pediatrics
Journal of Real Estate and Economics
Journal of Studies on Alcohol
Journal of Sustainable Tourism
Journal of the American College of Cardiology
Journal of the American Geriatrics Society
Journal of the American Medical Association
Journal of the American Public Health
    Association
Journal of the Royal Statistical Society
Journal of Tropical Ecology
Journal of Zoology, London
Kelley Blue Book
League of Conservation Voters
LH Research
Library Journal
Literary Digest
Limnology and Oceanography
Los Angeles Times
M Street Corporation
Marine Ecology Progress Series
Marine Mammal Science
Mediamark Research Inc.
Medical Biology and Etruscan Origins
Medical College of Wisconsin Eye Institute
Mellman Group for the Coalition to Reduce
    Nuclear Dangers
Merck Manual
Monthly Labor Review
Motor Vehicle Manufacturers Association
National Aeronautics and Space
    Administration
National Association for Gardening
National Association of Colleges and
    Employers
National Association of REALTORS®
National Basketball Association
National Center for Education Statistics
National Center for Health Statistics
National Council of the Churches of Christ
National Football League
National Gardening Association
National Geographic
National Governors Association
National Health and Nutrition Examination
    Survey
National Institute on Alcohol Abuse and
    Alcoholism
National Institute of Mental Health
National Institute on Drug Abuse
National Oceanic and Atmospheric
    Administration
National Opinion Research Center
National Safety Council
National Science Foundation
National Sporting Goods Association
National Survey of Salaries and Wages in
    Public Schools
New England Journal of Medicine
New York Times
Newsweek
Nielsen Media Research
Northwestern Endicott-Lindquist Report
Northwestern University Placement Center

Nutrition
Obstetrics & Gynecology
Office of Juvenile Justice and Delinquency
    Prevention
O'Neil Associates
Opinion Research Corporation
Organization for Economic Cooperation
    and Development
Origin of Species
Parade Magazine
Patent and Trademark Office
Payless ShoeSource
PC World
Pediatrics
Perspectives
Peterson's Annual Survey
Peterson's Guide to Four-Year Colleges, 2000
Philosophical Magazine
Physician's Handbook
Pollstar
Preventative Medicine
Pricewatch.com
Princeton Religion Research Center
Proceedings of the 6th Berkeley Symposium on
    Mathematics and Statistics, VI
Proceeding of the National Academy of
    Science USA
Professional Golf Association
Project Vote Smart
Public Broadcasting System
Public Health Reports
Public Interest Research Group
Quality Engineering
R. Jacobowitz, Ph.D. & G. Vishteh, M.D.
R. R. Bowker Company of New York
Radio Advertising Bureau of New York
Reader's Digest
Reader's Digest and Almanac Yearbook
Reader's Digest/Gallup Survey
Real Estate Research Corp.
Recording Industry Association of
    America, Inc.
Regional Markets, Vol 2/Households
Remote Sensing of Environment
Research Quarterly for Exercise and Sport
Research Resources, Inc.
Roper Starch Worldwide for ABC Global
    Kids Study
Rubber Age
Runner's World
Salt River Project
Scarborough Research
Science
Science News
Scientific American
Scientific Computing & Automation
Semiannual Wireless Survey
Social Forces
Sourcebook of Criminal Justice Statistics
Southwest Airlines
Sports Illustrated
Statistical Abstract of the United States, 1999
Statistical Abstract of the United States, 2002
Statistics of Income, Individual Income Tax
    Returns
STATS
Storm Prediction Center

Substance Abuse and Mental Health
    Services Administration
Survey of Current Business
Technometrics
TELENATION/Market Facts Inc.
Television Bureau of Advertising, Inc.
Tempe Daily News
The 2000 American Express Retail Index
The American Freshman
The American Midland Naturalist
The American Statistician
The Beer Institute
The Daily Courier
The Design and Analysis of Factorial
    Experiments
The Detection of Psychiatric Illness by
    Questionnaire
The Earth: Structure, Composition and
    Evolution
The History of Statistics
The Lancet
The Lobster Almanac
The Marathon: Physiological, Medical,
    Epidemiological and Psychological Studies
The Method of Statistics
The Morgan Horse
The Phoenix Gazette
The Wall Street Journal
The World Fact Book
Thomas Stanley, Georgia State University
Thoroughbred Times
TIME
TIMS
Trends in Television
Tropical Biodiversity
Truck Trader
U.N. Children's Fund
U.S. Air Force
U.S. Coast Guard
U.S. Congress, Joint Committee on Printing
U.S. News & World Report
United States Postal Office
University of Michigan Institute for Social
    Research
Urban Studies
USA TODAY
USA TODAY/CNN/Gallup Poll
USA WEEKEND
Vegetarian Journal
Vegetarian Resource Group
VentureOne Corporation
Vital Statistics of the United States
Wall Street Journal
Washington Post/ABC News Poll
Weatherwise
Webster's New World Dictionary
Western Journal of Medicine
Western North American Naturalist
Wichita Eagle
World Almanac, 1999
World of Wireless Communication
Zero Population Growth
Zogby American Poll

part I

Introduction

CHAPTER 1   **The Nature of Statistics**

The Nature of Statistics

GENERAL OBJECTIVES   What does the word *statistics* bring to mind? Most people immediately think of numerical facts or data, such as unemployment figures, farm prices, or the number of marriages and divorces. *Webster's New World Dictionary* gives two definitions of the word *statistics:*

1. facts or data of a numerical kind, assembled, classified, and tabulated so as to present significant information about a given subject.
2. [construed as sing.], the science of assembling, classifying, and tabulating such facts or data.

But statistics encompasses much more than these definitions convey. Not only do statisticians assemble, classify, and tabulate data, but they also analyze data for the purpose of making generalizations and decisions. For example, a political analyst can use data from a portion of the voting population to predict the political preferences of the entire voting population. And a city council can decide where to build a new airport runway based on environmental impact statements and demographic reports that include a variety of statistical data.

In this chapter, we introduce some basic terminology so that the various meanings of the word *statistics* will become clear to you. We also examine two primary ways of producing data, namely, through sampling and experimentation. We discuss sampling designs in Sections 1.3 and 1.4, and experimental designs in Section 1.5.

TABLE VIII (cont.) Values of F_α

dfn

dfd	α	1	2	3	4	5	6	7	8	9
9	0.10	3.36	3.01	2.81	2.69	2.61	2.55	2.51	2.47	2.44
	0.05	5.12	4.26	3.86	3.63	3.48	3.37	3.29	3.23	3.18
	0.025	7.21	5.71	5.08	4.72	4.48	4.32	4.20	4.10	4.03
	0.01	10.56	8.02	6.99	6.42	6.06	5.80	5.61	5.47	5.35
	0.005	13.61	10.11	8.72	7.96	7.47	7.13	6.88	6.69	6.54
10	0.10	3.29	2.92	2.73	2.61	2.52	2.46	2.41	2.38	2.35
	0.05	4.96	4.10	3.71	3.48	3.33	3.22	3.14	3.07	3.02
	0.025	6.94	5.46	4.83	4.47	4.24	4.07	3.95	3.85	3.78
	0.01	10.04	7.56	6.55	5.99	5.64	5.39	5.20	5.06	4.94
	0.005	12.83	9.43	8.08	7.34	6.87	6.54	6.30	6.12	5.97
11	0.10	3.23	2.86	2.66	2.54	2.45	2.39	2.34	2.30	2.27
	0.05	4.84	3.98	3.59	3.36	3.20	3.09	3.01	2.95	2.90
	0.025	6.72	5.26	4.63	4.28	4.04	3.88	3.76	3.66	3.59
	0.01	9.65	7.21	6.22	5.67	5.32	5.07	4.89	4.74	4.63
	0.005	12.23	8.91	7.60	6.88	6.42	6.10	5.86	5.68	5.54
12	0.10	3.18	2.81	2.61	2.48	2.39	2.33	2.28	2.24	2.21
	0.05	4.75	3.89	3.49	3.26	3.11	3.00	2.91	2.85	2.80
	0.025	6.55	5.10	4.47	4.12	3.89	3.73	3.61	3.51	3.44
	0.01	9.33	6.93	5.95	5.41	5.06	4.82	4.64	4.50	4.39
	0.005	11.75	8.51	7.23	6.52	6.07	5.76	5.52	5.35	5.20
13	0.10	3.14	2.76	2.56	2.43	2.35	2.28	2.23	2.20	2.16
	0.05	4.67	3.81	3.41	3.18	3.03	2.92	2.83	2.77	2.71
	0.025	6.41	4.97	4.35	4.00	3.77	3.60	3.48	3.39	3.31
	0.01	9.07	6.70	5.74	5.21	4.86	4.62	4.44	4.30	4.19
	0.005	11.37	8.19	6.93	6.23	5.79	5.48	5.25	5.08	4.94
14	0.10	3.10	2.73	2.52	2.39	2.31	2.24	2.19	2.15	2.12
	0.05	4.60	3.74	3.34	3.11	2.96	2.85	2.76	2.70	2.65
	0.025	6.30	4.86	4.24	3.89	3.66	3.50	3.38	3.29	3.21
	0.01	8.86	6.51	5.56	5.04	4.69	4.46	4.28	4.14	4.03
	0.005	11.06	7.92	6.68	6.00	5.56	5.26	5.03	4.86	4.72
15	0.10	3.07	2.70	2.49	2.36	2.27	2.21	2.16	2.12	2.09
	0.05	4.54	3.68	3.29	3.06	2.90	2.79	2.71	2.64	2.59
	0.025	6.20	4.77	4.15	3.80	3.58	3.41	3.29	3.20	3.12
	0.01	8.68	6.36	5.42	4.89	4.56	4.32	4.14	4.00	3.89
	0.005	10.80	7.70	6.48	5.80	5.37	5.07	4.85	4.67	4.54
16	0.10	3.05	2.67	2.46	2.33	2.24	2.18	2.13	2.09	2.06
	0.05	4.49	3.63	3.24	3.01	2.85	2.74	2.66	2.59	2.54
	0.025	6.12	4.69	4.08	3.73	3.50	3.34	3.22	3.12	3.05
	0.01	8.53	6.23	5.29	4.77	4.44	4.20	4.03	3.89	3.78
	0.005	10.58	7.51	6.30	5.64	5.21	4.91	4.69	4.52	4.38

TABLE VIII Values of F_α

dfn

dfd	α	1	2	3	4	5	6	7	8	9
1	0.10	39.86	49.50	53.59	55.83	57.24	58.20	58.91	59.44	59.86
	0.05	161.45	199.50	215.71	224.58	230.16	233.99	236.77	238.88	240.54
	0.025	647.79	799.50	864.16	899.58	921.85	937.11	948.22	956.66	963.28
	0.01	4052.2	4999.5	5403.4	5624.6	5763.6	5859.0	5928.4	5981.1	6022.5
	0.005	16211	20000	21615	22500	23056	23437	23715	23925	24091
2	0.10	8.53	9.00	9.16	9.24	9.29	9.33	9.35	9.37	9.38
	0.05	18.51	19.00	19.16	19.25	19.30	19.33	19.35	19.37	19.38
	0.025	38.51	39.00	39.17	39.25	39.30	39.33	39.36	39.37	39.39
	0.01	98.50	99.00	99.17	99.25	99.30	99.33	99.36	99.37	99.39
	0.005	198.50	199.00	199.17	199.25	199.30	199.33	199.36	199.37	199.39
3	0.10	5.54	5.46	5.39	5.34	5.31	5.28	5.27	5.25	5.24
	0.05	10.13	9.55	9.28	9.12	9.01	8.94	8.89	8.85	8.81
	0.025	17.44	16.04	15.44	15.10	14.88	14.73	14.62	14.54	14.47
	0.01	34.12	30.82	29.46	28.71	28.24	27.91	27.67	27.49	27.35
	0.005	55.55	49.80	47.47	46.19	45.39	44.84	44.43	44.13	43.88
4	0.10	4.54	4.32	4.19	4.11	4.05	4.01	3.98	3.95	3.94
	0.05	7.71	6.94	6.59	6.39	6.26	6.16	6.09	6.04	6.00
	0.025	12.22	10.65	9.98	9.60	9.36	9.20	9.07	8.98	8.90
	0.01	21.20	18.00	16.69	15.98	15.52	15.21	14.98	14.80	14.66
	0.005	31.33	26.28	24.26	23.15	22.46	21.97	21.62	21.35	21.14
5	0.10	4.06	3.78	3.62	3.52	3.45	3.40	3.37	3.34	3.32
	0.05	6.61	5.79	5.41	5.19	5.05	4.95	4.88	4.82	4.77
	0.025	10.01	8.43	7.76	7.39	7.15	6.98	6.85	6.76	6.68
	0.01	16.26	13.27	12.06	11.39	10.97	10.67	10.46	10.29	10.16
	0.005	22.78	18.31	16.53	15.56	14.94	14.51	14.20	13.96	13.77
6	0.10	3.78	3.46	3.29	3.18	3.11	3.05	3.01	2.98	2.96
	0.05	5.99	5.14	4.76	4.53	4.39	4.28	4.21	4.15	4.10
	0.025	8.81	7.26	6.60	6.23	5.99	5.82	5.70	5.60	5.52
	0.01	13.75	10.92	9.78	9.15	8.75	8.47	8.26	8.10	7.98
	0.005	18.63	14.54	12.92	12.03	11.46	11.07	10.79	10.57	10.39
7	0.10	3.59	3.26	3.07	2.96	2.88	2.83	2.78	2.75	2.72
	0.05	5.59	4.74	4.35	4.12	3.97	3.87	3.79	3.73	3.68
	0.025	8.07	6.54	5.89	5.52	5.29	5.12	4.99	4.90	4.82
	0.01	12.25	9.55	8.45	7.85	7.46	7.19	6.99	6.84	6.72
	0.005	16.24	12.40	10.88	10.05	9.52	9.16	8.89	8.68	8.51
8	0.10	3.46	3.11	2.92	2.81	2.73	2.67	2.62	2.59	2.56
	0.05	5.32	4.46	4.07	3.84	3.69	3.58	3.50	3.44	3.39
	0.025	7.57	6.06	5.42	5.05	4.82	4.65	4.53	4.43	4.36
	0.01	11.26	8.65	7.59	7.01	6.63	6.37	6.18	6.03	5.91
	0.005	14.69	11.04	9.60	8.81	8.30	7.95	7.69	7.50	7.34

NOTATION The following notation is used on this card:

n = sample size $\qquad$ σ = population stdev

$\overline{x}$ = sample mean $\qquad$ d = paired difference

s = sample stdev $\qquad$ $\hat{p}$ = sample proportion

Q_j = jth quartile $\qquad$ p = population proportion

N = population size $\qquad$ O = observed frequency

μ = population mean $\qquad$ E = expected frequency

CHAPTER 3   Descriptive Measures

- Sample mean: $\overline{x} = \dfrac{\Sigma x}{n}$

- Range: Range = Max − Min

- Sample standard deviation:

$$s = \sqrt{\frac{\Sigma(x - \overline{x})^2}{n - 1}} \quad \text{or} \quad s = \sqrt{\frac{\Sigma x^2 - (\Sigma x)^2/n}{n - 1}}$$

- Interquartile range: $\text{IQR} = Q_3 - Q_1$

- Lower limit = $Q_1 - 1.5 \cdot \text{IQR}$, $\qquad$ Upper limit = $Q_3 + 1.5 \cdot \text{IQR}$

- Population mean (mean of a variable): $\mu = \dfrac{\Sigma x}{N}$

- Population standard deviation (standard deviation of a variable):

$$\sigma = \sqrt{\frac{\Sigma(x - \mu)^2}{N}} \quad \text{or} \quad \sigma = \sqrt{\frac{\Sigma x^2}{N} - \mu^2}$$

- Standardized variable: $z = \dfrac{x - \mu}{\sigma}$

CHAPTER 4   Probability Concepts

- Probability for equally likely outcomes:

$$P(E) = \frac{f}{N},$$

where f denotes the number of ways event E can occur and N denotes the total number of outcomes possible.

- Special addition rule:

$$P(A \text{ or } B \text{ or } C \text{ or } \cdots) = P(A) + P(B) + P(C) + \cdots$$

($A, B, C, \ldots$ mutually exclusive)

- Complementation rule: $P(E) = 1 - P(\text{not } E)$

- General addition rule: $P(A \text{ or } B) = P(A) + P(B) - P(A \& B)$

- Conditional probability rule: $P(B \mid A) = \dfrac{P(A \& B)}{P(A)}$

- General multiplication rule: $P(A \& B) = P(A) \cdot P(B \mid A)$

- Special multiplication rule:

$$P(A \& B \& C \& \cdots) = P(A) \cdot P(B) \cdot P(C) \cdots$$

($A, B, C, \ldots$ independent)

- Rule of total probability:

$$P(B) = \sum_{j=1}^{k} P(A_j) \cdot P(B \mid A_j)$$

($A_1, A_2, \ldots, A_k$ mutually exclusive and exhaustive)

- Bayes's rule:

$$P(A_i \mid B) = \frac{P(A_i) \cdot P(B \mid A_i)}{\sum_{j=1}^{k} P(A_j) \cdot P(B \mid A_j)}$$

($A_1, A_2, \ldots, A_k$ mutually exclusive and exhaustive)

- Factorial: $k! = k(k - 1) \cdots 2 \cdot 1$

- Permutations rule: $_mP_r = \dfrac{m!}{(m - r)!}$

- Special permutations rule: $_mP_m = m!$

- Combinations rule: $_mC_r = \dfrac{m!}{r!\,(m - r)!}$

- Number of possible samples: $_NC_n = \dfrac{N!}{n!\,(N - n)!}$

CHAPTER 5   Discrete Random Variables

- Mean of a discrete random variable X: $\mu = \Sigma x P(X = x)$

- Standard deviation of a discrete random variable X:

$$\sigma = \sqrt{\Sigma(x - \mu)^2 P(X = x)} \quad \text{or} \quad \sigma = \sqrt{\Sigma x^2 P(X = x) - \mu^2}$$

- Factorial: $k! = k(k - 1) \cdots 2 \cdot 1$

- Binomial coefficient: $\dbinom{n}{x} = \dfrac{n!}{x!\,(n - x)!}$

- Binomial probability formula:

$$P(X = x) = \binom{n}{x} p^x (1 - p)^{n-x},$$

where n denotes the number of trials and p denotes the success probability.

- Mean of a binomial random variable: $\mu = np$

- Standard deviation of a binomial random variable: $\sigma = \sqrt{np(1 - p)}$

- Poisson probability formula: $P(X = x) = e^{-\lambda} \dfrac{\lambda^x}{x!}$

- Mean of a Poisson random variable: $\mu = \lambda$

- Standard deviation of a Poisson random variable: $\sigma = \sqrt{\lambda}$

CHAPTER 7   The Sampling Distribution of the Sample Mean

- Mean of the variable $\overline{x}$: $\mu_{\overline{x}} = \mu$

- Standard deviation of the variable $\overline{x}$: $\sigma_{\overline{x}} = \sigma/\sqrt{n}$

CHAPTER 8   Confidence Intervals for One Population Mean

- Standardized version of the variable $\bar{x}$:

$$z = \frac{\bar{x} - \mu}{\sigma/\sqrt{n}}$$

- z-interval for μ (σ known, normal population or large sample):

$$\bar{x} \pm z_{\alpha/2} \cdot \frac{\sigma}{\sqrt{n}}$$

- Margin of error for the estimate of μ: $E = z_{\alpha/2} \cdot \dfrac{\sigma}{\sqrt{n}}$

- Sample size for estimating μ:

$$n = \left(\frac{z_{\alpha/2} \cdot \sigma}{E} \right)^2,$$

rounded up to the nearest whole number.

- Studentized version of the variable $\bar{x}$:

$$t = \frac{\bar{x} - \mu}{s/\sqrt{n}}$$

- t-interval for μ (σ unknown, normal population or large sample):

$$\bar{x} \pm t_{\alpha/2} \cdot \frac{s}{\sqrt{n}}$$

with df $= n - 1$.

CHAPTER 9   Hypothesis Tests for One Population Mean

- z-test statistic for H_0: $\mu = \mu_0$ (σ known, normal population or large sample):

$$z = \frac{\bar{x} - \mu_0}{\sigma/\sqrt{n}}$$

- t-test statistic for H_0: $\mu = \mu_0$ (σ unknown, normal population or large sample):

$$t = \frac{\bar{x} - \mu_0}{s/\sqrt{n}}$$

with df $= n - 1$.

- Wilcoxon signed-rank test statistic for H_0: $\mu = \mu_0$ (symmetric population):

$$W = \text{sum of the positive ranks}$$

CHAPTER 10   Inferences for Two Population Means

- Pooled sample standard deviation:

$$s_p = \sqrt{\frac{(n_1 - 1)s_1^2 + (n_2 - 1)s_2^2}{n_1 + n_2 - 2}}$$

- Pooled t-test statistic for H_0: $\mu_1 = \mu_2$ (independent samples, normal populations or large samples, and equal population standard deviations):

$$t = \frac{\bar{x}_1 - \bar{x}_2}{s_p\sqrt{(1/n_1) + (1/n_2)}}$$

with df $= n_1 + n_2 - 2$.

- Pooled t-interval for $\mu_1 - \mu_2$ (independent samples, normal populations or large samples, and equal population standard deviations):

$$(\bar{x}_1 - \bar{x}_2) \pm t_{\alpha/2} \cdot s_p\sqrt{(1/n_1) + (1/n_2)}$$

with df $= n_1 + n_2 - 2$.

- Degrees of freedom for nonpooled-t procedures:

$$\Delta = \frac{\left[(s_1^2/n_1) + (s_2^2/n_2) \right]^2}{\dfrac{(s_1^2/n_1)^2}{n_1 - 1} + \dfrac{(s_2^2/n_2)^2}{n_2 - 1}},$$

rounded down to the nearest integer.

- Nonpooled t-test statistic for H_0: $\mu_1 = \mu_2$ (independent samples, and normal populations or large samples):

$$t = \frac{\bar{x}_1 - \bar{x}_2}{\sqrt{(s_1^2/n_1) + (s_2^2/n_2)}}$$

with df $= \Delta$.

- Nonpooled t-interval for $\mu_1 - \mu_2$ (independent samples, and normal populations or large samples):

$$(\bar{x}_1 - \bar{x}_2) \pm t_{\alpha/2} \cdot \sqrt{(s_1^2/n_1) + (s_2^2/n_2)}$$

with df $= \Delta$.

- Mann–Whitney test statistic for H_0: $\mu_1 = \mu_2$ (independent samples and same-shape populations):

$$M = \text{sum of the ranks for sample data from Population 1}$$

- Paired t-test statistic for H_0: $\mu_1 = \mu_2$ (paired sample, and normal differences or large sample):

$$t = \frac{\bar{d}}{s_d/\sqrt{n}}$$

with df $= n - 1$.

- Paired t-interval for $\mu_1 - \mu_2$ (paired sample, and normal differences or large sample):

$$\bar{d} \pm t_{\alpha/2} \cdot \frac{s_d}{\sqrt{n}}$$

with df $= n - 1$.

- Paired Wilcoxon signed-rank test statistic for H_0: $\mu_1 = \mu_2$ (paired sample and symmetric differences):

$$W = \text{sum of the positive ranks}$$

FORMULA/TABLE CARD FOR WEISS'S *INTRODUCTORY STATISTICS, SEVENTH EDITION*

Larry R. Griffey

TABLE I   Random numbers

Line number	00–09		10–19		20–29		30–39		40–49	
00	15544	80712	97742	21500	97081	42451	50623	56071	28882	28739
01	01011	21285	04729	39986	73150	31548	30168	76189	56996	19210
02	47435	53308	40718	29050	74858	64517	93573	51058	68501	42723
03	91312	75137	86274	59834	69844	19853	06917	17413	44474	86530
04	12775	08768	80791	16298	22934	09630	98862	39746	64623	32768
05	31466	43761	94872	92230	52367	13205	38634	55882	77518	36252
06	09300	43847	40881	51243	97810	18903	53914	31688	06220	40422
07	73582	13810	57784	72454	68997	72229	30340	08844	53924	89630
08	11092	81392	58189	22697	41063	09451	09789	00637	06450	85990
09	93322	98567	00116	35605	66790	52965	62877	21740	56476	49296
10	80134	12484	67089	08674	70753	90959	45842	59844	45214	36505
11	97888	31797	95037	84400	76041	96668	75920	68482	56855	97417
12	92612	27082	59459	69380	98654	20407	88151	56263	27126	63797
13	72744	45586	43279	44218	83638	05422	00995	70217	78925	39097
14	96256	70653	45285	26293	78305	80252	03625	40159	68760	84716
15	07851	47452	66742	83331	54701	06573	98169	37499	67756	68301
16	25594	41552	96475	56151	02089	33748	65289	89956	89559	33687
17	65358	15155	59374	80940	03411	94656	69440	47156	77115	99463
18	09402	31008	53424	21928	02198	61201	02457	87214	59750	51330
19	97424	90765	01634	37328	41243	33564	17884	94747	93650	77668

TABLE III   Normal scores

| Ordered position | n | | | | | | | | |
	5	6	7	8	9	10	11	12	13
1	−1.18	−1.28	−1.36	−1.43	−1.50	−1.55	−1.59	−1.64	−1.68
2	−0.50	−0.64	−0.76	−0.85	−0.93	−1.00	−1.06	−1.11	−1.16
3	0.00	−0.20	−0.35	−0.47	−0.57	−0.65	−0.73	−0.79	−0.85
4	0.50	0.20	0.00	−0.15	−0.27	−0.37	−0.46	−0.53	−0.60
5	1.18	0.64	0.35	0.15	0.00	−0.12	−0.22	−0.31	−0.39
6		1.28	0.76	0.47	0.27	0.12	0.00	−0.10	−0.19
7			1.36	0.85	0.57	0.37	0.22	0.10	0.00
8				1.43	0.93	0.65	0.46	0.31	0.19
9					1.50	1.00	0.73	0.53	0.39
10						1.55	1.06	0.79	0.60
11							1.59	1.11	0.85
12								1.64	1.16
13									1.68

TABLE VI   Values of M_α

| n_2 | α | n_1 | | | | | | | |
		3	4	5	6	7	8	9	10
3	0.10	14	20	27	36	45	55	66	78
	0.05	15	21	29	37	46	57	68	80
	0.025	—	22	30	38	48	58	70	82
	0.01	—	—	—	39	49	59	71	83
	0.005	—	—	—	—	—	60	72	85
4	0.10	16	23	31	40	49	60	72	85
	0.05	17	24	32	41	51	62	74	87
	0.025	18	25	33	43	53	64	76	89
	0.01	—	26	35	44	54	65	78	91
	0.005	—	—	—	45	55	66	79	93
5	0.10	18	26	34	44	54	65	78	91
	0.05	20	27	36	46	56	68	80	94
	0.025	21	28	37	47	58	70	83	96
	0.01	—	30	39	49	60	72	85	99
	0.005	—	—	40	50	61	73	86	101
6	0.10	21	29	38	48	59	71	84	98
	0.05	22	30	40	50	61	73	87	101
	0.025	23	32	41	52	63	76	89	103
	0.01	24	33	43	54	65	78	92	106
	0.005	—	34	44	55	67	80	94	108
7	0.10	23	31	41	52	63	76	89	104
	0.05	24	33	43	54	66	79	93	107
	0.025	26	35	45	56	68	81	95	110
	0.01	27	36	47	58	71	84	98	114
	0.005	—	37	48	60	72	86	101	116
8	0.10	25	34	44	56	68	81	95	110
	0.05	27	36	47	58	71	84	99	114
	0.025	28	38	49	61	73	87	102	117
	0.01	29	39	51	63	76	90	105	121
	0.005	30	40	52	65	78	92	108	124
9	0.10	27	37	48	60	72	86	101	116
	0.05	29	39	50	63	76	90	105	121
	0.025	31	41	53	65	78	93	108	124
	0.01	32	43	55	68	81	96	112	129
	0.005	33	44	56	70	84	99	114	131
10	0.10	29	40	51	64	77	91	106	123
	0.05	31	42	54	67	80	95	111	127
	0.025	33	44	56	69	83	98	114	131
	0.01	34	46	59	72	87	102	119	136
	0.005	36	48	61	74	89	105	121	139

TABLE VII   Values of χ_α^2

$\chi_{0.10}^2$	$\chi_{0.05}^2$	$\chi_{0.025}^2$	$\chi_{0.01}^2$	$\chi_{0.005}^2$	df
2.706	3.841	5.024	6.635	7.879	1
4.605	5.991	7.378	9.210	10.597	2
6.251	7.815	9.348	11.345	12.838	3
7.779	9.488	11.143	13.277	14.860	4
9.236	11.070	12.833	15.086	16.750	5
10.645	12.592	14.449	16.812	18.548	6
12.017	14.067	16.013	18.475	20.278	7
13.362	15.507	17.535	20.090	21.955	8
14.684	16.919	19.023	21.666	23.589	9
15.987	18.307	20.483	23.209	25.188	10
17.275	19.675	21.920	24.725	26.757	11
18.549	21.026	23.337	26.217	28.300	12
19.812	22.362	24.736	27.688	29.819	13
21.064	23.685	26.119	29.141	31.319	14
22.307	24.996	27.488	30.578	32.801	15
23.542	26.296	28.845	32.000	34.267	16
24.769	27.587	30.191	33.409	35.718	17
25.989	28.869	31.526	34.805	37.156	18
27.204	30.143	32.852	36.191	38.582	19
28.412	31.410	34.170	37.566	39.997	20
29.615	32.671	35.479	38.932	41.401	21
30.813	33.924	36.781	40.290	42.796	22
32.007	35.172	38.076	41.638	44.181	23
33.196	36.415	39.364	42.980	45.559	24
34.382	37.653	40.647	44.314	46.928	25
35.563	38.885	41.923	45.642	48.290	26
36.741	40.113	43.195	46.963	49.645	27
37.916	41.337	44.461	48.278	50.994	28
39.087	42.557	45.722	49.588	52.336	29
40.256	43.773	46.979	50.892	53.672	30
51.805	55.759	59.342	63.691	66.767	40
63.167	67.505	71.420	76.154	79.490	50
74.397	79.082	83.298	88.381	91.955	60
85.527	90.531	95.023	100.424	104.213	70
96.578	101.879	106.628	112.328	116.320	80
107.565	113.145	118.135	124.115	128.296	90
118.499	124.343	129.563	135.811	140.177	100

FORMULA/TABLE CARD FOR WEISS'S *INTRODUCTORY STATISTICS, SEVENTH EDITION*

Larry R. Griffey

TABLE IV   Values of t_α

df	$t_{0.10}$	$t_{0.05}$	$t_{0.025}$	$t_{0.01}$	$t_{0.005}$	df
1	3.078	6.314	12.706	31.821	63.657	1
2	1.886	2.920	4.303	6.965	9.925	2
3	1.638	2.353	3.182	4.541	5.841	3
4	1.533	2.132	2.776	3.747	4.604	4
5	1.476	2.015	2.571	3.365	4.032	5
6	1.440	1.943	2.447	3.143	3.707	6
7	1.415	1.895	2.365	2.998	3.499	7
8	1.397	1.860	2.306	2.896	3.355	8
9	1.383	1.833	2.262	2.821	3.250	9
10	1.372	1.812	2.228	2.764	3.169	10
11	1.363	1.796	2.201	2.718	3.106	11
12	1.356	1.782	2.179	2.681	3.055	12
13	1.350	1.771	2.160	2.650	3.012	13
14	1.345	1.761	2.145	2.624	2.977	14
15	1.341	1.753	2.131	2.602	2.947	15
16	1.337	1.746	2.120	2.583	2.921	16
17	1.333	1.740	2.110	2.567	2.898	17
18	1.330	1.734	2.101	2.552	2.878	18
19	1.328	1.729	2.093	2.539	2.861	19
20	1.325	1.725	2.086	2.528	2.845	20
21	1.323	1.721	2.080	2.518	2.831	21
22	1.321	1.717	2.074	2.508	2.819	22
23	1.319	1.714	2.069	2.500	2.807	23
24	1.318	1.711	2.064	2.492	2.797	24
25	1.316	1.708	2.060	2.485	2.787	25
26	1.315	1.706	2.056	2.479	2.779	26
27	1.314	1.703	2.052	2.473	2.771	27
28	1.313	1.701	2.048	2.467	2.763	28
29	1.311	1.699	2.045	2.462	2.756	29
30	1.310	1.697	2.042	2.457	2.750	30
31	1.309	1.696	2.040	2.453	2.744	31
32	1.309	1.694	2.037	2.449	2.738	32
33	1.308	1.692	2.035	2.445	2.733	33
34	1.307	1.691	2.032	2.441	2.728	34
35	1.306	1.690	2.030	2.438	2.724	35
36	1.306	1.688	2.028	2.434	2.719	36
37	1.305	1.687	2.026	2.431	2.715	37
38	1.304	1.686	2.024	2.429	2.712	38
39	1.304	1.685	2.023	2.426	2.708	39
40	1.303	1.684	2.021	2.423	2.704	40
41	1.303	1.683	2.020	2.421	2.701	41
42	1.302	1.682	2.018	2.418	2.698	42
43	1.302	1.681	2.017	2.416	2.695	43
44	1.301	1.680	2.015	2.414	2.692	44
45	1.301	1.679	2.014	2.412	2.690	45
46	1.300	1.679	2.013	2.410	2.687	46
47	1.300	1.678	2.012	2.408	2.685	47
48	1.299	1.677	2.011	2.407	2.682	48
49	1.299	1.677	2.010	2.405	2.680	49

TABLE IV   (cont.)   Values of t_α

df	$t_{0.10}$	$t_{0.05}$	$t_{0.025}$	$t_{0.01}$	$t_{0.005}$	df
50	1.299	1.676	2.009	2.403	2.678	50
51	1.298	1.675	2.008	2.402	2.676	51
52	1.298	1.675	2.007	2.400	2.674	52
53	1.298	1.674	2.006	2.399	2.672	53
54	1.297	1.674	2.005	2.397	2.670	54
55	1.297	1.673	2.004	2.396	2.668	55
56	1.297	1.673	2.003	2.395	2.667	56
57	1.297	1.672	2.002	2.394	2.665	57
58	1.296	1.672	2.002	2.392	2.663	58
59	1.296	1.671	2.001	2.391	2.662	59
60	1.296	1.671	2.000	2.390	2.660	60
61	1.296	1.670	2.000	2.389	2.659	61
62	1.295	1.670	1.999	2.388	2.657	62
63	1.295	1.669	1.998	2.387	2.656	63
64	1.295	1.669	1.998	2.386	2.655	64
65	1.295	1.669	1.997	2.385	2.654	65
66	1.295	1.668	1.997	2.384	2.652	66
67	1.294	1.668	1.996	2.383	2.651	67
68	1.294	1.668	1.995	2.382	2.650	68
69	1.294	1.667	1.995	2.382	2.649	69
70	1.294	1.667	1.994	2.381	2.648	70
71	1.294	1.667	1.994	2.380	2.647	71
72	1.293	1.666	1.993	2.379	2.646	72
73	1.293	1.666	1.993	2.379	2.645	73
74	1.293	1.666	1.993	2.378	2.644	74
75	1.293	1.665	1.992	2.377	2.643	75
80	1.292	1.664	1.990	2.374	2.639	80
85	1.292	1.663	1.988	2.371	2.635	85
90	1.291	1.662	1.987	2.368	2.632	90
95	1.291	1.661	1.985	2.366	2.629	95
100	1.290	1.660	1.984	2.364	2.626	100
200	1.286	1.653	1.972	2.345	2.601	200
300	1.284	1.650	1.968	2.339	2.592	300
400	1.284	1.649	1.966	2.336	2.588	400
500	1.283	1.648	1.965	2.334	2.586	500
600	1.283	1.647	1.964	2.333	2.584	600
700	1.283	1.647	1.963	2.332	2.583	700
800	1.283	1.647	1.963	2.331	2.582	800
900	1.282	1.647	1.963	2.330	2.581	900
1000	1.282	1.646	1.962	2.330	2.581	1000
2000	1.282	1.646	1.961	2.328	2.578	2000

$z_{0.10}$	$z_{0.05}$	$z_{0.025}$	$z_{0.01}$	$z_{0.005}$
1.282	1.645	1.960	2.326	2.576

TABLE V   Values of W_α

n	$W_{0.10}$	$W_{0.05}$	$W_{0.025}$	$W_{0.01}$	$W_{0.005}$	n
7	22	24	26	28	—	7
8	28	30	32	34	36	8
9	34	37	39	42	43	9
10	41	44	47	50	52	10
11	48	52	55	59	61	11
12	56	61	64	68	71	12
13	65	70	74	78	81	13
14	74	79	84	89	92	14
15	83	90	95	100	104	15
16	94	100	106	112	117	16
17	104	112	118	125	130	17
18	116	124	131	138	143	18
19	128	136	144	152	158	19
20	140	150	158	167	173	20

CHAPTER 11   Inferences for Population Standard Deviations

- χ^2-test statistic for H_0: $\sigma = \sigma_0$ (normal population):

$$\chi^2 = \frac{n-1}{\sigma_0^2}\, s^2$$

with df $= n - 1$.

- χ^2-interval for σ (normal population):

$$\sqrt{\frac{n-1}{\chi_{\alpha/2}^2}} \cdot s \qquad \text{to} \qquad \sqrt{\frac{n-1}{\chi_{1-\alpha/2}^2}} \cdot s$$

with df $= n - 1$.

- F-test statistic for H_0: $\sigma_1 = \sigma_2$ (independent samples and normal populations):

$$F = s_1^2 / s_2^2$$

with df $= (n_1 - 1, n_2 - 1)$.

- F-interval for σ_1/σ_2 (independent samples and normal populations):

$$\frac{1}{\sqrt{F_{\alpha/2}}} \cdot \frac{s_1}{s_2} \qquad \text{to} \qquad \frac{1}{\sqrt{F_{1-\alpha/2}}} \cdot \frac{s_1}{s_2}$$

with df $= (n_1 - 1, n_2 - 1)$.

CHAPTER 12   Inferences for Population Proportions

- Sample proportion:

$$\hat{p} = \frac{x}{n},$$

where x denotes the number of members in the sample that have the specified attribute.

- One-sample z-interval for p:

$$\hat{p} \pm z_{\alpha/2} \cdot \sqrt{\hat{p}(1 - \hat{p})/n}$$

(*Assumption:* both x and $n - x$ are 5 or greater)

- Margin of error for the estimate of p:

$$E = z_{\alpha/2} \cdot \sqrt{\hat{p}(1 - \hat{p})/n}$$

- Sample size for estimating p:

$$n = 0.25 \left(\frac{z_{\alpha/2}}{E} \right)^2 \qquad \text{or} \qquad n = \hat{p}_g(1 - \hat{p}_g)\left(\frac{z_{\alpha/2}}{E} \right)^2$$

rounded up to the nearest whole number (g = "educated guess")

- One-sample z-test statistic for H_0: $p = p_0$:

$$z = \frac{\hat{p} - p_0}{\sqrt{p_0(1 - p_0)/n}}$$

(*Assumption:* both np_0 and $n(1 - p_0)$ are 5 or greater)

- Pooled sample proportion: $\hat{p}_p = \dfrac{x_1 + x_2}{n_1 + n_2}$

- Two-sample z-test statistic for H_0: $p_1 = p_2$:

$$z = \frac{\hat{p}_1 - \hat{p}_2}{\sqrt{\hat{p}_p(1 - \hat{p}_p)}\sqrt{(1/n_1) + (1/n_2)}}$$

(*Assumptions:* independent samples; $x_1, n_1 - x_1, x_2, n_2 - x_2$ are all 5 or greater)

- Two-sample z-interval for $p_1 - p_2$:

$$(\hat{p}_1 - \hat{p}_2) \pm z_{\alpha/2} \cdot \sqrt{\hat{p}_1(1 - \hat{p}_1)/n_1 + \hat{p}_2(1 - \hat{p}_2)/n_2}$$

(*Assumptions:* independent samples; $x_1, n_1 - x_1, x_2, n_2 - x_2$ are all 5 or greater)

- Margin of error for the estimate of $p_1 - p_2$:

$$E = z_{\alpha/2} \cdot \sqrt{\hat{p}_1(1 - \hat{p}_1)/n_1 + \hat{p}_2(1 - \hat{p}_2)/n_2}$$

- Sample size for estimating $p_1 - p_2$:

$$n_1 = n_2 = 0.5 \left(\frac{z_{\alpha/2}}{E} \right)^2$$

or

$$n_1 = n_2 = \left(\hat{p}_{1g}(1 - \hat{p}_{1g}) + \hat{p}_{2g}(1 - \hat{p}_{2g}) \right) \left(\frac{z_{\alpha/2}}{E} \right)^2$$

rounded up to the nearest whole number (g = "educated guess")

CHAPTER 13   Chi-Square Procedures

- Expected frequencies for a chi-square goodness-of-fit test:

$$E = np$$

- Test statistic for a chi-square goodness-of-fit test:

$$\chi^2 = \Sigma(O - E)^2 / E$$

with df $= k - 1$, where k is the number of possible values for the variable under consideration.

- Expected frequencies for a chi-square independence test:

$$E = \frac{R \cdot C}{n}$$

where R = row total and C = column total.

- Test statistic for a chi-square independence test:

$$\chi^2 = \Sigma(O - E)^2 / E$$

with df $= (r - 1)(c - 1)$, where r and c are the number of possible values for the two variables under consideration.

CHAPTER 14   Descriptive Methods in Regression and Correlation

- S_{xx}, S_{xy}, and S_{yy}:

$$S_{xx} = \Sigma(x - \bar{x})^2 = \Sigma x^2 - (\Sigma x)^2/n$$

$$S_{xy} = \Sigma(x - \bar{x})(y - \bar{y}) = \Sigma xy - (\Sigma x)(\Sigma y)/n$$

$$S_{yy} = \Sigma(y - \bar{y})^2 = \Sigma y^2 - (\Sigma y)^2/n$$

- Regression equation: $\hat{y} = b_0 + b_1 x$, where

$$b_1 = \frac{S_{xy}}{S_{xx}} \qquad \text{and} \qquad b_0 = \frac{1}{n}(\Sigma y - b_1 \Sigma x) = \bar{y} - b_1 \bar{x}$$

- Total sum of squares: $SST = \Sigma(y - \overline{y})^2 = S_{yy}$

- Regression sum of squares: $SSR = \Sigma(\hat{y} - \overline{y})^2 = S_{xy}^2/S_{xx}$

- Error sum of squares: $SSE = \Sigma(y - \hat{y})^2 = S_{yy} - S_{xy}^2/S_{xx}$

- Regression identity: $SST = SSR + SSE$

- Coefficient of determination: $r^2 = \dfrac{SSR}{SST}$

- Linear correlation coefficient:

$$r = \frac{\frac{1}{n-1}\Sigma(x - \overline{x})(y - \overline{y})}{s_x s_y} \quad \text{or} \quad r = \frac{S_{xy}}{\sqrt{S_{xx}S_{yy}}}$$

CHAPTER 15   Inferential Methods in Regression and Correlation

- Population regression equation: $y = \beta_0 + \beta_1 x$

- Standard error of the estimate: $s_e = \sqrt{\dfrac{SSE}{n-2}}$

- Test statistic for H_0: $\beta_1 = 0$:

$$t = \frac{b_1}{s_e/\sqrt{S_{xx}}}$$

with df $= n - 2$.

- Confidence interval for β_1:

$$b_1 \pm t_{\alpha/2} \cdot \frac{s_e}{\sqrt{S_{xx}}}$$

with df $= n - 2$.

- Confidence interval for the conditional mean of the response variable corresponding to x_p:

$$\hat{y}_p \pm t_{\alpha/2} \cdot s_e \sqrt{\frac{1}{n} + \frac{(x_p - \Sigma x/n)^2}{S_{xx}}}$$

with df $= n - 2$.

- Prediction interval for an observed value of the response variable corresponding to x_p:

$$\hat{y}_p \pm t_{\alpha/2} \cdot s_e \sqrt{1 + \frac{1}{n} + \frac{(x_p - \Sigma x/n)^2}{S_{xx}}}$$

with df $= n - 2$.

- Test statistic for H_0: $\rho = 0$:

$$t = \frac{r}{\sqrt{\dfrac{1 - r^2}{n - 2}}}$$

with df $= n - 2$.

- Test statistic for a correlation test for normality:

$$R_p = \frac{\Sigma x w}{\sqrt{S_{xx}\,\Sigma w^2}}$$

where x and w denote observations of the variable and the corresponding normal scores, respectively.

CHAPTER 16   Analysis of Variance (ANOVA)

- Notation in one-way ANOVA:

$$k = \text{number of populations}$$
$$n = \text{total number of observations}$$
$$\overline{x} = \text{mean of all } n \text{ observations}$$
$$n_j = \text{size of sample from Population } j$$
$$\overline{x}_j = \text{mean of sample from Population } j$$
$$s_j^2 = \text{variance of sample from Population } j$$
$$T_j = \text{sum of sample data from Population } j$$

- Defining formulas for sums of squares in one-way ANOVA:

$$SST = \Sigma(x - \overline{x})^2$$
$$SSTR = \Sigma n_j(\overline{x}_j - \overline{x})^2$$
$$SSE = \Sigma(n_j - 1)s_j^2$$

- One-way ANOVA identity: $SST = SSTR + SSE$

- Computing formulas for sums of squares in one-way ANOVA:

$$SST = \Sigma x^2 - (\Sigma x)^2/n$$
$$SSTR = \Sigma(T_j^2/n_j) - (\Sigma x)^2/n$$
$$SSE = SST - SSTR$$

- Mean squares in one-way ANOVA:

$$MSTR = \frac{SSTR}{k - 1}, \qquad MSE = \frac{SSE}{n - k}$$

- Test statistic for one-way ANOVA (independent samples, normal populations, and equal population standard deviations):

$$F = \frac{MSTR}{MSE}$$

with df $= (k - 1, n - k)$.

- Confidence interval for $\mu_i - \mu_j$ in the Tukey multiple-comparison method (independent samples, normal populations, and equal population standard deviations):

$$(\overline{x}_i - \overline{x}_j) \pm \frac{q_\alpha}{\sqrt{2}} \cdot s\sqrt{(1/n_i) + (1/n_j)},$$

where $s = \sqrt{MSE}$ and q_α is obtained for a q-curve with parameters k and $n - k$.

- Test statistic for a Kruskal–Wallis test (independent samples, same-shape populations, all sample sizes 5 or greater):

$$H = \frac{SSTR}{SST/(n-1)} \quad \text{or} \quad H = \frac{12}{n(n+1)} \sum_{j=1}^{k} \frac{R_j^2}{n_j} - 3(n+1),$$

where $SSTR$ and SST are computed for the ranks of the data, and R_j denotes the sum of the ranks for the sample data from Population j. H is approximately chi-square with df $= k - 1$.

TABLE II (cont.) Areas under the standard normal curve

Second decimal place in z

z	0.00	0.01	0.02	0.03	0.04	0.05	0.06	0.07	0.08	0.09
0.0	0.5000	0.5040	0.5080	0.5120	0.5160	0.5199	0.5239	0.5279	0.5319	0.5359
0.1	0.5398	0.5438	0.5478	0.5517	0.5557	0.5596	0.5636	0.5675	0.5714	0.5753
0.2	0.5793	0.5832	0.5871	0.5910	0.5948	0.5987	0.6026	0.6064	0.6103	0.6141
0.3	0.6179	0.6217	0.6255	0.6293	0.6331	0.6368	0.6406	0.6443	0.6480	0.6517
0.4	0.6554	0.6591	0.6628	0.6664	0.6700	0.6736	0.6772	0.6808	0.6844	0.6879
0.5	0.6915	0.6950	0.6985	0.7019	0.7054	0.7088	0.7123	0.7157	0.7190	0.7224
0.6	0.7257	0.7291	0.7324	0.7357	0.7389	0.7422	0.7454	0.7486	0.7517	0.7549
0.7	0.7580	0.7611	0.7642	0.7673	0.7704	0.7734	0.7764	0.7794	0.7823	0.7852
0.8	0.7881	0.7910	0.7939	0.7967	0.7995	0.8023	0.8051	0.8078	0.8106	0.8133
0.9	0.8159	0.8186	0.8212	0.8238	0.8264	0.8289	0.8315	0.8340	0.8365	0.8389
1.0	0.8413	0.8438	0.8461	0.8485	0.8508	0.8531	0.8554	0.8577	0.8599	0.8621
1.1	0.8643	0.8665	0.8686	0.8708	0.8729	0.8749	0.8770	0.8790	0.8810	0.8830
1.2	0.8849	0.8869	0.8888	0.8907	0.8925	0.8944	0.8962	0.8980	0.8997	0.9015
1.3	0.9032	0.9049	0.9066	0.9082	0.9099	0.9115	0.9131	0.9147	0.9162	0.9177
1.4	0.9192	0.9207	0.9222	0.9236	0.9251	0.9265	0.9279	0.9292	0.9306	0.9319
1.5	0.9332	0.9345	0.9357	0.9370	0.9382	0.9394	0.9406	0.9418	0.9429	0.9441
1.6	0.9452	0.9463	0.9474	0.9484	0.9495	0.9505	0.9515	0.9525	0.9535	0.9545
1.7	0.9554	0.9564	0.9573	0.9582	0.9591	0.9599	0.9608	0.9616	0.9625	0.9633
1.8	0.9641	0.9649	0.9656	0.9664	0.9671	0.9678	0.9686	0.9693	0.9699	0.9706
1.9	0.9713	0.9719	0.9726	0.9732	0.9738	0.9744	0.9750	0.9756	0.9761	0.9767
2.0	0.9772	0.9778	0.9783	0.9788	0.9793	0.9798	0.9803	0.9808	0.9812	0.9817
2.1	0.9821	0.9826	0.9830	0.9834	0.9838	0.9842	0.9846	0.9850	0.9854	0.9857
2.2	0.9861	0.9864	0.9868	0.9871	0.9875	0.9878	0.9881	0.9884	0.9887	0.9890
2.3	0.9893	0.9896	0.9898	0.9901	0.9904	0.9906	0.9909	0.9911	0.9913	0.9916
2.4	0.9918	0.9920	0.9922	0.9925	0.9927	0.9929	0.9931	0.9932	0.9934	0.9936
2.5	0.9938	0.9940	0.9941	0.9943	0.9945	0.9946	0.9948	0.9949	0.9951	0.9952
2.6	0.9953	0.9955	0.9956	0.9957	0.9959	0.9960	0.9961	0.9962	0.9963	0.9964
2.7	0.9965	0.9966	0.9967	0.9968	0.9969	0.9970	0.9971	0.9972	0.9973	0.9974
2.8	0.9974	0.9975	0.9976	0.9977	0.9977	0.9978	0.9979	0.9979	0.9980	0.9981
2.9	0.9981	0.9982	0.9982	0.9983	0.9984	0.9984	0.9985	0.9985	0.9986	0.9986
3.0	0.9987	0.9987	0.9987	0.9988	0.9988	0.9989	0.9989	0.9989	0.9990	0.9990
3.1	0.9990	0.9991	0.9991	0.9991	0.9992	0.9992	0.9992	0.9992	0.9993	0.9993
3.2	0.9993	0.9993	0.9994	0.9994	0.9994	0.9994	0.9994	0.9995	0.9995	0.9995
3.3	0.9995	0.9995	0.9995	0.9996	0.9996	0.9996	0.9996	0.9996	0.9996	0.9997
3.4	0.9997	0.9997	0.9997	0.9997	0.9997	0.9997	0.9997	0.9997	0.9997	0.9998
3.5	0.9998	0.9998	0.9998	0.9998	0.9998	0.9998	0.9998	0.9998	0.9998	0.9998
3.6	0.9998	0.9998	0.9999	0.9999	0.9999	0.9999	0.9999	0.9999	0.9999	0.9999
3.7	0.9999	0.9999	0.9999	0.9999	0.9999	0.9999	0.9999	0.9999	0.9999	0.9999
3.8	0.9999	0.9999	0.9999	0.9999	0.9999	0.9999	0.9999	0.9999	0.9999	0.9999
3.9	1.0000†									

† For z ≥ 3.90, the areas are 1.0000 to four decimal places.

TABLE II Areas under the standard normal curve

Second decimal place in z

0.09	0.08	0.07	0.06	0.05	0.04	0.03	0.02	0.01	0.00	z
0.0001	0.0001	0.0001	0.0001	0.0001	0.0001	0.0001	0.0001	0.0001	0.0000†	−3.9
0.0001	0.0001	0.0001	0.0001	0.0001	0.0001	0.0001	0.0001	0.0001	0.0001	−3.8
0.0001	0.0001	0.0001	0.0001	0.0001	0.0001	0.0001	0.0001	0.0001	0.0001	−3.7
0.0001	0.0001	0.0001	0.0001	0.0001	0.0001	0.0001	0.0001	0.0002	0.0002	−3.6
0.0002	0.0002	0.0002	0.0002	0.0002	0.0002	0.0002	0.0002	0.0002	0.0002	−3.5
0.0002	0.0003	0.0003	0.0003	0.0003	0.0003	0.0003	0.0003	0.0003	0.0003	−3.4
0.0003	0.0004	0.0004	0.0004	0.0004	0.0004	0.0004	0.0005	0.0005	0.0005	−3.3
0.0005	0.0005	0.0005	0.0006	0.0006	0.0006	0.0006	0.0006	0.0007	0.0007	−3.2
0.0007	0.0007	0.0008	0.0008	0.0008	0.0008	0.0009	0.0009	0.0009	0.0010	−3.1
0.0010	0.0010	0.0011	0.0011	0.0011	0.0012	0.0012	0.0013	0.0013	0.0013	−3.0
0.0014	0.0014	0.0015	0.0015	0.0016	0.0016	0.0017	0.0018	0.0018	0.0019	−2.9
0.0019	0.0020	0.0021	0.0021	0.0022	0.0023	0.0023	0.0024	0.0025	0.0026	−2.8
0.0026	0.0027	0.0028	0.0029	0.0030	0.0031	0.0032	0.0033	0.0034	0.0035	−2.7
0.0036	0.0037	0.0038	0.0039	0.0040	0.0041	0.0043	0.0044	0.0045	0.0047	−2.6
0.0048	0.0049	0.0051	0.0052	0.0054	0.0055	0.0057	0.0059	0.0060	0.0062	−2.5
0.0064	0.0066	0.0068	0.0069	0.0071	0.0073	0.0075	0.0078	0.0080	0.0082	−2.4
0.0084	0.0087	0.0089	0.0091	0.0094	0.0096	0.0099	0.0102	0.0104	0.0107	−2.3
0.0110	0.0113	0.0116	0.0119	0.0122	0.0125	0.0129	0.0132	0.0136	0.0139	−2.2
0.0143	0.0146	0.0150	0.0154	0.0158	0.0162	0.0166	0.0170	0.0174	0.0179	−2.1
0.0183	0.0188	0.0192	0.0197	0.0202	0.0207	0.0212	0.0217	0.0222	0.0228	−2.0
0.0233	0.0239	0.0244	0.0250	0.0256	0.0262	0.0268	0.0274	0.0281	0.0287	−1.9
0.0294	0.0301	0.0307	0.0314	0.0322	0.0329	0.0336	0.0344	0.0351	0.0359	−1.8
0.0367	0.0375	0.0384	0.0392	0.0401	0.0409	0.0418	0.0427	0.0436	0.0446	−1.7
0.0455	0.0465	0.0475	0.0485	0.0495	0.0505	0.0516	0.0526	0.0537	0.0548	−1.6
0.0559	0.0571	0.0582	0.0594	0.0606	0.0618	0.0630	0.0643	0.0655	0.0668	−1.5
0.0681	0.0694	0.0708	0.0721	0.0735	0.0749	0.0764	0.0778	0.0793	0.0808	−1.4
0.0823	0.0838	0.0853	0.0869	0.0885	0.0901	0.0918	0.0934	0.0951	0.0968	−1.3
0.0985	0.1003	0.1020	0.1038	0.1056	0.1075	0.1093	0.1112	0.1131	0.1151	−1.2
0.1170	0.1190	0.1210	0.1230	0.1251	0.1271	0.1292	0.1314	0.1335	0.1357	−1.1
0.1379	0.1401	0.1423	0.1446	0.1469	0.1492	0.1515	0.1539	0.1562	0.1587	−1.0
0.1611	0.1635	0.1660	0.1685	0.1711	0.1736	0.1762	0.1788	0.1814	0.1841	−0.9
0.1867	0.1894	0.1922	0.1949	0.1977	0.2005	0.2033	0.2061	0.2090	0.2119	−0.8
0.2148	0.2177	0.2206	0.2236	0.2266	0.2296	0.2327	0.2358	0.2389	0.2420	−0.7
0.2451	0.2483	0.2514	0.2546	0.2578	0.2611	0.2643	0.2676	0.2709	0.2743	−0.6
0.2776	0.2810	0.2843	0.2877	0.2912	0.2946	0.2981	0.3015	0.3050	0.3085	−0.5
0.3121	0.3156	0.3192	0.3228	0.3264	0.3300	0.3336	0.3372	0.3409	0.3446	−0.4
0.3483	0.3520	0.3557	0.3594	0.3632	0.3669	0.3707	0.3745	0.3783	0.3821	−0.3
0.3859	0.3897	0.3936	0.3974	0.4013	0.4052	0.4090	0.4129	0.4168	0.4207	−0.2
0.4247	0.4286	0.4325	0.4364	0.4404	0.4443	0.4483	0.4522	0.4562	0.4602	−0.1
0.4641	0.4681	0.4721	0.4761	0.4801	0.4840	0.4880	0.4920	0.4960	0.5000	−0.0

† For z ≤ −3.90, the areas are 0.0000 to four decimal places.

Top Films of All Time

The American Film Institute (AFI) conducted a survey as part of a celebration of the 100th anniversary of cinema. AFI polled 1500 filmmakers, actors, critics, politicians, and film historians, asking them to pick their 100 favorite films from a list of 400. The films on the list were made between 1896 and 1996.

After tallying the responses, AFI compiled a list representing the top 100 films. *Citizen Kane,* made in 1941, finished in first place, followed by *Casablanca,* which was made in 1942. The following table shows the top 40 finishers in the poll.

Armed with the knowledge that you gain in this chapter, you will be asked to analyze further this AFI poll at the end of the chapter.

Rank	Film	Year	Rank	Film	Year
1	Citizen Kane	1941	21	The Grapes of Wrath	1940
2	Casablanca	1942	22	2001: A Space Odyssey	1968
3	The Godfather	1972	23	The Maltese Falcon	1941
4	Gone With the Wind	1939	24	Raging Bull	1980
5	Lawrence of Arabia	1962	25	E.T. The Extra-Terrestrial	1982
6	The Wizard of Oz	1939	26	Dr. Strangelove	1964
7	The Graduate	1967	27	Bonnie & Clyde	1967
8	On the Waterfront	1954	28	Apocalypse Now	1979
9	Schindler's List	1993	29	Mr. Smith Goes to Washington	1939
10	Singin' in the Rain	1952	30	The Treasure of the Sierra Madre	1948
11	It's a Wonderful Life	1946	31	Annie Hall	1977
12	Sunset Blvd.	1950	32	The Godfather, Part II	1974
13	The Bridge on the River Kwai	1957	33	High Noon	1952
14	Some Like It Hot	1959	34	To Kill a Mockingbird	1962
15	Star Wars	1977	35	It Happened One Night	1934
16	All About Eve	1950	36	Midnight Cowboy	1969
17	The African Queen	1951	37	The Best Years of Our Lives	1946
18	Psycho	1960	38	Double Indemnity	1944
19	Chinatown	1974	39	Doctor Zhivago	1965
20	One Flew Over the Cuckoo's Nest	1975	40	North by Northwest	1959

1.1   Two Kinds of Statistics

You probably already know something about statistics. If you read newspapers, surf the Web, watch the news on television, or follow sports, you see and hear the word *statistics* frequently. In this section, we use familiar examples such as baseball statistics and voter polls to introduce the two major types of statistics: **descriptive statistics** and **inferential statistics.** We also examine how to classify studies as either descriptive or inferential.

Descriptive Statistics

Each spring in the late 1940s, President Harry Truman officially opened the major league baseball season by throwing out the "first ball" at the opening game of the Washington Senators. Both President Truman and the Washington Senators had reason to be interested in statistics in 1948. We use the 1948 baseball season to illustrate the first major type of statistics, descriptive statistics, in Example 1.1.

Example 1.1   **Descriptive Statistics**

The 1948 Baseball Season   In 1948, the Washington Senators played 153 games, winning 56 and losing 97. They finished seventh in the American League and were led in hitting by Bud Stewart, whose batting average was .279. These and many other statistics were compiled by baseball statisticians who took the complete records for each game of the season and organized that large mass of information effectively and efficiently.

Although baseball fans take baseball statistics for granted, a great deal of time and effort is required to gather and organize them. Moreover, without such statistics, baseball would be much harder to understand. For instance, picture yourself trying to select the best hitter in the American League with only the official score sheets for each game. (More than 600 games were played in 1948; the best hitter was Ted Williams, who led the league with a batting average of .369.)

The work of baseball statisticians provides an excellent illustration of descriptive statistics. A formal definition of the term *descriptive statistics* follows.

DEFINITION 1.1   **Descriptive Statistics**

Descriptive statistics consists of methods for organizing and summarizing information.

Descriptive statistics includes the construction of graphs, charts, and tables and the calculation of various descriptive measures such as averages, measures of variation, and percentiles. We discuss descriptive statistics in detail in Chapters 2 and 3.

Inferential Statistics

We use the 1948 presidential election to introduce the other major type of statistics, inferential statistics, in Example 1.2.

Example 1.2   **Inferential Statistics**

The 1948 Presidential Election   In the fall of 1948, President Truman was also concerned about statistics. The *Gallup Poll* taken just prior to the election predicted that he would win only 44.5% of the vote and be defeated by the Republican nominee, Thomas E. Dewey. But this time the statisticians had predicted incorrectly. Truman won more than 49% of the vote and, with it, the presidency. The Gallup Organization modified some of its procedures and has correctly predicted the winner ever since.

Political polling provides an example of inferential statistics. Interviewing everyone of voting age in the United States on their voting preferences would be expensive and unrealistic. Statisticians who want to gauge the sentiment of the entire **population** of U.S. voters can afford to interview only a carefully chosen group of a few thousand voters. This group is called a **sample** of the population. Statisticians analyze the information obtained from a sample of the voting population to make inferences (draw conclusions) about the preferences of the entire voting population. Inferential statistics provides methods for making such inferences.

The terminology just introduced in the context of political polling is used in general in statistics. Specifically, the terms *population* and *sample* are defined as follows.

DEFINITION 1.2   **Population and Sample**

Population: The collection of all individuals or items under consideration in a statistical study.

Sample: That part of the population from which information is obtained.

Figure 1.1 depicts the relationship between a population and a sample from the population.

FIGURE 1.1
Relationship between population and sample

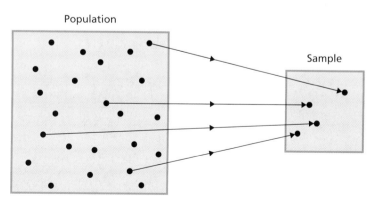

Now that we have discussed the terms *population* and *sample,* we can define *inferential statistics.*

DEFINITION 1.3

> **Inferential Statistics**
>
> *Inferential statistics* consists of methods for drawing and measuring the reliability of conclusions about a population based on information obtained from a sample of the population.

Descriptive statistics and inferential statistics are interrelated. You must almost always use techniques of descriptive statistics to organize and summarize the information obtained from a sample before carrying out an inferential analysis. Furthermore, the preliminary descriptive analysis of a sample often reveals features that lead you to the choice of (or to a reconsideration of the choice of) the appropriate inferential method.

Classifying Statistical Studies

As you proceed through this book, you will obtain a thorough understanding of the principles of descriptive and inferential statistics. At this point, you should be able to classify statistical studies as either descriptive or inferential. In doing so, you should consider the intent of the study.

If the intent of the study is to examine and explore the information obtained for its own intrinsic interest only, the study is descriptive. However, if the information is obtained from a sample of a population and the intent of the study is to use that information to draw conclusions about the population, the study is inferential.

Thus, a descriptive study may be performed on a sample as well as on a population. Only when an inference is made about the population, based on information obtained from the sample, does the study become inferential.

Examples 1.3 and 1.4 further illustrate the distinction between descriptive and inferential studies. In each example, we present the result of a statistical study and classify the study as either descriptive or inferential. Try to classify each study yourself before reading our explanation.

▍▍Example 1.3   **Classifying Statistical Studies**

The 1948 Presidential Election   Table 1.1 displays the voting results for the 1948 presidential election.

TABLE 1.1
Final results of the 1948 presidential election

Ticket	Votes	Percentage
Truman–Barkley(Democratic)	24,179,345	49.7
Dewey–Warren (Republican)	21,991,291	45.2
Thurmond–Wright (States Rights)	1,176,125	2.4
Wallace–Taylor (Progressive)	1,157,326	2.4
Thomas–Smith (Socialist)	139,572	0.3

Classification This study is descriptive. It is a summary of the votes cast by U.S. voters in the 1948 presidential election. No inferences are made.

Example 1.4 Classifying Statistical Studies

Testing Baseballs For the 101 years preceding 1977, baseballs used by the major leagues were purchased from the Spalding Company. In 1977, that company stopped manufacturing major league baseballs, and the major leagues arranged to buy their baseballs from the Rawlings Company.

Early in the 1977 season, pitchers began to complain that the Rawlings ball was "livelier" than the Spalding ball. They claimed it was harder, bounced farther and faster, and gave hitters an unfair advantage. There was some evidence for this claim: In the first 616 games of 1977, 1033 home runs were hit, compared to only 762 home runs hit in the first 616 games of 1976.

Sports Illustrated magazine sponsored a careful study of the liveliness question, and the results appeared in the article "They're Knocking the Stuffing Out of It" by Larry Keith (*Sports Illustrated*, June 13, 1977, pp. 23–27). In this study, an independent testing company randomly selected 85 baseballs from the current (1977) supplies of various major league teams. The bounce, weight, and hardness of the baseballs chosen were carefully measured. Those measurements were then compared with measurements obtained from similar tests on baseballs used in 1952, 1953, 1961, 1963, 1970, and 1973.

The conclusion, presented on page 24 of the *Sports Illustrated* issue, was that "…the 1977 Rawlings ball is livelier than the 1976 Spalding, but not as lively as it could be under big league rules, or as the ball has been in the past."

Classification This study is inferential. The independent testing company used a sample of 85 baseballs from the 1977 supplies of major league teams to make an inference about the population of all such baseballs. (An estimated 360,000 baseballs were used by the major leagues in 1977.)

The *Sports Illustrated* study also shows that it is often not feasible to obtain information for the entire population. Indeed, after the bounce and hardness tests, all of the baseballs sampled were taken to a butcher in Plainfield, New Jersey, to be sliced in half so that researchers could look inside them. Clearly, testing every baseball in this way would not have been practical.

The Development of Statistics

According to the *Dictionary of Scientific Biography*, "The word *Statistik*, first printed in 1672, meant *Staatswissenschaft*, or, rather, a science concerning the states. It was cultivated at the German universities, where it consisted of more or less systematically collecting 'state curiosities' rather than quantitative material."

The modern science of statistics is much broader than just collecting "state curiosities." Historically, descriptive statistics appeared first. Censuses were taken as long ago as Roman times. Over the centuries, records of such things as births, deaths, marriages, and taxes led naturally to the development of descriptive statistics.

Inferential statistics is a newer arrival. Major developments began to occur with the research of Karl Pearson (1857–1936) and Ronald Fisher (1890–1962), who published their findings in the early years of the twentieth century. Since the work of Pearson and Fisher, inferential statistics has evolved rapidly and is now applied in a myriad of fields.

Familiarity with statistics will help you make sense of many things you read in newspapers and magazines and on the Internet. For instance, in the description of the *Sports Illustrated* baseball test (Example 1.4), you may have questioned whether a sample of only 85 baseballs could be used to draw a conclusion about a population of some 360,000 baseballs. By the time you complete Chapter 9, you will understand why such inferences are not unreasonable.

EXERCISES 1.1

Statistical Concepts and Skills

1.1 Define the following terms:
a. Population   **b.** Sample

1.2 What are the two major types of statistics? Describe them in detail.

1.3 Identify some methods used in descriptive statistics.

1.4 Explain two ways in which descriptive statistics and inferential statistics are interrelated.

In Exercises 1.5–1.10, classify each of the studies as either descriptive or inferential.

1.5 TV Viewing Times. The A. C. Nielsen Company collects and publishes information on the television viewing habits of Americans. Data from a sample of Americans yielded the following estimates of average TV viewing time per week for all Americans. The times are in hours and minutes. [SOURCE: Nielsen Media Research, *Nielsen Report on Television*.]

Group (by age)		Time
Average all persons		30:14
Women	Total 18+	34:47
	18–24	28:54
	25–54	31:05
	55+	44:11
Men	Total 18+	30.41
	18–24	23:31
	25–54	28:44
	55+	38:47
Teens	12–17	21:50
Children	2–11	23:01

1.6 Professional Athlete Salaries. In the *Statistical Abstract of the United States*, average professional athletes' salaries in baseball, basketball, and football were compiled and compared for the years 1990 and 2000.

	Average salary ($1000)	
Sport	**1990**	**2000**
Baseball (MLB)	598	1,720
Basketball (NBA)	750	3,522
Football (NFL)	395	1,071

1.7 Causes of Death. The U.S. National Center for Health Statistics published the following rate estimates for the leading causes of death in 2000 in *Vital Statistics of the United States*. The estimates are based on the tenth revision of the *International Classification of Diseases*. Rates are per 100,000 population.

Cause	Rate
Diseases of heart	257.9
Malignant neoplasms	200.5
Cerebrovascular diseases	60.3
Chronic lower respiratory diseases	44.9
Accidents	34.0

1.8 Drug Use. The U.S. Substance Abuse and Mental Health Services Administration collects and publishes data on drug use, by type of drug and age group, in *National Household Survey on Drug Abuse*. The following table provides information for the years 1990 and 2000. The percentages shown are estimates obtained from national samples.

Type of drug	Percentage, 12 years old and over			
	Ever used		*Current user*	
	1990	**2000**	**1990**	**2000**
Marijuana	30.5	34.2	5.4	4.8
Cocaine	11.2	11.2	0.9	0.5
Inhalants	5.7	7.5	0.4	0.3
Hallucinogens	7.9	11.7	0.4	0.4
Heroin	0.8	1.2	—	0.1
Stimulants[1]	5.5	6.6	0.6	0.4
Sedatives[1]	2.8	3.2	0.2	0.1
Tranquilizers[1]	4.0	5.8	0.6	0.4
Pain relievers[1]	6.3	8.6	0.9	1.2
Alcohol	82.2	81.0	52.6	46.6

1 = Nonmedical use

Music type	Expenditure (%)
Rock	24.4
Pop	12.1
Rap/Hip hop	11.4
R&B/Urban	10.6
Country	10.5
Religious	6.7
Jazz	3.4
Classical	3.2
Soundtracks	1.4
New Age	1.0
Oldies	0.8
Children's	0.5
Other	7.9
Unknown	6.1

1.9 Dow Jones Industrial Averages. The following table provides the closing values of the Dow Jones Industrial Averages as of the end of December for the years 1997–2002. [SOURCE: Global Financial Data.]

Year	Closing value
1997	7,908.25
1998	9,181.43
1999	11,497.12
2000	10,786.85
2001	10,021.50
2002	8,341.63

1.10 The Music People Buy. Results of monthly telephone surveys yielded the percentage estimates of all music expenditures shown in the table at the top of the next column. These statistics were published in *2001 Consumer Profile*. [SOURCE: Recording Industry Association of America, Inc.]

Extending the Concepts and Skills

1.11 Organically Grown Produce. A *Newsweek* poll of a sample of Americans revealed that "84% of those surveyed would choose organically grown produce over produce grown using chemical fertilizers, pesticides, and herbicides."
a. Is the statement in quotes an inferential or a descriptive statement?
b. Based on the same information, what if the statement had been "84% of Americans would choose organically grown produce over produce grown using chemical fertilizers, pesticides, and herbicides"?

1.12 Ballistic Fingerprinting. In a press release dated October 22, 2002, ABCNews.com reported that "...73 percent of Americans...favor a law that would require every gun sold in the United States to be test-fired first, so law enforcement would have its fingerprint in case it were ever used in a crime."
a. Do you think that the statement in the press release is inferential or descriptive? Can you be sure?
b. Actually, ABCNews.com conducted a telephone survey of a random national sample of 1032 adults and determined that 73% of them favored a law that would require every gun sold in the United States to be test-fired first, so law enforcement would have its fingerprint in case it were ever used in a crime. How would you rephrase the statement in the press release to make clear that it is a descriptive statement? an inferential statement?

1.2   The Technology Center

Today, programs for conducting statistical and data analyses are available in dedicated statistical software packages, general-use spreadsheet software, and graphing calculators. In this book, we discuss three of the most popular technologies for doing statistics: Minitab, Excel, and the TI-83/84 Plus.[†]

For Excel, we mostly use Data Desk/XL (DDXL) from Data Description, Inc. This statistics add-in complements Excel's standard statistics capabilities; it is included on the WeissStats CD, which comes with your book. Further details of Minitab, Excel, the TI-83/84 Plus, and other statistical technologies are provided in supplements written specifically to accompany this book.

At the end of appropriate sections of the book, in subsections titled The Technology Center, we present and interpret output from the three technologies mentioned. The output from each technology addresses problems that were solved by hand earlier in the section. For this aspect of The Technology Center, you need neither a computer nor a graphing calculator, nor do you need any working knowledge of the technologies under discussion.

For those who want to learn how to obtain the output for one or more of the three technologies, step-by-step instructions are presented in subsections titled Obtaining the Output (Optional). When studying this material, you will get the best results by using your computer or graphing calculator to perform the steps described.

Each statistical technology has a slightly different method for data input and output. At this point, you should spend some time learning the method for each technology you want to study. The documentation or online help for your technology gives you the necessary details. For Minitab, Excel, and the TI-83/84 Plus, you can also find this information in the appropriate technology supplement to this book.

1.3   Simple Random Sampling

Throughout this book, we present examples of organizations or people conducting studies: A consumer group wants information about the gas mileage of a particular make of car, so it performs mileage tests on a sample of such cars and statistically analyzes the resulting data; or a teacher wants to know about the comparative merits of two teaching methods, so she tests those methods on two groups of students. This approach reflects a healthy attitude: To obtain information about a subject of interest, plan and conduct a study.

However, the possibility always exists that a study being considered has already been done. Repeating it would be a waste of time, energy, and money. Therefore, before a study is planned and conducted, a literature search should be made. Doing so does not require going through all the books in the library or making an extensive Internet search. Many information collection agencies specialize in finding studies on specific topics in specific areas.

> **What**
> *Does it Mean?*
>
> You can often avoid the effort and expense of a study if someone else has already done that study and published the results.

[†]For brevity, we write TI-83/84 Plus for TI-83 Plus and/or TI-84 Plus. Keystrokes and output remain the same from the TI-83 Plus to the TI-84 Plus. Thus, instructions and output given in the book apply to both calculators.

Census, Sampling, and Experimentation

If information required is not already available from a previous study, you can plan a new study to obtain the information. One method for acquiring information is to conduct a **census,** that is, obtain information on the entire population of interest. However, conducting a census is generally time consuming and costly, frequently impractical, and sometimes impossible.

Two methods other than a census for obtaining information are **sampling** and **experimentation.** In much of this book, we concentrate on sampling. However, we introduce experimentation in Section 1.5, discuss it sporadically throughout the text, and examine it in detail in the chapter *Design of Experiments and Analysis of Variance* on the WeissStats CD accompanying this book or on the Weiss Web site, www.aw-bc.com/weiss.

If sampling is deemed appropriate, you must then decide how to select the sample; that is, you must choose the method for obtaining a sample from the population. In making that choice, keep in mind that the sample will be used to draw conclusions about the entire population. Consequently, the sample should be a **representative sample,** that is, it should reflect as closely as possible the relevant characteristics of the population under consideration.

For instance, it would not make sense to use the average weight of a sample of professional football players to make an inference about the average weight of all adult males. Nor would it be reasonable to estimate the median income of California residents by sampling the incomes of Beverly Hills residents.

To see what can happen when a sample is not representative, consider the presidential election of 1936. Before the election, the *Literary Digest* magazine conducted an opinion poll of the voting population. Its survey team asked a sample of the voting population whether they would vote for Franklin D. Roosevelt, the Democratic candidate, or for Alfred Landon, the Republican candidate.

Based on the results of the survey, the magazine predicted an easy win for Landon. But when the actual election results were in, Roosevelt won by the greatest landslide in the history of presidential elections! What happened? Here are two reasons given for the failure of the poll.

- The sample was obtained from among people who owned a car or had a telephone. In 1936, that group included only the more well-to-do people, and historically such people tend to vote Republican.
- The response rate was low (less than 25% of those polled responded), and there was a nonresponse bias (a disproportionate number of those who responded to the poll were Landon supporters).

Whatever the reason for the poll's failure, the sample obtained by the *Literary Digest* obviously was not representative.

Most modern sampling procedures involve the use of **probability sampling.** In probability sampling, a random device, such as tossing a coin or consulting a table of random numbers, is used to decide which members of the population will constitute the sample instead of leaving such decisions to human judgment.

The use of probability sampling may still yield a nonrepresentative sample. However, probability sampling eliminates unintentional selection bias and permits the researcher to control the chance of obtaining a nonrepresentative sample. Furthermore, the use of probability sampling guarantees that the techniques of inferential statistics can be applied. In this section and the next, we examine the most important probability-sampling methods.

Simple Random Sampling

The inferential techniques considered in this book are intended for use with only one particular sampling procedure: **simple random sampling.**

DEFINITION 1.4

> **Simple Random Sampling; Simple Random Sample**
>
> *Simple random sampling:*   A sampling procedure for which each possible sample of a given size is equally likely to be the one obtained.
>
> *Simple random sample:*   A sample obtained by simple random sampling.

What
Does it Mean?

Simple random sampling corresponds to our intuitive notion of random selection by lot.

There are two types of simple random sampling. One is **simple random sampling with replacement,** whereby a member of the population can be selected more than once; the other is **simple random sampling without replacement,** whereby a member of the population can be selected at most once. *Unless we specify otherwise, assume that simple random sampling is done without replacement.*

In Example 1.5, we chose a very small population—the five top Oklahoma state officials—to illustrate simple random sampling. In practice, we would not sample from such a small population but would instead take a census. Using a small population here makes understanding the concept of simple random sampling easier.

▌▌Example 1.5   Simple Random Samples

TABLE 1.2
Five top Oklahoma state officials

Governor (G)
Lieutenant Governor (L)
Secretary of State (S)
Attorney General (A)
Treasurer (T)

Sampling Oklahoma State Officials   As reported by *The World Almanac*, the top five state officials of Oklahoma are as shown in Table 1.2. Consider these five officials a population of interest.

a. List the possible samples (without replacement) of two officials from this population of five officials.

b. Describe a method for obtaining a simple random sample of two officials from this population of five officials.

c. For the sampling method described in part (b), what are the chances that any particular sample of two officials will be the one selected?

d. Repeat parts (a)–(c) for samples of size 4.

Solution   For convenience, we use the letters in parentheses after the officials in Table 1.2 to represent them.

TABLE 1.3

The 10 possible samples of two officials

G, L	G, S	G, A	G, T
L, S	L, A	L, T	S, A
S, T	A, T		

TABLE 1.4

The five possible samples of four officials

G, L, S, A	G, L, S, T
G, L, A, T	G, S, A, T
L, S, A, T	

a. There are 10 possible samples of two officials from the population of five officials, as listed in Table 1.3.

b. To obtain a simple random sample of size 2 we could first write the letters that correspond to the five officials, G, L, S, A, and T, on separate pieces of paper. Next, we could place the five slips of paper in a box and shake it. Then, while blindfolded, we could pick two slips of paper.

c. The sampling procedure described in part (b) ensures that we are taking a simple random sample. Consequently, each of the possible samples of two officials is equally likely to be the one selected. There are 10 possible samples, so the chances are $\frac{1}{10}$ (1 in 10) that any particular sample of two officials will be the one selected.

d. There are five possible samples of four officials from the population of five officials, as listed in Table 1.4. In this case, a simple random sampling procedure, such as picking four slips of paper out of a box, gives each of the five possible samples in Table 1.4 a 1 in 5 chance of being the one selected.

Random-Number Tables

Obtaining a simple random sample by picking slips of paper out of a box is usually not practical, especially when the population to be sampled is large. But there are several practical procedures to get simple random samples. One common method is to use a **table of random numbers**—a table of randomly chosen digits. In Example 1.6, we explain how a table of random numbers can be used to obtain a simple random sample.

|||Example 1.6   Random-Number Tables

Sampling Student Opinions   Student questionnaires, known as "teacher evaluations," gained widespread use in the late 1960s and early 1970s. Generally, student evaluations of teachers are not done at final exam time. More commonly, professors hand out evaluation forms a week or so before the final.

That practice, however, poses several problems. On some days, less than 60% of the students registered for a class may actually attend. Moreover, because many of those who are present have preparations to make for other classes, they often complete their teacher evaluation forms in a hurry so that they can leave class early. A better method, therefore, might be to select a sample of students from the class and interview them individually. In this kind of situation, a simple random sample is appropriate.

During one semester, Professor Hassett wanted to sample the attitudes of the students taking college algebra at his school. He decided to interview 15 of the 728 students enrolled in the course. Professor Hassett had a registration list on which the 728 students were numbered 1–728, so he could obtain a simple random sample of 15 students by randomly selecting 15 numbers between 1 and 728. To do so, he used a table of random numbers. The random-number table used by Professor Hassett is presented as Table I in Appendix A. For ease of reference, we repeat it here as Table 1.5.

TABLE 1.5
Random numbers

Line number	00–09		10–19		20–29		30–39		40–49	
00	15544	80712	97742	21500	97081	42451	50623	56071	28882	28739
01	01011	21285	04729	39986	73150	31548	30168	76189	56996	19210
02	47435	53308	40718	29050	74858	64517	93573	51058	68501	42723
03	91312	75137	86274	59834	69844	19853	06917	17413	44474	86530
04	12775	08768	80791	16298	22934	09630	98862	39746	64623	32768
05	31466	43761	94872	92230	52367	13205	38634	55882	77518	36252
06	09300	43847	40881	51243	97810	18903	53914	31688	06220	40422
07	73582	13810	57784	72454	68997	72229	30340	08844	53924	89630
08	11092	81392	58189	22697	41063	09451	09789	00637	06450	85990
09	93322	98567	00116	35605	66790	52965	62877	21740	56476	49296
10	80134	12484	67089	08674	70753	90959	45842	59844	45214	36505
11	97888	31797	95037	84400	76041	96668	75920	68482	56855	97417
12	92612	27082	59459	69380	98654	20407	88151	56263	27126	63797
13	72744	45586	43279	44218	83638	05422	00995	70217	78925	39097
14	96256	70653	45285	26293	78305	80252	03625	40159	68760	84716
15	07851	47452	66742	83331	54701	06573	98169	37499	67756	68301
16	25594	41552	96475	56151	02089	33748	65289	89956	89559	33687
17	65358	15155	59374	80940	03411	94656	69440	47156	77115	99463
18	09402	31008	53424	21928	02198	61201	02457	87214	59750	51330
19	97424	90765	01634	37328	41243	33564	17884	94747	93650	77668

To select 15 random numbers between 1 and 728, we first pick a random starting point, say, by closing our eyes and placing a finger on Table 1.5. Then, beginning with the three digits under the finger, we go down the table and record the numbers as we go. Because we want numbers between 1 and 728 only, we discard the number 000 and numbers between 729 and 999. To avoid repetition, we also eliminate numbers that have occurred previously. If we have not found enough numbers by the time we reach the bottom of the table, we move over to the next column of three-digit numbers and go up.

Using this procedure, Professor Hassett obtained 069, circled in Table 1.5, as a starting point. Reading down from 069 to the bottom of Table 1.5 and then up the next column of three-digit numbers, he found the 15 random numbers displayed in Fig. 1.2 and in Table 1.6. Thus Professor Hassett interviewed the 15 students whose numbers on the registration list are shown in Table 1.6.

TABLE 1.6
Registration numbers of students interviewed

69	303	458	652	178
386	97	9	694	578
539	628	36	24	404

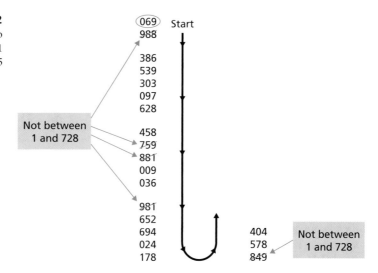

FIGURE 1.2

Procedure used by Professor Hassett to obtain 15 random numbers between 1 and 728 from Table 1.5

Simple random sampling is the basic type of probability sampling and is also the foundation for the more complex types of probability sampling. See Section 1.4 for more details.

The Technology Center

Nowadays, statistical software packages or graphing calculators, instead of random-number tables, are generally used to obtain random numbers. For instance, Minitab, Excel, and the TI-83/84 Plus all have built-in programs that can be utilized to generate random numbers within any specified range. See the technology manuals for details.

EXERCISES 1.3

Statistical Concepts and Skills

1.13 Explain why a census is often not the best way to obtain information about a population.

1.14 Identify two methods other than a census for obtaining information.

1.15 In sampling, why is obtaining a representative sample important?

1.16 Memorial Day Poll. An online poll conducted over Memorial Day Weekend (May 26–29, 2000) asked people what they were doing to observe Memorial Day. The choices were: (1) stay home and relax, (2) vacation outdoors over the weekend, or (3) visit a military cemetery. More than 22,000 people participated in the poll, with 86% selecting option 1. Discuss this poll with regard to its suitability.

1.17 Estimating Median Income. Explain why a sample of 30 dentists from Seattle taken to estimate the median income of all Seattle residents is not representative.

1.18 Provide a scenario of your own in which a sample is not representative.

1.19 Regarding probability sampling:
a. What is it?
b. Does probability sampling always yield a representative sample? Explain your answer.
c. Identify some advantages of probability sampling.

1.20 Regarding simple random sampling:
a. What is simple random sampling?
b. What is a simple random sample?
c. Identify two forms of simple random sampling and explain the difference between the two.

1.21 The inferential procedures discussed in this book are intended for use with only one particular sampling procedure. What sampling procedure is that?

1.22 Identify two methods for obtaining a simple random sample.

1.23 Oklahoma State Officials. The five top Oklahoma state officials are displayed in Table 1.2 on page 12. Use that table to solve the following problems.
a. List the 10 possible samples (without replacement) of size 3 that can be obtained from the population of five officials.
b. If a simple random sampling procedure is used to obtain a sample of three officials, what are the chances that it is the first sample on your list in part (a)? the second sample? the tenth sample?

1.24 Best-Selling Albums. *Billboard Online* provides data on the best-selling albums of all time. As of April, 2000, the top six best-selling albums of all time, by artist, are the Eagles (E), Michael Jackson (M), Pink Floyd (P), Led Zeppelin (L), Billy Joel (B), and Fleetwood Mac (F). [SOURCE: Record Industry Association of America, Inc.]
a. List the 15 possible samples (without replacement) of two artists that can be selected from the six. For brevity, use the initial provided.
b. Describe a procedure for taking a simple random sample of two artists from the six.
c. If a simple random sampling procedure is used to obtain two artists, what are the chances of selecting P and F? M and E?

1.25 Best-Selling Albums. Refer to Exercise 1.24.
a. List the 15 possible samples (without replacement) of four artists that can be selected from the six.
b. Describe a procedure for taking a simple random sample of four artists from the six.
c. If a simple random sampling procedure is used to obtain four artists, what are the chances of selecting E, F, L, and B? P, B, M, and F?

1.26 Best-Selling Albums. Refer to Exercise 1.24.
a. List the 20 possible samples (without replacement) of three artists that can be selected from the six.

b. Describe a procedure for taking a simple random sample of three artists from the six.
c. If a simple random sampling procedure is used to obtain three artists, what are the chances of selecting M, F, and L? P, L, and E?

1.27 The International 500. Each year, *Fortune Magazine* publishes an article titled "The International 500" that provides a ranking by sales of the top 500 firms outside the United States. Suppose that you want to examine various characteristics of successful firms. Further suppose that, for your study, you decide to take a simple random sample of 10 firms from *Fortune Magazine*'s list of "The International 500." Use Table I in Appendix A to obtain 10 random numbers that you can use to specify your sample.

1.28 Keno. In the game of keno, 20 balls are selected at random from 80 balls, numbered 1–80. Use Table I in Appendix A to simulate one game of keno by obtaining 20 random numbers between 1 and 80.

Extending the Concepts and Skills

1.29 Oklahoma State Officials. Refer to Exercise 1.23.
a. List the possible samples of size 1 that can be obtained from the population of five officials.
b. What is the difference between obtaining a simple random sample of size 1 and selecting one official at random?

1.30 Oklahoma State Officials. Refer to Exercise 1.23.
a. List the possible samples (without replacement) of size 5 that can be obtained from the population of five officials.
b. What is the difference between obtaining a simple random sample of size 5 and taking a census?

Using Technology

1.31 The International 500. Refer to Exercise 1.27. Use the technology of your choice to obtain 10 random numbers that you can use to specify your simple random sample of 10 firms.

1.32 Keno. Refer to Exercise 1.28. Use the technology of your choice to simulate one game of keno, that is, to obtain a simple random sample of 20 numbers between 1 and 80.

1.4 Other Sampling Designs

Simple random sampling is the most natural and easily understood method of probability sampling—it corresponds to our intuitive notion of random selection by lot. However, simple random sampling does have drawbacks. For instance, it may fail to provide sufficient coverage when information about subpopulations is required and may be impractical when the members of the population are widely scattered geographically.

In this section, we examine some commonly used sampling procedures that are often more appropriate than simple random sampling. Remember, though, the inferential procedures discussed in this book must be modified before they can be applied to data that are obtained by sampling procedures other than simple random sampling.

Systematic Random Sampling

One method that takes less effort to implement than simple random sampling is **systematic random sampling.** Consider Example 1.7.

Example 1.7   **Systematic Random Sampling**

Sampling Student Opinions   Let's return to the situation in Example 1.6 where Professor Hassett wanted to obtain a sample of 15 of the 728 students enrolled in college algebra at his school. Use systematic random sampling to obtain the sample.

TABLE 1.7

Numbers obtained by systematic random sampling

22	166	310	454	598
70	214	358	502	646
118	262	406	550	694

Solution   To begin, we divide the population size by the sample size and round the answer down to the nearest whole number: $\frac{728}{15} = 48$ (rounded down). Next, we select a number at random between 1 and 48 by using, say, a table of random numbers. Suppose that we do so and obtain the number 22. Then, we list every 48th number, starting at 22, until we have 15 numbers. This method yields the 15 numbers displayed in Table 1.7.

Had Professor Hassett used systematic random sampling to obtain his sample of students and had he obtained the number 22 as his starting point, he would have interviewed the 15 students whose numbers on the registration list are shown in Table 1.7.

As illustrated in Example 1.7, we use the following procedure to implement systematic random sampling.

PROCEDURE 1.1

Systematic Random Sampling

STEP 1 **Divide the population size by the sample size and round the result down to the nearest whole number, *m*.**

STEP 2 **Use a random-number table (or a similar device) to obtain a number, *k*, between 1 and *m*.**

STEP 3 **Select for the sample those members of the population that are numbered *k*, *k* + *m*, *k* + 2*m*, ...**

Systematic random sampling is not only easier to execute than simple random sampling, but it also usually provides results comparable to simple random sampling. The exception is the presence of some kind of cyclical pattern in the listing of the members of the population (e.g., male, female, male, female, …), a phenomenon that is relatively rare.

Cluster Sampling

Another sampling method is **cluster sampling.** It is particularly useful when the members of the population under consideration are widely scattered geographically. We illustrate this method in Example 1.8.

Example 1.8   Cluster Sampling

Bike Paths Survey   Several years ago, the city council of Tempe, Arizona, was being pressured by citizens' groups to install bike paths in the city. The council members wanted to be sure they had the support of a majority of the taxpayers, so they decided to poll the city's homeowners.

Their first attempt at surveying public opinion was a questionnaire mailed out with the city's 18,000 homeowner water bills. Unfortunately, this method did not work very well. Only 19.4% of the questionnaires were returned, and a large number of those had written comments that indicated they came from avid bicyclists or from people who strongly resented bicyclists. The questionnaire generally had not been returned by the average voter, and the city council realized that.

The city had an employee in the planning department with sample survey experience, so the council asked her to do a survey. She was given two assistants to help interview a representative sample of voters and was instructed to report back within 10 days.

The planner thought about taking a simple random sample of 300 voters, 100 interviews for herself and for each of her two assistants. However, the simple random sample plan created some time problems. The city was so spread out that an interviewer with a list of 100 voters randomly scattered around the city would have to drive an average of 18 minutes from one interview to the next. Doing so would require approximately 30 hours of driving time for each interviewer and could delay completion of the report. Obviously, simple random sampling would not do.

To save time, the planner decided to use cluster sampling. The residential portion of the city was divided into 947 blocks, each containing approximately 20 houses, as shown in Fig. 1.3.

FIGURE 1.3
A typical block of homes

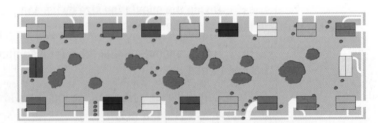

The planner numbered the blocks (clusters) on the city map from 1 to 947 and then used a table of random numbers to obtain a simple random sample of 15 of the 947 blocks. Each of the three interviewers was then assigned five of the 15 blocks obtained. This method gave each interviewer roughly 100 homes to visit but saved a great deal of travel time; an interviewer could work on a block for nearly a full day without having to drive to another neighborhood. The report was finished on time.

In the simplest case, as illustrated by Example 1.8, cluster sampling is implemented by using the following procedure.

PROCEDURE 1.2

Cluster Sampling

STEP 1 Divide the population into groups (clusters).

STEP 2 Obtain a simple random sample of the clusters.

STEP 3 Use all the members of the clusters obtained in Step 2 as the sample.

Although cluster sampling can save time and money, it does have disadvantages. Ideally, each cluster should mirror the entire population. However, that is often not the case, as members of a cluster are frequently more homogeneous than the members of the population as a whole. This situation can cause problems.

For instance, consider a simplified small town, as depicted in Fig. 1.4. The town council is thinking about building a town swimming pool. A planner for the town needs to sample voter opinion about using public funds to build the pool. Many upper-income and middle-income homeowners will probably say "No" because they own pools or can use a neighbor's. Many low-income voters will probably say "Yes" because they generally do not have access to pools.

FIGURE 1.4
Clusters for a small town

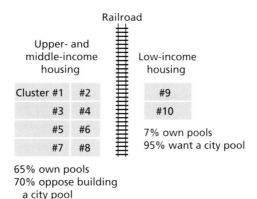

If the planner uses cluster sampling and interviews the voters of, say, three randomly selected clusters, there is a good chance that no low-income voters will be interviewed.[†] And if no low-income voters are interviewed, the results of the survey will be misleading. Suppose, for instance, that the planner obtained clusters #3, #5, and #8. Then his survey would show that only about 30% of the voters want a pool. But that is not true because more than 40% of the voters actually want a pool. The clusters that most strongly support the town swimming pool would not have been included in the survey.

In this hypothetical example, the town is so small that common sense indicates that a cluster sample may not be representative. However, in situations where there are hundreds of clusters, such problems may be quite difficult to detect.

Stratified Sampling

Another sampling method, known as **stratified sampling,** is often more reliable than cluster sampling. In stratified sampling the population is first divided into subpopulations, called **strata,** and then sampling is done from each stratum. Ideally, the members of each stratum should be homogeneous relative to the characteristic under consideration, as illustrated in Example 1.9.

▌▌ Example 1.9   Stratified Sampling

Town Swimming Pool   Consider again the town swimming pool situation. In stratified sampling, voters could be divided into three strata: upper income, middle income, and low income. A simple random sample could then be taken from each of the three strata.

This stratified sampling procedure would ensure that no income group is missed. It would also improve the precision of the statistical estimates (because the voters within each income group tend to be somewhat homogeneous) and would make it possible to estimate the separate opinions of each of the three strata.

           ▬

In stratified sampling, the strata are often sampled in proportion to their size, which is called **proportional allocation.** For instance, suppose that the strata consisting of the three income groups (upper, middle, and low) in Example 1.9 comprise, respectively, 10%, 70%, and 20% of the town. Then, for a sample size of, say, 50, the number of upper-income, middle-income, and low-income individuals sampled would be, respectively, 5 (10% of 50), 35 (70% of 50), and 10 (20% of 50).

The simplest type of stratified sampling, called **stratified random sampling with proportional allocation,** is implemented by using the following procedure.

[†]There are 120 possible three-cluster samples, and 56 of those contain neither of the low-income clusters, #9 and #10. In other words, 46.7% of the possible three-cluster samples contain neither of the low-income clusters.

> ### PROCEDURE 1.3
>
> ### Stratified Random Sampling with Proportional Allocation
>
> STEP 1 Divide the population into subpopulations (strata).
>
> STEP 2 From each stratum, obtain a simple random sample of size proportional to the size of the stratum; that is, the sample size for a stratum equals the total sample size times the stratum size divided by the population size.
>
> STEP 3 Use all the members obtained in Step 2 as the sample.

Multistage Sampling

Most large-scale surveys combine one or more of simple random sampling, systematic random sampling, cluster sampling, and stratified sampling in ways that can be quite complex. Such **multistage sampling** is used frequently by pollsters and government agencies.

For instance, the U.S. National Center for Health Statistics conducts surveys of the civilian noninstitutional U.S. population to obtain information on illnesses, injuries, and other health issues. Data collection is by a multistage probability sample of approximately 42,000 households. Information obtained from the surveys is published in the *National Health Interview Survey*.

EXERCISES 1.4

Statistical Concepts and Skills

1.33 The International 500. In Exercise 1.27 on page 16, you used simple random sampling to obtain a sample of 10 firms from *Fortune Magazine*'s list of "The International 500."
a. Use systematic random sampling to accomplish that same task.
b. Which method is easier: simple random sampling or systematic random sampling?
c. Does it seem reasonable to use systematic random sampling to obtain a representative sample? Explain your answer.

1.34 Keno. In the game of keno, 20 balls are selected at random from 80 balls, numbered 1–80. In Exercise 1.28 on page 16, you used simple random sampling to simulate one game of keno.
a. Use systematic random sampling to obtain a sample of 20 of the 80 balls.
b. Which method is easier: simple random sampling or systematic random sampling?
c. Does it seem reasonable to use systematic random sampling to simulate one game of keno? Explain your answer.

1.35 Sampling Dorm Residents. Students in the dormitories of a university in the state of New York live in clusters of four double rooms, called *suites*. There are 48 suites, with eight students per suite.
a. Describe a cluster sampling procedure for obtaining a sample of 24 dormitory residents.
b. Students typically choose friends from their classes as suitemates. With that in mind, do you think cluster sampling is a good procedure for obtaining a representative sample of dormitory residents? Explain your answer.
c. The university housing office has separate lists of dormitory residents by class level. The number of dormitory residents in each class level is as follows.

Class level	Number of dorm residents
Freshman	128
Sophomore	112
Junior	96
Senior	48

Use the table to design a procedure for obtaining a stratified sample of 24 dormitory residents. Use stratified random sampling with proportional allocation.

Extending the Concepts and Skills

1.36 In simple random sampling, all samples of a given size are equally likely. Is that true in systematic random sampling? Explain your answer.

1.37 White House Ethics. On June 27, 1996, an article appeared in *The Wall Street Journal* presenting the results of a nationwide poll regarding the White House procurement of FBI files on prominent Republicans and related ethical controversies. The article was headlined "White House Assertions on FBI Files Are Widely Rejected, Survey Shows." At the end of the article, the following explanation of the sampling procedure was given. Discuss the different aspects of sampling that appear in this explanation.

> The Wall Street Journal/NBC News poll was based on nationwide telephone interviews of 2,010 adults, including 1,637 registered voters, conducted Thursday to Tuesday by the polling organizations of Peter Hart and Robert Teeter. Questions related to politics were asked only of registered voters; questions related to economics and health were asked of all adults.
>
> The sample was drawn from 520 randomly selected geographic points in the continental U.S. Each region was represented in proportion to its population. Households were selected by a method that gave all telephone numbers, listed and unlisted, an equal chance of being included.
>
> One adult, 18 years or older, was selected from each household by a procedure to provide the correct number of male and female respondents.
>
> Chances are 19 of 20 that if all adults with telephones in the U.S. had been surveyed, the finding would differ from these poll results by no more than 2.2 percentage points in either direction among all adults and 2.5 among registered voters. Sample tolerances for subgroups are larger.

1.5 Experimental Designs

As we mentioned earlier, two methods for obtaining information other than a census are sampling and experimentation. In Sections 1.3 and 1.4, we discussed some of the basic principles and techniques of sampling. Now, we do the same for experimentation. To begin, we introduce some important terminology that further helps us differentiate among types of studies.

Observational Studies and Designed Experiments

Often the purpose of a statistical study is to investigate whether a relationship exists between two characteristics, such as smoking and lung cancer, height and weight, or educational attainment and annual income. For these kinds of studies, it is essential to distinguish between two types of procedures: observational studies and designed experiments.

In an **observational study,** researchers simply observe characteristics and take measurements, as in a sample survey. In a **designed experiment,** researchers impose treatments and controls and then observe characteristics and take measurements. Observational studies can reveal only *association*, whereas designed experiments can help establish *causation*. Examples 1.10 and 1.11 illustrate some major differences between observational studies and designed experiments.

Example 1.10 An Observational Study

Vasectomies and Prostate Cancer Approximately 450,000 vasectomies are performed each year in the United States. In this surgical procedure for contraception, the tube carrying sperm from the testicles is cut and tied.

Several studies have been conducted to analyze the relationship between vasectomies and prostate cancer. The results of one such study by E. Giovannucci et al. appeared in the paper "A Retrospective Cohort Study of Vasectomy and Prostate Cancer in U.S. Men" (*The Journal of the American Medical Association*, Vol. 269(7), pp. 878–882).

Dr. Edward Giovannucci, leader of the study and epidemiologist at Harvard-affiliated Brigham and Women's Hospital, said that "...we found 113 cases of prostate cancer among 22,000 men who had a vasectomy. This compares to a rate of 70 cases per 22,000 among men who didn't have a vasectomy."

The study shows about a 60% elevated risk of prostate cancer for men who have had a vasectomy, thereby revealing an association between vasectomy and prostate cancer. But does it establish causation: that having a vasectomy causes an increased risk of prostate cancer?

The answer is no, because the study is observational. The researchers simply observed two groups of men, one with vasectomies and the other without. Thus, although an association was established between vasectomy and prostate cancer, the association might be due to other factors (e.g., temperament) that make some men more likely to have vasectomies and also put them at greater risk of prostate cancer.

In the words of Dr. Stuart Howards, a urology professor at the University of Virginia Medical School who did not participate in the study, "...[these results] have to be considered seriously but do not prove that vasectomy causes prostate cancer."

▌▌Example 1.11   A Designed Experiment

Folic Acid and Birth Defects   For several years, evidence had been mounting that folic acid reduces major birth defects. An issue of *The Arizona Republic* reported on a Hungarian study that provided the strongest evidence to date. The results of the study, directed by Drs. Andrew E. Czeizel and Istvan Dudas of the National Institute of Hygiene in Budapest, were published in the paper "Prevention of the First Occurrence of Neural-Tube Defects by Periconceptional Vitamin Supplementation" (*The New England Journal of Medicine*, Vol. 327(26), p. 1832).

For the study, the doctors enrolled 4753 women prior to conception. The women were divided randomly into two groups. One group took daily multivitamins containing 0.8 mg of folic acid, whereas the other group received only trace elements. A drastic reduction in the rate of major birth defects occurred among the women who took folic acid: 13 per 1000 as compared to 23 per 1000 for those women who did not take folic acid.

In contrast to the observational study considered in Example 1.10, this is a designed experiment and does help establish causation. The researchers did not simply observe two groups of women but, instead, randomly assigned one group to take daily doses of folic acid and the other group to take only trace elements.

Principles of Experimental Design[†]

The study presented in Example 1.11 illustrates three basic principles of experimental design: **control, randomization,** and **replication.**

- *Control:* The doctors compared the rate of major birth defects for the women who took folic acid to that for the women who took only trace elements. This comparison controlled for such things as the *placebo effect* where subjects respond to the idea of a specific treatment rather than to the treatment itself.
- *Randomization:* The women were divided randomly into two groups to avoid unintentional selection bias in constituting the groups and thereby help eliminate the problem of potential confounding factors such as lifestyle and emotional state.
- *Replication:* A large number of women were recruited for the study to make it likely that the two groups created by randomization would be similar and also to increase the chances of detecting an effect due to the folic acid if such an effect exists.

In the folic acid study, both dosages of folic acid (0.8 mg and essentially none) are called *treatments* in the context of experimental design. Generally, each experimental condition is called a **treatment,** of which there may be several. Key Fact 1.1 summarizes our discussion about the principles of experimental design.

KEY FACT 1.1

Principles of Experimental Design

The following principles of experimental design enable a researcher to conclude that differences in the results of an experiment not reasonably attributable to chance are likely caused by the treatments.

- **Control:** Some method should be used to control for effects due to factors other than the ones of primary interest.
- **Randomization:** Subjects should be randomly divided into groups to avoid unintentional selection bias in constituting the groups, that is, to make the groups as similar as possible.
- **Replication:** A sufficient number of subjects should be used to ensure that randomization creates groups that resemble each other closely and to increase the chances of detecting differences among the treatments when such differences actually exist.

An important method of control is to compare several treatments. In fact, one of the most common experimental situations involves a specified treatment and *placebo,* an inert or innocuous medical substance. Technically, both the specified treatment and placebo are treatments. The group receiving the specified treatment is called the **treatment group,** and the group receiving placebo is called the **control group.** In the folic acid study, the women who took folic acid constituted the treatment group and those who took only trace elements constituted the control group.

[†]The remainder of this section should be covered if you plan to study the chapter *Design of Experiments and Analysis of Variance* on the WeissStats CD.

Terminology of Experimental Design

We now introduce some additional terminology used in experimental design. Each woman in the folic acid study is, in the language of experimental design, an **experimental unit,** or a **subject.** More generally, we have the following definition.

DEFINITION 1.5

Experimental Units; Subjects

In a designed experiment, the individuals or items on which the experiment is performed are called *experimental units.* When the experimental units are humans, the term *subject* is often used in place of experimental unit.

In the folic acid study, the researchers were interested in the effect of folic acid on major birth defects. Birth-defect classification (major or not) is the **response variable** for this study. The daily dosage of folic acid is called the **factor.** In this case, the factor has two **levels,** namely, 0.8 mg and essentially none.

When there is only one factor, as in the folic acid study, the treatments are the same as the levels of the factor. But, if there is more than one factor, each treatment is a combination of levels of the various factors. Example 1.12 presents an experiment in which there are two factors.

Example 1.12 Experimental Design

Weight Gain of Golden Torch Cacti   The Golden Torch Cactus (botanical name, *Trichocereus spachianus*), a columnar cactus native to Argentina, has excellent landscape potential. William Feldman and Frank Crosswhite, two researchers at the Boyce Thompson Southwestern Arboretum, conducted a thorough investigation of the optimal method for producing these cacti.

The researchers examined, among other things, the effects of a hydrophilic polymer and irrigation regime on weight gain. Hydrophilic polymers are used as soil additives to keep moisture in the root zone. For this study, the researchers chose Broadleaf P-4 polyacrylamide, abbreviated P4. The hydrophilic polymer was either used or not used, and five irrigation regimes were employed: none, light, medium, heavy, and very heavy. Identify the

a. experimental units.   b. response variable.   c. factors.

d. levels of each factor.   e. treatments.

Solution

a. The experimental units are the cacti used in the study.

b. The response variable is weight gain.

c. The factors are hydrophilic polymer and irrigation regime.

d. Hydrophilic polymer has two levels: with and without. Irrigation regime has five levels: none, light, medium, heavy, and very heavy.

e. Each treatment is a combination of a level of hydrophilic polymer and a level of irrigation regime. Table 1.8 depicts the treatments. In the table, we abbreviated "very heavy" as "Xheavy."

TABLE 1.8

Schematic for the 10 treatments in the cactus study

		Irrigation regime				
		None	Light	Medium	Heavy	Xheavy
Polymer	No P4	No water No P4 (Treatment 1)	Light water No P4 (Treatment 2)	Medium water No P4 (Treatment 3)	Heavy water No P4 (Treatment 4)	Xheavy water No P4 (Treatment 5)
	With P4	No water With P4 (Treatment 6)	Light water With P4 (Treatment 7)	Medium water With P4 (Treatment 8)	Heavy water With P4 (Treatment 9)	Xheavy water With P4 (Treatment 10)

Note that there are 10 different treatments for this experiment. ▬

We now present formal definitions of several important terms used in experimental design that we introduced earlier.

DEFINITION 1.6

Response Variable, Factors, Levels, and Treatments

Response variable: The characteristic of the experimental outcome that is to be measured or observed.

Factor: A variable whose effect on the response variable is of interest in the experiment.

Levels: The possible values of a factor.

Treatment: Each experimental condition. For one-factor experiments, the treatments are the levels of the single factor. For multifactor experiments, each treatment is a combination of levels of the factors.

Statistical Designs

Once we have chosen the treatments, we must decide how the experimental units are to be assigned to the treatments (or vice versa). The women in the folic acid study were randomly divided into two groups; one group received folic acid and the other only trace elements. In the cactus study, 40 cacti were divided randomly into 10 groups of four cacti each and then each group was assigned a different treatment from among the 10 depicted in Table 1.8. Both of these experiments involved the use of a **completely randomized design.**

DEFINITION 1.7

Completely Randomized Design

In a *completely randomized design,* all the experimental units are assigned randomly among all the treatments.

The completely randomized design is one of the most commonly used and simplest designs, but it is not always the best design. There are several alternatives to the completely randomized design.

For instance, in a **randomized block design,** experimental units that are similar in ways that are expected to affect the response variable are grouped in **blocks.** Then the random assignment of experimental units to the treatments is made block by block.

DEFINITION 1.8

Randomized Block Design

In a *randomized block design,* the experimental units are assigned randomly among all the treatments separately within each block.

In Example 1.13, we contrast completely randomized designs and randomized block designs.

Example 1.13 Statistical Designs

Golf Ball Driving Distances   Suppose that we want to compare the driving distances for five different brands of golf ball. For 40 golfers, discuss a method of comparison based on

a. a completely randomized design.   **b.** a randomized block design.

Solution   Here the experimental units are the golfers, the response variable is driving distance, the factor is brand of golf ball, and the levels (and treatments) are the five brands.

a. For a completely randomized design, we would randomly divide the 40 golfers into five groups of 8 golfers each and then randomly assign each group to drive a different brand of ball, as illustrated in Fig. 1.5.

FIGURE 1.5
Completely randomized design for golf ball experiment

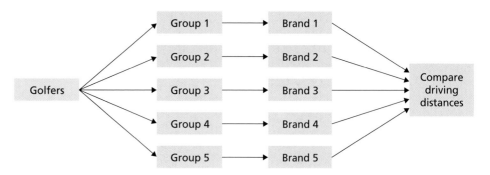

b. As driving distance is affected by gender, using a randomized block design, with blocking by gender, is probably a better approach. We could do so with 40 golfers, say, 20 men and 20 women. We would randomly divide the 20 men into five groups of 4 men each and then randomly assign each group of men to drive a different brand of ball, as shown in Fig. 1.6. Likewise, we would randomly divide the 20 women into five groups of 4 women each and then randomly assign each group of women to drive a different brand of ball, as also shown in Fig. 1.6.

FIGURE 1.6

Randomized block design for golf ball experiment

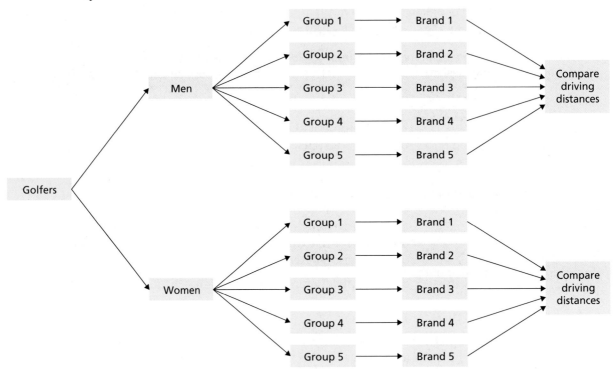

By blocking, we can isolate and remove the variation in driving distances between men and women and thereby make it easier to detect differences in driving distances among the five brands of golf ball, if such differences exist. Additionally, blocking permits us to analyze separately the differences in driving distances among the five brands for men and women.

As illustrated in Example 1.13, blocking can be used to isolate and remove systematic differences among blocks, thereby making differences among treatments easier to detect when such differences exist. Blocking also makes possible the separate analysis of treatment effects on each block.

In this section, we introduced some of the basic terminology and principles of experimental design. However, we have just scratched the surface of this vast and important topic to which entire courses and books are devoted. Further discussion of experimental design is provided in the chapter *Design of Experiments and Analysis of Variance* on the WeissStats CD accompanying this book or on the Weiss Web site, www.aw-bc.com/weiss.

EXERCISES 1.5

Statistical Concepts and Skills

1.38 Define
a. observational study.   b. designed experiment.

1.39 Fill in the following blank. Observational studies can reveal only association, whereas designed experiments can help establish _____.

*1.40 State and explain the significance of the three basic principles of experimental design.

1.41 Folic Acid and Birth Defects. In the folic acid study, 4753 women were enrolled by the doctors. Explain how a table of random numbers could be used to divide the women randomly into two groups, one of size 2376 and the other of size 2377.

1.42 Vasectomies and Prostate Cancer. Refer to the vasectomy/prostate cancer study discussed in Example 1.10 on page 22.
a. How could the study be modified to make it a designed experiment?
b. Comment on the feasibility of the designed experiment that you described in part (a).

1.43 The Salk Vaccine. In the 1940s and early 1950s, there was great public concern over epidemics of polio. In an attempt to alleviate this serious problem, Jonas Salk of the University of Pittsburgh developed a vaccine for polio. Various preliminary experiments indicated that the vaccine was safe and potentially effective. Nonetheless, a large-scale study was needed to determine whether the vaccine would truly work. A test involving nearly 2 million grade-school children was devised. All the children were inoculated, but only half received the Salk vaccine; the other half were given placebo, in this case an injection of salt dissolved in water. Neither the children nor the doctors performing the diagnoses knew which children belonged to which group. Instead, an evaluation center kept records of who received the Salk vaccine and who did not. The center found that

the incidence of polio was far less among the children inoculated with the Salk vaccine. From that information, the researchers concluded that the Salk vaccine would be effective in preventing polio for all U.S. schoolchildren and, consequently, it was then made available for general use. Was this investigation an observational study or a designed experiment? Justify your answer.

1.44 Do Left-Handers Die Earlier? According to a study published in the *Journal of the American Public Health Association*, left-handed people do not die at an earlier age than right-handed people, contrary to the conclusion of a highly publicized report done 2 years earlier. The investigation involved a 6-year study of 3800 people in East Boston older than age 65. Researchers at Harvard University and the National Institute of Aging found that the "lefties" and "righties" died at exactly the same rate. "There was no difference, period," said Dr. Jack Guralnik, an epidemiologist at the institute and one of the coauthors of the report. Was this investigation an observational study or a designed experiment? Justify your answer.

*1.45 In a designed experiment,
a. what are the experimental units?
b. if the experimental units are humans, what term is often used in place of experimental unit?

*1.46 Adverse Effects of Prozac. Prozac (fluoxetine hydrochloride), a product of Eli Lilly and Company, is used for the treatment of depression, obsessive–compulsive disorder (OCD), and bulimia nervosa. An issue of the magazine *Arthritis Today* contained an advertisement reporting on the "...treatment-emergent adverse events that occurred in 2% or more patients treated with Prozac and with incidence greater than placebo in the treatment of depression, OCD, or bulimia." In the study, 2444 patients took Prozac and 1331 patients were given placebo. Identify the
a. treatment group.   b. control group
c. treatments.

In Exercises 1.47–1.50, we present descriptions of designed experiments. In each case, identify the
a. experimental units.
b. response variable.
c. factor(s).
d. levels of each factor.
e. treatments.

*1.47 Lifetimes of Flashlight Batteries. To compare the lifetimes of four brands of flashlight battery, 20 flashlights were used. The 20 flashlights were randomly divided into four groups of 5 flashlights each. Then each group of flashlights was equipped with a different brand of battery.

*1.48 Storage of Perishable Items. Storage of perishable items is an important concern for many companies. One study examined the effects of storage time and storage temperature on the deterioration of a particular item. Three different storage temperatures and five different storage times were used.

*1.49 Increasing Unit Sales. Supermarkets are interested in strategies to increase temporarily the unit sales of a product. In one study, researchers compared the effect of display type and price on unit sales for a particular product. The following display types and pricing schemes were employed.

- Display types: normal display space interior to an aisle, normal display space at the end of an aisle, and enlarged display space.
- Pricing schemes: regular price, reduced price, and cost.

*1.50 Oat Yield and Manure. In a classic study, described by F. Yates in *The Design and Analysis of Factorial Experiments*, the effect on oat yield was compared for three different varieties of oats and four different concentrations of manure (0, 0.2, 0.4, and 0.6 cwt per acre).

*1.51 Lifetimes of Flashlight Batteries. Refer to Exercise 1.47. Did this study utilize a completely randomized design or a randomized block design? Explain your answer.

*1.52 Lifetimes of Flashlight Batteries. Refer to Exercise 1.47. Suppose that we compare the lifetimes of the four brands of flashlight battery by using 20 flashlights, five different brands of 4 flashlights each. Each of the 4 flashlights of a given brand uses a different brand of battery. Does this study utilize a completely randomized design or a randomized block design? Explain your answer.

Extending the Concepts and Skills

1.53 The Salk Vaccine. In Exercise 1.43, we discussed the Salk vaccine experiment. The experiment utilized a technique called *double-blinding* because neither the children nor the doctors involved knew which children had been given the vaccine and which had been given placebo. Explain the advantages of using double-blinding in the Salk vaccine experiment.

*1.54 In sampling from a population, state which type of sampling design corresponds to each of the following experimental designs:
a. completely randomized design.
b. randomized block design.

Using Technology

1.55 Folic Acid and Birth Defects. Use the technology of your choice to carry out the process described in Exercise 1.41.

CHAPTER REVIEW

You Should Be Able To

1. classify statistical studies as either descriptive or inferential.

2. identify the population and the sample in an inferential study.

3. explain what is meant by a representative sample.

4. describe simple random sampling.

5. use a table of random numbers to obtain a simple random sample.

*6. describe systematic random sampling, cluster sampling, and stratified sampling.

7. explain the difference between an observational study and a designed experiment.

8. classify a study concerning the relationship between two variables or characteristics as either an observational study or a designed experiment.

*9. state the three basic principles of experimental design.

*10. identify the treatment group and control group in a study.

*11. identify the experimental units, response variable, factor(s), levels of each factor, and treatments in a designed experiment.

*12. distinguish between a completely randomized design and a randomized block design.

Key Terms

blocks,* 27
census, 11
cluster sampling,* 18
completely randomized design,* 26
control,* 24
control group,* 24
descriptive statistics, 4
designed experiment, 22
experimental unit,* 25
experimentation, 11
factor,* 26
inferential statistics, 6
levels,* 26
multistage sampling,* 21

observational study, 22
population, 5
probability sampling, 11
proportional allocation,* 20
randomization,* 24
randomized block design,* 27
replication,* 24
representative sample, 11
response variable,* 26
sample, 5
sampling, 11
simple random sample, 12
simple random sampling, 12

simple random sampling
    with replacement, 12
simple random sampling
    without replacement, 12
strata,* 20
stratified random sampling with
    proportional allocation,* 20
stratified sampling,* 20
subject,* 25
systematic random sampling,* 17
table of random numbers, 13
treatment,* 26
treatment group,* 24

REVIEW TEST

Statistical Concepts and Skills

1. In a newspaper or magazine, or on the Internet, find an example of
a. a descriptive study.   **b.** an inferential study.

2. Almost any inferential study involves aspects of descriptive statistics. Explain why.

3. Baseball Scores. On June 29, 2003, the following baseball scores were printed in *The Daily Courier*. Is this study descriptive or inferential? Explain your answer.

Interleague Boxes	
Reds 5, Indians 4	White Sox 7, Cubs 6
Yankees 7, Mets 1	Marlins 10, Red Sox 9
Yankees 9, Mets 8	Expos 4, Blue Jays 2
Twins 5, Brewers 2	Giants 8, A's 7
Devil Rays 9, Braves 7	Phillies 9, Orioles 5
Cardinals 13, Royals 9	Angels 3, Dodgers 1
Padres 6, Mariners 0	Astros 2, Rangers 0

4. Serious Energy Situation. In a *USA TODAY/CNN Gallup Poll*, conducted in May of 2001, 94% of those surveyed said that the United States faced a serious energy situation, but, by 47% to 35%, they preferred an emphasis on conservation rather than on more production. Is this study descriptive or inferential? Explain your answer.

5. Before planning and conducting a study to obtain information, what should be done?

6. Explain the meaning of
a. a representative sample.
b. probability sampling.
c. simple random sampling.

7. Incomes of College Students' Parents. A researcher wants to estimate the average income of parents of college students. To accomplish that, he surveys a sample of 250 students at Yale. Is this a representative sample? Explain your answer.

8. Which of the following sampling procedures involve the use of probability sampling?
a. A college student is hired to interview a sample of voters in her town. She stays on campus and interviews 100 students in the cafeteria.
b. A pollster wants to interview 20 gas station managers in Baltimore. He posts a list of all such managers on his wall, closes his eyes, and tosses a dart at the list 20 times. He interviews the people whose names the dart hits.

9. On-Time Airlines. According to the U.S. Department of Transportation, the five airlines with the highest percentage of on-time arrivals for 2002 were American (AA), United (UA), U.S. Airways (US), America West (HP), and Continental (CO).
a. List the 10 possible samples (without replacement) of size 3 that can be obtained from the population of five airlines.
b. If a simple random sampling procedure is used to obtain a sample of three of these five airlines, what are the chances that it is the first sample on your list in part (a)? the second sample? the tenth sample?

10. Top North American Athletes. As part of ESPN's *SportsCenturyRetrospective*, a panel chosen by ESPN ranked the top 100 North American athletes of the twentieth century. For a class project, you are to obtain a simple random sample of 15 of these 100 athletes and briefly describe their athletic feats.

a. Explain how you can use Table I in Appendix A to obtain the simple random sample.
b. Starting at the three-digit number in line number 10 and column numbers 7–9 of Table I, read down the column, up the next, and so on, to find 15 numbers that you can use to identify the athletes to be considered.

*11. Describe each of the following sampling methods and indicate conditions under which each is appropriate.
a. Systematic random sampling
b. Cluster sampling
c. Stratified random sampling with proportional allocation

*12. Top North American Athletes. Refer to Problem 10.
a. Use systematic random sampling to obtain a sample of 15 athletes.
b. In this case, is systematic random sampling an appropriate alternative to simple random sampling? Explain your answer.

*13. Surveying the Faculty. The faculty of a college consists of 820 members. A new president has just been appointed. The president wants to get an idea of what the faculty considers the most important issues currently facing the school. She does not have time to interview all the faculty members and so decides to stratify the faculty by rank and use stratified random sampling with proportional allocation to obtain a sample of 40 faculty members. There are 205 full professors, 328 associate professors, 246 assistant professors, and 41 instructors.
a. How many faculty members of each rank should be selected for interviewing?
b. Use Table I in Appendix A to obtain the required sample. Explain your procedure in detail.

14. QuickVote. *TalkBack Live* conducts online surveys on various issues. The photo at the top of the next column shows the result of a *quickvote* taken on July 5, 2000, that asked whether a person would vote for a third-party candidate. Beneath the vote tally is a statement regarding the sampling procedure. Discuss this statement in light of what you have learned in this chapter.

15. Regarding observational studies and designed experiments,
a. describe each type of study.
b. with respect to possible conclusions, what important difference exists between the two types of study?

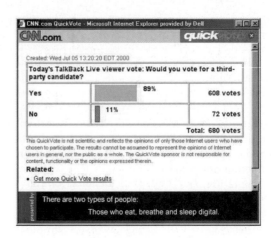

16. Persistent Poverty and IQ. An article appearing in an issue of *The Arizona Republic* reported on a study conducted by Greg Duncan of the University of Michigan. According to the report, "Persistent poverty during the first 5 years of life leaves children with IQs 9.1 points lower at age 5 than children who suffer no poverty during that period...." Is this an observational study or is it a designed experiment? Explain your answer.

17. AVONEX and MS. An issue of *Inside MS* contained an article describing AVONEX (Interferon beta-1a), a drug used in the treatment of relapsing forms of multiple sclerosis. Included in the article was a report on "...adverse events and selected laboratory abnormalities that occurred at an incidence of 2% or more among the 158 multiple sclerosis patients treated with 30 mcg of AVONEX once weekly by IM injection." In the study, 158 patients took AVONEX and 143 patients were given placebo.
a. Is this study observational or is it a designed experiment?
*b. Identify the treatment group, control group, and treatments.

*18. Identify and explain the significance of the three basic principles of experimental design.

*19. Doughnuts and Fat. A classic study, conducted in 1935 by B. Lowe, analyzed differences in the amount of fat absorbed by doughnuts in cooking with four different fats. For the experiment, 24 batches of doughnuts were randomly divided into four groups of 6 batches each. The four groups were then randomly assigned to the four fats. Identify the
a. experimental units.   b. response variable.
c. factor(s).   d. levels of each factor.
e. treatments.

*20. Plant Density and Tomato Yield. In the paper "Effects of Plant Density on Tomato Yields in Western Nigeria" (*Experimental Agriculture*, Vol. 12(1), pp. 43–47), B. Adelana reported on the effect of tomato variety and planting density on yield. Identify the

a. experimental units.   b. response variable.

c. factor(s).   d. levels of each factor.

e. treatments.

*21. Comparing Gas Mileages. An experiment is to be conducted to compare four different brands of gasoline for gas mileage.

a. Suppose that you randomly divide 24 cars into four groups of 6 cars each and then randomly assign the four groups to the four brands of gasoline, one group per brand. Is this experimental design a completely randomized design or a randomized block design? If it is the latter, what are the blocks?

b. Suppose, instead, that you use six different models of car whose varying characteristics (e.g., weight and horsepower) affect gas mileage. Four cars of each model are randomly assigned to the four different brands of gasoline. Is this experimental design a completely randomized design or a randomized block design? If it is the latter, what are the blocks?

c. Which design is better, the one in part (a) or the one in part (b)? Explain your answer.

22. *USA Today* **Polls.** The following explanation of *USA TODAY* polls and surveys was obtained from the *USA TODAY* Web site. Discuss the explanation in detail.

> USATODAY.com frequently publishes the results of both scientific opinion polls and online reader surveys. Sometimes the topics of these two very different types of public opinion sampling are similar but the results appear very different. It is important that readers understand the difference between the two.
>
> USA TODAY/CNN/Gallup polling is a scientific phone survey taken from a random sample of U.S. residents and weighted to reflect the population at large. This is a process that has been used and refined for more than 50 years. Scientific polling of this type has been used to predict the outcome of elections with considerable accuracy.
>
> Online surveys, such as USATODAY.com's "Quick Question," are not scientific and reflect the views of a self-selected slice of the population. People using the Internet and answering online surveys tend to have different demographics than the nation as a whole and as such, results will differ---sometimes dramatically---from scientific polling.
>
> USATODAY.com will clearly label results from the various types of surveys for the convenience of our readers.

23. Crosswords and Dementia. An article appearing in the *Los Angeles Times* on June 21, 2003, discussed a report by researchers from a recent issue of the *New England Journal of Medicine*. The article, titled "Crosswords Reduce Risk of Dementia," stated that "Elderly people who frequently read, do crossword puzzles, practice a musical instrument or play board games cut their risk of Alzheimer's and other forms of dementia by nearly two-thirds compared with people who seldom do such activities…" Comment on the statement in quotes, keeping in mind the type of study for which causation can be reasonably inferred.

Using Technology

24. Top North American Athletes. Refer to Problem 10. Use the technology of your choice to obtain 15 random numbers that can be used to specify a simple random sample of 15 athletes.

StatExplore in MyMathLab
Analyzing Data Online

Now, with MyMathLab, you can access StatExplore online statistical software. StatExplore offers an easy-to-use interface and is an alternative to commercial statistical software. It is customized for this textbook and is available for your use at no charge. To use StatExplore, you need a Java-capable Web browser, such as Internet Explorer or Netscape Navigator. You can access StatExplore from the student content area of your Weiss MyMathLab course.

At the end of each chapter of the book, we illustrate the use of StatExplore to perform a statistical analysis discussed in the chapter. For best results, you should implement the steps we present and thereby obtain the StatExplore output for yourself. You will also be asked to apply StatExplore further to conduct other statistical analyses examined in each chapter.

Most applications will require you to load data into the StatExplore data table (worksheet). Several methods are available for doing that. Here, we explain how to load data into the StatExplore data table from your Weiss MyMathLab course and from Excel files on the WeissStats CD. (*Note:* All appropriate data sets in this book are contained in Excel files on the WeissStats CD.)

EXAMPLE **Loading Data**

Table 1.9 provides data on age and price for a sample of 11 cars. Ages are in years; prices are in hundreds of dollars, rounded to the nearest hundred dollars. Load these data into the StatExplore data table.

TABLE 1.9
Age and price for a sample of 11 cars

Age (yr)	Price ($100)
5	85
4	103
6	70
5	82
5	89
5	98
6	66
6	95
2	169
7	70
7	48

SOLUTION The age and price data in Table 1.9 are in the file Tb01-09. To load these data into the StatExplore data

table from the StatExplore page in MyMathLab, simply click the link Tb01-09.

To load the age and price data into the StatExplore data table from the WeissStats CD, first place the WeissStats CD into your CD drive and then proceed as follows:

1 Choose **Data ➤ Load data ➤ from file**
2 Click the **My computer** link
3 Click the **Browse...** button
4 Navigate to the folder
   `Data Sets\Excel Files\Examples\Chapter 01`
   on the WeissStats CD and double click the
   file `Tb01-09.xls`
5 Click the **Load File** button

The age and price data are now available for performing any required statistical analysis in StatExplore.

To access StatExplore, go to the student content area of your Weiss MyMathLab course.

Internet Project

The Internet Projects Page for your book, *Introductory Statistics*, 7th ed., is a location on the Internet designed to help you understand statistics. The starting Internet address (URL) for the page is www.aw-bc.com/weiss. From this Web page, you can reach the Internet Projects Page. We suggest that you bookmark the Internet Projects Page for easy access in the future.

Each chapter in the book includes an Internet project that provides a set of simulations, demonstrations, and other activities as a supplement to those found in the text. The material comes from universities, individuals, and companies from all over the world.

The Titanic Disaster

In this first Internet project, you will become acquainted with the controversy and data surrounding the Titanic disaster. You will visit Web sites at several locations, viewing and thinking about the data and facts presented there. You will also see statistical demonstrations and animations that illustrate the concepts you have covered in class.

In 1912, the Titanic was the largest, most luxurious, and most technologically advanced liner in the world. At 11:40 P.M. on Sunday, April 14 of that year, the Titanic struck an iceberg. In $2\frac{1}{2}$ hours, the liner that many thought unsinkable went down. Only 705 of her 2224 passengers and crew were saved.

Some believe that the rescue procedures used that night unfairly favored the wealthier passengers. In this project, you will use statistical thinking to explore the data from the disaster to arrive at your own (statistically informed) conclusion.

Focusing on Data Analysis

The Focus Database

The file Focus.txt in the Focus Database folder of the WeissStats CD contains information on the undergraduate students at the University of Wisconsin - Eau Claire (UWEC). Those students will constitute the population of interest in the *Focusing on Data Analysis* sections that appear at the end of each chapter of the book.[†]

Thirteen variables are considered: sex (SEX), high school percentile (HSP), cumulative GPA (GPA), age (AGE), total earned credits (CREDITS), classification (CLASS), school/college (COLLEGE), primary major (MAJOR), residency (RESIDENCY), admission type (TYPE), ACT English score (ENGLISH), ACT math score (MATH), and ACT composite score (COMP). We call the database of information for those variables the *Focus database*.

Large data sets are almost always analyzed by computer, and that is how you should handle the data sets in the Focus database. We have supplied the Focus database in several file formats, namely, those used by some of the most common statistical technologies, including Minitab (.mtw), Excel (.xls), and SPSS (.sav).

If you use a different statistical technology, you should (1) input the file Focus.txt into that technology, (2) name the variables as indicated above parenthetically, and (3) save the worksheet to a file named Focus in the format suitable to your technology (i.e., with the appropriate file extension). Now, any time that you want to analyze the Focus database, you can simply retrieve your Focus worksheet.

[†]We have restricted attention to those undergraduate students at UWEC with complete records for all the variables under consideration.

Case Study Discussion

Top Films of All Time

At the beginning of this chapter, we discussed the results of a survey by the American Film Institute (AFI). Now that you have learned some of the basic terminology of statistics, we want you to examine that survey in greater detail.

Answer each of the following questions pertaining to the survey. In doing so, you may want to reread the description of the survey given on page 3.

a. Identify the population.
b. Identify the sample.
c. Is the sample representative of the population of all U.S. moviegoers? Explain your answer.
d. Consider the following statement: "Among the 1500 filmmakers, actors, critics, politicians, and film historians polled by AFI, the top-ranking film was *Citizen Kane.*" Is this statement descriptive or inferential? Explain your answer.
e. Suppose that the statement in part (d) is changed to: "Based on the AFI poll, *Citizen Kane* is the top-ranking film among all filmmakers, actors, critics, politicians, and film historians." Is this statement descriptive or inferential? Explain your answer.

Internet Resources: Visit the Weiss Web site www.aw-bc.com/weiss for additional discussion, exercises, and resources related to this case study.

Biography

FLORENCE NIGHTINGALE: Lady of the Lamp

Florence Nightingale (1820–1910), the founder of modern nursing, was born in Florence, Italy, into a wealthy English family. In 1849, over the objections of her parents, she entered the Institution of Protestant Deaconesses at Kaiserswerth, Germany, which "…trained country girls of good character to nurse the sick."

The Crimean War began in March, 1854, when England and France declared war on Russia. After serving as superintendent of the Institution for the Care of Sick Gentlewomen in London, Nightingale was appointed by the English Secretary of State at War, Sidney Herbert, to be in charge of 38 nurses who were to be stationed at military hospitals in Turkey.

Nightingale found the conditions in the hospitals appalling—overcrowded, filthy, and without sufficient facilities. In addition to the administrative duties she undertook to alleviate those conditions, she spent many hours tending patients. After 8:00 P.M. she allowed none of her nurses in the wards, but made rounds herself every night, a deed that earned her the epithet Lady of the Lamp.

Nightingale was an ardent believer in the power of statistics and used statistics extensively to gain an understanding of social and health issues. She lobbied to introduce statistics into the curriculum at Oxford and invented the coxcomb chart, a type of pie chart. Nightingale felt that charts and diagrams were a means of making statistical information understandable to people who would otherwise be unwilling to digest the dry numbers.

In May 1857, as a result of Nightingale's interviews with officials ranging from the Secretary of State to Queen Victoria herself, the Royal Commission on the Health of the Army was established. Under the auspices of the commission, the Army Medical

School was founded. In 1860, Nightingale used a fund set up by the public to honor her work in the Crimean War to create the Nightingale School for Nurses at St. Thomas's Hospital. During that same year, at the International Statistical Congress in London, she authored one of the three papers discussed in the Sanitary Section and also met Adolphe Quetelet (see Chapter 2 biography) who had greatly influenced her work.

After 1857, Nightingale lived as an invalid, although it has never been determined that she had any specific illness. In fact, many speculated that her invalidism was a stratagem she employed to devote herself to her work.

Nightingale was elected an Honorary Member of the American Statistical Association in 1874. In 1907, she was presented the Order of Merit for meritorious service by King Edward VII; she was the first woman to receive that award.

Florence Nightingale died in 1910. An offer of a national funeral and burial at Westminster Abbey was declined, and, according to her wishes, Nightingale was buried in the family plot in East Mellow, Hampshire, England.

Descriptive Statistics

Organizing Data

CHAPTER OUTLINE

GENERAL OBJECTIVES In Chapter 1, we introduced two major interrelated branches of statistics: *descriptive statistics* and *inferential statistics*. As you discovered there, descriptive statistics consists of methods for organizing and summarizing information clearly and effectively.

In this chapter, you begin your study of descriptive statistics. As a prerequisite, in Section 2.1, we show you how to classify data by type. Data type is often an important factor in selecting the correct statistical method, both in descriptive statistics and inferential statistics.

In Section 2.2, we explain how to group data so that they are easier to work with and understand. In Section 2.3, we demonstrate various classical ways to portray data graphically, thus providing a "picture" of the data. In Section 2.4, we introduce stem-and-leaf diagrams—one of an arsenal of statistical tools known collectively as *exploratory data analysis.*

In Section 2.5, we discuss the identification of the shape of a data set, an important aspect of both descriptive and inferential statistics. And, in Section 2.6, we present some tips for avoiding confusion when you read and interpret graphical displays.

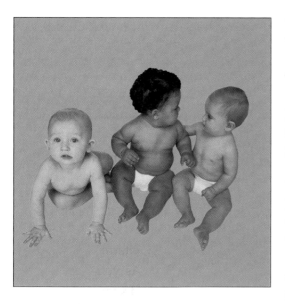

Preventing Infant Mortality

Infant mortality is concerned with infant deaths during the first year of life. Generally, the infant mortality rate (IMR) provides the number of such deaths per 1000 live births during a calendar year. In 1987, the U.S. Congress established the National Commission to Prevent Infant Mortality, whose charge is to create a national strategy for reducing the infant mortality rate of the United States.

From the *Statistical Abstract of the United States*, we obtained information on 2001 infant mortality rates for nations. If we rank those nations according to their infant mortality rates, the 30 with the lowest rates are as shown in the following table.

Rank	Country	IMR	Rank	Country	IMR
1	Sweden	3.5	16	Ireland	5.5
2	Singapore	3.6	17	United Kingdom	5.5
3	Finland	3.8	18	Czech Republic	5.6
4	Japan	3.9	19	Hong Kong	5.8
5	Norway	3.9	20	Italy	5.8
6	Netherlands	4.4	21	Portugal	5.9
7	Austria	4.4	22	New Zealand	6.3
8	France	4.5	23	Greece	6.4
9	Switzerland	4.5	24	**United States**	**6.8**
10	Belgium	4.7	25	Taiwan	6.9
11	Germany	4.7	26	Croatia	7.2
12	Spain	4.9	27	Cuba	7.4
13	Australia	5.0	28	South Korea	7.7
14	Canada	5.0	29	Israel	7.7
15	Denmark	5.0	30	Hungary	9.0

At the end of this chapter, you will be asked to revisit the information presented here and apply some of your newly learned statistical skills to identify and analyze various aspects of these infant mortality rates.

2.1 Variables and Data

A characteristic that varies from one person or thing to another is called a **variable.** Examples of variables for humans are height, weight, number of siblings, sex, marital status, and eye color. The first three of these variables yield numerical information and are examples of **quantitative variables;** the last three yield nonnumerical information and are examples of **qualitative variables,** also called **categorical variables.**

Quantitative variables can be classified as either *discrete* or *continuous*. A **discrete variable** is a variable whose possible values can be listed, even though the list may continue indefinitely. Mathematically, the numbers that make sense for values of a discrete variable form a finite or countably infinite set, usually some collection of whole numbers. "Number of siblings" is an example of a discrete variable. A discrete variable usually involves a count of something, such as the number of siblings a person has, the number of cars owned by a family, or the number of students in an introductory statistics class.

A **continuous variable** is a variable whose possible values form some interval of numbers—all numbers within the interval make sense for values of the variable. "Height" is an example of a continuous variable. Typically, a continuous variable involves a measurement of something, such as the height of a person, the weight of a newborn baby, or the length of time a car battery lasts.

The preceding discussion is summarized graphically in Fig. 2.1 and verbally in the following definition.

DEFINITION 2.1

What
Does it Mean?

A discrete variable usually involves a count of something, whereas a continuous variable usually involves a measurement of something.

Variables

Variable: A characteristic that varies from one person or thing to another.

Qualitative variable: A nonnumerically valued variable.[†]

Quantitative variable: A numerically valued variable.

Discrete variable: A quantitative variable whose possible values form a finite (or countably infinite) set of numbers.

Continuous variable: A quantitative variable whose possible values form some interval of numbers.

FIGURE 2.1
Types of variables

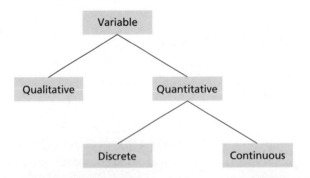

[†]Values of a qualitative variable are sometimes coded with numbers—for example, zip codes, which represent geographical locations. Note that it is not meaningful to do arithmetic with such numbers, in contrast to numbers obtained from a quantitative variable.

Observing the values of a variable for one or more people or things yields **data.** Thus the information collected, organized, and analyzed by statisticians is data. As for variables, data can be classified as **qualitative data, quantitative data, discrete data,** and **continuous data.**

DEFINITION 2.2

What
Does it Mean?

Data are classified according to the type of variable from which they were obtained.

Data

Data: Information obtained by observing values of a variable.

Qualitative data: Data obtained by observing values of a qualitative variable.

Quantitative data: Data obtained by observing values of a quantitative variable.

Discrete data: Data obtained by observing values of a discrete variable.

Continuous data: Data obtained by observing values of a continuous variable.

Each individual piece of data is called an **observation,** and the collection of all observations for a particular variable is called a **data set.**[†] We illustrate various types of variables and data in Examples 2.1–2.4.

Example 2.1   **Variables and Data**

The 107th Boston Marathon   At noon on April 21, 2003, more than 20,000 men and women set out to run from Hopkinton Center to the John Hancock Building in Boston. Their run, covering 26 miles and 385 yards, would be watched by thousands of people lining the streets leading into Boston and by millions more on television. It was the 107th running of the Boston Marathon.

A great deal of information was accumulated and recorded that afternoon by the Boston Athletic Association. Robert Cheruiyot of Kenya won the men's competition with a time of 2 hours, 10 minutes, and 11 seconds. The winner of the women's competition was Svetlana Zakharova of Russia, with a time of 2 hours, 25 minutes, and 20 seconds.

The Boston Marathon provides examples of different types of variables and data. The simplest type of variable is illustrated by the classification of each entrant as either male or female. "Sex" is a qualitative variable because its possible values (male or female) are nonnumerical. Thus, for instance, the information that Robert Cheruiyot is a male is qualitative data—data obtained by observing the value of the variable "sex" for Robert Cheruiyot.

Most racing fans are interested in the places of the finishers. "Place of finish" is a quantitative variable, which is also a discrete variable because it makes sense to talk only about first place, second place, and so on—there are only a finite number of possible finishing places. The information that, among the

---
[†]Sometimes *data set* is used to refer to all the data for all the variables under consideration.

women, Svetlana Zakharova and Lyubov Denisova finished first and second, respectively, and Jill Gaitenby and Esther Kiplagat finished ninth and tenth, respectively, is discrete, quantitative data—data obtained by observing the values of the variable "place of finish" for the four runners.

More can be learned about what happened in a race by looking at the times of the finishers. For instance, Svetlana Zakharova finished 1 minute and 31 seconds ahead of Lyubov Denisova, whereas Jill Gaitenby beat Esther Kiplagat by only 24 seconds. "Finishing time" is a quantitative variable, which is also a continuous variable because the finishing time of a runner can conceptually be any positive number. The information that Robert Cheruiyot ran his race in 2:10:11 and Svetlana Zakharova ran hers in 2:25:20 is continuous, quantitative data—data obtained by observing the values of the variable "finishing time" for Robert Cheruiyot and Svetlana Zakharova.

Example 2.2   Variables and Data

Human Blood Types   Human beings have one of four blood types: A, B, AB, or O. What kind of data do you receive when you are told your blood type?

Solution   Your blood type is qualitative data—data obtained by observing your value of the variable "blood type."

Example 2.3   Variables and Data

Household Size   The U.S. Bureau of the Census collects data on household size and publishes the information in *Current Population Reports*. What kind of data is the number of people in your household?

Solution   The number of people in your household is discrete, quantitative data—data obtained by observing the value of the variable "household size" for your particular household.

Example 2.4   Variables and Data

The World's Highest Waterfalls   The *Information Please Almanac* lists the world's highest waterfalls. The list shows that Angel Falls in Venezuela is 3281 feet high, or more than twice as high as Ribbon Falls in Yosemite, California, which is 1612 feet high. What kind of data are these heights?

Solution   The waterfall heights are continuous, quantitative data—data obtained by observing the values of the variable "height" for the two waterfalls.

Classification and the Choice of a Statistical Method

Some of the descriptive and inferential procedures that you will study are valid for only certain types of data. This limitation is one reason why you must be able to correctly classify data. Statisticians use other classifications besides the ones presented here. But the types we have discussed are sufficient for the majority of applications.

Data classification is difficult sometimes; even statisticians occasionally disagree over data type. For example, some classify amounts of money as discrete data, whereas others classify amounts of money as continuous data. In most cases, however, the appropriate classification of data is fairly clear and may help you choose the correct statistical method for analyzing the data.

EXERCISES 2.1

Statistical Concepts and Skills

2.1 Give an example, other than those presented in this section, of a
a. qualitative variable.
b. discrete, quantitative variable.
c. continuous, quantitative variable.

2.2 Explain the meaning of
a. qualitative variable.
b. discrete, quantitative variable.
c. continuous, quantitative variable.

2.3 Explain the meaning of
a. qualitative data.
b. discrete, quantitative data.
c. continuous, quantitative data.

2.4 Provide a reason why the classification of data is important.

2.5 Of the variables you have studied so far, which type yields nonnumerical data?

For each part of Exercises **2.6–2.9**, *classify the data as either qualitative or quantitative; if quantitative, further classify it as discrete or continuous. Also, identify the variable under consideration in each case.*

2.6 Doctor Disciplinary Actions. In the June 2003 issue of *Health Letter*, the Public Citizen Health Research Group ranked the states by serious doctor disciplinary actions per 1000 doctors for the year 2002. Here are data for the 10 states with the lowest rates.

State	No. of actions	No. of doctors	Actions per 1000 doctors
Hawaii	4	3,746	1.07
Delaware	3	2,219	1.35
Wisconsin	20	14,241	1.40
Tennessee	22	14,954	1.47
South Carolina	17	9,607	1.77
Maryland	39	21,883	1.78
North Carolina	43	20,851	2.06
Florida	93	44,747	2.08
Pennsylvania	82	39,052	2.10
Minnesota	30	14,218	2.11

Identify the type of data provided by the information in each of the following columns of the table.
a. first   **b.** second   **c.** third

2.7 How Hot Does It Get? The highest temperatures on record for selected cities are collected by the U.S. National Oceanic and Atmospheric Administration and published in *Comparative Climatic Data*. The following table displays data for years through 2000.

City	Rank	Highest temp. (°F)
Phoenix, AZ	1	122
Sacramento, CA	2	115
El Paso, TX	3	114
Omaha, NE	3	114
Dallas–Fort Worth	5	113
Wichita, KS	5	113

a. What type of data is presented in the second column of the table?

b. What type of data is provided in the third column of the table?

c. What type of data is provided by the information that Phoenix is in Arizona?

2.8 Earnings from the Crypt. Drawing on *Forbes'* 18 years of wealth-estimating experience, four ABCNews.com reporters calculated pretax earnings to the estates of deceased celebrities from licensing agreements and book and record sales for the 12-month period from June 2001 to June 2002. The following table shows those deceased celebrities with the top 13 earnings during that time period.

Rank	Name	Earnings ($millions)
1	Elvis Presley	37
2	Charles Schulz	28
3	John Lennon	20
4	Dale Earnhardt	20
5	Theodor Geisel	19
6	George Harrison	17
7	J.R.R. Tolkien	12
8	Bob Marley	10
9	Jimi Hendrix	8
10	Tupac Shakur	7
11	Marilyn Monroe	7
12	Jerry Garcia	5
13	Robert Ludlum	5

a. What type of data is presented in the first column of the table?

b. What type of data is provided by the information in the third column of the table?

2.9 Most Wired Cities. According to *Forrester Research*, in 2002, the 10 U.S. cities with the highest percentage of families using the Internet were as follows.

Rank	City	% families using Internet
1	San Francisco, CA	78.8
2	Austin–San Marcos, TX	76.5
3	Oxnard–Ventura Counties, CA	72.9
4	Albany–Schenectady–Troy, NY	72.7
5	Raleigh–Durham–Chapel Hill, NC	71.9
6	Wilmington–Newark, DE	70.2
7	Las Vegas, NV	70.1
8	Columbus, OH	69.1
9	Washington, DC	68.9
10	Salt Lake City–Ogden, UT	68.8

What type of data is provided by the information in the

a. first column of the table?

b. second column of the table?

c. third column of the table? (*Hint:* The possible ratios of positive whole numbers form a countably infinite set.)

Extending the Concepts and Skills

2.10 Ordinal data. Another important type of data is *ordinal data*, which is data about order or rank given on a scale such as 1, 2, 3, … or A, B, C, …. Following are several variables. Which, if any, yield ordinal data? Explain your answer.

a. Height   b. Weight   c. Age   d. Sex

e. Number of siblings

f. Religion

g. Place of birth

h. High school class rank

2.2 Grouping Data

The amount of data collected in real-world situations can sometimes be overwhelming. For example, a list of U.S. colleges and universities with information on enrollment, number of teachers, highest degree offered, and governing official can be found in *The World Almanac*. These data occupy 28 pages of small type!

By suitably organizing data, we can often make a large and complicated set of data more compact and easier to understand. In this section, we discuss **grouping**, which involves, as the term implies, putting data into groups rather than treating each observation individually. Grouping is one of the most common methods of organizing data.

We first consider the grouping of quantitative data, both continuous and discrete. Later in this section, we examine the grouping of qualitative data.

Grouping Quantitative Data

To begin, consider the grouping of quantitative data presented in Example 2.5.

Example 2.5 **Grouping Quantitative Data**

TABLE 2.1
Days to maturity for 40 short-term investments

70	64	99	55	64	89	87	65
62	38	67	70	60	69	78	39
75	56	71	51	99	68	95	86
57	53	47	50	55	81	80	98
51	36	63	66	85	79	83	70

TABLE 2.2
Classes and counts for the days-to-maturity data in Table 2.1

Days to maturity	Tally	Number of investments
30 ≺ 40	|||	3
40 ≺ 50			1
50 ≺ 60	Ⅷ |||	8
60 ≺ 70	Ⅷ Ⅷ	10
70 ≺ 80	Ⅷ ||	7
80 ≺ 90	Ⅷ ||	7
90 ≺ 100	||||	4
		40

What
Does it Mean?

Comparing Tables 2.1 and 2.2 clearly shows that grouping the data makes the data much easier to read and understand.

Days to Maturity for Short-Term Investments Table 2.1 displays the number of days to maturity for 40 short-term investments. The data are from *Barron's National Business and Financial Weekly*. Getting a clear picture of the data in Table 2.1 is difficult. By grouping the data into categories, or **classes,** we can make the data much easier to comprehend. The first step is to decide on the classes. One convenient way to group these data is by 10s.

Because the shortest maturity period is 36 days, our first class is for maturity periods from 30 days up to, but not including, 40 days. We use the symbol ≺ as a shorthand for "up to, but not including," so our first class is denoted 30 ≺ 40.[†] The longest maturity period is 99 days, so grouping by 10s results in the seven classes given in the first column of Table 2.2.

The final step for grouping the data is to determine the number of investments in each class. We do so by placing a tally mark for each investment from Table 2.1 on the appropriate line of Table 2.2. For instance, the first investment in Table 2.1 has a 70-day maturity period, calling for a tally mark on the line for the class 70 ≺ 80. The results of the tallying procedure are shown in the second column of Table 2.2. Now we count the tallies for each class and record each total in the third column of Table 2.2.

By simply glancing at Table 2.2, we can easily obtain various pieces of useful information. For instance, more investments are in the 60s range than in any other.

In Example 2.5, we used a commonsense approach to grouping data into classes. Some of that common sense can be used as guidelines for grouping. Three of the most important guidelines are the following.

1. *The number of classes should be small enough to provide an effective summary but large enough to display the relevant characteristics of the data.*

In Example 2.5, we used seven classes. Usually the number of classes should be between 5 and 20, but that is only a rule of thumb.

2. *Each observation must belong to one, and only one, class.*

Careless planning in Example 2.5 could have led to classes such as 30–40, 40–50, 50–60, and so on. Then, for instance, to which class would the investment

---

[†]The symbol ≺ is obtained by superimposing a less-than symbol over a dash. Using the notation 30 ≺ 40 is less awkward than using the notation 30–< 40 or 30–under 40.

What
Does it Mean?

Always keep in mind that the reason for grouping is to organize the data into a sensible number of classes to make the data more accessible and understandable.

with a 50-day maturity period belong? The classes in Table 2.2 do not cause such confusion; they cover all maturity periods and do not overlap.

3. *Whenever feasible, all classes should have the same width.*

The classes in Table 2.2 all have a width of 10 days. Among other things, choosing classes of equal width facilitates the graphical display of the data.

The list could go on, but for our purposes these three guidelines provide a solid basis for grouping data.

Frequency and Relative-Frequency Distributions

The number of observations that fall into a particular class is called the **frequency** (or **count**) of that class. For instance, in Table 2.2, the frequency of the class $50 \leqslant 60$ is 8 because eight investments are in the 50s days-to-maturity range. A table that provides all classes and their frequencies is called a **frequency distribution.** The first and third columns of Table 2.2 constitute a frequency distribution for the days-to-maturity data.

In addition to the frequency of a class, we are often interested in the **percentage** of a class. We find the percentage by first dividing the frequency of the class by the total number of observations and then multiplying the result by 100. From Table 2.2, the percentage of investments in the class $50 \leqslant 60$ is

What
Does it Mean?

A frequency distribution or relative-frequency distribution provides a table of the values of the observations and how often they occur.

$$\frac{8}{40} = 0.20 \quad \text{or} \quad 20\%.$$

Thus 20% of the investments have a number of days to maturity in the 50s.

The percentage of a class, expressed as a decimal, is called the **relative frequency** of the class. For the class $50 \leqslant 60$, the relative frequency is 0.20. A table that provides all classes and their relative frequencies is called a **relative-frequency distribution.** Table 2.3 displays a relative-frequency distribution for the days-to-maturity data. Note that the relative frequencies sum to 1 (100%).

Relative-frequency distributions are better than frequency distributions for comparing two data sets. The reason is that relative frequencies always fall between 0 and 1 and hence provide a standard for comparison. Two data sets that have identical frequency distributions have identical relative-frequency distributions. But two data sets that have identical relative-frequency distributions have identical frequency distributions only if both data sets have the same number of observations.

TABLE 2.3
Relative-frequency distribution for the days-to-maturity data in Table 2.1

Days to maturity	Relative frequency		
$30 \leqslant 40$	0.075	$\leftarrow$	3/40
$40 \leqslant 50$	0.025	$\leftarrow$	1/40
$50 \leqslant 60$	0.200	$\leftarrow$	8/40
$60 \leqslant 70$	0.250	$\leftarrow$	10/40
$70 \leqslant 80$	0.175	$\leftarrow$	7/40
$80 \leqslant 90$	0.175	$\leftarrow$	7/40
$90 \leqslant 100$	0.100	$\leftarrow$	4/40
	1.000		

Grouping Terminology

To be adept at grouping, you need to become familiar with and understand the various terms associated with grouping. We have already discussed several of these terms. To introduce some additional ones, let's return to the days-to-maturity data.

Consider, for example, the class $50 \leqslant 60$. The smallest maturity period that could go in this class is 50; this value is called the **lower cutpoint** of the class. The smallest maturity period that could go in the next higher class is 60; this value is called the **upper cutpoint** of the class and, of course, is also the lower cutpoint of the next higher class, $60 \leqslant 70$.

The number in the middle of the class $50 \leqslant 60$ is $(50 + 60)/2 = 55$ and is called the **midpoint** of the class. Midpoints provide single numbers for representing classes and are often used in graphical displays and for computing descriptive measures. The **width** of the class $50 \leqslant 60$, obtained by subtracting its lower cutpoint from its upper cutpoint, is $60 - 50 = 10$. We summarize the terminology of grouping as follows.

DEFINITION 2.3

Terms Used in Grouping

Classes: Categories for grouping data.

Frequency: The number of observations that fall in a class.

Frequency distribution: A listing of all classes and their frequencies.

Relative frequency: The ratio of the frequency of a class to the total number of observations.

Relative-frequency distribution: A listing of all classes and their relative frequencies.

Lower cutpoint: The smallest value that could go in a class.

Upper cutpoint: The smallest value that could go in the next higher class. The upper cutpoint of a class is the same as the lower cutpoint of the next higher class.

Midpoint: The middle of a class, obtained by taking the average of its lower and upper cutpoints.

Width: The difference between the upper and lower cutpoints of a class.

A table that provides the classes, frequencies, relative frequencies, and midpoints of a data set is called a **grouped-data table.** Table 2.4 is a grouped-data table for the days-to-maturity data. Construction of a grouped-data table from raw data is further illustrated in Example 2.6.

TABLE 2.4
Grouped-data table for the days-to-maturity data

Days to maturity	Frequency	Relative frequency	Midpoint
$30 \leqslant 40$	3	0.075	35
$40 \leqslant 50$	1	0.025	45
$50 \leqslant 60$	8	0.200	55
$60 \leqslant 70$	10	0.250	65
$70 \leqslant 80$	7	0.175	75
$80 \leqslant 90$	7	0.175	85
$90 \leqslant 100$	4	0.100	95
	40	1.000	

▌▌ Example 2.6 **Grouped-Data Tables**

Weights of 18–24-Year-Old Males The U.S. National Center for Health Statistics publishes data on weights and heights by age and sex in *Vital and Health Statistics*. The weights shown in Table 2.5, given to the nearest tenth of a pound, were obtained from a sample of 18–24-year-old males. Construct a grouped-data table for these weights. Use a class width of 20 and a first cutpoint of 120.

TABLE 2.5
Weights of 37 males, aged 18–24 years

129.2	185.3	218.1	182.5	142.8
155.2	170.0	151.3	187.5	145.6
167.3	161.0	178.7	165.0	172.5
191.1	150.7	187.0	173.7	178.2
161.7	170.1	165.8	214.6	136.7
278.8	175.6	188.7	132.1	158.5
146.4	209.1	175.4	182.0	173.6
149.9	158.6			

Solution Because we are to use a class width of 20 and a first cutpoint of 120, the first class will be 120 ≤ 140. In Table 2.5, the largest weight is 278.8 lb. Thus we choose the classes displayed in the first column of Table 2.6.

Applying the tallying procedure to the data in Table 2.5, we obtain the frequencies in the second column of Table 2.6. To illustrate some typical computations for the third and fourth columns of Table 2.6, we use the class 160 ≤ 180:

$$\text{relative frequency} = \frac{14}{37} = 0.378$$

and

$$\text{midpoint} = \frac{160 + 180}{2} = 170.$$

The other entries in the third and fourth columns are computed similarly.

TABLE 2.6
Grouped-data table for the weights of 37 males, aged 18–24 years

Weight (lb)	Frequency	Relative frequency	Midpoint
120 ≤ 140	3	0.081	130
140 ≤ 160	9	0.243	150
160 ≤ 180	14	0.378	170
180 ≤ 200	7	0.189	190
200 ≤ 220	3	0.081	210
220 ≤ 240	0	0.000	230
240 ≤ 260	0	0.000	250
260 ≤ 280	1	0.027	270
	37	0.999	

Recall that relative frequencies must always sum to 1. However, the sum of the relative frequencies in the third column of Table 2.6 is given as 0.999. The reason for this discrepancy is as follows. Each relative frequency is rounded to three decimal places and, in this case, the resulting sum differs from 1 by a little. Such a result is usually referred to as *rounding error* or *roundoff error*.

——■

An Alternate Method for Depicting Classes

An alternate method is often used to depict classes, especially when the data are expressed as whole numbers. For instance, the third class for the days-to-maturity data, $50 \lessdot 60$, is for maturity periods from 50 days up to, but not including, 60 days. In this case the maturity periods are expressed to the nearest whole day, so the third class can also be characterized as being for maturity periods from 50 days up to, and including, 59 days.

Thus, as an alternative to depicting the third class by $50 \lessdot 60$, we can use 50–59. The smallest maturity period that could go in this class is 50 and, in this context, is called the **lower limit** of the class. The largest maturity period that can go in this class is 59 and, in this context, is called the **upper limit** of the class.

When utilizing class limits to depict classes, we generally use *marks* instead of midpoints as representatives for the classes. The **mark** of a class is the average of its lower and upper limits. For example, the mark of the class 50–59 is $(50 + 59)/2 = 54.5$.

An alternate grouped-data table for the days-to-maturity data—one that uses class limits and class marks instead of class cutpoints and class midpoints— is displayed in Table 2.7. Compare Tables 2.4 and 2.7.

TABLE 2.7
Grouped-data table for the days-to-maturity data, using class limits and class marks

Days to maturity	Frequency	Relative frequency	Mark
30–39	3	0.075	34.5
40–49	1	0.025	44.5
50–59	8	0.200	54.5
60–69	10	0.250	64.5
70–79	7	0.175	74.5
80–89	7	0.175	84.5
90–99	4	0.100	94.5
	40	1.000	

The weight data in Table 2.5 are presented to one decimal place. Consequently, the alternate method for depicting the classes shown in the first column of Table 2.6 would be 120–139.9, 140–159.9, 160–179.9, and so on. The marks for these classes are, respectively, 129.95, 149.95, 169.95, and so on.

Both of these methods of depicting classes and their representatives are commonly used, and each has its advantages and disadvantages. We freely use both methods throughout this book. So, be sure that you have a clear understanding of the two approaches.

Single-Value Grouping

Up to this point, each class that we have used for grouping data represents a range of possible values. For instance, in Table 2.6, the first class, $120 \lessdot 140$, is for weights from 120 lb up to, but not including, 140 lb. In some cases, however, using classes that each represent a single possible value is more appropriate. Such classes are particularly suitable for discrete data in which there are only relatively few distinct observations. Consider, for instance, Example 2.7.

Example 2.7   Single-Value Grouping

TVs per Household   *Trends in Television*, published by the Television Bureau of Advertising, contains information on the number of television sets owned by U.S. households. Data on the number of TV sets per household for 50 randomly selected households are displayed in Table 2.8. Use classes based on a single value to construct a grouped-data table for these data.

TABLE 2.8

Number of TV sets in each of 50 randomly selected households

1	1	1	2	6	3	3	4	2	4
3	2	1	5	2	1	3	6	2	2
3	1	1	4	3	2	2	2	2	3
0	3	1	2	1	2	3	1	1	3
3	2	1	2	1	1	3	1	5	1

Solution   We first note that, because each class is to represent a single numerical value, the classes are 0, 1, 2, 3, 4, 5, and 6. These classes are displayed in the first column of Table 2.9.

Tallying the data in Table 2.8, we obtain the frequencies in the second column of Table 2.9. Dividing each frequency by the total number of observations, 50, we get the relative frequencies shown in the third column of Table 2.9. The table indicates, for example, that 14 of the 50 households, or 0.280 (28.0%), have two television sets.

Because each class is based on a single value, the midpoint (and mark) of each class is the same as the class. For instance, the midpoint (and mark) of the class 3 is 3. Therefore a midpoint (or mark) column in Table 2.9 is unnecessary, as such a column would be identical to the first column. In other words, Table 2.9 can serve as a grouped-data table.[†]

TABLE 2.9

Grouped-data table for number of TV sets

Number of TVs	Freq.	Relative freq.
0	1	0.020
1	16	0.320
2	14	0.280
3	12	0.240
4	3	0.060
5	2	0.040
6	2	0.040
	50	1.000

Grouping Qualitative Data

The concepts of cutpoints and midpoints are not appropriate for qualitative data. For instance, with data that categorize people as male or female, the classes are "male" and "female." To consider cutpoints or midpoints for such data makes no sense.

We can, of course, still group qualitative data and compute frequencies and relative frequencies for the classes. For qualitative data, the classes coincide with the observed values of the corresponding variable, as illustrated in Example 2.8.

Example 2.8   Grouping Qualitative Data

Political Party Affiliations   Professor Weiss asked his introductory statistics students to state their political party affiliations as Democratic (D), Republican (R), or Other (O). The responses are given in Table 2.10. Determine the frequency and relative-frequency distributions for these data.

TABLE 2.10

Political party affiliations of the students in introductory statistics

D	R	O	R	R	R	R
D	O	R	D	O	R	D
D	R	O	D	R	O	R
D	O	D	D	D	R	O
O	R	D	R	R	R	D

Solution   The classes for grouping the data are "Democratic," "Republican," and "Other." Tallying the data in Table 2.10, we obtain the frequency distribution displayed in the first two columns of Table 2.11.

[†]For single-value grouped data, the upper and lower limits of each class are also identical to the class. And, although it is generally unnecessary, we can also obtain the cutpoints. For the grouped data in Table 2.9, the cutpoints are -0.5, 0.5, 1.5, 2.5, 3.5, 4.5, 5.5, and 6.5.

TABLE 2.11
Frequency and relative-frequency
distributions for political party
affiliations

Party	Frequency	Relative frequency
Democratic	13	0.325
Republican	18	0.450
Other	9	0.225
	40	1.000

Dividing each frequency in the second column of Table 2.11 by the total number of students, 40, we get the relative frequencies in the third column. The first and third columns of Table 2.11 provide the relative-frequency distribution for the data.

INTERPRETATION   From Table 2.11, we see that, of the 40 students in the class, 13 are Democrats, 18 are Republicans, and 9 are Other. Equivalently, 32.5% are Democrats, 45.0% are Republicans, and 22.5% are Other.

The Technology Center

Grouping data can be tedious if done by hand. You can avoid the tedium by using a computer or graphing calculator. In this subsection, we present output and step-by-step instructions for grouping data into single-value classes. Refer to the technology manuals for other grouping methods.

Minitab and Excel have dedicated programs for single-value grouping of quantitative data or for grouping of qualitative (categorical) data. At the time of this writing, the TI-83/84 Plus does not have such a program. Example 2.9 illustrates the use of Minitab and Excel to group quantitative data. You can use the same steps to group qualitative data.

Example 2.9   Using Technology to Obtain a Grouped-Data Table

TVs per Household   Table 2.8 displays data on the number of TV sets per household for 50 randomly selected households. Use Minitab or Excel to obtain a grouped-data table for these data, based on single-value classes.

Solution   Printout 2.1 on the following page shows the output obtained by applying the grouping programs to the data on the number of TV sets per household. Compare the output in Printout 2.1 to the grouped-data table we obtained by hand in Table 2.9. Note, in particular, that both technologies use percents instead of relative frequencies.

Obtaining the Output (Optional)

Printout 2.1 provides output from Minitab and Excel for a grouped-data table based on single-value grouping. Here are detailed instructions for obtaining that output. First, we store the data from Table 2.8 in a column (Minitab) or range (Excel) named TVs. Then, we proceed as follows.

MINITAB	EXCEL	TI-83/84 PLUS
1 Choose **Stat ➤ Tables ➤ Tally Individual Variables...**	1 Choose **DDXL ➤ Tables**	SEE THE TI-83/84 PLUS MANUAL
2 Specify TVs in the **Variables** text box	2 Select **Frequency Table** from the **Function type** drop-down box	
3 Select the **Counts** and **Percents** check boxes	3 Specify TVs in the **Categorical Variable** text box	
4 Click **OK**	4 Click **OK**	

PRINTOUT 2.1

Grouped-data table output for the data on the number of TVs per household

MINITAB

Tally for Discrete Variables: TVs

TVs	Count	Percent
0	1	2.00
1	16	32.00
2	14	28.00
3	12	24.00
4	3	6.00
5	2	4.00
6	2	4.00
N=	50	

EXCEL

Summary of TVs

Total Cases	50	
Number of Categories	7	

TVs Frequency Table

Group	Count	%
0	1	2
1	16	32
2	14	28
3	12	24
4	3	6
5	2	4
6	2	4

EXERCISES 2.2

Statistical Concepts and Skills

2.11 Identify an important reason for grouping data.

2.12 Do the concepts of cutpoints and midpoints make sense for qualitative data? Explain your answer.

2.13 State three of the most important guidelines in choosing the classes for grouping a data set.

2.14 Explain the difference between
a. frequency and relative frequency.
b. percentage and relative frequency.

2.15 Are frequency distributions or relative-frequency distributions better for comparing two data sets? Explain your answer.

2.16 What are the four elements of a grouped-data table? Explain the meaning of each element.

2.17 With regard to grouping quantitative data into classes that each represent a range of possible values, we discussed two methods for depicting the classes. Identify the two methods and explain the relative advantages and disadvantages of each.

2.18 For grouping quantitative data, we examined three types of classes: (1) $a \ll b$, (2) a–b, and (3) single-value grouping. For each type of data given, decide which of these three types is usually best. Explain your answers.
a. Continuous data displayed to one or more decimal places
b. Discrete data in which there are relatively few distinct observations

2.19 When you group quantitative data into classes that each represent a single possible numerical value, why is it unnecessary to include a midpoint column in the grouped-data table?

2.20 Residential Energy Consumption. The U.S. Energy Information Administration collects data on residential energy consumption and expenditures. Results are published in the document *Residential Energy Consumption Survey: Consumption and Expenditures*. The following table gives one year's energy consumptions for a sample of 50 households in the South. Data are in millions of BTUs.

130	55	45	64	155	66	60	80	102	62
58	101	75	111	151	139	81	55	66	90
97	77	51	67	125	50	136	55	83	91
54	86	100	78	93	113	111	104	96	113
96	87	129	109	69	94	99	97	83	97

Use classes of equal width beginning with $40 \ll 50$ to construct a grouped-data table for these data.

2.21 Clocking the Cheetah. The Cheetah (*Acinonyx jubatus*) is the fastest land mammal on earth and is highly spe-

cialized to run down prey. According to the Cheetah Conservation of Southern Africa *Trade Environment Database*, the cheetah often exceeds speeds of 60 mph and has been clocked at speeds of more than 70 mph. The following table, based on information in the database, gives the speeds, in miles per hour, over a 1/4 mile for 35 cheetahs.

57.3	57.5	59.0	56.5	61.3	57.6	59.2
65.0	60.1	59.7	62.6	52.6	60.7	62.3
65.2	54.8	55.4	55.5	57.8	58.7	57.8
60.9	75.3	60.6	58.1	55.9	61.6	59.6
59.8	63.4	54.7	60.2	52.4	58.3	66.0

Use 52 as the first cutpoint and classes of equal width 2 to construct a grouped-data table for these speeds.

2.22 Residential Energy Consumption. Redo Exercise 2.20 by using the alternative method for grouping data based on class limits and class marks.

2.23 Clocking the Cheetah. Redo Exercise 2.21 by using the alternative method for grouping data based on class limits and class marks.

2.24 Household Size. The U.S. Bureau of the Census conducts nationwide surveys on characteristics of U.S. households and publishes the results in *Current Population Reports*. Following are data on the number of people per household for a sample of 40 households.

2	5	2	1	1	2	3	4
1	4	4	2	1	4	3	3
7	1	2	2	3	4	2	2
6	5	2	5	1	3	2	5
2	1	3	3	2	2	3	3

Construct a grouped-data table for these household sizes. Use classes based on a single value.

2.25 The Great White Shark. In a recent article entitled "Great White, Deep Trouble" (*National Geographic*,

Vol. 197(4), pp. 2–29), Peter Benchley—the author of *JAWS*—discussed various aspects of the Great White Shark (*Carcharodon carcharias*). The following table, based on information in that article, provides data on the number of pups borne in a lifetime by each of 80 Great White Shark females. Construct an appropriate grouped-data table for these data.

3	5	4	5	5	9	8	7
5	8	9	8	7	6	7	9
4	7	6	8	7	5	8	9
8	8	7	4	5	9	10	4
5	10	8	8	7	7	8	12
7	9	6	9	6	7	7	9
8	9	6	7	11	8	8	8
7	7	10	6	6	8	4	6
5	5	9	7	8	3	6	5
8	7	11	10	7	9	6	6

2.26 Road Rage. The report *Controlling Road Rage: A Literature Review and Pilot Study*, dated June 9, 1999, was prepared for the AAA Foundation for Traffic Safety by Daniel B. Rathbone, Ph.D., and Jorg C. Huckabee, MSCE. The authors discuss the results of a literature review and pilot study on how to prevent aggressive driving and road rage. As described in the study, road rage is criminal behavior by motorists characterized by uncontrolled anger that results in violence or threatened violence on the road. One of the goals of the study was to determine when road rage occurs most often. The following table provides the days on which 69 road rage incidents occurred. Construct a frequency distribution and a relative-frequency distribution for these data.

F	Sa	W	M	Tu	F	Th	M
Tu	F	Tu	F	Su	W	Th	F
Th	W	Th	Sa	W	W	F	F
Tu	Su	Tu	Th	W	Sa	Tu	Th
F	W	F	F	Su	F	Th	Tu
F	Tu	Tu	Tu	Sa	W	W	Sa
F	Sa	Th	W	F	Th	F	M
F	M	F	Su	W	Th	M	Tu
Sa	Th	F	Su	W			

2.27 All-Time Top TV Programs. According to *The World Almanac*, as of August 2002, the all-time top television programs by rating (percentage of TV-owning households tuned in to the program) are as follows. [SOURCE: Nielsen Media Research.]

Program	Telecast date	Network	Rating (%)	Audience (millions)
M*A*S*H (last episode)	02/28/83	CBS	60.2	50.2
Dallas (Who shot J.R.?)	11/21/80	CBS	53.3	41.5
Roots—Pt. 8	01/30/77	ABC	51.1	36.4
Super Bowl XVI	01/24/82	CBS	49.1	40.0
Super Bowl XVII	01/30/83	NBC	48.6	40.5
XVII Winter Olympics—2d Wed.	02/23/94	CBS	48.5	45.7
Super Bowl XX	01/26/86	NBC	48.3	41.5
Gone With the Wind—Pt. 1	11/07/76	NBC	47.7	34.0
Gone With the Wind—Pt. 2	11/08/76	NBC	47.4	33.8
Super Bowl XII	01/15/78	CBS	47.2	34.4
Super Bowl XIII	01/21/79	NBC	47.1	35.1
Bob Hope Christmas Show	01/15/70	NBC	46.6	27.3
Super Bowl XVIII	01/22/84	CBS	46.4	38.8
Super Bowl XIX	01/20/85	ABC	46.4	39.4
Super Bowl XIV	01/20/80	CBS	46.3	35.3
Super Bowl XXX	01/28/96	NBC	46.0	44.2
ABC Theatre (The Day After)	11/20/83	ABC	46.0	38.6
Roots—Pt. 6	01/28/77	ABC	45.9	32.7
The Fugitive	08/29/67	ABC	45.9	25.7
Super Bowl XXI	01/25/87	CBS	45.8	40.0

Construct frequency and relative-frequency distributions for the network data. (*Hint:* The classes are "ABC," "CBS," and "NBC.")

Extending the Concepts and Skills

The table at the top of the following column gives the closing Dow Jones Industrial Averages (Dow) on July 2, 2003. Use these data in Exercises 2.28–2.30.

2.28 Dow Closing Prices. The column headed "Last" in the Dow table gives the closing price per share, in dollars, for each of the Dow stocks. Construct a grouped-data table for these closing prices. Use equal-width classes and begin with the class 15 ≤ 25.

2.29 Dow Volume. The column headed "Volume" in the Dow table shows the number of shares sold, in hundreds, for each of the Dow stocks.

Symbol	Last ($)	Change ($)	Volume (100s)
AA	25.53	0.03	30,356
AXP	42.29	0.38	39,884
BA	34.80	0.15	39,868
C	44.40	0.59	130,909
CAT	55.74	1.00	21,686
DD	42.20	0.48	21,300
DIS	20.21	0.32	91,782
EK	27.25	−0.10	35,035
GE	28.61	−0.02	211,768
GM	35.94	0.20	58,825
HD	33.44	0.34	74,321
HON	27.82	0.82	33,345
HPQ	21.56	0.38	128,680
IBM	84.74	1.15	59,856
INTC	22.21	0.80	736,463
IP	36.89	0.82	33,801
JNJ	52.90	0.48	84,143
JPM	34.37	0.16	72,661
KO	46.38	−0.11	44,104
MCD	22.72	0.78	65,217
MMM	29.89	1.01	11,279
MO	46.59	0.40	56,377
MRK	61.72	0.25	66,453
MSFT	26.88	0.73	935,678
PG	90.11	0.15	32,565
SBC	26.60	0.57	78,069
T	19.88	0.70	46,032
UTX	71.90	0.61	18,895
WMT	55.73	1.38	100,099
XOM	36.28	0.10	109,873

a. Construct a grouped-data table for these sales volumes. Use the classes $1 \lessdot 2, 2 \lessdot 3, \ldots, 9 \lessdot 10$, and "10 & over," where the values are in millions of sales.

b. Why is there no midpoint for the last class?

2.30 Dow Gains and Losses. The column headed "Change" in the Dow table provides the difference, in dollars, between the closing price per share given in the second column of the table and the closing price per share on the previous trading day. Construct a grouped-data table for the changes, using classes of your choice. Explain your choice of classes.

2.31 Exam Scores. The exam scores for the students in an introductory statistics class are as follows.

88	82	89	70	85
63	100	86	67	39
90	96	76	34	81
64	75	84	89	96

a. Group these exam scores, using the classes 30–39, 40–49, 50–59, 60–69, 70–79, 80–89, and 90–100.

b. What are the widths of the classes?

c. If you wanted all the classes to have the same width, what classes would you use?

Contingency Tables. The methods presented in this section apply to grouping data obtained from observing values of one variable of a population. Such data are called *univariate data.* For instance, in Example 2.6 on page 50, we examined data obtained from observing the values of the variable "weight" for a sample of 18–24-year-old males; those data are univariate. We could have considered not only the weights of the males but also their heights. Then, we would have data on two variables, height and weight. Data obtained from observing values of two variables of a population are called *bivariate data.* Tables called *contingency tables* can be used to group bivariate data, as explained in Exercise 2.32.

2.32 Age and Sex. The following bivariate data on age (in years) and sex were obtained from the students in a freshman calculus course. The data show, for example, that the first student on the list is 21 years old and is a male.

Age	Sex	Age	Sex	Age	Sex	Age	Sex	Age	Sex
21	M	29	F	22	M	23	F	21	F
20	M	20	M	23	M	44	M	28	F
42	F	18	F	19	F	19	M	21	F
21	M	21	M	21	M	21	F	21	F
19	F	26	M	21	F	19	M	24	F
21	F	24	F	21	F	25	M	24	F
19	F	19	M	20	F	21	M	24	F
19	M	25	M	20	F	19	M	23	M
23	M	19	F	20	F	18	F	20	F
20	F	23	M	22	F	18	F	19	M

a. Group these data in the following contingency table. For the first student, place a tally mark in the box labeled by the "21–25" column and the "Male" row, as indicated. Tally the data for the other 49 students.

		Age (yr)			
		Under 21	21–25	Over 25	**Total**
Sex	Male		|		
	Female				
	Total				

b. Construct a table like the one in part (a) but with frequencies replacing the tally marks. Add the frequencies in each row and column of your table and record the sums in the proper "Total" boxes.

c. What do the row and column totals in your table in part (b) represent?

d. Add the row totals and add the column totals. Why are those two sums equal, and what does their common value represent?

e. Construct a table that shows the relative frequencies for the data. (*Hint:* Divide each frequency obtained in part (b) by the total of 50 students.)

f. Interpret the entries in your table in part (e) as percentages.

Using Technology

2.33 Clocking the Cheetah. Use the technology of your choice to obtain the grouped-data table for Exercise 2.21.

2.34 The Great White Shark. Use the technology of your choice to obtain the grouped-data table for Exercise 2.25.

2.35 All-Time Top TV Programs. Use the technology of your choice to obtain the frequency and relative-frequency distributions required in Exercise 2.27.

2.3 Graphs and Charts

Besides grouping, another method for organizing and summarizing data is to draw a picture of some kind. The old saying "a picture is worth a thousand words" has particular relevance in statistics—a graph or chart of a data set often provides the simplest and most efficient display.

In this section, we examine various techniques for organizing and summarizing data with graphs and charts. We first discuss histograms, beginning with an illustration of their use in Example 2.10.

Example 2.10 Histograms

Days to Maturity for Short-Term Investments Table 2.4 in Section 2.2 shows a grouped-data table for the number of days to maturity for 40 short-term investments. The first three columns of that table are repeated here in Table 2.12. Obtain graphical displays for these grouped data.

TABLE 2.12
Frequency and relative-frequency distributions for the days-to-maturity data

Days to maturity	Freq.	Relative freq.
30 < 40	3	0.075
40 < 50	1	0.025
50 < 60	8	0.200
60 < 70	10	0.250
70 < 80	7	0.175
80 < 90	7	0.175
90 < 100	4	0.100
	40	1.000

Solution One way to display these grouped data pictorially is to construct a graph, called a **frequency histogram,** that depicts the classes on the horizontal axis and the frequencies on the vertical axis. A frequency histogram for the days-to-maturity data is shown in Fig. 2.2(a).

Here are some important observations about Fig. 2.2(a).

- The height of each bar is equal to the frequency of the class it represents.
- The bar for each class extends from the lower cutpoint of the class to the upper cutpoint of the class.†

†This method is only one of several that can be used to depict the classes on the horizontal axis. Another common method is to use midpoints instead of cutpoints to label the horizontal axis. In that case, each bar is centered over the midpoint of the class it represents.

FIGURE 2.2

Days-to-maturity:
(a) frequency histogram;
(b) relative-frequency histogram

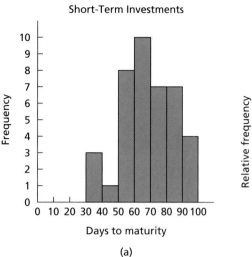

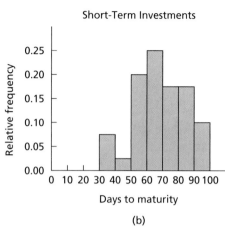

(a)

(b)

- Each axis of the frequency histogram has a label, and the frequency histogram as a whole has a title.

A frequency histogram displays the frequencies of the classes. To display the relative frequencies (or percentages), we use a **relative-frequency histogram,** which is similar to a frequency histogram. The only difference is that the height of each bar in a relative-frequency histogram is equal to the relative frequency of the class instead of the frequency of the class. A relative-frequency histogram for the days-to-maturity data is shown in Fig. 2.2(b).

Note that the shapes of the relative-frequency histogram in Fig. 2.2(b) and the frequency histogram in Fig. 2.2(a) are identical. The reason is that the frequencies and relative frequencies are proportional.

DEFINITION 2.4

What *Does it Mean?*

A histogram provides a graph of the values of the observations and how often they occur.

Frequency and Relative-Frequency Histograms

Frequency histogram: A graph that displays the classes on the horizontal axis and the frequencies of the classes on the vertical axis. The frequency of each class is represented by a vertical bar whose height is equal to the frequency of the class.

Relative-frequency histogram: A graph that displays the classes on the horizontal axis and the relative frequencies of the classes on the vertical axis. The relative frequency of each class is represented by a vertical bar whose height is equal to the relative frequency of the class.

For purposes of visually comparing the distributions of two data sets, relative-frequency histograms are better than frequency histograms. The same vertical scale is used for all relative-frequency histograms—a minimum of 0 and a maximum of 1—making direct comparison easy. In contrast, the vertical scale of a frequency histogram depends on the number of observations, making comparison more difficult.

Histograms for Single-Value Grouping

For the days-to-maturity data, each class represents a range of possible days to maturity, and in Figs. 2.2(a) and 2.2(b) the histogram bar for each class extends over that range. When data are grouped into classes based on a single value, the procedure is somewhat different. In that case, each bar is centered over the only possible value in the class, as illustrated in Example 2.11.

Example 2.11   **Histograms for Single-Value Grouped Data**

TABLE 2.13

Frequency and relative-frequency distributions for number of TV sets

Number of TVs	Freq.	Relative freq.
0	1	0.020
1	16	0.320
2	14	0.280
3	12	0.240
4	3	0.060
5	2	0.040
6	2	0.040
	50	1.000

TVs per Household   In Example 2.7, we considered data on the number of television sets per household for 50 randomly selected U.S. households. We used classes based on a single value to group those data. The frequency and relative-frequency distributions are given in Table 2.9 in Section 2.2 and are repeated here in Table 2.13. Construct a frequency histogram and a relative-frequency histogram for these grouped data.

Solution   For single-value grouping, we place the middle of each histogram bar directly over the single value represented by the class. Hence the frequency and relative-frequency histograms for the grouped data in Table 2.13 are those depicted in Figs. 2.3(a) and (b).

Note the symbol // on the horizontal axes in Figs. 2.3(a) and (b). This symbol indicates that the zero point on that axis is not in its usual position at the intersection of the horizontal and vertical axes. Whenever any such modification is made, whether on the horizontal axis or vertical axis, the symbol // or some similar symbol should be used to indicate that fact.

FIGURE 2.3

Number of TVs per household: (a) frequency histogram; (b) relative-frequency histogram

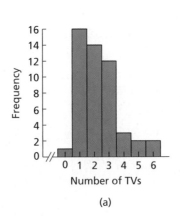

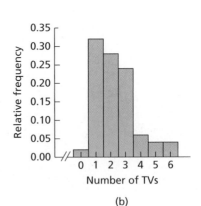

(a)   (b)

Dotplots

Another type of graphical display for quantitative data is the **dotplot.** Dotplots are particularly useful for showing the relative positions of the data in a data set or for comparing two or more data sets. We introduce dotplots in Example 2.12.

Example 2.12 Dotplots

TABLE 2.14
Prices, in dollars, of 16 DVD players

Prices of DVD Players One of Professor Weiss's sons was interested in adding a new DVD player to his home theater system. He decided to use the Internet to shop and went to *Price Watch U.S.A.* There he found 16 quotes on different brands and styles of DVD players. The prices, in dollars, are listed in Table 2.14. Construct a dotplot for these data.

210	219	214	197
224	219	199	199
208	209	215	199
212	212	219	210

Solution To construct a dotplot for the data in Table 2.14, we begin by drawing a horizontal axis that displays the possible prices. Then we record each price by placing a dot over the appropriate value on the horizontal axis. For instance, the first price is $210, which calls for a dot over the "210" on the horizontal axis. The dotplot for the data in Table 2.14 is shown in Fig. 2.4.

FIGURE 2.4
Dotplot for prices of DVD players

Prices of DVD Players
Price ($)

Note that dotplots are similar to histograms. In fact, when data are grouped in classes based on a single value, a dotplot and a frequency histogram are essentially identical. However, for single-value grouped data that involve decimals, dotplots are generally preferable to histograms because they are easier to construct and use.

Graphical Displays for Qualitative Data

Histograms and dotplots are designed for use with quantitative data. Qualitative data are portrayed with different techniques. Two common methods for displaying qualitative data graphically are *pie charts* and *bar graphs,* as illustrated in Example 2.13.

Example 2.13 Pie Charts and Bar Graphs

TABLE 2.15
Frequency and relative-frequency distributions for political party affiliations

Political Party Affiliations In Example 2.8, we obtained frequency and relative-frequency distributions for the political party affiliations of the students in Professor Weiss's introductory statistics class. We repeat those distributions in Table 2.15. Display the relative-frequency distribution of these qualitative data with a

Party	Freq.	Relative freq.
Democratic	13	0.325
Republican	18	0.450
Other	9	0.225
	40	1.000

a. pie chart.　　　　　　　　　　**b.** bar graph.

Solution

a. A **pie chart** is a disk divided into wedge-shaped pieces that are proportional to the relative frequencies. In this case, we need to divide a disk into three wedge-shaped pieces that comprise 32.5%, 45.0%, and 22.5% of the disk. We do so by using a protractor and the fact that there are 360° in a circle. Thus, for instance, the first piece of the disk is obtained by marking off 117° (32.5% of 360°). The pie chart for the relative-frequency distribution in Table 2.15 is shown in Fig. 2.5(a).

FIGURE 2.5
Political party affiliations:
(a) pie chart;
(b) bar graph

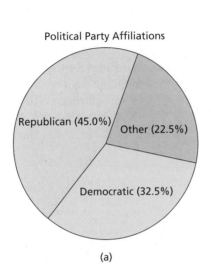

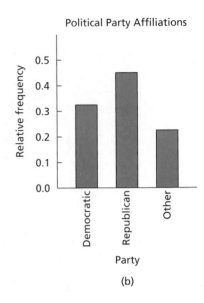

(a)   (b)

b. A **bar graph** is like a histogram. However, to avoid confusing bar graphs and histograms, we position the bars in a bar graph so that they do not touch each other. The bar graph for the relative-frequency distribution in Table 2.15 is shown in Fig. 2.5(b).

Histograms, dotplots, pie charts, and bar graphs are only a few of the countless ways that data can be portrayed pictorially. We will consider some additional graphical displays in the exercises for this section.

The **Technology Center**

Most statistical software packages and some graphing calculators have built-in programs for constructing histograms, dotplots, pie charts, and bar graphs. In this subsection, we present output and step-by-step instructions for obtaining a frequency histogram. Refer to the technology manuals for producing dotplots, pie charts, and bar graphs.

Example 2.14   **Using Technology to Obtain Histograms**

Days to Maturity for Short-Term Investments   Obtain a frequency histogram of the days-to-maturity data displayed in Table 2.1 on page 47 by using Minitab, Excel, or the TI-83/84 Plus.

Solution   Printout 2.2 shows the output obtained by applying the histogram programs to the days-to-maturity data.

PRINTOUT 2.2
Histogram output for the
days-to-maturity data

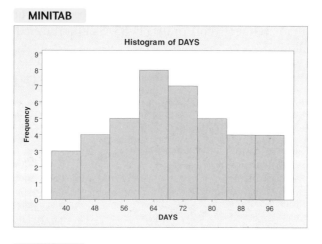

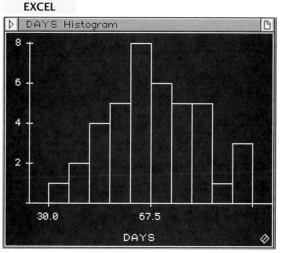

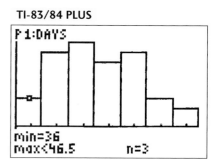

Some technologies require the user to specify the classes that are used for the construction of a histogram; others (e.g., Excel) automatically choose the classes; and still others (e.g., Minitab and the TI-83/84 Plus) give the user the choice of either specifying the classes or letting the program do it.

We obtained all three histograms in Printout 2.2 by letting the programs automatically choose the classes. This choice explains why the three histograms differ from each other and from the histogram we constructed by hand in Fig. 2.2(a) on page 59. Refer to the technology manuals for instructions to obtain histograms that are based on user-specified classes.

If the option is available, letting the technology automatically choose the classes for construction of a histogram is fast and easy. For most applications of histograms, the automatic method is sufficient for the required tasks.

Obtaining the Output (Optional)

Printout 2.2 provides output from Minitab, Excel, and the TI-83/84 Plus for a histogram of the days-to-maturity data. Here are detailed instructions for obtaining that output. First, we store the data from Table 2.1 in a column (Minitab), range (Excel), or list (TI-83/84 Plus) named DAYS. Then, we proceed as follows.

MINITAB
1 Choose **Graph ➤ Histogram…**
2 Select the **Simple** histogram and click **OK**
3 Specify DAYS in the **Graph variables** text box
4 Click **OK**

EXCEL
1 Choose **DDXL ➤ Charts and Plots**
2 Select **Histogram** from the **Function type** drop-down box
3 Specify DAYS in the **Quantitative Variable** text box
4 Click **OK**

TI-83/84 PLUS
1 Press **2nd ➤ STAT PLOT** and then press **ENTER** twice
2 Arrow to the third graph icon and press **ENTER**
3 Press the down-arrow key
4 Press **2nd ➤ LIST**
5 Arrow down to DAYS and press **ENTER**
6 Press **ZOOM** and then **9** (and then **TRACE**, if desired)

EXERCISES 2.3

Statistical Concepts and Skills

2.36 Explain the advantages and disadvantages of histograms relative to grouped-data tables.

2.37 Explain the difference between a frequency histogram and a relative-frequency histogram.

2.38 For data that are grouped in classes based on more than a single value, cutpoints are used on the horizontal axis of a histogram for depicting the classes. But, midpoints can also be used, in which case each bar is centered over the midpoint of the class it represents. Explain the advantages and disadvantages of each method.

2.39 In a bar graph, unlike in a histogram, the bars do not abut. Give a reason for that.

2.40 Some users of statistics prefer pie charts to bar graphs because people are accustomed to having the horizontal axis of a graph show order. For example, someone might infer from Fig. 2.5(b) on page 62 that "Republican" is less than "Other" because "Republican" is shown to the left of "Other" on the horizontal axis. Pie charts do not lead to such inferences. Give other advantages and disadvantages of each method.

2.41 DVD Players. Refer to Example 2.12 on page 61.
a. Explain why a frequency histogram of the DVD prices with classes based on a single value would be essentially identical to the dotplot shown in Fig. 2.4.

b. Would the dotplot and a frequency histogram be essentially identical if the classes for the histogram are not each based on a single value? Explain your answer.

2.42 Residential Energy Consumption. Shown in the following table are frequency and relative-frequency distributions for the data given in Exercise 2.20 on one year's energy consumptions for a sample of 50 households in the South.

Energy consumption (millions of BTU)	Frequency	Relative frequency
40 ≺ 50	1	0.02
50 ≺ 60	7	0.14
60 ≺ 70	7	0.14
70 ≺ 80	3	0.06
80 ≺ 90	6	0.12
90 ≺ 100	10	0.20
100 ≺ 110	5	0.10
110 ≺ 120	4	0.08
120 ≺ 130	2	0.04
130 ≺ 140	3	0.06
140 ≺ 150	0	0.00
150 ≺ 160	2	0.04

a. Construct a frequency histogram.
b. Construct a relative-frequency histogram.

2.43 Clocking the Cheetah. The following table provides frequency and relative-frequency distributions for the data presented in Exercise 2.21 on the speeds in miles per hour for a sample of 35 cheetahs.

Speed (mph)	Frequency	Relative frequency
52 < 54	2	0.057
54 < 56	5	0.143
56 < 58	6	0.171
58 < 60	8	0.229
60 < 62	7	0.200
62 < 64	3	0.086
64 < 66	2	0.057
66 < 68	1	0.029
68 < 70	0	0.000
70 < 72	0	0.000
72 < 74	0	0.000
74 < 76	1	0.029

a. Construct a frequency histogram.
b. Construct a relative-frequency histogram.

2.44 Household Size. The following table gives frequency and relative-frequency distributions for the data given in Exercise 2.24 on the number of people per household for a sample of 40 U.S. households.

Number of people	Frequency	Relative frequency
1	7	0.175
2	13	0.325
3	9	0.225
4	5	0.125
5	4	0.100
6	1	0.025
7	1	0.025

a. Construct a frequency histogram.
b. Construct a relative-frequency histogram.

2.45 The Great White Shark. The following table presents frequency and relative-frequency distributions for the data presented in Exercise 2.25 on the number of pups borne in a lifetime by each of 80 Great White Shark females.

Number of pups	Frequency	Relative frequency
3	2	0.025
4	5	0.063
5	10	0.125
6	11	0.138
7	17	0.213
8	17	0.213
9	11	0.138
10	4	0.050
11	2	0.025
12	1	0.013

a. Construct a frequency histogram.
b. Construct a relative-frequency histogram.

2.46 Exam Scores. Construct a dotplot for the following exam scores of the students in an introductory statistics class.

88	82	89	70	85
63	100	86	67	39
90	96	76	34	81
64	75	84	89	96

2.47 Ages of Trucks. The Motor Vehicle Manufacturers Association of the United States publishes information in *Motor Vehicle Facts and Figures* on the ages of cars and trucks currently in use. A sample of 37 trucks provided the ages, in years, displayed in the following table. Construct a dotplot for the ages.

8	12	14	16	15	5	11	13
4	12	12	15	12	3	10	9
11	3	18	4	9	11	17	
7	4	12	12	8	9	10	
9	9	1	7	6	9	7	

2.48 Road Rage. The following table shows frequency and relative-frequency distributions for the data given in Exercise 2.26 on the days on which 69 road rage incidents occurred.

Day	Frequency	Relative frequency
Su	5	0.072
M	5	0.072
Tu	11	0.159
W	12	0.174
Th	11	0.159
F	18	0.261
Sa	7	0.101

a. Draw a pie chart for the relative frequencies.
b. Construct a bar graph for the relative frequencies.

2.49 All-Time Top TV Programs. The following table displays frequency and relative-frequency distributions for the network data presented in Exercise 2.27 on the all-time top television programs by rating (percentage of TV-owning households tuned in to the program) as of August 2002.

Network	Frequency	Relative frequency
CBS	8	0.40
ABC	5	0.25
NBC	7	0.35

a. Draw a pie chart for the relative frequencies.
b. Construct a bar graph for the relative frequencies.

2.50 Adjusted Gross Incomes. The Internal Revenue Service (IRS) publishes data on adjusted gross incomes in *Statistics of Income, Individual Income Tax Returns.* The following relative-frequency histogram shows one year's individual income tax returns for adjusted gross incomes of less than $50,000.

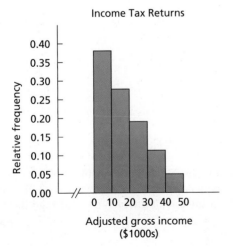

Use the histogram and the fact that adjusted gross incomes are expressed to the nearest whole dollar to answer each of the following questions.
a. Approximately what percentage of the individual income tax returns had an adjusted gross income between $10,000 and $19,999, inclusive?
b. Approximately what percentage had an adjusted gross income of less than $30,000?
c. The IRS reported that 89,928,000 individual income tax returns had an adjusted gross income of less than $50,000. Approximately how many had an adjusted gross income between $30,000 and $49,999, inclusive?

2.51 Cholesterol Levels. A pediatrician who tested the cholesterol levels of several young patients was alarmed to find that many had levels higher than 200 mg per 100 mL. The following relative-frequency histogram shows the readings for some patients who had high cholesterol levels.

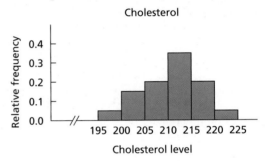

Use the graph to answer the following questions. Note that cholesterol levels are always expressed as whole numbers.
a. What percentage of the patients have cholesterol levels between 205 and 209, inclusive?
b. What percentage of the patients have levels of 215 or higher?
c. If the number of patients is 20, how many have levels between 210 and 214, inclusive?

Extending the Concepts and Skills

Relative-Frequency Polygons. Another graphical display commonly used is the relative-frequency polygon. In a *relative-frequency polygon,* a point is plotted above each class midpoint at a height equal to the relative frequency of the class. Then the points are connected with lines. For instance, the days-to-maturity data given in Table 2.4 on page 49 yield the following relative-frequency polygon.

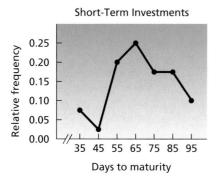

Short-Term Investments

2.52 Residential Energy Consumption. Construct a relative-frequency polygon for the energy-consumption data given in Exercise 2.42.

Ogives. Cumulative information can be portrayed using a graph called an *ogive* (ō′jīv). To construct an ogive, first make a table that displays cumulative frequencies and cumulative relative frequencies, as shown in Table 2.16 for the days-to-maturity data.

TABLE 2.16
Cumulative information for
days-to-maturity data

Less than	Cumulative frequency	Cumulative relative frequency
30	0	0.000
40	3	0.075
50	4	0.100
60	12	0.300
70	22	0.550
80	29	0.725
90	36	0.900
100	40	1.000

The first column of Table 2.16 gives the cutpoints of the classes and the second column gives the cumulative frequencies. A *cumulative frequency* is obtained by summing the frequencies of all classes representing values less than the specified cutpoint. For instance, by referring to Table 2.12 on page 58, we can find the cumulative frequency of investments with a maturity period of less than 50 days:

$$\text{Cumulative frequency} = 3 + 1 = 4.$$

That is, four of the investments have a maturity period of less than 50 days.

The third column of Table 2.16 gives the cumulative relative frequencies. A *cumulative relative frequency* is found by dividing the corresponding cumulative frequency by the total number of observations which, in this case, is 40. For instance, the cumulative relative frequency of investments with a maturity period of less than 50 days is

$$\text{cumulative relative frequency} = \frac{4}{40} = 0.100.$$

That is, 10% of the investments have a maturity period of less than 50 days.

Using Table 2.16, we can now construct an ogive for the days-to-maturity data. In an ogive, a point is plotted above each cutpoint at a height equal to the cumulative relative frequency. Then the points are connected with lines. The ogive for the days-to-maturity data is as follows.

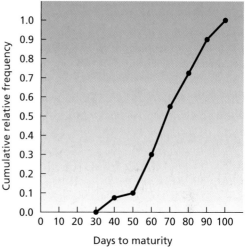

Short-Term Investments

2.53 Residential Energy Consumption. Refer to Exercise 2.42.
a. Construct a table similar to Table 2.16 for the energy-consumption data. Interpret your results.
b. Draw an ogive for the data.

Using Technology

2.54 Clocking the Cheetah. Use the technology of your choice to obtain a frequency histogram and a relative-frequency (or percent) histogram, as required in Exercise 2.43. The raw (ungrouped) data are given in Exercise 2.21 on page 55.

2.55 The Great White Shark. Use the technology of your choice to obtain the frequency histogram required in Exercise 2.45. The raw (ungrouped) data are given in Exercise 2.25 on page 56.

2.56 Ages of Trucks. Use the technology of your choice to obtain the dotplot required in Exercise 2.47.

2.57 All-Time Top TV Programs. Use the technology of your choice to obtain the pie chart and bar graph required in Exercise 2.49. The raw data are given in Exercise 2.27 on page 56.

2.4 Stem-and-Leaf Diagrams

New ways of displaying data are constantly being invented. One method, developed in the 1960s by the late Professor John Tukey of Princeton University, is called a *stem-and-leaf diagram,* or *stemplot.* This ingenious diagram is often easier to construct than either a frequency distribution or a histogram and generally displays more information. Returning once more to the days-to-maturity data, we illustrate stem-and-leaf diagrams in Example 2.15.

▌▐ Example 2.15 Stem-and-Leaf Diagrams

TABLE 2.17

Days to maturity for 40 short-term investments

70	64	99	55	64	89	87	65
62	38	67	70	60	69	78	39
75	56	71	51	99	68	95	86
57	53	47	50	55	81	80	98
51	36	63	66	85	79	83	70

Days to Maturity for Short-Term Investments The data on the number of days to maturity for 40 short-term investments are repeated in Table 2.17.

In Table 2.2 on page 47, we grouped these data by 10s, and in Fig. 2.2(a) on page 59, we portrayed the data graphically with a frequency histogram. Now we construct a stem-and-leaf diagram, which simultaneously groups the data and yields a graphical display similar to a histogram.

First, we select the leading digits from the data in Table 2.17, or 3, 4, ... , 9. Next, we list those leading digits in a column, as shown to the left of the vertical rule in Fig. 2.6(a) .

Then we write the final digit of each number from Table 2.17 to the right of the appropriate leading digit: The first investment has a maturity period of 70 days, which calls for a 0 to the right of the leading digit, 7. Reading down the first column of Table 2.17, we find that the second investment has a maturity period of 62 days, which calls for a 2 to the right of the leading digit, 6.

FIGURE 2.6

Diagrams for days-to-maturity data: (a) stem-and-leaf; (b) shaded stem-and-leaf; (c) ordered stem-and-leaf

Stems Leaves

	(a)		(b)		(c)
3	869	3	869	3	689
4	7	4	7	4	7
5	71635105	5	71635105	5	01135567
6	2473640985	6	2473640985	6	0234456789
7	0510980	7	0510980	7	0001589
8	5917036	8	5917036	8	0135679
9	9958	9	9958	9	5899

Continuing in this manner, we obtain the diagram displayed in Fig. 2.6(a). As indicated in the figure, the leading digits are called **stems** and the final digits **leaves;** the entire diagram is called a **stem-and-leaf diagram.**

The stem-and-leaf diagram for the days-to-maturity data is similar to a frequency histogram for those data because the length of the row of leaves for a class equals the frequency of the class. [Turn the stem-and-leaf diagram 90° counterclockwise, and compare it to the frequency histogram shown in Fig. 2.2(a) on page 59.]

By shading each row of leaves, as in Fig. 2.6(b), we get a diagram that looks even more like a frequency histogram of the data. It is called a **shaded stem-and-leaf diagram.** Because the numbers in it are still visible under the shading, a shaded stem-and-leaf diagram exhibits the raw (ungrouped) data in addition to providing a graphical display of a frequency distribution. Although a frequency histogram provides a graphical display of a frequency distribution, recovering the raw data from a frequency histogram generally is not possible.

Another form of stem-and-leaf diagram is called an **ordered stem-and-leaf diagram,** in which the leaves in each row are ordered from smallest to largest. Ordering makes the data easier to comprehend and facilitates computation of descriptive measures such as the median (discussed in Chapter 3). The ordered stem-and-leaf diagram for the days-to-maturity data is presented in Fig. 2.6(c).

In Example 2.16, we describe use of the stem-and-leaf diagram for three-digit numbers.

Example 2.16 Stem-and-Leaf Diagrams

Cholesterol Levels A pediatrician tested the cholesterol levels of several young patients and was alarmed to find that many had levels higher than 200 mg per 100 mL. The readings of 20 patients with high levels are presented in Table 2.18. Construct a stem-and-leaf diagram for these data.

TABLE 2.18
Cholesterol levels for 20 high-level patients

210	209	212	208
217	207	210	203
208	210	210	199
215	221	213	218
202	218	200	214

Solution Because these data are three-digit numbers, we use the first two digits as the stems and the third digit as the leaves. A stem-and-leaf diagram for the cholesterol levels is displayed in Fig. 2.7(a).

The stem-and-leaf diagram in Fig. 2.7(a) is only moderately helpful because there are so few stems. We can construct a better stem-and-leaf diagram by using two lines for each stem, with the first line for the leaf digits 0–4 and the second line for the leaf digits 5–9. This stem-and-leaf diagram is shown in Fig. 2.7(b).

FIGURE 2.7
Stem-and-leaf diagram for cholesterol levels:
(a) one line per stem;
(b) two lines per stem

```
                                        19 |
                                        19 | 9
                                        20 | 2 0 3
                                        20 | 8 9 7 8
           19 | 9                       21 | 0 0 2 0 0 3 4
           20 | 8 2 9 7 0 8 3           21 | 7 5 8 8
           21 | 0 7 5 0 8 2 0 0 3 8 4   22 | 1
           22 | 1                       22 |
                  (a)                          (b)
```

Although stem-and-leaf diagrams have several advantages over the more classical techniques for grouping and graphing, they do have some drawbacks. For instance, they are generally not useful with large data sets and can be awkward with data containing many digits; histograms are usually preferable to stem-and-leaf diagrams in such cases.

The Technology Center

Some technologies (e.g., Minitab) have built-in programs for constructing stem-and-leaf diagrams, but others (e.g., Excel and the TI-83/84 Plus) do not. Refer to the technology manuals for detailed instructions on obtaining stem-and-leaf diagrams—provided, of course, such plots are available in the technology under consideration. See Exercise 2.70 on page 72 for an example of a stem-and-leaf diagram generated by Minitab.

EXERCISES 2.4

Statistical Concepts and Skills

2.58 Discuss the relative advantages and disadvantages of stem-and-leaf diagrams and frequency histograms.

2.59 Suppose that you have a data set that contains a large number of observations. Which graphical display is generally preferable: a histogram or a stem-and-leaf diagram? Explain your answer.

2.60 Explain why the raw data generally cannot be recovered from a frequency histogram. Under what circumstances is recovering the raw data possible?

2.61 Suppose that you have constructed a stem-and-leaf diagram and discover that it is only moderately useful because there are too few stems. How can you remedy the problem?

2.62 Contents of Soft Drinks. A soft-drink bottler fills bottles with soda. For quality assurance purposes, filled bottles are sampled to ensure that they contain close to the content indicated on the label. A sample of 30 "one-liter" bottles of soda contain the amounts, in milliliters, shown in the following table.

1025	977	1018	975	977
990	959	957	1031	964
986	914	1010	988	1028
989	1001	984	974	1017
1060	1030	991	999	997
996	1014	946	995	987

Using the stems 91, 92, …, 106,
a. construct a stem-and-leaf diagram.
b. construct an ordered stem-and-leaf diagram.

2.63 Stressed-Out Bus Drivers. Frustrated passengers, congested streets, time schedules, and air and noise pollution are just some of the physical and social pressures that lead many urban bus drivers to retire prematurely with disabilities such as coronary heart disease and stomach disorders. An intervention program designed by the Stockholm Transit District was implemented to improve the work conditions of the city's bus drivers. Improvements were evaluated by Evans et al., who collected physiological and psychological data for bus drivers who drove the improved routes (intervention) and for drivers who were assigned the normal routes (control). Their findings were published in the article "Hassles on the Job: A Study of a Job Intervention with Urban Bus Drivers" (*Journal of Organizational Behavior*, Vol. 20, pp. 199–208). Following are data based on the results of the study for the heart rates, in beats per minute, of the control drivers.

74	52	67	63	77	57	80	77
53	76	54	73	54	60	77	63
60	68	64	66	71	66	55	71
84	63	73	59	68	64	82	

a. Construct a stem-and-leaf diagram having one line per stem.
b. Construct an ordered stem-and-leaf diagram having one line per stem.
c. Repeat parts (a) and (b) for two lines per stem.

2.64 High School Completion Rates. As reported by the U.S. Bureau of the Census in *Current Population Reports*, the percentage of adults in each state and the District of Columbia who have completed high school is as follows.

State	Percent	State	Percent	State	Percent
AL	78	KY	79	ND	86
AK	90	LA	81	OH	87
AZ	85	ME	89	OK	86
AR	82	MD	86	OR	88
CA	81	MA	85	PA	86
CO	90	MI	86	RI	81
CT	88	MN	91	SC	83
DE	86	MS	80	SD	92
DC	83	MO	87	TN	80
FL	84	MT	90	TX	79
GA	83	NE	90	UT	91
HI	87	NV	83	VT	90
ID	86	NH	88	VA	87
IL	86	NJ	87	WA	92
IN	85	NM	82	WV	77
IA	90	NY	83	WI	87
KS	88	NC	79	WY	90

Construct a stem-and-leaf diagram for the percentages with
a. one line per stem.   b. two lines per stem.
c. five lines per stem.
d. Which stem-and-leaf diagram do you consider most useful? Explain your answer.

2.65 Crime Rates. The U.S. Federal Bureau of Investigation published the following annual crime rates in *Crime in the United States*. Rates are per 1000 population.

State	Rate	State	Rate	State	Rate
AL	45	KY	30	ND	23
AK	42	LA	54	OH	40
AZ	58	ME	26	OK	46
AR	41	MD	48	OR	48
CA	37	MA	30	PA	30
CO	40	MI	41	RI	35
CT	32	MN	35	SC	52
DE	45	MS	40	SD	23
DC	73	MO	45	TN	49
FL	57	MT	35	TX	50
GA	48	NE	41	UT	45
HI	52	NV	43	VT	30
ID	32	NH	24	VA	30
IL	43	NJ	32	WA	51
IN	38	NM	55	WV	26
IA	32	NY	31	WI	32
KS	44	NC	49	WY	33

Construct a stem-and-leaf diagram for the crime rates with
a. one line per stem.   b. two lines per stem.
c. five lines per stem.
d. Which stem-and-leaf diagram do you consider most useful? Explain your answer.

Extending the Concepts and Skills

Further Stem-and-Leaf Techniques. We mentioned earlier that the use of stem-and-leaf diagrams can be awkward with data that contain many digits. In such cases, we can either round or truncate each observation to a suitable number of digits. Exercises 2.66 and 2.67 involve rounding and truncating numbers for use in stem-and-leaf diagrams.

2.66 Weights of Males. The U.S. National Center for Health Statistics publishes data on weights and heights by age and sex in *Vital and Health Statistics*. The following weights, to the nearest tenth of a pound, were obtained from a sample of 18–24-year-old males.

129.2	185.3	218.1	182.5	142.8
155.2	170.0	151.3	187.5	145.6
167.3	161.0	178.7	165.0	172.5
191.1	150.7	187.0	173.7	178.2
161.7	170.1	165.8	214.6	136.7
278.8	175.6	188.7	132.1	158.5
146.4	209.1	175.4	182.0	173.6
149.9	158.6			

a. Round each observation to the nearest pound and then construct a stem-and-leaf diagram of the rounded data.
b. Truncate each observation by dropping the decimal part and then construct a stem-and-leaf diagram of the truncated data.
c. Compare the stem-and-leaf diagrams that you obtained in parts (a) and (b).

2.67 Contents of Soft Drinks. Refer to Exercise 2.62.
a. Round each observation to the nearest 10 mL, drop the terminal 0s, and then obtain a stem-and-leaf diagram of the resulting data.
b. Truncate each observation by dropping the units digit and then construct a stem-and-leaf diagram of the truncated data.
c. Compare the stem-and-leaf diagrams that you obtained in parts (a) and (b) with each other and with the one obtained in Exercise 2.62.

Using Technology

2.68 High School Completion Rates. Use the technology of your choice to construct the stem-and-leaf diagrams required in Exercise 2.64.

2.69 Crime Rates. Use the technology of your choice to construct the stem-and-leaf diagrams required in Exercise 2.65.

2.70 NBA Leading Scorers. The National Basketball Association (NBA) provides information on average points per game for leading scorers on the Web site www.nba.com. A random sample of 24 NBA scoring leaders for the 2002–2003 regular season gave the following data on average points per game (PPG).

Player	PPG	Player	PPG
Tracy McGrady	32.1	Kobe Bryant	30.0
Allen Iverson	27.6	Paul Pierce	25.9
Dirk Nowitzki	25.1	Ray Allen	22.5
Allan Houston	22.5	Antawn Jamison	22.2
Shawn Marion	21.2	Steve Francis	21.0
Glenn Robinson	20.8	Jermaine O'Neal	20.8
Karl Malone	20.6	Michael Jordan	20.0
Richard Hamilton	19.7	Pau Gasol	19.0
Jason Kidd	18.7	Gilbert Arenas	18.3
Rashard Lewis	18.1	Matt Harpring	17.6
Cuttino Mobley	17.5	David Wesley	16.7
Kenyon Martin	16.7	Latrell Sprewell	16.4

The following Minitab printout displays a stem-and-leaf diagram for the data on average number of points per game. The second column gives the stems and the third column gives the leaves.

Stem-and-Leaf Display: PPG

```
Stem-and-leaf of PPG   N  = 24
Leaf Unit = 1.0


    5   1   66677
   10   1   88899
  (6)   2   000011
    8   2   222
    5   2   55
    3   2   7
    2   2
    2   3   0
    1   3   2
```

Did Minitab use rounding or truncation to obtain this stem-and-leaf diagram? Explain your answer.

2.5   Distribution Shapes; Symmetry and Skewness

In this section, we discuss distributions and their associated properties. To begin, we present a formal definition for the **distribution of a data set.**

DEFINITION 2.5

Distribution of a Data Set

The *distribution of a data set* is a table, graph, or formula that provides the values of the observations and how often they occur.

Up to now, we have portrayed distributions of data sets by frequency distributions, relative-frequency distributions, frequency histograms, relative-frequency histograms, dotplots, stem-and-leaf diagrams, pie charts, and bar graphs.

An important aspect of the distribution of a quantitative data set is its shape. Indeed, as we demonstrate in later chapters, the shape of a distribution frequently plays a role in determining the appropriate method of statistical analysis. To identify the shape of a distribution, the best approach usually is to use a smooth curve that approximates the overall shape.

For instance, Fig. 2.8 displays a relative-frequency histogram for the heights of the 3264 female students who attend a midwestern college. Also included in Fig. 2.8 is a smooth curve that approximates the overall shape of the distribution. Both the histogram and the smooth curve show that this distribution of

FIGURE 2.8
Relative-frequency histogram and
approximating smooth curve for the
distribution of heights

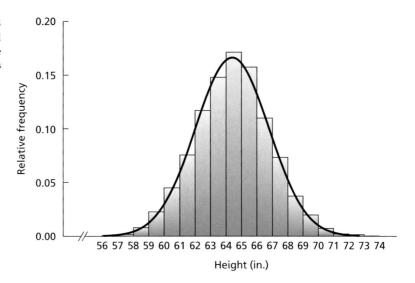

heights is bell shaped (or mound shaped), but the smooth curve makes seeing the shape a little easier.

Another advantage of using smooth curves to identify distribution shapes is that we need not worry about minor differences in shape. Instead we can concentrate on overall patterns, which, in turn, allows us to classify most distributions by designating relatively few shapes.

Distribution Shapes

Figure 2.9 displays some common distribution shapes: **bell shaped, triangular, uniform, reverse J shaped, J shaped, right skewed, left skewed, bimodal,** and **multimodal.** These shapes are idealized forms; in practice, distributions rarely have these exact shapes. Consequently, in identifying the shape of a distribu-

FIGURE 2.9
Common distribution shapes

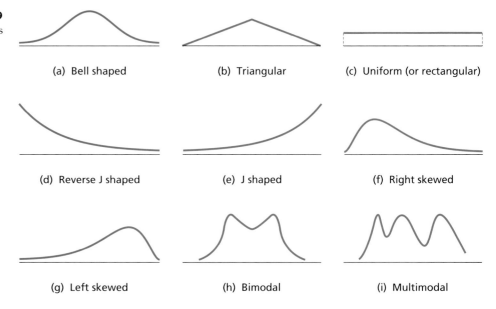

tion, exact conformance is not required, especially when we are considering small data sets. So, for example, we describe the distribution of heights displayed in Fig. 2.8 as bell shaped, even though the histogram does not form a perfect bell. In Example 2.17, we illustrate how to identify a distribution shape.

■▌▌ **Example 2.17**   **Identifying Distribution Shapes**

Household Size   The relative-frequency histogram for household size in the United States shown in Fig. 2.10(a) is based on data contained in *Current Population Reports*, a publication of the U.S. Bureau of the Census.[†] Identify the distribution shape for sizes of U.S. households.

FIGURE 2.10
Relative-frequency histogram for household size

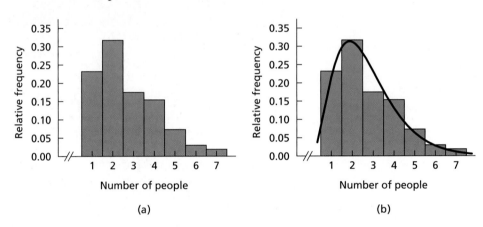

(a)   (b)

Solution   First, we draw a smooth curve through the histogram shown in Fig. 2.10(a) to get Fig. 2.10(b). Then, by referring to Fig. 2.9, we find that the distribution of household sizes is right skewed.

◼━

There are distribution shapes other than those represented in Fig. 2.9. However, the types shown in Fig. 2.9 comprise the most commonly encountered distribution shapes and suffice for our purposes in this book.

Modality

When considering the shape of a distribution, you should observe the number of peaks (highest points). A distribution is said to be **unimodal** if it has one peak; **bimodal** if it has two peaks; and **multimodal** if it has three or more peaks.

The distribution of heights in Fig. 2.8 is unimodal. More generally, we see from Fig. 2.9 that bell-shaped, triangular, reverse J-shaped, J-shaped, right-skewed, and left-skewed distributions are unimodal. Representations of bimodal and multimodal distributions are displayed in Figs. 2.9(h) and (i), respectively.[‡]

> **What**
> *Does it Mean?*
>
> Technically, a distribution is bimodal or multimodal only if the peaks are the same height. However, in practice, distributions with pronounced but not necessarily equal-height peaks are often called bimodal or multimodal.

[†]Actually, the class 7 portrayed in Fig. 2.10 is for seven or more people.

[‡]A uniform distribution has either no peaks or infinitely many peaks, depending on how you look at it. In any case, we do not classify a uniform distribution according to modality.

Symmetry and Skewness

Each of the three distributions in Figs. 2.9(a)–(c) can be divided into two pieces that are mirror images of one another. A distribution that has that property is called **symmetric.** Therefore bell-shaped, triangular, and uniform distributions are symmetric. The bimodal distribution pictured in Fig. 2.9(h) also is symmetric, but that is not always true for bimodal or multimodal distributions. Figure 2.9(i) shows an asymmetric multimodal distribution.

Again, when classifying distributions, we must be somewhat flexible. Thus exact symmetry is not required to classify a distribution as symmetric. For example, the distribution of heights in Fig. 2.8 is considered symmetric.

A unimodal distribution that is not symmetric is either right skewed, as in Fig. 2.9(f), or left skewed, as in Fig. 2.9(g). On the one hand, a right-skewed distribution rises to its peak rapidly and comes back toward the horizontal axis more slowly—its "right tail" is longer than its "left tail." On the other hand, a left-skewed distribution rises to its peak slowly and comes back toward the horizontal axis more rapidly—its "left tail" is longer than its "right tail." Note that reverse J-shaped distributions (Fig. 2.9(d)) and J-shaped distributions (Fig. 2.9(e)) are special types of right-skewed and left-skewed distributions, respectively.

Population and Sample Distributions

Recall that a variable is a characteristic that varies from one person or thing to another and that observing one or more values of a variable yields data. The data set obtained by observing the values of a variable for an entire population is called **population data** or **census data;** a data set obtained by observing the values of a variable for a sample of the population is called **sample data.** To distinguish their distributions, we use the terminology **population distribution** (or **distribution of the variable**) and **sample distribution.**

DEFINITION 2.6

Population and Sample Distributions; Distribution of a Variable

The distribution of population data is called the *population distribution,* or the *distribution of the variable.*

The distribution of sample data is called a *sample distribution.*

For a particular population and variable, sample distributions vary from sample to sample. However, there is only one population distribution, namely, the distribution of the variable under consideration on the population under consideration. Example 2.18 illustrates this point and some others as well.

Example 2.18 **Population and Sample Distributions**

Household Size In Example 2.17, we considered the distribution of household size for U.S. households. Here the variable is household size, and the population consists of all U.S. households. Figure 2.10(a) on page 74 provides a relative-frequency histogram of household size for the population of all U.S. households—this graph gives the population distribution or, equivalently, the distribution of the variable "household size." For ease in reference, we repeat Fig. 2.10(a) as Fig. 2.11(a).

FIGURE 2.11

Population distribution and six sample distributions for household size

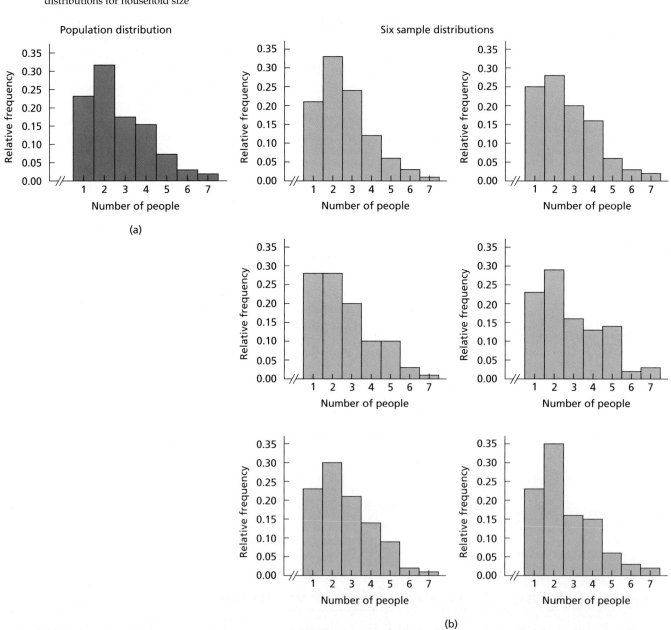

We obtained six simple random samples of 100 households each from the population of all U.S. households. Figure 2.11(b) shows relative-frequency histograms of household size for all six samples. Compare the six sample distributions in Fig. 2.11(b) to each other and to the population distribution in Fig. 2.11(a).

Solution As shown in Fig. 2.11(b), the distributions of the six samples, although similar, have definite differences. This result is not surprising because we would expect variation from one sample to another. Nonetheless, the overall shapes of the six sample distributions are roughly the same and also are similar in shape to the population distribution shown in Fig. 2.11(a)—all are right skewed.

—■

In practice, we usually do not know the population distribution. As Example 2.18 suggests, we can, under those circumstances, use the distribution of a simple random sample from the population to get a rough idea of the population distribution.

KEY FACT 2.1 **Population and Sample Distributions**

The distribution of a simple random sample from a population approximates the population distribution. In other words, if a simple random sample is taken from a population, the distribution of the observed values of the variable under consideration will approximate the distribution of the variable. The larger the sample, the better the approximation tends to be.

EXERCISES 2.5

Statistical Concepts and Skills

2.71 Explain the meaning of
a. distribution of a data set.   b. sample data.
c. population data.   d. census data.
e. sample distribution.   f. population distribution.
g. distribution of a variable.

2.72 Give two reasons why the use of smooth curves to describe shapes of distributions is helpful.

2.73 Suppose that a variable of a population has a bell-shaped distribution. If you take a large simple random sample from the population, roughly what shape would you expect the distribution of the sample to be? Explain your answer.

2.74 Suppose that a variable of a population has a reverse J-shaped distribution and that two simple random samples are taken from the population.

a. Would you expect the distributions of the two samples to have roughly the same shape? If so, what shape?
b. Would you expect some variation in shape for the distributions of the two samples? Explain your answer.

2.75 Identify and sketch three distribution shapes that are symmetric.

In each of Exercises 2.76–2.81, we have provided a graphical display of a data set. For each exercise,
a. *identify the overall shape of the distribution by referring to Fig. 2.9 on page 73.*
b. *state whether the distribution is (roughly) symmetric, right skewed, or left skewed.*

2.76 Children of U.S. Presidents. The *Information Please Almanac* provides the number of children of each of the U.S. presidents. A frequency histogram for number of children by president, through President George W. Bush, is as follows.

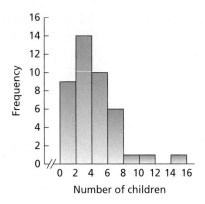

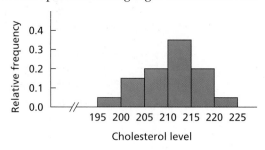

2.79 Cholesterol Levels. Refer to Exercise 2.51 on page 66. The following relative-frequency histogram shows the readings of some patients having high cholesterol levels.

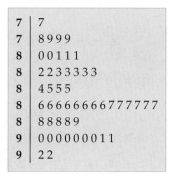

2.77 The Great White Shark. Refer to Exercise 2.25 on page 55. The following frequency histogram was obtained from data on the number of pups borne in a lifetime by each of 80 Great White Shark females.

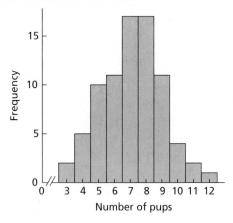

2.80 High School Completion Rates. Refer to Exercise 2.64 on page 70. The following stem-and-leaf diagram displays the completion rates.

```
7 | 7
7 | 8 9 9 9
8 | 0 0 1 1 1
8 | 2 2 3 3 3 3 3
8 | 4 5 5 5
8 | 6 6 6 6 6 6 6 6 7 7 7 7 7 7
8 | 8 8 8 8 9
9 | 0 0 0 0 0 0 0 0 1 1
9 | 2 2
```

2.81 Stays in Europe and the Mediterranean. The Bureau of Economic Analysis gathers information on the length of stay in Europe and the Mediterranean by U.S. travelers. Data are published in *Survey of Current Business*. The following stem-and-leaf diagram portrays the length of stay, in days, of a sample of 36 U.S. residents who traveled to Europe and the Mediterranean last year.

```
0 | 5 3 3 1 6 2 1 1 8 5 3
1 | 6 3 1 5 4 0 2 7 8 0 2 2 0
2 | 1 0 1 7 1
3 | 1 2
4 | 1 4 8
5 | 6
6 | 4
```

2.78 Adjusted Gross Incomes. Refer to Exercise 2.50 on page 66. The following relative-frequency histogram shows one year's individual income tax returns for adjusted gross incomes of less than $50,000.

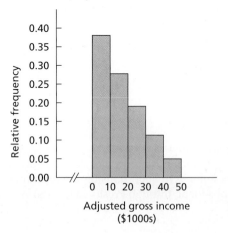

Extending the Concepts and Skills

2.82 Class Project: Number of Siblings. This exercise is a class project and works best in relatively large classes.

a. Determine the number of siblings for each student in the class.

b. Obtain a relative-frequency histogram of the number of siblings. Use single-value grouping.

c. Obtain a simple random sample of about one-third of the students in the class.

d. Determine the number of siblings for each student in the sample.

e. Obtain a relative-frequency histogram of the number of siblings for the sample. Use single-value grouping.

f. Repeat parts (c)–(e) three more times.

g. Compare the histograms for the samples to each other and to that for the entire population. Relate your observations to Key Fact 2.1.

2.83 Class Project: Random Digits. This exercise can be done individually or, better yet, as a class project.

a. Use a table of random numbers or a random-number generator to obtain 50 random integers between 0 and 9.

b. Without graphing the distribution of the 50 numbers you obtained, guess its shape. Explain your reasoning.

c. Construct a relative-frequency histogram based on single-value grouping for the 50 numbers that you obtained in part (a). Is its shape about what you expected?

d. If your answer to part (c) was "no," provide an explanation.

e. What would you do to make getting a "yes" answer to part (c) more plausible?

f. If you are doing this exercise as a class project, repeat parts (a)–(c) for 1000 random integers.

Using Technology

Simulation. For purposes of both understanding and research, simulating variables is often useful. Simulating a variable involves the use of a computer or statistical calculator to generate observations of the variable. In Exercises 2.84 and 2.85, the use of simulation will enhance your understanding of distribution shapes and the relation between population and sample distributions.

2.84 Random Digits. In this exercise, use technology to work Exercise 2.83, as follows:

a. Use the technology of your choice to obtain 50 random integers between 0 and 9.

b. Use the technology of your choice to get a relative-frequency histogram based on single-value grouping for the numbers that you obtained in part (a).

c. Repeat parts (a) and (b) five more times.

d. Are the shapes of the distributions that you obtained in parts (a)–(c) about what you expected?

e. Repeat parts (a)–(d), but generate 1000 random integers each time instead of 50.

2.85 Standard Normal Distribution. One of the most important distributions in statistics is the *standard normal distribution*. We discuss this distribution in detail in Chapter 6.

a. Use the technology of your choice to generate a sample of 3000 observations from a variable that has the standard normal distribution, that is, a normal distribution with mean 0 and standard deviation 1.

b. Use the technology of your choice to get a relative-frequency histogram for the 3000 observations that you obtained in part (a).

c. Based on the histogram you obtained in part (b), what shape does the standard normal distribution have? Explain your reasoning.

2.6 Misleading Graphs

Graphs and charts are frequently constructed in a manner that causes them to be misleading. Sometimes the misleading is intentional, and sometimes it is inadvertent. Regardless of intent, graphs and charts must be read and interpreted with a great deal of care. In this section, we examine some misleading graphs and charts, beginning with one in Example 2.19.

Example 2.19 Truncated Graphs

Unemployment Rates Figure 2.12(a) shows a bar graph from an article in a major metropolitan newspaper. The graph displays the unemployment rates in the United States from September of one year through March of the next year.

Because the bar for March is about one-fourth smaller than the bar for January, a quick look at Fig. 2.12(a) might lead you to conclude that the unem-

FIGURE 2.12

Unemployment rates:
(a) truncated graph;
(b) nontruncated graph

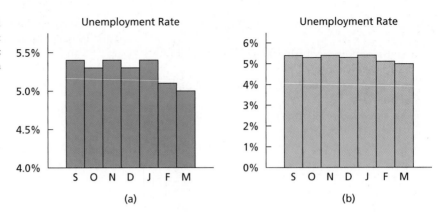

ployment rate dropped by roughly one-fourth between January and March. In reality, however, the unemployment rate dropped by less than one-thirteenth, from 5.4% to 5.0%. Consequently, you must analyze the graph more carefully to discover what it truly represents.

Figure 2.12(a) is an example of a **truncated graph** because the vertical axis, which should start at 0%, starts at 4% instead. Thus the part of the graph from 0% to 4% has been cut off, or truncated. This truncation causes the bars to be out of correct proportion and hence creates a misleading impression.

The graph would be even more deceptive if it started at 4.5%. To see this result, slide a piece of paper over the bottom of Fig. 2.12(a) so that the bars begin at 4.5%. By how much does it now appear that the unemployment rate dropped between January and March?

Although the truncated graph in Fig. 2.12(a) is potentially misleading, the truncation probably was done to present a picture of the "ups" and "downs" in the unemployment rate pattern rather than to mislead the reader intentionally.

A nontruncated version of Fig. 2.12(a) is shown in Fig. 2.12(b). Figure 2.12(b) provides a correct graphical display of the unemployment rate data, but the "ups" and "downs" are not so easy to spot as they are in the truncated graph in Fig. 2.12(a).

Truncated graphs have long been a target of statisticians, and many statistics books warn against their use. Nonetheless, as illustrated by Example 2.19, truncated graphs are still used today, even in reputable publications.

However, Example 2.19 also suggests that cutting off part of the vertical axis of a graph may be desirable. Doing so may allow relevant information, such as the "ups" and "downs" of the monthly unemployment rates, to be conveyed more easily. In such cases, though, a truncated graph should not be used. Instead, a special symbol, such as //, should be utilized to signify that the vertical axis has been modified.

The two graphs shown in Fig. 2.13 provide an excellent illustration. Both portray the number of new single-family homes sold per month over several months. The graph shown in Fig. 2.13(a) is truncated—most likely in an attempt to present a clear visual display of the variation in sales. The graph shown in Fig. 2.13(b) accomplishes the same result but is less subject to misinterpretation; you are aptly warned by the slashes that part of the vertical axis between 0 and 500 has been removed.

FIGURE 2.13
New single-family home sales

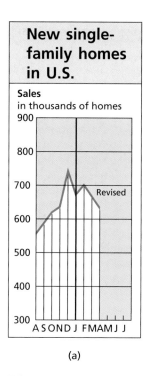

New single-family homes in U.S.

Sales
in thousands of homes

Revised

A S O N D J F M A M J J

(a)

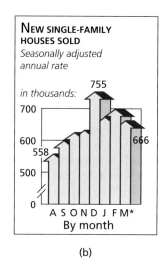

NEW SINGLE-FAMILY HOUSES SOLD
Seasonally adjusted annual rate

in thousands:

755

558

666

A S O N D J F M*
By month

(b)

SOURCES: Figure 2.13(a) reprinted by permission of Tribune Media Services. Figure 2.13(b) data from U.S. Department of Commerce and U.S. Department of Housing and Urban Development.

Improper Scaling

Misleading graphs and charts can also result from **improper scaling.** In Example 2.20, we show how that can happen.

▌Example 2.20   Improper Scaling

FIGURE 2.14
Pictogram for home building

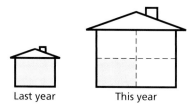

Last year   This year

Home Building   A developer is preparing a brochure to attract investors for a new shopping center to be built in an area of Denver, Colorado. The area is growing rapidly; this year twice as many homes will be built there as last year. To illustrate that fact, the developer draws a **pictogram,** as shown in Fig. 2.14.

The house on the left represents the number of homes built last year. Because the number of homes that will be built this year is double the number built last year, the developer makes the house on the right twice as tall and twice as wide as the house on the left. However, this scaling is improper as it gives the visual impression that four times as many homes will be built this year as last. Thus the developer's brochure may mislead the unwary investor.   ▬

Graphs and charts can be misleading in countless ways besides the two that we discussed. Many more examples of misleading graphs can be found in the entertaining and classic book *How to Lie with Statistics* by Darrell Huff (New York: Norton, 1955). The main purpose of this section has been to show you that graphs and charts should be constructed and read carefully.

EXERCISES 2.6

Statistical Concepts and Skills

2.86 Give one reason why constructing and reading graphs and charts carefully is important.

2.87 This exercise deals with truncated graphs.
a. What is a truncated graph?
b. Give a legitimate motive for truncating the axis of a graph.
c. If you have a legitimate motive for truncating the axis of a graph, how can you correctly obtain that objective without creating the possibility of misinterpretation?

2.88 In a current newspaper or magazine, find two examples of graphs that might be misleading. Explain why you think the graphs are potentially misleading.

2.89 Reading Skills. Each year the director of the reading program in a school district administers a standard test of reading skills. Then the director compares the average score for his district with the national average. Figure 2.15 was presented to the school board in the year 2002.

FIGURE 2.15
Average reading scores

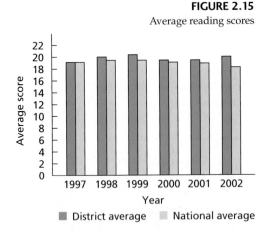

■ District average   ■ National average

a. Obtain a truncated version of Fig. 2.15 by sliding a piece of paper over the bottom of the graph so that the bars start at 16.
b. Repeat part (a) but have the bars start at 18.
c. What misleading impression about the year 2002 scores is given by the truncated graphs obtained in parts (a) and (b)?

2.90 America's Melting Pot. The following bar graph is based on a newspaper article entitled "Immigrants add seasoning to America's melting pot." [Used with permission from American Demographics, Ithaca, NY.]

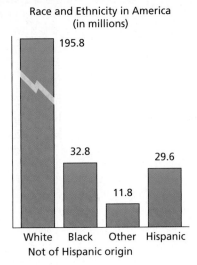

Race and Ethnicity in America
(in millions)

Data from Census Bureau July 1998 Estimates

a. Explain why a break is shown in the first bar.
b. Why was the graph constructed with a broken bar?
c. Is this graph potentially misleading? Explain your answer.

2.91 M2 Money Supply. The following bar graph, taken from *The Arizona Republic*, provides data on the M2 money supply over several months. M2 consists of cash in circulation, deposits in checking accounts, nonbank traveler's checks, accounts such as savings deposits, and money-market mutual funds.

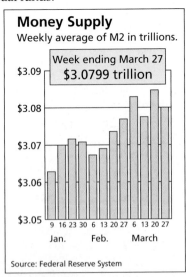

Money Supply
Weekly average of M2 in trillions.

Week ending March 27
$3.0799 trillion

Source: Federal Reserve System

a. What is wrong with the bar graph?
b. Construct a version of the bar graph with a nontruncated and unmodified vertical axis.

c. Construct a version of the bar graph in which the vertical axis is modified in an acceptable manner.

Extending the Concepts and Skills

2.92 Home Building. Refer to Example 2.20 on page 81. Suggest a way in which the developer can accurately illustrate that twice as many homes will be built in the area this year as last.

2.93 Marketing Golf Balls. A golf ball manufacturer has determined that a newly developed process results in a ball that lasts roughly twice as long as a ball produced by the current process. To illustrate this advance graphically, she designs a brochure showing a "new" ball having twice the radius of the "old" ball.

Old ball   New ball

a. What is wrong with this depiction?
b. How can the manufacturer accurately illustrate the fact that the "new" ball lasts twice as long as the "old" ball?

CHAPTER REVIEW

You Should Be Able To

1. classify variables and data as either qualitative or quantitative.

2. distinguish between discrete and continuous variables and data.

3. identify terms associated with the grouping of data.

4. group data into a frequency distribution and a relative-frequency distribution.

5. construct a grouped-data table.

6. draw a frequency histogram and a relative-frequency histogram.

7. construct a dotplot.

8. draw a pie chart and a bar graph.

9. construct stem-and-leaf, shaded stem-and-leaf, and ordered stem-and-leaf diagrams.

10. identify the shape and modality of the distribution of a data set.

11. specify whether a unimodal distribution is symmetric, right skewed, or left skewed.

12. understand the relationship between sample distributions and the population distribution (distribution of the variable under consideration).

13. identify and correct misleading graphs.

Key Terms

bar graph, *62*
bell shaped, *73*
bimodal, *73, 74*
categorical variables, *42*
census data, *75*
classes, *49*
continuous data, *43*
continuous variable, *42*
count, *48*
data, *43*
data set, *43*
discrete data, *43*
discrete variable, *42*
distribution of a data set, *72*
distribution of a variable, *75*

dotplot, *60*
frequency, *49*
frequency distribution, *49*
frequency histogram, *59*
grouped-data table, *49*
grouping, *46*
improper scaling, *81*
J shaped, *73*
leaves, *68*
left skewed, *73*
lower cutpoint, *49*
lower limit, *51*
mark, *51*
midpoint, *49*
multimodal, *73, 74*

observation, *43*
ordered stem-and-leaf diagram, *69*
percentage, *48*
pictogram, *81*
pie chart, *61*
population data, *75*
population distribution, *75*
qualitative data, *43*
qualitative variables, *42*
quantitative data, *43*
quantitative variables, *42*
relative frequency, *49*
relative-frequency distribution, *49*
relative-frequency histogram, *59*
reverse J shaped, *73*

REVIEW TEST

Statistical Concepts and Skills

1. This problem is about variables and data.
a. What is a variable?
b. Identify two main types of variables.
c. Identify the two types of quantitative variables.
d. What are data?
e. How is data type determined?

2. Explain why grouping data is important.

3. To which type of data do the concepts of cutpoints and midpoints not apply? Explain your answer.

4. A quantitative data set has been grouped into a grouped-data table with equal-width classes of width 8.
a. If the midpoint of the first class is 10, what are its lower and upper cutpoints?
b. What is the midpoint of the second class?
c. What are the lower and upper cutpoints of the third class?
d. Into which class would an observation of 22 go?

5. A quantitative data set has been grouped into a grouped-data table with equal-width classes.
a. If the lower and upper cutpoints of the first class are 5 and 15, respectively, what is the common class width?
b. What is the midpoint of the second class?
c. What are the lower and upper cutpoints of the third class?

6. When is the use of single-value grouping particularly appropriate?

7. In each of the following cases, explain the relative positioning of the bars in a histogram to the numbers that label the horizontal axis.
a. Cutpoints are used to label the horizontal axis.
b. Midpoints are used to label the horizontal axis.

8. Identify two main types of graphical displays that are used for qualitative data.

9. Which is preferable as a graphical display for a large, quantitative data set: a histogram or a stem-and-leaf diagram? Explain your answer.

10. Sketch the curve corresponding to each of the following distribution shapes.
a. Bell shaped   **b.** Right skewed
c. Reverse J shaped   **d.** Uniform

11. Make an educated guess as to the distribution shape of each of the following variables. Explain your answers.
a. Height of American adult males
b. Annual income of U.S. households
c. Age of full-time college students
d. Cumulative GPA of college seniors

12. A variable of a population has a left-skewed distribution.
a. If a large simple random sample is taken from the population, roughly what shape will the distribution of the sample have? Explain your answer.
b. If two simple random samples are taken from the population, would you expect the two sample distributions to have identical shapes? Explain your answer.
c. If two simple random samples are taken from the population, would you expect the two sample distributions to have similar shapes? If so, what shape would that be? Explain your answers.

13. Largest Hydroelectric Plants. The world's five largest hydroelectric plants, based on ultimate capacity, are as shown in the table at the top of the next column. Capacities are in megawatts. [SOURCE: T. W. Mermel, *Intl. Waterpower & Dam Construction Handbook*.]
a. What type of data is given in the first column of the table?
b. What type of data is given in the fourth column?
c. What type of data is given in the third column?

Rank	Name	Country	Capacity
1	Turukhansk	Russia	20,000
2	Three Gorges	China	18,200
3	Itaipu	Brazil/Para.	13,320
4	Grand Coulee	U.S.A.	10,830
5	Guri	Venezuela	10,300

14. Inauguration Ages. The ages at inauguration for the first 43 presidents of the United States (from George Washington to George W. Bush) are as follows.

President	Age at inaug.	President	Age at inaug.
G. Washington	57	B. Harrison	55
J. Adams	61	G. Cleveland	55
T. Jefferson	57	W. McKinley	54
J. Madison	57	T. Roosevelt	42
J. Monroe	58	W. Taft	51
J. Q. Adams	57	W. Wilson	56
A. Jackson	61	W. Harding	55
M. Van Buren	54	C. Coolidge	51
W. Harrison	68	H. Hoover	54
J. Tyler	51	F. Roosevelt	51
J. Polk	49	H. Truman	60
Z. Taylor	64	D. Eisenhower	62
M. Fillmore	50	J. Kennedy	43
F. Pierce	48	L. Johnson	55
J. Buchanan	65	R. Nixon	56
A. Lincoln	52	G. Ford	61
A. Johnson	56	J. Carter	52
U. Grant	46	R. Reagan	69
R. Hayes	54	G. Bush	64
J. Garfield	49	W. Clinton	46
C. Arthur	50	G. W. Bush	54
G. Cleveland	47		

a. Construct a grouped-data table for the inauguration ages in the preceding table. Use equal-width classes and begin with the class 40–44.
b. Identify the lower and upper cutpoints of the first class. (*Hint:* Be careful!)
c. Identify the common class width.
d. Draw a frequency histogram for the inauguration ages based on your grouping in part (a).

15. Inauguration Ages. Refer to Problem 14. Construct a dotplot for the ages at inauguration of the first 43 presidents of the United States.

16. Inauguration Ages. Refer to Problem 14. Construct an ordered stem-and-leaf diagram for the inauguration ages of the first 43 presidents of the United States.
a. Use one line per stem.
b. Use two lines per stem.
c. Which of the two stem-and-leaf diagrams that you just constructed corresponds to the frequency distribution of Problem 14(a)?

17. Busy Bank Tellers. The Prescott National Bank has six tellers available to serve customers. The data in the following table provide the number of busy tellers observed during 25 spot checks.

6	5	4	1	5
6	1	5	5	5
3	5	2	4	3
4	5	0	6	4
3	4	2	3	6

a. Construct a grouped-data table for these data. Use single-value grouping.
b. Draw a relative-frequency histogram for the data based on the grouping in part (a).

18. Student Class Levels. The class levels of the students in Professor Weiss's introductory statistics course are shown in the following table. The abbreviations Fr, So, Ju, and Se represent Freshman, Sophomore, Junior, and Senior, respectively.

Fr	So	Ju	So	Ju	Ju	Se	Ju
Se	So	Fr	Ju	So	Ju	So	Se
So	So	Se	So	So	Se	So	Fr
Ju	So	Ju	Fr	Fr	Ju	Ju	Fr
So	Se	Ju	Ju	So	So	So	Se

a. Obtain frequency and relative-frequency distributions for these data.
b. Draw a pie chart of the data that displays the percentage of students at each class level.
c. Draw a bar graph of the data that displays the relative frequency of students at each class level.

19. Dow Jones Annual Highs. According to *The World Almanac*, the highs for the Dow Jones Industrial Averages for 1967–2002 are as follows.

Year	High	Year	High
1967	943.08	1985	1553.10
1968	985.21	1986	1955.57
1969	968.85	1987	2722.42
1970	842.00	1988	2183.50
1971	950.82	1989	2791.41
1972	1036.27	1990	2999.75
1973	1051.70	1991	3168.83
1974	891.66	1992	3413.21
1975	881.81	1993	3794.33
1976	1014.79	1994	3978.36
1977	999.75	1995	5216.47
1978	907.74	1996	6560.91
1979	897.61	1997	8259.31
1980	1000.17	1998	9547.94
1981	1024.05	1999	11568.80
1982	1071.55	2000	11722.98
1983	1287.20	2001	11337.92
1984	1286.64	2002	10635.25

a. Construct a grouped-data table for the highs. Use classes of equal width and start with the class $0 \le 1000$.
b. Draw a relative-frequency histogram for the highs based on your result in part (a).

20. Identify the distribution shapes of each of the following data sets.
a. The inauguration ages of the first 43 presidents of the United States (from Problem 14).
b. The number of tellers busy with customers at Prescott National Bank during 25 spot checks (from Problem 17).

21. Draw a smooth curve that represents a symmetric tri-modal (three-peak) distribution.

22. Reshaping the Labor Force. The following graph is based on one that appeared in a newspaper article entitled "Hand that rocked cradle turns to work as women reshape U.S. labor force." The graph depicts the labor force participation rates for the years 1960, 1980, and 2000.

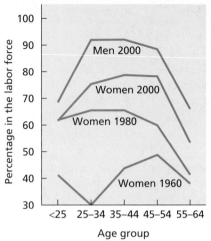

Working Men and Women by Age, 1960–2000

a. Cover the numbers on the vertical axis of the graph with a piece of paper.
b. Look at the 1960 and 2000 graphs for women, focusing on the 35–44-year-old age group. What impression does the graph convey regarding the ratio of the percentages of women in the labor force for 1960 and 2000?
c. Now remove the piece of paper from the graph. Use the vertical scale to find the actual ratio of the percentages of 35–44-year-old women in the labor force for 1960 and 2000.
d. Why is the graph potentially misleading?
e. What can be done to make the graph less potentially misleading?

Using Technology

23. Inauguration Ages. Refer to the age data in Problem 14. Use the technology of your choice to obtain a
a. frequency histogram of the data similar to the one found in Problem 14(d).
b. dotplot of the data.
c. stem-and-leaf diagram similar to the one constructed in Problem 16(b).

24. Student Class Levels. Refer to Problem 18. Use the technology of your choice to obtain a
a. pie chart of the data.
b. bar graph of the data.

StatExplore in MyMathLab
Analyzing Data Online

You can use StatExplore to perform all statistical analyses discussed in this chapter.
To illustrate, we show how to obtain a histogram and a stem-and-leaf diagram.

EXAMPLE   Histogram and Stem-and-Leaf

Days to Maturity for Short-Term Investments Table 2.1 on page 47 displays the number of days to maturity for 40 short-term investments. Use StatExplore to obtain a relative-frequency histogram and a stem-and-leaf diagram.

SOLUTION   To begin, we store the sample data from Table 2.1 in a column named DAYS. For the relative-frequency histogram, we proceed as follows:

> 1   Choose **Graphics ➤ Histogram**
> 2   Select the column DAYS
> 3   Click **Next →**
> 4   Click the arrow button at the right of the **Type** drop-down list box and select **Relative Frequency**
> 5   Click **Create Graph!**

The resulting output is shown in Printout 2.3.

PRINTOUT 2.3
StatExplore output for relative-frequency histogram

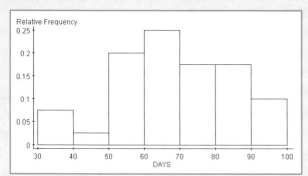

Note that StatExplore provides the option for choosing the classes of a histogram, but here we have just used the default classes.

To obtain a stem-and-leaf diagram for the days-to-maturity data, we proceed in the following way:

> 1   Choose **Graphics ➤ Stem and Leaf**
> 2   Select the column DAYS
> 3   Click **Create Graph!**

The resulting output is shown in Printout 2.4.

PRINTOUT 2.4
StatExplore output for stem and leaf

> **Variable: DAYS**
> 3 : 689
> 4 : 7
> 5 : 01135567
> 6 : 0234456789
> 7 : 0001589
> 8 : 0135679
> 9 : 5899

STATEXPLORE EXERCISES Solve the following problems by using StatExplore:

a. Obtain a frequency histogram and a relative-frequency histogram of the Cheetah speeds in Exercise 2.21 on page 55. Interpret your results.

b. Obtain a stem-and-leaf diagram for the heart rates in Exercise 2.63 on page 70. How many lines per stem did StatExplore use here?

c. Get the dotplot required in Example 2.12 on page 61. Compare the dotplot with the one we obtained by hand in Fig. 2.4 on page 61.

d. Determine the pie chart and bar graph required in Example 2.13 on page 61. (*Note:* The raw data for the political party affiliations are given in Table 2.10 on page 52.) Compare the pie chart and bar graph with those we obtained by hand in Fig. 2.5 on page 62.

To access StatExplore, go to the student content area of your Weiss MyMathLab course.

Internet Project

Simpson's Paradox

One of the most important and basic of all statistical questions is whether there is a relationship (or association) between variables. Sometimes two variables may appear to be related in a certain way, but when a third, hidden variable is taken into consideration, the association vanishes or even reverses direction! This behavior is known as *Simpson's paradox* and highlights the importance of looking for hidden or confounding variables.

In this Internet project, you will explore two data sets that illustrate Simpson's paradox. One data set is from a famous study of graduate admissions at a university and seems to show gender bias. The other data set comes from a recent study of healthcare for children in South Africa and appears to reveal a surprising association. In both cases, you will discover a hidden variable that significantly changes the apparent association.

Additionally, this project will provide an opportunity for further practice with the organization and display of data.

URL for access to Internet Projects Page: www.aw-bc.com/weiss

Focusing on Data Analysis

The Focus Database

Recall from Chapter 1 (see page 35) that the Focus database contains information on the undergraduate students at the University of Wisconsin - Eau Claire (UWEC). Statistical analyses for this database should be done with the technology of your choice.

  a. For each of the following variables, make an educated guess at its distribution shape: high school percentile (HSP), cumulative GPA (GPA), age (AGE), ACT English score (ENGLISH), ACT math score (MATH), and ACT composite score (COMP).
  b. Now obtain histograms for each of the variables in part (a) and compare your results with the educated guesses that you made in part (a). Which distributions are (roughly) symmetric?
  c. Obtain individual relative-frequency histograms for high school percentiles of males and females. Compare and discuss your results.
  d. Repeat part (c) for the variables cumulative GPA, age, ACT English score, ACT math score, and ACT composite score.
  e. Determine and interpret pie charts for the variables sex (SEX), classification (CLASS), residency (RESIDENCY), and admission type (TYPE).
  f. Repeat part (e) for bar graphs.
  g. Take five simple random samples (without replacement) of 250 students each. For each sample, obtain a histogram of the ages of the students in the sample and compare it with the histogram of ages for the entire population of students, as found in part (b).
  h. Repeat part (g) for high school percentile.
  i. Repeat part (g) for ACT composite score.
  j. What do the results of parts (g)–(i) illustrate?
  k. Take five simple random samples (without replacement) of 250 students each. For each sample, obtain a pie chart of the classification of the students in the sample and compare it with the pie chart of classification for the entire population of students, as found in part (e). What do the results of this exercise illustrate?

Case Study Discussion

Preventing Infant Mortality

Recall that the infant mortality rate of a nation represents the number of deaths of children under 1 year of age per 1000 live births in a calendar year. At the beginning of this chapter, we presented data on infant mortality for the 30 nations with the lowest rates. Refer to that data table on page 41 and solve each of the following problems.

a. What type of data is displayed in the second column of the table?

b. What type of data is given by the statement that the United States ranks 24th in infant mortality rate?

c. Construct a grouped-data table for the infant mortality rates. Use classes of equal width and start with the class $3.5 \leqslant 4.0$.

d. Construct a frequency histogram for the infant mortality rates based on your grouping in part (c).

e. Construct an ordered stem-and-leaf diagram for the infant mortality rates.

f. Use the technology of your choice to solve parts (c)–(e).

Internet Resources: Visit the Weiss Web site www.aw-bc.com/weiss for additional discussion, exercises, and resources related to this case study.

Biography

ADOLPHE QUETELET: On "The Average Man"

Lambert Adolphe Jacques Quetelet was born in Ghent, Belgium, on February 22, 1796. He attended school locally and, in 1819, received the first doctorate of science degree granted at the newly established University of Ghent. In that same year, he obtained a position as a professor of mathematics at the Brussels Athenaeum.

Quetelet was elected to the Belgian Royal Academy in 1820 and served as its secretary from 1834 until his death in 1874. He was founder and director of the Royal Observatory in Brussels, founder and a major contributor to the journal *Correspondance Mathématique et Physique*, and, according to Stephen M. Stigler in *The History of Statistics*, was "…active in the founding of more statistical organizations than any other individual in the nineteenth century." Among the organizations he established was the International Statistical Congress, initiated in 1853.

In 1835, Quetelet wrote a two-volume set titled *A Treatise on Man and the Development of His Faculties,* the publication in which he introduced his concept of the "average man" and that firmly established his international reputation as a statistician and sociologist. A review in the *Athenaeum* stated, "We consider the appearance of these volumes as forming an epoch in the literary history of civilization."

In 1855, Quetelet suffered a stroke that limited his work but not his popularity. He died on February 17, 1874. His funeral was attended by royalty and famous scientists from around the world. A monument to his memory was erected in Brussels in 1880.

3

Descriptive Measures

GENERAL OBJECTIVES   In Chapter 2, you began your study of descriptive statistics. There you learned how to organize data into tables and summarize data with graphs. Another method of summarizing data is to compute numbers, such as averages and percentiles, that describe the data set. Numbers that are used to describe data sets are called **descriptive measures.** In this chapter, we continue our discussion of descriptive statistics by examining some of the most commonly used descriptive measures.

In Section 3.1, we present *measures of center*—descriptive measures that indicate the center, or most typical value, in a data set. A particularly important measure of center is the mean; we discuss the mean of a sample in Section 3.2. Next, we examine *measures of variation*—descriptive measures that indicate the amount of variation or spread in a data set. In Section 3.3, we introduce two important measures of variation, the range and standard deviation.

The five-number summary, which we discuss in Section 3.4, includes descriptive measures that can be used to obtain both measures of center and measures of variation. That summary also provides the basis for a widely used graphical display, the boxplot.

In Section 3.5, we apply descriptive measures to populations. We also suggest how sample data can be used to provide estimates of descriptive measures of populations when census data are unavailable.

The Triple Crown

The Triple Crown is the most prestigious title that a 3-year-old thoroughbred racehorse can win. It is garnered by winning the Kentucky Derby, Preakness Stakes, and Belmont Stakes. These three races are open only to 3-year-olds; thus a horse has only one chance to win the Triple Crown.

Triple Crown was coined by sportswriter Charles Hatton in 1930. Sir Barton, in 1919, was the first horse to win all three races. Since then, only 10 horses have won the Triple Crown. Two trainers, James Fitzsimmons and Ben A. Jones, have trained two Triple Crown champions. Eddie Arcaro is the only jockey to ride two Triple Crown winners, on Whirlaway in 1941 and on Citation in 1948.

The Kentucky Derby is run on the first Saturday in May and has always been a $1\frac{1}{4}$ mile race. The Preakness Stakes is run 2 weeks later at a distance of $1\frac{3}{16}$ miles. Three weeks then elapse until the grueling $1\frac{1}{2}$ miles of the Belmont Stakes.

The following table provides the year, name, and times of the 11 Triple Crown winners. Note that, in 1919, the distances for the Preakness and Belmont were different from those for 1930 and thereafter. In this chapter, we demonstrate several additional techniques to help you analyze data. At the end of the chapter, after you have mastered those techniques, we ask you to apply them in analyzing the times of these Triple Crown winners.

		Time (minutes:seconds)		
Year	Horse	Kentucky Derby	Preakness Stakes	Belmont Stakes
1919	Sir Barton	2:09$\frac{4}{5}$	1:53 ($1\frac{1}{8}$ mi.)	2:17$\frac{2}{5}$ ($1\frac{3}{8}$ mi.)
1930	Gallant Fox	2:07$\frac{3}{5}$	2:00$\frac{3}{5}$	2:30$\frac{3}{5}$
1935	Omaha	2:05	1:58$\frac{2}{5}$	2:30$\frac{3}{5}$
1937	War Admiral	2:03$\frac{1}{5}$	1:58$\frac{2}{5}$	2:28$\frac{3}{5}$
1941	Whirlaway	2:01$\frac{2}{5}$	1:58$\frac{4}{5}$	2:31
1943	Count Fleet	2:04	1:57$\frac{2}{5}$	2:28$\frac{1}{5}$
1946	Assault	2:06$\frac{3}{5}$	2:01$\frac{2}{5}$	2:30$\frac{4}{5}$
1948	Citation	2:05$\frac{2}{5}$	2:02$\frac{2}{5}$	2:28$\frac{1}{5}$
1973	Secretariat	1:59$\frac{2}{5}$	1:54$\frac{2}{5}$	2:24
1977	Seattle Slew	2:02$\frac{1}{5}$	1:54$\frac{2}{5}$	2:29$\frac{3}{5}$
1978	Affirmed	2:01$\frac{2}{5}$	1:54$\frac{2}{5}$	2:26$\frac{4}{5}$

3.1   Measures of Center

Descriptive measures that indicate where the center or most typical value of a data set lies are called **measures of central tendency** or, more simply, **measures of center.** Measures of center are often referred to as *averages*.

In this section, we discuss the three most important measures of center: the *mean, median,* and *mode.* The mean and median apply only to quantitative data, whereas the mode can be used with either quantitative or qualitative (categorical) data.

The Mean

The most commonly used measure of center is the **mean.** When people speak of taking an average, they are most often referring to the mean.

DEFINITION 3.1

What❓ Does it Mean?

The mean of a data set is its arithmetic average.

Mean of a Data Set

The *mean* of a data set is the sum of the observations divided by the number of observations.

Example 3.1 illustrates calculation of the mean for data sets.

▌▌▌Example 3.1   The Mean

TABLE 3.1
Data Set I

$300	300	300	940	300
300	400	300	400	
450	800	450	1050	

TABLE 3.2
Data Set II

$300	300	940	450	400
400	300	300	1050	300

Weekly Salaries   Professor Hassett spent one summer working for a small mathematical consulting firm. The firm employed a few senior consultants, who made between $800 and $1050 per week; a few junior consultants, who made between $400 and $450 per week; and several clerical workers, who made $300 per week.

Because the first half of the summer was busier than the second half, more employees were required during the first half. Tables 3.1 and 3.2 display typical lists of weekly earnings for the two halves of the summer. Find the mean of each of the two data sets.

Solution   According to Definition 3.1, the mean of a data set is obtained by summing all the observations and then dividing that sum by the total number of observations. Data Set I has 13 observations. The sum of those observations is $6290, so

$$\text{Mean of Data Set I} = \frac{\$6290}{13} = \$483.85 \text{ (rounded to the nearest cent)}.$$

Similarly,

$$\text{Mean of Data Set II} = \frac{\$4740}{10} = \$474.00.$$

Thus the mean salary of the 13 employees in Data Set I is $483.85 and that of the 10 employees in Data Set II is $474.00.

INTERPRETATION The employees who worked in the first half of the summer earned more, on average (a mean salary of $483.85), than those who worked in the second half of the summer (a mean salary of $474.00).

—■

The Median

Another frequently used measure of center is the median. Essentially, the **median** of a data set is the number that divides the bottom 50% of the data from the top 50%. A more precise definition of the median follows.

DEFINITION 3.2

What *Does it Mean?*

The median of a data set is the middle value in its ordered list.

Median of a Data Set

Arrange the data in increasing order.

- If the number of observations is odd, then the *median* is the observation exactly in the middle of the ordered list.
- If the number of observations is even, then the *median* is the mean of the two middle observations in the ordered list.

In both cases, if we let n denote the number of observations, then the median is at position $(n + 1)/2$ in the ordered list.

Example 3.2 shows how to find the median of data sets.

║Example 3.2   **The Median**

Weekly Salaries   Consider again the two sets of salary data shown in Tables 3.1 and 3.2. Determine the median of each of the two data sets.

Solution   To find the median of Data Set I, we apply Definition 3.2. First, we arrange the data in increasing order:

300   300   300   300   300   300   **400**   400   450   450   800   940   1050

The number of observations in Data Set I is 13, which is an odd number. Because $n = 13$, we have $(n + 1)/2 = (13 + 1)/2 = 7$. Consequently, the median is the seventh observation in the ordered list, which is 400 (shown in boldface). The median salary of the 13 employees in Data Set I is $400.

To find the median of Data Set II, we again apply Definition 3.2. First, we arrange the data in increasing order:

300   300   300   300   **300**   **400**   400   450   940   1050

The number of observations in Data Set II is 10, which is an even number. Because $n = 10$, we have $(n + 1)/2 = (10 + 1)/2 = 5.5$. Consequently, the median is halfway between the fifth and sixth observations (shown in boldface) in the ordered list. In other words, the median salary of the 10 employees in Data Set II is $(300 + 400)/2 = \$350$.

INTERPRETATION Again, the analysis shows that the employees who worked in the first half of the summer tended to earn more (a median salary of $400) than those who worked in the second half of the summer (a median salary of $350).

To determine the median of a data set, you must first arrange the data in increasing order. Constructing a stem-and-leaf diagram as a preliminary step to ordering the data is often helpful.

The Mode

The final measure of center that we discuss here is the **mode.**

DEFINITION 3.3

What
Does it Mean?

The mode of a data set is its most frequently occurring value.

Mode of a Data Set

Obtain the frequency of occurrence of each value and note the greatest frequency.

- If the greatest frequency is 1 (i.e., no value occurs more than once), then the data set has no mode.
- If the greatest frequency is 2 or greater, then any value that occurs with that greatest frequency is called a *mode* of the data set.

To obtain the mode(s) of a data set, we first construct a frequency distribution for the data with classes based on a single value. The mode(s) can then be determined easily from the frequency distribution, as explained in Example 3.3.

|||Example 3.3   The Mode

TABLE 3.3
Frequency distribution for Data Set I, using single-value grouping

Salary	Frequency
300	6
400	2
450	2
800	1
940	1
1050	1

Weekly Salaries Determine the mode(s) of each of the two sets of salary data given in Tables 3.1 and 3.2 on page 92.

Solution First, we consider the salary data in Data Set I. Referring to Table 3.1, we obtain the frequency distribution of the data with classes based on a single value, as shown in Table 3.3.

From Table 3.3, we see that the greatest frequency of occurrence is 6, and 300 is the only value that occurs with that frequency. So the mode of the 13 salaries in Data Set I is $300.

Proceeding in the same way, we find that, for Data Set II, the greatest frequency of occurrence is 5 and that 300 is the only value that occurs with that frequency. So the mode of the 10 salaries in Data Set II is $300.

INTERPRETATION The most frequent salary was $300 both for the employees who worked in the first half of the summer and those who worked in the second half of the summer.

A data set can have more than one mode if there is more than one value that occurs with the greatest frequency. For instance, suppose that two of the clerical workers in Data Set I, who make $300 per week, were promoted to $400-per-week jobs. Then both the value 300 and the value 400 would occur with greatest frequency, 4. This new data set would thus have two modes, $300 and $400.

Comparison of the Mean, Median, and Mode

The mean, median, and mode of a data set are often different. Table 3.4 summarizes the definitions of these three measures of center and gives their values for Data Set I and Data Set II, which we computed in Examples 3.1–3.3.

TABLE 3.4
Means, medians, and modes of salaries in Data Set I and Data Set II

Measure of center	Definition	Data Set I	Data Set II
Mean	$\dfrac{\text{Sum of observations}}{\text{Number of observations}}$	$483.85	$474.00
Median	Middle value in ordered list	$400.00	$350.00
Mode	Most frequent value	$300.00	$300.00

In both Data Sets I and II, the mean is larger than the median. The reason is that the mean is strongly affected by the few large salaries in each data set. In general, the mean is sensitive to extreme (very large or very small) observations, whereas the median is not. Consequently, when the choice for the measure of center is between the mean and the median, the median is usually preferred for data sets that have extreme observations.

Figure 3.1 shows the relative positions of the mean and median for right-skewed, symmetric, and left-skewed distributions. Note that the mean is pulled in the direction of skewness, that is, in the direction of the extreme observations. For a right-skewed distribution, the mean is greater than the median; for a symmetric distribution, the mean and the median are equal; and, for a left-skewed distribution, the mean is less than the median.

FIGURE 3.1
Relative positions of the mean and median for (a) right-skewed, (b) symmetric, and (c) left-skewed distributions

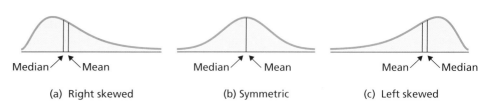

| (a) Right skewed | (b) Symmetric | (c) Left skewed |

A **resistant measure** is not sensitive to the influence of a few extreme observations. Thus the median is a resistant measure of center, whereas the mean is not. The resistance of the mean can be improved by using **trimmed means,** whereby a specified percentage of the smallest and largest observations are removed before computing the mean. In Exercise 3.16, we discuss trimmed means in more detail.

The mode for each of Data Sets I and II differs from both the mean and the median. Whereas the mean and the median are aimed at finding the center of a data set, the mode is really not—the value that occurs most frequently may not be near the center.

It should now be clear that the mean, median, and mode generally provide different information. There is no simple rule for deciding which measure of center to use in a given situation. Although skill in making such decisions is attained through practice, even experts may disagree about the most suitable measure of center for a particular data set. In Example 3.4, we discuss three data sets and suggest the most appropriate measure of center for each.

Example 3.4   Selecting an Appropriate Measure of Center

a. A student takes four exams in a biology class. His grades are 88, 75, 95, and 100. If asked for his average, which measure of center is the student likely to report?

b. The National Association of REALTORS publishes data on resale prices of U.S. homes. Which measure of center is most appropriate for such resale prices?

c. In the 2003 Boston Marathon, there were two categories of official finishers: male and female, of which there were 10,737 and 6,309, respectively. Which measure of center should be used here?

Solution

a. Chances are that the student would report the mean of his four exam scores, which is 89.5. The mean is probably the most suitable measure of center for the student to use because it takes into account the numerical value of each score and therefore indicates total overall performance.

b. The most appropriate measure of center for resale home prices is the median because it is aimed at finding the center of the data on resale home prices and because it is not strongly affected by the relatively few homes with extremely high resale prices. Thus the median provides a better indication of the "typical" resale price than either the mean or the mode.

c. The only suitable measure of center for these data is the mode, which is "male." Each observation in this data set is either "male" or "female." There is no way to compute a mean or median for such data. *Of the mean, median, and mode, the mode is the only measure of center that can be used for qualitative data.*

Many measures of center that appear in newspapers or that are reported by government agencies are medians, as is the case for household income and number of years of school completed. In an attempt to provide a clearer picture, some reports include both the mean and the median. For instance, the National Center for Health Statistics does so for daily intake of nutrients in the publication *Vital and Health Statistics*.

Population Mean and Sample Mean

Recall that a variable is a characteristic that varies from one person or thing to another and that observing one or more values of a variable yields data. The data set obtained by observing the values of a variable for an entire population is called *population data;* a data set obtained by observing the values of a variable for a sample of the population is called *sample data.*

The mean of population data is called the **population mean** or the **mean of the variable;** the mean of sample data is called a **sample mean.** The same terminology is used for the median and mode and, for that matter, any descriptive measure. Figure 3.2 shows the two ways in which the mean of a data set can be interpreted.

FIGURE 3.2

Possible interpretations for the mean of a data set

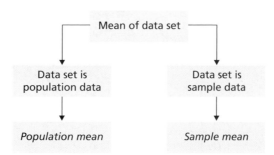

In Sections 3.2–3.4, we concentrate on descriptive measures of samples. Then, in Section 3.5, we discuss descriptive measures of populations and their relationship to descriptive measures of samples.

The **Technology Center**

For small data sets, such as the two sets of salary data in Tables 3.1 and 3.2, obtaining the mean, median, and mode by hand is easy. However, for even moderately large data sets, determining these descriptive measures by hand is tedious and prone to error.

All statistical software packages and most graphing calculators have built-in procedures for obtaining commonly used descriptive measures. Some offer individual descriptive measures, others provide related groups of descriptive measures, and still others supply both.

In Example 3.5, we present output for descriptive measures from three different technologies. Then we give step-by-step instructions for obtaining that output. Refer to the technology manuals for more details and other options.

Example 3.5 Using Technology to Obtain Descriptive Measures

Weekly Salaries Use Minitab, Excel, or the TI-83/84 Plus to obtain the mean and median of the salary data for Data Set I, displayed in Table 3.1 on page 92.

Solution Printout 3.1 shows output from Minitab, Excel, and the TI-83/84 Plus that provides several descriptive measures. At this point, we want to concentrate on the mean and median.

PRINTOUT 3.1
Output giving descriptive measures for
Data Set I

MINITAB

Descriptive Statistics: SETI

Variable	N	N*	Mean	SE Mean	StDev	Minimum	Q1	Median	Q3	Maximum
SETI	13	0	483.8	73.7	265.8	300.0	300.0	400.0	625.0	1050.0

EXCEL

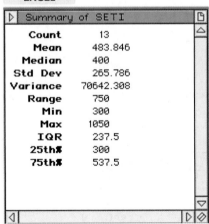

Count	13
Mean	483.846
Median	400
Std Dev	265.786
Variance	70642.308
Range	750
Min	300
Max	1050
IQR	237.5
25th%	300
75th%	537.5

TI-83/84 PLUS

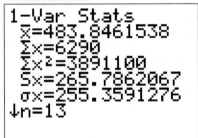

All three outputs shown in Printout 3.1 reveal that the mean and median of Data Set I are 483.8 and 400, respectively. *Note:* The TI-83/84 Plus uses $\bar{x}$ to denote the mean, a notation that we discuss in Section 3.2. ▬

Obtaining the Output (Optional)

Here are detailed instructions for obtaining descriptive measures for Data Set I by using Minitab, Excel, and the TI-83/84 Plus. First, store the data from Table 3.1 in a column (Minitab), range (Excel), or list (TI-83/84 Plus) named SETI. Then proceed as follows.

MINITAB	EXCEL	TI-83/84 PLUS
1 Choose **Stat ➤ Basic Statistics ➤ Display Descriptive Statistics. . .** 2 Specify SETI in the **Variables** text box 3 Click **OK**	1 Choose **DDXL ➤ Summaries** 2 Select **Summary of One Variable** from the **Function type** drop-down box 3 Specify SETI in the **Quantitative Variable** text box 4 Click **OK**	1 Press **STAT** 2 Arrow over to **CALC** 3 Press **1** 4 Press **2nd ➤ LIST** 5 Arrow down to SETI and press **ENTER** twice

EXERCISES 3.1

Statistical Concepts and Skills

3.1 Explain in detail the purpose of a measure of center.

3.2 Name and describe the three most important measures of center.

3.3 Of the mean, median, and mode, which is the only one appropriate for use with qualitative data?

3.4 True or false: The mean, median, and mode can all be used with quantitative data.

3.5 Consider the data set 1, 2, 3, 4, 5, 6, 7, 8, 9.
a. Obtain the mean and median of the data.
b. Replace the 9 in the data set by 99 and again compute the mean and median. Decide which measure of center works better here and explain your answer.
c. For the data set in part (b), the mean is neither central nor typical for the data. The lack of what property of the mean accounts for this result?

3.6 Complete the following statement: A descriptive measure is *resistant* if ....

3.7 Floor Space. The U.S. Department of Housing and Urban Development and the U.S. Bureau of the Census compile information on new, privately owned single-family houses. According to the document *Characteristics of New Housing*, in 2001 the mean floor space of such homes was 2324 sq ft and the median was 2103 sq ft. Which measure of center do you think is more appropriate? Justify your answer.

3.8 Net Worth. The Board of Governors of the Federal Reserve System publishes information on family net worth in the *Federal Reserve Bulletin*. In 2001, the mean net worth of families in the United States was $395.5 thousand and the median net worth was $86.1 thousand. Which measure of center do you think is more appropriate? Explain your answer.

In Exercises **3.9–3.12**, *determine the mean, median, and mode(s) for each of the data sets by hand, that is, using only basic calculator functions. For the mean and the median, round each answer to one more decimal place than that used for the observations.*

3.9 Amphibian Embryos. In a study of the effects of radiation on amphibian embryos titled "Shedding Light on Ultraviolet Radiation and Amphibian Embryos" (*BioScience*, Vol. 53, No. 6, pp. 551–561), L. Licht recorded the time it took for a sample of seven different species of frogs' and toads' eggs to hatch. The following table shows the times to hatch, in days.

6	7	11	6	5	5	11

3.10 Hurricanes. A recent article by D. Schaefer et al. (*Journal of Tropical Ecology*, Vol. 16, pp. 189–207) reported on a long-term study of the effects of hurricanes on tropical streams of the Luquillo Experimental Forest in Puerto Rico. The study shows that Hurricane Hugo had a significant impact on stream water chemistry. The following table shows a sample of 10 ammonia fluxes in the first year after Hugo. Data are in kilograms per hectare per year.

96	66	147	147	175
116	57	154	88	154

3.11 Tornado Touchdowns. Each year, tornadoes that touch down are recorded by the Storm Prediction Center and published in *Monthly Tornado Statistics*. The following table gives the number of tornadoes that touched down in the United States during each month of 2002. [SOURCE: National Oceanic and Atmospheric Administration.]

3	2	47	118	204	97
68	86	62	57	98	99

3.12 Technical Merit. In the 2002 Winter Olympics, Michelle Kwan competed in the Short Program ladies singles event. From nine judges, she received scores ranging from 1 (poor) to 6 (perfect). The following table provides the marks that the judges gave her on technical merit, found in an article by S. Berry (*Chance*, Vol. 15, No. 2, pp. 14–18).

5.8	5.7	5.9	5.7	5.5	5.7	5.7	5.7	5.6

3.13 All-Time Top TV Programs. According to *The World Almanac*, as of August 2002, the all-time top television programs by rating (percentage of TV-owning households tuned into the program) are as shown in the following table. [SOURCE: Nielsen Media Research.]

Program	Telecast date	Network	Rating (%)	Audience (millions)
M*A*S*H (last episode)	02/28/83	CBS	60.2	50.2
Dallas (Who shot J.R.?)	11/21/80	CBS	53.3	41.5
Roots—Pt. 8	01/30/77	ABC	51.1	36.4
Super Bowl XVI	01/24/82	CBS	49.1	40.0
Super Bowl XVII	01/30/83	NBC	48.6	40.5
XVII Winter Olympics—2d Wed.	02/23/94	CBS	48.5	45.7
Super Bowl XX	01/26/86	NBC	48.3	41.5
Gone With the Wind—Pt. 1	11/07/76	NBC	47.7	34.0
Gone With the Wind—Pt. 2	11/08/76	NBC	47.4	33.8
Super Bowl XII	01/15/78	CBS	47.2	34.4
Super Bowl XIII	01/21/79	NBC	47.1	35.1
Bob Hope Christmas Show	01/15/70	NBC	46.6	27.3
Super Bowl XVIII	01/22/84	CBS	46.4	38.8
Super Bowl XIX	01/20/85	ABC	46.4	39.4
Super Bowl XIV	01/20/80	CBS	46.3	35.3
Super Bowl XXX	01/28/96	NBC	46.0	44.2
ABC Theatre (The Day After)	11/20/83	ABC	46.0	38.6
Roots—Pt. 6	01/28/77	ABC	45.9	32.7
The Fugitive	08/29/67	ABC	45.9	25.7
Super Bowl XXI	01/25/87	CBS	45.8	40.0

a. Determine the mode of the network data.
b. Would it be appropriate to use either the mean or the median here? Explain your answer.

3.14 Road Rage. The report *Controlling Road Rage: A Literature Review and Pilot Study*, dated June 9, 1999, was prepared for the AAA Foundation for Traffic Safety by Daniel B. Rathbone, Ph.D., and Jorg C. Huckabee, MSCE. The authors discuss the results of a literature review and pilot study on how to prevent aggressive driving and road rage. As described in the report, *road rage* is criminal behavior by motorists characterized by uncontrolled anger that results in violence or threatened violence on the road. One of the goals of the study was to determine when road rage occurs most often. The following data provide the days on which 69 road-rage incidents occurred.

F	Sa	W	M	Tu	F	Th	M
Tu	F	Tu	F	Su	W	Th	F
Th	W	Th	Sa	W	W	F	F
Tu	Su	Tu	Th	W	Sa	Tu	Th
F	W	F	F	Su	F	Th	Tu
F	Tu	Tu	Tu	Sa	W	W	Sa
F	Sa	Th	W	F	Th	F	M
F	M	F	Su	W	Th	M	Tu
Sa	Th	F	Su	W			

a. Determine the mode of the road-rage data.
b. Would it be appropriate to use either the mean or the median here? Explain your answer.

Extending the Concepts and Skills

3.15 Food Choice. As you discovered earlier, *ordinal data* are data about order or rank given on a scale such as $1, 2, 3, \ldots$ or A, B, C, ... Most statisticians recommend using the median to indicate the center of an ordinal data set, but some researchers also use the mean. In the paper "Measurement of Ethical Food Choice Motives" (*Appetite*, Vol. 34, pp. 55–59), research psychologists M. Lindeman and M. Väänänen of the University of Helsinki published a study on the factors that most influence people's choice of food. One of the questions asked of the participants was how important, on a scale of 1 to 4 (1 = not at all important, 4 = very important), is ecological welfare in food choice motive, where ecological welfare includes animal welfare and environmental protection. Here are the ratings given by 14 of the participants.

2	4	1	2	4	3	3
2	2	1	2	4	2	3

a. Compute the mean of the data.
b. Compute the median of the data.
c. Decide which of the two measures of center is best.

3.16 Outliers and Trimmed Means. Some data sets contain *outliers*, observations that fall well outside the overall pattern of the data. (We discuss outliers in more detail in Section 3.4.) Suppose, for instance, that you are interested in the ability of high school algebra students to compute square roots. You decide to give a square-root exam to 10 of these students. Unfortunately, one of the students had a fight with his girlfriend and cannot concentrate—he gets a 0. The 10 scores are displayed in increasing order in the following table. The score of 0 is an outlier.

0	58	61	63	67	69	70	71	78	80

Statisticians have a systematic method for avoiding extreme observations and outliers when they calculate means. They compute *trimmed means*, in which high and low observations are deleted or "trimmed off" before the mean is calculated. For instance, to compute the 10% trimmed mean of the test-score data, we first delete both the bottom 10% and the top 10% of the ordered data, that is, 0 and 80. Then we calculate the mean of the remaining data. Thus the 10% trimmed mean of the test-score data is

$$\frac{58 + 61 + 63 + 67 + 69 + 70 + 71 + 78}{8} = 67.1.$$

The following table displays a set of scores for a 40-question algebra final exam.

2	15	16	16	19	21	21	25	26	27
4	15	16	17	20	21	24	25	27	28

a. Do any of the scores look like outliers?
b. Compute the usual mean of the data.

c. Compute the 5% trimmed mean of the data.
d. Compute the 10% trimmed mean of the data.
e. Compare the means you obtained in parts (b)–(d). Which of the three means provides the best measure of center for the data?

Using Technology

3.17 Tornado Touchdowns. Use the technology of your choice to determine the mean and median of the tornado data in Exercise 3.11.

3.18 Technical Merit. Use the technology of your choice to determine the mean and median of the technical-merit data in Exercise 3.12.

3.19 Body Temperature. A study by researchers at the University of Maryland addressed the question of whether the mean body temperature of humans is 98.6°F. The results of the study by P. Mackowiak, S. Wasserman, and M. Levine appeared in the article "A Critical Appraisal of 98.6°F, the Upper Limit of the Normal Body Temperature, and Other Legacies of Carl Reinhold August Wunderlich" (*Journal of the American Medical Association*, Vol. 268, pp. 1578–1580). The researchers obtained the body temperatures of 93 healthy humans. We have supplied these temperatures on the the WeissStats CD. Use the technology of your choice to determine the mean and median of the temperature data. Interpret your results.

3.20 Sex and Direction. In the paper "The Relation of Sex and Sense of Direction to Spatial Orientation in an Unfamiliar Environment" (*Journal of Environmental Psychology*, Vol. 20, pp. 17–28), Sholl et al. published the results of examining the sense of direction of 30 male and 30 female students. After being taken to an unfamiliar wooded park, the students were given a number of spatial orientation tests, including pointing to south, which tested their absolute frame of reference. Pointing to south was done by moving a pointer attached to a 360° protractor. We have supplied on the WeissStats CD the absolute pointing errors, in degrees, for both the male and female participants.
a. Use the technology of your choice to obtain the mean and median of each of the two data sets.
b. Use the results that you obtained in part (a) to compare the two data sets.

3.2 The Sample Mean

In this section, we discuss the sample mean in more detail. We also introduce some mathematical notation that is useful for expressing the formula for the sample mean and many other descriptive measures.

To begin, we note that in statistics, as in algebra, we use letters such as x, y, and z to denote variables. So, for instance, if we are studying heights and weights of college students, we might let x denote the variable "height" and y denote the variable "weight."

In Definition 3.1, we defined the mean of a data set in words: *The mean of a data set is the sum of the observations divided by the number of observations.* Using mathematical notation, we can express such definitions concisely. First, we introduce the mathematical notation for "sum of the observations" in Example 3.6.

Example 3.6   **Introducing Summation Notation**

Exam Scores   The exam scores for the student in Example 3.4(a) are 88, 75, 95, and 100.

a. Use mathematical notation to represent the individual exam scores.

b. Use summation notation to express the sum of the four exam scores.

Solution   Let x denote the variable "exam score."

a. We use the symbol x_i to represent the ith observation of the variable x. Thus, for the exam scores, we have

$$x_1 = \text{score on Exam 1} = 88;$$
$$x_2 = \text{score on Exam 2} = 75;$$
$$x_3 = \text{score on Exam 3} = 95;$$
$$x_4 = \text{score on Exam 4} = 100.$$

More simply, we can just write $x_1 = 88$, $x_2 = 75$, $x_3 = 95$, and $x_4 = 100$. The numbers 1, 2, 3, and 4 written below the xs are called **subscripts.**

b. Using the notation introduced in part (a), we can express the sum of the exam-score data symbolically as

$$x_1 + x_2 + x_3 + x_4.$$

We can use **summation notation** to obtain a shorthand description for this sum. This notation is the uppercase Greek letter Σ (sigma). That letter, which corresponds to the English letter S, is an abbreviation for the phrase "the sum of." So, in place of the lengthy expression $x_1 + x_2 + x_3 + x_4$, we can use Σx, read as "summation x" or "the sum of the observations of the variable x." For the exam-score data,

$$\Sigma x = x_1 + x_2 + x_3 + x_4 = 88 + 75 + 95 + 100 = 358.$$

INTERPRETATION   The grand total of the student's four exam scores is 358 points.

For clarity, we sometimes incorporate subscripts into the summation notation. We do so by writing Σx_i instead of Σx. The subscript i is a generic subscript. To be even more precise, we can use *indices* and write $\sum_{i=1}^{n} x_i$, read as "the sum of x_i from i equals 1 to n," where n denotes the number of observations.

Notation for a Sample Mean

Recall that a data set obtained by observing the values of a variable for a sample of a population is called *sample data* and that the mean of sample data is called a *sample mean*. The symbol used for a sample mean is a bar over the letter representing the variable.

So, for a variable x, a sample mean is denoted $\bar{x}$, read as "x bar." If we also use the letter n to denote the **sample size** or, equivalently, the number of observations, we can express the definition of a sample mean concisely.

DEFINITION 3.4

What Does it Mean?

A sample mean is the arithmetic average (mean) of sample data.

Sample Mean

For a variable x, the mean of the observations for a sample is called a *sample mean* and is denoted $\bar{x}$. Symbolically,

$$\bar{x} = \frac{\Sigma x}{n},$$

where n is the sample size.

We illustrate calculation of the sample mean in Example 3.7.

Example 3.7   **The Sample Mean**

AML and the Cost of Labor   Active Management of Labor (AML) was introduced in the 1960s to reduce the amount of time a woman spends in labor during the birth process. Recently, R. Rogers et al. conducted a study to determine whether AML also translates into a reduction in delivery cost to the patient. They reported their findings in the paper "Active Management of Labor: A Cost Analysis of a Randomized Controlled Trial" (*Western Journal of Medicine*, Vol. 172, pp. 240–243).

TABLE 3.5
Costs ($) of eight AML deliveries

Table 3.5 displays the costs, in dollars, of eight randomly sampled AML deliveries. Determine the sample mean of these delivery costs.

3141.09	2873.28
2115.64	1683.61
3470.08	1798.89
2539.48	3092.56

Solution   Let x denote the variable "cost" for AML deliveries. We want to obtain the mean, $\bar{x}$, of the eight observations of the variable x shown in Table 3.5. Summing those observations, we obtain $\Sigma x = 20{,}714.63$. Because the sample size (number of observations) is 8, we have $n = 8$. Thus

$$\bar{x} = \frac{\Sigma x}{n} = \frac{20{,}714.63}{8} = \$2{,}589.329.$$

INTERPRETATION   The mean cost of the sample of eight AML deliveries is $2589.329.

In Chapter 9, we return to the AML study and decide whether it provides sufficient evidence to conclude that, on average, AML reduces delivery cost to the patient.

Other Important Sums

We must often find sums other than the sum of the observations, Σx. One such sum is the sum of the squares of the observations, Σx^2. In Section 3.3, we need to obtain Σx, Σx^2, and various other sums. So that we can concentrate on the concepts presented there instead of the computations, we discuss computing those sums now, in Example 3.8.

▌▌ Example 3.8   Other Important Sums

Exam Scores   The exam-score data from Example 3.6 are repeated in the first column of Table 3.6. The remaining columns of the table contain some related quantities whose significance will become apparent in Section 3.3.

In Example 3.6, we found that the sum of the exam-score data is 358, a fact that we record at the bottom of the first column of Table 3.6. The second column of Table 3.6 displays the squares, x^2, of the exam scores. The sum of those squares is 32,394; that is, $\Sigma x^2 = 32,394$.

To obtain the third column of Table 3.6, we must first compute the mean, $\bar{x}$, of the four exam scores. Because $n = 4$ and $\Sigma x = 358$,

$$\bar{x} = \frac{\Sigma x}{n} = \frac{358}{4} = 89.5.$$

Subtracting 89.5 from each of the four exam scores in the first column of Table 3.6, we get the $x - \bar{x}$ values shown in the third column. The sum of those values is 0; that is, $\Sigma(x - \bar{x}) = 0$. The fourth column of Table 3.6 gives the squares, $(x - \bar{x})^2$, of the $x - \bar{x}$ values. The sum of those squares is 353; that is, $\Sigma(x - \bar{x})^2 = 353$.

TABLE 3.6
Exam-score data and related quantities

x	x^2	$x - \bar{x}$	$(x - \bar{x})^2$
88	7,744	−1.5	2.25
75	5,625	−14.5	210.25
95	9,025	5.5	30.25
100	10,000	10.5	110.25
358	32,394	0	353.00

EXERCISES 3.2

Statistical Concepts and Skills

3.21 Explain in your own words why mathematical notation is useful.

3.22 Explain what each symbol represents.
a. Σ **b.** n **c.** $\bar{x}$

3.23 For a given population, is the population mean a variable? What about a sample mean?

3.24 Let $x_1 = 1$, $x_2 = 7$, $x_3 = 4$, $x_4 = 5$, and $x_5 = 10$.
a. Compute Σx.   **b.** Find n.   **c.** Determine $\bar{x}$.

3.25 Let $x_1 = 12$, $x_2 = 8$, $x_3 = 9$, and $x_4 = 17$.
a. Compute Σx.   **b.** Find n.   **c.** Determine $\bar{x}$.

3.26 Honeymoons. Popular destinations for the newly-weds of today are the Caribbean and Hawaii. According to *Bride's Magazine*, a honeymoon, on average, lasts 9 days and costs $3657. A sample of 12 newlyweds reported the following lengths of stay of their honeymoons.

5	14	7	10	6	8
12	9	10	9	7	11

a. Compute Σx.   **b.** Find n.
c. Determine the sample mean. Round your answer to one more decimal place than that used for the observations.

3.27 Sleep. In 1908, W. S. Gosset published the article "The Probable Error of a Mean" (*Biometrika*, Vol. 6, pp. 1–25). It is in this pioneering paper, written under the pseudonym "Student," that Gosset introduced what later became known as Student's *t*-distribution, which we discuss in Chapter 8. Gosset used the following data set, which shows the additional sleep in hours obtained by a sample of 10 patients given laevohysocyamine hydrobromide.

1.9	0.8	1.1	0.1	−0.1
4.4	5.5	1.6	4.6	3.4

a. Compute Σx. **b.** Find n.
c. Determine the sample mean. Round your answer to one more decimal place than that used for the observations.

3.28 Acute Postoperative Days. Several neurosurgeons wanted to determine whether a dynamic system (Z-plate) reduced the number of acute postoperative days spent in the hospital relative to a static system (ALPS plate).

Ronald Jacobowitz, Ph.D., an Arizona State University professor, along with G. Vishteh, M.D., and other neurosurgeons, obtained the following data on the number of acute postoperative days in the hospital for the static system.

6	18	9	7	14	9

a. Compute $\bar{x}$.
b. Compute Σx^2, $\Sigma(x - \bar{x})$, and $\Sigma(x - \bar{x})^2$ by constructing a table similar to Table 3.6.

3.29 Fat Content. Using five packaged food items from your home, record the name of the item and the fat content per serving (in grams) as provided on the nutritional information label. For your fat-content data,
a. compute $\bar{x}$.
b. compute Σx^2, $\Sigma(x - \bar{x})$, and $\Sigma(x - \bar{x})^2$ by constructing a table similar to Table 3.6.

Extending the Concepts and Skills

3.30 Explain the difference between the quantities $(\Sigma x)^2$ and Σx^2. Construct an example to show that, in general, those two quantities are unequal.

3.31 Explain the difference between the quantities Σxy and $\Sigma x \Sigma y$. Provide an example to show that, in general, those two quantities are unequal.

3.32 For the exam-score data in Example 3.8, we found that $\Sigma(x - \bar{x}) = 0$. Explain why this result holds for any data set. (*Hint:* Write out the sum and use the fact that $\Sigma x = n\bar{x}$.)

3.3 Measures of Variation; The Sample Standard Deviation

Up to this point, we have discussed only descriptive measures of center, specifically, the mean, median, and mode. However, two data sets can have the same mean, median, or mode and yet still be quite different in other respects. For example, consider the heights of the five starting players on each of two men's college basketball teams, as shown in Fig. 3.3 at the top of the next page.

The two teams have the same mean heights, 75 inches (6′ 3″); the same median heights, 76 inches (6′ 4″); and the same modes, 76 inches (6′ 4″). Nonetheless, the two data sets clearly differ. In particular, the heights of the players on Team II vary much more than those on Team I. To describe that difference quantitatively, we use a descriptive measure that indicates the amount of variation, or spread, in a data set. Such descriptive measures are referred to as **measures of variation** or **measures of spread.**

Just as there are several different measures of center, there are also several different measures of variation. In this section, we examine two of the most frequently used measures of variation: the *range* and *sample standard deviation*. We begin with the range because it is the simplest to understand and compute.

FIGURE 3.3
Five starting players on each of two men's college basketball teams and their heights

Feet and inches	6'	6'1"	6'4"	6'4"	6'6"	5'7"	6'	6'4"	6'4"	7'
Inches	72	73	76	76	78	67	72	76	76	84

The Range

The contrast between the heights of the two teams in Fig. 3.3 becomes clear if we place the shortest player on each team next to the tallest, as in Fig. 3.4.

FIGURE 3.4
Shortest and tallest starting players on each of two men's college basketball teams and their heights

Feet and inches	6'	6'6"	5'7"	7'
Inches	72	78	67	84

The **range** of a data set is obtained by computing the difference between the maximum (largest) and minimum (smallest) observations. From Fig. 3.4,

$$\text{Team I: Range} = 78 - 72 = 6 \text{ inches,}$$
$$\text{Team II: Range} = 84 - 67 = 17 \text{ inches.}$$

INTERPRETATION The difference between the heights of the tallest and shortest players on Team I is 6 inches, whereas the difference between the heights of the tallest and shortest players on Team II is 17 inches.

DEFINITION 3.5

What *Does it Mean?*

The range of a data set is the difference between its largest and smallest values.

Range of a Data Set

The *range* of a data set is given by the formula

$$\text{Range} = \text{Max} - \text{Min,}$$

where Max and Min denote the maximum and minimum observations, respectively.

The range of a data set is quite easy to compute. However, when we use the range, a great deal of information is ignored: Only the largest and smallest observations are considered; the other observations are disregarded.

For that reason, two other measures of variation, the *standard deviation* and the *interquartile range,* are generally favored over the range. The standard deviation is the preferred measure of variation when the mean is used as the measure of center; the interquartile range is preferred when the median is used as the measure of center. We discuss the standard deviation in this section and consider the interquartile range in Section 3.4.

The Sample Standard Deviation

In contrast to the range, the standard deviation takes into account all the observations. The calculations required to determine a standard deviation are more involved than those needed to obtain a range. However, this problem is not serious because almost all computers and statistical calculators have built-in functions to do the necessary computations.

Roughly speaking, the **standard deviation** measures variation by indicating how far, on average, the observations are from the mean. For a data set with a large amount of variation, the observations will, on average, be far from the mean; hence the standard deviation will be large. For a data set with a small amount of variation, the observations will, on average, be close to the mean; consequently, the standard deviation will be small.

To compute the standard deviation of a data set, we need to know whether the set is population data or sample data. This information is necessary because the formulas for the standard deviations of sample data and population data differ slightly. In this section, we concentrate on the sample standard deviation. We discuss the population standard deviation in Section 3.5.

The first step in computing a sample standard deviation is to find how far each observation is from the mean, that is, the **deviations from the mean.** We show how to calculate them in Example 3.9.

❙❙ Example 3.9   **The Deviations From the Mean**

Heights of Starting Players   The heights, in inches, of the five starting players on Team I are 72, 73, 76, 76, and 78, as shown in Fig. 3.3. Find the deviations from the mean.

Solution   The mean height of the starting players on Team I is

$$\bar{x} = \frac{\Sigma x}{n} = \frac{72 + 73 + 76 + 76 + 78}{5} = \frac{375}{5} = 75 \text{ inches.}$$

To obtain the deviation from the mean for a particular observation, we subtract the mean from it; that is, we compute $x - \bar{x}$. For instance, the deviation from the mean for the height of 72 inches is $x - \bar{x} = 72 - 75 = -3$. The deviations from the mean for all five observations are given in the second column of Table 3.7 and are displayed graphically in Fig. 3.5.

■

TABLE 3.7

Deviations from the mean

Height x	Deviation from mean $x - \bar{x}$
72	−3
73	−2
76	1
76	1
78	3

FIGURE 3.5

Graphical display of the deviations from the mean (dots represent observations)

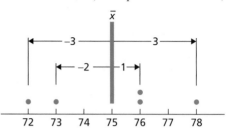

The second step in computing a sample standard deviation is to obtain a measure of the total deviation from the mean for all the observations. Although the quantities $x - \bar{x}$ represent deviations from the mean, adding them to get a total deviation from the mean is of no value because their sum, $\Sigma(x - \bar{x})$, always equals zero. Summing the data in the second column of Table 3.7 shows this result to be true for the height data of Team I but, in fact, it is true in general.

In computing a sample standard deviation, the deviations from the mean, $x - \bar{x}$, are squared to obtain quantities that do not sum to zero. The sum of the squared deviations from the mean, $\Sigma(x - \bar{x})^2$, is called the **sum of squared deviations** and provides a measure of total deviation from the mean for all the observations. We show how to calculate it in Example 3.10.

Example 3.10 The Sum of Squared Deviations

Heights of Starting Players   Compute the sum of squared deviations for the heights of the starting players on Team I.

Solution   To get Table 3.8, we added a column for $(x - \bar{x})^2$ to Table 3.7.

TABLE 3.8

Table for computing the sum of squared deviations for the heights of Team I

Height x	Deviation from mean $x - \bar{x}$	Squared deviation $(x - \bar{x})^2$
72	−3	9
73	−2	4
76	1	1
76	1	1
78	3	9
		24

From the third column of Table 3.8, $\Sigma(x - \bar{x})^2 = 24$. The sum of squared deviations is 24 inches^2.

The third step in computing a sample standard deviation is to take an average of the squared deviations. We do so by dividing the sum of squared deviations by $n - 1$, or 1 less than the sample size. The resulting quantity is called a **sample variance** and is denoted s_x^2 or, when no confusion can arise, s^2.

In symbols,

$$s^2 = \frac{\Sigma(x - \bar{x})^2}{n - 1}.$$

Note: If we divided by n instead of by $n - 1$, the sample variance would be the mean of the squared deviations. Although dividing by n seems more natural, we divide by $n - 1$ for the following reason. One of the main uses of the sample variance is to estimate the population variance (defined in Section 3.5). Division by n tends to underestimate the population variance, whereas division by $n - 1$ gives, on average, the correct value.

The computation of a sample variance is shown in Example 3.11.

▌▌Example 3.11   The Sample Variance

Heights of Starting Players   Obtain the sample variance of the heights of the starting players on Team I.

Solution   From Example 3.10, the sum of squared deviations is 24 inches². Because $n = 5$,

$$s^2 = \frac{\Sigma(x - \bar{x})^2}{n - 1} = \frac{24}{5 - 1} = 6.$$

The sample variance is 6 inches².   ▬

A sample variance is in units that are the square of the original units, the result of squaring the deviations from the mean. For instance, as we determined in Example 3.11, the sample variance of the heights of the players on Team I is 6 inches². Because descriptive measures should be expressed in the original units, the final step in computing a sample standard deviation is to take the square root of the sample variance. In other words, the **sample standard deviation,** denoted s_x or s, is

$$s = \sqrt{\frac{\Sigma(x - \bar{x})^2}{n - 1}},$$

which, for the heights of the players on Team I, is computed in Example 3.12.

▌▌Example 3.12   The Sample Standard Deviation

Heights of Starting Players   Determine the sample standard deviation of the heights of the starting players on Team I.

Solution   From Example 3.11, the sample variance is 6 inches². Thus the sample standard deviation is

$$s = \sqrt{\frac{\Sigma(x - \bar{x})^2}{n - 1}} = \sqrt{6} = 2.4 \text{ inches,}$$

rounded to the nearest tenth of an inch.

INTERPRETATION Roughly speaking, on average, the heights of the players on Team I vary from the mean height of 75 inches by 2.4 inches.

The following definition summarizes our discussion of the sample standard deviation.

DEFINITION 3.6

> **Sample Standard Deviation**
>
> For a variable x, the standard deviation of the observations for a sample is called a *sample standard deviation.* It is denoted s_x or, when no confusion will arise, simply s. We have
>
> $$s = \sqrt{\frac{\Sigma(x - \bar{x})^2}{n - 1}},$$
>
> where n is the sample size.

What
Does it Mean?

Roughly speaking, the sample standard deviation indicates how far, on average, the observations in the sample are from the mean of the sample.

We performed the computations required to obtain a sample standard deviation in four separate examples to explain the sample standard deviation and the calculations involved. Now that we have done that, we can present a simple procedure for computing a sample standard deviation, which we illustrate in Example 3.13.

STEP 1 Calculate the sample mean, $\bar{x}$.

STEP 2 Construct a table to obtain the sum of squared deviations, $\Sigma(x - \bar{x})^2$.

STEP 3 Apply Definition 3.6 to determine the sample standard deviation, s.

Example 3.13 **The Sample Standard Deviation**

Heights of Starting Players The heights, in inches, of the five starting players on Team II are 67, 72, 76, 76, and 84. Obtain the sample standard deviation of these heights.

TABLE 3.9

Table for computing the sum of squared deviations for the heights of Team II

x	$x - \bar{x}$	$(x - \bar{x})^2$
67	−8	64
72	−3	9
76	1	1
76	1	1
84	9	81
		156

Solution We apply the three-step procedure just described.

STEP 1 Calculate the sample mean, $\bar{x}$.

We have

$$\bar{x} = \frac{\Sigma x}{n} = \frac{67 + 72 + 76 + 76 + 84}{5} = \frac{375}{5} = 75 \text{ inches.}$$

STEP 2 Construct a table to obtain the sum of squared deviations, $\Sigma(x - \bar{x})^2$.

Table 3.9 provides columns for x, $x - \bar{x}$, and $(x - \bar{x})^2$. From the third column, $\Sigma(x - \bar{x})^2 = 156$ inches^2.

STEP 3 Apply Definition 3.6 to determine the sample standard deviation, *s*.

We have $n = 5$ and $\Sigma(x - \bar{x})^2 = 156$. Consequently, the sample standard deviation of the heights for Team II is

$$s = \sqrt{\frac{\Sigma(x - \bar{x})^2}{n - 1}} = \sqrt{\frac{156}{5 - 1}} = \sqrt{39} = 6.2 \text{ inches,}$$

rounded to the nearest tenth of an inch.

INTERPRETATION Roughly speaking, on average, the heights of the players on Team II vary from the mean height of 75 inches by 6.2 inches.

In Examples 3.12 and 3.13, we found that the sample standard deviations of the heights of the starting players on Teams I and II are 2.4 inches and 6.2 inches, respectively. Hence Team II, which has more variation in height than Team I, also has a larger standard deviation. That is the way a measure of variation is supposed to work.

KEY FACT 3.1

Variation and the Standard Deviation

The more variation there is in a data set, the larger is its standard deviation.

Key Fact 3.1 shows that the standard deviation satisfies the basic criterion for a measure of variation; in fact, the standard deviation is the most commonly used measure of variation. However, the standard deviation does have its drawbacks. For instance, it is not resistant: Its value can be strongly affected by a few extreme observations.

A Computing Formula for *s*

Next, we present an alternative formula for obtaining a sample standard deviation. Thus we need a name for the original formula given in Definition 3.6 to distinguish it from the alternative formula. Because the original formula was used to define the sample standard deviation, we call it the *defining formula* for *s*. The alternative formula for obtaining a sample standard deviation is given in Formula 3.1, which we call the *computing formula* for *s*.

FORMULA 3.1

Computing Formula for a Sample Standard Deviation

A sample standard deviation can be computed using the formula

$$s = \sqrt{\frac{\Sigma x^2 - (\Sigma x)^2/n}{n - 1}},$$

where *n* is the sample size.

The computing formula for s is equivalent to the defining formula—both formulas give the same answer, although differences owing to roundoff error are possible. However, the computing formula is usually faster and easier for doing calculations by hand and also reduces the chance for roundoff error.

Before illustrating the computing formula for s in Example 3.14, we need to comment on the similar-looking expressions, Σx^2 and $(\Sigma x)^2$, that occur in that formula. The expression Σx^2 represents the sum of the squares of the data; it is obtained by first squaring each observation and then summing those squared values. The expression $(\Sigma x)^2$ represents the square of the sum of the data; it is obtained by first summing the observations and then squaring that sum.

In the numerator of the computing formula, the division of $(\Sigma x)^2$ by n should be performed before the subtraction from Σx^2. In other words, first compute $(\Sigma x)^2/n$ and then subtract the result from Σx^2.

Example 3.14 Computing Formula for a Sample Standard Deviation

Heights of Starting Players   In Example 3.13, we obtained the sample standard deviation of the heights for the five starting players on Team II by using the defining formula for s. Obtain that sample standard deviation by using the computing formula.

TABLE 3.10

Table for computation of s, using the computing formula

x	x^2
67	4,489
72	5,184
76	5,776
76	5,776
84	7,056
375	28,281

Solution   To apply the computing formula for s, we need the sums Σx and Σx^2. They are determined in Table 3.10.

We know that $n = 5$ and the bottom row of Table 3.10 shows that $\Sigma x = 375$ and $\Sigma x^2 = 28,281$. Thus, by Formula 3.1,

$$s = \sqrt{\frac{\Sigma x^2 - (\Sigma x)^2/n}{n-1}} = \sqrt{\frac{28,281 - (375)^2/5}{5-1}}$$

$$= \sqrt{\frac{28,281 - 28,125}{4}} = \sqrt{\frac{156}{4}} = \sqrt{39} = 6.2 \text{ inches,}$$

rounded to the nearest tenth of an inch.

We have now obtained the sample standard deviation of the heights of the players on Team II in two ways—using the defining formula and using the computing formula. Both formulas give the same value, 6.2 inches, for the sample standard deviation. For these height data, either formula is relatively easy to apply. However, for most data sets—especially for those in which the mean is not a whole number—the computing formula is preferable.

Here is an important rule to remember when you use only basic calculator functions to obtain a sample standard deviation or any other descriptive measure.

Rounding Rule: Do not perform any rounding until the computation is complete; otherwise, substantial roundoff error can result.

Another common rounding rule is to round final answers that contain units to one more decimal place than the raw data. Although we usually abide by this convention, occasionally we vary from it for pedagogical reasons. In general, you should stick to this rounding rule as well.

Further Interpretation of the Standard Deviation

Again, the standard deviation is a measure of variation—the more variation there is in a data set, the larger is its standard deviation. Table 3.11 displays two data sets, each with 10 observations. A brief inspection of the table reveals that Data Set II has more variation than Data Set I.

TABLE 3.11
Data sets that have different variation

Data Set I	41	44	45	47	47	48	51	53	58	66
Data Set II	20	37	48	48	49	50	53	61	64	70

TABLE 3.12
Means and standard deviations of the data sets in Table 3.11

Data Set I	Data Set II
$\bar{x} = 50.0$	$\bar{x} = 50.0$
$s = 7.4$	$s = 14.2$

We computed the sample mean and sample standard deviation of each data set and summarized the results in Table 3.12. As expected, the standard deviation of Data Set II is larger than that of Data Set I.

To enable you to compare visually the variations in the two data sets, we produced the graphs shown in Figs. 3.6 and 3.7. On each graph, we marked the observations with dots. In addition, we located the sample mean, $\bar{x} = 50$, and measured intervals equal in length to the standard deviation: 7.4 for Data Set I and 14.2 for Data Set II.

FIGURE 3.6
Data Set I; $\bar{x} = 50$, $s = 7.4$

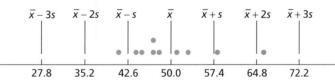

FIGURE 3.7
Data Set II; $\bar{x} = 50$, $s = 14.2$

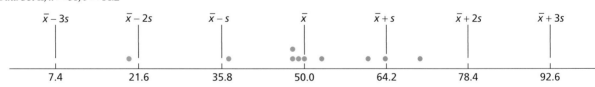

In Fig. 3.6, note that the horizontal position labeled $\bar{x} + 2s$ represents the number that is two standard deviations to the right of the mean, which in this case is

$$\bar{x} + 2s = 50.0 + 2 \cdot 7.4 = 50.0 + 14.8 = 64.8.^{\dagger}$$

[†]Recall that, for an expression of the form $a + b \cdot c$, the multiplication should be done before the addition. Thus $50.0 + 2 \cdot 7.4 = 50.0 + 14.8 = 64.8$. Similarly, for an expression of the form $a - b \cdot c$, the multiplication should be done before the subtraction.

Likewise, the horizontal position labeled $\bar{x} - 3s$ represents the number that is three standard deviations to the left of the mean, which in this case is

$$\bar{x} - 3s = 50.0 - 3 \cdot 7.4 = 50.0 - 22.2 = 27.8.$$

Figure 3.7 is interpreted in a similar manner.

The graphs shown in Figs. 3.6 and 3.7 vividly illustrate that Data Set II has more variation than Data Set I. They also show that for each data set all observations lie within a few standard deviations to either side of the mean. This result is no accident.

KEY FACT 3.2

Three-Standard-Deviations Rule

Almost all the observations in any data set lie within three standard deviations to either side of the mean.

A data set with a great deal of variation has a large standard deviation, so three standard deviations to either side of its mean will be extensive, as shown in Fig. 3.7. A data set with little variation has a small standard deviation, and hence three standard deviations to either side of its mean will be narrow, as shown in Fig. 3.6.

The three-standard-deviations rule is somewhat vague—what does "almost all" mean? It can be made more precise in several ways, two of which we now briefly describe. We can apply **Chebychev's rule,** which is valid for all data sets and implies, in particular, that at least 89% of the observations lie within three standard deviations to either side of the mean. If the distribution of the data set is approximately bell shaped, we can apply the **empirical rule** which implies, in particular, that roughly 99.7% of the observations lie within three standard deviations to either side of the mean.

In general, Chebychev's rule states that, for any data set and for any number $k > 1$, at least $100(1 - 1/k^2)\%$ of the observations lie within k standard deviations to either side of the mean. For more details about the empirical rule, refer to Section 6.3.

The Technology Center

In this section, we discussed two measures of variation: the range and the sample standard deviation. For learning purposes, calculating a few of these and other descriptive measures by hand is essential. However, in practice, because such calculations are usually tedious and prone to error, you should use a computer or statistical calculator to obtain descriptive measures whenever possible.

In Section 3.1, we described a method for using technology to determine simultaneously several descriptive measures. Printout 3.1 on page 98—the outputs from Minitab, Excel, and the TI-83/84 Plus—contains the sample standard deviation, labeled StDev, Std Dev, and Sx, respectively.

These outputs also contain the maximum and minimum observations. From these two descriptive measures, we can easily obtain the range by subtraction: Range = Max − Min. Refer to the technology manuals for more details and other options.

EXERCISES 3.3

Statistical Concepts and Skills

3.33 Explain the purpose of a measure of variation.

3.34 Why is the standard deviation preferable to the range as a measure of variation?

3.35 When you use the standard deviation as a measure of variation, what is the reference point?

3.36 The following dartboards represent darts thrown by two players, Tracey and Joan.

Tracey   Joan

For the variable "distance from the center," which player's board represents data with a smaller sample standard deviation? Explain your answer.

3.37 Consider the data set 1, 2, 3, 4, 5, 6, 7, 8, 9.
a. Use the defining formula to obtain the sample standard deviation.
b. Replace the 9 in the data set by 99 and again use the defining formula to compute the sample standard deviation.
c. Compare your answers in parts (a) and (b). The lack of what property of the standard deviation accounts for its extreme sensitivity to the change of 9 to 99?

3.38 Consider the following four data sets.

Data Set I		Data Set II		Data Set III		Data Set IV	
1	5	1	9	5	5	2	4
1	8	1	9	5	5	4	4
2	8	1	9	5	5	4	4
2	9	1	9	5	5	4	10
5	9	1	9	5	5	4	10

a. Compute the mean of each data set.

b. Although the four data sets have the same means, in what respect are they quite different?
c. Which data set appears to have the least variation? the greatest variation?
d. Compute the range of each data set.
e. Use the defining formula to compute the sample standard deviation of each data set.
f. From your answers to parts (d) and (e), which measure of variation better distinguishes the spread in the four data sets: the range or the standard deviation? Explain your answer.
g. Are your answers from parts (c) and (e) consistent?

3.39 IQ Scores. Below are 10 IQ scores.

110	122	132	107	101
97	115	91	125	142

Time each of the following calculations.
a. Use the defining formula to obtain the sample standard deviation of the 10 IQs.
b. Use the computing formula to obtain the sample standard deviation of the 10 IQs.
c. Did the computing formula save time? Explain why it did or did not.

3.40 Consider the data set 3, 3, 3, 3, 3, 3.
a. Guess the value of the sample standard deviation without calculating it. Explain your reasoning.
b. Use the defining formula to calculate the sample standard deviation.
c. Complete the following statement and explain your reasoning: If all observations in a data set are equal, the sample standard deviation is _____.
d. Complete the following statement and explain your reasoning: If the sample standard deviation of a data set is 0, then....

*In Exercises **3.41–3.44**, we repeat the essential data from Exercises 3.9–3.12 on pages 99–100. For each exercise, using only basic calculator functions, obtain the*
a. range of the data.
b. sample standard deviation of the data from the defining formula.
c. sample standard deviation of the data from the computing formula.
d. State which formula you found easier to use in obtaining s.

Note: *In parts (b) and (c), round your final answers to one more decimal place than that used for the data.*

3.41 Amphibian Embryos. For a sample of seven different species of frogs' and toads' eggs, the times to hatch, in days, were as follows.

6	7	11	6	5	5	11

3.42 Hurricanes. Hurricane Hugo had a significant impact on the water chemistry of streams in the Luquillo Experimental Forest in Puerto Rico. The following data are a sample of 10 ammonia fluxes in the first year after Hugo. Data are in kilograms per hectare per year.

96	66	147	147	175
116	57	154	88	154

3.43 Tornado Touchdowns. The following table gives the number of tornadoes that touched down in the United States during each month of the year 2002.

3	2	47	118	204	97
68	86	62	57	98	99

3.44 Technical Merit. The following table provides the marks that the judges gave Michelle Kwan on technical merit in the Short Program ladies singles event of the 2002 Winter Olympics.

5.8	5.7	5.9	5.7	5.5	5.7	5.7	5.7	5.6

Extending the Concepts and Skills

3.45 Outliers. In Exercise 3.16 on page 101, we discussed *outliers*, or observations that fall well outside the overall pattern of the data. The following table contains two data sets. Data Set II was obtained by removing the outliers from Data Set I.

Data Set I					Data Set II			
0	12	14	15	23	10	14	15	17
0	14	15	16	24	12	14	15	
10	14	15	17		14	15	16	

a. Compute the sample standard deviation of each of the two data sets.
b. Compute the range of each of the two data sets.
c. What effect do outliers have on variation? Explain your answer.

Grouped-Data Formulas. When data are grouped in a frequency distribution, we use the following formulas to obtain the sample mean and sample standard deviation.

Grouped-Data Formulas

$$\bar{x} = \frac{\Sigma x f}{n} \quad \text{and} \quad s = \sqrt{\frac{\Sigma (x - \bar{x})^2 f}{n - 1}},$$

where x denotes class midpoint, f denotes class frequency, and $n \, (= \Sigma f)$ denotes sample size.

In general, these formulas yield only approximations to the actual sample mean and sample standard deviation. We ask you to apply the grouped-data formulas in Exercises 3.46 and 3.47.

3.46 Weekly Salaries. In the following table, we repeat the salary data in Data Set II from Example 3.1.

300	300	940	450	400
400	300	300	1050	300

a. Use Definitions 3.4 and 3.6 on pages 103 and 110, respectively, to obtain the sample mean and sample standard deviation of this (ungrouped) data set.
b. A frequency distribution for Data Set II, based on single-value grouping, is presented in the first two columns of the table at the top of the next page. The third column of the table is for the xf-values, that is, class midpoint (which here is the same as the class) times class frequency. Complete the missing entries in the table and then use the grouped-data formula to obtain the sample mean.
c. Compare the answers that you obtained for the sample mean in parts (a) and (b). Explain why the grouped-data formula always yields the actual sample mean when the data are grouped in classes that are each based on a single value. (*Hint:* What does xf represent for each class?)

Salary x	Frequency f	Salary · Frequency xf
300	5	1500
400	2	
450	1	
940	1	
1050	1	
	10	

d. Construct a table similar to the one in part (b) but with columns for x, f, $x - \bar{x}$, $(x - \bar{x})^2$, and $(x - \bar{x})^2 f$. Use the table and the grouped-data formula to obtain the sample standard deviation.
e. Compare your answers for the sample standard deviation in parts (a) and (d). Explain why the grouped-data formula always yields the actual sample standard deviation when the data are grouped in classes that are each based on a single value.

3.47 Days to Maturity. Following is a grouped-data table for the days to maturity for 40 short-term investments, as found in *Barron's National Business and Financial Weekly*.

Days to maturity	Frequency f	Relative frequency	Midpoint x
30 ◂ 40	3	0.075	35
40 ◂ 50	1	0.025	45
50 ◂ 60	8	0.200	55
60 ◂ 70	10	0.250	65
70 ◂ 80	7	0.175	75
80 ◂ 90	7	0.175	85
90 ◂ 100	4	0.100	95
	40	1.000	

a. Use the grouped-data formulas to estimate the sample mean and sample standard deviation of the days-to-maturity data. Round your final answers to one decimal place.

b. The following table gives the raw days-to-maturity data.

70	64	99	55	64	89	87	65
62	38	67	70	60	69	78	39
75	56	71	51	99	68	95	86
57	53	47	50	55	81	80	98
51	36	63	66	85	79	83	70

Using Definitions 3.4 and 3.6 on pages 103 and 110, respectively, gives the true sample mean and sample standard deviation of the days-to-maturity data as 68.3 and 16.7, rounded to one decimal place. Compare these actual values of $\bar{x}$ and s to the estimates from part (a). Explain why the grouped-data formulas generally yield only approximations to the sample mean and sample standard deviation for non–single-value grouping.

Using Technology

3.48 Tornado Touchdowns. Use the technology of your choice to determine the range and sample standard deviation of the tornado occurrence data in Exercise 3.43.

3.49 Technical Merit. Use the technology of your choice to determine the range and sample standard deviation of the technical-merit data in Exercise 3.44.

3.50 Body Temperature. Refer to Exercise 3.19 on page 101. Use the technology of your choice to determine the range and sample standard deviation of the temperature data, which we have supplied on the WeissStats CD. Interpret your results.

3.51 Sex and Direction. Refer to Exercise 3.20 on page 101 and do the following:
a. Use the technology of your choice to obtain the range and sample standard deviation of each of the two data sets, which we have supplied on the WeissStats CD.
b. Use the results that you obtained in part (a) to compare the two data sets.

3.4 The Five-Number Summary; Boxplots

So far, we have focused on the mean and standard deviation to measure center and variation. We now examine several descriptive measures based on percentiles.

Unlike the mean and standard deviation, descriptive measures based on percentiles are *resistant*—they are not sensitive to the influence of a few extreme observations. For this reason, descriptive measures based on percentiles are often preferred over those based on the mean and standard deviation.

Quartiles

As you learned in Section 3.1, the median of a data set divides the data into two equal parts: the bottom 50% and the top 50%. The **percentiles** of a data set divide it into hundredths, or 100 equal parts. A data set has 99 percentiles, denoted $P_1, P_2, \ldots, P_{99}$. Roughly speaking, the first percentile, P_1, is the number that divides the bottom 1% of the data from the top 99%; the second percentile, P_2, is the number that divides the bottom 2% of the data from the top 98%; and so on. Note that the median is also the 50th percentile.

Certain percentiles are particularly important: the **deciles** divide a data set into tenths (10 equal parts), the **quintiles** divide a data set into fifths (five equal parts), and the **quartiles** divide a data set into quarters (four equal parts).

Quartiles are the most commonly used percentiles. A data set has three quartiles, which we denote Q_1, Q_2, and Q_3. Roughly speaking, the **first quartile, Q_1,** is the number that divides the bottom 25% of the data from the top 75%; the **second quartile, Q_2,** is the median, which, as you know, is the number that divides the bottom 50% of the data from the top 50%; and the **third quartile, Q_3,** is the number that divides the bottom 75% of the data from the top 25%. Note that the first and third quartiles are the 25th and 75th percentiles, respectively.

Figure 3.8 depicts the quartiles for uniform, bell-shaped, right-skewed, and left-skewed distributions.

FIGURE 3.8
Quartiles for (a) uniform, (b) bell-shaped, (c) right-skewed, and (d) left-skewed distributions

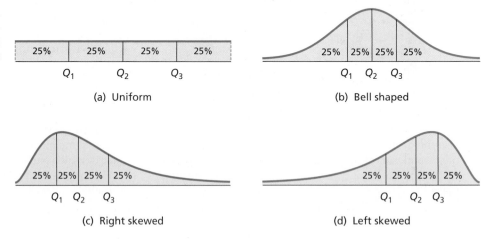

To determine the quartiles of a data set, we first obtain the set's median. Then we find the median of each of the two parts formed by the median of the entire data set. The resulting three numbers are the quartiles because they divide the data set into four parts that each contain (approximately) 25% of the data.

DEFINITION 3.7

Quartiles

Arrange the data in increasing order and determine the median.

- The *first quartile* is the median of the portion of the entire data set that lies at or below the median of the entire data set.
- The *second quartile* is the median of the entire data set.
- The *third quartile* is the median of the portion of the entire data set that lies at or above the median of the entire data set.

> **What**
> *Does it Mean?*
>
> The quartiles divide a data set into quarters (four equal parts).

Note: Not all statisticians define quartiles in exactly the same way. Other definitions may lead to different values but, in practice, the differences tend to be small with large data sets.

In Example 3.15, we demonstrate how to find quartiles.

▌▌▌Example 3.15   Quartiles

Weekly TV-Viewing Times   The A. C. Nielsen Company publishes data on the TV-viewing habits of Americans by various characteristics in *Nielsen Report on Television*. A sample of 20 people yielded the weekly viewing times, in hours, displayed in Table 3.13. Determine and interpret the quartiles for these data.

TABLE 3.13
Weekly TV-viewing times

25	41	27	32	43
66	35	31	15	5
34	26	32	38	16
30	38	30	20	21

Solution   To find the quartiles, we apply Definition 3.7. First, we arrange the data in Table 3.13 in increasing order:

5 15 16 20 21 25 26 27 30 **30 31** 32 32 34 35 38 38 41 43 66

Next, we determine the median of the entire data set. The number of observations is 20, so the position of the median is at $(20 + 1)/2 = 10.5$, halfway between the tenth and eleventh observations (shown in boldface) in the ordered list. Thus, the median of the entire data set is $(30 + 31)/2 = 30.5$.

The first quartile is the median of the portion of the entire data set that lies at or below the median of the entire data set. Referring to the ordered list of the entire data set and recalling that the median is 30.5, we see that the portion of the entire data set that lies at or below the median of the entire data set is

5 15 16 20 **21 25** 26 27 30 30

We note that this data set has 10 observations. Its median is therefore at position $(10 + 1)/2 = 5.5$, halfway between the fifth and sixth observations (shown in boldface) in the ordered list. Thus the median of this data set—and hence the first quartile—is $(21 + 25)/2 = 23$; that is, $Q_1 = 23$.

The second quartile is the median of the entire data set, or 30.5. Therefore, we have $Q_2 = 30.5$.

The third quartile is the median of the portion of the entire data set that lies at or above the median of the entire data set. Referring to the ordered list of the entire data set and recalling that the median is 30.5, we see that the portion of the entire data set that lies at or above the median of the entire data set is

31 32 32 34 **35 38** 38 41 43 66

We note that this data set has 10 observations. Its median is therefore at position $(10 + 1)/2 = 5.5$, halfway between the fifth and sixth observations (shown in boldface) in the ordered list. Thus the median of this data set—and hence the third quartile—is $(35 + 38)/2 = 36.5$; that is, $Q_3 = 36.5$.

In summary, the three quartiles for the TV-viewing times in Table 3.13 are $Q_1 = 23$ hours, $Q_2 = 30.5$ hours, and $Q_3 = 36.5$ hours.

INTERPRETATION We see that 25% of the TV-viewing times are less than 23 hours, 25% are between 23 hours and 30.5 hours, 25% are between 30.5 hours and 36.5 hours, and 25% are greater than 36.5 hours.

In Example 3.15, the number of observations is 20, which is even. To illustrate the determination of quartiles when the number of observations is odd, we consider the TV-viewing-time data again, but this time without the largest observation, 66. In this case, the ordered list of the entire data set is

5 15 16 20 21 25 26 27 30 **30** 31 32 32 34 35 38 38 41 43

The median of the entire data set (also the second quartile) is 30, shown in boldface. The first quartile is the median of the portion of the entire data set that lies at or below the median of the entire data set, or the median of the 10 observations from 5 through the boldfaced 30, which is $(21 + 25)/2 = 23$. The third quartile is the median of the portion of the entire data set that lies at or above the median of the entire data set, or the median of the 10 observations from the boldfaced 30 through 43, which is $(34 + 35)/2 = 34.5$. Thus, for this data set, we have $Q_1 = 23$ hours, $Q_2 = 30$ hours, and $Q_3 = 34.5$ hours.

The Interquartile Range

Next, we discuss the **interquartile range,** or **IQR.** Because quartiles are used to define the interquartile range, it is the preferred measure of variation when the median is used as the measure of center. Like the median, the interquartile range is a resistant measure.

DEFINITION 3.8

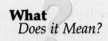

What Does it Mean?

Roughly speaking, the IQR gives the range of the middle 50% of the observations.

Interquartile Range

The *interquartile range,* or **IQR,** is the difference between the first and third quartiles; that is, $\text{IQR} = Q_3 - Q_1$.

In Example 3.16, we show how to obtain the interquartile range for the data on TV-viewing times.

Example 3.16   The Interquartile Range

Weekly TV-Viewing Times   Obtain the IQR for the TV-viewing-time data displayed in Table 3.13 on page 119.

Solution As we discovered in Example 3.15, the first and third quartiles are 23 and 36.5, respectively. Therefore the interquartile range is

$$Q_3 - Q_1 = 36.5 - 23 = 13.5 \text{ hours.}$$

In symbols, IQR = 13.5 hours.

INTERPRETATION The middle 50% of the TV-viewing times are spread out over a 13.5 hour interval, roughly.

The Five-Number Summary

From the three quartiles, we can obtain a measure of center (the median, Q_2) and measures of variation of the two middle quarters of the data, $Q_2 - Q_1$ for the second quarter and $Q_3 - Q_2$ for the third quarter. But the three quartiles don't tell us anything about the variation of the first and fourth quarters.

To gain that information, we need only include the minimum and maximum observations as well. Then the variation of the first quarter can be measured as the difference between the minimum and the first quartile, $Q_1 - \text{Min}$, and the variation of the fourth quarter can be measured as the difference between the third quartile and the maximum, $\text{Max} - Q_3$.

Thus the minimum, maximum, and quartiles together provide, among other things, information on center and variation. Written in increasing order, they comprise the **five-number summary** of a data set.

DEFINITION 3.9

Five-Number Summary

The *five-number summary* of a data set is Min, Q_1, Q_2, Q_3, Max.

What
Does it Mean?

The five-number summary of a data set consists of the minimum, maximum, and quartiles, written in increasing order.

In Example 3.17, we show how to obtain and interpret the five-number summary of a set of data.

Example 3.17 **The Five-Number Summary**

Weekly TV-Viewing Times Obtain and interpret the five-number summary for the TV-viewing-time data given in Table 3.13 on page 119.

Solution From the ordered list of the entire data set (see page 119), Min = 5 and Max = 66. Furthermore, as we showed earlier, $Q_1 = 23$, $Q_2 = 30.5$, and $Q_3 = 36.5$. Consequently, the five-number summary of the data on TV-viewing times is given by 5, 23, 30.5, 36.5, and 66 hours. From the five-number summary, the variations of the four quarters of the TV-viewing-time data are 18, 7.5, 6, and 29.5 hours, respectively.

INTERPRETATION There is less variation in the middle two quarters of the TV-viewing times than in the first and fourth quarters, and the fourth quarter has the greatest variation of all.

Outliers

In data analysis, the identification of **outliers,** or observations that fall well outside the overall pattern of the data, is important. An outlier requires special attention. It may be the result of a measurement or recording error, an observation from a different population, or an unusual extreme observation. Note that an extreme observation need not be an outlier; it may instead be an indication of skewness.

As an example of an outlier, consider the data set consisting of the individual wealths (in dollars) of all U.S. residents. For this data set, the wealth of Bill Gates is an outlier—in this case, an unusual extreme observation.

When an outlier is observed, you should always try to determine its cause. If an outlier is caused by a measurement or recording error, or for some other reason it clearly does not belong in the data set, the outlier can simply be removed. However, if no explanation for the outlier is apparent, the decision whether to retain it in the data set can often be a difficult judgment call.

We can use quartiles and the IQR to identify potential outliers, that is, as a diagnostic tool for spotting observations that may be outliers. To do so, we first define the **lower limit** and the **upper limit** of a data set.

DEFINITION 3.10

What *Does it Mean?*

The lower limit is the number that lies 1.5 IQRs below the first quartile; the upper limit is the number that lies 1.5 IQRs above the third quartile.

Lower and Upper Limits

The *lower limit* and *upper limit* of a data set are

$$\text{Lower limit} = Q_1 - 1.5 \cdot \text{IQR};$$
$$\text{Upper limit} = Q_3 + 1.5 \cdot \text{IQR}.$$

Observations that lie outside the lower and upper limits—either below the lower limit or above the upper limit—are potential outliers. Further data analysis should be done (using histograms, stem-and-leaf diagrams, or other methods) to find out whether such observations are truly outliers. In Example 3.18, we demonstrate this process.

Example 3.18 Outliers

Weekly TV-Viewing Times For the TV-viewing-time data in Table 3.13 on page 119,

a. obtain the lower and upper limits.

b. determine potential outliers, if any.

Solution

a. As before, $Q_1 = 23$, $Q_3 = 36.5$, and IQR $= 13.5$. Therefore

$$\text{Lower limit} = Q_1 - 1.5 \cdot \text{IQR} = 23 - 1.5 \cdot 13.5 = 2.75 \text{ hours};$$
$$\text{Upper limit} = Q_3 + 1.5 \cdot \text{IQR} = 36.5 + 1.5 \cdot 13.5 = 56.75 \text{ hours}.$$

These limits are portrayed graphically in Fig. 3.9.

FIGURE 3.9
Lower and upper limits for
TV-viewing times

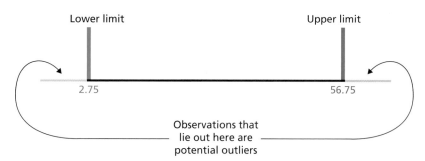

b. The ordered list of the entire data set on page 119 reveals one observation, 66, that lies outside the lower and upper limits—specifically, above the upper limit. Consequently, 66 is a potential outlier. A histogram and a stem-and-leaf diagram both indicate that the observation of 66 hours is truly an outlier.

INTERPRETATION The weekly viewing time of 66 hours lies outside the overall pattern of the other 19 viewing times in the data set.

Boxplots

A **boxplot,** also called a **box-and-whisker diagram,** is based on the five-number summary and can be used to provide a graphical display of the center and variation of a data set. These diagrams, like stem-and-leaf diagrams, were invented by Professor John Tukey.

Actually, two types of boxplots are in common use. One is simply called a boxplot; the other is called a **modified boxplot.** The main difference between the two types of boxplots is that potential outliers are plotted individually in a modified boxplot, but not in a boxplot. Thus, when outliers are of concern, modified boxplots are the preferred type of boxplot. Procedures 3.1 and 3.2 provide step-by-step instructions for constructing boxplots and modified boxplots, respectively.

PROCEDURE 3.1

To Construct a Boxplot

STEP 1 **Determine the five-number summary.**

STEP 2 **Draw a horizontal axis on which the numbers obtained in Step 1 can be located. Above this axis, mark the quartiles and the minimum and maximum with vertical lines.**

STEP 3 **Connect the quartiles to make a box and then connect the box to the minimum and maximum with lines.**

To construct a modified boxplot, we need to use the concept of *adjacent values.* The **adjacent values** of a data set are the most extreme observations that still lie within the lower and upper limits, that is, the most extreme observations that are not potential outliers. Note that, if a data set has no potential outliers, the adjacent values are just the minimum and maximum observations.

PROCEDURE 3.2

To Construct a Modified Boxplot

STEP 1 **Determine the quartiles.**

STEP 2 **Determine potential outliers and the adjacent values.**

STEP 3 **Draw a horizontal axis on which the numbers obtained in Steps 1 and 2 can be located. Above this axis, mark the quartiles and the adjacent values with vertical lines.**

STEP 4 **Connect the quartiles to make a box and then connect the box to the adjacent values with lines.**

STEP 5 **Plot each potential outlier with an asterisk.**

Note:

- If a data set has no potential outliers, its boxplot and modified boxplot are identical.
- In both types of boxplots, the two lines emanating from the box are called **whiskers.**
- Boxplots are frequently drawn vertically instead of horizontally.
- Symbols other than an asterisk are often used to plot potential outliers.

In Example 3.19, we demonstrate the use of Procedures 3.1 and 3.2 in obtaining a boxplot and a modified boxplot for a set of data.

Example 3.19 Boxplots and Modified Boxplots

Weekly TV-Viewing Times The weekly TV-viewing times for a sample of 20 people are displayed in Table 3.13 on page 119. Construct a boxplot and, if appropriate, a modified boxplot.

Solution To obtain a boxplot for the TV-viewing times, we apply the step-by-step method presented in Procedure 3.1.

STEP 1 **Determine the five-number summary.**

We have already obtained the five-number summary for the TV-viewing times: Min $= 5$, $Q_1 = 23$, $Q_2 = 30.5$, $Q_3 = 36.5$, and Max $= 66$.

FIGURE 3.10
(a) Boxplot for TV-viewing times;
(b) modified boxplot for TV-viewing
times

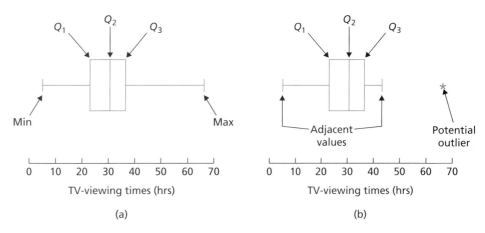

STEP 2 **Draw a horizontal axis on which the numbers obtained in Step 1 can be located. Above this axis, mark the quartiles and the minimum and maximum with vertical lines.**

See Fig. 3.10(a).

STEP 3 **Connect the quartiles to make a box and then connect the box to the minimum and maximum with lines.**

See Fig. 3.10(a).

Figure 3.10(a) is a boxplot for the data on TV-viewing times. Because the ends of the combined box are at the quartiles, the width of that box equals the interquartile range, IQR. Observe that the two boxes in the boxplot indicate the spread of the second and third quarters of the data and that the two whiskers indicate the spread of the first and fourth quarters.

INTERPRETATION There is less variation in the middle two quarters of the TV-viewing times than in the first and fourth quarters, and the fourth quarter has the greatest variation of all.

We discovered earlier that the TV-viewing times contain a potential outlier. Therefore it is appropriate to construct a modified boxplot because it will be different from the boxplot that we just obtained. To do so, we apply Procedure 3.2.

STEP 1 **Determine the quartiles.**

The quartiles for the TV-viewing times were found in Example 3.15: $Q_1 = 23$, $Q_2 = 30.5$, and $Q_3 = 36.5$.

STEP 2 **Determine potential outliers and the adjacent values.**

The TV-viewing times contain one potential outlier, 66. From the ordered list of the entire data set on page 119, the adjacent values are 5 and 43.

STEP 3 **Draw a horizontal axis on which the numbers obtained in Steps 1 and 2 can be located. Above this axis, mark the quartiles and the adjacent values with vertical lines.**

See Fig. 3.10(b).

STEP 4 **Connect the quartiles to make a box and then connect the box to the adjacent values with lines.**

See Fig. 3.10(b).

STEP 5 **Plot each potential outlier with an asterisk.**

As we noted in Step 2, this data set contains one potential outlier—namely, 66. It is plotted with an asterisk in Fig. 3.10(b).

Other Uses of Boxplots

Boxplots are especially suited to comparing two or more data sets. In doing so, you should use the same scale for all the boxplots. We introduce this use of boxplots in the exercises for this section and, later, when we discuss inferential statistics, we further apply this use of boxplots as an exploratory tool preliminary to inference.

We can also use a boxplot to identify the approximate shape of the distribution of a data set. Figure 3.11 displays some common distribution shapes and their corresponding boxplots. Study Fig. 3.11 carefully, noting especially how box width and whisker length relate to skewness and symmetry.

FIGURE 3.11
Distribution shapes and boxplots for (a) uniform, (b) bell-shaped, (c) right-skewed, and (d) left-skewed distributions

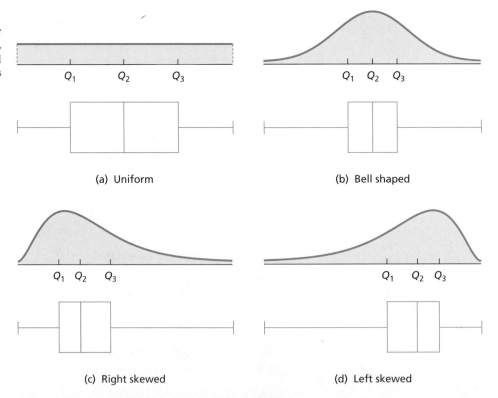

Employing boxplots to identify the shape of a distribution is most useful with large data sets. For small data sets, boxplots can be unreliable in identifying distribution shape; using a stem-and-leaf diagram or dotplot is generally a better way to ascertain distribution shape for a small data set.

The Technology Center

In Section 3.1, we discussed a method for using technology to determine simultaneously several descriptive measures. Printout 3.1 on page 98—the output from Minitab, Excel, and the TI-83/84 Plus—includes the five-number summary.

Minitab, Excel, and the TI-83/84 Plus have built-in programs for obtaining modified boxplots. Both Minitab and the TI-83/84 Plus also provide programs for boxplots. Here we discuss the use of these three technologies to get modified boxplots. For other options, refer to the technology manuals.

Example 3.20   Using Technology to Obtain a Modified Boxplot

Weekly TV-Viewing Times   Use Minitab, Excel, or the TI-83/84 Plus to obtain a modified boxplot for the TV-viewing times displayed in Table 3.13 on page 119.

Solution   Printout 3.2 shows the output obtained by applying the modified-boxplot programs to the TV-viewing-time data.

PRINTOUT 3.2
Modified-boxplot output for the TV-viewing times

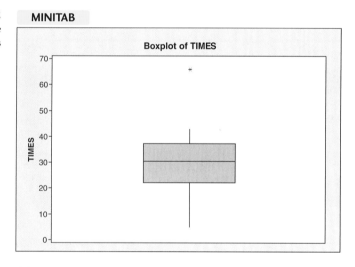

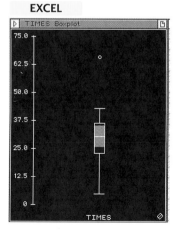

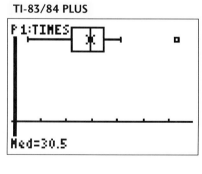

As we mentioned earlier and as Printout 3.2 shows, boxplots are frequently drawn vertically instead of horizontally. Compare the modified boxplots in Printout 3.2 to the one in Fig. 3.10(b) on page 125.

Obtaining the Output (Optional)

Here are detailed instructions for obtaining a modified boxplot for the TV-viewing times, using Minitab, Excel, and the TI-83/84 Plus. First, store the data from Table 3.13 on page 119 in a column (Minitab), range (Excel), or list (TI-83/84 Plus) named TIMES. Then proceed as follows.

MINITAB	EXCEL	TI-83/84 PLUS
1 Choose **Graph ➤ Boxplot...**	1 Choose **DDXL ➤ Charts and Plots**	1 Press **2nd ➤ STAT PLOT** and then press **ENTER** twice
2 Select the **Simple** boxplot and click **OK**	2 Select **Boxplot** from the **Function type** drop-down box	2 Arrow to the fourth graph icon and press **ENTER**
3 Specify TIMES in the **Graph variables** text box	3 Specify TIMES in the **Quantitative Variables** text box	3 Press the down-arrow key
4 Click **OK**	4 Click **OK**	4 Press **2nd ➤ LIST**
		5 Arrow down to TIMES and press **ENTER**
		6 Press **ZOOM** and then **9** (and then **TRACE**, if desired)

EXERCISES 3.4

Statistical Concepts and Skills

3.52 Identify by name three important groups of percentiles.

3.53 Identify an advantage that the median and interquartile range have over the mean and standard deviation, respectively.

3.54 Explain why the minimum and maximum observations are added to the three quartiles to describe better the variation in a data set.

3.55 Is an extreme observation necessarily an outlier? Explain your answer.

3.56 Under what conditions are boxplots useful for identifying the shape of a distribution?

3.57 Regarding the interquartile range,
a. what type of descriptive measure is it?
b. what does it measure?

3.58 Identify a use of the lower and upper limits.

3.59 When is a modified boxplot the same as a boxplot? Explain your answer.

3.60 Which measure of variation is preferred when
a. the mean is used as a measure of center?
b. the median is used as a measure of center?

In Exercises 3.61–3.64, use only basic calculator functions to determine the quartiles.

3.61 The Great Gretzky. Wayne Gretzky, a retired professional hockey player, played 20 seasons in the National Hockey League (NHL), from 1980 through 1999. S. Berry explored some of Gretzky's accomplishments in "A Statistician Reads the Sports Pages" (*Chance*, Vol. 16, No. 1, pp. 49–54). The following table shows the number of games in which Gretzky played during each of his 20 seasons in the NHL.

79	80	80	80	74
80	80	79	64	78
73	78	74	45	81
48	80	82	82	70

3.62 Parenting Grandparents. In the article "Grandchildren Raised by Grandparents, a Troubling Trend" (*California*

Agriculture, Vol. 55, No. 2, pp. 10–17), M. Blackburn considered the rates of children (under 18 years of age) living in California with grandparents as their primary caretakers. A sample of 14 California counties yielded the following percentages of children under 18 living with grandparents in 1990.

5.9	4.0	5.7	5.1	4.1	4.4	6.5
4.4	5.8	5.1	6.1	4.5	4.9	4.9

3.63 Hospital Stays. The U.S. National Center for Health Statistics compiles data on the length of stay by patients in short-term hospitals and publishes its findings in *Vital and Health Statistics*. A random sample of 21 patients yielded the following data on length of stay, in days.

4	4	12	18	9	6	12
3	6	15	7	3	55	1
10	13	5	7	1	23	9

3.64 Miles Driven. The U.S. Federal Highway Administration conducts studies on motor vehicle travel by type of vehicle. Results are published annually in *Highway Statistics*. A sample of 15 cars yields the following data on number of miles driven, in thousands, for last year.

13.2	13.3	11.9	15.7	11.3
12.2	16.7	10.7	3.3	13.6
14.8	9.6	11.6	8.7	15.0

In Exercises 3.65–3.68,
a. determine the interquartile range.
b. obtain the five-number summary.
c. identify potential outliers, if any.
d. construct and interpret a boxplot and, if appropriate, a modified boxplot.

3.65 The Great Gretzky. The number of games in which Wayne Gretzky played during each of his 20 seasons in the NHL, as given in Exercise 3.61.

3.66 Parenting Grandparents. The percentage of children under 18 years old living with grandparents in 1990 for each of 14 California counties, as given in Exercise 3.62.

3.67 Hospital Stays. The lengths of stay in short-term hospitals by 21 randomly selected patients, as given in Exercise 3.63.

3.68 Miles Driven. The number of thousands of miles 15 cars were driven last year, as given in Exercise 3.64.

3.69 Nicotine Patches. In the paper "The Smoking Cessation Efficacy of Varying Doses of Nicotine Patch Delivery Systems 4 to 5 Years Post-Quit Day" (*Preventative Medicine*, 28, pp. 113–118), D. Daughton et al. discussed the long-term effectiveness of transdermal nicotine patches on participants who had previously smoked at least 20 cigarettes per day. A sample of 15 participants in the Transdermal Nicotine Study Group (TNSG) reported that they now smoke the following number of cigarettes per day.

10	9	10	8	7
6	10	9	10	8
9	10	8	8	10

a. Determine the quartiles for these data.
b. Remark on the usefulness of quartiles with respect to this data set.

Extending the Concepts and Skills

3.70 Starting Salaries. Surveys are conducted by the Northwestern University Placement Center, Evanston, Illinois, on starting salaries for college graduates. Results of the surveys can be found in *The Northwestern Lindquist-Endicott Report*. The following diagram shows boxplots for the starting annual salaries, in thousands of dollars, obtained from samples of 32 computer-science graduates (top boxplot) and 35 accounting graduates (bottom boxplot). Use the boxplots to compare the starting salaries of the computer-science and accounting graduates sampled.

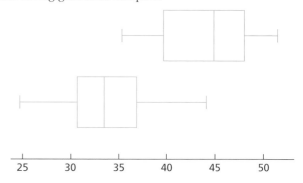

3.71 Obesity. Researchers in obesity wanted to compare the effectiveness of dieting with exercise against dieting without exercise. Seventy-three patients were randomly divided into two groups. Group 1, composed of 37 patients, was put on a program of dieting with exercise. Group 2, composed of 36 patients, dieted only. The results for weight loss, in pounds, after 2 months are summarized in the following boxplots. The top boxplot is for Group 1 and the bottom boxplot is for Group 2. Use the boxplots to compare the weight losses for the two groups.

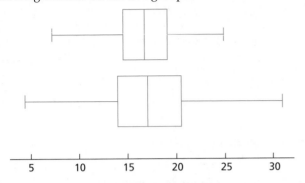

3.72 Each of the following boxplots was obtained from a very large data set. Use the boxplots to identify the approximate shape of the distribution of each data set.

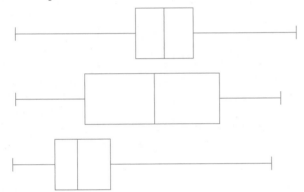

3.73 What can you say about the boxplot of a symmetric distribution?

Using Technology

3.74 The Great Gretzky. Refer to the data on the number of games in which Wayne Gretzky played during each of his 20 seasons in the NHL, as given in Exercise 3.61. Use the technology of your choice to
a. obtain a boxplot for the data.
b. determine the five-number summary of the data.

3.75 Parenting Grandparents. Refer to the data on the percentage of children under 18 years old living with grandparents in 1990 for each of 14 California counties, as given in Exercise 3.62. Use the technology of your choice to
a. obtain a boxplot for the data.
b. determine the five-number summary of the data.

3.76 Body Temperature. Refer to Exercise 3.19 on page 101. Use the technology of your choice to determine a boxplot and the five-number summary of the temperature data, which we have supplied on the WeissStats CD. Interpret your results.

3.77 Sex and Direction. Refer to Exercise 3.20 on page 101 and do the following:
a. Use the technology of your choice to obtain side-by-side boxplots of the two data sets, which we have supplied on the WeissStats CD.
b. Use the boxplots that you obtained in part (a) to compare the two data sets.

3.5 Descriptive Measures for Populations; Use of Samples

In this section, we discuss several descriptive measures for *population data*—the data obtained by observing the values of a variable for an entire population. Although, in reality, we often don't have access to population data, it is nonetheless helpful to become familiar with the notation and formulas used for descriptive measures of such data.

The Population Mean

Recall that, for a variable x and a sample of size n from a population, the sample mean is

$$\bar{x} = \frac{\Sigma x}{n}.$$

First, we sum the observations of the variable for the sample, and then we divide by the size of the sample.

TABLE 3.14
Notation used for a sample and for the population

	Size	Mean
Sample	n	$\bar{x}$
Population	N	μ

The mean of a finite population is obtained similarly: First, we sum all possible observations of the variable for the entire population, and then we divide by the size of the population. However, to distinguish the **population mean** from a sample mean, we use the Greek letter μ (pronounced "mew") to denote the population mean. We also use the uppercase English letter N to represent the size of the population.

Table 3.14 summarizes the notation that is used for both a sample and the population.

DEFINITION 3.11

What
Does it Mean?

A population mean (mean of a variable) is the arithmetic average (mean) of population data.

Population Mean (Mean of a Variable)

For a variable x, the mean of all possible observations for the entire population is called the *population mean* or *mean of the variable x*. It is denoted μ_x or, when no confusion will arise, simply μ. For a finite population,

$$\mu = \frac{\Sigma x}{N},$$

where N is the population size.

Example 3.21 illustrates calculation of the population mean.

Example 3.21   **The Population Mean**

U.S. Women's World Cup Soccer Team   From FIFA.com, the official Web site of the *Fédération Internationale de Football Association*, we obtained data for the players on the 2003 U.S. Women's World Cup soccer team, as shown in Table 3.15 on the next page. Heights are given in centimeters (cm) and weights in kilograms (kg). Obtain the population mean weight of these soccer players.

Solution   The variable under consideration here is weight, and the population of interest consists of the players on the 2003 U.S. Women's World Cup soccer team. The sum of the weights in the fourth column of Table 3.15 is 1251 kg. Because there are 20 players, $N = 20$. Consequently,

$$\mu = \frac{\Sigma x}{N} = \frac{1251}{20} = 62.55 \text{ kg}.$$

INTERPRETATION   The population mean weight of the players on the 2003 U.S. Women's World Cup soccer team is 62.55 kg.

Using a Sample Mean to Estimate a Population Mean

In inferential studies, we analyze sample data. Nonetheless, the objective is to describe the entire population. The reason for resorting to a sample is that its use is generally more practical. We illustrate this point in Example 3.22.

TABLE 3.15
U.S. Women's World Cup
soccer team, 2003

Name	Position	Height (cm)	Weight (kg)	College
Scurry, Briana	GK	175	68	UMass
Bivens, Kylie	DF	165	59	Santa Clara
Pearce, Christie	DF	165	61	Monmouth
Reddick, Cat	DF	170	68	UNC
Roberts, Tiffany	MF	161	51	UNC
Chastain, Brandi	DF	171	58	Santa Clara
Boxx, Shannon	MF	173	67	Notre Dame
MacMillan, Shannon	FW	164	61	Portland
Hamm, Mia	FW	163	61	UNC
Wagner, Aly	MF	165	61	Santa Clara
Foudy, Julie	MF	168	59	Stanford
Parlow, Cindy	FW	180	70	UNC
Lilly, Kristine	MF	163	57	UNC
Fawcett, Joy	DF	170	61	California
Sobrero, Kate	DF	170	61	Notre Dame
Milbrett, Tiffeny	FW	157	61	Portland
Slaton, Danielle	DF	168	66	Santa Clara
Mullinix, Siri	GK	173	64	UNC
Hucles, Angela	MF	168	64	Virginia
Wambach, Abby	FW	180	73	Florida

Example 3.22 A Use of a Sample Mean

Estimating Mean Household Income   The U.S. Bureau of the Census reports the mean (annual) income of U.S. households in the publication *Current Population Reports*.   To obtain the population data—the incomes of all U.S. households—would be extremely expensive and time consuming. It is also unnecessary because accurate estimates of the mean income of all U.S. households can be obtained from the mean income of a sample of such households. In reality, the Census Bureau samples 60,000 households from a total of more than 100 million U.S. households.

The variable under consideration here is household income; the population of interest is all U.S. households; the mean income of all U.S. households is the population mean, μ. The sample consists of the 60,000 households obtained by the Census Bureau; the mean income of those households is a sample mean, $\bar{x}$. Figure 3.12, displayed at the top of the following page, summarizes this discussion graphically.

After the sample has been taken, the Census Bureau can compute the sample mean income, $\bar{x}$, of the 60,000 households obtained. Using the value of $\bar{x}$, the Census Bureau can then estimate the population mean income, μ, of all U.S. households. We discuss these types of inferences in Chapter 8, where we examine confidence intervals for a population mean.

FIGURE 3.12
Population and sample for incomes of
U.S. households

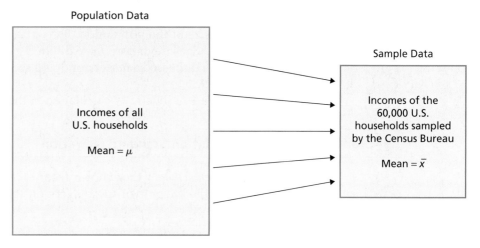

The Population Standard Deviation

Recall that, for a variable x and a sample of size n from a population, the sample standard deviation is

$$s = \sqrt{\frac{\Sigma (x - \bar{x})^2}{n - 1}}.$$

The standard deviation of a finite population is obtained in a similar, but slightly different, way. To distinguish the **population standard deviation** from a sample standard deviation, we use the Greek letter σ (pronounced "sigma") to denote the population standard deviation.

DEFINITION 3.12

What *Does it Mean?*

Roughly speaking, the population standard deviation indicates how far, on average, the observations in the population are from the mean of the population.

Population Standard Deviation (Standard Deviation of a Variable)

For a variable x, the standard deviation of all possible observations for the entire population is called the *population standard deviation* or *standard deviation of the variable x.* It is denoted σ_x or, when no confusion will arise, simply σ. For a finite population,

$$\sigma = \sqrt{\frac{\Sigma (x - \mu)^2}{N}},$$

where N is the population size. The population standard deviation can also be obtained from the computing formula

$$\sigma = \sqrt{\frac{\Sigma x^2}{N} - \mu^2}.$$

Note: Just as s^2 is called a sample variance, σ^2 is called the **population variance** (or **variance of the variable**).

In the defining formula for the population standard deviation, σ, we divide by N, the size of the population. In contrast, in the defining formula for the sample standard deviation, s, we divide by $n - 1$, or 1 less than the size of the sample. We discussed a reason for doing so in the note preceding Example 3.11 on page 109.

Example 3.23 illustrates calculation of the population standard deviation.

Example 3.23   The Population Standard Deviation

U.S. Women's World Cup Soccer Team   Obtain the population standard deviation of the weights of the players on the 2003 U.S. Women's World Cup soccer team.

Solution   We apply the computing formula given in Definition 3.12. To do so, we need the sum of the squares of the weights and the population mean weight, μ. From Example 3.21, $\mu = 62.55$ kg. Squaring each weight in Table 3.15 and adding the results yields $\Sigma x^2 = 78{,}737$. Recalling that there are 20 players, we have

$$\sigma = \sqrt{\frac{\Sigma x^2}{N} - \mu^2} = \sqrt{\frac{78{,}737}{20} - (62.55)^2} = 4.9 \text{ kg.}$$

INTERPRETATION   The population standard deviation of the weights of the players on the 2003 U.S. Women's World Cup soccer team is 4.9 kg. Roughly speaking, on average, the weights of the individual players fall 4.9 kg from their mean weight of 62.55 kg.

Using a Sample Standard Deviation to Estimate a Population Standard Deviation

We have shown that a sample mean can be used to estimate a population mean. Likewise, a sample standard deviation can be used to estimate a population standard deviation, as demonstrated in Example 3.24.

Example 3.24   A Use of a Sample Standard Deviation

Estimating Variation in Bolt Diameters   A hardware manufacturer produces "10-millimeter (mm)" bolts. The manufacturer knows that the diameters of the bolts produced vary somewhat from 10 mm and also from each other. But even if he is willing to accept some variation in bolt diameters, he cannot tolerate too much variation—if the variation is too large, too many of the bolts will be unusable (too narrow or too wide).

To evaluate the variation in bolt diameters, the manufacturer needs to know the population standard deviation, σ, of bolt diameters. Because, in this case, σ cannot be determined exactly (do you know why?), the manufacturer must use the standard deviation of the diameters of a sample of bolts to estimate σ. He decides to take a sample of 20 bolts.

The variable under consideration here is bolt diameter; the population of interest consists of all 10-mm bolts that have been or ever will be produced by the manufacturer; the standard deviation of the diameters of all such bolts is the population standard deviation, σ. The sample consists of the 20 bolts obtained by the manufacturer; the standard deviation of the diameters of those bolts is a sample standard deviation, s. Figure 3.13 summarizes this discussion graphically.

FIGURE 3.13

Population and sample for bolt diameters

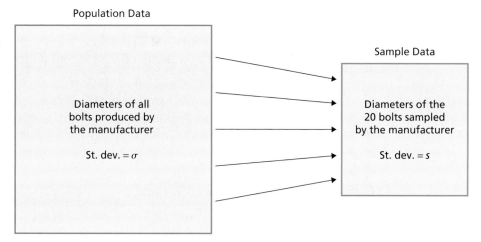

Population Data

Diameters of all bolts produced by the manufacturer

St. dev. $= \sigma$

Sample Data

Diameters of the 20 bolts sampled by the manufacturer

St. dev. $= s$

After the sample has been taken, the manufacturer can compute the sample standard deviation, s, of the diameters of the 20 bolts obtained. Using the value of s, he can then estimate the population standard deviation, σ, of the diameters of all bolts being produced. We examine such inferences in Chapter 11.

Parameter and Statistic

The following definition provides statistical terminology that helps us distinguish between a descriptive measure for a population—a **parameter**—and a descriptive measure for a sample—a **statistic.**

DEFINITION 3.13

Parameter and Statistic

Parameter: A descriptive measure for a population.

Statistic: A descriptive measure for a sample.

Thus, for example, μ and σ are parameters, whereas $\bar{x}$ and s are statistics.

Standardized Variables

We can associate with any variable x a new variable z, called the **standardized version** of x or the **standardized variable,** defined as follows.

DEFINITION 3.14

> **Standardized Variable**
>
> For a variable x, the variable
>
> $$z = \frac{x - \mu}{\sigma}$$
>
> is called the ***standardized version*** of x or the ***standardized variable*** corresponding to the variable x.

What Does it Mean?

The standardized version of a variable x is obtained by first subtracting from x its mean and then dividing by its standard deviation.

A standardized variable always has mean 0 and standard deviation 1. For this and other reasons, standardized variables play an important role in many aspects of statistical theory and practice. We present a few applications of standardized variables in this section; several others appear throughout the rest of the book. We begin with Example 3.25.

Example 3.25   Standardized Variables

TABLE 3.16

Possible observations of x and z

x	−1	3	3	3	5	5
z	−2	0	0	0	1	1

Understanding the Basics   Let's consider a simple variable x—namely, one with possible observations shown in the first row of Table 3.16.

a. Determine the standardized version of x.

b. Find the observed value of z corresponding to an observed value of x of 5.

c. Obtain all possible observations of z.

d. Find the mean and standard deviation of z using Definitions 3.11 and 3.12. Was it necessary to do these calculations to obtain the mean and standard deviation?

e. Obtain dotplots of the distributions of both x and z. Interpret the results.

Solution

a. Using Definitions 3.11 and 3.12, we find that the mean and standard deviation of x are $\mu = 3$ and $\sigma = 2$. Consequently, the standardized version of x is

$$z = \frac{x - 3}{2}.$$

b. The observed value of z corresponding to an observed value of x of 5 is

$$z = \frac{x - 3}{2} = \frac{5 - 3}{2} = 1.$$

c. Applying the formula $z = (x - 3)/2$ to each of the possible observations of the variable x shown in the first row of Table 3.16, we obtain the possible observations of the standardized variable z shown in the second row of Table 3.16.

d. From the second row of Table 3.16,

$$\mu_z = \frac{\Sigma z}{N} = \frac{0}{6} = 0$$

and

$$\sigma_z = \sqrt{\frac{\Sigma(z - \mu_z)^2}{N}} = \sqrt{\frac{6}{6}} = 1.$$

The results of these two computations illustrate something that we mentioned earlier: The mean of a standardized variable is always 0 and its standard deviation is always 1. Thus we really didn't need to perform the calculations to obtain the mean and standard deviation of the variable z.

e. Figures 3.14(a) and 3.14(b) show dotplots of the distributions of x and z, respectively.

FIGURE 3.14

Dotplots of the distributions of x and its standardized version z

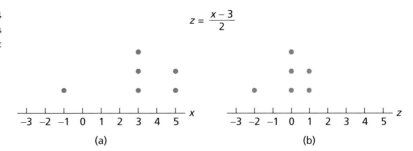

$$z = \frac{x - 3}{2}$$

(a)

(b)

INTERPRETATION The two dotplots in Fig. 3.14 illustrate visually something that we already know: Standardizing shifts a distribution so that the new mean is 0, and it changes the scale so that the new standard deviation is 1.

z-Scores

An important concept associated with standardized variables is that of the **z-score,** or **standard score,** which we now define.

DEFINITION 3.15

What *Does it Mean?*

The z-score of an observation tells us the number of standard deviations that the observation is from the mean, that is, how far the observation is from the mean in units of standard deviation.

z-Score

For an observed value of a variable x, the corresponding value of the standardized variable z is called the **z-score** of the observation. The term *standard score* is often used instead of *z-score*.

A negative *z*-score indicates that the observation is below (less than) the mean, whereas a positive *z*-score indicates that the observation is above (greater than) the mean. Example 3.26 illustrates calculation and interpretation of *z*-scores.

Example 3.26 z-Scores

U.S. Women's World Cup Soccer Team The weight data for the 2003 U.S. Women's World Cup soccer team is given in the fourth column of Table 3.15 on page 132. We determined earlier that the mean and standard deviation of the weights are 62.55 kg and 4.9 kg, respectively. So, in this case, the standardized variable is

$$z = \frac{x - 62.55}{4.9}.$$

a. Find and interpret the z-score of Tiffany Roberts's weight of 51 kg.

b. Find and interpret the z-score of Cindy Parlow's weight of 70 kg.

c. Construct a graph showing the results obtained in parts (a) and (b).

Solution

a. The z-score for Tiffany Roberts's weight of 51 kg is

$$z = \frac{x - 62.55}{4.9} = \frac{51 - 62.55}{4.9} = -2.36.$$

INTERPRETATION Tiffany Roberts's weight is 2.36 standard deviations below the mean.

b. The z-score for Cindy Parlow's weight of 70 kg is

$$z = \frac{x - 62.55}{4.9} = \frac{70 - 62.55}{4.9} = 1.52.$$

INTERPRETATION Cindy Parlow's weight is 1.52 standard deviations above the mean.

c. In Fig. 3.15, we marked Tiffany Roberts's weight of 51 kg with a color dot and Cindy Parlow's weight of 70 kg with a black dot. Additionally, we located the mean, $\mu = 62.55$ kg, and measured intervals equal in length to the standard deviation, $\sigma = 4.9$ kg.

FIGURE 3.15

Graph showing Tiffany Roberts's weight (color dot) and Cindy Parlow's weight (black dot)

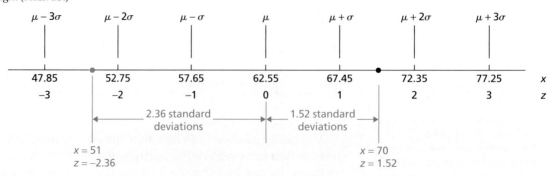

In Fig. 3.15, the numbers in the row labeled x represent weights in kilograms and the numbers in the row labeled z represent z-scores (i.e., number of standard deviations from the mean).

        ▆━■

The z-Score as a Measure of Relative Standing

The three-standard-deviations rule states that almost all the observations in any data set lie within three standard deviations to either side of the mean. Thus, for any variable, almost all possible observations have z-scores between -3 and 3.

Consequently, the z-score of an observation can be used as a measure of its relative standing among all the observations comprising a data set. A large positive z-score (i.e., a z-score roughly 3 or more) indicates that the observation is larger than most of the other observations; a large negative z-score (i.e., a z-score roughly -3 or less) indicates that the observation is smaller than most of the other observations; and a z-score near 0 indicates that the observation is located near the mean.

We can refine and make more precise the preceding statements by applying Chebychev's rule, which we discussed on page 114. And, if the distribution of the variable is roughly bell shaped, then, as we show in Chapter 6, we can do even better by applying the empirical rule.

We can also use z-scores to compare the relative standings of two observations from different populations. For example, we could use z-scores to compare the exam scores of two students in different sections of a beginning English course.

Other Descriptive Measures for Populations

Up to this point, we have concentrated on the mean and standard deviation in our discussion of descriptive measures for populations. The reason is that many of the classical inference procedures for center and variation concern those two parameters.

However, modern statistical analyses also rely heavily on descriptive measures based on percentiles. Quartiles, the IQR, and other descriptive measures based on percentiles are defined in the same way for (finite) populations as they are for samples. For simplicity and with one exception, we use the same notation for descriptive measures based on percentiles whether we are considering a sample or a population. The exception is: We use M to denote a sample median and η (eta) to denote a population median.

▌ EXERCISES 3.5

Statistical Concepts and Skills

3.78 Identify each quantity as a parameter or a statistic.
a. μ **b.** s **c.** $\bar{x}$ **d.** σ

3.79 Although, in practice, sample data are generally analyzed in inferential studies, what is the ultimate objective of such studies?

3.80 Microwave Popcorn. For a specific brand of microwave popcorn, what property is desirable for the population standard deviation of the cooking time? Explain your answer.

3.81 Complete the following sentences.
a. A standardized variable always has mean _____ and standard deviation _____.
b. The z-score corresponding to an observed value of a variable tells you _____.
c. A positive z-score indicates that the observation is _____ the mean, whereas a negative z-score indicates that the observation is _____ the mean.

3.82 Identify the statistic that is used to estimate
a. a population mean.
b. a population standard deviation.

3.83 Women's Soccer. Earlier in this section, we found that the population mean weight of the players on the 2003 U.S. Women's World Cup soccer team is 62.55 kg. In this context, is the number 62.55 a parameter or a statistic? Explain your answer.

3.84 Heights of Basketball Players. In Section 3.3, we analyzed the heights of the starting five players on each of two men's college basketball teams. The heights, in inches, of the players on Team II are 67, 72, 76, 76, and 84. Regarding the five players as a sample of all male starting college basketball players,
a. compute the sample mean height, $\bar{x}$.
b. compute the sample standard deviation, s.

Regarding the players now as a population,
c. compute the population mean height, μ.
d. compute the population standard deviation, σ.

Comparing your answers from parts (a) and (c) and from parts (b) and (d),
e. why are the values for $\bar{x}$ and μ equal?
f. why are the values for s and σ different?

3.85 Hurricane Hunters. The Air Force Reserve's 53rd Weather Reconnaissance Squadron, better known as the *Hurricane Hunters,* fly into the eye of tropical cyclones in their WC-130 Hercules aircraft to collect and report vital meteorological data for advance storm warnings. The data are relayed to the National Hurricane Center in Miami, Florida, for broadcasting emergency storm warnings on land. Over the Atlantic, 2002 was a busy year for the Hurricane Hunters with 12 tropical cyclones, 4 of which were hurricanes. The maximum winds were recorded for each cyclone, and are shown in the following table, where TS = tropical storm and Hu = hurricane. [SOURCE: *U.S. Air Force, National Oceanic and Atmospheric Administration.*]

Cyclone	Date	Max wind (knots)
TS Arthur	Jul 9–15	50
TS Bertha	Aug 4–9	43
TS Cristobal	Aug 5–8	45
TS Dolly	Aug 29–Sep 4	50
TS Edouard	Sep 1–6	55
TS Fay	Sep 5–11	50
Hu Gustav	Sep 8–12	85
TS Hanna	Sep 12–15	50
Hu Isidore	Sep 14–27	110
TS Josephine	Sep 17–19	50
Hu Kyle	Sep 20–Oct 12	75
Hu Lili	Sep 21–Oct 4	125

Using only basic calculator functions, obtain the following parameters for the population of maximum wind speeds. Use the appropriate mathematical notation for the parameters to express your answers.
a. Mean   **b.** Standard deviation
c. Median   **d.** Mode   **e.** IQR

3.86 Dart Doubles. The top two players in the 2001–2002 Professional Darts Corporation World Championship were Phil Taylor and Peter Manley. Taylor and Manley dominated the competition with a record number of doubles. A *double* is a throw that lands in either the outer ring of the dartboard or the outer ring of the bull's-eye. The following table provides the number of doubles thrown by each of the two players during the five rounds of competition, as found in *Chance* (Vol. 15, No. 3, pp. 48–55).

Taylor	21	18	18	19	13
Manley	5	24	20	26	14

a. Using only basic calculator functions, obtain the individual population means of the number of doubles.
b. Without doing any calculations, decide for which player the standard deviation of the number of doubles is smaller. Explain your answer.

c. Using only basic calculator functions, obtain the individual population standard deviations of the number of doubles.

d. Are your answers to parts (b) and (c) consistent? Explain your answer.

3.87 Doing Time. According to *Statistical Report*, published by the U.S. Bureau of Prisons, the mean time served by prisoners released from federal institutions for the first time is 16.3 months. Assume the standard deviation of the times served is 17.9 months. Let x denote time served by a prisoner released for the first time from a federal institution.

a. Find the standardized version of x.

b. Find the mean and standard deviation of the standardized variable.

c. Determine the z-scores for prison times served of 64.7 months and 4.2 months. Round your answers to two decimal places.

d. Interpret your answers in part (c).

e. Construct a graph similar to Fig. 3.15 on page 138 that depicts your results from parts (b) and (c).

3.88 Gestation Periods of Humans. Gestation periods of humans have a mean of 266 days and a standard deviation of 16 days. Let y denote the variable "gestation period" for humans.

a. Find the standardized variable corresponding to y.

b. What are the mean and standard deviation of the standardized variable?

c. Obtain the z-scores for gestation periods of 227 days and 315 days. Round your answers to two decimal places.

d. Interpret your answers in part (c).

e. Construct a graph similar to Fig. 3.15 on page 138 that shows your results from parts (b) and (c).

3.89 Exam Scores. Suppose that you take an exam with 400 possible points and are told that the mean score is 280 and that the standard deviation is 20. You are also told that you got 350. Did you do well on the exam? Explain your answer.

Extending the Concepts and Skills

3.90 SAT Scores. Each year, thousands of high school students bound for college take the Scholastic Assessment Test, or SAT. The test measures the verbal and mathematical abilities of prospective college students. Student scores are reported on a scale that ranges from a low of 200 to a high of 800. In one high school graduating class, the mean mathematics score on the SAT was 493 and the standard deviation was 105; the mean verbal score was 420 and the standard deviation was 98. A student in the graduating class scored 703 on the math and 665 on the verbal. Relative to the other students in the graduating class, on which test did the student do better? Explain your answer.

Population and Sample Standard Deviations. In Exercises 3.91 and 3.92, you examine the numerical relationship between the population standard deviation and the sample standard deviation computed from the same data. This relationship is helpful when the computer or statistical calculator being used has a built-in program for sample standard deviation but not for population standard deviation.

3.91 Consider the following three data sets.

Data Set 1		Data Set 2				Data Set 3				
2	4	7	5	5	3	4	7	8	9	7
7	3	9	8	6		4	5	3	4	5

a. Assuming that each of these data sets is sample data, compute the standard deviations. (Round your final answers to two decimal places.)

b. Assuming that each of these data sets is population data, compute the standard deviations. (Round your final answers to two decimal places.)

c. Using your results from parts (a) and (b), make an educated guess about the answer to the question: If both s and σ are computed for the same data set, will they tend to be closer together if the data set is large or if it is small?

3.92 Consider a data set with m observations. If the data are sample data, you compute the sample standard deviation, s, whereas if the data are population data, you compute the population standard deviation, σ.

a. Derive a mathematical formula that gives σ in terms of s when both are computed for the same data set. (*Hint:* First note that, numerically, the values of $\bar{x}$ and μ are identical. Consider the ratio of the defining formula for σ to the defining formula for s.)

b. Refer to the three data sets in Exercise 3.91. Verify that your formula in part (a) works for each of the three data sets.

c. Suppose that a data set consists of 15 observations. You compute the sample standard deviation of the data and obtain $s = 38.6$. Then you realize that the data are actually population data and that you should have obtained the population standard deviation instead. Use your formula from part (a) to obtain σ.

Using Technology

Note: Some statistical software packages and calculators have a built-in program for obtaining a sample standard deviation but do not have one for obtaining a population standard deviation. You can deal with this lack by using the following formula that expresses the population standard deviation in terms of the sample standard deviation when both are computed for the same data set:

$$\sigma = \sqrt{\frac{m-1}{m} \cdot s},$$

where m is the number of observations.

3.93 Women's Soccer. Refer to the heights of the 2003 U.S. Women's World Cup soccer team in the third column of Table 3.15 on page 132. Use the technology of your choice to obtain

a. the population mean height.
b. the population standard deviation of the heights. *Note:* Depending on the technology that you're using, you may need to refer to the formula given in the note preceding this exercise.

3.94 Hurricane Hunters. Refer to Exercise 3.85. Use the technology of your choice to obtain the population mean and population standard deviation of the maximum wind speeds. See the note in Exercise 3.93(b).

CHAPTER REVIEW

You Should Be Able To

1. use and understand the formulas in this chapter.

2. explain the purpose of a measure of center.

3. obtain and interpret the mean, the median, and the mode(s) of a data set.

4. choose an appropriate measure of center for a data set.

5. use and understand summation notation.

6. define, compute, and interpret a sample mean.

7. explain the purpose of a measure of variation.

8. define, compute, and interpret the range of a data set.

9. define, compute, and interpret a sample standard deviation.

10. define percentiles, deciles, and quartiles.

11. obtain and interpret the quartiles, IQR, and five-number summary of a data set.

12. obtain the lower and upper limits of a data set and identify potential outliers.

13. construct and interpret a boxplot and a modified boxplot.

14. use a boxplot to identify distribution shape for large data sets.

15. define the population mean (mean of a variable).

16. define the population standard deviation (standard deviation of a variable).

17. compute the population mean and population standard deviation of a finite population.

18. distinguish between a parameter and a statistic.

19. understand how and why statistics are used to estimate parameters.

20. obtain and interpret z-scores.

Key Terms

adjacent values, *124*
box-and-whisker diagram, *123*
boxplot, *123*
Chebychev's rule, *114*
deciles, *118*
descriptive measures, *90*
deviations from the mean, *107*
empirical rule, *114*
first quartile (Q_1), *119*
five-number summary, *121*
interquartile range (IQR), *120*
lower limit, *122*

mean, *92*
mean of a variable (μ), *97, 131*
measures of center, *92*
measures of central tendency, *92*
measures of spread, *105*
measures of variation, *105*
median, *93*
mode, *94*
modified boxplot, *123*
outliers, *122*
parameter, *135*
percentiles, *118*

population mean (μ), *97, 131*
population standard deviation (σ), *133*
population variance (σ^2), *133*
quartiles, *118*
quintiles, *118*
range, *106*
resistant measure, *96*
sample mean ($\bar{x}$), *97, 103*
sample size (n), *103*
sample standard deviation (s), *110*
sample variance (s^2), *108*

REVIEW TEST

Statistical Concepts and Skills

1. Define
a. descriptive measures.
b. measures of center.
c. measures of variation.

2. Identify the two most commonly used measures of center for quantitative data. Explain the relative advantages and disadvantages of each.

3. Among the measures of center discussed, which is the only one appropriate for qualitative data?

4. Identify the most appropriate measure of variation corresponding to each of the following measures of center.
a. Mean　　　　　　　　**b.** Median

5. Specify the mathematical symbol used for each of the following descriptive measures.
a. Sample mean
b. Sample standard deviation
c. Population mean
d. Population standard deviation

6. Data Set A has more variation than Data Set B. Decide which of the following statements are necessarily true.
a. Data Set A has a larger mean than Data Set B.
b. Data Set A has a larger standard deviation than Data Set B.

7. Complete the statement: Almost all the observations in any data set lie within _____ standard deviations to either side of the mean.

8. Regarding the five-number summary:
a. Identify its components.
b. How can it be employed to describe center and variation?
c. What graphical display is based on it?

9. Regarding outliers:
a. What is an outlier?
b. Explain how you can identify potential outliers, using only the first and third quartiles.

10. Regarding z-scores:
a. How is a z-score obtained?

b. What is the interpretation of a z-score?
c. An observation has a z-score of 2.9. Roughly speaking, what is the relative standing of the observation?

11. Party Time. An integral part of doing business in the dot-com culture is frequenting the party circuit centered in San Francisco. Here high-tech companies throw as many as five parties a night to recruit or retain talented workers in a highly competitive job market. With as many as 700 guests at a single party, the food and booze flow with an average alcohol cost per guest of $15–$18 and average food bill of $75–$150. A sample of a recent dot-com party yielded the following data on number of alcoholic drinks consumed per person. [SOURCE: *USA TODAY* Online, May 17, 2000]

4	4	1	0	5
1	1	2	4	3
1	5	3	0	2
2	2	1	2	4

a. Find the mean, median, and mode of these data.
b. Which measure of center do you think is best here? Explain your answer.

12. Duration of Marriages. The National Center for Health Statistics publishes information on the duration of marriages in *Vital Statistics of the United States*. Which measure of center is more appropriate for data on the duration of marriages, the mean or the median? Explain your answer.

13. Causes of Death. Death certificates provide data on the causes of death. Which of the three main measures of center is appropriate here? Explain your answer.

14. Road Patrol. In the paper "Injuries and Risk Factors in a 100-Mile (161-km) Infantry Road March" (*Preventative Medicine*, Vol. 28, pp. 167–173), Reynolds et al. reported on a study commissioned by the U.S. Army. The purpose of the study was to improve medical planning and identify risk factors during multiple-day road patrols by examining the acute effects of long-distance marches by light-infantry soldiers. Each soldier carried a standard U.S. Army rucksack, Meal-Ready-to-Eat packages, and other field equipment. A sample of 10 participating soldiers revealed the following data on total load mass, in kilograms.

48	50	45	49	44
47	37	54	40	43

a. Obtain the sample mean of these 10 load masses.
b. Obtain the range of the load masses.
c. Obtain the sample standard deviation of the load masses.

15. Millionaires. Dr. Thomas Stanley of Georgia State University has collected information on millionaires, including their ages, since 1973. A sample of 36 millionaires has a mean age of 58.5 years and a standard deviation of 13.4 years.
a. Complete the following graph.

$\bar{x} - 3s$ $\bar{x} - 2s$ $\bar{x} - s$ $\bar{x}$ $\bar{x} + s$ $\bar{x} + 2s$ $\bar{x} + 3s$

18.3   58.5   85.3

b. Fill in the blanks: Almost all the 36 millionaires are between _____ and _____ years old.

16. Millionaires. Refer to Problem 15. The ages of the 36 millionaires sampled are arranged in increasing order in the following table.

31	38	39	39	42	42	45	47	48
48	48	52	52	53	54	55	57	59
60	61	64	64	66	66	67	68	68
69	71	71	74	75	77	79	79	79

a. Determine the quartiles for the data.
b. Obtain and interpret the interquartile range.
c. Find and interpret the five-number summary.
d. Calculate the lower and upper limits.
e. Identify potential outliers, if any.
f. Construct and interpret a boxplot and, if appropriate, a modified boxplot.

17. UC Enrollment. According to *Information and Communications, University of California*, the Fall 2002 enrollment figures for undergraduates at the University of California campuses were as follows.

Campus	Enrollment (1000s)
Berkeley	23.7
Davis	22.8
Irvine	19.4
Los Angeles	24.9
Riverside	14.2
San Diego	19.1
Santa Barbara	17.7
Santa Cruz	12.9

a. Compute the population mean enrollment, μ, of the UC campuses. (Round your answer to two decimal places.)
b. Compute σ. (Round your answer to two decimal places.)
c. Letting x denote enrollment, specify the standardized variable, z, corresponding to x.
d. Without performing any calculations, give the mean and standard deviation of z. Explain your answers.
e. Construct dotplots for the distributions of both x and z. Interpret your graphs.
f. Obtain and interpret the z-scores for the enrollments at the Los Angeles and Riverside campuses.

18. Gasoline Prices. The U.S. Energy Information Administration reports figures on retail gasoline prices in the *Monthly Energy Review*. Data are obtained by sampling 10,000 gasoline service stations from a total of more than 185,000. For the 10,000 stations sampled in June 2003, the mean price per gallon for unleaded regular gasoline was $1.49.
a. Is the mean price given here a sample mean or a population mean? Explain your answer.
b. What letter or symbol would you use to designate the mean of $1.49?
c. Is the mean price given here a statistic or a parameter? Explain your answer.

Using Technology

19. Millionaires. Use the technology of your choice to determine the following statistics for the age data in Problem 16.
a. Mean
b. Median
c. Range
d. Sample standard deviation

20. Millionaires. Refer to the data in Problem 16. Use the technology of your choice to obtain
a. a (modified) boxplot for the data.
b. the five-number summary of the data.

21. Millionaires. Refer to Problem 16.
a. Printout 3.3 shows a boxplot generated by Minitab for the age data. Use the boxplot to discuss the variation in the data set.
b. Are there potential outliers in the data? If so, identify them approximately and provide a possible explanation for their cause.

PRINTOUT 3.3
Minitab output for Problem 21(a)

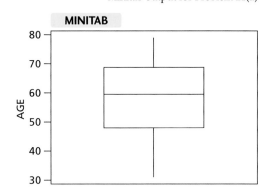

StatExplore in MyMathLab
Analyzing Data Online

You can use StatExplore to perform all statistical analyses discussed in this chapter. To illustrate, we show how to obtain the five-number summary and boxplots.

EXAMPLE Five-Number Summary; Boxplots

Weekly TV-Viewing Times Table 3.13 on page 119 provides data on the weekly TV-viewing times, in hours, for a sample of 20 people. Use StatExplore to obtain the five-number summary and a modified boxplot of the data.

SOLUTION To begin, we store the sample data from Table 3.13 in a column named TIMES. To determine the five-number summary of the sample of TV-viewing times, we proceed as follows:

> 1 Choose **Stat ➤ Summary Stats ➤ Columns**
> 2 Select the column TIMES
> 3 Click **Next →**
> 4 In the left pane of the **Statistics** list box, click on all items other than **Median, Min, Max, Q1,** and **Q3**
> 5 Click **Calculate**

The resulting output is shown in Printout 3.4.

PRINTOUT 3.4
StatExplore output for five-number summary

Summary statistics:					
Column	Median	Min	Max	Q1	Q3
TIMES	30.5	5	66	23	36.5

From Printout 3.4, we see that the five-number summary for the sample of TV-viewing times is 5, 23, 30.5, 36.5, and 66 hours.

To obtain a modified boxplot for the sample of TV-viewing times, we proceed in the following way, assuming again that the data are stored in a column named TIMES:

> 1 Choose **Graphics ➤ Boxplot**
> 2 Select the column TIMES
> 3 Click **Next →**
> 4 Select the **Use fences to identify outliers** check box
> 5 Click **Create Graph!**

The resulting output is shown in Printout 3.5.

PRINTOUT 3.5

StatExplore output for modified boxplot

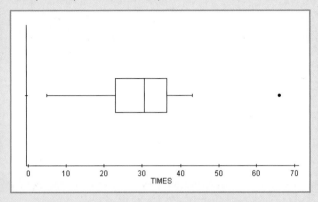

Compare the modified boxplot in Printout 3.5 with the one we drew by hand in Fig. 3.10(b) on page 125. (*Note:*

To obtain an ordinary boxplot, simply omit the third and fourth steps in the previous list of StatExplore instructions.)

STATEXPLORE EXERCISES Solve the following problems by using StatExplore:

a. Determine the five-number summary, a boxplot, and a modified boxplot for the miles-driven data in Exercise 3.64 on page 129. Interpret your results.

b. Find the mean, median, sample standard deviation, and range for the heights, in inches, of the players on Team I: 72, 73, 76, 76, and 78. (*Note:* Input the heights into the StatExplore data table by hand.) Interpret your results.

To access StatExplore, go to the student content area of your Weiss MyMathLab course.

Internet Project

Old Faithful Geyser and a Survey of Wages

In this Internet project, you are to explore two topics. The first involves the Old Faithful Geyser. Specifically, you will examine data on the duration of eruptions and the times between successive eruptions.

The second is a survey of wages. Wages provide an example in which the use of different descriptive measures is important to help in understanding and explaining data. The statistic you choose can make a striking difference in the conclusions you draw; in this case, the difference is between using the mean and the median.

In each part of the project, you will see how certain measures can help describe the data and how some measures work better than others. As usual, you will look at several different views of the data to understand better the true nature of the phenomenon that the data describes.

URL for access to Internet Projects Page: www.aw-bc.com/weiss

Focusing on Data Analysis

The Focus Database

Recall from Chapter 1 (see page 35) that the Focus database contains information on the undergraduate students at the University of Wisconsin - Eau Claire (UWEC). Statistical analyses for this database should be done with the technology of your choice.

a. Find the mean and median of the ACT English scores. Explain any difference between these two measures of center.

b. Find the five-number summary of the ACT English scores. Interpret your results.

c. Find the means and medians of the ACT English scores, individually for males and females. Use these measures of center to compare the two sets of scores.

d. Find the five-number summaries of the ACT English scores, individually for males and females. Use these statistics to compare the two sets of scores.

e. Repeat parts (a)–(d) for ACT math scores.

f. Determine the population mean and population standard deviation of the ages of the students at UWEC.

g. Take five simple random samples (without replacement) of 50 students each. For each sample, obtain the sample mean and sample standard deviation of the students' ages, and compare those statistics with the corresponding parameters. What do the results of this exercise illustrate?

Case Study Discussion

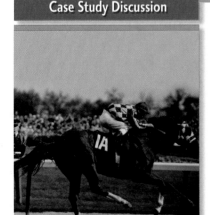

The Triple Crown

The table on page 91 displays the year, name, and times for each of the 11 Triple Crown winners. Do the following separately for the 11 times in each of the three races—the Kentucky Derby, Preakness Stakes, and Belmont Stakes.

a. Convert all times to speeds, in miles per hour, and use those data in the parts that follow. *Hint:* Divide each distance run by the running time in seconds and then multiply the result by 3600 (the number of seconds in an hour).

b. Determine the mean and median of the speeds. Explain any difference between these two measures of center.

c. Obtain the range and population standard deviation of the speeds.

d. Find and interpret the z-scores for the speeds of Citation and Secretariat.

e. Determine and interpret the quartiles of the speeds.

f. Find the lower and upper limits. Use them to identify potential outliers.

g. Construct a boxplot for the speeds and interpret your result in terms of the variation in the speeds.

h. Use the technology of your choice to solve parts (a)–(g).

Internet Resources: Visit the Weiss Web site www.aw-bc.com/weiss for additional discussion, exercises, and resources related to this case study.

Biography

JOHN TUKEY: A Pioneer of EDA

John Wilder Tukey was born on June 16, 1915, in New Bedford, Massachusetts. After earning bachelor's and master's degrees in chemistry from Brown University in 1936 and 1937, respectively, he enrolled in the mathematics program at Princeton University, where he received a master's degree in 1938 and a doctorate in 1939.

After graduating, Tukey was appointed Henry B. Fine Instructor in Mathematics at Princeton; 10 years later he was advanced to a full professorship. In 1965, Princeton established a department of statistics, and Tukey was named its first chairperson. In addition to his position at Princeton, he was a member of the Technical Staff at AT&T Bell Laboratories where he served as Associate Executive Director, Research in the Information Sciences Division, from 1945 until his retirement in 1985.

Tukey was among the leaders in the field of exploratory data analysis (EDA), which provides techniques such as stem-and-leaf diagrams for effectively investigating data. He also made fundamental contributions to the areas of robust estimation and time series analysis. Tukey wrote numerous books and more than 350 technical papers on mathematics, statistics, and other scientific subjects. Additionally, he coined the word *bit*, a contraction of *binary digit* (a unit of information, often as processed by a computer).

Tukey's participation in educational, public, and government service was most impressive. He was appointed to serve on the President's Science Advisory Committee by President Eisenhower; was chairperson of the committee that prepared "Restoring the Quality of our Environment" in 1965; helped develop the National Assessment of Educational Progress; and was a member of the Special Advisory Panel on the 1990 Census of the U.S. Department of Commerce, Bureau of the Census—to name only a few of his involvements.

Among many honors, Tukey received the National Medal of Science, the IEEE Medal of Honor, Princeton University's James Madison Medal, and Foreign Member, The Royal Society (London). He was the first recipient of the Samuel S. Wilks Award of the American Statistical Association. Until his death, Tukey remained on the faculty at Princeton as Donner Professor of Science, Emeritus; Professor of Statistics, Emeritus; and Senior Research Statistician. Tukey died of a heart attack on July 26, 2000, after a short illness. He was 85 years old.

part **III**

Probability, Random Variables, and Sampling Distributions

Probability Concepts

GENERAL OBJECTIVES Until now, we have concentrated on *descriptive statistics,* or methods for organizing and summarizing data. Another important aspect of this text is to present the fundamentals of *inferential statistics,* or methods of drawing conclusions about a population based on information obtained from a sample of the population.

Because inferential statistics involves using information obtained from part of a population (a sample) to draw conclusions about the entire population, we can never be certain that our conclusions are correct. And as uncertainty is inherent in inferential statistics, you need to become familiar with uncertainty before you can understand, develop, and apply the methods of inferential statistics.

The science of uncertainty is called **probability theory.** It enables you to evaluate and control the likelihood that a statistical inference is correct. More generally, probability theory provides the mathematical basis for inferential statistics. This chapter begins your study of probability.

The Powerball

The Powerball is a lottery that was introduced in 1992. It is now sold in Arizona, Colorado, Connecticut, Delaware, District of Columbia, Idaho, Indiana, Iowa, Kansas, Kentucky, Louisiana, Minnesota, Missouri, Montana, Nebraska, New Hampshire, New Mexico, Oregon, Pennsylvania, Rhode Island, South Carolina, South Dakota, Vermont, West Virginia, Wisconsin, and the Virgin Islands. Profits from Powerball tickets stay in the state where the ticket is sold and each state uses its own computer system to issue and validate Powerball tickets.

Although the Powerball is a multistate lottery game, it isn't the first. That distinction goes to Lotto*America, which was created in 1988 when Iowa and six other states joined forces to offer a game with a large jackpot. The more people who play, the bigger the jackpots tend to be, so multistate lotteries offer larger prizes than those of individual states.

A Powerball jackpot starts at $10 million and grows if no one wins it. Because the chance of winning a jackpot is small, the jackpot often grows to huge amounts, sometimes as much as $300 million. When there are multiple winners, the jackpot is divided equally among them. Drawings take place on Wednesday and Saturday evenings at 9:59 P.M. and can be watched during the 10:00 P.M. news.

To play the basic Powerball, a player first selects five numbers from the numbers 1–53 and then chooses a Powerball number, which can be any number between 1 and 42. A ticket costs $1. In the drawing, five white balls are drawn randomly from 53 white balls numbered 1–53; then one red Powerball is drawn randomly from 42 red balls numbered 1–42.

To win the jackpot, a ticket must match all the balls drawn; smaller prizes are awarded for matching some but not all the balls drawn. What are the chances of winning the jackpot? What are the chances of winning any prize at all? After studying probability, you will be able to answer these and similar questions. You will be asked to do so when you revisit the Powerball at the end of this chapter.

4.1 Probability Basics

Although most applications of probability theory to statistical inference involve large populations, the fundamental concepts of probability are most easily illustrated and explained with relatively small populations and games of chance. So, keep in mind that many of the examples in this chapter are designed expressly to demonstrate clearly the principles of probability.

The Equal-Likelihood Model

We discussed an important aspect of probability when we examined probability sampling in Chapter 1. In Example 4.1, we return to the illustration of simple random sampling given in Example 1.5 on page 12.

▐▐ Example 4.1   **Introducing Probability**

TABLE 4.1
Five top Oklahoma state officials

Governor (G)
Lieutenant Governor (L)
Secretary of State (S)
Attorney General (A)
Treasurer (T)

TABLE 4.2
The 10 possible samples of two officials

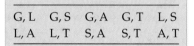

G, L	G, S	G, A	G, T	L, S
L, A	L, T	S, A	S, T	A, T

Oklahoma State Officials   As reported by *The World Almanac*, the top five state officials of Oklahoma are as shown in Table 4.1.

Suppose that we take a simple random sample without replacement of two officials from the five officials.

a. Find the probability that we obtain the governor and treasurer.

b. Find the probability that the attorney general is included in the sample.

Solution   For convenience, we use the letters in parentheses after the titles in Table 4.1 to represent the officials. As we discovered in Example 1.5, there are 10 possible samples of two officials from the population of five officials. They are listed in Table 4.2. If we take a simple random sample of size 2, each of the possible samples of two officials is equally likely to be the one selected.

a. Because there are 10 possible samples, the probability is $\frac{1}{10}$, or 0.1, of selecting the governor and treasurer (G, T). Another way of looking at this result is that 1 out of 10, or 10%, of the samples include both the governor and treasurer; hence the probability of obtaining such a sample is 10%, or 0.1. The same goes for any other two particular officials.

b. Table 4.2 shows that the attorney general (A) is included in 4 of the 10 possible samples of size 2. As each of the 10 possible samples is equally likely to be the one selected, the probability is $\frac{4}{10}$, or 0.4, that the attorney general is included in the sample. Another way of looking at this result is that 4 out of 10, or 40%, of the samples include the attorney general; hence the probability of obtaining such a sample is 40%, or 0.4.   ▬ ■

The essential idea in Example 4.1 is that when outcomes are equally likely, probabilities are nothing more than percentages (relative frequencies). In other words, we can use a simple formula, which we refer to as the *f/N* **rule,** to compute probabilities.

DEFINITION 4.1

What
Does it Mean?

For an experiment with equally likely outcomes, probabilities are identical to relative frequencies (or percentages).

Probability for Equally Likely Outcomes

Suppose that an experiment has N possible outcomes, all equally likely. Then the probability that a specified event occurs equals the number of ways, f, that the event can occur, divided by the total number of possible outcomes. In symbols,

Number of ways event can occur

$$\text{Probability of an event} = \frac{f}{N}.$$

Total number of possible outcomes

In stating Definition 4.1, we used the terms *experiment* and *event* in their intuitive sense. Basically, by an **experiment,** we mean an action whose outcome cannot be predicted with certainty. By an **event,** we mean some specified result that may or may not occur when an experiment is performed.

For instance, in Example 4.1 the experiment consists of taking a random sample of size 2 from the five officials. It has 10 possible outcomes ($N = 10$), all equally likely. In part (b), the event is that the sample obtained includes the attorney general, which can occur in four ways ($f = 4$); hence its probability equals

$$\frac{f}{N} = \frac{4}{10} = 0.4,$$

as we noted in Example 4.1(b). Examples 4.2 and 4.3 provide two additional illustrations of Definition 4.1. These examples further indicate the varied contexts under which the equal-likelihood model applies.

Example 4.2 **Probability for Equally Likely Outcomes**

TABLE 4.3
Frequency distribution of annual income for U.S. families

Income	Frequency (1000s)
Under $10,000	4,187
$10,000–$14,999	3,653
$15,000–$24,999	8,639
$25,000–$34,999	8,996
$35,000–$49,999	12,192
$50,000–$74,999	15,676
$75,000 & over	18,192
	71,535

Family Income The U.S. Bureau of the Census compiles data on family income and publishes its findings in *Current Population Reports*. Table 4.3 gives a frequency distribution of annual income for U.S. families.

A U.S. family is selected **at random,** meaning that each family is equally likely to be the one obtained (simple random sample of size 1). Determine the probability that the family selected has an annual income of

a. between $50,000 and $74,999, inclusive.

b. between $25,000 and $74,999, inclusive.

c. under $15,000.

Solution The second column of Table 4.3 shows that there are 71,535 thousand U.S. families; so $N = 71,535$ thousand.

a. The event in question is that the family selected makes between $50,000 and $74,999. Table 4.3 shows that the number of such families is 15,676 thousand, so $f = 15{,}676$ thousand. Applying the f/N rule, we find that the probability the family selected makes between $50,000 and $74,999 is

$$\frac{f}{N} = \frac{15{,}676}{71{,}535} = 0.219.$$

INTERPRETATION 21.9% of families in the United States make between $50,000 and $74,999, inclusive.

b. The event in question is that the family selected makes between $25,000 and $74,999. Table 4.3 reveals that the number of such families is $8{,}996 + 12{,}192 + 15{,}676$, or 36,864 thousand. Consequently, $f = 36{,}864$ thousand, so the required probability is

$$\frac{f}{N} = \frac{36{,}864}{71{,}535} = 0.515.$$

INTERPRETATION 51.5% of families in the United States make between $25,000 and $74,999, inclusive.

c. Proceeding as in parts (a) and (b), we find that the probability that the family selected makes under $15,000 is

$$\frac{f}{N} = \frac{4{,}187 + 3{,}653}{71{,}535} = \frac{7{,}840}{71{,}535} = 0.110.$$

INTERPRETATION 11.0% of families in the United States make less than $15,000.

Example 4.3   Probability for Equally Likely Outcomes

Dice   When two balanced dice are rolled, 36 equally likely outcomes are possible, as depicted in Fig. 4.1. Find the probability that

a. the sum of the dice is 11.

b. doubles are rolled; that is, both dice come up the same number.

FIGURE 4.1
Possible outcomes for rolling a pair
of dice

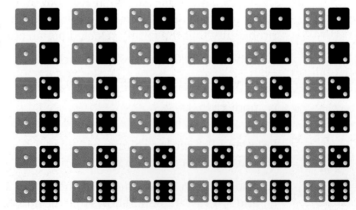

Solution For this experiment, $N = 36$.

a. The sum of the dice can be 11 in two ways, as is apparent from Fig. 4.1. Hence the probability that the sum of the dice is 11 equals $f/N = 2/36 = 0.056$.

> **INTERPRETATION** There is a 5.6% chance of a sum of 11 when two balanced dice are rolled.

b. Figure 4.1 also shows that doubles can be rolled in six ways. Consequently, the probability of rolling doubles equals $f/N = 6/36 = 0.167$.

> **INTERPRETATION** There is a 16.7% chance of doubles when two balanced dice are rolled.

— ■

The Meaning of Probability

Essentially, probability is a generalization of the concept of percentage. When we select a member at random from a finite population, as we did in Example 4.2, probability is nothing more than percentage. But, in general, how do we interpret probability? For instance, what do we mean when we say that

- the probability is 0.314 that the gestation period of a woman will exceed 9 months or
- the probability is 0.667 that the favorite in a horse race finishes in the money (first, second, or third place) or
- the probability is 0.40 that a traffic fatality involves an intoxicated or alcohol-impaired driver or nonoccupant?

Some probabilities are easy to interpret: A probability near 0 indicates that the event in question is very unlikely to occur when the experiment is performed, whereas a probability near 1 (100%) suggests that the event is quite likely to occur. To gain further insight into the meaning of probability, it is useful to consider the **frequentist interpretation of probability,** which construes the probability of an event to be the proportion of times it occurs in a large number of repetitions of the experiment.

Consider, for instance, the simple experiment of tossing a balanced coin once. Because the coin is balanced, we reason that there is a 50–50 chance the coin will land with heads facing up. Consequently, we attribute a probability of 0.5 to that event. The frequentist interpretation is that in a large number of tosses, the coin will land with heads facing up about half the time.

We used a computer to perform two simulations of tossing a balanced coin 100 times. The results are displayed in Fig. 4.2 at the top of the next page. Each graph shows the number of tosses of the coin versus the proportion of heads. Both graphs seem to corroborate the frequentist interpretation.

Although the frequentist interpretation is helpful for understanding the meaning of probability, it cannot be used as a definition of probability. One common way to define probabilities is to specify a **probability model**—a mathematical description of the experiment based on certain primary aspects and assumptions.

The **equal-likelihood model** discussed earlier in this section is an example of a probability model. Its primary aspect and assumption are that all possible

FIGURE 4.2

Two computer simulations of tossing a balanced coin 100 times

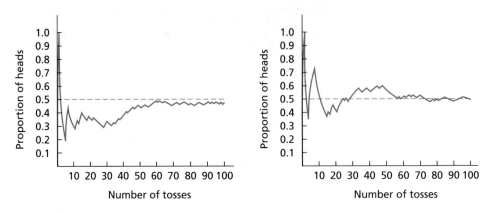

outcomes are equally likely to occur. We discuss other probability models later in this and subsequent chapters.

Basic Properties of Probabilities

Probabilities have some simple but basic properties, as listed in Key Fact 4.1.

KEY FACT 4.1

Basic Properties of Probabilities

Property 1: The probability of an event is always between 0 and 1, inclusive.

Property 2: The probability of an event that cannot occur is 0. (An event that cannot occur is called an **impossible event.**)

Property 3: The probability of an event that must occur is 1. (An event that must occur is called a **certain event.**)

Property 1 indicates that numbers such as 5 or −0.23 could not possibly be probabilities. Thus, if you calculate a probability and get an answer such as 5 or −0.23, you made an error. Example 4.4 illustrates Properties 2 and 3.

Example 4.4   Basic Properties of Probabilities

Dice   Let's return to Example 4.3, wherein two balanced dice are rolled. Determine the probability that

a. the sum of the dice is 1.

b. the sum of the dice is 12 or less.

Solution

a. Figure 4.1 on page 154 shows that the sum of the dice must be no less than 2. Thus the probability that the sum of the dice is 1 equals $f/N = 0/36 = 0$. This result illustrates Property 2 of Key Fact 4.1.

INTERPRETATION   Getting a sum of 1 when two balanced dice are rolled is impossible and hence has probability 0.

b. From Fig. 4.1, the sum of the dice is always 12 or less. Thus the probability of that event equals $f/N = 36/36 = 1$. This result illustrates Property 3 of Key Fact 4.1.

INTERPRETATION Getting a sum of 12 or less when two balanced dice are rolled is certain and hence has probability 1. ∎

EXERCISES 4.1

Statistical Concepts and Skills

4.1 Roughly speaking, what is an experiment? an event?

4.2 Concerning the equal-likelihood model of probability,
a. what is it?
b. how is the probability of an event found?

4.3 What is the difference between selecting a member at random from a finite population and taking a simple random sample of size 1?

4.4 If a member is selected at random from a finite population, probabilities are identical to _____.

4.5 State the frequentist interpretation of probability.

4.6 Interpret each of the following probability statements, using the frequentist interpretation of probability.
a. The probability is 0.487 that a newborn baby will be a girl.
b. The probability of a single ticket winning a prize in the Powerball lottery is 0.028.
c. If a balanced dime is tossed three times, the probability that it will come up heads all three times is 0.125.

4.7 Which of the following numbers could not possibly be probabilities? Justify your answer.
a. 0.462   **b.** −0.201   **c.** 1
d. $\frac{5}{6}$ **e.** 3.5   **f.** 0

4.8 Oklahoma State Officials. Refer to Table 4.1, presented on page 152.
a. List the possible samples without replacement of size 3 that can be obtained from the population of five officials. (*Hint:* There are 10 possible samples.)

If a simple random sample without replacement of three officials is taken from the five officials, determine the probability that
b. the governor, attorney general, and treasurer are obtained.
c. the governor and treasurer are included in the sample.
d. the governor is included in the sample.

In the following exercises, express your probability answers as a decimal rounded to three places.

4.9 Housing Units. The U.S. Bureau of the Census publishes data on housing units in *American Housing Survey in the United States*. The following table provides a frequency distribution for the number of rooms in U.S. housing units. The frequencies are in thousands.

Rooms	No. of units
1	471
2	1,470
3	11,715
4	23,468
5	24,476
6	21,327
7	13,782
8+	15,647

A U.S. housing unit is selected at random. Find the probability that the housing unit obtained has
a. four rooms.   **b.** more than four rooms.
c. one or two rooms.   **d.** fewer than one room.
e. one or more rooms.

4.10 Murder Victims. As reported by the Federal Bureau of Investigation in *Crime in the United States*, the age distribution of murder victims between 20 and 59 years old is as shown in the following table.

Age	Frequency
20–24	2916
25–29	2175
30–34	1842
35–39	1581
40–44	1213
45–49	888
50–54	540
55–59	372

A murder case in which the person murdered was between 20 and 59 years old is selected at random. Find the probability that the murder victim was

a. between 40 and 44 years old, inclusive.
b. at least 25 years old, that is, 25 years old or older.
c. between 45 and 59 years old, inclusive.
d. under 30 or over 54.

4.11 Occupations in Seoul. The population of Seoul was studied in an article by B. Lee and J. McDonald, "Determinants of Commuting Time and Distance for Seoul Residents: The Impact of Family Status on the Commuting of Women" (*Urban Studies*, Vol. 40, No. 7, pp. 1283–1302). The authors examined the different occupations for males and females in Seoul. Following is a frequency distribution of occupation type for males taking part in a survey. (*Note:* M = manufacturing, N = nonmanufacturing)

Occupation	Frequency
Administrative/M	2,197
Administrative/N	6,450
Technical/M	2,166
Technical/N	6,677
Clerk/M	1,640
Clerk/N	4,538
Production workers/M	5,721
Production workers/N	10,266
Service	9,274
Agriculture	159

If one of these males is selected at random, find the probability that his occupation is

a. service.   **b.** administrative.
c. manufacturing.   **d.** not manufacturing.

4.12 Nobel Prize Winners. The National Science Foundation collects data on Nobel Prize Laureates in the field of science and the date and location of their award-winning research. A frequency distribution for the number of winners, by country, for the years 1901–1999 are as follows.

Country	Winners
United States	199
United Kingdom	71
Germany	61
France	25
Soviet Union	10
Japan	4
Other Countries	89

Suppose that a recipient of a Nobel Prize in science between 1901–1999 is selected at random. Find the probability that the Nobel Laureate is from

a. Japan.
b. either France or Germany.
c. any country other than the United States.

4.13 Dice. Two balanced dice are rolled. Refer to Fig. 4.1 on page 154 and determine the probability that the sum of the dice is

a. 6.   **b.** even.
c. 7 or 11.   **d.** 2, 3, or 12.

4.14 Coin Tossing. A balanced dime is tossed three times. The possible outcomes can be represented as follows.

HHH	HTH	THH	TTH
HHT	HTT	THT	TTT

Here, for example, HHT means that the first two tosses come up heads and the third tails. Find the probability that

a. exactly two of the three tosses come up heads.
b. the last two tosses come up tails.
c. all three tosses come up the same.
d. the second toss comes up heads.

4.15 Housing Units. Refer to Exercise 4.9. Which, if any, of the events in parts (a)–(e) are certain? impossible?

Extending the Concepts and Skills

4.16 Explain what is wrong with the argument: When two balanced dice are rolled, the sum of the dice can be 2, 3, 4, 5, 6, 7, 8, 9, 10, 11, or 12, giving 11 possibilities. Therefore the probability is $\frac{1}{11}$ that the sum is 12.

4.17 Why can't the frequentist interpretation be used as a definition of probability?

Odds. Closely related to probabilities are *odds*. Newspapers, magazines, and other popular publications often express likelihood in terms of odds instead of probabilities, and odds are used much more than probabilities in gambling contexts. If the probability that an event occurs is p, the odds that the event occurs are p to $1 - p$. This fact is also expressed by saying that the odds are p to $1 - p$ *in favor of the event* or that the odds are $1 - p$ to p *against the event*. Conversely, if the odds in favor of an event are a to b (or, equivalently, the odds against it are b to a), the probability the event occurs is $a/(a + b)$. For example, if an event has probability 0.75 of occurring, the odds that the event occurs are 0.75 to 0.25, or 3 to 1; if the odds against an event are 3 to 2, the probability that the event occurs is $2/(2 + 3)$, or 0.4. We examine odds in Exercises 4.18–4.20.

4.18 Roulette. An American roulette wheel contains 38 numbers, of which 18 are red, 18 are black, and 2 are green. When the roulette wheel is spun, the ball is equally likely to land on any of the 38 numbers. For a bet on red, the house pays even odds (i.e., 1 to 1). What should the odds actually be to make the bet fair?

4.19 Cyber Affair. As found in *USA TODAY*, results of a survey by International Communications Research revealed that roughly 75% of adult women believe that a romantic relationship over the Internet while in an exclusive relationship in the real world is cheating. What are the odds against randomly selecting an adult female Internet user who believes that having a "cyber affair" is cheating?

4.20 The Triple Crown. Funny Cide, winner of both the 2003 Kentucky Derby and the 2003 Preakness Stakes, was the even-money (1-to-1 odds) favorite to win the 2003 Belmont Stakes and thereby capture the coveted Triple Crown of thoroughbred horseracing. The second favorite and actual winner of the 2003 Belmont Stakes, Empire Maker,

posted odds at 2 to 1 (against) to win the race. Based on the posted odds, determine the probability that the winner of the race would be
a. Funny Cide.
b. Empire Maker.

4.2 Events

Before continuing, we need to discuss events in greater detail. In Section 4.1, we used the word *event* intuitively. To be more precise, as used in probability, an event consists of a collection of outcomes, as illustrated in Example 4.5.

Example 4.5 Introducing Events

Playing Cards A deck of playing cards contains 52 cards, as displayed in Fig. 4.3. When we perform the experiment of randomly selecting one card from the deck, one of these 52 cards will be obtained. The collection of all 52 cards—the possible outcomes—is called the **sample space** for this experiment.

FIGURE 4.3
A deck of playing cards

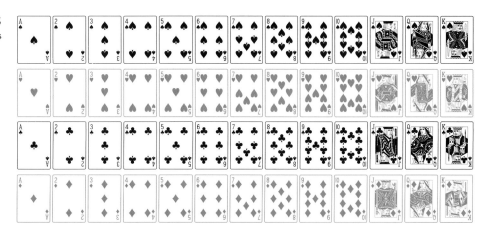

Many different events can be associated with this card-selection experiment. Let's consider four:

1. The event that the card selected is the king of hearts.
2. The event that the card selected is a king.
3. The event that the card selected is a heart.
4. The event that the card selected is a face card.

List the outcomes constituting each of these four events.

FIGURE 4.4
The event the king of hearts is selected

Solution The first event—that the card selected is the king of hearts—consists of the single outcome "king of hearts." This event is pictured in Fig. 4.4.

The second event—that the card selected is a king—consists of the four outcomes "king of spades," "king of hearts," "king of clubs," and "king of diamonds." This event is depicted in Fig. 4.5.

FIGURE 4.5
The event a king is selected

Thirteen outcomes make up the third event—that the card selected is a heart—namely, the outcomes "ace of hearts," "two of hearts,"..., "king of hearts." This event is shown in Fig. 4.6.

The fourth event—that the card selected is a face card—consists of 12 outcomes, namely, the 12 face cards shown in Fig. 4.7.

FIGURE 4.6
The event a heart is selected

FIGURE 4.7
The event a face card is selected

When the experiment of selecting a card from the deck is performed, an event *occurs* if it includes the card selected. For instance, if the card selected turns out to be the king of spades, the second and fourth events (Figs. 4.5 and 4.7) occur, whereas the first and third events (Figs. 4.4 and 4.6) do not.

The term *sample space* reflects the fact that, in statistics, the collection of possible outcomes often consists of the possible samples of a given size, as illustrated in Table 4.2 on page 152. The following definition summarizes the terminology discussed so far in this section.

DEFINITION 4.2

Sample Space and Event

Sample space: The collection of all possible outcomes for an experiment.

Event: A collection of outcomes for the experiment, that is, any subset of the sample space.

Notation and Graphical Displays for Events

For convenience, we use letters such as $A, B, C, D, \ldots$ to represent events. In the card-selection experiment of Example 4.5, for instance, we might let

$$A = \text{event the card selected is the king of hearts,}$$
$$B = \text{event the card selected is a king,}$$
$$C = \text{event the card selected is a heart, and}$$
$$D = \text{event the card selected is a face card.}$$

FIGURE 4.8
Venn diagram for event E

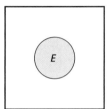

Graphical displays of events are useful for explaining and under-standing probability. **Venn diagrams,** named after English logician John Venn (1834–1923), are one of the best ways to portray events and relationships among events visually. The sample space is depicted as a rectangle, and the various events are drawn as disks (or other geometric shapes) inside the rectangle. In the simplest case, only one event is displayed, as shown in Fig. 4.8, with the colored portion representing the event.

Relationships Among Events

Each event E has a corresponding event defined by the condition that "E does not occur." That event is called the **complement** of E and is denoted **(not E).** Event (not E) consists of all outcomes not in E. A Venn diagram clarifies this idea, as shown in Fig. 4.9(a).

FIGURE 4.9
Venn diagrams for (a) event (not E), (b) event (A & B), and (c) event (A or B)

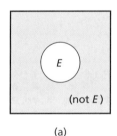

(a)

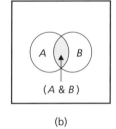

(b)

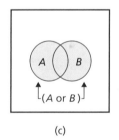

(c)

What
Does it Mean?

Event (not E) consists of all outcomes not in event E; event (A & B) consists of all outcomes common to event A and event B; event (A or B) consists of all outcomes either in event A or in event B or both.

With any two events, say, A and B, we can associate two new events. One new event is defined by the condition that "both event A and event B occur" and is denoted **(A & B).** Event (A & B) consists of all outcomes common to both event A and event B, as illustrated in Fig. 4.9(b).

The other new event associated with A and B is defined by the condition that "either event A or event B or both occur" or, equivalently, that "at least one of events A and B occurs." That event is denoted **(A or B)** and consists of all outcomes in either event A or event B or both, as Fig. 4.9(c) shows.

DEFINITION 4.3

Relationships Among Events

(not E): The event "E does not occur"

(A & B): The event "both A and B occur"

(A or B): The event "either A or B or both occur"

Note: Because the event "both *A* and *B* occur" is the same as the event "both *B* and *A* occur," event (*A* & *B*) is the same as event (*B* & *A*). Similarly, event (*A* or *B*) is the same as event (*B* or *A*).

In Example 4.6, we demonstrate relationships among events.

Example 4.6   Relationships Among Events

Playing Cards   For the experiment of randomly selecting one card from a deck of 52, let

> *A* = event the card selected is the king of hearts,
> *B* = event the card selected is a king,
> *C* = event the card selected is a heart, and
> *D* = event the card selected is a face card.

We showed the outcomes for each of those four events in Figs. 4.4–4.7, respectively, in Example 4.5. Determine the following events.

a. (not *D*)   **b.** (*B* & *C*)   **c.** (*B* or *C*)   **d.** (*C* & *D*)

Solution

a. (not *D*) is the event *D* does not occur—the event that a face card is not selected. Event (not *D*) consists of the 40 cards in the deck that are not face cards, as depicted in Fig. 4.10.

FIGURE 4.10
Event (not *D*)

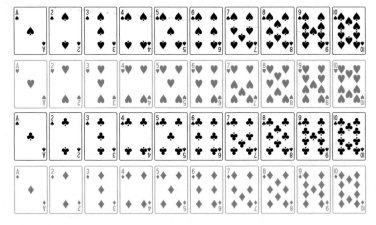

FIGURE 4.11
Event (*B* & *C*)

b. (*B* & *C*) is the event both *B* and *C* occur—the event that the card selected is both a king and a heart. This event can occur only if the card selected is the king of hearts. Consequently, (*B* & *C*) is the event that the card selected is the king of hearts and consists of the single outcome shown in Fig. 4.11.

Note: Event (*B* & *C*) is the same as event *A*, so we can write *A* = (*B* & *C*).

c. (*B* or *C*) is the event either *B* or *C* or both occur—the event that the card selected is either a king or a heart or both. Event (*B* or *C*) consists of 16 outcomes—namely, the 4 kings and the 12 nonking hearts—as illustrated in Fig. 4.12.

FIGURE 4.12
Event (*B* or *C*)

Note: Event (*B* or *C*) can occur in 16, not 17, ways because the outcome "king of hearts" is common to both event *B* and event *C*.

FIGURE 4.13
Event (*C* & *D*)

d. (*C* & *D*) is the event both *C* and *D* occur—the event that the card selected is both a heart and a face card. For that event to occur, the card selected must be the jack, queen, or king of hearts. Thus event (*C* & *D*) consists of the three outcomes displayed in Fig. 4.13. These three outcomes are those common to events *C* and *D*.

In Example 4.6, we described each of four events by listing their outcomes (Figs. 4.10–4.13). Sometimes, describing events verbally, as in Example 4.7, is more appropriate.

Example 4.7 Relationships Among Events

TABLE 4.4
Frequency distribution for students' ages

Age (yrs)	Frequency
17	1
18	1
19	9
20	7
21	7
22	5
23	3
24	4
26	1
35	1
36	1

Student Ages   A frequency distribution for the ages of the 40 students in Professor Weiss's introductory statistics class is presented in Table 4.4. One student is selected at random. Let

A = event the student selected is under 21,
B = event the student selected is over 30,
C = event the student selected is in his or her 20s, and
D = event the student selected is over 18.

Determine the following events.

a. (not *D*)   **b.** (*A* & *D*)   **c.** (*A* or *D*)   **d.** (*B* or *C*)

Solution

a. (not *D*) is the event *D* does not occur—the event that the student selected is not over 18, that is, is 18 or under. From Table 4.4, (not *D*) comprises the two students in the class who are 18 or under.

b. (*A* & *D*) is the event both *A* and *D* occur—the event that the student selected is both under 21 and over 18, that is, is either 19 or 20. Event (*A* & *D*) comprises the 16 students in the class who are 19 or 20.

c. (*A* or *D*) is the event either *A* or *D* or both occur—the event that the student selected is either under 21 or over 18 or both. But every student in the class is either under 21 or over 18. Consequently, event (*A* or *D*) comprises all 40 students in the class and is certain to occur.

d. (*B* or *C*) is the event either *B* or *C* or both occur—the event that the student selected is either over 30 or in his or her 20s. Table 4.4 shows that (*B* or *C*) comprises the 29 students in the class who are 20 or over.

Mutually Exclusive Events

Next, we introduce the concept of **mutually exclusive events.**

DEFINITION 4.4

> **Mutually Exclusive Events**
>
> Two or more events are *mutually exclusive events* if no two of them have outcomes in common.

What *Does it Mean?*

Events are mutually exclusive if at most one of them can occur when the experiment is performed.

The Venn diagrams shown in Fig. 4.14 portray the difference between two events that are mutually exclusive and two events that are not mutually exclusive. In Fig. 4.15, we show three mutually exclusive events and two cases of three events that are not mutually exclusive.

FIGURE 4.14
(a) Two mutually exclusive events;
(b) two non–mutually exclusive events

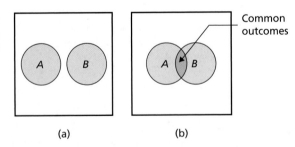

(a)   (b)

FIGURE 4.15
(a) Three mutually exclusive events;
(b) three non–mutually exclusive events;
(c) three non–mutually exclusive events

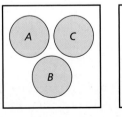   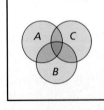   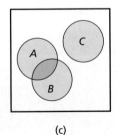

(a)   (b)   (c)

Example 4.8 Mutually Exclusive Events

Playing Cards For the experiment of randomly selecting one card from a deck of 52, let

$$C = \text{event the card selected is a heart,}$$
$$D = \text{event the card selected is a face card,}$$
$$E = \text{event the card selected is an ace,}$$
$$F = \text{event the card selected is an 8, and}$$
$$G = \text{event the card selected is a 10 or a jack.}$$

Which of the following collections of events are mutually exclusive?

a. C and D **b.** C and E **c.** D and E

d. D, E, and F **e.** D, E, F, and G

Solution

a. Event C and event D are not mutually exclusive because they have the common outcomes "king of hearts," "queen of hearts," and "jack of hearts." Both events occur if the card selected is the king, queen, or jack of hearts.

b. Event C and event E are not mutually exclusive because they have the common outcome "ace of hearts." Both events occur if the card selected is the ace of hearts.

c. Event D and event E are mutually exclusive because they have no common outcomes. They cannot both occur when the experiment is performed because selecting a card that is both a face card and an ace is impossible.

d. Events D, E, and F are mutually exclusive because no two of them can occur simultaneously.

e. Events D, E, F, and G are not mutually exclusive because event D and event G both occur if the card selected is a jack.

EXERCISES 4.2

Statistical Concepts and Skills

4.21 What type of graphical displays are useful for portraying events and relationships among events?

4.22 Construct a Venn diagram representing each event.
a. (not E) **b.** (A or B) **c.** (A & B)
d. (A & B & C) **e.** (A or B or C) **f.** ((not A) & B)

4.23 What does it mean for two events to be mutually exclusive? for three events?

4.24 Answer true or false to each statement, and give reasons for your answers.
a. If event A and event B are mutually exclusive, so are events A, B, and C for every event C.

b. If event A and event B are not mutually exclusive, neither are events A, B, and C for every event C.

4.25 Dice. When one die is rolled, the following six outcomes are possible:

List the outcomes constituting

$$A = \text{event the die comes up even,}$$
$$B = \text{event the die comes up 4 or more,}$$
$$C = \text{event the die comes up at most 2, and}$$
$$D = \text{event the die comes up 3.}$$

4.26 Horse Racing. In a horse race, the odds against winning are as shown in the following table. For example, the odds against winning are 8 to 1 for horse #1.

Horse	#1	#2	#3	#4	#5	#6	#7	#8
Odds	8	15	2	3	30	5	10	5

List the outcomes constituting

A = event one of the top two favorites wins (the top two favorites are the two horses with the lowest odds against winning),

B = event the winning horse's number is above 5,

C = event the winning horse's number is at most 3, that is, 3 or less, and

D = event one of the two long shots wins (the two long shots are the two horses with the highest odds against winning).

4.27 Dice. Refer to Exercise 4.25. For each of the following events, list the outcomes that constitute the event and describe the event in words.
a. (not A)   **b.** (A & B)   **c.** (B or C)

4.28 Horse Racing. Refer to Exercise 4.26. For each of the following events, list the outcomes that constitute the event and describe the event in words.
a. (not C)   **b.** (C & D)   **c.** (A or C)

4.29 Housing Units. The U.S. Bureau of the Census publishes data on housing units in *American Housing Survey in the United States*. The following table provides a frequency distribution for the number of rooms in U.S. housing units. The frequencies are in thousands.

Rooms	No. of units
1	471
2	1,470
3	11,715
4	23,468
5	24,476
6	21,327
7	13,782
8+	15,647

For a U.S. housing unit selected at random, let

A = event the unit has at most four rooms,

B = event the unit has at least two rooms,

C = event the unit has between five and seven rooms, inclusive, and

D = event the unit has more than seven rooms.

Describe each of the following events in words and determine the number of outcomes (housing units) that constitute each event.
a. (not A)   **b.** (A & B)   **c.** (C or D)

4.30 Protecting the Environment. A survey was conducted in Canada to ascertain public opinion about a major national park region in the Banff-Bow Valley. One question asked the amount that respondents would be willing to contribute per year to protect the environment in the Banff-Bow Valley region. The following frequency distribution was found in an article by J. Ritchie, S. Hudson, and S. Timur, titled "Public Reactions to Policy Recommendations from the Banff-Bow Valley Study" (*Journal of Sustainable Tourism*, Vol. 10, No. 4, pp. 295–308).

Contribution ($)	Frequency
0	85
1–50	116
51–100	59
101–200	29
201–300	5
301–500	7
501–1000	3

For a respondent selected at random, let

A = event that the respondent would be willing to contribute at least $101,

B = event that the respondent would not be willing to contribute more than $50,

C = event that the respondent would be willing to contribute between $1 and $200, and

D = event that the respondent would be willing to contribute at least $1.

Describe the following events in words and determine the number of outcomes (respondents) that make up each event.
a. (not D)   **b.** (A & B)   **c.** (C or A)   **d.** (B & D)

4.31 Dice. Refer to Exercise 4.25.
a. Are events A and B mutually exclusive?
b. Are events B and C mutually exclusive?

c. Are events *A*, *C*, and *D* mutually exclusive?
d. Are there three mutually exclusive events among *A*, *B*, *C*, and *D*? four?

4.32 Horse Racing. Each part of this exercise contains events from Exercise 4.26. In each case, decide whether the events are mutually exclusive.
a. *A* and *B*
b. *B* and *C*
c. *A*, *B*, and *C*
d. *A*, *B*, and *D*
e. *A*, *B*, *C*, and *D*

4.33 Housing Units. Refer to Exercise 4.29. Among the events *A*, *B*, *C*, and *D*, identify the collections of events that are mutually exclusive.

4.34 Protecting the Environment. Refer to Exercise 4.30. Among the events *A*, *B*, *C*, and *D*, identify the collections of events that are mutually exclusive.

4.35 Draw a Venn diagram portraying four mutually exclusive events.

Extending the Concepts and Skills

4.36 Construct a Venn diagram that portrays four events, *A*, *B*, *C*, and *D* that have the following properties: Events *A*, *B*, and *C* are mutually exclusive; events *A*, *B*, and *D* are mutually exclusive; no other three of the four events are mutually exclusive.

4.37 Suppose that *A*, *B*, and *C* are three events that cannot all occur simultaneously. Does this condition necessarily imply that *A*, *B*, and *C* are mutually exclusive? Justify your answer and illustrate it with a Venn diagram.

4.3 Some Rules of Probability

In this section, we discuss several rules of probability. Before beginning, however, we need to introduce an additional notation used in probability, which we do in Example 4.9.

 **Example 4.9   Probability Notation**

Dice   When a balanced die is rolled once, six equally likely outcomes are possible, as shown in Fig. 4.16. Use probability notation to express the probability that the die comes up an even number.

FIGURE 4.16
Sample space for rolling a die once

What *Does it Mean?*

Keep in mind that *A* refers to the event that the die comes up even, whereas *P(A)* refers to the probability of that event occurring.

Solution   The event that the die comes up an even number can occur in three ways—namely, if 2, 4, or 6 is rolled. Because $f/N = 3/6 = 0.5$, *the probability that the die comes up even is 0.5.* We want to express the italicized phrase using probability notation.

Let *A* denote the event that the die comes up even. We use the notation $P(A)$ to represent the probability that event *A* occurs. Hence we can rewrite the italicized statement simply as $P(A) = 0.5$, which is read "the probability of *A* is 0.5."

DEFINITION 4.5   Probability Notation

If *E* is an event, then $P(E)$ represents the probability that event *E* occurs. It is read "the probability of *E*."

The Special Addition Rule

The first rule of probability that we present is the **special addition rule,** which states that, for mutually exclusive events, the probability that one or another of the events occurs equals the sum of the individual probabilities. We use a Venn diagram to show the validity of the special addition rule. Figure 4.17 shows two mutually exclusive events, event A and event B.

FIGURE 4.17

Two mutually exclusive events

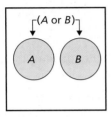

If you think of the colored regions shown in Fig. 4.17 as probabilities, the colored disk on the left is $P(A)$, the colored disk on the right is $P(B)$, and the total colored region is $P(A \text{ or } B)$. Because event A and event B are mutually exclusive, the total colored region equals the sum of the two colored disks; that is, $P(A \text{ or } B) = P(A) + P(B)$. We express this result and a more general result in Formula 4.1.

FORMULA 4.1

What
Does it Mean?

For mutually exclusive events, the probability that at least one occurs equals the sum of their individual probabilities.

The Special Addition Rule

If event A and event B are mutually exclusive, then

$$P(A \text{ or } B) = P(A) + P(B).$$

More generally, if events $A, B, C, \ldots$ are mutually exclusive, then

$$P(A \text{ or } B \text{ or } C \text{ or } \cdots) = P(A) + P(B) + P(C) + \cdots.$$

Example 4.10 illustrates use of the special addition rule.

Example 4.10   The Special Addition Rule

Size of Farms   The U.S. Bureau of the Census compiles information about farms and publishes its findings in the *Census of Agriculture*. From that publication, we present a relative-frequency distribution for the size of farms in the United States in the first two columns of Table 4.5.

In the third column of Table 4.5, we introduce events that correspond to the size classes. For example, if a farm is selected at random, D denotes the event that the farm has between 100 and 180 acres. The probabilities of the events in the third column of Table 4.5 equal the relative frequencies displayed in the second column. Thus, for instance, the probability is 0.156 that a randomly selected farm has between 100 and 180 acres: $P(D) = 0.156$.

Use Table 4.5 and the special addition rule to determine the probability that a randomly selected farm has between 100 and 500 acres.

TABLE 4.5

Size of farms in the United States

Size (acres)	Relative freq.	Event
Under 10	0.082	A
10 < 50	0.215	B
50 < 100	0.154	C
100 < 180	0.156	D
180 < 260	0.086	E
260 < 500	0.124	F
500 < 1000	0.092	G
1000 < 2000	0.054	H
2000 & over	0.037	I

Solution   The event that the farm selected has between 100 and 500 acres can be expressed as $(D \text{ or } E \text{ or } F)$. Events D, E, and F are mutually exclusive and hence by the special addition rule,

$$P(D \text{ or } E \text{ or } F) = P(D) + P(E) + P(F)$$
$$= 0.156 + 0.086 + 0.124 = 0.366.$$

The probability that a randomly selected U.S. farm has between 100 and 500 acres is 0.366.

INTERPRETATION 36.6% of U.S. farms have between 100 and 500 acres.

The Complementation Rule

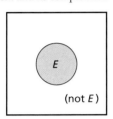

The second rule of probability that we discuss is the **complementation rule.** It states that the probability an event occurs equals 1 minus the probability the event does not occur. We use a Venn diagram to show the validity of this rule. Figure 4.18 shows an event, E, and its complement, (not E).

If you think of the regions shown in Fig. 4.18 as probabilities, the entire region enclosed by the rectangle is the probability of the sample space, or 1. Furthermore, the colored region is $P(E)$ and the uncolored region is $P(\text{not } E)$. Thus, $P(E) + P(\text{not } E) = 1$ or, equivalently, $P(E) = 1 - P(\text{not } E)$, which we express as the following formula.

FORMULA 4.2

The Complementation Rule

For any event E,
$$P(E) = 1 - P(\text{not } E).$$

What
Does it Mean?

The probability that an event occurs equals 1 minus the probability that it does not occur.

The complementation rule is useful because sometimes computing the probability that an event does not occur is easier than computing the probability that it does occur. In such cases, we can, because of the complementation rule, obtain the probability that the event occurs by first computing the probability that it does not occur and then subtracting the result from 1. Example 4.11 illustrates this approach.

Example 4.11 **The Complementation Rule**

Size of Farms The first two columns of Table 4.5 provide a relative-frequency distribution for the size of U.S. farms. For a randomly selected farm, find the probability that the farm has

a. less than 2000 acres.   **b.** 50 acres or more.

Solution

a. Let
$$J = \text{event the farm selected has less than 2000 acres.}$$

To determine $P(J)$, we apply the complementation rule because $P(\text{not } J)$ is easier to compute than $P(J)$. Note that (not J) is the event the farm obtained has 2000 or more acres, which is event I in Table 4.5. Thus $P(\text{not } J) = P(I) = 0.037$. Applying the complementation rule yields
$$P(J) = 1 - P(\text{not } J) = 1 - 0.037 = 0.963.$$

The probability that a randomly selected U.S. farm has less than 2000 acres is 0.963.

> **INTERPRETATION** 96.3% of U.S. farms have less than 2000 acres.

b. Let

$$K = \text{event the farm selected has 50 acres or more.}$$

We apply the complementation rule to find $P(K)$. Now, (not K) is the event the farm obtained has less than 50 acres. From Table 4.5, event (not K) is the same as event (A or B). Because event A and event B are mutually exclusive, the special addition rule implies that

$$P(\text{not } K) = P(A \text{ or } B) = P(A) + P(B) = 0.082 + 0.215 = 0.297.$$

Using this result and the complementation rule, we conclude that

$$P(K) = 1 - P(\text{not } K) = 1 - 0.297 = 0.703.$$

The probability that a randomly selected U.S. farm has 50 acres or more is 0.703.

> **INTERPRETATION** 70.3% of U.S. farms have at least 50 acres.

The General Addition Rule

The special addition rule (Formula 4.1) allows us to find the probability of event (A or B) from the probabilities of event A and event B, provided that event A and event B are mutually exclusive. For events that are not mutually exclusive, we must use a different rule—the *general addition rule*. To introduce it, we use the Venn diagram shown in Fig. 4.19.

If you think of the colored regions shown in Fig. 4.19 as probabilities, the colored disk on the left is $P(A)$, the colored disk on the right is $P(B)$, and the total colored region is $P(A \text{ or } B)$. To obtain the total colored region, $P(A \text{ or } B)$, we first sum the two colored disks, $P(A)$ and $P(B)$. When we do so, the common colored region, $P(A \& B)$, is counted twice. Thus, we must subtract $P(A \& B)$ from the sum. So, we see that $P(A \text{ or } B) = P(A) + P(B) - P(A \& B)$. This formula is the **general addition rule,** which we express as the following formula.

FIGURE 4.19
Non–mutually exclusive events

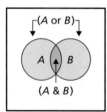

(A or B)

A B

(A & B)

FORMULA 4.3

The General Addition Rule

If A and B are any two events, then

$$P(A \text{ or } B) = P(A) + P(B) - P(A \& B).$$

What *Does it Mean?*

For any two events, the probability that at least one occurs equals the sum of their individual probabilities less the probability that both occur.

In Example 4.12, we compute the probability of selecting either a spade or a face card in two ways: first without using the general addition rule and then using it.

Example 4.12 The General Addition Rule

Playing Cards   Consider again the experiment of selecting one card at random from a deck of 52 playing cards. Find the probability that the card selected is either a spade or a face card

a. without using the general addition rule.

b. by using the general addition rule.

Solution

a. Let

$$E = \text{event the card selected is either a spade or a face card.}$$

Event E consists of 22 cards—namely, the 13 spades plus the other nine face cards that are not spades—as shown in Fig. 4.20. So, by the f/N rule,

$$P(E) = \frac{f}{N} = \frac{22}{52} = 0.423.$$

The probability that a randomly selected card is either a spade or a face card is 0.423.

FIGURE 4.20
Event E

b. To determine $P(E)$ by using the general addition rule, we first note that we can write $E = (C \text{ or } D)$, where

$$C = \text{event the card selected is a spade, and}$$
$$D = \text{event the card selected is a face card.}$$

Event C consists of the 13 spades, and event D consists of the 12 face cards. Also, event $(C \& D)$ consists of the three spades that are face cards—the jack, queen, and king of spades. Applying the general addition rule gives

$$P(E) = P(C \text{ or } D) = P(C) + P(D) - P(C \& D)$$
$$= \frac{13}{52} + \frac{12}{52} - \frac{3}{52} = 0.250 + 0.231 - 0.058 = 0.423,$$

which agrees with the answer obtained in part (a).

In Example 4.12, computing the probability was simpler without using the general addition rule. Frequently, however, the general addition rule is the easier or even the only way to compute a probability, as illustrated in Example 4.13.

Example 4.13 The General Addition Rule

Characteristics of People Arrested Data on people arrested are published by the U.S. Federal Bureau of Investigation in *Crime in the United States*. Records for one year showed that 79.6% of the people arrested were male, 18.3% were under 18 years of age, and 13.5% were males under 18 years of age. If a person arrested that year is selected at random, what is the probability that the person obtained is either male or under 18?

Solution Let

$$M = \text{event the person obtained is male, and}$$
$$E = \text{event the person obtained is under 18.}$$

The event that the person obtained is either male or under 18 can be expressed as (M or E). We want to determine $P(M \text{ or } E)$. From the percentage data provided, we know that $P(M) = 0.796$, $P(E) = 0.183$, and $P(M \text{ \& } E) = 0.135$. Applying the general addition rule, we conclude that

$$P(M \text{ or } E) = P(M) + P(E) - P(M \text{ \& } E)$$
$$= 0.796 + 0.183 - 0.135 = 0.844.$$

The probability that the person obtained is either male or under 18 is 0.844.

INTERPRETATION 84.4% of those arrested during the year in question were either male or under 18 years of age (or both).

The general addition rule is consistent with the special addition rule—if two events are mutually exclusive, both rules yield the same result.

There are also general addition rules for more than two events. For instance, the general addition rule for three events is

$$P(A \text{ or } B \text{ or } C) = P(A) + P(B) + P(C) - P(A \text{ \& } B) - P(A \text{ \& } C) - P(B \text{ \& } C) + P(A \text{ \& } B \text{ \& } C).$$

Try to write down the general addition rule for four events.

EXERCISES 4.3

Statistical Concepts and Skills

4.38 A Lottery. Suppose that you hold 20 out of a total of 500 tickets sold for a lottery. The grand-prize winner is determined by the random selection of one of the 500 tickets. Let G be the event that you win the grand prize. Find the probability that you win the grand prize. Express your answer in probability notation.

4.39 Ages of Senators. According to the *Congressional Directory*, the age distribution for senators in the 104th U.S. Congress is as follows.

Age (yrs)	No. of senators
Under 40	1
40–49	14
50–59	41
60–69	27
70 and over	17
	100

Suppose that a senator from the 104th U.S. Congress is selected at random. Let

A = event the senator is under 40,

B = event the senator is in his or her 40s,

C = event the senator is in his or her 50s, and

S = event the senator is under 60.

a. Use the table and the f/N rule to find $P(S)$.
b. Express event S in terms of events A, B, and C.
c. Determine $P(A)$, $P(B)$, and $P(C)$.
d. Compute $P(S)$, using the special addition rule and your answers from parts (b) and (c). Compare your answer with that in part (a).

4.40 Home Internet Access. The online publication *Cyber-Stats*, by Mediamark Research, Inc., reports Internet access and usage. The following is a percentage distribution of household income for households with home Internet access only.

Household income	Percentage	Event
Under $50,000	27.2	A
$50,000 < $75,000	27.3	B
$75,000 < $150,000	37.2	C
$150,000 or above	8.3	D

Suppose that a household with home Internet access only is selected at random. Let A denote the event the household has an income under $50,000, B denote the event the household has an income between $50,000 and $75,000, and so on (see the third column of the table). Apply the special addition rule to find the probability that the household obtained has an income
a. under $75,000. **b.** $50,000 or above.
c. between $50,000 and $150,000.
d. Interpret each of your answers in parts (a)–(c) in terms of percentages.

4.41 Oil Spills. The U.S. Coast Guard maintains a database of the number, source, and location of oil spills in U.S. navigable and territorial waters. The following is a probability distribution for location of oil spill events. [SOURCE: *Statistical Abstract of the United States*.]

Location	Probability
Atlantic Ocean	0.011
Pacific Ocean	0.059
Gulf of Mexico	0.271
Great Lakes	0.018
Other lakes	0.003
Rivers and canals	0.211
Bays and sounds	0.094
Harbors	0.099
Other	0.234

Apply the special addition rule to find the percentage of oil spills in U.S. navigable and territorial waters that
a. occur in an ocean.
b. occur in a lake or harbor.
c. do not occur in a lake, ocean, river, or canal.

4.42 Religion in America. R. Doyle diagrammed the religious preferences of Americans for the years 1940–2000 in an article titled "Religion in America," published in the February 2003 issue of *Scientific American*. Following is a relative-frequency distribution for the religious preference of Americans for the year 2000.

Preference	Relative frequency
Protestant	0.560
Catholic	0.250
Jewish	0.025
Other	0.015
None	0.150

Find the probability that the religious preference of a randomly selected American is
a. Catholic or Protestant.
b. not Jewish.
c. not Catholic, Protestant, or Jewish.

4.43 Ages of Senators. Refer to Exercise 4.39. Use the complementation rule to find the probability that a randomly selected senator in the 104th Congress is
a. 40 years old or older. **b.** under 60 years old.

4.44 Home Internet Access. Solve part (b) of Exercise 4.40 by using the complementation rule. Compare your work here to that in Exercise 4.40(b) where you used the special addition rule.

4.45 Craps. In the game of *craps*, a player rolls two balanced dice. Thirty-six equally likely outcomes are possible, as shown in Fig. 4.1 on page 154. Let

A = event the sum of the dice is 7,

B = event the sum of the dice is 11,

C = event the sum of the dice is 2,

D = event the sum of the dice is 3,

E = event the sum of the dice is 12,

F = event the sum of the dice is 8, and

G = event doubles are rolled.

a. Compute the probability of each of the seven events.
b. The player wins on the first roll if the sum of the dice is 7 or 11. Find the probability of that event by using the special addition rule and your answers from part (a).
c. The player loses on the first roll if the sum of the dice is 2, 3, or 12. Determine the probability of that event by using the special addition rule and your answers from part (a).
d. Compute the probability that either the sum of the dice is 8 or doubles are rolled, without using the general addition rule.
e. Compute the probability that either the sum of the dice is 8 or doubles are rolled by using the general addition rule and compare your answer to the one you obtained in part (d).

4.46 Gender and Divorce. According to the *Current Population Reports*, published by the Census Bureau, 51.8% of U.S. adults are female, 9.8% are divorced, and 5.6% are divorced females. For a U.S. adult selected at random, let

F = event the person is female, and

D = event the person is divorced.

a. Obtain $P(F)$, $P(D)$, and $P(F \& D)$.
b. Determine $P(F \text{ or } D)$ and interpret your answer in terms of percentages.
c. Find the probability that a randomly selected adult is male.

4.47 Let A and B be events such that $P(A) = \frac{1}{4}$, $P(B) = \frac{1}{3}$, and $P(A \text{ or } B) = \frac{1}{2}$.
a. Are events A and B mutually exclusive? Explain your answer.
b. Determine $P(A \& B)$.

4.48 Suppose that A and B are events such that $P(A) = \frac{1}{3}$, $P(A \text{ or } B) = \frac{1}{2}$, and $P(A \& B) = \frac{1}{10}$. Find $P(B)$.

Extending the Concepts and Skills

4.49 Suppose that A and B are mutually exclusive events.
a. Use the special addition rule to express $P(A \text{ or } B)$ in terms of $P(A)$ and $P(B)$.
b. Show that the general addition rule gives the same answer as that in part (a).

4.50 Secrets of Success. Gerald Kushel, Ed.D., was interviewed by *Bottom Line/Personal* on the secrets of successful people. To study success, Kushel questioned 1200 people, among whom were lawyers, artists, teachers, and students. He found that 15% enjoy neither their jobs nor their personal lives, 80% enjoy their jobs but not their personal lives, and 4% enjoy both their jobs and their personal lives. Determine the percentage of the 1200 people interviewed who
a. enjoy either their jobs or their personal lives.
b. enjoy their personal lives but not their jobs.

4.4 Contingency Tables; Joint and Marginal Probabilities[†]

In Section 2.2, we discussed grouping data obtained from one variable of a population into a frequency distribution. Data obtained by observing values of one variable of a population are called **univariate data.**

We often need to group and analyze data obtained from two variables of a population. Data obtained by observing values of two variables of a population are called **bivariate data,** and a frequency distribution for bivariate data is called a **contingency table** or **two-way table.** In Example 4.14, we introduce contingency tables.

[†]Sections 4.4–4.6 are recommended for those studying the binomial distribution (Section 5.3).

▌▌▌ Example 4.14   Introducing Contingency Tables

ASU Faculty   The *Arizona State University Statistical Summary* provides information on various characteristics of the ASU faculty. Data from one year on the variables age and rank of ASU faculty members yielded the contingency table shown in Table 4.6. Discuss and interpret the numbers in the table.

TABLE 4.6
Contingency table for age and rank of ASU faculty members

		Rank				
		Full professor R_1	Associate professor R_2	Assistant professor R_3	Instructor R_4	**Total**
Age	Under 30 A_1	2	3	57	6	68
	30–39 A_2	52	170	163	17	402
	40–49 A_3	156	125	61	6	348
	50–59 A_4	145	68	36	4	253
	60 & over A_5	75	15	3	0	93
	Total	430	381	320	33	1164

Solution   The small boxes inside the rectangle formed by the heavy lines are called **cells.** The number 2 in the upper left cell indicates that two faculty members are full professors under the age of 30. The number 170, diagonally below and to the right of the 2, shows that 170 faculty members are associate professors in their 30s.

The row total in the first row reveals that 68 ($2 + 3 + 57 + 6$) of the faculty members are under 30. Similarly, the column total in the third column shows that 320 of the faculty members are assistant professors. The number 1164 in the lower right corner gives the total number of faculty. That total can be found by summing either the row totals or the column totals; it can also be found by summing the frequencies in the 20 cells of the contingency table.   ▬■

Joint and Marginal Probabilities

We now use the age and rank data from Table 4.6 to introduce the concepts of *joint probabilities* and *marginal probabilities* in Example 4.15.

▌▌▌ Example 4.15   Joint and Marginal Probabilities

ASU Faculty   Suppose that an ASU faculty member is selected at random.

a. Identify the events associated with the subscripted letters that label the rows and columns of the contingency table shown in Table 4.6.

b. Identify the events associated with the cells of the contingency table.

c. Determine the probabilities of the events discussed in parts (a) and (b).

d. Summarize the results of part (c) in a table.

e. Discuss the relationship among the probabilities in the table obtained in part (d).

Solution

a. The subscripted letter A_1 that labels the first row of Table 4.6 represents the event that the faculty member obtained is under 30 years of age:

$$A_1 = \text{event the faculty member is under 30.}$$

Similarly, the subscripted letter R_2 that labels the second column represents the event that the faculty member obtained is an associate professor:

$$R_2 = \text{event the faculty member is an associate professor.}$$

Likewise, we can identify the remaining seven of the nine events associated with the subscripted letters that label the rows and columns. Note that the events A_1, A_2, A_3, A_4, and A_5 are mutually exclusive, as are the events R_1, R_2, R_3, and R_4.

b. In addition to considering events A_1 through A_5 and R_1 through R_4 separately, we can also consider them jointly. For instance, the event that the faculty member obtained is under 30 (event A_1) *and* is also an associate professor (event R_2) can be expressed as (A_1 & R_2):

$$(A_1 \text{ \& } R_2) = \text{event the faculty member is an associate professor under 30.}$$

The joint event (A_1 & R_2) is represented by the cell in the first row and second column of Table 4.6. That joint event is one of 20 different joint events—one for each cell of the contingency table—associated with this random experiment.

    Thinking of a contingency table as a Venn diagram can be useful. The Venn diagram corresponding to Table 4.6 is shown in Fig. 4.21, and it makes clear that the 20 joint events, (A_1 & R_1), (A_1 & R_2), . . . , (A_5 & R_4), are mutually exclusive.

FIGURE 4.21
Venn diagram corresponding to Table 4.6

	R_1	R_2	R_3	R_4
A_1	(A_1 & R_1)	(A_1 & R_2)	(A_1 & R_3)	(A_1 & R_4)
A_2	(A_2 & R_1)	(A_2 & R_2)	(A_2 & R_3)	(A_2 & R_4)
A_3	(A_3 & R_1)	(A_3 & R_2)	(A_3 & R_3)	(A_3 & R_4)
A_4	(A_4 & R_1)	(A_4 & R_2)	(A_4 & R_3)	(A_4 & R_4)
A_5	(A_5 & R_1)	(A_5 & R_2)	(A_5 & R_3)	(A_5 & R_4)

c. To determine the probabilities of the events discussed in parts (a) and (b), we begin by observing that the total number of faculty members is 1164, or, $N = 1164$. The probability that the faculty member selected is under 30 (event A_1) is found by first noting from Table 4.6 that $f = 68$ and then applying the f/N rule:

$$P(A_1) = \frac{f}{N} = \frac{68}{1164} = 0.058.$$

Similarly, we can find the probability that the faculty member obtained is an associate professor (event R_2):

$$P(R_2) = \frac{f}{N} = \frac{381}{1164} = 0.327.$$

Likewise, we can determine the probabilities of the remaining seven of the nine events associated with the subscripted letters that label the table's rows and columns. The nine probabilities considered here are often called **marginal probabilities** because they correspond to events represented in the margin of the contingency table.

We can also find probabilities for joint events, so-called **joint probabilities.** For instance, the probability that the faculty member obtained is an associate professor under 30 [event $(A_1 \,\&\, R_2)$] is

$$P(A_1 \,\&\, R_2) = \frac{f}{N} = \frac{3}{1164} = 0.003.$$

Similarly, we can determine the probabilities of the remaining 19 of the 20 joint events associated with the cells of the contingency table.

d. By referring to part (c), we can replace the joint frequency distribution in Table 4.6 with the **joint probability distribution** in Table 4.7, where probabilities are displayed instead of frequencies.

TABLE 4.7
Joint probability distribution corresponding to Table 4.6

		Rank				
		Full professor R_1	Associate professor R_2	Assistant professor R_3	Instructor R_4	$P(A_i)$
Age	Under 30 A_1	0.002	0.003	0.049	0.005	0.058
	30–39 A_2	0.045	0.146	0.140	0.015	0.345
	40–49 A_3	0.134	0.107	0.052	0.005	0.299
	50–59 A_4	0.125	0.058	0.031	0.003	0.217
	60 & over A_5	0.064	0.013	0.003	0.000	0.080
	$P(R_j)$	0.369	0.327	0.275	0.028	1.000

Note that in Table 4.7 the joint probabilities are displayed in the cells and the marginal probabilities in the margin. Also observe that the row and column labels "Total" in Table 4.6 have been changed in Table 4.7 to $P(R_j)$ and $P(A_i)$, respectively. The reason is that in Table 4.7 the last row gives the probabilities of events R_1 through R_4, and the last column gives the probabilities of events A_1 through A_5.

e. The sum of the joint probabilities in a row or column of a joint probability distribution equals the marginal probability in that row or column, with any observed discrepancy being due to roundoff error. For example, for the A_4 row of Table 4.7, the sum of the joint probabilities is

$$0.125 + 0.058 + 0.031 + 0.003 = 0.217,$$

which equals the marginal probability at the end of the A_4 row.

EXERCISES 4.4

Statistical Concepts and Skills

4.51 Identify three ways in which the total number of observations of bivariate data can be obtained from the frequencies in a contingency table.

4.52 Suppose that bivariate data are to be grouped into a contingency table. Determine the number of cells that the contingency table will have if the number of possible values for the two variables are
a. two and three.
b. four and three.
c. m and n.

4.53 Fill in the blanks.
a. Data obtained by observing values of one variable of a population are called _____ data.
b. Data obtained by observing values of two variables of a population are called _____ data.

4.54 Give an example of
a. univariate data.
b. bivariate data.

4.55 New England Patriots. The National Football League updates team rosters and posts them on its Web site at www.nfl.com. The following contingency table provides a cross-classification of players on the New England Patriots roster as of July 15, 2003, by weight and years of experience.

		\multicolumn{4}{c}{**Years of experience**}	
		Rookie Y_1	1–5 Y_2	6–10 Y_3	10+ Y_4	**Total**
Weight (lb)	Under 200 W_1	4	6	5	2	17
	200–300 W_2	15	20	17	3	55
	Over 300 W_3	6	11	1	1	19
	Total	25	37	23	6	91

a. How many cells are in this contingency table?
b. How many players are on the New England Patriots roster as of July 15, 2003?
c. How many players are rookies?
d. How many players weigh between 200 and 300 lb?
e. How many players are rookies who weigh between 200 and 300 lb?

4.56 Motor Vehicle Use. The U.S. Federal Highway Administration compiles information on motor vehicle use around the globe and publishes its findings in *Highway Statistics*. Following is a contingency table for the number of motor vehicles in use in North American countries by country and type of vehicle. Frequencies are in thousands.

	Country			
	U.S. C_1	Canada C_2	Mexico C_3	**Total**
Automobiles V_1	129,728	13,138	8,607	151,473
Motorcycles V_2	3,871	320	270	4,461
Trucks V_3	75,940	6,933	4,287	87,160
Total	209,539	20,391	13,164	243,094

(Vehicle type — row labels)

a. How many cells are in this contingency table?
b. How many vehicles are Canadian?
c. How many vehicles are motorcycles?
d. How many vehicles are Canadian motorcycles?
e. How many vehicles are either Canadian or motorcycles?
f. How many automobiles are Mexican?
g. How many vehicles are not automobiles?

4.57 Female Physicians. Characteristics of physicians are collected and recorded by the American Medical Association in *Physician Characteristics and Distribution in the U.S.* The following is a contingency table for female physicians in the United States, cross-classified by age and specialty.

	Age			
	Under 35 A_1	35–44 A_2	45 or over A_3	**Total**
Family practice S_1	5,842	6,797	2,828	15,467
Internal medicine S_2	11,799			30,501
Obstetrics/ gynecology S_3	4,454	4,857		
Plastic surgery S_4	85	262	131	478
Total	22,180		11,025	57,959

(Specialty — row labels)

For the female physicians in the United States whose specialty is one of those shown in the table,
a. fill in the five empty cells.
b. how many are between 35 and 44 years old?

c. how many are plastic surgeons under 35?
d. how many are either plastic surgeons or under 35?
e. how many are neither plastic surgeons nor under 35?
f. how many are not in family practice?

4.58 What does the general addition rule (Formula 4.3 on page 170) mean in the context of the probabilities in a joint probability distribution?

4.59 New England Patriots. Refer to Exercise 4.55.
a. For a randomly selected player on the New England Patriots, describe the events Y_3, W_2, and (W_1 & Y_2) in words.
b. Compute the probability of each event in part (a). Interpret your answers in terms of percentages.
c. Construct a joint probability distribution similar to that shown in Table 4.7 on page 177.
d. Verify that the sum of each row and column of joint probabilities equals the marginal probability in that row or column. (*Note:* Rounding may cause slight deviations.)

4.60 Motor Vehicle Use. Refer to Exercise 4.56.
a. For a randomly selected vehicle, describe the events C_1, V_3, and (C_1 & V_3) in words.
b. Compute the probability of each event in part (a).
c. Compute $P(C_1$ or $V_3)$, using the contingency table and the f/N rule.
d. Compute $P(C_1$ or $V_3)$, using the general addition rule and your answers from part (b).
e. Construct a joint probability distribution.

4.61 Female Physicians. Refer to Exercise 4.57. A female physician in the United States whose specialty is one of those shown in the table is selected at random.
a. Use the letters in the margins of the contingency table to represent each of the following three events: The physician obtained is (i) an internist, (ii) 45 or over, and (iii) in family practice and under 35.
b. Compute the probability of each event in part (a).
c. Construct a *joint percentage distribution,* a table similar to a joint probability distribution except with percentages instead of probabilities.

Extending the Concepts and Skills

4.62 Explain why the joint events in a contingency table are mutually exclusive.

4.63 In this exercise, you are asked to verify that the sum of the joint probabilities in a row or column of a joint probability distribution equals the marginal probability in that row or column. Consider the following joint probability distribution.

	C_1	$\cdots$	C_n	$P(R_i)$
R_1	$P(R_1 \,\&\, C_1)$	$\cdots$	$P(R_1 \,\&\, C_n)$	$P(R_1)$
.	.	$\cdots$	.	.
.	.	$\cdots$	.	.
.	.	$\cdots$	.	.
R_m	$P(R_m \,\&\, C_1)$	$\cdots$	$P(R_m \,\&\, C_n)$	$P(R_m)$
$P(C_j)$	$P(C_1)$	$\cdots$	$P(C_n)$	1

a. Explain why
$$R_1 = \big((R_1 \,\&\, C_1) \text{ or } \cdots \text{ or } (R_1 \,\&\, C_n)\big).$$

b. Why are the events $(R_1 \,\&\, C_1), \ldots, (R_1 \,\&\, C_n)$ mutually exclusive?

c. Explain why parts (a) and (b) imply that
$$P(R_1) = P(R_1 \,\&\, C_1) + \cdots + P(R_1 \,\&\, C_n).$$

This equation shows that the first row of joint probabilities sums to the marginal probability at the end of that row. A similar argument applies to any other row or column.

4.5   Conditional Probability

In this section, we introduce the concept of **conditional probability**.

DEFINITION 4.6

What
Does it Mean?

A conditional probability of an event is the probability that the event occurs under the assumption that another event occurs.

Conditional Probability

The probability that event B occurs given that event A occurs is called a *conditional probability*. It is denoted $P(B \mid A)$, which is read "the probability of B given A." We call A the *given event*.

Example 4.16 illustrates the calculation of conditional probabilities in the simple experiment of rolling a balanced die once.

▌▌Example 4.16   Conditional Probability

Rolling a Die   When a balanced die is rolled once, six equally likely outcomes are possible, as displayed in Fig. 4.22.

FIGURE 4.22
Sample space for rolling a die once

Let

F = event a 5 is rolled, and

O = event the die comes up odd.

Determine the following probabilities:

a. $P(F)$, the probability that a 5 is rolled.

b. $P(F \mid O)$, the conditional probability that a 5 is rolled, given that the die comes up odd.

c. $P(O \mid (\text{not } F))$, the conditional probability that the die comes up odd, given that a 5 is not rolled.

Solution

a. To obtain $P(F)$, the probability that a 5 is rolled, we proceed as usual. From Fig. 4.22 we see that six outcomes are possible. Also, event F can occur in only one way: if the die comes up 5. Thus the probability that a 5 is rolled is

$$P(F) = \frac{f}{N} = \frac{1}{6} = 0.167.$$

INTERPRETATION There is a 16.7% chance of rolling a 5.

FIGURE 4.23

Event O

b. Given that the die comes up odd, that is, that event O occurs, there are no longer six possible outcomes. There are only three, as Fig. 4.23 shows. Therefore the conditional probability that a 5 is rolled, given that the die comes up odd, is

$$P(F \mid O) = \frac{f}{N} = \frac{1}{3} = 0.333.$$

Comparison of this probability with the one obtained in part (a) shows that $P(F \mid O) \neq P(F)$; that is, the conditional probability that a 5 is rolled, given that the die comes up odd, is not the same as the (unconditional) probability that a 5 is rolled.

INTERPRETATION Given that the die comes up odd, there is a 33.3% chance of rolling a 5, compared with a 16.7% (unconditional) chance of rolling a 5. Knowing that the die comes up odd affects the chance of rolling a 5.

FIGURE 4.24

Event (not F)

c. Given that a 5 is not rolled, that is, that event (not F) occurs, the possible outcomes are the five shown in Fig. 4.24. Under these circumstances, event O (odd) can occur in two ways: if a 1 or a 3 is rolled. So the conditional probability that the die comes up odd, given that a 5 is not rolled, is

$$P\big(O \mid (\text{not } F)\big) = \frac{f}{N} = \frac{2}{5} = 0.4.$$

Compare this probability with the (unconditional) probability that the die comes up odd, which is 0.5. ▬

Conditional probability is often used to analyze bivariate data. In Section 4.4, we discussed contingency tables as a method for tabulating such data. We show in Example 4.17 how to obtain conditional probabilities for bivariate data directly from a contingency table.

Example 4.17 Conditional Probability

ASU Faculty Table 4.8 repeats the contingency table for age and rank of ASU faculty members. Suppose that an ASU faculty member is selected at random.

	Rank				
	Full professor R_1	Associate professor R_2	Assistant professor R_3	Instructor R_4	**Total**
Under 30 A_1	2	3	57	6	68
30–39 A_2	52	170	163	17	402
40–49 A_3	156	125	61	6	348
50–59 A_4	145	68	36	4	253
60 & over A_5	75	15	3	0	93
Total	430	381	320	33	1164

a. Determine the (unconditional) probability that the faculty member selected is in his or her 50s.

b. Determine the (conditional) probability that the faculty member selected is in his or her 50s given that an assistant professor is selected.

Solution

a. We are to determine the probability of event A_4. From Table 4.8, $N = 1164$, the total number of faculty members. Also, because 253 of the faculty members are in their 50s, we have $f = 253$. Therefore

$$P(A_4) = \frac{f}{N} = \frac{253}{1164} = 0.217.$$

INTERPRETATION   21.7% of the faculty are in their 50s.

b. We are to find the probability of event A_4, given that an assistant professor is selected (event R_3); in other words, we want to determine $P(A_4 \mid R_3)$. To obtain that probability, we restrict our attention to the assistant-professor column of Table 4.8. We have $N = 320$, the total number of assistant professors. Also, because 36 of the assistant professors are in their 50s, we have $f = 36$. Consequently,

$$P(A_4 \mid R_3) = \frac{f}{N} = \frac{36}{320} = 0.113.$$

INTERPRETATION   11.3% of the assistant professors are in their 50s.

The Conditional Probability Rule

In the previous two examples, we computed conditional probabilities *directly*, meaning that we first obtained the new sample space determined by the given event and then, using the new sample space, we calculated probabilities in the usual manner. Sometimes we cannot determine conditional probabilities directly but must instead compute them in terms of unconditional probabilities. Returning to the situation presented in Example 4.17, we obtain a formula for computing conditional probabilities in terms of unconditional probabilities in Example 4.18.

| **Example 4.18** | **Introducing the Conditional Probability Rule** |

ASU Faculty   In Example 4.17(b), we used a direct computation to determine the conditional probability that a faculty member is in his or her 50s (event A_4), given that an assistant professor is selected (event R_3). To do that, we restricted our attention to the R_3 column of Table 4.8 and obtained

$$P(A_4 \mid R_3) = \frac{36}{320} = 0.113.$$

Express the conditional probability $P(A_4 \mid R_3)$ in terms of unconditional probabilities.

Solution   First, we note that the number 36 in the numerator of the preceding fraction is the number of assistant professors in their 50s, that is, the number of ways event (R_3 & A_4) can occur. Next, we observe that the number 320 in the denominator of the preceding fraction is the total number of assistant professors, that is, the number of ways event R_3 can occur. Thus the numbers 36 and 320 are those used to compute the unconditional probabilities of events (R_3 & A_4) and R_3, respectively:

$$P(R_3 \text{ \& } A_4) = \frac{36}{1164} = 0.031 \quad \text{and} \quad P(R_3) = \frac{320}{1164} = 0.275.$$

From the previous three probabilities,

$$P(A_4 \mid R_3) = \frac{36}{320} = \frac{36/1164}{320/1164} = \frac{P(R_3 \text{ \& } A_4)}{P(R_3)}.$$

In other words, we can express the conditional probability $P(A_4 \mid R_3)$ in terms of the unconditional probabilities $P(R_3 \text{ \& } A_4)$ and $P(R_3)$ by using the formula

$$P(A_4 \mid R_3) = \frac{P(R_3 \text{ \& } A_4)}{P(R_3)}.$$

The general form of this formula is called the **conditional probability rule,** and we express it as the following formula.

FORMULA 4.4

The Conditional Probability Rule

If A and B are any two events with $P(A) > 0$, then

$$P(B \mid A) = \frac{P(A \, \& \, B)}{P(A)}.$$

What
Does it Mean?

The conditional probability of one event given another equals the probability that both events occur divided by the probability of the given event.

For the faculty member example, conditional probabilities can be obtained either directly or by applying the conditional probability rule. However, as Example 4.19 illustrates, the conditional probability rule is sometimes the only way that conditional probabilities can be determined.

Example 4.19 **The Conditional Probability Rule**

Marital Status and Gender Data on the marital status of U.S. adults can be found in the *Current Population Reports*, a publication of the U.S. Bureau of the Census. Table 4.9 provides a joint probability distribution for the marital status of U.S. adults by gender. We used "Single" to mean "Never married."

TABLE 4.9

Joint probability distribution of marital status and gender

		Marital status				
		Single M_1	Married M_2	Widowed M_3	Divorced M_4	$P(S_i)$
Gender	Male S_1	0.129	0.297	0.014	0.042	0.482
	Female S_2	0.106	0.300	0.056	0.056	0.518
	$P(M_j)$	0.235	0.597	0.070	0.098	1.000

A U.S. adult is selected at random.

a. Determine the probability that the adult selected is divorced, given that the adult selected is a male.

b. Determine the probability that the adult selected is a male, given that the adult selected is divorced.

Solution Unlike our previous work with contingency tables, we do not have frequency data here; rather, we have only probability (relative-frequency) data. Hence we cannot compute conditional probabilities directly; we must instead use the conditional probability rule.

a. We want $P(M_4 \mid S_1)$. Using the conditional probability rule and Table 4.9, we get

$$P(M_4 \mid S_1) = \frac{P(S_1 \, \& \, M_4)}{P(S_1)} = \frac{0.042}{0.482} = 0.087.$$

INTERPRETATION In the United States, 8.7% of adult males are divorced.

b. We want $P(S_1 \mid M_4)$. Using the conditional probability rule and Table 4.9, we get

$$P(S_1 \mid M_4) = \frac{P(M_4 \text{ \& } S_1)}{P(M_4)} = \frac{0.042}{0.098} = 0.429.$$

INTERPRETATION In the United States, 42.9% of divorced adults are males.

EXERCISES 4.5

Statistical Concepts and Skills

4.64 Give an example of the conditional probability of an event being the same as the unconditional probability of the event. (*Hint:* Consider the experiment of tossing a coin twice.)

4.65 Playing Cards. One card is selected at random from an ordinary deck of 52 playing cards. Let

A = event a face card is selected,

B = event a king is selected, and

C = event a heart is selected.

Find the following probabilities and express your results in words. Compute the conditional probabilities directly; do not use the conditional probability rule.
a. $P(B)$ **b.** $P(B \mid A)$
c. $P(B \mid C)$ **d.** $P(B \mid (\text{not } A))$
e. $P(A)$ **f.** $P(A \mid B)$
g. $P(A \mid C)$ **h.** $P(A \mid (\text{not } B))$

4.66 Housing Units. The U.S. Bureau of the Census publishes data on housing units in *American Housing Survey in the United States*. The following table provides a frequency distribution for the number of rooms in U.S. housing units. The frequencies are in thousands.

Rooms	No. of units
1	471
2	1,470
3	11,715
4	23,468
5	24,476
6	21,327
7	13,782
8+	15,647

Compute the following conditional probabilities directly; that is, do not use the conditional probability rule. For a U.S. housing unit selected at random, determine
a. the probability that the unit has exactly four rooms.
b. the conditional probability that the unit has exactly four rooms, given that it has at least two rooms.
c. the conditional probability that the unit has at most four rooms, given that it has at least two rooms.
d. Interpret your answers in parts (a)–(c) in terms of percentages.

4.67 New England Patriots. The National Football League updates team rosters and posts them on its Web site at www.nfl.com. The following contingency table provides a cross-classification of players on the New England Patriots roster as of July 15, 2003, by weight and years of experience.

		Years of experience				
		Rookie Y_1	1–5 Y_2	6–10 Y_3	10+ Y_4	**Total**
Weight (lb)	Under 200 W_1	4	6	5	2	17
	200–300 W_2	15	20	17	3	55
	Over 300 W_3	6	11	1	1	19
	Total	25	37	23	6	91

Compute the following conditional probabilities directly; that is, do not use the conditional probability rule. A player on the New England Patriots is selected at random. Find the probability that the player selected
a. is a rookie.   **b.** weighs over 300 pounds.
c. is a rookie, given that he weighs over 300 pounds.
d. weighs over 300 pounds, given that he is a rookie.
e. Interpret your answers in parts (a)–(d) in terms of percentages.

4.68 Shark Attacks. The *International Shark Attack File*, maintained by the American Elasmobranch Society and the Florida Museum of Natural History, is a compilation of all known shark attacks around the globe from the mid 1500s to the present. Following is a contingency table providing a cross-classification of worldwide reported shark attacks during the 1990s, by country and lethality of attack.

		Lethality		
		Fatal L_1	Nonfatal L_2	**Total**
Country	Australia C_1	9	56	65
	Brazil C_2	12	21	33
	South Africa C_3	8	57	65
	United States C_4	5	244	249
	Other C_5	36	92	128
	Total	70	470	540

a. Find $P(C_2)$. **b.** Find $P(C_2 \,\&\, L_1)$.
c. Obtain $P(L_1 \,|\, C_2)$ directly from the table.
d. Obtain $P(L_1 \,|\, C_2)$ by using the conditional probability rule and your answers from parts (a) and (b).
e. State your results in parts (a)–(c) in words.

4.69 Living Arrangements. As reported in the U.S. Census Bureau's *Current Population Reports*, the living arrangements by age of U.S. citizens 15 years of age and older are as shown in the following joint probability distribution.

		Living arrangement			
		Alone L_1	With spouse L_2	With others L_3	$P(A_i)$
Age (yrs)	15–24 A_1	0.006	0.017	0.154	0.177
	25–44 A_2	0.038	0.241	0.122	0.401
	45–64 A_3	0.035	0.187	0.047	0.269
	Over 64 A_4	0.047	0.084	0.022	0.153
	$P(L_j)$	0.126	0.529	0.345	1.000

A U.S. citizen 15 years of age or older is selected at random. Determine the probability that the person selected
a. lives with spouse.
b. is over 64.
c. lives with spouse and is over 64.
d. lives with spouse, given that the person is over 64.
e. is over 64, given that the person lives with spouse.
f. Interpret your answers in parts (a)–(e) in terms of percentages.

4.70 Property Crime. As reported by the Federal Bureau of Investigation in *Crime in the United States*, 4.9% of property crimes are committed in rural areas and 1.9% of property crimes are burglaries committed in rural areas. What percentage of property crimes committed in rural areas are burglaries?

4.71 Give an example of an experiment in which conditional probabilities
a. can be computed both directly and by using the conditional probability rule.
b. cannot be computed directly but only by using the conditional probability rule.

Extending the Concepts and Skills

4.72 New England Patriots. Refer to Exercise 4.67.
a. Construct a joint probability distribution.
b. Determine the probability distribution of weight for rookies; that is, construct a table showing the conditional probabilities that a rookie weighs under 200 pounds, between 200 and 300 pounds, and over 300 pounds.
c. Determine the probability distribution of years of experience for players who weigh over 300 pounds.
d. The probability distributions in parts (b) and (c) are examples of a *conditional probability distribution*. Determine two other conditional probability distributions for the data on weight and years of experience for the New England Patriots.

Correlation of Events.   An important application of conditional probability is to the concept of the *correlation of events*.   Event B is said to be *positively correlated* with event A if $P(B \mid A) > P(B)$; *negatively correlated* with event A if $P(B \mid A) < P(B)$; and *independent* of event A if $P(B \mid A) = P(B)$. You are asked to examine correlation of events in Exercises 4.73 and 4.74.

4.73 Let A and B be events, each with positive probability.
a. State in words what it means for event B to be positively correlated with event A; negatively correlated with event A; independent of event A.
b. Show that event B is positively correlated with event A if and only if event A is positively correlated with event B.
c. Show that event B is negatively correlated with event A if and only if event A is negatively correlated with event B.
d. Show that event B is independent of event A if and only if event A is independent of event B.

4.74 Drugs and Car Accidents. Suppose that it has been determined that "one-fourth of drivers at fault in a car accident use a certain drug."
a. Explain in words what it means to say that being the driver at fault in a car accident is positively correlated with use of the drug.
b. Under what condition on the percentage of drivers involved in car accidents who use the drug does the statement in quotes imply that being the driver at fault in a car accident is positively correlated with use of the drug? negatively correlated with use of the drug? independent of use of the drug? Explain your answers.
c. Suppose that, in fact, being the driver at fault in a car accident is positively correlated with use of the drug. Can you deduce that a cause-and-effect relationship exists between use of the drug and being the driver at fault in a car accident? Explain your answer.

4.6   The Multiplication Rule; Independence

The conditional probability rule is used to compute conditional probabilities in terms of unconditional probabilities. That is,

$$P(B \mid A) = \frac{P(A \ \& \ B)}{P(A)}.$$

Multiplying both sides of this equation by $P(A)$, we obtain a formula for computing joint probabilities in terms of marginal and conditional probabilities. It is called the **general multiplication rule,** and we express it as the following formula.

FORMULA 4.5

The General Multiplication Rule

If A and B are any two events, then

$$P(A \ \& \ B) = P(A) \cdot P(B \mid A).$$

What
Does it Mean?

For any two events, the probability that both occur equals the probability that a specified one occurs times the conditional probability of the other event, given the specified event.

The conditional probability rule and the general multiplication rule are simply variations of each other. On the one hand, when the joint and marginal probabilities are known or can be easily determined directly, we use the conditional probability rule to obtain conditional probabilities. On the other hand, when the marginal and conditional probabilities are known or can be easily determined directly, we use the general multiplication rule to obtain joint probabilities, as illustrated in Example 4.20.

Example 4.20   The General Multiplication Rule

U.S. Congress   The U.S. Congress's Joint Committee on Printing provides information on the composition of the Congress in the *Congressional Directory*.

For the 108th Congress, 18.7% of the members are senators and 48% of the senators are Democrats. What is the probability that a randomly selected member of the 108th Congress is a Democratic senator?

Solution Let

$$D = \text{event the member selected is a Democrat, and}$$
$$S = \text{event the member selected is a senator.}$$

The event that the member selected is a Democratic senator can be expressed as $(S \& D)$. We want to determine the probability of that event.

Because 18.7% of members are senators, $P(S) = 0.187$, and because 48% of senators are Democrats, $P(D \mid S) = 0.480$. Applying the general multiplication rule, we get

$$P(S \& D) = P(S) \cdot P(D \mid S) = 0.187 \cdot 0.480 = 0.090.$$

The probability that a randomly selected member of the 108th Congress is a Democratic senator is 0.090.

INTERPRETATION 9.0% of members of the 108th Congress are Democratic senators.

Another application of the general multiplication rule relates to sampling two or more members from a population. Example 4.21 provides an illustration.

Example 4.21 The General Multiplication Rule

Gender of Students In Professor Weiss's introductory statistics class, the number of males and females are as shown in the frequency distribution presented in Table 4.10. Two students are selected at random from the class. The first student selected is not returned to the class for possible reselection; that is, the sampling is without replacement. Find the probability that the first student selected is female and the second is male.

TABLE 4.10
Frequency distribution of males and females in Professor Weiss's introductory statistics class

Gender	Frequency
Male	17
Female	23
	40

Solution Let

$$F1 = \text{event the first student obtained is female, and}$$
$$M2 = \text{event the second student obtained is male.}$$

We want to determine $P(F1 \& M2)$. Using the general multiplication rule, we write

$$P(F1 \& M2) = P(F1) \cdot P(M2 \mid F1).$$

Computing the two probabilities on the right side of this equation is easy. To find $P(F1)$—the probability that the first student selected is female—we note from Table 4.10 that 23 of the 40 students are female, so

$$P(F1) = \frac{f}{N} = \frac{23}{40}.$$

Next, we find $P(M2 \mid F1)$—the conditional probability that the second student selected is male, given that the first one selected is female. Given that the first student selected is female, of the 39 students remaining in the class 17 are male, so

$$P(M2 \mid F1) = \frac{f}{N} = \frac{17}{39}.$$

Applying the general multiplication rule, we conclude that

$$P(F1 \ \& \ M2) = P(F1) \cdot P(M2 \mid F1) = \frac{23}{40} \cdot \frac{17}{39} = 0.251.$$

INTERPRETATION When two students are randomly selected from the class, the probability is 0.251 that the first student selected is female and the second is male.

You will find that drawing a **tree diagram** is often helpful when you are applying the general multiplication rule. An appropriate tree diagram for Example 4.21 is shown in Fig. 4.25.

FIGURE 4.25

Tree diagram for student selection problem

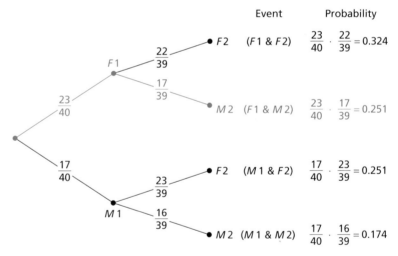

Each branch of the tree corresponds to one possibility for selecting two students at random from the class. For instance, the second branch of the tree, shown in color, corresponds to event ($F1 \ \& \ M2$)—the event that the first student selected is female (event $F1$) and the second is male (event $M2$).

Starting from the left on that branch, the number $\frac{23}{40}$ is the probability that the first student selected is female, $P(F1)$; the number $\frac{17}{39}$ is the conditional probability that the second student selected is male, given that the first student selected is female, $P(M2 \mid F1)$. The product of those two probabilities is, by the general multiplication rule, the probability that the first student selected is female and the second is male, $P(F1 \ \& \ M2)$. The second entry in the Probability column of Fig. 4.25 shows that this probability is 0.251, as we discovered at the end of Example 4.21.

There are also general multiplication rules for more than two events. For instance, the general multiplication rule for three events is

$$P(A \text{ \& } B \text{ \& } C) = P(A) \cdot P(B \mid A) \cdot P(C \mid (A \text{ \& } B)).$$

Try to write down the general multiplication rule for four events.

Independence

One of the most important concepts in probability is that of **statistical independence** of events. For two events, statistical independence or, more simply, **independence,** is defined as follows.

DEFINITION 4.7

Independent Events

Event B is said to be *independent* of event A if $P(B \mid A) = P(B)$.

What *Does it Mean?*

One event is independent of another event if knowing whether (or not) the latter event occurs does not affect the probability of the former event.

In Example 4.22, we show how to determine whether one event is independent of another event.

Example 4.22 **Independent Events**

Playing Cards   Consider again the experiment of randomly selecting one card from a deck of 52 playing cards. Let

$$F = \text{event a face card is selected,}$$
$$K = \text{event a king is selected, and}$$
$$H = \text{event a heart is selected.}$$

a. Determine whether event K is independent of event F.

b. Determine whether event K is independent of event H.

Solution   First we note that the unconditional probability that event K occurs is

$$P(K) = \frac{f}{N} = \frac{4}{52} = \frac{1}{13} = 0.077.$$

a. To determine whether event K is independent of event F, we must compute $P(K \mid F)$ and compare it to $P(K)$. If those two probabilities are equal, event K is independent of event F; otherwise, event K is not independent of event F. Now, given that event F occurs, 12 outcomes are possible (four jacks, four queens, and four kings), and event K can occur in four ways out of those 12 possibilities. Hence

$$P(K \mid F) = \frac{f}{N} = \frac{4}{12} = 0.333,$$

which does not equal $P(K)$; event K is not independent of event F.

INTERPRETATION We note that the lack of independence of event K (king) and event F (face card) stems from the fact that the percentage of kings among the face cards (33.3%) is not the same as the percentage of kings among all the cards (7.7%).

b. We need to compute $P(K \mid H)$ and compare it to $P(K)$. Given that event H occurs, 13 outcomes are possible (the 13 hearts), and event K can occur in one way out of those 13 possibilities. Therefore

$$P(K \mid H) = \frac{f}{N} = \frac{1}{13} = 0.077,$$

which equals $P(K)$; event K is independent of event H.

INTERPRETATION We note that the independence of event K (king) and event H (heart) stems from the fact that the percentage of kings among the hearts is the same as the percentage of kings among all the cards; namely, 7.7%.

If event B is independent of event A, then event A is also independent of event B. In such cases, we often say that event A and event B are independent, or that A and B are **independent events.** If two events are not independent, we say that they are **dependent events.** In Example 4.22, F and K are dependent events, whereas K and H are independent events.

The Special Multiplication Rule

Recall that the general multiplication rule states that, for any two events A and B,

$$P(A \text{ \& } B) = P(A) \cdot P(B \mid A).$$

If A and B are independent events, $P(B \mid A) = P(B)$. Thus, for the special case of independent events, we can replace the term $P(B \mid A)$ in the general multiplication rule by the term $P(B)$. Doing so yields the **special multiplication rule,** which we express as the following formula.

FORMULA 4.6

The Special Multiplication Rule (for Two Independent Events)

If A and B are independent events, then

$$P(A \text{ \& } B) = P(A) \cdot P(B),$$

and conversely, if $P(A \text{ \& } B) = P(A) \cdot P(B)$, then A and B are independent events.

What
Does it Mean?

Two events are independent if and only if the probability that both occur equals the product of their individual probabilities.

As in Example 4.22, we can use the definition of independence to decide whether two specified events are independent: If the two events are, say, A and B, we determine whether $P(B \mid A) = P(B)$. Alternatively, we can decide whether event A and event B are independent by using the special multiplication rule, that is, by determining whether $P(A \text{ \& } B) = P(A) \cdot P(B)$.

The definition of independence for three or more events is more complicated than that for two events. (See Exercise 4.87.) Nevertheless, the special multiplication rule still holds, as expressed in the following formula.

FORMULA 4.7

The Special Multiplication Rule

If events $A, B, C, \ldots$ are independent, then

$$P(A \ \& \ B \ \& \ C \ \& \ \cdots) = P(A) \cdot P(B) \cdot P(C) \cdots .$$

What *Does it Mean?*

For independent events, the probability that they all occur equals the product of their individual probabilities.

We can use the special multiplication rule to compute joint probabilities when we know or can reasonably assume that two or more events are independent. Example 4.23 illustrates this point.

Example 4.23 **The Special Multiplication Rule**

Roulette An American roulette wheel contains 38 numbers, of which 18 are red, 18 are black, and 2 are green. When the roulette wheel is spun, the ball is equally likely to land on any of the 38 numbers. In three plays at a roulette wheel, what is the probability that the ball will land on green the first time and on black the second and third times?

Solution First, we can reasonably assume that outcomes on successive plays at the wheel are independent. Now, we let

$G1$ = event the ball lands on green the first time,

$B2$ = event the ball lands on black the second time, and

$B3$ = event the ball lands on black the third time.

We want to determine $P(G1 \ \& \ B2 \ \& \ B3)$.

Because outcomes on successive plays at the wheel are independent, we know that event $G1$, event $B2$, and event $B3$ are independent. Applying the special multiplication rule, we conclude that

$$P(G1 \ \& \ B2 \ \& \ B3) = P(G1) \cdot P(B2) \cdot P(B3) = \frac{2}{38} \cdot \frac{18}{38} \cdot \frac{18}{38} = 0.012.$$

INTERPRETATION In three plays at a roulette wheel, there is a 1.2% chance that the ball will land on green the first time and on black the second and third times.

Mutually Exclusive Versus Independent Events

The terms *mutually exclusive* and *independent* refer to different concepts. Mutually exclusive events are those that cannot occur simultaneously; independent events are those for which the occurrence of some does not affect the probabilities of the others occurring.

In fact, if two or more events are mutually exclusive, the occurrence of one precludes the occurrence of the others. Two or more (nonimpossible) events cannot be both mutually exclusive and independent.

EXERCISES 4.6

Statistical Concepts and Skills

4.75 Internet Isolation. An article in *Science News* (Vol. 157, p. 135) reported on research concerning the effects of regular Internet usage. According to the article, 36% of Americans with Internet access are regular Internet users, meaning that they log on for at least 5 hours per week. Among regular Internet users, 25% say that the Web has reduced their social contact (e.g., talking with family and friends and going out on the town). Determine the probability that a randomly selected American with Internet access is a regular Internet user who feels that the Web has reduced his or her social contact. Interpret your answer in terms of percentages.

4.76 ESP Experiment. A person has agreed to participate in an ESP experiment. He is asked to randomly pick two numbers between 1 and 6. The second number must be different from the first. Let

H = event the first number picked is a 3, and

K = event the second number picked exceeds 4.

Determine

a. $P(H)$.   **b.** $P(K \mid H)$.   **c.** $P(H \,\&\, K)$.

Find the probability that both numbers picked are

d. less than 3.   **e.** greater than 3.

4.77 U.S. Governors. According to the National Governors Association, the political-party distribution of U.S. governors, as of 2003, is as follows.

Party	Frequency
Democratic	24
Republican	26

Two U.S. governors are selected at random without replacement.
a. Find the probability that the first is a Republican and the second a Democrat.
b. Find the probability that both are Republicans.
c. Draw a tree diagram for this problem similar to the one shown in Fig. 4.25 on page 189.
d. What is the probability that the two governors selected have the same political-party affiliation?
e. What is the probability that the two governors selected have different political-party affiliations?

4.78 Injured Americans. The U.S. National Center for Health Statistics compiles data on injuries and publishes the information in *Vital and Health Statistics*. A contingency table for injuries in the United States, by circumstance and gender, is as follows. Frequencies are in millions.

		Circumstance			
		Work C_1	Home C_2	Other C_3	Total
Gender	Male S_1	8.0	9.8	17.8	35.6
	Female S_2	1.3	11.6	12.9	25.8
	Total	9.3	21.4	30.7	61.4

a. Find $P(C_1)$.
b. Find $P(C_1 \mid S_2)$.
c. Are events C_1 and S_2 independent? Explain your answer.
d. Is the event that an injured person is male independent of the event that an injured person was hurt at home? Explain your answer.

4.79 Dice. When two balanced dice are rolled, 36 equally likely outcomes are possible, as depicted in Fig. 4.1 on page 154. Let

A = event the colored die comes up even,

B = event the black die comes up odd,

C = event the sum of the dice is 10, and

D = event the sum of the dice is even.

a. Compute $P(A)$, $P(B)$, $P(C)$, and $P(D)$.
b. Compute $P(B \mid A)$.
c. Are events A and B independent? Why or why not?
d. Compute $P(C \mid A)$.
e. Are events A and C independent? Why or why not?
f. Compute $P(D \mid A)$.
g. Are events A and D independent? Why or why not?

4.80 Living Arrangements. Refer to the joint probability distribution in Exercise 4.69 on page 186.
a. Determine $P(A_2)$, $P(L_1)$, and $P(A_2 \,\&\, L_1)$.
b. Use the special multiplication rule to determine whether events A_2 and L_1 are independent. Interpret your result.

4.81 Drawing Cards. Two cards are drawn at random from an ordinary deck of 52 cards. Determine the probability that both cards are aces if
a. the first card is replaced before the second card is drawn.
b. the first card is not replaced before the second card is drawn.

4.82 Yahtzee. In the game of Yahtzee, five balanced dice are rolled.
a. What is the probability of rolling all 2s?
b. What is the probability that all the dice come up the same number?

4.83 The Challenger Disaster. In a letter to the editor that appeared in the February 23, 1987, issue of *U.S. News and World Report*, a reader discussed the issue of space shuttle safety. Each "criticality 1" item must have 99.99% reliability, according to NASA standards, meaning that the probability of failure for such an item is 0.0001. Mission 25, the mission in which the Challenger exploded, had 748 "criticality 1" items. Determine the probability that
a. none of the "criticality 1" items would fail.
b. at least one "criticality 1" item would fail.
c. Interpret your answer in part (b) in words.

4.84 An Aging World. The growth of the elderly population in the world was studied in a joint effort by the U.S. Department of Commerce, the Economics and Statistics Administration, and the U.S. Census Bureau in *An Aging World: 2001*. The following table gives the percentages of elderly in three age groups for North America and Asia in the year 2000.

	65–74	75–79	80 or older
North America	6.6%	2.7%	3.3%
Asia	4.2%	0.8%	0.8%

Determine the following probabilities and express your answers to three significant digits (three digits following the last leading zero).
a. If three people are chosen at random from North America, what is the probability that all three are 80 years old or older?
b. If three people are chosen at random from Asia, what is the probability that all three are 80 years old or older?
c. If three people are chosen at random from North America, what is the probability that the first person is 65–74 years old, the second person is 75–79 years old, and the third person is 80 years old or older?

d. In doing your calculations in parts (a)–(c), are you assuming sampling with replacement or without replacement? Does it make much difference which of those two types of sampling is used? Explain your answers.

4.85 Activity Limitations. The National Center for Health Statistics compiles information on activity limitations. Results are published in *Vital and Health Statistics*. The data show that 13.6% of males and 14.4% of females have an activity limitation. Are gender and activity limitation statistically independent? Explain your answer.

Extending the Concepts and Skills

4.86 In this exercise, we examine further the concepts of independent events and mutually exclusive events.
a. If two events are mutually exclusive, determine their joint probability.
b. If two nonimpossible events are independent, explain why their joint probability is not 0.
c. Give an example of two events that are neither mutually exclusive nor independent.

4.87 Independence Extended. Three events, A, B, and C, are said to be independent if

$$P(A \& B) = P(A) \cdot P(B),$$
$$P(A \& C) = P(A) \cdot P(C),$$
$$P(B \& C) = P(B) \cdot P(C), \text{ and}$$
$$P(A \& B \& C) = P(A) \cdot P(B) \cdot P(C).$$

What is required for four events to be independent? Explain your definition in words.

4.88 Dice. When two balanced dice are rolled, 36 equally likely outcomes are possible, as illustrated in Fig. 4.1 on page 154. Let

A = event the colored die comes up even,
B = event the black die comes up even,
C = event the sum of the dice is even,
D = event the colored die comes up 1, 2, or 3,
E = event the colored die comes up 3, 4, or 5, and
F = event the sum of the dice is 5.

Apply the definition of independence for three events stated in Exercise 4.87 to solve each problem.
a. Are A, B, and C independent events?
b. Show that $P(D \& E \& F) = P(D) \cdot P(E) \cdot P(F)$ but that D, E, and F are not independent events.

4.89 Coin Tossing. When a balanced coin is tossed four times, 16 equally likely outcomes are possible, as shown in the following table.

HHHH	THHH	THHT	THTT
HHHT	HHTT	THTH	TTHT
HHTH	HTHT	TTHH	TTTH
HTHH	HTTH	HTTT	TTTT

Let

A = event the first toss is heads,

B = event the second toss is tails, and

C = event the last two tosses are heads.

Apply the definition of independence for three events stated in Exercise 4.87 to show that A, B, and C are independent events.

4.7 Bayes's Rule

In this section, we discuss a rule of probability developed by Thomas Bayes, an eighteenth-century clergyman. This rule is aptly called *Bayes's rule.*

One of the primary uses of Bayes's rule is to revise probabilities in accordance with newly acquired information. Such revised probabilities are actually conditional probabilities, and so, in some sense, we have already examined much of the material in this section. However, as you will see, application of Bayes's rule involves some new concepts and the use of some new techniques.

The Rule of Total Probability

In preparation for discussion of Bayes's rule, we need to study another rule of probability called the *rule of total probability.* First, we consider the concept of *exhaustive events.* Events $A_1, A_2, \ldots, A_k$ are said to be **exhaustive events** if one or more of them must occur.

For instance, the National Governors' Association classifies governors as Democrat, Republican, or Independent. Suppose that a governor is selected at random; let E_1, E_2, and E_3 denote the events that the governor selected is a Democrat, Republican, and Independent, respectively. Then events E_1, E_2, and E_3 are exhaustive because at least one of them must occur when a governor is selected—the governor selected must be a Democrat, Republican, or Independent.

The events E_1, E_2, and E_3 are not only exhaustive, but they are also mutually exclusive; a governor cannot have more than one political party affiliation at the same time. In general, if events are both exhaustive and mutually exclusive, exactly one of them must occur. This statement is true because at least one of the events must occur (the events are exhaustive) and at most one of the events can occur (the events are mutually exclusive).

An event and its complement are always mutually exclusive and exhaustive. Figure 4.26(a) on the next page portrays three events, A_1, A_2, and A_3, that are both mutually exclusive and exhaustive. Note that the three events do not overlap, indicating that they are mutually exclusive; furthermore, they fill out the entire region enclosed by the heavy rectangle (i.e., the sample space), indicating that they are exhaustive.

Now consider, say, three mutually exclusive and exhaustive events, A_1, A_2, and A_3, and any event B, as portrayed in Fig. 4.26(b). Note that event B com-

FIGURE 4.26
(a) Three mutually exclusive and exhaustive events; (b) an event B and three mutually exclusive and exhaustive events

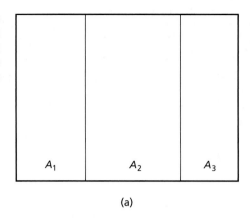

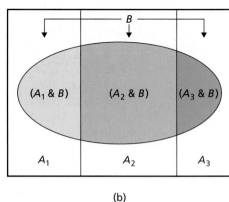

(a)

(b)

prises the mutually exclusive events $(A_1 \,\&\, B)$, $(A_2 \,\&\, B)$, and $(A_3 \,\&\, B)$, which are shown in color. This condition means that event B must occur in conjunction with exactly one of the events, A_1, A_2, or A_3.

If you think of the colored regions in Fig. 4.26(b) as probabilities, the total colored region is $P(B)$ and the three colored subregions are, from left to right, $P(A_1 \,\&\, B)$, $P(A_2 \,\&\, B)$, and $P(A_3 \,\&\, B)$. Because events $(A_1 \,\&\, B)$, $(A_2 \,\&\, B)$, and $(A_3 \,\&\, B)$ are mutually exclusive, the total colored region equals the sum of the three colored subregions; in other words,

$$P(B) = P(A_1 \,\&\, B) + P(A_2 \,\&\, B) + P(A_3 \,\&\, B).$$

Applying the general multiplication rule (Formula 4.5 on page 187) to each term on the right side of this equation, we obtain

$$P(B) = P(A_1) \cdot P(B \,|\, A_1) + P(A_2) \cdot P(B \,|\, A_2) + P(A_3) \cdot P(B \,|\, A_3).$$

This formula holds in general and is called the **rule of total probability,** which we express as Formula 4.8. It is also referred to as the **stratified sampling theorem** because of its importance in stratified sampling.

FORMULA 4.8

What *Does it Mean?*

Let $A_1, A_2, \ldots, A_k$ be mutually exclusive and exhaustive events. Then the probability of an event B can be obtained by multiplying the probability of each A_j by the conditional probability of B given A_j and then summing those products.

The Rule of Total Probability

Suppose that events $A_1, A_2, \ldots, A_k$ are mutually exclusive and exhaustive; that is, exactly one of the events must occur. Then for any event B,

$$P(B) = \sum_{j=1}^{k} P(A_j) \cdot P(B \,|\, A_j).$$

We apply the rule of total probability in Example 4.24.

Example 4.24 **The Rule of Total Probability**

U.S. Demographics The U.S. Bureau of the Census collects data on the resident population, by age and region of residence, and presents its findings in the *Current Population Reports*. The first two columns of Table 4.11 contain a

TABLE 4.11

Percentage distribution for region of residence and percentage of seniors in each region

Region	Percentage of U.S. population	Percentage seniors
Northeast	19.0	13.8
Midwest	23.1	13.0
South	35.5	12.8
West	22.4	11.1
	100.0	

percentage distribution for region of residence; the third column displays the percentage of seniors (age 65 or over) in each region. For instance, 19.0% of U.S. residents live in the Northeast, and 13.8% of those who live in the Northeast are seniors. Use Table 4.11 to determine the percentage of U.S. residents that are seniors.

Solution   To solve this problem, we first translate the information displayed in Table 4.11 into the language of probability. Suppose that a U.S. resident is selected at random. Let

$$S = \text{event the resident selected is a senior,}$$

and

$$R_1 = \text{event the resident selected lives in the Northeast,}$$
$$R_2 = \text{event the resident selected lives in the Midwest,}$$
$$R_3 = \text{event the resident selected lives in the South, and}$$
$$R_4 = \text{event the resident selected lives in the West.}$$

The percentages shown in the second and third columns of Table 4.11 translate into the probabilities displayed in Table 4.12.

TABLE 4.12

Probabilities derived from Table 4.11

$P(R_1) = 0.190$	$P(S \mid R_1) = 0.138$
$P(R_2) = 0.231$	$P(S \mid R_2) = 0.130$
$P(R_3) = 0.355$	$P(S \mid R_3) = 0.128$
$P(R_4) = 0.224$	$P(S \mid R_4) = 0.111$

The problem is to determine the percentage of U.S. residents that are seniors, or, in terms of probability, $P(S)$. Because a U.S. resident can reside in only one of the four regions at one time, events $R_1, R_2, R_3,$ and R_4 are mutually exclusive and exhaustive. Therefore, by the rule of total probability applied to the event S, and from Table 4.12 we have

$$P(S) = \sum_{j=1}^{4} P(R_j) \cdot P(S \mid R_j)$$
$$= 0.190 \cdot 0.138 + 0.231 \cdot 0.130 + 0.355 \cdot 0.128 + 0.224 \cdot 0.111$$
$$= 0.127.$$

A tree diagram for this calculation is shown in Fig. 4.27, where J represents the event that the resident selected is not a senior. We obtain $P(S)$ from the tree diagram by first multiplying the two probabilities on each branch of the tree that ends with S (the colored branches) and then summing all those products.

In any case, we see that $P(S) = 0.127$; the probability is 0.127 that a randomly selected U.S. resident is a senior.

INTERPRETATION   12.7% of U.S. residents are seniors.

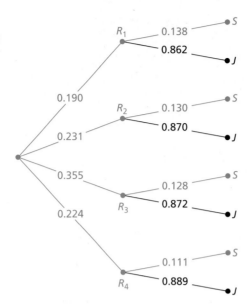

Bayes's Rule

Using the rule of total probability, we can derive Bayes's rule. For simplicity, let's consider three events, A_1, A_2, and A_3, that are mutually exclusive and exhaustive and let B be any event. For Bayes's rule we assume that the probabilities $P(A_1)$, $P(A_2)$, $P(A_3)$, $P(B \mid A_1)$, $P(B \mid A_2)$, and $P(B \mid A_3)$ are known. The problem is to use those six probabilities to determine the conditional probabilities $P(A_1 \mid B)$, $P(A_2 \mid B)$, and $P(A_3 \mid B)$.

We now show how to express $P(A_2 \mid B)$ in terms of the six known probabilities; $P(A_1 \mid B)$ and $P(A_3 \mid B)$ are handled similarly. First, we apply the conditional probability rule (Formula 4.4 on page 184), to write

$$P(A_2 \mid B) = \frac{P(B \ \& \ A_2)}{P(B)} = \frac{P(A_2 \ \& \ B)}{P(B)}. \tag{4.1}$$

Next, to the fraction on the right in Equation (4.1) we apply the general multiplication rule (Formula 4.5 on page 187) to the numerator and the rule of total probability to the denominator, giving

$$P(A_2 \ \& \ B) = P(A_2) \cdot P(B \mid A_2)$$

and

$$P(B) = P(A_1) \cdot P(B \mid A_1) + P(A_2) \cdot P(B \mid A_2) + P(A_3) \cdot P(B \mid A_3).$$

Substituting from the previous two equations into the fraction on the right in Equation (4.1), we obtain

$$P(A_2 \mid B) = \frac{P(A_2) \cdot P(B \mid A_2)}{P(A_1) \cdot P(B \mid A_1) + P(A_2) \cdot P(B \mid A_2) + P(A_3) \cdot P(B \mid A_3)}.$$

This formula holds in general and is called **Bayes's rule,** which we express as the following formula.

FORMULA 4.9   **Bayes's Rule**

Suppose that events $A_1, A_2, \ldots, A_k$ are mutually exclusive and exhaustive. Then for any event B,

$$P(A_i \mid B) = \frac{P(A_i) \cdot P(B \mid A_i)}{\sum_{j=1}^{k} P(A_j) \cdot P(B \mid A_j)},$$

where A_i can be any one of events $A_1, A_2, \ldots, A_k$.

Examples 4.25 and 4.26 are applications of Bayes's rule.

Example 4.25   **Bayes's Rule**

U.S. Demographics   From Table 4.11 on page 197, we know that 13.8% of Northeast residents are seniors. Now we ask: What percentage of seniors are Northeast residents?

Solution   The notation introduced at the beginning of the solution to Example 4.24 indicates that, in terms of probability, the problem is to find $P(R_1 \mid S)$—the probability that a U.S. resident lives in the Northeast, given that the resident is a senior. To obtain that conditional probability, we apply Bayes's rule and the probabilities shown in Table 4.12:

$$P(R_1 \mid S) = \frac{P(R_1) \cdot P(S \mid R_1)}{\sum_{j=1}^{4} P(R_j) \cdot P(S \mid R_j)}$$

$$= \frac{0.190 \cdot 0.138}{0.190 \cdot 0.138 + 0.231 \cdot 0.130 + 0.355 \cdot 0.128 + 0.224 \cdot 0.111}$$

$$= 0.207.$$

INTERPRETATION   20.7% of seniors are Northeast residents.

Example 4.26   **Bayes's Rule**

Smoking and Lung Disease   According to the Arizona Chapter of the American Lung Association, 7.0% of the population has lung disease. Of those having lung disease, 90.0% are smokers; of those not having lung disease, 25.3% are smokers. Determine the probability that a randomly selected smoker has lung disease.

Solution   Suppose that a person is selected at random. Let

$$S = \text{event the person selected is a smoker,}$$

and

$$L_1 = \text{event the person selected has no lung disease, and}$$
$$L_2 = \text{event the person selected has lung disease.}$$

Note that events L_1 and L_2 are complementary, which implies that they are mutually exclusive and exhaustive.

The data given in the statement of the problem indicate that $P(L_2) = 0.070$, $P(S \mid L_2) = 0.900$, and $P(S \mid L_1) = 0.253$. Also, $L_1 = ($not $L_2)$, so we can conclude that $P(L_1) = P$ (not $L_2) = 1 - P(L_2) = 1 - 0.070 = 0.930$. We summarize this information in Table 4.13.

The problem is to determine the probability that a randomly selected smoker has lung disease, $P(L_2 \mid S)$. Applying Bayes's rule to the probability data in Table 4.13, we obtain

$$P(L_2 \mid S) = \frac{P(L_2) \cdot P(S \mid L_2)}{P(L_1) \cdot P(S \mid L_1) + P(L_2) \cdot P(S \mid L_2)}$$

$$= \frac{0.070 \cdot 0.900}{0.930 \cdot 0.253 + 0.070 \cdot 0.900} = 0.211.$$

The probability is 0.211 that a randomly selected smoker has lung disease.

INTERPRETATION 21.1% of smokers have lung disease.

TABLE 4.13
Known probability information

$P(L_1) = 0.930$	$P(S \mid L_1) = 0.253$
$P(L_2) = 0.070$	$P(S \mid L_2) = 0.900$

Example 4.26 demonstrates that the rate of lung disease among smokers (21.1%) is more than three times the rate among the general population (7.0%). Using arguments similar to those in Example 4.26, we can show that the probability is 0.010 that a randomly selected nonsmoker has lung disease; in other words, 1.0% of nonsmokers have lung disease.

Hence the rate of lung disease among smokers (21.1%) is more than 20 times that among nonsmokers (1.0%). But, because this study is observational, we cannot conclude solely on the basis of this information that smoking causes lung disease; we can only infer that a strong positive association exists between smoking and lung disease.

Prior and Posterior Probabilities

Two important terms associated with Bayes's rule are *prior probability* and *posterior probability*. We introduce these terms by referring to Example 4.26.

From the information provided, we know that the probability is 0.070 that a randomly selected person has lung disease: $P(L_2) = 0.070$. This probability does not take into consideration whether the person is a smoker. It is therefore called a **prior probability** because it represents the probability that the person selected has lung disease *before* knowing whether the person is a smoker.

Now suppose that the person selected is found to be a smoker. On the basis of this additional information, we can revise the probability that the person has lung disease. We do so by determining the conditional probability that the person selected has lung disease, given that the person selected is a smoker: $P(L_2 \mid S) = 0.211$ (from Example 4.26). This revised probability is called a **posterior probability** because it represents the probability that the person selected has lung disease *after* we learn that the person is a smoker.

EXERCISES 4.7

Statistical Concepts and Skills

4.90 Explain why an event and its complement are always mutually exclusive and exhaustive.

4.91 Refer to Example 4.24 on page 196. In probability notation, the percentage of Midwest residents can be expressed as $P(R_2)$. Do the same for the percentage of
a. Southern residents.
b. Southern residents who are seniors.
c. seniors who are Southern residents.

4.92 Playing Golf. The National Sporting Goods Association collects and publishes data on participation in selected sports activities. For Americans 7 years old or older, 17.4% of males and 4.5% of females play golf. And, according to the U.S. Census Bureau's *Current Population Reports*, of Americans 7 years old or older, 48.6% are male and 51.4% are female. From among those who are 7 years old or older, one is selected at random. Find the probability that the person selected
a. plays golf.
b. plays golf, given that the person is a male.
c. is a female, given that the person plays golf.
d. Interpret your answers in parts (a)–(c) in terms of percentages.

4.93 Belief in Aliens. According to an Opinion Dynamics Poll published in *USA TODAY*, roughly 54% of U.S. men and 33% of U.S. women believe in aliens. Of U.S. adults, roughly 48% are men and 52% women.
a. What percentage of U.S. adults believe in aliens?
b. What percentage of U.S. women believe in aliens?
c. What percentage of U.S. adults that believe in aliens are women?

4.94 Moviegoers. A survey conducted by TELENATION/Market Facts, Inc., combined with information from the U.S. Census Bureau's *Current Population Reports*, yielded the following table. The first two columns provide a percentage distribution of adults by age group; the third column gives the percentage of people in each age group who go to the movies at least once a month—people whom we refer to as *moviegoers*.

Age	% adults	% moviegoers
18–24	12.7	83
25–34	20.7	54
35–44	22.0	43
45–54	16.5	37
55–64	10.9	27
65 & over	17.2	20

An adult is selected at random.
a. Find the probability that the adult selected is a moviegoer.
b. Find the probability that the adult selected is between 25 and 34 years old, given that he or she is a moviegoer.
c. Interpret your answers in parts (a) and (b) in terms of percentages.

4.95 Textbook Revision. Textbook publishers must estimate the sales of new (first-edition) books. The records of one major publishing company indicate that 10% of all new books sell more than projected, 30% sell close to the number projected, and 60% sell less than projected. Of those that sell more than projected, 70% are revised for a second edition, as are 50% of those that sell close to the number projected, and 20% of those that sell less than projected.
a. What percentage of books published by this publishing company go to a second edition?
b. What percentage of books published by this publishing company that go to a second edition sold less than projected in their first edition?

4.96 Suicide and Gender. The National Center for Health Statistics provides information on suicides by gender and method used. Data are published in *Vital Statistics of the United States*. One year in the United States, there were 31,284 suicides, of which 25,369 were males and 5,915 were females. The following table gives a relative-frequency distribution for method used by males and females who committed suicide.

Method used	Relative freq. for males	Relative freq. for females
Poisoning	0.133	0.364
Hanging/ strangulation	0.154	0.126
Firearms	0.651	0.408
Other	0.062	0.102

A suicide report is selected at random.
a. Find the probability that a firearm was used.
b. Determine the prior probability that the person who committed suicide was a female.
c. Determine the posterior probability that the person who committed suicide was a female, given that a firearm was used.
d. Interpret the probabilities obtained in parts (a)–(c) in terms of percentages.

4.97 Broken Eggs. At a grocery store, eggs come in cartons that hold a dozen eggs. Experience indicates that 78.5% of the cartons have no broken eggs, 19.2% have one broken egg, 2.2% have two broken eggs, 0.1% have three broken eggs, and that the percentage of cartons with four or more broken eggs is negligible. An egg selected at random from a carton is found to be broken. What is the probability that this egg is the only broken one in the carton?

Extending the Concepts and Skills

4.98 Medical Diagnostics. Medical tests are frequently used to decide whether a person has a particular disease. The *sensitivity* of a test is the probability that a person having the disease will test positive; the *specificity* of a test is the probability that a person not having the disease will test negative. A test for a certain disease has been used for many years. Experience with the test indicates that its sensitivity is 0.934 and that its specificity is 0.968. Furthermore, it is known that roughly 1 in 500 people has the disease.
a. Interpret the sensitivity and specificity of this test in terms of percentages.
b. Determine the probability that a person testing positive actually has the disease.

c. Interpret your answer from part (b) in terms of percentages.

4.99 Secrets of Success. Gerald Kushel, Ed.D., was interviewed by *Bottom Line/Personal* on the secrets of successful people. To study success, Kushel questioned 1200 people, among whom were lawyers, artists, teachers, and students. He found that 15% enjoy neither their jobs nor their personal lives, 80% enjoy their jobs but not their personal lives, and 4% enjoy both their jobs and their personal lives.
a. Determine the percentage of the people interviewed who enjoy their jobs.
b. What percentage of the people interviewed who enjoy their jobs also enjoy their personal lives?

4.100 Smoking and Lung Disease. Refer to Example 4.26 on page 199.
a. Determine the probability that a randomly selected non-smoker has lung disease.
b. Use the probability obtained in part (a) and the result of Example 4.26 to compare the rates of lung disease for smokers and nonsmokers.

4.101 Regarding the concepts of exhaustive and mutually exclusive,
a. draw a Venn diagram illustrating three events that are mutually exclusive but not exhaustive.
b. give an example of three events that are mutually exclusive but not exhaustive.
c. draw a Venn diagram illustrating three events that are exhaustive but not mutually exclusive.
d. give an example of three events that are exhaustive but not mutually exclusive.

4.8 Counting Rules

We often need to determine the number of ways something can happen—the number of possible outcomes for an experiment, the number of ways an event can occur, the number of ways a certain task can be performed, and so on. Sometimes, we can list the possibilities and then count them, but in most cases the number of possibilities is so large that doing so is impractical.

Therefore we need to develop techniques that do not rely on a direct listing for determining the number of ways something can happen. Such techniques are usually referred to as **counting rules.** In this section, we examine some widely used counting rules.

The Basic Counting Rule

One counting rule, called the **basic counting rule (BCR),** is fundamental to all the counting techniques we will discuss. We introduce the basic counting rule in Example 4.27.

▌▌Example 4.27 Introducing the Basic Counting Rule

Home Models and Elevations Robson Communities, Inc., builds new-home communities in several parts of Arizona. In Sun Lakes, it offers four models—the Shalimar, Palacia, Valencia, and Monterey—each in three different elevations, designated A, B, and C. How many choices are there for the selection of a home, including both model and elevation?

Solution We first use a tree diagram (see Fig. 4.28) to obtain systematically a direct listing of the possibilities. We use S for Shalimar, P for Palacia, V for Valencia, and M for Monterey.

FIGURE 4.28

Tree diagram for model and elevation possibilities

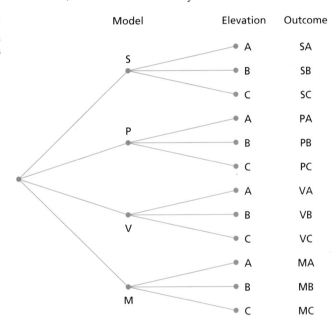

Each branch of the tree corresponds to one possibility for model and elevation. For instance, the first branch of the tree, ending in SA, corresponds to the Shalimar model with the A elevation. The total number of possibilities can be obtained by counting the number of branches at the end of the tree. Hence there are 12 choices for the selection of a home, including both model and elevation.

Although the tree-diagram approach for determining the number of possibilities is a direct listing, it provides a clue for obtaining the number of possibilities without resorting to a direct listing. Specifically, there are four possibilities for model, indicated by the four branches emanating from the starting point of the tree; corresponding to each possibility for model are three possibilities for elevation, indicated by the three subbranches emanating from the end of each model branch. Consequently, there are

$$\underbrace{3 + 3 + 3 + 3}_{4 \text{ times}} = 4 \cdot 3 = 12$$

possibilities altogether. Thus we can obtain the total number of possibilities by multiplying the number of possibilities for the model by the number of possibilities for the elevation.

The same multiplication principle applies regardless of the number of actions. We state this principle more precisely in the following key fact.

KEY FACT 4.2

The Basic Counting Rule (BCR)[†]

Suppose that r actions are to be performed in a definite order. Further suppose that there are m_1 possibilities for the first action and that corresponding to each of these possibilities are m_2 possibilities for the second action, and so on. Then there are $m_1 \cdot m_2 \cdots m_r$ possibilities altogether for the r actions.

What *Does it Mean?*

The total number of ways that several actions can occur equals the product of the individual number of ways for each action.

In Example 4.27 there are two actions ($r = 2$)—selecting a model and selecting an elevation. Because there are four possibilities for model, $m_1 = 4$; and because corresponding to each model are three possibilities for elevation, $m_2 = 3$. Therefore, by the BCR, the total number of possibilities, including both model and elevation, again is

$$m_1 \cdot m_2 = 4 \cdot 3 = 12.$$

Because the number of possibilities in the model/elevation problem is quite small, determining the number by a direct listing, as we did in the tree diagram shown in Fig. 4.28, is relatively simple. Even easier, however, is obtaining the number of possibilities by applying the BCR. Moreover, in problems having a large number of possibilities, a direct listing is not feasible and the BCR is the only practical way to proceed, as illustrated in Example 4.28.

Example 4.28 The Basic Counting Rule

License Plates The license plates of Arizona consist of three letters followed by three digits.

a. How many different license plates are possible?

b. How many possibilities are there for license plates on which no letter or digit is repeated?

Solution For both parts (a) and (b), we apply the BCR with six actions ($r = 6$).

a. There are 26 possibilities for the first letter, 26 for the second letter, and 26 for the third letter; there are 10 possibilities for the first digit, 10 for the second digit, and 10 for the third digit. Applying the BCR, we get

$$m_1 \cdot m_2 \cdot m_3 \cdot m_4 \cdot m_5 \cdot m_6 = 26 \cdot 26 \cdot 26 \cdot 10 \cdot 10 \cdot 10 = 17{,}576{,}000$$

possibilities altogether for different license plates. Obviously, obtaining the number of possibilities by a direct listing would not be practical—the tree diagram would have 17,576,000 branches!

[†]The basic counting rule is also known as the *basic principle of counting,* the *fundamental counting rule,* and the *multiplication rule.*

b. Again, there are 26 possibilities for the first letter. But for each possibility for the first letter, there are 25 corresponding possibilities for the second letter because the second letter cannot be the same as the first. And for each possibility for the first two letters, there are 24 corresponding possibilities for the third letter because the third letter cannot be the same as either the first or the second. Similarly, there are 10 possibilities for the first digit, 9 for the second digit, and 8 for the third digit. So by the BCR, there are

$$m_1 \cdot m_2 \cdot m_3 \cdot m_4 \cdot m_5 \cdot m_6 = 26 \cdot 25 \cdot 24 \cdot 10 \cdot 9 \cdot 8 = 11{,}232{,}000$$

possibilities for license plates on which no letter or digit is repeated.

■

Factorials

Before we continue our presentation of counting rules, we need to discuss factorials. **Factorials,** which are used extensively in mathematics and its applications, are defined as follows.

DEFINITION 4.8

What Does it Mean?

The factorial of a counting number is obtained by successively multiplying it by the next smaller integer until reaching 1.

Factorials

The product of the first k positive integers is called k *factorial* and is denoted $k!$. In symbols,

$$k! = k(k-1)\cdots 2 \cdot 1.$$

We also define $0! = 1$.

In Example 4.29, we demonstrate calculation of factorials.

▌▌▌Example 4.29 Factorials

Determine 3!, 4!, and 5!.

Solution Applying Definition 4.8, we obtain

$$3! = 3 \cdot 2 \cdot 1 = 6,$$
$$4! = 4 \cdot 3 \cdot 2 \cdot 1 = 24, \text{ and}$$
$$5! = 5 \cdot 4 \cdot 3 \cdot 2 \cdot 1 = 120,$$

as required.

■

Note, for instance, that $6! = 6 \cdot 5!$, $6! = 6 \cdot 5 \cdot 4!$, $6! = 6 \cdot 5 \cdot 4 \cdot 3!$, and so on. In general, if $j \le k$, then $k! = k(k-1)\cdots(k-j+1)(k-j)!$.

Permutations

A **permutation** of r objects from a collection of m objects is any *ordered* arrangement of r of the m objects. The number of possible permutations of r objects that can be formed from a collection of m objects is denoted $_mP_r$ (read "m permute r").[†] Example 4.30 provides a simple illustration of permutation.

▌▌Example 4.30   Introducing Permutations

Arrangement of Letters   Consider the collection of objects consisting of the five letters a, b, c, d, e.

a. List all possible permutations of three letters from this collection of five letters.

b. Use part (a) to determine the number of possible permutations of three letters that can be formed from the collection of five letters; that is, find $_5P_3$.

c. Use the BCR to determine the number of possible permutations of three letters that can be formed from the collection of five letters; that is, find $_5P_3$ by using the BCR.

Solution

a. The list of all possible permutations (ordered arrangements) of three letters from the first five letters of the English alphabet is shown in Table 4.14.

TABLE 4.14

Possible permutations of three letters from the collection of five letters

abc	abd	abe	acd	ace	ade	bcd	bce	bde	cde
acb	adb	aeb	adc	aec	aed	bdc	bec	bed	ced
bac	bad	bae	cad	cae	dae	cbd	cbe	dbe	dce
bca	bda	bea	cda	cea	dea	cdb	ceb	deb	dec
cab	dab	eab	dac	eac	ead	dbc	ebc	ebd	ecd
cba	dba	eba	dca	eca	eda	dcb	ecb	edb	edc

b. Table 4.14 indicates that there are 60 possible permutations of three letters from the collection of five letters; in other words, $_5P_3 = 60$.

c. There are five possibilities for the first letter, four possibilities for the second letter, and three possibilities for the third letter. Hence, by the BCR, there are

$$m_1 \cdot m_2 \cdot m_3 = 5 \cdot 4 \cdot 3 = 60$$

possibilities altogether, again giving $_5P_3 = 60$.   ◼

We can make two relevant observations from Example 4.30. First, listing all possible permutations under consideration is generally tedious or impractical.

[†]Other notations used for the number of possible permutations include P_r^m and $(m)_r$.

Second, listing all possible permutations is not necessary in order to determine how many there are—we can use the BCR to count them.

Part (c) of Example 4.30 reveals that we can use the BCR to obtain a general formula for $_mP_r$: $_mP_r = m(m-1)\cdots(m-r+1)$. Multiplying and dividing the right side of this formula by $(m-r)!$, we get the equivalent expression $_mP_r = m!/(m-r)!$. We summarize this discussion in the **permutations rule**, which we express in the following formula and apply in Examples 4.31 and 4.32.

FORMULA 4.10 **The Permutations Rule**

The number of possible permutations of r objects from a collection of m objects is given by the formula

$$_mP_r = \frac{m!}{(m-r)!}.$$

Example 4.31   The Permutations Rule

Exacta Wagering   In an exacta wager at the race track, a bettor picks the two horses that she thinks will finish first and second in a specified order. For a race with 12 entrants, determine the number of possible exacta wagers.

Solution   Selecting two horses from the 12 horses for an exacta wager is equivalent to specifying a permutation of two objects from a collection of 12 objects. The first object is the horse selected to finish in first place, and the second object is the horse selected to finish in second place.

Thus the number of possible exacta wagers is $_{12}P_2$—the number of possible permutations of two objects from a collection of 12 objects. Applying the permutations rule, with $m = 12$ and $r = 2$, we obtain

$$_{12}P_2 = \frac{12!}{(12-2)!} = \frac{12!}{10!} = \frac{12 \cdot 11 \cdot \cancel{10!}}{\cancel{10!}} = 12 \cdot 11 = 132.$$

INTERPRETATION   In a 12-horse race, there are 132 possible exacta wagers.

Example 4.32   The Permutations Rule

Arranging Books on a Shelf   A student has 10 books to arrange on a shelf of a bookcase. In how many ways can the 10 books be arranged?

Solution   Any particular arrangement of the 10 books on the shelf is a permutation of 10 objects from a collection of 10 objects. Hence we need to determine $_{10}P_{10}$, the number of possible permutations of 10 objects from a collection of 10 objects, more commonly expressed as the number of possible permutations of 10 objects among themselves. Applying the permutations rule, we get

$$_{10}P_{10} = \frac{10!}{(10-10)!} = \frac{10!}{0!} = \frac{10!}{1} = 10! = 3{,}628{,}800.$$

INTERPRETATION   There are 3,628,800 ways to arrange 10 books on a shelf.

Let's generalize Example 4.32 to find the number of possible permutations of m objects among themselves. Using the permutations rule, we conclude that

$$_{m}P_{m} = \frac{m!}{(m-m)!} = \frac{m!}{0!} = \frac{m!}{1} = m!,$$

which is called the **special permutations rule** and which we express in words in the following formula.

FORMULA 4.11   **The Special Permutations Rule**

The number of possible permutations of m objects among themselves is $m!$.

Combinations

A **combination** of r objects from a collection of m objects is any *unordered* arrangement of r of the m objects—in other words, any subset of r objects from the collection of m objects. Note that order matters in permutations but not in combinations.

The number of possible combinations of r objects that can be formed from a collection of m objects is denoted $_{m}C_{r}$ (read "m choose r").[†] In Example 4.33, we return to the situation in Example 4.30 to illustrate combinations.

Example 4.33   **Introducing Combinations**

Arrangement of Letters   Consider the collection of objects consisting of the five letters a, b, c, d, e.

a. List all possible combinations of three letters from this collection of five letters.

TABLE 4.15
Combinations

b. Use part (a) to determine the number of possible combinations of three letters that can be formed from the collection of five letters; that is, find $_{5}C_{3}$.

$\{a,b,c\}$	$\{a,b,d\}$	$\{a,b,e\}$	$\{a,c,d\}$
$\{a,c,e\}$	$\{a,d,e\}$	$\{b,c,d\}$	$\{b,c,e\}$
$\{b,d,e\}$	$\{c,d,e\}$		

Solution

a. The list of all possible combinations (unordered arrangements) of three letters from the first five letters of the English alphabet is shown in Table 4.15.

[†]Other notations used for the number of possible combinations include C_r^m and $\binom{m}{r}$.

b. Table 4.15 reveals that there are 10 possible combinations of three letters from the collection of five letters; in other words, $_5C_3 = 10$.

⎯ ∎

In Example 4.33, we obtained the number of possible combinations by a direct listing. We can avoid resorting to a direct listing by deriving a formula for determining the number of possible combinations. To do so, we return once more to the English letters example.

Look at the first combination in Table 4.15, $\{a, b, c\}$. By the special permutations rule, there are $3! = 6$ permutations of these three letters among themselves; they are *abc, acb, bac, bca, cab,* and *cba.* These six permutations are the ones displayed in the first column of Table 4.14 on page 206. Similarly, there are $3! = 6$ permutations of the three letters appearing as the second combination in Table 4.15, $\{a, b, d\}$. These six permutations are the ones displayed in the second column of Table 4.14. The same comments apply to the other eight combinations in Table 4.15.

Thus, for each combination of three letters from the collection of five letters, there are $3!$ corresponding permutations of three letters from the collection of five letters. Because any such permutation is accounted for in this way, there must be $3!$ times as many permutations as combinations. Equivalently, the number of possible combinations of three letters from the collection of five letters must equal the number of possible permutations of three letters from the collection of five letters divided by $3!$. Thus

$$_5C_3 = \frac{_5P_3}{3!} = \frac{5!/(5-3)!}{3!} = \frac{5!}{3!(5-3)!} = \frac{5 \cdot 4 \cdot \cancel{3!}}{\cancel{3!}\,2!} = \frac{5 \cdot 4}{2} = 10,$$

which is the number we obtained in Example 4.33 by a direct listing.

The same type of argument holds in general. Therefore we have the **combinations rule,** expressed in the following formula, for determining the number of possible combinations. We apply the combinations rule in Examples 4.34 and 4.35.

FORMULA 4.12   **The Combinations Rule**

The number of possible combinations of r objects from a collection of m objects is given by the formula

$$_mC_r = \frac{m!}{r!(m-r)!}.$$

Example 4.34   **The Combinations Rule**

CD-Club Introductory Offer   To recruit new members, a compact-disc (CD) club advertises a special introductory offer: A new member agrees to buy 1 CD at regular club prices and receives free any 4 CDs of his choice from a collection of 69 CDs. How many possibilities does the new member have for the selection of the 4 free CDs?

Solution Any particular selection of 4 CDs from 69 CDs is a combination of 4 objects from a collection of 69 objects. By the combinations rule, the number of possible selections is

$$_{69}C_4 = \frac{69!}{4!(69-4)!} = \frac{69!}{4!\,65!} = \frac{69 \cdot 68 \cdot 67 \cdot 66 \cdot \cancel{65!}}{4!\,\cancel{65!}} = 864{,}501.$$

INTERPRETATION There are 864,501 possibilities for the selection of 4 CDs from a collection of 69 CDs.

Example 4.35 The Combinations Rule

Sampling Students An economics professor is using a new method to teach a junior-level course with an enrollment of 42 students. The professor wants to conduct in-depth interviews with the students to get feedback on the new teaching method, but she does not want to interview all 42 of them. She decides to interview a sample of 5 students from the class. How many different samples are possible?

Solution A sample of 5 students from the class of 42 students can be considered a combination of 5 objects from a collection of 42 objects. By the combinations rule, the number of possible samples is

$$_{42}C_5 = \frac{42!}{5!(42-5)!} = \frac{42!}{5!\,37!} = 850{,}668.$$

INTERPRETATION There are 850,668 possible samples of 5 students from a class of 42 students.

Example 4.35 shows how to determine the number of possible samples of a specified size from a finite population. This method is so important that we record it as the following formula.

FORMULA 4.13 Number of Possible Samples

The number of possible samples of size n from a population of size N is $_NC_n$.

Applications to Probability

Suppose that an experiment has N equally likely possible outcomes. Then, according to the f/N rule, the probability that a specified event occurs equals the number of ways, f, that the event can occur divided by the total number of possible outcomes, N.

In the probability problems that we have considered so far, determining f and N has been easy, but that isn't always the case. We must often use counting rules to obtain the number of possible outcomes and the number of ways that the specified event can occur. Example 4.36 illustrates how counting rules can be applied to solve probability problems.

Example 4.36 Applying Counting Rules to Probability

Quality Assurance The quality assurance engineer of a television manufacturer inspects TVs in lots of 100. He selects 5 of the 100 TVs at random and inspects them thoroughly. Assuming that 6 of the 100 TVs in the current lot are defective, find the probability that exactly 2 of the 5 TVs selected by the engineer are defective.

Solution Because the engineer makes his selection at random, each of the possible outcomes is equally likely. Thus we can apply the f/N rule to obtain the required probability.

First, we determine the number of possible outcomes for the experiment. It is the number of ways that 5 TVs can be selected from the 100 TVs—the number of possible combinations of 5 objects from a collection of 100 objects. Applying the combinations rule yields

$$_{100}C_5 = \frac{100!}{5!(100-5)!} = \frac{100!}{5!\,95!} = 75{,}287{,}520,$$

or $N = 75{,}287{,}520$.

Next, we determine the number of ways the specified event can occur, that is, the number of outcomes in which exactly 2 of the 5 TVs selected are defective. To do so, we think of the 100 TVs as partitioned into two groups—namely, the defective TVs and the nondefective TVs, as shown in the top part of Fig. 4.29.

FIGURE 4.29

Calculating the number of outcomes in which exactly 2 of the 5 TVs selected are defective

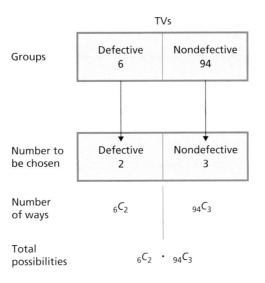

There are 6 TVs in the first group and 2 are to be selected, which can be done in

$$_6C_2 = \frac{6!}{2!(6-2)!} = \frac{6!}{2!\,4!} = 15$$

ways. There are 94 TVs in the second group and 3 are to be selected, which can be done in

$$_{94}C_3 = \frac{94!}{3!(94-3)!} = \frac{94!}{3!\,91!} = 134{,}044$$

ways. Consequently, by the BCR, there are a total of

$$_6C_2 \cdot {}_{94}C_3 = 15 \cdot 134{,}044 = 2{,}010{,}660$$

outcomes in which exactly 2 of the 5 TVs selected are defective, so $f = 2{,}010{,}660$. Figure 4.29 summarizes these calculations.

Applying the f/N rule, we now conclude that the probability that exactly 2 of the 5 TVs selected are defective is

$$\frac{f}{N} = \frac{2{,}010{,}660}{75{,}287{,}520} = 0.027.$$

INTERPRETATION   There is a 2.7% chance that exactly 2 of the 5 TVs selected by the engineer will be defective.

EXERCISES 4.8

Statistical Concepts and Skills

4.102 What are counting rules? Why are they important?

4.103 Why is the basic counting rule (BCR) often referred to as the multiplication rule?

4.104 Regarding permutations and combinations,
a. what is a permutation?
b. what is a combination?
c. what is the major distinction between the two?

4.105 Home Models and Elevations. Refer to Example 4.27 on page 203. Suppose that the developer discontinues the Shalimar model but provides an additional elevation choice, D, for each of the remaining three model choices.
a. Draw a tree diagram similar to the one shown in Fig. 4.28 depicting the possible choices for selection of a home, including both model and elevation.
b. Use the tree diagram in part (a) to determine the total number of choices for the selection of a home, including both model and elevation.
c. Use the BCR to determine the total number of choices for selection of a home, including both model and elevation.

4.106 Zip Codes. The author spoke with a representative of the U.S. Postal Service and obtained the following information about zip codes. A five-digit zip code consists of five digits, of which the first three give the sectional center and the last two the post office or delivery area. In addition

to the five-digit zip code, there is a trailing *plus four zip code*. The first two digits of the plus four zip code give the sector or several blocks and the last two the segment or side of the street. For the five-digit zip code, the first four digits can be any of the digits 0–9 and the fifth any of the digits 1–8. For the plus four zip code, the first three digits can be any of the digits 0–9 and the fourth any of the digits 1–9.
a. How many possible five-digit zip codes are there?
b. How many possible plus four zip codes are there?
c. How many possibilities are there in all, including both the five-digit zip code and the plus four zip code?

4.107 Technology Profiles. *Scientific Computing & Automation* magazine offers free subscriptions to the scientific community. The magazine does ask, however, that a person answer six questions: primary title, type of facility, area of work, brand of computer used, type of operating system in use, and type of instruments in use. Six choices are offered for the first question, eight for the second, five for the third, 19 for the fourth, 16 for the fifth, and 14 for the sixth. How many possibilities are there for answering all six questions?

4.108 Determine the value of each quantity.
a. $_7P_3$ b. $_5P_2$ c. $_8P_4$ d. $_6P_0$ e. $_9P_9$

4.109 Mutual Fund Investing. Investment firms usually have a large selection of mutual funds from which an investor can choose. One such firm has 30 mutual funds. Suppose that you plan to invest in four of these mutual funds, one during each quarter of next year. In how many different ways can you make these four investments?

4.110 Testing for ESP. An extrasensory perception (ESP) experiment is conducted by a psychologist. For part of the experiment, the psychologist takes 10 cards, numbered 1–10, and shuffles them. Then she looks at the cards one at a time. While she looks at each card, the subject writes down the number he thinks is on the card.

a. How many possibilities are there for the order in which the subject writes down the numbers?

b. If the subject has no ESP and is just guessing each time, what is the probability that he writes down the numbers in the correct order, that is, in the order that the cards are actually arranged?

c. Based on your result from part (b), what would you conclude if the subject writes down the numbers in the correct order? Explain your answer.

4.111 Determine the value of each quantity.

a. $_7C_3$ b. $_5C_2$ c. $_8C_4$ d. $_6C_0$ e. $_9C_9$

4.112 Poker. A poker hand consists of 5 cards dealt from an ordinary deck of 52 playing cards.

a. How many poker hands are possible?

b. How many different hands consisting of three kings and two queens are possible?

c. The hand in part (b) is an example of a full house: 3 cards of one denomination and 2 of another. How many different full houses are possible?

d. Calculate the probability of being dealt a full house.

4.113 Senate Committees. The U.S. Senate consists of 100 senators, two from each state. A committee consisting of 5 senators is to be formed.

a. How many different committees are possible?

b. How many are possible if no state may have more than 1 senator on the committee?

c. If the committee is selected at random from all 100 senators, what is the probability that no state will have both of its senators on the committee?

4.114 How many samples of size 5 are possible from a population of size 70?

4.115 Which Key? Suppose that you have a key ring with eight keys on it, one of which is your house key. Further suppose that you get home after dark and can't see the keys on the key ring. You randomly try one key at a time, being careful not to mix the keys that you've already tried with the ones you haven't. What is the probability that you get the right key

a. on the first try?   b. on the eighth try?

c. on or before the fifth try?

4.116 Quality Assurance. Refer to Example 4.36, which starts on page 211. Determine the probability that the number of defective TVs obtained by the engineer is

a. exactly one.   b. at most one.   c. at least one.

4.117 The Birthday Problem. A biology class has 38 students. Find the probability that at least 2 students in the class have the same birthday. For simplicity, assume that there are always 365 days in a year and that birth rates are constant throughout the year. (*Hint:* First, determine the probability that no 2 students have the same birthday and then apply the complementation rule.)

4.118 Lotto. A previous Arizona state lottery, called *Lotto*, is played as follows: The player selects six numbers from the numbers 1–42 and buys a ticket for $1. There are six winning numbers, which are selected at random from the numbers 1–42. To win a prize, a *Lotto* ticket must contain three or more of the winning numbers. A ticket with exactly three winning numbers is paid $2. The prize for a ticket with exactly four, five, or six winning numbers depends on sales and on how many other tickets were sold that have exactly four, five, or six winning numbers, respectively. If you buy one *Lotto* ticket, determine the probability that

a. you win the jackpot; that is, your six numbers are the same as the six winning numbers.

b. your ticket contains exactly four winning numbers.

c. you don't win a prize.

4.119 True–False Tests. A student takes a true–false test consisting of 15 questions. Assume that the student guesses at each question and find the probability that

a. the student gets at least one question correct.

b. the student gets a 60% or better on the exam.

Extending the Concepts and Skills

4.120 Political Studies. According to the Center for Political Studies at the University of Michigan, Ann Arbor, roughly 50% of U.S. adults are Democrats. Suppose that 10 U.S. adults are selected at random. Determine the approximate probability that

a. exactly 5 are Democrats.

b. 8 or more are Democrats.

4.121 The Birthday Problem. Refer to Exercise 4.117 but now assume that the class consists of N students. Determine the probability that at least two of the students have the same birthday.

4.122 Sampling Without Replacement. A simple random sample of size n is to be taken without replacement from a population of size N.

a. Determine the probability that any particular sample of size *n* is the one selected.

b. Determine the probability that any specified member of the population is included in the sample.

c. Determine the probability that any *k* specified members of the population are included in the sample.

Using Technology

4.123 The Birthday Problem. Use a computer or a programmable calculator and your answer from Exercise 4.121 to construct a table giving the probability that at least two of the students in the class have the same birthday, for $N = 2, 3, \ldots, 70$.

CHAPTER REVIEW

You Should Be Able To

1. use and understand the formulas in this chapter.

2. compute probabilities for experiments having equally likely outcomes.

3. interpret probabilities, using the frequentist interpretation of probability.

4. state and understand the basic properties of probability.

5. construct and interpret Venn diagrams.

6. find and describe (not *E*), (*A* & *B*), and (*A* or *B*).

7. determine whether two or more events are mutually exclusive.

8. understand and use probability notation.

9. state and apply the special addition rule.

10. state and apply the complementation rule.

11. state and apply the general addition rule.

*12. read and interpret contingency tables.

*13. construct a joint probability distribution.

*14. compute conditional probabilities both directly and by using the conditional probability rule.

*15. state and apply the general multiplication rule.

*16. state and apply the special multiplication rule.

*17. determine whether two events are independent.

*18. understand the difference between mutually exclusive events and independent events.

*19. determine whether two or more events are exhaustive.

*20. state and apply the rule of total probability.

*21. state and apply Bayes's rule.

*22. state and apply the basic counting rule (BCR).

*23. state and apply the permutations and combinations rules.

*24. apply counting rules to solve probability problems where appropriate.

Key Terms

(*A* & *B*), *161*
(*A* or *B*), *161*
at random, *153*
basic counting rule (BCR),* *204*
Bayes's rule,* *199*
bivariate data,* *174*
cells,* *175*
certain event, *156*
combination,* *208*
combinations rule,* *209*
complement, *161*
complementation rule, *169*
conditional probability,* *180*
conditional probability rule,* *184*
contingency table,* *174*

counting rules,* *202*
dependent events,* *191*
equal-likelihood model, *155*
event, *153, 160*
exhaustive events,* *195*
experiment, *153*
f/N rule, *152*
factorials,* *205*
frequentist interpretation of probability, *155*
general addition rule, *170*
general multiplication rule,* *187*
given event,* *180*
impossible event, *156*
independent events,* *190, 191*

joint probabilities,* *177*
joint probability distribution,* *177*
marginal probabilities,* *177*
mutually exclusive events, *164*
(not *E*), *161*
$P(B \mid A)$,* *180*
$P(E)$, *167*
permutation,* *206*
permutations rule,* *207*
posterior probability,* *200*
prior probability,* *200*
probability model, *155*
probability theory, *150*
rule of total probability,* *196*
sample space, *160*

REVIEW TEST

Statistical Concepts and Skills

1. Why is probability theory important to statistics?

2. Regarding the equal-likelihood model,
a. what is it?
b. how are probabilities computed?

3. What meaning is given to the probability of an event by the frequentist interpretation of probability?

4. Decide which of these numbers could not possibly be probabilities. Explain your answers.
a. 0.047 **b.** −0.047 **c.** 3.5 **d.** 1/3.5

5. Identify a commonly used graphical technique for portraying events and relationships among events.

6. What does it mean for two or more events to be mutually exclusive?

7. Suppose that E is an event. Use probability notation to represent
a. the probability that event E occurs.
b. the probability that event E occurs is 0.436.

8. Answer true or false to each statement and explain your answers.
a. For any two events, the probability that one or the other of the events occurs equals the sum of the two individual probabilities.
b. For any event, the probability that it occurs equals 1 minus the probability that it does not occur.

9. Identify one reason why the complementation rule is useful.

*10. Fill in the blanks.
a. Data obtained by observing values of one variable of a population are called _____ data.
b. Data obtained by observing values of two variables of a population are called _____ data.
c. A frequency distribution for bivariate data is called a _____.

*11. The sum of the joint probabilities in a row or column of a joint probability distribution equals the _____ probability in that row or column.

*12. Let A and B be events.
a. Use probability notation to represent the conditional probability that event B occurs, given that event A has occurred.
b. In part (a), which is the given event, A or B?

*13. Identify two possible ways in which conditional probabilities can be computed.

*14. What is the relationship between the joint probability and marginal probabilities of two independent events?

*15. If two or more events have the property that at least one of them must occur when the experiment is performed, the events are said to be _____.

*16. State the basic counting rule (BCR).

*17. For the first four letters in the English alphabet,
a. list the possible permutations of three letters from the four.
b. list the possible combinations of three letters from the four.
c. Use parts (a) and (b) to obtain $_4P_3$ and $_4C_3$.
d. Use the permutations and combinations rules to obtain $_4P_3$ and $_4C_3$. Compare your answers in parts (c) and (d).

18. Adjusted Gross Incomes. The U.S. Internal Revenue Service compiles data on income tax returns and summarizes its findings in *Statistics of Income*. The first two columns of Table 4.16 show a frequency distribution (number of returns) for adjusted gross income (AGI) from federal individual income tax returns, where K = thousand.

TABLE 4.16
Adjusted gross incomes

Adjusted gross income	Frequency (1000s)	Event	Probability
Under $10K	28,269	A	
$10K < 20K	24,568	B	
$20K < 30K	18,010	C	
$30K < 40K	12,967	D	
$40K < 50K	9,788	E	
$50K < 100K	21,635	F	
$100K & over	7,186	G	
	122,423		

A federal individual income tax return is selected at random.
a. Determine $P(A)$, the probability that the return selected shows an AGI under $10K.
b. Find the probability that the return selected shows an AGI between $30K and $100K (i.e., at least $30K but less than $100K).
c. Compute the probability of each of the seven events in the third column of Table 4.16, and record those probabilities in the fourth column.

19. Adjusted Gross Incomes. Refer to Problem 18. A federal individual income tax return is selected at random. Let

H = event the return shows an AGI between $20K and $100K,

I = event the return shows an AGI of less than $50K,

J = event the return shows an AGI of less than $100K, and

K = event the return shows an AGI of at least $50K.

Describe each of the following events in words and determine the number of outcomes (returns) that constitute each event.
a. (not J) b. (H & I)
c. (H or K) d. (H & K)

20. Adjusted Gross Incomes. For the following groups of events from Problem 19, determine which are mutually exclusive.
a. H and I
b. I and K
c. H and (not J)
d. H, (not J), and K

21. Adjusted Gross Incomes. Refer to Problems 18 and 19.
a. Use the second column of Table 4.16 and the f/N rule to compute the probability of each of the events H, I, J, and K.
b. Express each of the events H, I, J, and K in terms of the mutually exclusive events displayed in the third column of Table 4.16.

c. Compute the probability of each of the events H, I, J, and K, using your answers from part (b), the special addition rule, and the fourth column of Table 4.16, which you completed in Problem 18(c).

22. Adjusted Gross Incomes. Consider the events (not J), (H & I), (H or K), and (H & K) discussed in Problem 19.
a. Find the probability of each of those four events, using the f/N rule and your answers from Problem 19.
b. Compute $P(J)$, using the complementation rule and your answer for P (not J) from part (a).
c. In Problem 21(a), you found that $P(H) = 0.510$ and $P(K) = 0.235$; and, in part (a) of this problem, you found that $P(H \& K) = 0.177$. Using those probabilities and the general addition rule, find $P(H \text{ or } K)$.
d. Compare the answers that you obtained for $P(H \text{ or } K)$ in parts (a) and (c).

*23. **School Enrollment.** The U.S. National Center for Education Statistics publishes information about school enrollment in the *Digest of Education Statistics*. Table 4.17 provides a contingency table for enrollment in public and private schools by level. Frequencies are in thousands of students.

TABLE 4.17

Enrollment by level and type

| | | Type | | |
		Public T_1	Private T_2	**Total**
	Elementary L_1	33,903	4,640	38,543
Level	High school L_2	13,537	1,366	14,903
	College L_3	11,626	3,263	14,889
	Total	59,066	9,269	68,335

a. How many cells are in this contingency table?
b. How many students are in high school?
c. How many students attend public schools?
d. How many students attend private colleges?

*24. **School Enrollment.** Refer to the information given in Problem 23. A student is selected at random.
a. Describe the events L_3, T_1, and (T_1 & L_3) in words.
b. Find the probability of each event in part (a), and interpret your answers in terms of percentages.

c. Construct a joint probability distribution corresponding to Table 4.17.
d. Compute $P(T_1 \text{ or } L_3)$, using Table 4.17 and the f/N rule.
e. Compute $P(T_1 \text{ or } L_3)$, using the general addition rule and your answers from part (b).

*25. School Enrollment. Refer to the information given in Problem 23. A student is selected at random.
a. Find $P(L_3 \mid T_1)$ directly, using Table 4.17 and the f/N rule. Interpret the probability you obtain in terms of percentages.
b. Use the conditional probability rule and your answers from Problem 24(b) to find $P(L_3 \mid T_1)$.

*26. School Enrollment. Refer to the information given in Problem 23. A student is selected at random.
a. Use Table 4.17 to find $P(T_2)$ and $P(T_2 \mid L_2)$.
b. Are events L_2 and T_2 independent? Explain your answer in terms of percentages.
c. Are events L_2 and T_2 mutually exclusive?
d. Is the event that a student is in elementary school independent of the event that a student attends public school? Justify your answer.

*27. Public Programs. During one year, the College of Public Programs at Arizona State University awarded the following number of master's degrees.

Type of degree	Frequency
Master of arts	3
Master of public administration	28
Master of science	19

Two students who received such master's degrees are selected at random without replacement. Determine the probability that
a. the first student selected received a master of arts and the second a master of science.
b. both students selected received a master of public administration.
c. Construct a tree diagram for this problem similar to the one shown in Fig. 4.25 on page 189.

d. Find the probability that the two students selected received the same degree.

*28. Divorced Birds. Recent research by B. Hatchwell et al. on divorce rates among the long-tailed tit (*Aegithalos caudatus*) appeared in *Science News* (Vol. 157, No. 20, p. 317). Tracking birds in Yorkshire from one breeding season to the next, the researchers noted that 63% of pairs divorced and that "…compared with moms whose offspring had died, nearly twice the percentage of females that raised their youngsters to the fledgling stage moved out of the family flock and took mates elsewhere the next season—81% versus 43%." For the females in this study,
a. find the percentage whose offspring died. (*Hint:* You will need to use the rule of total probability and the complementation rule.)
b. find the percentage that divorced and whose offspring died.
c. find the percentage whose offspring died among those that divorced.

*29. Color Blindness. According to Maureen and Jay Neitz of the Medical College of Wisconsin Eye Institute, 9% of men are color blind. For four randomly selected men, find the probability that
a. none are color blind.
b. the first three are not color blind and the fourth is color blind.
c. exactly one of the four is color blind.

*30. Suppose that A and B are events such that $P(A) = 0.4$, $P(B) = 0.5$, and $P(A \& B) = 0.2$. Answer each question and explain your reasoning.
a. Are A and B mutually exclusive?
b. Are A and B independent?

*31. Alcohol and Accidents. The National Safety Council publishes information about automobile accidents in *Accident Facts*. The first two columns of the following table provide a percentage distribution of age group for drivers at fault in fatal crashes; the third column gives the percentage of such drivers in each age group with a blood alcohol content (BAC) of 0.10% or greater.

Age group (years)	Percentage of drivers	Percentage with BAC of 0.10% or greater
16–20	14.1	12.7
21–24	11.4	27.8
25–34	23.8	26.8
35–44	19.5	22.8
45–64	19.8	14.3
65 & over	11.4	5.0

Suppose that the report of an accident in which a fatality occurred is selected at random. Determine the probability that the driver at fault

a. had a BAC of 0.10% or greater, given that he or she was between 21 and 24 years old.

b. had a BAC of 0.10% or greater.

c. was between 21 and 24 years old, given that he or she had a BAC of 0.10% or greater.

d. Interpret your answers in parts (a)–(c) in terms of percentages.

e. Of the three probabilities in parts (a)–(c), which are prior and which are posterior?

*32. Quinella and Trifecta Wagering. In Example 4.31 on page 207, we considered exacta wagering in horse racing. Two similar wagers are the quinella and the trifecta. In a quinella wager, the bettor picks the two horses that he or she believes will finish first and second, but not in a specified order. In a trifecta wager, the bettor picks the three horses he or she thinks will finish first, second, and third in a specified order. For a 12-horse race,

a. how many different quinella wagers are there?

b. how many different trifecta wagers are there?

c. Repeat parts (a) and (b) for an 8-horse race.

*33. Bridge. A bridge hand consists of 13 cards dealt at random from an ordinary deck of 52 playing cards.

a. How many possible bridge hands are there?

b. Find the probability of being dealt a bridge hand that contains exactly two of the four aces.

c. Find the probability of being dealt an 8-4-1 distribution, that is, eight cards of one suit, four of another, and one of another.

d. Determine the probability of being dealt a 5-5-2-1 distribution.

e. Determine the probability of being dealt a hand void in a specified suit.

*34. Sweet Sixteen. In the NCAA basketball tournament, 64 teams compete in 63 games during six rounds of single-elimination bracket competition. During the "Sweet Sixteen" competition (the third round of the tournament), 16 teams compete in eight games. If you were to choose in advance of the tournament the eight teams that would win in the "Sweet Sixteen" competition and thus play in the fourth round of competition, how many different possibilities would you have?

*35. TVs and VCRs. According to *Trends in Television*, published by the Television Bureau of Advertising, Inc., 98.2% of (U.S.) households own a TV and 85.1% of TV households own a VCR.

a. Under what condition can you use the information provided to determine the percentage of households that own a VCR? Explain your reasoning.

b. Assuming that the condition you stated in part (a) actually holds, determine the percentage of households that own a VCR.

c. Assuming that the condition you stated in part (a) does not hold, what other piece of information would you need to find the percentage of households that own a VCR?

StatExplore in MyMathLab
Analyzing Data Online

You can use StatExplore to conduct various types of simulations. To illustrate, we show how to perform a simulation to corroborate the frequentist interpretation of probability, as discussed on page 155.

EXAMPLE Frequentist Interpretation

Gestation Periods The probability is 0.314 that the gestation period of a woman will exceed 9 months. Do the following by using StatExplore.

a. Simulate 500 times the experiment of observing whether a gestation period exceeds 9 months.

b. Obtain the proportion of times that the gestation periods exceed 9 months.

c. Interpret your result in part (b) in terms of the frequentist interpretation of probability.

SOLUTION

a. To perform the required simulation, we proceed as follows:

> 1 Choose **Data ➤ Simulate data ➤ Bernoulli**
> 2 Type 500 in the **Rows** text box
> 3 Type 1 in the **Columns** text box
> 4 Type 0.314 in the **p** text box
> 5 Click **Simulate**

In the StatExplore data table, the result of these five instructions is a column named Bernoulli1 that contains 1s and 0s, where a 1 indicates a gestation period that exceeds 9 months and a 0 indicates a gestation period that is 9 months or less.

b. Here, we need to determine the proportion of times that the gestation periods exceed 9 months, that is, the proportion of 1s that occur in the column named Bernoulli1. Several methods are available for doing that. One method is as follows:

> 1 Choose **Stat ➤ Tables ➤ Frequency**
> 2 Select the column Bernoulli1
> 3 Click **Calculate**

The resulting output is shown in Printout 4.1, from which we learn that the proportion of times the gestation periods exceed 9 months (the proportion of 1s that occur in the column Bernoulli1) is 0.308.

PRINTOUT 4.1
StatExplore output for Bernoulli1

Frequency table results for Bernoulli1:		
Bernoulli1	**Frequency**	**Relative Frequency**
0	346	0.692
1	154	0.308

c. The frequentist interpretation of probability is that, in a large number of gestation periods, the proportion that exceed 9 months (in this case, 0.308) will approximately equal the probability that a gestation period exceeds 9 months, which is 0.314. Thus, our simulation corroborates the frequentist interpretation of probability.

STATEXPLORE EXERCISES The probability is 0.667 that the favorite in a horse race will finish in the money (first, second, or third place). Use StatExplore to do the following:

a. Simulate 1000 times the experiment of observing whether the favorite in a horse race finishes in the money. Repeat the simulation four times. (*Hint:* Type 4 in the **Columns** text box.)

b. For each simulation, obtain the proportion of times that the favorite finishes in the money.

c. Interpret your result in part (b) in terms of the frequentist interpretation of probability.

> To access StatExplore, go to the student content area of your Weiss MyMathLab course.

Internet Project

The Space Shuttle Challenger

In this Internet project, you are to interactively explore the concepts of probability. The animations and demonstrations can help you understand how these concepts actually work.

On January 20, 1986, the 25th flight of the National Aeronautics and Space Administration's (NASA) space shuttle program took off. Just after liftoff, a puff of gray smoke could be seen coming from the right solid rocket booster. Seventy-three seconds into the flight, the space shuttle Challenger had climbed 10 miles and then exploded into a fireball, killing all aboard.

The cause of the explosion was found to be an O-ring failure in the right solid rocket booster. Cold weather was a contributing factor. In this project, you are to examine the data concerning temperature and its effect on the Challenger's O-rings.

When you studied the theory and concepts of probability in this chapter, you saw that everyday events follow probabilistic rules. The importance of learning how probability works is related to the significance of the events. As you will see in this project, sometimes those events have momentous consequences indeed.

URL for access to Internet Projects Page: www.aw-bc.com/weiss

**Focusing on
Data Analysis**

The Focus Database

Recall from Chapter 1 (see page 35) that the Focus database contains information on the undergraduate students at the University of Wisconsin - Eau Claire (UWEC). Statistical analyses for this database should be done with the technology of your choice.

a. Obtain a relative-frequency distribution for the classification (CLASS) data.
b. Using your answer from part (a), determine the probability that a randomly selected UWEC undergraduate student is a freshman.
c. Consider the experiment of selecting a UWEC undergraduate student at random and observing the classification of the student obtained. Simulate that experiment 1000 times. (*Hint:* The simulation is equivalent to taking a random sample of size 1000 with replacement.)
d. Referring to the simulation performed in part (c), in approximately what percentage of the 1000 experiments would you expect a freshman to be selected? Compare that percentage with the actual percentage of the 1000 experiments in which a freshman was selected.
e. Repeat parts (b)–(d) for sophomores; juniors; seniors.

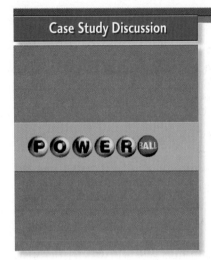

The Powerball

At the beginning of this chapter on page 151, we discussed the Powerball lottery and described some of its rules. Recall that, for a single ticket, a player first selects five numbers from the numbers 1–53 and then chooses a Powerball number, which can be any number between 1 and 42. A ticket costs $1. In the drawing, five white balls are drawn randomly from 53 white balls numbered 1–53, and one red Powerball is drawn randomly from 42 red balls numbered 1–42.

To win the jackpot, a ticket must match all the balls drawn. Prizes are also given for matching some but not all the balls drawn. Table 4.18 displays the number of matches, the prizes given, and the probabilities of winning.

Here are some things to note about Table 4.18.

- In the first column, an entry of the form $w + 1$ indicates w matches out of the five plus the Powerball; one of the form w indicates w matches out of the five and no Powerball.
- The prize amount for the jackpot depends on how recently it has been won, how many people win it, and choice of jackpot payment.
- Each probability in the third column is given to eight decimal places and can be obtained by using the techniques discussed in Section 4.8. If you studied that section, verify the probabilities in Table 4.18.

TABLE 4.18

Powerball winning combinations, prizes, and probabilities

Matches	Prize	Probability
5 + 1	Jackpot	0.00000001
5	$100,000	0.00000034
4 + 1	$5,000	0.00000199
4	$100	0.00008164
3 + 1	$100	0.00009359
3	$7	0.00383716
2 + 1	$7	0.00143503
1 + 1	$4	0.00807207
0 + 1	$3	0.01420684

Let E be an event having probability p. In independent repetitions of the experiment, it takes, on average, $1/p$ times until event E occurs. We apply this fact to the Powerball.

Suppose that you were to purchase one Powerball ticket per week. How long should you expect to wait before winning the jackpot? The third column of Table 4.18 shows that, for the jackpot, $p = 0.00000001$, and therefore we have $1/0.00000001 = 100,000,000$. So, if you purchased one Powerball ticket per week, you should expect to wait approximately 100 million weeks, or roughly 1.9 million years, before winning the jackpot.[†]

a. If you purchase one ticket, what is the probability that you win a prize?
b. If you purchase one ticket, what is the probability that you don't win a prize?
c. If you win a prize, what is the probability it is the $3 prize for having only the Powerball number?
d. If you were to buy one ticket per week, approximately how long should you expect to wait before getting a ticket with exactly three winning numbers and no Powerball?
e. If you were to buy one ticket per week, approximately how long should you expect to wait before winning a prize?

Internet Resources: Visit the Weiss Web site www.aw-bc.com/weiss for additional discussion, exercises, and resources related to this case study.

[†]The probability 0.00000001 for winning the jackpot is approximate. Using a more exact value, we find that you should expect to wait about 2.3 million years before winning the jackpot.

Biography

ANDREI KOLMOGOROV: Father of Modern Probability Theory

Andrei Nikolaevich Kolmogorov was born on April 25, 1903, in Tambov, Russia. At the age of 17, Kolmogorov entered Moscow State University, from which he graduated in 1925. His contributions to the world of mathematics, many of which appear in his numerous articles and books, encompass a formidable range of subjects.

Kolmogorov revolutionized probability theory with the introduction of the modern axiomatic approach to probability and by proving many of the fundamental theorems that are a consequence of that approach. He also developed two systems of partial differential equations, which bear his name. Those systems extended the development of probability theory and allowed its broader application to the fields of physics, chemistry, biology, and civil engineering.

In 1938, Kolmogorov published an extensive article entitled "Mathematics," which appeared in the first edition of the *Bolshaya Sovyetskaya Entsiklopediya* (Great Soviet Encyclopedia). In this article he discussed the development of mathematics from ancient to modern times and interpreted it in terms of dialectical materialism, the philosophy originated by Karl Marx and Friedrich Engels.

Kolmogorov became a member of the faculty at Moscow State University in 1925, at the age of 22. In 1931, he was promoted to professor; in 1933, he was appointed a director of the Institute of Mathematics of the university; and in 1937, he became Head of the University.

In addition to his work in higher mathematics, Kolmogorov was interested in the mathematical education of schoolchildren. He was chairman of the Commission for Mathematical Education under the Presidium of the Academy of Sciences of the U.S.S.R. During his tenure as chairman, he was instrumental in the development of a new mathematics training program that was introduced into Soviet schools.

Kolmogorov remained on the faculty at Moscow State University until his death in Moscow on October 20, 1987.

chapter 5

Discrete Random Variables

GENERAL OBJECTIVES In Chapters 2 and 3, we examined, among other things, variables and their distributions. Most of the variables we discussed in those chapters were variables of finite populations. But many variables are not of that type: the number of people waiting in line at a bank, the lifetime of an automobile tire, and the weight of a newborn baby, to name just three.

Probability theory enables us to extend concepts that apply to variables of finite populations—concepts such as relative-frequency distribution, mean, and standard deviation—to other types of variables. In doing so, we are led to the notion of a *random variable* and its *probability distribution.*

In this chapter, we discuss the fundamentals of discrete random variables and probability distributions and examine the concepts of the mean and standard deviation of a discrete random variable. Additionally, we describe in detail two of the most important discrete random variables: the binomial and Poisson.

Note: Those studying the normal approximation to the binomial distribution (Section 6.5) should study Sections 5.1–5.3.

Aces Wild on the Sixth at Oak Hill

A most amazing event occurred during the second round of the 1989 U.S. Open at Oak Hill in Pittsford, New York. Four golfers—Doug Weaver, Mark Wiebe, Jerry Pate, and Nick Price—made holes in one on the sixth hole. What are the chances for the occurrence of such a remarkable event?

An article appeared the next day in the *Boston Globe* that discussed the event in detail. To quote the article, "...for perspective, consider this: This is the 89th U.S. Open, and through the thousands and thousands and thousands of rounds played in the previous 88, there had been only 17 holes in one. Yet on this dark Friday morning, there were four holes in one on the same hole in less than two hours. Four times into a cup $4\frac{1}{2}$ inches in diameter from 160 yards away."

The article also reported odds estimates obtained from several different sources. These estimates varied considerably, from 1 in 10 million to 1 in 1,890,000,000,000,000 to 1 in 8.7 million to 1 in 332,000. After you have completed this chapter, you will be able to compute the odds for yourself.

5.1 Discrete Random Variables and Probability Distributions

In this section, we introduce discrete random variables and probability distributions. As you will discover, these concepts are natural extensions of the ideas of variables and relative-frequency distributions. Example 5.1 introduces random variables.

◼▮ **Example 5.1** **Introducing Random Variables**

TABLE 5.1
Grouped-data table for number of siblings for students in introductory statistics

Siblings x	Freq. f	Relative freq.
0	8	0.200
1	17	0.425
2	11	0.275
3	3	0.075
4	1	0.025
	40	1.000

Number of Siblings Professor Weiss asked his introductory statistics students to state how many siblings they have. Table 5.1 presents a grouped-data table for that information. The table shows, for instance, that 11 of the 40 students, or 27.5%, have two siblings. Discuss the "number of siblings" in the context of randomness.

Solution Because the "number of siblings" varies from student to student, it is a variable. Suppose now that a student is selected at random. Then the "number of siblings" of the student obtained is called a *random variable* because its value depends on chance—namely, on which student is selected. ▬■

Keeping Example 5.1 in mind, we now present the definition of a **random variable** as Definition 5.1.

DEFINITION 5.1

Random Variable

A *random variable* is a quantitative variable whose value depends on chance.

Example 5.1 shows how random variables arise naturally as quantitative variables of finite populations in the context of randomness. But random variables occur in many other ways. Four examples are

- the sum of the dice when a pair of fair dice are rolled,
- the number of puppies in a litter,
- the return on an investment, and
- the lifetime of a flashlight battery.

What Does it Mean?

The possible values of a discrete random variable can be listed, even though the list may continue indefinitely. A discrete random variable usually involves a count of something.

As you learned in Chapter 2, a *discrete variable* is a variable whose possible values form a finite or countably infinite set of numbers. The variable "number of siblings" in Example 5.1 is a discrete variable, its possible values being 0, 1, 2, 3, and 4. We use the adjective *discrete* for random variables in the same way that we do for variables—hence the term **discrete random variable.**

DEFINITION 5.2

Discrete Random Variable

A *discrete random variable* is a random variable whose possible values form a finite or countably infinite set of numbers.

Random-Variable Notation

Recall that we use letters near the end of the alphabet, such as x, y, and z, to denote variables. We also use such letters to represent random variables but, in this context, we usually make the letters uppercase. For instance, we could use x to denote the variable "number of siblings"; in the context of randomness, however, we would generally use X.

By utilizing random-variable notation, we can develop useful shorthands for discussing and analyzing random variables. For example, suppose that we use X to denote the number of siblings of a randomly selected student. Then we can represent the event that the student selected has, say, two siblings by $\{X = 2\}$, read "X equals two." And we can express the probability of that event as $P(X = 2)$, read "the probability that X equals two."

Probability Distributions and Histograms

Recall that the relative-frequency distribution of a discrete variable gives the possible values of the variable and the proportion of times each value occurs. Using the language of probability, we can extend the notions of relative-frequency distribution and relative-frequency histogram—concepts applying to variables of finite populations—to any discrete random variable. In doing so, we use the terms **probability distribution** and **probability histogram** as defined in Definition 5.3.

DEFINITION 5.3

Probability Distribution and Probability Histogram

Probability distribution: A listing of the possible values and corresponding probabilities of a discrete random variable, or a formula for the probabilities.

Probability histogram: A graph of the probability distribution that displays the possible values of a discrete random variable on the horizontal axis and the probabilities of those values on the vertical axis. The probability of each value is represented by a vertical bar whose height equals the probability.

What
Does it Mean?

The probability distribution and probability histogram of a discrete random variable show its possible values and their likelihood.

Example 5.2 illustrates probability distributions and probability histograms.

Example 5.2 **Probability Distributions and Histograms**

Number of Siblings   Refer to Example 5.1 and let X denote the number of siblings of a randomly selected student.

a. Determine the probability distribution of the random variable X.

b. Construct a probability histogram for the random variable X.

Solution

a. We want to determine the probability of each of the possible values of the random variable X. To obtain, for instance, $P(X = 2)$, the probability that

Siblings x	Probability $P(X = x)$
0	0.200
1	0.425
2	0.275
3	0.075
4	0.025
	1.000

the student selected has two siblings, we apply the f/N rule. From Table 5.1 on page 226, we find that

$$P(X = 2) = \frac{f}{N} = \frac{11}{40} = 0.275.$$

The other probabilities are found in the same way. Table 5.2 displays these probabilities and provides the probability distribution of the random variable X.

b. To construct a probability histogram for X, we plot its possible values on the horizontal axis and display the corresponding probabilities as vertical bars. Referring to Table 5.2, we get the probability histogram of the random variable X, as shown in Fig. 5.1.

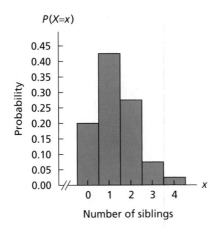

The probability histogram provides a quick and easy way to visualize how the probabilities are distributed.

⸻

The variable "number of siblings" is a variable of a finite population, so its probabilities are identical to relative frequencies. As a consequence, its probability distribution, given in the first and second columns of Table 5.2, is the same as its relative-frequency distribution, shown in the first and third columns of Table 5.1. Apart from labeling, the variable's probability histogram is identical to its relative-frequency histogram. These statements hold for any variable of a finite population.

Note also that the probabilities in the second column of Table 5.2 sum to 1, which is always the case for discrete random variables.

KEY FACT 5.1

Sum of the Probabilities of a Discrete Random Variable

For any discrete random variable X, we have $\Sigma P(X = x) = 1$.

What
Does it Mean?

The sum of the probabilities of the possible values of a discrete random variable equals 1.

Examples 5.3 and 5.4 provide additional illustrations of random-variable notation and probability distributions.

Example 5.3   Random Variables and Probability Distributions

TABLE 5.3

Frequency distribution for enrollment by grade level in U.S. public elementary schools

Grade level y	Frequency f
0	4,208
1	3,769
2	3,596
3	3,518
4	3,447
5	3,447
6	3,486
7	3,457
8	3,398
	32,326

TABLE 5.4

Probability distribution of the random variable Y, the grade level of a randomly selected elementary-school student

Grade level y	Probability $P(Y = y)$
0	0.130
1	0.117
2	0.111
3	0.109
4	0.107
5	0.107
6	0.108
7	0.107
8	0.105
	1.001

Elementary-School Enrollment   The U.S. National Center for Education Statistics compiles enrollment data on U.S. public schools and publishes the results in the *Digest of Education Statistics*. Table 5.3 displays a frequency distribution for the enrollment by grade level in public elementary schools, where $0 =$ kindergarten, $1 =$ first grade, and so on. Frequencies are in thousands of students.

For a randomly selected student in elementary school, let Y denote the grade level of the student obtained. Then Y is a discrete random variable whose possible values are $0, 1, 2, \ldots, 8$.

a. Use random-variable notation to represent the event that the student selected is in the fifth grade.

b. Determine $P(Y = 5)$ and express the result in terms of percentages.

c. Determine the probability distribution of Y.

Solution

a. The event that the student selected is in the fifth grade can be represented as $\{Y = 5\}$.

b. $P(Y = 5)$ is the probability that the student selected is in the fifth grade. Using Table 5.3 and the f/N rule, we get

$$P(Y = 5) = \frac{f}{N} = \frac{3{,}447}{32{,}326} = 0.107.$$

INTERPRETATION 10.7% of elementary-school students in the United States are in the fifth grade.

c. The probability distribution of Y is obtained by computing $P(Y = y)$ for $y = 0, 1, 2, \ldots, 8$. We have already done that for $y = 5$. The other probabilities are computed similarly and are displayed in Table 5.4.

Note: In Table 5.4, the sum of the probabilities is given as 1.001. However, Key Fact 5.1 states that the sum of the probabilities must be exactly 1. Our computation is off slightly because we rounded the probabilities for Y to three decimal places.

Once we have the probability distribution of a discrete random variable, we can easily determine any probability involving that random variable. The basic tool for accomplishing this is the special addition rule (Formula 4.1 on page 168). We illustrate this use of the special addition rule in part (e) of Example 5.4.

Example 5.4   Random Variables and Probability Distributions

Coin Tossing   When a balanced dime is tossed three times, eight equally likely outcomes are possible, as shown in Table 5.5. Here, for instance, HHT means that the first two tosses are heads and the third is tails. Let X denote the total number of heads obtained in the three tosses. Then X is a discrete random variable whose possible values are 0, 1, 2, and 3.

a. Use random-variable notation to represent the event that exactly two heads are tossed.

b. Determine $P(X = 2)$.

c. Find the probability distribution of X.

d. Use random-variable notation to represent the event that at most two heads are tossed.

e. Find $P(X \leq 2)$.

TABLE 5.5

Possible outcomes

HHH	HTH	THH	TTH
HHT	HTT	THT	TTT

Solution

a. The event that exactly two heads are tossed can be represented as $\{X = 2\}$.

b. $P(X = 2)$ is the probability that exactly two heads are tossed. Table 5.5 indicates there are three ways to get a total of two heads and that there are eight possible (equally likely) outcomes altogether. So, by the f/N rule,

$$P(X = 2) = \frac{f}{N} = \frac{3}{8} = 0.375.$$

The probability that exactly two heads are tossed is 0.375.

INTERPRETATION There is a 37.5% chance of obtaining exactly two heads in three tosses of a balanced dime.

c. The remaining probabilities for X are computed as in part (b) and are shown in Table 5.6.

d. The event that at most two heads are tossed can be represented as $\{X \leq 2\}$, read as "X is less than or equal to two."

e. $P(X \leq 2)$ is the probability that at most two heads are tossed. The event that at most two heads are tossed can be expressed as

$$\{X \leq 2\} = (\{X = 0\} \text{ or } \{X = 1\} \text{ or } \{X = 2\}).$$

Because the three events on the right are mutually exclusive, we use the special addition rule and Table 5.6 to conclude that

$$P(X \leq 2) = P(X = 0) + P(X = 1) + P(X = 2)$$
$$= 0.125 + 0.375 + 0.375 = 0.875.$$

The probability that at most two heads are tossed is 0.875.

INTERPRETATION There is an 87.5% chance of obtaining two or fewer heads in three tosses of a balanced dime.

TABLE 5.6

Probability distribution of the random variable X, the number of heads obtained in three tosses of a balanced dime

No. of heads x	Probability $P(X = x)$
0	0.125
1	0.375
2	0.375
3	0.125
	1.000

Interpretation of Probability Distributions

Recall that the frequentist interpretation of probability construes the probability of an event to be the proportion of times it occurs in a large number of (independent) repetitions of the experiment. Using that interpretation, we clarify the meaning of a probability distribution in Example 5.5.

Example 5.5   Interpreting a Probability Distribution

Coin Tossing   Consider once again the random variable X discussed in Example 5.4: the number of heads obtained in three tosses of a balanced dime. Suppose that we repeat the experiment of observing the number of heads obtained in three tosses of a balanced dime a large number of times. Then the proportion of those times in which, say, no heads are obtained (i.e., $X = 0$) should approximately equal the probability of that event [i.e., $P(X = 0)$]. The same statement holds for the other three possible values of the random variable X. Use simulation to verify these facts.

TABLE 5.7

Frequencies and proportions for the numbers of heads obtained in three tosses of a balanced dime for 1000 observations

No. of heads x	Freq. f	Proportion $f/1000$
0	136	0.136
1	377	0.377
2	368	0.368
3	119	0.119
	1000	1.000

Solution   Simulating a random variable means that we use a computer or statistical calculator to generate observations of the random variable. In this instance, we used a computer to simulate 1000 observations of the random variable X, the number of heads obtained in three tosses of a balanced dime.

Table 5.7 shows the frequencies and proportions for the numbers of heads obtained in the 1000 observations. For example, 136 of the 1000 observations resulted in no heads out of three tosses, which gives a proportion of 0.136.

As expected, the proportions in the third column of Table 5.7 are fairly close to the true probabilities in the second column of Table 5.6. This result is more easily seen if we compare the histogram for the proportions to the probability histogram of the random variable X, as shown in Fig. 5.2.

FIGURE 5.2

(a) Histogram of proportions for the numbers of heads obtained in three tosses of a balanced dime for 1000 observations; (b) probability histogram for the number of heads obtained in three tosses of a balanced dime

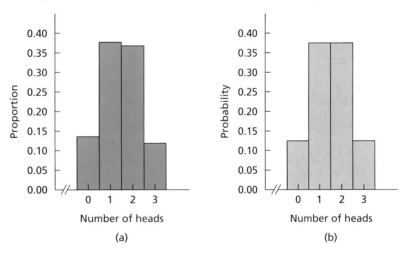

If we simulated, say, 10,000 observations instead of 1000, the proportions that would appear in the third column of Table 5.7 would most likely be even closer to the true probabilities listed in the second column of Table 5.6.

In light of Example 5.5, we can make the following statement concerning the interpretation of a probability distribution.

KEY FACT 5.2

Interpretation of a Probability Distribution

In a large number of independent observations of a random variable X, the proportion of times each possible value occurs will approximate the probability distribution of X; or, equivalently, a histogram of the proportions will approximate the probability histogram for X.

The Technology Center

Most statistical software packages and some graphing calculators have built-in procedures to simulate observations of specified random variables. For instance, both Minitab and Excel can be used to conduct the simulation described in Example 5.5. Refer to the technology manuals for details.

EXERCISES 5.1

Statistical Concepts and Skills

5.1 Fill in the blanks.

a. A relative-frequency distribution is to a variable as a _____ distribution is to a random variable.

b. A relative-frequency histogram is to a variable as a _____ histogram is to a random variable.

5.2 Provide an example (other than one discussed in the text) of a random variable that does not arise from a quantitative variable of a finite population in the context of randomness.

5.3 Let X denote the number of siblings of a randomly selected student. Explain the difference between $\{X = 3\}$ and $P(X = 3)$.

5.4 Fill in the blank. For a discrete random variable, the sum of the probabilities of its possible values equals _____.

5.5 Suppose that you make a large number of independent observations of a random variable and then construct a table giving the possible values of the random variable and the proportion of times each value occurs. What will this table resemble?

5.6 What rule of probability permits you to obtain any probability for a discrete random variable by simply knowing its probability distribution?

5.7 Space Shuttles. The National Aeronautics and Space Administration (NASA) compiles data on space-shuttle launches and publishes them on its Web site. The following table displays a frequency distribution for the number of crew members on each shuttle mission from April 1981 to July 2000.

Crew size	2	3	4	5	6	7	8
Frequency	4	1	2	36	18	33	2

Let X denote the crew size of a randomly selected shuttle mission between April 1981 and July 2000.

a. What are the possible values of the random variable X?

b. Use random-variable notation to represent the event that the shuttle mission obtained has a crew size of 7.

c. Find $P(X = 4)$; interpret in terms of percentages.

d. Obtain the probability distribution of X.
e. Construct a probability histogram for X.

5.8 Dice. When two balanced dice are rolled, 36 equally likely outcomes are possible, as depicted in Fig. 4.1 on page 154. Let Y denote the sum of the dice.
a. What are the possible values of the random variable Y?
b. Use random-variable notation to represent the event that the sum of the dice is 7.
c. Find $P(Y = 7)$.
d. Find the probability distribution of Y. Leave your probabilities in fraction form.
e. Construct a probability histogram for Y.

5.9 Busy Tellers. Prescott National Bank has six tellers available to serve customers. The number of tellers busy with customers at, say, 1:00 P.M. varies from day to day and depends on chance, so it is a random variable, say, X. Past records indicate that the probability distribution of X is as shown in the following table.

x	0	1	2	3	4	5	6
$P(X = x)$	0.029	0.049	0.078	0.155	0.212	0.262	0.215

For example, the probability is 0.262 that exactly five of the tellers will be busy with customers at 1:00 P.M.; that is, about 26.2% of the time, exactly five tellers are busy with customers at 1:00 P.M. Use random-variable notation to represent each of the following events. At 1:00 P.M.,
a. exactly four tellers are busy.
b. at least two tellers are busy.
c. fewer than five tellers are busy.
d. at least two but fewer than five tellers are busy.

Use the special addition rule and the probability distribution to determine
e. $P(X = 4)$.   **f.** $P(X \geq 2)$.
g. $P(X < 5)$.   **h.** $P(2 \leq X < 5)$.

5.10 Solar Eclipses. The *World Almanac* provides information on past and projected total solar eclipses from 1955–2015. Unlike total lunar eclipses, observing a total solar eclipse from Earth is rare because it can be seen along only a very narrow path and for only a short period of time.

a. Let X denote the duration, in minutes, of a total solar eclipse. Is X a discrete random variable? Explain your answer.
b. Let Y denote the duration, to the nearest minute, of a total solar eclipse. Is Y a discrete random variable? Explain your answer.

Extending the Concepts and Skills

5.11 Suppose that $P(Z > 1.96) = 0.025$. Find $P(Z \leq 1.96)$. (*Hint:* Use the complementation rule.)

5.12 Suppose that T and Z are random variables.
a. If $P(T > 2.02) = 0.05$ and $P(T < -2.02) = 0.05$, obtain $P(-2.02 \leq T \leq 2.02)$.
b. Suppose that $P(-1.64 \leq Z \leq 1.64) = 0.90$ and also that $P(Z > 1.64) = P(Z < -1.64)$. Find $P(Z > 1.64)$.

5.13 Let $c > 0$ and $0 \leq \alpha \leq 1$. Also let X, Y, and T be random variables.
a. If $P(X > c) = \alpha$, determine $P(X \leq c)$ in terms of α.
b. If $P(Y > c) = \alpha/2$ and $P(Y < -c) = P(Y > c)$, obtain $P(-c \leq Y \leq c)$ in terms of α.
c. Suppose that $P(-c \leq T \leq c) = 1 - \alpha$ and, moreover, that $P(T < -c) = P(T > c)$. Find $P(T > c)$ in terms of α.

Using Technology

5.14 Simulation. Refer to the probability distribution displayed in Table 5.6 on page 230.
a. Use the technology of your choice to repeat the simulation done in Example 5.5 on page 231.
b. Obtain the proportions for the number of heads in three tosses and compare it to the probability distribution in Table 5.6.
c. Obtain a histogram of the proportions and compare it to the probability histogram in Fig. 5.2(b) on page 231.
d. What do parts (b) and (c) illustrate?

5.2   The Mean and Standard Deviation of a Discrete Random Variable

In this section, we introduce the mean and standard deviation of a discrete random variable. As you will see, the mean and standard deviation of a discrete random variable are analogous to the population mean and population standard deviation.

Mean of a Discrete Random Variable

Recall that, for a variable x, the mean of all possible observations for the entire population is called the *population mean* or *mean of the variable x*. In Section 3.5, we gave a formula for the mean of a variable x:

$$\mu = \frac{\Sigma x}{N}.$$

Although this formula applies only to variables of finite populations, we can use it and the language of probability to extend the concept of the mean to any discrete variable. We show how to do so in Example 5.6.

▌▌▌Example 5.6   Introducing the Mean of a Discrete Random Variable

---

TABLE 5.8
Ages of eight students

19	20	20	19
21	27	20	21

TABLE 5.9
Probability distribution of X, the age of a randomly selected student

Age x	Probability $P(X = x)$		
19	0.250	←	2/8
20	0.375	←	3/8
21	0.250	←	2/8
27	0.125	←	1/8

Student Ages   Consider a population of eight students whose ages are those given in Table 5.8. The variable under consideration here is age, and the population consists of the eight students. Let X denote the age of a randomly selected student. In view of Table 5.8, the probability distribution of the random variable X is as shown in Table 5.9. Express the mean age of the eight students in terms of the probability distribution of the random variable X.

Solution   Referring first to Table 5.8 and then to Table 5.9, we get

$$\mu = \frac{\Sigma x}{N} = \frac{19 + 20 + 20 + 19 + 21 + 27 + 20 + 21}{8}$$

$$= \frac{\overbrace{19 + 19}^{2} + \overbrace{20 + 20 + 20}^{3} + \overbrace{21 + 21}^{2} + \overbrace{27}^{1}}{8}$$

$$= \frac{19 \cdot 2 + 20 \cdot 3 + 21 \cdot 2 + 27 \cdot 1}{8}$$

$$= 19 \cdot \frac{2}{8} + 20 \cdot \frac{3}{8} + 21 \cdot \frac{2}{8} + 27 \cdot \frac{1}{8}$$

$$= 19 \cdot P(X = 19) + 20 \cdot P(X = 20) + 21 \cdot P(X = 21) + 27 \cdot P(X = 27)$$

$$= \Sigma x P(X = x). \qquad \blacksquare$$

Example 5.6 shows that we can express the mean of a variable of a finite population in terms of the probability distribution of the corresponding random variable: $\mu = \Sigma x P(X = x)$. Because the expression on the right of this equation is meaningful for any discrete random variable, we can define the **mean of a discrete random variable** as follows, along with the synonyms **expected value** and **expectation** for *mean*.

DEFINITION 5.4

Mean of a Discrete Random Variable
The ***mean of a discrete random variable*** X is denoted μ_X or, when no confusion will arise, simply $\boldsymbol{\mu}$. It is defined by
$$\mu = \Sigma x P(X = x).$$
The terms ***expected value*** and ***expectation*** are commonly used in place of the term *mean*.

What
Does it Mean?

To obtain the mean of a discrete random variable, multiply each possible value by its probability and then add those products.

We now have a definition of *mean* consistent with that for variables of finite populations and applicable to any discrete random variable. In other words, we have extended the concept of population mean to any discrete variable.[†]

As you learned in Chapter 3, constructing appropriate tables provides an efficient way to compute descriptive measures by hand. In Example 5.7, we apply this technique to obtain the mean of a discrete random variable.

Example 5.7   **The Mean of a Discrete Random Variable**

TABLE 5.10

Table for computing the mean of the random variable X, the number of tellers busy with customers

x	$P(X = x)$	$xP(X = x)$
0	0.029	0.000
1	0.049	0.049
2	0.078	0.156
3	0.155	0.465
4	0.212	0.848
5	0.262	1.310
6	0.215	1.290
		4.118

Busy Tellers   Prescott National Bank has six tellers available to serve customers. The number of tellers busy with customers at, say, 1:00 P.M. varies from day to day and depends on chance; hence it is a random variable, say, X. Past records indicate that the probability distribution of X is as shown in the first two columns of Table 5.10. For instance, the probability is 0.262 that exactly five tellers will be busy with customers at 1:00 P.M. Find the mean of the random variable X.

Solution   To obtain the mean of the random variable X, we append a column for the product of x with $P(X = x)$, shown as the third column in Table 5.10, and apply Definition 5.4. Summing the entries in the third column, we find that

$$\mu = \Sigma x P(X = x) = 4.118.$$

INTERPRETATION   The mean number of tellers busy with customers is 4.118.

___

[†]We can also extend the concept of population mean to any continuous variable and, using integral calculus, develop a formula analogous to the one given in Definition 5.4 for discrete variables. We do not present the formula for the mean of a continuous variable because it is not needed in this book.

Interpretation of the Mean of a Random Variable

Recall that the mean of a variable of a finite population is the arithmetic average of all possible observations. A similar interpretation holds for the mean of a random variable.

For instance, in Example 5.7 the random variable X is the number of tellers busy with customers at 1:00 P.M., and the mean of that random variable is 4.118. Of course, there never will be a day when 4.118 tellers are busy with customers at 1:00 P.M. The mean of 4.118 simply indicates that, over many days, the average number of busy tellers at 1:00 P.M. will be about 4.118.

This interpretation holds in all cases. It is commonly known as the **law of averages** and in mathematical circles as the **law of large numbers.** We restate this interpretation as the following fact.

KEY FACT 5.3

What
Does it Mean?

The mean of a random variable can be considered the long-run-average value of the random variable in repeated independent observations.

Interpretation of the Mean of a Random Variable

In a large number of independent observations of a random variable X, the average value of those observations will approximately equal the mean, μ, of X. The larger the number of observations, the closer the average tends to be to μ.

We used a computer to simulate the number of busy tellers at 1:00 P.M. on 100 randomly selected days; that is, we obtained 100 independent observations of the random variable X. The data are displayed in Table 5.11.

TABLE 5.11
One hundred observations of the random variable X, the number of tellers busy with customers

5	3	5	3	4	3	4	3	6	5	6	4	5	4	3	5	4	5	6	3
4	1	6	5	3	6	3	5	5	4	6	4	1	6	5	3	3	6	4	5
3	4	2	5	5	6	5	4	6	2	4	5	4	6	4	5	5	3	4	6
1	5	4	6	4	4	4	5	6	2	5	4	5	1	3	3	6	4	6	4
5	6	5	5	3	2	4	6	6	1	5	1	3	6	5	3	5	4	3	6

The average value of the 100 observations in Table 5.11 is 4.25. This value is quite close to the mean, $\mu = 4.118$, of the random variable X. If we made, say, 1000 observations instead of 100, the average value of those 1000 observations would most likely be even closer to 4.118.

Figure 5.3(a) shows a plot of the average number of busy tellers versus the number of observations for the data in Table 5.11. The dashed line is at $\mu = 4.118$. Figure 5.3(b) depicts a plot for a different simulation of the number of busy tellers at 1:00 P.M. on 100 randomly selected days. Both plots suggest that, as the number of observations increases, the average number of busy tellers approaches the mean, $\mu = 4.118$, of the random variable X.

Standard Deviation of a Discrete Random Variable

Similar reasoning also lets us extend the concept of population standard deviation (standard deviation of a variable) to any discrete variable. When we do so, we have the **standard deviation of a discrete random variable,** which we define in Definition 5.5.

FIGURE 5.3
Graphs showing the average number of busy tellers versus the number of observations for two simulations of 100 observations each

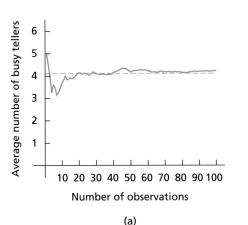

(a)

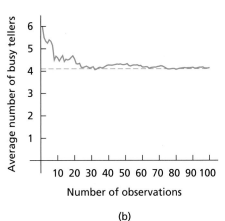

(b)

DEFINITION 5.5

<table>
<tr><td>

What
Does it Mean?

Roughly speaking, the standard deviation of a random variable X indicates how far, on average, an observed value of X is from its mean. In particular, the smaller the standard deviation of X, the more likely that an observed value of X will be close to its mean.

</td></tr>
</table>

Standard Deviation of a Discrete Random Variable

The *standard deviation of a discrete random variable* X is denoted σ_X or, when no confusion will arise, simply σ. It is defined as

$$\sigma = \sqrt{\Sigma(x - \mu)^2 P(X = x)}.$$

The standard deviation of a discrete random variable can also be obtained from the computing formula

$$\sigma = \sqrt{\Sigma x^2 P(X = x) - \mu^2}.$$

Note: The square of the standard deviation, σ^2, is called the **variance** of the random variable X.

In Example 5.8, we demonstrate computing the standard deviation of a discrete random variable.

Example 5.8   **The Standard Deviation of a Discrete Random Variable**

Busy Tellers   Refer to Example 5.7, where X denotes the number of tellers busy with customers at 1:00 P.M. Find the standard deviation of the random variable X.

Solution   We apply the computing formula given in Definition 5.5. To use that formula, we need the mean of X, which we obtained in Example 5.7, and columns for x^2 and $x^2 P(X = x)$, which are presented in the last two columns of Table 5.12 on the following page. From the final column of Table 5.12, we see that $\Sigma x^2 P(X = x) = 19.438$. Recalling that $\mu = 4.118$, we can now obtain the standard deviation of X:

$$\sigma = \sqrt{\Sigma x^2 P(X = x) - \mu^2} = \sqrt{19.438 - (4.118)^2} = 1.6,$$

rounded to one decimal place. The standard deviation of the number of tellers busy with customers is 1.6.

INTERPRETATION   Roughly speaking, on average, the number of busy tellers is 1.6 from the mean of 4.118 busy tellers.

TABLE 5.12
Table for computing the standard deviation of the random variable X, the number of tellers busy with customers

x	$P(X = x)$	x^2	$x^2 P(X = x)$
0	0.029	0	0.000
1	0.049	1	0.049
2	0.078	4	0.312
3	0.155	9	1.395
4	0.212	16	3.392
5	0.262	25	6.550
6	0.215	36	7.740
			19.438

EXERCISES 5.2

Statistical Concepts and Skills

5.15 What concept does the mean of a discrete random variable X generalize?

5.16 Comparing Investments. Suppose that the random variables X and Y represent the amount of return on two different investments. Further suppose that the mean of X equals the mean of Y but that the standard deviation of X is greater than the standard deviation of Y.
a. On average, is there a difference between the returns of the two investments? Explain your answer.
b. Which investment is more conservative? Why?

5.17 Space Shuttles. In Exercise 5.7 on page 232, you were asked to determine the probability distribution of the crew size, X, of a randomly selected shuttle mission between April 1981 and July 2000. That probability distribution is as follows.

x	2	3	4	5	6	7	8
$P(X = x)$	0.042	0.010	0.021	0.375	0.188	0.344	0.021

a. Find and interpret the mean of the random variable X.
b. Obtain the standard deviation of X by using one of the formulas given in Definition 5.5 on page 237.
c. Draw a probability histogram for the random variable; locate the mean; and show one, two, and three standard-deviation intervals.

5.18 Dice. In Exercise 5.8 on page 233, you were asked to determine the probability distribution of the sum of the dice, Y, when two balanced dice are rolled. That probability distribution is as follows.

y	2	3	4	5	6	7	8	9	10	11	12
$P(Y = y)$	$\frac{1}{36}$	$\frac{1}{18}$	$\frac{1}{12}$	$\frac{1}{9}$	$\frac{5}{36}$	$\frac{1}{6}$	$\frac{5}{36}$	$\frac{1}{9}$	$\frac{1}{12}$	$\frac{1}{18}$	$\frac{1}{36}$

a. Find and interpret the mean of Y.
b. Obtain the standard deviation of Y by using one of the formulas given in Definition 5.5 on page 237.
c. Draw a probability histogram for the random variable; locate the mean; and show one, two, and three standard-deviation intervals.

Expected Value. *As noted in Definition 5.4 on page 235, the mean of a random variable is also called its expected value. This terminology is especially useful in gambling and decision theory, as illustrated in Exercises 5.19 and 5.20.*

5.19 Roulette. An American roulette wheel contains 38 numbers: 18 are red, 18 are black, and 2 are green. When the roulette wheel is spun, the ball is equally likely to land on any of the 38 numbers. Suppose that you bet $1 on red. If the ball lands on a red number, you win $1; otherwise you lose your $1. Let X be the amount you win on your $1 bet. Then X is a random variable whose probability distribution is as follows.

x	1	−1
$P(X = x)$	0.474	0.526

a. Verify that the probability distribution shown in the table is correct.
b. Find the expected value of the random variable X.
c. On average, how much will you lose per play?

d. Approximately how much would you expect to lose if you bet $1 on red 100 times? 1000 times?

e. Is roulette a profitable game to play? Explain.

5.20 Evaluating Investments. An investor plans to put $50,000 in one of four investments. The return on each investment depends on whether next year's economy is strong or weak. The following table summarizes the possible payoffs, in dollars, for the four investments.

	Next year's economy	
	Strong	Weak
Certificate of deposit	6,000	6,000
Office complex	15,000	5,000
Land speculation	33,000	-17,000
Technical school	5,500	10,000

(row label: **Investment**)

Let V, W, X, and Y denote the payoffs for the certificate of deposit, office complex, land speculation, and technical school, respectively. Then V, W, X, and Y are random variables. Assume that next year's economy has a 40% chance of being strong and a 60% chance of being weak.

a. Find the probability distribution of each random variable V, W, X, and Y.

b. Determine the expected value of each random variable.

c. Which investment has the best expected payoff? Which has the worst?

d. Which investment would you select? Explain.

5.21 Equipment Breakdowns. A factory manager collected data on the number of equipment breakdowns per day. From those data, she derived the probability distribution shown in the following table, where W denotes the number of breakdowns on a given day.

w	0	1	2
$P(W = w)$	0.80	0.15	0.05

a. Determine μ_W and σ_W.

b. On average, how many breakdowns occur per day?

c. About how many breakdowns are expected during a 1-year period, assuming 250 work days per year?

Extending the Concepts and Skills

Properties of the Mean and Standard Deviation. In Exercises 5.22 and 5.23, you will develop some important properties of the mean and standard deviation of a random variable. Two of them relate the mean and standard deviation of the sum of two random variables to the individual means and standard deviations, respectively; two others relate the mean and standard deviation of a constant times a random variable to the constant and the mean and standard deviation of the random variable, respectively.

In developing these properties, you will need to use the concept of independent random variables. Two discrete random variables, X and Y, are said to be *independent random variables* if

$$P\left(\{X = x\} \& \{Y = y\}\right) = P(X = x) \cdot P(Y = y)$$

for all x and y—that is, if the joint probability distribution of X and Y equals the product of their marginal probability distributions. This condition is equivalent to requiring that events $\{X = x\}$ and $\{Y = y\}$ are independent for all x and y. A similar definition holds for independence of more than two discrete random variables.

5.22 Equipment Breakdowns. Refer to Exercise 5.21. Assume that the number of breakdowns on different days are independent of one another. Let X and Y denote the number of breakdowns on each of two consecutive days.

a. Complete the following joint probability distribution table.

			y		
		0	1	2	$P(X = x)$
x	0				
	1				
	2				
	$P(Y = y)$				

Hint: To obtain the joint probability in the first row, third column, use the definition of independence for discrete random variables and the table in Exercise 5.21:

$$P(\{X = 0\} \& \{Y = 2\}) = P(X = 0) \cdot P(Y = 2)$$
$$= 0.80 \cdot 0.05 = 0.04.$$

b. Use the joint probability distribution you obtained in part (a) to determine the probability distribution of the random variable $X + Y$, the total number of breakdowns in two days; that is, complete the following table.

u	0	1	2	3	4
$P(X + Y = u)$					

c. Use part (b) to find μ_{X+Y} and σ^2_{X+Y}.
d. Use part (c) to verify that the following equations hold for this example:

$$\mu_{X+Y} = \mu_X + \mu_Y \quad \text{and} \quad \sigma^2_{X+Y} = \sigma^2_X + \sigma^2_Y.$$

(*Note:* The mean and variance of X and Y are the same as that of W in Exercise 5.21.)

e. The equations in part (d) hold in general: If X and Y are any two random variables,

$$\mu_{X+Y} = \mu_X + \mu_Y.$$

In addition, if X and Y are independent,

$$\sigma^2_{X+Y} = \sigma^2_X + \sigma^2_Y.$$

Interpret these two equations in words.

5.23 Equipment Breakdowns. The factory manager in Exercise 5.21 estimates that each breakdown costs the company \$500 in repairs and loss of production. If W is the number of breakdowns in a day, then \$500W is the cost of breakdowns for that day.

a. Refer to the probability distribution shown in Exercise 5.21 and determine the probability distribution of the random variable 500W.

b. Determine the mean daily breakdown cost, μ_{500W}, by using your answer from part (a).
c. What is the relationship between μ_{500W} and μ_W? (*Note:* From Exercise 5.21, $\mu_W = 0.25$.)
d. Find σ_{500W} by using your answer from part (a).
e. What is the apparent relationship between σ_{500W} and σ_W? (*Note:* From Exercise 5.21, $\sigma_W = 0.536$.)
f. The results in parts (c) and (e) hold in general: If W is any random variable and c is a constant,

$$\mu_{cW} = c\mu_W \quad \text{and} \quad \sigma_{cW} = |c|\sigma_W.$$

Interpret these two equations in words.

Using Technology

5.24 Queueing Simulation. Benny's Barber Shop in Cleveland has five chairs for waiting customers. The number of customers waiting is a random variable Y with the following probability distribution.

y	0	1	2	3	4	5
$P(Y = y)$	0.424	0.161	0.134	0.111	0.093	0.077

a. Compute the mean number of customers waiting, μ.
b. In a large number of independent observations, how many customers will be waiting, on average?
c. Use the technology of your choice to simulate 100 observations of the number of customers waiting.
d. Obtain the mean of the observations in part (c) and compare it to μ.
e. What does part (d) illustrate?

5.3 The Binomial Distribution

Many applications of probability and statistics concern the repetition of an experiment. In such contexts, each repetition is called a **trial.** Of particular interest is the case when the experiment (each trial) has only two possible outcomes. Here are three examples.

- Testing the effectiveness of a drug: Several patients take the drug (the trials), and for each patient the drug is either effective or not effective (the two possible outcomes).
- Weekly sales of a car salesperson: The salesperson has several customers during the week (the trials), and for each customer the salesperson either makes a sale or does not make a sale (the two possible outcomes).
- Taste tests for colas: A number of people taste two different colas (the trials), and for each person the preference is either for the first cola or for the second cola (the two possible outcomes).

To analyze repeated trials of an experiment that has two possible outcomes requires knowledge of factorials, binomial coefficients, Bernoulli trials, and the binomial distribution. We begin with factorials.

Factorials

Factorials are defined as follows.

DEFINITION 5.6

Factorials

The product of the first k positive integers is called **k factorial** and is denoted $k!$. In symbols,

$$k! = k(k-1)\cdots 2 \cdot 1.$$

We also define $0! = 1$.

We illustrate the calculation of factorials in Example 5.9.

|| **Example 5.9**   **Factorials**

Doing the Calculations   Determine $3!$, $4!$, and $5!$.

Solution   Applying Definition 5.6 gives $3! = 3 \cdot 2 \cdot 1 = 6$, $4! = 4 \cdot 3 \cdot 2 \cdot 1 = 24$, and $5! = 5 \cdot 4 \cdot 3 \cdot 2 \cdot 1 = 120$.   ■

Note, for instance, that $6! = 6 \cdot 5!$, $6! = 6 \cdot 5 \cdot 4!$, $6! = 6 \cdot 5 \cdot 4 \cdot 3!$, and so on. In general, if $j \leq k$, then $k! = k(k-1)\cdots(k-j+1)(k-j)!$.

Binomial Coefficients

You may have already encountered binomial coefficients in algebra when you studied the binomial expansion, the expansion of $(a+b)^n$. Here is the definition of **binomial coefficients**.[†]

DEFINITION 5.7

Binomial Coefficients

If n is a positive integer and x is a nonnegative integer less than or equal to n, then the **binomial coefficient** $\binom{n}{x}$ is defined as

$$\binom{n}{x} = \frac{n!}{x!\,(n-x)!}.$$

[†]If you have read Section 4.8, you will note that the binomial coefficient $\binom{n}{x}$ equals the number of possible combinations of x objects from a collection of n objects.

Example 5.10   Binomial Coefficients

Doing the Calculations   Determine the value of each binomial coefficient.

a. $\binom{6}{1}$ b. $\binom{5}{3}$ c. $\binom{7}{3}$ d. $\binom{4}{4}$

Solution   We apply Definition 5.7.

a. $\binom{6}{1} = \dfrac{6!}{1!\,(6-1)!} = \dfrac{6!}{1!\,5!} = \dfrac{6 \cdot \cancel{5!}}{1!\,\cancel{5!}} = \dfrac{6}{1} = 6$

b. $\binom{5}{3} = \dfrac{5!}{3!\,(5-3)!} = \dfrac{5!}{3!\,2!} = \dfrac{5 \cdot 4 \cdot \cancel{3!}}{\cancel{3!}\,2!} = \dfrac{5 \cdot 4}{2} = 10$

c. $\binom{7}{3} = \dfrac{7!}{3!\,(7-3)!} = \dfrac{7!}{3!\,4!} = \dfrac{7 \cdot 6 \cdot 5 \cdot \cancel{4!}}{3!\,\cancel{4!}} = \dfrac{7 \cdot 6 \cdot 5}{6} = 35$

d. $\binom{4}{4} = \dfrac{4!}{4!\,(4-4)!} = \dfrac{4!}{4!\,0!} = \dfrac{\cancel{4!}}{\cancel{4!}\,0!} = \dfrac{1}{1} = 1$

Bernoulli Trials

Next we define **Bernoulli trials,** the trial outcomes—**success** and **failure**—and the **success probability.**

DEFINITION 5.8

Bernoulli Trials

Repeated trials of an experiment are called *Bernoulli trials* if the following three conditions are satisfied:

1. The experiment (each trial) has two possible outcomes, denoted generically *s,* for *success,* and *f,* for *failure.*
2. The trials are independent.
3. The probability of a success remains the same from trial to trial, called the *success probability* and denoted *p.*

Introducing the Binomial Distribution

The **binomial distribution** is the probability distribution for the number of successes in a sequence of Bernoulli trials. We introduce this concept in Example 5.11.

Example 5.11   Introducing the Binomial Distribution

Mortality   Mortality tables enable actuaries to obtain the probability that a person at any particular age will live a specified number of years. Such probabilities, in turn, permit the determination of life-insurance premiums, retirement pensions, annuity payments, and related items of importance to insurance companies and others.

According to tables provided by the U.S. National Center for Health Statistics in *Vital Statistics of the United States*, there is about an 80% chance that a person aged 20 will be alive at age 65. Suppose that three people aged 20 are selected at random.

a. Formulate the process of observing which people are alive at age 65 as a sequence of three Bernoulli trials.

b. Obtain the possible outcomes of the three Bernoulli trials.

c. Determine the probability of each outcome in part (b).

d. Find the probability that exactly two of the three people will be alive at age 65.

e. Obtain the probability distribution of the number of people of the three who are alive at age 65.

Solution

a. Each trial consists of observing whether a person currently aged 20 is alive at age 65 and has two possible outcomes: alive or dead. The trials are independent. If we let a success, *s*, correspond to being alive at age 65, the success probability is 0.8 (80%); that is, $p = 0.8$.

TABLE 5.13
Possible outcomes

sss	*ssf*	*sfs*	*sff*
fss	*fsf*	*ffs*	*fff*

b. The possible outcomes of the three Bernoulli trials are shown in Table 5.13 (*s* = success = alive, *f* = failure = dead). For instance, *ssf* represents the outcome that at age 65 the first two people are alive and the third is not.

c. As Table 5.13 indicates, eight outcomes are possible. However, because these eight outcomes are not equally likely, we cannot use the f/N rule to determine their probabilities; instead, we must proceed as follows. First of all, by part (a), the success probability equals 0.8, or

$$P(s) = p = 0.8.$$

Therefore the failure probability is

$$P(f) = 1 - p = 1 - 0.8 = 0.2.$$

TABLE 5.14
Outcomes and probabilities for observing whether each of three people is alive at age 65

Outcome	Probability
sss	$(0.8)(0.8)(0.8) = 0.512$
ssf	$(0.8)(0.8)(0.2) = 0.128$
sfs	$(0.8)(0.2)(0.8) = 0.128$
sff	$(0.8)(0.2)(0.2) = 0.032$
fss	$(0.2)(0.8)(0.8) = 0.128$
fsf	$(0.2)(0.8)(0.2) = 0.032$
ffs	$(0.2)(0.2)(0.8) = 0.032$
fff	$(0.2)(0.2)(0.2) = 0.008$

Now, because the trials are independent, we can apply the special multiplication rule (Formula 4.7 on page 192) to obtain the probability of each of the eight possible outcomes. For instance, the probability of the outcome *ssf* is

$$P(ssf) = P(s) \cdot P(s) \cdot P(f) = 0.8 \cdot 0.8 \cdot 0.2 = 0.128.$$

Similar computations yield the probabilities of the other seven possible outcomes. All eight possible outcomes and their probabilities are shown in Table 5.14. Note that outcomes containing the same number of successes have the same probability. For instance, three outcomes contain exactly two successes: *ssf*, *sfs*, and *fss*. Each of those three outcomes has the same probability: 0.128. The reason is that each probability is obtained by multiplying two success probabilities of 0.8 and one failure probability of 0.2.

A tree diagram is useful for organizing and summarizing the possible outcomes of this experiment and their probabilities. The tree diagram corresponding to Table 5.14 is presented in Fig. 5.4.

FIGURE 5.4

Tree diagram corresponding to Table 5.14

First person	Second person	Third person	Outcome	Probability

s — s 0.8 — s   *sss*   (0.8)(0.8)(0.8) = 0.512

0.8 — 0.2 — f   *ssf*   (0.8)(0.8)(0.2) = 0.128

s — 0.2 — f 0.8 — s   *sfs*   (0.8)(0.2)(0.8) = 0.128

0.8 — 0.2 — f   *sff*   (0.8)(0.2)(0.2) = 0.032

s 0.8 — s   *fss*   (0.2)(0.8)(0.8) = 0.128

0.2 — 0.8 — 0.2 — f   *fsf*   (0.2)(0.8)(0.2) = 0.032

f — 0.2 — f 0.8 — s   *ffs*   (0.2)(0.2)(0.8) = 0.032

0.8 — 0.2 — f   *fff*   (0.2)(0.2)(0.2) = 0.008

TABLE 5.15

Probability distribution of the random variable *X*, the number of people of three who are alive at age 65

Number alive x	Probability $P(X = x)$
0	0.008
1	0.096
2	0.384
3	0.512

d. Table 5.14 shows that the event that exactly two of the three people are alive at age 65 consists of three outcomes: *ssf*, *sfs*, and *fss*. It also shows that each of those three outcomes has the same probability, 0.128. So, by the special addition rule (Formula 4.1 on page 168), we have

$$P(\text{Exactly two will be alive}) = P(ssf) + P(sfs) + P(fss)$$
$$= \underbrace{0.128 + 0.128 + 0.128}_{3 \text{ times}} = 3 \cdot 0.128 = 0.384.$$

The probability that exactly two of the three people will be alive at age 65 is 0.384.

e. Let *X* denote the number of people of the three who are alive at age 65. In part (d), we found $P(X = 2)$. Proceeding in the same way, we can determine the remaining three probabilities: $P(X = 0)$, $P(X = 1)$, and $P(X = 3)$. The results are displayed in Table 5.15. A probability histogram for the distribution in Table 5.15 is given in Fig. 5.5. Note for future reference that the probability distribution is left skewed.

FIGURE 5.5

Probability histogram for the random variable *X*, the number of people of three who are alive at age 65

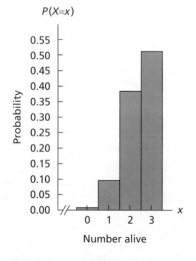

Number alive

The Binomial Probability Formula

Table 5.15 displays the probability distribution of the random variable "number alive at age 65" for three people currently aged 20. To obtain that probability distribution, we used a tabulation method (Table 5.14), which required a significant amount of work.

In most practical applications, the amount of work required would be considerably more and often would be prohibitive because the number of trials is generally much larger than three. For instance, for twenty 20-year-olds instead of three 20-year-olds, there would be over 1 million possible outcomes. In that case, the tabulation method certainly would not be feasible.

The good news is that there is a relatively simple formula for obtaining binomial probabilities. A first step in developing that formula is the following fact.

KEY FACT 5.4

> **Number of Outcomes Containing a Specified Number of Successes**
>
> In n Bernoulli trials, the number of outcomes that contain exactly x successes equals the binomial coefficient $\binom{n}{x}$.

What
Does it Mean?

There are $\binom{n}{x}$ ways of getting exactly x successes in n Bernoulli trials.

We won't stop to prove Key Fact 5.4, but let's quickly see whether it is consistent with the results obtained in Example 5.11. For instance, the direct listing in Table 5.13 shows that there are three outcomes in which exactly two of the three people are alive at age 65—namely, *ssf*, *sfs*, and *fss*. Using binomial coefficients, we can determine that fact without resorting to a direct listing. Applying Key Fact 5.4, we have

$$\begin{bmatrix} \text{Number of outcomes} \\ \text{comprising the event} \\ \text{exactly two alive} \end{bmatrix} = \binom{3}{2} = \frac{3!}{2!\,(3-2)!} = \frac{3!}{2!\,1!} = 3.$$

We can now develop a probability formula for the number of successes in Bernoulli trials. We indicate briefly how that formula is derived by referring to Example 5.11. For instance, to determine the probability that exactly two of the three people will be alive at age 65, $P(X = 2)$, we reason as follows:

1. Any particular outcome in which exactly two of the three people are alive at age 65 (e.g., *sfs*) has probability

$$\overset{\text{Two alive}}{\underset{\underset{\text{alive}}{\text{Probability}}}{(0.8)^2}} \cdot \overset{\text{One dead}}{\underset{\underset{\text{dead}}{\text{Probability}}}{(0.2)^1}} = 0.64 \cdot 0.2 = 0.128,$$

obtained by multiplying two success probabilities of 0.8 and one failure probability of 0.2.

2. By Key Fact 5.4, the number of outcomes in which exactly two of the three people are alive at age 65 is

$$\overset{\text{Number of trials}}{\underset{\text{Number alive}}{\binom{3}{2}}} = \frac{3!}{2!\,(3-2)!} = 3$$

3. By the special addition rule, the probability that exactly two of the three people will be alive at age 65 is

$$P(X = 2) = \binom{3}{2} \cdot (0.8)^2(0.2)^1 = 3 \cdot 0.128 = 0.384.$$

Of course, this result is the same as that obtained in Example 5.11(d). However, this time we determined the probability quickly and easily—no tabulation and no listing were required. More important, the reasoning we used applies to any sequence of Bernoulli trials and leads to the **binomial probability formula,** which we express in the following formula.

FORMULA 5.1

Binomial Probability Formula

Let X denote the total number of successes in n Bernoulli trials with success probability p. Then the probability distribution of the random variable X is given by

$$P(X = x) = \binom{n}{x} p^x (1 - p)^{n-x}, \qquad x = 0, 1, 2, ..., n.$$

The random variable X is called a **binomial random variable** and is said to have the **binomial distribution** with parameters n and p.

To determine a binomial probability formula in specific problems, having a well-organized strategy, such as the one presented in Procedure 5.1, is useful.

PROCEDURE 5.1

To Find a Binomial Probability Formula

Assumptions

1. n trials are to be performed.
2. Two outcomes, success or failure, are possible for each trial.
3. The trials are independent.
4. The success probability, p, remains the same from trial to trial.

STEP 1 Identify a success.

STEP 2 Determine p, the success probability.

STEP 3 Determine n, the number of trials.

STEP 4 The binomial probability formula for the number of successes, X, is

$$P(X = x) = \binom{n}{x} p^x (1 - p)^{n-x}.$$

In Example 5.12, we illustrate Procedure 5.1 by applying it to the random variable considered in Example 5.11.

Example 5.12 Obtaining Binomial Probabilities

Mortality According to tables provided by the U.S. National Center for Health Statistics in *Vital Statistics of the United States*, there is roughly an 80% chance that a person aged 20 will be alive at age 65. Suppose that three people aged 20 are selected at random. Find the probability that the number alive at age 65 will be

a. exactly two.   **b.** at most one.   **c.** at least one.

d. Determine the probability distribution of the number alive at age 65.

Solution Let X denote the number of people of the three who are alive at age 65. To solve parts (a)–(d), we first apply Procedure 5.1.

STEP 1 Identify a success.

A success is that a person currently aged 20 will be alive at age 65.

STEP 2 Determine p, the success probability.

The probability that a person currently aged 20 will be alive at age 65 is 80%, so $p = 0.8$.

STEP 3 Determine n, the number of trials.

The number of trials is the number of people in the study, which is three, so $n = 3$.

STEP 4 The binomial probability formula for the number of successes, X, is

$$P(X = x) = \binom{n}{x} p^x (1 - p)^{n-x}.$$

As $n = 3$ and $p = 0.8$, the formula becomes

$$P(X = x) = \binom{3}{x}(0.8)^x (0.2)^{3-x}.$$

Now that we have applied Procedure 5.1, solving parts (a)–(d) is relatively easy.

a. Applying the binomial probability formula with $x = 2$ yields

$$P(X = 2) = \binom{3}{2}(0.8)^2(0.2)^{3-2} = \frac{3!}{2!\,(3-2)!}(0.8)^2(0.2)^1 = 0.384.$$

INTERPRETATION Chances are 38.4% that exactly two of the three people will be alive at age 65.

b. The probability that at most one person will be alive at age 65 is

$$P(X \le 1) = P(X = 0) + P(X = 1)$$
$$= \binom{3}{0}(0.8)^0(0.2)^{3-0} + \binom{3}{1}(0.8)^1(0.2)^{3-1}$$
$$= 0.008 + 0.096 = 0.104.$$

INTERPRETATION Chances are 10.4% that one or fewer of the three people will be alive at age 65.

c. The probability that at least one person will be alive at age 65 is $P(X \ge 1)$, which we can obtain by first using the fact that

$$P(X \ge 1) = P(X = 1) + P(X = 2) + P(X = 3)$$

and then applying the binomial probability formula to calculate each of the three individual probabilities. However, using the complementation rule is easier:

$$P(X \geq 1) = 1 - P(X < 1) = 1 - P(X = 0)$$

$$= 1 - \binom{3}{0}(0.8)^0(0.2)^{3-0} = 1 - 0.008 = 0.992.$$

INTERPRETATION Chances are 99.2% that one or more of the three people will be alive at age 65.

d. To obtain the probability distribution of the random variable X, we need to use the binomial probability formula to compute $P(X = x)$, for $x = 0, 1, 2$, and 3. We have already done so for $x = 0, 1$, and 2 in parts (a) and (b). For $x = 3$, we have

$$P(X = 3) = \binom{3}{3}(0.8)^3(0.2)^{3-3} = (0.8)^3 = 0.512.$$

Thus the probability distribution of X is as shown in Table 5.15 on page 244. But this time we computed the probabilities quickly and easily by using the binomial probability formula. ∎

Shape of a Binomial Distribution

Figure 5.5 on page 244 shows that, for three people currently 20 years old, the probability distribution of the number who will be alive at age 65 is left skewed. The reason is that the success probability, $p = 0.8$, exceeds 0.5.

More generally, *a binomial distribution is right skewed if $p < 0.5$, is symmetric if $p = 0.5$, and is left skewed if $p > 0.5$.* Figure 5.6 illustrates these facts for three different binomial distributions with $n = 6$.

FIGURE 5.6

Probability histograms for binomial distributions with parameters $n = 6$ and (a) $p = 0.25$, (b) $p = 0.5$, and (c) $p = 0.75$

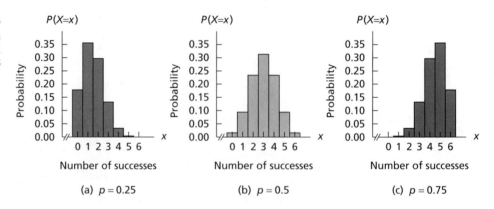

Mean and Standard Deviation of a Binomial Random Variable

In Section 5.2, we discussed the mean and standard deviation of a discrete random variable. We presented formulas to compute these parameters in Definition 5.4 on page 235 and Definition 5.5 on page 237.

Because these formulas apply to any discrete random variable, they work for a binomial random variable. Hence we can determine the mean and standard deviation of a binomial random variable by first using the binomial probability formula to obtain its probability distribution and then applying Definitions 5.4 and 5.5.

But there is an easier way. If we substitute the binomial probability formula into the formulas for the mean and standard deviation of a discrete random variable and then simplify mathematically, we obtain the following.

FORMULA 5.2

Mean and Standard Deviation of a Binomial Random Variable

The mean and standard deviation of a binomial random variable with parameters n and p are

$$\mu = np \quad \text{and} \quad \sigma = \sqrt{np(1-p)},$$

respectively.

What
Does it Mean?

To obtain the mean of a binomial random variable, multiply the number of trials by the success probability. To obtain the standard deviation, multiply the number of trials by the success probability by the failure probability, and take the square root of the result.

In Example 5.13, we apply the two formulas in Formula 5.2 to obtain the mean and standard deviation of the binomial random variable considered in the mortality illustration.

Example 5.13 **Mean and Standard Deviation of a Binomial Random Variable**

Mortality   For three randomly selected 20-year-olds, let X denote the number who are still alive at age 65. Obtain the mean and standard deviation of the random variable X.

Solution   As we previously found, X is a binomial random variable with parameters $n = 3$ and $p = 0.8$. Applying Formula 5.2, we get

$$\mu = np = 3 \cdot 0.8 = 2.4$$

and

$$\sigma = \sqrt{np(1-p)} = \sqrt{3 \cdot 0.8 \cdot 0.2} = 0.69.$$

INTERPRETATION   Of three randomly selected 20-year-olds, the number expected to be alive at age 65 is 2.4; that is, on average, 2.4 of every three 20-year-olds will still be alive at age 65.

Binomial Approximation to the Hypergeometric Distribution

Many statistical studies are concerned with the proportion (percentage) of members of a finite population that have a specified attribute. For instance, we might be interested in the proportion of U.S. adults who have Internet access. Here the population consists of all U.S. adults, and the specified attribute is "has Internet access." Or we might want to know the proportion of U.S. businesses that are minority owned. In this case, the population consists of all U.S. businesses, and the specified attribute is "minority owned."

Generally, the population under consideration is large, and it is therefore usually impractical and often impossible to determine the population proportion by taking a census; for instance, imagine trying to interview every U.S. adult to ascertain the proportion who have Internet access. So, in practice, we rely mostly on sampling and use the sample data to estimate the population proportion.

Suppose that a simple random sample of size n is taken from a population in which the proportion of members that have a specified attribute is p. Then a random variable of primary importance in the estimation of p is the number of members sampled that have the specified attribute, which we denote X. The exact probability distribution of X depends on whether the sampling is done with or without replacement.

If sampling is done with replacement, the sampling process constitutes Bernoulli trials: Each selection of a member from the population corresponds to a trial. A success occurs on a trial if the member selected in that trial has the specified attribute; otherwise, a failure occurs. The trials are independent because the sampling is done with replacement. The success probability remains the same from trial to trial—it always equals the proportion of the population that has the specified attribute. Therefore the random variable X has the binomial distribution with parameters n (the sample size) and p (the population proportion).

In reality, however, sampling is ordinarily done without replacement. Under these circumstances, the sampling process does not constitute Bernoulli trials because the trials are not independent and the success probability varies from trial to trial. In other words, the random variable X does not have a binomial distribution. Its distribution is important, however, and is referred to as a **hypergeometric distribution.**

We won't present the hypergeometric probability formula here because, in practice, a hypergeometric distribution can usually be approximated by a binomial distribution. The reason is that, if the sample size does not exceed 5% of the population size, there is little difference between sampling with and without replacement. We summarize the previous discussion as follows.

KEY FACT 5.5

What *Does it Mean?*

When a simple random sample is taken from a finite population, you can use a binomial distribution for the number of members obtained having a specified attribute, regardless of whether the sampling is with or without replacement, provided that, in the latter case, the sample size is small relative to the population size.

Sampling and the Binomial Distribution

Suppose that a simple random sample of size n is taken from a finite population in which the proportion of members that have a specified attribute is p. Then the number of members sampled that have the specified attribute

- has exactly a binomial distribution with parameters n and p if the sampling is done with replacement and
- has approximately a binomial distribution with parameters n and p if the sampling is done without replacement and the sample size does not exceed 5% of the population size.

For example, according to the U.S. Bureau of the Census publication *Current Population Reports*, 81.7% of U.S. adults have completed high school. Suppose that eight U.S. adults are to be randomly selected without replacement. Let X denote the number of those sampled that have completed high school. Then, as the sample size does not exceed 5% of the population size, the random variable X has approximately a binomial distribution with parameters $n = 8$ and $p = 0.817$.

Other Discrete Probability Distributions

The binomial distribution is the most important and most widely used discrete probability distribution. However, many other discrete probability distributions are often used in practice.

In addition to the hypergeometric distribution are the Poisson, discrete uniform, geometric, negative binomial, and multinomial distributions. We briefly discuss the Poisson, hypergeometric, and geometric distributions in the exercises of this section. We discuss the Poisson distribution in detail in Section 5.4.

The
Technology
Center

Almost all statistical technologies provide programs for determining binomial probabilities. In this subsection, we present output and step-by-step instructions to obtain individual binomial probabilities. For other options, refer to the technology manuals.

Example 5.14   **Using Technology to Obtain Binomial Probabilities**

Mortality   Consider once again the mortality illustration discussed in Example 5.12 on page 246. Use Minitab, Excel, or the TI-83/84 Plus to determine the probability that exactly two of the three people will be alive at age 65.

Solution   Recall that, of three randomly selected people aged 20, the number, X, who are alive at age 65 has a binomial distribution with parameters $n = 3$ and $p = 0.8$. We want the probability that exactly two of the three people will be alive at age 65—that is, $P(X = 2)$.

Printout 5.1 on the following page shows output obtained for this probability by applying the binomial probability programs from Minitab, Excel, and the TI-83/84 Plus. From any of these outputs, we find that the required probability is 0.384. This result, of course, agrees with the result that we obtained in part (a) of Example 5.12.

Obtaining the Output (Optional)

Printout 5.1 displays the probability that a binomial random variable with parameters $n = 3$ and $p = 0.8$ will be 2. Detailed instructions for obtaining that output are as follows.

MINITAB	EXCEL	TI-83/84 PLUS
1 Choose **Calc** ➤ **Probability Distributions** ➤ **Binomial...** 2 Select the **Probability** option button 3 Click in the **Number of trials** text box and type 3 4 Click in the **Probability of success** text box and type 0.8 5 Select the **Input constant** option button 6 Click in the **Input constant** text box and type 2 7 Click **OK**	1 Click **f_x** on the button bar 2 Select **Statistical** from the **Function category** list 3 Select **BINOMDIST** from the **Function name** list 4 Click **OK** 5 Type 2 in the **Number_s** text box 6 Click in the **Trials** text box and type 3 7 Click in the **Probability_s** text box and type 0.8 8 Click in the **Cumulative** text box and type FALSE	1 Press **2nd** ➤ **DISTR** 2 Arrow down to **binompdf(** and press **ENTER** 3 Type 3,0.8,2) and press **ENTER**

PRINTOUT 5.1

Output giving the probability that exactly two of the three people will be alive at age 65

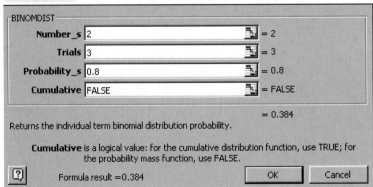

MINITAB

Probability Density Function

Binomial with n = 3 and p = 0.8

x   P(X = x)
2   0.384

TI-83/84 PLUS

```
binompdf(3,0.8,2
)
              .384
■
```

EXCEL

BINOMDIST

Number_s |2| = 2
Trials |3| = 3
Probability_s |0.8| = 0.8
Cumulative |FALSE| = FALSE

= 0.384

Returns the individual term binomial distribution probability.

Cumulative is a logical value: for the cumulative distribution function, use TRUE; for the probability mass function, use FALSE.

Formula result =0.384   OK   Cancel

EXERCISES 5.3

Statistical Concepts and Skills

5.25 Give two examples of Bernoulli trials other than those presented in the text.

5.26 What does the "bi" in "binomial" signify?

5.27 Compute 3!, 7!, 8!, and 9!.

5.28 Evaluate the following binomial coefficients.
a. $\binom{4}{1}$ **b.** $\binom{6}{2}$ **c.** $\binom{8}{3}$ **d.** $\binom{9}{6}$

5.29 Determine the value of each binomial coefficient.
a. $\binom{5}{3}$ **b.** $\binom{10}{0}$ **c.** $\binom{10}{10}$ **d.** $\binom{9}{5}$

5.30 Pinworm Infestation. Pinworm infestation, commonly found in children, can be treated with the drug pyrantel pamoate. According to the *Merck Manual*, the treatment is effective in 90% of cases. Suppose that three children with pinworm infestation are given pyrantel pamoate.
a. Considering a success in a given case to be "a cure," formulate the process of observing which children are cured

and which children are not cured as a sequence of three Bernoulli trials.

b. Construct a table similar to Table 5.14 on page 243 for the three cases. Display the probabilities to three decimal places.

c. Draw a tree diagram for this problem similar to the one shown in Fig. 5.4 on page 244.

d. List the outcomes in which exactly two of the three children are cured.

e. Find the probability of each outcome in part (d). Why are those probabilities all the same?

f. Use parts (d) and (e) to determine the probability that exactly two of the three children will be cured.

g. Without using the binomial probability formula, obtain the probability distribution of the random variable X, the number of children out of three who are cured.

5.31 Psychiatric Disorders. The National Institute of Mental Health reports that there is a 20% chance of an adult American suffering from a psychiatric disorder. Four randomly selected adult Americans are examined for psychiatric disorders.

a. If you let a success correspond to an adult American having a psychiatric disorder, what is the success probability, p? (*Note:* The use of the word *success* in Bernoulli trials need not reflect its usually positive connotation.)

b. Construct a table similar to Table 5.14 on page 243 for the four people examined. Display the probabilities to four decimal places.

c. Draw a tree diagram for this problem similar to the one shown in Fig. 5.4 on page 244.

d. List the outcomes in which exactly three of the four people examined have a psychiatric disorder.

e. Find the probability of each outcome in part (d). Why are those probabilities all the same?

f. Use parts (d) and (e) to determine the probability that exactly three of the four people examined have a psychiatric disorder.

g. Without using the binomial probability formula, obtain the probability distribution of the random variable Y, the number of adults out of four who have a psychiatric disorder.

5.32 Pinworm Infestation. Use Procedure 5.1 on page 246 to solve part (g) of Exercise 5.30.

5.33 Psychiatric Disorders. Use Procedure 5.1 on page 246 to solve part (g) of Exercise 5.31.

5.34 For each of the following probability histograms of binomial distributions, specify whether the success probability is less than, equal to, or greater than 0.5. Explain your answers.

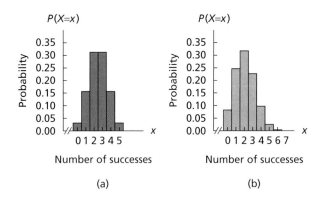

(a)   (b)

*In Exercises **5.35–5.37**, use Procedure 5.1 on page 246 to obtain the required probabilities. Express each probability answer as a decimal rounded to three places.*

5.35 Horse Racing. According to the *Daily Racing Form*, the probability is about 0.67 that the favorite in a horse race will finish in the money (first, second, or third place). In the next five races, what is the probability that the favorite finishes in the money

a. exactly twice?

b. exactly four times?

c. at least four times?

d. between two and four times, inclusive?

e. Determine the probability distribution of the random variable X, the number of times the favorite finishes in the money in the next five races.

f. Identify the probability distribution of X as right skewed, symmetric, or left skewed without consulting its probability distribution or drawing its probability histogram.

g. Draw a probability histogram for X.

h. Use your answer from part (e) and Definitions 5.4 and 5.5 on pages 235 and 237, respectively, to obtain the mean and standard deviation of the random variable X.

i. Use Formula 5.2 on page 249 to obtain the mean and standard deviation of the random variable X.

j. Interpret your answer for the mean in words.

5.36 Recidivism. In the May 2003 issue of *Scientific American*, R. Doyle examined rehabilitation of felons in the article, "Reducing Crime: Rehabilitation is Making a Comeback." One aspect of the article discussed recidivism of juvenile prisoners between 14 and 17 years old, indicating that 82% of those released in 1994 were rearrested within 3 years. Assuming that recidivism rate still applies today, solve the following problems for six newly released juvenile prisoners between 14 and 17 years old.

a. Determine the probability that the number rearrested within 3 years will be exactly four; at least four; at most five; between two and five, inclusive.

b. Determine the probability distribution of the random variable Y, the number of released prisoners of the six who are rearrested within 3 years.

c. Determine and interpret the mean of the random variable Y.

d. Obtain the standard deviation of Y.

e. If, in fact, exactly two of the six released juvenile prisoners are rearrested within 3 years, would you be inclined to conclude that the recidivism rate today has decreased from the 82% rate in 1994? Explain your reasoning. *Hint:* First consider the probability $P(Y \leq 2)$.

5.37 Teen Pregnancy. According to the periodical *Zero Population Growth* (Vol. 32, No. 2, p. 18), the United States leads the fully industrialized world in teen pregnancy rates, with 40% of U.S. females getting pregnant at least once before reaching the age of 20. Suppose that 10 U.S. 20-year-old females are selected at random. Find the probability that the number who have been pregnant at least once before the age of 20 will be

a. exactly five.

b. between three and five, inclusive.

c. less than 20% of those surveyed.

d. more than one.

Extending the Concepts and Skills

5.38 Roulette. A success, *s*, in Bernoulli trials is often derived from a collection of outcomes. For example, an American roulette wheel consists of 38 numbers, of which 18 are red, 18 are black, and 2 are green. When the roulette wheel is spun, the ball is equally likely to land on any one of the 38 numbers. If you are interested in which number the ball lands on, each play at the roulette wheel has 38 possible outcomes. Suppose, however, that you are betting on red. Then you are interested only in whether the ball lands on a red number. From this point of view, each play at the wheel has only two possible outcomes—either the ball lands on a red number or it doesn't. Hence successive bets on red constitute a sequence of Bernoulli trials with success probability $\frac{18}{38}$. In four plays at a roulette wheel, what is the probability that the ball lands on red

a. exactly twice?   **b.** at least once?

5.39 Lotto. A previous Arizona state lottery, called *Lotto*, is played as follows: The player selects six numbers from the numbers 1–42 and buys a ticket for $1. There are six winning numbers, which are selected at random from the numbers 1–42. To win a prize, a *Lotto* ticket must contain three or more of the winning numbers. A probability distribution for the number of winning numbers for a single ticket is shown in the following table.

Number of winning numbers	Probability
0	0.3713060
1	0.4311941
2	0.1684352
3	0.0272219
4	0.0018014
5	0.0000412
6	0.0000002

a. If you buy one *Lotto* ticket, determine the probability that you win a prize. Round your answer to three decimal places.

b. If you buy one *Lotto* ticket per week for a year, determine the probability that you win a prize at least once in the 52 tries.

5.40 Sickle Cell Anemia. Sickle cell anemia is an inherited blood disease that occurs primarily in blacks. In the United States, about 15 of every 10,000 black children have sickle cell anemia. The red blood cells of an affected person are abnormal; the result is severe chronic anemia (inability to carry the required amount of oxygen), which causes headaches, shortness of breath, jaundice, increased risk of pneumococcal pneumonia and gallstones, and other severe problems. Sickle cell anemia occurs in children who inherit an abnormal type of hemoglobin, called hemoglobin S, from both parents. If hemoglobin S is inherited from only one parent, the person is said to have sickle cell trait and is generally free from symptoms. There is a 50% chance that a person who has sickle cell trait will pass hemoglobin S to an offspring.

a. Obtain the probability that a child of two people who have sickle cell trait will have sickle cell anemia.

b. If two people who have sickle cell trait have five children, determine the probability that at least one of the children will have sickle cell anemia.

c. If two people who have sickle cell trait have five children, find the probability distribution of the number of those children who will have sickle cell anemia.

d. Construct a probability histogram for the probability distribution in part (c).

e. If two people who have sickle cell trait have five children, how many can they expect will have sickle cell anemia?

5.41 Sampling and the Binomial Distribution. Refer to the discussion on the binomial approximation to the hypergeometric distribution that begins on page 249.

a. If sampling is with replacement, explain why the trials are independent and the success probability remains the same from trial to trial—always the proportion of the population that has the specified attribute.

b. If sampling is without replacement, explain why the trials are not independent and the success probability varies from trial to trial.

5.42 The Hypergeometric Distribution. In this exercise, we discuss the *hypergeometric distribution* in more detail. When sampling is done without replacement from a finite population, the hypergeometric distribution is the exact probability distribution for the number of members sampled that have a specified attribute. The hypergeometric probability formula is

$$P(X = x) = \frac{\binom{Np}{x}\binom{N(1-p)}{n-x}}{\binom{N}{n}},$$

where X denotes the number of members sampled that have the specified attribute, N is the population size, n is the sample size, and p is the population proportion.

To illustrate, suppose that a customer purchases 4 fuses from a shipment of 250, of which 94% are not defective. Let a success correspond to a fuse that is not defective.
a. Determine N, n, and p.
b. Use the hypergeometric probability formula to find the probability distribution of the number of nondefective fuses the customer gets.
Key Fact 5.5 shows that a hypergeometric distribution can be approximated by a binomial distribution provided the sample size does not exceed 5% of the population size. In particular, you can use the binomial probability formula

$$P(X = x) = \binom{n}{x}p^x(1-p)^{n-x},$$

with $n = 4$ and $p = 0.94$, to approximate the probability distribution of the number of nondefective fuses that the customer gets.
c. Obtain the binomial distribution with parameters $n = 4$ and $p = 0.94$.
d. Compare the hypergeometric distribution that you obtained in part (b) with the binomial distribution that you obtained in part (c).

5.43 The Geometric Distribution. In this exercise, we discuss the *geometric distribution*, the probability distribution for the number of trials until the first success in Bernoulli trials. The geometric probability formula is

$$P(X = x) = p(1-p)^{x-1},$$

where X denotes the number of trials until the first success and p the success probability. Using the geometric proba-

bility formula and Definition 5.4 on page 235, we can show that the mean of the random variable X is $1/p$.

To illustrate, again consider the Arizona state lottery, *Lotto*, as described in Exercise 5.39. Suppose that you buy one *Lotto* ticket per week. Let X denote the number of weeks until you win a prize.
a. Find and interpret the probability formula for the random variable X. (*Note:* The appropriate success probability was obtained in Exercise 5.39(a).)
b. Compute the probability that the number of weeks until you win a prize is exactly 3; at most 3; at least 3.
c. On average, how long will it be until you win a prize?

5.44 The Poisson Distribution. Another important discrete probability distribution is the *Poisson distribution*, named in honor of the French mathematician and physicist Simeon Poisson (1781–1840). This probability distribution is often used to model the frequency with which a specified event occurs during a particular period of time. The Poisson probability formula is

$$P(X = x) = e^{-\lambda}\frac{\lambda^x}{x!},$$

where X is the number of times the event occurs and λ is a parameter equal to the mean of X. The number e is the base of natural logarithms and is approximately equal to 2.7183.

To illustrate, consider the following problem: Desert Samaritan Hospital, located in Mesa, Arizona, keeps records of emergency room traffic. Those records reveal that the number of patients who arrive between 6:00 P.M. and 7:00 P.M. has a Poisson distribution with parameter $\lambda = 6.9$. Determine the probability that, on a given day, the number of patients who arrive at the emergency room between 6:00 P.M. and 7:00 P.M. will be
a. exactly four.
b. at most two.
c. between four and 10, inclusive.

Using Technology

5.45 Horse Racing. Use the technology of your choice to obtain the required probability or probabilities in parts (a)–(e) of Exercise 5.35.

5.46 Recidivism. Use the technology of your choice to obtain the required probabilities in parts (a) and (b) of Exercise 5.36.

5.4   The Poisson Distribution

Another important discrete probability distribution is the **Poisson distribution,** named in honor of the French mathematician and physicist Simeon D. Poisson (1781–1840). The Poisson distribution is often used to model the frequency with which a specified event occurs during a particular period of time. For instance, we might apply the Poisson distribution when analyzing

- the number of patients who arrive at an emergency room between 6:00 P.M. and 7:00 P.M.,
- the number of telephone calls received per day at a switchboard, or
- the number of alpha particles emitted per minute by a radioactive substance.

Additionally, we might use the Poisson distribution to describe the probability distribution of the number of misprints in a book, the number of defective teeth a person has, or the number of bacterial colonies appearing on a petri dish smeared with a bacterial suspension.

The Poisson Probability Formula

Any particular Poisson distribution is identified by one parameter, usually denoted λ (the Greek letter lambda). Formula 5.3 provides the **Poisson probability formula,** the formula used to obtain probabilities for a random variable that has a Poisson distribution.

FORMULA 5.3

> **Poisson Probability Formula**
>
> Probabilities for a random variable X that has a Poisson distribution are given by the formula
>
> $$P(X = x) = e^{-\lambda}\frac{\lambda^x}{x!}, \qquad x = 0, 1, 2, \ldots,$$
>
> where λ is a positive real number and $e \approx 2.718$. (Most calculators have an e key.) The random variable X is called a *Poisson random variable* and is said to have the *Poisson distribution* with parameter λ.

A Poisson random variable has infinitely many possible values—namely, all whole numbers. Consequently, we cannot display all the probabilities for a Poisson random variable in a probability distribution table.

Example 5.15   The Poisson Distribution

Emergency Room Traffic   Desert Samaritan Hospital in Mesa, Arizona, keeps records of emergency room (ER) traffic. Those records indicate that the number of patients arriving between 6:00 P.M. and 7:00 P.M. has a Poisson distribution with parameter $\lambda = 6.9$. Determine the probability that, on a given day, the number of patients who arrive at the emergency room between 6:00 P.M. and 7:00 P.M. will be

a. exactly 4.

b. at most 2.

c. between 4 and 10, inclusive.

d. Obtain a table of probabilities for the random variable X, the number of patients arriving between 6:00 P.M. and 7:00 P.M. Stop when the probabilities become zero to three decimal places.

e. Use part (d) to construct a (partial) probability histogram for X.

f. Identify the shape of the probability distribution of X.

Solution The random variable X—the number of patients arriving between 6:00 P.M. and 7:00 P.M.—has a Poisson distribution with parameter $\lambda = 6.9$. Thus, by Formula 5.3, the probabilities for X are given by the Poisson probability formula,

$$P(X = x) = e^{-6.9}\, \frac{(6.9)^x}{x!}.$$

Using this formula, we can now solve the problems posed in parts (a)–(f).

a. Applying the Poisson probability formula with $x = 4$ gives

$$P(X = 4) = e^{-6.9}\, \frac{(6.9)^4}{4!} = e^{-6.9} \cdot \frac{2266.7121}{24} = 0.095.$$

INTERPRETATION Chances are 9.5% that exactly 4 patients will arrive at the ER between 6:00 P.M. and 7:00 P.M.

b. The probability of at most 2 arrivals is

$$P(X \le 2) = P(X = 0) + P(X = 1) + P(X = 2)$$

$$= e^{-6.9}\, \frac{(6.9)^0}{0!} + e^{-6.9}\, \frac{(6.9)^1}{1!} + e^{-6.9}\, \frac{(6.9)^2}{2!}$$

$$= e^{-6.9}\left(\frac{6.9^0}{0!} + \frac{6.9^1}{1!} + \frac{6.9^2}{2!} \right)$$

$$= e^{-6.9}\,(1 + 6.9 + 23.805) = e^{-6.9} \cdot 31.705 = 0.032.$$

INTERPRETATION Chances are only 3.2% that 2 or fewer patients will arrive at the ER between 6:00 P.M. and 7:00 P.M.

c. The probability of between 4 and 10 arrivals, inclusive, is

$$P(4 \le X \le 10) = P(X = 4) + P(X = 5) + \cdots + P(X = 10)$$

$$= e^{-6.9}\left(\frac{6.9^4}{4!} + \frac{6.9^5}{5!} + \cdots + \frac{6.9^{10}}{10!} \right) = 0.821.$$

INTERPRETATION Chances are 82.1% that between 4 and 10 patients will arrive at the ER between 6:00 P.M. and 7:00 P.M.

d. Proceeding as in part (a), we obtain a partial probability distribution of the random variable X, as shown in Table 5.16.

TABLE 5.16

Partial probability distribution of the random variable X, the number of patients arriving at the emergency room between 6:00 P.M. and 7:00 P.M.

Number arriving x	Probability $P(X = x)$	Number arriving x	Probability $P(X = x)$
0	0.001	10	0.068
1	0.007	11	0.043
2	0.024	12	0.025
3	0.055	13	0.013
4	0.095	14	0.006
5	0.131	15	0.003
6	0.151	16	0.001
7	0.149	17	0.001
8	0.128	18	0.000
9	0.098		

e. Using Table 5.16, we obtain a partial probability histogram for the random variable X, as depicted in Fig. 5.7.

FIGURE 5.7

Partial probability histogram for the random variable X, the number of patients arriving at the emergency room between 6:00 P.M. and 7:00 P.M.

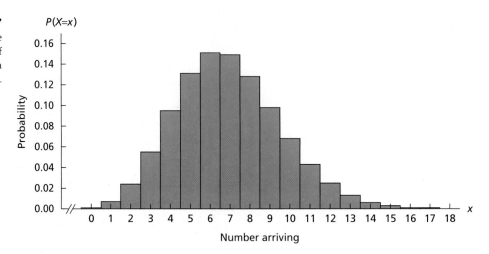

f. Figure 5.7 shows that the probability distribution of the random variable X is right skewed.

Shape of a Poisson Distribution

In part (f) of Example 5.15, we determined that the probability distribution of the number of patients arriving at the emergency room between 6:00 P.M. and 7:00 P.M. is right skewed. As a matter of fact: *All Poisson distributions are right skewed.*

Mean and Standard Deviation of a Poisson Random Variable

We can derive special formulas for the mean and standard deviation of a Poisson random variable. If we substitute the Poisson probability formula into the formulas for the mean and standard deviation of a discrete random variable and then simplify mathematically, we obtain the following formulas.

FORMULA 5.4

What *Does it Mean?*

The mean and standard deviation of a Poisson random variable are its parameter and square root of its parameter, respectively.

Mean and Standard Deviation of a Poisson Random Variable

The mean and standard deviation of a Poisson random variable with parameter λ are

$$\mu = \lambda \quad \text{and} \quad \sigma = \sqrt{\lambda},$$

respectively.

We apply Formula 5.4 in Example 5.16.

Example 5.16 Mean and Standard Deviation of a Poisson Random Variable

Emergency Room Traffic Let X denote the number of patients arriving at the emergency room of Desert Samaritan Hospital between 6:00 P.M. and 7:00 P.M.

a. Determine and interpret the mean of the random variable X.

b. Determine the standard deviation of X.

Solution As we know, X has the Poisson distribution with parameter $\lambda = 6.9$. So we apply Formula 5.4 to determine the mean and standard deviation of X.

a. The mean of X is $\mu = \lambda = 6.9$.

INTERPRETATION On average, 6.9 patients arrive at the emergency room between 6:00 P.M. and 7:00 P.M.

b. The standard deviation of X is $\sigma = \sqrt{\lambda} = \sqrt{6.9} = 2.6$.

Poisson Approximation to the Binomial Distribution

Recall that the binomial probability formula is

$$P(X = x) = \binom{n}{x} p^x (1 - p)^{n-x}.$$

We use this formula to obtain probabilities for the number of successes, X, in n Bernoulli trials with success probability p.

When n is large, the binomial probability formula can be awkward or impractical to use because of computational difficulties. These difficulties persist, even with sophisticated computers and calculators. Consequently, methods have been developed that permit us to approximate binomial probabilities with simpler formulas.

One of those methods involves use of the Poisson probability formula and applies when n is large and p is small. (As a rule of thumb, we require that $n \geq 100$ and $np \leq 10$.) In such cases, we can use a Poisson distribution to approximate the binomial distribution. And, as you might expect, the appropriate Poisson distribution is the one whose mean is the same as that of the binomial distribution; that is, $\lambda = np$. Specifically, we have Procedure 5.2.

PROCEDURE 5.2

To Approximate Binomial Probabilities Using a Poisson Probability Formula

STEP 1 **Determine *n*, the number of trials, and *p*, the success probability.**

STEP 2 **Determine whether *n* ≥ 100 and *np* ≤ 10. If they are not, do not use the Poisson approximation.**

STEP 3 **Use the Poisson probability formula,**

$$P(X = x) = e^{-np} \frac{(np)^x}{x!},$$

to approximate the required binomial probabilities.

Example 5.17 illustrates the use of Procedure 5.2 to approximate probabilities for the number of infant deaths in Sweden.

Example 5.17 Poisson Approximation to the Binomial

IMR in Sweden According to data obtained from the International Data Base and published in the *Statistical Abstract of the United States*, the infant mortality rate (IMR) in Sweden is 3.5 per 1000 live births. Use the Poisson approximation to determine the probability that, of 500 randomly selected live births, there are

a. no infant deaths.

b. at most three infant deaths.

Solution Let X denote the number of infant deaths out of 500 live births in Sweden. We apply Procedure 5.2 to approximate the required probabilities for X.

STEP 1 **Determine *n*, the number of trials, and *p*, the success probability.**

We have $n = 500$ (number of live births) and $p = \frac{3.5}{1000} = 0.0035$ (probability of an infant death).

STEP 2 Determine whether $n \geq 100$ and $np \leq 10$.

From Step 1, we conclude that $n = 500$ and $np = 500 \cdot 0.0035 = 1.75$. So $n \geq 100$ and $np \leq 10$.

STEP 3 Use the Poisson probability formula,

$$P(X = x) = e^{-np}\frac{(np)^x}{x!},$$

to approximate the required binomial probabilities.

As we noted in Step 2, $np = 1.75$. Thus the appropriate Poisson probability formula is

$$P(X = x) = e^{-1.75}\frac{(1.75)^x}{x!}.$$

a. The probability of no infant deaths in 500 live births is

$$P(X = 0) = e^{-1.75}\frac{(1.75)^0}{0!} = 0.174,$$

approximately.

INTERPRETATION Chances are about 17.4% that there will be no infant deaths in 500 live births.

b. The probability of at most three infant deaths in 500 live births is

$$P(X \leq 3) = P(X = 0) + P(X = 1) + P(X = 2) + P(X = 3)$$

$$= e^{-1.75}\left(\frac{1.75^0}{0!} + \frac{1.75^1}{1!} + \frac{1.75^2}{2!} + \frac{1.75^3}{3!}\right) = 0.899,$$

approximately.

INTERPRETATION Chances are about 89.9% that there will be three or fewer infant deaths in 500 live births.

We now refer to Example 5.17 to illustrate the accuracy of the Poisson approximation to the binomial distribution. We used a computer to obtain both the binomial distribution with parameters $n = 500$ and $p = 0.0035$ and the Poisson distribution with parameter $\lambda = np = 500 \cdot 0.0035 = 1.75$.

Table 5.17 at the top of the following page shows both distributions and exhibits how well the Poisson approximates the binomial. The probabilities are displayed to four decimal places to present a clear picture of the differences between the two distributions. Note that we stopped listing the probabilities once they became zero to four decimal places.

For large n and small p, using a computer instead of a Poisson approximation to obtain a required binomial distribution is not always possible—sometimes n is so large or p is so small that even a computer can't handle the computations. Nonetheless, the Poisson approximation remains easy to apply.

TABLE 5.17

Comparison of the binomial distribution with parameters $n = 500$ and $p = 0.0035$ to the Poisson distribution with parameter $\lambda = 1.75$

x	Binomial probability	Poisson approximation
0	0.1732	0.1738
1	0.3042	0.3041
2	0.2666	0.2661
3	0.1554	0.1552
4	0.0678	0.0679
5	0.0236	0.0238
6	0.0068	0.0069
7	0.0017	0.0017
8	0.0004	0.0004
9	0.0001	0.0001
10	0.0000	0.0000

The Technology Center

Most statistical technologies provide programs to determine Poisson probabilities. In this subsection, we present output and step-by-step instructions to obtain individual Poisson probabilities. For other options, refer to the technology manuals.

Example 5.18   **Using Technology to Obtain Poisson Probabilities**

Emergency Room Traffic   Consider again the illustration of emergency room traffic discussed in Example 5.15, which begins on page 256. Use Minitab, Excel, or the TI-83/84 Plus to determine the probability that exactly 4 patients will arrive at the emergency room between 6:00 P.M. and 7:00 P.M.

Solution   Recall that the number of patients, X, that arrive at the ER between 6:00 P.M. and 7:00 P.M. has a Poisson distribution with parameter $\lambda = 6.9$. We want the probability of exactly 4 arrivals, that is, $P(X = 4)$. Printout 5.2 shows output obtained for this probability by applying the Poisson probability programs from Minitab, Excel, and the TI-83/84 Plus.

From any of these outputs, we find that the required probability is 0.095 (to three decimal places). This result, of course, agrees with the result we obtained in part (a) of Example 5.15. ■

Obtaining the Output (Optional)

Printout 5.2 displays the probability that a Poisson random variable with parameter $\lambda = 6.9$ will be 4. Here are detailed instructions for obtaining that output.

MINITAB

1 Choose **Calc ➤ Probability Distributions ➤ Poisson...**
2 Select the **Probability** option button
3 Click in the **Mean** text box and type 6.9
4 Select the **Input constant** option button
5 Click in the **Input constant** text box and type 4
6 Click **OK**

EXCEL

1 Click **f$_x$** on the button bar
2 Select **Statistical** from the **Function category** list
3 Select **POISSON** from the **Function name** list
4 Click **OK**
5 Type 4 in the **X** text box
6 Click in the **Mean** text box and type 6.9
7 Click in the **Cumulative** text box and type FALSE

TI-83/84 PLUS

1 Press **2nd ➤ DISTR**
2 Arrow down to **poissonpdf(** and press **ENTER**
3 Type 6.9,4) and press **ENTER**

PRINTOUT 5.2
Output giving the probability that exactly 4 patients will arrive at the emergency room between 6:00 P.M. and 7:00 P.M.

MINITAB

Probability Density Function

Poisson with mean = 6.9

x   P(X = x)
4   0.0951816

TI-83/84 PLUS

poissonpdf(6.9,4
)
           .0951816428

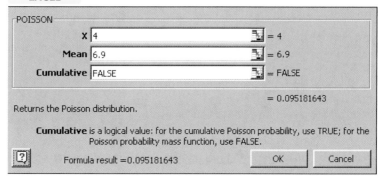

EXCEL

POISSON

X 4   = 4
Mean 6.9   = 6.9
Cumulative FALSE   = FALSE

                       = 0.095181643
Returns the Poisson distribution.

Cumulative is a logical value: for the cumulative Poisson probability, use TRUE; for the Poisson probability mass function, use FALSE.

Formula result = 0.095181643   OK   Cancel

EXERCISES 5.4

Statistical Concepts and Skills

5.47 Identify two uses of Poisson distributions.

5.48 Suppose that X has a Poisson distribution with parameter $\lambda = 3$. Determine
a. $P(X = 2)$.
b. $P(X \leq 3)$.
c. $P(X > 0)$. (*Hint:* Use the complementation rule.)
d. the mean of X.
e. the standard deviation of X.

5.49 Wars. A paper by L. F. Richardson, published in the *Journal of the Royal Statistical Society*, analyzed the distribution of wars in time. From the data, we find that the number of wars that begin during a given calendar year has approximately a Poisson distribution with parameter $\lambda = 0.7$. If a calendar year is selected at random, find the probability that the number of wars that begin during that calendar year will be
a. zero.
b. at most two.
c. between one and three, inclusive.

5.50 Motel Reservations. M. F. Driscoll and N. A. Weiss discussed the modeling and solution of problems concerning motel reservation networks in "An Application of Queuing Theory to Reservation Networks" (*TIMS*, Vol. 22, No. 5, pp. 540–546). They defined a Type 1 call to be a call from a motel's computer terminal to the national reservation center. For a certain motel, the number of Type 1 calls per hour has a Poisson distribution with parameter $\lambda = 1.7$. Determine the probability that the number of Type 1 calls made from this motel during a period of 1 hour will be
a. exactly one.
b. at most two.
c. at least two. (*Hint:* Use the complementation rule.)

5.51 Wars. Refer to Exercise 5.49. Let X denote the number of wars that begin during a randomly selected calendar year.
a. Find and interpret the mean of the random variable X.
b. Determine the standard deviation of X.

5.52 Motel Reservations. Refer to Exercise 5.50. Let X denote the number of Type 1 calls made by the motel during a 1-hour period.
a. Construct a table of probabilities for the random variable X. Compute the probabilities until they are zero to three decimal places.
b. Draw a histogram of the probabilities in part (a).

5.53 Fragile X Syndrome. The second leading genetic cause of mental retardation is Fragile X Syndrome, named for the fragile appearance of the tip of the X chromosome in affected individuals. One in 1500 males are affected worldwide, with no ethnic bias.
a. In a sample of 10,000 males, how many would you expect to have Fragile X Syndrome?
b. For a sample of 10,000 males, use the Poisson approximation to the binomial distribution to determine the probability that more than 7 of the males have Fragile X Syndrome; that at most 10 of the males have Fragile X Syndrome.

5.54 Holes in One. Refer to the Case Study at the beginning of the chapter. According to the experts, the odds against a PGA golfer making a hole in one are 3708 to 1; that is, the probability is $\frac{1}{3709}$. Use the Poisson approximation to the binomial distribution (Procedure 5.2) to determine the probability that at least four of the 155 golfers playing the second round would get a hole in one on the sixth hole.

5.55 A Yellow Lobster! As reported by the Associated Press, a veteran lobsterman recently hauled up a yellow lobster less than a quarter mile south of Princes Point in Harpswell Cove. Yellow lobsters are considerably rarer than blue lobsters and, according to *The Lobster Almanac*, roughly 1 in every 30 million lobsters hatched is yellow.

Apply the Poisson approximation to the binomial distribution to answer the following questions:
a. Of 100 million lobsters hatched, what is the probability that between 3 and 5, inclusive, are yellow?
b. Roughly how many lobsters must be hatched in order to be at least 90% sure that at least one is yellow?

Extending the Concepts and Skills

5.56 With regard to the use of a Poisson distribution to approximate binomial probabilities, on page 260 we stated that "...as you might expect, the appropriate Poisson distribution is the one whose mean is the same as that of the binomial distribution...." Explain why you might expect this result.

5.57 Roughly speaking, you can use the Poisson probability formula to approximate binomial probabilities when n is large and p is small (i.e., near 0). Explain how to use the Poisson probability formula to approximate binomial probabilities when n is large and p is large (i.e., near 1).

Using Technology

5.58 Wars. Use the technology of your choice to obtain the required probabilities in Exercise 5.49.

5.59 Motel Reservations. Use the technology of your choice to obtain the required probabilities in Exercise 5.50.

5.60 Fragile X Syndrome. In Exercise 5.53, you used the Poisson approximation to the binomial distribution with parameters $n = 10,000$ and $p = 1/1500$. Use the technology of your choice to obtain both the binomial distribution and its approximating Poisson distribution. Construct a table similar to Table 5.17 on page 262 to compare the two distributions.

CHAPTER REVIEW

You Should Be Able To

1. use and understand the formulas in this chapter.

2. determine the probability distribution of a discrete random variable.

3. construct a probability histogram.

4. describe events using random-variable notation, when appropriate.

5. use the frequentist interpretation of probability to understand the meaning of the probability distribution of a random variable.

6. find and interpret the mean and standard deviation of a discrete random variable.

7. compute factorials and binomial coefficients.

8. define and apply the concept of Bernoulli trials.

9. assign probabilities to the outcomes in a sequence of Bernoulli trials.

10. obtain binomial probabilities.

11. compute the mean and standard deviation of a binomial random variable.

12. obtain Poisson probabilities.

13. compute the mean and standard deviation of a Poisson random variable.

14. use the Poisson distribution to approximate binomial probabilities, when appropriate.

Key Terms

Bernoulli trials, *242*
binomial coefficients, *241*
binomial distribution, *242, 246*
binomial probability formula, *246*
binomial random variable, *246*
discrete random variable, *226*
expectation, *235*
expected value, *235*
factorials, *241*
failure, *242*

hypergeometric distribution, *250*
law of averages, *236*
law of large numbers, *236*
mean of a discrete random
    variable, *235*
Poisson distribution, *256*
Poisson probability formula, *256*
Poisson random variable, *256*
probability distribution, *227*
probability histogram, *227*

random variable, *226*
standard deviation of a discrete
    random variable, *237*
success, *242*
success probability, *242*
trial, *240*
variance of a discrete random
    variable, *237*

REVIEW TEST

Statistical Concepts and Skills

1. Fill in the blanks.
a. A _____ is a quantitative variable whose value depends on chance.
b. A discrete random variable is a random variable whose possible values form a _____ (or _____) set of numbers.

2. What does the probability distribution of a discrete random variable tell you?

3. How do you graphically portray the probability distribution of a discrete random variable?

4. If you sum the probabilities of the possible values of a discrete random variable, the result always equals _____.

5. A random variable X equals 2 with probability 0.386.
a. Use probability notation to express that fact.
b. If you make repeated independent observations of the random variable X, in approximately what percentage of those observations will you observe the value 2?
c. Roughly how many times would you expect to observe the value 2 in 50 observations? 500 observations?

6. A random variable X has mean 3.6. If you make a large number of repeated independent observations of the random variable X, the average value of those observations will be approximately _____.

7. Two random variables, X and Y, have standard deviations 2.4 and 3.6, respectively. Which one is more likely to take a value close to its mean? Explain your answer.

8. List the three requirements for repeated trials of an experiment to constitute Bernoulli trials.

9. What is the relationship between Bernoulli trials and the binomial distribution?

10. In 10 Bernoulli trials, how many outcomes contain exactly three successes?

11. Explain how the special formulas for the mean and standard deviation of a binomial or Poisson random variable are derived.

12. Suppose that a simple random sample of size n is taken from a finite population in which the proportion of members having a specified attribute is p. Let X be the number of members sampled that have the specified attribute.
a. If the sampling is done with replacement, identify the probability distribution of X.
b. If the sampling is done without replacement, identify the probability distribution of X.
c. Under what conditions is it acceptable to approximate the probability distribution in part (b) by the probability distribution in part (a)? Why is it acceptable?

13. ASU-Main Enrollment. According to the *Arizona State University Main Facts Book*, a frequency distribution for the number of undergraduate students attending the main campus one year, by class level, is as shown in the following table. Here, 1 = freshman, 2 = sophomore, 3 = junior, and 4 = senior.

Class level	1	2	3	4
No. of students	6,159	6,790	8,141	11,220

Let X denote the class level of a randomly selected ASU undergraduate.
a. What are the possible values of the random variable X?
b. Use random-variable notation to represent the event that the student selected is a junior (class-level 3).
c. Determine $P(X = 3)$ and interpret your answer in terms of percentages.
d. Determine the probability distribution of the random variable X.
e. Construct a probability histogram for the random variable X.

14. Busy Phone Lines. An accounting office has six incoming telephone lines. The probability distribution of the number of busy lines, Y, is as follows. Use random-variable notation to express each of the following events. The number of busy lines is
a. exactly four.   b. at least four.
c. between two and four, inclusive.
d. at least one.

y	$P(Y = y)$
0	0.052
1	0.154
2	0.232
3	0.240
4	0.174
5	0.105
6	0.043

Apply the special addition rule and the probability distribution to determine
e. $P(Y = 4)$.   f. $P(Y \geq 4)$.
g. $P(2 \leq Y \leq 4)$.   h. $P(Y \geq 1)$.

15. Busy Phone Lines. Refer to the probability distribution displayed in the table in Problem 14.
a. Find the mean of the random variable Y.
b. On average, how many lines are busy?
c. Compute the standard deviation of Y.
d. Construct a probability histogram for Y; locate the mean; and show one, two, and three standard deviation intervals.

16. Determine $0!$, $3!$, $4!$, and $7!$.

17. Determine the value of each binomial coefficient.
a. $\binom{8}{3}$ b. $\binom{8}{5}$ c. $\binom{6}{6}$ d. $\binom{10}{2}$ e. $\binom{40}{4}$ f. $\binom{100}{0}$

18. DUI Fatalities. According to *Reader's Digest*, there is a 40% chance that a traffic fatality involves an intoxicated or alcohol-impaired driver or nonoccupant. For brevity, the word *drinker* is used to designate an intoxicated or alcohol-impaired driver or nonoccupant. For a given traffic fatality, regard a success, s, to be that a drinker is involved.
a. Identify the success probability, p.
b. Construct a table showing the possible success–failure results and their probabilities for three traffic fatalities.
c. Draw a tree diagram for part (b).
d. List the outcomes in which exactly two of the three traffic fatalities involve a drinker.
e. Determine the probability of each of the outcomes in part (d). Explain why those probabilities are equal.
f. Find the probability that exactly two of the three traffic fatalities involve a drinker.
g. Without using the binomial probability formula, obtain the probability distribution of the random variable Y, the number of traffic fatalities of the three that involve a drinker.
h. Identify the probability distribution in part (g).

19. Booming Pet Business. The pet industry has undergone a surge in recent years, surpassing even the $20 billion a year toy industry. According to *U.S. News & World*

Report, 60% of U.S. households live with one or more pets. If four U.S. households are selected at random without replacement, determine the (approximate) probability that the number living with one or more pets will be

a. exactly three.　　　　**b.** at least three.

c. at most three.

d. Find the probability distribution of the random variable X, the number of U.S. households in a random sample of four that live with one or more pets.

e. Without referring to the probability distribution obtained in part (d) or constructing a probability histogram, decide whether the probability distribution is right skewed, symmetric, or left skewed. Explain your answer.

f. Draw a probability histogram for X.

g. Strictly speaking, why is the probability distribution that you obtained in part (d) only approximately correct? What is the exact distribution called?

h. Determine and interpret the mean of the random variable X.

i. Determine the standard deviation of X.

20. Following are two probability histograms of binomial distributions. For each, specify whether the success probability is less than, equal to, or greater than 0.5.

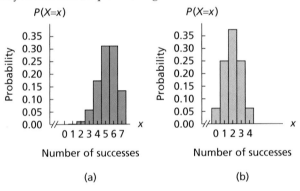

(a)　　　　　　　　　(b)

21. Wrong Number. A classic study by F. Thorndike on the number of calls to a wrong number appeared in the paper "Applications of Poisson's Probability Summation" (*Bell Systems Technical Journal*, Vol. 5, pp. 604–624). The study examined the number of calls to a wrong number from coin-box telephones in a large transportation terminal. According to the paper, the number of calls to a wrong number, X, in a 1-minute period has a Poisson distribution with parameter $\lambda = 1.75$. Determine the probability that during a 1-minute period the number of calls to a wrong number will be

a. exactly two.

b. between four and six, inclusive.

c. at least one.

d. Obtain a table of probabilities for X, stopping when the probabilities become zero to three decimal places.

e. Use part (d) to construct a partial probability histogram for the random variable X.

f. Identify the shape of the probability distribution of X. Is this shape typical of Poisson distributions?

22. Wrong Number. Refer to Problem 21.

a. Find and interpret the mean of the random variable X.

b. Determine the standard deviation of X.

23. Four of a Kind. The probability is about 0.00024 of being dealt four of a kind in a hand of five-card poker.

a. In 10,000 hands of five-card poker, roughly how many times would you expect to be dealt four of a kind?

b. Use the Poisson approximation to the binomial distribution to find the probability of being dealt four of a kind exactly twice in 10,000 hands of five-card poker.

c. Use the Poisson approximation to the binomial distribution to find the probability of being dealt four of a kind at least twice in 10,000 hands of five-card poker.

Using Technology

24. ASU-Main Enrollment. Refer to the probability distribution obtained in Problem 13(d).

a. Use the technology of your choice to simulate 2500 observations of the class level of a randomly selected undergraduate at ASU.

b. Obtain the proportions for the 2500 class levels simulated in part (a), and compare it to the probability distribution you obtained in Problem 13(d).

c. Construct a histogram of the proportions and compare it to the probability histogram you obtained in Problem 13(e).

d. What do parts (b) and (c) illustrate?

25. Busy Phone Lines. Refer to the probability distribution displayed in the table in Problem 14.

a. Use the technology of your choice to simulate 200 observations of the number of busy lines.

b. Obtain the mean of the observations in part (a), and compare it to μ, which you determined in Problem 15(a).

c. What is part (b) illustrating?

26. Booming Pet Business. Use the technology of your choice to obtain the required probability or probabilities in parts (a)–(d) of Problem 19.

27. Wrong Number. Use the technology of your choice to get the required probability or probabilities in parts (a)–(d) of Problem 21.

StatExplore in MyMathLab
Analyzing Data Online

You can use StatExplore to obtain probabilities for the most important probability distributions. To illustrate, we show how to find binomial probabilities considered earlier in Example 5.12 on page 246.

EXAMPLE Binomial Probabilities

Mortality There is an 80% chance that a person aged 20 will be alive at age 65. Suppose that three people aged 20 are selected at random. Use StatExplore to determine the probability that the number alive at age 65 will be exactly two; at most one.

SOLUTION We first note that, of three randomly selected 20-year-olds, the number alive at age 65 has the binomial distribution with parameters $n = 3$ and $p = 0.8$ (80%).

To obtain the probability that exactly two of the three 20-year-olds will be alive at age 65, we proceed as follows:

> 1 Choose **StatExplore ➤ Calculators ➤ Binomial**
> 2 Type 3 in the **n** text box
> 3 Type 0.8 in the **p** text box
> 4 Click the arrow button at the right of the **Prob(X** drop-down list box and select =
> 5 Type 2 in the **Prob(X** text box
> 6 Click **Compute**

The resulting output is shown in Printout 5.3. From this printout, we see that the probability that exactly two of the three 20-year-olds will be alive at age 65 is 0.384. (*Note:* The StatExplore output actually includes a graphic representation of the binomial probability in addition to the output shown in Printout 5.3.)

PRINTOUT 5.3
StatExplore output for probability of exactly two alive

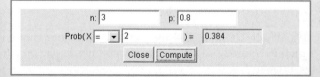

To obtain the probability that at most one of the three 20-year-olds will be alive at age 65, we proceed as in the previous list of instructions, except that in step 4 we select <= and in step 5 we type 1. The resulting output is shown in Printout 5.4. From this printout, we see that the probability that at most one (i.e., one or fewer) of the three 20-year-olds will be alive at age 65 is 0.104.

PRINTOUT 5.4
StatExplore output for probability of at most one alive

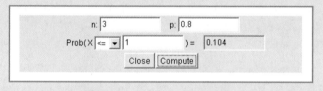

STATEXPLORE EXERCISES Solve the following problems by using StatExplore:
a. Obtain the probabilities required in parts (a), (c), and (d) of Exercise 5.37 on page 254.
b. Find the probability required in part (b) of Exercise 5.37. (*Hint:* The required probability can be obtained as the difference between two "at most" probabilities.)
c. Determine the Poisson probabilities that are required in parts (a)–(c) of Example 5.15 on page 256.

> To access StatExplore, go to the student content area of your Weiss MyMathLab course.

Internet Project

Racial Bias in Jury Selection

Everyone has biases—humans inevitably evaluate new circumstances based on their past experience and knowledge. These biases can effect people in systematic ways.

In this Internet project, you are to examine the possibility of racial bias in jury selection by looking at two actual U.S. court cases. In each case, an African-American was on trial and had a jury with no African-Americans seated. One trial was in Monroe County, Alabama, which has a population of about 24,000. The proportion of whites in that county is 0.58, or 58%; the proportion of blacks is 0.40, or 40%.

For the two court cases, the possibility of racial bias in the selection of the juries was raised on their appeal and, in both instances, the higher court determined that there was insufficient evidence of racial bias. The defendant in each court case was convicted and given the death sentence.

If 40% of the residents of a community are African-Americans, what is the chance of randomly selecting 12 people from the community and obtaining no African-Americans? This question is an oversimplification of the facts underlying the cases, but it is a part of the evidence that was considered in the aftermath of both convictions. In this project, you will look into the legal labyrinth of jury selection.

URL for access to Internet Projects Page: www.aw-bc.com/weiss

Focusing on Data Analysis

The Focus Database

Recall from Chapter 1 (see page 35) that the Focus database contains information on the undergraduate students at the University of Wisconsin - Eau Claire (UWEC). Statistical analyses for this database should be done with the technology of your choice.

a. Let X denote the age of a randomly selected undergraduate student at UWEC. Obtain the probability distribution of the random variable X. Display the probabilities to six decimal places.
b. Obtain a probability histogram or similar graphic for the random variable X.
c. Determine the mean and standard deviation of the random variable X.
d. Simulate 100 observations of the random variable X.
e. Roughly, what would you expect the average value of the 100 observations obtained in part (d) to be? Explain your reasoning.
f. In actuality, what is the average value of the 100 observations obtained in part (d)? Compare this value to the value you expected, as answered in part (e).
g. Consider the experiment of randomly selecting 10 UWEC undergraduates with replacement and observing the number of those selected who are 21 years old. Simulate that experiment 2000 times. (*Hint:* Simulate an appropriate binomial distribution.)
h. Referring to the simulation in part (g), in approximately what percentage of the 2000 experiments would you expect exactly 3 of the 10 students selected to be 21 years old? Compare that percentage to the actual percentage of the 2000 experiments in which exactly 3 of the 10 students selected are 21 years old.

Case Study Discussion

Aces Wild on the Sixth at Oak Hill

As we reported at the beginning of this chapter, on June 16, 1989, during the second round of the 1989 U.S. Open, four golfers—Doug Weaver, Mark Wiebe, Jerry Pate, and Nick Price—made holes in one on the sixth hole at Oak Hill in Pittsford, New York. Now that you have studied the material in this chapter, you can determine for yourself the likelihood of such an event.

According to the experts, the odds against a professional golfer making a hole in one are 3708-1; in other words, the probability is $\frac{1}{3709}$ that a professional golfer will make a hole in one. One hundred fifty-five golfers participated in the second round.

a. Determine the probability that at least 4 of the 155 golfers would get a hole in one on the sixth hole.
b. What assumptions did you make in solving part (a)? Do those assumptions seem reasonable to you? Explain your answer.
c. Use the technology of your choice to solve part (a).

Internet Resources: Visit the Weiss Web site www.aw-bc.com/weiss for additional discussion, exercises, and resources related to this case study.

Biography

JAMES BERNOULLI: Paving the Way for Probability Theory

James Bernoulli was born on December 27, 1654, in Basle, Switzerland. He was the first of the Bernoulli family of mathematicians; his younger brother John and various nephews and grandnephews were also renowned mathematicians. His father, Nicolaus Bernoulli (1623–1708), planned the ministry as James's career. James rebelled, however; to him, mathematics was much more interesting.

Although Bernoulli was schooled in theology, he studied mathematics on his own. He was especially fascinated with calculus. In a 1690 issue of the journal *Acta eruditorum,* Bernoulli used the word *integral* to describe the inverse of differential. The results of his studies of calculus and the catenary (the curve formed by a cord freely suspended between two fixed points) were soon applied to the building of suspension bridges.

Some of Bernoulli's most important work was published posthumously in *Ars Conjectandi* (The Art of Conjecturing) in 1713. This book contains his theory of permutations and combinations, the Bernoulli numbers, and his writings on probability, which include the weak law of large numbers for Bernoulli trials. *Ars Conjectandi* has been regarded as the beginning of the theory of probability.

Both James and his brother John were highly accomplished mathematicians. But rather than collaborating in their work, they were most often competing. James would publish a question inviting solutions in a professional journal. John would reply in the same journal with a solution, only to find that an ensuing issue would contain another article by James, telling him that he was wrong. In their later years, they communicated only in this manner.

Bernoulli began lecturing in natural philosophy and mechanics at the University of Basle in 1682 and became a Professor of Mathematics there in 1687. He remained at the university until his death of a "slow fever" on August 10, 1705.

The Normal Distribution

GENERAL OBJECTIVES In this chapter, we discuss the most important distribution in statistics—the *normal distribution.* As you will see, its importance lies in the fact that it appears again and again in both theory and practice.

A variable is said to be *normally distributed* or to have a *normal distribution* if its distribution has the shape of a normal curve, a special type of bell-shaped curve. In Section 6.1, we introduce normally distributed variables, show that percentages (or probabilities) for such a variable are equal to areas under its associated normal curve, and explain how all normal distributions can be converted to a single normal distribution—the *standard normal distribution.*

In Section 6.2, we demonstrate how to obtain areas under the *standard normal curve,* the normal curve corresponding to a variable that has the standard normal distribution. Then, in Section 6.3, we provide an efficient procedure for finding percentages (or probabilities) for any normally distributed variable from areas under the standard normal curve.

We present a method for graphically assessing whether a variable is normally distributed—*the normal probability plot*—in Section 6.4. Finally, in Section 6.5, we show how to approximate binomial probabilities with areas under a suitable normal curve.

Foot Length and Shoe Size for Women

According to research done on foot length of women, the mean length of women's feet is 9.58 inches with a standard deviation of 0.51 inch. Moreover, the distribution of foot length for women is bell shaped and, in fact, has the *normal distribution* shown in the graph below. (You will study normal distributions in this chapter.)

Knowing this distribution is useful to shoe manufacturers, shoe stores, and related merchants because it permits them to make informed decisions about shoe production, inventory, and so forth. In that regard, the following table provides a foot-length–to–shoe-size conversion, obtained from Payless ShoeSource.

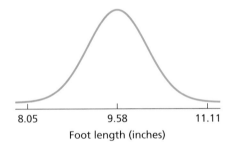

Foot length (inches)

Length (inches)	Size (USA)	Length (inches)	Size (USA)
8	3	10	9
$8\frac{3}{16}$	$3\frac{1}{2}$	$10\frac{3}{16}$	$9\frac{1}{2}$
$8\frac{5}{16}$	4	$10\frac{5}{16}$	10
$8\frac{8}{16}$	$4\frac{1}{2}$	$10\frac{8}{16}$	$10\frac{1}{2}$
$8\frac{11}{16}$	5	$10\frac{11}{16}$	11
$8\frac{13}{16}$	$5\frac{1}{2}$	$10\frac{13}{16}$	$11\frac{1}{2}$
9	6	11	12
$9\frac{3}{16}$	$6\frac{1}{2}$	$11\frac{3}{16}$	$12\frac{1}{2}$
$9\frac{5}{16}$	7	$11\frac{5}{16}$	13
$9\frac{8}{16}$	$7\frac{1}{2}$	$11\frac{8}{16}$	$13\frac{1}{2}$
$9\frac{11}{16}$	8	$11\frac{11}{16}$	14
$9\frac{13}{16}$	$8\frac{1}{2}$		

At the end of this chapter, you will be asked to answer several questions about foot length and shoe size for women, based on the information provided here.

6.1 Introducing Normally Distributed Variables

FIGURE 6.1
A normal curve

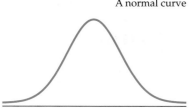

In everyday life, people deal with and use a wide variety of variables, many of which are intrinsically different. But some—such as aptitude-test scores, heights of women, and wheat yield—share an important characteristic: Their distributions have roughly the shape of a **normal curve,** that is, a special type of bell-shaped curve like the one shown in Fig. 6.1. Such a variable is called a **normally distributed variable** and is said to have a **normal distribution.** We summarize this discussion in Definition 6.1.

DEFINITION 6.1

Normally Distributed Variable

A variable is said to be a *normally distributed variable* or to have a *normal distribution* if its distribution has the shape of a normal curve.

What
Does it Mean?

In the last half of the nineteenth century, researchers discovered that it is quite usual, or "normal," for a variable to have a distribution shaped like that in Fig. 6.1. So, following the lead of noted British statistician Karl Pearson, such a distribution began to be referred to as a *normal distribution.*

Here is some important terminology associated with normal distributions.

- If a variable of a population is normally distributed and is the only variable under consideration, common practice is to say that the **population is normally distributed** or that it is a **normally distributed population.**
- In practice, a distribution is unlikely to have exactly the shape of a normal curve. If a variable's distribution is shaped roughly like a normal curve, we say that the variable is an **approximately normally distributed variable** or that it has **approximately a normal distribution.**

A normal distribution (and hence a normal curve) is completely determined by the mean and standard deviation; that is, two normally distributed variables having the same mean and standard deviation must have the same distribution. We often identify a normal curve by stating the corresponding mean and standard deviation and calling those the **parameters** of the normal curve.[†]

A normal distribution is symmetric about and centered at the mean of the variable, and its spread depends on the standard deviation of the variable—the larger the standard deviation, the flatter and more spread out is the distribution. Figure 6.2 displays three normal distributions.

FIGURE 6.2
Three normal distributions

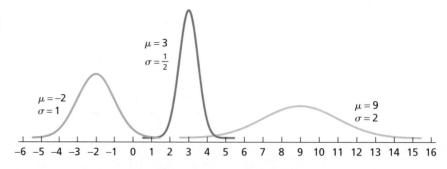

$\mu = 3$
$\sigma = \frac{1}{2}$

$\mu = -2$
$\sigma = 1$

$\mu = 9$
$\sigma = 2$

-6 -5 -4 -3 -2 -1 0 1 2 3 4 5 6 7 8 9 10 11 12 13 14 15 16

[†]The equation of the normal curve with parameters μ and σ is $y = e^{-(x-\mu)^2/2\sigma^2}/(\sqrt{2\pi}\sigma)$, where $e \approx 2.718$ and $\pi \approx 3.142$.

The three-standard-deviations rule, when applied to a variable, states that almost all the possible observations of the variable lie within three standard deviations to either side of the mean. This rule is illustrated by the three normal distributions in Fig. 6.2: Each normal curve is close to the horizontal axis outside the range of three standard deviations to either side of the mean.

For instance, the third normal distribution in Fig. 6.2 has mean $\mu = 9$ and standard deviation $\sigma = 2$. Three standard deviations below (to the left of) the mean is

$$\mu - 3\sigma = 9 - 3 \cdot 2 = 3,$$

and three standard deviations above (to the right of) the mean is

$$\mu + 3\sigma = 9 + 3 \cdot 2 = 15.$$

As shown in Fig. 6.2, the corresponding normal curve is close to the horizontal axis outside the range from 3 to 15.

In summary, the normal curve associated with a normal distribution is

- bell shaped,
- centered at μ, and
- close to the horizontal axis outside the range from $\mu - 3\sigma$ to $\mu + 3\sigma$,

as depicted in Figs. 6.2 and 6.3. This information helps us sketch a normal distribution.

FIGURE 6.3

Graph of generic normal distribution

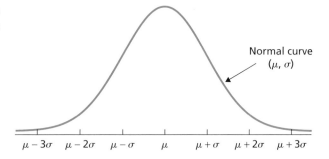

Example 6.1 illustrates a normally distributed variable and discusses some additional properties of such variables.

Example 6.1   A Normally Distributed Variable

Heights of Female College Students   A midwestern college has an enrollment of 3264 female students. Records show that the mean height of these students is 64.4 inches and that the standard deviation is 2.4 inches. Here the variable is height, and the population consists of the 3264 female students attending the college. Frequency and relative-frequency distributions for these heights are presented in Table 6.1 at the top of the next page. The table shows, for instance, that 7.35% (0.0735) of the students are between 67 and 68 inches tall.

a. Show that the variable "height" is approximately normally distributed for this population.

TABLE 6.1

Frequency and relative-frequency distributions for heights

Height (inches)	Freq. f	Rel. freq.
56 < 57	3	0.0009
57 < 58	6	0.0018
58 < 59	26	0.0080
59 < 60	74	0.0227
60 < 61	147	0.0450
61 < 62	247	0.0757
62 < 63	382	0.1170
63 < 64	483	0.1480
64 < 65	559	0.1713
65 < 66	514	0.1575
66 < 67	359	0.1100
67 < 68	240	0.0735
68 < 69	122	0.0374
69 < 70	65	0.0199
70 < 71	24	0.0074
71 < 72	7	0.0021
72 < 73	5	0.0015
73 < 74	1	0.0003
	3264	1.0000

b. Identify the normal curve associated with the variable "height" for this population.

c. Examine the relationship between the percentage of female students whose heights lie within a specified range and the corresponding area under the associated normal curve.

Solution

a. A relative-frequency histogram for the heights of the 3264 female students is presented in Fig. 6.4. It shows that the distribution of heights has roughly the shape of a normal curve and, consequently, that the variable "height" is approximately normally distributed for this population.

b. The associated normal curve is the one whose parameters are the same as the mean and standard deviation of the population, which are 64.4 and 2.4, respectively. Thus the normal curve associated with the variable "height" for this population of students is the one with parameters $\mu = 64.4$ and $\sigma = 2.4$. This normal curve is superimposed on the histogram in Fig. 6.4.

c. Consider, for instance, the percentage of female students who are between 67 and 68 inches tall. According to Table 6.1, the exact percentage is 7.35%, or 0.0735. Note that 0.0735 also equals the area of the cross-hatched bar in Fig. 6.4 because the bar has height 0.0735 and width 1. Now look at the area under the normal curve between 67 and 68, the area shaded in Fig. 6.4. This area approximates the area of the cross-hatched bar which, as we have noted, equals the percentage of students who are between 67 and 68 inches tall. Thus we can approximate the percentage of students who are between 67 and 68 inches tall by the area under the normal curve between 67 and 68. This same type of result holds in general.

FIGURE 6.4

Relative-frequency histogram for heights with superimposed normal curve

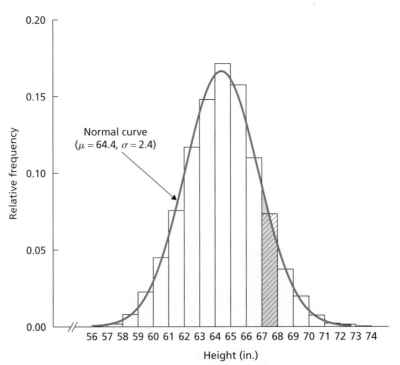

INTERPRETATION The percentage of female students whose heights lie within any specified range can be approximated by the corresponding area under the normal curve associated with the variable "height" for this population of female students.

In Key Fact 6.1, we summarize the important relationship established in Example 6.1 between percentages for a normally distributed variable and areas under its associated normal curve.

KEY FACT 6.1

Normally Distributed Variables and Normal-Curve Areas

For a normally distributed variable, the percentage of all possible observations that lie within any specified range equals the corresponding area under its associated normal curve, expressed as a percentage. This result holds approximately for a variable that is approximately normally distributed.

Note: For brevity, we often paraphrase the content of Key Fact 6.1 with the statement "percentages for a normally distributed variable are equal to areas under its associated normal curve."

Standardizing a Normally Distributed Variable

Let's summarize two essential facts about normally distributed variables and their associated normal curves.

- Once we know the mean and standard deviation of a normally distributed variable, we know its distribution and associated normal curve.
- Percentages for a normally distributed variable are equal to areas under its associated normal curve.

Consequently, once we know the mean and standard deviation of a normally distributed variable, we can obtain the percentage of all possible observations that lie within any specified range by determining the corresponding area under its associated normal curve. Now the question is: How do we find areas under a normal curve?

Conceptually, we need a table of areas for each normal curve. This, of course, is not possible because there are infinitely many different normal curves—one for each choice of μ and σ. The way out of this difficulty is standardizing, which transforms every normal distribution into one particular normal distribution, the **standard normal distribution**.

DEFINITION 6.2

FIGURE 6.5
Standard normal distribution

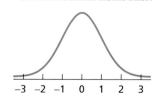

Standard Normal Distribution; Standard Normal Curve

A normally distributed variable having mean 0 and standard deviation 1 is said to have the *standard normal distribution*. Its associated normal curve is called the *standard normal curve*, which is shown in Fig. 6.5.

Recall from Chapter 3 (page 136) that we standardize a variable *x* by subtracting its mean and then dividing by its standard deviation. The resulting variable, $z = (x - \mu)/\sigma$, is called the *standardized version* of *x* or the *standardized variable* corresponding to *x*.

The standardized version of any variable has mean 0 and standard deviation 1. For a normally distributed variable, we can say even more: If a variable is normally distributed, then so is its standardized version. In other words, we have the following essential fact.

KEY FACT 6.2

What *Does it Mean?*

Subtracting from a normally distributed variable its mean and then dividing by its standard deviation results in a variable with the standard normal distribution.

Standardized Normally Distributed Variable

The standardized version of a normally distributed variable *x*,

$$z = \frac{x - \mu}{\sigma},$$

has the standard normal distribution.

We can interpret Key Fact 6.2 in several ways. Theoretically, it says that standardizing converts all normal distributions to the standard normal distribution, as depicted in Fig. 6.6.

FIGURE 6.6
Standardizing normal distributions

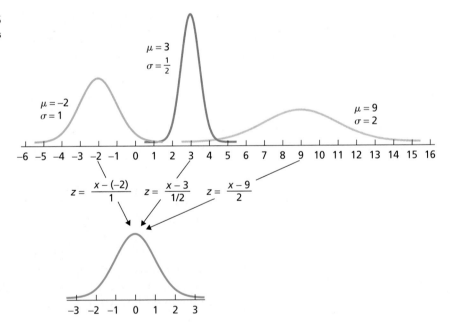

But we need a more practical interpretation of Key Fact 6.2. Let *x* be a normally distributed variable and let *a* and *b* be real numbers with $a < b$. The percentage of all possible observations of *x* that lie between *a* and *b* is the same as the percentage of all possible observations of *z* that lie between $(a - \mu)/\sigma$ and $(b - \mu)/\sigma$. And, in light of Key Fact 6.2, this latter percentage equals the area under the standard normal curve between $(a - \mu)/\sigma$ and $(b - \mu)/\sigma$. We summarize these ideas graphically in Fig. 6.7.

FIGURE 6.7
Finding percentages for a normally
distributed variable from areas under
the standard normal curve

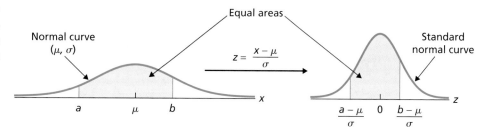

Consequently, for a normally distributed variable, we can find the percent-age of all possible observations that lie within any specified range by

1. expressing the range in terms of z-scores, and
2. obtaining the corresponding area under the standard normal curve.

You already know how to convert to z-scores. Therefore you need only learn how to find areas under the standard normal curve, which we demonstrate in Section 6.2.

Simulating a Normal Distribution

Both for purposes of understanding and research, simulating a variable is often useful. Doing so involves use of a computer or statistical calculator to gener-ate observations of the variable. In Example 6.2, we conduct and interpret a simulation of a normally distributed variable.

Example 6.2   **Simulating a Normally Distributed Variable**

Gestation Periods of Humans   Gestation periods of humans are normally dis-tributed with a mean of 266 days and a standard deviation of 16 days. Simulate 1000 human gestation periods, obtain a histogram of the simulated data, and interpret the results.

Solution   Here the variable x is gestation period and, for humans, it is normally distributed with mean $\mu = 266$ days and standard deviation $\sigma = 16$ days. We used a computer to simulate 1000 observations of the variable x for humans. A histogram for those observations is shown in Printout 6.1. For purposes

PRINTOUT 6.1
Histogram of 1000 simulated human
gestation periods with superimposed
normal curve

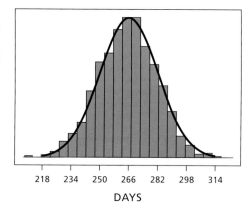

<table>
<tr><td>

What
Does it Mean?

When we simulate a normally distributed variable, a histogram of the observations will have roughly the shape of the normal curve associated with the variable.

</td><td>

of comparison, we have superimposed the normal curve associated with the variable—namely, the normal curve with parameters $\mu = 266$ and $\sigma = 16$.

As expected, the histogram in Printout 6.1 is shaped roughly like the normal curve associated with the variable. Because we have only sample data here, we would not expect the histogram to be shaped exactly like the normal curve. But, because the sample size is large, we would expect the histogram to be close to that shape, which it is.

If you do the simulation, you will (almost) certainly obtain different results than those depicted in Printout 6.1. However, your results should be similar.

</td></tr>
</table>

The Technology Center

Most statistical software packages and some graphing calculators have built-in procedures to simulate observations of normally distributed variables. In Example 6.2, we used Minitab but, for instance, Excel can also be used to conduct that simulation and obtain the histogram. Refer to the technology manuals for details.

EXERCISES 6.1

Statistical Concepts and Skills

6.1 A variable is approximately normally distributed. If you draw a histogram of the distribution of the variable, roughly what shape will it have?

6.2 Precisely what is meant by the statement that a population is normally distributed?

6.3 Two normally distributed variables have the same means and the same standard deviations. What can you say about their distributions? Explain your answer.

6.4 Which normal distribution has a wider spread: the one with mean 1 and standard deviation 2 or the one with mean 2 and standard deviation 1? Explain your answer.

6.5 Consider two normal distributions, one with mean -4 and standard deviation 3 and the other with mean 6 and standard deviation 3. Answer true or false to each statement and explain your answers.
a. The two normal distributions have the same shape.
b. The two normal distributions are centered at the same place.

6.6 True or false: The mean of a normal distribution has no effect on its shape. Explain your answer.

6.7 What are the parameters for a normal curve?

6.8 Sketch the normal distribution with
a. $\mu = 3$ and $\sigma = 3$. **b.** $\mu = 1$ and $\sigma = 3$.
c. $\mu = 3$ and $\sigma = 1$.

6.9 For a normally distributed variable, what is the relationship between the percentage of all possible observations that lie between 2 and 3 and the area under the associated normal curve between 2 and 3? What if the variable is only approximately normally distributed?

6.10 The area under a particular normal curve between 10 and 15 is 0.6874. A normally distributed variable has the same mean and standard deviation as the parameters for this normal curve. What percentage of all possible observations of the variable lie between 10 and 15? Explain your answer.

6.11 Female College Students. Refer to Example 6.1 on page 275.
a. Use the relative-frequency distribution in Table 6.1 to obtain the percentage of female students who are between 60 and 65 inches tall.
b. Use your answer from part (a) to estimate the area under the normal curve having parameters $\mu = 64.4$ and $\sigma = 2.4$ that lies between 60 and 65. Why do you get only an estimate of the true area?

6.12 Female College Students. Refer to Example 6.1 on page 275.
a. The area under the standard normal curve with parameters $\mu = 64.4$ and $\sigma = 2.4$ that lies to the left of 61 is 0.0783. Use this information to estimate the percentage of female students who are shorter than 61 inches.

b. Use the relative-frequency distribution in Table 6.1 to obtain the exact percentage of female students who are shorter than 61 inches.

c. Compare your answers from parts (a) and (b).

6.13 Giant Tarantulas. One of the larger species of tarantulas is the *Grammostola mollicoma*, whose common name is the Brazilian Giant Tawny Red. A tarantula has two body parts. The anterior part of the body is covered above by a shell, or carapace. From a recent article by F. Costa and F. Perez–Miles titled, "Reproductive Biology of Uruguayan Theraphosids" (*The Journal of Arachnology*, Vol. 30, No. 3, pp. 571–587), we find that the carapace length of the adult male *G. mollicoma* is normally distributed with a mean of 18.14 mm and a standard deviation of 1.76 mm. Let x denote carapace length for the adult male *G. mollicoma*.

a. Sketch the distribution of the variable x.

b. Obtain the standardized version, z, of x.

c. Identify and sketch the distribution of z.

d. The percentage of adult male *G. mollicoma* that have carapace lengths between 16 mm and 17 mm is equal to the area under the standard normal curve between _____ and _____.

e. The percentage of adult male *G. mollicoma* that have carapace lengths exceeding 19 mm is equal to the area under the standard normal curve that lies to the _____ of _____.

6.14 Serum Cholesterol Levels. According to the *National Health and Nutrition Examination Survey*, the serum (noncellular portion of blood) total cholesterol level of U.S. females 20 years old or older is normally distributed with a mean of 206 mg/dL (milligrams per deciliter) and a standard deviation of 44.7 mg/dL. Let x denote serum total cholesterol level for U.S. females 20 years old or older.

a. Sketch the distribution of the variable x.

b. Obtain the standardized version, z, of x.

c. Identify and sketch the distribution of z.

d. The percentage of U.S. females 20 years old or older who have a serum total cholesterol level between 150 mg/dL and 250 mg/dL is equal to the area under the standard normal curve between _____ and _____.

e. The percentage of U.S. females 20 years old or older who have a serum total cholesterol level below 220 mg/dL is equal to the area under the standard normal curve that lies to the _____ of _____.

Extending the Concepts and Skills

6.15 Chips Ahoy! 1,000 Chips Challenge. Students in an introductory statistics course at the U.S. Air Force Academy participated in Nabisco's "Chips Ahoy! 1,000 Chips Challenge" by confirming that there were at least 1000 chips in every 18-ounce bag of cookies that they examined. As part of their assignment, they concluded that the number of chips per bag is approximately normally distributed. Could the number of chips per bag be exactly normally distributed? Explain your answer. [SOURCE: Brad Warner and Jim Rutledge, "Checking the Chips Ahoy! Guarantee," *Chance*, Vol. 12(1), pp. 10–14]

6.16 Use the footnote on page 274 to write the equation of

a. the associated normal curve for a normally distributed variable with mean 5 and standard deviation 2.

b. the standard normal curve.

Using Technology

6.17 Using any technology available to you, graph the normal distribution with mean 5 and standard deviation 2.

6.18 Gestation Periods of Humans. Refer to the simulation of human gestation periods discussed in Example 6.2 on page 279.

a. Sketch the normal curve for human gestation periods.

b. Simulate 1000 human gestation periods. (*Note:* Users of the TI-83/84 Plus should simulate 500 human gestation periods.)

c. Approximately what values would you expect for the sample mean and sample standard deviation of the 1000 observations? Explain your answers.

d. Obtain the sample mean and sample standard deviation of the 1000 observations and compare your answers to your estimates in part (c).

e. Roughly what would you expect a histogram of the 1000 observations to look like? Explain your answer.

f. Obtain a histogram of the 1000 observations and compare your result to your expectation in part (e).

6.2 Areas Under the Standard Normal Curve

In Section 6.1, we introduced normally distributed variables. There we demonstrated, among other things, that we can obtain the percentage of all possible observations of a normally distributed variable that lie within any specified range by (1) expressing the range in terms of z-scores and (2) determining the corresponding area under the standard normal curve.

You already know how to convert to z-scores. Now, in this section, you will discover how to implement the second step—determining areas under the standard normal curve.

Basic Properties of the Standard Normal Curve

FIGURE 6.8

Standard normal distribution and standard normal curve

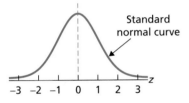

Before we show how to find areas under the standard normal curve, we need to discuss some of the basic properties of the standard normal curve. Recall that the standard normal curve is the curve associated with the standard normal distribution, that is, the normal distribution having mean 0 and standard deviation 1. Figure 6.8 again depicts the standard normal distribution and the standard normal curve.

In Section 6.1, we showed that a normal curve is bell shaped, is centered at μ, and is close to the horizontal axis outside the range from $\mu - 3\sigma$ to $\mu + 3\sigma$. Applied to the standard normal curve, these characteristics mean that it is bell shaped, is centered at 0, and is close to the horizontal axis outside the range from -3 to 3. Thus the standard normal curve is symmetric about 0. All of these properties are reflected in Fig. 6.8.

One property of the standard normal curve that is not obvious from Fig. 6.8 is that the total area under the curve is 1. This property is not unique to the standard normal curve; in fact, the total area under any curve that represents the distribution of a variable is equal to 1. Key Fact 6.3 summarizes our discussion about the properties of the standard normal curve.

KEY FACT 6.3

Basic Properties of the Standard Normal Curve

Property 1: The total area under the standard normal curve is 1.

Property 2: The standard normal curve extends indefinitely in both directions, approaching, but never touching, the horizontal axis as it does so.

Property 3: The standard normal curve is symmetric about 0; that is, the part of the curve to the left of the dashed line in Fig. 6.8 is the mirror image of the part of the curve to the right of it.

Property 4: Almost all the area under the standard normal curve lies between -3 and 3.

Because the standard normal curve is the associated normal curve for a standardized normally distributed variable, we labeled the horizontal axis in Fig. 6.8 with the letter z and refer to numbers on that axis as z-scores. For these reasons, the standard normal curve is sometimes called the **z-curve**.

Using the Standard Normal Table (Table II)

Because of the importance of areas under the standard normal curve, tables of those areas have been constructed. Such a table is Table II, which can be found inside the back cover of this book and in Appendix A.

A typical four-decimal-place number in the body of Table II gives the area under the standard normal curve that lies to the left of a specified z-score. The left page of Table II is for negative z-scores and the right page is for positive z-scores. In Example 6.3, we show how to find the area to the left of a z-score.

Example 6.3   **Finding the Area to the Left of a Specified z-Score**

Determine the area under the standard normal curve that lies to the left of 1.23, as shown in Fig. 6.9(a).

FIGURE 6.9
Finding the area under the standard normal curve to the left of 1.23

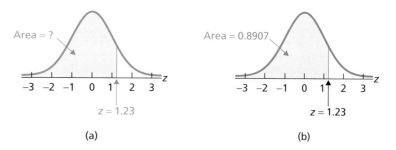

(a)   (b)

Solution   We use Table II, specifically the portion on the right page because 1.23 is positive. First, we go down the left-hand column, labeled z, to "1.2." Then, we go across that row to "0.03" in the top row. The number in the body of the table there, 0.8907, is the area under the standard normal curve that lies to the left of 1.23, as shown in Fig. 6.9(b).

Finding the area under the standard normal curve that lies to the left of a specified z-score is one important use of Table II. Two other important uses of that table are finding the area to the right of a specified z-score and finding the area between two specified z-scores. We illustrate these two uses in Examples 6.4 and 6.5, respectively.

Example 6.4   **Finding the Area to the Right of a Specified z-Score**

Determine the area under the standard normal curve that lies to the right of 0.76, as shown in Fig. 6.10(a) at the top of the following page.

Solution   Because the total area under the standard normal curve is 1 (Property 1 of Key Fact 6.3), the area to the right of 0.76 equals 1 minus the area to the left of 0.76. This latter area can be found in Table II, as explained in Example 6.3.

FIGURE 6.10

Finding the area under the standard
normal curve to the right of 0.76

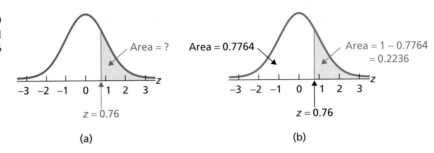

(a)   (b)

First, we go down the left-hand column, labeled z, to "0.7." Then, we go across that row to "0.06" in the top row. The number in the body of the table there, 0.7764, is the area under the standard normal curve that lies to the left of 0.76. Consequently, the area under the standard normal curve that lies to the right of 0.76 is $1 - 0.7764 = 0.2236$, as shown in Fig. 6.10(b).

Example 6.5   Finding the Area Between Two Specified z-Scores

Determine the area under the standard normal curve that lies between -0.68 and 1.82, as shown in Fig. 6.11(a).

FIGURE 6.11

Finding the area under the standard
normal curve that lies between -0.68
and 1.82

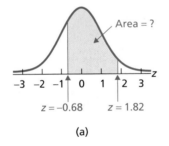
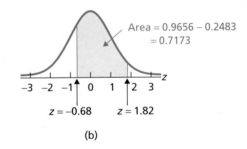

(a)   (b)

Solution The area under the standard normal curve that lies between -0.68 and 1.82 equals the area to the left of 1.82 minus the area to the left of -0.68. Table II shows that the area to the left of 1.82 is 0.9656 and that the area to the left of -0.68 is 0.2483. So the area under the standard normal curve that lies between -0.68 and 1.82 is $0.9656 - 0.2483 = 0.7173$, as depicted in Fig. 6.11(b).

The discussion presented in Examples 6.3–6.5 is summarized by the three graphs in Fig. 6.12. Each graph shows how Table II, which gives areas to the left of a specified z-score, can be used to obtain a required area.

Obtaining the area to the left of a specified z-score requires one table lookup, as shown in Fig. 6.12(a). Obtaining the area to the right of a specified z-score requires one table lookup and one subtraction (from 1), as shown in Fig. 6.12(b). Obtaining the area between two specified z-scores requires two table lookups and one subtraction, as shown in Fig. 6.12(c).

FIGURE 6.12

Using Table II to find the area under the standard normal curve that lies (a) to the left of a specified z-score, (b) to the right of a specified z-score, and (c) between two specified z-scores

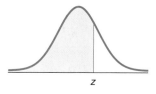

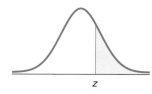

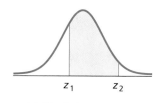

(a) Shaded area:
Area to left of z

(b) Shaded area:
$1 - $ (Area to left of z)

(c) Shaded area:
(Area to left of z_2)
$-$ (Area to left of z_1)

A Note Concerning Table II

The first area given in Table II is for $z = -3.90$. According to the table, the area under the standard normal curve that lies to the left of -3.90 is 0.0000. This entry does not mean that the area under the standard normal curve that lies to the left of -3.90 is exactly 0, but only that it is 0 to four decimal places (the area is 0.0000481 to seven decimal places). Indeed, as the standard normal curve extends indefinitely to the left without ever touching the axis, the area to the left of any z-score is greater than 0.

The last area given in Table II is for $z = 3.90$. According to the table, the area under the standard normal curve that lies to the left of 3.90 is 1.0000. This entry does not mean that the area under the standard normal curve that lies to the left of 3.90 is exactly 1, but only that it is 1 to four decimal places (the area is 0.9999519 to seven decimal places). Indeed, as the total area under the standard normal curve is exactly 1 and the curve extends indefinitely to the right without ever touching the axis, the area to the left of any z-score is less than 1.

Finding the z-score for a Specified Area

So far, we have used Table II to find areas under the standard normal curve to the left of a specified z-score, to the right of a specified z-score, and between two specified z-scores. Now, in Example 6.6, we show how to use Table II to find the z-score(s) corresponding to a specified area under the standard normal curve.

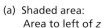Example 6.6   **Finding the z-Score Having a Specified Area to Its Left**

Determine the z-score having an area of 0.04 to its left under the standard normal curve, as depicted in Fig. 6.13(a).

FIGURE 6.13

Finding the z-score having an area of 0.04 to its left

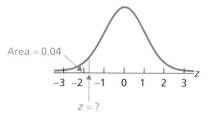

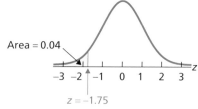

(a)

(b)

Solution We use Table II to obtain the *z*-score corresponding to the area 0.04. For ease of reference, we have reproduced a portion of Table II as Table 6.2.

TABLE 6.2
Areas under the standard normal curve

	Second decimal place in *z*									*z*
0.09	0.08	0.07	0.06	0.05	0.04	0.03	0.02	0.01	0.00	
.	.	.	.	.	.	.	.	.	.	.
.	.	.	.	.	.	.	.	.	.	.
					.					
0.0233	0.0239	0.0244	0.0250	0.0256	0.0262	0.0268	0.0274	0.0281	0.0287	−1.9
0.0294	0.0301	0.0307	0.0314	0.0322	0.0329	0.0336	0.0344	0.0351	0.0359	−1.8
0.0367	0.0375	0.0384	0.0392	0.0401	0.0409	0.0418	0.0427	0.0436	0.0446	−1.7
0.0455	0.0465	0.0475	0.0485	0.0495	0.0505	0.0516	0.0526	0.0537	0.0548	−1.6
0.0559	0.0571	0.0582	0.0594	0.0606	0.0618	0.0630	0.0643	0.0655	0.0668	−1.5
.	.	.	.	.	.	.	.	.	.	.
.	.	.	.	.	.	.	.	.	.	.
.	.	.	.	.	.	.	.	.	.	.

We search the body of Table 6.2 (or Table II) for the area 0.04. Because we find no such area, we use the area closest to 0.04, which is 0.0401. As we see from Table 6.2, the *z*-score corresponding to that area is −1.75. Thus the *z*-score having area 0.04 to its left under the standard normal curve is roughly −1.75, as depicted in Fig. 6.13(b).

In Example 6.6, we were to determine the *z*-score having an area of 0.04 to its left. Because we were unable to find an area entry of 0.04 in Table II, we selected the area closest to 0.04 and took the *z*-score corresponding to that area as an approximation of the required *z*-score.

This approach illustrates what we do in the most typical case: When no area entry in Table II equals the one desired, but there is one area entry closest to the one desired, we take the *z*-score corresponding to the closest area entry as an approximation of the required *z*-score.

Two other cases are possible. One is when an area entry in Table II equals the one desired; nothing more need be done. The other is when no area entry in Table II equals the one desired, but two area entries are equally closest to the one desired; in this case, we take the mean of the two corresponding *z*-scores as an approximation of the required *z*-score. Both of these cases are illustrated in Example 6.7, which we present momentarily.

Finding the *z*-score having a specified area to its right is often necessary. We have to make this determination so frequently that a special notation, z_α, is required.

DEFINITION 6.3

FIGURE 6.14
The z_α notation

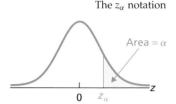

The z_α Notation

The symbol z_α is used to denote the z-score having an area of α (alpha) to its right under the standard normal curve, as illustrated in Fig. 6.14. Read "z_α" as "z sub α" or more simply as "z α."

We apply the z_α notation in Examples 6.7 and 6.8.

Example 6.7 **Finding z_α**

Use Table II to find

a. $z_{0.025}$. **b.** $z_{0.05}$.

Solution

a. $z_{0.025}$ is the z-score having an area of 0.025 to its right under the standard normal curve, as shown in Fig. 6.15(a).

FIGURE 6.15
Finding $z_{0.025}$

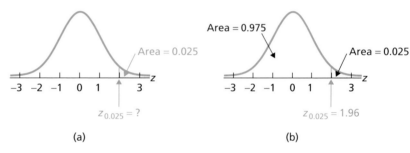

(a) (b)

Because the area under the standard normal curve to the right of $z_{0.025}$ is 0.025, the area to its left is $1 - 0.025 = 0.975$, as shown in Fig. 6.15(b). We search the body of Table II for the area 0.975 and find that entry. Its corresponding z-score is 1.96. Thus, $z_{0.025} = 1.96$, as shown in Fig. 6.15(b).

b. $z_{0.05}$ is the z-score having an area of 0.05 to its right under the standard normal curve, as shown in Fig. 6.16(a).

FIGURE 6.16
Finding $z_{0.05}$

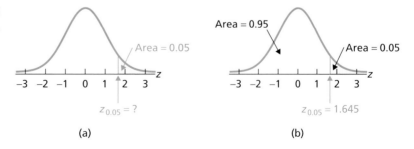

(a) (b)

Because the area under the standard normal curve to the right of $z_{0.05}$ is 0.05, the area to its left is $1 - 0.05 = 0.95$, as shown in Fig. 6.16(b). We search the body of Table II for the area 0.95 but find no such area. Instead, we find two areas are equally closest to 0.95—namely, 0.9495 and 0.9505. The z-scores corresponding to those two areas are 1.64 and 1.65, respectively. So our approximation of $z_{0.05}$ is the z-score halfway between 1.64 and 1.65; that is, $z_{0.05} = 1.645$, as shown in Fig. 6.16(b).

In Example 6.8, we show how to determine the two z-scores that divide the area under the standard normal curve into a specified middle area and two outside areas.

Example 6.8 | Finding the z-Scores for a Specified Area

Find the two z-scores that divide the area under the standard normal curve into a middle 0.95 area and two outside 0.025 areas, as depicted in Fig. 6.17(a).

FIGURE 6.17

Finding the two z-scores that divide the area under the standard normal curve into a middle 0.95 area and two outside 0.025 areas

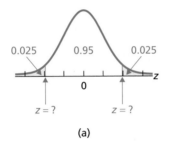
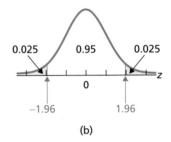

(a) (b)

Solution In Fig. 6.17(a), the area of the shaded region on the right is 0.025, which means that the z-score on the right is $z_{0.025}$. In Example 6.7(a), we found that $z_{0.025} = 1.96$. Thus the z-score on the right is 1.96. The standard normal curve is symmetric about 0, so the z-score on the left is -1.96. Therefore the two required z-scores are ± 1.96, as shown in Fig. 6.17(b).

Note: We could also solve Example 6.8 by first using Table II to find the z-score on the left in Fig. 6.17(a), which is -1.96, and then applying the symmetry property to obtain the z-score on the right, which is 1.96. Can you think of a third way to solve the problem?

EXERCISES 6.2

Statistical Concepts and Skills

6.19 Explain why being able to obtain areas under the standard normal curve is important.

6.20 With which normal distribution is the standard normal curve associated?

6.21 Without consulting Table II, explain why the area under the standard normal curve that lies to the right of 0 is 0.5.

6.22 According to Table II, the area under the standard normal curve that lies to the left of -2.08 is 0.0188. Without further consulting Table II, determine the area under the standard normal curve that lies to the right of 2.08. Explain your reasoning.

6.23 According to Table II, the area under the standard normal curve that lies to the left of 0.43 is 0.6664. Without further consulting Table II, determine the area under the standard normal curve that lies to the right of 0.43. Explain your reasoning.

6.24 According to Table II, the area under the standard normal curve that lies to the left of 1.96 is 0.975. Without further consulting Table II, determine the area under the standard normal curve that lies to the left of −1.96. Explain your reasoning.

6.25 Property 4 of Key Fact 6.3 states that most of the area under the standard normal curve lies between −3 and 3. Use Table II to determine precisely the percentage of the area under the standard normal curve that lies between −3 and 3.

6.26 Why is the standard normal curve sometimes referred to as the z-curve?

6.27 Explain how Table II is used to determine the area under the standard normal curve that lies
a. to the left of a specified z-score.
b. to the right of a specified z-score.
c. between two specified z-scores.

6.28 The area under the standard normal curve that lies to the left of a z-score is always strictly between _____ and _____.

Use Table II to obtain the areas under the standard normal curve required in Exercises **6.29–6.32**. *Sketch a standard normal curve and shade the area of interest in each problem.*

6.29 Determine the area under the standard normal curve that lies to the left of
a. 2.24.   **b.** −1.56.
c. 0.   **d.** −4.

6.30 Find the area under the standard normal curve that lies to the right of
a. −1.07.   **b.** 0.6.
c. 0.   **d.** 4.2.

6.31 Determine the area under the standard normal curve that lies between
a. −2.18 and 1.44.   **b.** −2 and −1.5.
c. 0.59 and 1.51.   **d.** 1.1 and 4.2.

6.32 Determine the area under the standard normal curve that lies
a. either to the left of −2.12 or to the right of 1.67.
b. either to the left of 0.63 or to the right of 1.54.

6.33 Use Table II to obtain each shaded area under the standard normal curve.

a.
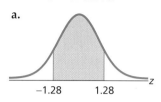

b.
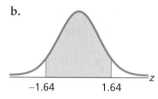

c.
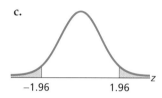

d.
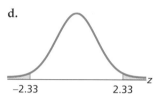

6.34 In each part, find the area under the standard normal curve that lies between the specified z-scores, sketch a standard normal curve, and shade the area of interest.
a. −1 and 1   **b.** −2 and 2   **c.** −3 and 3

6.35 The total area under the following standard normal curve is divided into eight regions.

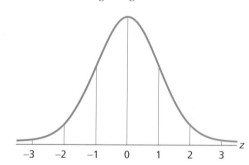

a. Determine the area of each region.
b. Complete the following table.

Region	Area	Percentage of total area
−∞ to −3	0.0013	0.13
−3 to −2		
−2 to −1		
−1 to 0		
0 to 1	0.3413	34.13
1 to 2		
2 to 3		
3 to ∞		
	1.0000	100.00

In Exercises 6.36–6.42, use Table II to obtain the required z-scores. Illustrate your work with graphs.

6.36 Obtain the z-score for which the area under the standard normal curve to its left is 0.025.

6.37 Find the z-score that has an area of 0.75 to its left under the standard normal curve.

6.38 Obtain the z-score that has an area of 0.95 to its right.

6.39 Determine $z_{0.33}$.

6.40 Find the following z-scores.
a. $z_{0.03}$ **b.** $z_{0.005}$

6.41 Determine the two z-scores that divide the area under the standard normal curve into a middle 0.90 area and two outside 0.05 areas.

6.42 Complete the following table.

$z_{0.10}$	$z_{0.05}$	$z_{0.025}$	$z_{0.01}$	$z_{0.005}$
1.28				

Extending the Concepts and Skills

6.43 In this section, we mentioned that the total area under any curve representing the distribution of a variable equals 1. Explain why.

6.44 Let $0 < \alpha < 1$. Determine the
a. z-score an area of α to its right in terms of z_α.
b. z-score having an area of α to its left in terms of z_α.
c. two z-scores that divide the area under the curve into a middle $1 - \alpha$ area and two outside areas of $\alpha/2$.
d. Draw graphs to illustrate your results in parts (a)–(c).

6.3   Working With Normally Distributed Variables

You have now learned everything required to obtain the percentage of all possible observations of a normally distributed variable that lie within any specified range. To do so, first express the range in terms of z-scores and then determine the corresponding area under the standard normal curve. More formally, use Procedure 6.1.

PROCEDURE 6.1

To Determine a Percentage or Probability for a Normally Distributed Variable

STEP 1   **Sketch the normal curve associated with the variable.**

STEP 2   **Shade the region of interest and mark the delimiting *x*-values.**

STEP 3   **Compute the *z*-scores for the delimiting *x*-values found in Step 2.**

STEP 4   **Use Table II to obtain the area under the standard normal curve delimited by the *z*-scores found in Step 3.**

FIGURE 6.18
Graphical portrayal of Procedure 6.1

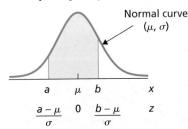

The steps in Procedure 6.1 are portrayed graphically in Fig. 6.18, with the specified range lying between the two numbers *a* and *b*. When the specified range is to the left (or right) of a specified number, it is represented similarly. However, there will be only one *x*-value, and the shaded region will be the area under the normal curve that lies to the left (or right) of that *x*-value.

Note: When computing z-scores in Step 3 of Procedure 6.1, round to two decimal places because that is the precision provided in Table II.

Example 6.9 Percentages for a Normally Distributed Variable

Intelligence Quotients Intelligence quotients (IQs) measured on the Stanford Revision of the Binet–Simon Intelligence Scale are normally distributed with a mean of 100 and a standard deviation of 16. Obtain the percentage of people who have IQs between 115 and 140.

Solution Here the variable is IQ, and the population consists of all people. Because IQs are normally distributed, we can determine the required percentage by applying Procedure 6.1.

STEP 1 Sketch the normal curve associated with the variable.

Here $\mu = 100$ and $\sigma = 16$. The normal curve associated with the variable is presented in Fig. 6.19. Note that the tick marks are 16 units apart; that is, the distance between successive tick marks is equal to the standard deviation.

FIGURE 6.19
Determination of the percentage of people having IQs between 115 and 140

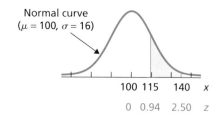

Normal curve
($\mu = 100$, $\sigma = 16$)

| | 100 | 115 | 140 | x |
| | 0 | 0.94 | 2.50 | z |

z-score computations:

Area to the left of *z*:

$$x = 115 \longrightarrow z = \frac{115 - 100}{16} = 0.94 \qquad 0.8264$$

$$x = 140 \longrightarrow z = \frac{140 - 100}{16} = 2.50 \qquad 0.9938$$

Shaded area = 0.9938 − 0.8264 = 0.1674

STEP 2 Shade the region of interest and mark the delimiting *x*-values.

See the shaded region and delimiting *x*-values in Fig. 6.19.

STEP 3 Compute the *z*-scores for the delimiting *x*-values found in Step 2.

We need to obtain the *z*-scores for the *x*-values 115 and 140:

$$x = 115 \quad \longrightarrow \quad z = \frac{115 - \mu}{\sigma} = \frac{115 - 100}{16} = 0.94,$$

and

$$x = 140 \quad \longrightarrow \quad z = \frac{140 - \mu}{\sigma} = \frac{140 - 100}{16} = 2.50.$$

These *z*-scores are marked beneath the *x*-values in Fig. 6.19.

STEP 4 Use Table II to obtain the area under the standard normal curve delimited by the *z*-scores found in Step 3.

The area under the standard normal curve to the left of 0.94 is 0.8264 and that to the left of 2.50 is 0.9938. Consequently, the required area, the area shaded in Fig. 6.19, is $0.9938 - 0.8264 = 0.1674$.

INTERPRETATION 16.74% of all people have IQs between 115 and 140. Equivalently, the probability is 0.1674 that a randomly selected person will have an IQ between 115 and 140. ▬■

Note: Procedure 6.1 can be accomplished efficiently by performing all four steps in a "picture," as illustrated in Fig. 6.19.

Visualizing a Normal Distribution

We now present a rule that permits us quickly and easily to visualize and obtain useful information about a normally distributed variable. The rule gives the percentages of all possible observations that lie within one, two, and three standard deviations to either side of the mean.

Recall that the *z*-score of an observation tells us how many standard deviations the observation is from the mean. Thus the percentage of all possible observations that lie within one standard deviation to either side of the mean equals the percentage of all observations whose *z*-scores lie between -1 and 1. For a normally distributed variable, that percentage is the same as the area under the standard normal curve between -1 and 1, which is 0.6826 or 68.26%. Proceeding in the same way, we get the **68.26-95.44-99.74 rule**, expressed as the following fact.

KEY FACT 6.4

68.26-95.44-99.74 Rule

Any normally distributed variable has the following properties.

Property 1: 68.26% of all possible observations lie within one standard deviation to either side of the mean, that is, between $\mu - \sigma$ and $\mu + \sigma$.

Property 2: 95.44% of all possible observations lie within two standard deviations to either side of the mean, that is, between $\mu - 2\sigma$ and $\mu + 2\sigma$.

Property 3: 99.74% of all possible observations lie within three standard deviations to either side of the mean, that is, between $\mu - 3\sigma$ and $\mu + 3\sigma$.

These properties are displayed graphically in Fig. 6.20.

FIGURE 6.20
68.26-95.44-99.74 rule

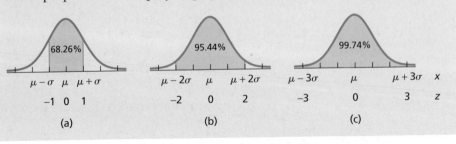

In Example 6.10, we apply the 68.26-95.44-99.74 rule to IQs.

▊▊ Example 6.10   68.26-95.44-99.74 Rule

Intelligence Quotients   Apply the 68.26-95.44-99.74 rule to IQs.

Solution Recall that IQs (measured on the Stanford Revision of the Binet–Simon Intelligence Scale) are normally distributed with a mean of 100 and a standard deviation of 16. In particular, we have $\mu = 100$ and $\sigma = 16$.

From Property 1 of the 68.26-95.44-99.74 rule, 68.26% of all people have IQs within one standard deviation to either side of the mean. One standard deviation below the mean is $\mu - \sigma = 100 - 16 = 84$; one standard deviation above the mean is $\mu + \sigma = 100 + 16 = 116$.

INTERPRETATION 68.26% of all people have IQs between 84 and 116, as illustrated in Fig. 6.21(a).

From Property 2 of the 68.26-95.44-99.74 rule, 95.44% of all people have IQs within two standard deviations to either side of the mean. Two standard deviations below the mean is $\mu - 2\sigma = 100 - 2 \cdot 16 = 100 - 32 = 68$; two standard deviations above the mean is $\mu + 2\sigma = 100 + 2 \cdot 16 = 100 + 32 = 132$.

INTERPRETATION 95.44% of all people have IQs between 68 and 132, as illustrated in Fig. 6.21(b).

From Property 3 of the 68.26-95.44-99.74 rule, 99.74% of all people have IQs within three standard deviations to either side of the mean. Three standard deviations below the mean is $\mu - 3\sigma = 100 - 3 \cdot 16 = 100 - 48 = 52$; three standard deviations above the mean is $\mu + 3\sigma = 100 + 3 \cdot 16 = 100 + 48 = 148$.

INTERPRETATION 99.74% of all people have IQs between 52 and 148, as illustrated in Fig. 6.21(c).

FIGURE 6.21
Graphical display of the
68.26-95.44-99.74 rule for IQs

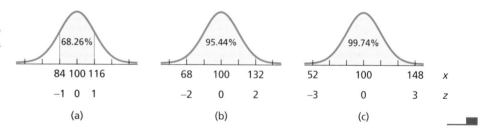

(a) (b) (c)

As illustrated in Example 6.10, the 68.26-95.44-99.74 rule allows us to obtain useful information about a normally distributed variable quickly and easily. Note, however, that similar facts are obtainable for any number of standard deviations. For instance, Table II reveals that, for any normally distributed variable, 86.64% of all possible observations lie within 1.5 standard deviations to either side of the mean.

Experience has shown that the 68.26-95.44-99.74 rule works reasonably well for any variable having approximately a bell-shaped distribution, regardless of whether the variable is normally distributed. This fact is referred to as the **empirical rule**, which we alluded to earlier in Chapter 3 (see page 114) in our discussion of the three-standard-deviations rule.

Finding the Observations for a Specified Percentage

Procedure 6.1 shows how to determine the percentage of all possible observations of a normally distributed variable that lie within any specified range. Frequently, however, we want to carry out the reverse procedure, that is, to find the observations corresponding to a specified percentage. Procedure 6.2 allows us to do that.

PROCEDURE 6.2

To Determine the Observations Corresponding to a Specified Percentage or Probability for a Normally Distributed Variable

STEP 1 Sketch the normal curve associated with the variable.

STEP 2 Shade the region of interest.

STEP 3 Use Table II to obtain the z-scores delimiting the region found in Step 2.

STEP 4 Obtain the x-values having the z-scores found in Step 3.

Among other things, we can use Procedure 6.2 to obtain quartiles, deciles, or any other percentile for a normally distributed variable. Example 6.11 shows how to find percentiles by this method.

▌▌Example 6.11   Obtaining Percentiles for a Normally Distributed Variable

Intelligence Quotients   Obtain and interpret the 90th percentile for IQs.

Solution The 90th percentile, P_{90}, is the IQ that is higher than those of 90% of all people. As IQs are normally distributed, we can determine the 90th percentile by applying Procedure 6.2.

STEP 1 Sketch the normal curve associated with the variable.

Here $\mu = 100$ and $\sigma = 16$. The normal curve associated with IQs is shown in Fig. 6.22.

STEP 2 Shade the region of interest.

See the shaded region in Fig. 6.22.

FIGURE 6.22
Finding the 90th percentile for IQs

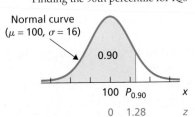

STEP 3 **Use Table II to obtain the *z*-scores delimiting the region found in Step 2.**

The *z*-score corresponding to P_{90} is the one having an area of 0.90 to its left under the standard normal curve. From Table II, that *z*-score is 1.28, approximately, as illustrated in Fig. 6.22.

STEP 4 **Obtain the *x*-values having the *z*-scores found in Step 3.**

We must find the *x*-value having the *z*-score 1.28—the IQ that is 1.28 standard deviations above the mean. It is $100 + 1.28 \cdot 16 = 100 + 20.48 = 120.48$.

INTERPRETATION The 90th percentile for IQs is 120.48. Thus, 90% of people have IQs below 120.48 and 10% have IQs above 120.48.

The Technology Center

Almost all statistical technologies provide programs to implement automatically the procedures that we discussed in this section—namely, to obtain for a normally distributed variable

- the percentage of all possible observations that lie within any specified range and
- the observations corresponding to a specified percentage.

In this subsection, we present output and step-by-step instructions to carry out these two procedures.

Example 6.12   **Using Technology to Obtain Normal Percentages**

Intelligence Quotients   Use Minitab, Excel, or the TI-83/84 Plus to determine the percentage of people who have IQs between 115 and 140.

Solution   Each of the three technologies provides a program for obtaining the area under the associated normal curve of a normally distributed variable that lies to the left of a specified value. Such an area corresponds to a **cumulative probability,** the probability that the variable will be less than or equal to the specified value.

Recall that IQs are normally distributed with a mean of 100 and a standard deviation of 16. We want to find the percentage of people who have IQs between 115 and 140. Printout 6.2 at the top of the next page shows output that can be easily used to obtain this percentage.

We get the required percentage from the Minitab output in Printout 6.2 by subtracting the two cumulative probabilities: $0.9938 - 0.8257 = 0.1681$, or 16.81%. Similarly, in Printout 6.2, Excel gives the required percentage as $0.99379032 - 0.825749307 = 0.168041013$, or 16.80%. Printout 6.2 shows that the TI-83/84 Plus does the subtraction for us and presents the required percentage as 0.1680410128, or 16.80%.

Note that the percentages obtained by the three technologies differ slightly from the percentage of 16.74% that we found in Example 6.9. The differences reflect the fact that the technologies retain more accuracy than we can get from Table II.

PRINTOUT 6.2

Output for obtaining the percentage of
people with IQs between 115 and 140

MINITAB

Cumulative Distribution Function

Normal with mean = 100 and standard deviation = 16

x	P(X <= x)
115	0.825749
140	0.993790

EXCEL

NORMDIST

X	115	= 115
Mean	100	= 100
Standard_dev	16	= 16
Cumulative	TRUE	= TRUE

= 0.825749307

Returns the normal cumulative distribution for the specified mean and standard deviation.

X is the value for which you want the distribution.

Formula result =0.825749307   OK   Cancel

TI-83/84 PLUS

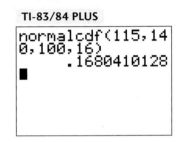
```
normalcdf(115,14
0,100,16)
          .1680410128
```

NOTE: Replacing 115 by 140 in the **X** text box yields 0.99379032.

Example 6.13   Using Technology to Obtain Normal Percentiles

Intelligence Quotients   Use Minitab, Excel, or the TI-83/84 Plus to find the
90th percentile for IQs.

Solution   Each technology provides a program for obtaining the observation that has a
specified area to its left under the associated normal curve of a normally distributed variable. Such an observation corresponds to an **inverse cumulative
probability,** the observation whose cumulative probability is the specified area.

Recall that IQs are normally distributed with a mean of 100 and a standard
deviation of 16. We want to find the 90th percentile for IQs, the IQ that is
higher than those of 90% of all people. Equivalently, we want to find the x-
value that has an area of 0.90 to its left under the normal curve with parameters
$\mu = 100$ and $\sigma = 16$. Printout 6.3 shows output obtained for this x-value by
applying the inverse cumulative probability programs from Minitab, Excel,
and the TI-83/84 Plus.

From any of the three outputs in Printout 6.3, the 90th percentile for IQs
is 120.5048. Note that this value differs slightly from the value of 120.48 that
we obtained in Example 6.11. The difference reflects the fact that the three
technologies retain more accuracy than we can get from Table II.

PRINTOUT 6.3
Output for obtaining the 90th percentile
for IQs

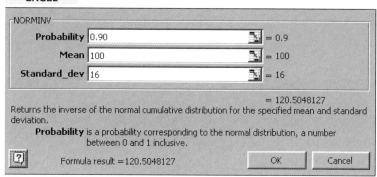

MINITAB

Inverse Cumulative Distribution Function

```
Normal with mean = 100 and standard deviation = 16

P( X <= x )        x
        0.9  120.505
```

EXCEL

TI-83/84 PLUS

```
invNorm(0.90,100
,16)
        120.5048251
```

Obtaining the Output (Optional)

Printout 6.2 provides output that can be easily used to determine the percentage of people who have IQs between 115 and 140. Here are detailed instructions for obtaining that output.

MINITAB	EXCEL	TI-83/84 PLUS
1 Store the delimiting IQs, 115 and 140, in a column named IQ 2 Choose **Calc ➤ Probability Distributions ➤ Normal...** 3 Select the **Cumulative probability** option button 4 Click in the **Mean** text box and type 100 5 Click in the **Standard deviation** text box and type 16 6 Select the **Input column** option button 7 Click in the **Input column** text box and specify IQ 8 Click **OK**	1 Click **fx** on the button bar 2 Select **Statistical** from the **Function category** list 3 Select **NORMDIST** from the **Function name** list 4 Click **OK** 5 Type 115 in the **X** text box 6 Click in the **Mean** text box and type 100 7 Click in the **Standard_dev** text box and type 16 8 Click in the **Cumulative** text box and type TRUE 9 To obtain the cumulative probability for 140, replace 115 by 140 in the **X** text box	1 Press **2nd ➤ DISTR** 2 Arrow down to **normalcdf(** and press **ENTER** 3 Type 115,140,100,16) and press **ENTER**

Printout 6.3 provides output that gives the 90th percentile for IQs. Here are detailed instructions for obtaining that output.

MINITAB	EXCEL	TI-83/84 PLUS
1 Choose **Calc ➤ Probability Distributions ➤ Normal...**	1 Click **f$_x$** on the button bar	1 Press **2nd ➤ DISTR**
2 Select the **Inverse cumulative probability** option button	2 Select **Statistical** from the **Function category** list	2 Arrow down to **invNorm(** and press **ENTER**
3 Click in the **Mean** text box and type <u>100</u>	3 Select **NORMINV** from the **Function name** list	3 Type <u>0.90,100,16)</u> and press **ENTER**
4 Click in the **Standard deviation** text box and type <u>16</u>	4 Click **OK**	
5 Select the **Input constant** option button	5 Type <u>0.90</u> in the **Probability** text box	
6 Click in the **Input constant** text box and type <u>0.90</u>	6 Click in the **Mean** text box and type <u>100</u>	
7 Click **OK**	7 Click in the **Standard_dev** text box and type <u>16</u>	

EXERCISES 6.3

Statistical Concepts and Skills

6.45 Briefly, for a normally distributed variable, how do you obtain the percentage of all possible observations that lie within a specified range?

6.46 Explain why the percentage of all possible observations of a normally distributed variable that lie within two standard deviations to either side of the mean equals the area under the standard normal curve between −2 and 2.

6.47 Drive for Show, Putt for Dough. An article by Scott M. Berry titled "Drive for Show and Putt for Dough" (*Chance*, Vol. 12(4), pp. 50–54) discussed driving distances of PGA players. The mean distance for tee shots on the 1999 men's PGA tour is 272.2 yards with a standard deviation of 8.12 yards. Assuming that the 1999 tee-shot distances are normally distributed, find the percentage of such tee shots that went
a. between 260 and 280 yards.
b. more than 300 yards.

6.48 Metastatic Carcinoid Tumors. A study of sizes of metastatic carcinoid tumors in the heart was conducted by U. Pandya et al. and published as the article, "Metastatic Carcinoid Tumor to the Heart: Echocardiographic-Pathologic Study of 11 Patients" (*Journal of the American College of Cardiology*, Vol. 40, pp. 1328–1332). Based on that study, we assume that lengths of metastatic carcinoid tumors in the heart are normally distributed with mean 1.8 cm and standard deviation 0.5 cm. Find the percentage of metastatic carcinoid tumors in the heart that
a. are between 1 cm and 2 cm long.
b. exceed 3 cm in length.

6.49 New York City 10 km Run. As reported in *Runner's World* magazine, the times of the finishers in the New York City 10 km run are normally distributed with a mean of 61 minutes and a standard deviation of 9 minutes. Let X be the time of a randomly selected finisher. Find
a. $P(X > 75)$.   **b.** $P(X < 50$ or $X > 70)$.

6.50 Polychaete Worms. *Opisthotrochopodus n. sp.* is a polychaete worm that inhabits deep sea hydrothermal vents along the Mid-Atlantic Ridge. According to an article by Van Dover et al. in *Marine Ecology Progress Series* (Vol. 181, pp. 201–214) the lengths of female polychaete worms are normally distributed with mean 6.1 mm and standard deviation 1.3 mm. Let X denote the length of a randomly selected female polychaete worm. Determine
a. $P(X \leq 3)$.   **b.** $P(5 < X < 7)$.

6.51 Gibbon Song Duration. A preliminary behavioral study of the Jingdong black gibbon, a primate endemic to the Wuliang Mountains in China, found that the mean song bout duration in the wet season is 12.59 minutes with a standard deviation of 5.31 minutes. [SOURCE: Lori K.

Sheeran, Zhang Yongzu, Frank E. Poirier, and Yang Dehua, "Preliminary Report on the Behavior of the Jingdong Black Gibbon (*Hylobates concolor jingdongensis*)," *Tropical Biodiversity*, Vol. 5(2), pp. 113-125] Assume that song bout is normally distributed and apply the 68.26-95.44-99.74 rule to determine the percentage of song bouts that have durations within

a. 1 standard deviation to either side of the mean.

b. 2 standard deviations to either side of the mean.

c. 3 standard deviations to either side of the mean.

6.52 Children Watching TV. The A. C. Nielsen Company reports in the *Nielsen Report on Television* that the mean weekly television viewing time for children aged 2–11 years is 24.50 hours. Assume that the weekly television viewing times of such children are normally distributed with a standard deviation of 6.23 hours and apply the 68.26-95.44-99.74 rule to fill in the blanks.

a. 68.26% of all such children watch between _____ and _____ hours of TV per week.

b. 95.44% of all such children watch between _____ and _____ hours of TV per week.

c. 99.74% of all such children watch between _____ and _____ hours of TV per week.

d. Draw graphs similar to those in Fig. 6.21 on page 293 to portray your results.

6.53 Drive for Show, Putt for Dough. Refer to Exercise 6.47.

a. Determine the quartiles of the driving distances.

b. Find the 95th percentile.

c. Obtain the third decile.

d. Interpret your answers for parts (a)–(c).

6.54 Metastatic Carcinoid Tumors. Refer to Exercise 6.48.

a. Determine the quartiles for lengths of metastatic carcinoid tumors in the heart.

b. Obtain the 20th percentile.

c. Find the seventh decile.

d. Interpret your answers for parts (a)–(c).

Extending the Concepts and Skills

6.55 For a normally distributed variable, fill in the blanks.

a. _____% of all possible observations lie within 1.96 standard deviations to either side of the mean.

b. _____% of all possible observations lie within 1.64 standard deviations to either side of the mean.

6.56 Emergency Room Traffic. Desert Samaritan Hospital, in Mesa, Arizona, keeps records of emergency room traffic. Those records reveal that the times between arriving patients have a mean of 8.7 minutes with a standard deviation of 8.7 minutes. Based solely on the values of these two parameters, explain why it is unreasonable to assume that the times between arriving patients is normally distributed or even approximately so.

6.57 Heights of Female Students. Refer to Example 6.1 on page 275. The heights of the 3264 female students attending a midwestern college are approximately normally distributed with mean 64.4 inches and standard deviation 2.4 inches. Thus we can use the normal distribution with $\mu = 64.4$ and $\sigma = 2.4$ to approximate the percentage of these students having heights within any specified range. In each part, (i) obtain the exact percentage from Table 6.1, (ii) use the normal distribution to approximate the percentage, and (iii) compare your answers.

a. The percentage of female students with heights between 62 and 63 inches.

b. The percentage of female students with heights between 65 and 70 inches.

6.58 Let $0 < \alpha < 1$. For a normally distributed variable, show that $100(1 - \alpha)\%$ of all possible observations lie within $z_{\alpha/2}$ standard deviations to either side of the mean, that is, between $\mu - z_{\alpha/2} \cdot \sigma$ and $\mu + z_{\alpha/2} \cdot \sigma$.

6.59 Let x be a normally distributed variable with mean μ and standard deviation σ.

a. Express the quartiles, $Q_1, Q_2,$ and Q_3, in terms of μ and σ.

b. Express the kth percentile, P_k, in terms of μ, σ, and k.

Using Technology

6.60 Drive for Show, Putt for Dough. Use the technology of your choice to solve Exercises 6.47 and 6.53.

6.61 Polychaete Worms. Use the technology of your choice to solve Exercise 6.50.

6.4 Assessing Normality; Normal Probability Plots

You have now seen how to work with normally distributed variables. For instance, you know how to determine the percentage of all possible observations that lie within any specified range and how to obtain the observations corresponding to a specified percentage.

Another problem involves deciding whether a variable is normally distributed, or at least approximately so, based on a sample of observations. Such decisions often play a major role in subsequent analyses—from percentage or percentile calculations to statistical inferences.

From Key Fact 2.1, if a simple random sample is taken from a population, the distribution of the observed values of a variable will approximate the distribution of the variable—and the larger the sample, the better the approximation tends to be. We can use this fact to help decide whether a variable is normally distributed.

On the one hand, if a variable is normally distributed, then, for a large sample, a histogram of the observations should be roughly bell shaped; for a very large sample, even moderate departures from a bell shape cast doubt on the normality of the variable. On the other hand, for a relatively small sample, ascertaining a clear shape in a histogram and, in particular, whether it is bell shaped is often difficult. These comments also hold for stem-and-leaf diagrams and dotplots.

Thus, for relatively small samples, a more sensitive graphical technique than the ones we have presented so far is required for assessing normality. Normal probability plots provide such a technique.

The idea behind a normal probability plot is simple: Compare the observed values of the variable to the observations expected for a normally distributed variable. More precisely, a **normal probability plot** is a plot of the observed values of the variable versus the **normal scores**—the observations expected for a variable having the standard normal distribution. If the variable is normally distributed, the normal probability plot should be roughly linear (i.e., fall roughly in a straight line) and vice versa.

When you use a normal probability plot to assess the normality of a variable, you must remember two things: (1) that the decision of whether a normal probability plot is roughly linear is a subjective one and (2) that you are using the observations of the variable for a sample to make a judgment about all possible observations of the variable (i.e., the distribution of the variable). With these considerations in mind, the following guidelines can be used to assess normality.

KEY FACT 6.5

Guidelines for Assessing Normality Using a Normal Probability Plot

To assess the normality of a variable using sample data, construct a normal probability plot.

- If the plot is roughly linear, accept as reasonable that the variable is approximately normally distributed.
- If the plot shows systematic deviations from linearity (e.g., if it displays significant curvature), conclude that the variable probably is not approximately normally distributed.

These guidelines should be interpreted loosely for small samples, but they can be interpreted rather strictly for large samples.

In practice, normal probability plots are obtained by computer. However, for purposes of understanding, constructing a few by hand is helpful. Table III in Appendix A provides the normal scores for sample sizes from 5 to 30. In Example 6.14, we explain how to use Table III to obtain a normal probability plot.

Example 6.14 Normal Probability Plots

TABLE 6.3
Adjusted gross incomes ($1000s)

9.7	93.1	33.0	21.2
81.4	51.1	43.5	10.6
12.8	7.8	18.1	12.7

Adjusted Gross Incomes   The Internal Revenue Service publishes data on federal individual income tax returns in *Statistics of Income, Individual Income Tax Returns*. A simple random sample of 12 returns from last year revealed the adjusted gross incomes, in thousands of dollars, shown in Table 6.3. Construct a normal probability plot for these data, and use the plot to assess the normality of adjusted gross incomes.

Solution   Here the variable is adjusted gross income, and the population consists of all last year's federal individual income tax returns. To construct a normal probability plot, we first arrange the data in increasing order and obtain the normal scores from Table III. The ordered data are shown in the first column of Table 6.4; the normal scores, from the $n = 12$ column of Table III, are shown in the second column of Table 6.4.

TABLE 6.4
Ordered data and normal scores

Adjusted gross income	Normal score
7.8	−1.64
9.7	−1.11
10.6	−0.79
12.7	−0.53
12.8	−0.31
18.1	−0.10
21.2	0.10
33.0	0.31
43.5	0.53
51.1	0.79
81.4	1.11
93.1	1.64

Next, we plot the points in Table 6.4, using the horizontal axis for the adjusted gross incomes and the vertical axis for the normal scores. For instance, the first point plotted has a horizontal coordinate of 7.8 and a vertical coordinate of −1.64.

Figure 6.23 shows all 12 points from Table 6.4, the normal probability plot for the sample of adjusted gross incomes. Note that the normal probability plot in Fig. 6.23 has significant curvature.

INTERPRETATION   In light of Key Fact 6.5, last year's adjusted gross incomes apparently are not (approximately) normally distributed.

FIGURE 6.23
Normal probability plot for the sample of adjusted gross incomes

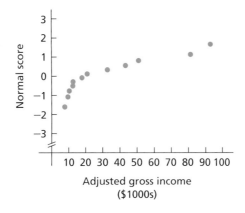

In some books and statistical technologies, you may encounter one or both of the following differences in normal probability plots:

- The vertical axis is used for the data and the horizontal axis for the normal scores.
- A probability or percent scale is used instead of normal scores.

Detecting Outliers with Normal Probability Plots

Recall that outliers are observations that fall well outside the overall pattern of the data. We can use normal probability plots to detect outliers, as explained in Example 6.15.

▌▌Example 6.15   Using Normal Probability Plots to Detect Outliers

TABLE 6.5

Sample of last year's chicken consumption (lb)

47	39	62	49	50	70
59	53	55	0	65	63
53	51	50	72	45	

Chicken Consumption   The U.S. Department of Agriculture publishes data on U.S. chicken consumption in *Food Consumption, Prices, and Expenditures*. Last year's chicken consumption, in pounds, for 17 randomly selected people are displayed in Table 6.5. A normal probability plot for these observations is presented in Fig. 6.24(a). Use the plot to discuss the distribution of chicken consumption and to detect any outliers.

FIGURE 6.24

Normal probability plots for chicken consumption: (a) original data; (b) data with outlier removed

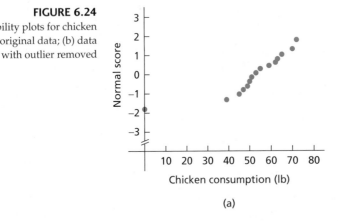

(a)

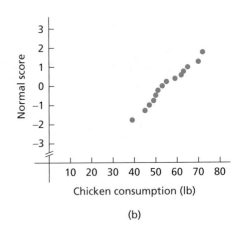

(b)

Solution   Figure 6.24(a) reveals that the normal probability plot falls roughly in a straight line except for the point corresponding to the consumption of 0 lb. That point falls well outside the overall pattern of the plot.

INTERPRETATION   The observation of 0 lb is an outlier, which might be a recording error or due to a person in the sample who does not eat chicken for some reason (e.g., a vegetarian).

If we remove the outlier 0 lb from the sample data and draw a normal proba-bility plot for the abridged data, Fig. 6.24(b) shows that this normal probability plot is quite linear.

INTERPRETATION It appears plausible that, among people who eat chicken, the amounts they consume annually are (approximately) normally distributed.

Note: There appear to be only 15 points (instead of 17) in Fig. 6.24(a). The reason is that we, as do many statistical software packages, use an averaging process to assign identical normal scores to identical observations; keep this fact in mind when you examine normal probability plots in figures and printouts. Thus, in Example 6.15, the two observations of 50 are assigned identical normal scores as are the two observations of 53. So, there really are 17 points in the graph, but only 15 are distinguishable because there are two sets of two identical points. An alternative procedure is to treat identical observations as being slightly different; then no averaging is required.

In this section, you learned how a normal probability plot for a sample of observations of a variable can be used as an aid for deciding whether the vari-able is (approximately) normally distributed. Although this visual assessment of normality is subjective, it is sufficient for most statistical analyses.

The Technology Center

Most statistical technologies have programs that automatically construct nor-mal probability plots. In this subsection, we present output and step-by-step instructions to implement such programs.

Example 6.16 **Using Technology to Obtain Normal Probability Plots**

Adjusted Gross Incomes Use Minitab, Excel, or the TI-83/84 Plus to obtain a normal probability plot for the adjusted gross incomes displayed in Table 6.3 on page 301.

Solution Printout 6.4 (next page) shows the output obtained by applying the normal-probability-plot programs to the data on adjusted gross incomes.

As we mentioned earlier and as you can see from the Excel output in Print-out 6.4, normal probability plots are sometimes drawn with the vertical axis used for the data and the horizontal axis for the normal scores. Compare the normal probability plots in Printout 6.4 to the one in Fig. 6.23 on page 301.

Obtaining the Output (Optional)

Here are detailed instructions for obtaining a normal probability plot for the sample of adjusted gross incomes using Minitab, Excel, and the TI-83/84 Plus. First, we store the data from Table 6.3 on page 301 in a column (Minitab), range (Excel), or list (TI-83/84 Plus) named AGI. Then, we proceed as follows.

MINITAB

1 Choose **Calc ➤ Calculator…**
2 Type <u>NSCORE</u> in the **Store result in variable** text box
3 Select **Normal scores** from the **Functions** list box
4 Specify AGI for **number** in the **Expression** text box
5 Click **OK**
6 Choose **Graph ➤ Scatterplot…**
7 Select the **Simple** scatterplot and click **OK**
8 Specify NSCORE in the **Y variables** text box for **1**
9 Click in the **X variables** text box for **1** and specify AGI
10 Click **OK**

EXCEL

1 Choose **DDXL ➤ Charts and Plots**
2 Select **Normal Probability Plot** from the **Function type** drop-down box
3 Specify AGI in the **Quantitative Variable** text box
4 Click **OK**

TI-83/84 PLUS

1 Press **2nd ➤ STAT PLOT** and then press **ENTER** twice
2 Arrow to the sixth graph icon and press **ENTER**
3 Press the down-arrow key
4 Press **2nd ➤ LIST**
5 Arrow down to AGI and press **ENTER**
6 Press **ZOOM** and then **9** (and then **TRACE**, if desired)

PRINTOUT 6.4

Normal-probability-plot output for the sample of adjusted gross incomes

MINITAB

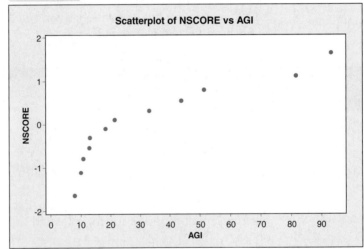

EXCEL

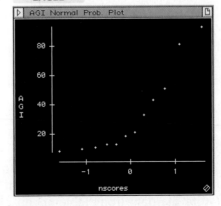

TI-83/84 PLUS

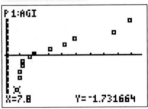

Note to Minitab users: Minitab has several programs for obtaining normal probability plots. In the instructions, we provided for getting a normal probability plot with normal scores on the vertical axis. A quicker procedure, but one that yields a plot with cumulative percents on the vertical axis, is to choose **Stat ➤ Basic Statistics ➤ Normality Test...,** specify AGI in the **Variable** text box, and click **OK**.

EXERCISES 6.4

Statistical Concepts and Skills

6.62 Explain why assessing the normality of a variable is often important.

6.63 Under what circumstances is using a normal probability plot to assess the normality of a variable usually better than using a histogram, stem-and-leaf diagram, or dotplot?

6.64 Regarding normal probability plots,
a. explain in detail what a normal probability plot is and how it is used to assess the normality of a variable.
b. how is a normal probability plot used to detect outliers?

In Exercises 6.65–6.68,
a. *use Table III in Appendix A to construct a normal probability plot of the given data. For simplicity, treat equal observations as being slightly different when obtaining normal scores.*
b. *use part (a) to identify any outliers.*
c. *use part (a) to assess the normality of the variable under consideration.*

6.65 Exam Scores. A sample of the final exam scores in a large introductory statistics course is as follows.

88	67	64	76	86
85	82	39	75	34
90	63	89	90	84
81	96	100	70	96

6.66 Cell Phone Rates. In the February 2003 issue of *Consumer Reports*, different cell phone providers and plans were compared. The monthly fees, in dollars, for a sample of the providers and plans are shown in the following table.

40	110	90	30	70
70	30	60	60	50
60	70	35	80	75

6.67 Thoroughbred Racing. The following table displays finishing times, in seconds, for the winners of fourteen 1-mile thoroughbred horse races, as found in two recent issues of *Thoroughbred Times*.

94.15	93.37	103.02	95.57	97.73	101.09	99.38
97.19	96.63	101.05	97.91	98.44	97.47	95.10

6.68 Beverage Expenditures. The Bureau of Labor Statistics publishes information on average annual expenditures by consumers in the *Consumer Expenditure Survey*. In 2000, the mean amount spent by consumers on nonalcoholic beverages was $250. A random sample of 12 consumers yielded the following data, in dollars, on last year's expenditures on nonalcoholic beverages.

361	176	184	265
259	281	240	273
259	249	194	258

Using Technology

6.69 Exam Scores. Use the technology of your choice to obtain a normal probability plot for the exam scores in Exercise 6.65.

6.70 Cell Phone Rates. Use the technology of your choice to obtain a normal probability plot for the cell phone rates in Exercise 6.66.

6.71 Fat Consumption in Vegetarians. A paper by Shao-Chun Lu et al., titled "LDL of Taiwanese Vegetarians are Less Oxidizable Than Those of Omnivores" (*Human Nutrition and Metabolism*, Vol. 130(6), pp. 1591–1596) compared fat consumption by vegetarians and omnivores. The following table displays the amount of fat consumed, in grams per day, for 28 vegetarians in the study.

20.5	31.4	35.7	52.8	27.0	40.3	45.7
19.7	32.5	33.5	58.5	30.1	61.4	33.3
35.3	54.7	54.1	56.7	35.9	58.8	25.7
66.3	35.9	35.7	47.1	38.7	16.4	42.0

a. Use the technology of your choice to obtain a normal probability plot of the data.
b. Identify outliers, if any.
c. Assess the normality of fat consumption for Taiwanese vegetarians.

6.72 Chips Ahoy! 1,000 Chips Challenge. Students in an introductory statistics course at the U.S. Air Force Academy participated in Nabisco's "Chips Ahoy! 1,000 Chips Challenge" by confirming that there were at least 1000 chips in every 18-ounce bag of cookies that they examined. As part of their assignment, they concluded that the number of chips per bag is approximately normally distributed. Their conclusion was based on the following data, which gives the number of chips per bag for 42 bags. Do you agree with the conclusion of the students? Explain your answer. [SOURCE: Brad Warner and Jim Rutledge, "Checking the Chips Ahoy! Guarantee," *Chance*, Vol. 12(1), pp. 10–14]

1200	1219	1103	1213	1258	1325	1295
1247	1098	1185	1087	1377	1363	1121
1279	1269	1199	1244	1294	1356	1137
1545	1135	1143	1215	1402	1419	1166
1132	1514	1270	1345	1214	1154	1307
1293	1546	1228	1239	1440	1219	1191

6.73 Finger Length of Criminals. In 1902, W. R. Macdonell published the article "On Criminal Anthropometry and the Identification of Criminals" (*Biometrika*, 1, pp. 177–227). Among other things, the author presented data on the left middle finger length, in centimeters. The following table provides the midpoints and frequencies of the finger length classes used.

Midpoint (cm)	Frequency	Midpoint (cm)	Frequency
9.5	1	11.6	691
9.8	4	11.9	509
10.1	24	12.2	306
10.4	67	12.5	131
10.7	193	12.8	63
11.0	417	13.1	16
11.3	575	13.4	3

Use these data and the technology of your choice to assess the normality of middle finger length of criminals by using
a. a histogram.
b. a normal probability plot. Explain your procedure and reasoning in detail.

6.74 Gestation Periods of Humans. For humans, gestation periods are normally distributed with a mean of 266 days and a standard deviation of 16 days.
a. Use the technology of your choice to simulate four random samples of 50 human gestation periods each.
b. Obtain a normal probability plot of each sample in part (a).
c. Are the normal probability plots in part (b) what you expected? Explain your answer.

6.75 Emergency Room Traffic. Desert Samaritan Hospital in Mesa, Arizona, keeps records of emergency room traffic. Those records reveal that the times between arriving patients have a special type of reverse J-shaped distribution called an *exponential distribution*. The records also show that the mean time between arriving patients is 8.7 minutes.
a. Use the technology of your choice to simulate four random samples of 75 interarrival times each.
b. Obtain a normal probability plot of each sample in part (a).
c. Are the normal probability plots in part (b) what you expected? Explain your answer.

6.5 Normal Approximation to the Binomial Distribution[†]

In this section, we demonstrate the approximation of binomial probabilities by using areas under a suitable normal curve. The development of the mathematical theory for doing so is credited to Abraham de Moivre (1667–1754) and Pierre-Simon Laplace (1749–1827). For more information on de Moivre and Laplace, see the biographies at the end of Chapters 12 and 7, respectively.

[†]Coverage of the binomial distribution (Section 5.3) is prerequisite to this section.

First, we need to review briefly the binomial distribution, which we discussed in detail in Section 5.3. Suppose that n identical independent success–failure experiments are performed, with the probability of success on any given trial being p. Let X denote the total number of successes in the n trials. Then, the probability distribution of the random variable X is given by the binomial probability formula,

$$P(X = x) = \binom{n}{x} p^x (1 - p)^{n-x}, \qquad x = 0, 1, 2, \ldots, n.$$

We say that X has the binomial distribution with parameters n and p.

You might be wondering why we would use normal-curve areas to approximate binomial probabilities when we can obtain them exactly with the binomial probability formula. Example 6.17 provides the reason.

Example 6.17   The Need to Approximate Binomial Probabilities

Mortality   Mortality tables enable actuaries to obtain the probability that a person at any particular age will live a specified number of years. This information, in turn, permits the determination of life-insurance premiums, pensions, annuity payments, and related items of importance to insurance companies and others.

According to tables provided by the U.S. National Center for Health Statistics in *Vital Statistics of the United States*, there is about an 80% chance that a person aged 20 will be alive at age 65. In Example 5.12 on page 246, we used the binomial probability formula to determine probabilities for the number of 20-year-olds of three who will be alive at age 65.

For most real-world problems, the number of people under investigation is much larger than three. Although, in principle, we can use the binomial probability formula to determine probabilities regardless of the number of people being considered, the practical realities dictate otherwise. Suppose, for instance, that 500 people aged 20 are selected at random. Find the probability that

a. exactly 390 of them will be alive at age 65.

b. between 375 and 425 of them, inclusive, will be alive at age 65.

Solution   Let X denote the number of people of the 500 who are alive at age 65. Then X has the binomial distribution with parameters $n = 500$ (the 500 people) and $p = 0.8$ (the probability a person aged 20 will be alive at age 65). Thus, in principle, probabilities for X can be determined exactly by using the binomial probability formula,

$$P(X = x) = \binom{500}{x} (0.8)^x (0.2)^{500-x}.$$

Let's apply that formula to the problems posed in parts (a) and (b).

a. The "answer" is

$$P(X = 390) = \binom{500}{390} (0.8)^{390} (0.2)^{110}.$$

However, to obtain the numerical value of the expression on the right of the preceding equation is not easy, even with a calculator. In performing the computations, we must be careful to avoid such pitfalls as making round-off errors and getting numbers so large or so small that they are outside the range of the calculator. Fortunately, we can sidestep the calculations altogether by using normal-curve areas.

b. The "answer" is

$$P(375 \leq X \leq 425) = P(X = 375) + P(X = 376) + \cdots + P(X = 425)$$

$$= \binom{500}{375}(0.8)^{375}(0.2)^{125} + \binom{500}{376}(0.8)^{376}(0.2)^{124}$$

$$+ \cdots + \binom{500}{425}(0.8)^{425}(0.2)^{75}.$$

Here we have the same computational difficulties as we did in part (a), except that we must evaluate 51 complex expressions instead of 1. Thus, although in theory we can use the binomial probability formula to determine the answer, doing so in practice is another matter. Surprising as it might seem, there is a way to use normal-curve areas to get the (approximate) answer—and it is easy!

We return to this problem momentarily and obtain the probabilities required in parts (a) and (b).

Example 6.17 makes clear why we often need to approximate binomial probabilities, even though the binomial probability formula is available for computing them exactly: *It is not practical to use the binomial probability formula when the number of trials, n, is very large.*

Under certain conditions on n and p, the distribution of a binomial random variable is (roughly) bell shaped. In such cases, we can approximate probabilities for the random variable by areas under a suitable normal curve. To show how, we present Example 6.18. For this example, calculating binomial probabilities exactly, using the binomial probability formula, is actually easy. However, for purposes of illustration, we also show how normal-curve areas approximate the binomial probabilities.

Example 6.18   Approximating Binomial Probabilities, Using Normal-Curve Areas

True–False Exams   A student is taking a true–false exam having 10 questions. Assume that the student guesses at all 10 questions and do the following.

a. Determine the probability that the student gets either 7 or 8 answers correct.

b. Approximate the probability obtained in part (a) by an area under a suitable normal curve.

Solution Use X to denote the number of correct answers by the student. Here X has the binomial distribution with parameters $n = 10$ (the 10 questions) and $p = 0.5$ (the probability of a correct guess).

a. Probabilities for X are given by the binomial probability formula,

$$P(X = x) = \binom{10}{x}(0.5)^x(1 - 0.5)^{10-x}.$$

Applying the formula, we obtain the probability distribution of X, as shown in Table 6.6. Using Table 6.6, we find that the probability the student gets either 7 or 8 answers correct is

$$P(X = 7 \text{ or } 8) = P(X = 7) + P(X = 8) = 0.1172 + 0.0439 = 0.1611.$$

b. To approximate the probability $P(X = 7 \text{ or } 8)$, we refer to Table 6.6 to obtain the probability histogram of the random variable X, which is displayed in Fig. 6.25.

Because the probability histogram in Fig. 6.25 is bell shaped, probabilities for X can be approximated by areas under a normal curve. As expected, the appropriate normal curve is the one whose parameters are the same as the mean and standard deviation of the random variable X. Because X has the binomial distribution with parameters $n = 10$ and $p = 0.5$, we can apply Formula 5.2 on page 249 to obtain easily the mean and standard deviation:

$$\mu = np = 10 \cdot 0.5 = 5$$

and

$$\sigma = \sqrt{np(1 - p)} = \sqrt{10 \cdot 0.5 \cdot (1 - 0.5)} = 1.58.$$

Therefore, the normal curve used here is the one with parameters $\mu = 5$ and $\sigma = 1.58$. That normal curve is superimposed on the probability histogram in Fig. 6.25.

TABLE 6.6

Probability distribution of the number of correct answers of 10 by the student

Number correct x	Probability $P(X = x)$
0	0.0010
1	0.0098
2	0.0439
3	0.1172
4	0.2051
5	0.2461
6	0.2051
7	0.1172
8	0.0439
9	0.0098
10	0.0010

FIGURE 6.25

Probability histogram for X with superimposed normal curve

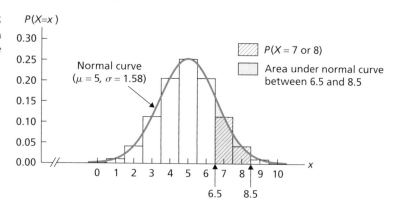

The probability $P(X = 7 \text{ or } 8)$ exactly equals the combined area of the corresponding bars of the histogram, the cross-hatched area in Fig. 6.25. Note that the cross-hatched area approximately equals the area under the normal curve between 6.5 and 8.5, the shaded area in Fig. 6.25.

Figure 6.25 makes clear why we consider the area under the normal curve between 6.5 and 8.5 instead of between 7 and 8. This adjustment is called the *correction for continuity.* It is required because we are approximating the distribution of a discrete variable by that of a continuous variable.

In any case, we see, at least graphically, that the probability $P(X = 7 \text{ or } 8)$ roughly equals the area under the normal curve with parameters $\mu = 5$ and $\sigma = 1.58$ that lies between 6.5 and 8.5. To compare these values quantitatively, we must obtain the normal-curve area. We do so in the usual way by converting to z-scores and then obtaining the corresponding area under the standard normal curve, as shown in Fig. 6.26.

FIGURE 6.26

Determination of the area under the normal curve with parameters $\mu = 5$ and $\sigma = 1.58$ that lies between 6.5 and 8.5

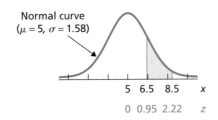

z-score computations:

$x = 6.5 \longrightarrow z = \dfrac{6.5 - 5}{1.58} = 0.95$ Area to the left of z:

                                                0.8289

$x = 8.5 \longrightarrow z = \dfrac{8.5 - 5}{1.58} = 2.22$ 0.9868

Shaded area = 0.9868 − 0.8289 = 0.1579

What
Does it Mean?

The normal-curve area provides an excellent approximation of the exact probability.

The last line in Fig. 6.26 shows that the area under the normal curve between 6.5 and 8.5 is 0.1579. This area is quite close to the exact value of $P(X = 7 \text{ or } 8)$, which, as we discovered in part (a), is 0.1611. ▬

As illustrated in Example 6.18, we can use normal-curve areas to approximate probabilities for binomial random variables that have bell-shaped distributions. Whether a particular binomial random variable has a bell-shaped distribution depends on its parameters, n and p. Figure 6.27 shows nine different binomial distributions.

As portrayed in Figs. 6.27(a) and 6.27(c), a binomial distribution with $p \neq 0.5$ is skewed. For small n, the skewness is enough to preclude using a normal approximation. However, as n increases, the skewness subsides and the binomial distribution becomes sufficiently bell shaped to permit a normal approximation. In contrast, as illustrated in Fig. 6.27(b), a binomial distribution with $p = 0.5$ is symmetric, regardless of the number of trials. Nonetheless, such a distribution will not be sufficiently bell shaped to permit a normal approximation if n is too small.

The customary rule of thumb for using the normal approximation is that *both np and n*$(1 - p)$ *are 5 or greater.* This restriction indicates, as suggested in Fig. 6.27, that the farther the success probability is from 0.5 (in either direction), the larger the number of trials must be to employ the normal approximation.

FIGURE 6.27

Nine different binomial distributions

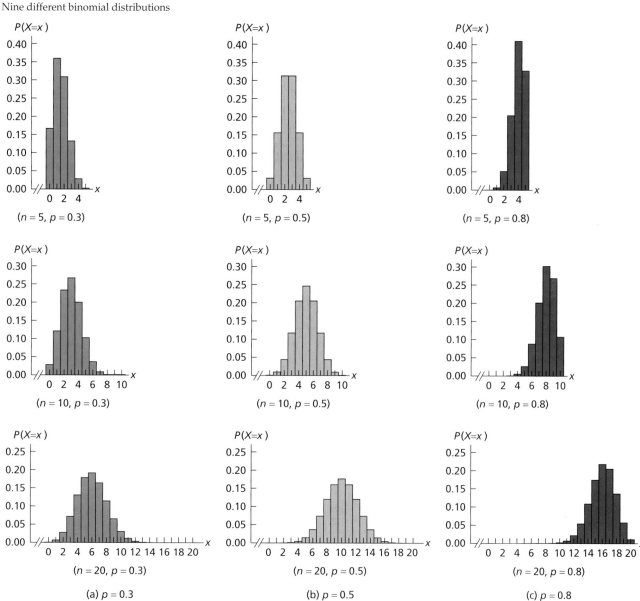

Precedure for Using the Normal Approximation to the Binomial Distribution

We can now write a general step-by-step method for approximating binomial probabilities by areas under a normal curve. We present this method as Procedure 6.3.

PROCEDURE 6.3

To Approximate Binomial Probabilities by Normal-Curve Areas

STEP 1 Determine *n*, the number of trials, and *p*, the success probability.

STEP 2 Determine whether both *np* and *n*(1 − *p*) are 5 or greater. If they are not, do not use the normal approximation.

STEP 3 Find μ and σ, using the formulas $\mu = np$ and $\sigma = \sqrt{np(1 - p)}$.

STEP 4 Make the correction for continuity and find the required area under the normal curve with parameters μ and σ.

Note: Step 4 of Procedure 6.3 requires the **correction for continuity**, as illustrated in Example 6.18. That is, when using normal-curve areas to approximate the probability that an observed value of a binomial random variable will be between two integers, inclusive, subtract 0.5 from the smaller integer and add 0.5 to the larger integer before finding the area under the normal curve.

In Example 6.19, we apply Procedure 6.3 to solve the mortality problems posed in Example 6.17.

Example 6.19   Normal Approximation to the Binomial

Mortality   The probability is 0.80 that a person aged 20 will be alive at age 65. Suppose that 500 people aged 20 are selected at random. Determine the probability that

a. exactly 390 of them will be alive at age 65.

b. between 375 and 425 of them, inclusive, will be alive at age 65.

Solution   We obtain the approximate values of the probabilities in parts (a) and (b) by applying Procedure 6.3.

STEP 1   Determine *n*, the number of trials, and *p*, the success probability.

We have $n = 500$ and $p = 0.8$.

STEP 2   Determine whether both *np* and *n*(1 − *p*) are 5 or greater.

From the values for *n* and *p* obtained in Step 1,

$$np = 500 \cdot 0.8 = 400 \quad \text{and} \quad n(1 - p) = 500 \cdot 0.2 = 100.$$

Both *np* and *n*(1 − *p*) are greater than 5, so we can apply the normal approximation.

STEP 3   Find μ and σ, using the formulas $\mu = np$ and $\sigma = \sqrt{np(1 - p)}$.

We get $\mu = 500 \cdot 0.8 = 400$ and $\sigma = \sqrt{500 \cdot 0.8 \cdot 0.2} = 8.94$.

STEP 4 Make the correction for continuity and find the required area under the normal curve with parameters μ and σ.

a. To make the correction for continuity, we subtract 0.5 from 390 and add 0.5 to 390. Thus we need to find the area under the normal curve with parameters $\mu = 400$ and $\sigma = 8.94$ that lies between 389.5 and 390.5. The required area, which is 0.0236, is obtained in Fig. 6.28. So, $P(X = 390) = 0.0236$, approximately.

INTERPRETATION The probability is about 0.0236 that exactly 390 of the 500 people selected will be alive at age 65.

FIGURE 6.28

Determination of the area under the normal curve with parameters $\mu = 400$ and $\sigma = 8.94$ that lies between 389.5 and 390.5

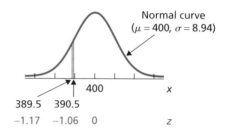

Normal curve ($\mu = 400$, $\sigma = 8.94$)

389.5   390.5

400   x

−1.17   −1.06   0 z

z-score computations:

$x = 389.5 \longrightarrow z = \dfrac{389.5 - 400}{8.94} = -1.17$

$x = 390.5 \longrightarrow z = \dfrac{390.5 - 400}{8.94} = -1.06$

Area to the left of z:

0.1210

0.1446

Shaded area $= 0.1446 - 0.1210 = 0.0236$

b. To make the correction for continuity, we subtract 0.5 from 375 and add 0.5 to 425. Thus we need to determine the area under the normal curve with parameters $\mu = 400$ and $\sigma = 8.94$ that lies between 374.5 and 425.5. We do so as in part (a) by converting to z-scores and then obtaining the corresponding area under the standard normal curve. We find the required area to be 0.9956. Consequently, $P(375 \leq X \leq 425) = 0.9956$, approximately.

INTERPRETATION The probability is roughly 0.9956 that between 375 and 425 of the 500 people selected will be alive at age 65.

—■

EXERCISES 6.5

Statistical Concepts and Skills

6.76 Why should you sometimes use normal-curve areas to approximate binomial probabilities even though you have a formula for computing them exactly?

6.77 The rule of thumb for using the normal approximation to the binomial is that both np and $n(1 - p)$ are 5 or greater. Why is this restriction necessary?

6.78 True–False Exams. Refer to Example 6.18 on page 308.
a. Use Table 6.6 to find the probability that the student guesses correctly on
    i. 4 or 5 questions, $P(X = 4 \text{ or } 5)$.
    ii. between 3 and 7 questions, $P(3 \leq X \leq 7)$.
b. Apply Procedure 6.3 to approximate the probabilities in part (a) by areas under a normal curve. Compare your answers.

6.79 True–False Exams. If, in Example 6.18, the true–false exam had 30 questions instead of 10, which normal curve would you use to approximate probabilities for the number of correct guesses?

In Exercises 6.80–6.83, apply Procedure 6.3 on page 312 to approximate the required binomial probabilities.

6.80 Airline Reservations. As reported by a spokesperson for Southwest Airlines, the no-show rate for reservations is 16%. In other words, the probability is 0.16 that a person making a reservation will not take the flight. For the next flight 42 people have reservations. What is the probability that
a. exactly 5 do not take the flight?
b. between 9 and 12, inclusive, do not take the flight?
c. at least 1 does not take the flight?
d. at most 2 do not take the flight?

6.81 Lightning-Induced Fatalities. As reported in the July/August 2003 issue of *Weatherwise*, according to the National Oceanic and Atmospheric Administration, people at ballparks and playgrounds are in more danger of being struck by lightning than those on golf courses. Of lightning-induced fatalities, 3.9% occur on golf courses. What is the probability that, of 250 randomly selected lightning-induced fatalities, the number occurring on golf courses is
a. exactly 4?
b. between 4 and 10, inclusive?
c. at least 10?

6.82 Teen Alcohol Use. The University of Michigan Institute for Social Research and National Institute on Drug Abuse publishes data on alcohol use in *Monitoring the Future*. According to the publication, 52.7% of all 12th graders in the United States have consumed alcohol in the past month. If 250 U.S. 12th graders are selected at random, find the probability that the number who have consumed alcohol in the past month is
a. exactly half of those sampled.
b. at least half of those sampled.
c. between 115 and 130, inclusive.

6.83 Mexican Households in Poverty. The U.N. Children's Fund defines a household living in poverty as one with an income below 50% of the national median. Mexico leads the world with 26% of households living in poverty. Of 300 randomly chosen Mexican households, what is the probability that the number living in poverty is
a. exactly 26% of those sampled?
b. at most 26% of those sampled?
c. at least 26% of those sampled?

Extending the Concepts and Skills

6.84 Roulette. An American roulette wheel consists of 38 numbers, of which 18 are red, 18 are black, and 2 are green. When the roulette wheel is spun, the ball is equally likely to land on each of the 38 numbers. A gambler is playing roulette and bets $10 on red each time. If the ball lands on a red number, the gambler wins $10; otherwise, the gambler loses $10. What is the probability that the gambler will be ahead after
a. 100 bets?
b. 1000 bets?
c. 5000 bets?
(*Hint:* The gambler will be ahead after a series of bets if and only if she has won more than half the bets.)

6.85 Flashlight Battery Lifetimes. A brand of flashlight battery has normally distributed lifetimes with a mean of 30 hours and a standard deviation of 5 hours. A supermarket purchases 500 of these batteries from the manufacturer. What is the probability that at least 80% of them will last longer than 25 hours?

6.86 Fragile X Syndrome. The second leading genetic cause of mental retardation is *Fragile X Syndrome*, named for the fragile appearance of the tip of the X chromosome in affected individuals. One in 1500 males are affected worldwide with no ethnic bias.
a. In a sample of 10,000 males, how many would you expect to have Fragile X Syndrome?
b. For a sample of 10,000 males, use the normal approximation to the binomial distribution to determine the probability that more than 7 of the males have Fragile X Syndrome; at most 10 of the males have Fragile X Syndrome.
c. The probabilities in part (b) were obtained in Exercise 5.53 on page 264 by using the Poisson approximation to the binomial distribution. Which estimates of the true binomial probabilities would you expect to be better, the ones using the normal approximation or those using the Poisson approximation? Explain your answer.

CHAPTER REVIEW

You Should Be Able To

1. use and understand the formulas in this chapter.

2. explain what it means for a variable to be normally distributed or approximately normally distributed.

3. explain the meaning of the parameters for a normal curve.

4. identify the basic properties of and sketch a normal curve.

5. identify the standard normal distribution and the standard normal curve.

6. use Table II to determine areas under the standard normal curve.

7. use Table II to determine the z-score(s) corresponding to a specified area under the standard normal curve.

8. use and understand the z_α notation.

9. determine a percentage or probability for a normally distributed variable.

10. state and apply the 68.26-95.44-99.74 rule.

11. determine the observations corresponding to a specified percentage or probability for a normally distributed variable.

12. explain how to assess the normality of a variable with a normal probability plot.

13. construct a normal probability plot with the aid of Table III.

14. use a normal probability plot to detect outliers.

*15. approximate binomial probabilities by normal-curve areas, when appropriate.

Key Terms

68.26-95.44-99.74 rule, *292*
approximately normally distributed
    variable, *274*
correction for continuity,* *312*
cumulative probability, *295*
empirical rule, *294*
inverse cumulative probability, *296*

normal curve, *274*
normal distribution, *274*
normal probability plot, *300*
normal scores, *300*
normally distributed population, *274*
normally distributed variable, *274*
parameters, *274*

standard normal curve, *277*
standard normal distribution, *277*
standardized normally distributed
    variable, *278*
z_α, *287*
z-curve, *282*

REVIEW TEST

Statistical Concepts and Skills

1. State two of the main reasons for studying the normal distribution.

2. Define
a. normally distributed variable.
b. normally distributed population.
c. parameters for a normal curve.

3. Answer true or false to each statement. Give reasons for your answers.
a. Two variables that have the same mean and standard deviation have the same distribution.
b. Two normally distributed variables that have the same mean and standard deviation have the same distribution.

4. Explain the relationship between percentages for a normally distributed variable and areas under the corresponding normal curve.

5. Identify the distribution of the standardized version of a normally distributed variable.

6. Answer true or false to each statement. Explain your answers.
a. Two normal distributions that have the same mean are centered at the same place, regardless of the relationship between their standard deviations.
b. Two normal distributions that have the same standard deviation have the same shape, regardless of the relationship between their means.

7. Consider the normal curves that have the parameters $\mu = 1.5$ and $\sigma = 3$; $\mu = 1.5$ and $\sigma = 6.2$; $\mu = -2.7$ and $\sigma = 3$; $\mu = 0$ and $\sigma = 1$.
a. Which curve has the largest spread?
b. Which curves are centered at the same place?
c. Which curves have the same shape?
d. Which curve is centered farthest to the left?
e. Which curve is the standard normal curve?

8. What key fact permits you to determine percentages for a normally distributed variable by first converting to z-scores and then determining the corresponding area under the standard normal curve?

9. Explain how to use Table II to determine the area under the standard normal curve that lies
a. to the left of a specified z-score.
b. to the right of a specified z-score.
c. between two specified z-scores.

10. Explain how to use Table II to determine the z-score that has a specified area to its
a. left under the standard normal curve.
b. right under the standard normal curve.

11. What does the symbol z_α signify?

12. State the 68.26-95.44-99.74 rule.

13. Roughly speaking, what are the normal scores corresponding to a sample of observations?

14. If you observe the values of a normally distributed variable for a sample, a normal probability plot should be roughly _____.

15. Sketch the normal curve having the parameters
a. $\mu = -1$ and $\sigma = 2$.   **b.** $\mu = 3$ and $\sigma = 2$.
c. $\mu = -1$ and $\sigma = 0.5$.

16. Forearm Length. In 1903, K. Pearson and A. Lee published a paper entitled "On the Laws of Inheritance in Man. I. Inheritance of Physical Characters" (*Biometrika*, Vol. 2, pp. 357–462). From information presented in that paper, forearm lengths of men, measured from the elbow to the middle fingertip, are (roughly) normally distributed with a mean of 18.8 inches and a standard deviation of 1.1 inches. Let x denote forearm length, in inches, for men.
a. Sketch the distribution of the variable x.
b. Obtain the standardized version, z, of x.
c. Identify and sketch the distribution of z.
d. The area under the normal curve with parameters 18.8 and 1.1 that lies between 17 and 20 is 0.8115. Determine the probability that a randomly selected man will have a forearm length between 17 inches and 20 inches.
e. The percentage of men who have forearm lengths less than 16 inches equals the area under the standard normal curve that lies to the _____ of _____.

17. According to Table II, the area under the standard normal curve that lies to the left of 1.05 is 0.8531. Without further reference to Table II, determine the area under the standard normal curve that lies
a. to the right of 1.05.
b. to the left of −1.05.
c. between −1.05 and 1.05.

18. Determine and sketch the area under the standard normal curve that lies
a. to the left of −3.02.
b. to the right of 0.61.
c. between 1.11 and 2.75.
d. between −2.06 and 5.02.
e. between −4.11 and −1.5.
f. either to the left of 1 or to the right of 3.

19. For the standard normal curve, find the z-score(s)
a. that has area 0.30 to its left.
b. that has area 0.10 to its right.
c. $z_{0.025}$, $z_{0.05}$, $z_{0.01}$, and $z_{0.005}$.
d. that divide the area under the curve into a middle 0.99 area and two outside 0.005 areas.

20. GRE Scores. Each year, thousands of college seniors take the Graduate Record Examination (GRE). The scores are transformed so they have a mean of 500 and a standard deviation of 100. Furthermore, the scores are known to be normally distributed. Determine the percentage of students that score
a. between 350 and 625.   **b.** 375 or greater.
c. below 750.

21. GRE Scores. Use the 68.26-95.44-99.74 rule to fill in the blanks with regard to Problem 20.
a. 68.26% of students who take the GRE score between _____ and _____.
b. 95.44% of students who take the GRE score between _____ and _____.
c. 99.74% of students who take the GRE score between _____ and _____.

22. GRE Scores. Refer to Problem 20. Solve the following problems and interpret your answers.
a. Obtain the quartiles for GRE scores.
b. Find the 99th percentile for GRE scores.

23. Costly Gas. According to the *AAA Daily Fuel Gauge Report*, the national average price for unleaded gasoline in July 2003 was $1.52. A random sample of 12 gas stations across the country yielded the following unleaded gas prices.

1.59	1.65	1.97	1.69
1.45	1.42	1.73	1.45
1.31	1.57	1.39	1.65

a. Use Table III to construct a normal probability plot for the data. For simplicity, treat equal observations as being slightly different when obtaining the normal scores.
b. Use part (a) to identify any outliers.
c. Use part (a) to assess normality.

*24. Diarrhea Vaccine. Acute rotavirus diarrhea is the leading cause of death among children under age 5, killing an estimated 4.5 million annually in developing countries. Scientists from Finland and Belgium claim that a new oral vaccine is 80% effective against rotavirus diarrhea. Assuming that the claim is correct, find the probability that, out of 1500 cases, the vaccine will be effective in
a. exactly 1225 cases.
b. at least 1175 cases.
c. between 1150 and 1250 cases, inclusive.

Using Technology

25. GRE Scores. Refer to Problem 20. Use the technology of your choice to do the following.
a. Sketch the normal curve for GRE scores.
b. Simulate 1000 GRE scores.

c. Approximately what values would you expect for the sample mean and sample standard deviation of the 1000 GRE scores obtained in part (b)? Explain your answers.
d. Determine the sample mean and sample standard deviation of the 1000 GRE scores obtained in part (b), and compare your answers to your answers in part (c).
e. Roughly what would you expect a histogram of the 1000 GRE scores obtained in part (b) to look like? Explain your answer.
f. Obtain a histogram of the 1000 GRE scores from part (b), and compare your result to your expectation in part (e).

26. GRE Scores. Use the technology of your choice to solve Problem 20. Comment on any discrepancies between the answers obtained here and those in Problem 20.

27. GRE Scores. Use the technology of your choice to solve Problem 22. Comment on any discrepancies between the answers obtained here and those in Problem 22.

28. Costly Gas. Use the technology of your choice to obtain a normal probability plot for the gasoline prices in Problem 23. Comment on any differences between the plot obtained here and the one in Problem 23.

StatExplore in MyMathLab
Analyzing Data Online

You can use StatExplore to perform all statistical analyses discussed in this chapter. To illustrate, we show how to obtain a normal probability plot. We note that StatExplore uses the terminology *QQ plot* instead of *normal probability plot*. (QQ is an abbreviation of *quantile–quantile*.)

EXAMPLE Normal Probability Plot

Adjusted Gross Incomes Table 6.3 on page 301 provides data on adjusted gross incomes, in thousands of dollars, for a simple random sample of 12 individual income tax returns from last year. Use StatExplore to obtain a normal probability plot of the data. Interpret the results.

SOLUTION   To begin, we store the sample data from Table 6.3 in a column named AGI. Then we proceed as follows:

    1 Choose **Graphics ➤ QQ Plot**
    2 Select the column AGI
    3 Click **Create Graph!**

The resulting output is shown in Printout 6.5.

PRINTOUT 6.5
StatExplore normal probability plot

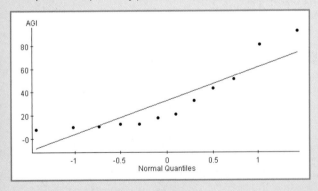

Note that StatExplore plots the sample data on the vertical axis and the normal quantiles (normal scores) on the horizontal axis. In any case, as the plot shows significant curvature (i.e., is far from linear), we conclude that last year's adjusted gross incomes are not normally distributed.

STATEXPLORE EXERCISES Solve the following problems by using StatExplore:

a. Obtain and interpret the normal probability plot required in Exercise 6.67 on page 305.
b. Determine the percentage required in Example 6.9 on page 291. (*Note:* Begin by choosing **StatExplore ➤ Calculators ➤ Normal.**)
c. Determine the IQ required in Example 6.11 on page 294.

To access StatExplore, go to the student content area of your Weiss MyMathLab course.

Internet Project

IQ of Boys and Girls

In this Internet project, you are to interact with several simulations and animations that show why and how the normal distribution is useful. You also are to explore a real data set containing the intelligence quotient (IQ) scores for boys and girls in primary school. In general, are boys and girls equally smart? Although most people believe there is no difference in IQ between the sexes, some believe there is.

IQ is a measure constructed to be normally distributed with a mean of 100 and a standard deviation of 16. Earlier in this chapter, we calculated probabilities for IQs based on this normal distribution model. In this Internet project, you will discover how well the model fits the real data.

URL for access to Internet Projects Page: www.aw-bc.com/weiss

Focusing on Data Analysis

The Focus Database

Recall from Chapter 1 (see page 35) that the Focus database contains information on the undergraduate students at the University of Wisconsin - Eau Claire (UWEC). Statistical analyses for this database should be done with the technology of your choice.

a. Obtain a histogram for each of the variables high school percentile, cumulative GPA, age, total earned credits, ACT English score, ACT math score, and ACT composite score.

b. Based on your results from part (a), which of the variables considered there appear to be approximately normally distributed?

c. Based on your results from part (a), which of the variables considered there appear to be far from normally distributed?

Foot Length and Shoe Size for Women

On page 273, we discussed foot length and shoe size for women. Now that you have studied the normal distribution, we ask you to solve several problems involving foot length and shoe size:

a. We mentioned on page 273 that the lengths of women's feet are normally distributed with a mean of 9.58 inches and a standard deviation of 0.51 inch. Based on that information, sketch the distribution of women's foot length and compare your result to the graph presented on page 273.
b. What percentage of women have foot lengths between 9 and 10 inches?
c. What percentage of women have foot lengths that exceed 11 inches?
d. Shoe manufacturers state that if a foot length is between two sizes, always move up to the larger size. Referring to the table on page 273, determine the percentage of women who wear size 8 shoes; size $11\frac{1}{2}$ shoes.
e. If an owner of a chain of shoe stores intends to purchase 10,000 pairs of women's shoes, roughly how many should he purchase of size 8? of size $11\frac{1}{2}$? Explain your reasoning.

Internet Resources: Visit the Weiss Web site www.aw-bc.com/weiss for additional discussion, exercises, and resources related to this case study.

Biography

CARL FRIEDRICH GAUSS: Child Prodigy

Carl Friedrich Gauss was born on April 30, 1777, in Brunswick, Germany, the only son in a poor, semiliterate peasant family; he taught himself to calculate before he could talk. At the age of 3, he pointed out an error in his father's calculations of wages. In addition to his arithmetic experimentation, he taught himself to read. At the age of 8, Gauss instantly solved the summing of all numbers from 1 to 100. His father was persuaded to allow him to stay in school and to study after school instead of working to help support the family.

Impressed by Gauss's brilliance, the Duke of Brunswick supported him monetarily from the ages of 14 to 30. This patronage permitted Gauss to pursue his studies exclusively. He conceived most of his mathematical discoveries by the time he was 17. Gauss was granted a doctorate in absentia from the university at Helmstedt; his doctoral thesis developed the concept of complex numbers and proved the fundamental theorem of algebra, which had previously been only partially established. Shortly thereafter, Gauss published his theory of numbers, which is considered one of the most brilliant achievements in mathematics.

Gauss made important discoveries in mathematics, physics, astronomy, and statistics. Two of his major contributions to statistics were the development of the least-squares method and fundamental work with the normal distribution, often called the *Gaussian distribution* in his honor.

In 1807, Gauss accepted the directorship of the observatory at the University of Göttingen which ended his dependence on the Duke of Brunswick. He remained there the rest of his life. In 1833, Gauss and a colleague, Wilhelm Weber, invented a working electric telegraph, 5 years before Samuel Morse. Gauss died in Göttingen in 1855.

The Sampling Distribution of the Sample Mean

GENERAL OBJECTIVES In the preceding chapters, you have studied sampling, descriptive statistics, probability, and the normal distribution. Now you will learn how these seemingly diverse topics can be integrated to lay the groundwork for inferential statistics.

In Section 7.1, we introduce the concepts of *sampling error* and *sampling distribution* and explain the essential role these concepts play in the design of inferential studies. The *sampling distribution* of a statistic is simply the distribution of the statistic, that is, the distribution of all possible observations of the statistic for samples of a given size from a population. In this chapter, we concentrate on the sampling distribution of the sample mean.

In Sections 7.2 and 7.3, we provide the required background for applying the sampling distribution of the sample mean. Specifically, in Section 7.2, we present formulas for the mean and standard deviation of the sample mean. Then, in Section 7.3, we indicate that, under certain general conditions, the sampling distribution of the sample mean is a normal distribution, or at least approximately so.

We apply this momentous fact in Chapters 8 and 9 to develop two important statistical-inference procedures: using the mean, $\bar{x}$, of a sample from a population to estimate and to draw conclusions about the mean, μ, of the entire population.

The Chesapeake and Ohio Freight Study

Can relatively small samples really provide results that are nearly as accurate as those obtained from a census? Although statisticians have shown mathematically that such is the case, a real study in which the results of a sample are compared with those of a census might make this assertion even more credible.

When a freight shipment travels over several railroads, the freight charge is divided among them according to prearranged agreements. With each shipment of freight, a document called a *waybill* is issued that provides information on the goods, route, and total charges. From the waybill of any particular freight shipment, the amount due each railroad can be calculated.

For a large number of shipments, the computations required for allocating the shares properly among the railroads are time consuming and costly. Consequently, if the division of total revenue among the railroads in question could be done accurately on the basis of a sample—as statisticians contend—considerable savings could be realized in accounting and clerical costs.

To convince themselves of the validity of the sampling approach, officials of the Chesapeake and Ohio Railroad Company (C&O) undertook a study of freight shipments that had traveled over its Pere Marquette district and another railroad during a 6-month period. The total number of waybills for that period was known (22,984), as was the total freight revenue.

Statistical theory was applied to determine the smallest number of waybills required to obtain an estimate of the total revenue due C&O with a prescribed accuracy. In all, 2072 of the 22,984 waybills, roughly 9%, were sampled. For each waybill in the sample, the necessary calculations were performed to find the amount of freight revenue for that shipment belonging to C&O. From those amounts the total revenue due C&O for all shipments was estimated to be $64,568.

How close was the estimate of $64,568, based on a sample of only 2072 waybills, to the total revenue actually due C&O for the 22,984 waybills? Take a guess! We'll discuss the answer at the end of this chapter.

7.1   Sampling Error; the Need for Sampling Distributions

We have already demonstrated that using a sample to acquire information about a population is often preferable to conducting a census, where data for the entire population are collected. Generally, sampling is less costly and can be done more quickly than a census; it is often the only practical way to gather information.

But now we need to deal with the following problem: Because a sample from a population provides data for only a portion of the entire population, we cannot expect the sample to yield perfectly accurate information about the population. Thus we should anticipate that a certain amount of error will result simply because we are sampling, which we call **sampling error** and define as follows.

DEFINITION 7.1

Sampling Error

Sampling error is the error resulting from using a sample to estimate a population characteristic.

Example 7.1 illustrates sampling error and the need for determining sampling distributions.

Example 7.1   **Sampling Error and the Need for Sampling Distributions**

Income Tax   The U.S. Internal Revenue Service (IRS) publishes annual figures on individual income tax returns in *Statistics of Income, Individual Income Tax Returns*. For the year 2001, the IRS reported that the mean tax of individual income tax returns was $9401. In actuality, the IRS reported the mean tax of a sample of 177,000 individual income tax returns from a total of more than 130 million such returns.

a. Identify the population under consideration.

b. Identify the variable under consideration.

c. Is the mean tax reported by the IRS a sample mean or the population mean?

d. Should we expect the mean tax, $\bar{x}$, of the 177,000 returns sampled by the IRS to be exactly the same as the mean tax, μ, of all individual income tax returns for 2001?

e. How can we answer questions about sampling error? For instance, is the sample mean tax, $\bar{x}$, reported by the IRS likely to be within $100 of the population mean tax, μ?

Solution

a. The population under consideration consists of all individual income tax returns for the year 2001.

b. The variable under consideration is "tax" (i.e., amount of income tax).

c. The mean tax reported is a sample mean, namely, the mean tax, $\bar{x}$, of the 177,000 returns sampled. It is not the population mean tax, μ, of all individual income tax returns for 2001.

d. We certainly cannot expect the mean tax, $\bar{x}$, of the 177,000 returns sampled by the IRS to be exactly the same as the mean tax, μ, of all individual income tax returns for 2001—some sampling error is to be anticipated.

e. To answer questions about the accuracy of estimating the mean tax of all individual income tax returns for 2001 by the mean tax of the 177,000 returns sampled, we need to know the distribution of all possible sample mean tax amounts (i.e., all possible $\bar{x}$-values) that could be obtained by sampling 177,000 individual income tax returns. That distribution is called the *sampling distribution of the sample mean*.

The distribution of a statistic (i.e., of all possible observations of the statistic for samples of a given size) is called the **sampling distribution** of the statistic. In this chapter, we concentrate on the **sampling distribution of the sample mean,** that is, of the statistic $\bar{x}$.

DEFINITION 7.2

What *Does it Mean?*

The sampling distribution of the sample mean is the distribution of all possible sample means for samples of a given size.

Sampling Distribution of the Sample Mean

For a variable x and a given sample size, the distribution of the variable $\bar{x}$ is called the *sampling distribution of the sample mean.*

In statistics, the following terms and phrases are synonymous.

- Sampling distribution of the sample mean
- Distribution of the variable $\bar{x}$
- Distribution of all possible sample means of a given sample size

Therefore, in this book, we use these three terms interchangeably.

Introducing the sampling distribution of the sample mean with an example that is both realistic and concrete is difficult because even for moderately large populations the number of possible samples is enormous, thus prohibiting an actual listing of the possibilities.[†] Consequently, in Example 7.2, we use an unrealistically small population to introduce the sampling distribution of the sample mean. Keep in mind, however, that populations are much larger in real-life applications.

Example 7.2 **Sampling Distribution of the Sample Mean**

TABLE 7.1
Heights, in inches, of the five starting players

Player	A	B	C	D	E
Height	76	78	79	81	86

Heights of Starting Players   Suppose that the population of interest consists of the five starting players on a men's basketball team whom, for convenience, we will call A, B, C, D, and E. Further suppose that the variable of interest is height, in inches. Table 7.1 lists the players and their heights.

a. Obtain the sampling distribution of the sample mean for samples of size 2.

[†]For example, the number of possible samples of size 50 from a population of size 10,000 is approximately equal to 3×10^{135}, a 3 followed by 135 zeros.

b. Make some observations about sampling error when the mean height of a random sample of two players is used to estimate the population mean height.[†]

c. Find the probability that, for a random sample of size 2, the sampling error made in estimating the population mean by the sample mean will be 1 inch or less; that is, determine the probability that $\bar{x}$ will be within 1 inch of μ.

Solution For future reference we first compute the population mean height:

$$\mu = \frac{\Sigma x}{N} = \frac{76 + 78 + 79 + 81 + 86}{5} = 80 \text{ inches.}$$

TABLE 7.2

Possible samples and sample means for samples of size 2

Sample	Heights	$\bar{x}$
A, B	76, 78	77.0
A, C	76, 79	77.5
A, D	76, 81	78.5
A, E	76, 86	81.0
B, C	78, 79	78.5
B, D	78, 81	79.5
B, E	78, 86	82.0
C, D	79, 81	80.0
C, E	79, 86	82.5
D, E	81, 86	83.5

a. The population under consideration here is so small that we can list the possibilities directly. There are 10 possible samples of size 2. The first column of Table 7.2 displays the 10 possible samples, the second column the corresponding heights (i.e., values of the variable "height"), and the third column the sample means. As a visual aid, we have also drawn a dotplot in Fig. 7.1 to portray the distribution of the sample means, that is, the sampling distribution of the sample mean for samples of size 2.

b. Referring to Table 7.2 or Fig. 7.1, we can make some simple but significant observations about sampling error when the mean height of a random sample of two players is used to estimate the population mean height. For instance, the mean height of the two players selected isn't likely to equal the population mean of 80 inches. In fact, only one of the 10 samples has a mean of 80 inches, the eighth sample in Table 7.2. Thus, in this case, the chances are only $\frac{1}{10}$, or 10%, that $\bar{x}$ will equal μ; some sampling error is likely.

c. Figure 7.1 shows that exactly 3 of the 10 samples have means within 1 inch of the population mean of 80 inches. So the probability is $\frac{3}{10}$, or 0.3, that the sampling error made in estimating μ by $\bar{x}$ will be 1 inch or less.

FIGURE 7.1

Dotplot for the sampling distribution of the sample mean for samples of size 2 ($n = 2$)

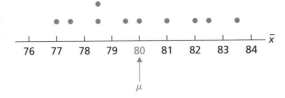

INTERPRETATION There is a 30% chance that the mean height of the two players selected will be within 1 inch of the population mean.

[†]As we mentioned in Section 1.3, the statistical-inference techniques considered in this book are intended for use only with simple random sampling. Therefore, unless otherwise specified, when we say *random sample*, we mean *simple random sample*. Furthermore, we assume that sampling is without replacement unless explicitly stated otherwise.

In Example 7.2, we determined the sampling distribution of the sample mean for samples of size 2. If we consider samples of another size—say, of size 4—we obtain a different sampling distribution of the sample mean, as demonstrated in Example 7.3.

Example 7.3 Sampling Distribution of the Sample Mean

Heights of Starting Players   The heights of the five starting players on a men's basketball team are presented in Table 7.1.

a. Obtain the sampling distribution of the sample mean for samples of size 4.

b. Make some observations about sampling error when the mean height of a random sample of four players is used to estimate the population mean height.

c. Find the probability that, for a random sample of size 4, the sampling error made in estimating the population mean by the sample mean will be 1 inch or less; that is, determine the probability that $\bar{x}$ will be within 1 inch of μ.

Solution

TABLE 7.3

Possible samples and sample means for samples of size 4

Sample	Heights	$\bar{x}$
A, B, C, D	76, 78, 79, 81	78.50
A, B, C, E	76, 78, 79, 86	79.75
A, B, D, E	76, 78, 81, 86	80.25
A, C, D, E	76, 79, 81, 86	80.50
B, C, D, E	78, 79, 81, 86	81.00

a. There are five possible samples of size 4. The first column of Table 7.3 displays the possible samples, the second column the corresponding heights (i.e., values of the variable "height"), and the third column the sample means. As a visual aid, we have also drawn a dotplot in Fig. 7.2 to portray the distribution of the sample means, that is, the sampling distribution of the sample mean for samples of size 4.

b. Referring to Table 7.3 or Fig. 7.2, we observe that none of the samples of size 4 has a mean equal to the population mean of 80 inches. Thus, when the mean height of a random sample of four players is used to estimate the population mean height, some sampling error is certain.

c. Figure 7.2 shows that exactly four of the five samples have means within 1 inch of the population mean of 80 inches. So the probability is $\frac{4}{5}$, or 0.8, that the sampling error made in estimating μ by $\bar{x}$ will be 1 inch or less.

FIGURE 7.2

Dotplot for the sampling distribution of the sample mean for samples of size 4 ($n = 4$)

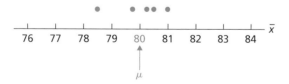

INTERPRETATION   There is an 80% chance that the mean height of the four players selected will be within 1 inch of the population mean.

Sample Size and Sampling Error

In Figs. 7.1 and 7.2, we drew dotplots for the sampling distributions of the sample mean for samples of sizes 2 and 4, respectively. Those two dotplots and dotplots for samples of sizes 1, 3, and 5 are displayed in Fig. 7.3.

FIGURE 7.3

Dotplots for the sampling distributions of the sample mean for samples of sizes 1, 2, 3, 4, and 5

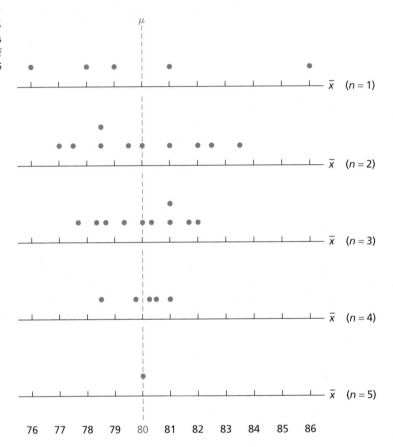

Figure 7.3 vividly shows that the possible sample means cluster more closely around the population mean as the sample size increases. This result, in turn, implies that sampling error tends to (although may not) be smaller for large samples than for small samples.

For example, Fig. 7.3 reveals that, for samples of size 1, two of five, or 40%, of the possible sample means lie within 1 inch of μ. For samples of size 2, three of ten, or 30%, of the possible sample means lie within 1 inch of μ. For samples of size 3, five of ten, or 50%, of the possible sample means lie within 1 inch of μ. For samples of size 4, four of five, or 80%, of the possible sample means lie within 1 inch of μ. And for samples of size 5, one of one, or 100%, of the possible sample means lie within 1 inch of μ.

Table 7.4 summarizes these results and also provides another sampling-error illustration easily obtained from Fig. 7.3.

TABLE 7.4

Sample size and sampling error illustrations for the heights of the basketball players

Sample size n	No. possible samples	No. within 1″ of μ	% within 1″ of μ	No. within 0.5″ of μ	% within 0.5″ of μ
1	5	2	40%	0	0%
2	10	3	30%	2	20%
3	10	5	50%	2	20%
4	5	4	80%	3	60%
5	1	1	100%	1	100%

More generally, we can make the following qualitative statement.

KEY FACT 7.1

Sample Size and Sampling Error

The larger the sample size, the smaller the sampling error tends to be in estimating a population mean, μ, by a sample mean, $\bar{x}$.

What *Does it Mean?*

The possible sample means cluster more closely around the population mean as the sample size increases.

What We Do in Practice

We have used the heights of a population of five basketball players to illustrate and explain the importance of the sampling distribution of the sample mean. For that small population, we can easily obtain the sampling distribution of the sample mean for any sample size by listing all of the possible sample means.

However, as we have noted, in practice, we usually deal with large populations. For such populations, obtaining the sampling distribution of the sample mean by a direct listing is not feasible. A more serious practical problem is that, in reality, we do not even know the population data (i.e., the distribution of the variable)—if we did, there would be no need to sample! Realistically then, we could not list the possible sample means to determine the sampling distribution of the sample mean even if we were willing to expend the effort.

So, what can we do in the usual case of a large and unknown population? Fortunately, mathematical relationships exist that allow us to determine, at least approximately, the sampling distribution of the sample mean for any specified sample size, which we discuss in Sections 7.2 and 7.3.

EXERCISES 7.1

Statistical Concepts and Skills

7.1 Why is sampling often preferable to conducting a census for the purpose of obtaining information about a population?

7.2 Why should you generally expect some error when estimating a parameter (e.g., a population mean) by a statistic (e.g., a sample mean)? What is this kind of error called?

Exercises 7.3–7.15 are intended solely to provide concrete illustrations of the sampling distribution of the sample mean. For that reason, the populations considered are unrealistically small. In each exercise, assume that sampling is without replacement.

7.3 NBA Champs. The winner of the 2002–2003 National Basketball Association (NBA) championship was the San Antonio Spurs. The following table provides the usual starting players and their positions and heights.

Player	Position	Height (in.)
Bruce Bowen (B)	Forward	79
Tim Duncan (D)	Forward	84
David Robinson (R)	Center	85
Emanuel Ginobli (G)	Guard	78
Tony Parker (P)	Guard	74

a. Find the population mean height of the five players.
b. For samples of size 2, construct a table similar to Table 7.2 on page 326. Use the letter in parentheses after each player's name to represent each player.
c. Draw a dotplot for the sampling distribution of the sample mean for samples of size 2 like the one shown in Fig. 7.1 on page 326.
d. For a random sample of size 2, what is the chance that the sample mean will equal the population mean? That is, determine $P(\bar{x} = \mu)$.
e. For a random sample of size 2, obtain the probability that the sampling error made in estimating the population mean by the sample mean will be 1 inch or less; that is, determine the probability that $\bar{x}$ will be within 1 inch of μ. Interpret your result in terms of percentages.

7.4 NBA Champs. Repeat parts (b)–(e) of Exercise 7.3 for samples of size 1.

7.5 NBA Champs. Repeat parts (b)–(e) of Exercise 7.3 for samples of size 3.

7.6 NBA Champs. Repeat parts (b)–(e) of Exercise 7.3 for samples of size 4.

7.7 NBA Champs. Repeat parts (b)–(e) of Exercise 7.3 for samples of size 5.

7.8 NBA Champs. This exercise requires that you have done Exercises 7.3–7.7.
a. Draw a graph similar to that shown in Fig. 7.3 on page 328 for sample sizes of 1, 2, 3, 4, and 5.
b. What does your graph in part (a) illustrate about the impact of increasing sample size on sampling error?
c. Construct a table similar to Table 7.4 on page 329 for some values of your choice.

7.9 World's Richest. Each year, *Forbes* magazine publishes a list of the world's richest people. In 2003, the six richest people, their countries, and their wealth (to the nearest billion dollars) are as shown in the following table. Consider these six people a population of interest.

Person	Country	Wealth ($billions)
William H. Gates III (G)	United States	41
Warren E. Buffett (B)	United States	31
Karl & Theo Albrecht (A)	Germany	26
Paul G. Allen (P)	United States	20
Prince Alwaleed Bin Talal Alsaud (T)	Saudi Arabia	18
Lawrence J. Ellison (E)	United States	17

a. Calculate the mean wealth, μ, of the six people.
b. For samples of size 2, construct a table similar to Table 7.2 on page 326. (There are 15 possible samples of size 2.)
c. Draw a dotplot for the sampling distribution of the sample mean for samples of size 2.
d. For a random sample of size 2, what is the chance that the sample mean will equal the population mean? That is, determine $P(\bar{x} = \mu)$.
e. For a random sample of size 2, determine the probability that the mean wealth of the two people obtained will be within 2 (i.e., $2 billion) of the population mean. Interpret your result in terms of percentages.

7.10 World's Richest. Repeat parts (b)–(e) of Exercise 7.9 for samples of size 1.

7.11 World's Richest. Repeat parts (b)–(e) of Exercise 7.9 for samples of size 3. (There are 20 possible samples.)

7.12 World's Richest. Repeat parts (b)–(e) of Exercise 7.9 for samples of size 4. (There are 15 possible samples.)

7.13 World's Richest. Repeat parts (b)–(e) of Exercise 7.9 for samples of size 5. (There are six possible samples.)

7.14 World's Richest. Repeat parts (b)–(e) of Exercise 7.9 for samples of size 6. What is the relationship between the only possible sample here and the population?

7.15 World's Richest. Explain what the dotplots in part (c) of Exercises 7.9–7.14 illustrate about the impact of increasing sample size on sampling error.

Extending the Concepts and Skills

7.16 Suppose that a sample is to be taken without replacement from a finite population of size N. If the sample size is the same as the population size,
a. how many possible samples are there?
b. what are the possible sample means?
c. what is the relationship between the only possible sample and the population?

7.17 Suppose that a random sample of size 1 is to be taken from a finite population of size N.
a. How many possible samples are there?
b. Identify the relationship between the possible sample means and the possible observations of the variable under consideration.
c. What is the difference between taking a random sample of size 1 from a population and selecting a member at random from the population?

7.2 The Mean and Standard Deviation of $\bar{x}$

In Section 7.1, we discussed the sampling distribution of the sample mean—the distribution of all possible sample means for any specified sample size or, equivalently, the distribution of the variable $\bar{x}$. We use the sampling distribution of the sample mean to make inferences about a population mean based on the mean of a sample from the population.

As we said earlier, we generally do not know the sampling distribution of the sample mean exactly. Fortunately, however, we can often approximate that sampling distribution by a normal distribution; that is, under certain conditions, the variable $\bar{x}$ is approximately normally distributed.

Recall that a variable is normally distributed if its distribution has the shape of a normal curve and that a normal distribution is determined by the mean and standard deviation. Hence a first step in learning how to approximate the sampling distribution of the sample mean by a normal distribution is to obtain the mean and standard deviation of the sample mean, that is, of the variable $\bar{x}$. We describe how to do that in this section.

To begin, let's review the notation used for the mean and standard deviation of a variable. Recall that the mean of a variable is denoted μ, subscripted if necessary with the letter representing the variable. So the mean of x is written as μ_x, the mean of y as μ_y, and so on. In particular, then, the mean of $\bar{x}$ is written as $\mu_{\bar{x}}$; similarly, the standard deviation of $\bar{x}$ is written as $\sigma_{\bar{x}}$.

The Mean of $\bar{x}$

There is a simple relationship between the mean of the variable $\bar{x}$ and the mean of the variable under consideration: They are equal, or $\mu_{\bar{x}} = \mu$. In other words, for any particular sample size, the mean of all possible sample means equals the population mean. This equality holds regardless of the size of the sample. In Example 7.4, we illustrate the relationship $\mu_{\bar{x}} = \mu$ by returning to the heights of the basketball players considered in Section 7.1.

Example 7.4 Mean of the Sample Mean

TABLE 7.5

Heights of the five starting players

Player	A	B	C	D	E
Height	76	78	79	81	86

Heights of Starting Players The heights, in inches, of the five starting players on a men's basketball team are repeated in Table 7.5. Here the population of interest consists of the five players, and the variable under consideration is height.

a. Determine the population mean, μ.

b. Obtain the mean, $\mu_{\bar{x}}$, of the variable $\bar{x}$ for samples of size 2. Verify that the relation $\mu_{\bar{x}} = \mu$ holds.

c. Repeat part (b) for samples of size 4.

Solution

a. To determine the population mean (i.e., the mean of the variable "height"), we apply Definition 3.11 on page 131 to the heights in Table 7.5:

$$\mu = \frac{\Sigma x}{N} = \frac{76 + 78 + 79 + 81 + 86}{5} = 80 \text{ inches.}$$

Thus the mean height of the five players is 80 inches.

b. To obtain the mean of the variable $\bar{x}$ for samples of size 2, we again apply Definition 3.11, but this time to $\bar{x}$. Referring to the third column of Table 7.2 on page 326, we get

$$\mu_{\bar{x}} = \frac{77.0 + 77.5 + \cdots + 83.5}{10} = 80 \text{ inches.}$$

By part (a), $\mu = 80$ inches. So, for samples of size 2, $\mu_{\bar{x}} = \mu$.

INTERPRETATION For samples of size 2, the mean of all possible sample means equals the population mean.

c. Proceeding as in part (b), but this time referring to the third column of Table 7.3 on page 327, we obtain the mean of the variable $\bar{x}$ for samples of size 4:

$$\mu_{\bar{x}} = \frac{78.50 + 79.75 + 80.25 + 80.50 + 81.00}{5} = 80 \text{ inches,}$$

which again is the same as μ.

INTERPRETATION For samples of size 4, the mean of all possible sample means equals the population mean.

For emphasis, we restate the relationship $\mu_{\bar{x}} = \mu$ as the following formula.

FORMULA 7.1

Mean of the Variable $\bar{x}$

For samples of size n, the mean of the variable $\bar{x}$ equals the mean of the variable under consideration. In symbols,

$$\mu_{\bar{x}} = \mu.$$

What
Does it Mean?

For each sample size, the mean of all possible sample means equals the population mean.

The Standard Deviation of $\bar{x}$

Next, we investigate the standard deviation of the variable $\bar{x}$. We want to discover any apparent relationship of the standard deviation of $\bar{x}$ to the standard deviation of the variable under consideration. To begin our investigation, in Example 7.5, we return to the basketball players.

Example 7.5

Standard Deviation of the Sample Mean

Heights of Starting Players   Refer to Table 7.5.

a. Determine the population standard deviation, σ.

b. Obtain the standard deviation, $\sigma_{\bar{x}}$, of the variable $\bar{x}$ for samples of size 2. Indicate any apparent relationship between $\sigma_{\bar{x}}$ and σ.

c. Repeat part (b) for samples of sizes 1, 3, 4, and 5.

d. Summarize and discuss the results obtained in parts (a)–(c).

Solution

a. To determine the population standard deviation (i.e., the standard deviation of the variable "height"), we apply Definition 3.12 on page 133 to the heights in Table 7.5. Recalling that $\mu = 80$ inches, we have

$$\sigma = \sqrt{\frac{\Sigma(x - \mu)^2}{N}}$$

$$= \sqrt{\frac{(76 - 80)^2 + (78 - 80)^2 + (79 - 80)^2 + (81 - 80)^2 + (86 - 80)^2}{5}}$$

$$= \sqrt{\frac{16 + 4 + 1 + 1 + 36}{5}} = \sqrt{11.6} = 3.41 \text{ inches.}$$

Thus the standard deviation of the heights of the five players is 3.41 inches.

b. To obtain the standard deviation of the variable $\bar{x}$ for samples of size 2, we again apply Definition 3.12, but this time to $\bar{x}$. Referring to the third column of Table 7.2 on page 326 and recalling that $\mu_{\bar{x}} = \mu = 80$ inches, we have

$$\sigma_{\bar{x}} = \sqrt{\frac{(77.0 - 80)^2 + (77.5 - 80)^2 + \cdots + (83.5 - 80)^2}{10}}$$

$$= \sqrt{\frac{9.00 + 6.25 + \cdots + 12.25}{10}} = \sqrt{4.35} = 2.09 \text{ inches,}$$

to two decimal places. Note that this result is not the same as the population standard deviation, which is $\sigma = 3.41$ inches. Also note that $\sigma_{\bar{x}}$ is smaller than σ.

c. Using the same procedure as in part (b), we compute $\sigma_{\bar{x}}$ for samples of sizes 1, 3, 4, and 5 and summarize the results in Table 7.6.

TABLE 7.6

The standard deviation of $\bar{x}$ for sample sizes 1, 2, 3, 4, and 5

Sample size n	Standard deviation of $\bar{x}$ $\sigma_{\bar{x}}$
1	3.41
2	2.09
3	1.39
4	0.85
5	0.00

d. Table 7.6 suggests that the standard deviation of $\bar{x}$ gets smaller as the sample size gets larger. We could have predicted this result from the dotplots shown in Fig. 7.3 on page 328 and the fact that the standard deviation of a variable measures the variation of its possible values.

INTERPRETATION Figure 7.3 indicates graphically that the variation, and hence the standard deviation, of all possible sample means decreases with increasing sample size. Table 7.6 indicates that same thing numerically.

▄

Example 7.5 provides evidence that the standard deviation of $\bar{x}$ gets smaller as the sample size gets larger; that is, the variation of all possible sample means decreases as the sample size increases. The question now is whether there is a formula that relates the standard deviation of $\bar{x}$ to the sample size and standard deviation of the population. The answer is yes! In fact, two different formulas express the precise relationship.

When sampling is done without replacement from a finite population, as in Example 7.5, the appropriate formula is

$$\sigma_{\bar{x}} = \sqrt{\frac{N - n}{N - 1}} \cdot \frac{\sigma}{\sqrt{n}},$$

where, as usual, n denotes the sample size and N the population size. When sampling is done with replacement from a finite population or when it is done from an infinite population, the appropriate formula is

$$\sigma_{\bar{x}} = \frac{\sigma}{\sqrt{n}}.$$

When the sample size is small relative to the population size, there is little difference between sampling with and without replacement.[†] So, in such cases, it is not surprising that the two formulas for $\sigma_{\bar{x}}$ yield almost the same numbers. In most practical applications, the sample size is in fact small relative to the population size, so with the understanding that the equality may be only approximate, we use the second formula, denoted Formula 7.2, exclusively in this book.

[†] As a rule of thumb, we say that the sample size is small relative to the population size if $n \leq 0.05N$, that is, if the size of the sample does not exceed 5% of the size of the population.

FORMULA 7.2

Standard Deviation of the Variable x̄

For samples of size n, the standard deviation of the variable $\bar{x}$ equals the standard deviation of the variable under consideration divided by the square root of the sample size. In symbols,

$$\sigma_{\bar{x}} = \frac{\sigma}{\sqrt{n}}.$$

Applying the Formulas

We have shown that simple formulas relate the mean and standard deviation of $\bar{x}$ to the mean and standard deviation of the population, namely, $\mu_{\bar{x}} = \mu$ and $\sigma_{\bar{x}} = \sigma/\sqrt{n}$ (at least approximately). We apply those formulas in Example 7.6.

Example 7.6 **Mean and Standard Deviation of the Sample Mean**

Living Space of Homes As reported by the U.S. Bureau of the Census in *Current Housing Reports*, the mean living space for single-family detached homes is 1742 square feet. The standard deviation is 568 square feet.

a. For samples of 25 single-family detached homes, determine the mean and standard deviation of the variable $\bar{x}$.

b. Repeat part (a) for a sample of size 500.

Solution Here the variable is living space, and the population consists of all single-family detached homes in the United States. From the given information, we know that $\mu = 1742$ sq. ft. and $\sigma = 568$ sq. ft.

a. The sample size is 25, so $\bar{x}$ denotes the mean living space of a sample of 25 single-family detached homes. We want to obtain the mean and standard deviation of all such possible sample means. Applying Formula 7.1 on page 333, we get

$$\mu_{\bar{x}} = \mu = 1742 \text{ sq. ft.}$$

As $n = 25$, we conclude from Formula 7.2 that

$$\sigma_{\bar{x}} = \frac{\sigma}{\sqrt{n}} = \frac{568}{\sqrt{25}} = 113.6 \text{ sq. ft.}$$

INTERPRETATION For samples of 25 single-family detached homes, the mean and standard deviation of all possible sample mean living spaces are 1742 sq. ft. and 113.6 sq. ft., respectively.

b. Proceeding as in part (a), we find that

$$\mu_{\bar{x}} = \mu = 1742 \text{ sq. ft.,}$$

and as $n = 500$,

$$\sigma_{\bar{x}} = \frac{\sigma}{\sqrt{n}} = \frac{568}{\sqrt{500}} = 25.4 \text{ sq. ft.}$$

INTERPRETATION For samples of 500 single-family detached homes, the mean and standard deviation of all possible sample mean living spaces are 1742 sq. ft. and 25.4 sq. ft., respectively.

Sample Size and Sampling Error (Revisited)

Key Fact 7.1 states that the possible sample means cluster more closely around the population mean as the sample size increases, and therefore the larger the sample size, the smaller the sampling error tends to be in estimating a population mean by a sample mean. Here is why that key fact is true.

- The larger the sample size, the smaller is the standard deviation of $\bar{x}$.
- The smaller the standard deviation of $\bar{x}$, the more closely the possible values of $\bar{x}$ (the possible sample means) cluster around the mean of $\bar{x}$.
- The mean of $\bar{x}$ equals the population mean.

Because the standard deviation of $\bar{x}$ determines the amount of sampling error to be expected when a population mean is estimated by a sample mean, it is often referred to as the **standard error of the sample mean.** In general, the standard deviation of a statistic used to estimate a parameter is called the **standard error (SE)** of the statistic.

EXERCISES 7.2

Statistical Concepts and Skills

7.18 Although, in general, you cannot know the sampling distribution of the sample mean exactly, by what distribution can you often approximate it?

7.19 Why is obtaining the mean and standard deviation of $\bar{x}$ a first step in approximating the sampling distribution of the sample mean by a normal distribution?

7.20 Does the sample size have an effect on the mean of all possible sample means? Explain your answer.

7.21 Does the sample size have an effect on the standard deviation of all possible sample means? Explain your answer.

7.22 Explain why increasing the sample size tends to result in a smaller sampling error when a sample mean is used to estimate a population mean.

7.23 What is another name for the standard deviation of the variable $\bar{x}$? What is the reason for that name?

7.24 In this section, we stated that, when the sample size is small relative to the population size, there is little difference between sampling with and without replacement. Explain in your own words why that statement is true.

Exercises 7.25–7.29 require that you have done Exercises 7.3–7.7.

7.25 NBA Champs. The winner of the 2002–2003 National Basketball Association (NBA) championship was the San Antonio Spurs. The following table provides the usual starting players and their positions and heights.

Player	Position	Height (in.)
Bruce Bowen (B)	Forward	79
Tim Duncan (D)	Forward	84
David Robinson (R)	Center	85
Emanuel Ginobli (G)	Guard	78
Tony Parker (P)	Guard	74

a. Determine the population mean height, μ, of the five players.

b. Consider samples of size 2 without replacement. Use your answer to Exercise 7.3(b) on page 330 and Definition 3.11 on page 131 to find the mean of the variable $\bar{x}$.

c. Find $\mu_{\bar{x}}$, using only the result of part (a).

7.26 NBA Champs. Repeat parts (b) and (c) of Exercise 7.25 for samples of size 1. For part (b), use your answer to Exercise 7.4(b).

7.27 NBA Champs. Repeat parts (b) and (c) of Exercise 7.25 for samples of size 3. For part (b), use your answer to Exercise 7.5(b).

7.28 NBA Champs. Repeat parts (b) and (c) of Exercise 7.25 for samples of size 4. For part (b), use your answer to Exercise 7.6(b).

7.29 NBA Champs. Repeat parts (b) and (c) of Exercise 7.25 for samples of size 5. For part (b), use your answer to Exercise 7.7(b).

7.30 Working at Home. According to the U.S. Bureau of Labor Statistics publication *News*, self-employed persons with home-based businesses work a mean of 23 hours per week at home with a standard deviation of 10 hours.

a. Identify the population and variable.

b. For samples of size 100, find the mean and standard deviation of all possible sample mean hours worked per week at home.

c. Repeat part (b) for samples of size 1000.

7.31 Baby Weight. The paper "Are Babies Normal?" by Traci Clemons and Marcello Pagano (*The American Statistician*, Vol. 53, No. 4, pp. 298–302) focused on babies born in 1991. According to the article, the mean birth weight is 3369 grams (7 pounds, 6.5 ounces) with a standard deviation of 581 grams.

a. Identify the population and variable.

b. For samples of size 200, find the mean and standard deviation of all possible sample mean weights.

c. Repeat part (b) for samples of size 400.

7.32 Menopause in Mexico. In the journal article "Age at Menopause in Puebla, Mexico" (*Human Biology*, Vol. 75, No. 2, pp. 205–206), authors L. Sievert and S. Hautaniemi compared the age of menopause for different populations. Menopause, the last menstrual period, is a universal phenomenon among females. According to the article, the mean age of menopause, surgical or natural, in Puebla, Mexico is 44.8 years with a standard deviation of 5.87 years. Let $\bar{x}$

denote the mean age of menopause for a sample of females in Puebla, Mexico.

a. For samples of size 40, find the mean and standard deviation of $\bar{x}$. Interpret your results in words.

b. Repeat part (a) with $n = 120$.

7.33 Mobile Homes. According to the U.S. Census Bureau publication *Current Construction Reports*, the mean price of new mobile homes is $51,300. The standard deviation of the prices is $7200. Let $\bar{x}$ denote the mean price of a sample of new mobile homes.

a. For samples of size 50, find the mean and standard deviation of $\bar{x}$. Interpret your results in words.

b. Repeat part (a) with $n = 100$.

7.34 Earthquakes. According to *The Earth: Structure, Composition and Evolution* (The Open University, S237), for earthquakes with a magnitude of 7.5 or greater on the Richter scale, the time between successive earthquakes has a mean of 437 days and a standard deviation of 399 days. Suppose that you observe a sample of four times between successive earthquakes that have a magnitude of 7.5 or greater on the Richter scale.

a. On average, what would you expect to be the mean of the four times?

b. How much variation would you expect from your answer in part (a)? (*Hint:* Use the three-standard-deviations rule.)

Extending the Concepts and Skills

7.35 Unbiased and Biased Estimators. A statistic is said to be an *unbiased estimator* of a parameter if the mean of all its possible values equals the parameter; otherwise, it is said to be a *biased estimator*. An unbiased estimator yields, on average, the correct value of the parameter, whereas a biased estimator does not.

a. Is the sample mean an unbiased estimator of the population mean? Explain your answer.

b. Is the sample median an unbiased estimator of the population median? (*Hint:* Refer to Example 7.2 on page 325. Consider samples of size 2.)

7.36 Class Project Simulation. This exercise can be done individually or, better yet, as a class project.

a. Use a random-number table or random-number generator to obtain a sample (with replacement) of four digits between 0 and 9. Do so a total of 50 times and compute the mean of each sample.

b. Theoretically, what are the mean and standard deviation of all possible sample means for samples of size 4?

c. Roughly what would you expect the mean and standard deviation of the 50 sample means you obtained in part (a) to be? Explain your answers.

d. Determine the mean and standard deviation of the 50 sample means you obtained in part (a).

e. Compare your answers in parts (c) and (d). Why are they different?

Using Technology

7.37 Gestation Periods of Humans. For humans, gestation periods are normally distributed with a mean of 266 days and a standard deviation of 16 days. Suppose that you observe the gestation periods for a sample of nine humans.

a. Theoretically, what are the mean and standard deviation of all possible sample means?

b. Use the technology of your choice to simulate 2000 samples of nine human gestation periods each.

c. Determine the mean of each of the 2000 samples you obtained in part (b).

d. Roughly what would you expect the mean and standard deviation of the 2000 sample means you obtained in part (c) to be? Explain your answers.

e. Determine the mean and standard deviation of the 2000 sample means you obtained in part (c).

f. Compare your answers in parts (d) and (e). Why are they different?

7.38 Emergency Room Traffic. Desert Samaritan Hospital in Mesa, Arizona, keeps records of emergency room traffic. Those records reveal that the times between arriving patients have a special type of reverse J-shaped distribution called an *exponential distribution*. They also indicate that the mean time between arriving patients is 8.7 minutes, as is the standard deviation. Suppose that you observe a sample of 10 interarrival times.

a. Theoretically, what are the mean and standard deviation of all possible sample means?

b. Use the technology of your choice to simulate 1000 samples of 10 interarrival times each.

c. Determine the mean of each of the 1000 samples you obtained in part (b).

d. Roughly what would you expect the mean and standard deviation of the 1000 sample means you obtained in part (c) to be? Explain your answers.

e. Determine the mean and standard deviation of the 1000 sample means you obtained in part (c).

f. Compare your answers in parts (d) and (e). Why are they different?

7.3 The Sampling Distribution of the Sample Mean

In Section 7.2, we took the first step in describing the sampling distribution of the sample mean, that is, the distribution of the variable $\bar{x}$. There, we showed that the mean and standard deviation of $\bar{x}$ can be expressed in terms of the sample size and the population mean and standard deviation: $\mu_{\bar{x}} = \mu$ and $\sigma_{\bar{x}} = \sigma/\sqrt{n}$.

In this section, we take the final step in describing the sampling distribution of the sample mean. In doing so, we distinguish between the case in which the variable under consideration is normally distributed and the case in which it may not be so.

Sampling Distribution of the Sample Mean for Normally Distributed Variables

Although it is by no means obvious, if the variable under consideration is normally distributed, so is the variable $\bar{x}$. The proof of this fact requires advanced mathematics, but we can make it plausible by simulation, as demonstrated in Example 7.7.

Example 7.7 — Sampling Distribution of the Sample Mean for a Normally Distributed Variable

Intelligence Quotients   Intelligence quotients (IQs) measured on the Stanford Revision of the Binet–Simon Intelligence Scale are normally distributed with a mean of 100 and a standard deviation of 16. For a sample size of, say, 4, use simulation to make plausible the fact that $\bar{x}$ is normally distributed, that is, that the possible sample mean IQs for samples of four people have a normal distribution.

PRINTOUT 7.1

Histogram of the sample means for 1000 samples of four IQs with superimposed normal curve

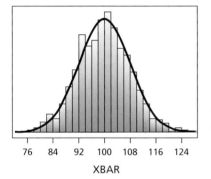

76   84   92   100   108   116   124

XBAR

Solution   First, we apply Formula 7.1 (page 333) and Formula 7.2 (page 335) to conclude that $\mu_{\bar{x}} = \mu = 100$ and $\sigma_{\bar{x}} = \sigma/\sqrt{n} = 16/\sqrt{4} = 8$; that is, the variable $\bar{x}$ has a mean of 100 and a standard deviation of 8.

We simulated 1000 samples of four IQs each, determined the sample mean ($\bar{x}$) of each of the 1000 samples, and obtained a histogram of the 1000 sample means. Printout 7.1 displays the histogram. For purposes of comparison, we superimposed on the histogram the normal distribution with a mean of 100 and a standard deviation of 8. Note that the histogram is shaped roughly like a normal curve—specifically, like the normal curve with parameters 100 and 8.

INTERPRETATION   The histogram in Printout 7.1 indicates that $\bar{x}$ is normally distributed, that is, that the possible sample mean IQs for samples of four people have a normal distribution.

Key Fact 7.2 summarizes our discussion of the sampling distribution of the sample mean when the variable under consideration is normally distributed.

KEY FACT 7.2

What Does it Mean?

For a normally distributed variable, the possible sample means for samples of a given size are also normally distributed.

Sampling Distribution of the Sample Mean for a Normally Distributed Variable

Suppose that a variable x of a population is normally distributed with mean μ and standard deviation σ. Then, for samples of size n, the variable $\bar{x}$ is also normally distributed and has mean μ and standard deviation $\sigma/\sqrt{n}$.

In Example 7.8, we compare the sampling distribution of the sample mean for two sample sizes to the normal distribution for IQs.

Example 7.8 — Sampling Distribution of the Sample Mean for a Normally Distributed Variable

Intelligence Quotients   Consider again the variable IQ, which is normally distributed with mean 100 and standard deviation 16. Obtain the sampling distribution of the sample mean for samples of size

a. 4.   **b.** 16.

Solution   The normal distribution for IQs is shown in Fig. 7.4(a). Because IQs are normally distributed, Key Fact 7.2 implies that, for any particular sample

FIGURE 7.4
(a) Normal distribution for IQs;
(b) sampling distribution of the sample
mean for $n = 4$; (c) sampling distribution
of the sample mean for $n = 16$

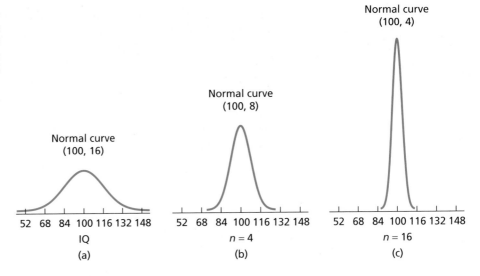

size n, the variable $\bar{x}$ is also normally distributed and has mean 100 and standard deviation $16/\sqrt{n}$.

a. For samples of size 4, we have $16/\sqrt{n} = 16/\sqrt{4} = 8$, and therefore the sampling distribution of the sample mean is a normal distribution with a mean of 100 and a standard deviation of 8. This normal distribution is shown in Fig. 7.4(b).

> **INTERPRETATION** The possible sample mean IQs for samples of four people have a normal distribution with mean 100 and standard deviation 8.

b. For samples of size 16, we have $16/\sqrt{n} = 16/\sqrt{16} = 4$, and therefore the sampling distribution of the sample mean is a normal distribution with a mean of 100 and a standard deviation of 4. This normal distribution is shown in Fig. 7.4(c).

> **INTERPRETATION** The possible sample mean IQs for samples of 16 people have a normal distribution with mean 100 and standard deviation 4.

The normal curves in Figs. 7.4(b) and 7.4(c) are drawn to scale so that you can compare them visually and observe two important things that you already know: Both curves are centered at the population mean ($\mu_{\bar{x}} = \mu$), and the spread becomes less extensive as the sample size increases ($\sigma_{\bar{x}} = \sigma/\sqrt{n}$).

Figure 7.4 also illustrates something else that you already know: The possible sample means cluster more closely around the population mean as the sample size increases, and therefore the larger the sample size, the smaller the sampling error tends to be in estimating a population mean by a sample mean.

Central Limit Theorem

According to Key Fact 7.2, if the variable x under consideration is normally distributed, so is the variable $\bar{x}$. Remarkably, that key fact holds approximately regardless of the distribution of x, provided only that the sample size is relatively large. This extraordinary fact, presented in the following key fact, is called the **central limit theorem** and is one of the most important theorems in statistics.

KEY FACT 7.3

What
Does it Mean?

For a large sample size, the possible sample means are approximately normally distributed, regardless of the distribution of the variable under consideration.

The Central Limit Theorem (CLT)

For a relatively large sample size, the variable $\bar{x}$ is approximately normally distributed, regardless of the distribution of the variable under consideration. The approximation becomes better with increasing sample size.

Roughly speaking, the farther the variable under consideration is from being normally distributed, the larger the sample size must be for a normal distribution to provide an adequate approximation to the distribution of $\bar{x}$. Usually, however, a sample size of 30 or more ($n \geq 30$) is large enough.

The proof of the central limit theorem is quite difficult. But we can make it plausible by simulation, as shown in Example 7.9.

Example 7.9 Checking the Plausibility of the CLT by Simulation

TABLE 7.7

Frequency distribution for U.S. household size

Number of people	Frequency (millions)
1	19.4
2	26.5
3	14.6
4	12.9
5	6.1
6	2.5
7	1.6

Household Size According to the U.S. Bureau of the Census publication *Current Population Reports*, a frequency distribution for the number of people per household in the United States is as displayed in Table 7.7. Frequencies are in millions of households.

Here, the variable under consideration is household size, and the population consists of all U.S. households. From Table 7.7, we find that the mean household size is $\mu = 2.685$ persons and the standard deviation is $\sigma = 1.47$ persons.

In Fig. 7.5, we present a relative-frequency histogram for household size, obtained from Table 7.7.

Note that the variable household size is far from being normally distributed; it is clearly right skewed. Nonetheless, according to the central limit theorem, the sampling distribution of the sample mean can be approximated by a normal distribution when the sample size is relatively large. Use simulation to make that fact plausible for a sample size of 30.

FIGURE 7.5

Relative-frequency histogram for household size

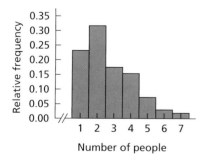

Number of people

Solution First, we apply Formula 7.1 (page 333) and Formula 7.2 (page 335) to conclude that, for samples of size 30,

$$\mu_{\bar{x}} = \mu = 2.685 \quad \text{and} \quad \sigma_{\bar{x}} = \sigma/\sqrt{n} = 1.47/\sqrt{30} = 0.27.$$

Thus the variable $\bar{x}$ has a mean of 2.685 and a standard deviation of 0.27.

We simulated 1000 samples of 30 household sizes each, determined the sample mean ($\bar{x}$) of each of the 1000 samples, and obtained a histogram of the 1000 sample means. Printout 7.2 at the top of the next page displays the histogram. For purposes of comparison, we superimposed on the histogram the normal distribution with a mean of 2.685 and a standard deviation of 0.27. Note that the histogram is shaped roughly like a normal curve—namely, like the normal curve with parameters 2.685 and 0.27.

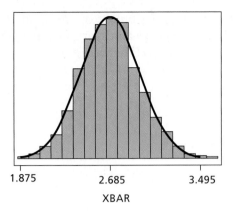

PRINTOUT 7.2
Histogram of the sample means for
1000 samples of 30 household sizes with
superimposed normal curve

1.875   2.685   3.495

XBAR

INTERPRETATION The histogram in Printout 7.2 indicates that $\bar{x}$ is approximately normally distributed, as guaranteed by the central limit theorem. Thus, for samples of 30 households, the possible sample mean household sizes have approximately a normal distribution.

The Sampling Distribution of the Sample Mean

We now summarize in Key Fact 7.4 the important facts that we have learned about the sampling distribution of the sample mean. We use Key Fact 7.4 frequently in the remainder of this book.

KEY FACT 7.4

What Does it Mean?

If either the variable under consideration is normally distributed or the sample size is large, then the possible sample means have, at least approximately, a normal distribution with mean μ and standard deviation $\sigma/\sqrt{n}$.

Sampling Distribution of the Sample Mean

Suppose that a variable x of a population has mean μ and standard deviation σ. Then, for samples of size n,

- the mean of $\bar{x}$ equals the population mean, or $\mu_{\bar{x}} = \mu$;
- the standard deviation of $\bar{x}$ equals the population standard deviation divided by the square root of the sample size, or $\sigma_{\bar{x}} = \sigma/\sqrt{n}$;
- if x is normally distributed, so is $\bar{x}$, regardless of sample size; and
- if the sample size is large, $\bar{x}$ is approximately normally distributed, regardless of the distribution of x.

The content of Key Fact 7.4 is illustrated graphically in Fig. 7.6 for normal, reverse-J-shaped, and uniform variables.

We know that, if the variable under consideration is normally distributed, so is the variable $\bar{x}$, regardless of sample size, as illustrated by Fig. 7.6(a). Additionally, we know that, if the sample size is large, the variable $\bar{x}$ is approximately normally distributed, regardless of the distribution of the variable under consideration. Figures 7.6(b) and 7.6(c) illustrate this fact for two nonnormal variables, one having a reverse-J-shaped distribution and the other having a uniform distribution.

FIGURE 7.6

Sampling distributions for (a) normal, (b) reverse-J-shaped, and (c) uniform variables

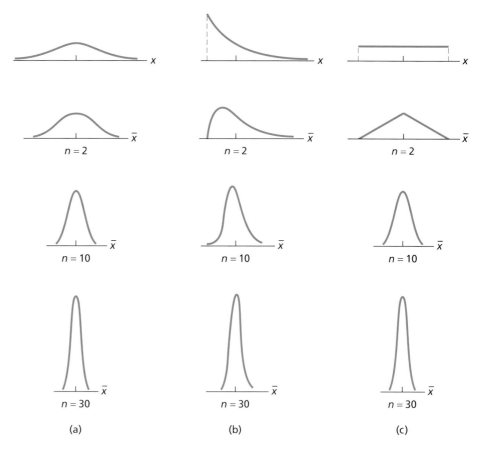

(a) (b) (c)

In each of these latter two cases, for samples of size 2, the variable $\bar{x}$ is far from being normally distributed; for samples of size 10, it is already somewhat normally distributed; and for samples of size 30, it is very close to being normally distributed.

EXERCISES 7.3

Statistical Concepts and Skills

7.39 A variable of a population has a mean of $\mu = 100$ and a standard deviation of $\sigma = 28$.
a. Identify the sampling distribution of the sample mean for samples of size 49.
b. In answering part (a), what assumptions did you make about the distribution of the variable?
c. Can you answer part (a) if the sample size is 16 instead of 49? Why or why not?

7.40 A variable of a population has a mean of $\mu = 35$ and a standard deviation of $\sigma = 42$.
a. If the variable is normally distributed, identify the sampling distribution of the sample mean for samples of size 9.

b. Can you answer part (a) if the distribution of the variable under consideration is unknown? Explain your answer.
c. Can you answer part (a) if the distribution of the variable under consideration is unknown but the sample size is 36 instead of 9? Why or why not?

7.41 A variable of a population is normally distributed with mean μ and standard deviation σ.
a. Identify the distribution of $\bar{x}$.
b. Does your answer to part (a) depend on the sample size? Explain your answer.
c. Identify the mean and the standard deviation of $\bar{x}$.
d. Does your answer to part (c) depend on the assumption that the variable under consideration is normally distributed? Why or why not?

7.42 A variable of a population has mean μ and standard deviation σ. For a large sample size n, answer the following questions.
a. Identify the distribution of $\bar{x}$.
b. Does your answer to part (a) depend on n being large? Explain your answer.
c. Identify the mean and the standard deviation of $\bar{x}$.
d. Does your answer to part (c) depend on the sample size being large? Why or why not?

7.43 Refer to Fig. 7.6.
a. Why are the four graphs in Fig. 7.6(a) all centered at the same place?
b. Why does the spread of the graphs diminish with increasing sample size? How does this result affect the sampling error when you estimate a population mean, μ, by a sample mean, $\bar{x}$?
c. Why are the graphs in Fig. 7.6(a) bell shaped?
d. Why do the graphs in Figs. 7.6(b) and (c) become bell shaped as the sample size increases?

7.44 According to the central limit theorem, for a relatively large sample size, the variable $\bar{x}$ is approximately normally distributed.
a. What rule of thumb is used for deciding whether the sample size is relatively large?
b. Roughly speaking, what property of the distribution of the variable under consideration determines how large the sample size must be for a normal distribution to provide an adequate approximation to the distribution of $\bar{x}$?

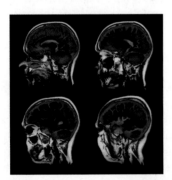

7.45 Brain Weights. In 1905, R. Pearl published the article "Biometrical Studies on Man. I. Variation and Correlation in Brain Weight" (*Biometrika*, Vol. 4, pp. 13–104). According to the study, brain weights of Swedish men are normally distributed with a mean of 1.40 kg and a standard deviation of 0.11 kg.
a. Determine the sampling distribution of the sample mean for samples of size 3. Interpret your answer in terms of the distribution of all possible sample mean brain weights for samples of three Swedish men.

b. Repeat part (a) for samples of size 12.
c. Construct graphs similar to those shown in Fig. 7.4 on page 340.

7.46 Teacher Salaries. Data on salaries in the public school system are published annually in *National Survey of Salaries and Wages in Public Schools*. The mean annual salary for classroom teachers is $43,658. Assume a standard deviation of $8000.
a. Determine the sampling distribution of the sample mean for samples of size 64. Interpret your answer in terms of the distribution of all possible sample mean salaries for samples of 64 teachers.
b. Repeat part (a) for samples of size 256.
c. Do you need to assume that public school teacher salaries are normally distributed to answer parts (a) and (b)? Explain your answer.

7.47 Loan Amounts. Ciochetti et al. studied mortgage loans in the article "A Proportional Hazards Model of Commercial Mortgage Default with Originator Bias" (*Journal of Real Estate and Economics*, Vol. 27, No. 1, pp. 5–23). According to the article, the loan amounts of loans originated by a large insurance-company lender have a mean of $6.74 million with a standard deviation of $15.37 million. The variable "loan amount" is known to have an extreme right-skewed distribution.
a. Using units of millions of dollars, determine the sampling distribution of the sample mean for samples of size 200. Interpret your result.
b. Repeat part (a) for samples of size 600.
c. Why can you still answer parts (a) and (b) when the distribution of loan amounts is not normal, but rather extremely right skewed?

7.48 New York City 10 km Run. As reported by *Runner's World* magazine, the times of the finishers in the New York City 10 km run are normally distributed with a mean of 61 minutes and a standard deviation of 9 minutes.
a. Determine the sampling distribution of the sample mean for samples of size 4.
b. Repeat part (a) for samples of size 9.
c. Construct graphs similar to those shown in Fig. 7.4 on page 340.

7.49 Brain Weights. Refer to Exercise 7.45.
a. Determine the percentage of all samples of three Swedish men that have mean brain weights within 0.1 kg of the population mean brain weight of 1.40 kg. Interpret your answer in terms of sampling error.
b. Repeat part (a) for samples of size 12.

7.50 Teacher Salaries. Refer to Exercise 7.46.
a. Determine the percentage of all samples of 64 public school teachers that have mean salaries within $1000 of

the population mean salary of $43,658. Interpret your answer in terms of sampling error.

b. Repeat part (a) for samples of size 256.

7.51 Loan Amounts. Refer to Exercise 7.47.

a. What is the probability that the sampling error made in estimating the population mean loan amount by that of a simple random sample of 200 loan amounts will be at most $1 million?

b. Repeat part (a) for samples of size 600.

7.52 New York City 10 km Run. Refer to Exercise 7.48.

a. What is the probability that the sampling error made in estimating the population mean finishing time by that of a random sample of four finishing times will be no more than 5 minutes?

b. Repeat part (a) for samples of size 9.

7.53 Air Conditioning Service Contracts. An air conditioning contractor is preparing to offer service contracts on the brand of compressor used in all of the units her company installs. Before she can work out the details, she must estimate how long those compressors last, on average. The contractor anticipated this need and has kept detailed records on the lifetimes of a random sample of 250 compressors. She plans to use the sample mean lifetime, $\bar{x}$, of those 250 compressors as her estimate for the population mean lifetime, μ, of all such compressors. If the lifetimes of this brand of compressor have a standard deviation of 40 months, what is the probability that the contractor's estimate will be within 5 months of the true mean?

Extending the Concepts and Skills

Use the 68.26-95.44–99.74 rule (page 292) to answer the questions posed in parts (a)–(c) of Exercises **7.54** *and* **7.55**.

7.54 A variable of a population is normally distributed with mean μ and standard deviation σ. For samples of size n, fill in the blanks. Justify your answers.

a. 68.26% of all possible samples have means that lie within _____ of the population mean, μ.

b. 95.44% of all possible samples have means that lie within _____ of the population mean, μ.

c. 99.74% of all possible samples have means that lie within _____ of the population mean, μ.

d. $100(1 - \alpha)\%$ of all possible samples have means that lie within _____ of the population mean, μ. (*Hint:* Draw a graph for the distribution of $\bar{x}$ and determine the z-scores dividing the area under the normal curve into a middle $1 - \alpha$ area and two outside areas of $\alpha/2$.)

7.55 A variable of a population has mean μ and standard deviation σ. For a large sample size n, fill in the blanks. Justify your answers.

a. Approximately _____% of all possible samples have means within $\sigma/\sqrt{n}$ of the population mean, μ.

b. Approximately _____% of all possible samples have means within $2\sigma/\sqrt{n}$ of the population mean, μ.

c. Approximately _____% of all possible samples have means within $3\sigma/\sqrt{n}$ of the population mean, μ.

d. Approximately _____% of all possible samples have means within $z_{\alpha/2}$ of the population mean, μ.

7.56 Testing for Content Accuracy. A brand of water-softener salt comes in packages marked "net weight 40 lb." The company that packages the salt claims that the bags contain an average of 40 lb of salt and that the standard deviation of the weights is 1.5 lb. Assume that the weights are normally distributed.

a. Obtain the probability that the weight of one randomly selected bag of water-softener salt will be 39 lb or less, if the company's claim is true.

b. Determine the probability that the mean weight of 10 randomly selected bags of water-softener salt will be 39 lb or less, if the company's claim is true.

c. If you bought one bag of water-softener salt and it weighed 39 lb, would you consider this evidence that the company's claim is incorrect? Explain your answer.

d. If you bought 10 bags of water-softener salt and their mean weight was 39 lb, would you consider this evidence that the company's claim is incorrect? Explain your answer.

Using Technology

7.57 Household Size. In Example 7.9 on page 341, we conducted a simulation to check the plausibility of the central limit theorem. The variable under consideration there is household size and the population consists of all U.S. households. A frequency distribution for household size of U.S. households is presented in Table 7.7.

a. Suppose that you simulate 1000 samples of four households each, determine the sample mean of each of the 1000 samples, and obtain a histogram of the 1000 sample means. Would you expect the histogram to be bell shaped? Explain your answer.

b. Carry out the tasks in part (a) and note the shape of the histogram.

c. Repeat parts (a) and (b) for samples of size 10.

d. Repeat parts (a) and (b) for samples of size 100.

7.58 Gestation Periods of Humans. For humans, gestation periods are normally distributed with a mean of 266 days and a standard deviation of 16 days. Suppose that you observe the gestation periods for a sample of nine humans.

a. Use the technology of your choice to simulate 2000 samples of nine human gestation periods each.

b. Find the sample mean of each of the 2000 samples.

c. Obtain the mean, the standard deviation, and a histogram of the 2000 sample means.

d. Theoretically, what are the mean, standard deviation, and distribution of all possible sample means for samples of size 9?

e. Compare your results from parts (c) and (d).

7.59 Emergency Room Traffic. A variable is said to have an *exponential distribution* or to be *exponentially distributed* if its distribution has the shape of an exponential curve, that is, a curve of the form $y = e^{-x/\mu}/\mu$ for $x > 0$, where μ is the mean of the variable. The standard deviation of such a variable also equals μ. At the emergency room at Desert Samaritan Hospital in Mesa, Arizona, the time from the arrival of one patient to the next, called an interarrival time, has an exponential distribution with a mean of 8.7 minutes.

a. Sketch the exponential curve for the distribution of the variable "interarrival time." Note that this variable is far from being normally distributed. What shape does its distribution have?

b. Use the technology of your choice to simulate 1000 samples of four interarrival times each.

c. Find the sample mean of each of the 1000 samples.

d. Determine the mean and standard deviation of the 1000 sample means.

e. Theoretically, what are the mean and the standard deviation of all possible sample means for samples of size 4? Compare your answers to those you obtained in part (d).

f. Obtain a histogram of the 1000 sample means. Is the histogram bell shaped? Would you necessarily expect it to be?

g. Repeat parts (b)–(f) for a sample size of 40.

CHAPTER REVIEW

You Should Be Able To

1. use and understand the formulas in this chapter.

2. define sampling error and explain the need for sampling distributions.

3. find the mean and standard deviation of the variable $\bar{x}$, given the mean and standard deviation of the population and the sample size.

4. state and apply the central limit theorem.

5. determine the sampling distribution of the sample mean when the variable under consideration is normally distributed.

6. determine the sampling distribution of the sample mean when the sample size is relatively large.

Key Terms

central limit theorem, *341*
sampling distribution, *325*

sampling distribution of the sample
    mean, *325*
sampling error, *324*

standard error (SE), *336*
standard error of the sample
    mean, *336*

REVIEW TEST

Statistical Concepts and Skills

1. Define sampling error.

2. What is the sampling distribution of a statistic? Why is it important?

3. Provide two synonyms for "the distribution of all possible sample means for samples of a given size."

4. Relative to the population mean, what happens to the possible sample means for samples of the same size as the sample size increases? Explain the relevance of this property in estimating a population mean by a sample mean.

5. Income Tax and the IRS. In 2001, the Internal Revenue Service (IRS) sampled approximately 177,000 tax returns to obtain estimates of various parameters. Data were published in *Statistics of Income, Individual Income Tax Returns*. According to that document, the mean income tax per return for the returns sampled was $9401.

a. Explain the meaning of sampling error in this context.

b. If, in reality, the population mean income tax per return in 2001 was $9489, how much sampling error was made in estimating that parameter by the sample mean of $9401?

c. If the IRS had sampled 250,000 returns instead of 177,000, would the sampling error necessarily have been smaller? Explain your answer.

d. In future surveys, how can the IRS increase the likelihood of small sampling error?

6. Officer Salaries. The following table gives the monthly salaries (in $1000s) of the six officers of a company.

Officer	A	B	C	D	E	F
Salary	8	12	16	20	24	28

a. Calculate the population mean monthly salary, μ.

There are 15 possible samples of size 4 from the population of six officers. They are listed in the first column of the following table.

Sample	Salaries	$\bar{x}$
A, B, C, D	8, 12, 16, 20	14
A, B, C, E	8, 12, 16, 24	15
A, B, C, F	8, 12, 16, 28	16
A, B, D, E	8, 12, 20, 24	16
A, B, D, F	8, 12, 20, 28	17
A, B, E, F	8, 12, 24, 28	18
A, C, D, E		
A, C, D, F		
A, C, E, F		
A, D, E, F		
B, C, D, E		
B, C, D, F		
B, C, E, F		
B, D, E, F		
C, D, E, F		

b. Complete the second and third columns of the table.
c. Complete the dotplot at the top of the next column for the sampling distribution of the sample mean for samples of size 4. Locate the population mean on the graph.
d. Obtain the probability that the mean salary of a random sample of four officers will be within 1 (i.e., $1000) of the population mean.

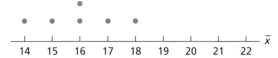

e. Use the answer you obtained in part (b) and Definition 3.11 on page 131 to find the mean of the variable $\bar{x}$. Interpret your answer.
f. Can you obtain the mean of the variable $\bar{x}$ without doing the calculation in part (e)? Explain your answer.

7. New Car Passion. The U.S. Bureau of Economic Analysis publishes data on sales of new motor vehicles in *Survey of Current Business*. In the year 2001, Americans spent an average of $21,605 for a new car. Assume a standard deviation of $10,200.
a. Identify the population and variable under consideration.
b. For samples of 50 new car sales in 2001, determine the mean and standard deviation of all possible sample mean prices.
c. Repeat part (b) for samples of size 100.
d. For samples of size 1000, answer the following question without doing any computations: Will the standard deviation of all possible sample mean prices be larger, smaller, or the same as that in part (c)? Explain your answer.

8. Hours Actually Worked. An article by Daniel Hecker in the *Monthly Labor Review* discussed the number of hours actually worked as opposed to the number of hours paid for. The study examined both full-time men and full-time women in 87 different occupations. According to the study, the mean number of hours (actually) worked by female marketing and advertising managers is $\mu = 45$ hours. Assuming a standard deviation of $\sigma = 7$ hours, decide whether each of the following statements is true or false or whether the information is insufficient to decide. Give a reason for each of your answers.
a. For a random sample of 196 female marketing and advertising managers, chances are roughly 95.44% that the sample mean number of hours worked will be between 31 hours and 59 hours.
b. 95.44% of all possible observations of the number of hours worked by female marketing and advertising managers lie between 31 hours and 59 hours.
c. For a random sample of 196 female marketing and advertising managers, chances are roughly 95.44% that the sample mean number of hours worked will be between 44 hours and 46 hours.

9. Hours Actually Worked. Repeat Problem 8, assuming that the number of hours worked by female marketing and advertising managers is normally distributed.

10. Antarctic Krill. In the Southern Ocean food web, the krill species *Euphausia superba* is the most important prey species for many marine predators, from seabirds to the largest whales. Body lengths of the species are normally distributed with a mean of 40 mm and a standard deviation of 12 mm. [SOURCE: K. Reid, J. Watkins, J. Croxall, E. Murphy, "Krill Population Dynamics at South Georgia 1991–1997 Based on Data From Predators and Nets," *Marine Ecology Progress Series*, Vol. 177, pp. 103–114]

a. Sketch the normal curve for the krill lengths.

b. Find the sampling distribution of the sample mean for samples of size 4. Draw a graph of the normal curve associated with $\bar{x}$.

c. Repeat part (b) for samples of size 9.

11. Antarctic Krill. Refer to Problem 10.

a. Determine the percentage of all samples of four krill that have mean lengths within 9 mm of the population mean length of 40 mm.

b. Obtain the probability that the mean length of four randomly selected krill will be within 9 mm of the population mean length of 40 mm.

c. Interpret the probability you obtained in part (b) in terms of sampling error.

d. Repeat parts (a)–(c) for samples of size 9.

12. The following graph shows the curve for a normally distributed variable. Superimposed are the curves for the sampling distributions of the sample mean for two different sample sizes.

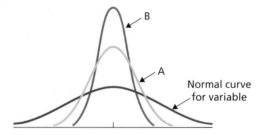

a. Explain why all three curves are centered at the same place.

b. Which curve corresponds to the larger sample size? Explain your answer.

c. Why is the spread of each curve different?

d. Which of the two sampling-distribution curves corresponds to the sample size that will tend to produce less sampling error? Explain your answer.

e. Why are the two sampling-distribution curves normal curves?

13. Blood Glucose Level. In the article "Drinking Glucose Improves Listening Span in Students Who Miss Breakfast" (*Educational Research*, Vol. 43, No. 2, pp. 201-207), authors N. Morris and P. Sarll explore the relationship between students who skip breakfast and their performance

on a number of cognitive tasks. According to their findings, blood glucose levels in the morning, after a 9-hour fast, have a mean of 4.60 mmol/l with a standard deviation of 0.16 mmol/l. (*Note:* mmol/l is an abbreviation of millimoles/liter, which is the world standard unit for measuring glucose in blood).

a. Determine the sampling distribution of the sample mean for samples of size 60.

b. Repeat part (a) for samples of size 120.

c. Must you assume that the blood glucose levels are normally distributed to answer parts (a) and (b)? Explain your answer.

14. Life Insurance in Force. The American Council of Life Insurance reports the mean life insurance in force per covered family in the *Life Insurance Fact Book*. Assume that the standard deviation of life insurance in force is $50,900.

a. Determine the probability that the sampling error made in estimating the population mean life insurance in force by that of a sample of 500 covered families will be $2000 or less.

b. Must you assume that life-insurance amounts are normally distributed in order to answer part (a)? What if the sample size is 20 instead of 500?

c. Repeat part (a) for a sample size of 5000.

15. Paint Durability. A paint manufacturer in Pittsburgh claims that his paint will last an average of 5 years. Assuming that paint life is normally distributed and has a standard deviation of 0.5 year, answer the following questions:

a. Suppose that you paint one house with the paint and that the paint lasts 4.5 years. Would you consider that evidence against the manufacturer's claim? (*Hint:* Assuming that the manufacturer's claim is correct, determine the probability that the paint life for a randomly selected house painted with the paint is 4.5 years or less.)

b. Suppose that you paint 10 houses with the paint and that the paint lasts an average of 4.5 years for the 10 houses. Would you consider that evidence against the manufacturer's claim?

c. Repeat part (b) if the paint lasts an average of 4.9 years for the 10 houses painted.

Using Technology

16. GRE Scores. Each year, thousands of college seniors take the Graduate Record Examination (GRE). The scores are transformed so that they have a mean of 500 and a standard deviation of 100. The scores are known to be normally distributed.

a. Use the technology of your choice to simulate 1000 samples of four GRE scores each.

b. Find the sample mean of each of the 1000 samples obtained in part (a).

c. Obtain the mean, the standard deviation, and a histogram of the 1000 sample means.
d. Theoretically, what are the mean, standard deviation, and distribution of all possible sample means for samples of size 4?
e. Compare your answers from parts (c) and (d).

17. Random Numbers. A variable is said to be *uniformly distributed* or to have a *uniform distribution* with parameters a and b if its distribution has the shape of the horizontal line segment $y = 1/(b - a)$, for $a < x < b$. The mean and standard deviation of such a variable are $(a + b)/2$ and $(b - a)/\sqrt{12}$, respectively. The basic random-number generator on a computer or calculator, which returns a number between 0 and 1, simulates a variable having a uniform distribution with parameters 0 and 1.

a. Sketch the distribution of a uniformly distributed variable with parameters 0 and 1. Observe from your sketch that such a variable is far from being normally distributed.
b. Use the technology of your choice to simulate 2000 samples of two random numbers between 0 and 1.
c. Find the sample mean of each of the 2000 samples obtained in part (b).
d. Determine the mean and standard deviation of the 2000 sample means.
e. Theoretically, what are the mean and the standard deviation of all possible sample means for samples of size 2? Compare your answers to those you obtained in part (d).
f. Obtain a histogram of the 2000 sample means. Is the histogram bell shaped? Would you expect it to be?
g. Repeat parts (b)–(f) for a sample size of 35.

StatExplore in MyMathLab
Analyzing Data Online

You can apply StatExplore to make the central limit theorem plausible through simulation. To illustrate, we use StatExplore to show the plausibility of the central limit theorem for a uniform variable.

EXAMPLE CLT Simulation

Uniform Variable A variable with a uniform distribution, as depicted in Fig. 2.9(c) on page 73, is far from normally distributed. Nonetheless, the central limit theorem assures that the distribution of sample means is approximately normal for relatively large sample sizes. Perform a StatExplore simulation to make this fact plausible for samples of size 30.

SOLUTION To begin, we simulate 1000 samples of size 30 from a uniform distribution as follows:

1 Choose **Data ➤ Simulate data ➤ Uniform**
2 Type 1000 in the **Rows** text box
3 Type 30 in the **Columns** text box
4 Click **Simulate**

The results of these four steps yield 1000 samples (the 1000 rows) of size 30 each (the 30 entries in each row) in the StatExplore data table. Next, we obtain the mean of each of the 1000 samples (i.e., of each of the 1000 rows) and store those means in another column in the data table, as follows:

1 Choose **Stat ➤ Summary Stats ➤ Rows**
2 Select the 30 columns labeled Uniform1 through Uniform30.
3 Click **Next →**
4 In the left pane of the **Statistics** list box, click on all items other than **Mean**
5 Select the **Store output in data table** check box
6 Click **Calculate**

The result of these six steps is a column named Row Mean in the StatExplore data table that contains the means of the 1000 samples of size 30. To visualize the distribution of these 1000 sample means, we obtain a histogram, as follows:

1 Choose **Graphics ➤ Histogram**
2 Select the column Row Mean
3 Click **Next →**
4 Type 0.01 in the **Binwidth** text box
5 Click **Create Graph!**

The resulting output is shown in Printout 7.3. As you can see from this printout, the 1000 sample means are roughly normally distributed, as expected in light of the central limit theorem.

PRINTOUT 7.3
StatExplore histogram of sample means

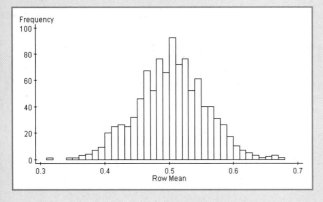

STATEXPLORE EXERCISES Using StatExplore, repeat the simulation done in the previous example for variables with the following distributions. (*Note:* You may have to experiment with the binwidth to get an appropriate histogram.)
a. standard normal distribution
b. exponential distribution with mean 1 (a special reverse-J-shaped distribution)
c. beta distribution with alpha = 1 and beta = 0.5 (a special J-shaped distribution)
d. chi-square distribution with DF = 5 (a special right-skewed distribution)
Now interpret your results.

To access StatExplore, go to the student content area of your Weiss MyMathLab course.

Internet Project

Simulations

The sampling distribution of the mean and the properties of this distribution are essential concepts in statistics. They are, at first, often difficult concepts to grasp, as well.

This Internet project uses several simulation applets to illustrate these concepts in a dynamic and interactive way that may help you understand them. Through these simulations, you are to explore the distribution of the sample mean, the mean and standard deviation of the sample mean, and the central limit theorem by using a variety of different population distributions.

URL for access to Internet Projects Page: www.aw-bc.com/weiss

**Focusing on
Data Analysis**

The Focus Database

Recall from Chapter 1 (see page 35) that the Focus database contains information on the undergraduate students at the University of Wisconsin - Eau Claire (UWEC). Suppose that you want to conduct extensive interviews with a simple random sample of 25 UWEC undergraduate students. Use the technology of your choice to obtain such a sample and the corresponding data for the thirteen variables in the Focus database.

Case Study Discussion

The Chesapeake and Ohio Freight Study

At the beginning of this chapter, we discussed a freight study commissioned by the Chesapeake and Ohio Railroad Company (C&O). A sample of 2072 waybills from a population of 22,984 waybills was used to estimate the total revenue due C&O. The estimate arrived at was $64,568.

Because all 22,984 waybills were available, a census could be taken to determine exactly the total revenue due C&O and thereby reveal the accuracy of the estimate obtained by sampling. The exact amount due C&O was found to be $64,651.

a. What percentage of the waybills constituted the sample?
b. What percentage error was made by using the sample to estimate the total revenue due C&O?
c. At the time, the cost of a complete examination was approximately $5000, whereas the cost of the sampling was only $1000. Knowing this information and your answers to parts (a) and (b), do you think that sampling was preferable to a census? Explain your answer.
d. In the study, the $83 error was against C&O. Would the error necessarily have to be that way?

Internet Resources: Visit the Weiss Web site www.aw-bc.com/weiss for additional discussion, exercises, and resources related to this case study.

Biography

PIERRE-SIMON LAPLACE: The Newton of France

Pierre-Simon Laplace was born on March 23, 1749, at Beaumount-en-Auge, Normandy, France, the son of a peasant farmer. His early schooling was at the military academy at Beaumount, where he developed his mathematical abilities. At the age of 18, he went to Paris. Within 2 years he was recommended for a professorship at the École Militaire by the French mathematician and philosopher Jean d'Alembert. (It is said that Laplace examined and passed Napoleon Bonaparte there in 1785.) In 1773 Laplace was granted membership in the Academy of Sciences.

Laplace held various positions in public life: He was president of the Bureau des Longitudes, professor at the École Normale, Minister of the Interior under Napoleon for six weeks (at which time he was replaced by Napoleon's brother), and Chancellor of the Senate; he was also made a marquis.

Laplace's professional interests were also varied. He published several volumes on celestial mechanics (which the Scottish geologist and mathematician John Playfair said were "the highest point to which man has yet ascended in the scale of intellectual attainment"), a book entitled *Théorie analytique des probabilités* (Analytic Theory of Probability), and other works on physics and mathematics. Laplace's primary contribution to the field of probability and statistics was the remarkable and all-important central limit theorem, which appeared in an 1809 publication and was read to the Academy of Sciences on April 9, 1810.

Astronomy was Laplace's major work; approximately half of his publications were concerned with the solar system and its gravitational interactions. These interactions were so complex that even Sir Isaac Newton had concluded "divine intervention was periodically required to preserve the system in equilibrium." Laplace, however, proved that planets' average angular velocities are invariable and periodic, and thus made the most important advance in physical astronomy since Newton.

When Laplace died in Paris on March 5, 1827, he was eulogized by the famous French mathematician and physicist Simeon Poisson as "the Newton of France."

part **IV**

Inferential Statistics

Confidence Intervals for One Population Mean

CHAPTER OUTLINE

GENERAL OBJECTIVES In this chapter, you begin your study of inferential statistics by examining methods for estimating the mean of a population. As you might suspect, the statistic used to estimate the population mean, μ, is the sample mean, $\bar{x}$. Because of sampling error, you cannot expect $\bar{x}$ to equal μ exactly. Thus, providing information about the accuracy of the estimate is important, which leads to a discussion of confidence intervals, the main topic of this chapter.

In Section 8.1, we provide the intuitive foundation for confidence intervals. Then, in Section 8.2, we present confidence intervals for one population mean when the population standard deviation, σ, is known. Although, in practice, σ is usually unknown, we first consider the case where σ is known to set the stage for handling the case where σ is unknown.

In Section 8.3, we investigate in detail how sample size affects the precision of estimating the population mean by a sample mean. In doing so, we introduce a salient quantity in that regard: the *margin of error*.

In Section 8.4, we discuss confidence intervals for one population when the population standard deviation is unknown. As a prerequisite to that topic, we introduce and describe one of the most important distributions in inferential statistics—*Student's t-distribution*.

The Chips Ahoy! 1,000 Chips Challenge

Nabisco, the maker of Chips Ahoy! cookies, challenged students across the nation to confirm the cookie maker's claim that there are [at least] 1000 chocolate chips in every 18-ounce bag of Chips Ahoy! cookies. According to the folks at Nabisco, a chocolate chip is defined as "...any distinct piece of chocolate that is baked into or on top of the cookie dough regardless of whether or not it is 100% whole." Students competed for $25,000 in scholarships and other prizes for participating in the Challenge.

As reported by Brad Warner and Jim Rutledge in the paper "Checking the Chips Ahoy! Guarantee" (*Chance*, Vol. 12(1), pp. 10–14), one such group that participated in the Challenge was an introductory statistics class at the United States Air Force Academy. With chocolate chips on their minds, cadets and faculty accepted the Challenge. Friends and families of the cadets sent 275 bags of Chips Ahoy! cookies from all over the country. From the 275 bags, 42 were randomly selected for the study, while the other bags were used to keep cadet morale high during counting.

For each of the 42 bags selected for the study, the cookies were first dissolved in water to separate the chips, and then the chips were counted. The following are the number of chips per bag for the 42 bags of Chips Ahoy! analyzed by the cadets in the introductory statistics class.

1200	1219	1103	1213	1258	1325	1295
1247	1098	1185	1087	1377	1363	1121
1279	1269	1199	1244	1294	1356	1137
1545	1135	1143	1215	1402	1419	1166
1132	1514	1270	1345	1214	1154	1307
1293	1546	1228	1239	1440	1219	1191

After studying confidence intervals in this chapter, you will be asked to analyze these data for the purpose of estimating the mean number of chips per bag for all bags of Chips Ahoy! cookies.

8.1   Estimating a Population Mean

A common problem in statistics is to obtain information about the mean, μ, of a population. For example, we might want to know

- the mean age of people in the civilian labor force,
- the mean cost of a wedding,
- the mean gas mileage of a new-model car, or
- the mean starting salary of liberal-arts graduates.

If the population is small, we can ordinarily determine μ exactly by first taking a census and then computing μ from the population data. But if the population is large, as it often is in practice, taking a census is generally impractical, extremely expensive, or impossible. Nonetheless, we can usually obtain sufficiently accurate information about μ by taking a sample from the population, as illustrated in Example 8.1.

Example 8.1   Point Estimate of a Population Mean

Prices of New Mobile Homes   The U.S. Bureau of the Census publishes annual price figures for new mobile homes in *Current Construction Reports*. The figures are obtained from sampling, not from a census. A simple random sample of 36 new mobile homes yielded the prices, in thousands of dollars, shown in Table 8.1. Use the data to estimate the population mean price, μ, of all new mobile homes.

TABLE 8.1

Prices ($1000s) of 36 randomly selected new mobile homes

53.8	54.4	45.2	42.9	49.9	48.2	41.6	58.9	48.6
53.1	59.4	49.7	43.7	52.7	47.7	41.5	35.3	58.9
35.9	42.5	57.2	45.1	50.3	50.0	41.9	37.3	39.7
42.0	62.7	62.8	46.6	60.5	43.9	56.4	49.8	63.9

Solution   We estimate the population mean price, μ, of all new mobile homes by the sample mean price, $\bar{x}$, of the 36 new mobile homes sampled. From Table 8.1,

$$\bar{x} = \frac{\Sigma x}{n} = \frac{1774.0}{36} = 49.28.$$

INTERPRETATION   Based on the sample data, we estimate the mean price, μ, of all new mobile homes to be approximately $49.28 thousand, that is, $49,280.

An estimate of this kind is called a *point estimate* for μ because it consists of a single number, or point.   ■

As indicated in the following definition, the term **point estimate** applies to the use of a statistic to estimate any parameter, not just a population mean.

DEFINITION 8.1

What
Does it Mean?

Roughly speaking, a point estimate of a parameter is our best guess for the value of the parameter based on sample data.

Point Estimate

A *point estimate* of a parameter is the value of a statistic used to estimate the parameter.

As you learned in Chapter 7, expecting a sample mean to exactly equal the population mean is unreasonable; some sampling error is to be anticipated. Therefore, in addition to reporting a point estimate for μ, we need to provide information that indicates the accuracy of the estimate. We do so by giving a **confidence-interval estimate** for μ. With a confidence-interval estimate for μ, we use the mean of a sample to construct an interval of numbers and state how confident we are that μ lies in that interval, called the **confidence level.**

DEFINITION 8.2

What
Does it Mean?

A confidence interval for a parameter provides a range of numbers along with a percentage confidence that the parameter lies in that range.

Confidence-Interval Estimate; Confidence Level

A *confidence-interval estimate* of a parameter consists of an interval of numbers obtained from a point estimate of the parameter and a percentage that specifies how confident we are that the parameter lies in the interval. The confidence percentage is called the *confidence level.*

Note: The term *confidence interval* is often abbreviated as CI.

In Example 8.2, we obtain a 95.44% confidence interval for the mean price of all new mobile homes. In doing so, we discuss in detail the logic behind determining confidence intervals. We present a general procedure based on this logic for obtaining confidence intervals at any prescribed confidence level in Section 8.2.

Example 8.2 | **Introducing Confidence Intervals**

Prices of New Mobile Homes We refer to Example 8.1 and continue to work in thousands of dollars. As a normal probability plot of the data in Table 8.1 shows, a reasonable presumption is that prices of new mobile homes are normally distributed. We assume that the population standard deviation of all such prices is $7.2 thousand, that is, $7200.[†]

a. Identify the distribution of the variable $\bar{x}$, that is, the sampling distribution of the sample mean for samples of size 36.

b. Use part (a) to show that 95.44% of all samples of 36 new mobile homes have the property that the interval from $\bar{x} - 2.4$ to $\bar{x} + 2.4$ contains μ.

c. Use part (b) and the sample data in Table 8.1 to obtain a 95.44% confidence interval for the mean price of all new mobile homes.

[†]We might know the population standard deviation from previous research or from a preliminary study of prices. We examine the more usual case, where σ is unknown, in Section 8.4.

Solution

a. Because $n = 36$, $\sigma = 7.2$, and prices of new mobile homes are normally distributed, Key Fact 7.4 on page 342 implies that

- $\mu_{\bar{x}} = \mu$ (which we don't know),
- $\sigma_{\bar{x}} = \sigma/\sqrt{n} = 7.2/\sqrt{36} = 1.2$, and
- $\bar{x}$ is normally distributed.

In other words, for samples of size 36, the variable $\bar{x}$ is normally distributed with mean μ and standard deviation 1.2.

b. The "95.44" part of the 68.26-95.44-99.74 rule states that, for a normally distributed variable, 95.44% of all possible observations lie within two standard deviations to either side of the mean. Applying this rule to the variable $\bar{x}$ and referring to part (a), we see that 95.44% of all samples of 36 new mobile homes have mean prices within $2 \cdot 1.2 = 2.4$ of μ. Or, equivalently, 95.44% of all samples of 36 new mobile homes have the property that the interval from $\bar{x} - 2.4$ to $\bar{x} + 2.4$ contains μ.

c. Because we are taking a simple random sample, each possible sample of size 36 is equally likely to be the one obtained. From part (b), 95.44% of all such samples have the property that the interval from $\bar{x} - 2.4$ to $\bar{x} + 2.4$ contains μ. Hence, chances are 95.44% that the sample we obtain has that property. Consequently, we can be 95.44% confident that the sample of 36 new mobile homes whose prices are shown in Table 8.1 has the property that the interval from $\bar{x} - 2.4$ to $\bar{x} + 2.4$ contains μ. For that sample, $\bar{x} = 49.28$, so

$$\bar{x} - 2.4 = 49.28 - 2.4 = 46.88 \quad \text{and} \quad \bar{x} + 2.4 = 49.28 + 2.4 = 51.68.$$

Thus our 95.44% confidence interval is from 46.88 to 51.68.

INTERPRETATION We can be 95.44% confident that the mean price, μ, of all new mobile homes is somewhere between \$46,880 and \$51,680.

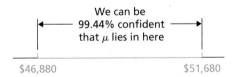

TABLE 8.2

Prices (\$1000s) of another sample of 36 randomly selected new mobile homes

59.0	58.1	47.2	39.0
61.5	49.8	42.0	61.7
51.7	39.2	52.6	51.3
54.9	44.4	55.1	51.8
50.1	46.6	52.5	50.7
48.5	47.3	48.1	54.0
65.2	55.2	54.0	46.2
58.1	40.9	52.1	50.1
58.0	54.8	50.3	63.9

Note: Although this or any other 95.44% confidence interval may or may not contain μ, we can be 95.44% confident that it does.

A confidence interval for a population mean depends on the sample mean, $\bar{x}$, which, in turn, depends on the sample selected. For example, suppose that the prices of the 36 new mobile homes sampled were as shown in Table 8.2 instead of as in Table 8.1. Then we would have $\bar{x} = 51.83$ so that

$$\bar{x} - 2.4 = 51.83 - 2.4 = 49.43 \quad \text{and} \quad \bar{x} + 2.4 = 51.83 + 2.4 = 54.23.$$

In this case, the 95.44% confidence interval for μ would be from 49.43 to 54.23. We could be 95.44% confident that the mean price, μ, of all new mobile homes is somewhere between \$49,430 and \$54,230.

Interpreting Confidence Intervals

Example 8.3 stresses the importance of interpreting a confidence interval correctly. It also illustrates that the population mean, μ, may or may not lie in the confidence interval obtained.

| Example 8.3 | **Interpreting Confidence Intervals** |

Prices of New Mobile Homes Consider again the prices of new mobile homes. As demonstrated in part (b) of Example 8.2, 95.44% of all samples of 36 new mobile homes have the property that the interval from $\bar{x} - 2.4$ to $\bar{x} + 2.4$ contains μ. In other words, if 36 new mobile homes are selected at random and their mean price, $\bar{x}$, is computed, the interval from

$$\bar{x} - 2.4 \quad \text{to} \quad \bar{x} + 2.4 \tag{8.1}$$

will be a 95.44% confidence interval for the mean price of all new mobile homes.

To illustrate that the mean price, μ, of all new mobile homes may or may not lie in the 95.44% confidence interval obtained, we used a computer to simulate 20 samples of 36 new mobile home prices each. For the simulation, we assumed that $\mu = 51$ (i.e., \$51 thousand) and $\sigma = 7.2$ (i.e., \$7.2 thousand). In reality, we don't know μ; we are assuming a value for μ to illustrate a point.

For each of the 20 samples of 36 new mobile home prices, we did three things: computed the sample mean price, $\bar{x}$; used Equation (8.1) to obtain the 95.44% confidence interval for μ based on the sample; and noted whether the population mean, $\mu = 51$, actually lies in the confidence interval.

Figure 8.1 at the top of the next page summarizes our results. For each sample, we have drawn a graph on the right-hand side of Fig. 8.1. The dot represents the sample mean, $\bar{x}$, in thousands of dollars, and the horizontal line represents the corresponding 95.44% confidence interval. Note that the population mean, μ, lies in the confidence interval only when the horizontal line crosses the dashed line.

Figure 8.1 reveals that μ lies in the 95.44% confidence interval in 19 of the 20 samples, that is, in 95% of the samples. If, instead of 20 samples, we simulated, say, 1000 samples, we would most likely find that the percentage of those 1000 samples for which μ lies in the 95.44% confidence interval would be even closer to 95.44%. Hence we can be 95.44% confident that any computed 95.44% confidence interval will contain μ.

      —▬

In Example 8.2, we obtained a 95.44% confidence interval for the mean price of all new mobile homes based on a sample of size 36. In doing so, we assumed the prices to be normally distributed, which we used to conclude that $\bar{x}$ is normally distributed.

If the prices are not normally distributed, then, because of the central limit theorem, we can still say that $\bar{x}$ is approximately normally distributed. The impact on the resulting 95.44% confidence interval would be that it is only approximately correct; that is, the true confidence level would only approximately equal 95.44%.

FIGURE 8.1

Twenty confidence intervals for the mean price of all new mobile homes, each based on a sample of 36 new mobile homes

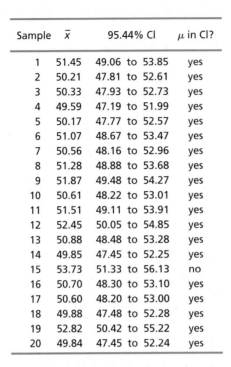

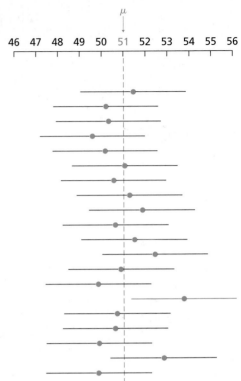

Sample	$\bar{x}$	95.44% CI	μ in CI?
1	51.45	49.06 to 53.85	yes
2	50.21	47.81 to 52.61	yes
3	50.33	47.93 to 52.73	yes
4	49.59	47.19 to 51.99	yes
5	50.17	47.77 to 52.57	yes
6	51.07	48.67 to 53.47	yes
7	50.56	48.16 to 52.96	yes
8	51.28	48.88 to 53.68	yes
9	51.87	49.48 to 54.27	yes
10	50.61	48.22 to 53.01	yes
11	51.51	49.11 to 53.91	yes
12	52.45	50.05 to 54.85	yes
13	50.88	48.48 to 53.28	yes
14	49.85	47.45 to 52.25	yes
15	53.73	51.33 to 56.13	no
16	50.70	48.30 to 53.10	yes
17	50.60	48.20 to 53.00	yes
18	49.88	47.48 to 52.28	yes
19	52.82	50.42 to 55.22	yes
20	49.84	47.45 to 52.24	yes

EXERCISES 8.1

Statistical Concepts and Skills

8.1 The value of a statistic used to estimate a parameter is called a _____ of the parameter.

8.2 What is a confidence-interval estimate of a parameter? Why is such an estimate superior to a point estimate?

8.3 **Wedding Costs.** According to *Bride's Magazine*, getting married these days can be expensive when the costs of the reception, engagement ring, bridal gown, pictures—just to name a few—are included. A simple random sample of

20 recent U.S. weddings yielded the following data on wedding costs, in dollars.

12,113	16,406	10,929	7,171	11,077
20,423	13,820	21,905	26,698	20,513
22,715	5,977	25,795	35,263	16,670
24,886	33,023	27,667	13,700	12,127

a. Use the data to obtain a point estimate for the population mean wedding cost, μ, of all recent U.S. weddings. (*Note:* The sum of the data is $378,878.)

b. Is your point estimate in part (a) likely to equal μ exactly? Explain your answer.

8.4 **Cottonmouth Litter Size.** A study by Blem and Blem, published in the *Journal of Herpetology* (Vol. 29, pp. 391–398), examined the reproductive characteristics of the eastern cottonmouth, a once widely distributed snake whose numbers have decreased recently due to encroachment by humans. A simple random sample of 44 female cottonmouths yielded the following data on number of young per litter.

5	12	7	7	6	8	12	9	7
4	9	6	12	7	5	6	10	3
10	8	8	12	5	6	10	11	3
8	4	5	7	6	11	7	6	8
8	14	8	7	11	7	5	4	

a. Use the data to obtain a point estimate for the mean number of young per litter, μ, of all female eastern cotton-mouths. (*Note:* $\Sigma x = 334$.)

b. Is your point estimate in part (a) likely to equal μ exactly? Explain your answer.

For Exercises 8.5–8.8, you may want to review Example 8.2, which begins on page 357.

8.5 Wedding Costs. Refer to Exercise 8.3. Assume that recent wedding costs in the United States are normally distributed with a standard deviation of $8100.

a. Determine a 95.44% confidence interval for the mean cost, μ, of all recent U.S. weddings.

b. Interpret your result in part (a).

c. Does the mean cost of all recent U.S. weddings lie in the confidence interval you obtained in part (a)? Explain your answer.

8.6 Cottonmouth Litter Size. Refer to Exercise 8.4. Assume that $\sigma = 2.4$.

a. Obtain an approximate 95.44% confidence interval for the mean number of young per litter of all female eastern cottonmouths.

b. Interpret your result in part (a).

c. Why is the 95.44% confidence interval that you obtained in part (a) not necessarily exact?

8.7 Fuel Tank Capacity. *Consumer Reports* provides information on new automobile models—including price, mileage ratings, engine size, body size, and indicators of features—in the data archive www.ConsumerReports.org. A simple random sample of 35 models for 2003 yielded the following data on fuel tank capacity, in gallons.

17.2	23.1	17.5	15.7	19.8	16.9	15.3
18.5	18.5	25.5	18.0	17.5	14.5	20.0
17.0	20.0	24.0	26.0	18.1	21.0	19.3
20.0	20.0	12.5	13.2	15.9	14.5	22.2
21.1	14.4	25.0	26.4	16.9	16.4	23.0

a. Find a point estimate for the mean fuel tank capacity of all 2003 automobile models. Interpret your answer in words. (*Note:* $\Sigma x = 664.9$ gallons.)

b. Determine a 95.44% confidence interval for the mean fuel tank capacity of all 2003 automobile models. Assume $\sigma = 3.50$ gallons.

c. How would you decide whether fuel tank capacities for 2003 automobile models are approximately normally distributed?

d. Must fuel tank capacities for 2003 automobile models be exactly normally distributed for the confidence interval that you obtained in part (b) to be approximately correct? Explain your answer.

8.8 Home Improvements. The *American Express Retail Index* provides information on budget amounts for home improvements. The following table displays the budgets, in dollars, of 45 randomly sampled home improvement jobs in the United States.

3179	1032	1822	4093	2285	1478	955	2773	514
3915	4800	3843	5265	2467	2353	4200	3146	551
2659	4660	3570	1598	2605	3643	2816	3125	3104
4503	2911	3605	2948	1421	1910	5145	4557	2026
2750	2069	3056	2550	631	4550	5069	2124	1573

a. Determine a point estimate for the population mean budget, μ, for such home improvement jobs. Interpret your answer in words. (*Note:* The sum of the data is $129,849.)

b. Obtain a 95.44% confidence interval for the population mean budget, μ, for such home improvement jobs and interpret your result in words. Assume that the population standard deviation of budgets for home improvement jobs is $1350.

c. How would you decide whether budgets for such home improvement jobs are approximately normally distributed?

d. Must the budgets for such home improvement jobs be exactly normally distributed for the confidence interval that you obtained in part (b) to be approximately correct? Explain your answer.

Extending the Concepts and Skills

8.9 New Mobile Homes. Refer to Examples 8.1 and 8.2. Use the data in Table 8.1 on page 356 to obtain a 99.74% confidence interval for the mean price of all new mobile homes. (*Hint:* Proceed as in Example 8.2 but use the "99.74" part of the 68.26-95.44-99.74 rule instead of the "95.44" part.)

8.10 New Mobile Homes. Refer to Examples 8.1 and 8.2. Use the data in Table 8.1 on page 356 to obtain a 68.26% confidence interval for the mean price of all new mobile homes. (*Hint:* Proceed as in Example 8.2 but use the "68.26" part of the 68.26-95.44-99.74 rule instead of the "95.44" part.)

8.2 Confidence Intervals for One Population Mean When σ Is Known

In Section 8.1, we showed how to find a 95.44% confidence interval for a population mean, that is, a confidence interval at a confidence level of 95.44%. In this section, we generalize the arguments used there to obtain a confidence interval for a population mean at any prescribed confidence level.

To begin, we introduce some general notation used with confidence intervals. Frequently, we want to write the confidence level in the form $1 - \alpha$, where α is a number between 0 and 1; that is, if the confidence level is expressed as a decimal, α is the number that must be subtracted from 1 to get the confidence level. To find α, we simply subtract the confidence level from 1. If the confidence level is 95.44%, then $\alpha = 1 - 0.9544 = 0.0456$; if the confidence level is 90%, then $\alpha = 1 - 0.90 = 0.10$; and so on.

Next, recall from Section 6.2 that the symbol z_α is used to denote the z-score that has area α to its right under the standard normal curve. So, for example, $z_{0.05}$ denotes the z-score that has area 0.05 to its right, $z_{0.025}$ denotes the z-score that has area 0.025 to its right, and $z_{\alpha/2}$ denotes the z-score that has area $\alpha/2$ to its right.

Obtaining Confidence Intervals for a Population Mean When σ Is Known

We now develop a simple step-by-step procedure for obtaining a confidence interval for a population mean when the population standard deviation is known. In doing so, we assume that the variable under consideration is normally distributed. But keep in mind that, because of the central limit theorem, the procedure also applies to obtaining an approximately correct confidence interval for a population mean when the sample size is large, regardless of the distribution of the variable.

The basis of our confidence-interval procedure is stated in Key Fact 7.4: If x is a normally distributed variable with mean μ and standard deviation σ, then, for samples of size n, the variable $\bar{x}$ is also normally distributed and has mean μ and standard deviation $\sigma/\sqrt{n}$. In Section 8.1, we used this fact and the "95.44" part of the 68.26-95.44-99.74 rule to conclude that 95.44% of all samples of size n have means within $2 \cdot \sigma/\sqrt{n}$ of μ, as depicted in Fig. 8.2(a).

FIGURE 8.2
(a) 95.44% of all samples have means within 2 standard deviations of μ;
(b) $100(1 - \alpha)$% of all samples have means within $z_{\alpha/2}$ standard deviations of μ

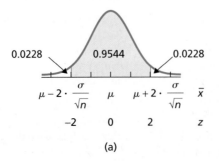

(a)

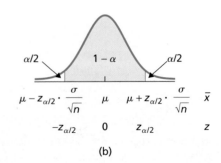

(b)

More generally, we can say that $100(1 - \alpha)\%$ of all samples of size n have means within $z_{\alpha/2} \cdot \sigma/\sqrt{n}$ of μ, as depicted in Fig. 8.2(b). Equivalently, we can say that $100(1 - \alpha)\%$ of all samples of size n have the property that the interval from

$$\bar{x} - z_{\alpha/2} \cdot \frac{\sigma}{\sqrt{n}} \quad \text{to} \quad \bar{x} + z_{\alpha/2} \cdot \frac{\sigma}{\sqrt{n}}$$

contains μ. Consequently, we have Procedure 8.1, sometimes referred to as the **one-sample z-interval procedure,** or, more briefly, as the **z-interval procedure.**

PROCEDURE 8.1

One-Sample z-Interval Procedure

Purpose To find a confidence interval for a population mean, μ

Assumptions
1. Simple random sample
2. Normal population or large sample
3. σ known

STEP 1 For a confidence level of $1 - \alpha$, use Table II to find $z_{\alpha/2}$.

STEP 2 The confidence interval for μ is from

$$\bar{x} - z_{\alpha/2} \cdot \frac{\sigma}{\sqrt{n}} \quad \text{to} \quad \bar{x} + z_{\alpha/2} \cdot \frac{\sigma}{\sqrt{n}},$$

where $z_{\alpha/2}$ is found in Step 1, n is the sample size, and $\bar{x}$ is computed from the sample data.

STEP 3 Interpret the confidence interval.

The confidence interval is exact for normal populations and is approximately correct for large samples from nonnormal populations.

Note: By saying that the confidence interval is *exact,* we mean that the true confidence level equals $1 - \alpha$; by saying that the confidence interval is *approximately correct,* we mean that the true confidence level only approximately equals $1 - \alpha$.

Before applying Procedure 8.1, we need to make several comments. In Chapter 6, we mentioned that, if a variable of a population is normally distributed and is the only variable under consideration, common statistical practice is to say that the population is normally distributed. In Assumption 2 of Procedure 8.1, we have abbreviated this statement as *normal population.* But remember, *normal population* really means that the variable under consideration is normally distributed on the population of interest. Similarly, *nonnormal population* means that the variable under consideration is not normally distributed on the population of interest.

Assumption 2 of Procedure 8.1 read in full means that the variable under consideration is normally distributed or that the sample size is large. Actually, the procedure works reasonably well even when the variable is not normally distributed and the sample size is small or moderate, provided the variable is not too far from being normally distributed. Procedures that are insensitive to departures from the assumptions on which they are based are called **robust procedures.** Thus the z-interval procedure is robust to moderate violations of the normality assumption.

When considering the z-interval procedure, you must also watch for outliers because their presence calls into question the normality assumption. Moreover, even for large samples, outliers can sometimes unduly affect a z-interval because the sample mean is not resistant to outliers. Key Fact 8.1 lists some general guidelines for use of the z-interval procedure.

KEY FACT 8.1

When to Use the One-Sample z-Interval Procedure[†]

- For small samples—say, of size less than 15—the z-interval procedure should be used only when the variable under consideration is normally distributed or very close to being so.
- For samples of moderate size—say, between 15 and 30—the z-interval procedure can be used unless the data contain outliers or the variable under consideration is far from being normally distributed.
- For large samples—say, of size 30 or more—the z-interval procedure can be used essentially without restriction. However, if outliers are present and their removal is not justified, the effect of the outliers on the confidence interval should be examined; that is, you should compare the confidence intervals obtained with and without the outliers. If the effect is substantial, using a different procedure or taking another sample probably is best.
- If outliers are present but their removal is justified and results in a data set for which the z-interval procedure is appropriate (as previously stated), the procedure can be used.

Key Fact 8.1 makes it clear that you should conduct preliminary data analyses before applying the z-interval procedure. Normal probability plots, boxplots, stem-and-leaf diagrams, and histograms are often useful in this regard. More generally, the following fundamental principle of data analysis is relevant to all inferential procedures.

[†]We can refine these guidelines further by considering the impact of skewness. Roughly speaking, the more skewed the distribution of the variable under consideration, the larger is the sample size required for the validity of the z-interval procedure. See, for instance, the paper "How Large Does n Have to be for Z and t Intervals?" by Dennis D. Boos and Jacqueline M. Hughes-Oliver (*The American Statistician*, Vol. 54, No. 2, pp. 121–128).

KEY FACT 8.2

A Fundamental Principle of Data Analysis

Before performing a statistical-inference procedure, examine the sample data. If any of the conditions required for using the procedure appear to be violated, do not apply the procedure. Instead use a different, more appropriate procedure, or, if you are unsure of one, consult a statistician.

> **What**
> *Does it Mean?*
>
> Always look at the sample data (by constructing a histogram, normal probability plot, boxplot, etc.) prior to performing a statistical-inference procedure to help check whether the procedure is appropriate.

Even for small samples, where graphical displays must be interpreted carefully, examining the data is still far better than not doing so. Even for very small samples, graphical displays can sometimes detect violations of assumptions required for inferential procedures. Remember to proceed cautiously when conducting graphical analyses of small samples, especially very small samples—say, of size 10 or less.

In Example 8.4, we apply Procedure 8.1, in the process checking the guidelines presented in Key Fact 8.1.

▌▌Example 8.4

The One-Sample z-Interval Procedure

Age of the Civilian Labor Force The U.S. Bureau of Labor Statistics collects information on the ages of people in the civilian labor force and publishes the results in *Employment and Earnings*. Fifty people in the civilian labor force are randomly selected; their ages are displayed in Table 8.3. Find a 95% confidence interval for the mean age, μ, of all people in the civilian labor force. Assume that the population standard deviation of the ages is 12.1 years.

TABLE 8.3

Ages of 50 randomly selected people in the civilian labor force

22	58	40	42	43
32	34	45	38	19
33	16	49	29	30
43	37	19	21	62
60	41	28	35	37
51	37	65	57	26
27	31	33	24	34
28	39	43	26	38
42	40	31	34	38
35	29	33	32	33

Solution We constructed (not shown) a normal probability plot, histogram, stem-and-leaf diagram, and boxplot for these data. The boxplot indicated potential outliers, but in view of the other three graphs, we concluded that, in fact, the data contain no outliers. As the sample size is 50, which is large, and the population standard deviation, σ, is known, we can apply Procedure 8.1 to obtain the required confidence interval.

STEP 1 **For a confidence level of $1 - \alpha$, use Table II to find $z_{\alpha/2}$.**

We want a 95% confidence interval, so $\alpha = 1 - 0.95 = 0.05$. From Table II, $z_{\alpha/2} = z_{0.05/2} = z_{0.025} = 1.96$.

STEP 2 **The confidence interval for μ is from**

$$\bar{x} - z_{\alpha/2} \cdot \frac{\sigma}{\sqrt{n}} \quad \text{to} \quad \bar{x} + z_{\alpha/2} \cdot \frac{\sigma}{\sqrt{n}}.$$

We have $\sigma = 12.1$, $n = 50$, and, from Step 1, $z_{\alpha/2} = 1.96$. To compute $\bar{x}$ for the data in Table 8.3, we apply the usual formula:

$$\bar{x} = \frac{\Sigma x}{n} = \frac{1819}{50} = 36.4,$$

to one decimal place. Consequently, a 95% confidence interval for μ is from

$$36.4 - 1.96 \cdot \frac{12.1}{\sqrt{50}} \quad \text{to} \quad 36.4 + 1.96 \cdot \frac{12.1}{\sqrt{50}},$$

or 33.0 to 39.8.

STEP 3 Interpret the confidence interval.

INTERPRETATION We can be 95% confident that the mean age, μ, of all people in the civilian labor force is somewhere between 33.0 years and 39.8 years.

Confidence and Precision

On the one hand, the confidence level of a confidence interval for a population mean, μ, signifies the confidence of the estimate, that is, the confidence we have that μ actually lies in the confidence interval. On the other hand, the length of the confidence interval indicates the precision of the estimate, that is, how well we have "pinned down" μ; long confidence intervals indicate poor precision, whereas short confidence intervals indicate good precision.

How does the confidence level affect the length of the confidence interval? To answer this question, let's return to Example 8.4, where we found a 95% confidence interval for the mean age, μ, of all people in the civilian labor force. The confidence level there is 0.95, and the confidence interval we computed is from 33.0 to 39.8 years. If we change the confidence level from 0.95 to, say, 0.90, then $z_{\alpha/2}$ changes from $z_{0.05/2} = z_{0.025} = 1.96$ to $z_{0.10/2} = z_{0.05} = 1.645$. The resulting confidence interval, using the same sample data (Table 8.3), is from

$$36.4 - 1.645 \cdot \frac{12.1}{\sqrt{50}} \quad \text{to} \quad 36.4 + 1.645 \cdot \frac{12.1}{\sqrt{50}},$$

or from 33.6 to 39.2 years. We show both the 90% and 95% confidence intervals in Fig. 8.3.

FIGURE 8.3
90% and 95% confidence intervals for μ, using the data in Table 8.3

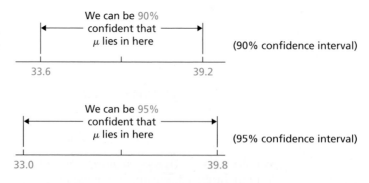

Thus, decreasing the confidence level decreases the length of the confidence interval, and vice versa, which makes sense: If we are willing to settle for less confidence that μ lies in our confidence interval, we can obtain a shorter interval. However, if we want to be more confident that μ lies in our confidence interval, we must settle for a greater interval. In other words, confidence and precision are related as stated in the following key fact.

KEY FACT 8.3   **Confidence and Precision**

For a fixed sample size, decreasing the confidence level improves the precision, and vice versa.

The **Technology Center**

Procedure 8.1 on page 363 provides a step-by-step method for obtaining a confidence interval for a population mean when the population standard deviation is known. Most statistical technologies have programs that automatically determine this type of confidence interval. In this subsection, we present output and step-by-step instructions to implement such programs.

Example 8.5   **Using Technology to Obtain a z-Interval**

Age of the Civilian Labor Force   Table 8.3 on page 365 displays the ages of 50 randomly selected people in the civilian labor force. Use Minitab, Excel, or the TI-83/84 Plus to determine a 95% confidence interval for the mean age, μ, of all people in the civilian labor force. Assume that the population standard deviation of the ages is 12.1 years.

Solution   Printout 8.1 on the following page shows the output obtained by applying the one-sample z-interval programs to the age data in Table 8.3.

As shown in the three outputs in Printout 8.1, the required 95% confidence interval is from 33.03 to 39.73. Hence we can be 95% confident that the mean age, μ, of all people in the civilian labor force is somewhere between 33.03 years and 39.73 years. Compare this confidence interval to the one obtained in Example 8.4. Can you explain the slight discrepancy?   ▬

Obtaining the Output (Optional)

Printout 8.1 provides output from Minitab, Excel, and the TI-83/84 Plus for a one-sample z-interval with a 95% confidence level based on the sample of ages presented in Table 8.3. The following are detailed instructions for obtaining that output. First, we store the age data in a column (Minitab), range (Excel), or list (TI-83/84 Plus) named AGE. Then, we proceed as follows.

MINITAB

1 Choose **Stat ➤ Basic Statistics ➤ 1-Sample Z...**
2 Select the **Samples in columns** option button
3 Click in the **Samples in columns** text box and specify AGE
4 Click in the **Standard deviation** text box and type 12.1
5 Click the **Options...** button
6 Type 95 in the **Confidence level** text box
7 Click the arrow button at the right of the **Alternative** drop-down list box and select **not equal**
8 Click **OK**
9 Click **OK**

EXCEL

1 Choose **DDXL ➤ Confidence Intervals**
2 Select **1 Var z Interval** from the **Function type** drop-down box
3 Specify AGE in the **Quantitative Variable** text box
4 Click **OK**
5 Click the **95%** button
6 Click in the **Type in the population standard deviation** text box and type 12.1
7 Click the **Compute Interval** button

TI-83/84 PLUS

1 Press **STAT**, arrow over to **TESTS**, and press **7**
2 Highlight **Data** and press **ENTER**
3 Press the down-arrow key, type 12.1 for σ, and press **ENTER**
4 Press **2nd ➤ LIST**
5 Arrow down to AGE and press **ENTER** three times
6 Type .95 for **C-Level** and press **ENTER** twice

PRINTOUT 8.1

One-sample z-interval output for the sample of ages

MINITAB

One-Sample Z: AGE

```
The assumed standard deviation = 12.1


Variable    N     Mean      StDev   SE Mean        95% CI
AGE         50  36.3800   11.0692   1.7112   (33.0261, 39.7339)
```

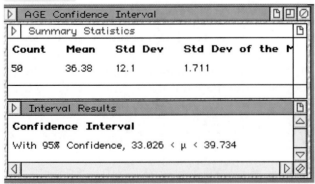

EXCEL

```
AGE Confidence Interval

Summary Statistics

Count   Mean    Std Dev   Std Dev of the M

50      36.38   12.1      1.711


Interval Results

Confidence Interval

With 95% Confidence, 33.026 < μ < 39.734
```

TI-83/84 PLUS

```
ZInterval
(33.026,39.734)
x̄=36.38
Sx=11.06915184
n=50
```

EXERCISES 8.2

Statistical Concepts and Skills

8.11 Find the confidence level and α for
a. a 90% confidence interval.
b. a 99% confidence interval.

8.12 What is meant by saying that a $1 - \alpha$ confidence interval is
a. exact?   **b.** approximately correct?

8.13 In developing Procedure 8.1, we assumed that the variable under consideration is normally distributed.
a. Explain why we needed that assumption.
b. Explain why the procedure yields an approximately correct confidence interval for large samples, regardless of the distribution of the variable under consideration.

8.14 For what is *normal population* an abbreviation?

8.15 Refer to Procedure 8.1.
a. Explain in detail the assumptions required for using the z-interval procedure.
b. How important is the normality assumption? Explain your answer.

8.16 What is meant by saying that a statistical procedure is robust?

8.17 In each part, assume that the population standard deviation is known. Decide whether use of the z-interval procedure to obtain a confidence interval for the population mean is reasonable. Explain your answers.
a. The variable under consideration is very close to being normally distributed, and the sample size is 10.
b. The variable under consideration is very close to being normally distributed, and the sample size is 75.
c. The sample data contain outliers, and the sample size is 20.
d. The sample data contain no outliers, the variable under consideration is roughly normally distributed, and the sample size is 20.
e. The distribution of the variable under consideration is highly skewed, and the sample size is 20.
f. The sample data contain no outliers, the sample size is 250, and the variable under consideration is far from being normally distributed.

8.18 Suppose that you have obtained data by taking a random sample from a population. Before performing a statistical inference, what should you do?

8.19 Suppose that you have obtained data by taking a random sample from a population and that you intend to find a confidence interval for the population mean, μ. Which confidence level, 95% or 99%, will result in the confidence interval's giving a more precise estimate of μ?

Preliminary data analyses indicate that you can reasonably apply the z-interval procedure (Procedure 8.1 on page 363) in Exercises 8.20–8.23.

8.20 Smelling Out the Enemy. Snakes deposit chemical trails as they travel through their habitats. These trails are often detected and recognized by lizards, which are potential prey. The ability to recognize their predators via tongue flicks can often mean life or death for lizards. Scientists from the University of Antwerp were interested in quantifying the responses of juveniles of the common lizard (*Lacerta vivipara*) to natural predator cues to determine whether the behavior is learned or congenital. Seventeen juvenile common lizards were exposed to the chemical cues of the viper snake. Their responses, in number of tongue flicks per 20 minutes, are presented in the following table. [SOURCE: Van Damme et al., "Responses of Naïve Lizards to Predator Chemical Cues," *Journal of Herpetology*, Vol. 29(1), pp. 38–43.]

425	510	629	236	654	200
276	501	811	332	424	674
676	694	710	662	633	

a. Find a 90% confidence interval for the mean number of tongue flicks per 20 minutes for all juvenile common lizards. Assume a population standard deviation of 190.0. (*Note:* The sum of the data is 9047.)
b. Interpret your answer from part (a).

8.21 Malnutrition and Poverty. R. Reifen et al. studied various nutritional measures of Ethiopian school children and published their findings in the paper "Ethiopian-Born and Native Israeli School Children Have Different Growth Patterns" (*Nutrition*, Vol. 19, pp. 427–431). The study, conducted in Azezo, North West Ethiopia, found that malnutrition is prevalent in primary and secondary school children because of economic poverty. The weights, in kilograms (kg), of 60 randomly selected male Ethiopian-born school children, ages 12–15 years old, are presented in the following table.

45.7	48.9	53.8	44.7	42.8	50.9	38.9	45.2	48.1	42.0
48.2	42.1	52.8	45.9	47.9	46.2	42.9	42.5	46.5	45.9
56.6	51.8	43.4	43.3	40.9	44.4	45.2	44.0	47.8	36.3
39.3	42.0	38.0	37.7	42.5	44.8	48.3	47.4	45.4	48.5
43.5	41.3	49.2	45.2	46.8	46.6	41.1	45.5	47.5	48.4
41.8	49.1	49.5	41.5	46.3	48.6	51.4	39.0	38.8	47.2

a. Find a 95% confidence interval for the mean weight of all male Ethiopian-born school children, ages 12–15 years old. Assume that the population standard deviation is 4.5 kg. (*Note:* The sum of the data is 2717.8 kg.)

b. Interpret your answer from part (a).

8.22 Political Prisoners. Ehlers, Maercker, and Boos studied various characteristics of political prisoners from the former East Germany and presented their findings in the paper "Posttraumatic Stress Disorder (PTSD) Following Political Imprisonment: The Role of Mental Defeat, Alienation, and Perceived Permanent Change" (*Journal of Abnormal Psychology*, Vol. 109, pp. 45–55). According to the article, the mean duration of imprisonment for 32 patients with chronic PTSD was 33.4 months. Assuming that $\sigma = 42$ months, determine a 95% confidence interval for the mean duration of imprisonment, μ, of all East German political prisoners with chronic PTSD. Interpret your answer in words.

8.23 Keep on Rolling. The Rolling Stones, a rock group formed in the 1960s, has toured extensively in support of new albums. *Pollstar* has collected data on the earnings from the Stones's North American tours. For 30 randomly selected Rolling Stones concerts, the mean gross earnings is $2.27 million. Assuming a population standard deviation gross earnings of $0.5 million, obtain a 99% confidence interval for the mean gross earnings of all Rolling Stones concerts. Interpret your answer in words.

8.24 Smelling Out the Enemy. Refer to Exercise 8.20.

a. Find a 99% confidence interval for μ.

b. Why is the confidence interval you found in part (a) longer than the one in Exercise 8.20?

c. Draw a graph similar to that shown in Fig. 8.3 on page 366 to display both confidence intervals.

d. Which confidence interval yields a more precise estimate of μ? Explain your answer.

8.25 Malnutrition and Poverty. Refer to Exercise 8.21.

a. Determine an 80% confidence interval for μ.

b. Why is the confidence interval you found in part (a) shorter than the one in Exercise 8.21?

c. Draw a graph similar to that shown in Fig. 8.3 on page 366 to display both confidence intervals.

d. Which confidence interval yields a more precise estimate of μ? Explain your answer.

Extending the Concepts and Skills

8.26 Family Size. The U.S. Bureau of the Census compiles data on family size and presents its findings in *Current Population Reports*. Suppose that 500 U.S. families are randomly selected to estimate the mean size, μ, of all U.S. families. Further suppose that the results are as shown in the following frequency distribution.

Size	2	3	4	5	6	7	8	9
Frequency	198	118	101	59	12	3	8	1

a. If the population standard deviation of family sizes is 1.3, determine a 95% confidence interval for the mean size, μ, of all U.S. families. (*Hint:* To find the sample mean, use the grouped-data formulas on page 116.)

b. Interpret your answer from part (a).

8.27 Key Fact 8.3 states that, for a fixed sample size, decreasing the confidence level improves the precision of the confidence-interval estimate of μ and vice versa.

a. Suppose that you want to increase the precision without reducing the level of confidence. What can you do?

b. Suppose that you want to increase the level of confidence without reducing the precision. What can you do?

Using Technology

*In Exercises **8.28** and **8.29**, use the technology of your choice to*

a. *obtain a normal probability plot, boxplot, histogram, and stem-and-leaf diagram of the data, and*

b. *construct the required confidence interval.*

c. *Justify the use of your procedure in part (b).*

8.28 Smelling out the Enemy. The data set and confidence interval in Exercise 8.20.

8.29 Malnutrition and Poverty. The data set and confidence interval in Exercise 8.21.

8.30 Ages of Diabetics. A research physician wants to estimate the average age of people with diabetes. She takes a random sample of 35 diabetics and obtains the following ages.

48	41	57	83	41	55	59
61	38	48	79	75	77	7
54	23	47	56	79	68	61
64	45	53	82	68	38	70
10	60	83	76	21	65	47

Use the technology of your choice to
a. find a 95% confidence interval for the mean age, μ, of people with diabetes. Assume that $\sigma = 21.2$ years.
b. obtain a normal probability plot, boxplot, histogram, and stem-and-leaf diagram of the data.
c. Remove the outliers (if any) from the data and then repeat part (a).
d. Comment on the advisability of using the z-interval procedure here.

8.31 Farm Children. A sociologist wants information on the number of children per farm family in her native state of Nebraska. Twenty-two randomly selected farm families have the following number of children.

1	2	1	2	0	2	1	5	4	1	0
2	1	3	1	0	1	0	1	8	0	1

Use the technology of your choice to
a. determine a 90% confidence interval for the mean number of children, μ, per farm family in Nebraska. Assume that $\sigma = 1.95$ children.
b. obtain a normal probability plot, boxplot, histogram, and stem-and-leaf diagram of the data.
c. Remove the outliers (if any) from the data and then repeat part (a).
d. Comment on the advisability of using the z-interval procedure here.

8.32 Gestation Periods of Humans. This exercise can be done individually or, better yet, as a class project. Gestation periods of humans are normally distributed with a mean of 266 days and a standard deviation of 16 days.
a. Simulate 100 samples of nine human gestation periods each.
b. For each sample in part (a), obtain a 95% confidence interval for the population mean gestation period.
c. For the 100 confidence intervals that you obtained in part (b), roughly how many would you expect to contain the population mean gestation period of 266 days?
d. For the 100 confidence intervals that you obtained in part (b), determine the number that contain the population mean gestation period of 266 days.
e. Compare your answers from parts (c) and (d) and comment on any observed difference.

8.3 Margin of Error

In this section, we examine in detail how sample size affects the precision of estimating a population mean by a sample mean. Recall Key Fact 7.1, which states that the larger the sample size, the smaller the sampling error tends to be. Now that you have studied confidence intervals, you can determine exactly how sample size affects the accuracy of the estimate. We begin by introducing the concept of the *margin of error* in Example 8.6.

⫿ Example 8.6   **Introducing Margin of Error**

Age of the Civilian Labor Force   In Example 8.4, we applied the one-sample z-interval procedure to the ages of a sample of 50 people in the civilian labor force to obtain a 95% confidence interval for the mean age, μ, of all people in the civilian labor force.

a. Discuss the precision with which $\bar{x}$ estimates μ.

b. By what quantity is the precision determined?

c. As demonstrated in Section 8.2, we can decrease the length of the confidence interval and thereby improve the precision of the estimate by decreasing the confidence level from 95% to some lower level. But suppose that we want to retain the same level of confidence and still improve the precision. How can we do so?

d. Explain why our answer to part (c) makes sense.

Solution Recalling first that $z_{\alpha/2} = z_{0.05/2} = z_{0.025} = 1.96$, $n = 50$, $\sigma = 12.1$, and $\bar{x} = 36.4$, we found that a 95% confidence interval for μ is from

$$\bar{x} - z_{\alpha/2} \cdot \frac{\sigma}{\sqrt{n}} \quad \text{to} \quad \bar{x} + z_{\alpha/2} \cdot \frac{\sigma}{\sqrt{n}}$$

or

$$36.4 - 1.96 \cdot \frac{12.1}{\sqrt{50}} \quad \text{to} \quad 36.4 + 1.96 \cdot \frac{12.1}{\sqrt{50}}$$

or

$$36.4 - 3.4 \quad \text{to} \quad 36.4 + 3.4$$

or

$$33.0 \quad \text{to} \quad 39.8.$$

We can be 95% confident that the mean age, μ, of all people in the civilian labor force is somewhere between 33.0 years and 39.8 years.

a. The confidence interval that we obtained provides a rather wide range for the possible values of μ. In other words, the precision of the estimate is poor.

b. Let's first look more closely at the confidence interval by displaying it graphically in Fig. 8.4.

FIGURE 8.4

95% confidence interval for the mean age, μ, of all people in the civilian labor force

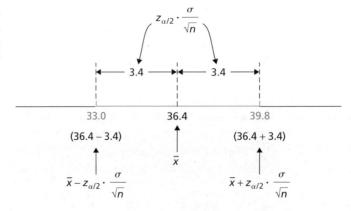

From Fig. 8.4 or the computations done at the beginning of the solution to this example, we find that the precision of the estimate is determined by the quantity

$$E = z_{\alpha/2} \cdot \frac{\sigma}{\sqrt{n}},$$

which is half the length of the confidence interval, or 3.4 in this case. The quantity E is called the **margin of error,** also known as the **maximum error of the estimate.** We use this terminology because, at the specified level of confidence (in this case, 95%), we are confident that our error in estimating μ by $\bar{x}$ is at most 3.4 years, as shown in Fig. 8.4. (*Note:* In newspapers and magazines, this fact is often expressed as "The poll has a margin of error of 3.4 years," or as "Theoretically, in 95 out of 100 such polls the margin of error will be 3.4 years.")

c. To improve the precision of the estimate, we need to decrease the margin of error, E. Because the sample size, n, occurs in the denominator of the formula for E, we can decrease E by increasing the sample size.

d. The answer to part (c) makes sense because we expect more precise information from larger samples.

In Example 8.6, we introduced some concepts and terminology important in confidence-interval analysis. We summarize that discussion in Definition 8.3 and in Key Fact 8.4.

DEFINITION 8.3

Margin of Error for the Estimate of μ

The *margin of error* for the estimate of μ is

$$E = z_{\alpha/2} \cdot \frac{\sigma}{\sqrt{n}}.$$

What *Does it Mean?*

The margin of error is equal to half the length of the confidence interval, as depicted in Fig. 8.5.

Figure 8.5 provides a graphical representation of the margin of error.

FIGURE 8.5

Margin of error, $E = z_{\alpha/2} \cdot \dfrac{\sigma}{\sqrt{n}}$

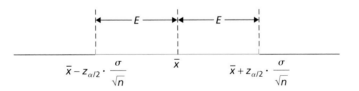

KEY FACT 8.4

Margin of Error, Precision, and Sample Size

The length of a confidence interval for a population mean, μ, and hence the precision with which $\bar{x}$ estimates μ, is determined by the margin of error, E. For a fixed confidence level, increasing the sample size improves the precision, and vice versa.

Determining the Required Sample Size

The margin of error and confidence level are often specified in advance. We must then determine the sample size required to meet the specifications. The formula for the required sample size can be obtained by solving for n in the formula for the margin of error, $E = z_{\alpha/2} \cdot \sigma/\sqrt{n}$. The result is Formula 8.1, which we apply in Example 8.7.

FORMULA 8.1

Sample Size for Estimating μ

The sample size required for a $(1 - \alpha)$-level confidence interval for μ with a specified margin of error, E, is given by the formula

$$n = \left(\frac{z_{\alpha/2} \cdot \sigma}{E}\right)^2,$$

rounded up to the nearest whole number.

Example 8.7   **Sample Size for Estimating μ**

Age of the Civilian Labor Force   Consider again the problem of estimating the mean age, μ, of all people in the civilian labor force.

a. Determine the sample size required to ensure that we can be 95% confident that μ is within 0.5 year of the estimate, $\bar{x}$. Recall that $\sigma = 12.1$ years.

b. Find a 95% confidence interval for μ if a sample of the size determined in part (a) has a mean age of 38.8 years.

Solution

a. To determine the required sample size, we apply Formula 8.1. In doing so, we must identify σ, E, and $z_{\alpha/2}$. We know that $\sigma = 12.1$ years; the margin of error, E, is specified at 0.5 year; and the confidence level is stipulated as 0.95, which means that $\alpha = 0.05$ and $z_{\alpha/2} = z_{0.05/2} = z_{0.025} = 1.96$. Thus the required sample size is

$$n = \left(\frac{z_{\alpha/2} \cdot \sigma}{E}\right)^2 = \left(\frac{1.96 \cdot 12.1}{0.5}\right)^2 = 2249.79,$$

which, rounded up to the nearest whole number, is 2250.

INTERPRETATION If 2250 people in the civilian labor force are randomly selected, we can be 95% confident that the mean age of all people in the civilian labor force is within 0.5 year of the mean age of the people in the sample.

b. We are to find a 95% confidence interval for the mean age of all people in the civilian labor force if a sample of the size determined in part (a) has a mean age of 38.8 years. Applying Procedure 8.1 with $\alpha = 0.05$, $\sigma = 12.1$, $\bar{x} = 38.8$, and $n = 2250$, we obtain the confidence interval

$$\bar{x} - z_{\alpha/2} \cdot \frac{\sigma}{\sqrt{n}} \quad \text{to} \quad \bar{x} + z_{\alpha/2} \cdot \frac{\sigma}{\sqrt{n}}$$

or

$$38.8 - 1.96 \cdot \frac{12.1}{\sqrt{2250}} \quad \text{to} \quad 38.8 + 1.96 \cdot \frac{12.1}{\sqrt{2250}}$$

or

$$38.8 - 0.5 \quad \text{to} \quad 38.8 + 0.5$$

or

$$38.3 \quad \text{to} \quad 39.3.$$

INTERPRETATION We can be 95% confident that the mean age, μ, of all people in the civilian labor force is somewhere between 38.3 years and 39.3 years.

Note: The sample size of 2250 was determined in part (a) of Example 8.7 to guarantee a margin of error of 0.5 year for a 95% confidence interval. Therefore, in light of Fig. 8.5 on page 373, we could have obtained the 95% confidence interval required in part (b) of Example 8.7 simply by computing

$$\bar{x} \pm E = 38.8 \pm 0.5.$$

Doing so would give the same confidence interval, 38.3 to 39.3, that we got in part (b) of Example 8.7 but with much less work. However, because the sample size is a rounded value, the simpler method we just used might have incorrectly yielded a slightly wider confidence interval. In practice, the simpler method is acceptable because, at worst, it provides a slightly conservative estimate.

The formula for finding the required sample size, Formula 8.1, involves the population standard deviation, σ. If σ is unknown, which is usually the case in practice, and we want to apply the formula, we must first estimate σ. One way to do so is to take a preliminary large sample, say, of size 30 or more. The sample standard deviation, s, of the sample obtained provides an estimate of σ and can be used in place of σ in Formula 8.1.

EXERCISES 8.3

Statistical Concepts and Skills

8.33 Discuss the relationship between the margin of error and the standard error of the mean.

8.34 Explain why the margin of error determines the precision with which a sample mean estimates a population mean.

8.35 In each part, explain the effect on the margin of error and hence the effect on the precision of estimating a population mean by a sample mean.
a. Increasing the confidence level while keeping the same sample size.
b. Increasing the sample size while keeping the same confidence level.

8.36 A confidence interval for a population mean has a margin of error of 3.4.
a. Determine the length of the confidence interval.
b. If the sample mean is 52.8, obtain the confidence interval.

8.37 A confidence interval for a population mean has length 20.
a. Determine the margin of error.
b. If the sample mean is 60, obtain the confidence interval.

8.38 Answer true or false to each statement concerning a confidence interval for a population mean. Give reasons for your answers.
a. The length of a confidence interval can be determined if you know only the margin of error.
b. The margin of error can be determined if you know only the length of the confidence interval.
c. The confidence interval can be obtained if you know only the margin of error.
d. The confidence interval can be obtained if you know only the margin of error and the sample mean.
e. The margin of error can be determined if you know only the confidence level.
f. The confidence level can be determined if you know only the margin of error.
g. The margin of error can be determined if you know only the confidence level, population standard deviation, and sample size.
h. The confidence level can be determined if you know only the margin of error, population standard deviation, and sample size.

8.39 Formula 8.1 provides a method for computing the sample size required to obtain a confidence interval with a specified confidence level and margin of error. The number resulting from the formula should be rounded up to the nearest whole number.
a. Why do you want a whole number?
b. Why do you round up instead of down?

8.40 Body Fat. J. McWhorter et al. of the College of Health Sciences at the University of Nevada, Las Vegas, studied physical therapy students during their graduate-school years. The researchers were interested in the fact that, although graduate physical-therapy students are taught the principles of fitness, some have difficulty finding the time to implement those principles. In the study, published as "An Evaluation of Physical Fitness Parameters for Graduate Students" (*Journal of American College Health*, Vol. 51, No. 1, pp. 32–37), a sample of 27 female graduate physical-therapy students had a mean of 22.46 percent body fat.
a. Assuming that percent body fat of female graduate physical-therapy students is normally distributed with

standard deviation 4.10 percent body fat, determine a 95% confidence interval for the mean percent body fat of all female graduate physical-therapy students.
b. Obtain the margin of error, E, for the confidence interval you found in part (a).
c. Explain the meaning of E in this context in terms of the accuracy of the estimate.
d. Determine the sample size required to have a margin of error of 1.55 percent body fat with a 99% confidence level.

8.41 Fuel Expenditures. In estimating the mean monthly fuel expenditure, μ, per household vehicle, the U.S. Energy Information Administration takes a sample of size 6841. Assuming that $\sigma = \$20.65$, determine the margin of error in estimating μ at the 95% level of confidence.

8.42 Smelling Out the Enemy. In Exercise 8.20, you found a 90% confidence interval for the mean number of tongue flicks per 20 minutes for all juvenile common lizards to be from 456.4 to 608.0. Obtain the margin of error by
a. taking half the length of the confidence interval.
b. using the formula in Definition 8.3 on page 373. (Recall that $n = 17$ and $\sigma = 190.0$.)

8.43 Malnutrition and Poverty. In Exercise 8.21, you found a 95% confidence interval for the mean weight of all male Ethiopian-born school children, ages 12–15 years old, to be from 44.16 kg to 46.44 kg. Obtain the margin of error by
a. taking half the length of the confidence interval.
b. using the formula in Definition 8.3 on page 373. (Recall that $n = 60$ and $\sigma = 4.5$ kg.)

8.44 Political Prisoners. In Exercise 8.22, you found a 95% confidence interval of 18.8 months to 48.0 months for the mean duration of imprisonment, μ, of all East German political prisoners with chronic PTSD.
a. Determine the margin of error, E.
b. Explain the meaning of E in this context in terms of the accuracy of the estimate.
c. Find the sample size required to have a margin of error of 12 months and a 99% confidence level. (Recall that $\sigma = 42$ months.)

d. Determine a 99% confidence interval for the mean duration of imprisonment, μ, if a sample of the size determined in part (c) has a mean of 36.2 months.

8.45 Keep on Rolling. In Exercise 8.23, you found a 99% confidence interval of $2.03 million to $2.51 million for the mean gross earnings of all Rolling Stones concerts.
a. Determine the margin of error, E.
b. Explain the meaning of E in this context in terms of the accuracy of the estimate.
c. Find the sample size required to have a margin of error of $0.1 million and a 95% confidence level. (Recall that $\sigma = 0.5$ million.)
d. Obtain a 95% confidence interval for the mean gross earnings if a sample of the size determined in part (c) has a mean of $2.35 million.

Extending the Concepts and Skills

8.46 Millionaires. Professor Thomas Stanley of Georgia State University has surveyed millionaires since 1973. Among other information, Professor Stanley obtains estimates for the mean age, μ, of all U.S. millionaires. Suppose that one year's study involved a simple random sample of 36 U.S. millionaires whose mean age was 58.53 years with a sample standard deviation of 13.36 years.

a. If, for next year's study, a confidence interval for μ is to have a margin of error of 2 years and a confidence level of 95%, determine the required sample size.
b. Why did you use the sample standard deviation, $s = 13.36$, in place of σ in your solution to part (a)? Why is it permissible to do so?

8.47 Corporate Farms. The U.S. Bureau of the Census estimates the mean value of the land and buildings per corporate farm. Those estimates are published in the *Census of Agriculture*. Suppose that an estimate, $\bar{x}$, is obtained and that the margin of error is $1000. Does this result imply that the true mean, μ, is within $1000 of the estimate? Explain your answer.

8.48 Suppose that a simple random sample is taken from a normal population having a standard deviation of 10 for the purpose of obtaining a 95% confidence interval for the mean of the population.
a. If the sample size is 4, obtain the margin of error.
b. Repeat part (a) for a sample size of 16.
c. Can you guess the margin of error for a sample size of 64? Explain your reasoning.

8.49 For a fixed confidence level, show that (approximately) quadrupling the sample size is necessary to halve the margin of error. (*Hint:* Use Formula 8.1.)

8.4 Confidence Intervals for One Population Mean When σ Is Unknown

In Section 8.2, you learned how to determine a confidence interval for a population mean, μ, when the population standard deviation, σ, is known. The basis of the procedure is in Key Fact 7.4: If x is a normally distributed variable with mean μ and standard deviation σ, then, for samples of size n, the variable $\bar{x}$ is also normally distributed and has mean μ and standard deviation $\sigma/\sqrt{n}$; equivalently, the **standardized version of $\bar{x}$,**

$$z = \frac{\bar{x} - \mu}{\sigma/\sqrt{n}}, \tag{8.2}$$

has the standard normal distribution.

But what if, as is usually the case in practice, the population standard deviation is unknown? Then we cannot base our confidence-interval procedure on the standardized version of $\bar{x}$. The best we can do is estimate the population standard deviation, σ, by the sample standard deviation, s; in other words, we replace σ by s in Equation (8.2) and base our confidence-interval procedure on the resulting variable,

$$t = \frac{\bar{x} - \mu}{s/\sqrt{n}}, \tag{8.3}$$

called the **studentized version of $\bar{x}$.**

Unlike the standardized version, the studentized version of $\bar{x}$ does not have a normal distribution. To get an idea of how their distributions differ, we used statistical software to simulate each variable for samples of size 4, assuming that $\mu = 15$ and $\sigma = 0.8$. (Any sample size, population mean, and population standard deviation will do.)

1. We simulated 5000 samples of size 4 each.
2. For each of the 5000 samples, we obtained the sample mean and sample standard deviation.
3. For each of the 5000 samples, we determined the observed values of both the standardized and studentized versions of $\bar{x}$, as given by Equations (8.2) and (8.3), respectively.
4. We obtained histograms of both the 5000 observed values of the standardized version of $\bar{x}$ and the 5000 observed values of the studentized version of $\bar{x}$, as shown in Printout 8.2.

PRINTOUT 8.2

Histograms of z (standardized version of $\bar{x}$) and t (studentized version of $\bar{x}$) for 5000 samples of size 4

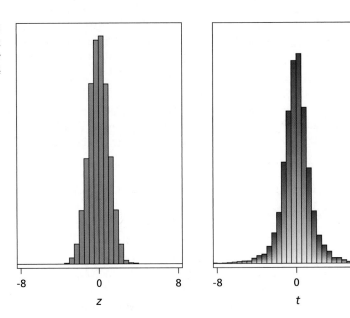

The two histograms shown in Printout 8.2 suggest that the distributions of the standardized version of $\bar{x}$—the variable z in Equation (8.2)—and the studentized version of $\bar{x}$—the variable t in Equation (8.3)—have things in common; both are bell shaped and symmetric about 0. But there is an important difference: The distribution of the studentized version has more spread than the standardized version. This difference is not surprising because the variation in the possible values of the standardized version is due solely to the variation of sample means, whereas that of the studentized version is due to the variation of both sample means and sample standard deviations.

As we know, the standardized version of $\bar{x}$ has the standard normal distribution. In 1908, William Gosset determined the distribution of the studentized version of $\bar{x}$, a distribution now called **Student's *t*-distribution** or, simply, the ***t*-distribution.** (See the Biography on page 395 for more on Gosset and the story of Student's *t*-distribution.)

t-Distributions and *t*-Curves

Actually, there is a different *t*-distribution for each sample size. We identify a particular *t*-distribution by giving its **degrees of freedom (df).** For the studentized version of $\bar{x}$, the number of degrees of freedom is 1 less than the sample size, which we indicate symbolically by **df = $n-1$.** Later, we encounter *t*-statistics other than the studentized version of $\bar{x}$ whose degrees of freedom are different. But for now, we summarize what we have presented about the studentized version of $\bar{x}$ as Key Fact 8.5.

KEY FACT 8.5

What Does it Mean?

For a normally distributed variable, the studentized version of the sample mean has the *t*-distribution with degrees of freedom 1 less than the sample size.

Studentized Version of the Sample Mean

Suppose that a variable x of a population is normally distributed with mean μ. Then, for samples of size n, the variable

$$t = \frac{\bar{x} - \mu}{s/\sqrt{n}}$$

has the *t*-distribution with $n-1$ degrees of freedom.

Like normally distributed variables, a variable having a *t*-distribution has an associated curve, called a ***t*-curve.** In this book, you don't need to know the equation of a *t*-curve, but you do need to understand the basic properties of a *t*-curve.

Although there is a different *t*-curve for each number of degrees of freedom, all *t*-curves are similar and resemble the standard normal curve. Figure 8.6 shows the standard normal curve and two *t*-curves. As illustrated in Fig. 8.6, *t*-curves have the properties delineated in Key Fact 8.6. Compare Properties 1–3 of *t*-curves with those of the standard normal curve, as delineated in Key Fact 6.3 on page 282.

FIGURE 8.6
Standard normal curve and two *t*-curves

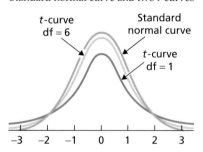

KEY FACT 8.6

Basic Properties of t-Curves

Property 1: The total area under a *t*-curve equals 1.

Property 2: A *t*-curve extends indefinitely in both directions, approaching, but never touching, the horizontal axis as it does so.

Property 3: A *t*-curve is symmetric about 0.

Property 4: As the number of degrees of freedom becomes larger, *t*-curves look increasingly like the standard normal curve.

Using the *t*-Table

Percentages (and probabilities) for a variable having a *t*-distribution equal areas under the variable's associated *t*-curve. For our purposes, one of which is obtaining confidence intervals for a population mean, we don't need a complete *t*-table for each *t*-curve; only certain areas will be important. Table IV, which appears in Appendix A and in abridged form inside the front cover, is sufficient for our purposes.

The two outside columns of Table IV, labeled df, display the number of degrees of freedom. As expected, the symbol t_α denotes the *t*-value having area α to its right under a *t*-curve. Thus the column headed $t_{0.10}$ contains *t*-values having area 0.10 to their right; the column headed $t_{0.05}$ contains *t*-values having area 0.05 to their right; and so on. We illustrate a use of Table IV in Example 8.8.

Example 8.8   Finding the t-Value Having a Specified Area to Its Right

For a *t*-curve with 13 degrees of freedom, determine $t_{0.05}$; that is, find the *t*-value having area 0.05 to its right, as shown in Fig. 8.7(a).

FIGURE 8.7
Finding the *t*-value having area 0.05 to its right

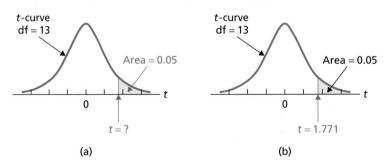

(a)   (b)

Solution   To find the *t*-value in question, we use Table IV. For ease of reference, we have repeated a portion of Table IV as Table 8.4.

TABLE 8.4
Values of t_α

df	$t_{0.10}$	$t_{0.05}$	$t_{0.025}$	$t_{0.01}$	$t_{0.005}$	df
.	.	.	.	.	.	.
.	.	.	.	.	.	.
.	.	.	.	.	.	.
12	1.356	1.782	2.179	2.681	3.055	12
13	1.350	1.771	2.160	2.650	3.012	13
14	1.345	1.761	2.145	2.624	2.977	14
15	1.341	1.753	2.131	2.602	2.947	15
.	.	.	.	.	.	.
.	.	.	.	.	.	.
.	.	.	.	.	.	.

The number of degrees of freedom is 13, so we first go down the outside columns, labeled df, to "13." Then we go across that row to the column headed $t_{0.05}$. The number in the body of the table there, 1.771, is the required *t*-value; that is, for a *t*-curve with df = 13, the *t*-value having area 0.05 to its right is $t_{0.05} = 1.771$, as shown in Fig. 8.7(b).

Note that Table IV in Appendix A contains degrees of freedom from 1 to 75 consecutively, but then contains only selected degrees of freedom. If you need a value of t_α with a number of degrees of freedom not included in Table IV, you have several options: Find a more detailed *t*-table, use technology, or use linear interpolation and Table IV. A less exact option, but one that usually suffices, is to use the degrees of freedom in Table IV closest to the one required.

As we noted earlier, *t*-curves look increasingly like the standard normal curve as the number of degrees of freedom gets larger. For degrees of freedom greater than 2000, a *t*-curve and the standard normal curve are virtually indistinguishable. Consequently, we stopped the *t*-table at df = 2000 and supplied the corresponding values of z_α beneath. These values can be used not only for the standard normal distribution, but also for any *t*-distribution having degrees of freedom greater than 2000.[†]

Obtaining Confidence Intervals for a Population Mean When σ Is Unknown

Having discussed *t*-distributions and *t*-curves, we can now develop a procedure for obtaining a confidence interval for a population mean when the population standard deviation is unknown. We proceed in essentially the same way as we did when the population standard deviation is known, except now we invoke a *t*-distribution instead of the standard normal distribution.

Hence we use $t_{\alpha/2}$ instead of $z_{\alpha/2}$ in the formula for the confidence interval. As a result, we have Procedure 8.2 (next page), which we refer to as the **one-sample *t*-interval procedure** or, simply, as the ***t*-interval procedure.**

Before applying Procedure 8.2, we need to make several comments. Although the *t*-interval procedure is based on the assumption that the variable under consideration is normally distributed (normal population), it also applies approximately for large samples, regardless of the distribution of the variable under consideration, as noted at the bottom of Procedure 8.2.

Actually, like the *z*-interval procedure, the *t*-interval procedure works reasonably well even when the variable under consideration is not normally distributed and the sample size is small or moderate, provided the variable is not too far from being normally distributed. In other words, the *t*-interval procedure is robust to moderate violations of the normality assumption.

When considering the *t*-interval procedure, we also have to watch for outliers. Again, the presence of outliers calls into question the normality assumption. In addition, even for large samples, outliers can sometimes unduly affect a *t*-interval because the sample mean and sample standard deviation are not resistant to outliers.

Guidelines for use of the *t*-interval procedure are the same as those for the *z*-interval procedure (Key Fact 8.1 on page 364). Remember always to examine the data before applying the *t*-interval procedure to ensure that its use is reasonable. Examples 8.9 and 8.10 illustrate use of Procedure 8.2.

[†]The values of z_α given at the bottom of Table IV are accurate to three decimal places and, because of that, some differ slightly from what you get by applying the method you learned for using Table II.

PROCEDURE 8.2

One-Sample *t*-Interval Procedure

Purpose To find a confidence interval for a population mean, μ

Assumptions

1. Simple random sample
2. Normal population or large sample
3. σ unknown

STEP 1 For a confidence level of 1 $-\alpha$, use Table IV to find $t_{\alpha/2}$ with df $= n - 1$, where n is the sample size.

STEP 2 The confidence interval for μ is from

$$\bar{x} - t_{\alpha/2} \cdot \frac{s}{\sqrt{n}} \quad \text{to} \quad \bar{x} + t_{\alpha/2} \cdot \frac{s}{\sqrt{n}},$$

where $t_{\alpha/2}$ is found in Step 1 and $\bar{x}$ and s are computed from the sample data.

STEP 3 Interpret the confidence interval.

The confidence interval is exact for normal populations and is approximately correct for large samples from nonnormal populations.

| Example 8.9 | The One-Sample t-Interval Procedure |

Pickpocket Offenses The U.S. Federal Bureau of Investigation (FBI) compiles data on robbery and property crimes and publishes the information in *Population-at-Risk Rates and Selected Crime Indicators*. A (simple random) sample of a recent year's pickpocket offenses yielded the values lost shown in Table 8.5. Use the data to obtain a 95% confidence interval for the mean value lost, μ, of all the year's pickpocket offenses.

TABLE 8.5

Value lost ($) for a sample of 25 pickpocket offenses

447	207	627	430	883
313	844	253	397	214
217	768	1064	26	587
833	277	805	653	549
649	554	570	223	443

Solution Because the sample size, $n = 25$, is moderate, we first need to consider questions of normality and outliers. (See the second bulleted item in Key Fact 8.1 on page 364.) To do that, we constructed a normal probability plot for the data in Table 8.5, as shown in Fig. 8.8.

The normal probability plot in Fig. 8.8 shows no outliers and falls roughly in a straight line. Thus we can apply Procedure 8.2 to obtain the required confidence interval.

STEP 1 For a confidence level of 1 $- \alpha$, use Table IV to find $t_{\alpha/2}$ with df $= n - 1$, where n is the sample size.

We want a 95% confidence interval, so $\alpha = 1 - 0.95 = 0.05$. For $n = 25$, df $= 25 - 1 = 24$. From Table IV, $t_{\alpha/2} = t_{0.05/2} = t_{0.025} = 2.064$.

FIGURE 8.8
Normal probability plot of the value-lost data in Table 8.5

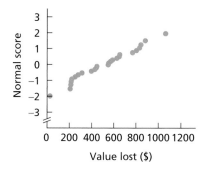

STEP 2 The confidence interval for μ is from

$$\bar{x} - t_{\alpha/2} \cdot \frac{s}{\sqrt{n}} \quad \text{to} \quad \bar{x} + t_{\alpha/2} \cdot \frac{s}{\sqrt{n}}.$$

From Step 1, we have $t_{\alpha/2} = 2.064$. Applying the usual formulas for $\bar{x}$ and s to the data in Table 8.5, we get $\bar{x} = 513.32$ and $s = 262.23$. Consequently, a 95% confidence interval for μ is from

$$513.32 - 2.064 \cdot \frac{262.23}{\sqrt{25}} \quad \text{to} \quad 513.32 + 2.064 \cdot \frac{262.23}{\sqrt{25}},$$

or 405.07 to 621.57.

STEP 3 Interpret the confidence interval.

INTERPRETATION We can be 95% confident that the mean value lost, μ, of all pickpocket offenses for the year in question is somewhere between \$405.07 and \$621.57.

Example 8.10 The One-Sample t-Interval Procedure

TABLE 8.6
Sample of year's chicken consumption (lb)

47	39	62	49	50	70
59	53	55	0	65	63
53	51	50	72	45	

Chicken Consumption The U.S. Department of Agriculture publishes data on chicken consumption in *Food Consumption, Prices, and Expenditures*. A recent year's chicken consumption, in pounds, for 17 randomly selected people is displayed in Table 8.6. Use the data to obtain a 90% confidence interval for the year's mean chicken consumption, μ.

Solution A normal probability plot of the data in Table 8.6 is displayed in Fig. 8.9(a). The plot reveals an outlier—the observation of 0 lb. Because the sample size is only moderate, applying Procedure 8.2 to the data in Table 8.6 is inappropriate.

FIGURE 8.9
Normal probability plots for chicken consumption: (a) original data and (b) data with outlier removed

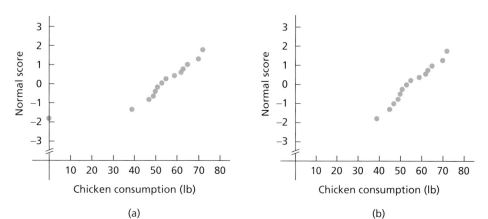

The outlier of 0 lb might be a recording error or it might reflect a person in the sample who does not eat chicken (e.g., a vegetarian). If we remove the outlier from the data, the normal probability plot for the abridged data shows no outliers and is roughly linear, as shown in Fig. 8.9(b).

Thus, if we are willing to take as our population of interest only people who eat chicken, we can use Procedure 8.2 to obtain a confidence interval. Applying that procedure to the sample data with the outlier removed yields a 90% confidence interval of 51.2 to 59.2.

INTERPRETATION We can be 90% confident that the year's mean chicken consumption, among people who eat chicken, is somewhere between 51.2 lb and 59.2 lb.

————————————◼

By making our population of interest in Example 8.10 only those people who eat chicken, we were justified in removing the outlier of 0 lb. Generally, an outlier should not be removed without careful consideration. *Simply removing an outlier because it is an outlier is unacceptable statistical practice.*

In Example 8.10, if we had been careless in our analysis by blindly finding a confidence interval without first examining the data, our result would have been invalid and misleading.

> **What** ?
> *Does it Mean?*
>
> Performing preliminary data analyses to check assumptions before applying inferential procedures is essential.

What If the Assumptions Are Not Satisfied?

We have now described two methods for obtaining a confidence interval for a population mean. If the population standard deviation is known, you can use the z-interval procedure (Procedure 8.1 on page 363); if the population standard deviation is unknown, you can use the t-interval procedure (Procedure 8.2 on page 382).

However, use of both procedures is based on an assumption: The variable under consideration should be approximately normally distributed or the sample size should be relatively large and, for small samples, both procedures should be avoided in the presence of outliers. (Again, refer to Key Fact 8.1 on page 364 for general guidelines.)

Suppose that you want to obtain a confidence interval for a population mean based on a small sample but that preliminary data analyses indicate either the presence of outliers or that the variable under consideration is far from normally distributed. As neither the z-interval procedure nor the t-interval procedure is appropriate, what can you do?

Under certain conditions, you can use a *nonparametric method*.[†] For example, if the variable under consideration has a symmetric distribution, you can use a nonparametric method called the *one-sample Wilcoxon confidence-interval procedure* to obtain a confidence interval for the population mean.

Most nonparametric methods do not require even approximate normality, are resistant to outliers and other extreme values, and can be applied regardless of sample size. However, parametric methods, such as the z-interval and t-interval procedures, tend to give more accurate results than nonparametric

---

[†]Recall that descriptive measures for a population, such as μ and σ, are called parameters. Technically, inferential methods concerned with parameters are called **parametric methods;** those that are not are called **nonparametric methods.** However, common practice is to refer to most methods that can be applied without assuming normality (regardless of sample size) as nonparametric. Thus the term *nonparametric method* as used in contemporary statistics is somewhat of a misnomer.

methods when the normality assumption and other requirements for their use are met.

We do not cover the one-sample Wilcoxon confidence-interval procedure in this book. But we do discuss several other nonparametric procedures, beginning in Chapter 9 with the Wilcoxon signed-rank test.

Example 8.11 illustrates the use of preliminary data analysis to decide which confidence-interval procedure to use.

Example 8.11   Choosing a Confidence Interval Procedure

Adjusted Gross Incomes   The U.S. Internal Revenue Service (IRS) publishes data on federal individual income tax returns in *Statistics of Income, Individual Income Tax Returns*. A sample of 12 returns from a recent year revealed the adjusted gross incomes, in thousands of dollars, shown in Table 8.7. Which procedure should be used to obtain a confidence interval for the mean adjusted gross income, μ, of all the year's individual income tax returns?

TABLE 8.7
Adjusted gross incomes ($1000)

9.7	93.1	33.0	21.2
81.4	51.1	43.5	10.6
12.8	7.8	18.1	12.7

Solution   Because the sample size is small ($n = 12$), we must first consider questions of normality and outliers. A normal probability plot of the sample data, as shown in Fig. 6.23 on page 301, suggests that adjusted gross incomes are far from being normally distributed. Consequently, neither the z-interval procedure nor the t-interval procedure should be used; instead, some nonparametric confidence interval procedure should be applied.

The Technology Center

Procedure 8.2 on page 382 provides a step-by-step method for obtaining a confidence interval for a population mean when the population standard deviation is unknown. Most statistical technologies have programs that automatically determine this type of confidence interval. In this subsection, we present output and step-by-step instructions to implement such programs.

Example 8.12   Using Technology to Obtain a t-Interval

Pickpocket Offenses   The values lost, in dollars, of 25 randomly selected pickpocket offenses from a recent year are displayed in Table 8.5 on page 382. Use Minitab, Excel, or the TI-83/84 Plus to obtain a 95% confidence interval for the mean value lost, μ, of all the year's pickpocket offenses.

Solution   Printout 8.3 at the top of the next page shows the output obtained by applying the one-sample t-interval programs to the value-lost data in Table 8.5.

As shown in the three outputs in Printout 8.3, the required 95% confidence interval is from 405.1 to 621.6. Hence we can be 95% confident that the mean value lost, μ, of all the year's pickpocket offenses is somewhere between $405.1 and $621.6. Compare this confidence interval to the one obtained in Example 8.9.

PRINTOUT 8.3

One-sample *t*-interval output for the
sample of values lost

MINITAB

One-Sample T: LOST

```
Variable    N      Mean    StDev   SE Mean        95% CI
LOST        25   513.320  262.231   52.446   (405.076, 621.564)
```

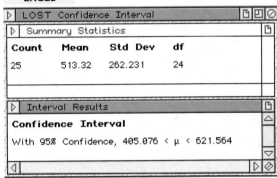

EXCEL

LOST Confidence Interval

Summary Statistics

Count	Mean	Std Dev	df
25	513.32	262.231	24

Interval Results

Confidence Interval

With 95% Confidence, 405.076 < μ < 621.564

TI-83/84 PLUS

```
TInterval
 (405.08,621.56)
 x̄=513.32
 Sx=262.2309288
 n=25
```

Obtaining the Output (Optional)

Printout 8.3 provides output from Minitab, Excel, and the TI-83/84 Plus for a one-sample *t*-interval with a 95% confidence level based on the sample of values lost presented in Table 8.5. The following are detailed instructions for obtaining that output. First, we store the values-lost data in a column (Minitab), range (Excel), or list (TI-83/84 Plus) named LOST. Then, we proceed as follows.

MINITAB

1 Choose **Stat ➤ Basic Statistics ➤ 1-Sample t...**
2 Select the **Samples in columns** option button
3 Click in the **Samples in columns** text box and specify LOST
4 Click the **Options...** button
5 Type 95 in the **Confidence level** text box
6 Click the arrow button at the right of the **Alternative** drop-down list box and select **not equal**
7 Click **OK**
8 Click **OK**

EXCEL

1 Choose **DDXL ➤ Confidence Intervals**
2 Select **1 Var t Interval** from the **Function type** drop-down box
3 Specify LOST in the **Quantitative Variable** text box
4 Click **OK**
5 Click the **95%** button
6 Click the **Compute Interval** button

TI-83/84 PLUS

1 Press **STAT**, arrow over to **TESTS**, and press **8**
2 Highlight **Data** and press **ENTER**
3 Press the down-arrow key
4 Press **2nd ➤ LIST**
5 Arrow down to LOST and press **ENTER** three times
6 Type .95 for **C-Level** and press **ENTER** twice

EXERCISES 8.4

Statistical Concepts and Skills

8.50 Explain the difference in the formulas for the standardized and studentized versions of $\bar{x}$.

8.51 Why do you need to consider the studentized version of $\bar{x}$ to develop a confidence-interval procedure for a population mean when the population standard deviation is unknown?

8.52 A variable has a mean of 100 and a standard deviation of 16. Four observations of this variable have a mean of 108 and a sample standard deviation of 12. Determine the observed value of the
a. standardized version of $\bar{x}$.
b. studentized version of $\bar{x}$.

8.53 A variable of a population has a normal distribution. Suppose that you want to find a confidence interval for the population mean.
a. If you know the population standard deviation, which procedure would you use?
b. If you do not know the population standard deviation, which procedure would you use?

8.54 Batting Averages. An issue of *Scientific American* reveals that the batting averages of major-league baseball players are normally distributed and have a mean of 0.270 and a standard deviation of 0.031. For samples of 20 batting averages, identify the distribution of each variable.
a. $\dfrac{\bar{x} - 0.270}{0.031/\sqrt{20}}$ **b.** $\dfrac{\bar{x} - 0.270}{s/\sqrt{20}}$

8.55 Explain why there is more variation in the possible values of the studentized version of $\bar{x}$ than in the possible values of the standardized version of $\bar{x}$.

8.56 Two *t*-curves have degrees of freedom 12 and 20, respectively. Which one more closely resembles the standard normal curve? Explain your answer.

8.57 For a *t*-curve with df = 6, use Table IV to find each *t*-value.
a. $t_{0.10}$ **b.** $t_{0.025}$ **c.** $t_{0.01}$

8.58 For a *t*-curve with df = 17, use Table IV to find each *t*-value.
a. $t_{0.05}$ **b.** $t_{0.025}$ **c.** $t_{0.005}$

8.59 For a *t*-curve with df = 21, find each *t*-value and illustrate your results graphically.
a. The *t*-value having area 0.10 to its right
b. $t_{0.01}$
c. The *t*-value having area 0.025 to its left (*Hint:* A *t*-curve is symmetric about 0.)

d. The two *t*-values that divide the area under the curve into a middle 0.90 area and two outside areas of 0.05

8.60 For a *t*-curve with df = 8, find each *t*-value and illustrate your results graphically.
a. The *t*-value having area 0.05 to its right
b. $t_{0.10}$
c. The *t*-value having area 0.01 to its left (*Hint:* A *t*-curve is symmetric about 0.)
d. The two *t*-values that divide the area under the curve into a middle 0.95 area and two outside 0.025 areas

8.61 A simple random sample of size 100 is taken from a population with unknown standard deviation. A normal probability plot of the data displays significant curvature but no outliers. Can you reasonably apply the *t*-interval procedure? Explain your answer.

8.62 A simple random sample of size 17 is taken from a population with unknown standard deviation. A normal probability plot of the data reveals an outlier but is otherwise roughly linear. Can you reasonably apply the *t*-interval procedure? Explain your answer.

Preliminary data analyses indicate the reasonableness of applying the t-interval procedure (Procedure 8.2 on page 382) in Exercises 8.63–8.66.

8.63 Family Fun? Taking the family to an amusement park has become increasingly costly according to the industry publication *Amusement Business*, which provides figures on the cost for a family of four to spend the day at one of America's amusement parks. A random sample of 25 families of four that attended amusement parks yielded the following costs, rounded to the nearest dollar.

122	166	171	148	135
173	137	163	119	144
164	153	162	140	142
158	130	167	173	186
92	170	126	163	172

a. Determine a 95% confidence interval for the mean cost of a family of four to spend the day at an American amusement park. (*Note:* $\bar{x} = \$151.04$; $s = \$22.01$.)

b. Interpret your answer from part (a).

8.64 Adrenomedullin and Pregnancy Loss. Adrenomedullin, a hormone found in the adrenal gland, participates in blood-pressure and heart-rate control. The level of adrenomedullin is raised in a variety of diseases, and medical complications, including recurrent pregnancy loss, can result. In an article by M. Nakatsuka et al. titled "Increased Plasma Adrenomedullin in Women With Recurrent Pregnancy Loss" (*Obstetrics & Gynecology*, Vol. 102, No. 2, pp. 319–324), the plasma levels of adrenomedullin for 38 women with recurrent pregnancy loss had a mean of 5.6 pmol/l and a sample standard deviation of 1.9 pmol/l, where pmol/l is an abbreviation of picomoles per liter.

a. Find a 90% confidence interval for the mean plasma level of adrenomedullin for all women with recurrent pregnancy loss.

b. Interpret your answer from part (a).

8.65 The Coruro's Burrow.   The subterranean coruro (*Spalacopus cyanus*) is a social rodent that lives in large colonies in underground burrows that can reach lengths of up to 600 meters. Zoologists Sabine Begall and Milton H. Gallardo studied the characteristics of the burrow systems of the subterranean coruro in central Chile and published their findings in the *Journal of Zoology, London*, (Vol. 251, pp. 53–60). A sample of 51 burrows had the following depths, in centimeters (cm).

15.1	16.0	18.3	18.8	13.9	15.8	14.2
12.3	11.8	12.1	17.9	16.6	16.5	16.0
12.8	14.7	15.9	13.9	17.2	12.2	18.2
16.9	13.3	14.4	15.0	12.1	11.0	16.7
17.4	8.2	19.3	17.4	15.3	15.6	19.7
14.5	12.5	12.8	13.3	16.8	17.5	14.0
14.9	16.7	12.0	15.0	16.2	9.7	15.4
18.9	14.9					

a. Obtain a 90% confidence interval for the mean depth of all subterranean coruro burrows. (*Note:* $\bar{x} = 15.05$ cm; $s = 2.50$ cm.)

b. Interpret your answer from part (a).

8.66 Sleep. In 1908, W. S. Gosset published the article "The Probable Error of a Mean" (*Biometrika*, Vol. 6, pp. 1–25). In this pioneering paper, written under the pseudonym "Student," Gosset introduced what later became known as Student's *t*-distribution. Gosset used the following data set, which gives the additional sleep in hours obtained by a sample of 10 patients using laevohysocyamine hydrobromide.

1.9	0.8	1.1	0.1	−0.1
4.4	5.5	1.6	4.6	3.4

a. Find a 95% confidence interval for the additional sleep that would be obtained on average for all people using laevohysocyamine hydrobromide. (*Note:* $\bar{x} = 2.33$ hr; $s = 2.002$ hr.)

b. Was the drug effective in increasing sleep? Explain your answer.

Extending the Concepts and Skills

8.67 Bicycle Commuting Times. A city planner working on bikeways designs a questionaire to obtain information about local bicycle commuters. One of the questions asks how long it takes the rider to pedal from home to his or her destination. A sample of local bicycle commuters yields the following times, in minutes.

22	19	24	31	29	29
21	15	27	23	37	31
30	26	16	26	12	
23	48	22	29	28	

a. Find a 90% confidence interval for the mean commuting time of all local bicycle commuters in the city. (*Note:* The sample mean and sample standard deviation of the data are 25.82 minutes and 7.71 minutes, respectively.)

b. Interpret your result in part (a).

c. Graphical analyses of the data indicate that the time of 48 minutes may be an outlier. Remove this potential outlier and repeat part (a). (*Note:* The sample mean and sample standard deviation of the abridged data are 24.76 and 6.05, respectively.)

d. Should you have used the procedure that you did in part (a)? Explain your answer.

8.68 Table IV in Appendix A contains degrees of freedom from 1 to 75 consecutively but then contains only selected degrees of freedom.

a. Why couldn't we provide entries for all possible degrees of freedom?

b. Why did we construct the table so that consecutive entries appear for smaller degrees of freedom but that only selected entries occur for larger degrees of freedom?

c. If you had only Table IV, what value would you use for $t_{0.05}$ with df = 87? with df = 125? with df = 650? with df = 3000? Explain your answers.

8.69 As we mentioned earlier in this section, we stopped the t-table at df $= 2000$ and supplied the corresponding values of z_α beneath. Explain why that makes sense.

8.70 A variable of a population has mean μ and standard deviation σ. For a sample of size n, under what conditions are the observed values of the studentized and standardized versions of $\bar{x}$ equal? Explain your answer.

8.71 Let $0 < \alpha < 1$. For a t-curve, determine
a. the t-value having area α to its right in terms of t_α.
b. the t-value having area α to its left in terms of t_α.
c. the two t-values that divide the area under the curve into a middle $1 - \alpha$ area and two outside $\alpha/2$ areas.
d. Draw graphs to illustrate your results in parts (a)–(c).

Using Technology

8.72 Forearm Length. In 1903, K. Pearson and A. Lee published a paper entitled "On the Laws of Inheritance in Man. I. Inheritance of Physical Characters" (*Biometrika*, Vol. 2, pp. 357–462). The article examined and presented data on forearm length, in inches, for a random sample of 140 men. The data are contained on the WeissStats CD.
a. Import the data into the technology of your choice.
b. Use the technology of your choice to obtain a normal probability plot, boxplot, and histogram of the data.
c. Is it reasonable to apply the t-interval procedure to the data? Explain your answer.
d. If you answered "yes" to part (c), obtain a 95% confidence interval for the mean forearm length of men. Interpret your result.

8.73 Family Fun? Refer to Exercise 8.63.
a. Use the technology of your choice to obtain a normal probability plot, boxplot, histogram, and stem-and-leaf diagram of the data.
b. Use the technology of your choice to construct the required confidence interval.
c. Justify the use of your procedure in part (b).

8.74 Calories in Soup. In the February 2003 issue of *Consumer Reports*, the quality of different vegetable, minestrone, and chicken noodle canned soups was evaluated. A sample of 14 such canned soups yielded the following data on the number of calories per serving. The serving size is typically 1 cup.

110	240	160	110	90	110	90
110	110	100	90	170	140	80

a. Use the technology of your choice to obtain a normal probability plot, a boxplot, a histogram, and a stem-and-leaf diagram for these data.
b. Is it reasonable to apply the one-sample t-interval procedure to these data in order to obtain a confidence interval for the mean number of calories per serving for all cans of vegetable, minestrone, and chicken noodle soups? Explain your answer.

8.75 Batting Averages. An issue of *Scientific American* reveals that the batting averages of major-league baseball players are normally distributed with mean .270 and standard deviation .031.
a. Simulate 2000 samples of five batting averages each.
b. Determine the sample mean and sample standard deviation of each of the 2000 samples.
c. For each of the 2000 samples, determine the observed value of the standardized version of $\bar{x}$.
d. Obtain a histogram of the 2000 observations in part (c).
e. Theoretically, what is the distribution of the standardized version of $\bar{x}$?
f. Compare your results from parts (d) and (e).
g. For each of the 2000 samples, determine the observed value of the studentized version of $\bar{x}$.
h. Obtain a histogram of the 2000 observations in part (g).
i. Theoretically, what is the distribution of the studentized version of $\bar{x}$?
j. Compare your results from parts (h) and (i).
k. Compare your histograms from parts (d) and (h). How and why do they differ?

CHAPTER REVIEW

You Should Be Able To

1. use and understand the formulas in this chapter.

2. obtain a point estimate for a population mean.

3. find and interpret a confidence interval for a population mean when the population standard deviation is known.

4. compute and interpret the margin of error for the estimate of μ.

5. understand the relationship between sample size, standard deviation, confidence level, and margin of error for a confidence interval for μ.

6. determine the sample size required for a specified confidence level and margin of error for the estimate of μ.

7. understand the difference between the standardized and studentized versions of $\bar{x}$.

8. state the basic properties of t-curves.

9. use Table IV to find $t_{\alpha/2}$ for df $= n - 1$ and selected values of α.

10. find and interpret a confidence interval for a population mean when the population standard deviation is unknown.

11. decide whether it is appropriate to use the z-interval procedure, t-interval procedure, or neither.

Key Terms

confidence-interval estimate, *357*
confidence level, *357*
degrees of freedom (df), *379*
margin of error (E), *373*
maximum error of the estimate, *373*
nonparametric methods, *384*
one-sample *t*-interval procedure, *382*

one-sample *z*-interval procedure, *363*
parametric methods, *384*
point estimate, *357*
robust procedures, *364*
standardized version of $\bar{x}$, *377*
studentized version of $\bar{x}$, *377*
Student's *t*-distribution, *378*

t_α, *380*
t-curve, *379*
t-distribution, *378*
t-interval procedure, *381*
z_α, *362*
z-interval procedure, *363*

REVIEW TEST

Statistical Concepts and Skills

1. Explain the difference between a point estimate of a parameter and a confidence-interval estimate of a parameter.

2. Answer true or false to the following statement and give a reason for your answer: If a 95% confidence interval for a population mean, μ, is from 33.8 to 39.0, the mean of the population must lie somewhere between 33.8 and 39.0.

3. Must the variable under consideration be normally distributed for you to use the z-interval procedure or t-interval procedure? Explain your answer.

4. If you obtained one thousand 95% confidence intervals for a population mean, μ, roughly how many of the intervals would actually contain μ?

5. Suppose that you have obtained a sample with the intent of performing a particular statistical-inference procedure. What should you do before applying the procedure to the sample data? Why?

6. Suppose that you intend to find a 95% confidence interval for a population mean by applying the one-sample z-interval procedure to a sample of size 100.
   a. What would happen to the precision of the estimate if you used a sample of size 50 instead but kept the same confidence level of 0.95?
   b. What would happen to the precision of the estimate if you changed the confidence level to 0.90 but kept the same sample size of 100?

7. A confidence interval for a population mean has a margin of error of 10.7.
   a. Obtain the length of the confidence interval.
   b. If the mean of the sample is 75.2, determine the confidence interval.

8. Suppose that you plan to apply the one-sample z-interval procedure to obtain a 90% confidence interval for a population mean, μ. You know that $\sigma = 12$ and that you are going to use a sample of size 9.
   a. What will be your margin of error?
   b. What else do you need to know in order to obtain the confidence interval?

9. A variable of a population has a mean of 266 and a standard deviation of 16. Ten observations of this variable have a mean of 262.1 and a sample standard deviation of 20.4. Obtain the observed value of the
   a. standardized version of $\bar{x}$.
   b. studentized version of $\bar{x}$.

10. **Baby Weight.** The paper "Are Babies Normal?" by Traci Clemons and Marcello Pagano (*The American Statistician*, Vol. 53, No. 4, pp. 298–302) focused on babies born in 1991. According to the article, for babies born within the "normal" gestational range of 37–43 weeks, birth weights are normally distributed with a mean of 3432 grams (7 pounds 9 ounces) and a standard deviation of 482 grams (1 pound 1 ounce). For samples of 15 such birth weights, identify the distribution of each variable.
   a. $\dfrac{\bar{x} - 3432}{482/\sqrt{15}}$
   b. $\dfrac{\bar{x} - 3432}{s/\sqrt{15}}$

11. The following figure shows the standard normal curve and two *t*-curves. Which of the two *t*-curves has the larger degrees of freedom? Explain your answer.

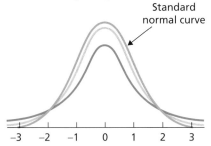

12. In each part of this problem, we have provided a scenario for a confidence interval. Decide whether the appropriate method for obtaining the confidence interval is the *z*-interval procedure, the *t*-interval procedure, or neither.

a. A random sample of size 17 is taken from a population. A normal probability plot of the sample data is found to be very close to linear (straight line). The population standard deviation is unknown.

b. A random sample of size 50 is taken from a population. A normal probability plot of the sample data is found to be roughly linear. The population standard deviation is known.

c. A random sample of size 25 is taken from a population. A normal probability plot of the sample data shows three outliers but is otherwise roughly linear. Checking reveals that the outliers are due to recording errors. The population standard deviation is known.

d. A random sample of size 20 is taken from a population. A normal probability plot of the sample data shows three outliers but is otherwise roughly linear. Removal of the outliers is questionable. The population standard deviation is unknown.

e. A random sample of size 128 is taken from a population. A normal probability plot of the sample data shows no outliers but has significant curvature. The population standard deviation is known.

f. A random sample of size 13 is taken from a population. A normal probability plot of the sample data shows no outliers but has significant curvature. The population standard deviation is unknown.

13. Millionaires. Dr. Thomas Stanley of Georgia State University has surveyed millionaires since 1973. Among other information, Stanley obtains estimates for the mean age, μ, of all U.S. millionaires. Suppose that 36 randomly selected U.S. millionaires are the following ages.

31	45	79	64	48	38	39	68	52
59	68	79	42	79	53	74	66	66
71	61	52	47	39	54	67	55	71
77	64	60	75	42	69	48	57	48

Determine a 95% confidence interval for the mean age, μ, of all U.S. millionaires. Assume that the standard deviation of ages of all U.S. millionaires is 13.0 years. (*Note:* The mean of the data is 58.53 years.)

14. Millionaires. From Problem 13, we know that "a 95% confidence interval for the mean age of all U.S. millionaires is from 54.3 years to 62.8 years." Decide which of the following sentences provide a correct interpretation of the statement in quotes. Justify your answers.

a. Ninety-five percent of all U.S. millionaires are between the ages of 54.3 years and 62.8 years.

b. There is a 95% chance that the mean age of all U.S. millionaires is between 54.3 years and 62.8 years.

c. We can be 95% confident that the mean age of all U.S. millionaires is between 54.3 years and 62.8 years.

d. The probability is 0.95 that the mean age of all U.S. millionaires is between 54.3 years and 62.8 years.

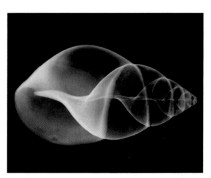

15. Sea Shell Morphology. In a 1903 paper, Abigail Camp Dimon discussed the effect of environment on the shape and form of two sea snail species, *Nassa obsoleta* and *Nassa trivittata*. One of the variables that Dimon considered was length of shell. She found the mean shell length of 461 randomly selected specimens of *N. trivittata* to be 11.9 mm. [SOURCE: "Quantitative Study of the Effect of Environment Upon the Forms of *Nassa obsoleta* and *Nassa trivittata* from Cold Spring Harbor, Long Island," *Biometrika*, Vol. 2, pp. 24–43.]

a. Assuming that $\sigma = 2.5$ mm, obtain a 90% confidence interval for the mean length, μ, of all *N. trivittata*.

b. Interpret your answer from part (a).

c. What properties should a normal probability plot of the data have for it to be permissible to apply the procedure that you used in part (a)?

16. Sea Shell Morphology. Refer to Problem 15.

a. Find the margin of error, E.

b. Explain the meaning of E as far as the accuracy of the estimate is concerned.

c. Determine the sample size required to have a margin of error of 0.1 mm and a 90% confidence level.

d. Find a 90% confidence interval for μ if a sample of the size determined in part (c) yields a mean of 12.0 mm.

17. For a *t*-curve with df $= 18$, obtain the *t*-value and illustrate your results graphically.

a. The *t*-value having area 0.025 to its right

b. $t_{0.05}$

c. The *t*-value having area 0.10 to its left

d. The two *t*-values that divide the area under the curve into a middle 0.99 area and two outside 0.005 areas

18. Children of Diabetic Mothers. A paper by Cho et al. in the May 2000 issue of *The Journal of Pediatrics* (Vol. 136(5), pp. 587–592) presented the results of research on various characteristics in children of diabetic mothers. Past studies have shown that maternal diabetes results in obesity, blood pressure, and glucose tolerance complications in the offspring. Following are the arterial blood pressures, in millimeters of mercury (mm Hg), for a random sample of 16 children of diabetic mothers.

81.6	84.1	87.6	82.8	82.0	88.9	86.7	96.4
84.6	104.9	90.8	94.0	69.4	78.9	75.2	91.0

a. A normal probability plot of these sample data shows no outliers and is roughly linear. Find a 95% confidence interval for the mean arterial blood pressure for all children of diabetic mothers. (*Note:* $\bar{x} = 86.2$ mm Hg; $s = 8.5$ mm Hg.)

b. Interpret your answer from part (a).

Using Technology

19. Millionaires. Use the technology of your choice to obtain the confidence interval required in Problem 13.

20. Children of Diabetic Mothers. Refer to Problem 18. Use the technology of your choice to

a. obtain and interpret a normal probability plot of the sample data.

b. find the required confidence interval.

21. Delaying Adulthood. The convict surgeonfish is a common tropical reef fish that has been found to delay metamorphosis into adult by extending its larval phase. This delay often leads to enhanced survivorship in the species by increasing the chances of finding suitable habitat. In the paper, "Delayed Metamorphosis of a Tropical Reef Fish (*Acanthurus triostegus*): A Field Experiment" (*Marine Ecology Progress Series*, Vol. 176, pp. 25–38), Mark I. McCormick published data that he obtained on the larval duration, in days, of 90 convict surgeonfish. The data are contained on the WeissStats CD.

a. Import the data into the technology of your choice.

b. Use the technology of your choice to obtain a normal probability plot, boxplot, and histogram of the data.

c. Is it reasonable to apply the *t*-interval procedure to the data? Explain your answer.

d. If you answered "yes" to part (c), obtain a 99% confidence interval for the mean larval duration of convict surgeonfish. Interpret your result.

22. Diamond Pricing. In a Singapore edition of *Business Times*, diamond pricing was explored. The price of a diamond is based on the diamond's weight, color, and clarity. A simple random sample of 18 one-half carat diamonds had the following prices, in dollars.

1676	1442	1995	1718	1826	2071	1947	1983	2146
1995	1876	2032	1988	2071	2234	2108	1941	2316

a. Use the technology of your choice to obtain a normal probability plot, a boxplot, a histogram, and a stem-and-leaf diagram for these data.

b. Is it reasonable to apply the one-sample *t*-interval procedure to these data in order to obtain a confidence interval for the mean price of all one-half carat diamonds? Explain your answer.

23. GRE Scores. Each year, thousands of college seniors take the Graduate Record Examination (GRE). The scores are transformed so they have a mean of 500 and a standard deviation of 100. Furthermore, the scores are known to be normally distributed.

a. Simulate 3000 samples of four GRE scores each.

b. Determine the sample mean and sample standard deviation of each of the 3000 samples.

c. For each of the 3000 samples, determine the observed value of the standardized version of $\bar{x}$.

d. Obtain a histogram of the 3000 observations in part (c).

e. Theoretically, what is the distribution of the standardized version of $\bar{x}$?

f. Compare your results from parts (d) and (e).

g. For each of the 3000 samples, determine the observed value of the studentized version of $\bar{x}$.

h. Obtain a histogram of the 3000 observations in part (g).

i. Theoretically, what is the distribution of the studentized version of $\bar{x}$?

j. Compare your results from parts (h) and (i).

k. Compare your histograms from parts (d) and (h). How and why do they differ?

StatExplore in MyMathLab
Analyzing Data Online

You can use StatExplore to perform all statistical analyses discussed in this chapter. To illustrate, we show how to perform a one-sample *t*-interval procedure.

EXAMPLE One-Sample t-Interval Procedure

Pickpocket Offenses Table 8.5 on page 382 displays the values lost, in dollars, of 25 randomly selected pickpocket offenses from a recent year. Use StatExplore to obtain a 95% confidence interval for the mean value lost, μ, of all the year's pickpocket offenses.

SOLUTION To obtain the required confidence interval by using StatExplore, we first store the sample data from Table 8.5 in a column named LOST. Then we proceed as follows:

1 Choose **Stat ➤ T statistics ➤ One sample**
2 Select the column LOST
3 Click **Next →**
4 Select the **Confidence Interval** option button
5 Click in **Level** text box and type <u>0.95</u>
6 Click **Calculate**

The resulting output is shown in Printout 8.4. From the printout, we see that the required 95% confidence interval is from 405.0764 (lower limit) to 621.5636 (upper limit). We can be 95% confident that the mean value lost, μ, of all pickpocket offenses for the year in question is somewhere between \$405.0764 and \$621.5636.

PRINTOUT 8.4
StatExplore output for one-sample *t*-interval

95% confidence interval results:
Parameter: mean of Variable

Variable	Sample Mean	Std. Err.	DF	L. Limit	U. Limit
LOST	513.32	52.446186	24	405.0764	621.5636

STATEXPLORE EXERCISES Solve the following problems by using StatExplore:

a. Obtain the confidence interval required in Exercise 8.63 on page 387.

b. Find the confidence interval required in Example 8.4 on page 365.

To access StatExplore, go to the student content area of your Weiss MyMathLab course.

Internet Project

Famous Data Sets

In 1797, British scientist Henry Cavendish performed a famous experiment to measure the density of the earth. In 1879, German–American physicist Albert Michelson performed an equally famous experiment to measure the velocity of light.

In both cases, the experiments were repeated many times, giving slightly different results each time. The variations were due, of course, to measurement errors and other factors beyond the control of the scientists. Thus these famous experiments were inherently statistical, and the fundamental problem in each case was to estimate the true value of the physical constant (the unknown parameter).

In this project, you are to study these experiments in some detail to understand their design and the possible sources of error. You are to compute a variety of confidence intervals for the physical constants. Because the true values of these constants are now known with great accuracy, you will be able to see how well your confidence intervals work.

URL for access to Internet Projects Page: www.aw-bc.com/weiss

Focusing on Data Analysis

The Focus Database

Recall from Chapter 1 (see page 35) that the Focus database contains information on the undergraduate students at the University of Wisconsin - Eau Claire (UWEC). Statistical analyses for this database should be done with the technology of your choice.

a. Obtain a simple random sample of size 50 of the high school percentiles (HSP) of UWEC undergraduate students.

b. Use your data from part (a) to determine a 95% confidence interval for the mean high school percentile of all UWEC undergraduate students. Interpret your result.

c. In practice, the (population) mean of the variable under consideration is unknown. However, in this case, we actually do have the population data. Obtain the mean high school percentile of all UWEC undergraduate students and note whether your confidence interval from part (b) contains that population mean. Must the confidence interval necessarily contain the population mean? Explain your answer.

d. Repeat parts (a)–(c) for the variables cumulative GPA, age, total earned credits, ACT English score, and ACT math score.

The following problems are intended as a class project.

e. Obtain 100 simple random samples of size 50 each of the ACT composite scores (COMP) of UWEC undergraduate students. (*Note:* If your class contains, say, 25 students, each student should obtain four simple random samples of size 50 each.)

f. For each sample in part (e), determine a 95% confidence interval for the mean ACT composite score of all UWEC undergraduate students.

g. Obtain the population mean ACT composite score of all UWEC undergraduate students.

h. Roughly how many of the 100 confidence intervals found in part (f) would you expect to contain the population mean ACT composite score? Explain your answer.

i. How many of the 100 confidence intervals found in part (f) actually contain the population mean ACT composite score? Compare this number to your answer in part (h).

The Chips Ahoy! 1,000 Chips Challenge

At the beginning of this chapter, on page 355, we presented data on the number of chocolate chips per bag for 42 bags of Chips Ahoy! cookies. These data were obtained by the students in an introductory statistics class at the United States Air Force Academy in response to the Chips Ahoy! 1,000 Chips Challenge sponsored by Nabisco, the makers of Chips Ahoy! cookies. Use the data collected by the students to answer the questions and conduct the analyses required in each part.

a. Obtain and interpret a point estimate for the mean number of chocolate chips per bag for all bags of Chips Ahoy! cookies. (*Note:* The sum of the data is 52,986.)
b. Construct and interpret a normal probability plot, boxplot, and histogram of the data.
c. Use the graphs in part (b) to identify outliers, if any.
d. Is it reasonable to use the one-sample *t*-interval procedure to obtain a confidence interval for the mean number of chocolate chips per bag for all bags of Chips Ahoy! cookies? Explain your answer.
e. Determine a 95% confidence interval for the mean number of chips per bag for all bags of Chips Ahoy! cookies, and interpret your result in words. (*Note:* $\bar{x} = 1261.6$; $s = 117.6$.)
f. Use the technology of your choice to solve parts (a), (b), and (e).

Internet Resources: Visit the Weiss Web site www.aw-bc.com/weiss for additional discussion, exercises, and resources related to this case study.

Biography

WILLIAM GOSSET: The "Student" in Student's t-Distribution

William Sealy Gosset was born in Canterbury, England, on June 13, 1876, the eldest son of Colonel Frederic Gosset and Agnes Sealy. He studied mathematics and chemistry at Winchester College and New College, Oxford, receiving a first-class degree in natural sciences in 1899.

After graduation Gosset began work with Arthur Guinness and Sons, a brewery in Dublin, Ireland. He saw the need for accurate statistical analyses of various brewing processes ranging from barley production to yeast fermentation, and pressed the firm to solicit mathematical advice. In 1906, the brewery sent him to work under Karl Pearson (see Biography in Chapter 13) at University College in London.

During the next few years, Gosset developed what has come to be known as Student's *t*-distribution. This distribution has proved to be fundamental in statistical analyses involving normal distributions. In particular, Student's *t*-distribution is used in performing inferences for a population mean when the population being sampled is (approximately) normally distributed and the population standard deviation is unknown. Although the statistical theory for large samples had been completed in the early 1800s, no small-sample theory was available before Gosset's work.

Because Guinness's brewery prohibited its employees from publishing any of their research, Gosset published his contributions to statistical theory under the pseudonym "Student"—thus the name "Student" in Student's *t*-distribution.

Gosset remained with Guinness his entire working life. In 1935, he moved to London to take charge of a new brewery. But his tenure there was short lived; he died in Beaconsfield, England, on October 16, 1937.

Hypothesis Tests for One Population Mean

GENERAL OBJECTIVES   In Chapter 8, we examined methods for obtaining confidence intervals for one population mean. We know that a confidence interval for a population mean, μ, is based on a sample mean, $\bar{x}$. Now we show how that statistic can be used to make decisions about hypothesized values of a population mean.

For example, suppose that we want to decide whether the mean prison sentence, μ, of all people imprisoned last year for drug offenses exceeds the 2000 mean of 75.5 months. To make that decision, we can take a random sample of people imprisoned last year for drug offenses, compute their sample mean sentence, $\bar{x}$, and then apply a statistical-inference technique called a *hypothesis test*.

In this chapter, we describe hypothesis tests for one population mean. In doing so, we consider three different procedures. The first two are called the *one-sample z-test* and the *one-sample t-test*, which are the hypothesis-test analogues of the one-sample z-interval and one-sample *t*-interval confidence-interval procedures, respectively, discussed in Chapter 8. The third is a nonparametric method called the *Wilcoxon signed-rank test*, which applies when the variable under consideration has a symmetric distribution.

We also examine two different approaches to hypothesis testing—namely, the critical-value approach and the *P*-value approach.

Sex and Sense of Direction

Many of you have been there, a classic scene: mom yelling at dad to turn left while dad decides to do just the opposite. Well, who made the right call? And, more generally, who has a better sense of direction, women or men?

Dr. Jeanne Sholl et al. considered these and other questions in a recent paper entitled "The Relation of Sex and Sense of Direction to Spatial Orientation in an Unfamiliar Environment" (*Journal of Environmental Psychology*, Vol. 20, pp. 17–28). These researchers defined sense of direction as "…the knowledge of the location and orientation of the body with respect to large stationary objects, or landmarks, attached to the surface of the earth."

In their study, the spatial orientation skills of 30 male students and 30 female students from Boston College were challenged in Houghton Garden Park, a wooded park near the BC campus in Newton, Massachusetts. Before driving to the park, the participants were asked to rate their own sense of direction as either good or poor.

In the park, students were instructed to point to predesignated target landmarks and also to the direction of south. Pointing was carried out by students moving a pointer attached to a 360° protractor; the angle of the pointing response was then recorded to the nearest degree. For the female students who had rated their sense of direction to be good, the following table provides the absolute pointing errors (in degrees) when they attempted to point south.

14	122	128	109	12
91	8	78	31	36
27	68	20	69	18

Based on these data, can you conclude that, in general, women who consider themselves to have a good sense of direction really do better, on average, than they would by just randomly guessing at the direction of south? To answer that question, you need to conduct a hypothesis test, which you will do after you study hypothesis testing in this chapter.

9.1   The Nature of Hypothesis Testing

We often use inferential statistics to make decisions or judgments about the value of a parameter, such as a population mean. For example, we might need to decide whether the mean weight, μ, of all bags of pretzels packaged by a particular company differs from the advertised weight of 454 grams (g); or, we might want to determine whether the mean age, μ, of all cars in use has increased from the 1995 mean of 8.5 years.

One of the most commonly used methods for making such decisions or judgments is to perform a **hypothesis test.** A **hypothesis** is a statement that something is true. For example, the statement "the mean weight of all bags of pretzels packaged differs from the advertised weight of 454 g" is a hypothesis.

Typically, a hypothesis test involves two hypotheses: the **null hypothesis** and the **alternative hypothesis** (or **research hypothesis**), which we define as follows.

DEFINITION 9.1

What *Does it Mean?*

Originally, the word *null* in *null hypothesis* stood for "no difference" or "the difference is null." Over the years, however, *null hypothesis* has come to mean simply a hypothesis to be tested. The problem in a hypothesis test is to decide whether the null hypothesis should be rejected in favor of the alternative hypothesis.

Null and Alternative Hypotheses

Null hypothesis: A hypothesis to be tested. We use the symbol H_0 to represent the null hypothesis.

Alternative hypothesis: A hypothesis to be considered as an alternative to the null hypothesis. We use the symbol H_a to represent the alternative hypothesis.

For instance, in the pretzel packaging example, the null hypothesis might be "the mean weight of all bags of pretzels packaged equals the advertised weight of 454 g," and the alternative hypothesis might be "the mean weight of all bags of pretzels packaged differs from the advertised weight of 454 g."

Choosing the Hypotheses

The first step in setting up a hypothesis test is to decide on the null hypothesis and the alternative hypothesis. The following are some guidelines for choosing these two hypotheses. Although the guidelines refer specifically to hypothesis tests for one population mean, μ, they apply to any hypothesis test concerning one parameter.

Null Hypothesis

In this book, the null hypothesis for a hypothesis test concerning a population mean, μ, always specifies a single value for that parameter. Hence the null hypothesis always takes the form $\mu = \mu_0$, where μ_0 is some number. In other words, an equals sign (=) should appear in the null hypothesis. We can therefore express the null hypothesis concisely as

$$H_0: \mu = \mu_0.$$

Alternative Hypothesis

The choice of the alternative hypothesis depends on and should reflect the purpose of the hypothesis test. Three choices are possible for the alternative hypothesis.

- If the primary concern is deciding whether a population mean, μ, is *different from* a specified value μ_0, the alternative hypothesis should be $\mu \neq \mu_0$. In other words, a does-not-equal sign ($\neq$) should appear in the alternative hypothesis. We express such an alternative hypothesis as

$$H_a: \mu \neq \mu_0.$$

A hypothesis test whose alternative hypothesis has this form is called a **two-tailed test.**

- If the primary concern is deciding whether a population mean, μ, is *less than* a specified value μ_0, the alternative hypothesis should be $\mu < \mu_0$. In other words, a less-than sign ($<$) should appear in the alternative hypothesis. We express such an alternative hypothesis as

$$H_a: \mu < \mu_0.$$

A hypothesis test whose alternative hypothesis has this form is called a **left-tailed test.**

- If the primary concern is deciding whether a population mean, μ, is *greater than* a specified value μ_0, the alternative hypothesis should be $\mu > \mu_0$. In other words, a greater-than sign ($>$) should appear in the alternative hypothesis. We express such an alternative hypothesis as

$$H_a: \mu > \mu_0.$$

A hypothesis test whose alternative hypothesis has this form is called a **right-tailed test.**

A hypothesis test is called a **one-tailed test** if it is either left tailed or right tailed, that is, if it is not two tailed. In Section 9.2, we explain the relevance of the term *tailed.* But for now let's consider Examples 9.1–9.3, which illustrate the preceding discussion.

Example 9.1   Choosing the Null and Alternative Hypotheses

Quality Assurance   A snack-food company produces a 454 g bag of pretzels. Although the actual net weights deviate slightly from 454 g and vary from one bag to another, the company insists that the mean net weight of the bags be kept at 454 g. Indeed, if the mean net weight is less than 454 g, the company will be shortchanging its customers; and if the mean net weight exceeds 454 g, the company will be unnecessarily overfilling the bags.

As part of its program, the quality assurance department periodically performs a hypothesis test to decide whether the packaging machine is working properly, that is, to decide whether the mean net weight of all bags packaged is 454 g.

a. Determine the null hypothesis for the hypothesis test.

b. Determine the alternative hypothesis for the hypothesis test.

c. Classify the hypothesis test as two tailed, left tailed, or right tailed.

Solution   Let μ denote the mean net weight of all bags packaged.

a. The null hypothesis for this hypothesis test is that the packaging machine is working properly, that is, that the mean net weight, μ, of all bags packaged *equals* 454 g. In symbols, H_0: $\mu = 454$ g.

b. The alternative hypothesis for this hypothesis test is that the packaging machine is not working properly, that is, that the mean net weight, μ, of all bags packaged is *different from* 454 g. In symbols, H_a: $\mu \neq 454$ g.

c. This hypothesis test is two tailed because a does-not-equal sign ($\neq$) appears in the alternative hypothesis.

                                                                              ▬■

Example 9.2   Choosing the Null and Alternative Hypotheses

Prices of History Books   The R. R. Bowker Company of New York collects information on the retail prices of books and publishes the data in *The Bowker Annual Library and Book Trade Almanac*. In 2000, the mean retail price of history books was $51.46. Suppose that we want to perform a hypothesis test to decide whether this year's mean retail price of history books has increased from the 2000 mean.

a. Determine the null hypothesis for the hypothesis test.

b. Determine the alternative hypothesis for the hypothesis test.

c. Classify the hypothesis test as two tailed, left tailed, or right tailed.

Solution   Let μ denote this year's mean retail price of history books.

a. The null hypothesis for this hypothesis test is that this year's mean retail price of history books *equals* the 2000 mean of $51.46; that is, H_0: $\mu = \$51.46$.

b. The alternative hypothesis for this hypothesis test is that this year's mean retail price of history books is *greater than* $51.46; that is, H_a: $\mu > \$51.46$.

c. This hypothesis test is right tailed because a greater-than sign ($>$) appears in the alternative hypothesis.

                                                                              ▬■

Example 9.3   Choosing the Null and Alternative Hypotheses

Poverty and Calcium   Calcium is the most abundant mineral in the body and also one of the most important. It works with phosphorus to build and maintain bones and teeth. According to the Food and Nutrition Board of the National Academy of Sciences, the recommended daily allowance (RDA) of calcium for adults is 800 milligrams (mg). Suppose that we want to perform a hypothesis test to decide whether the average person with an income below the poverty level gets less than the RDA of 800 mg.

a. Determine the null hypothesis for the hypothesis test.

b. Determine the alternative hypothesis for the hypothesis test.

c. Classify the hypothesis test as two tailed, left tailed, or right tailed.

Solution Let μ denote the mean calcium intake (per day) of all people with incomes below the poverty level.

a. The null hypothesis for this hypothesis test is that the mean calcium intake of all people with incomes below the poverty level *equals* the RDA of 800 mg per day; that is, H_0: $\mu = 800$ mg.

b. The alternative hypothesis for this hypothesis test is that the mean calcium intake of all people with incomes below the poverty level is *less than* the RDA of 800 mg per day; that is, H_a: $\mu < 800$ mg.

c. This hypothesis test is left tailed because a less-than sign ($<$) appears in the alternative hypothesis.

The Logic of Hypothesis Testing

After we have chosen appropriate null and alternative hypotheses for a hypothesis test, the next question is: How do we decide which of the two hypotheses is true; that is, how do we decide whether to reject the null hypothesis in favor of the alternative hypothesis? Very roughly, the procedure for deciding is as follows.

Basic Logic of Hypothesis Testing

Take a random sample from the population. If the sample data are consistent with the null hypothesis, do not reject the null hypothesis; if the sample data are inconsistent with the null hypothesis (in the direction of the alternative hypothesis), reject the null hypothesis and conclude that the alternative hypothesis is true.

In practice, of course, we must have a precise criterion for deciding whether to reject the null hypothesis. Example 9.4 illustrates how such a criterion can be devised for a two-tailed hypothesis test about a population mean. The example also introduces the logic and some of the terminology of hypothesis testing. Later in this chapter we provide general step-by-step procedures for performing hypothesis tests.

▌ **The Logic of Hypothesis Testing**

Quality Assurance   A company that produces snack foods uses a machine to package 454 g bags of pretzels. We assume that the net weights are normally distributed and that the population standard deviation of all such weights is 7.8 g.[†] A simple random sample of 25 bags of pretzels has the net weights, in grams, displayed in Table 9.1. Do the data provide sufficient evidence to conclude that the packaging machine is not working properly? We use the following steps to answer the question.

TABLE 9.1

Weights, in grams, of 25 randomly selected bags of pretzels

465	456	438	454	447
449	442	449	446	447
468	433	454	463	450
446	447	456	452	444
447	456	456	435	450

a. State the null and alternative hypotheses for the hypothesis test.

b. Discuss the logic behind carrying out the hypothesis test.

c. Identify the distribution of the variable $\bar{x}$, that is, the sampling distribution of the sample mean for samples of size 25.

d. Obtain a precise criterion for deciding whether to reject the null hypothesis in favor of the alternative hypothesis.

e. Apply the criterion in part (d) to the sample data and state the conclusion.

Solution   Let μ denote the mean net weight of all bags packaged.

a. The null and alternative hypotheses for the hypothesis test, as stated in Example 9.1, are

H_0: $\mu = 454$ g (the packaging machine is working properly)

H_a: $\mu \neq 454$ g (the packaging machine is not working properly).

b. Basically, the logic behind carrying out the hypothesis test is this: If the null hypothesis is true, that is, if $\mu = 454$ g, the mean weight, $\bar{x}$, of the sample of 25 bags of pretzels should approximately equal 454 g. We say "approximately equal" because we cannot expect a sample mean to equal exactly the population mean; some sampling error is to be anticipated. However, if the sample mean weight differs "too much" from 454 g, we would be inclined to reject the null hypothesis and conclude that the alternative hypothesis is true. As we show in part (d), we can use our knowledge of the sampling distribution of the sample mean to decide how much difference is "too much."

c. Because $n = 25$, $\sigma = 7.8$, and the weights are normally distributed, Key Fact 7.4 on page 342 implies that

- $\mu_{\bar{x}} = \mu$ (which we don't know),
- $\sigma_{\bar{x}} = \sigma/\sqrt{n} = 7.8/\sqrt{25} = 1.56$, and
- $\bar{x}$ is normally distributed.

In other words, for samples of size 25, the variable $\bar{x}$ is normally distributed with mean μ and standard deviation 1.56 g.

d. The "95.44" part of the 68.26-95.44-99.74 rule states that, for a normally distributed variable, 95.44% of all possible observations lie within two standard

---

[†]We might know the population standard deviation from previous research or from a preliminary study of net weights. In Section 9.6, we consider the more usual case of an unknown σ.

deviations to either side of the mean. Applying this part of the rule to the variable $\bar{x}$ and referring to part (c), we see that 95.44% of all samples of 25 bags of pretzels have mean weights within $2 \cdot 1.56 = 3.12$ g of μ. Or, equivalently, only 4.56% of all samples of 25 bags of pretzels have mean weights that are not within 3.12 g of μ, as illustrated in Fig. 9.1.

Thus, if the mean weight, $\bar{x}$, of the 25 bags of pretzels sampled is not within two standard deviations (3.12 g) of 454 g, we have evidence against the null hypothesis. Why? Because observing such a sample mean would occur by chance only 4.56% of the time if the null hypothesis, $\mu = 454$ g, is true.

In summary, then, we have obtained the following precise criterion for deciding whether to reject the null hypothesis. This criterion is portrayed graphically in Fig. 9.2(a).

If the mean weight, $\bar{x}$, of the 25 bags of pretzels sampled is more than two standard deviations (3.12 g) from 454 g, reject the null hypothesis, $\mu = 454$ g, and conclude that the alternative hypothesis, $\mu \neq 454$ g, is true. Otherwise, do not reject the null hypothesis.

FIGURE 9.1

95.44% of all samples of 25 bags of pretzels have mean weights within two standard deviations (3.12 g) of μ

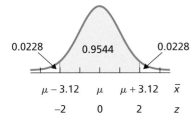

FIGURE 9.2

(a) Criterion for deciding whether to reject the null hypothesis; (b) normal curve associated with $\bar{x}$ if the null hypothesis is true, superimposed on the decision criterion

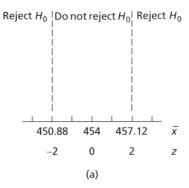

If the null hypothesis is true, the normal curve associated with $\bar{x}$ is the one with parameters 454 and 1.56; that normal curve is superimposed on Fig. 9.2(a) in Fig. 9.2(b).

FIGURE 9.3

Graph showing the number of standard deviations that the sample mean of 450 g is from the null-hypothesis population mean of 454 g

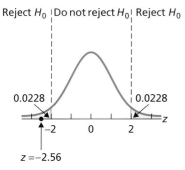

e. The mean weight, $\bar{x}$, of the sample of 25 bags of pretzels whose weights are given in Table 9.1 is 450 g. Therefore,

$$z = \frac{\bar{x} - 454}{1.56} = \frac{450 - 454}{1.56} = -2.56.$$

That is, the sample mean of 450 g is 2.56 standard deviations below the null-hypothesis population mean of 454 g, as shown in Fig. 9.3.

Because the mean weight of the 25 bags of pretzels sampled is more than two standard deviations from 454 g, we reject the null hypothesis, $\mu = 454$ g, and conclude that the alternative hypothesis, $\mu \neq 454$ g, is true.

INTERPRETATION The data provide sufficient evidence to conclude that the packaging machine is not working properly.

Example 9.4 contains all the elements of a hypothesis test, including the necessary theory. But don't worry too much about the details at this point. What you should understand now is (1) how to choose the null and alternative hypotheses for a hypothesis test, (2) how to classify a hypothesis test as two tailed, left tailed, or right tailed, and (3) the logic behind performing a hypothesis test.

EXERCISES 9.1

Statistical Concepts and Skills

9.1 Explain the meaning of the term *hypothesis* as used in inferential statistics.

9.2 What role does the decision criterion play in a hypothesis test?

9.3 Suppose that you want to perform a hypothesis test for a population mean μ.
a. Express the null hypothesis both in words and in symbolic form.
b. Express each of the three possible alternative hypotheses in words and in symbolic form.

9.4 Suppose that you are considering a hypothesis test for a population mean, μ. In each part, express the alternative hypothesis symbolically and identify the hypothesis test as two tailed, left tailed, or right tailed.
a. You want to decide whether the population mean is different from a specified value μ_0.
b. You want to decide whether the population mean is less than a specified value μ_0.
c. You want to decide whether the population mean is greater than a specified value μ_0.

In Exercises 9.5–9.10, hypothesis tests are proposed. For each hypothesis test,
a. determine the null hypothesis.
b. determine the alternative hypothesis.
c. classify the hypothesis test as two tailed, left tailed, or right tailed.

9.5 Toxic Mushrooms? Cadmium, a heavy metal, is toxic to animals. Mushrooms, however, are able to absorb and accumulate cadmium at high concentrations. The Czech and Slovak governments have set a safety limit for cadmium in dry vegetables at 0.5 part per million (ppm). M. Melgar et al. measured the cadmium levels in a random sample of the edible mushroom *Boletus pinicola* and published the results in the *Journal of Environmental Science and Health* (Vol. B33(4), pp. 439–455). A hypothesis test is to be performed to decide whether the mean cadmium level in *Boletus pinicola* mushrooms is greater than the government's recommended limit.

9.6 Serving Time. According to the Bureau of Crime Statistics and Research of Australia, as reported by the website www.agd.nsw.gov.au, the mean length of imprisonment for motor-vehicle theft offenders in Australia is 16.7 months. You want to perform a hypothesis test to decide whether the mean length of imprisonment for motor-vehicle theft offenders in Sydney differs from the national mean in Australia.

9.7 Cell Phones. The number of cell phone users has increased dramatically since 1987. This increase in cell phone use could bring a reduction in customer phone bills because of heightened competition. According to the *Semiannual Wireless Survey*, published by the Cellular Telecommunications & Internet Association, the mean local monthly bill for cell phone users in the United States was $47.37 in 2001. A hypothesis test is to be performed to determine whether last year's mean local monthly bill for cell phone users has decreased from the 2001 mean of $47.37.

9.8 Iron Deficiency? The Food and Nutrition Board of the National Academy of Sciences states that the recommended daily allowance (RDA) of iron for adult females under the age of 51 is 18 mg. A hypothesis test is to be performed to decide whether adult females under the age of 51 are, on average, getting less than the RDA of 18 mg of iron.

9.9 Body Temperature. A study by researchers at the University of Maryland addressed the question of whether the mean body temperature of humans is 98.6°F. The results of the study by P. Mackowiak, S. Wasserman, and M. Levine appeared in the article "A Critical Appraisal of 98.6°F, the Upper Limit of the Normal Body Temperature, and Other Legacies of Carl Reinhold August Wunderlich" (*Journal of the American Medical Association*, Vol. 268, pp. 1578–1580). Among other data, the researchers obtained the body temperatures of 93 healthy humans. Suppose that you want to use that data to decide whether the mean body temperature of healthy humans differs from 98.6°F.

9.10 Worker Fatigue. A study by M. Chen et al. titled "Heat Stress Evaluation and Worker Fatigue in a Steel Plant" (*American Industrial Hygiene Association*, Vol. 64, pp. 352–359) assessed fatigue in steel-plant workers due to heat stress. Among other things, the researchers monitored the heart rates of a random sample of 29 casting workers. A hypothesis test is to be conducted to decide whether the mean post-work heart rate of casting workers exceeds the normal resting heart rate of 72 beats per minute (bpm).

Extending the Concepts and Skills

9.11 Energy Use. The U.S. Energy Information Administration compiles data on energy consumption and publishes its findings in *Residential Energy Consumption Survey: Consumption and Expenditures*. One year, the mean energy consumed per U.S. household was 103.6 million British thermal units (BTU). For that same year, 20 randomly selected households in the West had the following energy consumptions, in millions of BTU.

104	84	72	95	69
80	78	74	76	81
82	61	94	65	100
70	65	83	76	84

Do the data provide sufficient evidence to conclude that the mean energy consumed by western households differed from that of all U.S. households? Assume that the

standard deviation of energy consumptions of all western households was 15 million BTU. Use the following steps to answer the question. You may want to refer to Example 9.4, which begins on page 402.
a. State the null and alternative hypotheses.
b. Discuss the logic of conducting the hypothesis test.
c. Identify the distribution of the variable $\bar{x}$, that is, the sampling distribution of the mean for samples of size 20.
d. Obtain a precise criterion for deciding whether to reject the null hypothesis in favor of the alternative hypothesis.
e. Apply the criterion in part (d) to the sample data and state your conclusion.

9.12 Quality Assurance. Refer to Example 9.4, which begins on page 402. Suppose that, in the solution to part (d), we use the "99.74" part of the 68.26-95.44-99.74 rule.
a. Determine the resulting decision criterion and portray it graphically, using a graph similar to the one shown in Fig. 9.2(a) on page 403.
b. Construct a graph similar to the one shown in Fig. 9.2(b) that illustrates the implications of the decision criterion in part (a) if in fact the null hypothesis is true.
c. Apply the criterion from part (a) to the sample data in Table 9.1 on page 402 and state your conclusion.

9.13 Quality Assurance. Refer to Example 9.4, which begins on page 402. In that example, we rejected the null hypothesis that the mean net weight of all bags packaged is 454 g in favor of the alternative hypothesis that the mean net weight of all bags packaged differs from 454 g. If in fact the null hypothesis is true, what is the chance of incorrectly rejecting it using a sample of 25 bags of pretzels? (*Hint:* Refer to Fig. 9.2(b).)

9.2 Terms, Errors, and Hypotheses

To explain fully the nature of hypothesis testing, we need some additional terms and concepts. In this section, we define several more terms used in hypothesis testing, discuss the two types of errors that can occur in a hypothesis test, and interpret the possible conclusions for a hypothesis test.

Some Additional Terminology

To introduce some additional terminology used in hypothesis testing, we refer to the pretzel packaging hypothesis test of Example 9.4 on page 402. Recall that the null and alternative hypotheses for that hypothesis test are

$$H_0: \mu = 454 \text{ g (the packaging machine is working properly)}$$
$$H_a: \mu \neq 454 \text{ g (the packaging machine is not working properly)},$$

where μ is the mean net weight of all bags of pretzels packaged.

FIGURE 9.4

Criterion used to decide whether to reject the null hypothesis

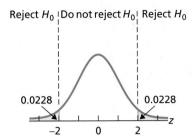

FIGURE 9.5

Rejection region, nonrejection region, and critical values for the pretzel packaging illustration

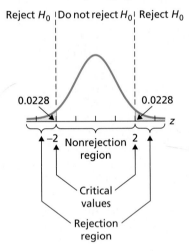

As a basis for deciding whether to reject the null hypothesis, in part (e) of Example 9.4 we utilized the variable

$$z = \frac{\bar{x} - \mu_0}{\sigma/\sqrt{n}} = \frac{\bar{x} - 454}{1.56},$$

which tells us how many standard deviations the sample mean is from the null hypothesis population mean of 454 g. That variable is called the **test statistic** for the hypothesis test.

Figure 9.3 includes a graph portraying the criterion used to decide whether the null hypothesis should be rejected. For ease of reference, we repeat that graph in Fig. 9.4.

The set of values for the test statistic that leads us to reject the null hypothesis is called the **rejection region.** In this case, the rejection region consists of all z-scores that lie either to the left of -2 or to the right of 2—that part of the horizontal axis under the shaded areas in Fig. 9.4.

The set of values for the test statistic that leads us not to reject the null hypothesis is called the **nonrejection region,** or **acceptance region.** In this case, the nonrejection region consists of all z-scores that lie between -2 and 2—that part of the horizontal axis under the unshaded area in Fig. 9.4.

The values of the test statistic that separate the rejection and nonrejection regions are called the **critical values.** In this case, the critical values are $z = \pm 2$, as shown in Fig. 9.4. We summarize the preceding discussion in Fig. 9.5.

The terminology introduced so far in this section, and defined formally as follows, applies to any hypothesis test, not just to hypothesis tests for a population mean.

DEFINITION 9.2

What
Does it Mean?

If the value of the test statistic falls in the rejection region, reject the null hypothesis; otherwise, do not reject the null hypothesis.

Test Statistic, Rejection Region, Nonrejection Region, and Critical Values

Test statistic: The statistic used as a basis for deciding whether the null hypothesis should be rejected.

Rejection region: The set of values for the test statistic that leads to rejection of the null hypothesis.

Nonrejection region: The set of values for the test statistic that leads to non-rejection of the null hypothesis.[†]

Critical values: The values of the test statistic that separate the rejection and nonrejection regions. A critical value is considered part of the rejection region.

For a two-tailed test, as in the pretzel packaging illustration, the null hypothesis is rejected when the test statistic is either too small or too large. Thus

---

[†]The reason that we prefer the term *nonrejection* to the term *acceptance* is explained in detail on page 410.

the rejection region for such a test consists of two parts: one on the left and one on the right, as shown in Fig. 9.5 and Fig. 9.6(a).

FIGURE 9.6
Graphical display of rejection regions for two-tailed, left-tailed, and right-tailed tests

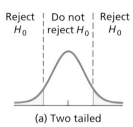

(a) Two tailed

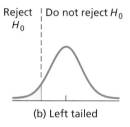

(b) Left tailed

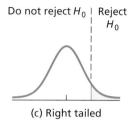

(c) Right tailed

For a left-tailed test, as in Example 9.3 on page 400 (the calcium intake illustration), the null hypothesis is rejected only when the test statistic is too small. Thus the rejection region for such a test consists of only one part, and that part is on the left, as shown in Fig. 9.6(b).

For a right-tailed test, as in Example 9.2 on page 400 (the history book illustration), the null hypothesis is rejected only when the test statistic is too large. Thus the rejection region for such a test consists of only one part, and that part is on the right, as shown in Fig. 9.6(c).

Table 9.2 and Fig. 9.6 summarize our discussion. Figure 9.6 shows why the term *tailed* is used: The rejection region is in both tails for a two-tailed test, in the left tail for a left-tailed test, and in the right tail for a right-tailed test.

TABLE 9.2
Rejection regions for two-tailed, left-tailed, and right-tailed tests

	Two-tailed test	**Left-tailed test**	**Right-tailed test**
Sign in H_a	$\neq$	$<$	$>$
Rejection region	Both sides	Left side	Right side

Type I and Type II Errors

Whenever we conduct a hypothesis test, any decision that we make may be incorrect. The reason is that partial information, obtained from a sample, is used to draw conclusions about the entire population.

There are two types of incorrect decisions—**Type I error** (rejection of a true null hypothesis) and **Type II error** (nonrejection of a false null hypothesis), as indicated in Table 9.3 and Definition 9.3.

TABLE 9.3
Correct and incorrect decisions for a hypothesis test

		H_0 is:	
		True	False
Decision:	Do not reject H_0	Correct decision	Type II error
	Reject H_0	Type I error	Correct decision

DEFINITION 9.3

Type I and Type II Errors

Type I error: Rejecting the null hypothesis when it is in fact true.

Type II error: Not rejecting the null hypothesis when it is in fact false.

Example 9.5 illustrates Type I and Type II errors and the consequences of these two types of errors.

Example 9.5 **Type I and Type II Errors**

Quality Assurance Consider once again the pretzel packaging hypothesis test. The null and alternative hypotheses are

H_0: $\mu = 454$ g (the packaging machine is working properly)

H_a: $\mu \neq 454$ g (the packaging machine is not working properly),

where μ is the mean net weight of all bags of pretzels packaged. Explain what each of the following would mean.

a. Type I error   **b.** Type II error   **c.** Correct decision

Recall from Example 9.4 that the results of sampling 25 bags of pretzels led to rejection of the null hypothesis, $\mu = 454$ g, that is, to the conclusion that $\mu \neq 454$ g. Classify that conclusion by error type or as a correct decision if

d. the mean net weight, μ, is in fact 454 g.

e. the mean net weight, μ, is in fact not 454 g.

Solution

a. A Type I error occurs when a true null hypothesis is rejected. In this case, a Type I error would occur if in fact $\mu = 454$ g but the results of the sampling lead to the conclusion that $\mu \neq 454$ g.

> **INTERPRETATION** A Type I error occurs if we conclude that the packaging machine is not working properly when in fact it is working properly.

b. A Type II error occurs when a false null hypothesis is not rejected. In this case, a Type II error would occur if in fact $\mu \neq 454$ g but the results of the sampling fail to lead to that conclusion.

> **INTERPRETATION** A Type II error occurs if we fail to conclude that the packaging machine is not working properly when in fact it is not working properly.

c. A correct decision can occur in either of two ways.

- A true null hypothesis is not rejected. That would happen if in fact $\mu = 454$ g and the results of the sampling do not lead to the rejection of that fact.

- A false null hypothesis is rejected. That would happen if in fact $\mu \neq 454$ g and the results of the sampling lead to that conclusion.

> **INTERPRETATION** A correct decision occurs if either we fail to conclude that the packaging machine is not working properly when in fact it is working properly, or we conclude that the packaging machine is not working properly when in fact it is not working properly.

d. If in fact $\mu = 454$ g, the null hypothesis is true. Consequently, by rejecting the null hypothesis, $\mu = 454$ g, a Type I error has been made—a true null hypothesis has been rejected.

e. If in fact $\mu \neq 454$ g, the null hypothesis is false. Consequently, by rejecting the null hypothesis, $\mu = 454$ g, a correct decision has been made—a false null hypothesis has been rejected.

———■

Probabilities of Type I and Type II Errors

Part of evaluating the effectiveness of a hypothesis test involves an analysis of the chances of making an incorrect decision. A Type I error occurs if the test statistic falls in the rejection region when in fact the null hypothesis is true. The probability of that happening, the **Type I error probability,** commonly called the **significance level** of the hypothesis test, is denoted α (the Greek letter alpha).

Figure 9.4 on page 406 shows the rejection and nonrejection regions for the pretzel packaging hypothesis test in Example 9.4. It also shows the normal curve for the test statistic

$$z = \frac{\bar{x} - 454}{1.56},$$

under the assumption that the null hypothesis, $\mu = 454$ g, is true. Figure 9.4 reveals that, if the null hypothesis is true, the probability is $0.0228 + 0.0228$, or 0.0456, that the test statistic, z, will fall in the rejection region. Thus, for this hypothesis test, the significance level is 0.0456; in symbols, $\alpha = 0.0456$.

> **INTERPRETATION** There is only a 4.56% chance of concluding that the packaging machine is not working properly when in fact it is working properly.

DEFINITION 9.4

Significance Level

The probability of making a Type I error, that is, of rejecting a true null hypothesis, is called the *significance level, α,* of a hypothesis test.

A Type II error occurs if the test statistic falls in the nonrejection region when in fact the null hypothesis is false. The probability of that happening, the **Type II error probability,** is denoted β (the Greek letter beta). It depends on the true value of μ. Calculation of Type II error probabilities is examined briefly in Exercise 9.29 and in detail in Section 9.4.

Ideally, both Type I and Type II errors should have small probabilities. Then the chance of making an incorrect decision would be small, regardless of whether the null hypothesis is true or false. As we demonstrate in Section 9.3, we can design a hypothesis test to have any specified significance level. So, for instance, if not rejecting a true null hypothesis is important, we should specify a small value for α. However, in making our choice for α, we must keep Key Fact 9.1 in mind.

KEY FACT 9.1

Relation Between Type I and Type II Error Probabilities

For a fixed sample size, the smaller we specify the significance level, α, the larger will be the probability, β, of not rejecting a false null hypothesis.

Consequently, we must always assess the risks involved in committing both types of errors and use that assessment as a method for balancing the Type I and Type II error probabilities.

Possible Conclusions for a Hypothesis Test

The significance level, α, is the probability of making a Type I error, that is, of rejecting a true null hypothesis. Therefore, if the hypothesis test is conducted at a small significance level (e.g., $\alpha = 0.05$), the chance of rejecting a true null hypothesis will be small. In this text, we generally specify a small significance level, so we can make the following statement concerning a hypothesis test: If we do reject the null hypothesis in a hypothesis test, we can be reasonably confident that the null hypothesis is false and therefore that the alternative hypothesis is true.

However, we usually do not know the probability, β, of making a Type II error, that is, of not rejecting a false null hypothesis. Consequently, if we do not reject the null hypothesis in a hypothesis test, we simply reserve judgment about which hypothesis is true. In other words, if we do not reject the null hypothesis, we conclude only that the data did not provide sufficient evidence to support the alternative hypothesis; we do not conclude that the data provided sufficient evidence to support the null hypothesis. Key Fact 9.2 summarizes this discussion.

KEY FACT 9.2

Possible Conclusions for a Hypothesis Test

Suppose that a hypothesis test is conducted at a small significance level.

- If the null hypothesis is rejected, we conclude that the alternative hypothesis is true.
- If the null hypothesis is not rejected, we conclude that the data do not provide sufficient evidence to support the alternative hypothesis.

When the null hypothesis is rejected in a hypothesis test performed at the significance level α, we frequently express that fact with the phrase "the test results are **statistically significant** at the α level." Similarly, when the null hypothesis is not rejected in a hypothesis test performed at the significance level α, we often express that fact with the phrase "the test results are **not statistically significant** at the α level."

EXERCISES 9.2

Statistical Concepts and Skills

9.14 Decide whether each statement is true or false. Explain your answers.

a. If it is important not to reject a true null hypothesis, the hypothesis test should be performed at a small significance level.

b. For a fixed sample size, decreasing the significance level of a hypothesis test results in an increase in the probability of making a Type II error.

9.15 Identify the two types of incorrect decisions in a hypothesis test. For each incorrect decision, what symbol is used to represent the probability of making that type of error?

Exercises **9.16–9.18** *contain graphs portraying the decision criterion for a hypothesis test for a population mean,* μ. *The null hypothesis for each test is* H_0: $\mu = \mu_0$; *the test statistic is*

$$z = \frac{\bar{x} - \mu_0}{\sigma/\sqrt{n}}.$$

The curve in each graph is the normal curve for the test statistic under the assumption that the null hypothesis is true. For each exercise, determine the
a. rejection region.   *b. nonrejection region.*
c. critical value(s).   *d. significance level.*
e. Construct a graph similar to that in Fig. 9.5 on page 406 that depicts your results from parts (a)–(d).
f. Identify the hypothesis test as two tailed, left tailed, or right tailed.

9.16 A graphical display of the decision criterion is:

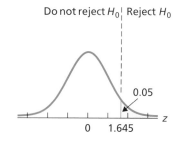

9.17 A graphical display of the decision criterion is:

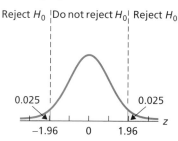

9.18 A graphical display of the decision criterion is:

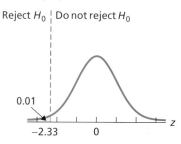

9.19 Toxic Mushrooms? The null and alternative hypotheses obtained in Exercise 9.5 on page 404 are

$$H_0: \mu = 0.5 \text{ ppm}$$
$$H_a: \mu > 0.5 \text{ ppm},$$

where μ is the mean cadmium level in *Boletus pinicola* mushrooms. Explain what each outcome would mean.
a. Type I error   **b.** Type II error   **c.** Correct decision

Now suppose that the results of carrying out the hypothesis test lead to nonrejection of the null hypothesis. Classify that conclusion by error type or as a correct decision if in fact the mean cadmium level in *Boletus pinicola* mushrooms
d. equals 0.5 ppm.   **e.** is greater than 0.5 ppm.

9.20 Serving Time. The null and alternative hypotheses obtained in Exercise 9.6 on page 404 are

$$H_0: \mu = 16.7 \text{ months}$$
$$H_a: \mu \neq 16.7 \text{ months},$$

where μ is the mean length of imprisonment for motor-vehicle theft offenders in Sydney, Australia. Explain what each outcome would mean.
a. Type I error   **b.** Type II error   **c.** Correct decision
Now suppose that the results of carrying out the hypothesis test lead to nonrejection of the null hypothesis. Classify that conclusion by error type or as a correct decision if in fact the mean length of imprisonment for motor-vehicle theft offenders in Sydney
d. equals 16.7 months.
e. does not equal 16.7 months.

9.21 Cell Phones. The null and alternative hypotheses obtained in Exercise 9.7 on page 404 are

$$H_0: \mu = \$47.37$$
$$H_a: \mu < \$47.37,$$

where μ is last year's mean local monthly bill for cell phone users. Explain what each outcome would mean.
a. Type I error   **b.** Type II error   **c.** Correct decision
Now suppose that the results of performing the hypothesis test lead to nonrejection of the null hypothesis. Classify that conclusion by error type or as a correct decision if in fact last year's mean local monthly bill for cell phone users
d. equals the 2001 mean of \$47.37.
e. is less than the 2001 mean of \$47.37.

9.22 Iron Deficiency? The null and alternative hypotheses obtained in Exercise 9.8 on page 404 are

$$H_0: \mu = 18 \text{ mg}$$
$$H_a: \mu < 18 \text{ mg},$$

where μ is the mean iron intake (per day) of all adult females under the age of 51. Explain what each outcome would mean.
a. Type I error   **b.** Type II error   **c.** Correct decision
Now suppose that the results of carrying out the hypothesis test lead to rejection of the null hypothesis, $\mu = 18$ mg, that is, to the conclusion that $\mu < 18$ mg. Classify that conclusion by error type or as a correct decision if in fact the mean iron intake, μ, of all adult females under the age of 51
d. equals the RDA of 18 mg per day.
e. is less than the RDA of 18 mg per day.

9.23 Body Temperature. The null and alternative hypotheses obtained in Exercise 9.9 on page 404 are

$$H_0: \mu = 98.6°F$$
$$H_a: \mu \neq 98.6°F,$$

where μ is the mean body temperature of all healthy humans. Explain what each outcome would mean.

a. Type I error   **b.** Type II error   **c.** Correct decision
Now suppose that the sample of temperatures leads to rejection of the null hypothesis. Classify that conclusion by error type or as a correct decision if in fact the mean body temperature, μ, of all healthy humans
d. is 98.6°F.   **e.** is not 98.6°F.

9.24 Worker Fatigue. The null and alternative hypotheses obtained in Exercise 9.10 on page 405 are

$$H_0: \mu = 72 \text{ bpm}$$
$$H_a: \mu > 72 \text{ bpm},$$

where μ is the mean post-work heart rate of casting workers. Explain what each outcome would mean.
a. Type I error   **b.** Type II error   **c.** Correct decision
Now suppose that the results of the sampling lead to rejection of the null hypothesis. Classify that conclusion by error type or as a correct decision if in fact the mean post-work heart rate of casting workers
d. is 72 bpm.   **e.** exceeds 72 bpm.

Extending the Concepts and Skills

9.25 Suppose that you choose the significance level of a hypothesis test to be 0.
a. What is the probability of a Type I error?
b. What is the probability of a Type II error?

9.26 Identify an exercise in this section for which it is important to have
a. a small α probability.   **b.** a small β probability.
c. both α and β probabilities small.

9.27 Approving Nuclear Reactors. Suppose that you are performing a statistical test to decide whether a nuclear reactor should be approved for use. Further suppose that failing to reject the null hypothesis corresponds to approval. What property would you want the Type II error probability, β, to have?

9.28 Guilty or Innocent? In the U.S. court system, a defendant is assumed innocent until proven guilty. Suppose that you regard a court trial as a hypothesis test with null and alternative hypotheses

$$H_0: \text{Defendant is innocent}$$
$$H_a: \text{Defendant is guilty.}$$

a. Explain the meaning of a Type I error.
b. Explain the meaning of a Type II error.
c. If you were the defendant, would you want α to be large or small? Explain your answer.
d. If you were the prosecuting attorney, would you want β to be large or small? Explain your answer.
e. What are the consequences to the court system if you make $\alpha = 0$? $\beta = 0$?

9.29 Type II Error Probabilities. For the pretzel packaging hypothesis test (Example 9.4, page 402), the null and alternative hypotheses are

$$H_0: \mu = 454 \text{ g (machine is working properly)}$$
$$H_a: \mu \neq 454 \text{ g (machine is not working properly)},$$

where μ is the mean net weight of all bags of pretzels packaged. Recall that the net weights are normally distributed with a standard deviation of 7.8 g. Figure 9.2(a) on page 403 portrays the decision criterion for a hypothesis test at the 4.56% significance level ($\alpha = 0.0456$) using a sample size of 25.

a. Identify the probability of a Type I error.
b. Assuming that the mean net weight being packaged is in fact 447 g, identify the distribution of the variable $\bar{x}$, that is, the sampling distribution of the sample mean for samples of size 25.

c. Use part (b) to determine the probability, β, of a Type II error if in fact the mean net weight being packaged is 447 g. (*Hint:* Referring to Fig. 9.2(a), note that β equals the percentage of all samples of 25 bags of pretzels whose mean weights are between 450.88 g and 457.12 g.)
d. Repeat parts (b) and (c) if in fact the mean net weight being packaged is 448 g, 449 g, 450 g, 451 g, 452 g, 453 g, 455 g, 456 g, 457 g, 458 g, 459 g, 460 g, and 461 g.
e. Use your answers from parts (b)–(d) to draw a graph of β versus the true value of μ. Interpret your graph.

Using Technology

9.30 Class Project: Quality Assurance. This exercise can be done individually or, better yet, as a class project. For the pretzel packaging hypothesis test in Example 9.4, the null and alternative hypotheses are

$$H_0: \mu = 454 \text{ g (machine is working properly)}$$
$$H_a: \mu \neq 454 \text{ g (machine is not working properly)},$$

where μ is the mean net weight of all bags of pretzels packaged. Recall that the net weights are normally distributed with a standard deviation of 7.8 g. Figure 9.4 on page 406 portrays the decision criterion for a test at the 4.56% significance level ($\alpha = 0.0456$). For a sample size of 25, the test statistic is

$$z = \frac{\bar{x} - \mu_0}{\sigma/\sqrt{n}} = \frac{\bar{x} - 454}{1.56}.$$

a. Assuming that $H_0: \mu = 454$ g is true, simulate 100 samples of 25 net weights each.
b. Determine the mean of each sample in part (a).
c. Use part (b) to determine the value of the test statistic, z, for each sample in part (a).
d. For the 100 samples obtained in part (a), roughly how many would you expect to lead to rejection of the null hypothesis? Explain your answer.
e. For the 100 samples obtained in part (a), determine the number that lead to rejection of the null hypothesis. (*Hint:* Refer to part (c) and Fig. 9.4.)
f. Compare your answers from parts (d) and (e) and comment on any observed difference.

9.3 Hypothesis Tests for One Population Mean When σ Is Known

In this section, we describe how to conduct a hypothesis test for a population mean at any prescribed significance level. You have already learned most of what you need to know in order to do that. What remains is to discuss how to obtain the critical value(s) when the significance level is specified in advance.

Recall that the significance level of a hypothesis test is the probability of rejecting a true null hypothesis or, equivalently, the probability that the test statistic will fall in the rejection region when the null hypothesis is true. With this fact in mind, we state Key Fact 9.3, which applies to any hypothesis test.

KEY FACT 9.3

Obtaining Critical Values

Suppose that a hypothesis test is to be performed at the significance level, α. Then the critical value(s) must be chosen so that, if the null hypothesis is true, the probability is α that the test statistic will fall in the rejection region.

Hypothesis Tests for a Population Mean When σ Is Known

We now develop a simple step-by-step procedure for performing a hypothesis test for a population mean when the population standard deviation is known. In doing so, we assume that the variable under consideration is normally distributed. But keep in mind that, because of the central limit theorem, the procedure will work reasonably well when the sample size is large, regardless of the distribution of the variable.

As you have seen, the null hypothesis for a hypothesis test concerning one population mean, μ, has the form H_0: $\mu = \mu_0$, where μ_0 is some number. Recall, also, that the test statistic for the hypothesis test is

$$z = \frac{\bar{x} - \mu_0}{\sigma/\sqrt{n}},$$

which tells you how many standard deviations the observed sample mean, $\bar{x}$, is from μ_0 (the value specified for the population mean in the null hypothesis).

The basis of the hypothesis testing procedure is in Key Fact 7.4: If x is a normally distributed variable with mean μ and standard deviation σ, then, for samples of size n, the variable $\bar{x}$ is also normally distributed and has mean μ and standard deviation $\sigma/\sqrt{n}$. This fact implies that, if the null hypothesis is true, the test statistic z has the standard normal distribution.

Consequently, in light of Key Fact 9.3, for a specified significance level, α, we need to choose the critical value(s) so that the area under the standard normal curve that lies above the rejection region equals α. In Example 9.6, we demonstrate how to choose these critical values.

Example 9.6 Obtaining the Critical Values

Determine the critical value(s) for a hypothesis test at the 5% significance level ($\alpha = 0.05$) if the test is

a. two tailed. **b.** left tailed. **c.** right tailed.

Solution As $\alpha = 0.05$, we need to choose the critical value(s) so that the area under the standard normal curve that lies above the rejection region equals 0.05.

a. For a two-tailed test, the rejection region is on both the left and right. So, in this case, the critical values are the two z-scores that divide the area under the standard normal curve into a middle 0.95 area and two outside areas of 0.025. In other words, the critical values are $\pm z_{0.025}$. From Table II in Appendix A, $\pm z_{0.025} = \pm 1.96$, as shown in Fig. 9.7(a).

FIGURE 9.7

Critical value(s) for a hypothesis test at the 5% significance level if the test is (a) two tailed, (b) left tailed, or (c) right tailed

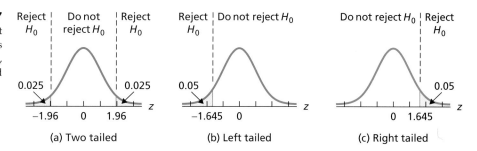

(a) Two tailed (b) Left tailed (c) Right tailed

b. For a left-tailed test, the rejection region is on the left. So, in this case, the critical value is the z-score having area 0.05 to its left under the standard normal curve, which is $-z_{0.05}$. From Table II, $-z_{0.05} = -1.645$, as shown in Fig. 9.7(b).

c. For a right-tailed test, the rejection region is on the right. So, in this case, the critical value is the z-score having area 0.05 to its right under the standard normal curve, which is $z_{0.05}$. From Table II, $z_{0.05} = 1.645$, as shown in Fig. 9.7(c).

By reasoning as we did in Example 9.6, we can obtain the critical value(s) for any specified significance level, α. As depicted in Fig. 9.8, for a two-tailed test, the critical values are $\pm z_{\alpha/2}$; for a left-tailed test, the critical value is $-z_\alpha$; and for a right-tailed test, the critical value is z_α.

FIGURE 9.8

Critical value(s) for a hypothesis test at the significance level α if the test is (a) two tailed, (b) left tailed, or (c) right tailed

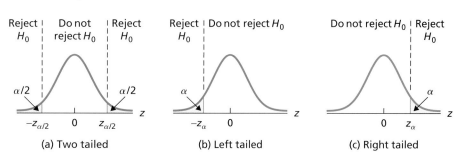

(a) Two tailed (b) Left tailed (c) Right tailed

TABLE 9.4

Some important values of z_α

$z_{0.10}$	$z_{0.05}$	$z_{0.025}$	$z_{0.01}$	$z_{0.005}$
1.28	1.645	1.96	2.33	2.575

The most commonly used significance levels are 0.10, 0.05, and 0.01. If we consider both one-tailed and two-tailed tests, these three significance levels give rise to five "tail areas." Using the standard-normal table, Table II, we obtained the value of z_α corresponding to each of those five tail areas, as displayed in Table 9.4.

Alternatively, these five values of z_α can be found at the bottom of the t-table, Table IV, where they are displayed to three decimal places. Can you explain the slight discrepancy between the values given for $z_{0.005}$ in the two tables?

The One-Sample z-Test

Procedure 9.1 (next page) provides a step-by-step method for performing a hypothesis test for a population mean when the population standard deviation is known. The procedure summarizes what we have done in this and the previous two sections. We often refer to Procedure 9.1 as the **one-sample z-test** or, more briefly, as the **z-test.**

One of the assumptions for using the z-test is that either the variable under consideration is normally distributed or the sample size is large. But, as with the z-interval procedure, the z-test is robust to moderate violations of the normality assumption. Thus the z-test works reasonably well even when the sample size is small or moderate and the variable is not normally distributed, provided the variable is not too far from being normally distributed.

Again, as with the z-interval procedure, you must watch for outliers when considering the z-test. Recall that the presence of outliers calls into question the normality assumption; and that, even for large samples, outliers can sometimes unduly affect a z-test because the sample mean is not resistant to outliers.

In Key Fact 9.4, we present guidelines for the use of Procedure 9.1.

KEY FACT 9.4

When to Use the One-Sample z-Test[†]

- For small samples—say, of size less than 15—the z-test should be used only when the variable under consideration is normally distributed or very close to being so.
- For samples of moderate size—say, between 15 and 30—the z-test can be used unless the data contain outliers or the variable under consideration is far from being normally distributed.
- For large samples—say, of size 30 or more—the z-test can be used essentially without restriction. However, if outliers are present and their removal is not justified, the effect of the outliers on the hypothesis test should be examined; that is, you should perform the hypothesis test, once with the outliers and once without them. If the conclusion remains the same either way, you may be content to take that as your conclusion and close the investigation. But if the conclusion is affected, you probably should make the more conservative conclusion, use a different procedure, or take another sample.
- If outliers are present but their removal is justified and results in a data set for which the z-test is appropriate (as previously stated), the procedure can be used.

Applying the One-Sample z-Test

Examples 9.7–9.9 illustrate use of the z-test, Procedure 9.1. We reexamine these examples in Section 9.5 when we discuss P-values.

[†]We can refine these guidelines further by considering the impact of skewness. Roughly speaking, the more skewed the distribution of the variable under consideration, the larger is the sample size required to use the z-test.

PROCEDURE 9.1

One-Sample *z*-Test (Critical-Value Approach)

Purpose To perform a hypothesis test for a population mean, μ

Assumptions

1. Simple random sample
2. Normal population or large sample
3. σ known

STEP 1 **The null hypothesis is H_0: $\mu = \mu_0$, and the alternative hypothesis is**

$$H_a\text{: } \mu \neq \mu_0 \quad \text{or} \quad H_a\text{: } \mu < \mu_0 \quad \text{or} \quad H_a\text{: } \mu > \mu_0$$
$$\text{(Two tailed)} \qquad \text{(Left tailed)} \qquad \text{(Right tailed)}$$

STEP 2 **Decide on the significance level, α.**

STEP 3 **Compute the value of the test statistic**

$$z = \frac{\bar{x} - \mu_0}{\sigma/\sqrt{n}}.$$

STEP 4 **The critical value(s) are**

$$\pm z_{\alpha/2} \quad \text{or} \quad -z_{\alpha} \quad \text{or} \quad z_{\alpha}$$
$$\text{(Two tailed)} \qquad \text{(Left tailed)} \qquad \text{(Right tailed)}$$

Use Table II to find the critical value(s).

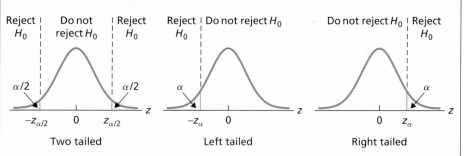

STEP 5 **If the value of the test statistic falls in the rejection region, reject H_0; otherwise, do not reject H_0.**

STEP 6 **Interpret the results of the hypothesis test.**

The hypothesis test is exact for normal populations and is approximately correct for large samples from nonnormal populations.

Note: By saying that the hypothesis test is *exact,* we mean that the true significance level equals α; by saying that it is *approximately correct,* we mean that the true significance level only approximately equals α.

▮▮ Example 9.7   The One-Sample z-Test

Prices of History Books   The R. R. Bowker Company of New York collects information on the retail prices of books and publishes its findings in *The Bowker Annual Library and Book Trade Almanac*. In 2000, the mean retail price of all history books was $51.46. This year's retail prices for 40 randomly selected history books are shown in Table 9.5. At the 1% significance level, do the data provide sufficient evidence to conclude that this year's mean retail price of all history books has increased from the 2000 mean of $51.46? Assume that the standard deviation of prices for this year's history books is $7.61.

TABLE 9.5

This year's prices ($) for 40 history books

56.00	46.25	47.34	53.99
53.71	47.88	54.82	55.73
51.00	61.70	47.03	62.68
47.80	50.89	52.36	50.95
51.28	50.94	60.70	72.38
47.70	56.16	52.33	51.70
53.80	50.90	63.74	52.87
41.08	64.93	57.44	54.09
75.37	56.48	69.04	42.71
53.76	72.17	61.26	42.65

Solution   We constructed (not shown) a normal probability plot, a histogram, a stem-and-leaf diagram, and a boxplot for these data. The boxplot indicated potential outliers, but in view of the other three graphs, we concluded that in fact the data contain no outliers. As the sample size is 40, which is large, and the population standard deviation is known, we can apply Procedure 9.1 to perform the required hypothesis test.

STEP 1 State the null and alternative hypotheses.

Let μ denote this year's mean retail price of all history books. We obtained the null and alternative hypotheses in Example 9.2 as

$$H_0: \mu = \$51.46 \text{ (mean price has not increased)}$$
$$H_a: \mu > \$51.46 \text{ (mean price has increased)}.$$

Note that the hypothesis test is right tailed because a greater-than sign (>) appears in the alternative hypothesis.

STEP 2 Decide on the significance level, α.

We are to perform the test at the 1% significance level, or $\alpha = 0.01$.

STEP 3 Compute the value of the test statistic

$$z = \frac{\bar{x} - \mu_0}{\sigma/\sqrt{n}}.$$

FIGURE 9.9

Criterion for deciding whether to reject the null hypothesis

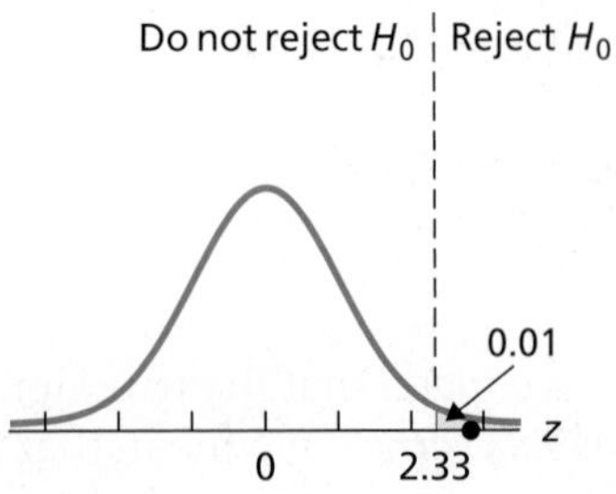

We have $\mu_0 = 51.46$, $\sigma = 7.61$, and $n = 40$. The mean of the sample data in Table 9.5 is $\bar{x} = 54.890$. Thus the value of the test statistic is

$$z = \frac{54.890 - 51.46}{7.61/\sqrt{40}} = 2.85.$$

This value of z is marked with a dot in Fig. 9.9.

STEP 4 The critical value for a right-tailed test is z_α. Use Table II to find the critical value.

As $\alpha = 0.01$, the critical value is $z_{0.01}$. From Table II (or Table 9.4 on page 415), $z_{0.01} = 2.33$, as shown in Fig. 9.9.

STEP 5 If the value of the test statistic falls in the rejection region, reject H_0; otherwise, do not reject H_0.

The value of the test statistic, found in Step 3, is $z = 2.85$. Figure 9.9 reveals that this value falls in the rejection region, so we reject H_0. The test results are statistically significant at the 1% level.

STEP 6 Interpret the results of the hypothesis test.

INTERPRETATION At the 1% significance level, the data provide sufficient evidence to conclude that this year's mean retail price of all history books has increased from the 2000 mean of $51.46. ∎

Example 9.8 The One-Sample z-Test

Poverty and Calcium Calcium is the most abundant mineral in the body and also one of the most important. It works with phosphorus to build and maintain bones and teeth. According to the Food and Nutrition Board of the National Academy of Sciences, the recommended daily allowance (RDA) of calcium for adults is 800 milligrams (mg).

A simple random sample of 18 people with incomes below the poverty level gives the daily calcium intakes shown in Table 9.6. At the 5% significance level, do the data provide sufficient evidence to conclude that the mean calcium intake of all people with incomes below the poverty level is less than the RDA of 800 mg? Assume that $\sigma = 188$ mg.

TABLE 9.6
Daily calcium intakes (mg) for 18 people with incomes below the poverty level

686	433	743	647	734	641
993	620	574	634	850	858
992	775	1113	672	879	609

Solution As the sample size, $n = 18$, is moderate, we first need to consider questions of normality and outliers. (See the second bulleted item in Key Fact 9.4 on page 416.) Hence we constructed a normal probability plot (not shown) for the data. The plot reveals no outliers and falls roughly in a straight line. Consequently, we can apply Procedure 9.1 to perform the required hypothesis test.

STEP 1 State the null and alternative hypotheses.

Let μ denote the mean calcium intake (per day) of all people with incomes below the poverty level. We obtained the null and alternative hypotheses in Example 9.3 as

H_0: $\mu = 800$ mg (mean calcium intake is not less than the RDA)
H_a: $\mu < 800$ mg (mean calcium intake is less than the RDA).

Note that the hypothesis test is left tailed because a less-than sign ($<$) appears in the alternative hypothesis.

STEP 2 Decide on the significance level, α.

We are to perform the test at the 5% significance level, or $\alpha = 0.05$.

STEP 3 Compute the value of the test statistic

$$z = \frac{\bar{x} - \mu_0}{\sigma/\sqrt{n}}.$$

We have $\mu_0 = 800$, $\sigma = 188$, and $n = 18$. From the data in Table 9.6, we find that $\bar{x} = 747.4$. Thus the value of the test statistic is

$$z = \frac{747.4 - 800}{188/\sqrt{18}} = -1.19.$$

This value of z is marked with a dot in Fig. 9.10.

FIGURE 9.10
Criterion for deciding whether to reject the null hypothesis

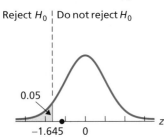

STEP 4 The critical value for a left-tailed test is $-z_\alpha$. Use Table II to find the critical value.

As $\alpha = 0.05$, the critical value is $-z_{0.05}$. From Table II (or Table 9.4 or Table IV), $z_{0.05} = 1.645$. Hence the critical value is $-z_{0.05} = -1.645$, as shown in Fig. 9.10.

STEP 5 If the value of the test statistic falls in the rejection region, reject H_0; otherwise, do not reject H_0.

The value of the test statistic, found in Step 3, is $z = -1.19$. Figure 9.10 reveals that this value does not fall in the rejection region, and so we do not reject H_0. The test results are not statistically significant at the 5% level.

STEP 6 Interpret the results of the hypothesis test.

INTERPRETATION At the 5% significance level, the data do not provide sufficient evidence to conclude that the mean calcium intake of all people with incomes below the poverty level is less than the RDA of 800 mg.

▌▌ Example 9.9 The One-Sample z-Test

Clocking the Cheetah The Cheetah (*Acinonyx jubatus*) is the fastest land mammal on earth and is highly specialized to run down prey. According to the Cheetah Conservation of Southern Africa *Trade Environment Database*, the cheetah often exceeds speeds of 60 miles per hour (mph) and has been clocked at speeds of more than 70 mph.

One common estimate of mean top speed for cheetahs is 60 mph. Table 9.7 gives the top speeds, in miles per hour, over a quarter mile for a sample of 35 cheetahs. At the 5% significance level, do the data provide sufficient evidence to conclude that the mean top speed of all cheetahs differs from 60 mph? Assume that the population standard deviation of top speeds is 3.2 mph.

TABLE 9.7
Top speeds, in miles per hour, for a sample of 35 cheetahs

57.3	57.5	59.0	56.5	61.3
57.6	59.2	65.0	60.1	59.7
62.6	52.6	60.7	62.3	65.2
54.8	55.4	55.5	57.8	58.7
57.8	60.9	75.3	60.6	58.1
55.9	61.6	59.6	59.8	63.4
54.7	60.2	52.4	58.3	66.0

Solution A frequency histogram for the data in Table 9.7, displayed in Fig. 9.11, suggests that the top speed of 75.3 mph (third entry in the fifth row) is an outlier. A stem-and-leaf diagram, a boxplot, and a normal probability plot further confirm that 75.3 is an outlier.

Thus, as suggested in the third bulleted item in Key Fact 9.4 (page 416), we first apply Procedure 9.1 to the full data set in Table 9.7 and then examine the effect on the test results when we remove the outlier, 75.3 mph.

FIGURE 9.11
Frequency histogram for the top
speeds in Table 9.7

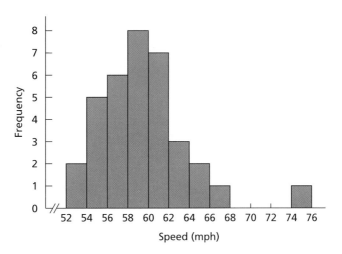

FIGURE 9.11
Frequency histogram for the top
speeds in Table 9.7

STEP 1 State the null and alternative hypotheses.

The null and alternative hypotheses are

$$H_0: \mu = 60 \text{ mph (mean top speed of cheetahs is 60 mph)}$$
$$H_a: \mu \neq 60 \text{ mph (mean top speed of cheetahs is not 60 mph)},$$

where μ denotes the mean top speed of all cheetahs. Note that the hypothesis test is two tailed because a does-not-equal sign ($\neq$) appears in the alternative hypothesis.

STEP 2 Decide on the significance level, α.

We are to perform the hypothesis test at the 5% significance level, or $\alpha = 0.05$.

STEP 3 Compute the value of the test statistic

$$z = \frac{\bar{x} - \mu_0}{\sigma/\sqrt{n}}.$$

We have $\mu_0 = 60$, $\sigma = 3.2$, and $n = 35$. From the data in Table 9.7, we find that $\bar{x} = 59.526$. Thus the value of the test statistic is

$$z = \frac{59.526 - 60}{3.2/\sqrt{35}} = -0.88.$$

This value of z is marked with a solid dot in Fig. 9.12.

FIGURE 9.12
Criterion for deciding whether to reject
the null hypothesis

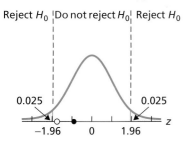

STEP 4 The critical values for a two-tailed test are $\pm z_{\alpha/2}$. Use Table II to find the critical values.

As $\alpha = 0.05$, we obtain from Table II (or Table 9.4 or Table IV) the critical values of $\pm z_{0.05/2} = \pm z_{0.025} = \pm 1.96$, as shown in Fig. 9.12.

STEP 5 If the value of the test statistic falls in the rejection region, reject H_0; otherwise, do not reject H_0.

From Step 3, the value of the test statistic is $z = -0.88$. This value does not fall in the rejection region shown in Fig. 9.12. Hence, we do not reject H_0. The test results are not statistically significant at the 5% level.

STEP 6 Interpret the results of the hypothesis test.

INTERPRETATION At the 5% significance level, the (complete) data do not provide sufficient evidence to conclude that the mean top speed of all cheetahs differs from 60 mph.

We have now completed the hypothesis test, using all 35 top speeds in Table 9.7. However, recall that the top speed of 75.3 mph is an outlier. For this problem, although we don't actually know whether removing this outlier is justified (a common situation), we can still remove it from the sample data and assess the effect on the hypothesis test.

Doing so, we find that the value of the test statistic for the abridged data is $z = -1.71$, which we have marked with a hollow dot in Fig. 9.12. This value still lies in the nonrejection region, although it is much closer to the rejection region than the value of the test statistic for the unabridged data, $z = -0.88$.

INTERPRETATION In this case, removing the outlier therefore does not affect the conclusion of the hypothesis test. We can probably accept that the mean top speed of all cheetahs is roughly 60 mph.

Statistical Significance Versus Practical Significance

Recall that the results of a hypothesis test are *statistically significant* if the null hypothesis is rejected at the chosen level of α. Statistical significance means that the data provide sufficient evidence to conclude that the truth is different from the stated null hypothesis. However, it does not necessarily mean that the difference is important in any practical sense.

For example, the manufacturer of a new car, the Orion, claims that a typical car gets 26 miles per gallon—that is, the mean gas mileage of all Orions is $\mu = 26$ mpg. Suppose that the mean gas mileage of a sample of 1000 Orions is 25.9 mpg. Assuming that the standard deviation of gas mileages for all Orions is 1.4 mpg, the value of the test statistic for a z-test of

$$H_0: \mu = 26 \text{ mpg (mean gas mileage is 26 mpg)}$$

$$H_a: \mu < 26 \text{ mpg (mean gas mileage is less than 26 mpg)}$$

is $z = -2.26$. This result is statistically significant at the 5% level (and even at the 1.19% level). Thus we can easily reject the null hypothesis, that is, the manufacturer's claim that the mean gas mileage of all Orions is 26 mpg.

Because the sample size, 1000, is so large, the sample mean, $\bar{x} = 25.9$ mpg, is probably nearly the same as the population mean. As a result, we rejected the manufacturer's claim because μ is about 25.9 mpg instead of 26 mpg. From a practical point of view, however, the difference between 25.9 mpg and 26 mpg is not important.

What
Does it Mean?

Statistical significance does not necessarily imply practical significance!

The Relation Between Hypothesis Tests and Confidence Intervals

Hypothesis tests and confidence intervals are closely related. Consider, for example, a two-tailed hypothesis test for a population mean at the significance level α. In this case, the null hypothesis will be rejected if and only if the value μ_0 given for the mean in the null hypothesis lies outside the $(1 - \alpha)$-level confidence interval for μ. Exercises 9.42 and 9.43 deal with this relation between hypothesis tests and confidence intervals in greater detail.

EXERCISES 9.3

Statistical Concepts and Skills

In Exercises **9.31–9.33,** *suppose that a hypothesis test is to be performed for a population mean,* μ, *with a null hypothesis of* H_0: $\mu = \mu_0$. *Further suppose that the test statistic to be used is*

$$z = \frac{\bar{x} - \mu_0}{\sigma/\sqrt{n}}.$$

For each exercise, obtain the required critical value(s) and draw a graph that illustrates your answers.

9.31 A left-tailed test with $\alpha = 0.05$.

9.32 A right-tailed test with $\alpha = 0.01$.

9.33 A two-tailed test with $\alpha = 0.05$.

9.34 Explain why considering outliers is important when you are conducting a one-sample z-test.

Preliminary data analyses indicate that applying the z-test (Procedure 9.1 on page 417) in Exercises **9.35–9.40** *is reasonable. Comment on the practical significance of all hypothesis tests whose results are statistically significant.*

9.35 Toxic Mushrooms? Refer to Exercise 9.5 on page 404. Here are the data obtained by the researchers.

0.24	0.59	0.62	0.16	0.77	1.33
0.92	0.19	0.33	0.25	0.59	0.32

At the 5% significance level, do the data provide sufficient evidence to conclude that the mean cadmium level in *Boletus pinicola* mushrooms is greater than the government's recommended limit of 0.5 ppm? Assume that the population standard deviation of cadmium levels in *Boletus pinicola* mushrooms is 0.37 ppm. (*Note:* The sum of the data is 6.31 ppm.)

9.36 Serving Time. From Exercise 9.6, the mean length of imprisonment for motor-vehicle theft offenders in Australia is 16.7 months. One hundred randomly selected motor-vehicle theft offenders in Sydney, Australia, had a mean length of imprisonment of 17.8 months. At the 5% significance level, do the data provide sufficient evidence to

conclude that the mean length of imprisonment for motor-vehicle theft offenders in Sydney differs from the national mean in Australia? Assume that the population standard deviation of the lengths of imprisonment for motor-vehicle theft offenders in Sydney is 6.0 months.

9.37 Cell Phones. Refer to Exercise 9.7 on page 404. Following are last year's local monthly bills, in dollars, for a random sample of 50 cell phone users.

32.88	31.09	16.13	104.20	47.65
116.96	61.97	43.74	13.52	15.37
30.09	14.62	27.10	28.80	42.04
16.93	61.31	15.50	32.48	22.43
44.67	35.64	19.95	28.95	46.54
28.08	88.95	23.45	31.60	48.32
64.68	50.48	46.24	35.60	24.53
16.17	16.41	41.80	24.74	77.21
28.04	16.56	49.35	44.82	15.12
57.67	13.48	81.16	51.62	126.84

At the 1% significance level, do the data provide sufficient evidence to conclude that last year's mean local monthly bill for cell phone users has decreased from the 2001 mean of $47.37? Assume that $\sigma = \$25$. (*Note:* The sum of the 50 cell phone bills is $2053.48.)

9.38 Iron Deficiency? Refer to Exercise 9.8 on page 404. The following iron intakes, in milligrams, were obtained during a 24-hour period for 45 randomly selected adult females under the age of 51.

15.0	18.1	14.4	14.6	10.9	18.1	18.2	18.3	15.0
16.0	12.6	16.6	20.7	19.8	11.6	12.8	15.6	11.0
15.3	9.4	19.5	18.3	14.5	16.6	11.5	16.4	12.5
14.6	11.9	12.5	18.6	13.1	12.1	10.7	17.3	12.4
17.0	6.3	16.8	12.5	16.3	14.7	12.7	16.3	11.5

At the 1% significance level, do the data suggest that adult females under the age of 51 are, on average, getting less than the RDA of 18 mg of iron? Assume that the population standard deviation is 4.2 mg. (*Note:* $\bar{x} = 14.68$ mg.)

9.39 Body Temperature. Refer to Exercise 9.9 on page 404. The researchers obtained the following body temperatures of 93 healthy humans.

98.0	97.6	98.8	98.0	98.8	98.8	97.6	98.6	98.6
98.8	98.0	98.2	98.0	98.0	97.0	97.2	98.2	98.1
98.2	98.5	98.5	99.0	98.0	97.0	97.3	97.3	98.1
97.8	99.0	97.6	97.4	98.0	97.4	98.0	98.6	98.6
98.4	97.0	98.4	99.0	98.0	99.4	97.8	98.2	99.2
99.0	97.7	98.2	98.2	98.8	98.1	98.5	97.2	98.5
99.2	98.3	98.7	98.8	98.6	98.0	99.1	97.2	97.6
97.9	98.8	98.6	98.6	99.3	97.8	98.7	99.3	97.8
98.4	97.7	98.3	97.7	97.1	98.4	98.6	97.4	96.7
96.9	98.4	98.2	98.6	97.0	97.4	98.4	97.4	96.8
98.2	97.4	98.0						

At the 1% significance level, do the data provide sufficient evidence to conclude that the mean body temperature of healthy humans differs from 98.6°F? Assume that $\sigma = 0.63$°F. (*Note:* The sum of the 93 temperatures is 9125.5°F.)

9.40 Worker Fatigue. From Exercise 9.10 on page 405, the normal resting heart rate is 72 beats per minute (bpm). A random sample of 29 casting workers had a mean post-work heart rate of 78.3 bpm. At the 5% significance level, do the data provide sufficient evidence to conclude that the mean post-work heart rate for casting workers exceeds the normal resting heart rate of 72 bpm? Assume that the population standard deviation of post-work heart rates for casting workers is 11.2 bpm.

Extending the Concepts and Skills

9.41 Each part of this exercise provides a scenario for a hypothesis test. Decide whether the z-test is an appropriate method for conducting the hypothesis test. Assume that the population standard deviation is known in each case.

a. Preliminary analyses reveal that the sample data contain no outliers but that the distribution of the variable under consideration is probably highly skewed. The sample size is 20.

b. A normal probability plot reveals an outlier but is otherwise roughly linear. It is determined that the outlier is a legitimate observation and should not be removed. The sample size is 15.

c. Preliminary analyses reveal that the sample data contain no outliers but that the distribution of the variable under consideration is probably mildly skewed. The sample size is 70.

In Exercises 9.42 and 9.43, you are to examine the relationship between hypothesis tests and confidence intervals for one population mean.

9.42 How Far People Drive. In 2000, the average car in the United States was driven 11.9 thousand miles, as reported by the U.S. Federal Highway Administration in *Highway Statistics*. A random sample of 500 cars revealed a mean of 11.7 thousand miles driven for last year. Let μ denote last year's mean distance driven for all cars.

a. Use Procedure 9.1 on page 417 to perform the hypothesis test

$$H_0: \mu = 11.9 \text{ thousand miles}$$
$$H_a: \mu \neq 11.9 \text{ thousand miles}$$

at the 5% significance level. Assume that last year's standard deviation of distances driven for all cars is 6.0 thousand miles.

b. Use Procedure 8.1 on page 363 to find a 95% confidence interval for μ.

c. Does the value of 11.9 thousand miles, hypothesized for the mean, μ, in the null hypothesis of part (a), lie within your confidence interval from part (b)?

d. Repeat parts (a)–(c) for a sample of 500 cars driven a mean of 12.5 thousand miles last year.

e. Based on your observations in parts (a)–(d), complete the following statements concerning the relationship between a two-tailed hypothesis test,

$$H_0: \mu = \mu_0$$
$$H_a: \mu \neq \mu_0,$$

at the significance level α and a $(1 - \alpha)$-level confidence interval for μ.

i. If μ_0 lies within the $(1 - \alpha)$-level confidence interval for μ, the null hypothesis *(will, will not)* be rejected.

ii. If μ_0 lies outside the $(1 - \alpha)$-level confidence interval for μ, the null hypothesis *(will, will not)* be rejected.

9.43 Hypothesis Tests and Confidence Intervals. In this exercise, you are to examine the general relationship between a two-tailed hypothesis test for a population mean and a confidence-interval estimate for a population mean.

a. Show that the inequalities

$$\bar{x} - z_{\alpha/2} \cdot \frac{\sigma}{\sqrt{n}} < \mu_0 < \bar{x} + z_{\alpha/2} \cdot \frac{\sigma}{\sqrt{n}}$$

are equivalent to

$$-z_{\alpha/2} < \frac{\bar{x} - \mu_0}{\sigma/\sqrt{n}} < z_{\alpha/2}.$$

b. Deduce the following fact from part (a): For a two-tailed hypothesis test,

$$H_0: \mu = \mu_0$$
$$H_a: \mu \neq \mu_0,$$

at the significance level α, the null hypothesis will not be rejected if μ_0 lies within the $(1 - \alpha)$-level confidence interval for μ, and conversely, the null hypothesis will be rejected if μ_0 does not lie within the $(1 - \alpha)$-level confidence interval for μ.

9.4   Type II Error Probabilities; Power

As you learned in Section 9.2, hypothesis tests do not always yield correct conclusions; they have built-in margins of error. An important part of planning a study is to take an advance look at the types of errors that can be made and the effects that those errors might have.

Recall that two types of errors are possible with hypothesis tests. One is a Type I error: rejecting a true null hypothesis. The other is a Type II error: not rejecting a false null hypothesis. Also recall that the probability of making a Type I error is called the significance level of the hypothesis test and is denoted α; that the probability of making a Type II error is denoted β.

In this section, we show how to compute Type II error probabilities. We also investigate the concept of the *power of a hypothesis test*. Although the discussion is limited to the one-sample z-test, the ideas apply to any hypothesis test.

Computing Type II Error Probabilities

The probability of making a Type II error depends on the sample size, the significance level, and the true value of the parameter under consideration. Example 9.10 explains how to compute the probability of making a Type II error for a one-sample z-test for a population mean.

Example 9.10   Computing Type II Error Probabilities

Questioning Gas Mileage Claims   The manufacturer of a new model car, the Orion, claims that a typical car gets 26 miles per gallon (mpg). A consumer advocacy group is skeptical of this claim and thinks that the mean gas mileage, μ, of all Orions may be less than 26 mpg. The group plans to perform the hypothesis test

$$H_0: \mu = 26 \text{ mpg (manufacturer's claim)}$$
$$H_a: \mu < 26 \text{ mpg (consumer group's conjecture),}$$

at the 5% significance level, using a sample of 30 Orions. Find the probability of making a Type II error if the true mean gas mileage of all Orions is

a. 25.8 mpg.   **b.** 25.0 mpg.

Assume that the gas mileages of Orions are normally distributed with a standard deviation of 1.4 mpg.

Solution The inference under consideration is a left-tailed hypothesis test for a population mean at the 5% significance level. The test statistic is

$$z = \frac{\bar{x} - \mu_0}{\sigma/\sqrt{n}} = \frac{\bar{x} - 26}{1.4/\sqrt{30}},$$

and the critical value is $-z_\alpha = -z_{0.05} = -1.645$. Thus the decision criterion for the hypothesis test is: If $z \le -1.645$, reject H_0; if $z > -1.645$, do not reject H_0.

Computing Type II error probabilities is somewhat simpler if the decision criterion is expressed in terms of $\bar{x}$ instead of z. To do that here, we must find the sample mean that is 1.645 standard deviations below the null hypothesis population mean of 26:

$$\bar{x} = 26 - 1.645 \cdot \frac{1.4}{\sqrt{30}} = 25.6.$$

FIGURE 9.13

Graphical display of decision criterion for the gas mileage illustration ($\alpha = 0.05, n = 30$)

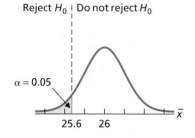

The decision criterion then can be expressed in terms of $\bar{x}$ as: If $\bar{x} \le 25.6$ mpg, reject H_0; if $\bar{x} > 25.6$ mpg, do not reject H_0. This decision criterion is portrayed graphically in Fig. 9.13.

a. If $\mu = 25.8$ mpg, then

- $\mu_{\bar{x}} = \mu = 25.8$,
- $\sigma_{\bar{x}} = \sigma/\sqrt{n} = 1.4/\sqrt{30} = 0.26$, and
- $\bar{x}$ is normally distributed.

Thus, the variable $\bar{x}$ is normally distributed with a mean of 25.8 mpg and a standard deviation of 0.26 mpg. The normal curve for $\bar{x}$ is shown in Fig. 9.14.

FIGURE 9.14

Determining the probability of a Type II error if $\mu = 25.8$ mpg

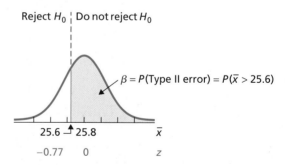

z-score computation:

$$\bar{x} = 25.6 \longrightarrow z = \frac{25.6 - 25.8}{0.26} = -0.77$$

Area to the left of z:

0.2206

Shaded area $= 1 - 0.2206 = 0.7794$

A Type II error occurs if we do not reject H_0, that is, if $\bar{x} > 25.6$ mpg. The probability of this happening equals the percentage of all samples whose means exceed 25.6 mpg, which we obtain in Fig. 9.14. From Fig. 9.14, if the true mean gas mileage of all Orions is 25.8 mpg, the probability of making a Type II error is 0.7794; that is, $\beta = 0.7794$.

INTERPRETATION There is roughly a 78% chance that the consumer group will fail to reject the manufacturer's claim that the mean gas mileage of all Orions is 26 mpg when in fact the true mean is 25.8 mpg.

Although this result is a rather high chance of error, we probably would not expect the hypothesis test to detect such a small difference in mean gas mileage (25.8 mpg as opposed to 26 mpg) with a sample size of only 30.

b. We proceed as we did in part (a) but this time assume that $\mu = 25.0$ mpg. Figure 9.15 shows the required computations.

FIGURE 9.15

Determining the probability of a Type II error if $\mu = 25.0$ mpg

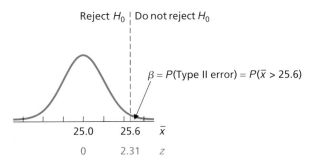

Reject H_0 | Do not reject H_0

$\beta = P(\text{Type II error}) = P(\bar{x} > 25.6)$

| 25.0 | 25.6 | $\bar{x}$ |
| 0 | 2.31 | z |

z-score computation:

Area to the left of z:

$\bar{x} = 25.6 \longrightarrow z = \dfrac{25.6 - 25.0}{0.26} = 2.31$ 0.9896

Shaded area $= 1 - 0.9896 = 0.0104$

From Fig. 9.15, if the true mean gas mileage of all Orions is 25.0 mpg, the probability of making a Type II error is 0.0104; that is, $\beta = 0.0104$.

INTERPRETATION There is only about a 1% chance that the consumer group will fail to reject the manufacturer's claim that the mean gas mileage of all Orions is 26 mpg when in fact the true mean is 25.0 mpg.

Combining figures such as Figs. 9.14 and 9.15 gives a better understanding of Type II error probabilities. In Fig. 9.16 at the top of the following page, we combined those two figures with two others. The Type II error probabilities for the two additional values of μ were obtained by using the same techniques as those in Example 9.10.

Figure 9.16 shows clearly that the farther the true mean is from the null hypothesis mean of 26 mpg, the smaller will be the probability of a Type II error. This result is hardly surprising: We would expect that a false null hypothesis is more likely to be detected when the true mean is far from the null hypothesis mean than when the true mean is close to the null hypothesis mean.

Power and Power Curves

As we mentioned earlier, part of evaluating the effectiveness of a hypothesis test involves an analysis of the chances of making an incorrect decision. Recall that the probability of making a Type I error is specified by the significance level and that the probability of making a Type II error depends on the true value of the parameter in question.

In modern statistical practice, analysts generally use the probability of not making a Type II error, called the **power,** to appraise the performance of a

FIGURE 9.16

Type II error probabilities for $\mu = 25.8$, 25.6, 25.3, and 25.0 ($\alpha = 0.05$, $n = 30$)

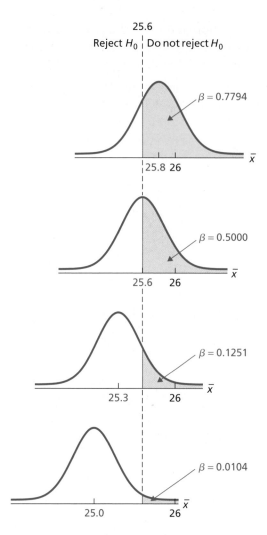

hypothesis test. Once we know the Type II error probability, β, obtaining the power is simple—we just subtract β from 1.

DEFINITION 9.5

Power

The *power* of a hypothesis test is the probability of not making a Type II error, that is, the probability of rejecting a false null hypothesis. We have

$$\text{Power} = 1 - P(\text{Type II error}) = 1 - \beta.$$

What?
Does it Mean?

The power of a hypothesis test is between 0 and 1 and measures the ability of the hypothesis test to detect a false null hypothesis. If the power is near 0, the hypothesis test is not very good at detecting a false null hypothesis; if the power is near 1, the hypothesis test is extremely good at detecting a false null hypothesis.

In reality, the true value of μ will be unknown, so constructing a table of powers for various values of μ is helpful in evaluating the effectiveness of the hypothesis test. For the gas mileage illustration, we have already obtained the Type II error probability, β, when the true mean is 25.8 mpg, 25.6 mpg, 25.3 mpg, and 25.0 mpg, as depicted in Fig. 9.16. Similar calculations yield the other β probabilities shown in the second column of Table 9.8. The third column of Table 9.8 shows the power that corresponds to each value of μ, obtained by subtracting β from 1.

TABLE 9.8

Selected Type II error probabilities and powers for the gas mileage illustration ($\alpha = 0.05$, $n = 30$)

True mean μ	P (Type II error) β	Power $1 - \beta$
25.9	0.8749	0.1251
25.8	0.7794	0.2206
25.7	0.6480	0.3520
25.6	0.5000	0.5000
25.5	0.3520	0.6480
25.4	0.2206	0.7794
25.3	0.1251	0.8749
25.2	0.0618	0.9382
25.1	0.0274	0.9726
25.0	0.0104	0.9896
24.9	0.0036	0.9964
24.8	0.0010	0.9990

We can use Table 9.8 to evaluate the overall effectiveness of the hypothesis test. We can also obtain from Table 9.8 a visual display of that effectiveness by plotting points of power against μ and then connecting the points with a smooth curve. The resulting curve is called a **power curve** and is shown in Fig. 9.17. In general, the closer a power curve is to 1 (i.e., the horizontal line 1 unit above the horizontal axis), the better the hypothesis test is at detecting a false null hypothesis.

FIGURE 9.17

Power curve for the gas mileage illustration ($\alpha = 0.05$, $n = 30$)

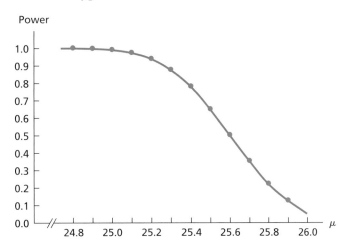

Sample Size and Power

Ideally, both Type I and Type II errors should have small probabilities. In terms of significance level and power, then, we want to specify a small significance level (close to 0) and yet have large power (close to 1).

As we noted in Section 9.2, for a fixed sample size, the smaller we specify the significance level, the larger will be the Type II error probability or, equivalently, the smaller will be the power. But there is a way that we can use a small significance level and have large power—namely, by employing a large sample. Example 9.11 provides an illustration.

Example 9.11   The Effect of Sample Size on Power

Questioning Gas Mileage Claims   Consider again the hypothesis test for the gas mileage illustration of Example 9.10,

$$H_0: \mu = 26 \text{ mpg (manufacturer's claim)}$$
$$H_a: \mu < 26 \text{ mpg (consumer group's conjecture)},$$

where μ is the mean gas mileage of all Orions. In Table 9.8, we presented selected powers when $\alpha = 0.05$ and $n = 30$. Now suppose that we keep the significance level at 0.05 but increase the sample size from 30 to 100.

a. Construct a table of powers similar to Table 9.8.

b. Use the table from part (a) to draw the power curve for $n = 100$ and compare it to the power curve drawn earlier for $n = 30$.

c. Interpret the results from parts (a) and (b).

Solution   The inference under consideration is a left-tailed hypothesis test for a population mean at the 5% significance level. The test statistic is

$$z = \frac{\bar{x} - \mu_0}{\sigma/\sqrt{n}} = \frac{\bar{x} - 26}{1.4/\sqrt{100}},$$

and the critical value is $-z_\alpha = -z_{0.05} = -1.645$. Thus the decision criterion for the hypothesis test is: If $z \leq -1.645$, reject H_0; if $z > -1.645$, do not reject H_0.

As we noted earlier, computing Type II error probabilities is somewhat simpler if the decision criterion is expressed in terms of $\bar{x}$ instead of z. To do that here, we must find the sample mean that is 1.645 standard deviations below the null hypothesis population mean of 26:

$$\bar{x} = 26 - 1.645 \cdot \frac{1.4}{\sqrt{100}} = 25.8.$$

The decision criterion then can be expressed in terms of $\bar{x}$ as: If $\bar{x} \leq 25.8$ mpg, reject H_0; if $\bar{x} > 25.8$ mpg, do not reject H_0. This decision criterion is portrayed graphically in Fig. 9.18.

a. Now that we have expressed the decision criterion in terms of $\bar{x}$, we can obtain Type II error probabilities by using the same techniques as in Example 9.10. We computed the Type II error probabilities that correspond to several values of μ, as shown in Table 9.9. The third column of Table 9.9 displays the powers.

b. Using Table 9.9, we can draw the power curve for the gas mileage illustration when $n = 100$, as shown in Fig. 9.19. For comparison purposes, we have also reproduced from Fig. 9.17 the power curve for $n = 30$.

c. **INTERPRETATION**   Comparing Tables 9.8 and 9.9 shows that each power is greater when $n = 100$ than when $n = 30$. Figure 9.19 displays that fact visually.

FIGURE 9.18

Graphical display of decision criterion for the gas mileage illustration ($\alpha = 0.05$, $n = 100$)

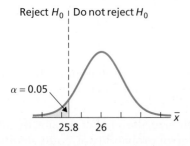

Reject H_0 | Do not reject H_0

$\alpha = 0.05$

25.8   26 $\bar{x}$

TABLE 9.9

Selected Type II error probabilities and powers for the gas mileage illustration ($\alpha = 0.05$, $n = 100$)

True mean μ	P (Type II error) β	Power $1 - \beta$
25.9	0.7611	0.2389
25.8	0.5000	0.5000
25.7	0.2389	0.7611
25.6	0.0764	0.9236
25.5	0.0162	0.9838
25.4	0.0021	0.9979
25.3	0.0002	0.9998
25.2	0.0000[†]	1.0000[‡]
25.1	0.0000	1.0000
25.0	0.0000	1.0000
24.9	0.0000	1.0000
24.8	0.0000	1.0000

[†]For $\mu \leq 25.2$, the β probabilities are 0 to four decimal places.
[‡]For $\mu \leq 25.2$, the powers are 1 to four decimal places.

FIGURE 9.19

Power curves for the gas mileage illustration when $n = 30$ and $n = 100$ ($\alpha = 0.05$)

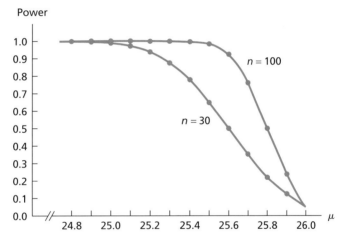

In Example 9.11, we found that increasing the sample size without changing the significance level increased the power. This relationship is true in general, so we express it as Key Fact 9.5.

KEY FACT 9.5

Sample Size and Power

For a fixed significance level, increasing the sample size increases the power.

What *Does it Mean?*

By using a sufficiently large sample size, we can obtain a hypothesis test with as much power as we want.

In practice, we need to keep in mind that larger sample sizes tend to increase the cost of a study. Consequently, we must balance, among other things, the cost of a large sample against the cost of possible errors.

As we have indicated, power is a useful way to evaluate the overall effectiveness of a hypothesis testing procedure. However, power can also be used to compare different procedures. For example, a researcher might decide between two hypothesis testing procedures on the basis of which test is more powerful for the situation under consideration.

The Technology Center

As we have shown, obtaining Type II error probabilities or powers is computationally intensive. Moreover, determining those quantities by hand can result in substantial roundoff error. Therefore, in practice, Type II error probabilities and powers are almost always calculated by computer. See the technology manuals for details.

EXERCISES 9.4

Statistical Concepts and Skills

9.44 Why don't hypothesis tests always yield correct decisions?

9.45 Define each term.
a. Type I error   **b.** Type II error   **c.** Significance level

9.46 What does the power of a hypothesis test tell you? How is it related to the probability of making a Type II error?

9.47 Why is it useful to obtain the power curve for a hypothesis test?

9.48 What happens to the power of a hypothesis test if the sample size is increased without changing the significance level? Explain your answer.

9.49 What happens to the power of a hypothesis test if the significance level is decreased without changing the sample size? Explain your answer.

9.50 Suppose that you must choose between two procedures for performing a hypothesis test—say, Procedure A and Procedure B. Further suppose that, for the same sample size and significance level, Procedure A has less power than Procedure B. Which procedure would you choose? Explain your answer.

In Exercises **9.51–9.56**, *we have given a hypothesis testing situation and (i) the population standard deviation σ, (ii) a significance level, (iii) a sample size, and (iv) some values of μ. For each exercise,*
a. express the decision criterion for the hypothesis test in terms of $\bar{x}$.
b. determine the probability of a Type I error.
c. construct a table similar to Table 9.8 on page 429 that provides the probability of a Type II error and the power for each of the given values of μ.
d. use the table obtained in part (c) to draw the power curve.

9.51 Toxic Mushrooms? The null and alternative hypotheses obtained in Exercise 9.5 on page 404 are

$$H_0: \mu = 0.5 \text{ ppm}$$
$$H_a: \mu > 0.5 \text{ ppm},$$

where μ is the mean cadmium level in *Boletus pinicola* mushrooms.
i. $\sigma = 0.37$ ii. $\alpha = 0.05$ iii. $n = 12$
iv. $\mu = 0.55, 0.60, 0.65, 0.70, 0.75, 0.80, 0.85$

9.52 Serving Time. The null and alternative hypotheses obtained in Exercise 9.6 on page 404 are

$$H_0: \mu = 16.7 \text{ months}$$
$$H_a: \mu \neq 16.7 \text{ months},$$

where μ is the mean length of imprisonment for motor-vehicle theft offenders in Sydney, Australia.
i. $\sigma = 6.0$ ii. $\alpha = 0.05$ iii. $n = 100$
iv. $\mu = 14.0, 14.5, 15.0, 15.5, 16.0, 16.5, 17.0, 17.5,$
      $18.0, 18.5, 19.0$

9.53 Cell Phones. The null and alternative hypotheses obtained in Exercise 9.7 on page 404 are

$$H_0: \mu = \$47.37$$
$$H_a: \mu < \$47.37,$$

where μ is last year's mean local monthly bill for cell phone users.
i. $\sigma = 25$ ii. $\alpha = 0.01$ iii. $n = 50$
iv. $\mu = 29, 32, 35, 38, 41, 44, 47$

9.54 Iron Deficiency? The null and alternative hypotheses obtained in Exercise 9.8 on page 404 are

$$H_0: \mu = 18 \text{ mg}$$
$$H_a: \mu < 18 \text{ mg},$$

where μ is the mean iron intake (per day) of all adult females under the age of 51.
i. $\sigma = 4.2$ ii. $\alpha = 0.01$ iii. $n = 45$
iv. $\mu = 15.50, 15.75, 16.00, 16.25, 16.50, 16.75,$
      $17.00, 17.25, 17.50, 17.75$

9.55 Body Temperature. The null and alternative hypotheses obtained in Exercise 9.9 on page 404 are

$$H_0: \mu = 98.6°\text{F}$$
$$H_a: \mu \neq 98.6°\text{F},$$

where μ is the mean body temperature of all healthy humans.
i. $\sigma = 0.63$ ii. $\alpha = 0.01$ iii. $n = 93$
iv. $\mu = 98.30, 98.35, 98.40, 98.45, 98.50, 98.55,$
      $98.65, 98.70, 98.75, 98.80, 98.85, 98.90$

9.56 Worker Fatigue. The null and alternative hypotheses obtained in Exercise 9.10 on page 405 are

$$H_0: \mu = 72 \text{ bpm}$$
$$H_a: \mu > 72 \text{ bpm},$$

where μ is the mean post-work heart rate of all casting workers.
i. $\sigma = 11.2$ ii. $\alpha = 0.05$ iii. $n = 29$
iv. $\mu = 73, 74, 75, 76, 77, 78, 79, 80$

9.57 Toxic Mushrooms? Repeat parts (a)–(d) of Exercise 9.51 for a sample size of 20. Compare your power curves for the two sample sizes and explain the principle being illustrated.

9.58 Body Temperature. Repeat parts (a)–(d) of Exercise 9.55 for a sample size of 150. Compare your power curves for the two sample sizes and explain the principle being illustrated.

Extending the Concepts and Skills

9.59 Consider a right-tailed hypothesis test for a population mean with null hypothesis $H_0: \mu = \mu_0$.

a. Draw the ideal power curve.
b. Explain what your curve in part (a) portrays.

9.60 Consider a left-tailed hypothesis test for a population mean with null hypothesis $H_0: \mu = \mu_0$.
a. Draw the ideal power curve.
b. Explain what your curve in part (a) portrays.

9.61 Consider a two-tailed hypothesis test for a population mean with null hypothesis $H_0: \mu = \mu_0$.
a. Draw the ideal power curve.
b. Explain what your curve in part (a) portrays.

Using Technology

9.62 Class Project: Questioning Gas Mileage. This exercise can be done individually or, better yet, as a class project. Refer to the gas mileage hypothesis test of Example 9.10 on page 425. Recall that the null and alternative hypotheses are

$$H_0: \mu = 26 \text{ mpg (manufacturer's claim)}$$
$$H_a: \mu < 26 \text{ mpg (consumer group's conjecture)},$$

where μ is the mean gas mileage of all Orions. Also recall that the mileages are normally distributed with a standard deviation of 1.4 mpg. Figure 9.13 on page 426 portrays the decision criterion for a test at the 5% significance level with a sample size of 30. Suppose that, in reality, the mean gas mileage of all Orions is 25.4 mpg.
a. Determine the probability of making a Type II error.
b. Simulate 100 samples of 30 gas mileages each.
c. Determine the mean of each sample in part (b).
d. For the 100 samples obtained in part (b), about how many would you expect to lead to nonrejection of the null hypothesis? Explain your answer.
e. For the 100 samples obtained in part (b), determine the number that lead to nonrejection of the null hypothesis.
f. Compare your answers from parts (d) and (e) and comment on any observed difference.

9.5 *P-Values*

In Section 9.3, we presented Procedure 9.1, a step-by-step method for performing a hypothesis test for a population mean when the population standard deviation is known. Step 4 of that procedure requires us to obtain critical values. Because of that requirement, Procedure 9.1 is said to use the **critical-value approach** to hypothesis testing.

    In this section, we discuss another approach to hypothesis testing, called the **P-value approach.** Very roughly speaking, the P-value indicates how likely (or unlikely) observation of the value obtained for the test statistic would be if the null hypothesis is true. In particular, a small *P*-value (close to 0) indicates that observation of the value obtained for the test statistic would be unlikely if the null hypothesis is true.

We can define the **P-value** (also called the **observed significance level** or the **probability value**) as the percentage of samples that would yield a value of the test statistic as extreme as or more extreme than that observed if the null hypothesis is true. But, more commonly, the *P*-value is defined in the language of probability as follows.

DEFINITION 9.6

P-Value

To obtain the *P-value* of a hypothesis test, we assume that the null hypothesis is true and compute the probability of observing a value of the test statistic as extreme as or more extreme than that observed. By *extreme* we mean "far from what we would expect to observe if the null hypothesis is true." We use the letter **P** to denote the *P*-value.

What *Does it Mean?*

Small *P*-values provide evidence against the null hypothesis; larger *P*-values do not. The smaller (closer to 0) the *P*-value, the stronger the evidence is against the null hypothesis.

In this section, we concentrate on *P*-values and the *P*-value approach for the one-sample *z*-test. But much of what we say here applies to *P*-values and the *P*-value approach for any hypothesis test.

Obtaining *P*-Values for a One-Sample *z*-Test

Recall that the test statistic for a one-sample *z*-test for a population mean with null hypothesis H_0: $\mu = \mu_0$ is

$$z = \frac{\bar{x} - \mu_0}{\sigma/\sqrt{n}}.$$

If the null hypothesis is true, this test statistic has the standard normal distribution, and its probabilities equal areas under the standard normal curve.

If we let z_0 denote the observed value of the test statistic z, we obtain the *P*-value as follows:

- *Two-tailed test:* The *P*-value is the probability of observing a value of the test statistic z at least as large in magnitude as the value actually observed, which is the area under the standard normal curve that lies outside the interval from $-|z_0|$ to $|z_0|$, as illustrated in Fig. 9.20(a).
- *Left-tailed test:* The *P*-value is the probability of observing a value of the test statistic z as small as or smaller than the value actually observed, which is the area under the standard normal curve that lies to the left of z_0, as illustrated in Fig. 9.20(b).
- *Right-tailed test:* The *P*-value is the probability of observing a value of the test statistic z as large as or larger than the value actually observed, which is the area under the standard normal curve that lies to the right of z_0, as illustrated in Fig. 9.20(c).

The best way to understand *P*-values is to examine several examples. In Examples 9.12 and 9.13, we obtain the *P*-values for two of the hypothesis tests we conducted in Section 9.3.

FIGURE 9.20

P-value for a *z*-test if the test is (a) two tailed, (b) left tailed, or (c) right tailed

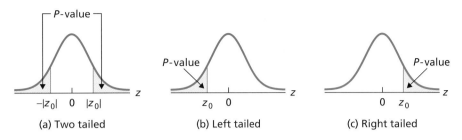

(a) Two tailed   (b) Left tailed   (c) Right tailed

‖ Example 9.12   P-Values

Prices of History Books   Consider again the history book hypothesis test of Example 9.7 where we wanted to decide whether this year's mean cost of all history books has increased from the 2000 mean of $51.46. Recall that the null and alternative hypotheses are

$$H_0: \mu = \$51.46 \text{ (mean price has not increased)}$$
$$H_a: \mu > \$51.46 \text{ (mean price has increased)},$$

where μ is this year's mean retail price of all history books. Note that the test is right tailed because a greater-than sign ($>$) appears in the alternative hypothesis.

Table 9.5 on page 418 displays this year's prices for 40 randomly selected history books; their mean price is $\bar{x} = \$54.890$. Using that sample mean and $\sigma = \$7.61$, we found the value of the test statistic to be 2.85. Obtain and interpret the *P*-value of the hypothesis test.

FIGURE 9.21

P-value for the history book hypothesis test

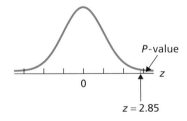

Solution   Because the hypothesis test is a right-tailed *z*-test, the *P*-value is the probability of observing a value of *z* of 2.85 or greater if the null hypothesis is true. That probability equals the area under the standard normal curve to the right of 2.85, the shaded area shown in Fig. 9.21. From Table II, we find that area to be $1 - 0.9978 = 0.0022$.

Therefore the *P*-value of this hypothesis test is 0.0022. If the null hypothesis is true, we would observe a value of the test statistic *z* of 2.85 or greater only about 2 times in 1000. In other words, if the null hypothesis is true, a random sample of 40 history books would have a mean of $54.89, or greater, about 0.2% of the time.

INTERPRETATION   The data provide very strong evidence against the null hypothesis.

‖ Example 9.13   P-Values

Clocking the Cheetah   In Example 9.9, we conducted a hypothesis test to decide whether the mean top speed of all cheetahs differs from 60 mph. Recall that the null and alternative hypotheses are

$$H_0: \mu = 60 \text{ mph (mean top speed of cheetahs is 60 mph)}$$
$$H_a: \mu \neq 60 \text{ mph (mean top speed of cheetahs is not 60 mph)},$$

where μ denotes the mean top speed of all cheetahs. Note that the hypothesis test is two tailed because a does-not-equal sign ($\neq$) appears in the alternative hypothesis.

Table 9.7 on page 420 shows the top speeds of a random sample of 35 cheetahs. Recall that the top speed of 75.3 mph is an outlier.

a. Obtain and interpret the *P*-value of the hypothesis test, using the unabridged data (i.e., including the outlier).

b. Obtain and interpret the *P*-value of the hypothesis test, using the abridged data (i.e., with the outlier removed).

c. Comment on the effect that removing the outlier has on the evidence against the null hypothesis.

Solution

a. In Example 9.9, we found the value of the test statistic for the unabridged data to be −0.88. Because the test is a two-tailed *z*-test, the *P*-value is the probability of observing a value of *z* of 0.88 or greater in magnitude if the null hypothesis is true. That probability, as depicted in Fig. 9.22(a), equals 0.3788.

Thus, if the null hypothesis is true, we would observe a value of the test statistic *z* of 0.88 or greater in magnitude more than 37 times in 100. In other words, if the null hypothesis is true, a random sample of 35 cheetahs would have a mean top speed at least as far from 60 mph as that of our unabridged sample more than 37% of the time.

INTERPRETATION The unabridged data do not provide evidence against the null hypothesis.

b. In Example 9.9, we found the value of the test statistic for the abridged data to be −1.71. In this case, the *P*-value, shown in Fig. 9.22(b), equals 0.0872. Thus, if the null hypothesis is true, we would observe a value of the test statistic *z* of 1.71 or greater in magnitude less than 9 times in 100. In other words, if the null hypothesis is true, a random sample of 34 cheetahs would have a mean top speed at least as far from 60 mph as that of our abridged sample less than 9% of the time.

INTERPRETATION The abridged data provide moderate evidence against the null hypothesis.

c. Parts (a) and (b) indicate that the strength of the evidence against the null hypothesis depends on whether the outlier is retained or removed. If the outlier is retained, there is virtually no evidence against the null hypothesis; if the outlier is removed, there is moderate evidence against the null hypothesis.

FIGURE 9.22

P-value for the cheetah speed hypothesis test (a) including the outlier, and (b) with the outlier removed

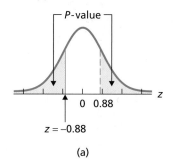

(a)

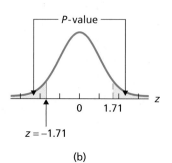

(b)

The *P*-Value Approach to Hypothesis Testing

The *P*-value can be interpreted as the *observed significance level* of a hypothesis test. To illustrate, suppose that the value of the test statistic for a right-tailed *z*-test turns out to be 1.88.   Then the *P*-value of the hypothesis test is 0.03 (actually 0.0301), as depicted by the shaded area in Fig. 9.23.

FIGURE 9.23

P-value as the observed
significance level

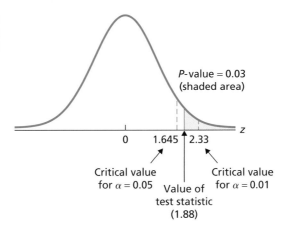

Figure 9.23 reveals that the null hypothesis would be rejected for a test at the 0.05 significance level but would not be rejected for a test at the 0.01 significance level.   In fact, Fig. 9.23 makes it clear that the *P*-value is precisely the smallest significance level at which the null hypothesis would be rejected, which we restate as Key Fact 9.6.

KEY FACT 9.6   **P-Value as the Observed Significance Level**

The *P*-value of a hypothesis test equals the smallest significance level at which the null hypothesis can be rejected, that is, the smallest significance level for which the observed sample data results in rejection of H_0.

In view of Key Fact 9.6, we have the following criterion, expressed as Key Fact 9.7, for deciding whether the null hypothesis should be rejected in favor of the alternative hypothesis.

KEY FACT 9.7   **Decision Criterion for a Hypothesis Test Using the P-Value**

If the *P*-value is less than or equal to the specified significance level, reject the null hypothesis; otherwise, do not reject the null hypothesis.

Key Fact 9.7 provides a foundation for the *P*-value approach to hypothesis testing.   For the one-sample *z*-test, then, we have Procedure 9.2 (next page).

In Example 9.8, we performed a one-sample *z*-test to decide whether the mean calcium intake of all people with incomes below the poverty level is less than the RDA of 800 mg.   To carry out that hypothesis test, we used Procedure 9.1 on page 417, which involves the critical-value approach.   We now perform that same test, using the *P*-value approach, in Example 9.14.

PROCEDURE 9.2

One-Sample z-Test (P-Value Approach)

Purpose To perform a hypothesis test for a population mean, μ

Assumptions

1. Simple random sample
2. Normal population or large sample
3. σ known

STEP 1 The null hypothesis is H_0: $\mu = \mu_0$, and the alternative hypothesis is

$$H_a: \mu \neq \mu_0 \quad \text{or} \quad H_a: \mu < \mu_0 \quad \text{or} \quad H_a: \mu > \mu_0$$
$$\text{(Two tailed)} \qquad \text{(Left tailed)} \qquad \text{(Right tailed)}$$

STEP 2 Decide on the significance level, α.

STEP 3 Compute the value of the test statistic

$$z = \frac{\bar{x} - \mu_0}{\sigma/\sqrt{n}}$$

and denote that value z_0.

STEP 4 Use Table II to obtain the P-value.

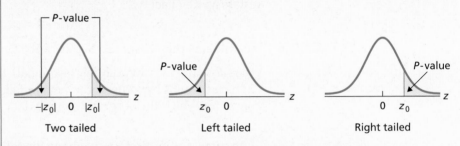

| Two tailed | Left tailed | Right tailed |

STEP 5 If $P \leq \alpha$, reject H_0; otherwise, do not reject H_0.

STEP 6 Interpret the results of the hypothesis test.

The hypothesis test is exact for normal populations and is approximately correct for large samples from nonnormal populations.

Example 9.14 The One-Sample z-Test (P-Value Approach)

Poverty and Calcium A simple random sample of 18 people with incomes below the poverty level gives the daily calcium intakes shown in Table 9.10. At the 5% significance level, do the data provide sufficient evidence to conclude that the mean calcium intake of all people with incomes below the poverty level is less than the RDA of 800 mg? Assume that $\sigma = 188$ mg.

TABLE 9.10

Daily calcium intakes (mg) for 18 people
with incomes below the poverty level

686	433	743	647	734	641
993	620	574	634	850	858
992	775	1113	672	879	609

Solution Recall that a normal probability plot of the data in Table 9.10 reveals no outliers and falls roughly in a straight line. So, for this moderate sample size, we can apply Procedure 9.2 to perform the hypothesis test.

STEP 1 State the null and alternative hypotheses.

Let μ denote the mean calcium intake (per day) of all people with incomes below the poverty level. The null and alternative hypotheses are

$$H_0: \mu = 800 \text{ mg (mean calcium intake is not less than the RDA)}$$
$$H_a: \mu < 800 \text{ mg (mean calcium intake is less than the RDA)}.$$

Note that the hypothesis test is left tailed because a less-than sign ($<$) appears in the alternative hypothesis.

STEP 2 Decide on the significance level, α.

We are to perform the test at the 5% significance level, or $\alpha = 0.05$.

STEP 3 Compute the value of the test statistic

$$z = \frac{\bar{x} - \mu_0}{\sigma/\sqrt{n}}.$$

We have $\mu_0 = 800$, $\sigma = 188$, and $n = 18$. From the data in Table 9.10, we find that $\bar{x} = 747.4$. Thus the value of the test statistic is

$$z = \frac{747.4 - 800}{188/\sqrt{18}} = -1.19.$$

FIGURE 9.24

Value of the test statistic and the *P*-value
for the calcium intake hypothesis test

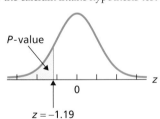

This value is shown in Fig. 9.24.

STEP 4 Use Table II to obtain the *P*-value.

The test is left tailed, so the *P*-value is the probability of observing a value of z of -1.19 or less if the null hypothesis is true. That probability equals the shaded area in Fig. 9.24, which by Table II is 0.1170. Hence $P = 0.1170$.

STEP 5 If $P \leq \alpha$, reject H_0; otherwise, do not reject H_0.

From Step 4, $P = 0.1170$. Because this *P*-value exceeds the specified significance level of 0.05, we do not reject H_0. The test results are not statistically significant at the 5% level.

STEP 6 Interpret the results of the hypothesis test.

INTERPRETATION At the 5% significance level, the data do not provide sufficient evidence to conclude that the mean calcium intake of all people with incomes below the poverty level is less than the RDA of 800 mg.

Comparison of the Critical-Value and *P*-Value Approaches

We have now discussed both the critical-value and *P*-value approaches to hypothesis testing, but we have done so only in terms of the one-sample *z*-test. For future reference, we now list the general steps involved in each approach in Table 9.11.

TABLE 9.11

Comparison of critical-value and *P*-value approaches

CRITICAL-VALUE APPROACH	*P*-VALUE APPROACH
STEP 1 State the null and alternative hypotheses.	STEP 1 State the null and alternative hypotheses.
STEP 2 Decide on the significance level, α.	STEP 2 Decide on the significance level, α.
STEP 3 Compute the value of the test statistic.	STEP 3 Compute the value of the test statistic.
STEP 4 Determine the critical value(s).	STEP 4 Determine the *P*-value.
STEP 5 If the value of the test statistic falls in the rejection region, reject H_0; otherwise, do not reject H_0.	STEP 5 If $P \le \alpha$, reject H_0; otherwise, do not reject H_0.
STEP 6 Interpret the result of the hypothesis test.	STEP 6 Interpret the result of the hypothesis test.

Using the *P*-Value to Assess the Evidence Against the Null Hypothesis

Key Fact 9.6 asserts that the *P*-value is the smallest significance level at which the null hypothesis can be rejected. Consequently, knowing the *P*-value allows us to assess significance at any level we desire. For instance, if the *P*-value of a hypothesis test is 0.03, the null hypothesis can be rejected at any significance level larger than 0.03 (e.g., $\alpha = 0.05$) and it cannot be rejected at any significance level smaller than 0.03 (e.g., $\alpha = 0.01$).

Knowing the *P*-value, we can also evaluate the strength of the evidence against the null hypothesis—the smaller the *P*-value, the stronger will be the evidence against the null hypothesis. Table 9.12 presents guidelines for interpreting the *P*-value of a hypothesis test.

In Example 9.12 (page 435) and Example 9.13 (page 435), we used the *P*-value to evaluate the strength of the evidence against the null hypothesis without reference to critical values or significance levels. This practice is common among researchers.

TABLE 9.12

Guidelines for using the *P*-value to assess the evidence against the null hypothesis

P-value	Evidence against H_0
$P > 0.10$	Weak or none
$0.05 < P \le 0.10$	Moderate
$0.01 < P \le 0.05$	Strong
$P \le 0.01$	Very strong

Hypothesis Tests Without Significance Levels: Many researchers do not explicitly refer to significance levels or critical values. Instead, they simply obtain the *P*-value of the hypothesis test and use it (or let the reader use it) to assess the strength of the evidence against the null hypothesis.

The Technology Center

Most statistical technologies have programs that automatically perform a one-sample z-test for a population mean. In this subsection, we present output and step-by-step instructions to implement such programs.

Example 9.15   Using Technology to Conduct a z-Test

Poverty and Calcium   Table 9.10 on page 439 displays the calcium intakes for a sample of 18 people with incomes below the poverty level. Use Minitab, Excel, or the TI-83/84 Plus to perform the hypothesis test in Example 9.14 on page 438.

Solution   Let μ denote the mean calcium intake (per day) of all people with incomes below the poverty level. You are to perform the hypothesis test

$$H_0\text{: } \mu = 800 \text{ mg (mean calcium intake is not less than the RDA)}$$
$$H_a\text{: } \mu < 800 \text{ mg (mean calcium intake is less than the RDA)}$$

at the 5% significance level ($\alpha = 0.05$). Note that the hypothesis test is left tailed because a less-than sign ($<$) appears in the alternative hypothesis. Also, recall that $\sigma = 188$ mg.

Printout 9.1 shows the output obtained by applying the one-sample z-test programs to the calcium intake data in Table 9.10.

PRINTOUT 9.1
One-sample z-test output for the sample of calcium intakes

MINITAB

One-Sample Z: CALCI

Test of mu = 800 vs < 800
The assumed standard deviation = 188

Variable	N	Mean	StDev	SE Mean	95% Upper Bound	Z	P
CALCI	18	747.389	171.989	44.312	820.276	-1.19	0.118

TI-83/84 PLUS

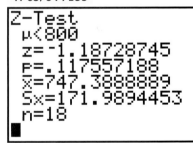

Using **Calculate**

EXCEL

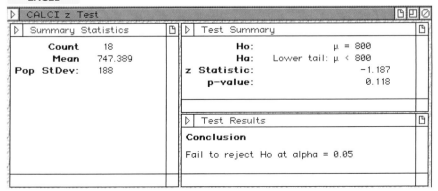

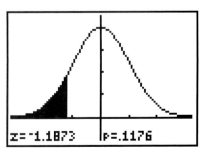

Using **Draw**

The three outputs in Printout 9.1 show that the *P*-value for the hypothesis test is 0.118. Because the *P*-value exceeds the specified significance level of 0.05, we do not reject H_0. The test results are not statistically significant at the 5% level; that is, at the 5% significance level, the data do not provide sufficient evidence to conclude that the mean calcium intake of all people with incomes below the poverty level is less than the RDA of 800 mg.

Obtaining the Output (Optional)

Printout 9.1 provides output from Minitab, Excel, and the TI-83/84 Plus for a one-sample *z*-test based on the sample of calcium intakes in Table 9.10. The following are detailed instructions for obtaining that output. First, we store the calcium intake data in a column (Minitab), range (Excel), or list (TI-83/84 Plus) named CALCI. Then, we proceed as follows.

MINITAB

1 Choose **Stat ➤ Basic Statistics ➤ 1-Sample Z...**
2 Select the **Samples in columns** option button
3 Click in the **Samples in columns** text box and specify CALCI
4 Click in the **Standard deviation** text box and type 188
5 Click in the **Test mean** text box and type 800
6 Click the **Options...** button
7 Click the arrow button at the right of the **Alternative** drop-down list box and select **less than**
8 Click **OK**
9 Click **OK**

EXCEL

1 Choose **DDXL ➤ Hypothesis Tests**
2 Select **1 Var z Test** from the **Function type** drop-down box
3 Specify CALCI in the **Quantitative Variable** text box
4 Click **OK**
5 Click the **Set μ0 and sd** button
6 Click in the **Hypothesized μ0** text box and type 800
7 Click in the **Population std dev** text box and type 188
8 Click **OK**
9 Click the **0.05** button
10 Click the $\mu < \mu$0 button
11 Click the **Compute** button

TI-83/84 PLUS

1 Press **STAT**, arrow over to **TESTS**, and press **1**
2 Highlight **Data** and press **ENTER**
3 Press the down-arrow key, type 800 for μ_0, and press **ENTER**
4 Type 188 for σ and press **ENTER**
5 Press **2nd ➤ LIST**
6 Arrow down to CALCI and press **ENTER** three times
7 Highlight $< \mu_0$ and press **ENTER**
8 Press the down-arrow key, highlight **Calculate** or **Draw**, and press **ENTER**

EXERCISES 9.5

Statistical Concepts and Skills

9.63 State two reasons why including the *P*-value is prudent when you are reporting the results of a hypothesis test.

9.64 What is the *P*-value of a hypothesis test? When does it provide evidence against the null hypothesis?

9.65 We presented two different approaches to hypothesis testing. Identify and compare these two approaches.

9.66 Explain how the *P*-value is obtained for a one-sample *z*-test in case the hypothesis test is
a. left tailed.   **b.** right tailed.   **c.** two tailed.

9.67 True or false: The *P*-value is the smallest significance level for which the observed sample data results in rejection of the null hypothesis.

9.68 In each part, we have given the significance level and *P*-value for a hypothesis test. For each case, decide whether the null hypothesis should be rejected.
a. $\alpha = 0.05$, $P = 0.06$ **b.** $\alpha = 0.10$, $P = 0.06$
c. $\alpha = 0.06$, $P = 0.06$

9.69 Which provides stronger evidence against the null hypothesis, a *P*-value of 0.02 or a *P*-value of 0.03? Explain your answer.

9.70 In each part, we have given the *P*-value for a hypothesis test. For each case, determine the strength of the evidence against the null hypothesis.
a. $P = 0.06$ **b.** $P = 0.35$
c. $P = 0.027$ **d.** $P = 0.004$

In Exercises **9.71**–**9.74**, *we have given the value obtained for the test statistic*

$$z = \frac{\bar{x} - \mu_0}{\sigma / \sqrt{n}}$$

in a one-sample z-test for a population mean. We have also specified whether the test is two tailed, left tailed, or right tailed. Determine the P-value in each case.

9.71 Right-tailed test:
a. $z = 2.03$ **b.** $z = -0.31$

9.72 Left-tailed test:
a. $z = -1.84$ **b.** $z = 1.25$

9.73 Two-tailed test:
a. $z = 3.08$ **b.** $z = -2.42$

9.74 Right-tailed test:
a. $z = 1.24$ **b.** $z = -0.69$

In Exercises 9.35–9.40 of Section 9.3, you were asked to use Procedure 9.1, which employs the critical-value approach to hypothesis testing, to perform a one-sample z-test for a population mean. Now, in Exercises **9.75**–**9.80**, *you are asked to use Procedure 9.2 on page 438, which employs the P-value approach to hypothesis testing, to perform those same hypothesis tests. In addition, use Table 9.12 on page 440 to assess the strength of the evidence against the null hypotheses.*

9.75 Toxic Mushrooms? Refer to Exercise 9.5 on page 404. Here are the data obtained by the researchers.

0.24	0.59	0.62	0.16	0.77	1.33
0.92	0.19	0.33	0.25	0.59	0.32

At the 5% significance level, do the data provide sufficient evidence to conclude that the mean cadmium level in *Boletus pinicola* mushrooms is greater than the government's recommended limit of 0.5 ppm? Assume that the population standard deviation of cadmium levels in *Boletus pinicola* mushrooms is 0.37 ppm. (*Note:* The sum of the data is 6.31 ppm.)

9.76 Serving Time. From Exercise 9.6, the mean length of imprisonment for motor-vehicle theft offenders in Australia is 16.7 months. One hundred randomly selected motor-vehicle theft offenders in Sydney, Australia, had a mean

length of imprisonment of 17.8 months. At the 5% significance level, do the data provide sufficient evidence to conclude that the mean length of imprisonment for motor-vehicle theft offenders in Sydney differs from the national mean in Australia? Assume that the population standard deviation of the lengths of imprisonment for motor-vehicle theft offenders in Sydney is 6.0 months.

9.77 Cell Phones. Refer to Exercise 9.7 on page 404. Following are last year's local monthly bills, in dollars, for a random sample of 50 cell phone users.

32.88	31.09	16.13	104.20	47.65
116.96	61.97	43.74	13.52	15.37
30.09	14.62	27.10	28.80	42.04
16.93	61.31	15.50	32.48	22.43
44.67	35.64	19.95	28.95	46.54
28.08	88.95	23.45	31.60	48.32
64.68	50.48	46.24	35.60	24.53
16.17	16.41	41.80	24.74	77.21
28.04	16.56	49.35	44.82	15.12
57.67	13.48	81.16	51.62	126.84

At the 1% significance level, do the data provide sufficient evidence to conclude that last year's mean local monthly bill for cell phone users has decreased from the 2001 mean of $47.37? Assume that $\sigma = \$25$. (*Note:* The sum of the 50 cell phone bills is $2053.48.)

9.78 Iron Deficiency? Refer to Exercise 9.8 on page 404. The following iron intakes, in milligrams, were obtained during a 24-hour period for 45 randomly selected adult females under the age of 51.

15.0	18.1	14.4	14.6	10.9	18.1	18.2	18.3	15.0
16.0	12.6	16.6	20.7	19.8	11.6	12.8	15.6	11.0
15.3	9.4	19.5	18.3	14.5	16.6	11.5	16.4	12.5
14.6	11.9	12.5	18.6	13.1	12.1	10.7	17.3	12.4
17.0	6.3	16.8	12.5	16.3	14.7	12.7	16.3	11.5

At the 1% significance level, do the data suggest that adult females under the age of 51 are, on average, getting less than the RDA of 18 mg of iron? Assume that the population standard deviation is 4.2 mg. (*Note:* $\bar{x} = 14.68$ mg.)

9.79 Body Temperature. Refer to Exercise 9.9 on page 404. The researchers obtained the following body temperatures of 93 healthy humans.

98.0	97.6	98.8	98.0	98.8	98.8	97.6	98.6	98.6
98.8	98.0	98.2	98.0	98.0	97.0	97.2	98.2	98.1
98.2	98.5	98.5	99.0	98.0	97.0	97.3	97.3	98.1
97.8	99.0	97.6	97.4	98.0	97.4	98.0	98.6	98.6
98.4	97.0	98.4	99.0	98.0	99.4	97.8	98.2	99.2
99.0	97.7	98.2	98.2	98.8	98.1	98.5	97.2	98.5
99.2	98.3	98.7	98.8	98.6	98.0	99.1	97.2	97.6
97.9	98.8	98.6	98.6	99.3	97.8	98.7	99.3	97.8
98.4	97.7	98.3	97.7	97.1	98.4	98.6	97.4	96.7
96.9	98.4	98.2	98.6	97.0	97.4	98.4	97.4	96.8
98.2	97.4	98.0						

At the 1% significance level, do the data provide sufficient evidence to conclude that the mean body temperature of healthy humans differs from 98.6°F? Assume that $\sigma = 0.63$°F. (*Note:* The sum of the 93 temperatures is 9125.5°F.)

9.80 Worker Fatigue. From Exercise 9.10 on page 405, the normal resting heart rate is 72 beats per minute (bpm). A random sample of 29 casting workers had a mean post-work heart rate of 78.3 bpm. At the 5% significance level, do the data provide sufficient evidence to conclude that the mean post-work heart rate for casting workers exceeds the normal resting heart rate of 72 bpm? Assume that the population standard deviation of post-work heart rates for casting workers is 11.2 bpm.

Extending the Concepts and Skills

9.81 Consider a one-sample z-test for a population mean. Denote z_0 as the observed value of the test statistic z. If the test is right tailed, then the P-value can be expressed as $P(z \geq z_0)$. Determine the corresponding expression for the P-value if the test is
a. left tailed.   **b.** two tailed.

9.82 The symbol $\Phi(z)$ is often used to denote the area under the standard normal curve that lies to the left of a specified value of z. Consider a one-sample z-test for a population mean. Denote z_0 as the observed value of the test statistic z. Express the P-value of the hypothesis test in terms of Φ if the test is
a. left tailed.   **b.** right tailed.   **c.** two tailed.

9.83 Obtaining the P-value. Let x denote the test statistic for a hypothesis test and x_0 its observed value. Then the P-value of the hypothesis test equals
a. $P(x \geq x_0)$ for a right-tailed test,
b. $P(x \leq x_0)$ for a left-tailed test,

c. $2 \cdot \min\{P(x \leq x_0), P(x \geq x_0)\}$ for a two-tailed test, where the probabilities are computed under the assumption that the null hypothesis is true. Suppose that you are considering a one-sample z-test for a population mean. Verify that the probability expressions in parts (a)–(c) are equivalent to those obtained in Exercise 9.81.

9.84 Discuss the relative advantages and disadvantages of using the P-value approach to hypothesis testing instead of the critical-value approach.

Using Technology

9.85 Toxic Mushrooms? Use the technology of your choice to perform the hypothesis test in Exercise 9.75.

9.86 Iron Deficiency? Use the technology of your choice to perform the hypothesis test in Exercise 9.78.

9.87 An issue of *Habitat World*, the publication of Habitat for Humanity International, contains an article on housing affordability. Included in the article is a table of fair market rents (FMR) for two-bedroom units, by state, obtained from the National Low Income Housing Coalition. According to the table, the FMR for Maine is $590. A sample of 32 randomly selected two-bedroom units in Maine yielded the following data on monthly rents.

289	597	648	669	745	577	626	661
657	595	604	739	598	545	696	450
521	669	656	565	610	503	589	472
675	586	663	609	560	507	643	749

At the 5% significance level, do the data provide sufficient evidence to conclude that the mean monthly rent for two-bedroom units in Maine differs from the FMR of $590? Assume that the standard deviation of monthly rents for two-bedroom units in Maine is $73.10.
a. Use the technology of your choice to identify potential outliers, if any.
b. Use the technology of your choice to perform the required hypothesis test, using the unabridged sample data.
c. Remove observations that are potential outliers, if any, and perform the required hypothesis test, using the abridged sample data.
d. Comment on the effect that removing the potential outliers has on the hypothesis test.
e. State your conclusion regarding the hypothesis test and explain your answer.

9.6 Hypothesis Tests for One Population Mean When σ Is Unknown

In Section 9.3, you learned how to perform a hypothesis test for one population mean when the population standard deviation, σ, is known. However, as we have mentioned, the population standard deviation is usually not known.

To develop a hypothesis-testing procedure for a population mean when σ is unknown, we begin by recalling Key Fact 8.5: If a variable x of a population is normally distributed with mean μ, then, for samples of size n, the studentized version of $\bar{x}$,

$$t = \frac{\bar{x} - \mu}{s/\sqrt{n}},$$

has the t-distribution with $n - 1$ degrees of freedom.

Because of Key Fact 8.5, we can perform a hypothesis test for a population mean when the population standard deviation is unknown by proceeding in essentially the same way as when it is known. The only difference is that we invoke a t-distribution instead of the standard normal distribution—that is, we employ the variable

$$t = \frac{\bar{x} - \mu_0}{s/\sqrt{n}}$$

as our test statistic and use the t-table, Table IV, to obtain the critical value(s) or P-value. We refer to this hypothesis-testing procedure as the **one-sample t-test** or, simply, as the **t-test.**

P-Values for a t-Test

Before presenting a step-by-step procedure for conducting a (one-sample) t-test, we need to discuss P-values for such a test. P-values for a t-test are obtained in a manner similar to that for a z-test.

As we have just shown, the test statistic for a one-sample t-test for a population mean with null hypothesis H_0: $\mu = \mu_0$ is

$$t = \frac{\bar{x} - \mu_0}{s/\sqrt{n}}.$$

If the null hypothesis is true, this test statistic has the t-distribution with $n - 1$ degrees of freedom, and its probabilities equal areas under the t-curve with df $= n - 1$.

If we let t_0 be the observed value of the test statistic t, we obtain the P-value as follows:

- *Two-tailed test:* The P-value is the probability of observing a value of the test statistic t at least as large in magnitude as the value actually observed, which is the area under the t-curve that lies outside the interval from $-|t_0|$ to $|t_0|$, as shown in Fig. 9.25(a) on the next page.
- *Left-tailed test:* The P-value is the probability of observing a value of the test statistic t as small as or smaller than the value actually observed, which is the area under the t-curve that lies to the left of t_0, as shown in Fig. 9.25(b).

- *Right-tailed test:* The P-value is the probability of observing a value of the test statistic t as large as or larger than the value actually observed, which is the area under the t-curve that lies to the right of t_0, as shown in Fig. 9.25(c).

FIGURE 9.25

P-value for a t-test if the test is (a) two tailed, (b) left tailed, or (c) right tailed

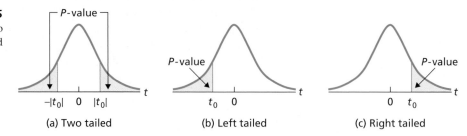

(a) Two tailed (b) Left tailed (c) Right tailed

Estimating the P-Value of a t-Test

To obtain the exact P-value of a t-test, we need to use a computer with statistical software or a statistical calculator. However, we can use t-tables, such as Table IV, to estimate the P-value of a t-test, and an estimate of the P-value is usually sufficient for deciding whether to reject the null hypothesis.

For instance, consider a right-tailed t-test with $n = 15$, $\alpha = 0.05$, and a value of the test statistic of $t = 3.458$. For df $= 15 - 1 = 14$, the t-value 3.458 is larger than any t-value in Table IV, the largest one being $t_{0.005} = 2.977$ (which means that the area under the t-curve that lies to the right of 2.977 equals 0.005). This fact, in turn, implies that the area to the right of 3.458 is less than 0.005; in other words, $P < 0.005$. Because the P-value is less than the designated significance level of 0.05, we reject H_0.

Example 9.16 provides two more illustrations of how Table IV can be used to estimate the P-value of a t-test.

Example 9.16 Using Table IV to Estimate the P-Value of a t-Test

Use Table IV to estimate the P-value of each t-test.

a. Left-tailed test, $n = 12$, and $t = -1.938$

b. Two-tailed test, $n = 25$, and $t = -0.895$

Solution

a. Because the test is left tailed, the P-value is the area under the t-curve with df $= 12 - 1 = 11$ that lies to the left of -1.938, as depicted in Fig. 9.26(a).

A t-curve is symmetric about 0, so the area to the left of -1.938 equals the area to the right of 1.938, and we can use Table IV to estimate this latter area. Concentrating on the df $= 11$ row of Table IV, we search for the two t-values that straddle 1.938; we find that they are $t_{0.05} = 1.796$ and $t_{0.025} = 2.201$. These values imply that the area under the t-curve that lies to the right of 1.938 is somewhere between 0.025 and 0.05, as depicted in Fig. 9.26(b).

Consequently, the area under the t-curve that lies to the left of -1.938 is also somewhere between 0.025 and 0.05, which means that $0.025 < P < 0.05$.

FIGURE 9.26

Estimating the *P*-value of a left-tailed *t*-test with a sample size of 12 and test statistic $t = -1.938$

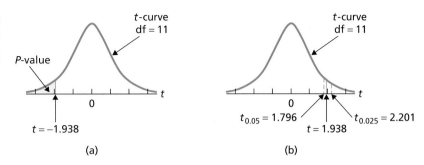

Hence we can reject H_0 at any significance level of 0.05 or larger, and we cannot reject H_0 at any significance level of 0.025 or smaller. For significance levels between 0.025 and 0.05, Table IV is not sufficiently detailed to help us to decide whether to reject H_0.[†]

b. Because the test is two tailed, the *P*-value is the area under the *t*-curve with $\text{df} = 25 - 1 = 24$ that lies either to the left of -0.895 or to the right of 0.895, as depicted in Fig. 9.27(a).

FIGURE 9.27

Estimating the *P*-value of a two-tailed *t*-test with a sample size of 25 and test statistic $t = -0.895$

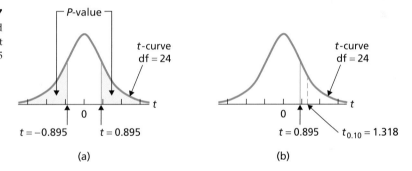

Because a *t*-curve is symmetric about 0, the area to the left of -0.895 and the area to the right of 0.895 are equal. Concentrating on the $\text{df} = 24$ row of Table IV, we find that 0.895 is smaller than any *t*-value in the table, the smallest being $t_{0.10} = 1.318$. This fact implies that the area under the *t*-curve that lies to the right of 0.895 is greater than 0.10, as depicted in Fig. 9.27(b).

Consequently, the area under the *t*-curve that lies either to the left of -0.895 or to the right of 0.895 is greater than 0.20, which means that $P > 0.20$. Hence we cannot reject H_0 at any significance level of 0.20 or smaller. For significance levels larger than 0.20, Table IV is not sufficiently detailed to help us to decide whether to reject H_0.

The One-Sample *t*-Test

We now present Procedure 9.3, a step-by-step method for performing a one-sample *t*-test. As you can see, Procedure 9.3 includes both the critical-value approach for a one-sample *t*-test (stay left) and the *P*-value approach for a one-sample *t*-test (stay right).

[†]This latter case provides an example of a *P*-value estimate that is not good enough. In such cases, use a computer with statistical software or a statistical calculator to obtain the exact *P*-value.

| PROCEDURE 9.3 | One-Sample *t*-Test |

Purpose To perform a hypothesis test for a population mean, μ

Assumptions
1. Simple random sample
2. Normal population or large sample
3. σ unknown

STEP 1 **The null hypothesis is H_0: $\mu = \mu_0$, and the alternative hypothesis is**

$$H_a: \mu \neq \mu_0 \quad \text{or} \quad H_a: \mu < \mu_0 \quad \text{or} \quad H_a: \mu > \mu_0$$
$$\text{(Two tailed)} \qquad \text{(Left tailed)} \qquad \text{(Right tailed)}$$

STEP 2 **Decide on the significance level, α.**

STEP 3 **Compute the value of the test statistic**

$$t = \frac{\bar{x} - \mu_0}{s/\sqrt{n}}$$

and denote that value t_0.

| CRITICAL-VALUE APPROACH | P-VALUE APPROACH |

STEP 4 **The critical value(s) are**

$$\pm t_{\alpha/2} \quad \text{or} \quad -t_{\alpha} \quad \text{or} \quad t_{\alpha}$$
$$\text{(Two tailed)} \qquad \text{(Left tailed)} \qquad \text{(Right tailed)}$$

with df = $n - 1$. Use Table IV to find the critical value(s).

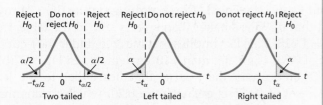

STEP 5 **If the value of the test statistic falls in the rejection region, reject H_0; otherwise, do not reject H_0.**

STEP 4 **The t-statistic has df = $n - 1$. Use Table IV to estimate the P-value, or obtain it exactly by using technology.**

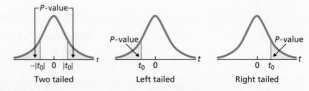

STEP 5 **If $P \leq \alpha$, reject H_0; otherwise, do not reject H_0.**

STEP 6 **Interpret the results of the hypothesis test.**

The hypothesis test is exact for normal populations and is approximately correct for large samples from nonnormal populations.

Before applying Procedure 9.3, we need to make several comments. Although the *t*-test is derived based on the assumption that the variable under consideration is normally distributed (normal population), it also applies approximately for large samples regardless of the distribution of the variable under consideration, as noted at the end of Procedure 9.3.

Actually, like the *z*-test, the *t*-test works reasonably well even when the variable under consideration is not normally distributed and the sample size is small or moderate, provided the variable is not too far from being normally distributed. In other words, the *t*-test is robust to moderate violations of the normality assumption.

When considering the *t*-test, you must also watch for outliers. Again, the presence of outliers calls into question the normality assumption. Moreover, even for large samples, outliers can sometimes unduly affect a *t*-test because the sample mean and sample standard deviation are not resistant to outliers.

Guidelines for use of the *t*-test are the same as those given for the *z*-test in Key Fact 9.4 on page 416. Remember, always look at the data before applying the *t*-test to ensure that it is reasonable to use it. Example 9.17 illustrates use of Procedure 9.3.

▌▌ Example 9.17   The One-Sample t-Test

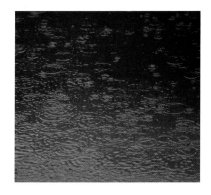

TABLE 9.13
pH levels for 15 lakes

7.2	7.3	6.1	6.9	6.6
7.3	6.3	5.5	6.3	6.5
5.7	6.9	6.7	7.9	5.8

Acid Rain and Lake Acidity   Acid rain from the burning of fossil fuels has caused many of the lakes around the world to become acidic. The biology in these lakes often collapses because of the rapid and unfavorable changes in water chemistry. A lake is classified as nonacidic if it has a pH greater than 6.

Aldo Marchetto and Andrea Lami measured the pH of high mountain lakes in the Southern Alps and reported their findings in the paper "Reconstruction of pH by Chrysophycean Scales in Some Lakes of the Southern Alps" (*Hydrobiologia*, Vol. 274, pp. 83–90). Table 9.13 shows the pH levels obtained by the researchers for 15 lakes. At the 5% significance level, do the data provide sufficient evidence to conclude that, on average, high mountain lakes in the Southern Alps are nonacidic?

Solution   A normal probability plot (not shown) of the data in Table 9.13 reveals no outliers and is quite linear. Consequently, we can apply Procedure 9.3 to conduct the required hypothesis test.

STEP 1   State the null and alternative hypotheses.

Let μ denote the mean pH level of all high mountain lakes in the Southern Alps. Then the null and alternative hypotheses are

$$H_0: \mu = 6 \text{ (mean pH level is not greater than 6)}$$
$$H_a: \mu > 6 \text{ (mean pH level is greater than 6).}$$

Note that the hypothesis test is right tailed because a greater-than sign ($>$) appears in the alternative hypothesis.

STEP 2   Decide on the significance level, α.

We are to perform the test at the 5% significance level, so $\alpha = 0.05$.

STEP 3 Compute the value of the test statistic

$$t = \frac{\bar{x} - \mu_0}{s/\sqrt{n}}.$$

We have $\mu_0 = 6$ and $n = 15$ and calculate the mean and standard deviation of the sample data in Table 9.13 as 6.6 and 0.672, respectively. Hence the value of the test statistic is

$$t = \frac{6.6 - 6}{0.672/\sqrt{15}} = 3.458.$$

CRITICAL-VALUE APPROACH	P-VALUE APPROACH

CRITICAL-VALUE APPROACH

STEP 4 The critical value for a right-tailed test is t_α, with df $= n - 1$. Use Table IV to find the critical value.

We have $n = 15$ and $\alpha = 0.05$. Table IV shows that for df $= 15 - 1 = 14$, $t_{0.05} = 1.761$, as shown in Fig. 9.28A.

FIGURE 9.28A

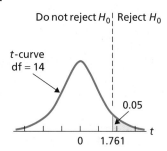

STEP 5 If the value of the test statistic falls in the rejection region, reject H_0; otherwise, do not reject H_0.

The value of the test statistic, found in Step 3, is $t = 3.458$. Figure 9.28A reveals that it falls in the rejection region. Consequently, we reject H_0. The test results are statistically significant at the 5% level.

P-VALUE APPROACH

STEP 4 The t-statistic has df $= n - 1$. Use Table IV to estimate the P-value, or obtain it exactly by using technology.

From Step 3, the value of the test statistic is $t = 3.458$. The test is right tailed, so the P-value is the probability of observing a value of t of 3.458 or greater if the null hypothesis is true. That probability equals the shaded area in Fig. 9.28B.

FIGURE 9.28B

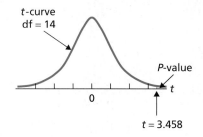

We have $n = 15$, and so df $= 15 - 1 = 14$. From Fig. 9.28B and Table IV, $P < 0.005$. (Using technology, we obtain $P = 0.00192$.)

STEP 5 If $P \leq \alpha$, reject H_0; otherwise, do not reject H_0.

From Step 4, $P < 0.005$. Because the P-value is less than the specified significance level of 0.05, we reject H_0. The test results are statistically significant at the 5% level and (see Table 9.12 on page 440) provide very strong evidence against the null hypothesis.

STEP 6 Interpret the results of the hypothesis test.

INTERPRETATION At the 5% significance level, the data provide sufficient evidence to conclude that, on average, high mountain lakes in the Southern Alps are nonacidic.

The Technology Center

Most statistical technologies have programs that automatically perform a one-sample *t*-test for a population mean. In this subsection, we present output and step-by-step instructions to implement such programs.

Example 9.18   **Using Technology to Conduct a t-Test**

Acid Rain and Lake Acidity   Table 9.13 on page 449 displays the pH levels of a sample of 15 lakes in the Southern Alps. Use Minitab, Excel, or the TI-83/84 Plus to perform the hypothesis test in Example 9.17 on page 449.

Solution   Let μ denote the mean pH level of all high mountain lakes in the Southern Alps. The problem is to perform the hypothesis test

$$H_0: \mu = 6 \text{ (mean pH level is not greater than 6)}$$
$$H_a: \mu > 6 \text{ (mean pH level is greater than 6)}$$

at the 5% significance level ($\alpha = 0.05$). Note that the hypothesis test is right tailed because a greater-than sign ($>$) appears in the alternative hypothesis.

Printout 9.2 on the next page shows the output obtained by applying the one-sample *t*-test programs to the pH data contained in Table 9.13.

From the outputs in Printout 9.2, the *P*-value for the hypothesis test is 0.002. The *P*-value is less than the specified significance level of 0.05, so we reject H_0. The test results are statistically significant at the 5% level. That is, at the 5% significance level, the data provide sufficient evidence to conclude that, on average, high mountain lakes in the Southern Alps are nonacidic.

Obtaining the Output (Optional)

Printout 9.2 provides output from Minitab, Excel, and the TI-83/84 Plus for a one-sample *t*-test based on the sample of pH levels presented in Table 9.13. The following are detailed instructions for obtaining that output. First, we store the pH data in a column (Minitab), range (Excel), or list (TI-83/84 Plus) named PH. Then, we proceed as follows.

MINITAB	**EXCEL**	**TI-83/84 PLUS**
1 Choose **Stat ➤ Basic Statistics ➤ 1-Sample t...**	1 Choose **DDXL ➤ Hypothesis Tests**	1 Press **STAT**, arrow over to **TESTS**, and press **2**
2 Select the **Samples in columns** option button	2 Select **1 Var t Test** from the **Function type** drop-down box	2 Highlight **Data** and press **ENTER**
3 Click in the **Samples in columns** text box and specify PH	3 Specify PH in the **Quantitative Variable** text box	3 Press the down-arrow key, type <u>6</u> for μ_0, and press **ENTER**
4 Click in the **Test mean** text box and type <u>6</u>	4 Click **OK**	4 Press **2nd ➤ LIST**
5 Click the **Options...** button	5 Click the **Set μ0** button and type <u>6</u>	5 Arrow down to PH and press **ENTER** three times
6 Click the arrow button at the right of the **Alternative** drop-down list box and select **greater than**	6 Click **OK**	6 Highlight $> \mu_0$ and press **ENTER**
7 Click **OK**	7 Click the **0.05** button	7 Press the down-arrow key, highlight **Calculate** or **Draw**, and press **ENTER**
8 Click **OK**	8 Click the $\mu > \mu$0 button	
	9 Click the **Compute** button	

PRINTOUT 9.2
One-sample *t*-test output for the
sample of pH levels

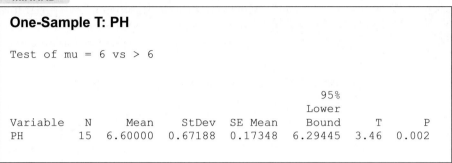

MINITAB

One-Sample T: PH

Test of mu = 6 vs > 6

					95% Lower		
Variable	N	Mean	StDev	SE Mean	Bound	T	P
PH	15	6.60000	0.67188	0.17348	6.29445	3.46	0.002

EXCEL

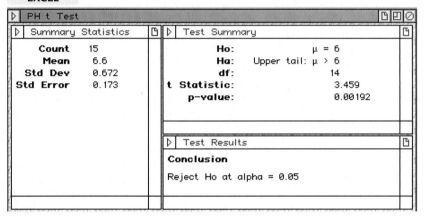

PH t Test	
Summary Statistics	**Test Summary**
Count 15	**Ho:** μ = 6
Mean 6.6	**Ha:** Upper tail: μ > 6
Std Dev 0.672	**df:** 14
Std Error 0.173	**t Statistic:** 3.459
	p-value: 0.00192
	Test Results
	Conclusion
	Reject Ho at alpha = 0.05

TI-83/84 PLUS

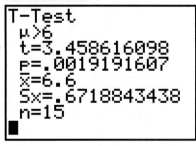

```
T-Test
 μ>6
 t=3.458616098
 p=.0019191607
 x̄=6.6
 Sx=.6718843438
 n=15
```

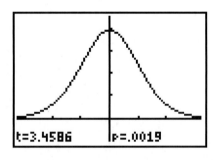

t=3.4586   p=.0019

Using **Calculate**   Using **Draw**

EXERCISES 9.6

Statistical Concepts and Skills

9.88 What is the difference in assumptions between the one-sample *t*-test and the one-sample *z*-test?

Preliminary data analyses indicate that you can reasonably use a t-test to conduct each of the hypothesis tests required in Exercises 9.89–9.94. Perform each t-test, using either the critical-value approach or the P-value approach. Comment on the practical significance of those tests whose results are statistically significant.

9.89 Apparel and Services. According to the document *Consumer Expenditures*, a publication of the U.S. Bureau of Labor Statistics, the average consumer unit spent $1856 on apparel and services in 2000. That same year, 36 consumer units in the Midwest had the following annual expenditures, in dollars, on apparel and services.

2347	1773	2199	2299	2252	2163
1437	1667	2309	1969	1597	1100
1751	2009	1546	1771	2023	1910
1981	1407	1524	1729	1726	1869
1194	1824	2245	1403	1970	2424
1682	1818	1450	1130	1610	1480

At the 5% significance level, do the data provide sufficient evidence to conclude that the 2000 mean annual expenditure on apparel and services for consumer units in the Midwest differed from the national mean of $1856? (*Note:* The sample mean and sample standard deviation of the data are $1794.11 and $351.69, respectively.)

9.90 Stressed-Out Bus Drivers. Previous studies have shown that urban bus drivers have an extremely stressful job, and a large proportion of drivers retire prematurely with disabilities due to occupational stress. These stresses come from a combination of physical and social sources such as traffic congestion, incessant time pressure, and unruly passengers. A recent paper, "Hassles on the Job: A Study of a Job Intervention With Urban Bus Drivers" by G. Evans et al. (*Journal of Organizational Behavior*, Vol. 20, pp. 199–208), examined the effects of an intervention program to improve the conditions of urban bus drivers. Among other variables, the researchers monitored diastolic blood pressure of bus drivers in downtown Stockholm, Sweden. The following data, in millimeters of mercury (mm Hg), are based on the blood pressures obtained prior to intervention for the 41 bus drivers in the study.

95	58	99	81	85	81	84
77	79	100	83	89	73	79
80	70	90	74	79	76	65
91	88	83	90	93	69	89
95	80	70	66	90	94	93
77	81	75	95	73	63	

At the 10% significance level, do the data provide sufficient evidence to conclude that the mean diastolic blood pressure of bus drivers in Stockholm, Sweden exceeds the normal diastolic blood pressure of 80 mm Hg? (*Note:* The sample mean and sample standard deviation of the data are 81.76 mm Hg and 10.32 mm Hg, respectively.)

9.91 Brewery Effluent and Crops. Because many industrial wastes contain nutrients that enhance crop growth,

efforts are being made for environmental purposes to use such wastes on agricultural soils. Two researchers, Mohammad Ajmal and Ahsan Ullah Khan, reported their findings on experiments with brewery wastes used for agricultural purposes in the article "Effects of Brewery Effluent on Agricultural Soil and Crop Plants" (*Environmental Pollution (Series A)*, 33, pp. 341–351). The researchers studied the physico-chemical properties of effluent from Mohan Meakin Breweries Ltd. (MMBL), Ghazibad, UP, India, and "…its effects on the physico-chemical characteristics of agricultural soil, seed germination pattern, and the growth of two common crop plants." They assessed the impact of using different concentrations of the effluent: 25%, 50%, 75%, and 100%. The following data, based on the results of the study, provide the percentages of limestone in the soil obtained by using 100% effluent.

2.41	2.31	2.54	2.28	2.72
2.60	2.51	2.51	2.42	2.70

Do the data provide sufficient evidence to conclude, at the 1% level of significance, that the mean available limestone in soil treated with 100% MMBL effluent exceeds 2.30%, the percentage ordinarily found? (*Note:* $\bar{x} = 2.5$ and $s = 0.149$.)

9.92 TV Viewing. According to *Communications Industry Forecast*, published by Veronis Suhler Stevenson of New York, NY, the average person watched 4.47 hours of television per day in 2000. A random sample of 40 people gave the following number of hours of television watched per day for last year.

2.3	9.2	3.3	8.0	7.0	5.3	2.1	9.0
2.7	4.7	2.6	4.2	3.4	6.4	4.7	2.2
2.9	5.2	3.5	5.8	2.9	3.3	3.7	7.8
5.3	5.4	5.5	1.6	5.4	7.5	2.3	5.2
3.3	2.6	8.9	4.5	0.0	3.8	2.6	8.5

At the 10% significance level, do the data provide sufficient evidence to conclude that the amount of television watched per day last year by the average person differed from that in 2000? (*Note:* $\bar{x} = 4.615$ hours and $s = 2.277$ hours.)

9.93 Ankle Brachial Index. The Ankle Brachial Index (ABI) compares the blood pressure of a patient's arm to the blood pressure of the patient's leg. The ABI can be an indicator of different diseases, including arterial diseases. A healthy (or normal) ABI is 0.9 or greater. In a study by M. McDermott et al. titled "Sex Differences in Peripheral Arterial Disease: Leg Symptoms and Physical Functioning" (*Journal of the American Geriatrics Society*, Vol. 51, No. 2, pp. 222–228), the researchers obtained the ABI of 187 women with peripheral arterial disease. The results were a mean ABI of 0.64 with a standard deviation of 0.15. At the 5% significance level, do the data provide sufficient evidence to conclude that women with peripheral arterial disease have an unhealthy ABI?

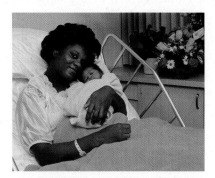

9.94 Active Management of Labor. Having a baby in a U.S. hospital costs $2528, on average. Active management of labor (AML) is a group of interventions designed to help reduce the length of labor and the rate of cesarean deliveries. Physicians from the Department of Obstetrics and Gynecology at the University of New Mexico Health Sciences Center were interested in determining whether AML would also translate into a reduced cost for delivery. The results of their study can be found in Rogers et al., "Active Management of Labor: A Cost Analysis of a Randomized Controlled Trial" (*Western Journal of Medicine*, Vol. 172, pp. 240–243). According to the article, 200 AML deliveries had a mean cost of $2480 with a standard deviation of $766. At the 5% significance level, do the data provide sufficient evidence to conclude that, on average, AML reduces the cost of having a baby in a U.S. hospital?

Extending the Concepts and Skills

9.95 Beef Consumption. According to *Food Consumption, Prices, and Expenditures*, published by the U.S. Department of Agriculture, the mean consumption of beef per person in 2000 was 64 lb (boneless, trimmed weight). A sample of 40 people taken this year yielded the following data, in pounds, on last year's beef consumption.

77	65	57	54	68	79	56	0
50	49	51	56	56	78	63	72
0	62	74	61	61	60	56	37
76	77	67	67	62	89	56	75
69	73	75	62	8	71	20	47

a. Use the sample data to decide, at the 5% significance level, whether last year's mean beef consumption is less than the 2000 mean of 64 lb. (*Note:* The mean and standard deviation of the sample data are 58.40 lb and 20.42 lb, respectively.)

b. The sample data contain four potential outliers: 0, 0, 8, and 20. Remove those four observations and repeat the hypothesis test in part (a). (*Note:* The mean and standard deviation of the abridged sample data are 64.11 lb and 11.02 lb, respectively.)

c. Compare your results in parts (a) and (b).

d. Assuming that the four potential outliers are not recording errors, comment on the advisability of removing them from the sample data before performing the hypothesis test.

e. What action would you take regarding this hypothesis test?

9.96 Suppose that you want to perform a hypothesis test for a population mean based on a small sample but that preliminary data analyses indicate either the presence of outliers or that the variable under consideration is far from normally distributed.

a. Is either the *z*-test or *t*-test appropriate?

b. If not, what type of procedure might be appropriate?

Using Technology

9.97 Apparel and Services. Refer to Exercise 9.89. Use the technology of your choice to

a. obtain a normal probability plot, a boxplot, a histogram, and a stem-and-leaf diagram of the data.

b. perform the required hypothesis test.

c. Is the use of your procedure in part (b) justified? Explain your answer.

9.98 Stressed-Out Bus Drivers. Refer to Exercise 9.90. Use the technology of your choice to

a. obtain a normal probability plot, a boxplot, a histogram, and a stem-and-leaf diagram of the data.

b. perform the required hypothesis test.

c. Is the use of your procedure in part (b) justified? Explain your answer.

9.99 Beef Consumption. Refer to Exercise 9.95. Use the technology of your choice to

a. identify the potential outliers.

b. perform the hypothesis test considered in part (a) of Exercise 9.95.

c. perform the hypothesis test considered in part (b) of Exercise 9.95.

d. Compare the results obtained in parts (b) and (c).

9.7   The Wilcoxon Signed-Rank Test[†]

Up to this point, we have presented two methods for performing a hypothesis test for a population mean. If the population standard deviation is known, we can use the z-test; if the population standard deviation is unknown, we can use the t-test.

But both procedures require another assumption for their use: The variable under consideration should be approximately normally distributed or the sample size should be relatively large. For small samples, both procedures should be avoided in the presence of outliers.

In this section, we describe a third method for performing a hypothesis test for a population mean—the **one-sample Wilcoxon signed-rank test** or, simply, the **Wilcoxon signed-rank test.** This test, which is sometimes more appropriate than either the z-test or the t-test, is an example of a nonparametric method.

What Is a Nonparametric Method?

Recall that descriptive measures for population data, such as μ and σ, are called parameters. Technically, inferential methods concerned with parameters are called **parametric methods;** those that are not are called **nonparametric methods.** However, common statistical practice is to refer to most methods that can be applied without assuming normality as nonparametric. Thus the term *nonparametric method* as used in contemporary statistics is somewhat of a misnomer.

Nonparametric methods have both advantages and disadvantages. On the one hand, they usually entail fewer and simpler computations than parametric methods and are resistant to outliers and other extreme values. On the other hand, they are not as powerful as parametric methods, such as the z-test and t-test, when the requirements for use of parametric methods are met.

The Logic Behind the Wilcoxon Signed-Rank Test

The Wilcoxon signed-rank test is based on the assumption that the variable under consideration has a *symmetric distribution*—one that can be divided into two pieces that are mirror images of each other—but does not require that its distribution be normal or have any other specific shape. Thus, for example, the Wilcoxon signed-rank test applies to a variable that has a normal, triangular, uniform, or symmetric bimodal distribution, but not to one that has a right-skewed or left-skewed distribution. Example 9.19 explains the reasoning behind the Wilcoxon signed-rank test.

Example 9.19   **Introducing the Wilcoxon Signed-Rank Test**

Weekly Food Costs   The U.S. Department of Agriculture publishes information about food costs in *Agricultural Research Service*. According to that document, a typical U.S. family of four spends about $157 per week on food. Ten

---

[†]This section begins the coverage of nonparametric statistics.

randomly selected Kansas families of four have the weekly food costs shown in Table 9.14. Do the data provide sufficient evidence to conclude that the mean weekly food cost for Kansas families of four is less than the national mean of $157?

TABLE 9.14

Sample of weekly food costs ($)

143	169	149	135	161
138	152	150	141	159

Solution Let μ denote the mean weekly food cost for all Kansas families of four. We want to perform the hypothesis test

H_0: $\mu = \$157$ (mean weekly food cost is not less than $157)

H_a: $\mu < \$157$ (mean weekly food cost is less than $157).

FIGURE 9.29

Stem-and-leaf diagram of sample data in Table 9.14

13	8 5
14	3 1
14	9
15	2 0
15	9
16	1
16	9

As we said, a condition for the use of the Wilcoxon signed-rank test is that the variable under consideration have a symmetric distribution. If the weekly food costs for Kansas families of four have a symmetric distribution, a graphical display of the sample data should be roughly symmetric.

Figure 9.29 shows a stem-and-leaf diagram of the sample data in Table 9.14. The diagram is roughly symmetric and so does not reveal any obvious violations of the symmetry condition.[†] We therefore apply the Wilcoxon signed-rank test to carry out the hypothesis test.

To begin, we rank the data in Table 9.14 according to distance and direction from the null hypothesis mean, $\mu_0 = \$157$. The steps for doing so are presented in Table 9.15.

TABLE 9.15

Steps for ranking the data in Table 9.14 according to distance and direction from the null hypothesis mean

Cost ($) x	Difference $D = x - 157$	$|D|$	Rank of $|D|$	Signed rank R
143	−14	14	7	−7
138	−19	19	9	−9
169	12	12	6	6
152	−5	5	3	−3
149	−8	8	5	−5
150	−7	7	4	−4
135	−22	22	10	−10
141	−16	16	8	−8
161	4	4	2	2
159	2	2	1	1

STEP 1   *Subtract μ_0 from x.*

STEP 2   *Make each difference positive by taking absolute values.*

STEP 3   *Rank the absolute differences in order from smallest (1) to largest (10).*

STEP 4   *Give each rank the same sign as the sign in Column 2.*

[†]For ease in explaining the Wilcoxon signed-rank test, we have chosen an example in which the sample size is very small. This selection, however, makes it difficult to effectively check the symmetry condition. In general, we must proceed cautiously when dealing with very small samples.

The absolute differences, $|D|$, displayed in the third column, identify how far each observation is from 157. The ranks of those absolute differences, displayed in the fourth column, show which observations are closer to 157 and which are farther away. The signed ranks, R, displayed in the last column, indicate additionally whether an observation is greater than 157 $(+)$ or less than 157 $(-)$. Figure 9.30 depicts the information for the second and third rows of Table 9.15.

FIGURE 9.30
Meaning of signed ranks for the observations 138 and 169

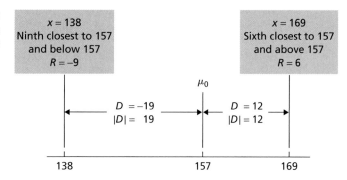

The reasoning behind the Wilcoxon signed-rank test is: If the null hypothesis, $\mu = \$157$, is true, then, because the distribution of weekly food costs is symmetric, we expect the sum of the positive ranks and the sum of the negative ranks to be roughly the same in magnitude. For the sample size of 10, the sum of all the ranks must be $1 + 2 + \cdots + 10 = 55$ and half of 55 is 27.5.

Thus, if the null hypothesis is true, we expect the sum of the positive ranks (and the sum of the negative ranks) to be roughly 27.5. To put it another way, if the sum of the positive ranks is too much smaller than 27.5, we take this result as evidence that the null hypothesis is false and conclude that the mean weekly food cost is less than $157. From the last column of Table 9.15, the sum of the positive ranks, denoted W, equals $6 + 2 + 1 = 9$. This value is quite a bit smaller than 27.5, or what we would expect if the mean is in fact $157.

The question now is: Can the difference between the observed and expected values of W be reasonably attributed to sampling error or does it indicate that the mean weekly food cost for Kansas families of four is actually less than $157? To answer that question, we need a table of critical values for W, which we present as Table V in Appendix A. We discuss that table and then return to complete the hypothesis test.

Using the Wilcoxon Signed-Rank Table

Table V in Appendix A provides critical values for a Wilcoxon signed-rank test.[†]
Recall that *a critical value from Table V is to be included as part of the rejection region.*

The two outside columns of Table V give the sample size, n. As expected, the symbol W_α denotes the W-value that has area (percentage, probability) α to its right. Thus the column headed $W_{0.10}$ contains W-values having area 0.10 to their right, the column headed $W_{0.05}$ contains W-values having area 0.05 to their right, and so on. The entries in these columns provide critical values for a right-tailed Wilcoxon signed-rank test.

[†]Actually, the significance levels presented in Table V are only approximate, but they are considered close enough to use in practice.

The distribution of the variable W is symmetric about $n(n+1)/4$. This characteristic, in turn, implies that the W-value that has area α to its left—or, equivalently, area $1 - \alpha$ to its right—equals $n(n+1)/2$ minus the W-value that has area α to its right. In symbols,

$$W_{1-\alpha} = n(n+1)/2 - W_{\alpha}. \qquad (9.1)$$

Similarly,

$$W_{1-\alpha/2} = n(n+1)/2 - W_{\alpha/2}. \qquad (9.2)$$

Equations (9.1) and (9.2) show that we can also obtain from Table V the critical value(s) for a left-tailed Wilcoxon signed-rank test and a two-tailed Wilcoxon signed-rank test, respectively. Example 9.20 illustrates use of Table V.

Example 9.20   Using the Wilcoxon Signed-Rank Table, Table V

Refer to Example 9.19, which begins on page 455. Determine the critical value, rejection region, and nonrejection region for the hypothesis test if it is to be performed at the 5% significance level.

Solution   The hypothesis test in Example 9.19 is left tailed and the sample size is 10. We want a 5% significance level, or $\alpha = 0.05$.

To find the critical value, we first go down the outside columns, labeled n, of Table V to "10." Then we go across that row to the column headed $W_{0.05}$. The number in the body of the table there, 44, is the critical value for a right-tailed test. Now, applying Equation (9.1) and recalling that $n = 10$, we find that the critical value for a left-tailed test at the 5% significance level is

$$W_{1-0.05} = W_{0.95} = 10(10+1)/2 - W_{0.05} = 55 - 44 = 11.$$

Thus, for the hypothesis test in Example 9.19 performed at the 5% significance level, the critical value is $W = 11$, the rejection region consists of all W-values less than or equal to 11, and the nonrejection region consists of all W-values greater than 11. Therefore we reject H_0 if $W \le 11$ and do not reject H_0 if $W > 11$, as shown in Fig. 9.31.

FIGURE 9.31
Rejection and nonrejection regions for a left-tailed Wilcoxon signed-rank test with $\alpha = 0.05$ and $n = 10$

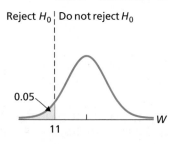

Reject H_0 ┆ Do not reject H_0

0.05

11

W

Note: Although the variable W is discrete, we drew its "histogram" in Fig. 9.31 in the shape of a normal curve. This approach is not only convenient, but is also acceptable, because W is close to normally distributed except for very small sample sizes. We use this graphical convention throughout this section.

We can also use Table V to estimate the P-value of a Wilcoxon signed-rank test, but to do so is awkward and tedious. Hence we assume that the user of the P-value approach to hypothesis testing for a Wilcoxon signed-rank test has access to statistical software.

Performing the Wilcoxon Signed-Rank Test

Procedure 9.4 provides a step-by-step method for performing a Wilcoxon signed-rank test. Note that we often use the phrase "symmetric population" to indicate that the variable under consideration has a symmetric distribution.

PROCEDURE 9.4   Wilcoxon Signed-Rank Test

Purpose To perform a hypothesis test for a population mean, μ

Assumptions

1. Simple random sample
2. Symmetric population

STEP 1 The null hypothesis is H_0: $\mu = \mu_0$, and the alternative hypothesis is

$$H_a: \mu \neq \mu_0 \quad \text{or} \quad H_a: \mu < \mu_0 \quad \text{or} \quad H_a: \mu > \mu_0$$
$$\text{(Two tailed)} \qquad \text{(Left tailed)} \qquad \text{(Right tailed)}$$

STEP 2 Decide on the significance level, α.

STEP 3 Construct a work table of the following form.

Observation x	Difference $D = x - \mu_0$	$|D|$	Rank of $|D|$	Signed rank R
.	.	.	.	.
.	.	.	.	.
.	.	.	.	.

STEP 4 Compute the value of the test statistic

$$W = \text{sum of the positive ranks}$$

and denote that value W_0.

CRITICAL-VALUE APPROACH

STEP 5 The critical value(s) are

$$W_{1-\alpha/2} \text{ and } W_{\alpha/2} \quad \text{or} \quad W_{1-\alpha} \quad \text{or} \quad W_{\alpha}$$
$$\text{(Two tailed)} \qquad \text{(Left tailed)} \qquad \text{(Right tailed)}$$

Use Table V and, if necessary, Equation (9.1) or (9.2) on page 458 to find the critical value(s).

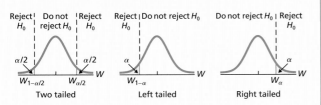

STEP 6 If the value of the test statistic falls in the rejection region, reject H_0; otherwise, do not reject H_0.

P-VALUE APPROACH

STEP 5 Obtain the *P*-value by using technology.

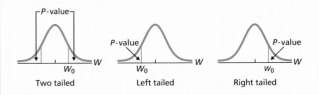

STEP 6 If $P \leq \alpha$, reject H_0; otherwise, do not reject H_0.

STEP 7 Interpret the results of the hypothesis test.

Example 9.21   The Wilcoxon Signed-Rank Test

Weekly Food Costs   To apply Procedure 9.4, let's complete the hypothesis test of Example 9.19. A random sample of 10 Kansas families of four yielded the data on weekly food costs shown in Table 9.14 on page 456. At the 5% significance level, do the data provide sufficient evidence to conclude that the mean weekly food cost for Kansas families of four is less than the national mean of $157?

Solution   We apply Procedure 9.4.

STEP 1 State the null and alternative hypotheses.

Let μ denote the mean weekly food cost for all Kansas families of four. Then the null and alternative hypotheses are

$$H_0: \mu = \$157 \text{ (mean weekly food cost is not less than \$157)}$$
$$H_a: \mu < \$157 \text{ (mean weekly food cost is less than \$157).}$$

Note that the hypothesis test is left tailed because a less-than sign ($<$) appears in the alternative hypothesis.

STEP 2 Decide on the significance level, α.

The test is to be performed at the 5% significance level, or $\alpha = 0.05$.

STEP 3 Construct a work table.

We have already done so (see Table 9.15 on page 456).

STEP 4 Compute the value of the test statistic

$$W = \text{sum of the positive ranks.}$$

The last column of Table 9.15 shows that the sum of the positive ranks equals

$$W = 6 + 2 + 1 = 9.$$

CRITICAL-VALUE APPROACH	P-VALUE APPROACH
STEP 5 The critical value for a left-tailed test is $W_{1-\alpha}$. Use Table V and Equation (9.1) to find the critical value.	**STEP 5 Obtain the P-value by using technology.**
We have already determined the critical value in Example 9.20; it is 11, as shown in Fig. 9.31 on page 458.	Using technology, we find that the P-value for the hypothesis test is $P = 0.033$.
STEP 6 If the value of the test statistic falls in the rejection region, reject H_0; otherwise, do not reject H_0.	**STEP 6 If $P \le \alpha$, reject H_0; otherwise, do not reject H_0.**
The value of the test statistic is $W = 9$, as found in Step 4, which falls in the rejection region shown in Fig. 9.31. Thus we reject H_0. The test results are statistically significant at the 5% level.	From Step 5, $P = 0.033$. Because the P-value is less than the specified significance level of 0.05, we reject H_0. The test results are statistically significant at the 5% level and (see Table 9.12 on page 440) provide strong evidence against the null hypothesis.

STEP 7 Interpret the results of the hypothesis test.

INTERPRETATION At the 5% significance level, the data provide sufficient evidence to conclude that the mean weekly food cost for Kansas families of four is less than the national mean of $157.

As mentioned earlier, one advantage of nonparametric methods is that they are resistant to outliers. We can illustrate that advantage for the Wilcoxon signed-rank test by referring to Example 9.21.

The stem-and-leaf diagram depicted in Fig. 9.29 on page 456 shows that the sample data presented in Table 9.15 contain no outliers. The smallest observation, and also the farthest from the null hypothesis mean of 157, is 135. Replacing 135 by, say, 85, introduces an outlier but has no effect on the value of the test statistic and hence none on the hypothesis test itself. (Why is that so?)

Note: The following points may be relevant when performing a Wilcoxon signed-rank test:

- If an observation equals μ_0 (the value for the mean in the null hypothesis), that observation should be removed and the sample size reduced by 1.
- If two or more absolute differences are tied, each should be assigned the mean of the ranks they would have had if there were no ties. For example, if two absolute differences are tied for second place, each should be assigned rank $(2 + 3)/2 = 2.5$, and rank 4 should be assigned to the next largest absolute difference, which really is fourth. If three absolute differences are tied for fifth place, each should be assigned rank $(5 + 6 + 7)/3 = 6$, and rank 8 should be assigned to the next largest absolute difference.

Because the mean and median of a symmetric distribution are identical, a Wilcoxon signed-rank test can be used to perform a hypothesis test for a population median, η, as well as for a population mean, μ. To use Procedure 9.4 to carry out a hypothesis test for a population median, simply replace μ by η and μ_0 by η_0. In some of the exercises at the end of this section, you will be asked to use the Wilcoxon signed-rank test to perform hypothesis tests for a population median.

Comparing the Wilcoxon Signed-Rank Test and the *t*-Test

As you learned in Section 9.6, a *t*-test can be used to conduct a hypothesis test for a population mean when the variable under consideration is normally distributed. Because a normally distributed variable necessarily has a symmetric distribution, we can also use the Wilcoxon signed-rank test to perform such a hypothesis test.

The question now becomes: If we want to perform a hypothesis test for a population mean and we know that the variable under consideration is normally distributed, should we use the *t*-test or the Wilcoxon signed-rank test? As you might expect, we should use the *t*-test. For a normally distributed variable, the *t*-test is more powerful than the Wilcoxon signed-rank test because it is designed expressly for such variables; surprisingly, however, the *t*-test is not much more powerful than the Wilcoxon signed-rank test.

However, if the variable under consideration has a symmetric distribution but is not normally distributed, the Wilcoxon signed-rank test is usually more powerful than the *t*-test and is often considerably more powerful. In summary, we have Key Fact 9.8.

KEY FACT 9.8

Wilcoxon Signed-Rank Test Versus the t-Test

Suppose that a hypothesis test is to be performed for a population mean. When deciding between the *t*-test and the Wilcoxon signed-rank test, follow these guidelines:

- If you are reasonably sure that the variable under consideration is normally distributed, use the *t*-test.
- If you are not reasonably sure that the variable under consideration is normally distributed but are reasonably sure that it has a symmetric distribution, use the Wilcoxon signed-rank test.

The Technology Center

Some statistical technologies have dedicated programs that automatically perform a Wilcoxon signed-rank test for one population mean, but others do not. For instance, Minitab currently has such a program, but neither Excel nor the TI-83/84 Plus does. In this subsection, we present output and step-by-step instructions to implement Minitab's Wilcoxon signed-rank test program.

For a statistical technology that does not have a dedicated program for conducting a particular type of hypothesis test, such as the Wilcoxon signed-rank test, the macro capabilities of the statistical technology can often be used to write a program. Refer to the technology manuals for more details.

As we said earlier, a Wilcoxon signed-rank test can be used to perform a hypothesis test for a population median, η, as well as for a population mean, μ. Many statistical technologies, including Minitab, present the output of that procedure in terms of the median, but that output can also be interpreted in terms of the mean.

Example 9.22 **Using Technology to Conduct a Wilcoxon Signed-Rank Test**

Weekly Food Costs Table 9.14 on page 456 displays the weekly food costs for 10 Kansas families of four. Use Minitab to perform the hypothesis test in Example 9.21 on page 460.

Solution Let μ denote the mean weekly food cost for all Kansas families of four. The problem is to perform the hypothesis test

$$H_0: \mu = \$157 \text{ (mean weekly food cost is not less than \$157)}$$
$$H_a: \mu < \$157 \text{ (mean weekly food cost is less than \$157)}$$

at the 5% significance level. Note that the hypothesis test is left tailed because a less-than sign ($<$) appears in the alternative hypothesis.

Printout 9.3 shows the output obtained by applying Minitab's Wilcoxon signed-rank test program to the weekly-food-cost data in Table 9.14.

PRINTOUT 9.3

Wilcoxon signed-rank test output for the sample of weekly food costs

MINITAB

Wilcoxon Signed Rank Test: COST

Test of median = 157.0 versus median < 157.0

	N	N for Test	Wilcoxon Statistic	P	Estimated Median
COST	10	10	9.0	0.033	149.5

From the Minitab output in Printout 9.3, the *P*-value for the hypothesis test is 0.033. Because the *P*-value is less than the specified significance level of 0.05, we reject H_0. At the 5% significance level, the data provide sufficient evidence to conclude that the mean weekly food cost for Kansas families of four is less than the national mean of $157.

Obtaining the Output (Optional)

Printout 9.3 provides output from Minitab for a Wilcoxon signed-rank test based on the sample of food costs in Table 9.14. The following are detailed instructions for obtaining that output. First, we store the food-cost data in a column named COST. Then, we proceed as follows.

MINITAB	**EXCEL**	**TI-83/84 PLUS**
1 Choose **Stat ➤ Nonparametrics ➤ 1-Sample Wilcoxon…** 2 Specify COST in the **Variables** text box 3 Select the **Test median** option button 4 Click in the **Test median** text box and type 157 5 Click the arrow button at the right of the **Alternative** drop-down list box and select **less than** 6 Click **OK**	SEE THE EXCEL MANUAL	SEE THE TI-83/84 PLUS MANUAL

EXERCISES 9.7

Statistical Concepts and Skills

9.100 Technically, what is a *nonparametric method?* In current statistical practice, how is that term used?

9.101 Discuss advantages and disadvantages of nonparametric methods relative to parametric methods.

9.102 What assumption must be met in order to use the Wilcoxon signed-rank test?

9.103 We mentioned that if, in a Wilcoxon signed-rank test, an observation equals μ_0 (the value given for the mean in the null hypothesis), that observation should be removed and the sample size reduced by 1. Why does that need to be done?

9.104 Suppose that you want to perform a hypothesis test for a population mean. Assume that the population standard deviation is unknown and that the sample size is relatively small. In each part, we have given the distribution shape of the variable under consideration. Decide whether you would use the *t*-test, the Wilcoxon signed-rank test, or neither.
a. Uniform **b.** Normal
c. Reverse J shaped

9.105 Suppose that you want to perform a hypothesis test for a population mean. Assume that the population standard deviation is unknown and that the sample size is relatively small. In each part, we have given the distribution shape of the variable under consideration. Decide whether you would use the *t*-test, the Wilcoxon signed-rank test, or neither.
a. Triangular **b.** Symmetric bimodal
c. Left skewed

9.106 The Wilcoxon signed-rank test can be used to perform a hypothesis test for a population median, η, as well as for a population mean, μ. Why is that so?

In each of Exercises 9.107–9.110, use the Wilcoxon signed-rank test to perform the required hypothesis test.

9.107 How Old People Are. In 2000, the median age of U.S. residents was 35.7 years, as reported by the U.S. Bureau of the Census in *Current Population Reports*. A random sample of 10 U.S. residents taken this year yielded the following ages, in years.

42	62	14	57	36
45	49	39	11	26

At the 1% significance level, do the data provide sufficient evidence to conclude that the median age of today's U.S. residents has increased from the 2000 median age of 35.7 years?

9.108 Beverage Expenditures. The U.S. Bureau of Labor Statistics publishes information on average annual expenditures by consumers in *Consumer Expenditures*. In 2000, the mean amount spent per consumer unit on nonalcoholic beverages was $250. A random sample of 12 consumer units yielded the following data, in dollars, on last year's expenditures on nonalcoholic beverages.

395	210	218	299	293	315
274	307	293	283	228	292

At the 5% significance level, do the data provide sufficient evidence to conclude that last year's mean amount spent by consumers on nonalcoholic beverages has increased from the 2000 mean of $250?

9.109 Pricing Pickups. The *Kelley Blue Book* provides data on retail and trade-in values for used cars and trucks. The retail value represents the price a dealer might charge after preparing the vehicle for sale. A 1996 Nissan XE King Cab pickup has a 2003 *Kelley Blue Book* retail value of $6735. We obtained the following asking prices, in dollars, for a sample of 1996 Nissan XE King Cab pickups for sale in Phoenix, Arizona.

5500	8250	4497	6088	4900
9377	6088	3877	6640	3900

At the 10% significance level, do the data provide sufficient evidence to conclude that the mean asking price for 1996 XE King Cab pickups in Phoenix is less than the 2003 *Kelley Blue Book* value?

9.110 Birth Weights. The National Center for Health Statistics reports in *Vital Statistics of the United States* that the median birth weight of U.S. babies was 7.4 lb in 2000. A random sample of this year's births provided the following weights, in pounds.

8.6	7.4	5.3	13.8	7.8	5.7	9.2
8.8	8.2	9.2	5.6	6.0	11.6	7.2

Can we conclude that this year's median birth weight differs from that in 2000? Use a significance level of 0.05.

9.111 Brewery Effluent and Crops. Refer to Exercise 9.91 on page 453. Two researchers, Mohammad Ajmal and Ahsan Ullah Khan, reported their findings on experiments with brewery wastes used for agricultural purposes in the article

"Effects of Brewery Effluent on Agricultural Soil and Crop Plants" (*Environmental Pollution (Series A)*, 33, pp. 341–351). The following data, based on the results of the study, provide the percentages of limestone in the soil obtained by using 100% effluent.

2.41	2.31	2.54	2.28	2.72
2.60	2.51	2.51	2.42	2.70

a. Can you conclude that the mean available limestone in soil treated with 100% MMBL effluent exceeds 2.30%, the percentage ordinarily found? Perform a Wilcoxon signed-rank test at the 1% significance level.

b. The hypothesis test considered in part (a) was done in Exercise 9.91 with a *t*-test. The assumption in that exercise is that the percentage of limestone in the soil obtained by using 100% effluent is normally distributed. If that is the case, why is it permissible to perform a Wilcoxon signed-rank test for the mean available limestone in soil treated with 100% MMBL effluent?

9.112 Ethical Food Choice Motives. In the paper "Measurement of Ethical Food Choice Motives" (*Appetite*, Vol. 34, pp. 55–59), research psychologists M. Lindeman and M. Väänänen of the University of Helsinki published a study on the factors that most influence peoples' choice of food. One of the questions asked of the participants was how important, on a scale of 1 to 4 (1 = not at all important, 4 = very important), is ecological welfare in food choice motive, where ecological welfare includes animal welfare and environmental protection. Following are the responses of a random sample of 18 Helsinkians.

3	2	2	3	3	3	2	2	3
3	3	1	3	4	2	1	3	1

At the 5% significance level, do the data provide sufficient evidence to conclude that, on average, Helsinkians respond with an ecological welfare food choice motive greater than 2?
a. Use the Wilcoxon signed-rank test.
b. Use the *t*-test.
c. Compare the results of the two tests.

9.113 Checking Advertised Contents. A manufacturer of liquid soap produces a bottle with an advertised content of 310 milliliters (mL). Sixteen bottles are randomly selected and found to have the following contents, in mL.

297	318	306	300	311	303	291	298
322	307	312	300	315	296	309	311

A normal probability plot of the data indicates that you can assume the contents are normally distributed. Let μ denote the mean content of all bottles produced. To decide whether the mean content is less than advertised, perform the hypothesis test

$$H_0: \mu = 310 \text{ mL}$$
$$H_a: \mu < 310 \text{ mL}$$

at the 5% significance level.
a. Use the *t*-test.
b. Use the Wilcoxon signed-rank test.
c. If the mean content is in fact less than 310 mL, how do you explain the discrepancy between the two tests?

9.114 Education of Jail Inmates. Twenty years ago, the U.S. Bureau of Justice Statistics reported in *Profile of Jail Inmates* that the median educational attainment of jail inmates was 10.2 years. Ten current inmates are randomly selected and found to have the following educational attainments, in years.

14	10	5	6	8
10	10	8	9	9

Assume that educational attainments of current jail inmates have a symmetric, nonnormal distribution. At the 10% significance level, do the data provide sufficient evidence to conclude that this year's median educational attainment has changed from what it was 20 years ago?
a. Use the *t*-test.
b. Use the Wilcoxon signed-rank test.
c. If this year's median educational attainment has in fact changed from what it was 20 years ago, how do you explain the discrepancy between the two tests?

Extending the Concepts and Skills

9.115 How Long Do Marriages Last? The National Center for Health Statistics publishes data on the duration of marriages in *Vital Statistics of the United States*. Ten years ago, the median duration of a marriage was 7.2 years. Suppose that you take a random sample of 50 divorce certificates from last year and record the marriage durations for the purpose of deciding whether the average marriage duration has decreased. Which test would give the better results, the Wilcoxon signed-rank test or the *t*-test? Explain your answer.

9.116 The Census Form. The U.S. Bureau of the Census estimates that the *U.S. Census Form* takes the average household 14 minutes to complete. To check that claim, completion times are recorded for 36 randomly selected households. Which test would give the better results, the

Wilcoxon signed-rank test or the *t*-test? Explain your answer.

9.117 Waiting for the Train. A commuter train arrives punctually at a station every half hour. Each morning, a commuter named John leaves his house and casually strolls to the train station. John thinks that he is unlucky and that he waits longer for the train on average than he should.
a. Assuming that John is not unlucky, how long should he expect to wait for the train, on average?
b. Assuming that John is not unlucky, identify the distribution of the times he waits for the train.
c. The following is a sample of the times, in minutes, that John waited for the train.

24	20	3	19	28	22
26	4	11	5	16	24

Use the Wilcoxon signed-rank test to decide, at the 10% significance level, whether the data provide sufficient evidence to conclude that, on average, John waits more than 15 minutes for the train.
d. Explain why the Wilcoxon signed-rank test is appropriate here.
e. Is the Wilcoxon signed-rank test more appropriate here than the *t*-test? Explain your answer.

Normal Approximation for *W*. The table of critical values for the Wilcoxon signed-rank test, Table V, stops at $n = 20$. For larger samples, a normal approximation can be used. In fact, the normal approximation works well even for sample sizes as small as 10.

Normal Approximation for *W*

Suppose that the variable under consideration has a symmetric distribution. Then, for samples of size n,

- $\mu_W = n(n + 1)/4$
- $\sigma_W = \sqrt{n(n + 1)(2n + 1)/24}$, and
- W is approximately normally distributed for $n \geq 10$.

Thus, for samples of size 10 or more, the standardized variable

$$z = \frac{W - n(n + 1)/4}{\sqrt{n(n + 1)(2n + 1)/24}}$$

has approximately the standard normal distribution.

9.118 Large Sample Wilcoxon Signed-Rank Test. Formulate a hypothesis-testing procedure for a Wilcoxon signed-rank test that uses the test statistic z given in the preceding box.

9.119 Birth Weights. Refer to Exercise 9.110.
a. Use the procedure you formulated in Exercise 9.118 to perform the hypothesis test in Exercise 9.110.
b. Compare your result in part (a) to the one you obtained in Exercise 9.110, where the normal approximation was not used.

9.120 The Distribution of *W*. In this exercise, you are to obtain the distribution of the variable W for samples of size 3 so that you can see how the critical values for the Wilcoxon signed-rank test are derived.
a. The rows of the following table give all possible signs for the signed ranks in a Wilcoxon signed-rank test with $n = 3$. For instance, the first row covers the possibility that all three observations are greater than μ_0 and thus have positive sign ranks. Fill in the empty column with values of W. (*Hint:* The first entry is 6 and the last is 0.)

Rank			
1	2	3	*W*
+	+	+	
+	+	−	
+	−	+	
+	−	−	
−	+	+	
−	+	−	
−	−	+	
−	−	−	

b. If the null hypothesis H_0: $\mu = \mu_0$ is true, what percentages of samples will match any particular row of the table? (*Hint:* The answer is the same for all rows.)
c. Use the answer from part (b) to obtain the distribution of W for samples of size 3.
d. Draw a relative-frequency histogram of the distribution obtained in part (c).
e. Use your histogram from part (d) to find the critical value for a left-tailed Wilcoxon signed-rank test with a sample size of 3 and a significance level of 0.125.

9.121 The Distribution of *W*. Repeat Exercise 9.120 for samples of size 4. (*Hint:* The table will have 16 rows.)

Using Technology

9.122 Easy Hole at the British Open? The Old Course at St. Andrews in Scotland is home of the British Open, one of the major tournaments in professional golf. The *Hole O'Cross Out*, known by both European and U.S. professional golfers as one of the friendliest holes at St. Andrews, is the fifth hole, a 514-yard, par 5 hole with an open fairway and a large green. As one reporter for PGATOUR.com put it: "If players think before they drive, they will easily walk away with birdies and pars." Following is a sample of scores on the *Hole O'Cross Out* posted by 156 golf professionals.

3	4	4	4	5	5	5	6	3	4	4	4	5	5	5	6
3	4	4	4	5	5	5	6	3	4	4	4	5	5	5	6
3	4	4	4	5	5	5	6	3	4	4	4	5	5	5	6
4	4	4	5	5	5	5	6	4	4	4	5	5	5	5	6
4	4	4	5	5	5	5	6	4	4	4	5	5	5	5	6
4	4	4	5	5	5	5	6	4	4	4	5	5	5	5	6
4	4	4	5	5	5	5	6	4	4	4	5	5	5	5	6
4	4	4	5	5	5	5	7	4	4	4	5	5	5	5	7
4	4	4	5	5	5	5	4	4	4	5	5	5	5	4	4
4	5	5	5	6	4	4	4	5	5	5	6				

Use the technology of your choice to decide whether, on average, professional golfers score better than par on the *Hole O'Cross Out*. Perform the required hypothesis test at the 0.01 level of significance.
a. Employ the Wilcoxon signed-rank test.
b. Employ the *t*-test.
c. Compare your results from parts (a) and (b).

9.123 Pricing Pickups. Refer to Exercise 9.109. Use the technology of your choice to perform the required hypothesis test, using the Wilcoxon signed-rank test.

9.124 Birth Weights. Refer to Exercise 9.110. Use the technology of your choice to perform the required hypothesis test, using the Wilcoxon signed-rank test.

9.8 Which Procedure Should Be Used?[†]

In this chapter, you have learned three procedures for performing a hypothesis test for one population mean: the *z*-test, the *t*-test, and the Wilcoxon signed-rank test.

The *z*-test and *t*-test are designed to be used when the variable under consideration has a normal distribution. In such cases, the *z*-test applies when the population standard deviation is known and the *t*-test applies when the population standard deviation is unknown.

Recall that both the *z*-test and the *t*-test are approximately correct when the sample size is large, regardless of the distribution of the variable under consideration. Moreover, these two tests should be used cautiously when outliers are present. Refer to Key Fact 9.4 on page 416 for guidelines covering use of the *z*-test and *t*-test.

Recall further that the Wilcoxon signed-rank test is designed to be used when the variable under consideration has a symmetric distribution. And, unlike the *z*-test and *t*-test, the Wilcoxon signed-rank test is resistant to outliers.

We summarize the three procedures in Table 9.16. Each row of the table gives the type of test, the conditions required for using the test, the test statistic, and the procedure to use. Note that we used the abbreviations "normal population" for "the variable under consideration is normally distributed," "*W*-test" for "Wilcoxon signed-rank test," and "symmetric population" for "the variable under consideration has a symmetric distribution."

In selecting the correct procedure, keep in mind that the best choice is the procedure expressly designed for the type of distribution under consideration, if such a procedure exists, and that the *z*-test and *t*-test are only approximately correct for large samples from nonnormal populations.

For instance, suppose that the variable under consideration is normally distributed and that the population standard deviation is known. Then both the *z*-test and Wilcoxon signed-rank test apply. The *z*-test applies because the vari-

---
[†]All previous sections in this chapter, including the material on the Wilcoxon signed-rank test, are prerequisite to this section.

TABLE 9.16

Summary of hypothesis-testing procedures for one population mean, μ. The null hypothesis for all tests is $H_0: \mu = \mu_0$

Type	Assumptions	Test statistic	Procedure to use
z-test	1. Simple random sample 2. Normal population or large sample 3. σ known	$z = \dfrac{\bar{x} - \mu_0}{\sigma/\sqrt{n}}$	9.1 (page 417) 9.2 (page 438)
t-test	1. Simple random sample 2. Normal population or large sample 3. σ unknown	$t = \dfrac{\bar{x} - \mu_0}{s/\sqrt{n}}$ (df $= n - 1$)	9.3 (page 448)
W-test	1. Simple random sample 2. Symmetric population	$W =$ sum of positive ranks	9.4 (page 459)

able under consideration is normally distributed and σ is known; the W-test applies because a normal distribution is symmetric. But the correct procedure is the z-test because it is designed specifically for variables that have a normal distribution.

The flowchart shown in Fig. 9.32 provides an organized strategy for choosing the correct hypothesis testing procedure for a population mean. It is based on the preceding discussion.

FIGURE 9.32

Flowchart for choosing the correct hypothesis testing procedure for one population mean

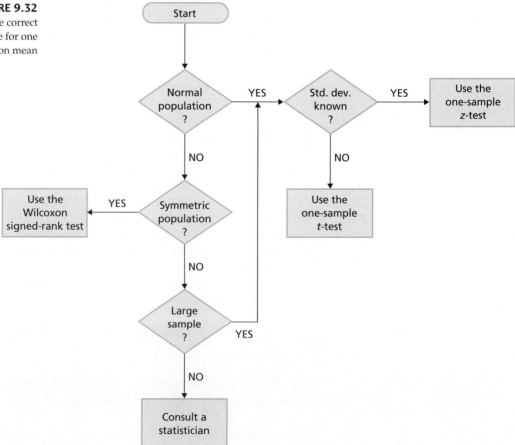

In practice, you need to look at the sample data to ascertain the type of distribution before selecting the appropriate procedure. We recommend using a normal probability plot and either a stem-and-leaf diagram (for small or moderate-size samples) or a histogram (for moderate-size or large samples). We demonstrate choosing the correct procedure in Example 9.23.

Example 9.23 Choosing the Correct Hypothesis Testing Procedure

TABLE 9.17
2000 entertainment expenditures ($) for a sample of consumer units in the West

1864	2741	2130	3153	3319
2324	1822	2579	2534	1982
2028	785	1580	1353	1591
2220	2001	2357	2352	2149
2321	2667	4009	2093	1855
1585	751	2245	2112	2528
1877	1512	2350	2581	2115
2442	2146	913	2617	2724
2048	1978	1701	1784	2102
2197	2213	2093	2283	2400

Entertainment Expenditure According to *Consumer Expenditure Survey*, published by the U.S. Bureau of Labor Statistics, the 2000 mean expenditure for entertainment was $1863 per consumer unit. For that same year, a sample of 50 consumer units in the West yielded the annual expenditures for entertainment shown in Table 9.17.

Suppose that we want to use the sample data in Table 9.17 to decide whether the 2000 mean entertainment expenditure by consumer units in the West exceeded that of the nation as a whole. Then we want to perform the hypothesis test

H_0: $\mu = \$1863$ (mean expenditure in the West is not greater)

H_a: $\mu > \$1863$ (mean expenditure in the West is greater),

where μ is the 2000 mean entertainment expenditure by consumer units in the West. Which procedure should be used to perform the hypothesis test?

Solution We begin by drawing a normal probability plot and a histogram of the sample data in Table 9.17, as shown in Fig. 9.33.

FIGURE 9.33
(a) Normal probability plot and (b) histogram of expenditure data in Table 9.17

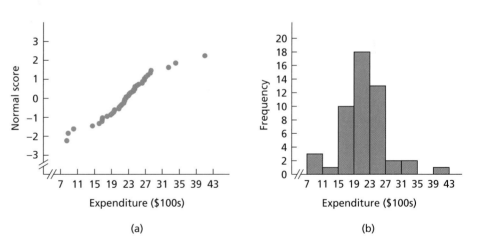

(a)   (b)

Next, we consult the flowchart shown in Fig. 9.32 and the graphs presented in Fig. 9.33 to decide which procedure to use. The first question we must answer is whether the variable under consideration is normally distributed. The normal probability plot in Fig. 9.33(a) shows some curvature and/or the presence of outliers; so the answer to the first question is "No."

This result leads to the next question: Does the variable under consideration have a symmetric distribution? The histogram shown in Fig. 9.33(b) suggests that we can reasonably assume that the answer to that question is "Yes."

The "Yes" answer to the preceding question leads us to the box in Fig. 9.32 that states: Use the Wilcoxon signed-rank test.

INTERPRETATION An appropriate procedure for carrying out the hypothesis test is the Wilcoxon signed-rank test.

EXERCISES 9.8

Statistical Concepts and Skills

9.125 In this chapter, we presented three procedures for conducting a hypothesis test for one population mean.
a. Identify the three procedures by name.
b. List the assumptions for using each procedure.
c. Identify the test statistic for each procedure.

9.126 Suppose that you want to perform a hypothesis test for a population mean. Assume that the variable under consideration is normally distributed and that the population standard deviation is unknown.
a. Can the t-test be used to perform the hypothesis test? Explain your answer.
b. Can the Wilcoxon signed-rank test be used to perform the hypothesis test? Explain your answer.
c. Which procedure is preferable, the t-test or the Wilcoxon signed-rank test? Explain your answer.

9.127 Suppose that you want to perform a hypothesis test for a population mean. Assume that the variable under consideration has a symmetric nonnormal distribution and that the population standard deviation is unknown. Further assume that the sample size is large and that no outliers are present in the sample data.
a. Can the t-test be used to perform the hypothesis test? Explain your answer.
b. Can the Wilcoxon signed-rank test be used to perform the hypothesis test? Explain your answer.
c. Which procedure is preferable, the t-test or the Wilcoxon signed-rank test? Explain your answer.

9.128 Suppose that you want to perform a hypothesis test for a population mean. Assume that the variable under consideration has a highly skewed distribution and that the population standard deviation is known. Further assume that the sample size is large and that no outliers are present in the sample data.
a. Can the z-test be used to perform the hypothesis test? Explain your answer.

b. Can the Wilcoxon signed-rank test be used to perform the hypothesis test? Explain your answer.

In Exercises 9.129–9.136, we have provided a normal probability plot and either a stem-and-leaf diagram or a frequency histogram for a set of sample data. The intent is to employ the sample data to perform a hypothesis test for the mean of the population from which the data were obtained. In each case, consult the graphs provided and the flowchart in Fig. 9.32 to decide which procedure should be used.

9.129 The normal probability plot and stem-and-leaf diagram of the data are shown in Fig. 9.34; σ is known.

9.130 The normal probability plot and histogram of the data are shown in Fig. 9.35; σ is known.

9.131 The normal probability plot and histogram of the data are shown in Fig. 9.36; σ is unknown.

9.132 The normal probability plot and stem-and-leaf diagram of the data are shown in Fig. 9.37; σ is unknown.

9.133 The normal probability plot and stem-and-leaf diagram of the data are shown in Fig. 9.38; σ is unknown.

9.134 The normal probability plot and stem-and-leaf diagram of the data are shown in Fig. 9.39; σ is unknown. (*Note:* The decimal parts of the observations were removed before the stem-and-leaf diagram was constructed.)

9.135 The normal probability plot and stem-and-leaf diagram of the data are shown in Fig. 9.40; σ is known.

9.136 The normal probability plot and stem-and-leaf diagram of the data are shown in Fig. 9.41; σ is known.

FIGURE 9.34
Normal probability plot and
stem-and-leaf diagram for Exercise 9.129

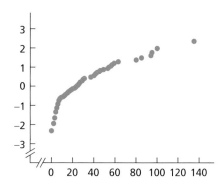
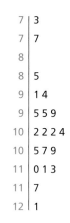

```
 7 | 3
 7 | 7
 8 |
 8 | 5
 9 | 1 4
 9 | 5 5 9
10 | 2 2 2 4
10 | 5 7 9
11 | 0 1 3
11 | 7
12 | 1
```

FIGURE 9.35
Normal probability plot and histogram
for Exercise 9.130

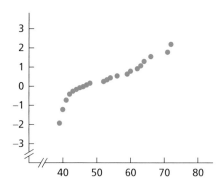

FIGURE 9.36
Normal probability plot and histogram
for Exercise 9.131

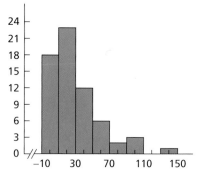
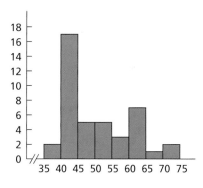

FIGURE 9.37
Normal probability plot and
stem-and-leaf diagram for Exercise 9.132

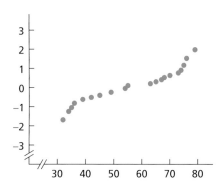

```
 3 | 2 2 4
 3 | 5 6 6 9
 4 | 2
 4 | 5 9 9
 5 | 4 4
 5 | 5
 6 | 3
 6 | 5 7 8
 7 | 0 3 4
 7 | 5 5 6 9
```

FIGURE 9.38
Normal probability plot and
stem-and-leaf diagram for Exercise 9.133

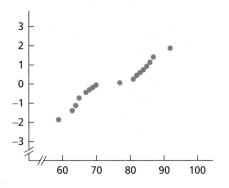

```
5 | 9
6 | 3 4
6 | 5 5 5 7 8 9
7 | 0
7 | 7
8 | 1 1 2 3 4
8 | 5 6 7
9 | 2
```

FIGURE 9.39
Normal probability plot and
stem-and-leaf diagram for Exercise 9.134

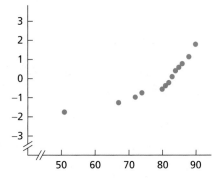

```
0 | 6
0 |
1 | 1
1 | 3 3 3
1 | 5 5
1 | 6 7
1 | 8 8 8
2 |
2 | 2
```

FIGURE 9.40
Normal probability plot and
stem-and-leaf diagram for Exercise 9.135

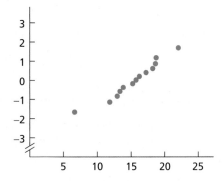

```
5 | 1
5 |
6 |
6 | 7
7 | 2 4
7 |
8 | 0 1 2 3 3 3 4
8 | 5 6 8 8
9 | 0
```

FIGURE 9.41
Normal probability plot and
stem-and-leaf diagram for Exercise 9.136

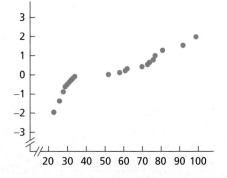

```
2 | 3 6 6 8 8 8 9
3 | 0 1 2 3 4
4 |
5 | 2 8
6 | 1 2
7 | 0 3 4 6 7 7
8 | 1
9 | 2 9
```

CHAPTER REVIEW

You Should Be Able To

1. use and understand the formulas in this chapter.

2. define the terms associated with hypothesis testing.

3. choose the null and alternative hypotheses for a hypothesis test.

4. explain the logic behind hypothesis testing.

5. identify the test statistic, rejection region, nonrejection region, and critical value(s) for a hypothesis test.

6. define and apply the concepts of Type I and Type II errors.

7. state and interpret the possible conclusions for a hypothesis test.

8. obtain the critical value(s) for a specified significance level.

9. perform a hypothesis test for a population mean when the population standard deviation is known.

*10. compute Type II error probabilities.

*11. calculate the power of a hypothesis test.

*12. draw a power curve.

*13. understand the relationship between sample size, significance level, and power.

14. obtain the *P*-value of a hypothesis test.

15. state and apply the steps for performing a hypothesis test, using the critical-value approach to hypothesis testing.

16. state and apply the steps for performing a hypothesis test, using the *P*-value approach to hypothesis testing.

17. perform a hypothesis test for a population mean when the population standard deviation is unknown.

*18. perform a hypothesis test for a population mean when the variable under consideration has a symmetric distribution.

*19. decide which procedure should be used to perform a hypothesis test for a population mean.

Key Terms

acceptance region, *406*
alternative hypothesis, *398*
critical-value approach to hypothesis
    testing, *433*
critical values, *406*
hypothesis, *398*
hypothesis test, *398*
left-tailed test, *399*
nonparametric methods,* *455*
nonrejection region, *406*
not statistically significant, *411*
null hypothesis, *398*
observed significance level, *434*
one-sample *t*-test, *445, 448*

one-sample Wilcoxon signed-rank
    test,* *455*
one-sample *z*-test, *417, 438*
one-tailed test, *399*
P-value (*P*), *434*
P-value approach to hypothesis
    testing, *433*
parametric methods,* *455*
power,* *428*
power curve,* *429*
probability value, *434*
rejection region, *406*
research hypothesis, *398*
right-tailed test, *399*

significance level (α), *409*
statistically significant, *411*
t-test, *445*
test statistic, *406*
two-tailed test, *399*
Type I error, *408*
Type I error probability (α), *409*
Type II error, *408*
Type II error probability (β), *409*
W_α,* *457*
Wilcoxon signed-rank test,* *455, 459*
z-test, *416*

REVIEW TEST

Statistical Concepts and Skills

1. Explain the meaning of each term.
   a. Null hypothesis
   b. Alternative hypothesis
   c. Test statistic
   d. Rejection region
   e. Nonrejection region
   f. Critical value(s)

2. The following statement appeared on a box of Tide laundry detergent: "Individual packages of Tide may weigh slightly more or less than the marked weight due to normal variations incurred with high speed packaging machines, but each day's production of Tide will average slightly above the marked weight."

a. Explain in statistical terms what the statement means.
b. Describe in words a hypothesis test for checking the statement.
c. Suppose that the marked weight is 76 ounces. State in words the null and alternative hypotheses for the hypothesis test. Then express those hypotheses in statistical terminology.

3. Regarding a hypothesis test,
a. what is the procedure, generally, for deciding whether the null hypothesis should be rejected?
b. how can the procedure identified in part (a) be made objective and precise?

4. There are three possible alternative hypotheses in a hypothesis test for a population mean. Identify them and explain when each is used.

5. Two types of incorrect decisions can be made in a hypothesis test: a Type I error and a Type II error.
a. Explain the meaning of each type of error.
b. Identify the letter used to represent the probability of each type of error.
c. If the null hypothesis is in fact true, only one type of error is possible. Which type is that? Explain your answer.
d. If you fail to reject the null hypothesis, only one type of error is possible. Which type is that? Explain your answer.

6. Suppose that you want to conduct a right-tailed hypothesis test at the 5% significance level. How must the critical value be chosen?

7. In each part, we have identified a hypothesis testing procedure for a population mean. State the assumptions required and the test statistic used in each case.
a. One-sample *t*-test   **b.** One-sample *z*-test
*c. One-sample Wilcoxon signed-rank test

8. What is meant when we say that a hypothesis test is
a. exact?   **b.** approximately correct?

9. Discuss the difference between statistical significance and practical significance.

10. For a fixed sample size, what happens to the probability of a Type II error if the significance level is decreased from 0.05 to 0.01?

*11. Regarding the power of a hypothesis test,
a. what does it represent?
b. what happens to the power of a hypothesis test if the significance level is kept at 0.01 while the sample size is increased from 50 to 100?

12. Regarding the *P*-value of a hypothesis test,
a. what is the *P*-value of a hypothesis test?
b. answer true or false: A *P*-value of 0.02 provides more evidence against the null hypothesis than a *P*-value of 0.03. Explain your answer.
c. answer true or false: A *P*-value of 0.74 provides essentially no evidence against the null hypothesis. Explain your answer.
d. explain why the *P*-value of a hypothesis test is also referred to as the observed significance level.

13. Discuss the differences between the critical-value and *P*-value approaches to hypothesis testing.

*14. Identify two advantages of nonparametric methods over parametric methods. When is a parametric procedure preferred? Explain your answer.

15. Cheese Consumption. The U.S. Department of Agriculture reports in *Food Consumption, Prices, and Expenditures* that the average American consumed 30.0 lb of cheese in 2000. Cheese consumption has increased steadily since 1960 when the average American ate only 8.3 lb of cheese annually. Suppose that you want to decide whether last year's mean cheese consumption is greater than the 2000 mean.
a. Identify the null hypothesis.
b. Identify the alternative hypothesis.
c. Classify the hypothesis test as two tailed, left tailed, or right tailed.

16. The following graph portrays the decision criterion for a hypothesis test about a population mean, μ. The null hypothesis for the test is $H_0: \mu = \mu_0$, and the test statistic is

$$z = \frac{\bar{x} - \mu_0}{\sigma/\sqrt{n}}.$$

The curve shown in the graph reveals the implications of the decision criterion if in fact the null hypothesis is true.

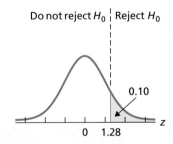

Determine the
a. rejection region.   **b.** nonrejection region.
c. critical value(s).   **d.** significance level.
e. Draw a graph that depicts the answers you obtained in parts (a)–(d).

f. Classify the hypothesis test as two tailed, left tailed, or right tailed.

17. Cheese Consumption. The null and alternative hypotheses for the hypothesis test in Problem 15 are

$$H_0: \mu = 30.0 \text{ lb (mean has not increased)}$$
$$H_a: \mu > 30.0 \text{ lb (mean has increased)},$$

where μ is last year's mean cheese consumption for all Americans. Explain what each of the following would mean.
a. Type I error   **b.** Type II error   **c.** Correct decision

Now suppose that the results of carrying out the hypothesis test lead to nonrejection of the null hypothesis. Classify that decision by error type or as a correct decision if in fact last year's mean cheese consumption
d. has not increased from the 2000 mean of 30.0 lb.
e. has increased from the 2000 mean of 30.0 lb.

***18. Cheese Consumption.** Refer to Problem 15. Suppose that you decide to use a z-test with a significance level of 0.10 and a sample size of 35. Assume that $\sigma = 6.9$ lb.
a. Determine the probability of a Type I error.
b. If last year's mean cheese consumption was 30.5 lb, identify the distribution of the variable $\bar{x}$, that is, the sampling distribution of the mean for samples of size 35.
c. Use part (b) to determine the probability, β, of a Type II error if in fact last year's mean cheese consumption was 30.5 lb.
d. Repeat parts (b) and (c) if in fact last year's mean cheese consumption was 31.0 lb, 31.5 lb, 32.0 lb, 32.5 lb, 33.0 lb, 33.5 lb, and 34.0 lb.
e. Use your answers from parts (c) and (d) to construct a table of selected Type II error probabilities and powers similar to Table 9.8 on page 429.
f. Use your answer from part (e) to construct the power curve.

Using a sample size of 60 instead of 35, repeat
g. part (b).   **h.** part (c).   **i.** part (d).
j. part (e).   **k.** part (f).
l. Compare your power curves for the two sample sizes and explain the principle being illustrated.

19. Cheese Consumption. Refer to Problem 15. The following table provides last year's cheese consumption, in pounds, for 35 randomly selected Americans.

42	25	29	34	38	36	30
29	28	32	24	43	22	38
32	28	41	20	35	24	29
40	29	22	33	23	27	32
33	33	32	18	40	32	25

a. At the 10% significance level, do the data provide sufficient evidence to conclude that last year's mean cheese consumption for all Americans has increased over the 2000 mean? Assume that $\sigma = 6.9$ lb. For your hypothesis test, use a z-test and the critical-value approach. (*Note:* The sum of the data is 1078 lb.)
b. Given the conclusion in part (a), if an error has been made, what type must it be? Explain your answer.

20. Cheese Consumption. Refer to Problem 19.
a. Repeat the hypothesis test, using the P-value approach to hypothesis testing.
b. Use Table 9.12 on page 440 to assess the strength of the evidence against the null hypothesis.

21. Purse Snatching. According to *Crime in the United States*, a publication of the FBI, the mean value lost to purse snatching was $356 in 2000. For last year, 12 randomly selected purse-snatching offenses yielded the following values lost, to the nearest dollar.

231	446	296	386	189	293
261	250	229	372	290	454

Use a t-test with either the critical-value approach or the P-value approach to decide, at the 5% significance level, whether last year's mean value lost to purse snatching has decreased from the 2000 mean. The mean and standard deviation of the data are $308.1 and $86.9, respectively.

***22. Purse Snatching.** Refer to Problem 21.
a. Perform the required hypothesis test, using the Wilcoxon signed-rank test.
b. In performing the hypothesis test in part (a), what assumption did you make about the distribution of last year's values lost to purse snatching?
c. In Problem 21, we used the t-test to perform the hypothesis test. The assumption in that problem is that last year's values lost to purse snatching are normally distributed. If that assumption is true, why is it permissible to perform a Wilcoxon signed-rank test for the mean value lost?

*23. Purse Snatching. Refer to Problems 21 and 22. If in fact last year's values lost to purse snatching are normally distributed, which is the preferred procedure for performing the hypothesis test—the *t*-test or the Wilcoxon signed-rank test? Explain your answer.

24. Betting the Spreads. College basketball, and particularly the NCAA basketball tournament, is a popular venue for gambling, from novices in office betting pools to the high roller. To encourage uniform betting across teams, Las Vegas oddsmakers assign a point spread to each game. The *point spread* is the oddsmakers' prediction for the number of points by which the favored team will win. If you bet on the favorite, you win the bet provided the favorite wins by more than the point spread; otherwise, you lose the bet. Is the point spread a good measure of the relative ability of the two teams? Hal S. Stern and Barbara Mock addressed this question in the paper "College Basketball Upsets: Will a 16-Seed Ever Beat a 1-Seed?" (*Chance*, Vol. 11(1), pp. 27–31). They obtained the difference between the actual margin of victory and the point spread, called the *point-spread error*, for 2109 college basketball games. The mean point-spread error was found to be −0.2 point with a standard deviation of 10.9 points. For a particular game, a point-spread error of 0 indicates that the point spread was a perfect estimate of the two teams' relative abilities.
a. If, on average, the oddsmakers are estimating correctly, what is the (population) mean point-spread error?
b. Use the data to decide, at the 5% significance level, whether the (population) mean point-spread error differs from 0.
c. Interpret your answer in part (b).

Problems 25 and 26 each include a normal probability plot and either a frequency histogram or a stem-and-leaf diagram for a set of sample data. The intent is to use the sample data to perform a hypothesis test for the mean of the population from which the data were obtained. In each case, consult the graphs provided to decide whether to use the z-test, the t-test, or neither. Explain your answer.

25. The normal probability plot and histogram of the data are depicted in Fig. 9.42; σ is known.

26. The normal probability plot and stem-and-leaf diagram of the data are depicted in Fig. 9.43; σ is unknown.

*27. Refer to Problems 25 and 26.
a. In each case, consult the appropriate graphs to decide whether using the Wilcoxon signed-rank test is reasonable for performing a hypothesis test for the mean of the population from which the data were obtained. Give reasons for your answers.

b. For each case where using either the *z*-test or the *t*-test is reasonable and where using the Wilcoxon signed-rank test is also appropriate, decide which test is preferable. Give reasons for your answers.

Using Technology

28. Cheese Consumption. Refer to Problems 19 and 20. Use the technology of your choice to
a. obtain a normal probability plot of the data.
b. perform the required hypothesis test.
c. Justify the use of your procedure in part (b).

29. Purse Snatching. Refer to Problem 21. Use the technology of your choice to
a. obtain a normal probability plot of the data.
b. perform the required hypothesis test.
c. Justify the use of your procedure in part (b).

*30. Purse Snatching. Use the technology of your choice to carry out the hypothesis test considered in part (a) of Problem 22.

*31. Nursing-Home Costs. The cost of staying in a nursing home in the United States is rising dramatically, as reported in the August 5, 2003 issue of *The Wall Street Journal*. In May 2002, the average cost of a private room in a nursing home was $168 per day. For August 2003, a random sample of 11 nursing homes yielded the following daily costs, in dollars, for a private room in a nursing home.

73	199	192	181	182	250
159	182	208	129	282	

Use the technology of your choice to do the following.
a. Apply the *t*-test to decide at the 10% significance level whether the average cost for a private room in a nursing home in August 2003 exceeded that in May 2002.
b. Repeat part (a) by using the Wilcoxon signed-rank test.
c. Obtain a normal probability plot, a boxplot, a stem-and-leaf diagram, and a histogram of the sample data.
d. Discuss the discrepancy in results between the *t*-test and the Wilcoxon signed-rank test.

32. Beer Drinking. According to *The Beer Institute*, the mean annual consumption of beer per person in the United States is 22.0 gallons (roughly 235 twelve-ounce bottles). A random sample of 300 Washington D.C. residents yielded the annual beer consumptions provided on the WeissStats CD.
a. Obtain a histogram of the data.

FIGURE 9.42

Normal probability plot and histogram for Problem 25

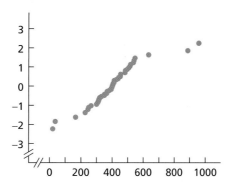
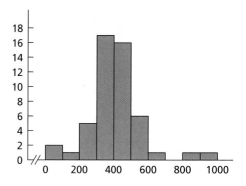

FIGURE 9.43

Normal probability plot and stem-and-leaf diagram for Problem 26

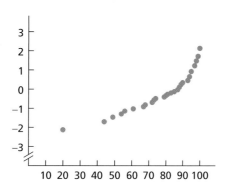
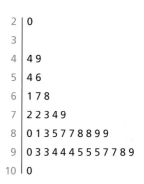

b. Does your histogram in part (a) indicate any outliers?

c. At the 1% significance level, do the data provide sufficient evidence to conclude that the mean annual consumption of beer per person for the nation's capital exceeds the national mean? (*Note:* See the third bulleted item in Key Fact 9.4 on page 416.)

StatExplore in MyMathLab
Analyzing Data Online

You can use StatExplore to perform all statistical analyses discussed in this chapter. To illustrate, we show how to perform a one-sample *t*-test.

EXAMPLE One-Sample t-Test

Acid Rain and Lake Acidity Table 9.13 on page 449 displays the pH levels of a sample of 15 lakes in the Southern Alps. Use StatExplore to decide, at the 5% significance level, whether the data provide sufficient evidence to conclude that, on average, high mountain lakes in the Southern Alps are nonacidic (i.e., have a pH level greater than 6).

SOLUTION Let μ denote the mean pH level of all high mountain lakes in the Southern Alps. The problem is to perform the hypothesis test

$H_0: \mu = 6$ (mean pH level is not greater than 6)

$H_a: \mu > 6$ (mean pH level is greater than 6)

at the 5% significance level. Note that the hypothesis test is right tailed because a greater-than sign ($>$) appears in the alternative hypothesis.

To carry out the hypothesis test by using StatExplore, we first store the sample data from Table 9.13 in a column named PH. Then, we proceed as follows.

1 Choose **Stat ➤ T statistics ➤ One sample**
2 Select the column PH
3 Click **Next →**
4 Select the **Hypothesis Test** option button
5 Click in the **Null: mean** = text box and type <u>6</u>
6 Click the arrow button at the right of the **Alternative** drop-down list box and select >
7 Click **Calculate**

The resulting output is shown in Printout 9.4. From Printout 9.4, the *P*-value for the hypothesis test is seen to be 0.0019. As the *P*-value is less than the specified significance level of 0.05, we reject H_0. At the 5% significance level, the data provide sufficient evidence to conclude that, on average, high mountain lakes in the Southern Alps are nonacidic.

PRINTOUT 9.4
StatExplore output for one-sample *t*-test

Hypothesis test results:
Parameter: mean of Variable
H0 : Parameter = 6
HA : Parameter > 6

Variable	Sample Mean	Std. Err.	DF	T-Stat	P-value
PH	6.6	0.1734798	14	3.458616	0.0019

STATEXPLORE EXERCISES Solve the following problems by using StatExplore.

a. Conduct the one-sample *t*-test required in Exercise 9.92 on page 453.
b. Carry out the one-sample *z*-test required in Example 9.14 on page 438.
*c. Perform the Wilcoxon signed-rank test required in Example 9.21 on page 460.

To access StatExplore, go to the student content area of your Weiss MyMathLab course.

Internet Project

The Ozone Hole

Ozone is a molecule made up of three atoms of oxygen. It plays many roles in the ecosystem, but one of the most important for every life form on Earth is the protection it provides from the harmful effects of the Sun's ultraviolet light.

For years, humans have used chemicals that only recently have been discovered to be harmful to the protective ozone layer. Humankind is beginning to see the possible effects of that damage. For example, the depletion of ozone in the stratosphere may increase the rate of skin cancer and cataracts, and it can harm crops and ocean life.

In this Internet project, you are to explore data collected worldwide on the decrease of ozone in the Earth's atmosphere. You also have the opportunity to investigate the possible consequences of ozone depletion: the amount of solar ultraviolet light reaching the Earth's surface and the effects it may have on humans.

URL for access to Internet Projects Page: www.aw-bc.com/weiss

Focusing on Data Analysis

The Focus Database

Recall from Chapter 1 (see page 35) that the Focus database contains information on the undergraduate students at the University of Wisconsin - Eau Claire (UWEC). Statistical analyses for this database should be done with the technology of your choice.

a. Obtain a simple random sample of size 50 of the ACT composite scores (COMP) of UWEC undergraduate students.
b. Use your data from part (a) to decide, at the 5% significance level, whether the mean ACT composite score of UWEC undergraduates exceeds the national mean of 21.0 points. Interpret your result.
c. In practice, the (population) mean of the variable under consideration is unknown. However, in this case, we actually do have the population data. Obtain the mean ACT composite score of all UWEC undergraduate students and note whether the decision concerning the hypothesis test in part (b) was correct.

The following problems are intended as a class project.

d. Obtain 100 simple random samples of size 50 each of the ages of UWEC undergraduate students. (*Note:* If your class contains, say, 25 students, each student should obtain four simple random samples of size 50 each.)
e. The population mean age of all UWEC undergraduates is 20.75 years. For each sample in part (d), perform a hypothesis test at the 10% significance level to decide whether the mean age of UWEC undergraduates differs from 20.75 years.
f. Roughly how many of the 100 hypothesis tests conducted in part (e) would you expect to result in rejection of the null hypothesis? Explain your answer.
g. How many of the 100 hypothesis tests conducted in part (e) actually result in rejection of the null hypothesis? Compare this number to your answer in part (f).

Case Study Discussion

Sex and Sense of Direction

At the beginning of this chapter, we discussed research by Jeanne Sholl et al. on the relationship between sex and sense of direction. Recall that, in their study, the spatial orientation skills of 30 male students and 30 female students from Boston College were challenged in Houghton Garden Park, a wooded park near the BC campus in Newton, Massachusetts. Before driving out to the park, the participants were asked to rate their own sense of direction as either good or poor.

In the park, students were instructed to point to predesignated target landmarks and also to the direction of south. Pointing was carried out by moving a pointer attached to a 360° protractor; the angle of the pointing response was then recorded to the nearest degree. For the female students who had rated their sense of direction to be good, the table on page 397 provides the absolute pointing errors (in degrees) when they attempted to point south.

a. If, on average, women who consider themselves to have a good sense of direction do no better than they would by just randomly guessing at the direction of south, what would be their mean absolute pointing error?

b. At the 1% significance level, do the data provide sufficient evidence to conclude that women who consider themselves to have a good sense of direction really do better, on average, than they would by just randomly guessing at the direction of south? Use a one-sample *t*-test.

c. Obtain a normal probability plot, boxplot, and stem-and-leaf diagram of the data. Based on these plots, is use of the *t*-test reasonable? Explain your answer.

d. Use the technology of your choice to perform the data analyses in parts (b) and (c).

*e. Solve part (b) by using the Wilcoxon signed-rank test.

*f. Based on the plots you obtained in part (c), is use of the Wilcoxon signed-rank test reasonable? Explain your answer.

*g. Use the technology of your choice to perform the Wilcoxon signed-rank test of part (e).

Internet Resources: Visit the Weiss Web site www.aw-bc.com/weiss for additional discussion, exercises, and resources related to this case study.

Biography

JERZY NEYMAN: A Principal Founder of Modern Statistical Theory

Jerzy Neyman was born on April 16, 1894, in Bendery, Russia. His father, Czeslaw, was a member of the Polish nobility, a lawyer, a judge, and an amateur archaeologist. Because Russian authorities prohibited the family from living in Poland, Jerzy Neyman grew up in various cities in Russia. He entered the university in Kharkov in 1912. At Kharkov he was at first interested in physics, but, because of his clumsiness in the laboratory, he decided to pursue mathematics.

After World War I, when Russia was at war with Poland over borders, Neyman was jailed as an enemy alien. In 1921, as a result of a prisoner exchange, he went to Poland for the first time. In 1924, he received his doctorate from the University of Warsaw. Between 1924 and 1934, Neyman worked with Karl Pearson (see Biography in Chapter 13) and his son Egon Pearson and held a position at the University of Kraków. In 1934, Neyman took a position in Karl Pearson's statistical laboratory at University College in London. He stayed in England, where he worked with Egon Pearson until 1938, at which time he accepted an offer to join the faculty at the University of California at Berkeley.

When the United States entered World War II, Neyman set aside development of a statistics program and did war work. After the war ended, Neyman organized a symposium to celebrate its end and "the return to theoretical research." That symposium, held in August 1945, and succeeding ones, held every 5 years until 1970, were instrumental in establishing Berkeley as a preeminent statistical center.

Neyman was a principal founder of the theory of modern statistics. His work on hypothesis testing, confidence intervals, and survey sampling transformed both the theory and the practice of statistics. His achievements were acknowledged by the granting of many honors and awards, including election to the United States National Academy of Sciences, the Guy Medal in Gold of the Royal Statistical Society, and the United States National Medal of Science.

Neyman remained active until his death of heart failure on August 5, 1981, at the age of 87, in Oakland, California.

10

Inferences for Two Population Means

GENERAL OBJECTIVES In Chapters 8 and 9, you learned how to obtain confidence intervals and perform hypothesis tests for one population mean. Frequently, however, inferential statistics is used to compare the means of two or more populations.

For example, we might want to perform a hypothesis test to decide whether the mean age of buyers of new domestic cars is greater than the mean age of buyers of new imported cars; or, we might want to find a confidence interval for the difference between the two mean ages.

Broadly speaking, in this chapter we examine two types of inferential procedures for comparing the means of two populations. The first type applies when the samples from the two populations are *independent*, meaning that the sample selected from one of the populations has no effect or bearing on the sample selected from the other population.

The second type of inferential procedure for comparing the means of two populations applies when the samples from the two populations are *paired*. A paired sample may be appropriate when there is a natural pairing of the members of the two populations such as husband and wife.

Breast Milk and IQ

Considerable controversy exists over whether long-term neurodevelopment is affected by nutritional factors in early life. Five researchers summarized their findings on that question for preterm babies in the publication "Breast Milk and Subsequent Intelligence Quotient in Children Born Preterm" (*The Lancet*, 339, pp. 261–264). Their study was a continuation of work begun in January, 1982.

Previously, these five researchers had showed that a mother's decision to provide breast milk for preterm infants is associated with higher developmental scores for the children at age 18 months. In the article mentioned, they analyzed IQ data on the same children at age $7\frac{1}{2}$–8 years. IQ was measured for 300 children, using an abbreviated form of the Weschler Intelligence Scale for Children (revised Anglicized version: WISC-R UK).

The mothers of the children in the study had chosen whether to provide their infants with breast milk within 72 hours of delivery; 90 did not and 210 did. Of those 210 who chose to provide their infants with breast milk, 193 succeeded and 17 did not.

The children whose mothers declined to provide breast milk were designated by the researchers as Group I; those whose mothers had chosen but were unable to provide breast milk were designated as Group IIa; and those whose mothers had chosen and were able to provide breast milk were designated as Group IIb. The following table displays statistics for all three groups.

Group	Sample size	Mean IQ	St. Dev.
I	90	92.8	15.2
IIa	17	94.8	19.0
IIb	193	103.7	15.3

After studying the inferential methods discussed in this chapter, you will be able to conduct statistical analyses to examine how breast feeding affects subsequent IQ for children age $7\frac{1}{2}$–8 years who were born preterm.

10.1 The Sampling Distribution of the Difference Between Two Sample Means for Independent Samples

In this section, we lay the groundwork for making statistical inferences to compare the means of two populations. The methods that we first consider require not only that the samples selected from the two populations be simple random samples but also that they be **independent samples.** That is, the sample selected from one of the populations has no effect or bearing on the sample selected from the other population.

With **independent simple random samples**, each possible pair of samples—one from one population and one from the other—is equally likely to be the pair of samples selected. Example 10.1 provides an unrealistically simple illustration of independent samples, but it will help you understand the concept.

Example 10.1 **Introducing Independent Random Samples**

Males and Females Let's consider two small populations, one consisting of three men and the other of four women, as shown in the following figure.

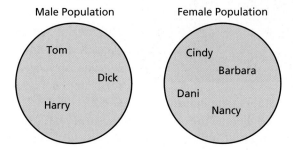

Suppose that we take a sample of size 2 from the male population and a sample of size 3 from the female population.

a. List the possible pairs of independent samples.

b. If the samples are selected at random, determine the chance that any particular pair of samples will be the pair of independent samples obtained.

Solution For convenience, we use the first letter of each name as an abbreviation for the actual name.

a. The possible samples of size 2 from the male population are listed on the left in Table 10.1. The possible samples of size 3 from the female population are listed on the right in Table 10.1.

To obtain the possible pairs of independent samples, we list each possible male sample of size 2 with each possible female sample of size 3, as shown in Table 10.2. There are 12 possible pairs of independent samples of two men and three women.

b. For independent simple random samples, each of the 12 possible pairs of samples shown in Table 10.2 is equally likely to be the pair selected. Therefore the chances are $\frac{1}{12}$ (1 in 12) that any particular pair of samples will be the one obtained.

TABLE 10.1	**TABLE 10.2**
Possible samples of size 2 from the male population and possible samples of size 3 from the female population	Possible pairs of independent samples of two men and three women

Male sample of size 2	Female sample of size 3
T, D	C, B, D
T, H	C, B, N
D, H	C, D, N
	B, D, N

Male sample of size 2	Female sample of size 3
T, D	C, B, D
T, D	C, B, N
T, D	C, D, N
T, D	B, D, N
T, H	C, B, D
T, H	C, B, N
T, H	C, D, N
T, H	B, D, N
D, H	C, B, D
D, H	C, B, N
D, H	C, D, N
D, H	B, D, N

The purpose of Example 10.1 is to provide a concrete illustration of independent samples and to emphasize that, for independent simple random samples of any given sizes, each possible pair of independent samples is equally likely to be the one selected. In practice, we neither obtain the number of possible pairs of independent samples nor explicitly compute the chance of selecting a particular pair of independent samples. But these concepts underlie the methods we do use.

Note: Recall that, when we say *random sample,* we mean *simple random sample* unless specifically stated otherwise. Likewise, when we say *independent random samples,* we mean *independent simple random samples,* unless specifically stated otherwise.

Comparing Two Population Means, Using Independent Samples

We can now examine the process for comparing the means of two populations based on independent samples. Example 10.2 introduces the pertinent ideas.

Example 10.2 **Comparing Two Population Means, Using Independent Samples**

Faculty Salaries   The American Association of University Professors (AAUP) conducts salary studies of college professors and publishes its findings in *AAUP Annual Report on the Economic Status of the Profession.* Suppose that we want to decide whether the mean salaries of college faculty in public and private institutions are different.

a. Formulate the problem statistically by posing it as a hypothesis test.

b. Explain the basic idea for carrying out the hypothesis test.

c. Suppose that 30 faculty members from public institutions and 35 faculty members from private institutions are randomly and independently selected and that the salaries of these faculty members are as displayed in Table 10.3, in thousands of dollars rounded to the nearest hundred. Discuss the use of these data to make a decision concerning the hypothesis test.

TABLE 10.3

Annual salaries ($1000s) for 30 faculty members in public institutions and 35 faculty members in private institutions

Sample 1 (public institutions)						Sample 2 (private institutions)					
41.2	97.0	107.4	31.6	114.4	70.6	99.9	59.0	70.1	125.5	44.7	109.2	69.9
63.8	48.4	42.0	61.2	31.4	63.0	60.8	108.0	75.6	58.5	53.4	52.2	83.0
65.2	83.8	91.2	86.4	49.2	88.8	63.1	84.6	68.6	85.3	73.3	38.1	78.1
36.2	22.8	40.8	47.2	58.2	48.2	80.5	74.6	59.4	54.2	66.3	104.5	34.2
67.2	95.2	51.6	71.4	81.0	78.0	88.2	31.8	78.0	104.3	99.6	63.0	69.3

Solution

a. To formulate the problem statistically, we first note that we have one variable (salary) and two populations (faculty in public institutions and faculty in private institutions). Let the two populations in question be designated Populations 1 and 2, respectively:

$$\text{Population 1: Faculty in public institutions}$$
$$\text{Population 2: Faculty in private institutions}$$

Next, we denote the means of the variable "salary" for the two populations μ_1 and μ_2, respectively:

$$\mu_1 = \text{mean salary of faculty in public institutions;}$$
$$\mu_2 = \text{mean salary of faculty in private institutions.}$$

Then, we can state the hypothesis test we want to perform as

$$H_0: \ \mu_1 = \mu_2 \ (\text{mean salaries are the same})$$
$$H_a: \ \mu_1 \neq \mu_2 \ (\text{mean salaries are different}).$$

b. Roughly speaking, we can carry out the hypothesis test as follows.

1. Independently and randomly take a sample of faculty members from public institutions (Population 1) and a sample of faculty members from private institutions (Population 2).
2. Compute the mean salary, $\bar{x}_1$, of the sample of faculty members from public institutions and the mean salary, $\bar{x}_2$, of the sample of faculty members from private institutions.
3. Reject the null hypothesis if the sample means, $\bar{x}_1$ and $\bar{x}_2$, differ by too much; otherwise, do not reject the null hypothesis.

This process is depicted in Fig. 10.1.

c. The means of the two samples in Table 10.3 are

$$\bar{x}_1 = \frac{\Sigma x}{n_1} = \frac{1934.4}{30} = 64.48 \quad \text{and} \quad \bar{x}_2 = \frac{\Sigma x}{n_2} = \frac{2568.8}{35} = 73.39.$$

FIGURE 10.1

Process for comparing two population means, using independent samples

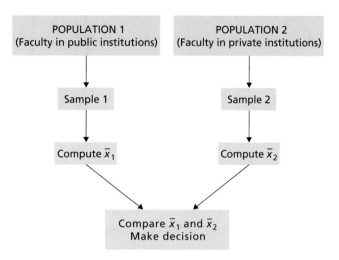

The question now is: Can the difference of 8.91 ($8910) between these two sample means be reasonably attributed to sampling error or is the difference large enough to indicate that the two populations have different means? To answer that question, we need to know the distribution of the difference between two sample means—the *sampling distribution of the difference between two sample means.* We examine that sampling distribution in this section and complete the hypothesis test in the next section.

The Sampling Distribution of the Difference Between Two Sample Means for Independent Samples

First, we need to discuss the notation used for parameters and statistics when we are analyzing two populations. Let's call the two populations under consideration Population 1 and Population 2. Then, as indicated in Example 10.2, we use a subscript 1 when referring to parameters or statistics for Population 1 and a subscript 2 when referring to them for Population 2. We present the notation in Table 10.4.

TABLE 10.4

Notation for parameters and statistics when considering two populations

	Population 1	**Population 2**
Population mean	μ_1	μ_2
Population std. dev.	σ_1	σ_2
Sample mean	$\bar{x}_1$	$\bar{x}_2$
Sample std. dev.	s_1	s_2
Sample size	n_1	n_2

Armed with this notation, we describe in Key Fact 10.1 the **sampling distribution of the difference between two sample means.** In doing so, we assume that the variable under consideration is normally distributed on each population. But keep in mind that, because of the central limit theorem, consequences of Key Fact 10.1 will still apply approximately for large samples regardless of distribution type.

Understanding Key Fact 10.1 is aided by recalling Key Fact 7.2: Suppose that a variable x of a population is normally distributed and has mean μ and standard deviation σ. Then, for samples of size n,

- $\mu_{\bar{x}} = \mu$,
- $\sigma_{\bar{x}} = \sigma/\sqrt{n}$, and
- $\bar{x}$ is normally distributed.

KEY FACT 10.1

The Sampling Distribution of the Difference Between Two Sample Means for Independent Samples

Suppose that x is a normally distributed variable on each of two populations. Then, for independent samples of sizes n_1 and n_2 from the two populations,

- $\mu_{\bar{x}_1 - \bar{x}_2} = \mu_1 - \mu_2$,
- $\sigma_{\bar{x}_1 - \bar{x}_2} = \sqrt{(\sigma_1^2/n_1) + (\sigma_2^2/n_2)}$, and
- $\bar{x}_1 - \bar{x}_2$ is normally distributed.

In words, the first bulleted item says that the mean of all possible differences between the two sample means equals the difference between the two population means. The second bulleted item indicates that the standard deviation of all possible differences between the two sample means equals the square root of the sum of the population variances each divided by the corresponding sample size.

The formulas for the mean and standard deviation of $\bar{x}_1 - \bar{x}_2$ given in the first and second bulleted items, respectively, hold regardless of the distributions of the variable on the two populations. The assumption that the variable is normally distributed on each of the two populations is needed only to conclude that $\bar{x}_1 - \bar{x}_2$ is normally distributed (third bulleted item) and, as we noted, that too holds approximately for large samples, regardless of distribution type.

The Two-Sample z-Procedures

Under the conditions of Key Fact 10.1, the standardized version of $\bar{x}_1 - \bar{x}_2$,

$$z = \frac{(\bar{x}_1 - \bar{x}_2) - (\mu_1 - \mu_2)}{\sqrt{(\sigma_1^2/n_1) + (\sigma_2^2/n_2)}},$$

has the standard normal distribution. Using that fact, we can develop hypothesis testing and confidence-interval procedures for comparing two population means when the population standard deviations are known. These procedures are often called the *two-sample z-test* and the *two-sample z-interval procedure*, respectively, or, collectively, as the *two-sample z-procedures*.

Because population standard deviations are usually unknown, however, we won't discuss the two-sample z-procedures in detail. Instead we relegate them to the exercises and concentrate on the more useful two-sample *t*-procedures, which apply in the case of unknown population standard deviations. In Sections 10.2 and 10.3, we examine the two-sample t-procedures, specifically, the pooled t-procedure and the nonpooled t-procedure, respectively.

The Relation Between Hypothesis Tests and Confidence Intervals

Hypothesis tests and confidence intervals are closely related. Consider, for example, a two-tailed hypothesis test for comparing two population means at the significance level α. In this case, the null hypothesis will be rejected if and only if the $(1 - \alpha)$-level confidence interval for $\mu_1 - \mu_2$ does not contain 0. Exercises 10.11–10.13 deal with this relation between hypothesis tests and confidence intervals in greater detail.

EXERCISES 10.1

Statistical Concepts and Skills

10.1 Age of Car Buyers. In the introduction to this chapter, we mentioned comparing the mean age of buyers of new domestic cars to the mean age of buyers of new imported cars.
a. Identify the variable that would be considered.
b. Identify the two populations that would be considered.
c. Suppose that we want to perform a hypothesis test to decide whether the mean age of buyers of new domestic cars is greater than the mean age of buyers of new imported cars. State the null and alternative hypotheses for the hypothesis test.

10.2 Spending at the Mall. An issue of *USA TODAY* compared the mean amounts spent by teens and adults at shopping malls.
a. Identify the variable under consideration.
b. Identify the two populations.
c. Suppose that we want to perform a hypothesis test to decide whether the mean amount spent by teens is less than the mean amount spent by adults. State the null and alternative hypotheses for the hypothesis test.

10.3 Give an example of interest to you for comparing two population means. Identify the variable under consideration and the two populations.

10.4 Define the phrase *independent samples*.

10.5 Consider the quantities μ_1, σ_1, $\bar{x}_1$, s_1, μ_2, σ_2, $\bar{x}_2$, and s_2.
a. Which quantities represent parameters and which represent statistics?
b. Which quantities are fixed numbers and which are variables?

10.6 Discuss the basic strategy for performing a hypothesis test to compare the means of two populations, based on independent samples.

10.7 Why do you need to know the sampling distribution of the difference between two sample means in order to perform a hypothesis test to compare two population means?

10.8 Identify the assumption for using the two-sample z-procedures that renders those procedures generally impractical.

10.9 A variable of two populations has a mean of 40 and a standard deviation of 12 for one of the populations and a mean of 40 and a standard deviation of 6 for the other population.
a. For independent samples of sizes 9 and 4, respectively, find the mean and standard deviation of $\bar{x}_1 - \bar{x}_2$.
b. Must the variable under consideration be normally distributed on each of the two populations for you to answer part (a)? Explain your answer.
c. Can you conclude that the variable $\bar{x}_1 - \bar{x}_2$ is normally distributed? Explain your answer.

10.10 A variable of two populations has a mean of 40 and a standard deviation of 12 for one of the populations and a mean of 40 and a standard deviation of 6 for the other population. Moreover, the variable is normally distributed on each of the two populations.
a. For independent samples of sizes 9 and 4, respectively, determine the mean and standard deviation of $\bar{x}_1 - \bar{x}_2$.
b. Can you conclude that the variable $\bar{x}_1 - \bar{x}_2$ is normally distributed? Explain your answer.
c. Determine the percentage of all pairs of independent samples of sizes 9 and 4, respectively, from the two populations that have the property that the difference between the sample means is between -10 and 10.

Extending the Concepts and Skills

The Two-Sample z-Procedures. Using Key Fact 10.1, we can derive the following two-sample z-procedures.

Two-Sample z-Procedures

Purpose To perform a hypothesis test or obtain a confidence interval to compare two population means, μ_1 and μ_2

Assumptions

1. Simple random samples
2. Independent samples
3. Normal populations or large samples
4. Known population standard deviations

- **Two-sample z-test.** The test statistic for a hypothesis test with null hypothesis H_0: $\mu_1 = \mu_2$ (population means are equal) is

$$z = \frac{\bar{x}_1 - \bar{x}_2}{\sqrt{(\sigma_1^2/n_1) + (\sigma_2^2/n_2)}}.$$

- **Two-sample z-interval procedure.** The endpoints of a $(1 - \alpha)$-level confidence interval for the difference, $\mu_1 - \mu_2$, between the two population means are

$$(\bar{x}_1 - \bar{x}_2) \pm z_{\alpha/2} \cdot \sqrt{(\sigma_1^2/n_1) + (\sigma_2^2/n_2)}.$$

---

*Exercises **10.11** and **10.12** illustrate the use of the two-sample z-procedures.*

10.11 Starting Salaries. The National Association of Colleges and Employers (NACE) conducts surveys of salary offers to new college graduates and publishes the results in *Salary Survey*. The following table gives the starting annual salaries obtained from independent random samples of 35 liberal-arts graduates and 32 criminal-justice graduates. Data are in thousands of dollars.

Liberal arts					Criminal justice			
32.0	28.4	33.2	30.8	27.1	28.1	26.2	28.9	26.2
26.0	27.1	27.1	30.0	31.8	28.3	29.5	30.5	25.5
31.3	27.4	31.3	32.0	31.1	26.7	29.8	28.4	27.1
29.5	28.3	30.4	28.5		28.8	28.4	27.9	29.8
30.9	30.0	29.5	30.6		28.1	26.8	24.6	28.4
31.5	28.2	26.7	34.2		29.1	26.7	28.5	28.7
31.0	29.2	30.7	31.3		29.3	28.3	27.4	24.4
31.4	31.9	32.6	30.2		28.5	25.0	27.9	27.4

a. At the 5% significance level, can you conclude that the mean starting salaries of liberal-arts graduates and criminal-justice graduates differ? Assume that the population standard deviations of starting salaries are 1.82 ($1820) for liberal-arts graduates and 1.73 ($1730) for criminal-justice graduates. (*Note:* The sum of the liberal-arts data is 1053.2, and the sum of the criminal-justice data is 889.2.)

b. Determine a 95% confidence interval for the difference between the mean starting annual salaries of liberal-arts graduates and criminal-justice graduates.

10.12 Hospital Stays. The U.S. National Center for Health Statistics compiles data on the length of stay by patients in short-term hospitals and publishes its findings in *Vital and Health Statistics*. Independent random samples of 39 male patients and 35 female patients gave the following data on length of stay, in days.

Male					Female				
4	4	12	18	9	14	7	15	1	12
6	12	10	3	6	1	3	7	21	4
15	7	3	13	1	1	5	4	4	3
2	10	13	5	7	5	18	12	5	1
1	23	9	2	1	7	7	2	15	4
17	2	24	11	14	9	10	7	3	6
6	2	1	8	1	5	9	6	2	14
3	19	3	1						

a. At the 5% significance level, do the data provide sufficient evidence to conclude that, on average, the lengths of stay in short-term hospitals by males and females differ? Assume that $\sigma_1 = 5.4$ days and $\sigma_2 = 4.6$ days. (*Note:* The sum of the male data is 308 days, and the sum of the female data is 249 days.)

b. Determine a 95% confidence interval for the difference between the mean lengths of stay in short-term hospitals by males and females.

10.13 Hypothesis Tests and Confidence Intervals. Use the results obtained in Exercises 10.11 and 10.12 to complete the statement: A hypothesis test of H_0: $\mu_1 = \mu_2$ versus H_a: $\mu_1 \neq \mu_2$ at the significance level α will lead to rejection of the null hypothesis if and only if the number _____ does not lie in the $(1 - \alpha)$-level confidence interval for $\mu_1 - \mu_2$.

Using Technology

10.14 To obtain the sampling distribution of the difference between two sample means for independent samples, as stated in Key Fact 10.1 on page 488, we need to know that, for independent observations, the difference of two normally distributed variables is also a normally distributed variable. In this exercise, you are to perform a computer simulation to make that fact plausible.

a. Simulate 2000 observations from a normally distributed variable with a mean of 100 and a standard deviation of 16.

b. Repeat part (a) for a normally distributed variable with a mean of 120 and a standard deviation of 12.

c. Determine the difference between each pair of observations in parts (a) and (b).

d. Obtain a histogram of the 2000 differences found in part (c). Why is the histogram bell shaped?

10.15 In this exercise, you are to perform a computer simulation to illustrate the sampling distribution of the difference between two sample means for independent samples, Key Fact 10.1.

a. Simulate 1000 samples of size 12 from a normally distributed variable with a mean of 640 and a standard deviation of 70. Obtain the sample mean of each of the 1000 samples.

b. Simulate 1000 samples of size 15 from a normally distributed variable with a mean of 715 and a standard deviation of 150. Obtain the sample mean of each of the 1000 samples.

c. Obtain the difference, $\bar{x}_1 - \bar{x}_2$, for each of the 1000 pairs of sample means obtained in parts (a) and (b).

d. Obtain the mean, the standard deviation, and a histogram of the 1000 differences found in part (c).

e. Theoretically, what are the mean, standard deviation, and distribution of all possible differences, $\bar{x}_1 - \bar{x}_2$?

f. Compare your answers from parts (d) and (e).

10.2 Inferences for Two Population Means, Using Independent Samples: Standard Deviations Assumed Equal

In Section 10.1, we took the first steps required to develop inferential procedures based on independent samples for comparing the means of two populations. Armed with that information, we can derive two inferential methods. One requires that the two populations have equal standard deviations; the other does not. We develop the first method in this section and the second method in Section 10.3.

Hypothesis Tests for the Means of Two Populations With Equal Standard Deviations, Using Independent Samples

We now develop a procedure for performing a hypothesis test based on independent samples to compare the means of two populations with equal, but unknown, standard deviations. Our immediate goal is to find a test statistic for such a test. In doing so, we assume that the variable under consideration is normally distributed on each population. As we demonstrate later, the resulting hypothesis-testing procedure is approximately correct for large samples, regardless of the type of distribution.

Let's use σ to denote the common standard deviation of the two populations. We know from Key Fact 10.1 on page 488 that, for independent samples, the standardized version of $\bar{x}_1 - \bar{x}_2$,

$$z = \frac{(\bar{x}_1 - \bar{x}_2) - (\mu_1 - \mu_2)}{\sqrt{(\sigma_1^2/n_1) + (\sigma_2^2/n_2)}},$$

has the standard normal distribution. Replacing σ_1 and σ_2 in the previous displayed equation with their common value σ and using some algebra, we obtain the variable

$$z = \frac{(\bar{x}_1 - \bar{x}_2) - (\mu_1 - \mu_2)}{\sigma\sqrt{(1/n_1) + (1/n_2)}}. \tag{10.1}$$

However, we cannot use this variable as a basis for the required test statistic because σ is unknown.

Consequently, we need to use sample information to estimate the unknown population standard deviation, σ. We do so by first obtaining an estimate of the unknown population variance, σ^2. The best way to do that is to regard the sample variances, s_1^2 and s_2^2, as two estimates of σ^2 and then **pool** those estimates by weighting them according to sample size (actually by degrees of freedom). Thus our estimate of σ^2 is

$$s_p^2 = \frac{(n_1 - 1)s_1^2 + (n_2 - 1)s_2^2}{n_1 + n_2 - 2}$$

and hence that of σ is

$$s_p = \sqrt{\frac{(n_1 - 1)s_1^2 + (n_2 - 1)s_2^2}{n_1 + n_2 - 2}}.$$

The subscript "p" stands for "pooled," and the quantity s_p is called the **pooled sample standard deviation.**

Replacing σ in Equation (10.1) with its estimate, s_p, we get the variable

$$\frac{(\bar{x}_1 - \bar{x}_2) - (\mu_1 - \mu_2)}{s_p\sqrt{(1/n_1) + (1/n_2)}},$$

which we can use as a basis for the required test statistic. Unlike the variable in Equation (10.1), this one does not have the standard normal distribution. However, its distribution is one with which you are already familiar—a t-distribution—which we describe in Key Fact 10.2.

KEY FACT 10.2   **Distribution of the Pooled t-Statistic**

Suppose that x is a normally distributed variable on each of two populations and that the population standard deviations are equal. Then, for independent samples of sizes n_1 and n_2 from the two populations, the variable

$$t = \frac{(\bar{x}_1 - \bar{x}_2) - (\mu_1 - \mu_2)}{s_p\sqrt{(1/n_1) + (1/n_2)}}$$

has the t-distribution with df $= n_1 + n_2 - 2$.

In light of Key Fact 10.2, for a hypothesis test that has null hypothesis H_0: $\mu_1 = \mu_2$ (population means are equal), we can use the variable

$$t = \frac{\bar{x}_1 - \bar{x}_2}{s_p\sqrt{(1/n_1) + (1/n_2)}}$$

as the test statistic and obtain the critical value(s) or P-value from the t-table, Table IV in Appendix A. We refer to this hypothesis-testing procedure as the **pooled t-test.** Procedure 10.1 provides a step-by-step method for performing a pooled t-test by using either the critical-value approach (stay left) or the P-value approach (stay right).

PROCEDURE 10.1   Pooled *t*-Test

Purpose To perform a hypothesis test to compare two population means, μ_1 and μ_2

Assumptions

1. Simple random samples
2. Independent samples
3. Normal populations or large samples
4. Equal population standard deviations

STEP 1 The null hypothesis is H_0: $\mu_1 = \mu_2$, and the alternative hypothesis is

$$H_a: \mu_1 \neq \mu_2 \quad \text{or} \quad H_a: \mu_1 < \mu_2 \quad \text{or} \quad H_a: \mu_1 > \mu_2$$
$$\text{(Two tailed)} \qquad \text{(Left tailed)} \qquad \text{(Right tailed)}$$

STEP 2 Decide on the significance level, α.

STEP 3 Compute the value of the test statistic

$$t = \frac{\bar{x}_1 - \bar{x}_2}{s_p \sqrt{(1/n_1) + (1/n_2)}},$$

where

$$s_p = \sqrt{\frac{(n_1 - 1)s_1^2 + (n_2 - 1)s_2^2}{n_1 + n_2 - 2}}.$$

Denote the value of the test statistic t_0.

CRITICAL-VALUE APPROACH	P-VALUE APPROACH

CRITICAL-VALUE APPROACH

STEP 4 The critical value(s) are

$$\pm t_{\alpha/2} \quad \text{or} \quad -t_\alpha \quad \text{or} \quad t_\alpha$$
$$\text{(Two tailed)} \qquad \text{(Left tailed)} \qquad \text{(Right tailed)}$$

with df $= n_1 + n_2 - 2$. Use Table IV to find the critical value(s).

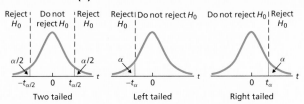

STEP 5 If the value of the test statistic falls in the rejection region, reject H_0; otherwise, do not reject H_0.

P-VALUE APPROACH

STEP 4 The *t*-statistic has df $= n_1 + n_2 - 2$. Use Table IV to estimate the *P*-value, or obtain it exactly by using technology.

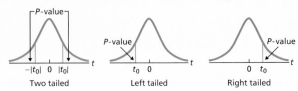

STEP 5 If $P \leq \alpha$, reject H_0; otherwise, do not reject H_0.

STEP 6 Interpret the results of the hypothesis test.

The hypothesis test is exact for normal populations and is approximately correct for large samples from nonnormal populations.

Before we apply the pooled t-test, several comments are in order. In Step 3 of Procedure 10.1, we need to calculate the pooled sample standard deviation, s_p. The pooled sample standard deviation always lies between the two sample standard deviations, s_1 and s_2, which is useful as a check when s_p is calculated by hand.

Regarding Assumption 3 for the pooled t-test, although the pooled t-test was derived under the condition that the variable under consideration is normally distributed on each of the two populations (normal populations), it also applies approximately for large samples regardless of distribution type, as noted at the bottom of Procedure 10.1. Actually, the pooled t-test works reasonably well even for small samples or samples of moderate size from nonnormal populations provided the populations are not too nonnormal. In other words, the pooled t-test is robust to moderate violations of the normality assumption.

The pooled t-test is also robust to moderate violations of Assumption 4 (equal population standard deviations) provided the sample sizes are roughly equal. We will have more to say about the robustness of the pooled t-test at the end of Section 10.3.

As before, we can check normality with normal probability plots. The equal-standard-deviations assumption is more difficult to check, especially when the sample sizes are small. We recommend checking it by informally comparing the standard deviations of the two samples and by comparing their stem-and-leaf diagrams, histograms, or boxplots. Be sure to use the same scales for each pair of graphs.

The equal-standard-deviations assumption is sometimes checked by performing a formal hypothesis test, called an F-test for the equality of two standard deviations. We don't recommend this procedure because, although the pooled t-test is robust to moderate violations of normality, the F-test is extremely nonrobust to such violations. As the noted statistician George E. P. Box remarked: "To make a preliminary test on variances [standard deviations] is rather like putting to sea in a rowing boat to find out whether conditions are sufficiently calm for an ocean liner to leave port!"

When considering the pooled t-test, you must also watch for outliers. Again, the presence of outliers calls into question the normality assumption. Moreover, even for large samples, outliers can sometimes unduly affect a pooled t-test because the sample mean and sample standard deviation are not resistant to them.

We also note that the pooled t-test is frequently the procedure of choice when comparing two means with a designed experiment. Indeed, the assumption of equal population standard deviations is often reasonable in such cases, for instance, when the null hypothesis of equal means actually implies that there is no difference at all between the two treatments.

Example 10.3 illustrates use of the pooled t-test.

▌▌Example 10.3   The Pooled t-Test

Faculty Salaries   We now return to the salary problem posed in Example 10.2. Recall that we want to perform a hypothesis test to decide whether there is a difference between the mean salaries of faculty in public institutions and private institutions.

Independent random samples of 30 faculty members in public institutions and 35 faculty members in private institutions yielded the data in Table 10.5. At the 5% significance level, do the data provide sufficient evidence to conclude that mean salaries for faculty in public and private institutions differ?

TABLE 10.5

Annual salaries ($1000s) for 30 faculty members in public institutions and 35 faculty members in private institutions

Sample 1 (public institutions)						Sample 2 (private institutions)					
41.2	97.0	107.4	31.6	114.4	70.6	99.9	59.0	70.1	125.5	44.7	109.2	69.9
63.8	48.4	42.0	61.2	31.4	63.0	60.8	108.0	75.6	58.5	53.4	52.2	83.0
65.2	83.8	91.2	86.4	49.2	88.8	63.1	84.6	68.6	85.3	73.3	38.1	78.1
36.2	22.8	40.8	47.2	58.2	48.2	80.5	74.6	59.4	54.2	66.3	104.5	34.2
67.2	95.2	51.6	71.4	81.0	78.0	88.2	31.8	78.0	104.3	99.6	63.0	69.3

TABLE 10.6

Summary statistics for the samples in Table 10.5

Public institutions	Private institutions
$\bar{x}_1 = 64.48$	$\bar{x}_2 = 73.39$
$s_1 = 23.95$	$s_2 = 22.26$
$n_1 = 30$	$n_2 = 35$

Solution First, we present in Table 10.6 the required summary statistics for the two samples in Table 10.5. These statistics are obtained in the usual way.

Next, we check the four conditions required for using the pooled *t*-test. As the samples are independent simple random samples, Assumptions 1 and 2 are satisfied. To check Assumption 3, we first note that the sample sizes are large. Thus, we need be concerned only with the presence of outliers. Careful graphical analyses (not shown) suggest no outliers for either sample, so we can consider Assumption 3 satisfied. From Table 10.6, the sample standard deviations are 23.95 and 22.26; these are certainly close enough for us to consider Assumption 4 satisfied.

The preceding paragraph suggests that the pooled *t*-test can be used to carry out the hypothesis test. We apply Procedure 10.1.

STEP 1 **State the null and alternative hypotheses.**

The null and alternative hypotheses are

$$H_0: \mu_1 = \mu_2 \text{ (mean salaries are the same)}$$
$$H_a: \mu_1 \neq \mu_2 \text{ (mean salaries are different)},$$

where μ_1 and μ_2 are the mean salaries of all faculty in public and private institutions, respectively. Note that the hypothesis test is two tailed because a does-not-equal sign ($\neq$) appears in the alternative hypothesis.

STEP 2 **Decide on the significance level, α.**

The test is to be performed at the 5% significance level, or $\alpha = 0.05$.

STEP 3 **Compute the value of the test statistic**

$$t = \frac{\bar{x}_1 - \bar{x}_2}{s_p \sqrt{(1/n_1) + (1/n_2)}}$$

where

$$s_p = \sqrt{\frac{(n_1 - 1)s_1^2 + (n_2 - 1)s_2^2}{n_1 + n_2 - 2}}.$$

We first determine the pooled sample standard deviation, s_p. Using the summary statistics in Table 10.6, we find that

$$s_p = \sqrt{\frac{(30-1)\cdot(23.95)^2 + (35-1)\cdot(22.26)^2}{30+35-2}} = 23.05.$$

Referring again to Table 10.6, we obtain the value of the test statistic:

$$t = \frac{\bar{x}_1 - \bar{x}_2}{s_p\sqrt{(1/n_1)+(1/n_2)}} = \frac{64.48 - 73.39}{23.05\sqrt{(1/30)+(1/35)}} = -1.554.$$

CRITICAL-VALUE APPROACH	P-VALUE APPROACH

CRITICAL-VALUE APPROACH

STEP 4 **The critical values for a two-tailed test are $\pm t_{\alpha/2}$ with df $= n_1 + n_2 - 2$. Use Table IV to find the critical values.**

From Table 10.6, $n_1 = 30$ and $n_2 = 35$, so df $= 30 + 35 - 2 = 63$. Also, from Step 2, we have $\alpha = 0.05$. In Table IV with df $= 63$, we find that the critical values are $\pm t_{\alpha/2} = \pm t_{0.05/2} = \pm t_{0.025} = \pm 1.998$, as shown in Fig. 10.2A.

FIGURE 10.2A

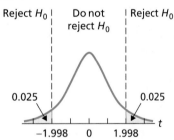

STEP 5 **If the value of the test statistic falls in the rejection region, reject H_0; otherwise, do not reject H_0.**

From Step 3, the value of the test statistic is $t = -1.554$, which does not fall in the rejection region (see Fig. 10.2A). Thus we do not reject H_0. The test results are not statistically significant at the 5% level.

P-VALUE APPROACH

STEP 4 **The t-statistic has df $= n_1 + n_2 - 2$. Use Table IV to estimate the P-value or obtain it exactly by using technology.**

From Step 3, the value of the test statistic is $t = -1.554$. As the test is two tailed, the P-value is the probability of observing a value of t of 1.554 or greater in magnitude if the null hypothesis is true. That probability equals the shaded area shown in Fig. 10.2B.

FIGURE 10.2B

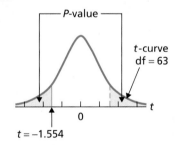

From Table 10.6, $n_1 = 30$ and $n_2 = 35$, so df $= 30 + 35 - 2 = 63$. Referring to Fig. 10.2B and to Table IV with df $= 63$, we find that $0.10 < P < 0.20$. (Using technology, we obtain $P = 0.125$.)

STEP 5 **If $P \leq \alpha$, reject H_0; otherwise, do not reject H_0.**

From Step 4, $0.10 < P < 0.20$. Because the P-value exceeds the specified significance level of 0.05, we do not reject H_0. The test results are not statistically significant at the 5% level and (see Table 9.12 on page 440) provide at most weak evidence against the null hypothesis.

STEP 6 **Interpret the results of the hypothesis test.**

INTERPRETATION At the 5% significance level, the data do not provide sufficient evidence to conclude that a difference exists between the mean salaries of faculty in public and private institutions. ▬■

Confidence Intervals for the Difference Between the Means of Two Populations With Equal Standard Deviations

We can also use Key Fact 10.2 on page 492 to derive a confidence-interval procedure, Procedure 10.2, for the difference between two population means, which we often refer to as the **pooled *t*-interval procedure.** We demonstrate its use in Example 10.4.

PROCEDURE 10.2

Pooled *t*-Interval Procedure

Purpose To find a confidence interval for the difference between two population means, μ_1 and μ_2

Assumptions
1. Simple random samples
2. Independent samples
3. Normal populations or large samples
4. Equal population standard deviations

STEP 1 **For a confidence level of 1 − α, use Table IV to find $t_{\alpha/2}$ with df = $n_1 + n_2 - 2$.**

STEP 2 **The endpoints of the confidence interval for $\mu_1 - \mu_2$ are**

$$(\bar{x}_1 - \bar{x}_2) \pm t_{\alpha/2} \cdot s_p \sqrt{(1/n_1) + (1/n_2)}.$$

STEP 3 **Interpret the confidence interval.**

The confidence interval is exact for normal populations and is approximately correct for large samples from nonnormal populations.

Example 10.4   The Pooled t-Interval Procedure

Faculty Salaries   Obtain a 95% confidence interval for the difference, $\mu_1 - \mu_2$, between the mean salaries of faculty in public and private institutions.

Solution We apply Procedure 10.2.

STEP 1 **For a confidence level of $1 - \alpha$, use Table IV to find $t_{\alpha/2}$ with df $= n_1 + n_2 - 2$.**

For a 95% confidence interval, $\alpha = 0.05$. From Table 10.6, $n_1 = 30$ and $n_2 = 35$, so df $= n_1 + n_2 - 2 = 30 + 35 - 2 = 63$. In Table IV, we find that with df $= 63$, $t_{\alpha/2} = t_{0.05/2} = t_{0.025} = 1.998$.

STEP 2 **The endpoints of the confidence interval for $\mu_1 - \mu_2$ are**

$$(\bar{x}_1 - \bar{x}_2) \pm t_{\alpha/2} \cdot s_p \sqrt{(1/n_1) + (1/n_2)}.$$

From Step 1, $t_{\alpha/2} = 1.998$. Also, $n_1 = 30$, $n_2 = 35$, and, from Example 10.3, we know that $\bar{x}_1 = 64.48$, $\bar{x}_2 = 73.39$, and $s_p = 23.05$. Hence the endpoints of the confidence interval for $\mu_1 - \mu_2$ are

$$(64.48 - 73.39) \pm 1.998 \cdot 23.05 \sqrt{(1/30) + (1/35)},$$

or -8.91 ± 11.46. Thus the 95% confidence interval is from -20.37 to 2.55.

STEP 3 **Interpret the confidence interval.**

INTERPRETATION   We can be 95% confident that the difference between the mean salaries of faculty in public and private institutions is somewhere between $-\$20{,}370$ and $\$2{,}550$.

The
Technology
Center

Most statistical technologies have programs that automatically perform pooled *t*-procedures. In this subsection, we present output and step-by-step instructions to implement such programs.

Example 10.5   **Using Technology to Conduct Pooled t-Procedures**

Faculty Salaries   Table 10.5 on page 495 displays the annual salaries, in thousands of dollars, for independent samples of 30 faculty members in public institutions and 35 faculty members in private institutions. Use Minitab, Excel, or the TI-83/84 Plus to perform the hypothesis test in Example 10.3 and obtain the confidence interval required in Example 10.4.

Solution   Let μ_1 and μ_2 denote the mean salaries of all faculty in public and private institutions, respectively. The task in Example 10.3 is to perform the hypothesis test

$$H_0\text{: } \mu_1 = \mu_2 \text{ (mean salaries are the same)}$$
$$H_a\text{: } \mu_1 \neq \mu_2 \text{ (mean salaries are different)}$$

at the 5% significance level; the task in Example 10.4 is to obtain a 95% confidence interval for $\mu_1 - \mu_2$.

Printout 10.1 on pages 499 and 500 shows the output obtained by applying the pooled *t*-procedures to the salary data in Table 10.5.

PRINTOUT 10.1

Pooled *t*-procedures output for the
salary data

MINITAB

Two-Sample T-Test and CI: PUBL, PRIV

```
Two-sample T for PUBL vs PRIV

        N  Mean  StDev  SE Mean
PUBL   30  64.5   24.0      4.4
PRIV   35  73.4   22.3      3.8


Difference = mu (PUBL) - mu (PRIV)
Estimate for difference:  -8.91429
95% CI for difference:  (-20.37737, 2.54880)
T-Test of difference = 0 (vs not =): T-Value = -1.55  P-Value = 0.125  DF = 63
Both use Pooled StDev = 23.0553
```

EXCEL

2 Sample t Test Results

Test Results		Test Summary
Conclusion		**Test:** Pooled t Test
Fail to reject Ho at alpha = 0.05		**Ho:** $\mu 1 - \mu 2 = 0$
		Ha: 2-tailed: $\mu 1 - \mu 2 \neq 0$
		df: 63

Summary Statistics
Diff Std Error
-8.914 5.736

t Statistic:	-1.554
p-value:	0.125

PUBL Summary
n Mean Std Dev
30 64.48 23.953

PRIV Summary
n Mean Std Dev
35 73.394 22.261

Using **2 Var t Test**

Confidence Interval

Interval Results
Confidence Interval
With 95% Confidence, -20.377 < $\mu 1 - \mu 2$ < 2.549

Interval Summary
Diff Std Err df t*
-8.914 5.736 63 1.998

Using **2 Var t Interval**

PRINTOUT 10.1 (cont.)

Pooled *t*-procedures output for the salary data

TI-83/84 PLUS

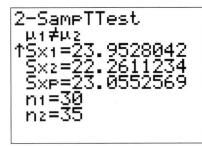

Using **2-SampTTest**   Using **2-SampTInt**

The outputs in Printout 10.1 reveal that the *P*-value for the hypothesis test is 0.125. Because the *P*-value exceeds the specified significance level of 0.05, we do not reject H_0. The outputs in Printout 10.1 also show that a 95% confidence interval for the difference between the means is from -20.38 to 2.55.

Obtaining the Output (Optional)

Printout 10.1 provides output from Minitab, Excel, and the TI-83/84 Plus for pooled *t*-procedures based on the samples of salaries in Table 10.5. The following are detailed instructions for obtaining that output. First, we store the two samples of salary data in columns (Minitab), ranges (Excel), or lists (TI-83/84 Plus) named PUBL and PRIV. Then, we proceed as follows.[†]

[†]Although Minitab simultaneously performs a hypothesis test and obtains a confidence interval, the type of confidence interval Minitab finds depends on the type of hypothesis test. Specifically, Minitab computes a two-sided confidence interval for a two-tailed test and a one-sided confidence interval for a one-tailed test. In this book, we consider only two-sided confidence intervals. To perform a one-tailed hypothesis test and obtain a two-sided confidence interval, we must apply Minitab's pooled *t*-procedure twice: once for the one-tailed hypothesis test and once for the confidence interval specifying a two-tailed hypothesis test.

MINITAB	EXCEL	TI-83/84 PLUS
1 Choose **Stat ➤ Basic Statistics ➤ 2-Sample t...**	FOR THE HYPOTHESIS TEST:	FOR THE HYPOTHESIS TEST:

MINITAB

1 Choose **Stat ➤ Basic Statistics ➤ 2-Sample t...**
2 Select the **Samples in different columns** option button
3 Click in the **First** text box and specify PUBL
4 Click in the **Second** text box and specify PRIV
5 Select the **Assume equal variances** check box
6 Click the **Options...** button
7 Click in the **Confidence level** text box and type 95
8 Click in the **Test difference** text box and type 0
9 Click the arrow button at the right of the **Alternative** drop-down list box and select **not equal**
10 Click **OK**
11 Click **OK**

EXCEL

FOR THE HYPOTHESIS TEST:
1 Choose **DDXL ➤ Hypothesis Tests**
2 Select **2 Var t Test** from the **Function type** drop-down box
3 Specify PUBL in the **1st Quantitative Variable** text box
4 Specify PRIV in the **2nd Quantitative Variable** text box
5 Click **OK**
6 Click the **Pooled** button
7 Click the **Set difference** button, type 0, and click **OK**
8 Click the **0.05** button
9 Click the $\mu1 - \mu2 \neq$ **diff** button
10 Click the **Compute** button

FOR THE CI:
1 Exit to Excel
2 Choose **DDXL ➤ Confidence Intervals**
3 Select **2 Var t Interval** from the **Function type** drop-down box
4 Specify PUBL in the **1st Quantitative Variable** text box
5 Specify PRIV in the **2nd Quantitative Variable** text box
6 Click **OK**
7 Click the **Pooled** button
8 Click the **95%** button
9 Click the **Compute Interval** button

TI-83/84 PLUS

FOR THE HYPOTHESIS TEST:
1 Press **STAT**, arrow over to **TESTS**, and press **4**
2 Highlight **Data** and press **ENTER**
3 Press the down-arrow key
4 Press **2nd ➤ LIST**, arrow down to PUBL, and press **ENTER** twice
5 Press **2nd ➤ LIST**, arrow down to PRIV, and press **ENTER** four times
6 Highlight $\neq \mu2$ and press **ENTER**
7 Press the down-arrow key, highlight **Yes**, and press **ENTER**
8 Press the down-arrow key, highlight **Calculate**, and press **ENTER**

FOR THE CI:
1 Press **STAT**, arrow over to **TESTS**, and press **0**
2 Highlight **Data** and press **ENTER**
3 Press the down-arrow key
4 Press **2nd ➤ LIST**, arrow down to PUBL, and press **ENTER** twice
5 Press **2nd ➤ LIST**, arrow down to PRIV, and press **ENTER** four times
6 Type .95 for **C-Level** and press **ENTER**
7 Highlight **Yes**, and press **ENTER**
8 Press the down-arrow key and press **ENTER**

EXERCISES 10.2

Statistical Concepts and Skills

10.16 Regarding the four conditions required for using the pooled *t*-procedures:
a. what are they?
b. how important is each condition?

10.17 Explain why s_p is called the pooled sample standard deviation.

10.18 Independent random samples are taken from two populations with the intent of performing a hypothesis test to compare their means. Preliminary data analyses indicate that the variable under consideration is normally distributed on each population. The following table provides summary statistics for the two samples.

Sample 1	Sample 2
$\bar{x}_1 = 468.3$	$\bar{x}_2 = 394.6$
$s_1 = 38.2$	$s_2 = 84.7$
$n_1 = 6$	$n_2 = 14$

Is the use of the pooled *t*-test on these data reasonable? Explain your answer.

Preliminary data analyses indicate that you can reasonably consider the assumptions for using pooled t-procedures satisfied in Exercises 10.19–10.24. For each exercise, perform the required hypothesis test by using either the critical-value approach or the P-value approach.

10.19 Doing Time. The U.S. Bureau of Prisons publishes data in *Statistical Report* on the times served by prisoners released from federal institutions for the first time. Independent random samples of released prisoners in the fraud and firearms offense categories yielded the following information on time served, in months.

Fraud		Firearms	
3.6	17.9	25.5	23.8
5.3	5.9	10.4	17.9
10.7	7.0	18.4	21.9
8.5	13.9	19.6	13.3
11.8	16.6	20.9	16.1

At the 5% significance level, do the data provide sufficient evidence to conclude that the mean time served for fraud is less than that for firearms offenses? (*Note:* $\bar{x}_1 = 10.12$, $s_1 = 4.90$, $\bar{x}_2 = 18.78$, and $s_2 = 4.64$.)

10.20 Sex and Direction. In the paper "The Relation of Sex and Sense of Direction to Spatial Orientation in an Unfamiliar Environment" (*Journal of Environmental Psychology*, Vol. 20, pp. 17–28), Sholl et al. published the results of examining the sense of direction of 30 male and 30 female students. After being taken to an unfamiliar wooded park, the students were given some spatial orientation tests, including pointing to south, which tested their absolute frame of reference. The students pointed by moving a pointer attached to a 360° protractor. Following are the absolute pointing errors, in degrees, of the participants.

Male					Female				
13	130	39	33	10	14	8	20	3	138
13	68	18	3	11	122	78	69	111	3
38	23	60	5	9	128	31	18	35	111
59	5	86	22	70	109	36	27	32	35
58	3	167	15	30	12	27	8	3	80
8	20	67	26	19	91	68	66	176	15

At the 1% significance level, do the data provide sufficient evidence to conclude that, on average, males have a better sense of direction and, in particular, a better frame of reference than females? (*Note:* $\bar{x}_1 = 37.6$, $s_1 = 38.5$, $\bar{x}_2 = 55.8$, and $s_2 = 48.3$.)

10.21 Fortified Juice and PTH. V. Tangpricha et al. did a study to determine whether fortifying orange juice with Vitamin D would result in changes in the blood levels of five biochemical variables. One of those variables was the concentration of parathyroid hormone (PTH), measured in picograms/milliliter (pg/mL). The researchers published their results in the paper "Fortification of Orange Juice with Vitamin D: a Novel Approach for Enhancing Vitamin D Nutritional Health" (*American Journal of Clinical Nutrition*, Vol. 77, pp. 1478–1483). A double-blind experiment was used in which 14 subjects drank 240 ml per day of orange juice fortified with 1000 IU of Vitamin D and 12 subjects drank 240 ml per day of unfortified orange juice. Concentration levels were recorded at the beginning of the experiment and again at the end of 12 weeks. The following data, based on the results of the study, provide the decrease (negative values indicate increase) in PTH levels for those drinking the fortified juice and for those drinking the unfortified juice.

Fortified				Unfortified		
−7.7	11.2	65.8	−45.6	65.1	0.0	40.0
−4.8	26.4	55.9	−15.5	−48.8	15.0	8.8
34.4	−5.0	−2.2		13.5	−6.1	29.4
−20.1	−40.2	73.5		−20.5	−48.4	−28.7

At the 5% significance level, do the data provide sufficient evidence to conclude that drinking fortified orange juice reduces PTH level more than drinking unfortified orange juice? (*Note:* The mean and standard deviation for the data on fortified juice are 9.0 pg/mL and 37.4 pg/mL, respectively, and for the data on unfortified juice, they are 1.6 pg/mL and 34.6 pg/mL, respectively.)

10.22 Driving Distances. Data on household vehicle miles of travel (VMT) are compiled annually by the Federal Highway Administration and are published in *National Personal Transportation Survey, Summary of Travel Trends*. Independent random samples of 15 midwestern households and 14 southern households provided the following data on last year's VMT, in thousands of miles.

Midwest			South		
16.2	12.9	17.3	22.2	19.2	9.3
14.6	18.6	10.8	24.6	20.2	15.8
11.2	16.6	16.6	18.0	12.2	20.1
24.4	20.3	20.9	16.0	17.5	18.2
9.6	15.1	18.3	22.8	11.5	

At the 5% significance level, does there appear to be a difference in last year's mean VMT for midwestern and southern households? (*Note:* $\bar{x}_1 = 16.23$, $s_1 = 4.06$, $\bar{x}_2 = 17.69$, and $s_2 = 4.42$.)

10.23 Vegetarians and Omnivores. Philosophical and health issues are prompting an increasing number of Taiwanese to switch to a vegetarian lifestyle. A study by Lu et al., published in the *Journal of Nutrition* (Vol. 130, pp. 1591–1596), compared the daily intake of nutrients by vegetarians and omnivores living in Taiwan. Among the nutrients considered was protein. Too little protein stunts growth and interferes with all bodily functions; too much protein puts a strain on the kidneys, can cause diarrhea and dehydration, and can leach calcium from bones and teeth. Independent random samples of 51 female vegetarians and 53 female omnivores yielded the following summary statistics, in grams, on daily protein intake.

Vegetarians	Omnivores
$\bar{x}_1 = 39.04$	$\bar{x}_2 = 49.92$
$s_1 = 18.82$	$s_2 = 18.97$
$n_1 = 51$	$n_2 = 53$

Do the data provide sufficient evidence to conclude that the mean daily protein intakes of female vegetarians and female omnivores differ? Perform the required hypothesis test at the 1% significance level.

10.24 Offspring of Diabetic Mothers. Previous research indicates that children borne by diabetic mothers may suffer from obesity, high blood pressure, and glucose intolerance. Independent random samples of adolescent offspring of diabetic mothers (ODM) and nondiabetic mothers (ONM) were taken by Cho et al. and evaluated for potential differences in vital measurements, including blood pressure and glucose tolerance. The study was published in *The Journal of Pediatrics* (Vol. 136(5), pp. 587–592). The following summary statistics are for the systolic blood pressures, in mm Hg, of the 99 ODM participants and the 80 ONM participants.

ODM	ONM
$\bar{x}_1 = 118$	$\bar{x}_2 = 110$
$s_1 = 12.04$	$s_2 = 11.25$
$n_1 = 99$	$n_2 = 80$

At the 1% significance level, do the data provide sufficient evidence to conclude that the mean systolic blood pressure of ODM children exceeds that of ONM children?

In Exercises 10.25–10.28, apply Procedure 10.2 on page 497 to obtain the required confidence interval.

10.25 Fortified Juice and PTH. Refer to Exercise 10.21.
a. Obtain a 90% confidence interval for the difference between the mean reductions in PTH levels for fortified and unfortified orange juice.
b. Interpret your result from part (a).

10.26 Driving Distances. Refer to Exercise 10.22.
a. Determine a 95% confidence interval for the difference between last year's mean VMTs by midwestern and southern households.
b. Interpret your result from part (a).

10.27 Vegetarians and Omnivores. Refer to Exercise 10.23.
a. Obtain a 99% confidence interval for the difference between the mean daily protein intakes of female vegetarians and female omnivores.
b. Interpret your answer in part (a).

10.28 Offspring of Diabetic Mothers. Refer to Exercise 10.24.
a. Determine a 98% confidence interval for the difference between the mean systolic blood pressures of ODM and ONM children.
b. Interpret your answer in part (a).

Extending the Concepts and Skills

10.29 In this section, we introduced the pooled *t*-test, which provides a method for comparing two population means. In deriving the pooled *t*-test, we stated that the variable

$$z = \frac{(\bar{x}_1 - \bar{x}_2) - (\mu_1 - \mu_2)}{\sigma\sqrt{(1/n_1) + (1/n_2)}}$$

cannot be used as a basis for the required test statistic because σ is unknown. Why can't that variable be used as a basis for the required test statistic?

10.30 The formula for the pooled variance, s_p^2, is given on page 492. Show that, if the sample sizes, n_1 and n_2, are equal, then s_p^2 is the mean of s_1^2 and s_2^2.

Using Technology

10.31 Vegetarians and Omnivores. Refer to Exercises 10.23 and 10.27. The raw data are on the WeissStats CD. Use the technology of your choice to
a. obtain normal probability plots, boxplots, and the standard deviations for the two samples.
b. perform the required hypothesis test and obtain the desired confidence interval.
c. Is your procedure in part (b) justified? Explain your answer.

10.32 Offspring of Diabetic Mothers. Refer to Exercises 10.24 and 10.28. We have supplied the raw data on the WeissStats CD. Use the technology of your choice to
a. obtain normal probability plots, boxplots, and the standard deviations for the two samples.
b. perform the required hypothesis test and obtain the desired confidence interval.
c. Is your procedure in part (b) justified? Explain your answer.

10.33 Blood Cholesterol and Heart Disease. Numerous studies have shown that high blood cholesterol leads to artery clogging and subsequent heart disease. One such study by Scott et al. was published in the paper "Plasma Lipids as Collateral Risk Factors in Coronary Artery Disease: A Study of 371 Males With Chest Pain" (*Journal of Chronic Diseases*, Vol. 31, pp. 337–345). The research compared the plasma cholesterol concentrations of independent random samples of patients with and without evidence of heart disease. Evidence of heart disease was based on the degree of narrowing in the arteries. The raw data, presented in milligrams/deciliter (mg/dl), are provided on the WeissStats CD.

a. Obtain normal probability plots, boxplots, and the standard deviations of the two samples.
b. At the 1% significance level, do the data provide sufficient evidence to conclude that a difference exists between the mean blood cholesterol concentrations of male patients with and without evidence of heart disease? Use the pooled *t*-test.
c. Determine and interpret a 99% confidence interval for the difference between the mean blood cholesterol concentrations of male patients with and without evidence of heart disease. Apply the pooled *t*-interval procedure.
d. Are the procedures in parts (b) and (c) justified? Explain your answer.

10.34 In this exercise, you are to perform a computer simulation to illustrate the distribution of the pooled *t*-statistic, given in Key Fact 10.2 on page 492.
a. Simulate 1000 random samples of size 4 from a normally distributed variable with a mean of 100 and a standard deviation of 16. Then obtain the sample mean and sample standard deviation of each of the 1000 samples.
b. Simulate 1000 random samples of size 3 from a normally distributed variable with a mean of 110 and a standard deviation of 16. Then obtain the sample mean and sample standard deviation of each of the 1000 samples.
c. Determine the value of the pooled *t*-statistic for each of the 1000 pairs of samples obtained in parts (a) and (b).
d. Obtain a histogram of the 1000 values found in part (c).
e. Theoretically, what is the distribution of all possible values of the pooled *t*-statistic?
f. Compare your results from parts (d) and (e).

10.3 Inferences for Two Population Means, Using Independent Samples: Standard Deviations Not Assumed Equal

In Section 10.2, we examined methods based on independent samples for performing inferences to compare the means of two populations. The methods discussed, called pooled *t*-procedures, require that the standard deviations of the two populations be equal.

In this section, we develop inferential procedures based on independent samples to compare the means of two populations that do not require the population standard deviations to be equal, even though they may be. As before, we assume that the population standard deviations are unknown, because that is usually the case in practice.

For our derivation, we also assume that the variable under consideration is normally distributed on each population. As we demonstrate later, the resulting inferential procedures are approximately correct for large samples, regardless of distribution type.

Hypothesis Tests for the Means of Two Populations, Using Independent Samples

We begin by finding a test statistic. We know from Key Fact 10.1 on page 488 that, for independent samples, the standardized version of $\bar{x}_1 - \bar{x}_2$,

$$z = \frac{(\bar{x}_1 - \bar{x}_2) - (\mu_1 - \mu_2)}{\sqrt{(\sigma_1^2/n_1) + (\sigma_2^2/n_2)}},$$

has the standard normal distribution. We are assuming that the population standard deviations, σ_1 and σ_2, are unknown, so we cannot use this variable as a basis for the required test statistic. We therefore replace σ_1 and σ_2 with their sample estimates, s_1 and s_2, and obtain the variable

$$\frac{(\bar{x}_1 - \bar{x}_2) - (\mu_1 - \mu_2)}{\sqrt{(s_1^2/n_1) + (s_2^2/n_2)}},$$

which we can use as a basis for the required test statistic. This variable does not have the standard normal distribution, but it does have roughly a t-distribution, as indicated in Key Fact 10.3.

KEY FACT 10.3

Distribution of the Nonpooled t-Statistic

Suppose that x is a normally distributed variable on each of two populations. Then, for independent samples of sizes n_1 and n_2 from the two populations, the variable

$$t = \frac{(\bar{x}_1 - \bar{x}_2) - (\mu_1 - \mu_2)}{\sqrt{(s_1^2/n_1) + (s_2^2/n_2)}}$$

has approximately a t-distribution. The degrees of freedom used is obtained from the sample data. It is denoted Δ and given by

$$\Delta = \frac{\left[(s_1^2/n_1) + (s_2^2/n_2)\right]^2}{\dfrac{\left(s_1^2/n_1\right)^2}{n_1 - 1} + \dfrac{\left(s_2^2/n_2\right)^2}{n_2 - 1}},$$

rounded down to the nearest integer.

In light of Key Fact 10.3, for a hypothesis test that has null hypothesis $H_0: \mu_1 = \mu_2$, we can use the variable

$$t = \frac{\bar{x}_1 - \bar{x}_2}{\sqrt{(s_1^2/n_1) + (s_2^2/n_2)}}$$

as the test statistic and obtain the critical value(s) or P-value from the t-table, Table IV. We refer to this hypothesis-testing procedure as the **nonpooled t-test**. Procedure 10.3 on the next page provides a step-by-step method for performing a nonpooled t-test by using either the critical-value approach or the P-value approach.

Regarding Assumption 3 for the nonpooled t-test, although the nonpooled t-test was derived under the condition that the variable under consideration is normally distributed on each of the two populations (normal populations), it also applies approximately for large samples regardless of distribution type. Actually, the nonpooled t-test works reasonably well even for small samples or samples of moderate size from nonnormal populations provided the populations are not too nonnormal. In other words, the nonpooled t-test is robust to moderate violations of the normality assumption.

PROCEDURE 10.3	Nonpooled t-Test

Purpose To perform a hypothesis test to compare two population means, μ_1 and μ_2

Assumptions

1. Simple random samples
2. Independent samples
3. Normal populations or large samples

STEP 1 **The null hypothesis is H_0: $\mu_1 = \mu_2$, and the alternative hypothesis is**

$$H_a: \mu_1 \neq \mu_2 \quad \text{or} \quad H_a: \mu_1 < \mu_2 \quad \text{or} \quad H_a: \mu_1 > \mu_2$$
$$\text{(Two tailed)} \qquad \text{(Left tailed)} \qquad \text{(Right tailed)}$$

STEP 2 **Decide on the significance level, α.**

STEP 3 **Compute the value of the test statistic**

$$t = \frac{\bar{x}_1 - \bar{x}_2}{\sqrt{(s_1^2/n_1) + (s_2^2/n_2)}}.$$

Denote the value of the test statistic t_0.

CRITICAL-VALUE APPROACH

STEP 4 **The critical value(s) are**

$$\pm t_{\alpha/2} \quad \text{or} \quad -t_\alpha \quad \text{or} \quad t_\alpha$$
$$\text{(Two tailed)} \qquad \text{(Left tailed)} \qquad \text{(Right tailed)}$$

with df = Δ, where

$$\Delta = \frac{\left[(s_1^2/n_1) + (s_2^2/n_2)\right]^2}{\dfrac{\left(s_1^2/n_1\right)^2}{n_1 - 1} + \dfrac{\left(s_2^2/n_2\right)^2}{n_2 - 1}},$$

rounded down to the nearest integer. Use Table IV to find the critical value(s).

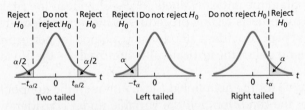

STEP 5 **If the value of the test statistic falls in the rejection region, reject H_0; otherwise, do not reject H_0.**

P-VALUE APPROACH

STEP 4 **The t-statistic has df = Δ, where**

$$\Delta = \frac{\left[(s_1^2/n_1) + (s_2^2/n_2)\right]^2}{\dfrac{\left(s_1^2/n_1\right)^2}{n_1 - 1} + \dfrac{\left(s_2^2/n_2\right)^2}{n_2 - 1}},$$

rounded down to the nearest integer. Use Table IV to estimate the P-value, or obtain it exactly by using technology.

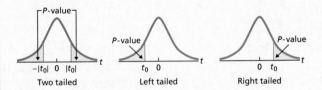

STEP 5 **If $P \leq \alpha$, reject H_0; otherwise, do not reject H_0.**

STEP 6 **Interpret the results of the hypothesis test.**

When considering the nonpooled *t*-test, you must also watch for outliers. Again, the presence of outliers calls into question the normality assumption. Moreover, even for large samples, outliers can sometimes unduly affect a non-pooled *t*-test because the sample mean and sample standard deviation are not resistant to them.

We also note that the nonpooled *t*-test can be used as a procedure for comparing two means with a designed experiment. When the null hypothesis of equal means doesn't necessarily imply that there is no difference at all between the two treatments, the nonpooled *t*-test is often preferable to the pooled *t*-test.

Example 10.6 illustrates use of the nonpooled *t*-test.

Example 10.6   The Nonpooled t-Test

Neurosurgery Operative Times   Several neurosurgeons wanted to determine whether a dynamic system (Z-plate) reduced the operative time relative to a static system (ALPS plate). R. Jacobowitz, Ph.D., an ASU professor, along with G. Vishteh, M.D., and other neurosurgeons, obtained the data displayed in Table 10.7 on operative times, in minutes, for the two systems. At the 1% significance level, do the data provide sufficient evidence to conclude that the mean operative time is less with the dynamic system than with the static system?

TABLE 10.7
Operative times, in minutes, for dynamic and static systems

Dynamic							Static		
370	360	510	445	295	315	490	430	445	455
345	450	505	335	280	325	500	455	490	535

TABLE 10.8
Summary statistics for the samples in Table 10.7

Dynamic	Static
$\bar{x}_1 = 394.6$	$\bar{x}_2 = 468.3$
$s_1 = 84.7$	$s_2 = 38.2$
$n_1 = 14$	$n_2 = 6$

Solution   First, we present in Table 10.8 the required summary statistics for the two samples in Table 10.7. These statistics are obtained in the usual way.

Next, we check the three conditions required for using the nonpooled *t*-test. These data were obtained from a randomized comparative experiment, a type of designed experiment. Therefore, we can consider Assumptions 1 and 2 satisfied. Boxplots and normal probability plots (not shown) of the two samples in Table 10.7 reveal no outliers and, keeping in mind that the nonpooled *t*-test is robust to moderate violations of normality, show that we can consider Assumption 3 satisfied. We can therefore apply the nonpooled *t*-test, Procedure 10.3, to carry out the hypothesis test.

STEP 1   State the null and alternative hypotheses.

Let μ_1 and μ_2 denote the mean operative times for the dynamic and static systems, respectively. Then the null and alternative hypotheses are

H_0: $\mu_1 = \mu_2$ (mean dynamic time is not less than mean static time)
H_a: $\mu_1 < \mu_2$ (mean dynamic time is less than mean static time).

Note that the hypothesis test is left tailed because a less-than sign ($<$) appears in the alternative hypothesis.

STEP 2 Decide on the significance level, α.

The test is to be performed at the 1% significance level, or $\alpha = 0.01$.

STEP 3 Compute the value of the test statistic

$$t = \frac{\bar{x}_1 - \bar{x}_2}{\sqrt{(s_1^2/n_1) + (s_2^2/n_2)}}.$$

Referring to Table 10.8, we get

$$t = \frac{394.6 - 468.3}{\sqrt{(84.7^2/14) + (38.2^2/6)}} = -2.681.$$

CRITICAL-VALUE APPROACH	**P-VALUE APPROACH**

CRITICAL-VALUE APPROACH

STEP 4 The critical value for a left-tailed test is $-t_\alpha$ with df $= \Delta$. Use Table IV to find the critical value.

From Step 2, $\alpha = 0.01$. Also, from Table 10.8, we see that

$$\mathrm{df} = \Delta = \frac{\left[(84.7^2/14) + (38.2^2/6)\right]^2}{\dfrac{(84.7^2/14)^2}{14-1} + \dfrac{(38.2^2/6)^2}{6-1}},$$

which equals 17 when rounded down. From Table IV with df $= 17$, we find that the critical value is $-t_\alpha = -t_{0.01} = -2.567$, as shown in Fig. 10.3A.

FIGURE 10.3A

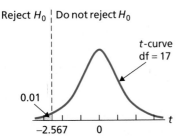

STEP 5 If the value of the test statistic falls in the rejection region, reject H_0; otherwise, do not reject H_0.

From Step 3, the value of the test statistic is $t = -2.681$, which, as we see from Fig. 10.3A, falls in the rejection region. Thus we reject H_0. The test results are statistically significant at the 1% level.

P-VALUE APPROACH

STEP 4 The t-statistic has df $= \Delta$. Use Table IV to estimate the P-value or obtain it exactly by using technology.

From Step 3, the value of the test statistic is $t = -2.681$. The test is left tailed, so the P-value is the probability of observing a value of t of -2.681 or less if the null hypothesis is true. That probability equals the shaded area shown in Fig. 10.3B.

FIGURE 10.3B

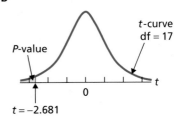

From Table 10.8, we find that

$$\mathrm{df} = \Delta = \frac{\left[(84.7^2/14) + (38.2^2/6)\right]^2}{\dfrac{(84.7^2/14)^2}{14-1} + \dfrac{(38.2^2/6)^2}{6-1}},$$

which equals 17 when rounded down. Referring to Fig. 10.3B and Table IV with df $= 17$, we find that $0.005 < P < 0.01$. (Using technology, $P = 0.0079$.)

STEP 5 If $P \leq \alpha$, reject H_0; otherwise, do not reject H_0.

From Step 4, $0.005 < P < 0.01$. Because the P-value is less than the specified significance level of 0.01, we reject H_0. The test results are statistically significant at the 1% level and (see Table 9.12 on page 440) provide very strong evidence against the null hypothesis.

STEP 6   Interpret the results of the hypothesis test.

INTERPRETATION   At the 1% significance level, the data provide sufficient evidence to conclude that the mean operative time is less with the dynamic system than with the static system.   ▬

Confidence Intervals for the Difference Between the Means of Two Populations, Using Independent Samples

We can also use Key Fact 10.3 on page 505 to derive a confidence-interval procedure, Procedure 10.4, for the difference between two means, which we often refer to as the **nonpooled** *t***-interval procedure.** We demonstrate its use in Example 10.7.

PROCEDURE 10.4

Nonpooled *t*-Interval Procedure

Purpose To find a confidence interval for the difference between two population means, μ_1 and μ_2

Assumptions

1. Simple random samples
2. Independent samples
3. Normal populations or large samples

STEP 1   For a confidence level of 1 − α, use Table IV to find $t_{\alpha/2}$ with df = Δ, where

$$\Delta = \frac{\left[\left(s_1^2/n_1 \right) + \left(s_2^2/n_2 \right) \right]^2}{\dfrac{\left(s_1^2/n_1 \right)^2}{n_1 - 1} + \dfrac{\left(s_2^2/n_2 \right)^2}{n_2 - 1}},$$

rounded down to the nearest integer.

STEP 2   The endpoints of the confidence interval for $\mu_1 - \mu_2$ are

$$(\bar{x}_1 - \bar{x}_2) \pm t_{\alpha/2} \cdot \sqrt{\left(s_1^2/n_1 \right) + \left(s_2^2/n_2 \right)}.$$

STEP 3   Interpret the confidence interval.

▌▌Example 10.7   The Nonpooled t-Interval Procedure

Neurosurgery Operative Times   Use the sample data in Table 10.7 on page 507 to obtain a 99% confidence interval for the difference, $\mu_1 - \mu_2$, between the mean operative times of the dynamic and static systems.

Solution   We apply Procedure 10.4.

STEP 1 **For a confidence level of 1 − α, use Table IV to find $t_{\alpha/2}$ with df = Δ.**

For a 99% confidence interval, $\alpha = 0.01$. From Example 10.6, df = 17. In Table IV, with df = 17, $t_{\alpha/2} = t_{0.01/2} = t_{0.005} = 2.898$.

STEP 2 **The endpoints of the confidence interval for $\mu_1 - \mu_2$ are**

$$(\bar{x}_1 - \bar{x}_2) \pm t_{\alpha/2} \cdot \sqrt{(s_1^2/n_1) + (s_2^2/n_2)}.$$

From Step 1, $t_{\alpha/2} = 2.898$. Referring to Table 10.8 on page 507, we conclude that the endpoints of the confidence interval for $\mu_1 - \mu_2$ are

$$(394.6 - 468.3) \pm 2.898 \cdot \sqrt{(84.7^2/14) + (38.2^2/6)}$$

or −153.4 to 6.0.

STEP 3 **Interpret the confidence interval.**

INTERPRETATION We can be 99% confident that the difference between the mean operative times of the dynamic and static systems is somewhere between −153.4 minutes and 6.0 minutes.

Pooled Versus Nonpooled *t*-Tests

Suppose that we want to perform a hypothesis test based on independent simple random samples to compare the means of two populations. Further suppose that either the variable under consideration is normally distributed on each of the two populations or the sample sizes are large. Then two tests are candidates for the job: the pooled *t*-test (Procedure 10.1 of Section 10.2) or the nonpooled *t*-test (Procedure 10.3 of this section).

In theory, the pooled *t*-test requires that the population standard deviations be equal. What if the pooled *t*-test is used when in fact the population standard deviations are not equal? The answer to this question depends on several factors. If the population standard deviations are unequal, but not too unequal, and the sample sizes are nearly the same, using the pooled *t*-test will not cause serious difficulties. If the population standard deviations are quite different, however, using the pooled *t*-test can result in a significantly larger Type I error probability than the one specified (i.e., α).

In contrast, the nonpooled *t*-test does not require that the population standard deviations be equal; it applies whether or not they are equal. Then why use the pooled *t*-test at all? The reason is that, if the population standard deviations are equal or nearly so, then, on average, the pooled *t*-test is slightly more powerful; that is, the probability of making a Type II error is somewhat smaller.

As emphasized in Key Fact 10.4, the pooled *t*-test should be used only when the two populations have nearly equal standard deviations; otherwise, the nonpooled *t*-test should be applied. Similar remarks apply to the pooled *t*-interval and nonpooled *t*-interval procedures.

KEY FACT 10.4   **Choosing Between a Pooled and Nonpooled Procedure**

Suppose that you want to use independent simple random samples to compare the means of two populations. When you are deciding between a pooled *t*-procedure and a nonpooled *t*-procedure, follow these guidelines: If you are reasonably sure that the populations have nearly equal standard deviations, use a pooled *t*-procedure; otherwise, use a nonpooled *t*-procedure.

The **Technology Center**

Most statistical technologies have programs that automatically perform nonpooled *t*-procedures. In this subsection, we present output and step-by-step instructions to implement such programs.

Example 10.8   **Using Technology to Conduct Nonpooled t-Procedures**

Neurosurgery Operative Times   Table 10.7 on page 507 displays samples of neurosurgery operative times, in minutes, for dynamic and static systems. Use Minitab, Excel, or the TI-83/84 Plus to perform the hypothesis test in Example 10.6 and obtain the confidence interval required in Example 10.7.

Solution   Let μ_1 and μ_2 denote, respectively, the mean operative times of the dynamic and static systems. The task in Example 10.6 is to perform the hypothesis test

H_0: $\mu_1 = \mu_2$ (mean dynamic time is not less than mean static time)

H_a: $\mu_1 < \mu_2$ (mean dynamic time is less than mean static time)

at the 1% significance level; the task in Example 10.7 is to obtain a 99% confidence interval for $\mu_1 - \mu_2$.

Printout 10.2, presented on pages 512 and 513, shows the output obtained by applying the nonpooled *t*-procedures to the data on operative times displayed in Table 10.7.

The outputs in Printout 10.2 reveal that the *P*-value for the hypothesis test is 0.008, rounded to three decimal places. As the *P*-value is less than the specified significance level of 0.01, we reject H_0. The outputs in Printout 10.2 also show that a 99% confidence interval for the difference between the means is from -153 to 6, rounded to the nearest integer.

Note that, if we round the endpoints of the confidence interval to one or more decimal places, there are discrepancies among the confidence intervals provided by the three technologies. This is due to the fact that some statistical technologies round degrees of freedom, whereas others do not round degrees of freedom.

PRINTOUT 10.2

Nonpooled *t*-procedures output for the
operative-time data

MINITAB

Two-Sample T-Test and CI: DYNA, STAT   [FOR THE HYPOTHESIS TEST]

Two-sample T for DYNA vs STAT

```
                          SE
        N    Mean  StDev  Mean
DYNA   14   394.6   84.7   23
STAT    6   468.3   38.2   16
```

Difference = mu (DYNA) - mu (STAT)
Estimate for difference: -73.6905
99% upper bound for difference: -3.1200
T-Test of difference = 0 (vs <): T-Value = -2.68 P-Value = 0.008 DF = 17

Two-Sample T-Test and CI: DYNA, STAT   [FOR THE CONFIDENCE INTERVAL]

Two-sample T for DYNA vs STAT

```
                          SE
        N    Mean  StDev  Mean
DYNA   14   394.6   84.7   23
STAT    6   468.3   38.2   16
```

Difference = mu (DYNA) - mu (STAT)
Estimate for difference: -73.6905
99% CI for difference: (-153.3690, 5.9881)
T-Test of difference = 0 (vs not =): T-Value = -2.68 P-Value = 0.016 DF = 17

EXCEL

2 Sample t Test Results		

Test Results		Test Summary
Conclusion		Test: 2-Sample t Test
Reject Ho at alpha = 0.01		Ho: $\mu 1 - \mu 2 = 0$
		Ha: Lower tail: $\mu 1 - \mu 2 < 0$
		df: 17

Summary Statistics		t Statistic: -2.68
Diff Std Error		p-value: 0.0079
-73.69 27.492		

DYNA Summary		STAT Summary
n Mean Std Dev		**n Mean Std Dev**
14 394.643 84.75		6 468.333 38.166

Using **2 Var t Test**

PRINTOUT 10.2 (cont.)

Nonpooled *t*-procedures output for the operative-time data

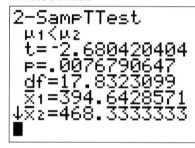

Using **2 Var t Interval**

TI-83/84 PLUS

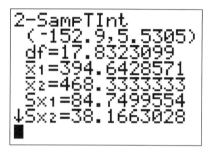

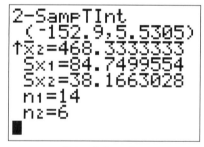

Using **2-SampTTest**   Using **2-SampTInt**

Obtaining the Output (Optional)

Printout 10.2 provides output from Minitab, Excel, and the TI-83/84 Plus for nonpooled *t*-procedures based on the samples of operative times in Table 10.7. The following are detailed instructions for obtaining that output. First, we store the two samples of operative times in columns (Minitab), ranges (Excel), or lists (TI-83/84 Plus) named DYNA and STAT. Then, we proceed as follows.[†]

[†]Although Minitab simultaneously performs a hypothesis test and obtains a confidence interval, the type of confidence interval Minitab finds depends on the type of hypothesis test. Specifically, Minitab computes a two-sided confidence interval for a two-tailed test and a one-sided confidence interval for a one-tailed test. In this book, we consider only two-sided confidence intervals. To perform a one-tailed hypothesis test and obtain a two-sided confidence interval, we must apply Minitab's nonpooled *t*-procedure twice: once for the one-tailed hypothesis test and once for the confidence interval specifying a two-tailed hypothesis test.

MINITAB	EXCEL	TI-83/84 PLUS
FOR THE HYPOTHESIS TEST:	**FOR THE HYPOTHESIS TEST:**	**FOR THE HYPOTHESIS TEST:**

MINITAB

FOR THE HYPOTHESIS TEST:
1 Choose **Stat ➤ Basic Statistics ➤ 2-Sample t...**
2 Select the **Samples in different columns** option button
3 Click in the **First** text box and specify DYNA
4 Click in the **Second** text box and specify STAT
5 Deselect the **Assume equal variances** check box
6 Click the **Options...** button
7 Click in the **Confidence level** text box and type 99
8 Click in the **Test difference** text box and type 0
9 Click the arrow button at the right of the **Alternative** drop-down list box and select **less than**
10 Click **OK**
11 Click **OK**

FOR THE CI:
1 Choose **Edit ➤ Edit Last Dialog**
2 Click the **Options...** button
3 Click the arrow button at the right of the **Alternative** drop-down list box and select **not equal**
4 Click **OK**
5 Click **OK**

EXCEL

FOR THE HYPOTHESIS TEST:
1 Choose **DDXL ➤ Hypothesis Tests**
2 Select **2 Var t Test** from the **Function type** drop-down box
3 Specify DYNA in the **1st Quantitative Variable** text box
4 Specify STAT in the **2nd Quantitative Variable** text box
5 Click **OK**
6 Click the **2-sample** button
7 Click the **Set difference** button, type 0, and click **OK**
8 Click the **0.01** button
9 Click the $\mu 1 - \mu 2 <$ **diff** button
10 Click the **Compute** button

FOR THE CI:
1 Exit to Excel
2 Choose **DDXL ➤ Confidence Intervals**
3 Select **2 Var t Interval** from the **Function type** drop-down box
4 Specify DYNA in the **1st Quantitative Variable** text box
5 Specify STAT in the **2nd Quantitative Variable** text box
6 Click **OK**
7 Click the **2-sample** button
8 Click the **99%** button
9 Click the **Compute Interval** button

TI-83/84 PLUS

FOR THE HYPOTHESIS TEST:
1 Press **STAT**, arrow over to **TESTS**, and press **4**
2 Highlight **Data** and press **ENTER**
3 Press the down-arrow key
4 Press **2nd ➤ LIST**, arrow down to DYNA, and press **ENTER** twice
5 Press **2nd ➤ LIST**, arrow down to STAT, and press **ENTER** four times
6 Highlight $< \mu 2$ and press **ENTER**
7 Press the down-arrow key, highlight **No**, and press **ENTER**
8 Press the down-arrow key, highlight **Calculate**, and press **ENTER**

FOR THE CI:
1 Press **STAT**, arrow over to **TESTS**, and press **0**
2 Highlight **Data** and press **ENTER**
3 Press the down-arrow key
4 Press **2nd ➤ LIST**, arrow down to DYNA, and press **ENTER** twice
5 Press **2nd ➤ LIST**, arrow down to STAT, and press **ENTER** four times
6 Type .99 for **C-Level** and press **ENTER**
7 Highlight **No** and press **ENTER**
8 Press the down-arrow key and press **ENTER**

EXERCISES 10.3

Statistical Concepts and Skills

10.35 Neurosurgery Operative Times. Refer to Example 10.6 on page 507. Explain why using the nonpooled *t*-test is more appropriate than using the pooled *t*-test.

10.36 Suppose that you know that a variable is normally distributed on each of two populations. Further suppose that you want to perform a hypothesis test based on independent random samples to compare the two population means. In each case, decide whether you would use the pooled or nonpooled *t*-test and give a reason for your answer.
a. You know that the population standard deviations are equal.
b. You know that the population standard deviations are not equal.

c. The sample standard deviations are 23.6 and 25.2, and each sample size is 25.
d. The sample standard deviations are 23.6 and 59.2.

10.37 What is the difference in assumptions between the pooled and nonpooled *t*-procedures?

10.38 Discuss the relative advantages and disadvantages of using pooled and nonpooled *t*-procedures.

Preliminary data analyses indicate that you can reasonably use nonpooled t-procedures in Exercises 10.39–10.44. For each exercise, apply a nonpooled t-test to perform the required hypothesis test, using either the critical-value approach or the P-value approach.

10.39 Political Prisoners. According to the American Psychiatric Association, posttraumatic stress disorder (PTSD) is a common psychological consequence of traumatic events that involve a threat to life or physical integrity. During the Cold War, some 200,000 people in East Germany were imprisoned for political reasons. Many were subjected to physical and psychological torture during their imprisonment, resulting in PTSD. Ehlers, Maercker, and Boos studied various characteristics of political prisoners from the former East Germany and presented their findings in the paper "Posttraumatic Stress Disorder (PTSD) Following Political Imprisonment: The Role of Mental Defeat, Alienation, and Perceived Permanent Change" (*Journal of Abnormal Psychology*, Vol. 109, pp. 45–55). The researchers randomly and independently selected 32 former prisoners diagnosed with chronic PTSD and 20 former prisoners that were diagnosed with PTSD after release from prison but had since recovered (remitted). The ages, in years, at arrest yielded the following summary statistics.

Chronic	Remitted
$\bar{x}_1 = 25.8$	$\bar{x}_2 = 22.1$
$s_1 = 9.2$	$s_2 = 5.7$
$n_1 = 32$	$n_2 = 20$

At the 10% significance level, is there sufficient evidence to conclude that a difference exists in the mean age at arrest of East German prisoners with chronic PTSD and remitted PTSD?

10.40 Nitrogen and Seagrass. The seagrass *Thalassia testudinum* is an integral part of the Texas coastal ecosystem. Essential to the growth of *T. testudinum* is ammonium. Researchers Kun-Seop Lee and Kenneth H. Dunton of the Marine Science Institute of the University of Texas at Austin noticed that the seagrass beds in Corpus Christi Bay (CCB) were taller and thicker than those in Lower Laguna Madre (LLM). They compared the sediment ammonium concentrations in the two locations and published their findings in *Marine Ecology Progress Series* (Vol. 196, pp. 39–48). Following are the summary statistics on sediment ammonium concentrations, in micromoles, obtained by the researchers.

CCB	LLM
$\bar{x}_1 = 115.1$	$\bar{x}_2 = 24.3$
$s_1 = 79.4$	$s_2 = 10.5$
$n_1 = 51$	$n_2 = 19$

At the 1% significance level, is there sufficient evidence to conclude that the mean sediment ammonium concentration in CCB exceeds that in LLM?

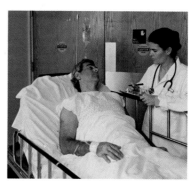

10.41 Acute Postoperative Days. Refer to Example 10.6 on page 507. The researchers also obtained the following data on the number of acute postoperative days in the hospital using the dynamic and static systems.

Dynamic							Static		
7	5	8	8	6	7	7	6	18	9
9	10	7	7	7	7	8	7	14	9

At the 5% significance level, do the data provide sufficient evidence to conclude that the mean number of acute postoperative days in the hospital is smaller with the dynamic system than with the static system? (*Note:* $\bar{x}_1 = 7.36$, $s_1 = 1.22$, $\bar{x}_2 = 10.50$, and $s_2 = 4.59$.)

10.42 Stressed-Out Bus Drivers. Frustrated passengers, congested streets, time schedules, and air and noise pollution are just some of the physical and social pressures that lead many urban bus drivers to retire prematurely with disabilities such as coronary heart disease and stomach disorders. An intervention program designed by the Stockholm Transit District was implemented to improve the work conditions of the city's bus drivers. Improvements were evaluated by Evans et al. who collected physiological and psychological data for bus drivers who drove on the improved routes (intervention) and for drivers who were assigned the normal routes (control). Their findings were published in the article "Hassles on the Job: a Study of a Job Intervention With Urban Bus Drivers" (*Journal of Organizational Behavior* Vol. 20, pp. 199–208). Following are data, based on the results of the study, for the heart rates, in beats per minute, of the intervention and control drivers.

Intervention		Control						
68	66	74	52	67	63	77	57	80
74	58	77	53	76	54	73	54	
69	63	60	77	63	60	68	64	
68	73	66	71	66	55	71	84	
64	76	63	73	59	68	64	82	

a. At the 5% significance level, do the data provide sufficient evidence to conclude that the intervention program reduces mean heart rate of urban bus drivers in Stockholm? (*Note:* $\bar{x}_1 = 67.90$, $s_1 = 5.49$, $\bar{x}_2 = 66.81$, and $s_2 = 9.04$.)

b. Can you provide an explanation for the somewhat surprising results of the study?

c. Is the study a designed experiment or an observational study? Explain your answer.

10.43 Schizophrenia and Dopamine. Previous research has suggested that changes in the activity of dopamine, a neurotransmitter in the brain, may be a causative factor for schizophrenia. In the paper "Schizophrenia: Dopamine b-Hydroxylase Activity and Treatment Response" (*Science*, Vol. 216, pp. 1423–1425), Sternberg et al. published the results of their study in which they examined 25 schizophrenic patients who had been classified as either psychotic or not psychotic by hospital staff. The activity of dopamine was measured in each patient by using the enzyme dopamine b-hydroxylase to assess differences in dopamine activity between the two groups. The following are the data, in nanomoles per milliliter-hour per milligram (nmol/ml-h/mg).

Psychotic		Not psychotic		
0.0150	0.0222	0.0104	0.0230	0.0145
0.0204	0.0275	0.0200	0.0116	0.0180
0.0306	0.0270	0.0210	0.0252	0.0154
0.0320	0.0226	0.0105	0.0130	0.0170
0.0208	0.0245	0.0112	0.0200	0.0156

At the 1% significance level, do the data suggest that dopamine activity is higher, on average, in psychotic patients? (*Note:* $\bar{x}_1 = 0.02426$, $s_1 = 0.00514$, $\bar{x}_2 = 0.01643$, and $s_2 = 0.00470$.)

10.44 Wing Length. D. Cristol et al. published results of their studies of two subspecies of dark-eyed juncos in the article "Migratory Dark-Eyed Juncos, *Junco Hyemalis*, Have Better Spatial Memory and Denser Hippocampal Neurons than Nonmigratory Conspecifics" (*Animal Behaviour*, Vol. 66, pp. 317–328). One of the subspecies migrates each year and the other does not migrate. Several physical characteristics of 14 birds of each subspecies were measured, one of which was wing length. The following data, based on results obtained by the researchers, provide the wing lengths, in millimeters (mm), for the samples of two subspecies.

a. At the 1% significance level, do the data provide sufficient evidence to conclude that the mean wing lengths for the two subspecies are different? (*Note:* The mean and standard deviation for the migratory-bird data

Migratory			Nonmigratory		
84.5	81.0	82.6	82.1	82.4	83.9
82.8	84.5	81.2	87.1	84.6	85.1
80.5	82.1	82.3	86.3	86.6	83.9
80.1	83.4	81.7	84.2	84.3	86.2
83.0	79.7		87.8	84.1	

are 82.1 mm and 1.501 mm, respectively, and that for the nonmigratory-bird data are 84.9 mm and 1.698 mm, respectively.)

b. Would it be reasonable to use a pooled *t*-test here? Explain your answer.

c. If your answer to part (b) was *yes*, then perform a pooled *t*-test to answer the question in part (a) and compare your results to that found in part (a) by using a nonpooled *t*-test.

In Exercises **10.45–10.48**, *apply Procedure 10.4 on page 509 to obtain the required confidence interval.*

10.45 Political Prisoners. Refer to Exercise 10.39.

a. Determine a 90% confidence interval for the difference, $\mu_1 - \mu_2$, between the mean ages at arrest of East German prisoners with chronic PTSD and remitted PTSD.

b. Interpret your answer in words.

10.46 Nitrogen and Seagrass. Refer to Exercise 10.40.

a. Determine a 98% confidence interval for the difference, $\mu_1 - \mu_2$, between the mean sediment ammonium concentrations in CCB and LLM.

b. Interpret your answer in words.

10.47 Acute Postoperative Days. Refer to Exercise 10.41.

a. Find a 90% confidence interval for the difference between the mean numbers of acute postoperative days in the hospital with the dynamic and static systems.

b. Interpret your answer in words.

10.48 Stressed-Out Bus Drivers. Refer to Exercise 10.42.

a. Find a 90% confidence interval for the difference between the mean heart rates of urban bus drivers in Stockholm in the two environments.

b. Interpret your answer in words.

Extending the Concepts and Skills

10.49 Acute Postoperative Days. In Exercise 10.41, you conducted a nonpooled *t*-test to decide whether the mean number of acute postoperative days spent in the hospital is smaller with the dynamic system than with the static system.

a. Using a pooled *t*-test, repeat that hypothesis test.

b. Compare your decisions with the pooled and nonpooled *t*-tests.

c. Which test do you think is more appropriate, the pooled or nonpooled *t*-test? Explain your answer.

10.50 Neurosurgery Operative Times. In Example 10.6 on page 507, we conducted a nonpooled *t*-test to decide whether the mean operative time is less with the dynamic system than with the static system.

a. Using a pooled *t*-test, repeat that hypothesis test.

b. Compare your decisions with the pooled and nonpooled *t*-tests.

c. Which test do you think is more appropriate, the pooled or nonpooled *t*-test? Explain your answer.

10.51 Each pair of graphs in Fig. 10.4 shows the distributions of a variable on two populations. Suppose that, in each case, you want to perform a small-sample hypothesis test based on independent simple random samples to compare the means of the two populations. In each case, decide whether the pooled *t*-test, nonpooled *t*-test, or neither should be used. Explain your answers.

10.52 Tukey's Quick Test. In this exercise, we examine an alternative method, conceived by the late Professor John Tukey, for performing a two-tailed hypothesis test for two population means based on independent random samples. To apply this procedure, one of the samples must contain the largest observation (high group) and the other sample must contain the smallest observation (low group). Here are the steps for performing Tukey's quick test.

Step 1 Count the number of observations in the high group that are greater than or equal to the largest observation in the low group. Count ties as $1/2$.

Step 2 Count the number of observations in the low group that are less than or equal to the smallest observation in the high group. Count ties as $1/2$.

Step 3 Add the two counts obtained in Steps 1 and 2, and denote the sum c.

Step 4 Reject the null hypothesis at the 5% significance level if and only if $c \geq 7$; reject it at the 1% significance level if and only if $c \geq 10$; and reject it at the 0.1% significance level if and only if $c \geq 13$.

a. Can Tukey's quick test be applied to Exercise 10.22 on page 502? Explain your answer.

b. If your answer to part (a) was *yes*, apply Tukey's quick test and compare your result to that found in Exercise 10.22, where a *t*-test was used.

c. Can Tukey's quick test be applied to Exercise 10.44? Explain your answer.

d. If your answer to part (c) was *yes*, apply Tukey's quick test and compare your result to that found in Exercise 10.44, where a *t*-test was used.

For more details about Tukey's quick test, see John Tukey's paper "A Quick, Compact, Two-Sample Test to Duckworth's Specifications" (*Technometrics*, Vol. 1, No. 1, pp. 31–48).

Using Technology

10.53 Schizophrenia and Dopamine. Refer to Exercise 10.43. Use the technology of your choice to

a. obtain boxplots and normal probability plots for the two samples.

b. perform the required hypothesis test.

c. obtain a 99% confidence interval for the difference between the two population means.

d. Justify the use of your procedure in parts (b) and (c).

10.54 Wing Length. Refer to Exercise 10.44. Use the technology of your choice to

a. obtain boxplots and normal probability plots for the two samples.

b. perform the required hypothesis test.

c. obtain a 99% confidence interval for the difference between the two population means.

d. Justify the use of your procedure in parts (b) and (c).

FIGURE 10.4
Figure for Exercise 10.51

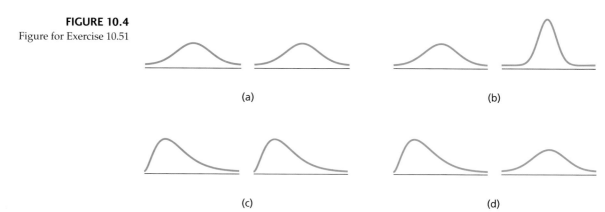

(a)   (b)

(c)   (d)

10.4 The Mann–Whitney Test[†]

We have now developed two procedures for performing a hypothesis test to compare the means of two populations—namely, the pooled and nonpooled *t*-tests. Both tests require (1) simple random samples, (2) independent samples, and (3) normal populations or large samples. The pooled *t*-test also requires that the population standard deviations be equal.

Recall that the shape of a normal distribution is determined by its standard deviation. In other words, two normal distributions have the same shape if and only if they have equal standard deviations. Consequently, the pooled *t*-test applies when the two distributions (one for each population) of the variable under consideration are normal and have the same shape; the nonpooled *t*-test applies when the two distributions of the variable under consideration are normal, even if they don't have the same shape.

Another procedure for performing a hypothesis test based on independent simple random samples to compare the means of two populations is the **Mann–Whitney test.** This nonparametric test, introduced by Wilcoxon and further developed by Mann and Whitney, is also commonly referred to as the **Wilcoxon rank-sum test** or the **Mann–Whitney–Wilcoxon test.**

The Mann–Whitney test applies when the two distributions of the variable under consideration have the same shape, but it does not require that they be normal or have any other specific shape. Figure 10.5 summarizes our discussion graphically.

FIGURE 10.5

Appropriate procedure for comparing two population means based on independent simple random samples

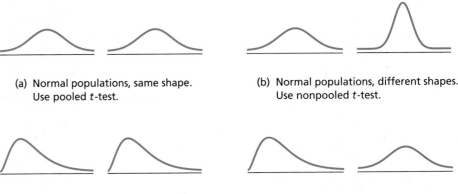

(a) Normal populations, same shape. Use pooled *t*-test.

(b) Normal populations, different shapes. Use nonpooled *t*-test.

(c) Nonnormal populations, same shape. Use Mann-Whitney test.

(d) Not both normal populations, different shapes. Use nonpooled *t*-test for large samples; otherwise, consult a statistician.

In Example 10.9, we introduce and explain the reasoning behind the Mann–Whitney test.

Example 10.9 Introducing the Mann–Whitney Test

Computer Training A nationwide shipping firm purchased a new computer system to track its current shipments, pickups, and deliveries. The system was linked to computer terminals in all the firm's regional offices where office

[†]This section continues the coverage of nonparametric statistics.

personnel could type requests for information on the location of shipments and get answers immediately on display screens.

Management had to devise a training program to teach the staff how to use the computer terminals and decided to hire a technical writer to compose a short self-study manual for that purpose. The manual was designed so that a person could read it and be ready to use the computer terminal in 2 hours.

Some employees were able to apply the procedures outlined in the manual in very little time; others took considerably longer. Someone suggested that the reason for this difference in comprehension times might be that some employees had previous experience with computers but that others did not. To test this suggestion, independent samples of employees with and without computer experience were randomly selected.

The times, in minutes, required for these employees to comprehend the manual are displayed in Table 10.9. At the 5% significance level, do the data provide sufficient evidence to conclude that the mean comprehension time for all employees without computer experience exceeds the mean comprehension time for all employees with computer experience?

Solution Let μ_1 and μ_2 denote the mean comprehension times for all employees without computer experience and with computer experience, respectively. Then the null and alternative hypotheses are

H_0: $\mu_1 = \mu_2$ (mean time for inexperienced employees is not greater)

H_a: $\mu_1 > \mu_2$ (mean time for inexperienced employees is greater).

Recall that use of the Mann–Whitney test requires that the two distributions of the variable under consideration have the same shape. If the comprehension-time distributions for employees without and with computer experience have the same shape, the distributions of the two samples in Table 10.9 should have roughly the same shape.

To check this, we constructed in Fig. 10.6 a **back-to-back stem-and-leaf diagram** of the two samples in Table 10.9. In such a diagram, the leaves for the first sample are on the left, the stems are in the middle, and the leaves for the second sample are on the right. The stem-and-leaf diagrams in Fig. 10.6 have roughly the same shape and so do not reveal any obvious violations of the same-shape condition.[†]

To apply the Mann–Whitney test, we first rank all the data from both samples combined. (Referring to Fig. 10.6 is helpful in ranking the data.) The results of ranking the data are depicted in Table 10.10 at the top of the next page, which shows, for instance, that the first employee without computer experience had the ninth shortest comprehension time among all 15 employees in the two samples combined.

The idea behind the Mann–Whitney test is a simple one: If the sum of the ranks for the sample of employees without experience is too large, we take that as evidence that the mean comprehension time for all employees without experience exceeds that for all employees with experience (i.e., we reject the null hypothesis and conclude that the alternative hypothesis is true).

TABLE 10.9

Times, in minutes, required to comprehend the self-study manual

Without experience	With experience
139	142
118	109
164	130
151	107
182	155
140	88
134	95
	104

FIGURE 10.6

Back-to-back stem-and-leaf diagram of the two comprehension-time samples in Table 10.9

Without experience		With experience
	8	8
	9	5
	10	9 7 4
8	11	
	12	
4 9	13	0
0	14	2
1	15	5
4	16	
	17	
2	18	

[†]For ease in explaining the Mann–Whitney test, we have chosen an example in which the sample sizes are very small. However, very small sample sizes makes effectively checking the same-shape condition difficult. In general, we must proceed cautiously when dealing with very small samples.

TABLE 10.10
Results of ranking the combined data
from Table 10.9

Without experience	Overall rank	With experience	Overall rank
139	9	142	11
118	6	109	5
164	14	130	7
151	12	107	4
182	15	155	13
140	10	88	1
134	8	95	2
		104	3

From Table 10.10, the sum of the ranks for the sample of employees without experience, denoted M, is

$$9 + 6 + 14 + 12 + 15 + 10 + 8 = 74.$$

To decide whether $M = 74$ is large enough to warrant rejection of the null hypothesis, we need a table of critical values for M, which we present as Table VI in Appendix A. We discuss that table and then return to complete the hypothesis test.

Using the Mann–Whitney Table

Table VI in Appendix A provides critical values for a Mann–Whitney test.[†] Note that *a critical value from Table VI is to be included as part of the rejection region.*

The size of the sample from Population 2 is given in the left-most column of Table VI, the values of α in the next column, and the size of the sample from Population 1 along the top. As expected, the symbol M_α denotes the M-value that has area (percentage, probability) α to its right. Thus the entries in Table VI provide critical values for a right-tailed Mann–Whitney test.

The distribution of the variable M is symmetric about $n_1(n_1 + n_2 + 1)/2$. This characteristic, in turn, implies that the M-value that has area α to its left—or, equivalently, area $1 - \alpha$ to its right—equals $n_1(n_1 + n_2 + 1)$ minus the M-value that has area α to its right. In symbols,

$$M_{1-\alpha} = n_1(n_1 + n_2 + 1) - M_\alpha. \tag{10.2}$$

Similarly,

$$M_{1-\alpha/2} = n_1(n_1 + n_2 + 1) - M_{\alpha/2}. \tag{10.3}$$

Equations (10.2) and (10.3) show that we can also obtain from Table VI the critical value(s) for a left-tailed Mann–Whitney test and a two-tailed Mann–Whitney test, respectively. Example 10.10 illustrates use of Table VI.

[†]Actually, the significance levels presented in Table VI are only approximate, but they are considered close enough for use in practice.

Example 10.10   Using the Mann–Whitney Table, Table VI

Refer to Example 10.9 on page 518. Determine the critical value, rejection region, and nonrejection region for a hypothesis test to be performed at the 5% significance level.

Solution The hypothesis test in Example 10.9 is right tailed, and the sample sizes are $n_1 = 7$ and $n_2 = 8$. We want a 5% significance level, or $\alpha = 0.05$.

To determine the critical value, we first go down the left-most column of Table VI to "8." Next, we concentrate on the row for α labeled 0.05, going across it to the column headed "7." The number in the body of the table there, 71, is the critical value for a right-tailed test.

Thus, for the hypothesis test in Example 10.9 performed at the 5% significance level, the critical value is $M = 71$, the rejection region consists of all M-values greater than or equal to 71, and the nonrejection region consists of all M-values less than 71. Therefore we reject H_0 if $M \geq 71$ and do not reject H_0 if $M < 71$, as shown in Fig. 10.7.

FIGURE 10.7

Rejection and nonrejection regions for a right-tailed Mann–Whitney test with $\alpha = 0.05$, $n_1 = 7$, and $n_2 = 8$.

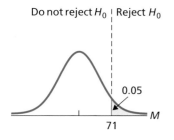

Note: Although the variable M is discrete, we have drawn its "histogram" in Fig. 10.7 in the shape of a normal curve. This method is not only convenient, but it is also acceptable, because M is close to being normally distributed except for very small sample sizes. We use this graphical convention throughout this section.

We can also use Table VI to estimate the P-value of a Mann–Whitney test. However, doing so is awkward and tedious, so we assume that the user of the P-value approach to hypothesis testing for a Mann–Whitney test has access to statistical software.

Performing the Mann–Whitney Test

Procedure 10.5 at the top of the next page provides a step-by-step method for performing a Mann–Whitney test. In stating the assumptions for the test we use, for brevity, the phrase "same-shape populations" to indicate that the two distributions (one for each population) of the variable under consideration have the same shape.

The Mann–Whitney test can be used to compare two population medians as well as two population means. We state Procedure 10.5 in terms of population means, but to apply it to population medians, we would simply replace μ_1 with η_1 and μ_2 with η_2.

Note: When there are ties in the sample data, ranks are assigned in the same way as in the Wilcoxon signed-rank test. Namely, if two or more observations are tied, each is assigned the mean of the ranks they would have had if there were no ties.

Example 10.11, which begins at the top of page 523, illustrates the use of Procedure 10.5.

PROCEDURE 10.5	Mann–Whitney Test

Purpose To perform a hypothesis test to compare two population means, μ_1 and μ_2

Assumptions

1. Simple random samples
2. Independent samples
3. Same-shape populations

STEP 1 The null hypothesis is H_0: $\mu_1 = \mu_2$, and the alternative hypothesis is

$$H_a: \mu_1 \neq \mu_2 \quad \text{or} \quad H_a: \mu_1 < \mu_2 \quad \text{or} \quad H_a: \mu_1 > \mu_2$$
$$\text{(Two tailed)} \qquad \text{(Left tailed)} \qquad \text{(Right tailed)}$$

STEP 2 Decide on the significance level, α.

STEP 3 Construct a work table of the following form.

Sample from Population 1	Overall rank	Sample from Population 2	Overall rank
.	.	.	.
.	.	.	.
.	.	.	.

STEP 4 Compute the value of the test statistic

$$M = \text{sum of the ranks for sample data from Population 1}$$

and denote that value M_0.

CRITICAL-VALUE APPROACH	**P-VALUE APPROACH**

CRITICAL-VALUE APPROACH

STEP 5 The critical value(s) are

$$M_{1-\alpha/2} \text{ and } M_{\alpha/2} \quad \text{or} \quad M_{1-\alpha} \quad \text{or} \quad M_{\alpha}$$
$$\text{(Two tailed)} \qquad \text{(Left tailed)} \qquad \text{(Right tailed)}$$

Use Table VI and, if necessary, Equation (10.2) or (10.3) on page 520 to find the critical value(s).

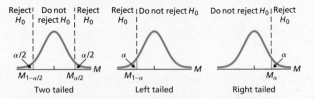

STEP 6 If the value of the test statistic falls in the rejection region, reject H_0; otherwise, do not reject H_0.

P-VALUE APPROACH

STEP 5 Obtain the *P*-value by using technology.

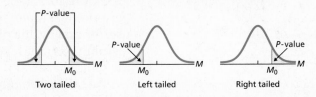

STEP 6 If $P \leq \alpha$, reject H_0; otherwise, do not reject H_0.

STEP 7 Interpret the results of the hypothesis test.

Example 10.11   The Mann–Whitney Test

Computer Training   To apply Procedure 10.5, let's complete the hypothesis test of Example 10.9. Independent samples of employees with and without computer experience were randomly selected. The employees obtained were timed to see how long it would take them to comprehend a self-study manual that explained how to use a computer to track their company's products.

The times, in minutes, are given in Table 10.9 on page 519. At the 5% significance level, do the data provide sufficient evidence to conclude that the mean comprehension time for employees without computer experience exceeds that for employees with computer experience?

Solution   We apply Procedure 10.5.

STEP 1   State the null and alternative hypotheses.

Let μ_1 and μ_2 denote the mean comprehension times for all employees without and with computer experience, respectively. Then the null and alternative hypotheses are

H_0: $\mu_1 = \mu_2$ (mean time for inexperienced employees is not greater)

H_a: $\mu_1 > \mu_2$ (mean time for inexperienced employees is greater).

Note that the hypothesis test is right tailed because a greater-than sign ($>$) appears in the alternative hypothesis.

STEP 2   Decide on the significance level, α.

We are to perform the hypothesis test at the 5% significance level; consequently, we have $\alpha = 0.05$.

STEP 3   Construct a work table.

We have already done so as Table 10.10 on page 520, which we repeat here.

Without experience	Overall rank	With experience	Overall rank
139	9	142	11
118	6	109	5
164	14	130	7
151	12	107	4
182	15	155	13
140	10	88	1
134	8	95	2
		104	3

STEP 4   Compute the value of the test statistic

$$M = \text{sum of the ranks for sample data from Population 1.}$$

From the second column of the work table,

$$M = 9 + 6 + 14 + 12 + 15 + 10 + 8 = 74.$$

CRITICAL-VALUE APPROACH	P-VALUE APPROACH
STEP 5 The critical value for a right-tailed test is M_α. Use Table VI to find the critical value. We have already determined the critical value in Example 10.10; it is 71, as shown in Fig. 10.7 on page 521. **STEP 6** If the value of the test statistic falls in the rejection region, reject H_0; otherwise, do not reject H_0. From Step 4, the value of the test statistic is $M = 74$. Figure 10.7 shows that this value falls in the rejection region. Thus we reject H_0. The test results are statistically significant at the 5% level.	**STEP 5** Obtain the *P*-value by using technology. Using technology, we find that the *P*-value for the hypothesis test is $P = 0.0214$. **STEP 6** If $P \le \alpha$, reject H_0; otherwise, do not reject H_0. From Step 5, $P = 0.0214$. Because the *P*-value is less than the specified significance level of 0.05, we reject H_0. The test results are statistically significant at the 5% level and (see Table 9.12 on page 440) provide strong evidence against the null hypothesis.

STEP 7 Interpret the results of the hypothesis test.

INTERPRETATION At the 5% significance level, the data provide sufficient evidence to conclude that the mean comprehension time for employees without computer experience exceeds that for employees with computer experience. Evidently, those with computer experience can, on average, comprehend the training manual more quickly than those without.

Comparing the Mann–Whitney Test and the Pooled *t*-Test

In Section 10.2, you learned how to perform a pooled *t*-test to compare two population means when the variable under consideration is normally distributed on each of the two populations and the population standard deviations are equal. As two normal distributions that have equal standard deviations have the same shape, we can also use the Mann–Whitney test to perform such a hypothesis test.

Which test is better under these circumstances? As you might expect, it is the pooled *t*-test because that test is designed expressly for normal populations; under conditions of normality, the pooled *t*-test is more powerful than the Mann–Whitney test. What is somewhat surprising is that the pooled *t*-test is not much more powerful than the Mann–Whitney test.

However, if the two distributions of the variable under consideration have the same shape but are not normal, the Mann–Whitney test is usually more powerful than the pooled *t*-test, often considerably so. We summarize this discussion in Key Fact 10.5.

KEY FACT 10.5 **The Mann–Whitney Test Versus the Pooled t-Test**

Suppose that the distributions of a variable of two populations have the same shape and that you want to compare, using independent simple random samples, the two population means. When you are deciding between the pooled *t*-test and the Mann–Whitney test, follow these guidelines: If you are reasonably sure that the two distributions are normal, use the pooled *t*-test; otherwise, use the Mann–Whitney test.

The Technology Center

Some statistical technologies have dedicated programs that automatically perform a Mann–Whitney test, but others do not. For instance, Minitab currently has such a program, but neither Excel nor the TI-83/84 Plus do. In this subsection, we present output and step-by-step instructions to implement Minitab's Mann–Whitney test program.

For a statistical technology that does not have a dedicated program for conducting a particular type of hypothesis test, such as the Mann–Whitney test, you can often use its macro capabilities to write a program. Refer to the technology manuals for more details.

As we said earlier, a Mann–Whitney test can be used to perform a hypothesis test to compare two population medians as well as two population means. Many statistical technologies, including Minitab, present the output of that procedure in terms of medians, but that output can also be interpreted in terms of means.

Example 10.12 **Using Technology to Conduct a Mann–Whitney Test**

Computer Training Use Minitab to perform the hypothesis test in Example 10.11 on page 523.

Solution Let μ_1 and μ_2 denote the mean times for comprehension of the self-study manual for all employees without and with computer experience, respectively. The problem is to use the Mann–Whitney procedure to perform the hypothesis test

H_0: $\mu_1 = \mu_2$ (mean time for inexperienced employees is not greater)

H_a: $\mu_1 > \mu_2$ (mean time for inexperienced employees is greater)

at the 5% significance level. Note that the hypothesis test is right tailed because a greater-than sign ($>$) appears in the alternative hypothesis.

Printout 10.3 shows the output obtained by applying Minitab's Mann–Whitney test program to the samples of comprehension times displayed in Table 10.9 on page 519.

PRINTOUT 10.3
Mann–Whitney test output for the samples of comprehension times

MINITAB

Mann-Whitney Test and CI: WOUT, WITH

```
        N  Median
WOUT  7  140.00
WITH  8  108.00


Point estimate for ETA1-ETA2 is 31.50
95.7 Percent CI for ETA1-ETA2 is (3.99,56.00)
W = 74.0
Test of ETA1 = ETA2 vs ETA1 > ETA2 is significant at 0.0214
```

The Minitab output in Printout 10.3 reveals that the *P*-value for the hypothesis test is 0.0214. Because the *P*-value is less than the specified significance level of 0.05, we reject H_0. At the 5% significance level, the data provide sufficient evidence to conclude that the mean comprehension time for employees without computer experience exceeds that for employees with computer experience.

Obtaining the Output (Optional)

Printout 10.3 provides output from Minitab for a Mann–Whitney test based on the samples of comprehension times in Table 10.9. The following are detailed instructions for obtaining that output. First, we store the comprehension-time data in columns named WOUT and WITH. Then, we proceed as follows.

MINITAB	EXCEL	TI-83/84 PLUS
1 Choose **Stat ➤ Nonparametrics ➤ Mann-Whitney...** 2 Specify WOUT in the **First Sample** text box 3 Specify WITH in the **Second Sample** text box 4 Click the arrow button at the right of the **Alternative** drop-down list box and select **greater than** 5 Click **OK**	SEE THE EXCEL MANUAL	SEE THE TI-83/84 PLUS MANUAL

EXERCISES 10.4

Statistical Concepts and Skills

10.55 Why do two normal distributions that have equal standard deviations have the same shape?

10.56 State the conditions that are required for using the Mann–Whitney test.

10.57 Suppose that, for two populations, the distributions of the variable under consideration have the same shape. Further suppose that you want to perform a hypothesis test based on independent random samples to compare the two population means. In each case, decide whether you would use the pooled *t*-test or the Mann–Whitney test and give a reason for your answer. You know that the distributions of the variable are
a. normal. **b.** not normal.

10.58 Part of conducting a Mann–Whitney test involves ranking all the data from both samples combined. Explain how to deal with ties.

In Exercises 10.59–10.62, use the Mann–Whitney test to carry out the required hypothesis tests.

10.59 Math and Chemistry. A college chemistry instructor was concerned about the detrimental effects of poor mathematics background on her students. She randomly selected 15 students and divided them according to math background. Their semester averages were the following.

Fewer than 2 years of high school algebra		Two or more years of high school algebra		
58	61	84	92	75
81	64	67	83	81
74	43	65	52	74

At the 5% significance level, do the data provide sufficient evidence to conclude that, in this teacher's chemistry courses, students with fewer than 2 years of high school algebra have a lower mean semester average than those with 2 or more years?

10.60 Picoplankton in the Bay. Picoplankton are micron-sized, single-cell algae that are an integral component of aquatic ecosystems, both in estuarine and open ocean waters. A study by Ning et al. (*Limnology and Oceanography*, Vol. 45(3), pp. 695–702) examined the spatial and temporal dynamics of picoplankton populations in the diverse estuarine environment of San Francisco Bay. Oceanographers classify the Bay into three spatial regions: North, Central, and South. The North Bay is strongly influenced by the Sacramento–San Joaquin Rivers. The South Bay is a semi-enclosed lagoon that receives constant nutrient inputs from the dense human populations surrounding it. The Central Bay receives a mix of inputs from the North and South Bays and the Pacific Ocean. Independent samples of picoplankton in the North and South Bays yielded the following data on concentration in units of 10^7 cells per liter.

North	16.2	11.2	24.8	36.4	15.0	23.6	12.1
South	9.8	18.7	26.0	7.4	15.0		

At the 5% significance level, do the data provide sufficient evidence to conclude that the mean concentrations of the picoplankton populations differ between the North and South Bays?

10.61 College Libraries. The National Center for Education Statistics surveys college libraries to obtain information on the number of volumes held. Results of the surveys are published in the *Digest of Education Statistics* and *Academic Libraries*. Independent random samples of public and private colleges yielded the following data on number of volumes held, in thousands.

Public	79	41	516	15	24	411	265
Private	139	603	113	27	67	500	

At the 5% significance level, can you conclude that the median number of volumes held by public colleges is less than that held by private colleges?

10.62 Weekly Earnings. The U.S. Bureau of Labor Statistics publishes data on weekly earnings of full-time wage and salary workers in *Employment and Earnings*. Independent random samples of male and female workers gave the following data on weekly earnings, in dollars.

Men		Women	
920	571	2073	353
2617	411	2188	369
1884	401	589	1176
382	812	370	1440
		505	407

At the 5% significance level, do the data provide sufficient evidence to conclude that the median weekly earnings of male full-time wage and salary workers exceeds the median weekly earnings of female full-time wage and salary workers?

10.63 Doing Time. The U.S. Bureau of Prisons publishes data in *Statistical Report* on the times served by prisoners released from federal institutions for the first time. Independent random samples of released prisoners in the fraud and firearms offense categories yielded the following information on time served, in months.

Fraud		Firearms	
3.6	17.9	25.5	23.8
5.3	5.9	10.4	17.9
10.7	7.0	18.4	21.9
8.5	13.9	19.6	13.3
11.8	16.6	20.9	16.1

a. Do the data provide sufficient evidence to conclude that the mean time served for fraud is less than that for firearms offenses? Perform a Mann–Whitney test at a significance level of 0.05.
b. The hypothesis test in part (a) was done in Exercise 10.19 with the pooled *t*-test. The assumption there is that times served for both offense categories are normally distributed and have equal standard deviations. If that in fact is true, why can you use a Mann–Whitney test to compare the means? Is the pooled *t*-test or the Mann–Whitney test better in this case? Explain your answers.

Extending the Concepts and Skills

10.64 Suppose that you want to perform a hypothesis test based on independent random samples to compare the means of two populations. For each part, decide whether you would use the pooled *t*-test, the nonpooled *t*-test, the Mann–Whitney test, or none of these tests if preliminary data analyses of the samples suggest that the two distributions of the variable under consideration are

a. normal but do not have the same shape.
b. not normal but have the same shape.
c. not normal and do not have the same shape; both sample sizes are large.

10.65 Suppose that you want to perform a hypothesis test based on independent random samples to compare the means of two populations. For each part, decide whether you would use the pooled *t*-test, the nonpooled *t*-test, the Mann–Whitney test, or none of these tests if preliminary data analyses of the samples suggest that the two distributions of the variable under consideration are

a. normal and have the same shape.
b. not normal and do not have the same shape; one of the sample sizes is large and the other is small.
c. different, one being normal and the other not; both sample sizes are large.

10.66 Suppose that you want to perform a hypothesis test based on independent random samples to compare the means of two populations. You know that the two distributions of the variable under consideration have the same shape and may be normal. You take the two samples and find that the data for one of the samples contain outliers. Which procedure would you use? Explain your answer.

Normal Approximation for M. The table of critical values for the Mann–Whitney test, Table VI, stops at $n_1 = 10$ and $n_2 = 10$. For larger samples, a normal approximation can be used.

Normal Approximation for M

Suppose that the two distributions of the variable under consideration have the same shape. Then, for samples of sizes n_1 and n_2,

- $\mu_M = n_1(n_1 + n_2 + 1)/2$,
- $\sigma_M = \sqrt{n_1 n_2 (n_1 + n_2 + 1)/12}$, and
- M is approximately normally distributed for $n_1 \geq 10$ and $n_2 \geq 10$.

Thus, for sample sizes of 10 or more, the standardized variable

$$z = \frac{M - n_1(n_1 + n_2 + 1)/2}{\sqrt{n_1 n_2 (n_1 + n_2 + 1)/12}}$$

has approximately the standard normal distribution.

10.67 Large Sample Mann–Whitney Test. Formulate a hypothesis testing procedure for a Mann–Whitney test that uses the test statistic *z* given in the preceding box.

10.68 Doing Time. Refer to Exercise 10.63.

a. Use your procedure from Exercise 10.67 to perform the hypothesis test.
b. Compare your result in part (a) to the one you obtained in Exercise 10.63(a), where you didn't use the normal approximation.

10.69 The Distribution of M. In this exercise, you are to obtain the distribution of the variable *M* when the sample sizes are both 3. Doing so enables you to see how the critical values for the Mann–Whitney test are derived. All possible ranks for the data are displayed in the following table; the letter *A* stands for a member from Population 1, and the letter *B* stands for a member from Population 2.

Rank						
1	2	3	4	5	6	**M**
A	*A*	*A*	*B*	*B*	*B*	6
A	*A*	*B*	*A*	*B*	*B*	7
A	*A*	*B*	*B*	*A*	*B*	8
.	.	.	.	.	.	.
.	.	.	.	.	.	.
.	.	.	.	.	.	.
B	*B*	*A*	*B*	*A*	*A*	14
B	*B*	*B*	*A*	*A*	*A*	15

a. Complete the table. (*Hint:* There are 20 rows.)
b. If the null hypothesis, $H_0: \mu_1 = \mu_2$, is true, what percentages of samples will match any given row of the table? (*Hint:* The answer is the same for all rows.)
c. Use the answer from part (b) to obtain the distribution of *M* when $n_1 = 3$ and $n_2 = 3$.
d. Draw a relative-frequency histogram of the distribution obtained in part (c).
e. Use your histogram from part (d) to obtain the entry in Table VI for $n_1 = 3$, $n_2 = 3$, and $\alpha = 0.10$.

Using Technology

10.70 Sex and Direction. See Exercise 10.20 (Section 10.2, page 502) for background information. For this exercise, however, the absolute pointing errors, in degrees, of the participants are provided on the WeissStats CD.

a. Use the Mann–Whitney test to decide whether, on average, males have a better sense of direction and, in particular, a better frame of reference than females. Perform the test with $\alpha = 0.01$.

b. Obtain boxplots and normal probability plots for both samples.

c. In Exercise 10.20, you used the pooled t-test to conduct the hypothesis test. Based on your graphs in part (b), which test is more appropriate, the pooled t-test or the Mann–Whitney test? Explain your answer.

10.71 Formaldehyde Exposure. One use of the chemical formaldehyde is to preserve animal specimens. In the article, "Exposure to Formaldehyde Among Animal Health Students" (*American Industrial Hygiene Association Journal*, Vol. 63, pp. 647–650), A. Dufresne et al. examined student exposure to formaldehyde. In the course of their lab work, 18 students at each of two animal health training centers were exposed to formaldehyde. Testing equipment recorded the total amount of formaldehyde, in milligrams per milliliter (mg/mL), to which each student was exposed. The results are presented in the following table.

Lab 1			Lab 2		
0.08	0.08	0.11	0.28	1.00	0.26
0.40	0.17	0.53	0.40	1.10	0.40
0.08	0.14	0.43	0.41	0.62	0.53
0.16	0.08	0.76	0.53	1.20	0.37
0.20	0.16	0.41	0.96	0.58	0.53
0.44	0.08	0.19	1.10	0.53	0.58

a. Obtain normal probability plots of the two samples.

b. Based on your results from part (a), given the choice between using a pooled t-test or a Mann–Whitney test, which would you choose? Explain your answer.

c. Use the test that you chose in part (b) to decide whether the data provide sufficient evidence to conclude that there is a difference in median formaldehyde exposure in the two labs. Perform the required hypothesis test at the 5% significance level.

10.72 Weekly Earnings. Refer to Exercise 10.62.

a. Use the technology of your choice to obtain normal probability plots and boxplots for the two samples.

b. Is it reasonable to use the pooled t-test to perform the hypothesis test required in Exercise 10.62? Explain your answer.

c. Is it reasonable to use the Mann–Whitney test to perform the hypothesis test required in Exercise 10.62? Explain your answer.

d. Use the technology of your choice to perform the Mann–Whitney test required in Exercise 10.62.

10.73 Transformations. Often data do not satisfy the conditions for use of any of the standard hypothesis-testing proce-

dures that we have discussed—the pooled t-test, nonpooled t-test, or Mann–Whitney test. However, by making a suitable *transformation*, you can often obtain data that do satisfy the assumptions of one or more of these standard tests.

In the paper "A Bayesian Analysis of a Multiplicative Treatment Effect in Weather Modification," Simpson, Alsen, and Eden presented the results of a study on cloud seeding with silver nitrate (*Technometrics*, Vol 17, pp. 161–166). The following table gives the rainfall amounts, in acre-feet, for unseeded and seeded clouds.

Unseeded			Seeded		
1202.6	87.0	26.1	2745.6	274.7	115.3
830.1	81.2	24.4	1697.8	274.7	92.4
372.4	68.5	21.7	1656.0	255.0	40.6
345.5	47.3	17.3	978.0	242.5	32.7
321.2	41.1	11.5	703.4	200.7	31.4
244.3	36.6	4.9	489.1	198.6	17.5
163.0	29.0	4.9	430.0	129.6	7.7
147.8	28.6	1.0	334.1	119.0	4.1
95.0	26.3		302.8	118.3	

Suppose that you want to perform a hypothesis test to decide whether cloud seeding with silver nitrate increases rainfall.

a. Obtain boxplots and normal probability plots for both samples.

b. Is use of the pooled t-test appropriate? Why or why not?

c. Is use of the nonpooled t-test appropriate? Why or why not?

d. Is use of the Mann–Whitney test appropriate? Why or why not?

e. Now transform each sample by taking logarithms. That is, for each observation, x, obtain $\log x$.

f. Obtain boxplots and normal probability plots for both transformed samples.

g. Is use of the pooled t-test on the transformed data appropriate? Why or why not?

h. Is use of the nonpooled t-test on the transformed data appropriate? Why or why not?

i. Is use of the Mann–Whitney test on the transformed data appropriate? Why or why not?

j. Which of the three procedures would you use to conduct the hypothesis test for the transformed data? Explain your answer.

k. Use the test you designated in part (j) to conduct the hypothesis test for the transformed data.

l. What conclusions can you draw?

10.5 Inferences for Two Population Means, Using Paired Samples

So far, the methods we have presented for comparing the means of two populations are based on independent samples. In this section and Section 10.6, we examine methods based on **paired samples** for comparing the means of two populations. A paired sample may be appropriate when there is a natural pairing of the members of the two populations.

Each pair in a paired sample consists of a member of one population and that member's corresponding member in the other population. With a **simple random paired sample,** each possible paired sample is equally likely to be the one selected. Example 10.13 provides an unrealistically simple illustration of paired samples, but it will help you understand the concept.

▋▋Example 10.13 **Introducing Random Paired Samples**

Husbands and Wives Let's consider two small populations, one consisting of five married women and the other of their five husbands, as shown in the following figure. The arrows in the figure indicate that Elizabeth and Karim are married, Carol and Harold are married, and so on. The married couples constitute the pairs for these two populations.

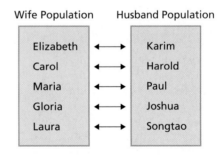

TABLE 10.11
Possible paired samples of size 3 from the wife and husband populations

Suppose that we take a paired sample of size 3 (i.e., a sample of three pairs) from these two populations.

Paired sample
(E, K), (C, H), (M, P)
(E, K), (C, H), (G, J)
(E, K), (C, H), (L, S)
(E, K), (M, P), (G, J)
(E, K), (M, P), (L, S)
(E, K), (G, J), (L, S)
(C, H), (M, P), (G, J)
(C, H), (M, P), (L, S)
(C, H), (G, J), (L, S)
(M, P), (G, J), (L, S)

a. List the possible paired samples.

b. If a paired sample is selected at random (simple random paired sample), find the chance that any particular paired sample will be the one obtained.

Solution For convenience, we use the first letter of each name as an abbreviation for the name and use parentheses to designate a wife–husband pair. For example, (E, K) represents the couple Elizabeth and Karim.

a. There are 10 possible paired samples of size 3, as displayed in Table 10.11.

b. For a simple random paired sample of size 3, each of the 10 possible paired samples listed in Table 10.11 is equally likely to be the one selected. Therefore the chances are $\frac{1}{10}$ (1 in 10) that any particular paired sample of size 3 will be the one obtained.

The purpose of Example 10.13 is to provide a concrete illustration of paired samples and to emphasize that, for simple random paired samples of any given size, each possible paired sample is equally likely to be the one selected. In practice, we neither obtain the number of possible paired samples nor explicitly compute the chance of selecting a particular paired sample. However, these concepts underlie the methods we do use.

Comparing Two Population Means, Using a Paired Sample

We are now ready to examine a process for comparing the means of two populations based on a paired sample. In Example 10.14, we introduce the ideas underlying this process.

Example 10.14   Comparing Two Means, Using a Paired Sample

Gasoline Additive   Suppose that we want to decide whether a newly developed gasoline additive increases gas mileage.

a. Formulate the problem statistically by posing it as a hypothesis test.

b. Explain the basic idea for carrying out the hypothesis test.

c. Suppose that 10 cars are selected at random and that the cars sampled are driven both with and without the additive, yielding a paired sample of size 10. Further suppose that the resulting gas mileages, in miles per gallon (mpg), are as displayed in the second and third columns of Table 10.12. Discuss the use of these data to make a decision concerning the hypothesis test.

TABLE 10.12
Gas mileages, with and without additive, for 10 randomly selected cars

Car	Gas mileage with additive	Gas mileage w/o additive	Paired difference d
1	25.7	24.9	0.8
2	20.0	18.8	1.2
3	28.4	27.7	0.7
4	13.7	13.0	0.7
5	18.8	17.8	1.0
6	12.5	11.3	1.2
7	28.4	27.8	0.6
8	8.1	8.2	−0.1
9	23.1	23.1	0.0
10	10.4	9.9	0.5
			6.6

Solution

a. To formulate the problem statistically, we first note that we have one variable—gas mileage—and two populations:

> Population 1: All cars when the additive is used
>
> Population 2: All cars when the additive is not used

Let μ_1 and μ_2 denote the means of the variable "gas mileage" for Population 1 and Population 2, respectively:

> μ_1 = mean gas mileage of all cars when the additive is used;
>
> μ_2 = mean gas mileage of all cars when the additive is not used.

We want to perform the hypothesis test

> H_0: $\mu_1 = \mu_2$ (mean gas mileage with additive is not greater)
>
> H_a: $\mu_1 > \mu_2$ (mean gas mileage with additive is greater).

b. Independent samples could be used to carry out the hypothesis test: Take independent simple random samples of, say, 10 cars each; have one group driven with the additive (sample from Population 1) and the other group driven without the additive (sample from Population 2); and then apply a pooled or nonpooled t-test to the gas mileage data obtained.

   However, in this case, a paired sample is probably more appropriate. Here, a pair consists of a car driven with the additive and the same car driven without the additive. The variable we analyze is the difference between the gas mileages of a car driven with and without the additive.

   By using a paired sample, we can remove extraneous sources of variation, in this case, the variation in the gas mileages of cars. The sampling error thus made in estimating the difference between the population means will generally be smaller. As a result, we are more likely to detect differences between the population means when such differences exist.

c. The last column of Table 10.12 contains the difference, d, between the gas mileages with and without the additive for each of the 10 cars sampled. We refer to each difference as a **paired difference** because it is the difference of a pair of observations. For example, the first car got 25.7 mpg with the additive and 24.9 mpg without the additive, giving a paired difference of $d = 25.7 - 24.9 = 0.8$ mpg, an increase in gas mileage of 0.8 mpg with the additive.

   If the null hypothesis is true, the paired differences of the gas mileages for the cars sampled should average about zero; that is, the sample mean $\bar{d}$ of the paired differences should be roughly zero. If $\bar{d}$ is too much greater than zero, we would take this as evidence that the null hypothesis is false.

   From the last column of Table 10.12, we find that the sample mean of the paired differences is

$$\bar{d} = \frac{\Sigma d}{n} = \frac{6.6}{10} = 0.66 \text{ mpg,}$$

a mean increase in gas mileage of 0.66 mpg when the additive is used. The question now is: Can this mean increase in gas mileage be reasonably attributed to sampling error or is it large enough to indicate that, on average, the additive improves gas mileage (i.e., $\mu_1 > \mu_2$)? To answer that question, we need to know the distribution of the variable $\bar{d}$. We discuss that distribution and then return to solve the gas mileage problem.

The Paired *t*-Statistic

Suppose that x is a variable on each of two populations whose members can be paired. For each pair, we let d denote the difference between the values of the variable x on the members of the pair. We call d the **paired-difference variable.**

It can be shown that the mean of the paired differences equals the difference between the two population means. In symbols,

$$\mu_d = \mu_1 - \mu_2.$$

Furthermore, if d is normally distributed, we can apply this equation and our knowledge of the studentized version of a sample mean (Key Fact 8.5 on page 379) to obtain Key Fact 10.6.

KEY FACT 10.6

Distribution of the Paired t-Statistic

Suppose that x is a variable on each of two populations whose members can be paired. Further suppose that the paired-difference variable d is normally distributed. Then, for paired samples of size n, the variable

$$t = \frac{\bar{d} - (\mu_1 - \mu_2)}{s_d/\sqrt{n}}$$

has the *t*-distribution with df $= n - 1$.

Note: We use the phrase **normal differences** as an abbreviation of "the paired-difference variable is normally distributed."

Hypothesis Tests for the Means of Two Populations, Using a Paired Sample

We now present a hypothesis-testing procedure based on a paired sample for comparing the means of two populations when the paired-difference variable is normally distributed. In light of Key Fact 10.6, for a hypothesis test with null hypothesis H_0: $\mu_1 = \mu_2$, we can use the variable

$$t = \frac{\bar{d}}{s_d/\sqrt{n}}$$

as the test statistic and obtain the critical value(s) or *P*-value from the *t*-table, Table IV.

We refer to this hypothesis-testing procedure as the **paired *t*-test**. Note that the paired *t*-test is just the one-sample *t*-test applied to the paired-difference variable with null hypothesis H_0: $\mu_d = 0$. Procedure 10.6 provides a step-by-step method for performing a paired *t*-test by using either the critical-value approach or the *P*-value approach.

PROCEDURE 10.6 | Paired *t*-Test

Purpose To perform a hypothesis test to compare two population means, μ_1 and μ_2

Assumptions

1. Simple random paired sample
2. Normal differences or large sample

STEP 1 The null hypothesis is H_0: $\mu_1 = \mu_2$, and the alternative hypothesis is

$$H_a:\ \mu_1 \neq \mu_2 \qquad \text{or} \qquad H_a:\ \mu_1 < \mu_2 \qquad \text{or} \qquad H_a:\ \mu_1 > \mu_2$$
$$\text{(Two tailed)} \qquad\qquad \text{(Left tailed)} \qquad\qquad \text{(Right tailed)}$$

STEP 2 Decide on the significance level, α.

STEP 3 Calculate the paired differences of the sample pairs.

STEP 4 Compute the value of the test statistic

$$t = \frac{\overline{d}}{s_d/\sqrt{n}}$$

and denote that value t_0.

CRITICAL-VALUE APPROACH	P-VALUE APPROACH

CRITICAL-VALUE APPROACH

STEP 5 The critical value(s) are

$$\pm t_{\alpha/2} \qquad \text{or} \qquad -t_\alpha \qquad \text{or} \qquad t_\alpha$$
$$\text{(Two tailed)} \qquad \text{(Left tailed)} \qquad \text{(Right tailed)}$$

with df $= n - 1$. Use Table IV to find the critical value(s).

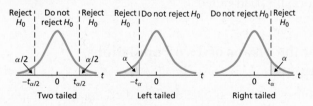

STEP 6 If the value of the test statistic falls in the rejection region, reject H_0; otherwise, do not reject H_0.

P-VALUE APPROACH

STEP 5 The *t*-statistic has df $= n - 1$. Use Table IV to estimate the *P*-value, or obtain it exactly by using technology.

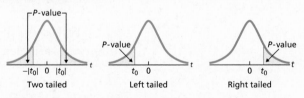

STEP 6 If $P \leq \alpha$, reject H_0; otherwise, do not reject H_0.

STEP 7 Interpret the results of the hypothesis test.

The hypothesis test is exact when the paired-difference variable is normally distributed (normal differences) and is approximately correct for large samples when the paired-difference variable is not normally distributed (nonnormal differences).

Before we apply the paired *t*-test, we need to discuss the assumptions for its use. Assumption 1 (simple random paired sample) is essential; the sample must be a simple random paired sample or the procedure does not apply. Thus, for instance, do not apply the paired *t*-test to independent samples and, conversely, do not apply a pooled *t*-test or a nonpooled *t*-test to a paired sample.

Regarding Assumption 2, although the paired *t*-test is based on the assumption that the paired-difference variable is normally distributed, it also applies approximately for large samples regardless of distribution type, as noted at the bottom of Procedure 10.6. And, like the one-sample *t*-test, the paired *t*-test works reasonably well even for small samples or samples of moderate size when the paired-difference variable is not normally distributed, provided that variable is not too far from being normally distributed.

When considering the paired *t*-test, you must also watch for outliers in the sample of paired differences. Again, the presence of outliers calls into question the normality assumption. Moreover, even for large samples, outliers can sometimes unduly affect a paired *t*-test because the sample mean and sample standard deviation are not resistant to them.

Practical guidelines for the use of the paired *t*-test are the same as those given for the one-sample *z*-test in Key Fact 9.4 on page 416 when applied to paired differences. Always look at the sample of paired differences before applying the paired *t*-test to ensure that its use is reasonable.

Finally, we emphasize that the normality assumption in Assumption 2 refers to the paired-difference variable. The two distributions of the variable under consideration need not be normally distributed.

Example 10.15 illustrates use of Procedure 10.6.

Example 10.15   The Paired t-Test

Gasoline Additive   We now return to the gas mileage problem posed in Example 10.14. The gas mileages of 10 randomly selected cars, both with and without a new gasoline additive, are displayed in the second and third columns of Table 10.12 on page 531. At the 5% significance level, do the data provide sufficient evidence to conclude that, on average, the gasoline additive improves gas mileage?

Solution   To begin, we check the two conditions required for using the paired *t*-test. We have a simple random paired sample—each pair consists of a car driven both with and without the additive. So, Assumption 1 is satisfied.

Because the sample size, $n = 10$, is small, we need to consider questions of normality and outliers. (See the first bulleted item in Key Fact 9.4 on page 416.) To do so, we constructed a normal probability plot (not shown) for the sample of paired differences contained in the last column of Table 10.12. The normal probability plot reveals no outliers and is quite linear, so we can consider Assumption 2 satisfied. Hence we can apply the paired *t*-test to perform the required hypothesis test.

STEP 1   State the null and alternative hypotheses.

Let μ_1 denote the mean gas mileage of all cars when the additive is used and μ_2 denote the mean gas mileage of all cars when the additive is not used. Then the null and alternative hypotheses are

H_0: $\mu_1 = \mu_2$ (mean gas mileage with additive is not greater)

H_a: $\mu_1 > \mu_2$ (mean gas mileage with additive is greater).

Note that the hypothesis test is right tailed because a greater-than sign ($>$) appears in the alternative hypothesis.

STEP 2   Decide on the significance level, α.

The test is to be performed at the 5% significance level, or $\alpha = 0.05$.

STEP 3   Calculate the paired differences of the sample pairs.

The paired differences are shown in the last column of the following table.

Car	Gas mileage with additive	Gas mileage w/o additive	Paired difference d
1	25.7	24.9	0.8
2	20.0	18.8	1.2
3	28.4	27.7	0.7
4	13.7	13.0	0.7
5	18.8	17.8	1.0
6	12.5	11.3	1.2
7	28.4	27.8	0.6
8	8.1	8.2	−0.1
9	23.1	23.1	0.0
10	10.4	9.9	0.5

STEP 4   Compute the value of the test statistic

$$t = \frac{\bar{d}}{s_d/\sqrt{n}}.$$

We first need to determine the sample mean and sample standard deviation of the paired differences. We do so in the usual manner:

$$\bar{d} = \frac{\Sigma d}{n} = \frac{6.6}{10} = 0.66$$

and

$$s_d = \sqrt{\frac{\Sigma d^2 - (\Sigma d)^2/n}{n-1}} = \sqrt{\frac{6.12 - (6.6)^2/10}{10-1}} = 0.443.$$

Consequently, the value of the test statistic is

$$t = \frac{\bar{d}}{s_d/\sqrt{n}} = \frac{0.66}{0.443/\sqrt{10}} = 4.711.$$

CRITICAL-VALUE APPROACH	P-VALUE APPROACH

STEP 5 The critical value for a right-tailed test is t_α with df = $n - 1$. Use Table IV to find the critical value.

We have $n = 10$ and $\alpha = 0.05$. Table IV reveals that, for df = $10 - 1 = 9$, we have $t_{0.05} = 1.833$, as shown in Fig. 10.8A.

FIGURE 10.8A

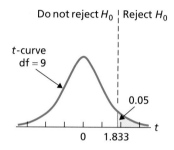

STEP 6 If the value of the test statistic falls in the rejection region, reject H_0; otherwise, do not reject H_0.

From Step 4, the value of the test statistic is $t = 4.711$, which falls in the rejection region, as shown in Fig. 10.8A. Hence we reject H_0. The test results are statistically significant at the 5% level.

STEP 5 The t-statistic has df = $n - 1$. Use Table IV to estimate the P-value or obtain it exactly by using technology.

From Step 4, the value of the test statistic is $t = 4.711$. The test is right tailed, so the P-value is the probability of observing a value of t of 4.711 or greater if the null hypothesis is true. That probability equals the shaded area shown in Fig. 10.8B.

FIGURE 10.8B

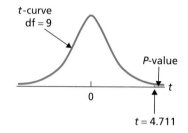

We have $n = 10$, so df = $10 - 1 = 9$. Referring to Fig. 10.8B and to Table IV with df = 9, we find that $P < 0.005$. (Using technology, we determined that $P = 0.000549$.)

STEP 6 If $P \leq \alpha$, reject H_0; otherwise, do not reject H_0.

From Step 5, $P < 0.005$. Because the P-value is less than the specified significance level of 0.05, we reject H_0. The test results are statistically significant at the 5% level and (see Table 9.12 on page 440) provide very strong evidence against the null hypothesis.

STEP 7 Interpret the results of the hypothesis test.

INTERPRETATION At the 5% significance level, the data provide sufficient evidence to conclude that the mean gas mileage of all cars when the additive is used is greater than the mean gas mileage of all cars when the additive is not used. Evidently, the additive is effective in increasing gas mileage.

Confidence Intervals for the Difference Between the Means of Two Populations, Using a Paired Sample

We can also use Key Fact 10.6 on page 533 to derive a confidence-interval procedure for the difference between two population means. We often refer to that confidence-interval procedure as the **paired *t*-interval procedure** and present it here as Procedure 10.7.

PROCEDURE 10.7

Paired *t*-Interval Procedure

Purpose To find a confidence interval for the difference between two population means, μ_1 and μ_2.

Assumptions
1. Simple random paired sample
2. Normal differences or large sample

STEP 1 For a confidence level of $1 - \alpha$, use Table IV to find $t_{\alpha/2}$ with df $= n - 1$.

STEP 2 The endpoints of the confidence interval for $\mu_1 - \mu_2$ are

$$\bar{d} \pm t_{\alpha/2} \cdot \frac{s_d}{\sqrt{n}}.$$

STEP 3 Interpret the confidence interval.

The confidence interval is exact when the paired-difference variable is normally distributed (normal differences) and is approximately correct for large samples when the paired-difference variable is not normally distributed (nonnormal differences).

Example 10.16   The Paired t-Interval Procedure

Gasoline Additive   Use the sample data in Table 10.12 on page 531 to obtain a 90% confidence interval for the difference, $\mu_1 - \mu_2$, between the mean gas mileage of all cars when the additive is used and the mean gas mileage of all cars when the additive is not used.

Solution   We apply Procedure 10.7.

STEP 1 For a confidence level of $1 - \alpha$, use Table IV to find $t_{\alpha/2}$ with df $= n - 1$.

For a 90% confidence interval, $\alpha = 0.10$. From Table IV, we determine that, for df $= n - 1 = 10 - 1 = 9$, we have $t_{\alpha/2} = t_{0.10/2} = t_{0.05} = 1.833$.

STEP 2 **The endpoints of the confidence interval for $\mu_1 - \mu_2$ are**

$$\bar{d} \pm t_{\alpha/2} \cdot \frac{s_d}{\sqrt{n}}.$$

From Step 1, $t_{\alpha/2} = 1.833$. Also, $n = 10$ and, from Example 10.15, we know that $\bar{d} = 0.66$ and $s_d = 0.443$. Consequently, the endpoints of the confidence interval for $\mu_1 - \mu_2$ are

$$0.66 \pm 1.833 \cdot \frac{0.443}{\sqrt{10}},$$

or 0.40 to 0.92.

STEP 3 **Interpret the confidence interval.**

INTERPRETATION We can be 90% confident that the difference between the mean gas mileage of all cars when the additive is used and the mean gas mileage of all cars when the additive is not used is somewhere between 0.40 mpg and 0.92 mpg. In particular, we can be confident that, on average, the additive increases gas mileage by at least 0.40 mpg.

The Technology Center

Most statistical technologies have programs that automatically perform paired t-procedures. In this subsection, we present output and step-by-step instructions to implement such programs.

Example 10.17 **Using Technology to Conduct Paired t-Procedures**

Gasoline Additive Table 10.12 on page 531 displays the gas mileages of 10 randomly selected cars, both with and without a new gasoline additive. Use Minitab, Excel, or the TI-83/84 Plus to perform the hypothesis test in Example 10.15 and obtain the confidence interval required in Example 10.16.

Solution   Let μ_1 denote the mean gas mileage of all cars when the additive is used and let μ_2 denote the mean gas mileage of all cars when the additive is not used. The task in Example 10.15 is to perform the hypothesis test

H_0: $\mu_1 = \mu_2$ (mean gas mileage with additive is not greater)

H_a: $\mu_1 > \mu_2$ (mean gas mileage with additive is greater),

at the 5% significance level; the task in Example 10.16 is to obtain a 90% confidence interval for $\mu_1 - \mu_2$.

Printout 10.4 on pages 540 and 541 shows the output obtained by applying the paired t-procedures to the mileage data presented in Table 10.12.

The outputs in Printout 10.4 reveal that the P-value for the hypothesis test is less than 0.001. Because the P-value is less than the significance level of 0.05, we reject H_0. The outputs in Printout 10.4 also show that a 90% confidence interval for the difference between the means is from 0.403 to 0.917.

PRINTOUT 10.4

Paired *t*-procedures output for the
mileage data

MINITAB

Paired T-Test and CI: WITH, WOUT   [FOR THE HYPOTHESIS TEST]

```
Paired T for WITH - WOUT

                 N      Mean     StDev   SE Mean
WITH            10   18.9100    7.4721    2.3629
WOUT            10   18.2500    7.4188    2.3460
Difference      10   0.660000  0.442719  0.140000


90% lower bound for mean difference: 0.466376
T-Test of mean difference = 0 (vs > 0): T-Value = 4.71   P-Value = 0.001
```

Paired T-Test and CI: WITH, WOUT   [FOR THE CONFIDENCE INTERVAL]

```
Paired T for WITH - WOUT

                 N      Mean     StDev   SE Mean
WITH            10   18.9100    7.4721    2.3629
WOUT            10   18.2500    7.4188    2.3460
Difference      10   0.660000  0.442719  0.140000


90% CI for mean difference: (0.403364, 0.916636)
T-Test of mean difference = 0 (vs not = 0): T-Value = 4.71   P-Value = 0.001
```

TI-83/84 PLUS

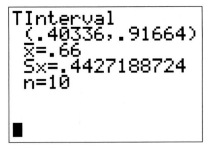

Using **T-Test**   Using **TInterval**

PRINTOUT 10.4 (cont.)
Paired *t*-procedures output for the
mileage data

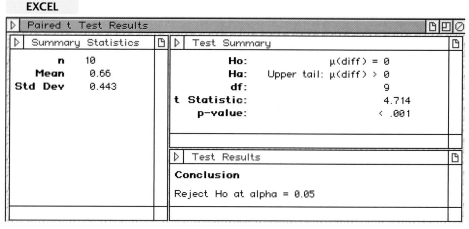

Using **Paired t Test**

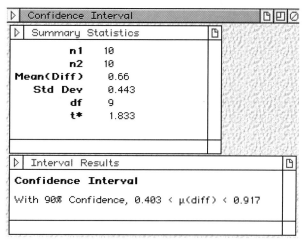

Using **Paired t Interval**

Obtaining the Output (Optional)

Printout 10.4 provides output from Minitab, Excel, and the TI-83/84 Plus for paired *t*-procedures based on the mileages (with and without additive) in Table 10.12. The following are detailed instructions for obtaining that output. First, we store the mileages in columns (Minitab), ranges (Excel), or lists (TI-83/84 Plus) named WITH and WOUT. Then, we proceed as follows.[†]

[†]Minitab computes a two-sided confidence interval for a two-tailed test and a one-sided confidence interval for a one-tailed test. To perform a one-tailed hypothesis test and obtain a two-sided confidence interval, we must apply Minitab's paired *t*-procedure twice: once for the one-tailed hypothesis test and once for the confidence interval specifying a two-tailed hypothesis test.

MINITAB	EXCEL	TI-83/84 PLUS

MINITAB

FOR THE HYPOTHESIS TEST:
1 Choose **Stat ➤ Basic Statistics ➤ Paired t...**
2 Select the **Samples in columns** option button
3 Click in the **First sample** text box and specify WITH
4 Click in the **Second sample** text box and specify WOUT
5 Click the **Options...** button
6 Click in the **Confidence level** text box and type 90
7 Click in the **Test mean** text box and type 0
8 Click the arrow button at the right of the **Alternative** drop-down list box and select **greater than**
9 Click **OK**
10 Click **OK**

FOR THE CI:
1 Choose **Edit ➤ Edit Last Dialog**
2 Click the **Options...** button
3 Click the arrow button at the right of the **Alternative** drop-down list box and select **not equal**
4 Click **OK**
5 Click **OK**

EXCEL

FOR THE HYPOTHESIS TEST:
1 Choose **DDXL ➤ Hypothesis Tests**
2 Select **Paired t Test** from the **Function type** drop-down box
3 Specify WITH in the **1st Quantitative Variable** text box
4 Specify WOUT in the **2nd Quantitative Variable** text box
5 Click **OK**
6 Click the **Set μ(diff)0** button, type 0, and click **OK**
7 Click the **0.05** button
8 Click the **μ(diff) > μ(diff)0** button
9 Click the **Compute** button

FOR THE CI:
1 Exit to Excel
2 Choose **DDXL ➤ Confidence Intervals**
3 Select **Paired t Interval** from the **Function type** drop-down box
4 Specify WITH in the **1st Quantitative Variable** text box
5 Specify WOUT in the **2nd Quantitative Variable** text box
6 Click **OK**
7 Click the **90%** button
8 Click the **Compute Interval** button

TI-83/84 PLUS

FOR THE PAIRED DIFFERENCES:
1 Press **2nd ➤ LIST**, arrow down to WITH, and press **ENTER**
2 Press −
3 Press **2nd ➤ LIST**, arrow down to WOUT, and press **ENTER**
4 Press **STO ▶**
5 Press **2nd ➤ A-LOCK**, type DIFF, and press **ENTER**

FOR THE HYPOTHESIS TEST:
1 Press **STAT**, arrow over to **TESTS**, and press **2**
2 Highlight **Data** and press **ENTER**
3 Press the down-arrow key, type 0 for μ_0, and press **ENTER**
4 Press **2nd ➤ LIST**, arrow down to DIFF, and press **ENTER** three times
5 Highlight > μ_0 and press **ENTER**
6 Press the down-arrow key, highlight **Calculate**, and press **ENTER**

FOR THE CI:
1 Press **STAT**, arrow over to **TESTS**, and press **8**
2 Highlight **Data** and press **ENTER**
3 Press the down-arrow key
4 Press **2nd ➤ LIST**, arrow down to DIFF, and press **ENTER** three times
5 Type .9 for **C-Level** and press **ENTER** twice

EXERCISES 10.5

Statistical Concepts and Skills

10.74 State one possible advantage of using paired samples instead of independent samples.

10.75 Ages of Married People. The U.S. Bureau of the Census publishes information on the ages of married people in *Current Population Reports*. Suppose that you want to use a paired sample to compare the mean ages of married men and married women. Identify
a. the variable under consideration.
b. the two populations.
c. the pairs.
d. the paired-difference variable.

10.76 TV Viewing by Married People. The A. C. Nielsen Company collects data on the TV viewing habits of Americans and publishes the information in *Nielsen Report on Television*. Suppose that you want to use a paired sample to compare the mean viewing times of married men and married women. Identify
a. the variable under consideration.
b. the two populations.
c. the pairs.
d. the paired-difference variable.

10.77 State the two conditions required for performing a paired *t*-procedure. How important are those conditions?

10.78 Provide an example (different from the ones considered in this section) of a procedure based on a paired sample being more appropriate than one based on independent samples.

Preliminary data analyses indicate that use of a paired t-test is reasonable in Exercises **10.79–10.84**. *Perform each hypothesis test by using either the critical-value approach or the P-value approach.*

10.79 Zea Mays. Charles Darwin, author of *Origin of Species*, investigated the effect of cross-fertilization on the heights of plants. In one study he planted 15 pairs of Zea mays plants. Each pair consisted of one cross-fertilized plant and one self-fertilized plant grown in the same pot. The following table gives the height differences, in eighths of an inch, for the 15 pairs. Each difference is obtained by subtracting the height of the self-fertilized plant from that of the cross-fertilized plant.

49	−67	8	16	6
23	28	41	14	29
56	24	75	60	−48

a. Identify the variable under consideration.
b. Identify the two populations.
c. Identify the paired-difference variable.
d. Are the numbers in the table paired differences? Why or why not?
e. At the 5% significance level, do the data provide sufficient evidence to conclude that the mean heights of cross-fertilized and self-fertilized Zea mays differ? (*Note:* $\bar{d} = 20.93$ and $s_d = 37.74$.)
f. Repeat part (e) at the 1% significance level.

10.80 Sleep. In 1908, W. S. Gosset published "The Probable Error of a Mean" (*Biometrika*, Vol. 6, pp. 1–25). In this pioneering paper, published under the pseudonym

"Student," he introduced what later became known as Student's *t*-distribution. Gosset used the following data set, which gives the additional sleep in hours obtained by 10 patients who used laevohysocyamine hydrobromide.

1.9	0.8	1.1	0.1	−0.1
4.4	5.5	1.6	4.6	3.4

a. Identify the variable under consideration.
b. Identify the two populations.
c. Identify the paired-difference variable.
d. Are the numbers in the table paired differences? Why or why not?
e. At the 5% significance level, do the data provide sufficient evidence to conclude that laevohysocyamine hydrobromide is effective in increasing sleep? (*Note:* $\bar{d} = 2.33$ and $s_d = 2.002$.)
f. Repeat part (e) at the 1% significance level.

10.81 Anorexia Treatment. Anorexia nervosa is a serious eating disorder, particularly among young women. The following data provide the weights, in pounds, of 17 anorexic young women before and after receiving a family therapy treatment for anorexia nervosa. [SOURCE: Hand et al. (ed.) *A Handbook of Small Data Sets*, London: Chapman & Hall, 1994. Raw data from B. Everitt (personal communication).]

Before	After	Before	After	Before	After
83.3	94.3	76.9	76.8	82.1	95.5
86.0	91.5	94.2	101.6	77.6	90.7
82.5	91.9	73.4	94.9	83.5	92.5
86.7	100.3	80.5	75.2	89.9	93.8
79.6	76.7	81.6	77.8	86.0	91.7
87.3	98.0	83.8	95.2		

Does family therapy appear to be effective in helping anorexic young women gain weight? Perform the appropriate hypothesis test at the 5% significance level.

10.82 Measuring Treadwear. Stichler, Richey, and Mandel compared two methods of measuring treadwear in their paper "Measurement of Treadwear of Commercial Tires" (*Rubber Age*, Vol. 73:2). Eleven tires were each measured for treadwear by two methods, one based on weight and the other on groove wear. The following are the data, in thousands of miles.

Weight method	Groove method	Weight method	Groove method
30.5	28.7	24.5	16.1
30.9	25.9	20.9	19.9
31.9	23.3	18.9	15.2
30.4	23.1	13.7	11.5
27.3	23.7	11.4	11.2
20.4	20.9		

At the 5% significance level, do the data provide sufficient evidence to conclude that, on average, the two measurement methods give different results?

10.83 Glaucoma and Corneal Thickness. Glaucoma is a leading cause of blindness in the United States. N. Ehlers measured the corneal thickness of eight patients who had glaucoma in one eye but not in the other. The results of the study were published as the paper "On Corneal Thickness and Intraocular Pressure, II" (*Acta Opthalmologica*, Vol. 48, pp. 1107–1112). The following are the data on corneal thickness, in microns.

Patient	Normal	Glaucoma
1	484	488
2	478	478
3	492	480
4	444	426
5	436	440
6	398	410
7	464	458
8	476	460

At the 10% significance level, do the data provide sufficient evidence to conclude that mean corneal thickness is greater in normal eyes than in eyes with glaucoma?

10.84 Fortified Orange Juice. V. Tangpricha et al. conducted a study to determine whether fortifying orange juice with Vitamin D would increase serum 25-hydroxyvitamin D [25(OH)D] concentration in the blood. The researchers reported their findings in the paper "Fortification of Orange Juice with Vitamin D: a Novel Approach for Enhancing Vitamin D Nutritional Health" (*American Journal of Clinical Nutrition*, Vol. 77, pp. 1478–1483). A double-blind experiment was used in which 14 subjects drank 240 ml per day of orange juice fortified with 1000 IU of Vitamin D and 12 subjects drank 240 ml per day of unfortified orange juice. Concentration levels were recorded at the beginning of the experiment and again at the end of 12 weeks. The following data, based on the results of the study, provide the before and after serum 25(OH)D concentrations in the blood, in nanomoles per liter (nmo/L), for the group that drank the fortified juice.

Before	After	Before	After
8.6	33.8	3.9	75.0
32.3	137.0	1.5	83.3
60.7	110.6	18.1	71.5
20.4	52.7	100.9	142.0
39.4	110.5	84.3	171.4
15.7	39.1	32.3	52.1
58.3	124.1	41.7	112.9

At the 1% significance level, do the data provide sufficient evidence to conclude that drinking fortified orange juice increases the serum 25(OH)D concentration in the blood? (*Note:* The mean and standard deviation of the paired differences are −56.99 nmo/L and 26.20 nmo/L, respectively.)

*In Exercises **10.85–10.88**, use Procedure 10.7 on page 538 to obtain the required confidence interval.*

10.85 Zea Mays. Refer to Exercise 10.79.
a. Determine a 95% confidence interval for the difference between the mean heights of cross-fertilized and self-fertilized Zea mays. Interpret your answer.
b. Repeat part (a) for a 99% confidence level.

10.86 Sleep. Refer to Exercise 10.80.
a. Determine a 90% confidence interval for the additional sleep that would be obtained, on average, by using laevo-hysocyamine hydrobromide. Interpret your answer.
b. Repeat part (a) for a 98% confidence level.

10.87 Anorexia Treatment. Refer to Exercise 10.81. Determine a 90% confidence interval for the weight gain that would be obtained, on average, by using the family therapy treatment. Interpret your answer.

10.88 Measuring Treadwear. Refer to Exercise 10.82. Determine a 95% confidence interval for the mean difference in measurement by the weight and groove methods. Interpret your answer.

Extending the Concepts and Skills

10.89 Explain exactly how a paired *t*-test can be formulated as a one-sample *t*-test. (*Hint:* Work solely with the paired-difference variable.)

10.90 Faculty Salaries. In Example 10.3 on page 494, we performed a hypothesis test based on independent samples to decide whether mean salaries differ for faculty in public and private institutions. Now you are to perform that same hypothesis test based on a paired sample. Pairs are formed by matching faculty in public and private institutions by rank and specialty. A random sample of 30 pairs yields the following annual salaries, in thousands of dollars.

Public	Private	Public	Private	Public	Private
85.1	91.9	41.9	51.2	42.5	46.2
55.3	64.0	36.3	50.2	87.8	97.8
88.7	95.4	91.3	102.6	78.5	85.0
78.5	87.6	63.8	66.5	53.1	67.4
59.8	63.8	67.9	70.4	84.0	85.8
102.6	100.3	88.9	94.2	54.0	60.0
53.2	57.5	76.7	84.0	77.3	86.8
51.3	57.4	45.0	53.4	62.1	73.6
28.9	32.9	67.5	77.3	56.2	61.8
58.2	67.8	68.2	84.4	60.8	69.2

a. Do the data provide sufficient evidence to conclude that mean salaries differ for faculty in public and private institutions? Perform the required hypothesis test at the 5% significance level. (*Note:* $\bar{d} = -7.367$ and $s_d = 3.992$.)
b. Compare your result in part (a) to the one obtained in Example 10.3.
c. Which test is more appropriate? Explain your answer.
d. Find a 95% confidence interval for the difference between the mean salaries of faculty in public and private institutions.
e. Compare your result in part (d) to the one obtained in Example 10.4 on page 497.

10.91 A hypothesis test, based on a paired sample, is to be performed to compare the means of two populations. The sample of 15 paired differences contains an outlier but otherwise is approximately bell shaped. Assuming that removal of the outlier would not be legitimate, would use of the paired *t*-test or a nonparametric test be better? Explain your answer.

10.92 Gasoline Additive. This exercise shows what can happen when a hypothesis-testing procedure designed for use with independent samples is applied to perform a hypothesis test on a paired sample. In Example 10.15 on page 535, we applied the paired *t*-test to decide whether a gasoline additive is effective in increasing gas mileage. Specifically, if we let μ_1 and μ_2 denote the mean gas mileages of all cars when the additive is and is not used, respectively,

the null and alternative hypotheses are

$$H_0: \mu_1 = \mu_2$$
$$H_a: \mu_1 > \mu_2.$$

a. Apply the nonpooled *t*-test to the sample data in the second and third columns of Table 10.12 on page 531 to perform the hypothesis test. Use $\alpha = 0.05$.
b. Why is performing the hypothesis test the way you did in part (a) inappropriate?
c. Compare your result in part (a) to the one obtained in Example 10.15.

Using Technology

10.93 Anorexia Treatment. Refer to both Exercises 10.81 and 10.87. Use the technology of your choice to
a. obtain a normal probability plot of the paired differences.
b. perform the required hypothesis test and obtain the desired confidence interval.
c. Justify the use of your procedure in part (b).

10.94 Measuring Treadwear. Refer to both Exercises 10.82 and 10.88. Use the technology of your choice to
a. obtain a normal probability plot of the paired differences.
b. perform the required hypothesis test and obtain the desired confidence interval.
c. Justify the use of your procedure in part (b).

10.95 Tobacco Mosaic Virus. To assess the effects of two different strains of the tobacco mosaic virus, W. Youden and H. Beale randomly selected eight tobacco leaves. Half of each leaf was subjected to one of the strains of tobacco mosaic virus and the other half to the other strain. The researchers then counted the number of local lesions apparent on each half of each leaf. The results of their study were published in the paper "A Statistical Study of the Local Lesion Method for Estimating Tobacco Mosaic Virus" (*Contributions to Boyce Thompson Institute*, Vol. 6, p. 437). The following are the data.

Leaf	Virus 1	Virus 2
1	31	18
2	20	17
3	18	14
4	17	11
5	9	10
6	8	7
7	10	5
8	7	6

Suppose that you want to perform a hypothesis test to determine whether a difference exists between the mean number of local lesions resulting from the two viral strains. Conduct preliminary graphical data analyses to decide whether applying the paired *t*-test is reasonable. Explain your decision.

10.96 Improving Car Emissions? The makers of the MAG-NETIZER Engine Energizer System (EES) claim that it improves gas mileage and reduces emissions in automobiles by using magnetic free energy to increase the amount of oxygen in the fuel for greater combustion efficiency. Following are test results, performed under International and U.S. Government agency standards, on a random sample of 14 vehicles. The data give the carbon monoxide (CO) levels, in parts per million, of each vehicle tested, both before installation of EES and after installation. [SOURCE: *Global Source Marketing*.]

Before	After	Before	After
1.60	0.15	2.60	1.60
0.30	0.20	0.15	0.06
3.80	2.80	0.06	0.16
6.20	3.60	0.60	0.35
3.60	1.00	0.03	0.01
1.50	0.50	0.10	0.00
2.00	1.60	0.19	0.00

Suppose that you want to perform a hypothesis test to determine whether, on average, EES reduces CO emissions. Conduct preliminary graphical data analyses to decide whether applying the paired *t*-test is reasonable. Explain your decision.

10.6   The Paired Wilcoxon Signed-Rank Test[†]

In Section 10.5, we discussed the paired *t*-procedures, which provide methods based on paired samples for comparing two population means. An assumption for use of those procedures is that the paired-difference variable is (approximately) normally distributed or that the sample size is large. In cases of a paired sample with small or moderate sample size and where the distribution of the paired-difference variable is far from normal, a paired *t*-procedure is inappropriate and a nonparametric procedure should be used instead.

For instance, if the distribution of the paired-difference variable is symmetric (but not necessarily normal), we can perform a hypothesis test to compare the means of the two populations by applying the Wilcoxon signed-rank test (Procedure 9.4 on page 459) to the sample of paired differences. In this context, the Wilcoxon signed-rank test is called the **paired Wilcoxon signed-rank test.**

Recall that the critical value for a right-tailed Wilcoxon test can be obtained directly from Table V. To obtain the critical value(s) for a left-tailed or two-tailed Wilcoxon test, we refer to Table V and use the relations

$$W_{1-\alpha} = n(n+1)/2 - W_\alpha \tag{10.4}$$

or

$$W_{1-\alpha/2} = n(n+1)/2 - W_{\alpha/2}, \tag{10.5}$$

respectively. We can also use Table V to estimate the *P*-value of a Wilcoxon test. However, as before, doing so is awkward and tedious, so we assume that the user of the *P*-value approach to hypothesis testing for a paired Wilcoxon signed-rank test has access to statistical software.

Procedure 10.8 provides the steps for performing a paired Wilcoxon signed-rank test. For convenience, we use the phrase **symmetric differences** as shorthand for "the paired-difference variable has a symmetric distribution."

[†]This section continues the coverage of nonparametric statistics.

| PROCEDURE 10.8 | Paired Wilcoxon Signed-Rank Test |

Purpose To perform a hypothesis test to compare two population means, μ_1 and μ_2

Assumptions

1. Simple random paired sample
2. Symmetric differences

STEP 1 The null hypothesis is H_0: $\mu_1 = \mu_2$, and the alternative hypothesis is

$$H_a: \mu_1 \neq \mu_2 \quad \text{or} \quad H_a: \mu_1 < \mu_2 \quad \text{or} \quad H_a: \mu_1 > \mu_2$$
$$\text{(Two tailed)} \qquad \text{(Left tailed)} \qquad \text{(Right tailed)}$$

STEP 2 Decide on the significance level, α.

STEP 3 Calculate the paired differences of the sample pairs.

STEP 4 Discard all paired differences that equal 0 and reduce the sample size accordingly.

STEP 5 Construct a work table to obtain the signed ranks of the paired differences.

STEP 6 Compute the value of the test statistic

$$W = \text{sum of the positive ranks}$$

and denote that value W_0.

| CRITICAL-VALUE APPROACH | P-VALUE APPROACH |

STEP 7 The critical value(s) are

$$W_{1-\alpha/2} \text{ and } W_{\alpha/2} \quad \text{or} \quad W_{1-\alpha} \quad \text{or} \quad W_{\alpha}$$
$$\text{(Two tailed)} \qquad \text{(Left tailed)} \qquad \text{(Right tailed)}$$

Use Table V and, if necessary, Equation (10.4) or (10.5) on page 546 to find the critical value(s).

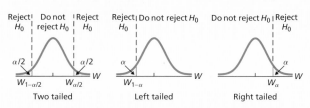

STEP 8 If the value of the test statistic falls in the rejection region, reject H_0; otherwise, do not reject H_0.

STEP 7 Obtain the *P*-value by using technology.

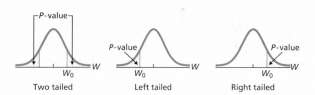

STEP 8 If $P \leq \alpha$, reject H_0; otherwise, do not reject H_0.

STEP 9 Interpret the results of the hypothesis test.

In Example 10.15, we used a paired *t*-test to decide whether, on average, a gasoline additive improves gas mileage. In Example 10.18, we use the paired Wilcoxon signed-rank test to perform that hypothesis test.

▌▌Example 10.18   The Paired Wilcoxon Signed-Rank Test

Gasoline Additive   The gas mileages of 10 randomly selected cars, both with and without a new gasoline additive, are displayed in the second and third columns of Table 10.13. At the 5% significance level, do the data provide sufficient evidence to conclude that, on average, the gasoline additive improves gas mileage? Use the paired Wilcoxon signed-rank test.

TABLE 10.13

Gas mileages, with and without additive, for 10 randomly selected cars

Car	Gas mileage with additive	Gas mileage w/o additive	Paired difference d
1	25.7	24.9	0.8
2	20.0	18.8	1.2
3	28.4	27.7	0.7
4	13.7	13.0	0.7
5	18.8	17.8	1.0
6	12.5	11.3	1.2
7	28.4	27.8	0.6
8	8.1	8.2	−0.1
9	23.1	23.1	0.0
10	10.4	9.9	0.5

Solution   We apply Procedure 10.8.

STEP 1 State the null and alternative hypotheses.

Let μ_1 denote the mean gas mileage of all cars when the additive is used and μ_2 denote the mean gas mileage of all cars when the additive is not used. Then the null and alternative hypotheses are

$$H_0: \ \mu_1 = \mu_2 \ \text{(mean gas mileage with additive is not greater)}$$
$$H_a: \ \mu_1 > \mu_2 \ \text{(mean gas mileage with additive is greater)}.$$

Note that the hypothesis test is right tailed because a greater-than sign (>) appears in the alternative hypothesis.

STEP 2 Decide on the significance level, α.

The test is to be performed at the 5% significance level, or $\alpha = 0.05$.

STEP 3 Calculate the paired differences of the sample pairs.

We have already done so in the last column of Table 10.13.

STEP 4 Discard all paired differences that equal 0 and reduce the sample size accordingly.

The last column of Table 10.13 contains one paired difference that equals 0. Discarding it, we now have a sample of size 9.

STEP 5 Construct a work table to obtain the signed ranks of the paired differences.

Paired difference d	$\|d\|$	Rank of $\|d\|$	Signed rank R
0.8	0.8	6	6
1.2	1.2	8.5	8.5
0.7	0.7	4.5	4.5
0.7	0.7	4.5	4.5
1.0	1.0	7	7
1.2	1.2	8.5	8.5
0.6	0.6	3	3
−0.1	0.1	1	−1
0.5	0.5	2	2

STEP 6 Compute the value of the test statistic

$$W = \text{sum of the positive ranks.}$$

From the last column of the work table, the sum of the positive ranks is

$$W = 6 + 8.5 + 4.5 + 4.5 + 7 + 8.5 + 3 + 2 = 44.$$

CRITICAL-VALUE APPROACH	P-VALUE APPROACH

CRITICAL-VALUE APPROACH

STEP 7 The critical value for a right-tailed test is W_α. Use Table V to find the critical value.

From Table V with $n = 9$ and $\alpha = 0.05$, the critical value is $W_{0.05} = 37$, as shown in Fig. 10.9A.

FIGURE 10.9A

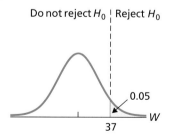

STEP 8 If the value of the test statistic falls in the rejection region, reject H_0; otherwise, do not reject H_0.

The value of the test statistic is $W = 44$, as found in Step 6. This value falls in the rejection region, as revealed in Fig. 10.9A. Thus we reject H_0. The test results are statistically significant at the 5% level.

P-VALUE APPROACH

STEP 7 Obtain the *P*-value by using technology.

Using technology, we find that the *P*-value for the hypothesis test is $P = 0.006$, as shown in Fig. 10.9B.

FIGURE 10.9B

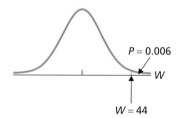

STEP 8 If $P \leq \alpha$, reject H_0; otherwise, do not reject H_0.

From Step 7, $P = 0.006$. Because the *P*-value is less than the specified significance level of 0.05, we reject H_0. The test results are statistically significant at the 5% level and (see Table 9.12 on page 440) provide very strong evidence against the null hypothesis.

STEP 9 Interpret the results of the hypothesis test.

INTERPRETATION   At the 5% significance level, the data provide sufficient evidence to conclude that the mean gas mileage of all cars when the additive is used is greater than the mean gas mileage of all cars when the additive is not used. Evidently, the additive is effective in increasing gas mileage.

Comparing the Paired Wilcoxon Signed-Rank Test and the Paired *t*-Test

As we demonstrated in Section 10.5, a paired *t*-test can be used to conduct a hypothesis test to compare two population means when we have a paired sample and the paired-difference variable is normally distributed. A normally distributed variable necessarily has a symmetric distribution, so we can also use the paired Wilcoxon signed-rank test to perform such a hypothesis test.

So now the question is: If we know that the paired-difference variable is normally distributed, should we use the paired *t*-test or the paired Wilcoxon signed-rank test? As you might expect, the answer is the paired *t*-test. For a normally distributed paired-difference variable, the paired *t*-test is more powerful than the paired Wilcoxon signed-rank test because it is designed expressly for such paired-difference variables. Surprisingly though, the paired *t*-test is not much more powerful than the paired Wilcoxon signed-rank test.

However, if the paired-difference variable has a symmetric distribution but is not normally distributed, the paired Wilcoxon signed-rank test is usually more powerful than the paired *t*-test and is often considerably more powerful. We summarize this discussion in Key Fact 10.7.

KEY FACT 10.7

Paired Wilcoxon Signed-Rank Test Versus the Paired t-Test

Suppose that a hypothesis test based on a paired sample is to be performed to compare the means of two populations. When deciding between the paired *t*-test and the paired Wilcoxon signed-rank test, you should follow these guidelines:

- If you are reasonably sure that the paired-difference variable is normally distributed, use the paired *t*-test.
- If you are not reasonably sure that the paired-difference variable is normally distributed but are reasonably sure that it has a symmetric distribution, use the paired Wilcoxon signed-rank test.

The Technology Center   Some statistical technologies have dedicated programs that automatically perform a paired Wilcoxon signed-rank test, but others do not. Currently, none of Minitab, Excel, and the TI-83/84 Plus have such a program.

Following are some possible options for a statistical technology that does not have a dedicated program for conducting a paired Wilcoxon signed-rank test:

- If the statistical technology has a dedicated program for a (one-sample) Wilcoxon signed-rank test, you can perform a paired Wilcoxon signed-rank test by applying the (one-sample) Wilcoxon signed-rank test program to the sample of paired differences with null hypothesis H_0: $\mu_d = 0$.
- If the statistical technology has appropriate macro capabilities, you can use them to write a program for performing a paired Wilcoxon signed-rank test.

Refer to the technology manuals for more details.

A paired Wilcoxon signed-rank test can be used to perform a hypothesis test to compare two population medians as well as two population means. Many statistical technologies present the output of that procedure in terms of medians, but that output can also be interpreted in terms of means.

EXERCISES 10.6

Statistical Concepts and Skills

10.97 Suppose that you want to perform a hypothesis test based on a simple random paired sample to compare the means of two populations and that you know that the paired-difference variable is normally distributed. Answer each question and explain your answers.
a. Is it acceptable to use the paired t-test?
b. Is it acceptable to use the paired Wilcoxon signed-rank test?
c. Which test is preferable, the paired t-test or the paired Wilcoxon signed-rank test?

10.98 Suppose that you want to perform a hypothesis test based on a simple random paired sample to compare the means of two populations and you know that the paired-difference variable has a symmetric distribution that is far from normal.
a. Is use of the paired t-test acceptable if the sample size is small or moderate? Why or why not?
b. Is use of the paired t-test acceptable if the sample size is large? Why or why not?
c. Is use of the paired Wilcoxon signed-rank test acceptable? Why or why not?
d. If both the paired t-test and the paired Wilcoxon signed-rank test are acceptable, which test is preferable? Explain your answer.

Exercises **10.99–10.104** *repeat Exercises 10.79–10.84 of Section 10.5. There, you applied the paired t-test to solve each problem. Now you are to solve each problem by applying the paired Wilcoxon signed-rank test.*

10.99 Zea Mays. Refer to Exercise 10.79 on page 543.
a. At the 5% significance level, do the data provide sufficient evidence to conclude that the mean heights of cross-fertilized and self-fertilized Zea mays differ?
b. Repeat part (a) at the 1% significance level.

10.100 Sleep. Refer to Exercise 10.80 on page 543.
a. At the 5% significance level, do the data provide sufficient evidence to conclude that laevohysocyamine hydrobromide is effective in increasing sleep?
b. Repeat part (a) at the 1% significance level.

10.101 Anorexia Treatment. Anorexia nervosa is a serious eating disorder, particularly among young women. The following data provide the weights, in pounds, of 17 anorexic young women before and after receiving a family therapy treatment for anorexia nervosa. [SOURCE: Hand et al., ed., *A Handbook of Small Data Sets*, London: Chapman & Hall, 1994. Raw data from B. Everitt (personal communication).]

Before	After	Before	After	Before	After
83.3	94.3	76.9	76.8	82.1	95.5
86.0	91.5	94.2	101.6	77.6	90.7
82.5	91.9	73.4	94.9	83.5	92.5
86.7	100.3	80.5	75.2	89.9	93.8
79.6	76.7	81.6	77.8	86.0	91.7
87.3	98.0	83.8	95.2		

Does family therapy appear to be effective in helping anorexic young women gain weight? Perform the appropriate hypothesis test at the 5% significance level.

10.102 Measuring Treadwear. Stichler, Richey, and Mandel compared two methods of measuring treadwear in their paper "Measurement of Treadwear of Commercial Tires" (*Rubber Age*, 73:2). Eleven tires were each measured for treadwear by two methods, one based on weight and the other on groove wear. The following are the data, in thousands of miles.

Weight method	Groove method	Weight method	Groove method
30.5	28.7	24.5	16.1
30.9	25.9	20.9	19.9
31.9	23.3	18.9	15.2
30.4	23.1	13.7	11.5
27.3	23.7	11.4	11.2
20.4	20.9		

At the 5% significance level, do the data provide sufficient evidence to conclude that, on average, the two measurement methods give different results?

10.103 Glaucoma and Corneal Thickness. Refer to Exercise 10.83 on page 544. At the 10% significance level, do the data provide sufficient evidence to conclude that mean corneal thickness is greater in normal eyes than in eyes with glaucoma?

10.104 Fortified Orange Juice. Refer to Exercise 10.84 on page 544. At the 1% significance level, do the data provide sufficient evidence to conclude that drinking fortified orange juice increases the serum 25(OH)D concentration in the blood?

Extending the Concepts and Skills

10.105 A hypothesis test based on a simple random paired sample is to be performed to compare the means of two populations. The sample of 15 paired differences contains an outlier but otherwise is approximately bell shaped. Assuming that removing the outlier is not legitimate, which test is better to use—the paired t-test or the paired Wilcoxon signed-rank test? Explain your answer.

10.106 Suppose that you want to perform a hypothesis test based on a simple random paired sample to compare the means of two populations. For each part, decide whether you would use the paired t-test, the paired Wilcoxon signed-rank test, or neither of these tests. Preliminary data analyses of the sample of paired differences suggest that the distribution of the paired-difference variable is
a. approximately normal.
b. highly skewed; the sample size is 20.
c. symmetric bimodal.

10.107 Suppose that you want to perform a hypothesis test based on a simple random paired sample to compare the means of two populations. For each part, decide whether you would use the paired t-test, the paired Wilcoxon signed-rank test, or neither of these tests. Preliminary data analyses of the sample of paired differences suggest that the distribution of the paired-difference variable is
a. uniform.
b. neither symmetric nor normal; the sample size is 132.
c. moderately skewed but otherwise approximately bell shaped.

10.108 Explain why the paired Wilcoxon signed-rank test is simply a Wilcoxon signed-rank test on the sample of paired differences with null hypothesis H_0: $\mu_d = 0$.

Using Technology

10.109 Anorexia Treatment. Refer to Exercise 10.101. Use the technology of your choice to
a. obtain a boxplot and stem-and-leaf diagram of the paired differences.
b. perform the required hypothesis test.
c. Justify the use of your procedure in part (b).

10.110 Measuring Treadwear. Refer to Exercise 10.102. Use the technology of your choice to
a. obtain a boxplot and stem-and-leaf diagram of the paired differences.
b. perform the required hypothesis test.
c. Justify the use of your procedure in part (b).

10.111 Tobacco Mosaic Virus. To assess the effects of two different strains of the tobacco mosaic virus, W. Youden and H. Beale randomly selected eight tobacco leaves. Half of each leaf was subjected to one of the strains of tobacco mosaic virus and the other half to the other strain. The researchers then counted the number of local lesions apparent on each half of each leaf. The results of their study were published in the paper "A Statistical Study of the Local

Lesion Method for Estimating Tobacco Mosaic Virus" (*Contributions to Boyce Thompson Institute*, Vol. 6, p. 437). The following are the data.

Leaf	Virus 1	Virus 2
1	31	18
2	20	17
3	18	14
4	17	11
5	9	10
6	8	7
7	10	5
8	7	6

Suppose that you want to perform a hypothesis test to determine whether a difference exists between the mean number of local lesions resulting from the two viral strains. Conduct preliminary graphical data analyses to decide whether applying the paired Wilcoxon signed-rank test is reasonable. Explain your decision.

10.112 Improving Car Emissions? The makers of the MAGNETIZER Engine Energizer System (EES) claim that it improves gas mileage and reduces emissions in automobiles by using magnetic free energy to increase the amount of oxygen in the fuel for greater combustion efficiency. Following are test results, performed under International and U.S. Government agency standards, on a random sample of 14 vehicles. The data give the carbon monoxide (CO) levels, in parts per million, of each vehicle tested, both before installation of EES and after installation. [SOURCE: *Global Source Marketing*.]

Before	After	Before	After
1.60	0.15	2.60	1.60
0.30	0.20	0.15	0.06
3.80	2.80	0.06	0.16
6.20	3.60	0.60	0.35
3.60	1.00	0.03	0.01
1.50	0.50	0.10	0.00
2.00	1.60	0.19	0.00

Suppose that you want to perform a hypothesis test to determine whether, on average, EES reduces CO emissions. Conduct preliminary graphical data analyses to decide whether applying the paired Wilcoxon signed-rank test is reasonable. Explain your decision.

10.7 Which Procedure Should Be Used?[†]

In this chapter, we developed several inferential procedures for comparing the means of two populations. Table 10.14 at the top of the next page summarizes the hypothesis-testing procedures; a similar table can be constructed for confidence-interval procedures.

Each row of Table 10.14 gives the type of test, the conditions required for using the test, the test statistic, and the procedure to use. For brevity, we have written "paired *W*-test" instead of "paired Wilcoxon signed-rank test." And, as previously, we have used the following abbreviations:

- *normal populations*—the two distributions of the variable under consideration are normally distributed;
- *same-shape populations*—the two distributions of the variable under consideration have the same shape;
- *normal differences*—the paired-difference variable is normally distributed;
- *symmetric differences*—the paired-difference variable has a symmetric distribution.

In selecting the correct procedure, keep in mind that the best choice is the procedure expressly designed for the types of distributions under consideration, if such a procedure exists, and that the three *t*-tests are only approximately correct for large samples from nonnormal populations.

[†]All previous sections in this chapter, including the material on the Mann–Whitney test and paired Wilcoxon signed-rank test, are prerequisite to this section.

TABLE 10.14

Summary of hypothesis testing procedures for comparing two population means. The null hypothesis for all tests is H_0: $\mu_1 = \mu_2$

Type	Assumptions	Test statistic	Procedure to use
Pooled t-test	1. Simple random samples 2. Independent samples 3. Normal populations or large samples 4. Equal population standard deviations	$t = \dfrac{\bar{x}_1 - \bar{x}_2}{s_p\sqrt{(1/n_1) + (1/n_2)}}$ † $(\text{df} = n_1 + n_2 - 2)$	10.1 (page 493)
Nonpooled t-test	1. Simple random samples 2. Independent samples 3. Normal populations or large samples	$t = \dfrac{\bar{x}_1 - \bar{x}_2}{\sqrt{(s_1^2/n_1) + (s_2^2/n_2)}}$ ‡	10.3 (page 506)
Mann–Whitney test	1. Simple random samples 2. Independent samples 3. Same-shape populations	M = sum of the ranks for sample data from Population 1	10.5 (page 522)
Paired t-test	1. Simple random paired sample 2. Normal differences or large sample	$t = \dfrac{\bar{d}}{s_d/\sqrt{n}}$ $(\text{df} = n - 1)$	10.6 (page 534)
Paired W-test	1. Simple random paired sample 2. Symmetric differences	W = sum of positive ranks	10.8 (page 547)

† $s_p = \sqrt{\dfrac{(n_1 - 1)s_1^2 + (n_2 - 1)s_2^2}{n_1 + n_2 - 2}}$ ‡ $\text{df} = \dfrac{[(s_1^2/n_1) + (s_2^2/n_2)]^2}{\dfrac{(s_1^2/n_1)^2}{n_1 - 1} + \dfrac{(s_2^2/n_2)^2}{n_2 - 1}}$

For instance, suppose that independent simple random samples are taken from two populations with equal standard deviations and that the two distributions (one for each population) of the variable under consideration are normally distributed. Then the pooled t-test, nonpooled t-test, and Mann–Whitney test are all applicable. But the correct procedure is the pooled t-test because that test is designed specifically for use with independent samples from two normally distributed populations that have equal standard deviations.

The flowchart shown in Fig. 10.10 provides an organized strategy for choosing the correct hypothesis-testing procedure for comparing two population means. It is based on the preceding discussion.

In practice, you need to look at the sample data to ascertain distribution type before selecting the appropriate procedure. We recommend using normal probability plots and either stem-and-leaf diagrams (for small samples or samples of moderate size) or histograms (for samples of moderate size or large samples); boxplots can also be quite helpful, especially for samples of moderate size or large samples. We demonstrate choosing the correct procedure in Example 10.19.

FIGURE 10.10

Flowchart for choosing the correct hypothesis testing procedure for comparing two population means

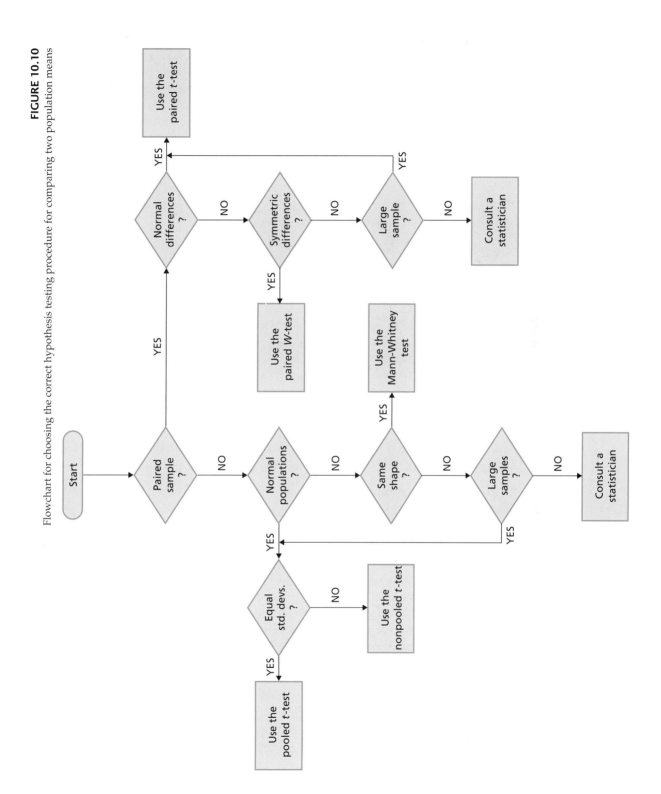

Example 10.19 Choosing the Correct Hypothesis-Testing Procedure

Skinfold Thickness A study entitled "Body Composition of Elite Class Distance Runners" was conducted by M. L. Pollock et al. to determine whether elite distance runners actually are thinner than other people. Their results were published in *The Marathon: Physiological, Medical, Epidemiological, and Psychological Studies*, P. Milvey (ed.), New York: New York Academy of Sciences, p. 366.

The researchers measured skinfold thickness, an indirect indicator of body fat, of runners and nonrunners in the same age group. The data presented in Table 10.15 are based on the skinfold-thickness measurements on the thighs of the people sampled. Data are in millimeters.

TABLE 10.15

Skinfold thickness (mm) for independent samples of elite runners and others

Runners			Others			
7.3	6.7	8.7	24.0	19.9	7.5	18.4
3.0	5.1	8.8	28.0	29.4	20.3	19.0
7.8	3.8	6.2	9.3	18.1	22.8	24.2
5.4	6.4	6.3	9.6	19.4	16.3	16.3
3.7	7.5	4.6	12.4	5.2	12.2	15.6

Suppose that we want to use the sample data displayed in Table 10.15 to decide whether elite runners have smaller skinfold thickness, on average, than other people. Let μ_1 denote the mean skinfold thickness of elite runners and let μ_2 denote the mean skinfold thickness of others. Then we want to perform the hypothesis test

$$H_0: \ \mu_1 = \mu_2 \ \text{(mean skinfold thickness is not smaller)}$$
$$H_a: \ \mu_1 < \mu_2 \ \text{(mean skinfold thickness is smaller)}.$$

Which procedure should we use to perform the hypothesis test?

Solution To decide which procedure to use, we refer to the flowchart in Fig. 10.10. The first question we must answer is: Do we have a paired sample? The samples are not paired, so the answer to the first question is "No."

This result leads to the question: Are the populations normal? To answer this question, we constructed the normal probability plots shown in Fig. 10.11. The plots are quite linear and thereby indicate that a reasonable presumption is that skinfold thickness is approximately normally distributed for both elite runners and others. Hence the answer to the second question is "Yes."

Next, we must answer the question: Are the population standard deviations equal? The standard deviations of the two samples in Table 10.15 are 1.80 mm and 6.61 mm, respectively. These statistics suggest that the population standard deviations are not equal.

We can also see that they aren't equal by looking at a back-to-back stem-and-leaf diagram, histograms, or boxplots. Figure 10.12 displays boxplots for the two samples presented in Table 10.15. The vast difference in the spreads of the two plots again suggests that the standard deviations of skinfold thickness differ for elite runners and others. Thus the answer to the third question is "No."

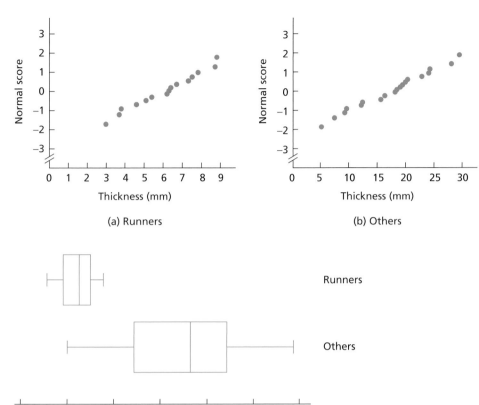

FIGURE 10.11
Normal probability plots of the sample data for (a) elite runners and (b) others

(a) Runners

(b) Others

FIGURE 10.12
Boxplots of the sample data for elite runners and others

Runners

Others

The "No" answer to the preceding question leads us to the box in Fig. 10.10 that states: Use the nonpooled *t*-test. Therefore, we should use Procedure 10.3 to conduct the hypothesis test.

EXERCISES 10.7

Statistical Concepts and Skills

10.113 We considered three hypothesis-testing procedures based on independent simple random samples to compare the means of two populations with unknown standard deviations.
a. Identify the three procedures by name.
b. List the conditions for using each procedure.
c. Identify the test statistic for each procedure.

10.114 We examined two hypothesis-testing procedures based on a simple random paired sample to compare the means of two populations.
a. Identify the two procedures by name.
b. List the conditions for using each procedure.
c. Identify the test statistic for each procedure.

10.115 Suppose that you want to perform a hypothesis test based on independent simple random samples to compare the means of two populations. Assume that the variable under consideration is normally distributed on each of the two populations and that the population standard deviations are equal.
a. Identify the procedures discussed in this chapter that could be used to carry out the hypothesis test, that is, the procedures whose assumptions are satisfied.
b. Among the procedures that you identified in part (a), which is the best one to use? Explain your answer.

10.116 Suppose that you want to perform a hypothesis test based on independent simple random samples to compare the means of two populations. Assume that the variable under consideration is normally distributed on each of the two populations and that the population standard deviations are unequal.

a. Identify the procedures discussed in this chapter that could be used to carry out the hypothesis test, that is, the procedures whose assumptions are satisfied.
b. Among the procedures that you identified in part (a), which is the best one to use? Explain your answer.

10.117 Suppose that you want to perform a hypothesis test based on independent simple random samples to compare the means of two populations. Assume that the two distributions of the variable under consideration have the same shape but are not normally distributed and that the sample sizes are both large.

a. Identify the procedures discussed in this chapter that could be used to carry out the hypothesis test, that is, the procedures whose assumptions are satisfied.
b. Among the procedures that you identified in part (a), which is the best one to use? Explain your answer.

10.118 Suppose that you want to perform a hypothesis test based on a simple random paired sample to compare the means of two populations. Assume that the paired-difference variable is normally distributed.

a. Identify the procedures discussed in this chapter that could be used to carry out the hypothesis test, that is, the procedures whose assumptions are satisfied.
b. Among the procedures that you identified in part (a), which is the best one to use? Explain your answer.

10.119 Suppose that you want to perform a hypothesis test based on a simple random paired sample to compare the means of two populations. Assume that the paired-difference variable has a nonnormal symmetric distribution and that the sample size is large.

a. Identify the procedures discussed in this chapter that could be used to carry out the hypothesis test, that is, the procedures whose assumptions are satisfied.
b. Among the procedures that you identified in part (a), which is the best one to use? Explain your answer.

*In Exercises **10.120–10.125**, we provide a type of sampling (independent or paired), sample size(s), and a figure showing the results of preliminary data analyses on the sample(s). For independent samples, the graphs are for the two samples; for a paired sample, the graphs are for the paired differences. The intent is to employ the sample data to perform a hypothesis test to compare the means of the two populations from which the data were obtained. In each case, use the information provided and the flowchart shown in Fig. 10.10 on page 555 to decide which procedure should be applied.*

10.120 Paired; $n = 75$; Fig. 10.13

10.121 Independent; $n_1 = 25$ and $n_2 = 20$; Fig. 10.14

10.122 Independent; $n_1 = 17$ and $n_2 = 17$; Fig. 10.15

10.123 Independent; $n_1 = 40$ and $n_2 = 45$; Fig. 10.16

10.124 Independent; $n_1 = 20$ and $n_2 = 15$; Fig. 10.17

10.125 Paired; $n = 18$; Fig. 10.18

FIGURE 10.13
Results of preliminary data analyses
in Exercise 10.120

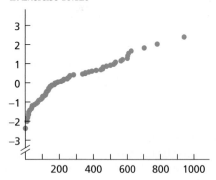

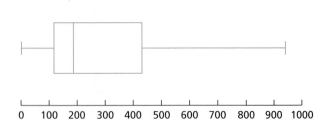

FIGURE 10.14
Results of preliminary data analyses
in Exercise 10.121

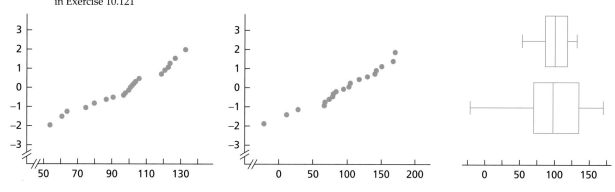

FIGURE 10.15
Results of preliminary data analyses
in Exercise 10.122

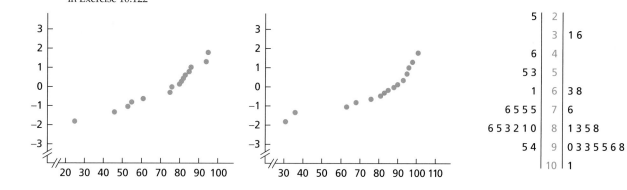

FIGURE 10.16
Results of preliminary data analyses
in Exercise 10.123

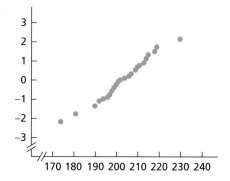
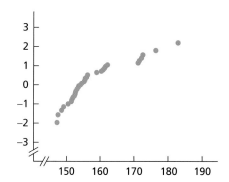

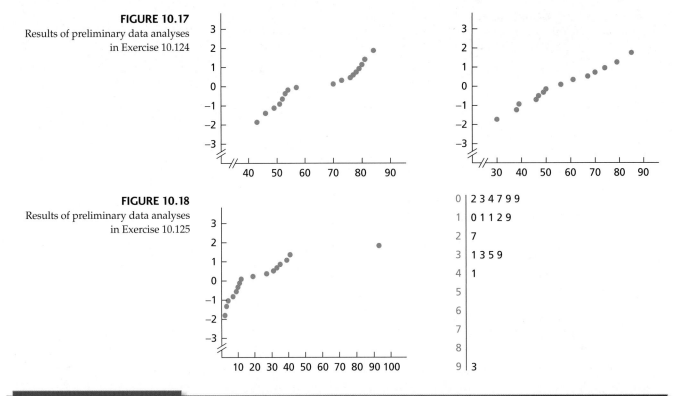

FIGURE 10.17
Results of preliminary data analyses in Exercise 10.124

FIGURE 10.18
Results of preliminary data analyses in Exercise 10.125

CHAPTER REVIEW

You Should Be Able To

1. use and understand the formulas in this chapter.

2. perform inferences based on independent simple random samples to compare the means of two populations when the population standard deviations are unknown but are assumed to be equal.

3. perform inferences based on independent simple random samples to compare the means of two populations when the population standard deviations are unknown but are not assumed to be equal.

*4. perform a hypothesis test based on independent simple random samples to compare the means of two populations when the distributions of the variable under consideration have the same shape.

5. perform inferences based on a simple random paired sample to compare the means of two populations.

*6. perform a hypothesis test based on a simple random paired sample to compare the means of two populations when the paired-difference variable has a symmetric distribution.

*7. decide which procedure should be used to perform an inference to compare the means of two populations.

Key Terms

back-to-back stem-and-leaf diagram,* 519
independent samples, 484
independent simple random samples, 484
M_{α},* 520
Mann–Whitney test,* 518, 522
Mann–Whitney–Wilcoxon test,* 518
nonpooled t-interval procedure, 509
nonpooled t-test, 505, 506

normal differences, 533
paired difference, 532
paired-difference variable, 533
paired sample, 530
paired t-interval procedure, 538
paired t-test, 533, 534
paired Wilcoxon signed-rank test,* 546, 547
pool, 492

pooled sample standard deviation (s_p), 492
pooled t-interval procedure, 497
pooled t-test, 492, 493
sampling distribution of the difference between two sample means, 488
simple random paired sample, 530
symmetric differences,* 546
Wilcoxon rank-sum test,* 518

REVIEW TEST

Statistical Concepts and Skills

1. Discuss the basic strategy for comparing the means of two populations based on independent simple random samples.

2. Discuss the basic strategy for comparing the means of two populations based on a simple random paired sample.

3. Regarding the pooled and nonpooled *t*-procedures,
a. what is the difference in assumptions between the two procedures?
b. how important is the assumption of independent simple random samples for these procedures?
c. how important is the normality assumption for these procedures?
d. Suppose that the variable under consideration is normally distributed on each of the two populations and that you are going to use independent simple random samples to compare the population means. Fill in the blank and explain your answer: Unless you are quite sure that the _____ are equal, the nonpooled *t*-procedures should be used instead of the pooled *t*-procedures.

*4. Suppose that independent simple random samples are taken from two populations to compare their means. Further suppose that the two distributions of the variable under consideration have the same shape.
a. Would the nonpooled *t*-test ever be the procedure of choice in these circumstances? Explain your answer.
b. Under what conditions would the pooled *t*-test be preferable to the Mann–Whitney test? Explain your answer.

5. Explain one possible advantage of using a paired sample instead of independent samples.

*6. Suppose that a simple random paired sample is taken from two populations to compare their means. Further suppose that the distribution of the paired-difference variable has a symmetric distribution. Under what conditions would the paired *t*-test be preferable to the paired Wilcoxon signed-rank test? Explain your answer.

7. Grip and Leg Strength. In the paper, "Sex Differences in Static Strength and Fatigability in Three Different Muscle Groups" (*Research Quarterly for Exercise and Sport*, Vol 61(3), pp. 238–242), J. Misner et al. published results of a study on grip and leg strength of males and females. The following data, in newtons, is based on their measurements of right-leg strength.

Male			Female		
2632	1796	2256	1344	1351	1369
2235	2298	1917	2479	1573	1665
1105	1926	2644	1791	1866	1544
1569	3129	2167	2359	1694	2799
1977			1868	2098	

Preliminary data analyses indicate that you can reasonably presume leg strength is normally distributed for both males and females and that the standard deviations of leg strength are approximately equal.
a. At the 5% significance level, do the data provide sufficient evidence to conclude that mean right-leg strength of males exceeds that of females? (*Note:* $\bar{x}_1 = 2127$, $s_1 = 513$, $\bar{x}_2 = 1843$, and $s_2 = 446$.)
b. Estimate the *P*-value of the hypothesis test, and use that estimate and Table 9.12 on page 440 to assess the strength of the evidence against the null hypothesis.

8. Grip and Leg Strength. Refer to Problem 7. Determine a 90% confidence interval for the difference between the mean right-leg strengths of males and females. Interpret your result.

9. Cottonmouth Litter Size. A study published by Blem and Blem in the *Journal of Herpetology* (Vol. 29, pp. 391–398) examined the reproductive characteristics of the eastern cottonmouth. The data in the table at the top of the next column, based on the results of the researchers' study, give the number of young per litter for 24 female cottonmouths in Florida and 44 female cottonmouths in Virginia.

Preliminary data analyses indicate that you can reasonably presume that litter sizes of cottonmouths in both states are approximately normally distributed. At the 1% signifi-

Florida			Virginia					
8	6	7	5	12	7	7	6	8
7	4	3	12	9	7	4	9	6
1	7	5	12	7	5	6	10	3
6	6	5	10	8	8	12	5	6
6	8	5	10	11	3	8	4	5
5	7	4	7	6	11	7	6	8
6	6	5	8	14	8	7	11	7
5	5	4	5	4				

Pair	Program 1	Program 2
1	1114	1032
2	996	1148
3	979	1074
4	1125	1076
5	910	959
6	1056	1094
7	1091	1091
8	1053	1096
9	996	1032
10	894	1012

cance level, do the data provide sufficient evidence to conclude that, on average, the number of young per litter of cottonmouths in Florida is less than that in Virginia? Do not assume that the population standard deviations are equal. (*Note:* $\bar{x}_1 = 5.46$, $s_1 = 1.59$, $\bar{x}_2 = 7.59$, and $s_2 = 2.68$.)

10. Cottonmouth Litter Size. Refer to Problem 9. Find a 98% confidence interval for the difference between the mean litter sizes of cottonmouths in Florida and Virginia. Interpret your result.

*11. Home Prices in LA and NY. The *National Association of REALTORS* compiles and publishes home prices of existing single-family homes in cities across the United States. Independent simple random samples of 10 homes each in New York City and Los Angeles yielded the following data on home prices in thousands of dollars.

New York		Los Angeles	
235.8	209.5	245.5	248.8
222.4	250.1	238.0	235.6
484.1	246.1	398.8	212.5
272.8	340.1	209.6	443.9
237.1	279.1	323.9	215.0

At the 5% significance level, can you conclude that the mean costs for existing single-family homes differ in New York City and Los Angeles? (*Note:* Preliminary data analyses suggest that you can reasonably presume that the cost distributions for the two cities have roughly the same shape but that those distributions are right skewed.)

12. Speed Reading. To compare the effectiveness of two speed-reading programs, 10 pairs of people were randomly selected. Each pair consisted of people whose then current reading speeds were essentially identical. From each pair, one person was randomly selected to take Program 1; the other person to take Program 2. After the 10 pairs of people completed the speed-reading programs, their reading speeds, in words per minute, were as follows.

At the 10% significance level, can you conclude that there is a difference in effectiveness of the two speed-reading programs? (*Note:* A normal probability plot of the paired differences suggests that it is reasonable to presume that the paired-difference variable is approximately normally distributed.)

13. Speed Reading. Refer to Problem 12. Find a 90% confidence interval for the difference in mean reading speeds for the two programs.

*14. Fiber Density. In the article "Comparison of Fiber Counting by TV Screen and Eyepieces of Phase Contrast Microscopy" (*American Industrial Hygiene Association Journal*, Vol. 63, pp. 756–761), I. Moa, H. Yeh, and M. Chen reported on determining fiber density by two different methods. The fiber density of 10 samples with varying fiber density was obtained by using both an eyepiece method and a TV-screen method. The results, in fibers per square millimeter, are presented in the following table.

Sample ID	Eyepiece	TV Screen
1	182.2	177.8
2	118.5	116.6
3	100.0	92.4
4	161.3	145.0
5	42.7	38.9
6	299.1	226.3
7	547.8	514.6
8	437.3	458.1
9	174.4	159.2
10	85.4	86.6

Use the paired Wilcoxon signed-rank test to decide whether, on average, the eyepiece method gives a greater fiber-density reading than the TV-screen method. Perform the required hypothesis test at the 5% significance level.

Using Technology

15. Grip and Leg Strength. Refer to Problems 7 and 8. Use the technology of your choice to

a. obtain normal probability plots, stem-and-leaf diagrams, boxplots, and the standard deviations of the two samples.

b. perform the required hypothesis test and obtain the desired confidence interval.

c. Justify the use of your procedures in part (b).

16. Cottonmouth Litter Size. Refer to Problems 9 and 10. Use the technology of your choice to

a. obtain normal probability plots, stem-and-leaf diagrams, boxplots, and the standard deviations of the two samples.

b. perform the required hypothesis test and obtain the desired confidence interval.

c. Justify the use of your procedures in part (b).

*17. Home Prices in LA and NY. Refer to Problem 11. Use the technology of your choice to

a. obtain normal probability plots, stem-and-leaf diagrams, boxplots, and the standard deviations of the two samples.

b. perform the required hypothesis test.

c. Justify the use of your procedure in part (b).

18. Speed Reading. Refer to Problems 12 and 13. Use the technology of your choice to

a. obtain a normal probability plot, stem-and-leaf diagram, and boxplot of the paired differences.

b. perform the required hypothesis test and obtain the desired confidence interval.

c. Justify the use of your procedures in part (b).

*19. Fiber Density. Refer to Problem 14. Use the technology of your choice to do the following:

a. Apply the paired t-test to decide whether, on average, the eyepiece method gives a greater fiber-density reading than the TV-screen method. Perform the required hypothesis test at the 5% significance level.

b. Apply the paired Wilcoxon signed-rank test to decide whether, on average, the eyepiece method gives a greater fiber-density reading than the TV-screen method. Perform the required hypothesis test at the 5% significance level.

c. Compare your results from parts (a) and (b).

d. Obtain a normal probability plot, a boxplot, and a stem-and-leaf diagram of the paired differences.

e. Based on your results from part (d), which test do you think is more appropriate, the paired t-test or the Wilcoxon signed-rank test? Explain your answer.

StatExplore in MyMathLab
Analyzing Data Online

You can use StatExplore to perform all statistical analyses discussed in this chapter. To illustrate, we show how to perform pooled t-procedures.

EXAMPLE Pooled t-Procedures

Faculty Salaries Table 10.5 on page 495 displays the annual salaries, in thousands of dollars, for independent simple random samples of 30 faculty members from public institutions and 35 faculty members from private institutions. Use StatExplore to do the following:

a. Decide, at the 5% significance level, whether the data provide sufficient evidence to conclude that mean salaries for faculty in public and private institutions differ.

b. Obtain a 95% confidence interval for the difference between the mean salaries of faculty in public and private institutions.

SOLUTION Let μ_1 and μ_2 denote the mean salaries of all faculty in public and private institutions, respectively.

a. Here the problem is to perform the hypothesis test

$$H_0: \mu_1 = \mu_2 \text{ (mean salaries are the same)}$$

$$H_a: \mu_1 \neq \mu_2 \text{ (mean salaries are different)}$$

at the 5% significance level. Note that the hypothesis test is two tailed because a does-not-equal sign ($\neq$) appears in the alternative hypothesis.

To carry out the hypothesis test by using StatExplore, we first store the two samples of salary data from Table 10.5 in columns named PUBLIC and PRIVATE, respectively. Then, we proceed as follows:

1 Choose **Stat ➤ T statistics ➤ Two sample**
2 Click the arrow button at the right of the **Sample 1 in** drop-down list box and select PUBLIC
3 Click the arrow button at the right of the **Sample 2 in** drop-down list box and select PRIVATE
4 Select the **Pool variances** check box
5 Click **Next →**
6 Select the **Hypothesis Test** option button
7 Click in the **Null: mean diff. =** text box and type <u>0</u>
8 Click the arrow button at the right of the **Alternative** drop-down list box and select **not =**
9 Click **Calculate**

The resulting output is shown in Printout 10.5. From Printout 10.5, we see that the *P*-value for the hypothesis test is 0.1252. As the *P*-value exceeds the specified significance level of 0.05, we do not reject H_0. At the 5% significance level, the data do not provide sufficient evidence to conclude that mean salaries for faculty in public and private institutions differ.

PRINTOUT 10.5
StatExplore output for pooled *t*-test

Hypothesis test results:
m1 = mean of PUBLIC
m2 = mean of PRIVATE
Parameter: m1 - m2
H0 : Parameter = 0
HA : Parameter not = 0
(with pooled variances)

Difference	Sample Mean	Std. Err.	DF	T-Stat	P-value
m1 − m2	-8.914286	5.736302	63	-1.5540127	0.1252

b. To determine a 95% confidence interval for the difference between the mean salaries of faculty in public and private institutions, we proceed as follows, again assuming that the two samples of salary data from Table 10.5 have been stored in columns named PUBLIC and PRIVATE:

1 Choose **Stat ➤ T statistics ➤ Two sample**
2 Click the arrow button at the right of the **Sample 1 in** drop-down list box and select PUBLIC
3 Click the arrow button at the right of the **Sample 2 in** drop-down list box and select PRIVATE
4 Select the **Pool variances** check box
5 Click **Next →**
6 Select the **Confidence Interval** option button
7 Click in the **Level** text box and type <u>0.95</u>
8 Click **Calculate**

The resulting output is shown in Printout 10.6. From Printout 10.6, we see that a 95% confidence interval is from −20.37737 (lower limit) to 2.5487988 (upper limit). We can be 95% confident that the difference between the mean salaries of faculty in public and private institutions is somewhere between −$20,377.37 and $2,548.7988.

PRINTOUT 10.6
StatExplore output for pooled *t*-interval

95% confidence interval results:
m1 = mean of PUBLIC
m2 = mean of PRIVATE
Parameter: m1 - m2
(with pooled variances)

Difference	Sample Mean	Std. Err.	DF	L. Limit	U. Limit
m1 - m2	-8.914286	5.736302	63	-20.37737	2.5487988

STATEXPLORE EXERCISES Solve the following problems by using StatExplore.

a. Conduct the pooled *t*-test required in Exercise 10.21 on page 502.

b. Obtain the pooled *t*-interval required in Exercise 10.25 on page 503.

c. Carry out the nonpooled *t*-test required in Example 10.6 on page 507.

d. Obtain the nonpooled *t*-interval required in Example 10.7 on page 509.

*e. Carry out the Mann–Whitney hypothesis test required in Example 10.11 on page 523.

f. Conduct the paired *t*-test required in Example 10.15 on page 535.

g. Obtain the paired *t*-interval required in Example 10.16 on page 538.

*h. Perform the paired Wilcoxon signed-rank test required in Example 10.18 on page 548. (*Note:* StatExplore does not have a dedicated program to conduct a paired Wilcoxon signed-rank test. To obtain the paired differences, choose **Data ➤ Transform data** and select the function Y - X. Then apply the one-sample Wilcoxon signed-rank test to the paired differences.)

To access StatExplore, go to the student content area of your Weiss MyMathLab course.

Internet Project

Women Workers and Equality in the United States

The rise of women in the U.S. labor force has been dramatic. According to the Women's Bureau of the U.S. Department of Labor, women workers made up 46% of the U.S. labor force in 1994. The Bureau predicts that this figure will increase to 48% by 2005.

Although these numbers reflect progress for women in the United States, problems remain. More women are working than in the past, but are they compensated the same as men? Some advocates for equality claim that most of the women in the U.S. work force have jobs that tend to pay less than male occupations. Additionally, these advocates suggest that such women have fewer opportunities for advancement. In this Internet project, you are to explore relevant data and come to your own conclusion.

URL for access to Internet Projects Page: www.aw-bc.com/weiss

Focusing on Data Analysis

The Focus Database

Recall from Chapter 1 (see page 35) that the Focus database contains information on the undergraduate students at the University of Wisconsin - Eau Claire (UWEC). Statistical analyses for this database should be done with the technology of your choice.

  a. Obtain independent simple random samples of size 50 each of the ACT composite scores (COMP) of male and female UWEC undergraduate students.
  b. Obtain normal probability plots, boxplots, and the sample standard deviations of the ACT composite scores of the sampled males and the sampled females.
  c. At the 5% significance level, do the data from part (a) provide sufficient evidence to conclude that mean ACT composite scores differ for male and female UWEC undergraduates? Justify the use of the procedure you chose to carry out the hypothesis test.
  d. Use the data from part (a) to determine a 95% confidence interval for the difference between the mean ACT composite scores of male and female UWEC undergraduates. Interpret your result.
  e. Repeat parts (a)–(d) for cumulative GPA (GPA).
  f. Obtain independent simple random samples of size 50 each of the ACT English scores (ENGLISH) of male and female UWEC undergraduate students.
  g. Obtain normal probability plots, boxplots, and the sample standard deviations of the ACT English scores of the sampled males and the sampled females.
  h. At the 5% significance level, do the data from part (f) provide sufficient evidence to conclude that the mean ACT English score of male UWEC undergraduates is less than that of female UWEC undergraduates? Justify the use of the procedure you chose to carry out the hypothesis test.
  i. Use the data from part (f) to determine a 90% confidence interval for the difference between the mean ACT English scores of male and female UWEC undergraduates. Interpret your result.
  j. Obtain independent simple random samples of size 50 each of the ACT math scores (MATH) of male and female UWEC undergraduate students.
  k. Obtain normal probability plots, boxplots, and the sample standard deviations of the ACT math scores of the sampled males and the sampled females.

l. At the 5% significance level, do the data from part (j) provide sufficient evidence to conclude that the mean ACT math score of male UWEC undergraduates exceeds that of female UWEC undergraduates? Justify the use of the procedure you chose to carry out the hypothesis test.

m. Use the data from part (j) to determine a 90% confidence interval for the difference between the mean ACT math scores of male and female UWEC undergraduates. Interpret your result.

Case Study Discussion

Breast Milk and IQ

On page 483 of this chapter, we presented data obtained by five researchers studying the effect of breast feeding on subsequent IQ of preterm babies at age $7\frac{1}{2}$–8 years. Three categories were considered: children whose mothers declined to provide breast milk (Group I); those whose mothers had chosen but were unable to provide breast milk (Group IIa); and those whose mothers had chosen and were able to provide breast milk (Group IIb).

Presuming that IQs are normally distributed in all three categories, solve each of the following problems:

a. Do the data provide sufficient evidence to conclude that, for children age $7\frac{1}{2}$–8 years who are born preterm, a difference exists in mean IQ between those whose mothers decline to provide breast milk and those whose mothers choose but are unable to provide breast milk? Perform the required hypothesis test at the 5% significance level.

b. Do the data provide sufficient evidence to conclude that, for children age $7\frac{1}{2}$–8 years who are born preterm, the mean IQ of those whose mothers decline to provide breast milk is less than that of those whose mothers choose and are able to provide breast milk? Perform the required hypothesis test at the 5% significance level.

c. Do the data provide sufficient evidence to conclude that, for children age $7\frac{1}{2}$–8 years who are born preterm, the mean IQ of those whose mothers choose but are unable to provide breast milk is less than that of those whose mothers choose and are able to provide breast milk? Perform the required hypothesis test at the 5% significance level.

d. Is this study observational, or is it a designed experiment? Explain your answer.

e. Based on your answers in parts (a)–(d), what conclusions do you draw?

The researchers also adjusted the data for such factors as social class, mother's education, and infant's sex, and still reached the same conclusions: "…preterm babies whose mothers provided breast milk had a substantial advantage in subsequent IQ at $7\frac{1}{2}$–8 years over those who did not…." However, the researchers emphasized that they could not exclude the possibility that their findings could be explained by differences in parental behavior or genetic potential between the groups.

Internet Resources: Visit the Weiss Web site www.aw-bc.com/weiss for additional discussion, exercises, and resources related to this case study.

Biography   *GERTRUDE COX: Spreading the Gospel According to St. Gertrude*

Gertrude Mary Cox was born on January 13, 1900, in Dayton, Iowa, the daughter of John and Emmaline Cox. She graduated from Perry High School, Perry, Iowa, in 1918. Between 1918 and 1925, she prepared to become a deaconess in the Methodist Episcopal Church.

In 1929 and 1931, Cox received a B.S. and an M.S., respectively, from Iowa State College in Ames. Her work there was directed by George W. Snedecor, and her degree was the first master's degree in statistics given by the department of mathematics at Iowa State.

From 1931 to 1933, Cox studied psychological statistics at the University of California at Berkeley. Snedecor meanwhile had established a new Statistical Laboratory at Iowa State, and in 1933 he asked her to be his assistant. This position launched her internationally influential career in statistics. Cox worked in the lab until becoming an Iowa State assistant professor in 1939.

In 1940, the committee in charge of filling a newly created position as head of the department of experimental statistics at North Carolina State College in Raleigh asked Snedecor for recommendations; he first named several male statisticians, then wrote, "…but if you would consider a woman for this position I would recommend Gertrude Cox of my staff." They did consider a woman and Cox accepted their offer.

In 1945, Cox organized and became director of the Institute of Statistics, which combined the teaching of statistics at the University of North Carolina and North Carolina State. Work conferences that Cox organized established the Institute as an international center for statistics. She also developed statistical programs throughout the South, referred to as "spreading the gospel according to St. Gertrude."

Cox's area of expertise was experimental design. She, with W. G. Cochran, wrote *Experimental Designs* (1950), recognized as the classic textbook on design and analysis of replicated experiments.

From 1960–1964, Cox was director of the Statistics Section of the Research Triangle Institute in Durham, North Carolina. She then retired, working only as a consultant. She died of leukemia on October 17, 1978, in Durham.

chapter 11

Inferences for Population Standard Deviations

GENERAL OBJECTIVES To this point in the study of inferential statistics, we have concentrated on inferences for population means. Another important class of inferences consists of those for population standard deviations (or variances).

For example, in Chapter 9, we discussed the problem of deciding whether the mean net weight of bags of pretzels being packaged by a machine equals the advertised weight of 454 grams. This decision involves a hypothesis test for a population mean.

We should also be concerned with the variation in weights from bag to bag. If the variation is too large, many bags will contain either considerably more or considerably less than they should. To investigate the variation, we can perform a hypothesis test or construct a confidence interval for the standard deviation of the weights. These are inferences for one population standard deviation.

Additionally, we might want to compare two different machines for packaging the pretzels to see whether one provides a smaller variation in weights than the other. We could do so by using inferences for two population standard deviations.

In this chapter, we discuss inferences for one population standard deviation and for two population standard deviations.

Speaker Woofer Driver Manufacturing

Speaker driver manufacturing is an important industry in many countries. In Taiwan, for example, more than 100 companies or factories produce and supply parts and driver units for speakers.

An essential component in driver units is the rubber edge, which affects aspects of sound quality such as musical image and clarity. And an important characteristic of the rubber edge is weight. Generally, each process for manufacturing rubber edges calls for a *production weight specification* that consists of a lower specification limit (LSL), a target weight (T), and an upper specification limit (USL). The actual (population) mean and standard deviation of the weights of the rubber edges produced are called, respectively, the process mean (μ) and process standard deviation (σ).

Several process capability indexes are used to measure process quality relative to the production weight specification and the process mean and standard deviation. An on-target process ($\mu = T$) is called *super* if (USL − LSL)/(6σ) > 2 or, equivalently, if σ < (USL − LSL)/12.

W. L. Pearn and K. S. Chen investigated five rubber-edge manufacturing processes at Bopro—a company located in Taipei, Taiwan—for process capability. The following table, adapted from their paper "Multiprocess Performance Analysis: A Case Study" (*Quality Engineering*, Vol. 10(1), pp. 1–8), provides weight data for one of the processes.

17.59	17.63	17.68	17.57	17.70	17.77	17.54	17.65	17.49	17.60
17.61	17.72	17.68	17.69	17.57	17.66	17.55	17.80	17.67	17.53
17.71	17.54	17.68	17.75	17.46	17.82	17.62	17.53	17.47	17.50
17.74	17.65	17.68	17.68	17.71	17.64	17.65	17.62	17.56	17.60
17.51	17.70	17.47	17.57	17.55	17.63	17.44	17.60	17.63	17.59
17.69	17.53	17.59	17.57	17.49	17.52	17.71	17.56	17.49	17.58

In this chapter, we introduce inferences for population standard deviations. After you have completed the chapter, you will be asked to use the data in the preceding table to ascertain the process capability.

11.1 Inferences for One Population Standard Deviation

Recall that standard deviation is a measure of the variation (spread) of a data set. A data set with a great deal of variation will have a large standard deviation, whereas one with little variation will have a small standard deviation.

Also recall that, for a variable x, the standard deviation of all possible observations for the entire population is called the *population standard deviation* or *standard deviation of the variable x*. It is denoted σ_x or, when no confusion will arise, simply σ.

Suppose that we want to obtain information about a population standard deviation. If the population is small, we can often determine σ exactly by first taking a census and then computing σ from the population data. However, if the population is large, which is usually the case, a census is generally not feasible, and we must use inferential methods to obtain the required information about σ.

In this section, we describe how to perform hypothesis tests and construct confidence intervals for the standard deviation of a normally distributed variable. Such inferences are based on a distribution called the *chi-square distribution*. We begin by discussing it.

The Chi-Square Distribution

A variable is said to have a **chi-square distribution** if its distribution has the shape of a special type of right-skewed curve, called a **chi-square (χ^2) curve.** Actually, there are infinitely many chi-square distributions, and we identify the chi-square distribution (and χ^2-curve) in question by stating its number of degrees of freedom, just as we did for t-distributions. Figure 11.1 on the next page shows three χ^2-curves and illustrates some basic properties of χ^2-curves, which we present in Key Fact 11.1.

KEY FACT 11.1

Basic Properties of χ^2-Curves

Property 1: The total area under a χ^2-curve equals 1.

Property 2: A χ^2-curve starts at 0 on the horizontal axis and extends indefinitely to the right, approaching, but never touching, the horizontal axis as it does so.

Property 3: A χ^2-curve is right skewed.

Property 4: As the number of degrees of freedom becomes larger, χ^2-curves look increasingly like normal curves.

Percentages (and probabilities) for a variable having a chi-square distribution equal areas under its associated χ^2-curve. To perform a hypothesis test or construct a confidence interval for a population standard deviation, we need to know how to find the χ^2-value that corresponds to a specified area under a χ^2-curve. Table VII in Appendix A provides χ^2-values corresponding to several areas for various degrees of freedom.

FIGURE 11.1
χ^2-curves for df = 5, 10, and 19

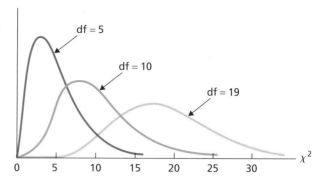

The χ^2-table (Table VII) is similar to the t-table (Table IV). The two outside columns of Table VII, labeled df, display the number of degrees of freedom. As expected, the symbol χ_α^2 denotes the χ^2-value having area α to its right under a χ^2-curve. Thus the column headed $\chi_{0.995}^2$ contains χ^2-values having area 0.995 to their right; the column headed $\chi_{0.99}^2$ contains χ^2-values having area 0.99 to their right; and so on. In Examples 11.1–11.3, we explain how to use Table VII.

Example 11.1   Finding the χ^2-Value Having a Specified Area to Its Right

For a χ^2-curve with 12 degrees of freedom, find $\chi_{0.025}^2$; that is, find the χ^2-value having area 0.025 to its right, as shown in Fig. 11.2(a).

FIGURE 11.2
Finding the χ^2-value having area 0.025 to its right

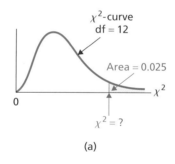

(a)

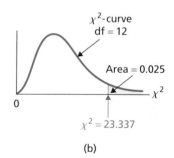

(b)

Solution   To find this χ^2-value, we use Table VII. As the number of degrees of freedom is 12, we first go down the outside columns, labeled df, to "12." Then we go across that row to the column headed $\chi_{0.025}^2$. The number in the body of the table there, 23.337, is the required χ^2-value; that is, for a χ^2-curve with df = 12, the χ^2-value having area 0.025 to its right is $\chi_{0.025}^2 = 23.337$, as shown in Fig. 11.2(b).

Example 11.2   Finding the χ^2-Value Having a Specified Area to Its Left

Determine the χ^2-value having area 0.05 to its left for a χ^2-curve with df = 7, as depicted in Fig. 11.3(a).

FIGURE 11.3
Finding the χ^2-value having
area 0.05 to its left

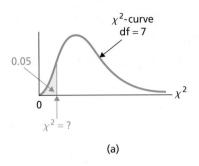

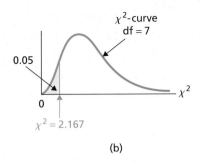

(a)   (b)

Solution As the total area under a χ^2-curve equals 1 (Property 1, Key Fact 11.1), the unshaded area in Fig. 11.3(a) must equal $1 - 0.05 = 0.95$. Thus the area to the right of the required χ^2-value is 0.95, so the required χ^2-value is $\chi^2_{0.95}$. From Table VII with df $= 7$, $\chi^2_{0.95} = 2.167$. Consequently, for a χ^2-curve with df $= 7$, the χ^2-value having area 0.05 to its left is 2.167, as depicted in Fig. 11.3(b).

Example 11.3 **Finding the χ^2-Values for a Specified Area**

For a χ^2-curve with df $= 20$, determine the two χ^2-values that divide the area under the curve into a middle 0.95 area and two outside 0.025 areas, as shown in Fig. 11.4(a).

FIGURE 11.4
Finding the two χ^2-values that divide
the area under the curve into a middle
0.95 area and two outside 0.025 areas

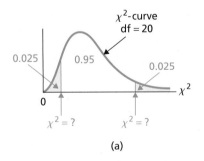

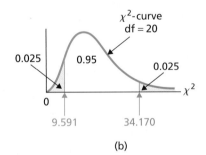

(a)   (b)

Solution First, we obtain the χ^2-value on the right in Fig. 11.4(a). As the shaded area on the right is 0.025, the χ^2-value on the right is $\chi^2_{0.025}$. From Table VII with df $= 20$, $\chi^2_{0.025} = 34.170$.

Next, we obtain the χ^2-value on the left in Fig. 11.4(a). Because the area to the left of that χ^2-value is 0.025, the area to its right is $1 - 0.025 = 0.975$. Hence the χ^2-value on the left is $\chi^2_{0.975}$, which, by Table VII, equals 9.591 for df $= 20$.

Consequently, for a χ^2-curve with df $= 20$, the two χ^2-values that divide the area under the curve into a middle 0.95 area and two outside 0.025 areas are 9.591 and 34.170, as shown in Fig. 11.4(b).

The Logic Behind Hypothesis Tests for One Population Standard Deviation

We illustrate the logic behind hypothesis tests for one population standard deviation in Example 11.4.

▌▎Example 11.4 **Hypothesis Tests for a Population Standard Deviation**

Quality Assurance A hardware manufacturer produces 10 mm bolts. The manufacturer knows that the diameters of the bolts produced vary somewhat from 10 mm and also from each other. But even if he is willing to accept some variation in bolt diameters, he cannot tolerate too much variation—if the variation is too large, too many of the bolts produced will be unusable.

Hence the manufacturer must make sure that the standard deviation, σ, of the bolt diameters is not unduly large. It has been determined that an acceptable standard deviation for the bolt diameters is one that is less than 0.09 mm.[†]

a. Formulate statistically the problem of deciding whether the standard deviation of bolt diameters is less than 0.09 mm by posing it as a hypothesis test.

b. Explain the basic idea for carrying out the hypothesis test.

TABLE 11.1
Diameters (in millimeters) of
12 randomly selected bolts

c. The diameters of 12 bolts randomly selected by the manufacturer are presented in Table 11.1. Discuss the use of these data to make a decision concerning the hypothesis test.

10.05	10.00	10.02	9.97
10.07	10.03	9.98	10.10
9.95	9.99	10.00	10.08

Solution

a. We want to perform the hypothesis test

$$H_0: \ \sigma = 0.09 \text{ mm (too much variation)}$$
$$H_a: \ \sigma < 0.09 \text{ mm (not too much variation)}.$$

If the null hypothesis can be rejected, the manufacturer can be confident that the variation in bolt diameters is acceptable.[‡]

b. Roughly speaking, the hypothesis test can be carried out in the following manner.

1. Take a random sample of bolts.
2. Compute the standard deviation, s, of the diameters of the bolts sampled.
3. If s is too much smaller than 0.09 mm, reject the null hypothesis in favor of the alternative hypothesis; otherwise, do not reject the null hypothesis.

c. The sample standard deviation of the bolt diameters in Table 11.1 is

$$s = \sqrt{\frac{\Sigma x^2 - (\Sigma x)^2/n}{n-1}} = \sqrt{\frac{1204.8290 - (120.24)^2/12}{11}} = 0.047 \text{ mm}.$$

[†]See Exercise 11.28 for an explanation of how that information could be obtained.

[‡]Instead, we could take the null hypothesis to be $H_0: \ \sigma = 0.09$ mm (not too much variation) and the alternative hypothesis to be $H_a: \ \sigma > 0.09$ mm (too much variation). Then rejection of the null hypothesis would indicate that the variation in bolt diameters is unacceptable.

Is this value of s too much smaller than 0.09 mm, suggesting that the null hypothesis be rejected? Or can the difference between $s = 0.047$ mm and the null hypothesis value of $\sigma = 0.09$ mm be attributed to sampling error? To answer these questions, we need to know the distribution of the variable s, that is, the distribution of all possible sample standard deviations that could be obtained by sampling 12 bolts. We examine that distribution and then return to complete the hypothesis test.

       ━■

The Sampling Distribution of the Sample Standard Deviation

Recall that, to perform a hypothesis test for the mean, μ, of a normally distributed variable, we use the variable

$$t = \frac{\bar{x} - \mu_0}{s/\sqrt{n}}$$

as the test statistic, not simply the variable $\bar{x}$. Similarly, when performing a hypothesis test for the standard deviation, σ, of a normally distributed variable, we do not utilize the variable s as the test statistic. Rather, we use a modified version of that variable:

$$\chi^2 = \frac{n-1}{\sigma_0^2}\, s^2.$$

This variable has a chi-square distribution, as emphasized in Key Fact 11.2 and illustrated in Example 11.5.

KEY FACT 11.2

The Sampling Distribution of the Sample Standard Deviation[†]

Suppose that a variable of a population is normally distributed with standard deviation σ. Then, for samples of size n, the variable

$$\chi^2 = \frac{n-1}{\sigma^2}\, s^2$$

has the chi-square distribution with $n-1$ degrees of freedom.

▌▐ **Example 11.5** **The Sampling Distribution of the Sample Standard Deviation**

Quality Assurance   Refer to Example 11.4. Suppose that, in reality, the bolt diameters are normally distributed with a mean of 10 mm and a standard deviation of 0.09 mm. Then, according to Key Fact 11.2, for samples of size 12, the variable χ^2 has a chi-square distribution with 11 degrees of freedom. Use simulation to make that fact plausible.

[†]Strictly speaking, the sampling distribution presented here is not the sampling distribution of the sample standard deviation but is the sampling distribution of a multiple of the sample variance (square of the sample standard deviation).

Solution We first simulated 1000 samples of 12 bolt diameters each, that is, 1000 samples of 12 observations each of a normally distributed variable with a mean of 10 and a standard deviation of 0.09. Then, for each of the 1000 samples, we determined the sample standard deviation, s, and obtained the value of the variable

$$\chi^2 = \frac{n-1}{\sigma^2}s^2 = \frac{11}{(0.09)^2}s^2.$$

PRINTOUT 11.1

Histogram of χ^2 for 1000 samples of 12 bolt diameters with the χ^2-curve for the sampling distribution superimposed

A histogram of those 1000 values of χ^2 is shown in Printout 11.1. As expected, the histogram is shaped like the superimposed χ^2-curve with df = 11.

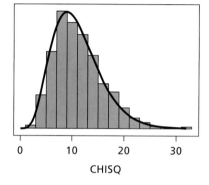

CHISQ

Hypothesis Tests for a Population Standard Deviation

In light of Key Fact 11.2, for a hypothesis test with null hypothesis H_0: $\sigma = \sigma_0$, we can use the variable

$$\chi^2 = \frac{n-1}{\sigma_0^2}s^2$$

as the test statistic and obtain the critical value(s) from the χ^2-table, Table VII. We refer to this hypothesis-testing procedure as the **χ^2-test for one population standard deviation**.

Procedure 11.1 on the next page provides a step-by-step method for performing a χ^2-test for one population standard deviation by using either the critical-value approach or the P-value approach. For the P-value approach, we could use Table VII to estimate the P-value. However, because doing so can be awkward and tedious, we assume that the user of the P-value approach has access to statistical software.

Unlike the z-tests and t-tests for one and two population means, the χ^2-test for one population standard deviation is not robust to moderate violations of the normality assumption. In fact, it is so nonrobust that many statisticians advise against its use unless there is considerable evidence that the variable under consideration is normally distributed or very nearly so.

Consequently, before applying Procedure 11.1, you should construct a normal probability plot. If the plot creates any doubt about the normality of the variable under consideration, you should not use Procedure 11.1.

Example 11.6 The χ^2-Test for a Population Standard Deviation

Quality Assurance We can now complete the hypothesis test proposed in Example 11.4. Recall that a hardware manufacturer needs to decide whether the standard deviation of bolt diameters is less than 0.09 mm. He randomly samples 12 bolts and measures their diameters. The results are displayed in Table 11.1 on page 573. At the 5% significance level, do the data provide sufficient evidence to conclude that the standard deviation of the diameters of all 10 mm bolts produced by the manufacturer is less than 0.09 mm?

| **PROCEDURE 11.1** | χ^2-Test for One Population Standard Deviation |

Purpose To perform a hypothesis test for a population standard deviation, σ

Assumptions

1. Simple random sample
2. Normal population

STEP 1 The null hypothesis is H_0: $\sigma = \sigma_0$, and the alternative hypothesis is

$$H_a: \sigma \neq \sigma_0 \quad \text{or} \quad H_a: \sigma < \sigma_0 \quad \text{or} \quad H_a: \sigma > \sigma_0$$
$$\text{(Two tailed)} \qquad \text{(Left tailed)} \qquad \text{(Right tailed)}$$

STEP 2 Decide on the significance level, α.

STEP 3 Compute the value of the test statistic

$$\chi^2 = \frac{n-1}{\sigma_0^2}s^2$$

and denote that value χ_0^2.

| **CRITICAL-VALUE APPROACH** | **P-VALUE APPROACH** |

STEP 4 The critical value(s) are

$$\chi_{1-\alpha/2}^2 \text{ and } \chi_{\alpha/2}^2 \quad \text{or} \quad \chi_{1-\alpha}^2 \quad \text{or} \quad \chi_\alpha^2$$
$$\text{(Two tailed)} \qquad \text{(Left tailed)} \qquad \text{(Right tailed)}$$

with df $= n - 1$. Use Table VII to find the critical value(s).

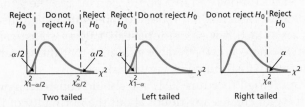

STEP 5 If the value of the test statistic falls in the rejection region, reject H_0; otherwise, do not reject H_0.

STEP 4 The χ^2-statistic has df $= n - 1$. Obtain the P-value by using technology.

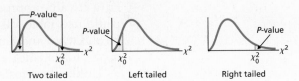

STEP 5 If $P \leq \alpha$, reject H_0; otherwise, do not reject H_0.

STEP 6 Interpret the results of the hypothesis test.

Solution To begin, we construct a normal probability plot for the data in Table 11.1, which is quite linear, as shown in Fig. 11.5. These and other data on bolt diameters collected previously by the manufacturer indicate that we can reasonably presume that the diameters of 10 mm bolts produced by this manufacturer are normally distributed. Thus we can apply Procedure 11.1 to perform the required hypothesis test.

FIGURE 11.5

Normal probability plot for the sample of bolt diameters in Table 11.1

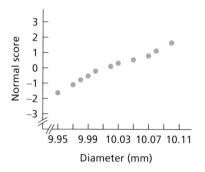

STEP 1 State the null and alternative hypotheses.

Let σ denote the population standard deviation of bolt diameters. Then the null and alternative hypotheses are

$$H_0: \ \sigma = 0.09 \text{ mm (too much variation)}$$
$$H_a: \ \sigma < 0.09 \text{ mm (not too much variation).}$$

Note that the hypothesis test is left tailed because a less-than sign ($<$) appears in the alternative hypothesis.

STEP 2 Decide on the significance level, α.

The test is to be performed at the 5% level of significance, or $\alpha = 0.05$.

STEP 3 Compute the value of the test statistic

$$\chi^2 = \frac{n-1}{\sigma_0^2} s^2.$$

First, we obtain the sample variance, s^2. From Table 11.1,

$$s^2 = \frac{\Sigma x^2 - (\Sigma x)^2 / n}{n-1} = \frac{1204.8290 - (120.24)^2 / 12}{11} = 0.0022.$$

As $n = 12$ and $\sigma_0 = 0.09$, the value of the test statistic is

$$\chi^2 = \frac{n-1}{\sigma_0^2} s^2 = \frac{12-1}{(0.09)^2} \cdot 0.0022 = 2.988.$$

CRITICAL-VALUE APPROACH	P-VALUE APPROACH

CRITICAL-VALUE APPROACH

STEP 4 The critical value for a left-tailed test is $\chi^2_{1-\alpha}$, with df = $n-1$. Use Table VII to find the critical value.

We have $\alpha = 0.05$. Also, $n = 12$, so df = $12 - 1 = 11$. In Table VII, we find that the critical value is $\chi^2_{1-\alpha} = \chi^2_{1-0.05} = \chi^2_{0.95} = 4.575$, as shown in Fig. 11.6A.

FIGURE 11.6A

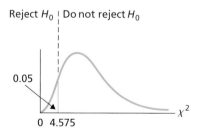

Reject H_0 | Do not reject H_0

0.05

0   4.575

STEP 5 If the value of the test statistic falls in the rejection region, reject H_0; otherwise, do not reject H_0.

From Step 3, the value of the test statistic is $\chi^2 = 2.988$, which falls in the rejection region shown in Fig. 11.6A. Thus we reject H_0. The test results are statistically significant at the 5% level.

P-VALUE APPROACH

STEP 4 The χ^2-statistic has df = $n-1$. Obtain the P-value by using technology.

For $n = 12$, df = $12 - 1 = 11$. Using technology, we find that the P-value for the hypothesis test is $P = 0.00911$, as shown in Fig. 11.6B.

FIGURE 11.6B

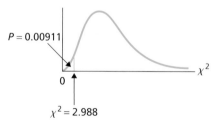

$P = 0.00911$

0

$\chi^2 = 2.988$

STEP 5 If $P \leq \alpha$, reject H_0; otherwise, do not reject H_0.

From Step 4, $P = 0.0091$. Because the P-value is less than the specified significance level of 0.05, we reject H_0. The test results are statistically significant at the 5% level and (see Table 9.12 on page 440) provide very strong evidence against the null hypothesis.

STEP 6 **Interpret the results of the hypothesis test.**

INTERPRETATION At the 5% significance level, the data provide sufficient evidence to conclude that the standard deviation, σ, of the diameters of all 10 mm bolts produced by the manufacturer is less than 0.09 mm. Evidently, the variation in bolt diameters is not too large. ▬■

Confidence Intervals for a Population Standard Deviation

Using Key Fact 11.2 on page 574, we can also obtain a confidence-interval procedure, Procedure 11.2, for a population standard deviation. This procedure is often called the χ^2-**interval procedure for a population standard deviation.**

PROCEDURE 11.2

χ^2-Interval Procedure

Purpose To find a confidence interval for a population standard deviation, σ

Assumptions
1. Simple random sample
2. Normal population

STEP 1 **For a confidence level of $1 - \alpha$, use Table VII to find $\chi^2_{1-\alpha/2}$ and $\chi^2_{\alpha/2}$ with df = $n - 1$.**

STEP 2 **The confidence interval for σ is from**

$$\sqrt{\frac{n-1}{\chi^2_{\alpha/2}}} \cdot s \quad \text{to} \quad \sqrt{\frac{n-1}{\chi^2_{1-\alpha/2}}} \cdot s,$$

where $\chi^2_{1-\alpha/2}$ and $\chi^2_{\alpha/2}$ are found in Step 1, n is the sample size, and s is computed from the sample data obtained.

STEP 3 **Interpret the confidence interval.**

Like the χ^2-test for one population standard deviation, the χ^2-interval procedure is not at all robust to violations of the normality assumption: Misleading information may be obtained if the procedure is used on data from a variable that is not normally distributed. In other words, the χ^2-interval procedure should not be used unless there is considerable evidence that the variable under consideration is normally distributed or very nearly so. Example 11.7 illustrates application of this procedure.

▌▌ **Example 11.7** **The χ^2-Interval Procedure**

Quality Assurance Use the sample data in Table 11.1 on page 573 to determine a 95% confidence interval for the standard deviation, σ, of the diameters of all 10 mm bolts produced by the manufacturer.

Solution   As we discovered in Example 11.6, we can reasonably presume that the diameters of 10 mm bolts produced by the manufacturer are normally distributed. Hence we can apply Procedure 11.2 to obtain the required confidence interval.

STEP 1 For a confidence level of $1 - \alpha$, use Table VII to find $\chi^2_{1-\alpha/2}$ and $\chi^2_{\alpha/2}$ with df $= n - 1$.

For a 95% confidence interval, the confidence level is $0.95 = 1 - 0.05$, and so $\alpha = 0.05$. Also, for $n = 12$, df $= 11$. In Table VII, we find that

$$\chi^2_{1-\alpha/2} = \chi^2_{1-0.05/2} = \chi^2_{0.975} = 3.816$$

and

$$\chi^2_{\alpha/2} = \chi^2_{0.05/2} = \chi^2_{0.025} = 21.920.$$

STEP 2 The confidence interval for σ is from

$$\sqrt{\frac{n-1}{\chi^2_{\alpha/2}}} \cdot s \quad \text{to} \quad \sqrt{\frac{n-1}{\chi^2_{1-\alpha/2}}} \cdot s.$$

We have $n = 12$, and from Step 1, $\chi^2_{1-\alpha/2} = 3.816$ and $\chi^2_{\alpha/2} = 21.920$. Also, we found in Example 11.4 that $s = 0.047$ mm. So, a 95% confidence interval for σ is from

$$\sqrt{\frac{12-1}{21.920}} \cdot 0.047 \quad \text{to} \quad \sqrt{\frac{12-1}{3.816}} \cdot 0.047,$$

or 0.033 to 0.080.

STEP 3 Interpret the confidence interval.

INTERPRETATION   We can be 95% confident that the standard deviation, σ, of the diameters of all 10 mm bolts produced by the manufacturer is somewhere between 0.033 mm and 0.080 mm.

The Technology Center

Some statistical technologies have dedicated programs that automatically perform a χ^2-test and obtain a χ^2-interval for one population standard deviation, but others do not. For a statistical technology that does not have such dedicated programs, you can often use its macro capabilities to write them. Refer to the technology manuals for more details.

EXERCISES 11.1

Statistical Concepts and Skills

11.1 What is meant by saying that a variable has a chi-square distribution?

11.2 How are different chi-square distributions identified?

11.3 Two χ^2-curves have degrees of freedom 12 and 20, respectively. Which curve more closely resembles a normal curve? Explain your answer.

11.4 The t-table has entries for areas of 0.10, 0.05, 0.025, 0.01, and 0.005. In contrast, the χ^2-table has entries for those areas and for 0.995, 0.99, 0.975, 0.95, and 0.90. Explain why the t-values corresponding to these additional areas can be obtained from the existing t-table, but must be provided explicitly in the χ^2-table.

*In Exercises **11.5–11.12**, use Table VII to find the required χ^2-values. Illustrate your work graphically.*

11.5 For a χ^2-curve with 19 degrees of freedom, find the χ^2-value that has area
a. 0.025 to its right.   **b.** 0.95 to its right.

11.6 For a χ^2-curve with 22 degrees of freedom, find the χ^2-value that has area
a. 0.01 to its right.   **b.** 0.995 to its right.

11.7 For a χ^2-curve with df $= 10$, determine
a. $\chi^2_{0.05}$.   **b.** $\chi^2_{0.975}$.

11.8 For a χ^2-curve with df $= 4$, determine
a. $\chi^2_{0.005}$.   **b.** $\chi^2_{0.99}$.

11.9 Consider a χ^2-curve with df $= 8$. Obtain the χ^2-value that has area
a. 0.01 to its left.   **b.** 0.95 to its left.

11.10 Consider a χ^2-curve with df $= 16$. Obtain the χ^2-value that has area
a. 0.025 to its left.   **b.** 0.975 to its left.

11.11 Determine the two χ^2-values that divide the area under the curve into a middle 0.95 area and two outside 0.025 areas for a χ^2-curve with
a. df $= 5$.   **b.** df $= 26$.

11.12 Determine the two χ^2-values that divide the area under the curve into a middle 0.90 area and two outside 0.05 areas for a χ^2-curve with
a. df $= 11$.   **b.** df $= 28$.

11.13 When you use chi-square procedures to make inferences about a population standard deviation, why should the variable under consideration be normally distributed or nearly so?

11.14 Give two situations in which making an inference about a population standard deviation would be important.

*Preliminary data analyses and other information indicate that you can reasonably assume that the variables under consideration in Exercises **11.15–11.20** are normally distributed. In each case, use either the critical-value approach or the P-value approach to perform the required hypothesis test.*

11.15 Body Temperature. A study by researchers at the University of Maryland addressed the question of whether the mean body temperature of humans is 98.6°F. The results of the study by P. Mackowiak, S. Wasserman, and M. Levine appeared in the article "A Critical Appraisal of 98.6°F, the

Upper Limit of the Normal Body Temperature, and Other Legacies of Carl Reinhold August Wunderlich" (*Journal of the American Medical Association*, Vol. 268, pp. 1578–1580). Among other data, the researchers obtained the following body temperatures of 93 healthy humans.

98.0	97.6	98.8	98.0	98.8	98.8	97.6	98.6	98.6
98.8	98.0	98.2	98.0	98.0	97.0	97.2	98.2	98.1
98.2	98.5	98.5	99.0	98.0	97.0	97.3	97.3	98.1
97.8	99.0	97.6	97.4	98.0	97.4	98.0	98.6	98.6
98.4	97.0	98.4	99.0	98.0	99.4	97.8	98.2	99.2
99.0	97.7	98.2	98.2	98.8	98.1	98.5	97.2	98.5
99.2	98.3	98.7	98.8	98.6	98.0	99.1	97.2	97.6
97.9	98.8	98.6	98.6	99.3	97.8	98.7	99.3	97.8
98.4	97.7	98.3	97.7	97.1	98.4	98.6	97.4	96.7
96.9	98.4	98.2	98.6	97.0	97.4	98.4	97.4	96.8
98.2	97.4	98.0						

In Exercise 9.39, you were asked to use these data to decide whether mean body temperature of healthy humans differs from 98.6°F. There, you were to assume that the population standard deviation of body temperatures for healthy humans is 0.63°F. At the 5% significance level, do the data provide sufficient evidence to conclude that the population standard deviation of body temperatures for healthy humans differs from 0.63°F? (*Note:* The sample standard deviation of the 93 temperatures is 0.647°F.)

11.16 EPA Gas Mileage Estimates. Gas mileage estimates for cars and light-duty trucks are determined and published by the Environmental Protection Agency (EPA). According to the EPA, "…the mileages obtained by most drivers will be within plus or minus 15 percent of the [EPA] estimates…." The mileage estimate given for one 2000 model is 23 mpg on the highway. If the EPA claim is true, the standard deviation of mileages should be about $0.15 \cdot 23/3 = 1.15$ mpg. A random sample of 12 cars of this model yields the following highway mileages.

24.1	23.3	22.5	23.2
22.3	21.1	21.4	23.4
23.5	22.8	24.5	24.3

At the 5% significance level, do the data suggest that the standard deviation of highway mileages for all 1998 cars of this model is different from 1.15 mpg? (*Note:* $s = 1.071$.)

11.17 Process Capability. R. Morris and E. Watson studied various aspects of process capability in the paper "Determining Process Capability in a Chemical Batch Process" (*Quality Engineering*, Vol. 10(2), pp. 389–396). In one part of the study, the researchers compared the variability in product of a particular piece of equipment to a known analytic capability to decide whether product consistency could be improved. The following data were obtained for 10 batches of product.

30.1	30.7	30.2	29.3	31.0
29.6	30.4	31.2	28.8	29.8

At the 1% significance level, do the data provide sufficient evidence to conclude that the process variation for this piece of equipment exceeds the analytic capability of 0.27? (*Note:* $s = 0.756$.)

11.18 Premade Pizza. Homestyle Pizza of Camp Verde, Arizona, provides baking instructions for its premade pizzas. According to the instructions, the average baking time is 12 to 18 minutes. If the times are normally distributed, the standard deviation of the times should be approximately 1 minute. A random sample of 15 pizzas yielded the following baking times to the nearest tenth of a minute.

15.4	15.1	14.0	15.8	16.0
13.7	15.6	11.6	14.8	12.8
17.6	15.1	16.4	13.1	15.3

At the 1% significance level, do the data provide sufficient evidence to conclude that the standard deviation of baking times exceeds 1 minute? (*Note:* The sample standard deviation of the 15 baking times is 1.54 minutes.)

11.19 Dispensing Coffee. A coffee machine is supposed to dispense 6 fluid ounces (fl oz) of coffee into a paper cup. In reality, the amounts dispensed vary from cup to cup. However, if the machine is working properly, most of the cups will contain within 10% of the advertised 6 fl oz. In other words, the standard deviation of the amounts dispensed should be less than 0.2 fl oz. A random sample of 15 cups provided the following data, in fluid ounces.

5.90	5.82	6.20	6.09	5.93
6.18	5.99	5.79	6.28	6.16
6.00	5.85	6.13	6.09	6.18

At the 5% significance level, do the data provide sufficient evidence to conclude that the standard deviation of the amounts being dispensed is less than 0.2 fl oz? (*Note:* $s = 0.154$.)

11.20 Counting Production. In Issue 10 of *STATS* from Iowa State University, data were published from an exper-

iment that examined the effects of machine adjustment on bolt production. An electronic counter records the number of bolts passing it on a conveyer belt and stops the run when the count reaches a preset number. The following data give the times, in seconds, needed to count 20 bolts for eight different runs.

10.78	9.39	9.84	13.94
12.33	7.32	7.91	15.58

Do the data provide sufficient evidence to conclude that the standard deviation in the time needed to count 20 bolts exceeds 2 seconds? Use $\alpha = 0.05$. (*Note:* The sample standard deviation of the eight times is 2.8875 seconds.)

*In Exercises **11.21–11.24**, use Procedure 11.2 on page 578 to obtain the required confidence interval.*

11.21 Body Temperature. Refer to Exercise 11.15. Determine a 95% confidence interval for the population standard deviation of body temperatures for healthy humans.

11.22 EPA Gas Mileage Estimates. Refer to Exercise 11.16. Find a 95% confidence interval for the standard deviation of highway gas mileages for all cars of the model and year in question.

11.23 Process Capability. Refer to Exercise 11.17. Determine a 98% confidence interval for the process variation of the piece of equipment under consideration.

11.24 Premade Pizza. Refer to Exercise 11.18. Obtain a 98% confidence interval for the standard deviation of baking times.

Extending the Concepts and Skills

11.25 Dispensing Coffee. Refer to Exercise 11.19. Why is it important that the standard deviation of the amounts of coffee being dispensed not be too large?

11.26 EPA Gas Mileage Estimates. Refer to Exercises 11.16 and 11.22. Why is it useful to know the standard deviation of the gas mileages as well as the mean gas mileage?

11.27 Quality Assurance. Refer to Example 11.4 on page 573. In that example, we chose the null hypothesis to be $H_0: \sigma = 0.09$ (too much variation) and the alternative hypothesis to be $H_a: \sigma < 0.09$ (not too much variation). Alternatively, we could take the null hypothesis to be $H_0: \sigma = 0.09$ (not too much variation) and the alternative hypothesis to be $H_a: \sigma > 0.09$ (too much variation). Explain the advantages and disadvantages of the two different choices of null and alternative hypotheses.

11.28 Quality Assurance. In Example 11.4 on page 573, we assumed that an acceptable standard deviation, σ, for the bolt diameters is less than 0.09 mm. We now help you show how such information might be obtained. Let's suppose that the manufacturer has set the tolerance specifications for the 10 mm bolts at ± 0.3 mm; that is, a bolt's diameter is considered satisfactory if it is between 9.7 mm and 10.3 mm. Further suppose that the manufacturer has decided that less than 0.1% (1 out of 1000) of the bolts produced should be defective.

a. Let X denote the diameter of a randomly selected bolt. Show that the manufacturer's production criteria can be expressed mathematically as $P(9.7 \leq X \leq 10.3) > 0.999$.

b. Draw a normal-curve figure that illustrates the equation $P(9.7 \leq X \leq 10.3) = 0.999$. Include both an x-axis and a z-axis. Assume that $\mu = 10$ mm.

c. Deduce from your figure in part (b) that the manufacturer's production criteria are equivalent to the condition that $0.3/\sigma > z_{0.0005}$.

d. Use part (c) to conclude that the manufacturer's production criteria are equivalent to requiring that the standard deviation of bolt diameters be less than 0.09 mm, that is, $\sigma < 0.09$ mm.

Using Technology

11.29 Body Temperature. Refer to Exercises 11.15 and 11.21. Use the technology of your choice to

a. obtain both a normal probability plot and a boxplot of the data.

b. perform the required hypothesis test and obtain the desired confidence interval.

c. Justify the use of your procedures in part (b).

11.30 EPA Gas Mileage Estimates. Refer to Exercises 11.16 and 11.22. Use the technology of your choice to

a. obtain both a normal probability plot and a boxplot of the data.

b. perform the required hypothesis test and obtain the desired confidence interval.

c. Justify the use of your procedures in part (b).

11.31 Intelligence Quotients. Measured on the Stanford Revision of the Binet–Simon Intelligence Scale, intelligence quotients (IQs) are known to be normally distributed with a mean of 100 and a standard deviation of 16. Use the technology of your choice to:

a. Simulate 1000 samples of four IQs each.

b. Determine the sample standard deviation of each of the 1000 samples.

c. Obtain the following quantity for each of the 1000 samples.

$$\frac{n-1}{\sigma^2} s^2 = \frac{4-1}{16^2} s^2$$

d. Obtain a histogram of the 1000 values found in part (c).

e. Theoretically, what is the distribution of the variable in part (c)?

f. Compare your answers from parts (d) and (e).

11.2 Inferences for Two Population Standard Deviations, Using Independent Samples

In Section 11.1, we discussed hypothesis tests and confidence intervals for one population standard deviation. We now introduce hypothesis tests and confidence intervals for two population standard deviations. More precisely, we examine inferences to compare the standard deviations of one variable of two different populations. Such inferences are based on a distribution called the *F-distribution*, named in honor of Sir Ronald Fisher. (See the Biography on page 846 for more on Fisher.)

The *F*-Distribution

A variable is said to have an **F-distribution** if its distribution has the shape of a special type of right-skewed curve, called an **F-curve**. Actually, there are infinitely many F-distributions, and we identify the F-distribution (and F-curve) in question by stating its number of degrees of freedom, just as we did for t-distributions and chi-square distributions.

But an *F*-distribution has two numbers of degrees of freedom instead of one. Figure 11.7 depicts two different *F*-curves; one has df = (10, 2), and the other has df = (9, 50).

FIGURE 11.7

Two different *F*-curves

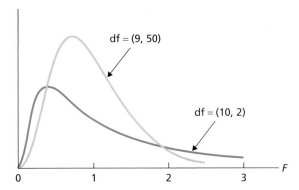

The first number of degrees of freedom for an *F*-curve is called the **degrees of freedom for the numerator** and the second the **degrees of freedom for the denominator.** (The reason for this terminology will become clear shortly.) Thus, for the *F*-curve in Fig. 11.7 with df = (10, 2), we have

$$df = (10, 2)$$

Degrees of freedom↗   ↖Degrees of freedom
for the numerator   for the denominator

Some basic properties of *F*-curves are presented in Key Fact 11.3.

KEY FACT 11.3

Basic Properties of F-Curves

Property 1: The total area under an *F*-curve equals 1.

Property 2: An *F*-curve starts at 0 on the horizontal axis and extends indefinitely to the right, approaching, but never touching, the horizontal axis as it does so.

Property 3: An *F*-curve is right skewed.

Percentages (and probabilities) for a variable having an *F*-distribution equal areas under its associated *F*-curve. To perform a hypothesis test or construct a confidence interval for comparing two population standard deviations, we need to know how to find the *F*-value that corresponds to a specified area under an *F*-curve. The symbol F_α is used to denote the *F*-value having area α to its right.

Table VIII in Appendix A provides *F*-values having areas 0.005, 0.01, 0.025, 0.05, and 0.10 to their right for various degrees of freedom. The degrees of freedom for the denominator (dfd) are displayed in the outside columns of the table; the values of α are in the next columns; and the degrees of freedom for the numerator (dfn) are along the top. In Examples 11.8–11.10, we show how to use Table VIII.

Example 11.8   Finding the F-Value Having a Specified Area to Its Right

For an F-curve with df $= (4, 12)$, find $F_{0.05}$; that is, find the F-value having area 0.05 to its right, as shown in Fig. 11.8(a).

FIGURE 11.8

Finding the F-value having area 0.05 to its right

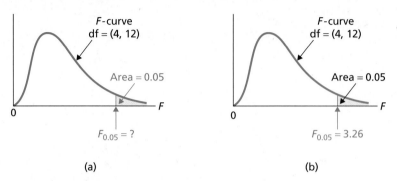

(a)   (b)

Solution   To obtain the F-value, we use Table VIII. In this case, $\alpha = 0.05$, the degrees of freedom for the numerator is 4, and the degrees of freedom for the denominator is 12.

We first go down the dfd column to "12." Next, we concentrate on the row for α labeled 0.05. Then we go across that row to the column headed "4." The number in the body of the table there, 3.26, is the required F-value; that is, for an F-curve with df $= (4, 12)$, the F-value having area 0.05 to its right is 3.26: $F_{0.05} = 3.26$, as shown in Fig. 11.8(b).

In many statistical analyses that involve the F-distribution, we also need to determine F-values having areas 0.005, 0.01, 0.025, 0.05, and 0.10 to their left. Although such F-values aren't available directly from Table VIII, we can obtain them indirectly from the table by using Key Fact 11.4, which we also illustrate in Examples 11.9 and 11.10.

KEY FACT 11.4

Reciprocal Property of F-Curves

For an F-curve with df $= (\nu_1, \nu_2)$, the F-value having area α to its left equals the reciprocal of the F-value having area α to its right for an F-curve with df $= (\nu_2, \nu_1)$.

Example 11.9   Finding the F-Value Having a Specified Area to Its Left

For an F-curve with df $= (60, 8)$, find the F-value having area 0.05 to its left.

Solution   We apply Key Fact 11.4. Accordingly, the required F-value is the reciprocal of the F-value having area 0.05 to its right for an F-curve with df $= (8, 60)$. From Table VIII, this latter F-value equals 2.10. Consequently, the required F-value is $\frac{1}{2.10}$, or 0.48, as shown in Fig. 11.9.

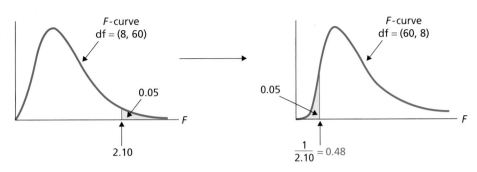

FIGURE 11.9
Finding the *F*-value having
area 0.05 to its left

Example 11.10   Finding the F-Values for a Specified Area

For an *F*-curve with df = (9, 8), determine the two *F*-values that divide the area under the curve into a middle 0.95 area and two outside 0.025 areas, as shown in Fig. 11.10(a).

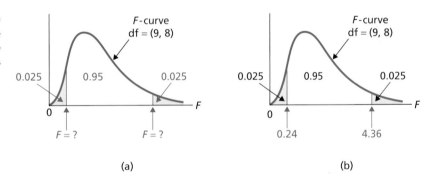

FIGURE 11.10
Finding the two *F*-values that divide the
area under the curve into a middle
0.95 area and two outside 0.025 areas

Solution   First, we obtain the *F*-value on the right in Fig. 11.10(a), which has area 0.025 to its right. In other words, the *F*-value on the right is $F_{0.025}$. From Table VIII with df = (9, 8), $F_{0.025} = 4.36$.

Next, we obtain the *F*-value on the left in Fig. 11.10(a). By Key Fact 11.4, that *F*-value is the reciprocal of the *F*-value having area 0.025 to its right for an *F*-curve with df = (8, 9). From Table VIII, we find that this latter *F*-value equals 4.10. Thus the *F*-value on the left in Fig. 11.10(a) is $\frac{1}{4.10}$, or 0.24.

Consequently, for an *F*-curve with df = (9, 8), the two *F*-values that divide the area under the curve into a middle 0.95 area and two outside 0.025 areas are 0.24 and 4.36, as shown in Fig. 11.10(b).

The Logic Behind Hypothesis Tests for Comparing Two Population Standard Deviations

We illustrate the logic behind hypothesis tests for comparing two population standard deviations in Example 11.11.

▌▌Example 11.11   Hypothesis Tests for Two Population Standard Deviations

Elmendorf Tear Strength   Variation within a method used for testing a product is an essential factor in deciding whether the method should be employed. Indeed, when the variation of such a test is high, ascertaining the true quality of a product is difficult.

In the article "Using Repeatability and Reproducibility Studies to Evaluate a Destructive Test Method" (*Quality Engineering*, Vol. 10(2), pp. 283–290), A. Phillips et al. studied the variability of the Elmendorf tear test. That test is used to evaluate material strength for fiberglass shingles, paper quality, and other manufactured products.

In one aspect of the study, the researchers independently and randomly obtained data on Elmendorf tear strength of three different vinyl floor coverings. Table 11.2 provides the data, in grams, for two of the three vinyl floor coverings.

Suppose that we want to decide whether the standard deviations of tear strength differ between the two vinyl floor coverings.

TABLE 11.2

Results of Elmendorf tear test on two different vinyl floor coverings (data in grams)

Brand A		Brand B	
2288	2384	2592	2384
2368	2304	2512	2432
2528	2240	2576	2112
2144	2208	2176	2288
2160	2112	2304	2752

a. Formulate the problem statistically by posing it as a hypothesis test.

b. Explain the basic idea for carrying out the hypothesis test.

c. Discuss the use of the data in Table 11.2 to make a decision concerning the hypothesis test.

Solution

a. We want to perform the hypothesis test

$$H_0: \sigma_1 = \sigma_2 \text{ (standard deviations of tear strength are the same)}$$
$$H_a: \sigma_1 \neq \sigma_2 \text{ (standard deviations of tear strength are different)}$$

where σ_1 and σ_2 denote the population standard deviations of tear strength for Brand A and Brand B, respectively.

b. We can carry out the hypothesis test by comparing the sample standard deviations, s_1 and s_2, of the two sets of sample data presented in Table 11.2. Such a comparison could be made in any of several ways. Here, we make it by looking at the square of the ratio of s_1 to s_2, or equivalently, the quotient of the sample variances. That statistic is called the **F-statistic.**

If the population standard deviations, σ_1 and σ_2, are equal, the sample standard deviations, s_1 and s_2, should be roughly the same. This property, in turn, implies that the value of the *F*-statistic should be close to 1. When the value of the *F*-statistic differs from 1 by too much, it provides evidence against the null hypothesis of equal population standard deviations.

c. For the data in Table 11.2, we have $s_1 = 128.3$ g and $s_2 = 199.7$ g. Thus the value of the *F*-statistic is

$$F = \frac{s_1^2}{s_2^2} = \frac{128.3^2}{199.7^2} = 0.413.$$

Does this value of *F* differ from 1 by enough to conclude that the null hypothesis of equal population standard deviations is false? To answer that question, we need to know the distribution of the *F*-statistic. We discuss that distribution and then return to complete the hypothesis test. ■

The Distribution of the *F*-Statistic

To perform hypothesis tests and obtain confidence intervals for two population standard deviations, we need Key Fact 11.5, which we illustrate in Example 11.12.

KEY FACT 11.5

Distribution of the F-Statistic for Comparing Two Population Standard Deviations

Suppose that the variable under consideration is normally distributed on each of two populations. Then, for independent samples of sizes n_1 and n_2 from the two populations, the variable

$$F = \frac{s_1^2/\sigma_1^2}{s_2^2/\sigma_2^2}$$

has the *F*-distribution with df $= (n_1 - 1, n_2 - 1)$.

▌▌Example 11.12   The Distribution of the F-Statistic

Elmendorf Tear Strength Suppose that the Elmendorf tear strengths for Brand A vinyl floor covering are normally distributed with mean 2275 g and standard deviation 132 g, and that those for Brand B vinyl floor covering are normally distributed with mean 2405 g and standard deviation 194 g. Then, for independent random samples of sizes 10 and 10 from Brand A and Brand B, the variable *F* in Key Fact 11.5 has the *F*-distribution with df $= (9, 9)$. Use simulation to make that fact plausible.

PRINTOUT 11.2

Histogram of *F* for 1000 independent samples with the corresponding *F*-curve superimposed

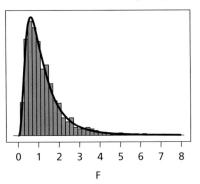

Solution We first simulated 1000 samples of 10 tear strengths each for Brand A vinyl floor covering, that is, 1000 samples of 10 observations each of a normally distributed variable with mean 2275 and standard deviation 132. Next, we simulated 1000 samples of 10 tear strengths each for Brand B vinyl floor covering, that is, 1000 samples of 10 observations each of a normally distributed variable with mean 2405 and standard deviation 194. Then, for each of the 1000 pairs of samples from the two brands, we determined the sample standard deviations, s_1 and s_2, and obtained the value of the variable

$$F = \frac{s_1^2/\sigma_1^2}{s_2^2/\sigma_2^2} = \frac{s_1^2/132^2}{s_2^2/194^2}.$$

As expected, the histogram of those 1000 values of *F*, shown in Printout 11.2, is shaped like the superimposed *F*-curve with df $= (9, 9)$. ■

Hypothesis Tests for Two Population Standard Deviations

In light of Key Fact 11.5, for a hypothesis test with null hypothesis H_0: $\sigma_1 = \sigma_2$ (population standard deviations are equal), we can use the variable

$$F = \frac{s_1^2}{s_2^2}$$

as the test statistic and obtain the critical value(s) from the F-table, Table VIII. We refer to this hypothesis-testing procedure as the **F-test for two population standard deviations,** also known as the *F-test for two population variances.*

Procedure 11.3 on the next page provides a step-by-step method for performing an F-test for two population standard deviations by using either the critical-value approach or the P-value approach. For the P-value approach, we could use Table VIII to estimate the P-value. However, because doing so can be awkward and tedious, we assume that the user of the P-value approach has access to statistical software.

Unlike the z-tests and t-tests for one and two population means, the F-test for two population standard deviations is not robust to moderate violations of the normality assumption. In fact, it is so nonrobust that many statisticians advise against its use unless there is considerable evidence that the variable under consideration is normally distributed, or very nearly so, on each population.

Consequently, before applying Procedure 11.3, you should construct a normal probability plot of each sample. If either plot creates any doubt about the normality of the variable under consideration, then you should not use Procedure 11.3. Example 11.13 illustrates application of the procedure.

▌▌ Example 11.13 The F-Test for Two Population Standard Deviations

Elmendorf Tear Strength We can now complete the hypothesis test proposed in Example 11.11. Independent random samples of two vinyl floor coverings yield the data on Elmendorf tear strength shown in Table 11.2 on page 586. At the 5% significance level, do the data provide sufficient evidence to conclude that the population standard deviations of tear strength differ for the two vinyl floor coverings?

Solution To begin, we construct normal probability plots (not shown) for the two samples in Table 11.2. The plots suggest that we can reasonably presume that tear strength is normally distributed for each brand of vinyl flooring. Hence we can apply Procedure 11.3 to perform the required hypothesis test.

STEP 1 State the null and alternative hypotheses.

Let σ_1 and σ_2 denote the population standard deviations of tear strength for Brand A and Brand B, respectively. Then the null and alternative hypotheses are

H_0: $\sigma_1 = \sigma_2$ (standard deviations of tear strength are the same)

H_a: $\sigma_1 \neq \sigma_2$ (standard deviations of tear strength are different).

| PROCEDURE 11.3 | *F*-Test for Two Population Standard Deviations |

Purpose To perform a hypothesis test to compare two population standard deviations, σ_1 and σ_2

Assumptions

1. Simple random samples
2. Independent samples
3. Normal populations

STEP 1 The null hypothesis is H_0: $\sigma_1 = \sigma_2$, and the alternative hypothesis is

$$H_a: \sigma_1 \neq \sigma_2 \quad \text{or} \quad H_a: \sigma_1 < \sigma_2 \quad \text{or} \quad H_a: \sigma_1 > \sigma_2$$
$$\text{(Two tailed)} \qquad \text{(Left tailed)} \qquad \text{(Right tailed)}$$

STEP 2 Decide on the significance level, α.

STEP 3 Compute the value of the test statistic

$$F = \frac{s_1^2}{s_2^2}$$

and denote that value F_0.

| CRITICAL-VALUE APPROACH | P-VALUE APPROACH |

STEP 4 The critical value(s) are

$$F_{1-\alpha/2} \text{ and } F_{\alpha/2} \quad \text{or} \quad F_{1-\alpha} \quad \text{or} \quad F_\alpha$$
$$\text{(Two tailed)} \qquad \text{(Left tailed)} \qquad \text{(Right tailed)}$$

with df $= (n_1 - 1, n_2 - 1)$. Use Table VIII to find the critical value(s).

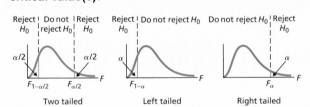

STEP 5 If the value of the test statistic falls in the rejection region, reject H_0; otherwise, do not reject H_0.

STEP 4 The *F*-statistic has df $= (n_1 - 1, n_2 - 1)$. Obtain the *P*-value by using technology.

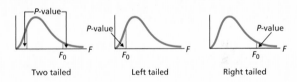

STEP 5 If $P \leq \alpha$, reject H_0; otherwise, do not reject H_0.

STEP 6 Interpret the results of the hypothesis test.

Note that the hypothesis test is two tailed because a does-not-equal sign ($\neq$) appears in the alternative hypothesis.

STEP 2 Decide on the significance level, α.

The test is to be performed at the 5% level of significance, or $\alpha = 0.05$.

STEP 3 Compute the value of the test statistic $F = s_1^2/s_2^2$.

We computed the value of the test statistic at the end of Example 11.11, where we found that $F = 0.413$.

CRITICAL-VALUE APPROACH	P-VALUE APPROACH

CRITICAL-VALUE APPROACH

STEP 4 The critical values for a two-tailed test are $F_{1-\alpha/2}$ and $F_{\alpha/2}$ with df $= (n_1 - 1, n_2 - 1)$. Use Table VIII to find the critical values.

We have $\alpha = 0.05$. Also, $n_1 = 10$ and $n_2 = 10$, so df $= (9, 9)$. Therefore the critical values are $F_{1-\alpha/2} = F_{1-0.05/2} = F_{0.975}$ and $F_{\alpha/2} = F_{0.05/2} = F_{0.025}$. From Table VIII, $F_{0.025} = 4.03$. To obtain $F_{0.975}$, we first note that it is the F-value having area 0.025 to its left. Applying the reciprocal property of F-curves (see page 584), we conclude that $F_{0.975}$ equals the reciprocal of the F-value having area 0.025 to its right for an F-curve with df $= (9, 9)$. (We switched the degrees of freedom, but as they are the same, the difference isn't apparent.) Thus $F_{0.975} = \frac{1}{4.03} = 0.25$. Figure 11.11A summarizes our results.

FIGURE 11.11A

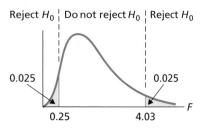

Reject H_0 | Do not reject H_0 | Reject H_0

0.025   0.025

0.25   4.03 F

STEP 5 If the value of the test statistic falls in the rejection region, reject H_0; otherwise, do not reject H_0.

From Step 3, the value of the test statistic is $F = 0.413$. This value does not fall in the rejection region shown in Fig. 11.11A. Consequently, we do not reject H_0. The test results are not statistically significant at the 5% level.

P-VALUE APPROACH

STEP 4 The F-statistic has df $= (n_1 - 1, n_2 - 1)$. Obtain the P-value by using technology.

We have $n_1 = 10$ and $n_2 = 10$, so df $= (9, 9)$. Using technology, we find that the P-value for the hypothesis test is $P = 0.2039$, as depicted in Fig. 11.11B.

FIGURE 11.11B

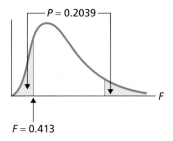

$P = 0.2039$

$F = 0.413$

STEP 5 If $P \leq \alpha$, reject H_0; otherwise, do not reject H_0.

From Step 4, $P = 0.2039$. Because the P-value exceeds the specified significance level of 0.05, we do not reject H_0. The test results are not statistically significant at the 5% level and (see Table 9.12 on page 440) provide at most weak evidence against the null hypothesis.

STEP 6 Interpret the results of the hypothesis test.

INTERPRETATION At the 5% significance level, the data do not provide sufficient evidence to conclude that the population standard deviations of tear strength differ for the two vinyl floor coverings.

Confidence Intervals for Two Population Standard Deviations

Using Key Fact 11.5 on page 587, we can also obtain a confidence-interval procedure, Procedure 11.4, for the ratio of two population standard deviations. This procedure is sometimes called the **F-interval procedure for two population standard deviations.**

PROCEDURE 11.4

F-Interval Procedure

Purpose To find a confidence interval for the ratio of two population standard deviations, σ_1 and σ_2

Assumptions

1. Simple random samples
2. Independent samples
3. Normal populations

STEP 1 For a confidence level of $1 - \alpha$, use Table VIII to find $F_{1-\alpha/2}$ and $F_{\alpha/2}$ with df $= (n_1 - 1, n_2 - 1)$.

STEP 2 The confidence interval for σ_1/σ_2 is from

$$\frac{1}{\sqrt{F_{\alpha/2}}} \cdot \frac{s_1}{s_2} \quad \text{to} \quad \frac{1}{\sqrt{F_{1-\alpha/2}}} \cdot \frac{s_1}{s_2},$$

where $F_{1-\alpha/2}$ and $F_{\alpha/2}$ are found in Step 1, n_1 and n_2 are the sample sizes, and s_1 and s_2 are computed from the sample data obtained.

STEP 3 Interpret the confidence interval.

Like the *F*-test for two population standard deviations, the *F*-interval procedure is not at all robust to violations of the normality assumption: Misleading information may be obtained if the procedure is used on data from a variable that is not normally distributed on both populations. In other words, the *F*-interval procedure should not be used unless there is considerable evidence that the variable under consideration is normally distributed, or very nearly so, on each population. We apply Procedure 11.4 in Example 11.14.

▌▌Example 11.14   The F-Interval Procedure

Elmendorf Tear Strength   Use the sample data in Table 11.2 on page 586 to determine a 95% confidence interval for the ratio, σ_1/σ_2, of the standard deviations of tear strength for Brand A and Brand B vinyl floor coverings.

Solution As found in Example 11.13, we can reasonably presume that tear strengths are normally distributed for both Brand A and Brand B vinyl floor coverings. Consequently, we can apply Procedure 11.4 to obtain the required confidence interval.

STEP 1 For a confidence level of $1 - \alpha$, use Table VIII to find $F_{1-\alpha/2}$ and $F_{\alpha/2}$ with df $= (n_1 - 1, n_2 - 1)$.

We want to obtain a 95% confidence interval; consequently, $\alpha = 0.05$. Hence we need to find $F_{0.975}$ and $F_{0.025}$ for df $= (n_1 - 1, n_2 - 1) = (9, 9)$. We did so earlier (Example 11.13, Step 4 of the critical-value approach), where we determined that $F_{0.975} = 0.25$ and $F_{0.025} = 4.03$.

STEP 2 The confidence interval for σ_1/σ_2 is from

$$\frac{1}{\sqrt{F_{\alpha/2}}} \cdot \frac{s_1}{s_2} \quad \text{to} \quad \frac{1}{\sqrt{F_{1-\alpha/2}}} \cdot \frac{s_1}{s_2}.$$

For the data in Table 11.2, $s_1 = 128.3$ g and $s_2 = 199.7$ g. From Step 1, we know that $F_{0.975} = 0.25$ and $F_{0.025} = 4.03$. Consequently, the required 95% confidence interval is from

$$\frac{1}{\sqrt{4.03}} \cdot \frac{128.3}{199.7} \quad \text{to} \quad \frac{1}{\sqrt{0.25}} \cdot \frac{128.3}{199.7},$$

or 0.32 to 1.28.

STEP 3 Interpret the confidence interval.

INTERPRETATION We can be 95% confident that the ratio of the population standard deviations of tear strength for Brand A and Brand B vinyl floor coverings is somewhere between 0.32 and 1.28.   ▬

The Technology Center

Some statistical technologies have dedicated programs that automatically perform an *F*-test and obtain an *F*-interval for two population standard deviations, but others do not. For a statistical technology that does not have such dedicated programs, you can often use its macro capabilities to write them. Refer to the technology manuals for more details.

EXERCISES 11.2

Statistical Concepts and Skills

11.32 What is meant by saying that a variable has an F-distribution?

11.33 How is an F-distribution and its corresponding F-curve identified?

11.34 How many numbers of degrees of freedom does an F-curve have? What are those degrees of freedom called?

11.35 What symbol is used to denote the F-value having area 0.05 to its right? 0.025 to its right? α to its right?

11.36 Using the F_α-notation, identify the F-value having area 0.975 to its left.

11.37 An F-curve has df $= (12, 7)$. What is the number of degrees of freedom for the
a. numerator?   **b.** denominator?

11.38 An F-curve has df $= (8, 19)$. What is the number of degrees of freedom for the
a. denominator?   **b.** numerator?

*In Exercises **11.39–11.46**, use Table VIII and, if necessary, the reciprocal property of F-curves to find the required F-values. Illustrate your work graphically.*

11.39 An F-curve has df $= (24, 30)$. In each case, find the F-value that has the specified area to its right.
a. 0.05   **b.** 0.01   **c.** 0.025

11.40 An F-curve has df $= (12, 5)$. In each case, find the F-value that has the specified area to its right.
a. 0.01   **b.** 0.05   **c.** 0.005

11.41 For an F-curve with df $= (20, 21)$, find
a. $F_{0.01}$.   **b.** $F_{0.05}$.   **c.** $F_{0.10}$.

11.42 For an F-curve with df $= (6, 10)$, find
a. $F_{0.05}$.   **b.** $F_{0.01}$.   **c.** $F_{0.025}$.

11.43 Consider an F-curve with df $= (6, 8)$. Obtain the F-value that has area
a. 0.01 to its left.   **b.** 0.95 to its left.

11.44 Consider an F-curve with df $= (15, 5)$. Obtain the F-value that has area
a. 0.025 to its left.   **b.** 0.975 to its left.

11.45 Determine the two F-values that divide the area under the curve into a middle 0.95 area and two outside 0.025 areas for an F-curve with
a. df $= (7, 4)$.   **b.** df $= (12, 20)$.

11.46 Determine the two F-values that divide the area under the curve into a middle 0.90 area and two outside 0.05 areas for an F-curve with
a. df $= (10, 8)$.   **b.** df $= (12, 12)$.

11.47 In using F-procedures to make inferences for two population standard deviations, why should the distributions (one for each population) of the variable under consideration be normally distributed or nearly so?

11.48 Give two situations in which comparing two population standard deviations would be important.

*Preliminary data analyses and other information indicate that you can reasonably assume that, in Exercises **11.49–11.52**, the variable under consideration is normally distributed on both populations. For each exercise, use either the critical-value approach or the P-value approach to perform the required hypothesis test.*

11.49 Algebra Exam Scores. One year at Arizona State University, the algebra course director decided to experiment with a new teaching method that might reduce variability in final-exam scores by eliminating lower scores. The director randomly divided the algebra students who were registered for class at 9:40 A.M. into two groups. One of the groups, called the control group, was taught the usual algebra course; the other group, called the experimental group, was taught by the new teaching method. Both classes covered the same material, took the same unit quizzes, and took the same final exam at the same time. The final-exam scores (out of 40 possible) for the two groups are shown in the following table.

Control						Experimental			
36	35	35	33	32	32	36	35	35	31
31	29	29	28	28	28	30	29	27	27
27	27	27	26	26	25	26	23	21	21
24	24	24	23	20	20	35	32	28	28
19	19	18	18	18	17	25	23	21	19
17	16	15	15	15	15				
14	11	10	9	4					

Do the data provide sufficient evidence to conclude that there is less variation among final-exam scores when the new teaching method is used? Perform an F-test at the 5% significance level. (*Note:* $s_1 = 7.813$ and $s_2 = 5.286$.)

11.50 Pulmonary Hypertension. In the paper "Persistent Pulmonary Hypertension of the Neonate and Asymmetric Growth Restriction" (*Obstetrics & Gynecology*, Vol. 91, No. 3, pp. 336–341), M. Williams et al. reported on a study of characteristics of neonates. Infants treated for pulmonary hypertension, called the PH group, were compared with those not so treated, called the control group. One of the characteristics measured was head circumference. The following data, in centimeters (cm) are based on the results obtained by the researchers.

PH		Control				
33.9	35.1	35.2	35.6	36.7	35.1	36.0
33.4	34.5	33.4	31.3	33.5	35.8	36.3
37.9	31.3	34.3	33.1	32.4	35.1	33.6
32.5	32.9	31.8	34.1	35.2	34.8	34.5
36.3	34.2	31.6	31.9	31.9	32.8	34.0

Do the data provide sufficient evidence to conclude that variation in head circumference differs among neonates treated for pulmonary hypertension and those not so treated? Perform an *F*-test at the 5% significance level. (*Note:* $s_1 = 1.907$ and $s_2 = 1.594$.)

11.51 Chronic Hemodialysis and Anxiety. Patients who undergo chronic hemodialysis often experience severe anxiety. Videotapes of progressive relaxation exercises were shown to one group of patients and neutral videotapes to another group. Then both groups took the State-Trait Anxiety Inventory, a psychiatric questionnaire used to measure anxiety, where higher scores correspond to higher anxiety. In the paper "The Effectiveness of Progressive Relaxation in Chronic Hemodialysis Patients" (*Journal of Chronic Diseases*, 35(10)), R. Alarcon et al. presented the results of the study. The following data are based on those results.

Relaxation tapes				Neutral tapes			
30	41	28	14	36	44	47	45
40	36	38	24	50	54	54	45
61	36	24	45	50	46	28	35
38	43	32	28	42	35	32	43
37	34	20	23	41	33	35	36
34	47	25	31	32	17	45	
39	14	43	40	24	46		
29	21	40					

Do the data provide sufficient evidence to conclude that variation in anxiety-test scores differs between patients who are shown videotapes of progressive relaxation exercises and those who are shown neutral videotapes? Perform an

F-test at the 10% significance level. (*Note:* $s_1 = 10.154$ and $s_2 = 9.197$.)

11.52 Whiskey Prices. During the 1960s, liquor stores either were run by the state as a monopoly or were privately owned. Independent samples of state-run and privately owned liquor stores yielded the following prices, in dollars, for a fifth of Seagram's 7 Crown Whiskey. [SOURCE: Julian L. Simon and Peter Bruce, "Resampling: A Tool for Everyday Statistical Work," *Chance*, Vol. 4(1), pp. 23–32.]

State run				Privately owned			
4.65	4.11	4.20	3.80	4.82	4.85	4.80	4.85
4.74	4.10	5.05	4.00	4.54	5.20	4.90	4.29
4.55	4.15	4.55	4.19	4.95	4.95	4.75	4.79
4.50	4.00	4.20	4.75	5.29	4.85	4.29	4.85
				4.75	5.10	5.25	4.95
				4.75	4.55	4.79	
				4.89	4.50	5.30	

At the 10% significance level, do the data provide sufficient evidence to conclude that there was more price variation in state-run stores than in privately owned stores? (*Note:* $s_1 = 0.344$ and $s_2 = 0.264$.)

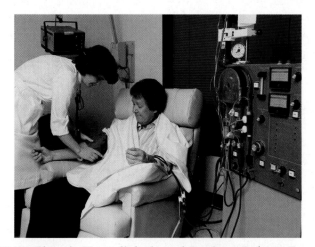

11.53 Chronic Hemodialysis and Anxiety. Refer to Exercise 11.51. Find a 90% confidence interval for the ratio of the population standard deviations of scores for patients who are shown videotapes of progressive relaxation exercises and those who are shown neutral videotapes. Interpret your result.

11.54 Pulmonary Hypertension. Refer to Exercise 11.50. Find a 95% confidence interval for the ratio of the population standard deviations of head circumferences for neonates treated for pulmonary hypertension and those not so treated. Interpret your result.

Extending the Concepts and Skills

11.55 Elmendorf Tear Strength. Refer to Example 11.13 on page 588. Use Table VIII to show that the *P*-value for the hypothesis test exceeds 0.20.

11.56 Because of space restrictions, the numbers of degrees of freedom in Table VIII are not consecutive. For instance, the degrees of freedom for the numerator skips from 24 to 30. If you had only Table VIII and you needed to find $F_{0.05}$ for $df = (25, 20)$, how would you do it?

Estimating F-values From Table VIII. One solution to Exercise 11.56 is to use linear interpolation as follows: For $df = (24, 20)$, we have $F_{0.05} = 2.08$; and for $df = (30, 20)$, we have $F_{0.05} = 2.04$. As 25 is $\frac{1}{6}$ of the way between 24 and 30, we estimate that for an *F*-curve with $df = (25, 20)$,

$$F_{0.05} = 2.08 + \frac{1}{6} \cdot (2.04 - 2.08) = 2.07.$$

In solving Exercises 11.57 and 11.58, use linear interpolation as required.

11.57 Algebra Exam Scores. Refer to Exercise 11.49. Obtain a 90% confidence interval for the ratio of the population standard deviations of final-exam scores for students taught by the conventional method and those taught by the new method. Interpret your result.

11.58 Whiskey Prices. Refer to Exercise 11.52. Find an 80% confidence interval for the ratio of the population standard deviations of whiskey prices for state-run stores and privately owned stores. Interpret your result.

Using Technology

11.59 Chronic Hemodialysis and Anxiety. Refer to Exercises 11.51 and 11.53. Use the technology of your choice to
a. obtain a normal probability plot and boxplot of each sample.
b. perform the required hypothesis test and obtain the desired confidence interval.
c. Is use of the procedures that you employed in part (b) reasonable? Explain your answer.

11.60 Pulmonary Hypertension. Refer to Exercises 11.50 and 11.54. Use the technology of your choice to
a. get a normal probability plot and boxplot of each sample.
b. perform the required hypothesis test and obtain the desired confidence interval.
c. Is use of the procedures that you employed in part (b) reasonable? Explain your answer.

11.61 Elmendorf Tear Strength. Refer to Example 11.13 on page 588. Use the technology of your choice to determine the exact *P*-value for the hypothesis test.

11.62 The Etruscans. Anthropologists are still trying to unravel the mystery of the origins of the Etruscan empire, a highly advanced Italic civilization formed around the eighth century B.C. in central Italy. Were they native to the Italian peninsula or, as many aspects of their civilization suggest, did they migrate from the East by land or sea? The maximum head breadth, in millimeters, of 70 modern Italian male skulls and 84 preserved Etruscan male skulls were analyzed to help researchers decide whether the Etruscans were native to Italy. The resulting data can be found on the WeissStats CD. [SOURCE: N. A. Barnicot and D. R. Brothwell. "The Evaluation of Metrical Data in the Comparison of Ancient and Modern Bones." In *Medical Biology and Etruscan Origins*, G. E. W. Wolstenholme and C. M. O'Connor, eds., Little, Brown & Co., 1959.] Use the technology of your choice to solve parts (a) and (b).
a. Perform an *F*-test at the 5% significance level to decide whether the data provide sufficient evidence to conclude that variation in skull measurements differ between the two populations.
b. Obtain a normal probability plot for each sample.
c. In light of your plots in part (b), does conducting the test you did in part (a) seem reasonable? Explain your answer.

11.63 Active Management of Labor. Active management of labor (AML) is a group of interventions designed to help reduce the length of labor and the rate of cesarean deliveries. Physicians from the Department of Obstetrics and Gynecology at the University of New Mexico Health Sciences Center were interested in determining whether AML would affect the cost of delivery. The results of their study can be found in Rogers et al., "Active Management of Labor: A Cost Analysis of a Randomized Controlled Trial" (*Western Journal of Medicine*, Vol. 172, pp. 240–243). Data, based on the researchers' findings, on the cost of cesarean deliveries for independent random samples of those using AML and those using standard hospital protocols are provided on the WeissStats CD. Use the technology of your choice to solve parts (a) and (b).
a. Do the data provide sufficient evidence to conclude that the variation in cost between the two methods differs? Perform an *F*-test at the 10% significance level.
b. Obtain a normal probability plot for each sample.
c. In light of your plots in part (b), does conducting the test you did in part (a) seem reasonable? Explain your answer.

11.64 Use the technology of your choice to conduct the simulation discussed in Example 11.12 on page 587.

CHAPTER REVIEW

You Should Be Able To

1. use and understand the formulas in this chapter.

2. state the basic properties of χ^2-curves.

3. use the chi-square table, Table VII.

4. perform a hypothesis test for a population standard deviation when the variable under consideration is normally distributed.

5. obtain a confidence interval for a population standard deviation when the variable under consideration is normally distributed.

6. state the basic properties of F-curves.

7. apply the reciprocal property of F-curves.

8. use the F-table, Table VIII.

9. perform a hypothesis test to compare two population standard deviations when the variable under consideration is normally distributed on both populations.

10. obtain a confidence interval for the ratio of two population standard deviations when the variable under consideration is normally distributed on both populations.

Key Terms

χ^2_α, *571*
χ^2-interval procedure for one
   population standard
   deviation, *578*
χ^2-test for one population standard
   deviation, *575, 576*
chi-square (χ^2) curve, *570*

chi-square distribution, *570*
degrees of freedom for the
   denominator, *583*
degrees of freedom for the
   numerator, *583*
F_α, *583*
F-curve, *582*

F-distribution, *582*
F-interval procedure for two
   population standard
   deviations, *591*
F-statistic, *586*
F-test for two population standard
   deviations, *588, 589*

REVIEW TEST

Statistical Concepts and Skills

1. What distribution is used in this chapter to make inferences for one population standard deviation?

2. Fill in the blanks.
a. A χ^2-curve is _____ skewed.
b. A χ^2-curve looks increasingly like a _____ curve as the number of degrees of freedom becomes larger.

3. When you use the χ^2-test or χ^2-interval procedure to conduct inferences for one population standard deviation, what assumption must be met by the variable under consideration? How important is that assumption?

4. Consider a χ^2-curve with 17 degrees of freedom. Use Table VII to determine
a. $\chi^2_{0.99}$.
b. $\chi^2_{0.01}$.
c. the χ^2-value having area 0.05 to its right.
d. the χ^2-value having area 0.05 to its left.
e. the two χ^2-values that divide the area under the curve into a middle 0.95 area and two outside 0.025 areas.

5. What distribution is used in this chapter to make inferences for two population standard deviations?

6. Fill in the blanks:
a. An F-curve is _____ skewed.
b. For an F-curve with df = (14, 5), the F-value having area 0.05 to its left equals the _____ of the F-value having area 0.05 to its right for an F-curve with df = (_____, _____).
c. The observed value of a variable having an F-distribution must be greater than or equal to _____.

7. When you use the F-test or F-interval procedure to conduct inferences for two population standard deviations, what assumption must be met by the variable under consideration? How important is that assumption?

8. Consider an F-curve with df = (4, 8). Use Table VIII to determine
a. $F_{0.01}$.
b. $F_{0.99}$.
c. the F-value having area 0.05 to its right.

d. the *F*-value having area 0.05 to its left.

e. the two *F*-values that divide the area under the curve into a middle 0.95 area and two outside 0.025 areas.

9. Intelligence Quotients. IQs measured on the Stanford Revision of the Binet–Simon Intelligence Scale are supposed to have a standard deviation of 16 points. Twenty-five randomly selected people were given the IQ test; here are the data that were obtained.

91	96	106	116	97
102	96	124	115	121
95	111	105	101	86
88	129	112	82	98
104	118	127	66	102

Preliminary data analyses and other information indicate the reasonableness of presuming that IQs measured on the Stanford Revision of the Binet–Simon Intelligence Scale are normally distributed.

a. Do the data provide sufficient evidence to conclude that IQs measured on this scale have a standard deviation different from 16 points? Perform the required hypothesis test at the 10% significance level. (*Note: s = 15.006.*)

b. How crucial is the normality assumption for the hypothesis test you performed in part (a)? Explain your answer.

10. Intelligence Quotients. Refer to Problem 9. Determine a 90% confidence interval for the standard deviation of IQs measured on the Stanford Revision of the Binet–Simon Intelligence Scale.

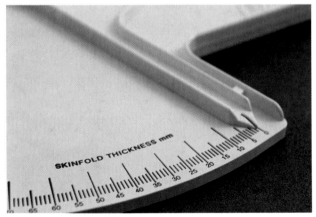

11. Skinfold Thickness. A study entitled "Body Composition of Elite Class Distance Runners" was conducted by M. L. Pollock et al. to decide whether elite distance runners are actually thinner than other people. Their results were published in *The Marathon: Physiological, Medical, Epidemiological, and Psychological Studies*, P. Milvey (ed.), New York: New York Academy of Sciences, 1977, p. 366. The researchers measured the skinfold thickness, an indirect indicator of body fat, of runners and nonrunners in the same age group. The data, in millimeters (mm), shown in the following table are based on the skinfold thickness measurements on the thighs of the people sampled.

Runners			Others			
7.3	6.7	8.7	24.0	19.9	7.5	18.4
3.0	5.1	8.8	28.0	29.4	20.3	19.0
7.8	3.8	6.2	9.3	18.1	22.8	24.2
5.4	6.4	6.3	9.6	19.4	16.3	16.3
3.7	7.5	4.6	12.4	5.2	12.2	15.6

a. For an *F*-test to compare the standard deviations of skinfold thickness of runners and others, identify the appropriate *F*-distribution.

b. At the 1% significance level, do the data provide sufficient evidence to conclude that runners have less variability in skinfold thickness than others? (*Note: $s_1 = 1.798$ and $s_2 = 6.606$ and, if you are using the critical-value approach, use the fact that, for an *F*-curve with df $= (14, 19)$, $F_{0.99} = 0.28$.*)

c. What assumption about skinfold thickness did you make in carrying out the hypothesis test in part (b)? How would you check that assumption?

d. In addition to the assumption on skinfold thickness discussed in part (c), what other assumption is required for performing the *F*-test?

12. Skinfold Thickness. Refer to Problem 11. We used technology to determine that, for an *F*-curve having df $= (14, 19)$, $F_{0.01} = 3.19$.

a. Find a 98% confidence interval for the ratio of the standard deviations of skinfold thickness for runners and for others.

b. Interpret your answer for part (a).

Using Technology

13. Intelligence Quotients. Use the technology of your choice to perform the required hypothesis test in Problem 9 and to obtain the desired confidence interval in Problem 10.

14. Skinfold Thickness. Use the technology of your choice to perform the required hypothesis test in Problem 11 and to obtain the desired confidence interval in Problem 12.

15. Sex and Direction. In the paper "The Relation of Sex and Sense of Direction to Spatial Orientation in an Unfamiliar Environment" (*Journal of Environmental Psychology*, Vol. 20, pp. 17–28), Sholl et al. published the results of examining the sense of direction of 30 male and 30 female students. After being taken to an unfamiliar wooded park, the students were given a number of spatial orientation tests, including pointing to south, which tested their absolute frame of reference. To point south, the students moved a pointer attached to a 360° protractor. The absolute pointing errors, in degrees, for students who rated themselves with a good sense of direction (GSOD) and those who rated themselves with a poor sense of direction (PSOD) are provided on the WeissStats CD. Can you reasonably apply the *F*-test to compare the variation in pointing errors between people who rate themselves with a good sense of direction and those who rate themselves with a poor sense of direction? Explain your answer.

16. Microwave Popcorn. Two brands of microwave popcorn, Brand A and Brand B, are compared for consistency in popping time. Following are the popping times, in seconds, for 30 bags of each brand.

Brand A			Brand B		
146	152	149	147	145	153
152	150	147	150	142	164
153	158	158	151	158	140
150	143	143	139	145	155
146	150	158	165	143	151
152	148	154	153	157	144
154	149	145	154	155	146
141	150	149	158	138	151
157	141	152	146	147	160
149	145	158	153	147	147

Use the technology of your choice to decide, at the 5% significance level, whether the data provide sufficient evidence to conclude that Brand A has a more consistent popping time than Brand B. Explain your reasoning in detail.

StatExplore in MyMathLab
Analyzing Data Online

You can use StatExplore to perform all statistical analyses discussed in this chapter.
To illustrate, we show how to perform an *F*-test and an *F*-interval procedure.

EXAMPLE   **F-Test and F-Interval Procedure**

Elmendorf Tear Strength   Table 11.2 on page 586 displays the Elmendorf tear strengths for independent random samples of two brands of vinyl floor coverings, Brand A and Brand B. Use StatExplore to do the following:

a. Decide, at the 5% significance level, whether the data provide sufficient evidence to conclude that the population standard deviations of tear strength differ for Brand A and Brand B.

b. Find a 95% confidence interval for the ratio of the standard deviations of tear strength for Brand A and Brand B.

SOLUTION   Let σ_1 and σ_2 denote the population standard deviations of tear strength for Brands A and B, respectively.

a. Here the problem is to perform the hypothesis test

H_0: $\sigma_1 = \sigma_2$ (standard deviations are the same)

H_a: $\sigma_1 \neq \sigma_2$ (standard deviations are different)

at the 5% significance level. Note that the hypothesis test is two tailed because a does-not-equal sign ($\neq$) appears in the alternative hypothesis.

To carry out the hypothesis test by using StatExplore, we first store the two samples of tear-strength data from Table 11.2 in columns named BRAND A and BRAND B, respectively. Then, we proceed as follows:

1   Choose **Stat ➤ Variance ➤ Two sample**
2   Click the arrow button at the right of the **Sample 1 in** drop-down list box and select BRAND A
3   Click the arrow button at the right of the **Sample 2 in** drop-down list box and select BRAND B
4   Click **Next →**
5   Select the **Hypothesis Test** option button
6   Click in the **Null: variance ratio =** text box and type <u>1</u>
7   Click the arrow button at the right of the **Alternative** drop-down list box and select **not =**
8   Click **Calculate**

The resulting output is shown in Printout 11.3. From Printout 11.3, the *P*-value for the hypothesis test is 0.2039. As the *P*-value exceeds the specified significance level of 0.05, we do not reject H_0. At the 5% significance level, the data do not provide sufficient evidence to conclude that the population standard deviations of tear strength differ for Brand A and Brand B.

PRINTOUT 11.3
StatExplore output for *F*-test

Hypothesis test results:
v1 = variance of BRAND A
v2 = variance of BRAND B
Parameter: v1/v2
H0 : Parameter = 1
HA : Parameter not = 1

Ratio	n1	n2	Sample Ratio	F-Stat	P-value
v1/v2	10	10	0.41302797	0.41302797	0.2039

b. To determine a 95% confidence interval for the ratio of the population standard deviations of tear strength for Brand A and Brand B, we proceed as follows, again assuming that the two samples of tear-strength data from Table 11.2 have been stored in columns named BRAND A and BRAND B, respectively.

1   Choose **Stat ➤ Variance ➤ Two sample**
2   Click the arrow button at the right of the **Sample 1 in** drop-down list box and select BRAND A
3   Click the arrow button at the right of the **Sample 2 in** drop-down list box and select BRAND B
4   Click **Next →**
5   Select the **Confidence Interval** option button
6   Click in the **Level** text box and type <u>0.95</u>
7   Click **Calculate**

The resulting output is shown in Printout 11.4. From Printout 11.4, a 95% confidence interval for the ratio of the population *variances* is from 0.10259031 (lower limit) to 1.6628482 (upper limit). Taking square roots, we deduce that we can be 95% confident that the ratio of the population standard deviations of tear strength for Brand A and Brand B is somewhere between 0.32 and 1.29.

PRINTOUT 11.4
StatExplore output for *F*-interval

95% confidence interval results:
v1 = variance of BRAND A
v2 = variance of BRAND B
Parameter: v1/v2

Ratio	n1	n2	Sample Ratio	L. Limit	U. Limit
v1/v2	10	10	0.41302797	0.10259031	1.6628482

STATEXPLORE EXERCISES Solve the following problems by using StatExplore.
a. Conduct the hypothesis test required in Exercise 11.49 on page 593.
b. Obtain the confidence interval required in Exercise 11.53 on page 594.
c. Carry out the hypothesis test required in Example 11.6 on page 575.
d. Find the confidence interval required in Example 11.7 on page 578.

To access StatExplore, go to the student content area of your Weiss MyMathLab course.

Internet Project

Firearm-Related Deaths for Men and Women

The Centers for Disease Control and Prevention (CDC) had previously estimated that, by 2003, firearms would become the number one cause of product-related deaths in the United States.

In this Internet project, you will see striking differences in firearm-related deaths for males and females. For instance, males are almost six times more likely to be killed by firearms than are females.

But the question posed in this project concerns the amount of variation in age at firearm-related death for males and females. You are to estimate the standard deviation for each population and perform a hypothesis test to decide whether the two population standard deviations differ.

A more interesting question—which statistics alone cannot answer—is what do the standard deviations tell us in this context? Are there, for example, social or psychological reasons for one group to have a larger amount of variation than the other?

URL for access to Internet Projects Page: www.aw-bc.com/weiss

Focusing on Data Analysis

The Focus Database

Recall from Chapter 1 (see page 35) that the Focus database contains information on the undergraduate students at the University of Wisconsin - Eau Claire (UWEC). Statistical analyses for this database should be done with the technology of your choice.

a. Obtain a random sample of size 50 of the ACT composite scores (COMP) of UWEC undergraduate students.

b. At the 5% significance level, do the data from part (a) provide sufficient evidence to conclude that the standard deviation of ACT composite scores of all UWEC undergraduates differs from 3 points?

c. Use the sample data from part (a) to determine a 95% confidence interval for the standard deviation of ACT composite scores of all UWEC undergraduates. Interpret your result.

d. Obtain a normal probability plot and a boxplot of the ACT composite scores of the sampled UWEC undergraduates.

e. Based on your results from part (d), do you think that performing the inferences in parts (b) and (c) is reasonable? Explain your answer.

f. Obtain independent simple random samples of size 50 each of the ACT English scores (ENGLISH) and the ACT math scores (MATH) of UWEC undergraduate students.

g. At the 5% significance level, do the data from part (f) provide sufficient evidence to conclude that the standard deviations of ACT English scores and ACT math scores differ for UWEC undergraduates?

h. Use the sample data from part (f) to determine a 95% confidence interval for the ratio of the standard deviation of ACT English scores to the standard deviation of ACT math scores for UWEC undergraduates.

i. Obtain normal probability plots and boxplots of the ACT English scores and the ACT math scores of the sampled UWEC undergraduates.

j. Based on your results from part (i), do you think that performing the inferences in parts (g) and (h) is reasonable? Explain your answer.

Case Study Discussion

Speaker Woofer Driver Manufacturing

On page 569, at the beginning of this chapter, we discussed rubber-edge manufacturing for speaker woofer drivers and a criterion for classifying process capability.

Recall that each process for manufacturing rubber edges requires a production weight specification that consists of a lower specification limit (LSL), a target weight (T), and an upper specification limit (USL). The actual mean and standard deviation of the weights of the rubber edges being produced are called the process mean (μ) and process standard deviation (σ), respectively. A process that is on target ($\mu = T$) is called *super* if $\sigma < (\text{USL} - \text{LSL})/12$.

The table on page 569 provides data on rubber-edge weight for a sample of 60 observations. Use those data and the procedures discussed in this chapter to solve the following problems:

a. Obtain a 99% confidence interval for the process standard deviation.
b. The process under consideration is known to be on target and its production weight specification is LSL $= 16.72$, $T = 17.60$, and USL $= 18.48$. Do the data provide sufficient evidence to conclude that the process is super? Perform the required hypothesis test at the 1% significance level.
c. Obtain a normal probability plot of the data.
d. Based on your plot in part (c), was conducting the inferences that you did in parts (a) and (b) reasonable? Explain your answer.

Internet Resources: Visit the Weiss Web site www.aw-bc.com/weiss for additional discussion, exercises, and resources related to this case study.

Biography   *W. EDWARDS DEMING: Transforming Industry with SQC*

William Edwards Deming was born on October 14, 1900, in Sioux City, Iowa. Shortly after his birth, his father secured homestead land and moved the family first to Cody, Wyoming, and then to Powell, Wyoming.

Deming obtained a B.S. in physics at the University of Wyoming in 1921, a Master's degree in physics and mathematics at the University of Colorado in 1924, and a doctorate in mathematical physics at Yale University in 1928.

While working for various federal agencies during the next decade, Deming became an expert on sampling and quality control. In 1939, he accepted the position of head mathematician and advisor in sampling at the United States Census Bureau. Deming began the use of sampling at the Census Bureau, and, expanding the work of Walter A. Shewhart (later known as the father of statistical quality control, or SQC), also applied statistical methods of quality control to provide reliability and quality to the nonmanufacturing environment.

In 1946, Deming left the Census Bureau, joined the Graduate School of Business Administration at New York University, and offered his services to the private sector as a consultant in statistical studies. It was in this latter capacity that Deming transformed industry in Japan. Deming began his long association with Japanese businesses in 1947 when the U.S. War Department engaged him to instruct Japanese industrialists in statistical quality control methods. The reputation of Japan's goods changed from definitely shoddy to amazingly excellent over the next two decades as the businessmen of Japan implemented Deming's teachings.

More than 30 years passed before Deming's methods gained widespread recognition by the business community in the United States. Finally, in 1980, as the result of the NBC white paper, *If Japan Can, Why Can't We?*, in which Deming's role was publicized, executives of major corporations (among them, Ford Motor Company) contracted with Deming to improve the quality of U.S. goods.

Deming maintained an intense work schedule throughout his eighties, giving four-day managerial seminars, teaching classes at NYU, sponsoring clinics for statisticians, and consulting with businesses internationally. His last book, *The New Economics,* was published in 1993. Dr. Deming died at his home in Washington, D.C., on December 20, 1993.

Inferences for Population Proportions

GENERAL OBJECTIVES In Chapters 8–10, we discussed methods for finding confidence intervals and performing hypothesis tests for one or two population means. Now we describe how to conduct those inferences for one or two population proportions.

A *population proportion* is the proportion (percentage) of a population that has a specified attribute. For example, if the population under consideration consists of all Americans and the specified attribute is "retired," the population proportion is the proportion of all Americans who are retired.

In Section 12.1, we first introduce the required notation and terminology for performing proportion inferences and then discuss confidence intervals for one population proportion. Next, in Section 12.2, we examine a method for conducting a hypothesis test for one population proportion.

In Section 12.3, we investigate how to perform a hypothesis test to compare two population proportions and how to construct a confidence interval for the difference between two population proportions.

Hormone-Replacement Therapy and Dementia

Initial studies of hormone-replacement therapy for post-menopausal women suggested remarkable benefits. More recently, however, researchers have reported adverse effects from such therapy. For instance, two articles in the May 28, 2003 issue of the *Journal of the American Medical Association* indicate that use of estrogen and progestin increases the risk of dementia and stroke in otherwise healthy women.

As reported in the May 31, 2003 issue of *Science News*, these results all stem from the Women's Health Initiative, a nationwide study beginning in the early 1990s. More than 27,000 women were involved in the component of the study dealing with hormone-replacement therapy. Three years before the scheduled completion of this component of the study, researchers halted their investigation because they discovered that "…postmenopausal women taking estrogen–progestin supplements had a greater risk of circulatory disorders, breast cancer, and stroke than women getting inert pills did."

The researchers had also monitored 4532 healthy women over the age of 65 for signs of Alzheimer's disease or other dementia. These women were randomly divided into two equal-size groups of 2266 women each. One group was assigned to take the hormone-replacement therapy and the other group received a placebo.

Over 5 years of treatment, 40 of the 2266 women who received the hormone-replacement therapy were diagnosed with dementia, compared with 21 of the 2266 women getting a placebo, as reported by Sally A. Shumaker of Wake Forest University, a coauthor of the study. It surprised the researchers that the women receiving hormones had a greater incidence of dementia because earlier research, as reported in the Feb. 17, 2001 issue of *Science News*, had suggested the opposite result.

One important question is whether the results of this designed experiment provide sufficient evidence to conclude that, indeed, healthy women over 65 years old who take hormone-replacement therapy are at greater risk for dementia than those who do not. After you study this chapter, you will be asked to answer that question.

12.1   Confidence Intervals for One Population Proportion

Many statistical studies are concerned with obtaining the proportion (percentage) of a population that has a specified attribute. For example, we might be interested in

- the percentage of U.S. adults who have health insurance,
- the percentage of cars in the United States that are imports,
- the percentage of U.S. adults who favor stricter clean air health standards, or
- the percentage of Canadian women in the labor force.

In the first case, the population consists of all U.S. adults and the specified attribute is "has health insurance." For the second case, the population consists of all cars in the United States and the specified attribute is "is an import." The population in the third case is all U.S. adults and the specified attribute is "favors stricter clean air health standards." And, in the fourth case, the population consists of all Canadian women and the specified attribute is "is in the labor force."

Generally, the population under consideration is large, and determining the population proportion by taking a census is therefore usually impractical and often impossible; for instance, imagine trying to interview every U.S. adult for the purpose of ascertaining the proportion who have health insurance. Thus, in practice, we mostly rely on sampling and use the sample data to make inferences about the population proportion. In Example 12.1, we introduce proportion notation and terminology.

Example 12.1   Proportion Notation and Terminology

Playing Hooky From Work   Many employers are concerned about employees who call in sick when in fact they are not ill. A survey commissioned by the Hilton Hotels Corporation investigated this issue. One question asked of the people taking part in the survey was whether they call in sick at least once a year when they simply need time to relax. For brevity, we use the phrase *play hooky* to refer to that practice.

In the survey, 1010 randomly selected U.S. employees were polled. The proportion of the 1010 employees sampled who play hooky was used to estimate the proportion of all U.S. employees who play hooky. Discuss the statistical notation and terminology used in this and similar studies on proportions.

Solution   We use p to denote the proportion of all U.S. employees who play hooky; it represents the **population proportion** and is the parameter whose value is to be estimated. The proportion of the 1010 U.S. employees sampled who play hooky is designated $\hat{p}$ (read "p hat") and represents a **sample proportion;** it is the statistic used to estimate the unknown population proportion, p.

Although unknown, the population proportion, p, is a fixed number. In contrast, the sample proportion, $\hat{p}$, is a variable; its value varies from sample to sample. For instance, if 202 of the 1010 employees sampled play hooky, then

$$\hat{p} = \frac{202}{1010} = 0.2,$$

that is, 20.0% of the employees sampled play hooky. Whereas, if 184 of the 1010 employees sampled play hooky, then

$$\hat{p} = \frac{184}{1010} = 0.182,$$

that is, 18.2% of the employees sampled play hooky.

These two calculations also reveal how to compute a sample proportion: Divide the number of employees sampled who play hooky, denoted x, by the total number of employees sampled, n. In symbols, $\hat{p} = x/n$. ▄

In Example 12.1, we introduced some notation and terminology used when we make inferences about a population proportion. In general, we have the following definitions.

DEFINITION 12.1

Population Proportion and Sample Proportion

Consider a population in which each member either has or does not have a specified attribute. Then we use the following notation and terminology.

Population proportion, p: The proportion (percentage) of the entire population that has the specified attribute.

Sample proportion, $\hat{p}$: The proportion (percentage) of a sample from the population that has the specified attribute.

In Example 12.1, the population consists of all U.S. employees, and the specified attribute is "plays hooky." The population proportion, p, is the proportion of all U.S. employees who play hooky; and a sample proportion, $\hat{p}$, is the proportion of employees sampled who play hooky. As demonstrated in Example 12.1, we compute a sample proportion as indicated in Formula 12.1.

FORMULA 12.1

Sample Proportion

A sample proportion, $\hat{p}$, is computed by using the formula

$$\hat{p} = \frac{x}{n},$$

where x denotes the number of members in the sample that have the specified attribute and, as usual, n denotes the sample size.

What
Does it Mean?

A sample proportion is obtained by dividing the number of members sampled that have the specified attribute by the total number of members sampled.

Note: For convenience, we sometimes refer to x (the number of members in the sample that have the specified attribute) as the **number of successes** and to $n - x$ (the number of members in the sample that do not have the specified attribute) as the **number of failures.** But remember that, in this context, the words *success* and *failure* need not have the ordinary meanings of the words.

Before proceeding, let's draw some parallels between proportions and means. Table 12.1 shows the correspondence between the notation for means and the notation for proportions.

Recall that a sample mean, $\bar{x}$, can be used to make inferences about a population mean, μ. Similarly, a sample proportion, $\hat{p}$, can be used to make inferences about a population proportion, p.

TABLE 12.1

Correspondence between notations for means and proportions

	Parameter	**Statistic**
Means	μ	$\bar{x}$
Proportions	p	$\hat{p}$

The Sampling Distribution of the Sample Proportion

To make inferences about a population mean, μ, we must know the sampling distribution of the sample mean, that is, the distribution of the variable $\bar{x}$. The same is true for proportions: To make inferences about a population proportion, p, we need to know the **sampling distribution of the sample proportion,** that is, the distribution of the variable $\hat{p}$.

Because a proportion can always be regarded as a mean, we can use our knowledge of the sampling distribution of the sample mean to derive the sampling distribution of the sample proportion. (See Exercise 12.33 for details.) In practice, the sample size usually is large, so we concentrate on that case, as presented in Key Fact 12.1.

KEY FACT 12.1

What
Does it Mean?

If *n* is large, the possible sample proportions for samples of size *n* have approximately a normal distribution with mean *p* and standard deviation $\sqrt{p(1-p)/n}$.

The Sampling Distribution of the Sample Proportion

For samples of size n,

- the mean of $\hat{p}$ equals the population proportion: $\mu_{\hat{p}} = p$;
- the standard deviation of $\hat{p}$ equals the square root of the product of the population proportion and one minus the population proportion divided by the sample size: $\sigma_{\hat{p}} = \sqrt{p(1-p)/n}$; and
- $\hat{p}$ is approximately normally distributed for large n.

The accuracy of the normal approximation depends on n and p. If p is close to 0.5, the approximation is quite accurate, even for moderate n. The farther p is from 0.5, the larger n must be for the approximation to be accurate. As a rule of thumb, we use the normal approximation when *np and n(1 − p) are both 5 or greater.*[†] In this chapter, when we say that n is large, we mean that np and $n(1 - p)$ are both 5 or greater.

[†]Another commonly used rule of thumb is that np and $n(1 - p)$ are both 10 or greater; still another is that $np(1 - p)$ is 25 or greater. However, our rule of thumb, which is less conservative than either of those two, is consistent with the conditions required for performing a chi-square goodness-of-fit test (discussed in Chapter 13).

We can make Key Fact 12.1 plausible through simulation. We do so in Example 12.2 by returning to the situation described in Example 12.1.

Example 12.2 Sampling Distribution of the Sample Proportion

Playing Hooky From Work Suppose that, in reality, 19.1% of U.S. employees play hooky; that is, that the population proportion is $p = 0.191$. Then, according to Key Fact 12.1, for samples of size 1010, the variable $\hat{p}$ is approximately normally distributed with mean $\mu_{\hat{p}} = p = 0.191$ and standard deviation

$$\sigma_{\hat{p}} = \sqrt{\frac{p(1-p)}{n}} = \sqrt{\frac{0.191(1-0.191)}{1010}} = 0.012.$$

Use simulation to make this fact plausible.

Solution First we simulated 2000 samples of 1010 U.S. employees each. Next we determined, for each of those 2000 samples, the sample proportion ($\hat{p}$) of employees sampled who play hooky. Then we obtained a histogram of the 2000 sample proportions, as shown in Printout 12.1. For purposes of comparison, we superimposed on the histogram the normal distribution with a mean of 0.191 and a standard deviation of 0.012.

PRINTOUT 12.1

Histogram of the sample proportions for 2000 samples of size 1010 with the approximating normal curve for $\hat{p}$ superimposed

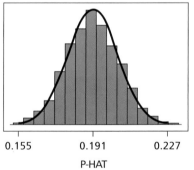

0.155   0.191   0.227

P-HAT

INTERPRETATION The histogram shown in Printout 12.1 is shaped roughly like a normal curve, specifically like the normal curve with parameters 0.191 and 0.012. This result makes it plausible that $\hat{p}$ is approximately normally distributed with mean 0.191 and standard deviation 0.012, as proposed in Key Fact 12.1.

Large-Sample Confidence Intervals for a Population Proportion

Procedure 12.1 on the next page provides a step-by-step method for obtaining a confidence interval for a population proportion. We often refer to this procedure as the **one-sample z-interval procedure** for a population proportion or, simply, as the **z-interval procedure** for a population proportion. The one-sample z-interval procedure for a population proportion is based on Key Fact 12.1 and is derived in a way similar to the one-sample z-interval procedure for a population mean (Procedure 8.1 on page 363).

Note: As stated in Assumption 2 of Procedure 12.1, a condition for using that procedure is that "the number of successes, x, and the number of failures, $n - x$, are both 5 or greater." We can restate this condition as "$n\hat{p}$ and $n(1 - \hat{p})$ are both 5 or greater," which, for an unknown p, corresponds to the rule of thumb for using the normal approximation given after Key Fact 12.1.

In our first application of Procedure 12.1, we obtain a confidence interval for the proportion of all U.S. employees who play hooky. We use the information obtained in the poll commissioned by the Hilton Hotels Corporation.

PROCEDURE 12.1

One-Sample z-Interval Procedure

Purpose To find a confidence interval for a population proportion, p

Assumptions
1. Simple random sample
2. The number of successes, x, and the number of failures, $n - x$, are both 5 or greater.

STEP 1 For a confidence level of $1 - \alpha$, use Table II to find $z_{\alpha/2}$.

STEP 2 The confidence interval for p is from

$$\hat{p} - z_{\alpha/2} \cdot \sqrt{\hat{p}(1 - \hat{p})/n} \quad \text{to} \quad \hat{p} + z_{\alpha/2} \cdot \sqrt{\hat{p}(1 - \hat{p})/n},$$

where $z_{\alpha/2}$ is found in Step 1, n is the sample size, and $\hat{p} = x/n$ is the sample proportion.

STEP 3 Interpret the confidence interval.

Example 12.3   The One-Sample z-Interval Procedure

Playing Hooky From Work   A poll was taken of 1010 U.S. employees. The employees sampled were asked whether they "play hooky," that is, call in sick at least once a year when they simply need time to relax; 202 responded "yes." Use these data to obtain a 95% confidence interval for the proportion, p, of all U.S. employees who play hooky.

Solution   The attribute in question is "plays hooky," the sample size is 1010, and the number of employees sampled who play hooky is 202. We have $n = 1010$. Also, $x = 202$ and $n - x = 1010 - 202 = 808$, both of which are 5 or greater. We can therefore apply Procedure 12.1 to obtain the required confidence interval.

STEP 1 For a confidence level of $1 - \alpha$, use Table II to find $z_{\alpha/2}$.

We want a 95% confidence interval, which means that $\alpha = 0.05$. In Table II or at the bottom of Table IV, we find that $z_{\alpha/2} = z_{0.05/2} = z_{0.025} = 1.96$.

STEP 2 The confidence interval for p is from

$$\hat{p} - z_{\alpha/2} \cdot \sqrt{\hat{p}(1 - \hat{p})/n} \quad \text{to} \quad \hat{p} + z_{\alpha/2} \cdot \sqrt{\hat{p}(1 - \hat{p})/n}.$$

We have $n = 1010$ and, from Step 1, $z_{\alpha/2} = 1.96$. Also, because 202 of the 1010 employees sampled play hooky, $\hat{p} = x/n = 202/1010 = 0.2$. Consequently, a 95% confidence interval for p is from

$$0.2 - 1.96 \cdot \sqrt{(0.2)(1 - 0.2)/1010} \quad \text{to} \quad 0.2 + 1.96 \cdot \sqrt{(0.2)(1 - 0.2)/1010},$$

or

$$0.2 - 0.025 \quad \text{to} \quad 0.2 + 0.025,$$

or 0.175 to 0.225.

STEP 3 Interpret the confidence interval.

INTERPRETATION We can be 95% confident that the percentage of all U.S. employees who play hooky is somewhere between 17.5% and 22.5%.

■

Margin of Error

In Section 8.3, we discussed the margin of error in estimating a population mean by a sample mean. In general, the **margin of error** of an estimator represents the precision with which it estimates the parameter in question. The confidence-interval formula in Step 2 of Procedure 12.1 indicates that the margin of error, E, in estimating a population proportion by a sample proportion is $z_{\alpha/2} \cdot \sqrt{\hat{p}(1-\hat{p})/n}$.

DEFINITION 12.2

> **Margin of Error for the Estimate of p**
>
> The *margin of error* for the estimate of p is
>
> $$E = z_{\alpha/2} \cdot \sqrt{\hat{p}(1-\hat{p})/n}.$$

What *Does it Mean?*

The margin of error is equal to half the length of the confidence interval. It represents the precision with which a sample proportion estimates the population proportion at the specified confidence level.

In Example 12.3, the margin of error is

$$E = z_{\alpha/2} \cdot \sqrt{\hat{p}(1-\hat{p})/n} = 1.96 \cdot \sqrt{(0.2)(1-0.2)/1010} = 0.025,$$

which can also be obtained by taking one-half the length of the confidence interval: $(0.225 - 0.175)/2 = 0.025$. Therefore we can be 95% confident that the error in estimating the proportion p of all U.S. employees who play hooky by the proportion, 0.2, of those in the sample who play hooky is at most 0.025, that is, plus or minus 2.5 percentage points.

On the one hand, given a confidence interval, we can find the margin of error by taking half the length of the confidence interval. On the other hand, given the sample proportion and the margin of error, we can determine the confidence interval—its endpoints are $\hat{p} \pm E$.

Most newspaper and magazine polls provide the sample proportion and the margin of error associated with a 95% confidence interval. For example, a survey of U.S. women conducted by Gallup for the CNBC cable network stated, "…36% of those polled believe their gender will hurt them; the margin of error for the poll is plus or minus 4 percentage points."

Translated into our terminology, $\hat{p} = 0.36$ and $E = 0.04$. Thus the confidence interval has endpoints $\hat{p} \pm E = 0.36 \pm 0.04$, or 0.32 to 0.40. As a result, we can be 95% confident that the percentage of all U.S. women who believe that their gender will hurt them is somewhere between 32% and 40%.

Determining the Required Sample Size

The margin of error and confidence level of a confidence interval are often specified in advance. We must then determine the sample size required to meet those specifications. If we solve for n in the formula for the margin of error, we obtain

$$n = \hat{p}(1 - \hat{p})\left(\frac{z_{\alpha/2}}{E}\right)^2. \tag{12.1}$$

This formula cannot be used to obtain the required sample size because the sample proportion, $\hat{p}$, is not known prior to sampling.

There are two ways around this problem. To begin, we examine the graph of $\hat{p}(1 - \hat{p})$ versus $\hat{p}$ shown in Fig. 12.1, which reveals that the largest $\hat{p}(1 - \hat{p})$ can be is 0.25, or when $\hat{p} = 0.5$. The farther $\hat{p}$ is from 0.5, the smaller will be the value of $\hat{p}(1 - \hat{p})$.

FIGURE 12.1
Graph of $\hat{p}(1 - \hat{p})$ versus $\hat{p}$

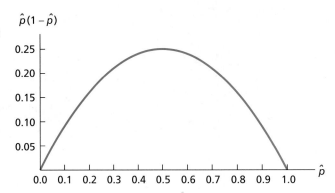

Because the largest possible value of $\hat{p}(1 - \hat{p})$ is 0.25, the most conservative approach for determining sample size is to use that value in Equation (12.1). The sample size obtained then will generally be larger than necessary and the margin of error less than required. Nonetheless, this approach guarantees that the specifications will be met or bettered.

However, because sampling tends to be time consuming and expensive, we usually do not want to take a larger sample than necessary. If we can make an educated guess for the observed value of $\hat{p}$—say, from a previous study or theoretical considerations—we can use that guess to obtain a more realistic sample size.

In this same vein, if we have in mind a likely range for the observed value of $\hat{p}$, then, in light of Fig. 12.1, we should take as our educated guess for $\hat{p}$ the value in the range closest to 0.5. But, in either case, we should be aware that, if the observed value of $\hat{p}$ is closer to 0.5 than is our educated guess, the margin of error will be larger than desired.

Formula 12.2 summarizes this discussion, and we apply it in Example 12.4.

FORMULA 12.2   **Sample Size for Estimating p**

A $(1 - \alpha)$-level confidence interval for a population proportion that has a margin of error of at most E can be obtained by choosing

$$n = 0.25 \left(\frac{z_{\alpha/2}}{E}\right)^2,$$

rounded up to the nearest whole number. If you can make an educated guess, $\hat{p}_g$ (g for guess), for the observed value of $\hat{p}$, then you should instead choose

$$n = \hat{p}_g(1 - \hat{p}_g) \left(\frac{z_{\alpha/2}}{E}\right)^2,$$

rounded up to the nearest whole number.

Example 12.4   Sample Size for Estimating p

Playing Hooky From Work   Consider again the problem of estimating the proportion of all U.S. employees who play hooky.

a. Obtain a sample size that will ensure a margin of error of at most 0.01 for a 95% confidence interval.

b. Find a 95% confidence interval for p if, for a sample of the size determined in part (a), the proportion of those who play hooky is 0.194.

c. Determine the margin of error for the estimate in part (b) and compare it to the margin of error specified in part (a).

d. Repeat parts (a)–(c) if the proportion of those sampled who play hooky can reasonably be presumed to be somewhere between 0.1 and 0.3.

e. Compare the results obtained in parts (a)–(c) with those obtained in part (d).

Solution

a. We apply the first displayed equation in Formula 12.2. To do so, we must identify $z_{\alpha/2}$ and the margin of error, E. The confidence level is stipulated to be 0.95, so $z_{\alpha/2} = z_{0.05/2} = z_{0.025} = 1.96$, and the margin of error is specified at 0.01. Thus a sample size that will ensure a margin of error of at most 0.01 for a 95% confidence interval is

$$n = 0.25 \left(\frac{z_{\alpha/2}}{E}\right)^2 = 0.25 \left(\frac{1.96}{0.01}\right)^2 = 9604.$$

INTERPRETATION   If we take a sample of 9604 U.S. employees, the margin of error for our estimate of the proportion of all U.S. employees who play hooky will be 0.01 or less, that is, at most plus or minus 1 percentage point.

b. We find, by applying Procedure 12.1 (page 610) with $\alpha = 0.05$, $n = 9604$, and $\hat{p} = 0.194$, that a 95% confidence interval for p has endpoints

$$0.194 \pm 1.96 \cdot \sqrt{(0.194)(1 - 0.194)/9604},$$

or 0.194 ± 0.008, or 0.186 to 0.202.

INTERPRETATION   Based on a sample of 9604 U.S. employees, we can be 95% confident that the percentage of all U.S. employees who play hooky is somewhere between 18.6% and 20.2%.

c. The margin of error for the estimate in part (b) is 0.008. Not surprisingly, this is less than the margin of error of 0.01 specified in part (a).

d. If we can reasonably presume that the proportion of those sampled who play hooky will be somewhere between 0.1 and 0.3, we use the second displayed equation in Formula 12.2, with $\hat{p}_g = 0.3$ (the value in the range closest to 0.5), to obtain the sample size:

$$n = \hat{p}_g(1 - \hat{p}_g)\left(\frac{z_{\alpha/2}}{E}\right)^2 = (0.3)(1 - 0.3)\left(\frac{1.96}{0.01}\right)^2 = 8068 \text{ (rounded up)}.$$

Applying Procedure 12.1 with $\alpha = 0.05$, $n = 8068$, and $\hat{p} = 0.194$, we find that a 95% confidence interval for p has endpoints

$$0.194 \pm 1.96 \cdot \sqrt{(0.194)(1 - 0.194)/8068},$$

or 0.194 ± 0.009, or 0.185 to 0.203.

INTERPRETATION   Based on a sample of 8068 U.S. employees, we can be 95% confident that the percentage of all U.S. employees who play hooky is somewhere between 18.5% and 20.3%. The margin of error for the estimate is 0.009.

e. By using the educated guess for $\hat{p}$ in part (d), we reduced the required sample size by more than 1500 (from 9604 to 8068). Moreover, only 0.1% (0.001) of precision was lost—the margin of error rose from 0.008 to 0.009. The risk of using the guess 0.3 for $\hat{p}$ is that, if the observed value of $\hat{p}$ had turned out to be larger than 0.3 (but smaller than 0.7), the achieved margin of error would have exceeded the specified 0.01.

The Technology Center   Most statistical technologies have programs that automatically obtain a confidence interval for a population proportion. In this subsection, we present output and step-by-step instructions to implement such programs.

Example 12.5   **Using Technology to Obtain a z-Interval**

Playing Hooky From Work   In Example 12.3, we applied Procedure 12.1 to determine a 95% confidence interval for the proportion of all U.S. employees who play hooky. Use Minitab, Excel, or the TI-83/84 Plus to obtain that confidence interval.

Solution   We first recall that the sample size is 1010 and that the number of employees sampled who play hooky is 202. Printout 12.2 shows the output obtained by applying the one-sample z-interval programs to the data.

PRINTOUT 12.2
One-sample z-interval output for the data on playing hooky from work

MINITAB

Test and CI for One Proportion

Test of p = 0.5 vs p not = 0.5

Sample	X	N	Sample p	95% CI	Z-Value	P-Value
1	202	1010	0.200000	(0.175331, 0.224669)	-19.07	0.000

EXCEL

▷	x Confidence Interval	🗗🗖⊘

▷	Summary Statistics	🗋
n	1010	
p-hat	0.2	
Std Err	0.0126	
z*	1.96	

▷	Interval Results	🗋
Confidence Interval		
With 95% Confidence, 0.175 < p < 0.225		

TI-83/84 PLUS

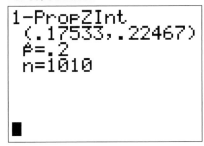

```
1-PropZInt
 (.17533,.22467)
 p̂=.2
 n=1010
■
```

The outputs in Printout 12.2 indicate that the required 95% confidence interval is from 0.175 to 0.225. Thus, we can be 95% confident that the percentage of all U.S. employees who play hooky is somewhere between 17.5% and 22.5%.

Obtaining the Output (Optional)

Printout 12.2 provides output from Minitab, Excel, and the TI-83/84 Plus for a 95% confidence interval for the proportion of all U.S. employees who play hooky. The following are detailed instructions for obtaining that output.

MINITAB
1 Choose **Stat ➤ Basic Statistics ➤ 1 Proportion...**
2 Select the **Summarized data** option button
3 Click in the **Number of trials** text box and type 1010
4 Click in the **Number of events** text box and type 202
5 Click the **Options...** button
6 Click in the **Confidence level** text box and type 95
7 Select the **Use test and interval based on normal distribution** check box
8 Click **OK**
9 Click **OK**

EXCEL
1 Store the sample size, 1010, and the number of successes, 202, in ranges named n and x, respectively
2 Choose **DDXL ➤ Confidence Intervals**
3 Select **Summ 1 Var Prop Interval** from the **Function type** drop-down list box
4 Specify x in the **Num Successes** text box
5 Specify n in the **Num Trials** text box
6 Click **OK**
7 Click the **95%** button
8 Click the **Compute Interval** button

TI-83/84 PLUS
1 Press **STAT**, arrow over to **TESTS**, and press **ALPHA ➤ A**
2 Type 202 for **x** and press **ENTER**
3 Type 1010 for **n** and press **ENTER**
4 Type .95 for **C-Level** and press **ENTER** twice

EXERCISES 12.1

Statistical Concepts and Skills

12.1 In a newspaper or magazine of your choice, find a statistical study that contains an estimated population proportion.

12.2 Why is statistical inference generally used to obtain information about a population proportion?

12.3 Is a population proportion a parameter or a statistic? What about a sample proportion? Explain your answers.

12.4 Answer the following questions about the basic notation and terminology for proportions.
a. What is a population proportion?
b. What symbol is used for a population proportion?
c. What is a sample proportion?
d. What symbol is used for a sample proportion?
e. For what is the phrase "number of successes" an abbreviation? What symbol is used for the number of successes?
f. For what is the phrase "number of failures" an abbreviation?
g. Explain the relationships among the sample proportion, the number of successes, and the sample size.

12.5 This exercise involves the use of an unrealistically small population to provide a concrete illustration for the exact distribution of a sample proportion. A population consists of three men and two women. The first names of the men are Jose, Pete, and Carlo; the first names of the women are Gail and Frances. Suppose that the specified attribute is "female."
a. Determine the population proportion, p.

b. The first column of the following table provides the possible samples of size 2, where each person is represented by the first letter of his or her first name; the second column gives the number of successes—the number of females obtained—for each sample, and the third column shows the sample proportion. Complete the table.

Sample	Number of females x	Sample proportion $\hat{p}$
J, G	1	0.5
J, P	0	0.0
J, C	0	0.0
J, F	1	0.5
G, P		
G, C		
G, F		
P, C		
P, F		
C, F		

c. Construct a dotplot for the sampling distribution of the proportion for samples of size 2. Mark the position of the population proportion on the dotplot.
d. Use the third column of the table to obtain the mean of the variable $\hat{p}$.
e. Compare your answers from parts (a) and (d). Why are they the same?

12.6 Repeat parts (b)–(e) of Exercise 12.5 for samples of size 1.

12.7 Repeat parts (b)–(e) of Exercise 12.5 for samples of size 3. (There are 10 possible samples.)

12.8 Repeat parts (b)–(e) of Exercise 12.5 for samples of size 4. (There are five possible samples.)

12.9 Repeat parts (b)–(e) of Exercise 12.5 for samples of size 5.

12.10 Prerequisite to this exercise are Exercises 12.5–12.9. What do your graphs in parts (c) of those exercises illustrate about the impact of increasing sample size on sampling error? Explain your answer.

12.11 NBA Draft Picks. Since 1966, 45% of the No. 1 draft picks in the National Basketball Association have been centers.
a. Identify the population.
b. Identify the specified attribute.
c. Is the proportion 0.45 (45%) a population proportion or a sample proportion? Explain your answer.

12.12 Staying Single. According to an article in *Time* magazine, women are staying single longer these days, by choice. In 1963, 83% of women in the United States between the ages of 25 and 54 were married, compared to 68% in 2000. For 2000,
a. identify the population.
b. identify the specified attribute.
c. Under what circumstances is the proportion 0.68 a population proportion? a sample proportion? Explain your answers.

12.13 Random Drug Testing. A *Harris Poll* conducted in June 2000, asked Americans whether states should be allowed to conduct random drug tests on elected officials. Of 21,355 respondents, 79% said "yes."
a. Determine the margin of error for a 99% confidence interval.
b. Without doing any calculations, indicate whether the margin of error is larger or smaller for a 90% confidence interval. Explain your answer.

12.14 Genetic Binge Eating. According to an article in the March 22, 2003 issue of *Science News*, binge eating has been associated with a mutation of the gene for a brain protein called melanocortin 4 receptor (MC4R). In one study, Fritz F. Horber of the Hirslanden Clinic in Zurich and his colleagues genetically analyzed the blood of 469 obese people and found that 24 carried a mutated MC4R gene. Suppose that you want to estimate the proportion of all obese people who carry a mutated MC4R gene.
a. Determine the margin of error for a 90% confidence interval.
b. Without doing any calculations, indicate whether the margin of error is larger or smaller for a 95% confidence interval. Explain your answer.

12.15 In each of parts (a)–(f), we have given a likely range for the observed value of a sample proportion $\hat{p}$. Based on the given range, identify the educated guess that should be used for the observed value of $\hat{p}$ to calculate the required sample size for a prescribed confidence level and margin of error.
a. 0.2 to 0.4   b. 0.4 to 0.7   c. 0.7 or greater
d. 0.2 or less   e. 0.4 or greater f. 0.7 or less
g. In each of parts (a)–(f), which observed values of the sample proportion will yield a larger margin of error than the one specified if the educated guess is used for the sample size computation?

In Exercises 12.16–12.19, apply Procedure 12.1 on page 610 to obtain the required confidence interval. Be sure to check the condition for using that procedure.

12.16 Asthmatics and Sulfites. Studies are performed to estimate the percentage of the nation's 10 million asthmatics who are allergic to sulfites. In one survey, 38 of 500 randomly selected U.S. asthmatics were found to be allergic to sulfites.
a. Find a 95% confidence interval for the proportion, p, of all U.S. asthmatics who are allergic to sulfites.
b. Interpret your result from part (a).

12.17 Drinking Habits. A *Reader's Digest/Gallup Survey* on the drinking habits of Americans estimated the percentage of adults across the country who drink beer, wine, or hard liquor, at least occasionally. Of the 1516 adults interviewed, 985 said that they drank.
a. Determine a 95% confidence interval for the proportion, p, of all Americans who drink beer, wine, or hard liquor, at least occasionally.
b. Interpret your result from part (a).

12.18 Factory Farming Funk. The U.S. Environmental Protection Agency recently reported that confined animal feeding operations (CAFOs) dump 2 trillion pounds of waste into the environment annually, contaminating the ground water in 17 states and polluting more than 35,000 miles of our nation's rivers. In a recent survey of 1000 registered voters by Snell, Perry and Associates, 80% favored the creation of standards to limit such pollution and, in general, viewed CAFOs unfavorably.

a. Find a 99% confidence interval for the percentage of all registered voters who favor the creation of standards on CAFO pollution and, in general, view CAFOs unfavorably.

b. Interpret your answer in part (a).

12.19 The Nipah Virus. From fall 1998 through mid 1999, Malaysia was the site of an encephalitis outbreak caused by the Nipah virus, a paramyxovirus that appears to spread from pigs to workers on pig farms. As reported by Goh et al. in the *New England Journal of Medicine* (Vol. 342(17), p. 1229), neurologists from the University of Malaysia found that, among 94 patients infected with the Nipah virus, 30 died from encephalitis.

a. Find a 90% confidence interval for the percentage of Malaysians infected with the Nipah virus who will die from encephalitis.

b. Interpret your answer in part (a).

12.20 Literate Adults. Suppose that you have been hired to estimate the percentage of adults in your state who are literate. You take a random sample of 100 adults and find that 96 are literate. You then obtain a 95% confidence interval of

$$0.96 \pm 1.96 \cdot \sqrt{(0.96)(0.04)/100},$$

or 0.922 to 0.998. From it you conclude that you can be 95% confident that the percentage of all adults in your state who are literate is somewhere between 92.2% and 99.8%. Is anything wrong with this reasoning?

12.21 IMR in Singapore. Recall that the infant mortality rate (IMR) is the number of infant deaths per 1000 live births. Suppose that you have been commissioned to estimate the IMR in Singapore. From a random sample of 1109 live births in Singapore, you find that 0.361% of them resulted in infant deaths. You next find a 90% confidence interval:

$$0.00361 \pm 1.645 \cdot \sqrt{(0.00361)(0.99639)/1109},$$

or 0.000647 to 0.00657. You then conclude that, "I can be 90% confident that the IMR in Singapore is somewhere between 0.647 and 6.57." How did you do?

12.22 Warming to Russia. A recent *ABCNEWS Poll* found that Americans now have relatively warm feelings toward Russia, a former adversary. The poll, conducted by telephone among a random sample of 1043 adults, found that 647 of those sampled consider the two countries friends. The margin of error for the poll was plus or minus 2.9 percentage points (for a 0.95 confidence level). Use this information to obtain a 95% confidence interval for the percentage of all Americans who consider the two countries friends.

12.23 Online Tax Returns. According to the U.S. Internal Revenue Service, among people entitled to tax refunds, those who file online receive their refunds twice as fast as paper filers. A study conducted by ICR of Media, Pennsylvania, found that 57% of those polled said that they are not worried about the privacy of their financial information when filing their tax returns online. The telephone survey of 1002 people from March 30, 2000, to April 5, 2000, had a margin of error of plus or minus 3 percentage points (for a 0.95 confidence level). Use this information to obtain a 95% confidence interval for the percentage of all people who are not worried about the privacy of their financial information when filing their tax returns online.

12.24 Asthmatics and Sulfites. Refer to Exercise 12.16.

a. Determine the margin of error for the estimate of p.

b. Obtain a sample size that will ensure a margin of error of at most 0.01 for a 95% confidence interval without making a guess for the observed value of $\hat{p}$.

c. Find a 95% confidence interval for p if, for a sample of the size determined in part (b), the proportion of asthmatics sampled who are allergic to sulfites is 0.071.

d. Determine the margin of error for the estimate in part (c) and compare it to the margin of error specified in part (b).

e. Repeat parts (b)–(d) if you can reasonably presume that the proportion of asthmatics sampled who are allergic to sulfites will be at most 0.10.

f. Compare the results you obtained in parts (b)–(d) with those obtained in part (e).

12.25 Drinking Habits. Refer to Exercise 12.17.

a. Find the margin of error for the estimate of p.

b. Obtain a sample size that will ensure a margin of error of at most 0.02 for a 95% confidence interval without making a guess for the observed value of $\hat{p}$.

c. Find a 95% confidence interval for p if, for a sample of the size determined in part (b), 63% of those sampled drink alcoholic beverages.

d. Determine the margin of error for the estimate in part (c) and compare it to the margin of error specified in part (b).

e. Repeat parts (b)–(d) if you can reasonably presume that the percentage of adults sampled who drink alcoholic beverages will be at least 60%.

f. Compare the results you obtained in parts (b)–(d) with those obtained in part (e).

12.26 Factory Farming Funk. Refer to Exercise 12.18.

a. Determine the margin of error for the estimate of the percentage.

b. Obtain a sample size that will ensure a margin of error of at most 1.5 percentage points for a 99% confidence interval without making a guess for the observed value of $\hat{p}$.

c. Find a 99% confidence interval for p if, for a sample of the size determined in part (b), 82.2% of the registered voters sampled favor the creation of standards on CAFO pollution and, in general, view CAFOs unfavorably.

d. Determine the margin of error for the estimate in part (c) and compare it to the margin of error specified in part (b).

e. Repeat parts (b)–(d) if you can reasonably presume that the percentage of registered voters sampled who favor the creation of standards on CAFO pollution and, in general, view CAFOs unfavorably will be between 75% and 85%.

f. Compare the results you obtained in parts (b)–(d) with those obtained in part (e).

12.27 The Nipah Virus. Refer to Exercise 12.19.

a. Find the margin of error for the estimate of the percentage.

b. Obtain a sample size that will ensure a margin of error of at most 5 percentage points for a 90% confidence interval without making a guess for the observed value of $\hat{p}$.

c. Find a 90% confidence interval for p if, for a sample of the size determined in part (b), 28.8% of the sampled Malaysians infected with the Nipah virus die from encephalitis.

d. Determine the margin of error for the estimate in part (c) and compare it to the margin of error specified in part (b).

e. Repeat parts (b)–(d) if you can reasonably presume that the percentage of sampled Malaysians infected with the Nipah virus who will die from encephalitis will be between 25% and 40%.

f. Compare the results you obtained in parts (b)–(d) with those obtained in part (e).

12.28 Product Response Rate. A company manufactures goods that are sold exclusively by mail order. The director of market research needed to test market a new product. She planned to send brochures to a random sample of households and use the proportion of orders obtained as an estimate of the true proportion, known as the *product response rate*. The results of the market research were to be utilized as a primary source for advance production planning, so the director wanted the figures she presented to be as accurate as possible. Specifically, she wanted to be 95% confident that the estimate of the product response rate would be accurate to within 1%.

a. Without making any assumptions, determine the sample size required.

b. Historically, product response rates for products sold by this company have ranged from 0.5% to 4.9%. If the director had been willing to assume that the sample prod-

uct response rate for this product would also fall in that range, find the required sample size.

c. Compare the results from parts (a) and (b).

d. Discuss the possible consequences if the assumption made in part (b) turns out to be incorrect.

12.29 Indicted Governor. On Thursday, June 13, 1996, then-Arizona Governor Fife Symington was indicted on 23 counts of fraud and extortion. Just hours after the federal prosecutors announced the indictment, several polls were conducted of Arizonans asking whether they thought Symington should resign. One poll, conducted by Research Resources, Inc., which appeared in the *Phoenix Gazette*, revealed that 58% of Arizonans felt that Symington should resign; it had a margin of error of plus or minus 4.9 percentage points. Another poll, conducted by Phoenix-based Behavior Research Center and appearing in the *Tempe Daily News*, reported that 54% of Arizonans felt that Symington should resign; it had a margin of error of plus or minus 4.4 percentage points. Can the conclusions of both polls be correct? Explain your answer.

Extending the Concepts and Skills

12.30 What important theorem in statistics implies that, for a large sample size, the possible sample proportions of that size have approximately a normal distribution?

12.31 In discussing the sample size required for obtaining a confidence interval with a prescribed confidence level and margin of error, we made the following statement: "If we have in mind a likely range for the observed value of $\hat{p}$, then, in light of Fig. 12.1, we should take as our educated guess for $\hat{p}$ the value in the range closest to 0.5." Explain why.

12.32 In discussing the sample size required for obtaining a confidence interval with a prescribed confidence level and margin of error, we made the following statement: "…we should be aware that, if the observed value of $\hat{p}$ is closer to 0.5 than is our educated guess, the margin of error will be larger than desired." Explain why.

12.33 Consider a population in which the proportion of members having a specified attribute is p. Let y be the variable whose value is 1 if a member has the specified attribute and 0 if a member does not.

a. If the size of the population is N, how many members of the population have the specified attribute?

b. Use part (a) and Definition 3.11 on page 131 to show that $\mu_y = p$.

c. Use part (b) and the computing formula in Definition 3.12 on page 133 to show that $\sigma_y = \sqrt{p(1-p)}$.

d. Explain why $\bar{y} = \hat{p}$.

e. Use parts (b)–(d) and Key Fact 7.4 on page 342, to justify Key Fact 12.1.

Using Technology

12.34 Asthmatics and Sulfites. Refer to Exercise 12.16.
a. Use the technology of your choice to obtain the confidence interval required in part (a) of Exercise 12.16.
b. Compare your answer to the one you obtained in Exercise 12.16. Explain any discrepancy that you observe.

12.35 Drinking Habits. Refer to Exercise 12.17.
a. Use the technology of your choice to obtain the confidence interval required in part (a) of Exercise 12.17.
b. Compare your answer to the one you obtained in Exercise 12.17. Explain any discrepancy that you observe.

12.2 Hypothesis Tests for One Population Proportion

In Section 12.1, we showed how to obtain confidence intervals for a population proportion. Now we demonstrate how to perform hypothesis tests for a population proportion. This procedure is actually a special case of the one-sample z-test for a population mean (Procedure 9.1, page 417).

From Key Fact 12.1 on page 608, we deduce that, for large n, the standardized version of $\hat{p}$,

$$z = \frac{\hat{p} - p}{\sqrt{p(1-p)/n}},$$

has approximately the standard normal distribution. Consequently, to perform a large-sample hypothesis test with null hypothesis $H_0: p = p_0$, we can use the variable

$$z = \frac{\hat{p} - p_0}{\sqrt{p_0(1-p_0)/n}}$$

as the test statistic and obtain the critical value(s) or P-value from the standard normal table, Table II. We refer to this hypothesis-testing procedure as the **one-sample z-test** for a population proportion or, simply, the **z-test** for a population proportion.

Procedure 12.2 on the next page provides a step-by-step method for performing a one-sample z-test for a population proportion by using either the critical-value approach or the P-value approach. We apply Procedure 12.2 in Example 12.6.

▌▌Example 12.6 The One-Sample z-Test

Gun Control   One of the more controversial issues in the United States is gun control; there are many avid proponents and opponents of banning handgun sales. In a survey conducted by Louis Harris of LH Research, 1250 U.S. adults were polled regarding their views on banning handgun sales. Of those sampled, 650 favored a ban. At the 5% significance level, do the data provide sufficient evidence to conclude that a majority of U.S. adults (i.e., more than 50%) favor banning handgun sales?

Solution   Because $n = 1250$ and $p_0 = 0.50$ (50%), we have

$$np_0 = 1250 \cdot 0.50 = 625 \quad \text{and} \quad n(1-p_0) = 1250 \cdot (1-0.50) = 625.$$

As both np_0 and $n(1-p_0)$ are 5 or greater, we can apply Procedure 12.2.

PROCEDURE 12.2 One-Sample z-Test

Purpose To perform a hypothesis test for a population proportion, p

Assumptions

1. Simple random sample

2. Both np_0 and $n(1 - p_0)$ are 5 or greater

STEP 1 The null hypothesis is H_0: $p = p_0$, and the alternative hypothesis is

$$H_a: p \neq p_0 \quad \text{or} \quad H_a: p < p_0 \quad \text{or} \quad H_a: p > p_0$$
$$\text{(Two tailed)} \qquad \text{(Left tailed)} \qquad \text{(Right tailed)}$$

STEP 2 Decide on the significance level, α.

STEP 3 Compute the value of the test statistic

$$z = \frac{\hat{p} - p_0}{\sqrt{p_0(1 - p_0)/n}}$$

and denote that value z_0.

CRITICAL-VALUE APPROACH	P-VALUE APPROACH

STEP 4 The critical value(s) are

$$\pm z_{\alpha/2} \quad \text{or} \quad -z_\alpha \quad \text{or} \quad z_\alpha$$
$$\text{(Two tailed)} \qquad \text{(Left tailed)} \qquad \text{(Right tailed)}$$

Use Table II to find the critical value(s).

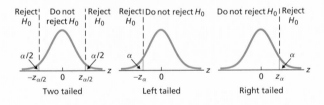

STEP 5 If the value of the test statistic falls in the rejection region, reject H_0; otherwise, do not reject H_0.

STEP 4 Use Table II to obtain the P-value.

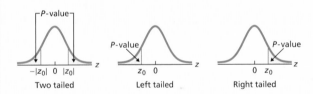

STEP 5 If $P \leq \alpha$, reject H_0; otherwise, do not reject H_0.

STEP 6 Interpret the results of the hypothesis test.

STEP 1 State the null and alternative hypotheses.

Let p denote the proportion of all U.S. adults who favor banning handgun sales. Then the null and alternative hypotheses are

$$H_0: \ p = 0.50 \text{ (it is not true that a majority favor a ban)}$$
$$H_a: \ p > 0.50 \text{ (a majority favor a ban)}.$$

The hypothesis test is right tailed because a greater-than sign ($>$) appears in the alternative hypothesis.

STEP 2 Decide on the significance level, α.

We are to perform the hypothesis test at the 5% significance level; consequently, we have $\alpha = 0.05$.

STEP 3 Compute the value of the test statistic

$$z = \frac{\hat{p} - p_0}{\sqrt{p_0(1 - p_0)/n}}.$$

We have $n = 1250$ and $p_0 = 0.50$. The number of U.S. adults surveyed who favor banning handgun sales is 650. Therefore the proportion of those surveyed who favor a ban is $\hat{p} = x/n = 650/1250 = 0.520$ (52.0%). Consequently, the value of the test statistic is

$$z = \frac{0.520 - 0.50}{\sqrt{(0.50)(1 - 0.50)/1250}} = 1.41.$$

CRITICAL-VALUE APPROACH	P-VALUE APPROACH

CRITICAL-VALUE APPROACH

STEP 4 The critical value for a right-tailed test is z_α. Use Table II to find the critical value.

For $\alpha = 0.05$, the critical value is $z_{0.05} = 1.645$, as shown in Fig. 12.2A.

FIGURE 12.2A

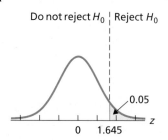

Do not reject H_0 | Reject H_0

0.05

0   1.645

STEP 5 If the value of the test statistic falls in the rejection region, reject H_0; otherwise, do not reject H_0.

From Step 3, the value of the test statistic is $z = 1.41$, which, as we see from Fig. 12.2A, does not fall in the rejection region. Thus we do not reject H_0. The test results are not statistically significant at the 5% level.

P-VALUE APPROACH

STEP 4 Use Table II to obtain the P-value.

From Step 3, the value of the test statistic is $z = 1.41$. The test is right tailed, so the P-value is the probability of observing a value of z of 1.41 or greater if the null hypothesis is true. That probability equals the shaded area in Fig. 12.2B, which by Table II is 0.0793, or $P = 0.0793$.

FIGURE 12.2B

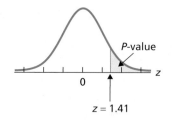

P-value

0

$z = 1.41$

STEP 5 If $P \leq \alpha$, reject H_0; otherwise, do not reject H_0.

From Step 4, $P = 0.0793$. Because the P-value exceeds the specified significance level of 0.05, we do not reject H_0. The test results are not statistically significant at the 5% level but (see Table 9.12 on page 440) do provide moderate evidence against the null hypothesis.

STEP 6 Interpret the results of the hypothesis test.

INTERPRETATION At the 5% significance level, the data do not provide sufficient evidence to conclude that a majority of U.S. adults favor banning handgun sales.

Example 12.6 provides a good illustration of how statistical results are sometimes misstated. The newspaper article that featured the survey had the headline "Fear prompts 52% in U.S. to back pistol-sale ban, poll says." In fact, the poll says no such thing. It says only that 52% of those *sampled* back a pistol-sale ban. And, as we have demonstrated, at the 5% significance level the poll does not provide sufficient evidence to conclude that a majority of U.S. adults back a pistol-sale ban.

Most statistical technologies have programs that automatically perform a hypothesis test for a population proportion. In this subsection, we present output and step-by-step instructions to implement such programs.

Example 12.7   **Using Technology to Conduct a z-Test**

Gun Control   Use Minitab, Excel, or the TI-83/84 Plus to perform the hypothesis test considered in Example 12.6 on page 620.

Solution   Let p denote the proportion of all U.S. adults who favor banning handgun sales. The task is to perform the hypothesis test

$$H_0:\ p = 0.50 \text{ (it is not true that a majority favor a ban)}$$
$$H_a:\ p > 0.50 \text{ (a majority favor a ban)}$$

at the 5% significance level. Note that the hypothesis test is right tailed because a greater-than sign ($>$) appears in the alternative hypothesis. Also, recall that 1250 U.S. adults were polled and that 650 of those adults favored banning handgun sales. Printout 12.3 on the next page shows the output obtained by applying the one-sample z-test programs to the data.

The outputs in Printout 12.3 show that the P-value for the hypothesis test is 0.079 (to three decimal places). The P-value exceeds the specified significance level of 0.05, so we do not reject H_0. The test results are not statistically significant at the 5% level; that is, at the 5% significance level, the data do not provide sufficient evidence to conclude that a majority of U.S. adults favor banning handgun sales.

Obtaining the Output (Optional)

Printout 12.3 provides output from Minitab, Excel, and the TI-83/84 Plus for a one-sample z-test to decide whether a majority of U.S. adults favor banning handgun sales. The following are detailed instructions for obtaining that output.

MINITAB

1 Choose **Stat ➤ Basic Statistics ➤ 1 Proportion...**
2 Select the **Summarized data** option button
3 Click in the **Number of trials** text box and type <u>1250</u>
4 Click in the **Number of events** text box and type <u>650</u>
5 Click the **Options...** button
6 Click in the **Test proportion** text box and type <u>0.50</u>
7 Click the arrow button at the right of the **Alternative** drop-down list box and select **greater than**
8 Select the **Use test and interval based on normal distribution** check box
9 Click **OK**
10 Click **OK**

EXCEL

1 Store the sample size, 1250, and the number of successes, 650, in ranges named n and x, respectively
2 Choose **DDXL ➤ Hypothesis Tests**
3 Select **Summ 1 Var Prop Test** from the **Function type** drop-down list box
4 Specify x in the **Num Successes** text box
5 Specify n in the **Num Trials** text box
6 Click **OK**
7 Click the **Set p0** button
8 Click in the **Hypothesized Population Proportion** text box and type <u>0.50</u>
9 Click **OK**
10 Click the **.05** button
11 Click the **p > p0** button
12 Click the **Compute** button

TI-83/84 PLUS

1 Press **STAT**, arrow over to **TESTS**, and press **5**
2 Type <u>0.50</u> for p_0 and press **ENTER**
3 Type <u>650</u> for **x** and press **ENTER**
4 Type <u>1250</u> for **n** and press **ENTER**
5 Highlight **> p_0** and press **ENTER**
6 Press the down-arrow key, highlight **Calculate** or **Draw**, and press **ENTER**

PRINTOUT 12.3
One-sample z-test output for the data on banning handguns

MINITAB

Test and CI for One Proportion

Test of p = 0.5 vs p > 0.5

Sample	X	N	Sample p	95% Lower Bound	Z-Value	P-Value
1	650	1250	0.520000	0.496757	1.41	0.079

TI-83/84 PLUS

```
1-PropZTest
 prop>.5
 z=1.414213562
 P=.0786496525
 p=.52
 n=1250
■
```

Using **Calculate**

EXCEL

```
▷ x Proportion Test                              ▢▢▧

 ▷ │ Summary Statistics │▢  ▷│ Test Summary        │▢
         n    1250              p0:           0.5
      p-hat   0.52              Ho:       p = 0.5
    Std Dev   0.0141            Ha:  Upper tail: p > 0.5
                            z Statistic:      1.414
                              p-value:        0.0786

                          ▷│ Test Results         │▢
                          Conclusion
                          Fail to reject Ho at alpha =  0.05
```

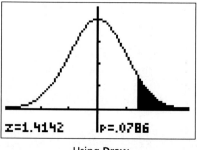

Using **Draw**

EXERCISES 12.2

Statistical Concepts and Skills

12.36 Of what procedure is Procedure 12.2 a special case? Why do you think that is so?

In Exercises 12.37–12.42, use either the critical-value approach or the P-value approach to perform each hypothesis test. Comment on the practical significance of all tests whose results are statistically significant.

12.37 Generation Y Online. People who were born between 1978 and 1983 are sometimes classified by demographers as belonging to Generation Y. According to a recent Forrester Research survey published in *American Demographics* (Vol. 22(1), p. 12), of 850 Generation Y Web users, 459 reported using the Internet to download music.
a. Determine the sample proportion.
b. At the 5% significance level, do the data provide sufficient evidence to conclude that a majority of Generation Y Web users use the Internet to download music?

12.38 Christmas Presents. The *Arizona Republic* conducted a telephone poll of 758 Arizona adults who celebrate Christmas. The question asked was, "In your family, do you open presents on Christmas Eve or Christmas Day?" Of those surveyed, 394 said they wait until Christmas Day.
a. Determine the sample proportion.
b. At the 5% significance level, do the data provide sufficient evidence to conclude that a majority of Arizona families who celebrate Christmas wait until Christmas Day to open their presents?

12.39 Marijuana and Hashish. The U.S. Substance Abuse and Mental Health Services Administration conducts surveys on drug use by type of drug and age group. Results are published in *National Household Survey on Drug Abuse*. According to that publication, 13.6% of 18–25-year-olds were current users of marijuana or hashish in 2000. A recent poll of 1283 randomly selected 18–25-year-olds revealed that 205 currently use marijuana or hashish. At the 10% significance level, do the data provide sufficient evidence to conclude that the percentage of 18–25-year-olds

who currently use marijuana or hashish has changed from the 2000 percentage of 13.6%?

12.40 Families in Poverty. In 2000, 8.6% of all U.S. families had incomes below the poverty level, as reported by the Census Bureau in *Current Population Reports*. During that same year, of 400 randomly selected families whose householder had at least a Bachelor's degree, 9 had incomes below the poverty level. At the 1% significance level, do the data provide sufficient evidence to conclude that, in 2000, the percentage of families that earned incomes below the poverty level was lower among those whose householders had at least a Bachelor's degree than among all U.S. families?

12.41 Labor-Union Support. Labor Day was created by the U.S. labor movement over 100 years ago. It was subsequently adopted by most states as an official holiday. In a recent poll by the Gallup Organization, 1003 randomly selected adults were asked whether they approve of labor unions; 65% said yes.
a. In 1936, about 72% of Americans approved of labor unions. At the 5% significance level, do the data provide sufficient evidence to conclude that the percentage of Americans who approve of labor unions now has decreased since 1936?
b. In 1963, roughly 67% of Americans approved of labor unions. At the 5% significance level, do the data provide sufficient evidence to conclude that the percentage of Americans who approve of labor unions now has decreased since 1963?

12.42 An Edge in Roulette? Of the 38 numbers on an American roulette wheel, 18 are red, 18 are black, and 2 are green. If the wheel is balanced, the probability of the ball landing on red is $\frac{18}{38} = 0.474$. A gambler has been studying a roulette wheel. If the wheel is out of balance, he can improve his odds of winning. The gambler observes 200 spins of the wheel and finds that the ball lands on red 93 times. At the 10% significance level, do the data provide sufficient evidence to conclude that the ball is not landing on red the correct percentage of the time for a balanced wheel?

Using Technology

12.43 Marijuana and Hashish. Use the technology of your choice to perform the hypothesis test in Exercise 12.39.

12.44 Families in Poverty. Use the technology of your choice to perform the hypothesis test in Exercise 12.40.

12.3 Inferences for Two Population Proportions, Using Independent Samples

In Sections 12.1 and 12.2, you studied inferences for one population proportion. Now we examine inferences for comparing two population proportions. In this case, we have two populations and one specified attribute; the problem is to compare the proportion of one population that has the specified attribute to the proportion of the other population that has the specified attribute. We begin by discussing hypothesis testing in Example 12.8.

Example 12.8 **Hypothesis Tests for Two Population Proportions**

Eating Out Vegetarian A *Zogby International* poll of 1181 U.S. adults was conducted to gauge the demand for vegetarian meals in restaurants. The study was commissioned by the Vegetarian Resource Group and was published in the *Vegetarian Journal*.

In the survey, samples of 747 U.S. men and 434 U.S. women were independently and randomly taken. Of those sampled, 276 men and 195 women said that they sometimes order a dish without meat, fish, or fowl when they eat out.

Suppose that we want to use the data to decide whether, in the United States, the percentage of men who sometimes order a dish without meat, fish, or fowl is smaller than the percentage of women who sometimes order a dish without meat, fish, or fowl.

a. Formulate the problem statistically by posing it as a hypothesis test.

b. Explain the basic idea for carrying out the hypothesis test.

c. Discuss the use of the data to make a decision concerning the hypothesis test.

Solution

a. The specified attribute is "sometimes orders a dish without meat, fish, or fowl," which we abbreviate as "sometimes orders veg." The two populations are

> Population 1: All U.S. men
> Population 2: All U.S. women.

Let p_1 and p_2 denote the population proportions for the two populations:

> p_1 = proportion of all U.S. men who sometimes order veg
> p_2 = proportion of all U.S. women who sometimes order veg.

We want to perform the hypothesis test

> H_0: $p_1 = p_2$ (percentage for men is not less than that for women)
> H_a: $p_1 < p_2$ (percentage for men is less than that for women).

b. Roughly speaking, we can carry out the hypothesis test as follows:

    1. Compute the proportion of the men sampled who sometimes order veg, $\hat{p}_1$, and compute the proportion of the women sampled who sometimes order veg, $\hat{p}_2$.

    2. If $\hat{p}_1$ is too much smaller than $\hat{p}_2$, reject H_0; otherwise, do not reject H_0.

c. The first step in part (b) is easy. Because 276 of the 747 men sampled sometimes order veg and 195 of the 434 women sampled sometimes order veg,

$$\hat{p}_1 = \frac{x_1}{n_1} = \frac{276}{747} = 0.369 \ (36.9\%)$$

and

$$\hat{p}_2 = \frac{x_2}{n_2} = \frac{195}{434} = 0.449 \ (44.9\%).$$

For the second step in part (b), we must decide whether the sample proportion $\hat{p}_1 = 0.369$ is less than the sample proportion $\hat{p}_2 = 0.449$ by a sufficient amount to warrant rejecting the null hypothesis in favor of the alternative hypothesis. In other words, we need to decide whether the difference between the two sample proportions can reasonably be attributed to sampling error or whether it indicates that the percentage of men who sometimes order veg is less than the percentage of women who sometimes order veg.

To make that decision, we need to know the distribution of the difference between two sample proportions. We discuss that sampling distribution and then complete the hypothesis test.   ▪

The Sampling Distribution of the Difference Between Two Sample Proportions for Large and Independent Samples

To begin our discussion of the sampling distribution of the difference between two sample proportions, we summarize the required notation in Table 12.2.

TABLE 12.2

Notation for parameters and statistics when two population proportions are being considered

	Population 1	**Population 2**
Population proportion	p_1	p_2
Sample size	n_1	n_2
Number of successes	x_1	x_2
Sample proportion	$\hat{p}_1$	$\hat{p}_2$

Recall that the *number of successes* refers to the number of members sampled that have the specified attribute. Consequently, we compute the sample proportions by using the formulas

$$\hat{p}_1 = \frac{x_1}{n_1} \quad \text{and} \quad \hat{p}_2 = \frac{x_2}{n_2}.$$

Armed with the notation in Table 12.2, we now describe, in Key Fact 12.2, the **sampling distribution of the difference between two sample proportions.**

Recalling Key Fact 12.1 on page 608, which gives the sampling distribution of one sample proportion, helps.

KEY FACT 12.2

> ### The Sampling Distribution of the Difference Between Two Sample Proportions for Independent Samples
>
> For independent samples of sizes n_1 and n_2 from the two populations,
>
> - $\mu_{\hat{p}_1 - \hat{p}_2} = p_1 - p_2$,
> - $\sigma_{\hat{p}_1 - \hat{p}_2} = \sqrt{p_1(1-p_1)/n_1 + p_2(1-p_2)/n_2}$, and
> - $\hat{p}_1 - \hat{p}_2$ is approximately normally distributed for large n_1 and n_2.

What
Does it Mean?

For large independent samples, the possible differences between two sample proportions have approximately a normal distribution with mean $p_1 - p_2$ and standard deviation $\sqrt{p_1(1-p_1)/n_1 + p_2(1-p_2)/n_2}$.

Key Fact 12.2 provides the necessary basis for deriving inferential procedures to compare two population proportions.

Large-Sample Hypothesis Tests for Two Population Proportions, Using Independent Samples

We now develop a hypothesis-testing procedure for comparing two population proportions. Our immediate goal is to identify a variable that we can use as the test statistic. From Key Fact 12.2, we know that, for large, independent samples, the standardized variable

$$z = \frac{(\hat{p}_1 - \hat{p}_2) - (p_1 - p_2)}{\sqrt{p_1(1-p_1)/n_1 + p_2(1-p_2)/n_2}} \tag{12.2}$$

has approximately the standard normal distribution.

The null hypothesis for a hypothesis test to compare two population proportions is

$$H_0: p_1 = p_2 \text{ (population proportions are equal)}.$$

If the null hypothesis is true, then $p_1 - p_2 = 0$ and, consequently, the variable in Equation (12.2) becomes

$$z = \frac{\hat{p}_1 - \hat{p}_2}{\sqrt{p(1-p)/n_1 + p(1-p)/n_2}}, \tag{12.3}$$

where p denotes the common value of p_1 and p_2. Factoring $p(1-p)$ out of the denominator of Equation (12.3) yields the variable

$$z = \frac{\hat{p}_1 - \hat{p}_2}{\sqrt{p(1-p)}\sqrt{(1/n_1) + (1/n_2)}}. \tag{12.4}$$

However, because p is unknown, we cannot use this variable as the test statistic.

Consequently, we must estimate p by using sample information. The best estimate of p is obtained by pooling the data to get the proportion of successes in both samples combined; that is, we estimate p by

$$\hat{p}_p = \frac{x_1 + x_2}{n_1 + n_2}.$$

We call $\hat{p}_p$ the **pooled sample proportion**.

Replacing p in Equation (12.4) with its estimate $\hat{p}_p$ yields the variable

$$\frac{\hat{p}_1 - \hat{p}_2}{\sqrt{\hat{p}_p(1 - \hat{p}_p)}\sqrt{(1/n_1) + (1/n_2)}},$$

which can be used as the test statistic and, like the variable in Equation (12.4), has approximately the standard normal distribution for large samples if the null hypothesis is true. Therefore we have Procedure 12.3 (on the next page), which we call the **two-sample z-test** for two population proportions. In Example 12.9, we demonstrate use of that procedure.

Example 12.9   The Two-Sample z-Test

Eating Out Vegetarian   We now return to the problem posed in Example 12.8. Independent random samples of 747 U.S. men and 434 U.S. women were taken. Of those sampled, 276 men and 195 women said that they sometimes order a dish without meat, fish, or fowl when they eat out. At the 5% significance level, do the data provide sufficient evidence to conclude that, in the United States, the percentage of men who sometimes order a dish without meat, fish, or fowl is smaller than the percentage of women who sometimes order a dish without meat, fish, or fowl?

Solution   We first recall that the specified attribute is "sometimes orders a dish without meat, fish, or fowl," abbreviated as "sometimes orders veg." We apply Procedure 12.3, noting first that the assumptions for its use are satisfied.

STEP 1   State the null and alternative hypotheses.

Let p_1 and p_2 denote the proportions of all U.S. men and all U.S. women who sometimes order veg, respectively. Then the null and alternative hypotheses are

H_0: $p_1 = p_2$ (percentage for men is not less than that for women)

H_a: $p_1 < p_2$ (percentage for men is less than that for women).

The hypothesis test is left tailed because a less-than sign ($<$) appears in the alternative hypothesis.

STEP 2   Decide on the significance level, α.

The test is to be performed at the 5% significance level, or $\alpha = 0.05$.

STEP 3   Compute the value of the test statistic

$$z = \frac{\hat{p}_1 - \hat{p}_2}{\sqrt{\hat{p}_p(1 - \hat{p}_p)}\sqrt{(1/n_1) + (1/n_2)}},$$

where $\hat{p}_p = (x_1 + x_2)/(n_1 + n_2)$.

We first obtain $\hat{p}_1$, $\hat{p}_2$, and $\hat{p}_p$. As 276 of the 747 men sampled and 195 of the 434 women sampled sometimes order veg, $x_1 = 276$, $n_1 = 747$, $x_2 = 195$,

(continued on page 631)

PROCEDURE 12.3	Two-Sample z-Test

Purpose To perform a hypothesis test to compare two population proportions, p_1 and p_2

Assumptions

1. Simple random samples
2. Independent samples
3. x_1, $n_1 - x_1$, x_2, and $n_2 - x_2$ are all 5 or greater

STEP 1 The null hypothesis is H_0: $p_1 = p_2$, and the alternative hypothesis is

$$H_a: p_1 \neq p_2 \quad \text{or} \quad H_a: p_1 < p_2 \quad \text{or} \quad H_a: p_1 > p_2$$
$$\text{(Two tailed)} \qquad \text{(Left tailed)} \qquad \text{(Right tailed)}$$

STEP 2 Decide on the significance level, α.

STEP 3 Compute the value of the test statistic

$$z = \frac{\hat{p}_1 - \hat{p}_2}{\sqrt{\hat{p}_p(1 - \hat{p}_p)}\sqrt{(1/n_1) + (1/n_2)}},$$

where $\hat{p}_p = (x_1 + x_2)/(n_1 + n_2)$. Denote the value of the test statistic z_0.

CRITICAL-VALUE APPROACH	**P-VALUE APPROACH**

CRITICAL-VALUE APPROACH

STEP 4 The critical value(s) are

$$\pm z_{\alpha/2} \quad \text{or} \quad -z_\alpha \quad \text{or} \quad z_\alpha$$
$$\text{(Two tailed)} \qquad \text{(Left tailed)} \qquad \text{(Right tailed)}$$

Use Table II to find the critical value(s).

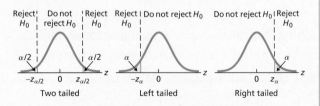

STEP 5 If the value of the test statistic falls in the rejection region, reject H_0; otherwise, do not reject H_0.

P-VALUE APPROACH

STEP 4 Use Table II to obtain the *P*-value.

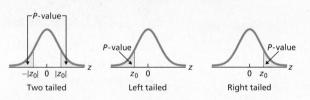

STEP 5 If $P \leq \alpha$, reject H_0; otherwise, do not reject H_0.

STEP 6 Interpret the results of the hypothesis test.

Note: Procedure 12.3 and its confidence-interval counterpart (Procedure 12.4 on page 632) also apply to designed experiments with two treatments.

and $n_2 = 434$. Therefore,

$$\hat{p}_1 = \frac{x_1}{n_1} = \frac{276}{747} = 0.369, \qquad \hat{p}_2 = \frac{x_2}{n_2} = \frac{195}{434} = 0.449,$$

and

$$\hat{p}_p = \frac{x_1 + x_2}{n_1 + n_2} = \frac{276 + 195}{747 + 434} = \frac{471}{1181} = 0.399.$$

Consequently, the value of the test statistic is

$$z = \frac{\hat{p}_1 - \hat{p}_2}{\sqrt{\hat{p}_p(1 - \hat{p}_p)}\sqrt{(1/n_1) + (1/n_2)}}$$

$$= \frac{0.369 - 0.449}{\sqrt{(0.399)(1 - 0.399)}\sqrt{(1/747) + (1/434)}} = -2.71.$$

CRITICAL-VALUE APPROACH	P-VALUE APPROACH

STEP 4 **The critical value for a left-tailed test is** $-z_\alpha$. **Use Table II to find the critical value.**

For $\alpha = 0.05$, we find from Table II that the critical value is $-z_{0.05} = -1.645$, as depicted in Fig. 12.3A.

FIGURE 12.3A

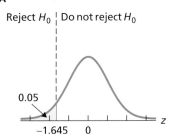

STEP 5 **If the value of the test statistic falls in the rejection region, reject** H_0; **otherwise, do not reject** H_0.

From Step 3, the value of the test statistic is $z = -2.71$, which, as we see from Fig. 12.3A, falls in the rejection region. Thus we reject H_0. The test results are statistically significant at the 5% level.

STEP 4 **Use Table II to obtain the** *P*-value.

From Step 3, the value of the test statistic is $z = -2.71$. As the test is left tailed, the *P*-value is the probability of observing a value of z of -2.71 or less if the null hypothesis is true. That probability equals the shaded area in Fig. 12.3B, which by Table II is 0.0034, or $P = 0.0034$.

FIGURE 12.3B

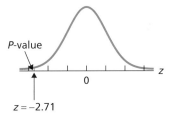

STEP 5 **If** $P \le \alpha$, **reject** H_0; **otherwise, do not reject** H_0.

From Step 4, $P = 0.0034$. Because the *P*-value is less than the specified significance level of 0.05, we reject H_0. The test results are statistically significant at the 5% level and (see Table 9.12 on page 440) provide very strong evidence against the null hypothesis.

STEP 6 **Interpret the results of the hypothesis test.**

INTERPRETATION At the 5% significance level, the data provide sufficient evidence to conclude that, in the United States, the percentage of men who sometimes order a dish without meat, fish, or fowl is smaller than the percentage of women who sometimes order a dish without meat, fish, or fowl.

Large-Sample Confidence Intervals for the Difference Between Two Population Proportions

We can also use Key Fact 12.2 on page 628 to derive a confidence-interval procedure, Procedure 12.4, for the difference between two population proportions, which we call the **two-sample z-interval procedure** for two population proportions. We apply the procedure in Example 12.10.

PROCEDURE 12.4

Two-Sample z-Interval Procedure

Purpose To find a confidence interval for the difference between two population proportions, p_1 and p_2

Assumptions

1. Simple random samples
2. Independent samples
3. x_1, $n_1 - x_1$, x_2, and $n_2 - x_2$ are all 5 or greater

STEP 1 For a confidence level of $1 - \alpha$, use Table II to find $z_{\alpha/2}$.

STEP 2 The endpoints of the confidence interval for $p_1 - p_2$ are

$$(\hat{p}_1 - \hat{p}_2) \pm z_{\alpha/2} \cdot \sqrt{\hat{p}_1(1 - \hat{p}_1)/n_1 + \hat{p}_2(1 - \hat{p}_2)/n_2}.$$

STEP 3 Interpret the confidence interval.

Example 12.10   The Two-Sample z-Interval Procedure

Eating Out Vegetarian   Refer to Example 12.9 and obtain a 90% confidence interval for the difference, $p_1 - p_2$, between the proportions of U.S. men and U.S. women who sometimes order a dish without meat, fish, or fowl.

Solution We apply Procedure 12.4, noting first that the conditions for its use are met.

STEP 1 For a confidence level of $1 - \alpha$, use Table II to find $z_{\alpha/2}$.

For a 90% confidence interval, we have $\alpha = 0.10$. From Table II, we determine that $z_{\alpha/2} = z_{0.10/2} = z_{0.05} = 1.645$.

STEP 2 The endpoints of the confidence interval for $p_1 - p_2$ are

$$(\hat{p}_1 - \hat{p}_2) \pm z_{\alpha/2} \cdot \sqrt{\hat{p}_1(1 - \hat{p}_1)/n_1 + \hat{p}_2(1 - \hat{p}_2)/n_2}.$$

From Step 1, $z_{\alpha/2} = 1.645$. As we found in Example 12.9, $\hat{p}_1 = 0.369$, $n_1 = 747$, $\hat{p}_2 = 0.449$, and $n_2 = 434$. Therefore the endpoints of the 90% confidence interval for $p_1 - p_2$ are

$$(0.369 - 0.449) \pm 1.645 \cdot \sqrt{(0.369)(1 - 0.369)/747 + (0.449)(1 - 0.449)/434},$$

or -0.080 ± 0.049, or -0.129 to -0.031.

STEP 3 Interpret the confidence interval.

INTERPRETATION We can be 90% confident that, in the United States, the difference between the proportions of men and women who sometimes order a dish without meat, fish, or fowl is somewhere between -0.129 and -0.031. In other words, we can be 90% confident that the percentage of U.S. women who sometimes order veg exceeds the percentage of U.S. men who sometimes order veg by somewhere between 3.1 and 12.9 percentage points.

Margin of Error and Sample Size

We can obtain the **margin of error** in estimating the difference between two population proportions by referring to Step 2 of Procedure 12.4. The formula for the margin of error is presented in Formula 12.3.

FORMULA 12.3

What Does it Mean?

The margin of error equals half the length of the confidence interval and represents the precision with which the difference between the sample proportions estimates the difference between the population proportions at the specified confidence level.

Margin of Error for the Estimate of $p_1 - p_2$

The margin of error for the estimate of $p_1 - p_2$ is

$$E = z_{\alpha/2} \cdot \sqrt{\hat{p}_1(1 - \hat{p}_1)/n_1 + \hat{p}_2(1 - \hat{p}_2)/n_2}.$$

From the formula for the margin of error, we can determine the sample sizes required to obtain a confidence interval with a specified confidence level and margin of error. Formula 12.4 provides two methods for obtaining the required samples sizes.

FORMULA 12.4

Sample Size for Estimating $p_1 - p_2$

A $(1 - \alpha)$-level confidence interval for the difference between two population proportions that has a margin of error of at most E can be obtained by choosing

$$n_1 = n_2 = 0.5 \left(\frac{z_{\alpha/2}}{E}\right)^2,$$

rounded up to the nearest whole number. If you can make educated guesses, $\hat{p}_{1g}$ and $\hat{p}_{2g}$, for the observed values of $\hat{p}_1$ and $\hat{p}_2$, you should instead choose

$$n_1 = n_2 = \left(\hat{p}_{1g}(1 - \hat{p}_{1g}) + \hat{p}_{2g}(1 - \hat{p}_{2g})\right)\left(\frac{z_{\alpha/2}}{E}\right)^2,$$

rounded up to the nearest whole number.

The first displayed formula in Formula 12.4 provides sample sizes that ensure obtaining a $(1 - \alpha)$-level confidence interval with a margin of error of at most E, but it may yield sample sizes that are unnecessarily large. The second displayed formula in Formula 12.4 yields smaller sample sizes, but it should not be used unless the guesses for the sample proportions are considered reasonably accurate.

If likely ranges for the observed values of the two sample proportions are known, the values in the ranges closest to 0.5 should be taken as the educated guesses. For further discussion of these ideas and for applications of Formulas 12.3 and 12.4, see Exercise 12.60.

The Technology Center

Most statistical technologies have programs that automatically perform a hypothesis test for comparing two population proportions and obtain a confidence interval for the difference between two population proportions. In this subsection, we present output and step-by-step instructions to implement such programs.

Example 12.11 **Using Technology to Conduct Two-Sample z-Procedures**

Eating Out Vegetarian Use Minitab, Excel, or the TI-83/84 Plus to perform the hypothesis test in Example 12.9 on page 629 and obtain the confidence interval required in Example 12.10 on page 632.

Solution Let p_1 and p_2 denote the proportions of all U.S. men and all U.S. women who sometimes order veg, respectively. The task in Example 12.9 is to perform the hypothesis test

H_0: $p_1 = p_2$ (percentage for men is not less than that for women)

H_a: $p_1 < p_2$ (percentage for men is less than that for women)

at the 5% significance level; the task in Example 12.10 is to obtain a 90% confidence interval for $p_1 - p_2$.

We recall that the sample sizes for the men and women are 747 and 434, respectively, and that the numbers of men and women sampled who sometimes order veg are 276 and 195, respectively. Printout 12.4 on pages 635 and 636 shows the output obtained by applying the two-sample-z programs to the data.

The outputs in Printout 12.4 reveal that the P-value for the hypothesis test is 0.003 (to three decimal places). The P-value is less than the specified significance level of 0.05, so we reject H_0. The outputs in Printout 12.4 also show that a 90% confidence interval for the difference between the proportions is from -0.129 to -0.031.

Obtaining the Output (Optional)

Printout 12.4 provides output from Minitab, Excel, and the TI-83/84 Plus for two-sample z-procedures, based on the sample data for the study on sometimes ordering vegetarian. Detailed instructions for obtaining that output are presented at the top of page 637.

PRINTOUT 12.4

Two-sample-z output for the study on ordering vegetarian

MINITAB

Test and CI for Two Proportions   [FOR THE HYPOTHESIS TEST]

```
Sample    X     N    Sample p
1       276   747   0.369478
2       195   434   0.449309


Difference = p (1) - p (2)
Estimate for difference:  -0.0798308
90% upper bound for difference:  -0.0417711
Test for difference = 0 (vs < 0):  Z = -2.70   P-Value = 0.003
```

Test and CI for Two Proportions   [FOR THE CONFIDENCE INTERVAL]

```
Sample    X     N    Sample p
1       276   747   0.369478
2       195   434   0.449309


Difference = p (1) - p (2)
Estimate for difference:  -0.0798308
90% CI for difference:  (-0.128680, -0.0309817)
Test for difference = 0 (vs not = 0):  Z = -2.70   P-Value = 0.007
```

EXCEL

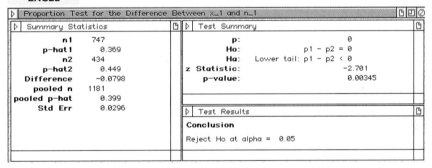

Using **Summ 2 Var Prop Test**

PRINTOUT 12.4 (cont.)

Two-sample-z output for the study on ordering vegetarian

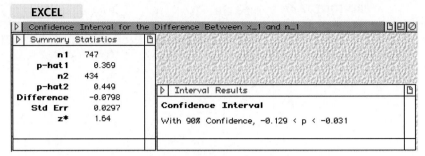

Using **Summ 2 Var Prop Interval**

TI-83/84 PLUS

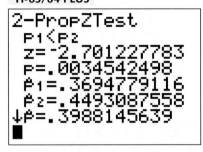

Using **2-PropZInt**

Using **2-PropZTest**

MINITAB	EXCEL	TI-83/84 PLUS

MINITAB

FOR THE HYPOTHESIS TEST:
1 Choose **Stat ➤ Basic Statistics ➤ 2 Proportions...**
2 Select the **Summarized data** option button
3 Click in the **Trials** text box for **First** and type 747
4 Click in the **Events** text box for **First** and type 276
5 Click in the **Trials** text box for **Second** and type 434
6 Click in the **Events** text box for **Second** and type 195
7 Click the **Options...** button
8 Click in the **Confidence level** text box and type 90
9 Click in the **Test difference** text box and type 0
10 Click the arrow button at the right of the **Alternative** drop-down list box and select **less than**
11 Select the **Use pooled estimate of p for test** check box
12 Click **OK**
13 Click **OK**

FOR THE CI:
1 Choose **Edit ➤ Edit Last Dialog**
2 Click the **Options...** button
3 Click the arrow button at the right of the **Alternative** drop-down list box and select **not equal**
4 Click **OK**
5 Click **OK**

EXCEL

Store the sample sizes, 747 and 434, in ranges named n_1 and n_2, respectively, and store the numbers of successes, 276 and 195, in ranges named x_1 and x_2, respectively.

FOR THE HYPOTHESIS TEST:
1 Choose **DDXL ➤ Hypothesis tests**
2 Select **Summ 2 Var Prop Test** from the **Function type** drop-down list box
3 Specify x_1 in the **Num Successes 1** text box, n_1 in the **Num Trials 1** text box, x_2 in the **Num Successes 2** text box, and n_2 in the **Num Trials 2** text box
4 Click **OK**
5 Click the **Set p** button
6 Click in the **Specify p** text box and type 0
7 Click **OK**
8 Click the **.05** button
9 Click the **p1 − p2 < p** button
10 Click the **Compute** button

FOR THE CI:
1 Choose **DDXL ➤ Confidence Intervals**
2 Select **Summ 2 Var Prop Interval** from the **Function type** drop-down list box
3 Specify x_1 in the **Num Successes 1** text box, n_1 in the **Num Trials 1** text box, x_2 in the **Num Successes 2** text box, and n_2 in the **Num Trials 2** text box
4 Click **OK**
5 Click the **90%** button
6 Click the **Compute Interval** button

TI-83/84 PLUS

FOR THE HYPOTHESIS TEST:
1 Press **STAT**, arrow over to **TESTS**, and press **6**
2 Type 276 for **x1** and press **ENTER**
3 Type 747 for **n1** and press **ENTER**
4 Type 195 for **x2** and press **ENTER**
5 Type 434 for **n2** and press **ENTER**
6 Highlight **<p2** and press **ENTER**
7 Press the down-arrow key, highlight **Calculate**, and press **ENTER**

FOR THE CI:
1 Press **STAT**, arrow over to **TESTS**, and press **ALPHA ➤ B**
2 Type 276 for **x1** and press **ENTER**
3 Type 747 for **n1** and press **ENTER**
4 Type 195 for **x2** and press **ENTER**
5 Type 434 for **n2** and press **ENTER**
6 Type .90 for **C-Level** and press **ENTER** twice

EXERCISES 12.3

Statistical Concepts and Skills

12.45 Explain the basic idea for performing a hypothesis test, based on independent samples, to compare two population proportions.

12.46 Kids Attending Church. A Roper Starch Worldwide for A.B.C. Global Kids Study conducted surveys in various countries to estimate the percentage of children who attend church at least once a week. Two of the countries in the survey were the United States and Germany. Considering these two countries only,
a. identify the specified attribute.
b. identify the two populations.

c. What are the two population proportions under consideration?

12.47 Sunscreen Use. Industry Research polled teenagers on sunscreen use. The survey revealed that 46% of teenage girls and 30% of teenage boys regularly use sunscreen before going out in the sun.
a. Identify the specified attribute.
b. Identify the two populations.
c. Are the proportions 0.46 (46%) and 0.30 (30%) sample proportions or population proportions? Explain your answer.

12.48 Consider a hypothesis test for two population proportions with the null hypothesis $H_0: p_1 = p_2$. What parameter is being estimated by the
a. sample proportion $\hat{p}_1$?
b. sample proportion $\hat{p}_2$?
c. pooled sample proportion $\hat{p}_p$?

12.49 Of the quantities p_1, p_2, x_1, x_2, $\hat{p}_1$, $\hat{p}_2$, and $\hat{p}_p$,
a. which represent parameters and which represent statistics?
b. which are fixed numbers and which are variables?

For Exercises **12.50**–**12.55**, *use either the critical-value approach or the P-value approach to perform the required hypothesis test.*

12.50 Vasectomies and Prostate Cancer. Approximately 450,000 vasectomies are performed each year in the United States. In this surgical procedure for contraception, the tube carrying sperm from the testicles is cut and tied. Several studies have been conducted to analyze the relationship between vasectomies and prostate cancer. The results of one such study by E. Giovannucci et al. appeared in the paper "A Retrospective Cohort Study of Vasectomy and Prostate Cancer in U.S. Men" (*Journal of the American Medical Association*, Vol. 269(7), pp. 878–882). Of 21,300 men who had not had a vasectomy, 69 were found to have prostate cancer; of 22,000 men who had had a vasectomy, 113 were found to have prostate cancer.
a. At the 1% significance level, do the data provide sufficient evidence to conclude that men who have had a vasectomy are at greater risk of having prostate cancer?
b. Is this study a designed experiment or an observational study? Explain your answer.
c. In view of your answers to parts (a) and (b), could you reasonably conclude that having a vasectomy causes an increased risk of prostate cancer? Explain your answer.

12.51 Folic Acid and Birth Defects. For several years, evidence had been mounting that folic acid reduces major birth defects. An issue of the *Arizona Republic* reported on a Hungarian study that provided the strongest evidence to date. The results of the study, directed by Drs. Andrew E. Czeizel and Istvan Dudas of the National Institute of Hygiene in Budapest, were published in the paper "Prevention of the First Occurrence of Neural-Tube Defects by Periconceptional Vitamin Supplementation" (*New England Journal of Medicine*, Vol. 327(26), p. 1832). For the study, the doctors enrolled 4753 women prior to conception. The women were divided randomly into two groups. One group, consisting of 2701 women, took daily multivitamins containing 0.8 mg of folic acid; the other group, consisting of 2052 women, received only trace elements. Major birth defects occurred

in 35 cases when the women took folic acid and in 47 cases when the women did not.
a. At the 1% significance level, do the data provide sufficient evidence to conclude that women who take folic acid are at lesser risk of having children with major birth defects?
b. Is this study a designed experiment or an observational study? Explain your answer.
c. In view of your answers to parts (a) and (b), could you reasonably conclude that taking folic acid causes a reduction in major birth defects? Explain your answer.

12.52 Racial Crossover. In the paper "The Racial Crossover in Comorbidity, Disability, and Mortality," (*Demography*, Vol. 37(3), pp. 267–283), Nan E. Johnson investigated the health of independent random samples of white and African-American elderly (aged 70 or older). Of the 4989 white elderly surveyed, 529 had at least one stroke, whereas 103 of the 906 African-American elderly surveyed reported at least one stroke. At the 5% significance level, do the data suggest that there is a difference in stroke incidence between white and African-American elderly?

12.53 Buckling Up. Response Insurance collects data on seat-belt use among U.S. drivers. Of 1000 drivers 25–34 years old, 27% said that they buckle up, whereas 330 of 1100 drivers 45–64 years old said that they did. At the 10% significance level, do the data suggest that there is a difference in seat-belt use between drivers 25–34 years old and those 45–64 years old? [SOURCE: *USA TODAY Online*.]

12.54 Ballistic Fingerprinting. Guns make unique markings on bullets they fire and their shell casings. These markings are referred to as *ballistic fingerprints*. A recent *ABCNEWS Poll* examined the opinions of Americans on the enactment of a law "…that would require every gun sold in the United States to be test-fired first, so law enforcement would have its fingerprint in case it were ever used in a crime." The following problem is based on the results of that poll. Independent simple random samples were taken of 495 men and 537 women. When asked whether they support a ballistic fingerprinting law, 307 of the men and 446 of the women said "yes." At the 1% significance level, do the data provide sufficient evidence to conclude that men tend to favor ballistic fingerprinting less than women?

12.55 Overweight by Degree. An adult is considered overweight if he or she has a body mass index (BMI) of at least 25. BMI is a measure that adjusts body weight for height, and is based on definitions provided in the *Dietary Guideline for Americans*, published by the U.S. Department of Agriculture and the U.S. Department of Health and Human Services. Of 750 randomly selected adults whose highest degree is a bachelors, 386 are overweight; and of 500 randomly selected adults with a graduate degree, 237 are overweight.

a. What assumptions are required for using the two-sample z-test here?

b. Apply the two-sample z-test to determine, at the 5% significance level, whether the percentage who are overweight is greater for those whose highest degree is a bachelors than for those with a graduate degree.

c. Repeat part (b) at the 10% significance level.

In Exercises 12.56–12.59, apply Procedure 12.4 on page 632 to obtain the required confidence interval.

12.56 Vasectomies and Prostate Cancer. Refer to Exercise 12.50.

a. Determine a 98% confidence interval for the difference between the prostate cancer rates of men who have had a vasectomy and those who have not.

b. Interpret your answer from part (a).

12.57 Folic Acid and Birth Defects. Refer to Exercise 12.51.

a. Determine a 98% confidence interval for the difference between the rates of major birth defects for babies born to women who have taken folic acid and those born to women who have not.

b. Interpret your answer from part (a).

12.58 Racial Crossover. Refer to Exercise 12.52. Determine and interpret a 95% confidence interval for the difference between the stroke incidences of white and African-American elderly.

12.59 Buckling Up. Refer to Exercise 12.53. Determine and interpret a 90% confidence interval for the difference between the proportions of seat-belt users for drivers in the age groups 25–34 years and 45–64 years.

Extending the Concepts and Skills

12.60 Eating Out Vegetarian. In this exercise, apply Formulas 12.3 and 12.4 on page 633 to the study on ordering vegetarian considered in Examples 12.8–12.10.

a. Obtain the margin of error for the estimate of the difference between the proportions of men and women who sometimes order veg by taking half the length of the confidence interval found in Example 12.10 on page 632. Interpret your answer in words.

b. Obtain the margin of error for the estimate of the difference between the proportions of men and women who sometimes order veg by applying Formula 12.3.

c. Without making a guess for the observed values of the sample proportions, obtain the common sample size that will ensure a margin of error of at most 0.01 for a 90% confidence interval.

d. Find a 90% confidence interval for $p_1 - p_2$ if, for samples of the size determined in part (c), 38.3% of the men and 43.7% of the women sometimes order veg.

e. Determine the margin of error for the estimate in part (d) and compare it to the required margin of error specified in part (c).

f. Repeat parts (c)–(e) if you can reasonably presume that at most 41% of the men sampled and at most 49% of the women sampled will be people who sometimes order veg.

g. Compare the results obtained in parts (c)–(e) to those obtained in part (f).

12.61 Identify the formula in this section that shows that the difference between the two sample proportions is an unbiased estimator of the difference between the two population proportions.

Using Technology

12.62 Racial Crossover. Refer to Exercises 12.52 and 12.58. Use the technology of your choice to conduct the required hypothesis test and obtain the desired confidence interval.

12.63 Folic Acid and Birth Defects. Refer to Exercises 12.51 and 12.57. Use the technology of your choice to conduct the required hypothesis test and obtain the desired confidence interval.

CHAPTER REVIEW

You Should Be Able To

1. use and understand the formulas in this chapter.

2. find a large-sample confidence interval for a population proportion.

3. compute the margin of error for the estimate of a population proportion.

4. understand the relationship between the sample size, confidence level, and margin of error for a confidence interval for a population proportion.

5. determine the sample size required for a specified confidence level and margin of error for the estimate of a population proportion.

6. perform a large-sample hypothesis test for a population proportion.

7. perform large-sample inferences (hypothesis tests and confidence intervals) to compare two population proportions.

8. understand the relationship between the sample sizes, confidence level, and margin of error for a confidence interval for the difference between two population proportions.

9. determine the sample sizes required for a specified confidence level and margin of error for the estimate of the difference between two population proportions.

Key Terms

margin of error, *611, 633*
number of failures, *608*
number of successes, *607*
one-sample z-interval procedure, *610*
one-sample z-test, *620, 621*
pooled sample proportion ($\hat{p}_p$), *628*
population proportion (p), *607*

sample proportion ($\hat{p}$), *607*
sampling distribution of the difference between two sample proportions, *628*
sampling distribution of the sample proportion, *608*

two-sample z-interval procedure, *632*
two-sample z-test, *629, 630*
z-interval procedure, *609*
z-test, *620*

REVIEW TEST

Statistical Concepts and Skills

1. Medical Marijuana? A *Harris Poll* was recently conducted to estimate the proportion of Americans who feel that marijuana should be legalized for medicinal use in patients with cancer and other painful and terminal diseases. Identify the
a. specified attribute.   b. population.
c. population proportion.
d. According to the poll, 80% of the 83,957 respondents said that marijuana should be legalized for medicinal use. Is the proportion 0.80 (80%) a sample proportion or a population proportion? Explain your answer.

2. Why is a sample proportion generally used to estimate a population proportion instead of obtaining the population proportion directly?

3. Explain what each phrase means in the context of inferences for a population proportion.
a. Number of successes
b. Number of failures

4. Fill in the blanks.
a. The mean of all possible sample proportions equals the _____.
b. For large samples, the possible sample proportions have approximately a _____ distribution.
c. A rule of thumb for using a normal distribution to approximate the distribution of all possible sample proportions is that both _____ and _____ are _____ or greater.

5. What does the margin of error for the estimate of a population proportion tell you?

6. Holiday Blues. A poll was conducted by Opinion Research Corporation to estimate the proportions of men and women who get the "holiday blues." Identify the
a. specified attribute.   b. two populations.
c. two population proportions.
d. two sample proportions.
e. According to the poll, 34% of men and 44% of women get the "holiday blues." Are the proportions 0.34 and 0.44 sample proportions or population proportions? Explain your answer.

7. Suppose that you are using independent samples to compare two population proportions. Fill in the blanks.

a. The mean of all possible differences between the two sample proportions equals the _____.

b. For large samples, the possible differences between the two sample proportions have approximately a _____ distribution.

8. Smallpox Vaccine. In December of 2002, *ABCNEWS.com* published the results of a poll that asked U.S. adults whether they would get a smallpox shot if it were available. Sampling, data collection, and tabulation were done by TNS Intersearch of Horsham, PA. When the risk of the vaccine was described in detail, 4 in 10 of those surveyed said they would take the smallpox shot. According to the article, "the results have a three-point margin of error" (for a 0.95 confidence level). Use the information provided to obtain a 95% confidence interval for the percentage of all U.S. adults who would take a smallpox shot, knowing the risk of the vaccine.

9. Suppose that you want to obtain a 95% confidence interval based on independent samples for the difference between two population proportions and that you want a margin of error of at most 0.01.

a. Without making an educated guess for the observed sample proportions, find the required common sample size.

b. Suppose that, from past experience, you are quite sure that the two sample proportions will be 0.75 or greater. What common sample size should you use?

10. Getting a Job. The National Association of Colleges and Employers sponsors the *Graduating Student and Alumni Survey*. Part of the survey gauges student optimism in landing a job after graduation. According to the survey results, published in the May 2000, issue of *American Demographics*, among the 1218 respondents, 733 said that they expected difficulty finding a job. Use these data to obtain and interpret a 95% confidence interval for the proportion of students who expect difficulty finding a job.

11. Getting a Job. Refer to Problem 10.

a. Find the margin of error for the estimate of p.

b. Obtain a sample size that will ensure a margin of error of at most 0.02 for a 95% confidence interval without making a guess for the observed value of $\hat{p}$.

c. Find a 95% confidence interval for p if, for a sample of the size determined in part (b), 58.7% of those surveyed say that they expect difficulty finding a job.

d. Determine the margin of error for the estimate in part (c) and compare it to the required margin of error specified in part (b).

e. Repeat parts (b)–(d) if you can reasonably presume that the percentage of those surveyed who say that they expect difficulty finding a job will be at least 56%.

f. Compare the results obtained in parts (b)–(d) with those obtained in part (e).

12. Justice in the Courts? In an issue of *Parade Magazine*, the editors reported on a national survey on law and order. One question asked of the 2512 U.S. adults who took part was whether they believed that juries "almost always" convict the guilty and free the innocent. Only 578 said that they did. Do the data provide sufficient evidence to conclude that less than one in four Americans believe that juries "almost always" convict the guilty and free the innocent?

a. Use either the critical-value approach or the *P*-value approach to perform the appropriate hypothesis test at the 5% significance level.

b. Assess the strength of the evidence against the null hypothesis by referring to Table 9.12 on page 440.

13. Height and Breast Cancer. An article published in an issue of the *Annals of Epidemiology* discussed the relationship between height and breast cancer. The study by the National Cancer Institute, which took 5 years and involved more than 1500 women with breast cancer and 2000 women without breast cancer, revealed a trend between height and breast cancer: "...taller women have a 50 to 80 percent greater risk of getting breast cancer than women who are closer to 5 feet tall." But Christine Swanson, a nutritionist who was involved with the study, added that "...height may be associated with the culprit, ...but no one really knows" the exact relationship between height and the risk of breast cancer.

a. Classify this study as either an observational study or a designed experiment. Explain your answer.

b. Interpret the statement made by Christine Swanson in light of your answer to part (a).

14. Views on the Economy. State and local governments often poll their constituents about their views on the economy. In two polls, taken approximately 1 year apart, O'Neil Associates asked 600 Maricopa County, Arizona, residents whether they thought the state's economy would improve over the next 2 years. In the first poll, 48% said "yes"; in the second poll, 60% said "yes." Do the data provide sufficient evidence to conclude that the percentage of Maricopa

County residents who thought the state's economy would improve over the next 2 years was less during the time of the first poll than during the time of the second?

a. Use either the critical-value approach or the *P*-value approach to perform the appropriate hypothesis test at the 1% significance level.

b. Assess the strength of the evidence against the null hypothesis by referring to Table 9.12 on page 440.

15. Views on the Economy. Refer to Problem 14.

a. Determine a 98% confidence interval for the difference, $p_1 - p_2$, between the proportions of Maricopa County residents who thought that the state's economy would improve over the next 2 years during the time of the first poll and during the time of the second poll.

b. Interpret your answer from part (a).

16. Views on the Economy. Refer to Problems 14 and 15.

a. Take half the length of the confidence interval found in Problem 15(a) to obtain the margin of error for the estimate of the difference between the two population proportions. Interpret your result in words.

b. Solve part (a) by applying Formula 12.3 on page 633.

c. Obtain the common sample size that will ensure a margin of error of at most 0.03 for a 98% confidence interval without making a guess for the observed values of the sample proportions.

d. Find a 98% confidence interval for $p_1 - p_2$ if, for samples of the size determined in part (c), the sample proportions are 0.475 and 0.603, respectively.

e. Determine the margin of error for the estimate in part (d) and compare it to the required margin of error specified in part (c).

Using Technology

17. Getting a Job. Use the technology of your choice to obtain the confidence interval required in Problem 10.

18. Justice in the Courts? Use the technology of your choice to conduct the hypothesis test required in Problem 12.

19. Views on the Economy. Use the technology of your choice to perform the required hypothesis test in Problem 14 and to obtain the desired confidence interval in Problem 15.

20. Finasteride and Prostate Cancer. The results of a major study to examine the effect of finasteride in reducing the risk of prostate cancer appeared on June 24, 2003, on the web site of the *New England Journal of Medicine*. The study, known as the Prostate Cancer Prevention Trial (PCPT), was sponsored by the U.S. Public Health Service and the National Cancer Institute. In the PCPT trial, 18,832 men, 55 years old or older with normal physical exams and prostate-specific antigen (PSA) levels of 3.0 nanograms per milliliter or lower, were randomly assigned to receive 5 mg of finasteride daily or a placebo. At 7 years, of the 9060 men included in the final analysis, 4368 had taken finasteride and 4692 had received a placebo. For those who took finasteride, 803 cases of prostate cancer were diagnosed, compared with 1147 cases for those who took a placebo. Use the technology of your choice to decide, at the 1% significance level, whether finasteride reduces the risk of prostate cancer. (*Note:* As reported in the August 2003 issue of the *Public Citizen's Health Research Group Newsletter*, most of the aforementioned detected cancers were "low-grade cancers of little clinical significance." Moreover, the risk of high-grade cancers was determined to be elevated for those taking finasteride.)

StatExplore in MyMathLab
Analyzing Data Online

You can use StatExplore to perform all statistical analyses discussed in this chapter. To illustrate, we show how to perform a one-sample z-interval procedure for a population proportion.

EXAMPLE   One-Sample z-Interval Procedure

Playing Hooky From Work A poll, commissioned by the Hilton Hotels Corporation, was taken of 1010 U.S. employees. The employees sampled were asked whether they "play hooky," that is, call in sick at least once a year when they simply need time to relax; 202 responded *yes*. Use StatExplore to obtain a 95% confidence interval for the proportion, *p*, of all U.S. employees who play hooky.

SOLUTION   To obtain the required confidence interval by using StatExplore, we proceed as follows:

> 1 Choose **Stat ➤ Proportions ➤ One sample ➤ with summary**
> 2 Click in the **Number of successes** text box and type 202
> 3 Click in the **Number of observations** text box and type 1010
> 4 Click **Next →**
> 5 Select the **Confidence Interval** option button
> 6 Click in the **Level** text box and type 0.95
> 7 Click **Calculate**

The resulting output is shown in Printout 12.5. From Printout 12.5, the required 95% confidence interval is from 0.17533123 (lower limit) to 0.22466877 (upper limit). We can be 95% confident that the percentage of all U.S. employees who play hooky is somewhere between 17.5% and 22.5%.

PRINTOUT 12.5
StatExplore output for one-sample z-interval

95% confidence interval results:
p = proportion of successes for population
Parameter: p

Proportion	Count	Total	Sample Prop.	Std. Err.	L. Limit	U. Limit
p	202	1010	0.2	0.012586336	0.17533123	0.22466877

STATEXPLORE EXERCISES Solve the following problems by using StatExplore.

a. Determine the confidence interval required in Exercise 12.17(a) on page 617.

b. Conduct the hypothesis test required in Example 12.6 on page 620.

c. Perform the hypothesis test required in Example 12.9 on page 629.

d. Determine the confidence interval required in Example 12.10 on page 632.

> To access StatExplore, go to the student content area of your Weiss MyMathLab course.

Internet Project

The Salk Polio Vaccine Trial

The Salk vaccine trial is one of the most famous statistical studies ever conducted. This field trial was designed to determine whether a new vaccine, developed by Dr. Jonas Salk of the University of Pittsburgh, would be effective in preventing poliomyelitis, a disease also known as infantile paralysis.

In the late 1940s and early 1950s, polio epidemics tended to come in cyclic waves of ever-increasing severity; children were particularly vulnerable. Thus, in 1954, when the trial was conducted, polio was one of the most feared of all diseases. In this Internet project, you will explore a number of interesting issues about the design of this experiment and competing experiments, and you will analyze the results.

URL for access to Internet Projects Page: www.aw-bc.com/weiss

**Focusing on
Data Analysis**

The Focus Database

Recall from Chapter 1 (see page 35) that the Focus database contains information on the undergraduate students at the University of Wisconsin - Eau Claire (UWEC). Statistical analyses for this database should be done with the technology of your choice.

a. Obtain a simple random sample of size 50 of the sexes of UWEC undergraduate students.

b. At the 5% significance level, do the data from part (a) provide sufficient evidence to conclude that more than half of UWEC undergraduates are females?

c. Use the data from part (a) to determine a 90% confidence interval for the percentage of female UWEC undergraduates. Interpret your result.

d. Obtain independent simple random samples of size 50 each of the sexes of resident and nonresident UWEC undergraduates.

e. At the 5% significance level, do the data from part (d) provide sufficient evidence to conclude that a difference exists in the percentages of females among resident and nonresident UWEC undergraduates?

f. Use the data from part (d) to determine a 95% confidence interval for the difference between the percentages of females among resident and nonresident UWEC undergraduates. Interpret your result.

g. Repeat parts (d)–(f) for samples of size 100 instead of 50.

h. Obtain the population proportions of females among resident and nonresident UWEC undergraduates.

i. If the results of your hypothesis tests in parts (e) and (g) differ, explain why.

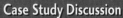

Hormone-Replacement Therapy and Dementia

At the beginning of this chapter, we discussed a study that newly examined the relationship between hormone-replacement therapy (in the form of estrogen and progestin) and dementia. Researchers randomly divided 4532 healthy women over the age of 65 into two equal-size groups. One group received the hormone-replacement therapy and the other group a placebo. Over 5 years, 40 of the women receiving the hormone-replacement therapy were diagnosed with dementia, compared with 21 of those getting a placebo. Suppose that you want to use these data to decide whether healthy women over 65 years old who take hormone-replacement therapy are at greater risk for dementia than those who do not.

a. Identify the two populations of interest.
b. Identify the specified attribute.
c. Determine the two sample proportions.
d. At the 5% significance level, do the data provide sufficient evidence to conclude that healthy women over 65 years old who take hormone-replacement therapy are at greater risk for dementia than those who do not?
e. Referring to part (d), use Table 9.12 on page 440 to assess the strength of the evidence against the null hypothesis and hence in favor of the alternative hypothesis that healthy women over 65 years old who take hormone-replacement therapy are at greater risk for dementia than those who do not.

Internet Resources: Visit the Weiss Web site www.aw-bc.com/weiss for additional discussion, exercises, and resources related to this case study.

Biography

ABRAHAM DE MOIVRE: Paving the Way for Proportion Inferences

Abraham de Moivre was born in Vitry-le-Francois, France, on May 26, 1667, the son of a country surgeon. He was educated in the Catholic school in his village and at the Protestant Academy at Sedan. In 1684, he went to Paris to study under Jacques Ozanam.

In late 1685, de Moivre, a French Huguenot (Protestant), was imprisoned in Paris because of his religion. (In October, 1685, Louis XIV revoked an edict that had allowed Protestantism in addition to the Catholicism favored by the French Court.) The duration of his incarceration is unclear, but de Moivre was probably jailed 1 to 3 years. In any case, upon his release he fled to London where he began tutoring students in mathematics.

In London, de Moivre mastered Sir Isaac Newton's *Principia* and became a close friend of Newton's and of Edmond Halley's, an English astronomer (in whose honor, incidentally, Halley's Comet is named). In Newton's later years, he would refuse to take new students, saying, "Go to Mr. de Moivre; he knows these things better than I do."

De Moivre's contributions to probability theory, mathematics, and statistics range from the definition of statistical independence to analytical trigonometric formulas to his major discovery: the normal approximation to the binomial distribution—of monumental importance in its own right, precursor to the central limit theorem, and fundamental to proportion inferences. The definition of statistical independence appeared in *The Doctrine of Chances*, published in 1718 and dedicated to Newton; the normal approximation to the binomial distribution was contained in a Latin pamphlet published in 1733. Many of his other papers were published in *Philosophical Transactions of the Royal Society*.

De Moivre also did research on the analysis of mortality statistics and the theory of annuities. In 1725, the first edition of his *Annuities on Lives*, in which he derived annuity formulas and addressed other annuity problems, was published.

De Moivre was elected to the Royal Society in 1697, to the Berlin Academy of Sciences in 1735, and to the Paris Academy in 1754. But, despite his obvious talents as a mathematician and his many champions, he was never able to obtain a position in any of England's universities. Instead, he had to rely on his meager earnings as a tutor in mathematics and a consultant on gambling and insurance, supplemented by the sales of his books. De Moivre died in London on November 27, 1754.

chapter 13

Chi-Square Procedures

CHAPTER OUTLINE

GENERAL OBJECTIVES The statistical-inference techniques presented so far have dealt exclusively with hypothesis tests and confidence intervals for population parameters, such as population means and population proportions. In this chapter, we consider two widely used inferential procedures that are not concerned with population parameters. These two procedures are often referred to as **chi-square procedures** because they rely on a distribution called the *chi-square distribution.*

To prepare for examination of chi-square procedures, we discuss the chi-square distribution in Section 13.1. In Section 13.2, we present the chi-square goodness-of-fit test, a hypothesis test that can be used to make inferences about the distribution of a variable. For instance, we could apply that test to a sample of university students to decide whether the political preference distribution of all university students differs from that of the population as a whole.

In Section 13.3, as a preliminary to the study of a second chi-square procedure, we discuss contingency tables—frequency distributions for bivariate data—and related topics. Then, in Section 13.4, we present the chi-square independence test, a hypothesis test used to decide whether an association exists between two characteristics of a population. For instance, we could apply that test to a sample of U.S. adults to decide whether an association exists between annual income and educational level for all U.S. adults.

Road Rage

The report *Controlling Road Rage: A Literature Review and Pilot Study* was prepared for the AAA Foundation for Traffic Safety by Daniel B. Rathbone, Ph.D., and Jorg C. Huckabee, MSCE. The authors discuss the results of a literature review and pilot study on how to prevent aggressive driving and road rage.

Road rage is defined as "...an incident in which an angry or impatient motorist or passenger intentionally injures or kills another motorist, passenger, or pedestrian, or attempts or threatens to injure or kill another motorist, passenger, or pedestrian."

One of the goals of the study was to determine when road rage occurs most often. The following table provides the days on which 69 road-rage incidents occurred. For this sample of 69 road-rage incidents, more incidents occurred on Friday than any other day. This information, of course, is a descriptive statistic and, by itself, does not imply that Friday is the mode for all road-rage incidents.

F	Sa	W	M	Tu	F	Th	M
Tu	F	Tu	F	Su	W	Th	F
Th	W	Th	Sa	W	W	F	F
Tu	Su	Tu	Th	W	Sa	Tu	Th
F	W	F	F	Su	F	Th	Tu
F	Tu	Tu	Tu	Sa	W	W	Sa
F	Sa	Th	W	F	Th	F	M
F	M	F	Su	W	Th	M	Tu
Sa	Th	F	Su	W			

Using the data to make inferences about all road-rage incidents would be more informative. For instance, do the data provide sufficient evidence to conclude that some days are more likely than others for the occurrence of road-rage incidents? We can answer this question by applying one of the chi-square procedures discussed in this chapter—namely, the chi-square goodness-of-fit test. At the end of this chapter, you are asked to apply that test to the road-rage data and obtain the answer to this question.

13.1 The Chi-Square Distribution

The statistical-inference procedures discussed in this chapter rely on a distribution called the **chi-square distribution.** A variable is said to have a chi-square distribution if its distribution has the shape of a special type of right-skewed curve, called a **chi-square (χ^2) curve.**

Actually, there are infinitely many chi-square distributions, and we identify the chi-square distribution (and χ^2-curve) in question by stating its number of degrees of freedom, just as we did for t-distributions. Figure 13.1 shows three χ^2-curves and illustrates some basic properties of χ^2-curves, which are presented in Key Fact 13.1.

FIGURE 13.1
χ^2-curves for df $= 5$, 10, and 19

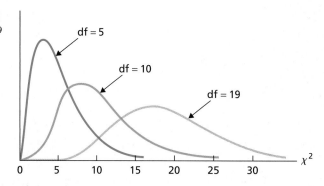

KEY FACT 13.1

Basic Properties of χ^2-Curves

Property 1: The total area under a χ^2-curve equals 1.

Property 2: A χ^2-curve starts at 0 on the horizontal axis and extends indefinitely to the right, approaching, but never touching, the horizontal axis.

Property 3: A χ^2-curve is right skewed.

Property 4: As the number of degrees of freedom becomes larger, χ^2-curves look increasingly like normal curves.

Using the χ^2-Table

Percentages (and probabilities) for a variable that has a chi-square distribution equal areas under its associated χ^2-curve. To perform a chi-square test, we need to know how to find the χ^2-value that has a specified area to its right. Table VII in Appendix A provides χ^2-values that correspond to several areas for various degrees of freedom.

The χ^2-table (Table VII) is similar to the t-table (Table IV). The two outside columns of Table VII, labeled df, display the number of degrees of freedom. As expected, the symbol χ^2_α denotes the χ^2-value that has area α to its right under a χ^2-curve. Thus the column headed $\chi^2_{0.995}$ contains χ^2-values that have area 0.995 to their right; the column headed $\chi^2_{0.99}$ contains χ^2-values that have area 0.99 to their right; and so on. We illustrate a use of Table VII in Example 13.1.

Example 13.1   Finding the χ^2-Value Having a Specified Area to Its Right

For a χ^2-curve with 12 degrees of freedom, find $\chi^2_{0.025}$; that is, find the χ^2-value that has area 0.025 to its right, as shown in Fig. 13.2(a).

FIGURE 13.2
Finding the χ^2-value that has area 0.025 to its right

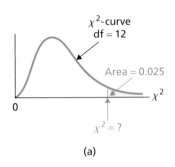

(a)

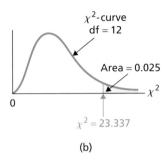

(b)

Solution To find this χ^2-value, we use Table VII. As the number of degrees of freedom is 12, we first go down the outside columns, labeled df, to "12." Then we go across that row to the column headed $\chi^2_{0.025}$. The number in the body of the table there, 23.337, is the required χ^2-value; that is, for a χ^2-curve with df = 12, the χ^2-value having area 0.025 to its right is $\chi^2_{0.025}$ = 23.337, as shown in Figure 13.2(b).

EXERCISES 13.1

Statistical Concepts and Skills

13.1 What is meant by saying that a variable has a chi-square distribution?

13.2 How do you identify different chi-square distributions?

13.3 Consider two χ^2-curves with degrees of freedom 12 and 20, respectively. Which one more closely resembles a normal curve? Explain your answer.

13.4 The *t*-table has entries for areas of 0.10, 0.05, 0.025, 0.01, and 0.005. In contrast, the χ^2-table has entries for those areas and for 0.995, 0.99, 0.975, 0.95, and 0.90. Explain why the *t*-values corresponding to these additional areas can be obtained from the existing *t*-table, but must be provided explicitly in the χ^2-table.

In Exercises 13.5–13.8, use Table VII to determine the required χ^2-values. Illustrate your work graphically.

13.5 For a χ^2-curve with 19 degrees of freedom, find the χ^2-value that has area
a. 0.025 to its right.   **b.** 0.95 to its right.

13.6 For a χ^2-curve with 22 degrees of freedom, find the χ^2-value that has area
a. 0.01 to its right.   **b.** 0.995 to its right.

13.7 For a χ^2-curve with df = 10, determine
a. $\chi^2_{0.05}$.   **b.** $\chi^2_{0.975}$.

13.8 For a χ^2-curve with df = 4, determine
a. $\chi^2_{0.005}$.   **b.** $\chi^2_{0.99}$.

Extending the Concepts and Skills

13.9 Explain how you would use Table VII to find the χ^2-value that has area 0.05 to its left. Obtain this χ^2-value for a χ^2-curve with df = 26.

13.10 Explain how you would use Table VII to find the two χ^2-values that divide the area under a χ^2-curve into a middle 0.95 area and two outside 0.025 areas. Find these two χ^2-values for a χ^2-curve with df = 14.

13.2 Chi-Square Goodness-of-Fit Test

The first chi-square procedure that we discuss is called the **chi-square goodness-of-fit test.** We can use this procedure to perform a hypothesis test about the distribution of a qualitative (categorical) variable or a discrete quantitative variable that has only finitely many possible values.[†] In Example 13.2, we introduce and explain the reasoning behind the chi-square goodness-of-fit test.

▊▊ Example 13.2 **Introduces the Chi-Square Goodness-of-Fit Test**

TABLE 13.1

Distribution of violent crimes in the United States, 2000

Type of violent crime	Relative frequency
Murder	0.011
Forcible rape	0.063
Robbery	0.286
Agg. assault	0.640
	1.000

TABLE 13.2

Sample results for 500 randomly selected violent-crime reports from last year

Type of violent crime	Frequency
Murder	3
Forcible rape	37
Robbery	154
Agg. assault	306
	500

Violent Crimes The U.S. Federal Bureau of Investigation (FBI) compiles data on crimes and crime rates and publishes the information in *Crime in the United States*. A violent crime is classified by the FBI as murder, forcible rape, robbery, or aggravated assault. Table 13.1 provides a relative-frequency distribution for (reported) violent crimes in 2000. For instance, in 2000, 28.6% of violent crimes were robberies.

A random sample of 500 violent-crime reports from last year yielded the frequency distribution shown in Table 13.2. Suppose that we want to use the data in Tables 13.1 and 13.2 to decide whether last year's distribution of violent crimes has changed from the 2000 distribution.

a. Formulate the problem statistically by posing it as a hypothesis test.

b. Explain the basic idea for carrying out the hypothesis test.

c. Discuss the details for making a decision concerning the hypothesis test.

Solution

a. The population is last year's (reported) violent crimes. The variable under consideration is "type of violent crime," and its possible values are murder, forcible rape, robbery, and aggravated assault. We want to perform the hypothesis test

> H_0: Last year's violent-crime distribution is the same as the 2000 distribution.
>
> H_a: Last year's violent-crime distribution is different from the 2000 distribution.

b. The basic idea behind the chi-square goodness-of-fit test is to compare the **observed frequencies** in the second column of Table 13.2 to the frequencies that would be expected—the **expected frequencies**—if last year's violent-crime distribution is the same as the 2000 distribution. If the observed and expected frequencies match fairly well, we do not reject the null hypothesis; otherwise, we reject the null hypothesis.

c. To transform the basic idea for carrying out the hypothesis test, as explained in part (b), into a precise procedure, we need to answer two questions:

[†]Actually, the chi-square goodness-of-fit test can be applied to any variable whose possible values have been grouped into a finite number of categories.

1. What frequencies should we expect from a random sample of 500 violent-crime reports from last year if last year's violent-crime distribution is the same as the 2000 distribution?
2. How do we decide whether the observed frequencies match reasonably well with those that we would expect?

The first question is easy to answer. If last year's violent-crime distribution is the same as the 2000 distribution, then, for instance, 28.6% of last year's violent crimes were robberies, as shown in Table 13.1. Therefore, in a random sample of 500 violent-crime reports from last year, we would expect about 28.6% of the 500, or 143, to be robberies. In general, we compute each expected frequency, denoted E, by using the formula

$$E = np,$$

where n is the sample size and p is the appropriate relative frequency from the second column of Table 13.1. For instance, the expected frequency for robberies is

$$E = np = 500 \cdot 0.286 = 143,$$

as we have already demonstrated. Calculations of the expected frequencies for all four types of violent crime are shown in Table 13.3.

TABLE 13.3

Expected frequencies if last year's violent-crime distribution is the same as the 2000 distribution

Type of violent crime	Relative frequency p	Expected frequency $np = E$
Murder	0.011	$500 \cdot 0.011 = 5.5$
Forcible rape	0.063	$500 \cdot 0.063 = 31.5$
Robbery	0.286	$500 \cdot 0.286 = 143.0$
Agg. assault	0.640	$500 \cdot 0.640 = 320.0$

The third column of Table 13.3 provides the answer to the first question. It gives the frequencies that we would expect if last year's violent-crime distribution is the same as the 2000 distribution.

The second question—whether the observed frequencies match reasonably well with the expected frequencies—is harder to answer. We need to calculate a number that measures how good the fit is.

The second column of Table 13.4 (next page) repeats the observed frequencies from the second column of Table 13.2. The third column of Table 13.4 lists the expected frequencies from the third column of Table 13.3.

To measure how well the observed and expected frequencies match, we look at the differences, $O - E$, displayed in the fourth column of Table 13.4. Summing these differences to obtain a "total difference" isn't very useful because the sum is 0. Instead, we square each difference (fifth column) and then divide by the corresponding expected frequency. Doing so gives the values $(O - E)^2 / E$, called **chi-square subtotals**, shown in the sixth column. The sum of the chi-square subtotals,

$$\Sigma(O - E)^2 / E = 3.555,$$

TABLE 13.4

Calculating the goodness of fit

Type of violent crime x	Observed frequency O	Expected frequency E	Difference $O - E$	Square of difference $(O - E)^2$	Chi-square subtotal $(O - E)^2/E$
Murder	3	5.5	−2.5	6.25	1.136
Forcible rape	37	31.5	5.5	30.25	0.960
Robbery	154	143.0	11.0	121.00	0.846
Agg. assault	306	320.0	−14.0	196.00	0.613
	500	500.0	0		3.555

is the statistic used to measure how well (or poorly) the observed and expected frequencies match.

If the null hypothesis is true, the observed and expected frequencies should be roughly equal, resulting in a small value of the test statistic, $\Sigma(O - E)^2/E$. In other words, large values of $\Sigma(O - E)^2/E$ provide evidence against the null hypothesis.

As we have seen, $\Sigma(O - E)^2/E = 3.555$. Can this value be reasonably attributed to sampling error, or is it large enough to suggest that the null hypothesis is false? To answer this question, we need to know the distribution of the test statistic $\Sigma(O - E)^2/E$, which is described in Key Fact 13.2.

KEY FACT 13.2

What *Does it Mean?*

To obtain a chi-square subtotal, square the difference between an observed and expected frequency and divide the result by the expected frequency. Adding the chi-square subtotals gives the χ^2-statistic, which has approximately a chi-square distribution.

Distribution of the χ^2-Statistic for a Chi-Square Goodness-of-Fit Test

For a chi-square goodness-of-fit test, the test statistic

$$\chi^2 = \Sigma(O - E)^2/E$$

has approximately a chi-square distribution if the null hypothesis is true. The number of degrees of freedom is 1 less than the number of possible values for the variable under consideration.

Procedure for the Chi-Square Goodness-of-Fit Test

In light of Key Fact 13.2, we now present, in Procedure 13.1, a step-by-step method for conducting a chi-square goodness-of-fit test by using either the critical-value approach or the *P*-value approach. Because the null hypothesis is rejected only when the test statistic is too large, a chi-square goodness-of-fit test is always right tailed.

Note: Regarding Assumptions 1 and 2 of Procedure 13.1, in many texts the rule given is that all expected frequencies be 5 or greater. Research by the noted statistician W. G. Cochran shows that the "rule of 5" is too restrictive.

| PROCEDURE 13.1 | Chi-Square Goodness-of-Fit Test |

Purpose To perform a hypothesis test for the distribution of a variable

Assumptions
1. All expected frequencies are 1 or greater
2. At most 20% of the expected frequencies are less than 5
3. Simple random sample

STEP 1 **The null and alternative hypotheses are**

> H_0: The variable has the specified distribution.
> H_a: The variable does not have the specified distribution.

STEP 2 **Calculate the expected frequency for each possible value of the variable by using the formula $E = np$, where n is the sample size and p is the relative frequency (or probability) given for the value in the null hypothesis.**

STEP 3 **Determine whether the expected frequencies satisfy Assumptions 1 and 2. If they do not, this procedure should not be used.**

STEP 4 **Decide on the significance level, α.**

STEP 5 **Compute the value of the test statistic**

$$\chi^2 = \Sigma(O - E)^2/E,$$

where O and E represent observed and expected frequencies, respectively. Denote the value of the test statistic χ_0^2.

CRITICAL-VALUE APPROACH

STEP 6 **The critical value is χ_α^2 with df $= k - 1$, where k is the number of possible values for the variable. Use Table VII to find the critical value.**

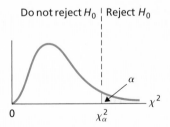

STEP 7 **If the value of the test statistic falls in the rejection region, reject H_0; otherwise, do not reject H_0.**

P-VALUE APPROACH

STEP 6 **The χ^2-statistic has df $= k - 1$, where k is the number of possible values for the variable. Use Table VII to estimate the P-value, or obtain it exactly by using technology.**

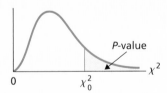

STEP 7 **If $P \le \alpha$, reject H_0; otherwise, do not reject H_0.**

STEP 8 **Interpret the results of the hypothesis test.**

Example 13.3 The Chi-Square Goodness-of-Fit Test

TABLE 13.5

Distribution of violent crimes in
the United States, 2000

Type of violent crime	Relative frequency
Murder	0.011
Forcible rape	0.063
Robbery	0.286
Agg. assault	0.640

Violent Crimes We can now complete the hypothesis test introduced in Example 13.2. Table 13.5 repeats the relative-frequency distribution for violent crimes in the United States in 2000.

A random sample of 500 violent-crime reports from last year yielded the frequency distribution shown in Table 13.6. At the 5% significance level, do the data provide sufficient evidence to conclude that last year's violent-crime distribution is different from the 2000 distribution?

Solution We apply Procedure 13.1.

STEP 1 State the null and alternative hypotheses.

The null and alternative hypotheses are

H_0: Last year's violent-crime distribution is the same as the 2000 distribution.

H_a: Last year's violent-crime distribution is different from the 2000 distribution.

TABLE 13.6

Sample results for 500 randomly selected
violent-crime reports from last year

Type of violent crime	Observed frequency
Murder	3
Forcible rape	37
Robbery	154
Agg. assault	306

STEP 2 Calculate the expected frequency for each possible value of the variable by using the formula $E = np$, where n is the sample size and p is the relative frequency given for the value in the null hypothesis.

We have $n = 500$, and the relative frequencies for the null hypothesis are shown in the second column of Table 13.5. The required calculations are summarized in Table 13.3 on page 653.

STEP 3 Determine whether the expected frequencies satisfy Assumptions 1 and 2.

1. Are all expected frequencies 1 or greater? Yes, as shown in Table 13.3.
2. Are at most 20% of the expected frequencies less than 5? Yes, in fact, none of the expected frequencies are less than 5, as shown in Table 13.3.

STEP 4 Decide on the significance level, α.

We are to perform the hypothesis test at the 5% significance level and, consequently, we have $\alpha = 0.05$.

STEP 5 Compute the value of the test statistic

$$\chi^2 = \Sigma(O - E)^2/E,$$

where O and E represent observed and expected frequencies, respectively.

The observed frequencies are displayed in the second column of Table 13.6. The calculated value of the test statistic, displayed at the bottom of the last column in Table 13.4 on page 654, is

$$\chi^2 = \Sigma(O - E)^2/E = 3.555.$$

CRITICAL-VALUE APPROACH	*P*-VALUE APPROACH

CRITICAL-VALUE APPROACH

STEP 6 The critical value is χ^2_α with df = $k - 1$, where k is the number of possible values for the variable. Use Table VII to find the critical value.

From Step 4, $\alpha = 0.05$. The variable under consideration is "type of violent crime." There are four types of violent crime, so $k = 4$. In Table VII, we find that, for df = $k - 1 = 4 - 1 = 3$, $\chi^2_{0.05} = 7.815$, as shown in Fig. 13.3A.

FIGURE 13.3A

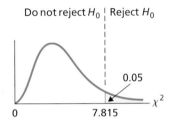

STEP 7 **If the value of the test statistic falls in the rejection region, reject H_0; otherwise, do not reject H_0.**

From Step 5, the value of the test statistic is $\chi^2 = 3.555$. Because it does not fall in the rejection region, as shown in Fig. 13.3A, we do not reject H_0. The test results are not statistically significant at the 5% level.

***P*-VALUE APPROACH**

STEP 6 The χ^2-statistic has df = $k - 1$, where k is the number of possible values for the variable. Use Table VII to estimate the *P*-value, or obtain it exactly by using technology.

From Step 5, the value of the test statistic is $\chi^2 = 3.555$. The test is right tailed, so the *P*-value is the probability of observing a value of χ^2 of 3.555 or greater if the null hypothesis is true. That probability equals the shaded area in Fig. 13.3B.

FIGURE 13.3B

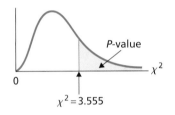

The variable under consideration is "type of violent crime." Because there are four types of violent crime, $k = 4$. Referring to Fig. 13.3B and to Table VII with df = $k - 1 = 4 - 1 = 3$, we find that $P > 0.10$. (Using technology, we obtain $P = 0.314$.)

STEP 7 **If $P \leq \alpha$, reject H_0; otherwise, do not reject H_0.**

From Step 6, $P > 0.10$. Because the *P*-value exceeds the specified significance level of 0.05, we do not reject H_0. The test results are not statistically significant at the 5% level and (see Table 9.12 on page 440) provide essentially no evidence against the null hypothesis.

STEP 8 **Interpret the results of the hypothesis test.**

INTERPRETATION At the 5% significance level, the data do not provide sufficient evidence to conclude that last year's violent-crime distribution differs from the 2000 distribution.

The Technology Center

Some statistical technologies have dedicated programs that automatically perform a chi-square goodness-of-fit test, but others do not. For a statistical technology that does not have a dedicated program for a chi-square goodness-of-fit test, you can often use the macro capabilities of the statistical technology to write a program. Refer to the technology manuals for more details.

EXERCISES 13.2

Statistical Concepts and Skills

13.11 Why is the phrase "goodness of fit" used to describe the type of hypothesis test considered in this section?

13.12 Are the observed frequencies variables? What about the expected frequencies? Explain your answers.

13.13 In each part of this exercise, we have given the relative frequencies for the null hypothesis of a chi-square goodness-of-fit test and the sample size. In each case, decide whether Assumptions 1 and 2 for using that test are satisfied.
a. Sample size: $n = 100$.
Relative frequencies: 0.65, 0.30, 0.05.
b. Sample size: $n = 50$.
Relative frequencies: 0.65, 0.30, 0.05.
c. Sample size: $n = 50$.
Relative frequencies: 0.20, 0.20, 0.25, 0.30, 0.05.
d. Sample size: $n = 50$.
Relative frequencies: 0.22, 0.21, 0.25, 0.30, 0.02.
e. Sample size: $n = 50$.
Relative frequencies: 0.22, 0.22, 0.25, 0.30, 0.01.
f. Sample size: $n = 100$.
Relative frequencies: 0.44, 0.25, 0.30, 0.01.

13.14 Primary Heating Fuel. According to *Current Housing Reports*, published by the U.S. Bureau of the Census, the primary heating fuel for all occupied housing units is distributed as follows.

Primary heating fuel	Percentage
Utility gas	51.5
Fuel oil, kerosene	9.8
Electricity	30.7
Bottled, tank, or LPG	5.7
Wood and other fuel	1.9
None	0.4

Suppose that you want to determine whether the distribution of primary heating fuel for occupied housing units built after 2000 differs from that of all occupied housing units. To decide, you take a random sample of housing units built after 2000 and obtain a frequency distribution of their primary heating fuel.
a. Identify the population and variable under consideration here.
b. For each of the following sample sizes, determine whether conducting a chi-square goodness-of-fit test is appropriate and explain your answers: 200; 250; 300.

c. Strictly speaking, what is the smallest sample size for which conducting a chi-square goodness-of-fit test is appropriate?

In Exercises 13.15–13.18, use either the critical-value approach or the P-value approach to perform the required hypothesis test.

13.15 Freshmen Politics. The Higher Education Research Institute of the University of California, Los Angeles, publishes information on characteristics of incoming college freshmen in *The American Freshman*. In 2000, 27.7% of incoming freshmen characterized their political views as liberal, 51.9% as moderate, and 20.4% as conservative. For this year, a random sample of 500 incoming college freshmen yielded the following frequency distribution for political views.

Political view	Frequency
Liberal	160
Moderate	246
Conservative	94

a. Identify the population and variable under consideration here.
b. At the 5% significance level, do the data provide sufficient evidence to conclude that this year's distribution of political views for incoming college freshmen has changed from the 2000 distribution?
c. Repeat part (b), using a significance level of 10%.

13.16 Car Sales. The American Automobile Manufacturers Association compiles data on U.S. car sales by type of car. Following is the 1990 distribution, as reported in the *World Almanac*.

Type of car	Small	Midsize	Large	Luxury
Percentage	32.8	44.8	9.4	13.0

A random sample of last year's U.S. car sales yielded the following data.

Type of car	Small	Midsize	Large	Luxury
Frequency	133	249	47	71

a. Identify the population and variable under consideration here.

b. At the 5% significance level, do the data provide sufficient evidence to conclude that last year's type-of-car distribution for U.S. car sales differs from the 1990 distribution?

13.17 M&M Colors. Observing that the proportion of blue M&Ms in his bowl of candy appeared to be less than that of the other colors, Ronald D. Fricker, Jr., decided to compare the color distribution in randomly chosen bags of M&Ms to the theoretical distribution reported by M&M/MARS consumer affairs. Fricker published his findings in the article "The Mysterious Case of the Blue M&Ms" (*Chance*, Vol. 9(4), pp. 19–22). The following is the theoretical distribution.

Color	Percentage
Brown	30
Yellow	20
Red	20
Orange	10
Green	10
Blue	10

For his study, Fricker bought three bags of M&Ms from local stores and counted the number of each color. The average number of each color in the three bags was distributed as follows.

Color	Frequency
Brown	152
Yellow	114
Red	106
Orange	51
Green	43
Blue	43

Do the data provide sufficient evidence to conclude that the color distribution of M&Ms differs from that reported by M&M/MARS consumer affairs? Use $\alpha = 0.05$.

13.18 An Edge in Roulette? An American roulette wheel contains 18 red numbers, 18 black numbers, and 2 green numbers. The following table shows the frequency with which the ball landed on each color in 200 trials.

Number	Red	Black	Green
Frequency	88	102	10

At the 5% significance level, do the data suggest that the wheel is out of balance?

Extending the Concepts and Skills

13.19 Table 13.4 on page 654 showed the calculated sums of the observed frequencies, the expected frequencies, and their differences. Strictly speaking, those sums are not needed. However, they serve as a check for computational errors.

a. In general, what common value should the sum of the observed frequencies and the sum of the expected frequencies equal? Explain your answer.

b. Fill in the blank. The sum of the differences between each observed and expected frequency should equal _____.

c. Suppose that you are conducting a chi-square goodness-of-fit test. If the sum of the expected frequencies does not equal the sample size, what do you conclude?

d. Suppose that you are conducting a chi-square goodness-of-fit test. If the sum of the expected frequencies equals the sample size, can you conclude that you made no error in calculating the expected frequencies? Explain your answer.

13.20 The chi-square goodness-of-fit test provides a method for performing a hypothesis test about the distribution of a variable that has k possible values. If the number of possible values is 2, that is, $k = 2$, the chi-square goodness-of-fit test is equivalent to a procedure that you studied earlier.

a. Which procedure is that? Explain your answer.

b. Suppose that you want to perform a hypothesis test to decide whether the proportion of a population that has a specified attribute is different from p_0. Discuss the method for performing such a test if you use (1) the one-sample z-test for a population proportion (page 621) or (2) the chi-square goodness-of-fit test.

Using Technology

13.21 Freshmen Politics. Use the technology of your choice to perform the hypothesis test in Exercise 13.15.

13.22 Car Sales. Use the technology of your choice to perform the hypothesis test in Exercise 13.16.

13.23 Credit Card Marketing. According to market research by Brittain Associates, published in an issue of *American Demographics*, the income distribution of adult Internet users closely mirrors that of credit card applicants. That is exactly what many major credit card issuers want to hear because they hope to replace direct mail marketing with more efficient Web-based marketing. Following is an income distribution for credit card applicants.

Income ($1000)	< 30	30 < 50	50 < 70	70+
Percentage	28	33	21	18

A random sample of 109 adult Internet users yielded the following income distribution.

Income ($1000)	< 30	30 < 50	50 < 70	70+
Frequency	25	29	26	29

Use the technology of your choice to decide, at the 5% significance level, whether the data do not support the claim by Brittain Associates. What if you use a 10% significance level?

13.24 Migrating Women. In the article "Waves of Rural Brides: Female Marriage Migration in China" (*Annals of the Association of American Geographers*, 88(2), pp. 227–251),

C. Fan and Y. Huang report on the reasons that women in China migrate within the country to new places of residence. The percentages for reasons given by 15–29-year-old women for migrating within the same province are presented in the second column of the following table. For a random sample of 500 women in the same age group who migrated to a different province, the number giving each of the reasons is recorded in the third column of the table.

Reason	Intraprovincial migrants (%)	Interprovincial migrants
Job transfer	4.8	20
Job assignment	7.2	23
Industry/business	17.8	108
Study/training	16.9	47
Help from friends/ relatives	6.2	43
Joining family	6.8	45
Marriage	36.8	205
Other	3.5	9

Use the technology of your choice to decide, at the 1% significance level, whether the data provide sufficient evidence to conclude that the distribution of reasons for migration between provinces is different from that for migration within provinces.

13.3 Contingency Tables; Association

The next chi-square procedure that we present is the **chi-square independence test,** which is used to decide, based on sample data, whether two variables of a population are statistically related. Before we can do that, however, we need to discuss two prerequisite concepts: *contingency tables* and *association*.

Contingency Tables

In Section 2.2, you learned how to group data from one variable into a frequency distribution. Recall that data obtained by observing values of one variable of a population are called **univariate data.**

Now, we show how to simultaneously group data from two variables into a frequency distribution. Data obtained by observing values of two variables of a population are called **bivariate data,** and a frequency distribution for bivariate data is called a **contingency table** or **two-way table.** In Example 13.4, we introduce contingency tables.

▌▌Example 13.4   Introducing Contingency Tables

Political Party and Class Level   In Example 2.8 on page 52, we considered data on political party affiliation for the students in Professor Weiss's introductory statistics course. These are univariate data obtained by observing values of the single variable "political party affiliation."

Now, we simultaneously consider data on political party affiliation and class level for the students in Professor Weiss's introductory statistics course, as shown in Table 13.7. These are bivariate data, obtained by observing values of the two variables "political party affiliation" and "class level." The task is to group these bivariate data into a contingency table.

TABLE 13.7

Political party affiliation and class level for students in introductory statistics

Student	Political party	Class level	Student	Political party	Class level
1	Democratic	Freshman	21	Democratic	Junior
2	Other	Junior	22	Democratic	Senior
3	Democratic	Senior	23	Republican	Freshman
4	Other	Sophomore	24	Democratic	Sophomore
5	Democratic	Sophomore	25	Democratic	Senior
6	Republican	Sophomore	26	Republican	Sophomore
7	Republican	Junior	27	Republican	Junior
8	Other	Freshman	28	Other	Junior
9	Other	Sophomore	29	Other	Junior
10	Republican	Sophomore	30	Democratic	Sophomore
11	Republican	Sophomore	31	Republican	Sophomore
12	Republican	Junior	32	Democratic	Junior
13	Republican	Sophomore	33	Republican	Junior
14	Democratic	Junior	34	Other	Senior
15	Republican	Sophomore	35	Other	Sophomore
16	Republican	Senior	36	Republican	Freshman
17	Democratic	Sophomore	37	Republican	Freshman
18	Democratic	Junior	38	Republican	Freshman
19	Other	Senior	39	Democratic	Junior
20	Republican	Sophomore	40	Republican	Senior

Solution   A contingency table must provide for each possible pair of values for the two variables. In this case, the contingency table has the form shown in Table 13.8 at the top of the next page. The small boxes inside the rectangle formed by the heavy lines are called **cells,** which hold the frequencies.

To group the bivariate data in Table 13.7 into the contingency table, we need to determine how many students fall in each cell. We do so by going through the data in Table 13.7 and placing a tally mark in the appropriate cell of Table 13.8 for each student. For instance, the first student is both a Democrat and a freshman, so this calls for a tally mark in the upper left cell of Table 13.8. The results of the tallying procedure are shown in Table 13.8.

We now count the tallies in each cell to determine the frequencies. Replacing the tallies in Table 13.8 by the frequencies, we obtain the contingency table for the bivariate data in Table 13.7, as shown in Table 13.9.

TABLE 13.8

Form of contingency table for political party affiliation and class level

		Class level				
		Freshman	Sophomore	Junior	Senior	**Total**
Party	Democratic	|	||||	ИН	|||	
	Republican	||||	ИН |||	||||	||	
	Other	|	|||	|||	||	
	Total					

TABLE 13.9

Contingency table for political party affiliation and class level

		Class level				
		Freshman	Sophomore	Junior	Senior	**Total**
Party	Democratic	1	4	5	3	13
	Republican	4	8	4	2	18
	Other	1	3	3	2	9
	Total	6	15	12	7	40

The number 1 in the upper left cell of Table 13.9 indicates that one student in the course is both a Democrat and a freshman. The number 8, diagonally below and to the right of the 1, shows that eight students in the course are both Republicans and sophomores.

The total in the first row indicates that 13 $(1 + 4 + 5 + 3)$ of the students are Democrats. Similarly, the total in the third column shows that 12 of the students are juniors. The number 40 in the lower right corner of the table gives the total number of students in the course. That total can be found by summing either the row totals or the column totals; it can also be found by summing the frequencies in the 12 cells.

Grouping bivariate data into a contingency table by hand, as we did in Example 13.4, is useful for purposes of understanding. However, in practice, computers are almost always used to accomplish such tasks.

Association

Next, we need to discuss the concept of **association** for two variables. We do so for variables that are either categorical or quantitative with only finitely many possible values. Roughly speaking, there is an association between two variables of a population if knowing the value of one of the variables imparts information about the value of the other variable. In Example 13.5, we introduce the concept of association.

Example 13.5 Introduces Association

Political Party and Class Level In Example 13.4, we presented data on political party affiliation and class level for the students in Professor Weiss's introductory statistics course. Consider those students a population of interest.

a. Obtain the distribution of political party affiliation within each class level.

b. Use the result of part (a) to decide whether there is an association between the variables "political party affiliation" and "class level."

c. What would it mean if there were no association between the variables "political party affiliation" and "class level"?

d. Explain how a segmented bar graph represents whether there is an association between the variables "political party affiliation" and "class level."

e. Discuss another method for deciding whether there is an association between the variables "political party affiliation" and "class level."

Solution

a. To obtain the distribution of political party affiliation within each class level, divide each entry in a column of the contingency table in Table 13.9 by its column total. The results are shown in Table 13.10.

TABLE 13.10
Conditional distributions of political party affiliation by class level

		Class level				
		Freshman	Sophomore	Junior	Senior	**Total**
Party	Democratic	0.167	0.267	0.417	0.429	0.325
	Republican	0.667	0.533	0.333	0.286	0.450
	Other	0.167	0.200	0.250	0.286	0.225
	Total	1.000	1.000	1.000	1.000	1.000

The first column of Table 13.10 gives the distribution of political party affiliation for freshman: 16.7% are Democrats, 66.7% are Republicans, and 16.7% are Other. This distribution is called the **conditional distribution** of the variable "political party affiliation" corresponding to the value "freshman" of the variable "class level"; or, more simply, the conditional distribution of political party affiliation for freshmen.

Similarly, the second, third, and fourth columns give the conditional distributions of political party affiliation for sophomores, juniors, and seniors, respectively. The "Total" column provides the (unconditional) distribution of political party affiliation for the entire population which, in this context, is called the **marginal distribution** of the variable "political party affiliation." This distribution is the same as the one we found in Example 2.8 (Table 2.11 on page 53).

b. Table 13.10 shows that there is an association between the variables "political party affiliation" and "class level" because knowing the value of the variable "class level" imparts information about the value of the variable "political party affiliation." For instance, as shown in Table 13.10, if we are given no information about the class level of a student in the course, there is a 32.5% chance that the student is a Democrat. But, if we know that the student is a junior, there is a 41.7% chance that the student is a Democrat.

c. If there were no association between the variables "political party affiliation" and "class level," the four conditional distributions of political party affiliation would be the same as each other and as the marginal distribution of political party affiliation; in other words, all five columns of Table 13.10 would be identical.

d. A **segmented bar graph** is helpful for understanding the concept of association. The first four bars of the segmented bar graph in Fig. 13.4 provide the conditional distributions of political party affiliation for freshmen, sophomores, juniors, and seniors, respectively, and the fifth bar gives the marginal distribution of political party affiliation. This segmented bar graph is obtained by referring to Table 13.10.

FIGURE 13.4

Segmented bar graph for the conditional distributions and marginal distribution of political party affiliation

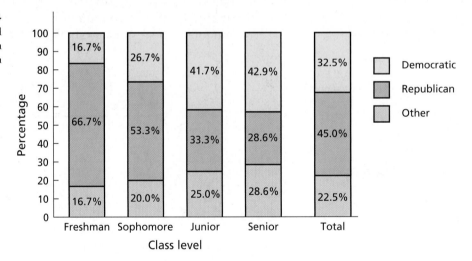

If there were no association between political party affiliation and class level, the four bars displaying the conditional distributions of political party affiliation would be the same as each other and as the bar displaying the marginal distribution of political party affiliation; in other words, all five bars in Fig. 13.4 would be identical. That there is in fact an association between political party affiliation and class level is reflected in the segmented bar graph by nonidentical bars.

e. Alternatively, we could decide whether there is an association between the two variables by obtaining the conditional distribution of class level within each political party affiliation. The conclusion regarding association (or nonassociation) will be the same, regardless of which variable's conditional distributions are obtained.

Keeping in mind the terminology introduced in Example 13.5, we now define the concept of *association* for two variables.

DEFINITION 13.1

What
Does it Mean?

Roughly speaking, two variables of a population are associated if knowing the value of one variable imparts information about the value of the other variable.

Association

We say that there is an **association** between two variables of a population if the conditional distributions of one variable given the other are not identical.

The phrase **statistically dependent variables** is also used to express the fact that there is an association between two variables. Similarly, the phrase **statistically independent variables** is often used to indicate that there is no association between two variables.

The Technology Center

Some statistical technologies have dedicated programs that automatically group bivariate data into a contingency table and obtain conditional and marginal distributions as well. For a statistical technology that does not have this kind of dedicated program, you can often use the macro capabilities of the statistical technology to write a program. Refer to the technology manuals for more details.

EXERCISES 13.3

Statistical Concepts and Skills

13.25 Provide an example of univariate data; of bivariate data.

13.26 Identify the type of table that is used to group bivariate data.

13.27 What are the small boxes inside the heavy lines of a contingency table called?

13.28 Suppose that bivariate data are to be grouped into a contingency table. Determine the number of cells that the contingency table will have if the number of possible values for the two variables are
a. two and three. **b.** four and three. **c.** m and n.

13.29 Identify three ways in which the total number of observations of bivariate data can be obtained from the frequencies in a contingency table.

13.30 U.S. Voting. Congressional Quarterly, Inc., provides information about the popular vote cast for president in the United States, by political party and region, in *America Votes*.

According to that publication, in the 1996 presidential election, 40.7% of those voting voted for the Republican candidate, whereas 45.2% of those voting who live in the South did so. For that presidential election, is there an association between the variables "party of presidential candidate voted for" and "region of residence" for those who voted? Explain your answer.

13.31 Physicians in Residency. According to the article "Women Physicians Statistics" (*Journal of the American Medical Association*, Vol. 290, No. 9, pp. 1234–1236), in 2002, 8.2% of male physicians in residency specialized in family practice and 12.2% of female physicians in residency specialized in family practice. Is there an association between the variables "gender" and "specialty" for physicians in residency in 2002? Explain your answer.

Table 13.11 at the top of the next column provides data on sex, class level, and college for the students in one section of the course Introduction to Computer Science during one semester at Arizona State University. In the table, we have used the abbreviations BUS for Business, ENG for Engineering and Applied Sciences, and LIB for Liberal Arts and Sciences.

TABLE 13.11

Sex, class level, and college for students in introduction to computer science

Sex	Class	College	Sex	Class	College
M	Junior	ENG	F	Soph	BUS
M	Soph	ENG	F	Junior	ENG
F	Senior	BUS	M	Junior	LIB
F	Junior	BUS	F	Junior	BUS
M	Junior	ENG	M	Soph	BUS
F	Junior	LIB	M	Junior	BUS
M	Senior	LIB	M	Soph	ENG
M	Soph	ENG	M	Junior	ENG
M	Junior	ENG	M	Junior	ENG
M	Soph	ENG	M	Soph	LIB
F	Soph	BUS	F	Senior	ENG
F	Junior	BUS	F	Senior	BUS
M	Junior	ENG			

In Exercises **13.32–13.34**, *use the data in Table 13.11.*

13.32 Sex and Class Level. Refer to Table 13.11. Consider the variables "sex" and "class level."
a. Group the bivariate data for these two variables into a contingency table.
b. Determine the conditional distribution of sex within each class level and the marginal distribution of sex.
c. Determine the conditional distribution of class level within each sex and the marginal distribution of class level.
d. Is there an association between the variables "sex" and "class level" for this population? Explain your answer.

13.33 Sex and College. Refer to Table 13.11. Consider the variables "sex" and "college."
a. Group the bivariate data for these two variables into a contingency table.
b. Determine the conditional distribution of sex within each college and the marginal distribution of sex.
c. Determine the conditional distribution of college within each sex and the marginal distribution of college.
d. Is there an association between the variables "sex" and "college" for this population? Explain your answer.

13.34 Class level and College. Refer to Table 13.11. Consider the variables "class level" and "college."
a. Group the bivariate data for these two variables into a contingency table.
b. Determine the conditional distribution of class level within each college and the marginal distribution of class level.

c. Determine the conditional distribution of college within each class level and the marginal distribution of college.
d. Is there an association between the variables "class level" and "college" for this population? Explain your answer.

Table 13.12 provides hypothetical data on political party affiliation and class level for the students in a night-school course.

TABLE 13.12

Political party affiliation and class level for the students in a night-school course (hypothetical data)

Party	Class	Party	Class	Party	Class
Rep	Jun	Rep	Soph	Rep	Jun
Dem	Soph	Other	Jun	Rep	Soph
Dem	Jun	Dem	Soph	Rep	Soph
Other	Jun	Rep	Soph	Rep	Fresh
Dem	Jun	Dem	Sen	Rep	Soph
Dem	Fresh	Rep	Jun	Rep	Jun
Dem	Soph	Dem	Jun	Rep	Sen
Dem	Sen	Dem	Jun	Rep	Jun
Other	Sen	Rep	Sen	Dem	Soph
Dem	Fresh	Rep	Fresh	Rep	Jun
Rep	Jun	Rep	Jun	Other	Jun
Rep	Jun	Dem	Jun	Dem	Jun
Dem	Sen	Rep	Sen	Other	Soph
Rep	Jun	Rep	Sen	Rep	Sen
Dem	Sen	Rep	Sen	Other	Soph
Rep	Jun	Dem	Soph	Rep	Soph
Rep	Soph	Other	Fresh	Other	Soph
Rep	Fresh	Rep	Soph	Other	Sen
Rep	Jun	Other	Jun	Rep	Soph
Dem	Soph	Dem	Jun	Dem	Jun

In Exercises **13.35** *and* **13.36**, *use the data in Table 13.12.*

13.35 Party and Class. Refer to Table 13.12.
a. Group the bivariate data for the two variables into a contingency table.
b. Determine the conditional distribution of political party affiliation within each class level.
c. Is there an association between the variables "political party affiliation" and "class level" for this population of night-school students? Explain your answer.
d. Without doing any further calculation, determine the marginal distribution of political party affiliation.
e. Without doing further calculation, respond true or false to the following statement and explain your answer:

"The conditional distributions of class level within political party affiliations are identical to each other and to the marginal distribution of class level."

13.36 Party and Class. Refer to Table 13.12.
a. If you have not done Exercise 13.35, group the bivariate data for the two variables into a contingency table.
b. Determine the conditional distribution of class level within each political party affiliation.
c. Is there an association between the variables "political party affiliation" and "class level" for this population of night-school students? Explain your answer.
d. Without doing any further calculation, determine the marginal distribution of class level.
e. Without doing further calculation, respond true or false to the following statement and explain your answer: "The conditional distributions of political party affiliation within class levels are identical to each other and to the marginal distribution of political party affiliation."

13.37 AIDS Cases. Acquired immunodeficiency syndrome (AIDS) is a specific group of diseases or conditions that indicate severe immunosuppression related to infection with the human immunodeficiency virus (HIV). According to the Centers for Disease Control (*HIV/AIDS Surveillance Report*, Vol. 13, No. 1), the number of AIDS cases in the United States through June 2001, by gender and race/ethnicity, is as shown in the following contingency table. Frequencies are in thousands, rounded to the nearest thousand.

		Gender		
		Male	Female	**Total**
Race/Ethnicity	White	307		337
	Black	221	81	302
	Hispanic	118	27	
	Other	8	1	9
	Total		139	

a. How many cells does this contingency table have?
b. Fill in the missing entries.
c. What was the total number of AIDS cases in the United States through June 2001?
d. How many AIDS cases were Hispanics?
e. How many AIDS cases were males?
f. How many AIDS cases were white females?

13.38 Vehicles in Use. As reported by the Motor Vehicle Manufacturers Association of the United States in *Motor Vehicle Facts and Figures*, the number of cars and trucks in use by age are as shown in the following contingency table. Frequencies are in millions.

		Type		
		Car	Truck	**Total**
Age (yr)	Under 6	46.2	27.8	74.0
	6–8	26.9		40.0
	9–11	23.3	10.7	
	12 & over	26.8	18.6	45.4
	Total	123.2		

a. How many cells does this contingency table have?
b. Fill in the missing entries.
c. What is the total number of cars and trucks in use?
d. How many vehicles are trucks?
e. How many vehicles are between 6 and 8 years old?
f. How many vehicles are trucks that are between 9 and 11 years old?

13.39 Education of Prisoners. In the article "Education and Correctional Populations" (*Bureau of Justice Statistics Special Report*, January 2003, NCJ 195670), C. Harlow examined the educational attainment of prisoners by type of prison facility. The following contingency table was adapted from Table 1 of the article. Frequencies are in thousands, rounded to the nearest hundred.

		Prison facility			
		State	Federal	Local	**Total**
Educational attainment	8th grade or less	149.9	10.6	66.0	226.5
	Some high school	269.1	12.9	168.2	450.2
	GED	300.8	20.1	71.0	391.9
	High school diploma	216.4	24.0	130.4	370.8
	Postsecondary	95.0	14.0	51.9	160.9
	College grad or more	25.3	7.2	16.1	48.6
	Total	1056.5	88.8	503.6	1648.9

How many prisoners
a. are in state facilities?
b. have at least a college education?
c. are in federal facilities and have at most an 8th grade education?

d. are in federal facilities or have at most an 8th grade education?

e. in local facilities have a postsecondary educational attainment?

f. with a postsecondary educational attainment are in local facilities?

g. are not in federal facilities?

13.40 U.S. Hospitals. The American Hospital Association publishes information about U.S. hospitals and nursing homes in *Hospital Statistics*. The following contingency table provides a cross classification of U.S. hospitals and nursing homes by type of facility and number of beds.

	Number of beds			
Facility		24 or fewer	25–74	75 or more	**Total**
	General	260	1586	3557	5403
	Psychiatric	24	242	471	737
	Chronic	1	3	22	26
	Tuberculosis	0	2	2	4
	Other	25	177	208	410
	Total	310	2010	4260	6580

In the following questions, the term *hospital* refers to either a hospital or nursing home.

a. How many hospitals have at least 75 beds?

b. How many hospitals are psychiatric facilities?

c. How many hospitals are psychiatric facilities with at least 75 beds?

d. How many hospitals either are psychiatric facilities or have at least 75 beds?

e. How many general facilities have between 25 and 74 beds?

f. How many hospitals with between 25 and 74 beds are chronic facilities?

g. How many hospitals have more than 24 beds?

13.41 AIDS Cases. Refer to Exercise 13.37. Here, we abbreviate "race/ethnicity" by "race." For AIDS cases in the United States through June 2001, answer the following questions:

a. Find and interpret the conditional distribution of gender by race.

b. Find and interpret the marginal distribution of gender.

c. Is there an association between the variables "gender" and "race"? Explain your answer.

d. What percentage of AIDS cases were females?

e. What percentage of AIDS cases among whites were females?

f. Without doing further calculations, respond true or false to the following statement and explain your answer: "The conditional distributions of race by gender are not identical."

g. Find and interpret the marginal distribution of race and the conditional distributions of race by gender.

13.42 Vehicles in Use. Refer to Exercise 13.38. Here, the term "vehicle" refers to either a U.S. car or truck currently in use.

a. Determine the conditional distribution of age group for each type of vehicle.

b. Determine the marginal distribution of age group for vehicles.

c. Is there an association between the variables "type" and "age group" for vehicles? Explain your answer.

d. Find the percentage of vehicles under 6 years old.

e. Find the percentage of cars under 6 years old.

f. Without doing any further calculations, respond true or false to the following statement and explain your answer: "The conditional distributions of type of vehicle within age groups are not identical."

g. Determine and interpret the marginal distribution of type of vehicle and the conditional distributions of type of vehicle within age groups.

13.43 Education of Prisoners. Refer to Exercise 13.39.

a. Find the conditional distribution of educational attainment within each type of prison facility.

b. Is there an association between educational attainment and type of prison facility for prisoners? Explain your answer.

c. Determine the marginal distribution of educational attainment for prisoners.

d. Construct a segmented bar graph for the conditional distributions of educational attainment and marginal distribution of educational attainment that you obtained in parts (a) and (c), respectively. Interpret the graph in light of your answer to part (b).

e. Without doing any further calculations, respond true or false to the following statement and explain your answer: "The conditional distributions of facility type within educational attainment categories are identical."

f. Determine the marginal distribution of facility type and the conditional distributions of facility type within educational attainment categories.

g. Find the percentage of prisoners who are in federal facilities.

h. Find the percentage of prisoners with at most an 8th grade education who are in federal facilities.

i. Find the percentage of prisoners in federal facilities who have at most an 8th grade education.

13.44 U.S. Hospitals. Refer to Exercise 13.40.
a. Determine the conditional distribution of number of beds within each facility type.
b. Is there an association between facility type and number of beds for U.S. hospitals? Explain your answer.
c. Determine the marginal distribution of number of beds for U.S. hospitals.
d. Construct a segmented bar graph for the conditional distributions and marginal distribution of number of beds. Interpret the graph in light of your answer to part (b).
e. Without doing any further calculations, respond true or false to the following statement and explain your answer: "The conditional distributions of facility type within number-of-beds categories are identical."
f. Obtain the marginal distribution of facility type and the conditional distributions of facility type within number-of-beds categories.
g. What percentage of hospitals are general facilities?
h. What percentage of hospitals that have at least 75 beds are general facilities?
i. What percentage of general facilities have at least 75 beds?

Extending the Concepts and Skills

13.45 In this exercise, you are to consider two variables, x and y, defined on a hypothetical population. Following are the conditional distributions of the variable y corresponding to each value of the variable x.

			x		
		A	B	C	**Total**
	0	0.316	0.316	0.316	
	1	0.422	0.422	0.422	
y	2	0.211	0.211	0.211	
	3	0.047	0.047	0.047	
	4	0.004	0.004	0.004	
	Total	1.000	1.000	1.000	

a. Is there an association between the variables x and y? Explain your answer.
b. Determine the marginal distribution of y.
c. Can you determine the marginal distribution of x? Explain your answer.

13.46 Age and Sex. The U.S. Bureau of the Census publishes census data on the resident population of the United States in *Current Population Reports*. According to that document, 7.2% of male residents are in the age group 20–24 years.
a. If there were no association between age group and sex, what percentage of the resident population would be in the age group 20–24 years? Explain your answer.
b. If there were no association between age group and sex, what percentage of female residents would be in the age group 20–24 years? Explain your answer.
c. There are about 145.0 million female residents of the United States. If there were no association between age group and sex, how many female residents would there be in the age group 20–24 years?
d. In fact there are some 9.6 million female residents in the age group 20–24 years. Given this number and your answer to part (c), what do you conclude?

Using Technology

13.47 Sex, Class Level, and College. Use the technology of your choice to carry out parts (a)–(c) of
a. Exercise 13.32.
b. Exercise 13.33.
c. Exercise 13.34.

13.48 Party and Class. Use the technology of your choice to carry out
a. parts (a) and (b) of Exercise 13.35.
b. part (b) of Exercise 13.36.

13.4 Chi-Square Independence Test

In Section 13.3, you learned how to determine whether there is an association between two variables of a population if you have the bivariate data for the entire population. However, because, in most cases, data for an entire population are not available, you must usually apply inferential methods to decide whether an association exists between two variables.

One of the most commonly used procedures for making such decisions is the **chi-square independence test.** In Example 13.6, we introduce and explain the reasoning behind the chi-square independence test.

Example 13.6 Introducing the Chi-Square Independence Test

Marital Status and Drinking   A national survey was conducted to obtain information on the alcohol consumption patterns of U.S. adults by marital status. A random sample of 1772 residents, 18 years old and older, yielded the data displayed in Table 13.13.[†] For instance, of the 1772 adults sampled, 1173 are married, 590 abstain, and 411 are married and abstain.

TABLE 13.13

Contingency table of marital status and alcohol consumption for 1772 randomly selected U.S. adults

		Drinks per month			
		Abstain	1–60	Over 60	**Total**
Marital status	Single	67	213	74	354
	Married	411	633	129	1173
	Widowed	85	51	7	143
	Divorced	27	60	15	102
	Total	590	957	225	1772

Suppose that we want to use the data in Table 13.13 to decide whether there is an association between marital status and alcohol consumption.

a. Formulate the problem statistically by posing it as a hypothesis test.

b. Explain the basic idea for carrying out the hypothesis test.

c. Develop a formula for computing the expected frequencies.

d. Construct a table that provides both the observed frequencies in Table 13.13 and the expected frequencies.

e. Discuss the details for making a decision concerning the hypothesis test.

Solution

a. For a chi-square independence test, the null hypothesis is that the two variables are not associated and the alternative hypothesis is that the two variables are associated. Thus, we want to perform the hypothesis test

H_0: Marital status and alcohol consumption are not associated.

H_a: Marital status and alcohol consumption are associated.

[†]Adapted from "Alcohol Use and Alcohol Problems among U.S. Adults: Results of the 1979 National Survey" by W. B. Clark and L. Midanik. In National Institute on Alcohol Abuse and Alcoholism, *Alcohol and Health Monograph No. 1, Alcohol Consumption and Related Problems.* DHHS Pub. No. (ADM) 82–1190.

b. The idea behind the chi-square independence test is to compare the observed frequencies in Table 13.13 with the frequencies that would be expected if the null hypothesis of nonassociation is true. The test statistic for making the comparison is the same one used for the goodness-of-fit test, $\chi^2 = \Sigma(O - E)^2/E$, where O represents observed frequency and E represents expected frequency.

c. To develop a formula for computing the expected frequencies, consider, for instance, the cell of Table 13.13 corresponding to "Married *and* Abstain," the cell in the second row and first column. We note that the population proportion of all adults who abstain can be estimated by the sample proportion of the 1772 adults sampled who abstain, that is, by

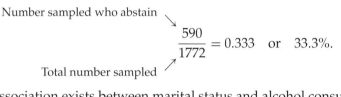

$$\frac{590}{1772} = 0.333 \quad \text{or} \quad 33.3\%.$$

If no association exists between marital status and alcohol consumption (i.e., if H_0 is true), then the proportion of married adults who abstain is the same as the proportion of all adults who abstain. Thus, if H_0 is true, the sample proportion 590/1772 is also an estimate of the population proportion of married adults who abstain. This result, in turn, implies that of the 1173 married adults sampled, we would expect about

$$\frac{590}{1772} \cdot 1173 = 390.6$$

to abstain from alcohol.

Let's rewrite the left side of this expected-frequency computation in a slightly different way. By using algebra and referring to Table 13.13, we obtain

$$\text{Expected frequency} = \frac{590}{1772} \cdot 1173 = \frac{1173 \cdot 590}{1772}$$

$$= \frac{(\text{Row total}) \cdot (\text{Column total})}{\text{Sample size}}.$$

If we let R denote "Row total" and C denote "Column total," we can express this equation compactly as

$$E = \frac{R \cdot C}{n}, \tag{13.1}$$

where, as usual, E denotes expected frequency and n denotes sample size.

d. Using Equation (13.1), we can obtain the expected frequencies for all 12 cells in Table 13.13. We have already done that for the cell in the second row and first column. For the cell in the upper right corner of the table, we get

$$E = \frac{R \cdot C}{n} = \frac{354 \cdot 225}{1772} = 44.9.$$

Similar computations give the expected frequencies for the remaining cells.

In Table 13.14, we have modified Table 13.13 by placing the expected frequency for each cell beneath the corresponding observed frequency. Table 13.14 shows, for instance, that of the adults sampled, 74 were observed to be single and consume more than 60 drinks per month, whereas if there is no association between marital status and alcohol consumption, the expected frequency is 44.9.

TABLE 13.14

Observed and expected frequencies for marital status and alcohol consumption (expected frequencies printed below observed frequencies)

Marital status		Drinks per month			
		Abstain	1–60	Over 60	**Total**
	Single	67 117.9	213 191.2	74 44.9	354
	Married	411 390.6	633 633.5	129 148.9	1173
	Widowed	85 47.6	51 77.2	7 18.2	143
	Divorced	27 34.0	60 55.1	15 13.0	102
	Total	590	957	225	1772

e. If the null hypothesis of nonassociation is true, the observed and expected frequencies should be approximately equal, which would result in a relatively small value of the test statistic, $\chi^2 = \Sigma (O - E)^2 / E$. Consequently, if χ^2 is too large, we reject the null hypothesis and conclude that an association exists between marital status and alcohol consumption. From Table 13.14, we find that

$$\begin{aligned}
\chi^2 &= \Sigma (O - E)^2 / E \\
&= (67 - 117.9)^2/117.9 + (213 - 191.2)^2/191.2 + (74 - 44.9)^2/44.9 \\
&\quad + (411 - 390.6)^2/390.6 + (633 - 633.5)^2/633.5 + (129 - 148.9)^2/148.9 \\
&\quad + (85 - 47.6)^2/47.6 + (51 - 77.2)^2/77.2 + (7 - 18.2)^2/18.2 \\
&\quad + (27 - 34.0)^2/34.0 + (60 - 55.1)^2/55.1 + (15 - 13.0)^2/13.0 \\
&= 21.952 + 2.489 + 18.776 + 1.070 + 0.000 + 2.670 \\
&\quad + 29.358 + 8.908 + 6.856 + 1.427 + 0.438 + 0.324 \\
&= 94.269.^{\dagger}
\end{aligned}$$

Can this value be reasonably attributed to sampling error, or is it large enough to indicate that marital status and alcohol consumption are associated? Before we can answer that question, we must know the distribution of the χ^2-statistic, which is described in Key Fact 13.3. ▬

---

†Although we have displayed the expected frequencies to one decimal place and the chi-square subtotals to three decimal places, the calculations were made at full calculator accuracy.

KEY FACT 13.3

Distribution of the χ^2-Statistic for a Chi-Square Independence Test

For a chi-square independence test, the test statistic

$$\chi^2 = \Sigma(O - E)^2/E$$

has approximately a chi-square distribution if the null hypothesis of nonassociation is true. The number of degrees of freedom is $(r - 1)(c - 1)$, where r and c are the number of possible values for the two variables under consideration.

What *Does it Mean?*

To obtain a chi-square subtotal, square the difference between an observed and expected frequency and divide the result by the expected frequency. Adding the chi-square subtotals gives the χ^2-statistic, which has approximately a chi-square distribution.

Procedure for the Chi-Square Independence Test

In light of Key Fact 13.3, we now present, in Procedure 13.2 on the next page, a step-by-step method for conducting a chi-square independence test by using either the critical-value approach or the P-value approach. Because the null hypothesis is rejected only when the test statistic is too large, a chi-square independence test is always right tailed.

Example 13.7 **The Chi-Square Independence Test**

Marital Status and Drinking A random sample of 1772 U.S. adults yielded the data on marital status and alcohol consumption displayed in Table 13.13 on page 670. At the 5% significance level, do the data provide sufficient evidence to conclude that an association exists between marital status and alcohol consumption?

Solution We apply Procedure 13.2.

STEP 1 State the null and alternative hypotheses.

The null and alternative hypotheses are

H_0: Marital status and alcohol consumption are not associated.

H_a: Marital status and alcohol consumption are associated.

STEP 2 Calculate the expected frequencies by using the formula $E = RC/n$, where R = row total, C = column total, and n = sample size. Place each expected frequency below its corresponding observed frequency in the contingency table.

We did so earlier and displayed the results in Table 13.14 on page 672.

STEP 3 Determine whether the expected frequencies satisfy Assumptions 1 and 2.

1. Are all expected frequencies 1 or greater? Yes, as shown in Table 13.14.
2. Are at most 20% of the expected frequencies less than 5? Yes, in fact, none of the expected frequencies are less than 5, as shown in Table 13.14.

STEP 4 Decide on the significance level, α.

The test is to be performed at the 5% significance level, or $\alpha = 0.05$.

PROCEDURE 13.2	Chi-Square Independence Test

Purpose To perform a hypothesis test to decide whether two variables are associated

Assumptions
1. All expected frequencies are 1 or greater
2. At most 20% of the expected frequencies are less than 5
3. Simple random sample

STEP 1 The null and alternative hypotheses are

H_0: The two variables are not associated.

H_a: The two variables are associated.

STEP 2 Calculate the expected frequencies by using the formula $E = RC/n$, where $R =$ row total, $C =$ column total, and $n =$ sample size. Place each expected frequency below its corresponding observed frequency in the contingency table.

STEP 3 Determine whether the expected frequencies satisfy Assumptions 1 and 2. If they do not, this procedure should not be used.

STEP 4 Decide on the significance level, α.

STEP 5 Compute the value of the test statistic

$$\chi^2 = \Sigma(O - E)^2/E,$$

where O and E represent observed and expected frequencies, respectively. Denote the value of the test statistic χ_0^2.

CRITICAL-VALUE APPROACH	P-VALUE APPROACH

STEP 6 The critical value is χ_α^2 with df $=$ $(r-1)(c-1)$, where r and c are the number of possible values for the two variables. Use Table VII to find the critical value.

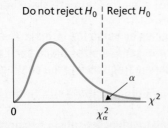

STEP 7 If the value of the test statistic falls in the rejection region, reject H_0; otherwise, do not reject H_0.

STEP 6 The χ^2-statistic has df $= (r-1)(c-1)$, where r and c are the number of possible values for the two variables. Use Table VII to estimate the P-value, or obtain it exactly by using technology.

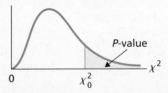

STEP 7 If $P \leq \alpha$, reject H_0; otherwise, do not reject H_0.

STEP 8 Interpret the results of the hypothesis test.

STEP 5 **Compute the value of the test statistic**

$$\chi^2 = \Sigma(O - E)^2/E,$$

where *O* and *E* represent observed and expected frequencies, respectively.

The observed and expected frequencies are displayed in Table 13.14. Using them, we compute the value of the test statistic:

$$\chi^2 = (67 - 117.9)^2/117.9 + (213 - 191.2)^2/191.2 + \cdots + (15 - 13.0)^2/13.0$$
$$= 21.952 + 2.489 + \cdots + 0.324$$
$$= 94.269.$$

CRITICAL-VALUE APPROACH	P-VALUE APPROACH

CRITICAL-VALUE APPROACH

STEP 6 **The critical value is χ^2_α with df = $(r - 1)(c - 1)$, where *r* and *c* are the number of possible values for the two variables. Use Table VII to find the critial value.**

The number of marital status categories is four, and the number of drinks per month categories is three. Hence $r = 4, c = 3$, and

$$\text{df} = (r - 1)(c - 1) = 3 \cdot 2 = 6.$$

For $\alpha = 0.05$, Table VII reveals that the critical value is $\chi^2_{0.05} = 12.592$, as depicted in Fig. 13.5A.

FIGURE 13.5A

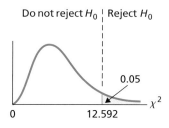

Do not reject H_0 | Reject H_0

0.05

0   12.592 χ^2

STEP 7 **If the value of the test statistic falls in the rejection region, reject H_0; otherwise, do not reject H_0.**

From Step 5, we see that the value of the test statistic is $\chi^2 = 94.269$, which falls in the rejection region, as shown in Fig. 13.5A. Thus we reject H_0. The test results are statistically significant at the 5% level.

P-VALUE APPROACH

STEP 6 **The χ^2-statistic has df = $(r - 1)(c - 1)$, where *r* and *c* are the number of possible values for the two variables. Use Table VII to estimate the P-value, or obtain it exactly by using technology.**

From Step 5, we see that the value of the test statistic is $\chi^2 = 94.269$. Because the test is right tailed, the P-value is the probability of observing a value of χ^2 of 94.269 or greater if the null hypothesis is true. That probability equals the shaded area shown in Fig. 13.5B.

FIGURE 13.5B

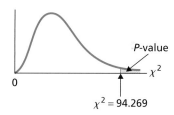

P-value

0

χ^2

$\chi^2 = 94.269$

The number of marital status categories is four and the number of drinks per month categories is three. Hence $r = 4, c = 3$, and

$$\text{df} = (r - 1)(c - 1) = 3 \cdot 2 = 6.$$

From Fig. 13.5B and Table VII with df = 6, we find that $P < 0.005$. (Using technology, we determined that $P = 0.000$ to three decimal places.)

STEP 7 **If $P \leq \alpha$, reject H_0; otherwise, do not reject H_0.**

From Step 6, $P < 0.005$. Because the P-value is less than the specified significance level of 0.05, we reject H_0. The test results are statistically significant at the 5% level and (see Table 9.12 on page 440) provide very strong evidence against the null hypothesis.

STEP 8 **Interpret the results of the hypothesis test.**

INTERPRETATION At the 5% significance level, the data provide sufficient evidence to conclude that there is an association between marital status and alcohol consumption.

Concerning the Assumptions

In Procedure 13.2, we made two assumptions about expected frequencies:

1. All expected frequencies are 1 or greater.
2. At most 20% of the expected frequencies are less than 5.

What can we do if one or both of these assumptions are violated? Three approaches are possible. We can combine rows or columns to increase the expected frequencies in those cells in which they are too small; we can eliminate certain rows or columns in which the small expected frequencies occur; or we can increase the sample size.

Association and Causation

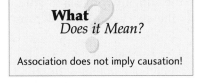

What *Does it Mean?*

Association does not imply causation!

The chi-square independence test is used to decide whether an association exists between two variables of a population—the null hypothesis is that the two variables are not associated, and the alternative hypothesis is that they are associated. If the null hypothesis is rejected, we can conclude that the two variables are associated, but not that they are causally related.

For instance, in Example 13.7, we rejected the null hypothesis of nonassociation for the variables marital status and alcohol consumption. In other words, knowing the marital status of a person imparts information about the alcohol consumption of that person, and vice versa. It does not necessarily mean, for instance, that being single causes a person to drink more.

Although we must keep in mind that association does not imply causation, we must also note that, if two variables are not associated, there is no point in looking for a causal relationship. In other words, association is a necessary but not sufficient condition for causation.

The **Technology** **Center**

Most statistical technologies have programs that automatically perform a chi-square independence test. In this subsection, we present output and step-by-step instructions to implement such programs for Minitab and the TI-83/84 Plus. Currently, Excel's program has some technical problems, and DDXL requires raw data (as opposed to contingency-table data); ways to work around these problems are discussed in the *Excel Manual*.

Example 13.8 **Using Technology to Perform an Independence Test**

Marital Status and Drinking Use Minitab or the TI-83/84 Plus to perform the chi-square independence test considered in Example 13.7 on page 673.

Solution   We want to perform the hypothesis test

H_0: Marital status and alcohol consumption are not associated.

H_a: Marital status and alcohol consumption are associated.

at the 5% significance level. Printout 13.1 shows the output obtained by applying the programs for a chi-square independence test to the data presented in Table 13.13.

PRINTOUT 13.1

Chi-square independence test output for the data on marital status and alcohol consumption

MINITAB

Chi-Square Test: ABSTAIN, 1-60, OVER 60

```
Expected counts are printed below observed counts
Chi-Square contributions are printed below expected counts

            ABSTAIN     1-60   OVER 60   Total
     1           67      213        74     354
            117.87   191.18     44.95
            21.952    2.489    18.776

     2          411      633       129    1173
            390.56   633.50    148.94
             1.070    0.000     2.670

     3           85       51         7     143
             47.61    77.23     18.16
            29.358    8.908     6.856

     4           27       60        15     102
             33.96    55.09     12.95
             1.427    0.438     0.324

Total          590      957       225    1772

Chi-Sq = 94.269, DF = 6, P-Value = 0.000
```

TI-83/84 PLUS

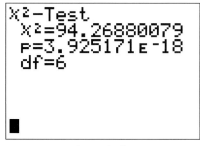

Using **Calculate**

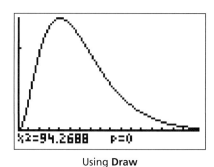

Using **Draw**

The outputs in Printout 13.1 reveal that the *P*-value for the hypothesis test is 0.000 to three decimal places. Because the *P*-value is less than the specified significance level of 0.05, we reject H_0. The test results are statistically significant at the 5% level. That is, at the 5% significance level, the data provide sufficient evidence to conclude that there is an association between marital status and alcohol consumption.

Obtaining the Output (Optional)

Printout 13.1 provides output from Minitab and the TI-83/84 Plus for a chi-square independence test based on the sample data displayed in Table 13.13 on page 670. The following are detailed instructions for obtaining that output.

MINITAB	EXCEL	TI-83/84 PLUS
1 Store the cell data from Table 13.13 in columns named ABSTAIN, 1-60, and OVER 60.	SEE THE EXCEL MANUAL	1 Press **2nd ➤ MATRIX**, arrow over to **EDIT**, and press **1**
2 Choose **Stat ➤ Tables ➤ Chi-Square Test (Table in Worksheet)...**		2 Type <u>4</u> and press **ENTER**
		3 Type <u>3</u> and press **ENTER**
3 Specify ABSTAIN, '1-60', and 'OVER 60' in the **Columns containing the table** text box		4 Enter the cell data from Table 13.13, pressing **ENTER** after each entry
		5 Press **STAT**, arrow over to **TESTS**, and press **ALPHA ➤ C**
4 Click **OK**		6 Press **2nd ➤ MATRIX**, press **1**, and press **ENTER**
		7 Press **2nd ➤ MATRIX**, press **2**, and press **ENTER**
		8 Highlight **Calculate** or **Draw**, and press **ENTER**

EXERCISES 13.4

Statistical Concepts and Skills

13.49 To decide whether two variables of a population are associated, we usually need to resort to inferential methods such as the chi-square independence test. Why?

13.50 Step 1 of Procedure 13.2 gives generic statements for the null and alternative hypotheses of a chi-square independence test. Use the terms *statistically dependent* and *statistically independent,* introduced on page 665, to restate those hypotheses.

13.51 In Example 13.6, we made the following statement: If no association exists between marital status and alcohol consumption, the proportion of married adults who abstain is the same as the proportion of all adults who abstain. Explain why that statement is true.

13.52 Explain why a chi-square independence test is always right tailed.

13.53 A chi-square independence test is to be conducted to decide whether an association exists between two variables of a population. One variable has six possible values and the other variable has four. What is the degrees of freedom for the χ^2-statistic?

13.54 Education and Salary. Studies have shown that a positive association exists between educational level and annual salary; in other words, people with more education tend to make more money.
a. Does this finding mean that more education *causes* a person to make more money? Explain your answer.
b. Do you think there is a causal relationship between educational level and annual salary? Explain your answer.

13.55 We stated earlier that, if two variables are not associated, there is no point in looking for a causal relationship. Why is that so?

13.56 Identify three techniques that can be tried as a remedy when one or more of the expected-frequency assumptions for a chi-square independence test are violated.

In Exercises **13.57–13.61**, use either the critical-value approach or the P-value approach to perform a chi-square independence test, provided the conditions for using the test are met.

13.57 Siskel and Ebert. In the late Gene Siskel and Roger Ebert's TV show *Sneak Preview,* the two Chicago movie critics reviewed the week's new movie releases and then rated them thumbs up (positive), mixed, or thumbs down (negative). These two critics often saw the merits of a movie differently. But, in general, were the ratings given by Siskel and Ebert associated? The answer to this question was the focus of the paper "Evaluating Agreement and Disagreement Among Movie Reviewers" by Alan Agresti and Larry Winner that appeared in *Chance* (Vol. 10(2), pp. 10–14). Following is a contingency table that summarizes the ratings by Siskel and Ebert for 160 movies.

		Ebert's rating			
		Thumbs down	Mixed	Thumbs up	**Total**
Siskel's rating	Thumbs down	24	8	13	45
	Mixed	8	13	11	32
	Thumbs up	10	9	64	83
	Total	42	30	88	160

At the 1% significance level, do the data provide sufficient evidence to conclude that an association exists between the ratings of Siskel and Ebert?

13.58 Diabetes in Native Americans. Preventable chronic diseases are increasing rapidly in Native American populations, particularly diabetes. Gilliland et al. examined the diabetes issue in the paper, "Preventative Health Care among Rural American Indians in New Mexico" (*Preventative Medicine*, Vol. 28, pp. 194–202). Following is a contingency table showing cross classification of educational attainment and diabetic state for a sample of 1273 Native Americans.

		Diabetic state		
		Diabetes	No diabetes	**Total**
Education	Less than HS	33	218	251
	HS grad	25	389	414
	Some college	20	393	413
	College grad	17	178	195
	Total	95	1178	1273

At the 1% significance level, do the data provide sufficient evidence to conclude that an association exists between educational level and diabetic state for Native Americans?

13.59 Learning at Home. M. Stuart et al. studied various aspects of grade-school children and their mothers and reported their findings in the article "Learning to Read at Home and at School" (*British Journal of Educational Psychology*, 68(1), pp. 3–14). The researchers gave a questionnaire to parents of 66 children in kindergarten through second grade. Two social-class groups, middle and working, were identified based on the mother's occupation.

a. One of the questions dealt with the children's knowledge of nursery rhymes. The following data were obtained.

		Nursery-rhyme knowledge		
		A few	Some	Lots
Social class	Middle	4	13	15
	Working	5	11	18

Are Assumptions 1 and 2 satisfied for a chi-square independence test? If so, conduct the test and interpret your results. Use $\alpha = 0.01$.

b. Another question dealt with whether the parents played "I Spy" games with their children. The following data were obtained.

		Frequency of games		
		Never	Sometimes	Often
Social class	Middle	2	8	22
	Working	11	10	13

Are Assumptions 1 and 2 satisfied for a chi-square independence test? If so, conduct the test and interpret your results. Use $\alpha = 0.01$.

13.60 Thoughts of Suicide. A study reported by D. Goldberg in *The Detection of Psychiatric Illness by Questionnaire* (Oxford University, London, p. 126, 1972) examined the relationship between mental-health classification and thoughts of suicide. The mental health of each person in a sample of 295 was classified as normal, mild psychiatric illness, or severe psychiatric illness. Also, each person was asked, "Have you recently found that the idea of taking your own life kept coming into your mind?" Following are the results.

| | Mental health | | | |
	Normal	Mild illness	Severe illness	Total
Definitely not	90	43	34	167
Don't think so	5	18	8	31
Crossed my mind	3	21	21	45
Definitely yes	1	15	36	52
Total	99	97	99	295

(Response)

Is there evidence that an association exists between response to the suicide question and mental-health classification? Perform the required hypothesis test at the 5% significance level.

13.61 Lawyers. The American Bar Foundation publishes information on the characteristics of lawyers in *The Lawyer Statistical Report*. The following contingency table cross classifies 307 randomly selected U.S. lawyers by status in practice and the size of the city in which they practice.

| | Size of city | | | |
	Less than 250,000	250,000–499,999	500,000 or more	Total
Government	12	4	14	30
Judicial	8	1	2	11
Private practice	122	31	69	222
Salaried	19	7	18	44
Total	161	43	103	307

(Status in practice)

At the 5% significance level, do the data provide sufficient evidence to conclude that size of city and status in practice are statistically dependent for U.S. lawyers?

Extending the Concepts and Skills

13.62 Lawyers. In Exercise 13.61, you couldn't perform the chi-square independence test because the assumptions regarding expected frequencies were not met. As mentioned in the text, three approaches are available for remedying the situation: (1) combine rows or columns; (2) eliminate rows or columns; or (3) increase the sample size.
a. Combine the first two rows of the contingency table in Exercise 13.61 to form a new contingency table.
b. Use the table obtained in part (a) to perform the hypothesis test required in Exercise 13.61, if possible.
c. Eliminate the second row of the contingency table in Exercise 13.61 to form a new contingency table.
d. Use the table obtained in part (c) to perform the hypothesis test required in Exercise 13.61, if possible.

Using Technology

13.63 Siskel and Ebert. Use the technology of your choice to perform the hypothesis test in Exercise 13.57.

13.64 Diabetes in Native Americans. Use the technology of your choice to perform the hypothesis test in Exercise 13.58.

13.65 Job Satisfaction. A *CNN/USA TODAY* poll conducted by Gallup asked a sample of employed Americans the following question: "Which do you enjoy more, the hours when you are on your job, or the hours when you are not on your job?" The responses to this question were cross tabulated against several characteristics, among which were sex, age, type of community, amount of education, income, and type of employer. The following table summarizes the poll's results.

	On the job	Off the job	Don't know
Male	77	263	28
Female	77	215	27
18–29 years	33	136	5
30–49 years	77	274	25
50–64 years	35	56	19
65 and older	9	11	5
Urban	62	197	14
Suburban	47	171	25
Rural	42	109	15
Postgraduate	23	51	10
College graduate	41	126	18
Some college	20	108	5
No college	68	192	21
Under $20,000	40	80	11
$20,000–$29,999	32	116	7
$30,000–$49,999	41	131	8
$50,000 and over	34	138	22
Private	67	326	27
Government	23	82	11
Self	61	69	14

Use the data to decide, at the 5% significance level, whether an association exists between each of the following pairs of variables.

a. sex and response (to the question)
b. age and response
c. type of community and response
d. amount of education and response
e. income and response
f. type of employer and response

CHAPTER REVIEW

You Should Be Able To

1. use and understand the formulas in this chapter.

2. identify the basic properties of χ^2-curves.

3. use the chi-square table, Table VII.

4. explain the reasoning behind the chi-square goodness-of-fit test.

5. perform a chi-square goodness-of-fit test.

6. group bivariate data into a contingency table.

7. find and graph marginal and conditional distributions.

8. decide whether an association exists between two variables of a population, given bivariate data for the entire population.

9. explain the reasoning behind the chi-square independence test.

10. perform a chi-square independence test to decide whether an association exists between two variables of a population, given bivariate data for a sample of the population.

Key Terms

association, *665*
bivariate data, *660*
cells, *661*
χ^2_α, *650*
chi-square (χ^2) curve, *650*
chi-square distribution, *650*
chi-square goodness-of-fit test, *652, 655*

chi-square independence test, *670, 674*
chi-square procedures, *648*
chi-square subtotals, *653*
conditional distribution, *663*
contingency table, *660*
expected frequencies, *652*
marginal distribution, *663*

observed frequencies, *652*
segmented bar graph, *664*
statistically dependent variables, *665*
statistically independent
    variables, *665*
two-way table, *660*
univariate data, *660*

REVIEW TEST

Statistical Concepts and Skills

1. How do you distinguish among the infinitely many chi-square distributions and corresponding χ^2-curves?

2. Regarding a χ^2-curve,
a. at what point on the horizontal axis does the curve begin?
b. what shape does it have?
c. As the number of degrees of freedom increases, a χ^2-curve begins to look like another type of curve. What type of curve is that?

3. Recall that the number of degrees of freedom for the *t*-distribution used in a one-sample *t*-test for a population mean depends on the sample size.
a. Is that true for the chi-square distribution used in a chi-square goodness-of-fit test? Explain your answer.
b. Is that true for the chi-square distribution used in a chi-square independence test? Explain your answer.

4. Explain why a chi-square goodness-of-fit test or a chi-square independence test is always right tailed.

5. If the observed and expected frequencies for a chi-square goodness-of-fit test or a chi-square independence test matched perfectly, what would be the value of the test statistic?

6. Regarding the expected-frequency assumptions for a chi-square goodness-of-fit test or a chi-square independence test,
a. state them.
b. how important are they?

7. Race and Region. T. G. Exter's book *Regional Markets, Vol. 2/Households* (Ithaca, NY: New Strategist Publications, Inc.) provides data on U.S. households by region of the country. One table in the book cross classifies households by race (of the householder) and region of residence. The table shows that 7.8% of all U.S. households are Hispanic.
a. If there were no association between race and region of residence, what percentage of Midwest households would be Hispanic?
b. There are 24.7 million Midwest households. If there were no association between race and region of residence, how many Midwest households would be Hispanic?
c. In fact, there are 645 thousand Midwest Hispanic households. Given this information and your answer to part (b), what can you conclude?

8. Suppose that you have bivariate data for an entire population.
a. How would you decide whether there is an association between the two variables under consideration?
b. Assuming that you make no calculation mistakes, could your conclusion be in error? Explain your answer.

9. Suppose that you have bivariate data for a sample of a population.
a. How would you decide whether there is an association between the two variables under consideration?
b. Assuming that you make no calculation mistakes, could your conclusion be in error? Explain your answer.

10. Consider a χ^2-curve with 17 degrees of freedom. Use Table VII to determine
a. $\chi^2_{0.99}$.
b. $\chi^2_{0.01}$.
c. the χ^2-value that has area 0.05 to its right.
d. the χ^2-value that has area 0.05 to its left.
e. the two χ^2-values that divide the area under the curve into a middle 0.95 area and two outside 0.025 areas.

11. Educational Attainment. The U.S. Bureau of the Census compiles census data on educational attainment of Americans. From the document *Current Population Reports*, we obtained the 2000 percent distribution of educational attainment for U.S. adults 25 years old and older. Here is that distribution.

Highest level	Percentage
Not HS graduate	15.8
HS graduate	33.2
Some college	17.6
Associate's degree	7.8
Bachelor's degree	17.0
Advanced degree	8.6

A random sample of 500 U.S. adults (25 years old and older) taken this year gave the following frequency distribution.

Highest level	Frequency
Not HS graduate	84
HS graduate	160
Some college	88
Associate's degree	32
Bachelor's degree	87
Advanced degree	49

a. Use either the critical-value approach or *P*-value approach to decide, at the 5% significance level, whether this year's distribution of educational attainment differs from the 2000 distribution.
b. Estimate the *P*-value of the hypothesis test and use that estimate to assess the evidence against the null hypothesis.

12. U.S. Governors. From the *Statistical Abstract of the United States*, we obtained the data on state and U.S. region displayed in the first, second, fourth, and fifth columns of the following table. And from the National Governors Association (www.nga.org), we got the data on each governor's political party affiliation, as of 2003, shown in the third and sixth columns of the table.

State	Region	Party	State	Region	Party
AL	SO	Rep	MT	WE	Rep
AK	WE	Rep	NE	MW	Rep
AZ	WE	Dem	NV	WE	Rep
AR	SO	Rep	NH	NE	Rep
CA	WE	Dem	NJ	NE	Dem
CO	WE	Rep	NM	WE	Dem
CT	NE	Rep	NY	NE	Rep
DE	SO	Dem	NC	SO	Dem
FL	SO	Rep	ND	MW	Rep
GA	SO	Rep	OH	MW	Rep
HI	WE	Rep	OK	SO	Dem
ID	WE	Rep	OR	WE	Dem
IL	MW	Dem	PA	NE	Dem
IN	MW	Dem	RI	NE	Rep
IA	MW	Dem	SC	SO	Rep
KS	MW	Dem	SD	MW	Rep
KY	SO	Dem	TN	SO	Dem
LA	SO	Rep	TX	SO	Rep
ME	NE	Dem	UT	WE	Rep
MD	SO	Rep	VT	NE	Rep
MA	NE	Rep	VA	SO	Dem
MI	MW	Dem	WA	WE	Dem
MN	MW	Rep	WV	SO	Dem
MS	SO	Dem	WI	MW	Dem
MO	MW	Dem	WY	WE	Dem

a. Group the bivariate data for the variables "region" and "party of governor" into a contingency table.
b. Determine the conditional distributions of region by party and the marginal distribution of region.
c. Determine the conditional distributions of party by region and the marginal distribution of party.
d. Is there an association between the variables "region" and "party of governor" for the states of the United States? Explain your answer.
e. What percentage of states have Republican governors?
f. If there were no association between region and party of governor, what percentage of Southern states would have Republican governors?
g. In reality, what percentage of Southern states have Republican governors?

h. What percentage of states are in the South?
i. If there were no association between region and party of governor, what percentage of states with Republican governors would be in the South?
j. In reality, what percentage of states with Republican governors are in the South?

13. U.S. Hospitals. From data in *Hospital Statistics*, published by the American Hospital Association, we obtained the following contingency table for U.S. hospitals and nursing homes, by type of facility and type of control. We used the abbreviations Gov for Government, Prop for Proprietary, and NP for nonprofit.

		Control			
		Gov	Prop	NP	**Total**
Facility	General	1697	660	3046	5403
	Psychiatric	266	358	113	737
	Chronic	21	1	4	26
	Tuberculosis	3	0	1	4
	Other	59	148	203	410
	Total	2046	1167	3367	6580

In the following questions, the term *hospital* refers to either a hospital or nursing home:
a. How many hospitals are government controlled?
b. How many hospitals are psychiatric facilities?
c. How many hospitals are government controlled psychiatric facilities?
d. How many general facilities are nonprofit?
e. How many hospitals are not under proprietary control?
f. How many hospitals are either general facilities or under proprietary control?

14. U.S. Hospitals. Refer to Problem 13.

a. Obtain the conditional distribution of control type within each facility type.

b. Is there an association between facility type and control type for U.S. hospitals? Explain your answer.

c. Determine the marginal distribution of control type for U.S. hospitals.

d. Construct a segmented bar graph for the conditional distributions and marginal distribution of control type. Interpret the graph in light of your answer to part (b).

e. Without doing any further calculations, respond true or false to the following statement and explain your answer: "The conditional distributions of facility type within control types are identical."

f. Determine the marginal distribution of facility type and the conditional distributions of facility type within control types.

g. What percentage of hospitals are under proprietary control?

h. What percentage of psychiatric hospitals are under proprietary control?

i. What percentage of hospitals under proprietary control are psychiatric hospitals?

15. Hodgkin's Disease. Hodgkin's disease is a malignant, progressive, sometimes fatal disease of unknown cause, and characterized by enlargement of the lymph nodes, spleen, and liver. The following contingency table summarizes data collected during a study by Hancock et al. (*Journal of Clinical Oncology*, Vol. 5(4), pp. 283–297) of 538 patients with Hodgkin's disease. The table cross classifies the histological types of patients and their responses to treatment three months prior to the study.

		Response			
		Positive	Partial	None	**Total**
Histological type	Lymphocyte depletion	18	10	44	72
	Lymphocyte predominance	74	18	12	104
	Mixed cellularity	154	54	58	266
	Nodular sclerosis	68	16	12	96
	Total	314	98	126	538

At the 1% significance level, do the data provide sufficient evidence to conclude that histological type and treatment response are statistically dependent?

Using Technology

16. Educational Attainment. Use the technology of your choice to conduct the required chi-square goodness-of-fit test in Problem 11.

17. U.S. Governors. Use the technology of your choice to solve parts (a)–(c) of Problem 12.

18. Hodgkin's Disease. Use the technology of your choice to perform the hypothesis test in Problem 15.

StatExplore in MyMathLab
Analyzing Data Online

You can use StatExplore to perform all statistical analyses discussed in this chapter.
To illustrate, we show how to perform a chi-square independence test.

EXAMPLE Chi-Square Independence Test

Marital Status and Drinking Table 13.13 on page 670 provides a contingency table of marital status and alcohol consumption for 1772 randomly selected U.S. adults. Use StatExplore to decide, at the 5% significance level, whether the data provide sufficient evidence to conclude that an association exists between marital status and alcohol consumption for U.S. adults.

SOLUTION   The problem is to perform the hypothesis test

H_0: Marital status and alcohol consumption are not associated

H_a: Marital status and alcohol consumption are associated

at the 5% significance level.

To carry out the hypothesis test by using StatExplore, we store the marital-status categories in a column named MARITAL STATUS and store the cell data from Table 13.13 in columns named Abstain, 1-60, and Over 60. Then we proceed as follows.

1 Choose **Stat ➤ Tables ➤ Contingency ➤ with summary**
2 Select Abstain, 1-60, and Over 60 for the **Select columns for table** text box
3 Click the arrow button at the right of the **Row labels in** drop-down list box and select MARITAL STATUS
4 Click in the **Column variable** text box and type `DRINKS PER MONTH`
5 Click **Calculate**

The resulting output is shown in Printout 13.2. From Printout 13.2, the *P*-value for the hypothesis test is less than 0.0001. As the *P*-value is less than the specified significance level of 0.05, we reject H_0. At the 5% significance level, the data provide sufficient evidence to conclude

that there is an association between marital status and alcohol consumption for U.S. adults.

PRINTOUT 13.2
StatExplore output for chi-square independence test

Contingency table results:
Rows: MARITAL STATUS
Columns: DRINKS PER MONTH

	Abstain	1-60	Over 60	Total
Single	67	213	74	354
Married	411	633	129	1173
Widowed	85	51	7	143
Divorced	27	60	15	102
Total	590	957	225	1772

Statistic	DF	Value	P-value
Chi-square	6	94.2688	<0.0001

STATEXPLORE EXERCISES Solve the following problems by using StatExplore.

a. Perform the chi-square goodness-of-fit test required in Example 13.3 on page 656.

b. Group the bivariate data on political party affiliation and class level from Table 13.7 on page 661 into a contingency table. (*Hint:* Start by choosing **Stat ➤ Tables ➤ Contingency ➤ with data.**)

c. Conduct the chi-square independence test required in Exercise 13.57 on page 679.

To access StatExplore, go to the student content area of your Weiss MyMathLab course.

Internet Project

Sex and the Death Penalty

In this Internet project, you are to examine data for men and women on death row to determine whether sex differences exist in the death-row population. Currently, there are 49 women on death row, which accounts for only 1.5% of the total death-row population.

In general, both the death-sentencing rate and the death-row population are very small for women compared to those for men. Additionally, actual execution of female offenders is rare, with only 533 documented instances. Now you can explore these differences yourself in an attempt to understand the many social aspects underlying the numbers.

URL for access to Internet Projects Page: www.aw-bc.com/weiss

Focusing on Data Analysis

The Focus Database

Recall from Chapter 1 (see page 35) that the Focus database contains information on the undergraduate students at the University of Wisconsin - Eau Claire (UWEC). Statistical analyses for this database should be done with the technology of your choice. In the following problems, omit the one foreign student from the analysis.

a. Obtain the sexes (SEX) and classifications (CLASS) of a simple random sample of 200 UWEC undergraduate students.
b. At the 5% significance level, do the data from part (a) provide sufficient evidence to conclude that an association exists between sex and classification for UWEC undergraduates? Interpret your results.
c. Was it reasonable to apply the chi-square independence test in part (b)? Explain your answer.

Repeat parts (a)–(c) for each of the following pairs of variables:

d. sex and residency
e. sex and college
f. classification and residency
g. classification and college
h. college and residency
i. Apply the following coding scheme to the cumulative GPAs of UWEC undergraduates and store the GPA category codes in a column named GPACAT.

Cumulative GPA	GPA category
Less than 2.50	1
2.50–2.99	2
3.00–3.49	3
At least 3.50	4

j. Obtain the sexes and GPA categories (GPACAT) of a simple random sample of 200 UWEC undergraduate students.
k. At the 5% significance level, do the data from part (j) provide sufficient evidence to conclude that an association exists between sex and GPA category for UWEC undergraduates? Interpret your results.
l. Was it reasonable to apply the chi-square independence test in part (k)? Explain your answer.

Repeat parts (j)–(l) for each of the following pairs of variables:

m. GPA category and residency
n. GPA category and college
o. GPA category and classification

Road Rage

At the beginning of this chapter, we discussed the report *Controlling Road Rage: A Literature Review and Pilot Study*, prepared for the AAA Foundation for Traffic Safety by Daniel B. Rathbone, Ph.D., and Jorg C. Huckabee, MSCE. The authors examined the results of a literature review and pilot study on how to prevent aggressive driving and road rage.

One aspect of the study was to investigate road rage as a function of the day of the week. The table on page 649 indicates the day of the week on which road rage occurred for a random sample of 69 road-rage incidents. Use those data to decide, at the 5% significance level, whether road-rage incidents are more likely to occur on some days than on others.

Internet Resources: Visit the Weiss Web site www.aw-bc.com/weiss for additional discussion, exercises, and resources related to this case study.

Biography   *KARL PEARSON: The Founding Developer of Chi-Square Tests*

Karl Pearson was born on March 27, 1857, in London, the second son of William Pearson, a prominent lawyer, and his wife, Fanny Smith. Karl Pearson's early education took place at home. At the age of 9, he was sent to University College School in London, where he remained for the next 7 years. Because of ill health, Pearson was then privately tutored for a year. He received a scholarship at King's College, Cambridge, in 1875. There he earned a B.A. (with honors) in mathematics in 1879 and an M.A. in law in 1882. He then studied physics and metaphysics in Heidelberg, Germany.

In addition to his expertise in mathematics, law, physics, and metaphysics, Pearson was competent in literature and knowledgeable about German history, folklore, and philosophy. He was also considered somewhat of a political radical because of his interest in the ideas of Karl Marx and the rights of women.

In 1884, Pearson was appointed Goldsmid professor of applied mathematics and mechanics at University College; from 1891–1894, he was also a lecturer in geometry at Gresham College, London. In 1911, he gave up the Goldsmid chair to become the first Galton professor of Eugenics at University College. Pearson was elected to the Royal Society—a prestigious association of scientists—in 1896 and was awarded the society's Darwin Medal in 1898.

Pearson really began his pioneering work in statistics in 1893, mainly through an association with Walter Weldon (a zoology professor at University College), Francis Edgeworth (a professor of logic at University College), and Sir Francis Galton (see the Chapter 15 Biography). An analysis of published data on roulette wheels at Monte Carlo led to Pearson's discovery of the chi-square goodness-of-fit test. He also coined the term *standard deviations*, introduced his amazingly diverse skew curves, and developed the most widely used measure of correlation, the correlation coefficient.

Pearson, Weldon, and Galton cofounded the statistical journal *Biometrika*, of which Pearson was editor (1901–1936) and a major contributor. Pearson retired from University College in 1933. He died in London on April 27, 1936.

part V

Regression, Correlation, and ANOVA

14

Descriptive Methods in Regression and Correlation

GENERAL OBJECTIVES We often want to know whether two or more variables are related and, if they are, how they are related. In Sections 13.3 and 13.4, we examined relationships between two variables where each variable is either a qualitative (categorical) variable or a quantitative variable whose possible values constitute or have been grouped into a finite number of categories. In this chapter, we discuss relationships between two quantitative variables.

Some commonly used methods for examining the relationship between two or more quantitative variables and for making predictions are *linear regression* and *correlation*. We discuss descriptive methods in linear regression and correlation in this chapter and consider inferential methods in Chapter 15.

In preparation for our discussion of linear regression, we review linear equations with one independent variable in Section 14.1. In Section 14.2, we explain how to determine the *regression equation*, the equation of the line that best fits a set of data points.

In Section 14.3, we examine the coefficient of determination, a descriptive measure of the utility of the regression equation for making predictions. In Section 14.4, we discuss the linear correlation coefficient, which provides a descriptive measure of the strength of the linear relationship between two quantitative variables.

Shoe Size and Height

Most of us have heard that tall people generally have larger feet than short people. Is that really true and, if so, what is the precise relationship between height and foot length? To examine the relationship, Professor Dennis Young obtained data on shoe size and height for a sample of students at Arizona State University. We have displayed the results obtained by Professor Young in the table below, where height is measured in inches.

At the end of this chapter, after you have studied the fundamentals of descriptive methods in regression and correlation, you will be asked to analyze these data to determine the relationship between shoe size and height and to ascertain the strength of that relationship. In particular, you will discover how shoe size can be used to predict height.

Shoe size	Height	Sex	Shoe size	Height	Sex
6.5	66.0	F	13.0	77.0	M
9.0	68.0	F	11.5	72.0	M
8.5	64.5	F	8.5	59.0	F
8.5	65.0	F	5.0	62.0	F
10.5	70.0	M	10.0	72.0	M
7.0	64.0	F	6.5	66.0	F
9.5	70.0	F	7.5	64.0	F
9.0	71.0	F	8.5	67.0	M
13.0	72.0	M	10.5	73.0	M
7.5	64.0	F	8.5	69.0	F
10.5	74.5	M	10.5	72.0	M
8.5	67.0	F	11.0	70.0	M
12.0	71.0	M	9.0	69.0	M
10.5	71.0	M	13.0	70.0	M

14.1 Linear Equations with One Independent Variable

As an aid to understanding linear regression, we first need to review linear equations with one independent variable. The general form of a **linear equation** with one independent variable can be written as

$$y = b_0 + b_1 x,$$

where b_0 and b_1 are constants (fixed numbers), x is the independent variable, and y is the dependent variable.[†]

The graph of a linear equation with one independent variable is a **straight line;** furthermore, any nonvertical straight line can be represented by such an equation. Examples of linear equations with one independent variable are $y = 4 + 0.2x$, $y = -1.5 - 2x$, and $y = -3.4 + 1.8x$. The straight-line graphs of these three linear equations are depicted in Fig. 14.1.

FIGURE 14.1

Straight-line graphs of three linear equations

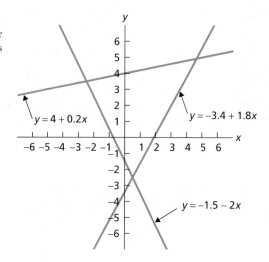

Linear equations with one independent variable occur frequently in applications of mathematics to many different fields, including the management, life, and social sciences, as well as the physical and mathematical sciences. In Examples 14.1 and 14.2, we illustrate the use of linear equations in a simple business application.

Example 14.1 Linear Equations

Word-Processing Costs CJ² Business Services does word processing as one of its basic functions. Its rate is \$20/hr plus a \$25 disk charge. The total cost to a customer depends, of course, on the number of hours needed to complete

---

[†]You may be familiar with the form $y = mx + b$ instead of the form $y = b_0 + b_1 x$. In statistics, the latter form is preferred because it allows a smoother transition to multiple regression, in which there is more than one independent variable. Material on multiple regression is provided in the chapters *Multiple Regression Analysis* and *Model Building in Regression* on the WeissStats CD accompanying this book or on the Weiss Web site.

the job. Find the equation that expresses the total cost in terms of the number of hours required to complete the job.

Solution Let x denote the number of hours required to complete the job, and let y denote the total cost to the customer. Because the rate for word processing is \$20/hr, a job that takes x hours will cost \$20x plus the \$25 disk charge. Hence the total cost, y, of a job that takes x hours is $y = 25 + 20x$.
█

The equation, $y = 25 + 20x$, for the total cost of a word processing job is a linear equation; here $b_0 = 25$ and $b_1 = 20$. Using the equation, we can find the exact cost for a job once we know the number of hours required. For instance, a job that takes 5 hours will cost $y = 25 + 20 \cdot 5 = \$125$; a job that takes 7.5 hours will cost $y = 25 + 20 \cdot 7.5 = \$175$. Table 14.1 displays these costs and a few others.

As we have already mentioned, a linear equation, such as $y = 25 + 20x$, has a straight-line graph. We can obtain the graph of $y = 25 + 20x$ by plotting the points displayed in Table 14.1 and connecting them with a straight line, as shown in Fig. 14.2.

The graph in Fig. 14.2 is useful for quickly estimating cost. For example, a glance at the graph shows that a 10-hour job will cost somewhere between \$200 and \$300. The exact cost is $y = 25 + 20 \cdot 10 = \$225$.

TABLE 14.1

Times and costs for five word processing jobs

Time (hr) x	Cost (\$) y
5.0	125
7.5	175
15.0	325
20.0	425
22.5	475

FIGURE 14.2

Graph of $y = 25 + 20x$, obtained from the points displayed in Table 14.1

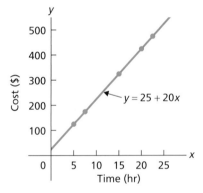

Intercept and Slope

For a linear equation $y = b_0 + b_1 x$, the numbers b_0 and b_1 have an important geometric interpretation. The number b_0 is the y-value where the straight-line graph of the linear equation intersects the y-axis. The number b_1 measures the steepness of the straight line; more precisely, b_1 indicates how much the y-value on the straight line changes when the x-value increases by 1 unit. Figure 14.3 illustrates these relationships.

FIGURE 14.3

Graph of $y = b_0 + b_1 x$

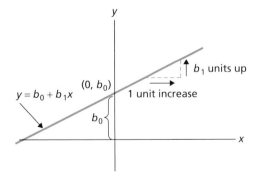

Because of the geometric interpretation of the numbers b_0 and b_1, they have special names that reflect that interpretation: the **y-intercept** and the **slope** of the line, respectively.

DEFINITION 14.1

What Does it Mean?

For a straight line, the y-intercept is where it intersects the y-axis and the slope measures its steepness.

y-Intercept and Slope

For a linear equation $y = b_0 + b_1 x$, the number b_0 is called the **y-intercept** and the number b_1 is called the **slope**.

Example 14.2 applies the concepts of y-intercept and slope, as defined in Definition 14.1, to the illustration of word-processing costs.

Example 14.2 y-Intercept and Slope

Word-Processing Costs In Example 14.1, we obtained the linear equation that expresses the total cost, y, of a word processing job in terms of the number of hours, x, required to complete the job. The equation is $y = 25 + 20x$.

a. Find the y-intercept and slope of that linear equation.

b. Interpret the y-intercept and slope in terms of the graph of the equation.

c. Interpret the y-intercept and slope in terms of word processing costs.

Solution

a. The y-intercept for the equation is $b_0 = 25$, and the slope is $b_1 = 20$.

b. The y-intercept $b_0 = 25$ is the y-value where the straight line $y = 25 + 20x$ intersects the y-axis. The slope $b_1 = 20$ indicates that the y-value increases by 20 units for every increase in x of 1 unit, as shown in Fig. 14.4.

FIGURE 14.4
Graph of $y = 25 + 20x$

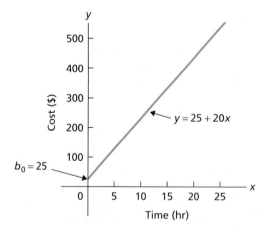

c. In terms of word-processing costs, the y-intercept $b_0 = 25$ represents the total cost of a job that takes 0 hours. In other words, the y-intercept of $25 is a fixed cost that is charged no matter how long the job takes. The slope $b_1 = 20$ represents the cost per hour of $20; it is the amount that the total cost, y, goes up for every increase of 1 hour in the time, x, required to complete the job.

A straight line is determined by any two distinct points that lie on the line. Thus the straight-line graph of a linear equation, $y = b_0 + b_1 x$, can be obtained by first substituting two different x-values into the equation to get two distinct points and then connecting those two points with a straight line.

For example, to graph the linear equation $y = 5 - 3x$, we can use the x-values 1 and 3 (or any other two x-values). The y-values corresponding to those two x-values are $y = 5 - 3 \cdot 1 = 2$ and $y = 5 - 3 \cdot 3 = -4$, respectively. Therefore the graph of the linear equation $y = 5 - 3x$ is the straight line that passes through the two points $(1, 2)$ and $(3, -4)$, as depicted in Fig. 14.5.

FIGURE 14.5
Graph of $y = 5 - 3x$

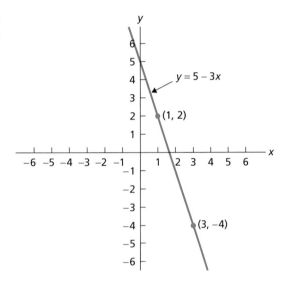

Note that the line in Fig. 14.5 slopes downward—the y-values decrease as x increases—because the slope of the line is negative: $b_1 = -3 < 0$. Now look at the line in Fig. 14.4, the graph of the linear equation $y = 25 + 20x$. That line slopes upward—the y-values increase as x increases—because the slope of the line is positive: $b_1 = 20 > 0$. In general, we have Key Fact 14.1.

KEY FACT 14.1

Graphical Interpretation of Slope

The straight-line graph of the linear equation $y = b_0 + b_1 x$ slopes upward if $b_1 > 0$, slopes downward if $b_1 < 0$, and is horizontal if $b_1 = 0$, as shown in Fig. 14.6.

FIGURE 14.6
Graphical interpretation of slope

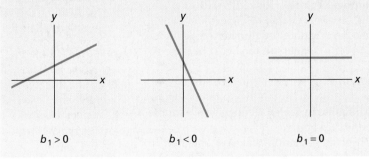

EXERCISES 14.1

Statistical Concepts and Skills

14.1 Regarding linear equations with one independent variable, answer the following questions:
a. What is the general form of such an equation?
b. In your expression in part (a), which letters represent constants and which represent variables?
c. In your expression in part (a), which letter represents the independent variable and which represents the dependent variable?

14.2 Fill in the blank. The graph of a linear equation with one independent variable is a _____.

14.3 Consider the linear equation $y = b_0 + b_1 x$.
a. Identify and give the geometric interpretation of b_0.
b. Identify and give the geometric interpretation of b_1.

14.4 Answer true or false to each statement and explain your answers.
a. The straight-line graph of a linear equation slopes upward unless the slope is 0.
b. The value of the y-intercept has no effect on the direction that the straight-line graph of a linear equation slopes.

14.5 Rental-Car Costs. In September 2003, the Avis Rent-A-Car rate for renting a Buick LeSabre in Mobile, Alabama, was $68.22 per day plus 25¢ per mile. For a 1-day rental, let x denote the number of miles driven and let y denote the total cost, in dollars.
a. Obtain the equation that expresses y in terms of x.
b. Find b_0 and b_1.
c. Construct a table similar to Table 14.1 on page 695 for the x-values 50, 100, and 250 miles.
d. Draw the graph of the equation that you obtained in part (a) by plotting the points from part (c) and connecting them with a straight line.
e. Apply the graph from part (d) to estimate visually the cost of driving the car 150 miles. Then calculate that cost exactly by using the equation from part (a).

14.6 Air-Conditioning Repairs. Richard's Heating and Cooling in Prescott, Arizona, charges $55 per hour plus a $30 service charge. Let x denote the number of hours required for a job and let y denote the total cost to the customer.
a. Obtain the equation that expresses y in terms of x.
b. Find b_0 and b_1.
c. Construct a table similar to Table 14.1 on page 695 for the x-values 0.5, 1, and 2.25 hours.
d. Draw the graph of the equation that you obtained in part (a) by plotting the points from part (c) and connecting them with a straight line.

e. Apply the graph from part (d) to estimate visually the cost of a job that takes 1.75 hours. Then calculate that cost exactly by using the equation from part (a).

14.7 Measuring Temperature. The two most commonly used scales for measuring temperature are the Fahrenheit and Celsius scales. If you let y denote Fahrenheit temperature and x denote Celsius temperature, you can express the relationship between those two scales with the linear equation $y = 32 + 1.8x$.
a. Determine b_0 and b_1.
b. Find the Fahrenheit temperatures corresponding to the Celsius temperatures $-40°$, $0°$, $20°$, and $100°$.
c. Graph the linear equation $y = 32 + 1.8x$, using the four points found in part (b).
d. Apply the graph obtained in part (c) to estimate visually the Fahrenheit temperature corresponding to a Celsius temperature of $28°$. Then calculate that temperature exactly by using the linear equation $y = 32 + 1.8x$.

14.8 A Law of Physics. A ball is thrown straight up in the air with an initial velocity of 64 feet per second (ft/sec). According to the laws of physics, if you let y denote the velocity of the ball after x seconds, $y = 64 - 32x$.
a. Determine b_0 and b_1 for this linear equation.
b. Determine the velocity of the ball after 1, 2, 3, and 4 sec.
c. Graph the linear equation $y = 64 - 32x$, using the four points obtained in part (b).
d. Use the graph from part (c) to estimate visually the velocity of the ball after 1.5 sec. Then calculate that velocity exactly by using the linear equation $y = 64 - 32x$.

In Exercises 14.9–14.12,
a. find the y-intercept and slope of the specified linear equation.
b. explain what the y-intercept and slope represent in terms of the graph of the equation.
c. explain what the y-intercept and slope represent in terms relating to the application.

14.9 Rental-Car Costs. $y = 68.22 + 0.25x$ (from Exercise 14.5)

14.10 Air-Conditioning Repairs. $y = 30 + 55x$ (from Exercise 14.6)

14.11 Measuring Temperature. $y = 32 + 1.8x$ (from Exercise 14.7)

14.12 A Law of Physics. $y = 64 - 32x$ (from Exercise 14.8)

In Exercises **14.13–14.22**, *we give linear equations. For each equation,*
a. find the y-intercept and slope.
b. determine whether the line slopes upward, slopes downward, or is horizontal, without graphing the equation.
c. use two points to graph the equation.

14.13 $y = 3 + 4x$

14.14 $y = -1 + 2x$

14.15 $y = 6 - 7x$

14.16 $y = -8 - 4x$

14.17 $y = 0.5x - 2$

14.18 $y = -0.75x - 5$

14.19 $y = 2$

14.20 $y = -3x$

14.21 $y = 1.5x$

14.22 $y = -3$

In Exercises **14.23–14.30**, *we identify the y-intercepts and slopes, respectively, of straight lines. For each line,*

a. determine whether it slopes upward, slopes downward, or is horizontal, without graphing the equation.
b. find its equation.
c. use two points to graph the equation.

14.23 5 and 2

14.24 −3 and 4

14.25 −2 and −3

14.26 0.4 and 1

14.27 0 and −0.5

14.28 −1.5 and 0

14.29 3 and 0

14.30 0 and 3

Extending the Concepts and Skills

14.31 In this section, we stated that any nonvertical straight line can be described by an equation of the form $y = b_0 + b_1 x$.
a. Explain why a vertical straight line can't be expressed in this form.
b. What is the form of the equation of a vertical straight line?
c. Does a vertical straight line have a slope? Explain your answer.

14.2 The Regression Equation

In Examples 14.1 and 14.2, we discussed the linear equation $y = 25 + 20x$, which expresses the total cost, y, of a word processing job in terms of the time in hours, x, required to complete it. Given the amount of time required, x, we can use the equation to determine the *exact* cost of the job, y.

Real-life applications are frequently not so simple as the word processing example, in which one variable (cost) can be predicted exactly in terms of another variable (time required). Rather, we must often be content with rough predictions. For instance, we cannot predict the exact price, y, of a particular make and model of car just by knowing its age, x. Indeed, even for a fixed age, say, 3 years old, price varies from car to car. We must be content with making a rough prediction for the price of a 3-year-old car of the particular make and model or with an estimate of the mean price of all such 3-year-old cars.

Table 14.2 displays data on age and price for a sample of cars of a particular make and model. We refer to the car as the Orion, but the data, obtained from the *Asian Import* edition of the *Auto Trader* magazine, is for a real car. Ages are in years; prices are in hundreds of dollars, rounded to the nearest hundred dollars.

Plotting the data helps us visualize any apparent relationships between age and price. Such a plot is called a **scatter diagram** (or **scatterplot**). The scatter diagram for the data in Table 14.2 is depicted in Fig. 14.7 on the next page.

Clearly, the data points do not lie on a straight line, but they appear to cluster about a straight line. We want to fit a straight line to the data points and use that line to predict the price of an Orion based on its age.

Because we could draw many different straight lines through the cluster of data points, we need a method to choose the "best" line. The method, called

TABLE 14.2
Age and price data for a sample of 11 Orions

Car	Age (yr) x	Price ($100) y
1	5	85
2	4	103
3	6	70
4	5	82
5	5	89
6	5	98
7	6	66
8	6	95
9	2	169
10	7	70
11	7	48

FIGURE 14.7

Scatter diagram for the age and price data of Orions from Table 14.2

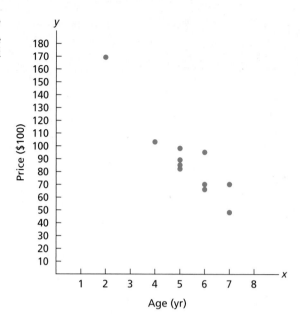

the *least-squares criterion,* is based on an analysis of the errors made in using a straight line to fit the data points. To introduce the least-squares criterion, we use a very simple data set in Example 14.3. We return to the Orion data shortly.

Example 14.3 Introducing the Least-Squares Criterion

TABLE 14.3

Four data points

x	y
1	1
1	2
2	2
4	6

Consider the problem of fitting a straight line to the four data points in Table 14.3, whose scatter diagram is depicted in Fig. 14.8. Many (in fact, infinitely many) straight lines can be fit to those four data points. Two possibilities are shown in Figs. 14.9(a) and 14.9(b).

To avoid confusion, we use $\hat{y}$ to denote the y-value predicted by a straight line for a value of x. For instance, the y-value predicted by Line A for $x = 2$ is

$$\hat{y} = 0.50 + 1.25 \cdot 2 = 3,$$

and the y-value predicted by Line B for $x = 2$ is

$$\hat{y} = -0.25 + 1.50 \cdot 2 = 2.75.$$

FIGURE 14.8

Scatter diagram for the data points in Table 14.3

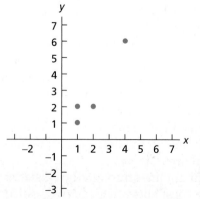

To measure quantitatively how well a line fits the data, we first consider the errors, e, made in using the line to predict the y-values of the data points. For instance, as we have just demonstrated, Line A predicts a y-value of $\hat{y} = 3$ when $x = 2$. The actual y-value for $x = 2$ is $y = 2$ (see Table 14.3). So, the error made in using Line A to predict the y-value of the data point $(2, 2)$ is

$$e = y - \hat{y} = 2 - 3 = -1,$$

as seen in Fig. 14.9(a). In general, an error, e, is the signed vertical distance from the line to a data point.

The fourth column of Table 14.4(a), shows the errors made by Line A for all four data points; the fourth column of Table 14.4(b) shows the same for Line B.

To decide which line, Line A or Line B, fits the data better, we first compute the sum of the squared errors, Σe^2, in the final column of Table 14.4(a) and Table 14.4(b). The line having the smaller sum of squared errors, in this case

FIGURE 14.9
Two possible straight-line fits to the data points in Table 14.3

Line *A*: $y = 0.50 + 1.25x$

Line *B*: $y = -0.25 + 1.50x$

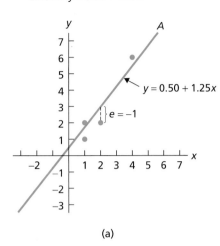

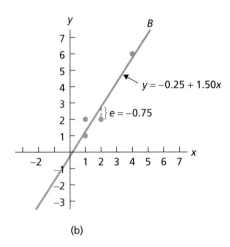

(a)

(b)

TABLE 14.4
Determining how well the data points in Table 14.3 are fit by (a) Line *A* and (b) Line *B*

Line *A*: $y = 0.50 + 1.25x$					Line *B*: $y = -0.25 + 1.50x$				
x	*y*	$\hat{y}$	*e*	e^2	*x*	*y*	$\hat{y}$	*e*	e^2
1	1	1.75	−0.75	0.5625	1	1	1.25	−0.25	0.0625
1	2	1.75	0.25	0.0625	1	2	1.25	0.75	0.5625
2	2	3.00	−1.00	1.0000	2	2	2.75	−0.75	0.5625
4	6	5.50	0.50	0.2500	4	6	5.75	0.25	0.0625
				1.8750					1.2500

(a)

(b)

Line *B*, is the one that fits the data better. And, among all straight lines, the *least-squares criterion* is that the line having the smallest sum of squared errors is the one that fits the data best.

◼

With Example 14.3 in mind, we now state in Key Fact 14.2 the **least-squares criterion** for the straight line that best fits a set of data points. Then, in Definition 14.2, we present the terminology used for that line and for its corresponding equation, namely, **regression line** and **regression equation**.

KEY FACT 14.2   **Least-Squares Criterion**

The straight line that best fits a set of data points is the one having the smallest possible sum of squared errors.

DEFINITION 14.2   **Regression Line and Regression Equation**

Regression line: The straight line that best fits a set of data points according to the least-squares criterion.

Regression equation: The equation of the regression line.

Although the least-squares criterion states the property that the regression line for a set of data points must satisfy, it does not tell us how to find that line. This task is accomplished by Formula 14.1, which provides formulas for obtaining the regression line. In preparation, we introduce, in Definition 14.3, some notation that will be used throughout our study of regression and correlation.

DEFINITION 14.3

Notation Used in Regression and Correlation

We define S_{xx}, S_{xy}, and S_{yy} by $S_{xx} = \Sigma(x - \bar{x})^2$, $S_{xy} = \Sigma(x - \bar{x})(y - \bar{y})$, and $S_{yy} = \Sigma(y - \bar{y})^2$. For hand computations, these three quantities are most easily obtained by using the following computing formulas:

$$S_{xx} = \Sigma x^2 - (\Sigma x)^2/n, \quad S_{xy} = \Sigma xy - (\Sigma x)(\Sigma y)/n, \quad S_{yy} = \Sigma y^2 - (\Sigma y)^2/n.$$

FORMULA 14.1

Regression Equation

The regression equation for a set of n data points is $\hat{y} = b_0 + b_1 x$, where

$$b_1 = \frac{S_{xy}}{S_{xx}} \quad \text{and} \quad b_0 = \frac{1}{n}(\Sigma y - b_1 \Sigma x) = \bar{y} - b_1 \bar{x}.$$

Example 14.4 illustrates determination and use of the regression equation.

Example 14.4 The Regression Equation

TABLE 14.5

Table for computing the regression equation for the Orion data

Age and Price of Orions Table 14.2 displays data on age and price for a sample of 11 Orions. We repeat that data in the first two columns of Table 14.5.

a. Determine the regression equation for the data.

b. Graph the regression equation and the data points.

c. Describe the apparent relationship between age and price of Orions.

d. Interpret the slope of the regression line in terms of prices for Orions.

e. Use the regression equation to predict the price of a 3-year-old Orion and a 4-year-old Orion.

Age (yr) x	Price ($100) y	xy	x^2
5	85	425	25
4	103	412	16
6	70	420	36
5	82	410	25
5	89	445	25
5	98	490	25
6	66	396	36
6	95	570	36
2	169	338	4
7	70	490	49
7	48	336	49
58	975	4732	326

Solution

a. To determine the regression equation, we first need to compute b_1 and b_0 by using Formula 14.1. We did so by constructing a table of values for x (age), y (price), xy, x^2, and their sums in Table 14.5.

The slope of the regression line therefore is

$$b_1 = \frac{S_{xy}}{S_{xx}} = \frac{\Sigma xy - (\Sigma x)(\Sigma y)/n}{\Sigma x^2 - (\Sigma x)^2/n} = \frac{4732 - (58)(975)/11}{326 - (58)^2/11} = -20.26.$$

The y-intercept is

$$b_0 = \frac{1}{n}(\Sigma y - b_1 \Sigma x) = \frac{1}{11}[975 - (-20.26) \cdot 58] = 195.47.$$

So the regression equation is $\hat{y} = 195.47 - 20.26x$.

Note: The usual warnings about rounding apply. When computing the slope, b_1, of the regression line, do not round until the computation is finished. And when computing the y-intercept, b_0, do not use the rounded value of b_1; instead, keep full calculator accuracy.

b. To graph the regression equation, we need to substitute two different x-values in the regression equation to obtain two distinct points. Let's use the x-values 2 and 8. The corresponding y-values are

$$\hat{y} = 195.47 - 20.26 \cdot 2 = 154.95 \quad \text{and} \quad \hat{y} = 195.47 - 20.26 \cdot 8 = 33.39.$$

Hence the regression line passes through the two points $(2, 154.95)$ and $(8, 33.39)$. In Fig. 14.10, we plotted these two points with hollow dots. Drawing a straight line through the two hollow dots yields the regression line, the graph of the regression equation.

FIGURE 14.10

Regression line and data points for Orion data

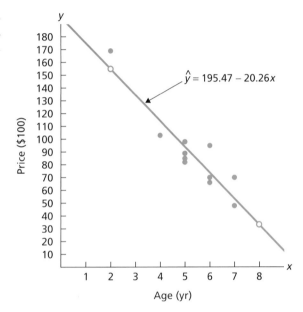

Also shown in Fig. 14.10 are the data points from the first two columns of Table 14.5. The regression line in Fig. 14.10 thus is the straight line that best fits the data points according to the least-squares criterion; that is, it is the straight line that has the smallest possible sum of squared errors.

c. Because the slope of the regression line is negative, price tends to decrease as age increases—no particular surprise.

d. Because x represents age, in years, and y represents price, in hundreds of dollars, the slope of -20.26 indicates that Orions depreciate an estimated $2026 per year, at least in the 2- to 7-year-old range.

e. For a 3-year-old Orion, $x = 3$, and the regression equation yields the predicted price of

$$\hat{y} = 195.47 - 20.26 \cdot 3 = 134.69.$$

INTERPRETATION   The estimated price of a 3-year-old Orion is $13,469.

Similarly, the predicted price for a 4-year-old Orion is

$$\hat{y} = 195.47 - 20.26 \cdot 4 = 114.43.$$

INTERPRETATION The estimated price of a 4-year-old Orion is $11,443.

We discuss questions concerning the accuracy and reliability of such predictions later in this chapter and also in Chapter 15.

Predictor Variable and Response Variable

For a linear equation $y = b_0 + b_1 x$, y is the dependent variable and x is the independent variable. However, in the context of regression analysis, we more customarily call y the **response variable** and x the **predictor variable** or **explanatory variable** (because it is used to predict or explain the values of the response variable). For the Orion example, then, age is the predictor variable and price is the response variable.

Extrapolation

If a scatter diagram indicates a linear relationship between two variables, we can reasonably use the regression equation to make predictions for values of the predictor variable within the range of the observed values of the predictor variable. However, to do so for values of the predictor variable outside that range may not be reasonable because the linear relationship between the variables may not hold there.

Using the regression equation to make predictions for values of the predictor variable outside the range of the observed values of the predictor variable is called **extrapolation.** Grossly incorrect predictions can result from extrapolation.

The Orion example provides an excellent illustration of extrapolation leading to grossly incorrect predictions. The regression equation for the sample of Orions is $\hat{y} = 195.47 - 20.26x$, and the observed ages (values of the predictor variable) range from 2 to 7 years old.

Suppose that we extrapolate by using the regression equation to predict the price of an 11-year-old Orion. The predicted price is

$$\hat{y} = 195.47 - 20.26 \cdot 11 = -27.39,$$

or −$2739. Clearly, this result is ridiculous—no one is going to pay us $2739 to take away their 11-year-old Orion.

Consequently, although the relationship between age and price of Orions appears to be linear in the range from 2 to 7 years old, it is definitely not so in the range from 2 to 11 years old. Figure 14.11 summarizes the discussion on extrapolation as it applies to age and price of Orions.

To help avoid extrapolation, some researchers include the range of the observed values of the predictor variable with the regression equation. For the Orion example, we would write

$$\hat{y} = 195.47 - 20.26x, \qquad 2 \leq x \leq 7.$$

FIGURE 14.11

Extrapolation in the Orion example

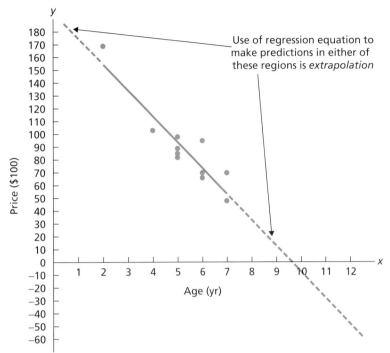

Writing the regression equation in this way makes clear that using it to predict price for ages outside the range from 2 to 7 years old is extrapolation.

Outliers and Influential Observations

Recall that an outlier is an observation that lies outside the overall pattern of the data. In the context of regression, an **outlier** is a data point that lies far from the regression line, relative to the other data points. Figure 14.10 on page 703 shows that the Orion data have no outliers.

An outlier can sometimes have a significant effect on a regression analysis. Thus, as usual, we need to identify outliers and remove them from the analysis when appropriate—for example, if we find the outlier to be a measurement or recording error.

We must also watch for influential observations. In regression analysis, an **influential observation** is a data point whose removal causes the regression equation (and line) to change considerably. A data point separated in the *x*-direction from the other data points is often an influential observation because the regression line is "pulled" toward such a data point without counteraction by other data points.

As with an outlier, we must try to determine the reason for an influential observation. If an influential observation is due to a measurement or recording error or for some other reason it clearly does not belong in the data set, it can be removed without further consideration. However, if no explanation for the influential observation is apparent, the decision whether to retain it is often difficult and calls for a judgment by the researcher.

For the Orion data, Fig. 14.10 on page 703 (or Table 14.5 on page 702) shows that the data point $(2, 169)$ is potentially an influential observation because the age of 2 years appears separated from the other observed ages. When we remove that data point and recalculate the regression equation, the result is $\hat{y} = 160.33 - 14.24x$. Figure 14.12 reveals that this equation differs markedly from the regression equation, $\hat{y} = 195.47 - 20.26x$, based on the full data set. The data point $(2, 169)$ is indeed an influential observation.

FIGURE 14.12

Regression lines with and without the influential observation removed

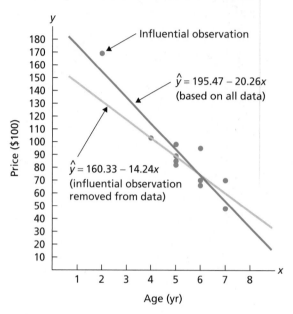

The influential observation $(2, 169)$ is not a recording error; it is a legitimate data point. Nonetheless, we may need either to remove it—thus limiting the analysis to Orions between 4 and 7 years old—or to obtain additional data on 2-year-old (and 3-year-old) Orions so that the regression analysis is not so dependent on one data point.

We added data for one 2-year-old and three 3-year-old Orions and obtained the regression equation $\hat{y} = 193.63 - 19.93x$. This regression equation differs little from our original regression equation, $\hat{y} = 195.47 - 20.26x$. Therefore we could justify using the original regression equation to analyze the relationship between age and price of Orions between 2 and 7 years of age, even though the corresponding data set contains an influential observation.

An outlier may or may not be an influential observation; and an influential observation may or may not be an outlier. Many statistical software packages identify potential outliers and influential observations.

A Warning on the Use of Linear Regression

The idea behind finding a regression line is based on the assumption that the data points are scattered about a straight line.[†] Frequently, however, the data points are scattered about a curve instead of a straight line, as depicted in Fig. 14.13(a).

[†]We discuss this assumption in detail and make it more precise in Section 15.1.

FIGURE 14.13
(a) Data points scattered about a curve;
(b) inappropriate straight line fit to
the data points

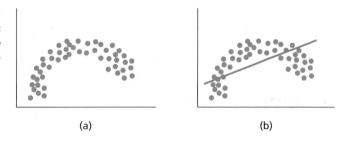

(a)   (b)

One can still compute the values of b_0 and b_1 to obtain a regression line for these data points. The result, however, will yield an inappropriate fit by a straight line, as shown in Fig. 14.13(b), when in fact a curve should be used. For instance, using the regression line would lead us to predict that y-values in Fig. 14.13(a) will keep increasing when they have actually begun to decrease.

Key Fact 14.3 summarizes the criterion for finding a regression line.

KEY FACT 14.3

Criterion for Finding a Regression Line

Before finding a regression line for a set of data points, draw a scatter diagram. If the data points do not appear to be scattered about a straight line, do not determine a regression line.

Techniques are available for fitting curves to data points showing a curved pattern, such as the data points plotted in Fig. 14.13(a). We discuss those techniques, referred to as **curvilinear regression,** in the chapter *Model Building in Regression* on the WeissStats CD accompanying this book or on the Weiss Web site.

The Technology Center

Almost all statistical technologies have programs that automatically obtain a scatter diagram and a regression line for a set of data points. In this subsection, we present output and step-by-step instructions to carry out these two procedures.

Again, before determining a regression line, we should look at a scatter diagram of the data to determine whether the data points appear to be scattered about a straight line. Consider Example 14.5.

Example 14.5

Using Technology to Obtain a Scatter Diagram

Age and Price of Orions Use Minitab, Excel, or the TI-83/84 Plus to obtain a scatter diagram for the age and price data presented in Table 14.2 on page 699.

Solution

Printout 14.1 at the top of the next page shows output obtained by applying the scatter diagram programs to the age and price data presented in Table 14.2.

These outputs show that the data points are scattered about a straight line and hence that we can reasonably find a regression line for these data. Compare the scatter diagrams in Printout 14.1 to the scatter diagram we drew by hand in Fig. 14.7 on page 700.

PRINTOUT 14.1
Scatter diagrams for the
age and price data
of 11 Orions

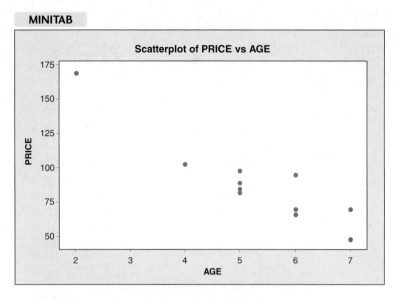

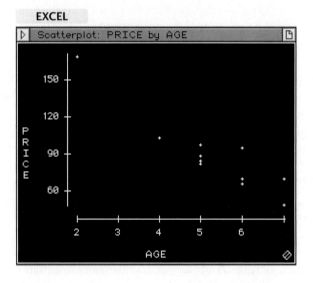

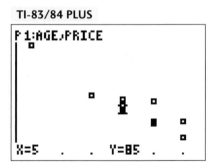

Now consider Example 14.6.

Example 14.6   Using Technology to Obtain a Regression Line

Age and Price of Orions   Use Minitab, Excel, or the TI-83/84 Plus to obtain the regression equation for the age and price data displayed in Table 14.2 on page 699.

Solution   Printout 14.2 shows output obtained by applying the regression programs to the age and price data displayed in Table 14.2.

PRINTOUT 14.2

Regression output for the age and price data of 11 Orions

MINITAB

Regression Analysis: PRICE versus AGE

```
The regression equation is
PRICE = 195 - 20.3 AGE


Predictor      Coef   SE Coef      T      P
Constant     195.47     15.24  12.83  0.000
AGE          -20.261     2.800  -7.24  0.000


S = 12.5766    R-Sq = 85.3%   R-Sq(adj) = 83.7%


Analysis of Variance

Source          DF       SS       MS       F       P
Regression       1   8285.0   8285.0   52.38  0.000
Residual Error   9   1423.5    158.2
Total           10   9708.5
```

EXCEL

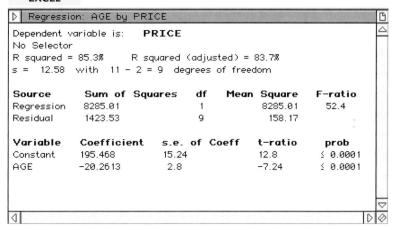

```
▷  Regression: AGE by PRICE

Dependent variable is:    PRICE
No Selector
R squared = 85.3%      R squared (adjusted) = 83.7%
s =  12.58  with  11 - 2 = 9  degrees of freedom

Source       Sum of Squares   df   Mean Square   F-ratio
Regression    8285.01          1      8285.01     52.4
Residual      1423.53          9       158.17

Variable     Coefficient   s.e. of Coeff   t-ratio     prob
Constant     195.468        15.24          12.8      ≤ 0.0001
AGE          -20.2613        2.8           -7.24     ≤ 0.0001
```

TI-83/84 PLUS

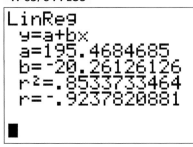

```
LinReg
 y=a+bx
 a=195.4684685
 b=-20.26126126
 r²=.8533733464
 r=-.9237820881
```

The outputs in Printout 14.2 reveal that the regression equation for the age and price data is $\hat{y} = 195.47 - 20.26x$. (In the Minitab and Excel outputs, the y-intercept and slope of the regression line are displayed as the first and second entries in the column labeled `Coef` in Minitab and the column labeled `Coefficient` in Excel, respectively.)

We can also use Minitab, Excel, or the TI-83/84 Plus to obtain a scatter diagram of the data with the regression line superimposed. We show output from the three technologies in Printout 14.3. Compare the scatter diagrams and superimposed regression lines in Printout 14.3 to those we drew by hand in Fig. 14.10 on page 703.

PRINTOUT 14.3

Scatter diagrams with superimposed regression lines for the age and price data of 11 Orions

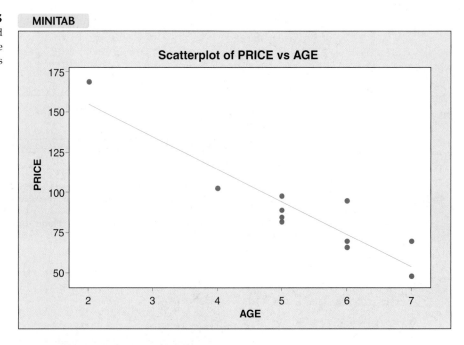

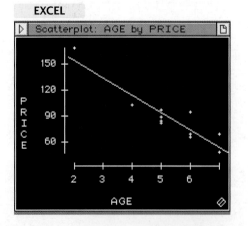

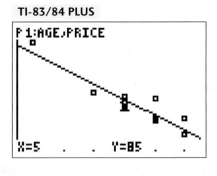

Obtaining the Output (Optional)

Printout 14.1 provides output from Minitab, Excel, and the TI-83/84 Plus for a scatter diagram for the age and price data of 11 Orions given in Table 14.2 on page 699. The following are detailed instructions for obtaining that output. First, we store the age and price data in columns (Minitab), ranges (Excel), or lists (TI-83/84 Plus) named AGE and PRICE, respectively. Then, we proceed as follows.

MINITAB	**EXCEL**	**TI-83/84 PLUS**
1 Choose **Graph ➤ Scatterplot...**	1 Choose **DDXL ➤ Charts and Plots**	1 Press **2nd ➤ STAT PLOT** and then press **ENTER** twice
2 Select the **Simple** scatterplot and click **OK**	2 Select **Scatterplot** from the **Function type** drop-down list box	2 Arrow to the first graph icon and press **ENTER**
3 Specify PRICE in the **Y variables** text box	3 Specify AGE in the **x-Axis Variable** text box	3 Press the down-arrow key
4 Specify AGE in the **X variables** text box	4 Specify PRICE in the **y-Axis Variable** text box	4 Press **2nd ➤ LIST**, arrow down to AGE, and press **ENTER** twice
5 Click **OK**	5 Click **OK**	5 Press **2nd ➤ LIST**, arrow down to PRICE, and press **ENTER** twice
		6 Press **ZOOM** and then **9** (and then **TRACE**, if desired)

Printout 14.2 provides output from Minitab, Excel, and the TI-83/84 Plus for the regression equation for the age and price data in Table 14.2. The following are detailed instructions for obtaining that output. First, we store the age and price data in columns (Minitab), ranges (Excel), or lists (TI-83/84 Plus) named AGE and PRICE, respectively. Then, we proceed as follows.

MINITAB	**EXCEL**	**TI-83/84 PLUS**
1 Choose **Stat ➤ Regression ➤ Regression...**	1 Choose **DDXL ➤ Regression**	1 Press **2nd ➤ CATALOG** and then press **D**
2 Specify PRICE in the **Response** text box	2 Select **Simple Regression** from the **Function type** drop-down list box	2 Arrow down to **DiagnosticOn** and press **Enter** twice
3 Specify AGE in the **Predictors** text box	3 Specify PRICE in the **Response Variable** text box	3 Press **STAT**, arrow over to **CALC**, and press **8**
4 Click the **Results...** button	4 Specify AGE in the **Explanatory Variable** text box	4 Press **2nd ➤ LIST**, arrow down to AGE, and press **ENTER**
5 Select the **Regression equation, table of coefficients, s, R-squared, and basic analysis of variance** option button	5 Click **OK**	5 Press **, ➤ 2nd ➤ LIST**, arrow down to PRICE, and press **ENTER**
6 Click **OK**		6 Press **, ➤ VARS**, arrow over to **Y-VARS**, and press **ENTER** three times
7 Click **OK**		

To obtain the Minitab output in Printout 14.3, proceed as indicated in the instructions for obtaining a scatter diagram except, in the second step, select the **With Regression** scatterplot instead of the **Simple** scatterplot. The Excel output in Printout 14.3 is part of the DDXL output that results from applying the steps in the Excel instructions for doing a regression. To obtain the TI-83/84 Plus output in Printout 14.3, simply press **GRAPH** and then **TRACE** after executing the steps in the TI-83/84 Plus instructions for doing a regression.

EXERCISES 14.2

Statistical Concepts and Skills

14.32 Regarding a scatter diagram,
a. identify one of its uses.
b. what property should it have to obtain a regression line for the data?

14.33 Regarding the criterion used to decide on the line that best fits a set of data points,
a. what is that criterion called?
b. specifically, what is the criterion?

14.34 Regarding the line that best fits a set of data points,
a. what is that line called?
b. what is the equation of that line called?

14.35 Regarding the two variables under consideration in a regression analysis,
a. what is the dependent variable called?
b. what is the independent variable called?

14.36 Using the regression equation to make predictions for values of the predictor variable outside the range of the observed values of the predictor variable is called _____.

14.37 Fill in the blanks.
a. In the context of regression, an _____ is a data point that lies far from the regression line, relative to the other data points.
b. In regression analysis, an _____ is a data point whose removal causes the regression equation to change considerably.

*In Exercises **14.38** and **14.39**,*
a. graph the linear equations and data points.
b. construct tables for x, y, ŷ, e, and e² similar to Table 14.4 on page 701.
c. determine which line fits the set of data points better, according to the least-squares criterion.

14.38 Line A: $y = 1.5 + 0.5x$
Line B: $y = 1.125 + 0.375x$

x	1	1	5	5
y	1	3	2	4

14.39 Line A: $y = 3 - 0.6x$
Line B: $y = 4 - x$

x	0	2	2	5	6
y	4	2	0	-2	1

*For Exercises **14.40**–**14.45**, be sure to save your work. You will need it in later sections.*

14.40 Refer to Exercise 14.38.
a. Find the regression equation for the data points.
b. Graph the regression equation and the data points.

14.41 Refer to Exercise 14.39.
a. Find the regression equation for the data points.
b. Graph the regression equation and the data points.

14.42 Tax Efficiency. *Tax efficiency* is a measure—ranging from 0 to 100—of how much tax due to capital gains stock or mutual funds investors pay on their investments each year; the higher the tax efficiency, the lower is the tax. The paper "At the Mercy of the Manager" (*Financial Planning*, Vol. 30(5), pp. 54–56) by Craig Israelsen examined the relationship between investments in mutual fund portfolios and their associated tax efficiencies. The following table shows percentage of investments in energy securities (x) and tax efficiency (y) for 10 mutual fund portfolios.

x	3.1	3.2	3.7	4.3	4.0	5.5	6.7	7.4	7.4	10.6
y	98.1	94.7	92.0	89.8	87.5	85.0	82.0	77.8	72.1	53.5

a. Determine the regression equation for the data.
b. Graph the regression equation and the data points.
c. Describe the apparent relationship between percent of investments in energy securities and tax efficiency for mutual fund portfolios.
d. What does the slope of the regression line represent in terms of percent of investments in energy securities and tax efficiency for mutual fund portfolios?
e. Use the regression equation that you obtained in part (a) to predict the tax efficiency of a mutual fund portfolio with 5.0% of its investments in energy securities; with 7.4% of its investments in energy securities.
f. Identify the predictor and response variables.
g. Identify outliers and potential influential observations.

14.43 Corvette Prices. The *Kelley Blue Book* provides information on wholesale and retail prices of cars. Following are age and price data for 10 randomly selected Corvettes between 1 and 6 years old. Here, x denotes age, in years, and y denotes price, in hundreds of dollars.

x	6	6	6	2	2	5	4	5	1	4
y	270	260	275	405	364	295	335	308	405	305

a. Determine the regression equation for the data.
b. Graph the regression equation and the data points.
c. Describe the apparent relationship between age and price for Corvettes.
d. What does the slope of the regression line represent in terms of Corvette age and price?
e. Use the regression equation that you obtained in part (a) to predict the price of a 2-year-old Corvette; a 3-year-old Corvette.
f. Identify the predictor and response variables.
g. Identify outliers and potential influential observations.

14.44 Custom Homes. Hanna Properties specializes in custom-home resales in the Equestrian Estates, an exclusive subdivision in Phoenix, Arizona. A random sample of nine custom homes currently listed for sale provided the following information on size and price. Here, x denotes size, in hundreds of square feet, rounded to the nearest hundred, and y denotes price, in thousands of dollars, rounded to the nearest thousand.

x	26	27	33	29	29	34	30	40	22
y	290	305	325	327	356	411	488	554	246

a. Determine the regression equation for the data.
b. Graph the regression equation and the data points.
c. Describe the apparent relationship between square footage and price for custom homes in the Equestrian Estates.
d. What does the slope of the regression line represent in terms of size and price of custom homes in the Equestrian Estates?
e. Use the regression equation determined in part (a) to predict the price of a custom home in the Equestrian Estates that has 2600 sq ft.
f. Identify the predictor and response variables.
g. Identify outliers and potential influential observations.

14.45 Plant Emissions. Plants emit gases that trigger the ripening of fruit, attract pollinators, and cue other physiological responses. N. G. Agelopolous, K. Chamberlain, and J. A. Pickett examined factors that affect the emission of volatile compounds by the potato plant *Solanum tuberosom* and published their findings in the *Journal of Chemical Ecology* (Vol. 26(2), pp. 497–511). The volatile compounds analyzed were hydrocarbons used by other plants and animals. Following are data on plant weight (x), in grams, and quantity of volatile compounds emitted (y), in hundreds of nanograms, for 11 potato plants.

x	57	85	57	65	52	67	62	80	77	53	68
y	8.0	22.0	10.5	22.5	12.0	11.5	7.5	13.0	16.5	21.0	12.0

a. Determine the regression equation for the data.
b. Graph the regression equation and the data points.
c. Describe the apparent relationship between potato plant weight and quantity of volatile compounds emitted.
d. What does the slope of the regression line represent in terms of potato plant weight and quantity of volatile compounds emitted?
e. Use the regression equation determined in part (a) to predict the quantity of volatile compounds emitted by a potato plant that weighs 75 grams.
f. Identify the predictor and response variables.
g. Identify outliers and potential influential observations.

14.46 For which of the following sets of data points can you reasonably determine a regression line? Explain your answer.

14.47 For which of the following sets of data points can you reasonably determine a regression line? Explain your answer.

14.48 Tax Efficiency. In Exercise 14.42, you determined a regression equation that relates the variables percentage of investments in energy securities and tax efficiency for mutual fund portfolios.
a. Should that regression equation be used to predict the tax efficiency of a mutual fund portfolio with 6.4% of its investments in energy securities? with 15% of its investments in energy securities? Explain your answers.
b. For which percentages of investments in energy securities is use of the regression equation to predict tax efficiency reasonable?

14.49 Corvette Prices. In Exercise 14.43, you determined a regression equation that can be used to predict the price of a Corvette, given its age.
a. Should that regression equation be used to predict the price of a 4-year-old Corvette? a 10-year-old Corvette? Explain your answers.
b. For which ages is use of the regression equation to predict price reasonable?

14.50 Palm Beach Fiasco. The 2000 U.S. presidential election brought great controversy to the election process. Many voters in Palm Beach, Florida, claimed that they were confused by the ballot format and may have accidentally voted for Pat Buchanan when they intended to vote for Al Gore. Professors Greg D. Adams of Carnegie Mellon University and Chris Fastnow of Chatham College compiled and analyzed data on election votes in Florida, by county, for both 1996 and 2000. What conclusions would you draw from the following scatter diagrams constructed by the researchers? Explain your answers.

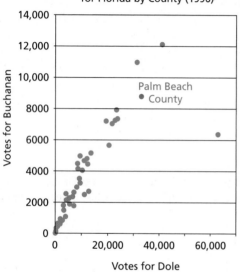

Republican Presidential Primary Election Results for Florida by County (1996)

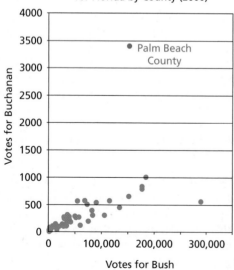

Presidential Election Results for Florida by County (2000)

Source: Prof. Greg D. Adams, Dept. of Social & Decision Sciences, Carnegie Mellon University, and Prof. Chris Fastnow, Director, Center for Women in Politics in Pennsylvania, Chatham College

14.51 Wasp Mating Systems. In the paper "Mating System and Sex Allocation in the Gregarious Parasitoid *Cotesia glomerata*" (*Animal Behaviour*, Vol. 66, pp. 259–264), H. Gu and S. Dorn reported on various aspects of the mating system and sex allocation strategy of the wasp *C. glomerata*. One part of the study involved the investigation of the percentage of male wasps dispersing before mating in relation to the brood sex ratio (proportion of males). Following are the data obtained by the researchers.

Sex ratio	% dispersing	Sex ratio	% dispersing
0.143	24.9	0.657	27.0
0.106	29.9	0.667	32.9
0.157	27.9	0.623	38.0
0.294	12.8	0.680	36.9
0.387	20.9	0.722	37.0
0.446	18.9	0.737	33.0
0.550	20.9	0.758	34.0
0.455	21.8	0.787	36.0
0.447	23.9	0.810	42.0
0.502	27.9	0.836	42.0
0.554	29.0	0.900	53.3

a. Obtain a scatter diagram for the data.
b. Is it reasonable to find a regression line for the data? Explain your answer.

Extending the Concepts and Skills

14.52 Sample Covariance. For n pairs of observations from two variables, x and y, the *sample covariance*, s_{xy}, is defined by

$$s_{xy} = \frac{\Sigma(x - \bar{x})(y - \bar{y})}{n - 1}. \qquad (14.1)$$

a. Determine the sample covariance of the data points in Exercise 14.39.

The sample covariance can be used as an alternative method for obtaining the slope and y-intercept of the regression line for a set of data points. The formulas are

$$b_1 = s_{xy}/s_x^2 \quad \text{and} \quad b_0 = \bar{y} - b_1\bar{x}, \qquad (14.2)$$

where s_x denotes the sample standard deviation of the x-values.

b. Use Equations (14.1) and (14.2) to find the regression equation for the data points in Exercise 14.39. Compare your answer to the answer you obtained in part (a) of Exercise 14.41.

Using Technology

In Exercises 14.53–14.58, use the technology of your choice to
a. obtain a scatter diagram for the data.
b. decide whether finding a regression line for the data is reasonable. If so, then also do parts (c) and (d).
c. determine and interpret the regression equation for the data.
d. identify potential outliers and influential observations.

14.53 Batting and Scoring. Is the number of runs a baseball team scores in a season related to its team batting average? ESPN compiles end-of-season statistics for Major League Baseball and maintains them on its Web site. The following table provides season team batting averages and total runs scored for a sample of major league baseball teams.

Average	Runs	Average	Runs
.294	968	.267	793
.278	938	.265	792
.278	925	.256	764
.270	887	.254	752
.274	825	.246	740
.271	810	.266	738
.263	807	.262	731
.257	798	.251	708

14.54 PCBs and Pelicans. Polychlorinated biphenyls (PCBs), industrial pollutants, are known to be a great danger to natural ecosystems. In a study by R. W. Risebrough

titled "Effects of Environmental Pollutants Upon Animals Other Than Man" (*Proceedings of the 6th Berkeley Symposium on Mathematics and Statistics, VI*, University of California Press, pp. 443–463), 60 Anacapa pelican eggs were collected and measured for their shell thickness, in millimeters (mm), and concentration of PCBs, in parts per million (ppm). The data are on the WeissStats CD.

14.55 More Money More Beer? Does a higher state per capita income equate to a higher per capita beer consumption? Data downloaded from *The Beer Institute Online* on per capita income, in dollars, and per capita beer consumption, in gallons, for the 50 states and Washington, D.C., are provided on the WeissStats CD.

14.56 Gas Guzzlers. The magazine *Consumer Reports* publishes information on automobile gas mileage and variables that affect gas mileage. In the April 1999 issue, data on gas mileage (in miles per gallon) and engine displacement (in liters) were published for 121 vehicles. Those data are available on the WeissStats CD.

14.57 Top Wealth Managers. The September 15, 2003 issue of *BARRON'S* presented information on top wealth managers in the United States, based on individual clients with accounts of $1 million or more. Data were given for various variables, two of which were number of private client managers and private client assets. Those data are provided on the WeissStats CD, where private client assets are in billions of dollars.

14.58 Shortleaf Pines. The ability to estimate the volume of a tree based on a simple measurement, such as the tree's diameter, is important to the lumber industry, ecologists, and conservationists. Data on volume, in cubic feet, and diameter at breast height, in inches, for 70 shortleaf pines was reported in C. Bruce and F. X. Schumacher's *Forest Mensuration* (New York: McGraw-Hill, 1935) and analyzed by A. C. Akinson in the article "Transforming Both Sides of a Tree" (*The American Statistician*, Vol. 48, pp. 307–312). The data are presented on the WeissStats CD.

14.3 The Coefficient of Determination

In Example 14.4, we determined the regression equation, $\hat{y} = 195.47 - 20.26x$, for data on age and price of a sample of 11 Orions, where x represents age, in years, and $\hat{y}$ represents predicted price, in hundreds of dollars. We also applied the regression equation to predict the price of a 4-year-old Orion,

$$\hat{y} = 195.47 - 20.26 \cdot 4 = 114.43,$$

or $11,443. But how valuable are such predictions? Is the regression equation useful for predicting price, or could we do just as well by ignoring age?

In general, the utility of a regression equation for making predictions can be evaluated in several ways. One way is to determine the percentage of variation in the observed values of the response variable explained by the regression (or predictor variable). To illustrate, we return to the data on age and price of Orions in Example 14.7.

■■■ Example 14.7 Introduces the Coefficient of Determination

Age and Price of Orions   The scatter diagram and regression line for the age and price data of 11 Orions are repeated in Fig. 14.14.

FIGURE 14.14

Scatter diagram and regression line for Orion data

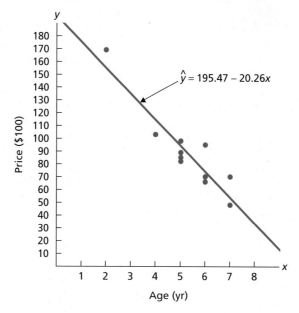

$$\hat{y} = 195.47 - 20.26x$$

The scatter diagram reveals that the prices of the 11 Orions vary widely, ranging from a low of 48 ($4800) to a high of 169 ($16,900). But Fig. 14.14 also shows that much of the price variation is "explained" by the regression (or age); that is, the regression line, with age as the predictor variable, predicts a sizeable portion of the type of variation found in the prices.

To describe quantitatively how much of the variation in the observed prices is explained by the regression, we need to define two measures of variation: (1) the total variation in the observed prices and (2) the amount of variation in the observed prices explained by the regression.

As a measure of total variation in the observed prices, we use the sum of squared deviations of the observed prices from the mean price. This measure is called the **total sum of squares, *SST***; in symbols, $SST = \Sigma(y - \bar{y})^2$. If we divide SST by $n - 1$, we get the sample variance of the observed prices, so SST really is a measure of total variation.

To compute the total sum of squares, we must first find the sample mean price:

$$\bar{y} = \frac{\Sigma y}{n} = \frac{975}{11} = 88.64.$$

Now, using Table 14.6, we get the total sum of squares for the Orion price data.[†]

TABLE 14.6

Table for computing *SST* for the Orion price data

Age (yr) x	Price ($100) y	$y - \bar{y}$	$(y - \bar{y})^2$
5	85	−3.64	13.2
4	103	14.36	206.3
6	70	−18.64	347.3
5	82	−6.64	44.0
5	89	0.36	0.1
5	98	9.36	87.7
6	66	−22.64	512.4
6	95	6.36	40.5
2	169	80.36	6458.3
7	70	−18.64	347.3
7	48	−40.64	1651.3
	975		9708.5

From the final column of Table 14.6,

$$SST = \Sigma(y - \bar{y})^2 = 9708.5,$$

which is the measure of total variation in the observed prices.

To obtain the amount of variation in the observed prices explained by the regression, let's first look at a particular observed price—say, $y = 98$—corresponding to the data point $(5, 98)$. In Fig. 14.15 at the top of the next page, we magnified the portion of Fig. 14.14 that shows that data point and included the mean of the observed prices ($\bar{y} = 88.64$) and the predicted price for a 5-year-old Orion ($\hat{y} = 94.16$).

The total variation in the observed prices is based on the deviation of each observed price from the mean price, $y - \bar{y}$. As illustrated in Fig. 14.15, each such deviation can be decomposed into two parts: the deviation explained by the regression line, $\hat{y} - \bar{y}$, and the remaining unexplained deviation, $y - \hat{y}$. Hence the amount of variation (squared deviation) in the observed prices explained by the regression is $\Sigma(\hat{y} - \bar{y})^2$. This measure is called the **regression sum of squares, *SSR*.**

To compute *SSR*, we need the predicted prices, $\hat{y}$, and the mean of the observed prices, $\bar{y}$. We have already computed the mean of the observed prices. Each predicted price is obtained by substituting the age of the Orion in question for x in the regression equation $\hat{y} = 195.47 - 20.26x$. The third column of Table 14.7 shows the predicted prices for all 11 Orions.

[†]Values in Table 14.6 and all other tables in this section are displayed to various numbers of decimal places, but computations were done with full calculator accuracy.

FIGURE 14.15

Magnification of a portion of Fig. 14.14 showing only the data point (5, 98)

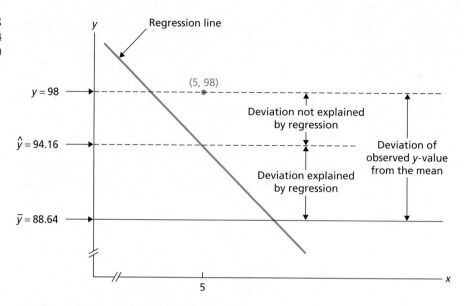

TABLE 14.7

Table for computing *SSR* for the Orion data

Age (yr) x	Price ($100) y	$\hat{y}$	$\hat{y} - \bar{y}$	$(\hat{y} - \bar{y})^2$
5	85	94.16	5.53	30.5
4	103	114.42	25.79	665.0
6	70	73.90	−14.74	217.1
5	82	94.16	5.53	30.5
5	89	94.16	5.53	30.5
5	98	94.16	5.53	30.5
6	66	73.90	−14.74	217.1
6	95	73.90	−14.74	217.1
2	169	154.95	66.31	4397.0
7	70	53.64	−35.00	1224.8
7	48	53.64	−35.00	1224.8
				8285.0

Recalling that $\bar{y} = 88.64$, we construct the fourth column of Table 14.7. We then calculate the entries for the fifth column and obtain the regression sum of squares, *SSR*:

$$SSR = \Sigma(\hat{y} - \bar{y})^2 = 8285.0.$$

This measure is the amount of variation in the observed prices explained by the regression.

Using the values of *SST* and *SSR*, we can now determine the percentage of variation in the observed prices explained by the regression, which is called the **coefficient of determination,** denoted r^2. We have

$$r^2 = \frac{SSR}{SST} = \frac{8285.0}{9708.5} = 0.853 \quad (85.3\%).$$

INTERPRETATION Evidently, age is quite useful for predicting price because 85.3% of the variation in the observed prices is explained by the regression of price on age. ___■

Now let's consider the remaining deviation portrayed in Fig. 14.15—the deviation not explained by the regression, $y - \hat{y}$. The amount of variation (squared deviation) in the observed prices not explained by the regression is $\Sigma(y - \hat{y})^2$, called the **error sum of squares, SSE.**

To compute *SSE*, we need the observed prices, y, and the predicted prices, $\hat{y}$. Both quantities are displayed in Table 14.7 and are repeated in the second and third columns of Table 14.8.

TABLE 14.8

Table for computing *SSE* for the Orion data

Age (yr) x	Price ($100) y	$\hat{y}$	$y - \hat{y}$	$(y - \hat{y})^2$
5	85	94.16	−9.16	83.9
4	103	114.42	−11.42	130.5
6	70	73.90	−3.90	15.2
5	82	94.16	−12.16	147.9
5	89	94.16	−5.16	26.6
5	98	94.16	3.84	14.7
6	66	73.90	−7.90	62.4
6	95	73.90	21.10	445.2
2	169	154.95	14.05	197.5
7	70	53.64	16.36	267.7
7	48	53.64	−5.64	31.8
				1423.5

From the final column of Table 14.8, we obtain the error sum of squares:

$$SSE = \Sigma(y - \hat{y})^2 = 1423.5.$$

This measure is the amount of variation in the observed prices not explained by the regression. Because the regression line is the line that best fits the data according to the least squares criterion, *SSE* is also the smallest possible sum of squared errors among all straight lines.

We now summarize our discussion of the three sums of squares and the coefficient of determination in the following two definitions.

DEFINITION 14.4

Sums of Squares in Regression

Total sum of squares, SST: The variation in the observed values of the response variable: $SST = \Sigma(y - \bar{y})^2$.

Regression sum of squares, SSR: The variation in the observed values of the response variable explained by the regression: $SSR = \Sigma(\hat{y} - \bar{y})^2$.

Error sum of squares, SSE: The variation in the observed values of the response variable not explained by the regression: $SSE = \Sigma(y - \hat{y})^2$.

DEFINITION 14.5

Coefficient of Determination

The *coefficient of determination, r^2,* is the proportion of variation in the observed values of the response variable explained by the regression:

$$r^2 = \frac{SSR}{SST}.$$

The Regression Identity

For the Orion data, we have determined that $SST = 9708.5$, $SSR = 8285.0$, and $SSE = 1423.5$. As $9708.5 = 8285.0 + 1423.5$, we see that $SST = SSR + SSE$. This equation is always true and is called the **regression identity,** which we state as Key Fact 14.4.

KEY FACT 14.4

Regression Identity

The total sum of squares equals the regression sum of squares plus the error sum of squares: $SST = SSR + SSE$.

Because of the regression identity, we can also express the coefficient of determination in terms of the total sum of squares and the error sum of squares:

$$r^2 = \frac{SSR}{SST} = \frac{SST - SSE}{SST} = 1 - \frac{SSE}{SST}.$$

This formula shows that we can also interpret the coefficient of determination as the percentage reduction obtained in the total squared error by using the regression equation instead of the mean, $\bar{y}$, to predict the observed values of the response variable. We examine this interpretation in Exercise 14.69.

Computing Formulas for the Sums of Squares

Calculating the three sums of squares—SST, SSR, and SSE—with the defining formulas is time consuming and can lead to significant roundoff error unless full accuracy is retained. For those reasons, we usually use computing formulas or a computer to obtain the sums of squares. Here are the computing formulas.

FORMULA 14.2

Computing Formulas for the Sums of Squares

The three sums of squares, SST, SSR, and SSE, can be obtained by using the following computing formulas:

$$\text{Total sum of squares: } SST = S_{yy}$$

$$\text{Regression sum of squares: } SSR = S_{xy}^2/S_{xx}$$

$$\text{Error sum of squares: } SSE = S_{yy} - S_{xy}^2/S_{xx}$$

The formulas for S_{yy}, S_{xy}, and S_{xx} are given in Definition 14.3 on page 702.

Example 14.8 Computing Formulas for the Sums of Squares

Age and Price of Orions The age and price data for a sample of 11 Orions are repeated in the first two columns of Table 14.9. Use the computing formulas in Formula 14.2 to determine the three sums of squares.

Solution To apply the computing formulas, we need a table of values for x (age), y (price), xy, x^2, y^2, and their sums, as shown in Table 14.9.

TABLE 14.9
Table for obtaining the three sums of squares for the Orion data by using the computing formulas

Age (yr) x	Price ($100) y	xy	x^2	y^2
5	85	425	25	7,225
4	103	412	16	10,609
6	70	420	36	4,900
5	82	410	25	6,724
5	89	445	25	7,921
5	98	490	25	9,604
6	66	396	36	4,356
6	95	570	36	9,025
2	169	338	4	28,561
7	70	490	49	4,900
7	48	336	49	2,304
58	975	4732	326	96,129

Using the last row of Table 14.9 and Formula 14.2, we can now obtain the three sums of squares for the Orion data. The total sum of squares is

$$SST = S_{yy} = \Sigma y^2 - (\Sigma y)^2/n = 96{,}129 - (975)^2/11 = 9708.5;$$

the regression sum of squares is

$$SSR = \frac{S_{xy}^2}{S_{xx}} = \frac{\left[\Sigma xy - (\Sigma x)(\Sigma y)/n\right]^2}{\Sigma x^2 - (\Sigma x)^2/n} = \frac{\left[4732 - (58)(975)/11\right]^2}{326 - (58)^2/11} = 8285.0;$$

and, from the two preceding results, the error sum of squares is

$$SSE = S_{yy} - \frac{S_{xy}^2}{S_{xx}} = 9708.5 - 8285.0 = 1423.5.$$

The values that we obtained for the three sums of squares by applying the computing formulas are, of course, the same as the values we found earlier by using the defining formulas. However, when the computing formulas are used, the computations are much simpler and less subject to roundoff error.

The Technology Center

Most statistical technologies have programs to compute the coefficient of determination, r^2, and the three sums of squares, *SST*, *SSR*, and *SSE*. In fact, many statistical technologies present those four statistics as part of the output for a regression equation. In Example 14.9, we concentrate on the coefficient of determination. Refer to the technology manuals for a discussion of the three sums of squares.

Example 14.9 Using Technology to Obtain a Coefficient of Determination

Age and Price of Orions The age and price data for a sample of 11 Orions is given in Table 14.2 on page 699. Use Minitab, Excel, or the TI-83/84 Plus to obtain the coefficient of determination, r^2, for those data.

Solution In Section 14.2, we used Minitab, Excel, and the TI-83/84 Plus to find the regression equation for the age and price data. The results are shown in Printout 14.2 on page 709. The outputs in Printout 14.2 also display the coefficient of determination: `R-Sq = 85.3%` (Minitab), `R Squared = 85.3%` (Excel), and $r^2 = .8533733464$ (TI-83/84 Plus). Thus, to three decimal places, $r^2 = 0.853$.

EXERCISES 14.3

Statistical Concepts and Skills

14.59 In this section, we introduced a descriptive measure of the utility of the regression equation for making predictions.
a. Identify the term and symbol for that descriptive measure.
b. Provide an interpretation of that descriptive measure.

14.60 Fill in the blanks.
a. A measure of total variation in the observed values of the response variable is the _____. The mathematical abbreviation for it is _____.
b. A measure of the amount of variation in the observed values of the response variable explained by the regression is the _____. The mathematical abbreviation for it is _____.
c. A measure of the amount of variation in the observed values of the response variable not explained by the regression is the _____. The mathematical abbreviation for it is _____.

14.61 For a particular regression analysis, $SST = 8291.0$ and $SSR = 7626.6$.
a. Obtain and interpret the coefficient of determination.
b. Determine SSE.

In Exercises **14.62** and **14.63**, we repeat the data from Exercises 14.38 and 14.39, respectively. We also provide the regression

equations for those data from Exercises 14.40 and 14.41, respectively. For each exercise here,
a. compute the three sums of squares, SST, SSR, and SSE, using the defining formulas (page 719).
b. verify the regression identity, SST = SSR + SSE.
c. compute the coefficient of determination.
d. determine the percentage of variation in the observed values of the response variable that is explained by the regression.
e. state how useful the regression equation appears to be for making predictions. (Answers for this part may vary, owing to differing interpretations.)

14.62 Following are the data from Exercise 14.38.

x	1	1	5	5
y	1	3	2	4

The regression equation is $\hat{y} = 1.75 + 0.25x$.

14.63 Following are the data from Exercise 14.39.

x	0	2	2	5	6
y	4	2	0	-2	1

The regression equation is $\hat{y} = 2.875 - 0.625x$.

*For Exercises **14.64–14.67**,*
a. compute SST, SSR, and SSE, using Formula 14.2 on page 720.
b. compute the coefficient of determination, r^2.
c. determine the percentage of variation in the observed values of the response variable explained by the regression, and interpret your answer.
d. state how useful the regression equation appears to be for making predictions.

14.64 Tax Efficiency. Following are the data on percentage of investments in energy securities and tax efficiency from Exercise 14.42.

x	3.1	3.2	3.7	4.3	4.0	5.5	6.7	7.4	7.4	10.6
y	98.1	94.7	92.0	89.8	87.5	85.0	82.0	77.8	72.1	53.5

14.65 Corvette Prices. Following are the age and price data for Corvettes from Exercise 14.43:

x	6	6	6	2	2	5	4	5	1	4
y	270	260	275	405	364	295	335	308	405	305

14.66 Custom Homes. Following are the size and price data for custom homes from Exercise 14.44.

x	26	27	33	29	29	34	30	40	22
y	290	305	325	327	356	411	488	554	246

14.67 Plant Emissions. Following are the data on plant weight and quantity of volatile emissions from Exercise 14.45.

x	57	85	57	65	52	67	62	80	77	53	68
y	8.0	22.0	10.5	22.5	12.0	11.5	7.5	13.0	16.5	21.0	12.0

Extending the Concepts and Skills

14.68 What can you say about SSE, SSR, and the utility of the regression equation for making predictions if
a. $r^2 = 1$? **b.** $r^2 = 0$?

14.69 On page 720, we noted that, because of the regression identity, we can also express the coefficient of determination in terms of the total sum of squares and the error sum of squares as

$$r^2 = 1 - \frac{SSE}{SST}.$$

a. Explain why this formula shows that the coefficient of determination can also be interpreted as the percentage reduction obtained in the total squared error by using the regression equation instead of the mean, $\bar{y}$, to predict the observed values of the response variable.
b. Refer to Exercise 14.65. What percentage reduction is obtained in the total squared error by using the regression equation instead of the mean of the observed prices to predict the observed prices?

Using Technology

*In Exercises **14.70–14.75**, use the technology of your choice to*
a. obtain the coefficient of determination.
b. determine the percentage of variation in the observed values of the response variable explained by the regression and interpret your answer.
c. state how useful the regression equation appears to be for making predictions.

14.70 Batting and Scoring. Following are the data on season team batting average and total runs scored for a sample of major league baseball teams from Exercise 14.53.

Average	Runs	Average	Runs
.294	968	.267	793
.278	938	.265	792
.278	925	.256	764
.270	887	.254	752
.274	825	.246	740
.271	810	.266	738
.263	807	.262	731
.257	798	.251	708

14.71 Body Fat. In the paper "Total Body Composition by Dual-Photon (^{153}Gd) Absorptiometry" (*American Journal of Clinical Nutrition*, Vol. 40, pp. 834–839), R. B. Mazess et al. studied methods for quantifying body composition. Eighteen randomly selected adults were measured for percentage of body fat, using dual-photon absorptiometry. Each adult's age and percentage of body fat are shown in the following table.

Age	% Fat	Age	% Fat	Age	% Fat
23	9.5	45	27.4	56	32.5
23	27.9	49	25.2	57	30.3
27	7.8	50	31.1	58	33.0
27	17.8	53	34.7	58	33.8
39	31.4	53	42.0	60	41.1
41	25.9	54	29.1	61	34.5

14.72 PCBs and Pelicans. The data for shell thickness and concentration of PCBs for 60 Anacapa pelican eggs from Exercise 14.54 are on the WeissStats CD.

14.73 More Money More Beer? The data for per capita income and per capita beer consumption for the 50 states and Washington, D.C., from Exercise 14.55 are on the WeissStats CD.

14.74 Gas Guzzlers. The data for gas mileage and engine displacement for 121 vehicles from Exercise 14.56 are on the WeissStats CD.

14.75 Estriol Level and Birth Weight.   J. Greene and J. Touchstone conducted a study on the relationship between the estriol levels of pregnant women and the birth weights of their children. Their findings, "Urinary Tract Estriol: An Index of Placental Function," were published in the *American Journal of Obstetrics and Gynecology* (Vol. 85(1), pp. 1–9). The data from the study are provided on the WeissStats CD, where estriol levels are in mg/24 hr and birth weights are in hectograms.

14.4 Linear Correlation

We often hear statements pertaining to the correlation or lack of correlation between two variables: "There is a positive correlation between advertising expenditures and sales" or "IQ and alcohol consumption are uncorrelated." In this section, we explain the meaning of such statements.

Several statistics can be used to measure the correlation between two variables. The statistic most commonly used is the **linear correlation coefficient, *r*,** which is also referred to as the **Pearson product moment correlation coefficient,** in honor of its developer, Karl Pearson. The following are the defining and computing formulas for the linear correlation coefficient.

DEFINITION 14.6

What
Does it Mean?

The linear correlation coefficient is a descriptive measure of the strength of the linear (straight-line) relationship between two variables.

Linear Correlation Coefficient

The *linear correlation coefficient, r,* of n data points is defined by

$$r = \frac{\frac{1}{n-1}\Sigma(x-\bar{x})(y-\bar{y})}{s_x s_y}.$$

It can also be obtained from the computing formula

$$r = \frac{S_{xy}}{\sqrt{S_{xx}S_{yy}}},$$

where S_{xx}, S_{xy}, and S_{yy} are given in Definition 14.3 on page 702.

The computing formula presented in Definition 14.6 is almost always pre-ferred for hand calculations, but the defining formula reveals the meaning and basic properties of the linear correlation coefficient. For instance, because of the division by the sample standard deviations, s_x and s_y, in the defining formula for r, we can conclude that *r is independent of the choice of units and always lies between −1 and 1.*

Understanding the Linear Correlation Coefficient

We now discuss some other important properties of the linear correlation co-efficient, r. Keep in mind that r measures the strength of the *linear* relationship between two variables and that the following properties of r are meaningful only when the data points are scattered about a straight line.

- *r reflects the slope of the scatter diagram.* The linear correlation coefficient is positive when the scatter diagram shows a positive slope and is negative when the scatter diagram shows a negative slope. To demonstrate why this property is true, we refer to the defining formula in Definition 14.6 and to Fig. 14.16, where we have drawn a coordinate system with a second set of axes centered at point $(\bar{x}, \bar{y})$.

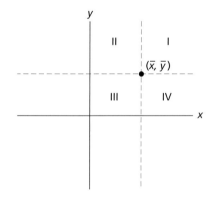

FIGURE 14.16

Coordinate system with a second set of axes centered at $(\bar{x}, \bar{y})$

  If the scatter diagram shows a positive slope, the data points, on average, will lie either in Region I or Region III. For such a data point, the deviations from the means, $x - \bar{x}$ and $y - \bar{y}$, will either both be positive or both be nega-tive. This condition implies that, on average, the product $(x - \bar{x})(y - \bar{y})$ will be positive and consequently that the correlation coefficient will be positive.

  If the scatter diagram shows a negative slope, the data points, on aver-age, will lie either in Region II or Region IV. For such a data point, one of the deviations from the mean will be positive and the other negative. This con-dition implies that, on average, the product $(x - \bar{x})(y - \bar{y})$ will be negative and consequently that the correlation coefficient will be negative.

- *The magnitude of r indicates the strength of the linear relationship.* A value of r close to −1 or to 1 indicates a strong linear relationship between the vari-ables and that the variable x is a good linear predictor of the variable y (i.e., the regression equation is extremely useful for making predictions). A value of r near 0 indicates at most a weak linear relationship between the variables and that the variable x is a poor linear predictor of the variable y (i.e., the regression equation is either useless or not very useful for making predictions).

- *The sign of r suggests the type of linear relationship.* A positive value of r suggests that the variables are **positively linearly correlated variables,** meaning that y tends to increase linearly as x increases, with the tendency being greater the closer that r is to 1. A negative value of r suggests that the variables are **negatively linearly correlated variables,** meaning that y tends to decrease linearly as x increases, with the tendency being greater the closer that r is to −1.

- *The sign of r and the sign of the slope of the regression line are identical.* If r is positive, so is the slope of the regression line (i.e., the regression line slopes upward); if r is negative, so is the slope of the regression line (i.e., the regression line slopes downward).

To graphically portray the meaning of the linear correlation coefficient, we present various degrees of linear correlation in Fig. 14.17.

FIGURE 14.17
Various degrees of linear correlation

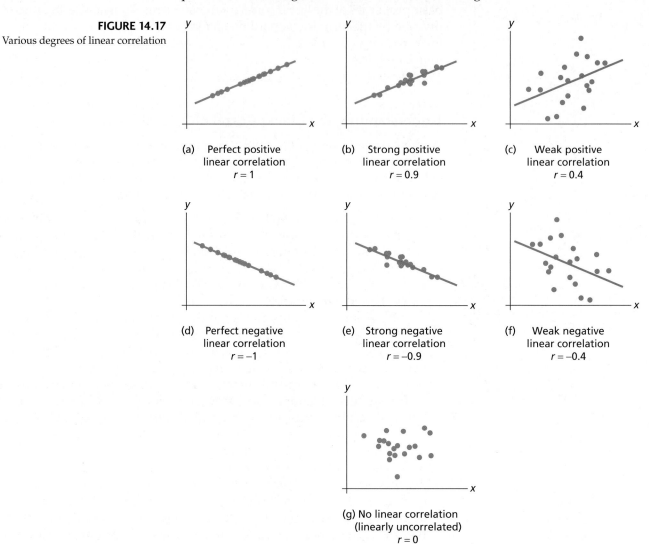

(a) Perfect positive
linear correlation
$r = 1$

(b) Strong positive
linear correlation
$r = 0.9$

(c) Weak positive
linear correlation
$r = 0.4$

(d) Perfect negative
linear correlation
$r = -1$

(e) Strong negative
linear correlation
$r = -0.9$

(f) Weak negative
linear correlation
$r = -0.4$

(g) No linear correlation
(linearly uncorrelated)
$r = 0$

If r is close to ± 1, the data points are clustered closely about the regression line, as shown in Fig. 14.17(b) and (e). If r is farther from ± 1, the data points are more widely scattered about the regression line, as shown in Fig. 14.17(c) and (f). And, if r is near 0, the data points are essentially scattered about a horizontal line (i.e., the slope of the regression line is near 0), as shown in Fig. 14.17(g), indicating at most a weak linear relationship between the variables.

Computing and Interpreting the Linear Correlation Coefficient

We demonstrate how to compute and interpret the linear correlation coefficient of a set of data points by returning to the data on age and price for a sample of Orions in Example 14.10.

Example 14.10 The Linear Correlation Coefficient

Age and Price of Orions The age and price data for a sample of 11 Orions are repeated in the first two columns of Table 14.10.

TABLE 14.10

Table for obtaining the linear correlation coefficient for the Orion data by using the computing formula

Age (yr) x	Price ($100) y	xy	x^2	y^2
5	85	425	25	7,225
4	103	412	16	10,609
6	70	420	36	4,900
5	82	410	25	6,724
5	89	445	25	7,921
5	98	490	25	9,604
6	66	396	36	4,356
6	95	570	36	9,025
2	169	338	4	28,561
7	70	490	49	4,900
7	48	336	49	2,304
58	975	4732	326	96,129

a. Compute the linear correlation coefficient, r, of the data.

b. Interpret the value of r obtained in part (a) in terms of the linear relationship between the variables age and price of Orions.

c. Discuss the graphical implications of the value of r.

Solution First recall that the scatter diagram shown in Fig. 14.7 on page 700 indicates that the data points are scattered about a straight line. Hence it is meaningful to obtain the linear correlation coefficient of these data.

a. We apply the computing formula in Definition 14.6 on page 724 to obtain the linear correlation coefficient. To do so, we need a table of values for x, y, xy, x^2, y^2, and their sums, as shown in Table 14.10. Referring to the last row of Table 14.10, we get

$$ r = \frac{S_{xy}}{\sqrt{S_{xx}S_{yy}}} = \frac{\Sigma xy - (\Sigma x)(\Sigma y)/n}{\sqrt{\left[\Sigma x^2 - (\Sigma x)^2/n\right]\left[\Sigma y^2 - (\Sigma y)^2/n\right]}} $$

$$ = \frac{4732 - (58)(975)/11}{\sqrt{\left[326 - (58)^2/11\right]\left[96{,}129 - (975)^2/11\right]}} = -0.924. $$

b.   **INTERPRETATION** The linear correlation coefficient, $r = -0.924$, suggests a strong negative linear correlation between age and price of Orions. In particular, it indicates that as age increases there is a strong tendency for price to decrease, which is not surprising. It also implies that the regression equation, $\hat{y} = 195.47 - 20.26x$, is extremely useful for making predictions.

c. Because the correlation coefficient, $r = -0.924$, is quite close to -1, the data points should be clustered rather closely about the regression line. Figure 14.14 on page 716 shows that to be the case.

_____ ■

Relationship Between the Correlation Coefficient and the Coefficient of Determination

In Section 14.3, we discussed the coefficient of determination, r^2, a descriptive measure of the utility of the regression equation for making predictions. In this section, we introduced the linear correlation coefficient, r, as a descriptive measure of the strength of the linear relationship between two variables.

We expect the strength of the linear relationship also to indicate the usefulness of the regression equation for making predictions. In other words, there should be a relationship between the linear correlation coefficient and the coefficient of determination—and there is. The relationship is precisely the one suggested by the notation used, as stated in Key Fact 14.5.

KEY FACT 14.5

Relationship Between the Correlation Coefficient and the Coefficient of Determination

The coefficient of determination is the square of the linear correlation coefficient.

In Example 14.10, we found that the linear correlation coefficient for the data on age and price of a sample of 11 Orions is $r = -0.924$. From this result and Key Fact 14.5, we can easily obtain the coefficient of determination: $r^2 = (-0.924)^2 = 0.854$. As expected, this value is the same (except for roundoff error) as the value we found for r^2 on page 718 by using the defining formula $r^2 = SSR/SST$. In general, we can obtain the coefficient of determination either by using the defining formula or by first finding the linear correlation coefficient and then squaring the result.

Likewise, we can obtain the linear correlation coefficient either by using Definition 14.6 or from the coefficient of determination, provided we also know the direction of the regression line. Specifically, the square root of the coefficient of determination gives the magnitude of the linear correlation coefficient; the sign of the linear correlation coefficient is the same as that of the slope of the regression line.

Warnings on the Use of the Linear Correlation Coefficient

As we mentioned in Section 14.2, an assumption for finding the regression line for a set of data points is that the data points are scattered about a straight line. Similarly, because the linear correlation coefficient is used to describe the strength of the *linear* relationship between two variables, it should be used as a descriptive measure only when a scatter diagram indicates that the data points are scattered about a straight line.

For instance, in general, we cannot say that a value of r near 0 implies that there is no relationship between the two variables under consideration, nor can we say that a value of r near ± 1 implies that a linear relationship exists between the two variables. Such statements are meaningful only when a scatter diagram indicates that the data points are scattered about a straight line. See Exercise 14.91 for more on these issues.

When using the linear correlation coefficient, you must also watch for outliers and influential observations. Such data points can sometimes unduly affect r because sample means and sample standard deviations are not resistant to outliers and other extreme values.

Correlation and Causation

Two variables may have a high correlation without being causally related. For example, Table 14.11 displays data on total pari-mutuel turnover (money wagered) at U.S. racetracks and college enrollment for five randomly selected years. [SOURCE: *National Association of State Racing Commissioners* and *U.S. National Center for Education Statistics*.]

TABLE 14.11

Pari-mutuel turnover and college enrollment for five randomly selected years

Pari-mutuel turnover ($millions) x	College enrollment (thousands) y
5,977	8,581
7,862	11,185
10,029	11,260
11,677	12,372
11,888	12,426

The linear correlation coefficient of the data points in Table 14.11 is $r = 0.931$, suggesting a strong positive linear correlation between pari-mutuel wagering and college enrollment. But this result doesn't mean that a causal relationship exists between the two variables, such as that when people go to racetracks they are somehow inspired to go to college. On the contrary, we can only infer that the two variables have a strong tendency to increase (or decrease) simultaneously and that total pari-mutuel turnover is a good predictor of college enrollment.

Two variables may be strongly correlated because they are both associated with other variables, called **lurking variables,** that cause changes in the two variables under consideration. For example, a study showed that teachers' salaries and the dollar amount of liquor sales are positively linearly correlated. A possible explanation for this curious fact might be that both variables are tied to other variables, such as the rate of inflation, that pull them along together.

The Technology Center

Almost all statistical technologies have programs that automatically obtain the linear correlation coefficient for a set of data points. Here, we present output and step-by-step instructions for obtaining a linear correlation coefficient.

Example 14.11   **Using Technology to Obtain a Linear Correlation Coefficient**

Age and Price of Orions   Use Minitab, Excel, or the TI-83/84 Plus to obtain the linear correlation coefficient of the age and price data displayed in Table 14.10 on page 727, for a sample of 11 Orions.

Solution   Printout 14.4 shows output obtained by applying the linear correlation coefficient programs to the age and price data.

PRINTOUT 14.4
Linear correlation coefficient for the age and price data of 11 Orions

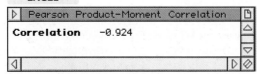

MINITAB

Correlations: AGE, PRICE

Pearson correlation of AGE and PRICE = -0.924
P-Value = 0.000

EXCEL

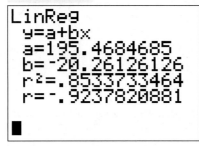

Pearson Product-Moment Correlation	
Correlation	-0.924

TI-83/84 PLUS

LinReg
 y=a+bx
 a=195.4684685
 b=-20.26126126
 r²=.8533733464
 r=-.9237820881
■

These outputs show that the linear correlation coefficient for the age and price data is -0.924.

We can also obtain the linear correlation coefficient for the age and price data from the output of the linear regression programs shown in Printout 14.2 on page 709. For the TI-83/84 Plus, r is the last item in the output. For Minitab and Excel, we find r by first taking the square root of the coefficient of determination, r^2, and then using the sign of the slope of the regression line, b_1, for r. From the Minitab and Excel outputs in Printout 14.2, $r^2 = 0.853$ and $b_1 = -20.261$. Therefore $r = -\sqrt{r^2} = -\sqrt{0.853} = -0.924$.

Obtaining the Output (Optional)

Printout 14.4 provides output from Minitab, Excel, and the TI-83/84 Plus for the linear correlation coefficient of the age and price data for a sample of 11 Orions, displayed in Table 14.10 on page 727. The following are detailed instructions for obtaining that output. First, we store the age and price data in columns (Minitab), ranges (Excel), or lists (TI-83/84 Plus) named AGE and PRICE, respectively. Then, we proceed as follows.

MINITAB

1 Choose **Stat ➤ Basic Statistics ➤ Correlation...**
2 Specify AGE and PRICE in the **Variables** text box
3 Click **OK**

EXCEL

1 Choose **DDXL ➤ Regression**
2 Select **Correlation** from the **Function type** drop-down list box
3 Specify AGE in the **x-Axis Quantitative Variable** text box
4 Specify PRICE in the **y-Axis Quantitative Variable** text box
5 Click **OK**

TI-83/84 PLUS

1 Press **2nd ➤ CATALOG** and then press **D**
2 Arrow down to **DiagnosticOn** and press **Enter** twice
3 Press **STAT**, arrow over to **CALC**, and press **8**
4 Press **2nd ➤ LIST**, arrow down to AGE, and press **ENTER**
5 Press **, ➤ 2nd ➤ LIST**, arrow down to PRICE, and press **ENTER** twice

---

EXERCISES 14.4

Statistical Concepts and Skills

14.76 What is one purpose of the linear correlation coefficient?

14.77 The linear correlation coefficient is also known by another name. What is it?

14.78 Fill in the blanks.
a. The symbol used for the linear correlation coefficient is _____.
b. A value of r close to ± 1 indicates that there is a _____ linear relationship between the variables.
c. A value of r close to _____ indicates that there is either no linear relationship between the variables or a weak one.

14.79 Fill in the blanks.
a. A value of r close to _____ indicates that the regression equation is extremely useful for making predictions.
b. A value of r close to 0 indicates that the regression equation is either useless or _____ for making predictions.

14.80 Fill in the blanks.
a. If y tends to increase linearly as x increases, the variables are _____ linearly correlated.

b. If y tends to decrease linearly as x increases, the variables are _____ linearly correlated.
c. If there is no linear relationship between x and y, the variables are linearly _____.

14.81 Answer true or false to the following statement and provide a reason for your answer: If there is a very strong positive correlation between two variables, a causal relationship exists between the two variables.

14.82 The linear correlation coefficient of a set of data points is 0.846.
a. Is the slope of the regression line positive or negative? Explain your answer.
b. Determine the coefficient of determination.

14.83 The coefficient of determination of a set of data points is 0.709 and the slope of the regression line is −3.58. Determine the linear correlation coefficient of the data.

*In Exercises **14.84** and **14.85**, we repeat data from exercises in Section 14.2. For each exercise here, obtain the linear correlation coefficient by using the defining formula in Definition 14.6 on page 724.*

14.84 Following are the data from Exercise 14.38.

x	1	1	5	5
y	1	3	2	4

14.85 Following are the data from Exercise 14.39.

x	0	2	2	5	6
y	4	2	0	−2	1

*In Exercises **14.86–14.89**, we repeat data from exercises in Section 14.2. For each exercise here,*

a. *obtain the linear correlation coefficient by using the computing formula in Definition 14.6 on page 724.*
b. *interpret the value of r in terms of the linear relationship between the two variables in question.*
c. *discuss the graphical interpretation of the value of r and verify that it is consistent with the graph you obtained in the corresponding exercise in Section 14.2.*
d. *square r and compare the result with the value of the coefficient of determination you obtained in the corresponding exercise in Section 14.3.*

14.86 Tax Efficiency. Following are the data on percentage of investments in energy securities and tax efficiency from Exercise 14.42.

x	3.1	3.2	3.7	4.3	4.0	5.5	6.7	7.4	7.4	10.6
y	98.1	94.7	92.0	89.8	87.5	85.0	82.0	77.8	72.1	53.5

Refer to the graph you obtained in Exercise 14.42 and your answer to Exercise 14.64(b).

14.87 Corvette Prices. Following are the age and price data for Corvettes from Exercise 14.43.

x	6	6	6	2	2	5	4	5	1	4
y	270	260	275	405	364	295	335	308	405	305

Refer to the graph you obtained in Exercise 14.43 and your answer to Exercise 14.65(b).

14.88 Custom Homes. Following are the size and price data for custom homes from Exercise 14.44.

x	26	27	33	29	29	34	30	40	22
y	290	305	325	327	356	411	488	554	246

Refer to the graph you obtained in Exercise 14.44 and your answer to Exercise 14.66(b).

14.89 Plant Emissions. Following are the data on plant weight and quantity of volatile emissions from Exercise 14.45.

x	57	85	57	65	52	67	62	80	77	53	68
y	8.0	22.0	10.5	22.5	12.0	11.5	7.5	13.0	16.5	21.0	12.0

Refer to the graph you obtained in Exercise 14.45 and your answer to Exercise 14.67(b).

14.90 Height and Score. A random sample of 10 students was taken from an introductory statistics class. The following data were obtained, where x denotes height, in inches, and y denotes score on the final exam.

x	71	68	71	65	66	68	68	64	62	65
y	87	96	66	71	71	55	83	67	86	60

a. What sort of value of r would you expect to find for these data? Explain your answer.
b. Compute r.

14.91 Consider the following set of data points.

x	−3	−2	−1	0	1	2	3
y	9	4	1	0	1	4	9

a. Compute the linear correlation coefficient, r.
b. Can you conclude from your answer in part (a) that the variables x and y are unrelated? Explain your answer.
c. Draw a scatter diagram for the data.
d. Is use of the linear correlation coefficient as a descriptive measure for the data appropriate? Explain your answer.
e. Show that the data are related by the quadratic equation $y = x^2$. Graph that equation and the data points.

Now consider the following set of data points.

x	−3	−2	−1	0	1	2	3
y	−27	−8	−1	0	1	8	27

f. Compute the linear correlation coefficient, r.
g. Can you conclude from your answer in part (f) that the variables x and y are linearly related? Explain your answer.

h. Draw a scatter diagram for the data.

i. Is use of the linear correlation coefficient as a descriptive measure for the data appropriate? Explain your answer.

j. Show that the data are related by the cubic equation $y = x^3$. Graph that equation and the data points.

14.92 Determine whether r is positive, negative, or zero for each of the following data sets.

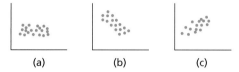

| (a) | (b) | (c) |

Extending the Concepts and Skills

14.93 The coefficient of determination of a set of data points is 0.716.

a. Can you determine the linear correlation coefficient? If yes, obtain it. If no, why not?

b. Can you determine whether the slope of the regression line is positive or negative? Why or why not?

c. If we tell you that the slope of the regression line is negative, can you determine the linear correlation coefficient? If yes, obtain it. If no, why not?

d. If we tell you that the slope of the regression line is positive, can you determine the linear correlation coefficient? If yes, obtain it. If no, why not?

14.94 Country Music Blues. A Knight-Ridder News Service article in the November 27, 1992, issue of the *Wichita Eagle* discussed a study on the relationship between country music and suicide. The results of the study, coauthored by Steven Stack and James Gundlach, appeared as the paper "The Effect of Country Music on Suicide" (*Social Forces*, Vol. 71(1), pp. 211–218). According to the article, "…analysis of 49 metropolitan areas shows that the greater the airtime devoted to country music, the greater the white suicide rate." (Suicide rates in the black population were found to be uncorrelated with the amount of country music airtime.)

a. Use the terminology introduced in this section to describe the statement quoted above.

b. One of the conclusions stated in the journal article was that country music "nurtures a suicidal mood" by dwelling on marital status and alienation from work. Is this conclusion warranted solely on the basis of the positive correlation found between airtime devoted to country music and white suicide rate? Explain your answer.

Using Technology

*In Exercises **14.95–14.101**, use the technology of your choice to a. obtain the linear correlation coefficient.*

b. interpret the value of r in terms of the linear relationship between the two variables in question.

14.95 Batting and Scoring. Following are the data on season team batting average and total runs scored for a sample of major league baseball teams from Exercise 14.53.

Average	Runs	Average	Runs
.294	968	.267	793
.278	938	.265	792
.278	925	.256	764
.270	887	.254	752
.274	825	.246	740
.271	810	.266	738
.263	807	.262	731
.257	798	.251	708

14.96 Body Fat. Following are the age and body fat data for 18 randomly selected adults from Exercise 14.71.

Age	% Fat	Age	% Fat	Age	% Fat
23	9.5	45	27.4	56	32.5
23	27.9	49	25.2	57	30.3
27	7.8	50	31.1	58	33.0
27	17.8	53	34.7	58	33.8
39	31.4	53	42.0	60	41.1
41	25.9	54	29.1	61	34.5

14.97 PCBs and Pelicans. The data on shell thickness and concentration of PCBs for 60 Anacapa pelican eggs from Exercise 14.54 are on the WeissStats CD.

14.98 More Money More Beer? The data for per capita income and per capita beer consumption for the 50 states and Washington, D.C., from Exercise 14.55 are on the WeissStats CD.

14.99 Gas Guzzlers. The data for gas mileage and engine displacement for 121 vehicles from Exercise 14.56 are on the WeissStats CD.

14.100 Estriol Level and Birth Weight. The data for estriol levels of pregnant women and birth weights of their children from Exercise 14.75 are on the WeissStats CD.

14.101 Fiber Density. In the article "Comparison of Fiber Counting by TV Screen and Eyepieces of Phase Contrast

Microscopy" (*American Industrial Hygiene Association Journal*, Vol. 63, pp. 756–761), I. Moa, H. Yeh, and M. Chen reported on determining fiber density by two different methods. Twenty samples of varying fiber density were each counted by 10 viewers by means of an eyepiece method and a television-screen method to determine the relationship between the counts done by each method. The results, in fibers per square millimeter, are presented in the table shown at the right.

Eyepiece	TV screen	Eyepiece	TV screen
3.5	1.8	56.3	55.4
4.5	7.4	89.2	115.8
6.8	3.2	105.1	92.4
7.9	8.8	116.1	182.4
9.3	11.5	145.2	135.7
13.8	5.6	201.5	261.4
18.3	23.9	224.8	89.6
20.8	46.6	250.5	194.7
23.1	21.7	265.0	314.9
49.5	84.6	307.9	300.4

CHAPTER REVIEW

You Should Be Able To

1. use and understand the formulas in this chapter.

2. define and apply the concepts related to linear equations with one independent variable.

3. explain the least-squares criterion.

4. obtain and graph the regression equation for a set of data points, interpret the slope of the regression line, and use the regression equation to make predictions.

5. define and use the terminology *predictor variable* and *response variable*.

6. understand the concept of extrapolation.

7. identify outliers and influential observations.

8. know when obtaining a regression line for a set of data points is appropriate.

9. calculate and interpret the three sums of squares, SST, SSE, and SSR, and the coefficient of determination, r^2.

10. determine and interpret the linear correlation coefficient, r.

11. explain and apply the relationship between the linear correlation coefficient and the coefficient of determination.

Key Terms

coefficient of determination (r^2), 720
curvilinear regression, 707
error sum of squares (SSE), 719
explanatory variable, 704
extrapolation, 704
influential observation, 705
least-squares criterion, 701
linear correlation coefficient (r), 724
linear equation, 694
lurking variables, 729

negatively linearly correlated
    variables, 725
outlier, 705
Pearson product moment correlation
    coefficient, 724
positively linearly correlated
    variables, 725
predictor variable, 704
regression equation, 701, 702
regression identity, 720

regression line, 701
regression sum of squares (SSR), 719
response variable, 704
scatter diagram, 699
scatterplot, 699
slope, 696
straight line, 694
total sum of squares (SST), 719
y-intercept, 696

REVIEW TEST

Statistical Concepts and Skills

1. For a linear equation $y = b_0 + b_1 x$, identify the
a. independent variable.   b. dependent variable.
c. slope.   d. y-intercept.

2. Consider the linear equation $y = 4 - 3x$.
a. At what y-value does its graph intersect the y-axis?
b. At what x-value does its graph intersect the y-axis?
c. What is its slope?

d. By how much does the *y*-value on the line change when the *x*-value increases by 1 unit?

e. By how much does the *y*-value on the line change when the *x*-value decreases by 2 units?

3. Answer true or false to each statement and explain your answers.

a. The *y*-intercept of a straight line has no effect on the steepness of the line.

b. A horizontal line has no slope.

c. If a line has a positive slope, *y*-values on the line decrease as the *x*-values decrease.

4. What kind of plot is useful for deciding whether finding a regression line for a set of data points is reasonable?

5. Identify one use of a regression equation.

6. Regarding the variables in a regression analysis,

a. what is the independent variable called?

b. what is the dependent variable called?

7. Fill in the blanks.

a. Based on the least-squares criterion, the line that best fits a set of data points is the one having the _____ possible sum of squared errors.

b. The line that best fits a set of data points according to the least-squares criterion is called the _____ line.

c. Using a regression equation to make predictions for values of the predictor variable outside the range of the observed values of the predictor variable is called _____.

8. In the context of regression analysis, what is an

a. outlier?   **b.** influential observation?

9. Identify a use of the coefficient of determination as a descriptive measure.

10. For each of the sums of squares in regression, state its name and what it measures.

a. *SST*   **b.** *SSR*   **c.** *SSE*

11. Fill in the blanks.

a. One use of the linear correlation coefficient is as a descriptive measure of the strength of the _____ relationship between two variables.

b. A positive linear relationship between two variables means that one variable tends to increase linearly as the other _____.

c. A value of *r* close to −1 suggests a strong _____ linear relationship between the variables.

d. A value of *r* close to _____ suggests at most a weak linear relationship between the variables.

12. Answer true or false to the following statement and explain your answer: A strong correlation between two variables doesn't necessarily mean that they're causally related.

13. Equipment Depreciation. A small company has purchased a microcomputer system for $7200 and plans to depreciate the value of the equipment by $1200 per year for 6 years. Let *x* denote the age of the equipment, in years, and *y* denote the value of the equipment, in hundreds of dollars.

a. Find the equation that expresses *y* in terms of *x*.

b. Find the *y*-intercept, b_0, and slope, b_1, of the linear equation in part (a).

c. Without graphing the equation in part (a), decide whether the line slopes upward, slopes downward, or is horizontal.

d. Find the value of the computer equipment after 2 years; after 5 years.

e. Obtain the graph of the equation in part (a) by plotting the points from part (d) and connecting them with a straight line.

f. Use the graph from part (e) to visually estimate the value of the equipment after 4 years. Then calculate that value exactly, using the equation from part (a).

14. Graduation Rates. Graduation rate—the percentage of entering freshmen, attending full time and graduating within 5 years—and what influences it have become a concern in U.S. colleges and universities. *U.S. News and World Report*'s "College Guide" provides data on graduation rates for colleges and universities as a function of the percentage of freshmen in the top 10% of their high school class, total spending per student, and student-to-faculty ratio. A random sample of 10 universities gave the following data on student-to-faculty ratio (S/F ratio) and graduation rate (grad rate).

S/F ratio *x*	Grad rate *y*	S/F ratio *x*	Grad rate *y*
16	45	17	46
20	55	17	50
17	70	17	66
19	50	10	26
22	47	18	60

a. Draw a scatter diagram of the data.

b. Is finding a regression line for the data reasonable? Explain your answer.

c. Determine the regression equation for the data and draw its graph on the scatter diagram you drew in part (a).

d. Describe the apparent relationship between student-to-faculty ratio and graduation rate.

e. What does the slope of the regression line represent in terms of student-to-faculty ratio and graduation rate?

f. Use the regression equation to predict the graduation rate of a university having a student-to-faculty ratio of 17.

g. Identify outliers and potential influential observations.

15. Graduation Rates. Refer to Problem 14.
a. Determine *SST*, *SSR*, and *SSE* by using the computing formulas.
b. Obtain the coefficient of determination.
c. Obtain the percentage of the total variation in the observed graduation rates that is explained by student-to-faculty ratio (i.e., by the regression line).
d. State how useful the regression equation appears to be for making predictions.

16. Graduation Rates. Refer to Problem 14.
a. Compute the linear correlation coefficient, *r*.
b. Interpret your answer from part (a) in terms of the linear relationship between student-to-faculty ratio and graduation rate.
c. Discuss the graphical implications of the value of the linear correlation coefficient, *r*.
d. Use your answer from part (a) to obtain the coefficient of determination.

Using Technology

17. Graduation Rates. Refer to Problem 14. Use the technology of your choice to
a. obtain a scatter diagram of the data.
b. determine the regression equation for the data.
c. find the coefficient of determination, r^2, and the three sums of squares, *SST*, *SSR*, and *SSE*.
d. identify potential outliers and influential observations.
e. obtain the regression equation with the potential influential observation removed.
f. Is the potential influential observation actually an influential observation? That is, does its removal markedly change the regression equation?

18. Graduation Rates. Use the technology of your choice to obtain the linear correlation coefficient of the data in Problem 14.

19. Fat Consumption and Prostate Cancer. Researchers have asked whether there is a relationship between nutrition and cancer, and many studies have shown that there is. In fact, one of the conclusions of a study by B. Reddy et al., "Nutrition and Its Relationship to Cancer" (*Advances in Cancer Research*, Vol. 32, pp. 237–345), was that "...none of the risk factors for cancer is probably more significant than diet and nutrition." One dietary factor that has been studied for its relationship with prostate cancer is fat consumption. On the WeissStats CD, you will find data on fat consumption and prostate cancer death rate for nations of the world. The data were obtained from a graph—adapted from information in the article mentioned—in John Robbins's classic book *Diet for a New America* (Walpole, NH: Stillpoint, 1987). Use the technology of your choice to solve the following problems:
a. Obtain a scatter diagram for the data on fat consumption and prostate cancer death rate. What does the scatter diagram tell you?
b. Does obtaining a regression equation for the data appear reasonable? Explain your answer.
c. Find the regression equation for the data, using fat consumption as the predictor variable.
d. Interpret the slope of the regression line.
e. Compute the correlation coefficient of the data and interpret your result.
f. Identify outliers and potential influential observations, if any.

20. Exotic Plants. In the article "Effects of Human Population, Area, and Time on Non-native Plant and Fish Diversity in the United States" (*Biological Conservation*, Vol. 100, No. 2, pp. 243–252), M. McKinney investigated the relationship of various factors on the number of exotic plants in each state. On the WeissStats CD, you will find the data on population (in millions), area (in thousands of square miles), and number of exotic plants for each state. Use the technology of your choice to obtain the correlation coefficient between
a. population and area.
b. population and number of exotic plants.
c. area and number of exotic plants.
d. Interpret and explain the results you obtained in parts (a)–(c).

21. Payroll and Success. Steve Galbraith, Morgan Stanley's Chief Investment Officer for U.S. Equity Research, examined several financial aspects of baseball in the article "Finding the Financial Equivalent of a Walk" (*Perspectives*, October 2003, pp. 1–3). The article includes a table containing the 2003 payrolls (in millions of dollars) and winning percentages (through August 3) of major league baseball teams. We have reproduced the data on the WeissStats CD. Use the technology of your choice to do the following:
a. Obtain a scatter diagram of the data.
b. Determine and interpret the correlation coefficient of the data.
c. Find the regression equation of the data with payroll as the predictor variable and winning percentage as the response variable. Interpret your result.
d. Determine and interpret the coefficient of determination.
e. Predict the winning percentage of a team with a payroll of $50 million; $100 million.

StatExplore in MyMathLab
Analyzing Data Online

You can use StatExplore to perform all statistical analyses discussed in this chapter. To illustrate, we show how to obtain a scatter diagram and conduct a regression analysis.

EXAMPLE Scatter Diagram

Age and Price of Orions Table 14.2 on page 699 displays data on age and price for a sample of 11 Orions. Use StatExplore to obtain a scatter diagram for these data.

SOLUTION To obtain a scatter diagram for the data, we first store the age and price data in columns named AGE and PRICE. Then we proceed as follows.

> 1 Choose **Graphics ➤ Scatter Plot**
> 2 Select AGE for the **X variable**
> 3 Select PRICE for the **Y variable**
> 4 Click **Create Graph!**

The resulting output is shown in Printout 14.5. The scatter diagram in Printout 14.5 indicates that the data points are scattered roughly about a straight line and, hence, that finding a regression line for these data is reasonable.

PRINTOUT 14.5
StatExplore output for scatter diagram

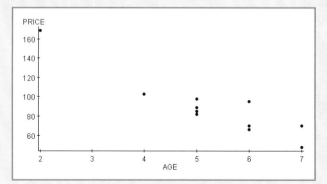

EXAMPLE Regression Analysis

Age and Price of Orions For the age and price data for the sample of 11 Orions, use StatExplore to obtain the

a. regression equation.
b. coefficient of determination.
c. three sums of squares.

SOLUTION We first store the age and price data in columns named AGE and PRICE. Then we proceed as follows.

> 1 Choose **Stat ➤ Regression ➤ Simple Linear**
> 2 Select AGE for the **X-Variable**
> 3 Select PRICE for the **Y-Variable**
> 4 Click **Calculate**

The resulting output is shown in Printout 14.6.

PRINTOUT 14.6
StatExplore output for regression analysis

Simple linear regression results:
Dependent Variable: PRICE
Independent Variable: AGE
PRICE = 195.46848 - 20.261261 AGE
Sample size: 11
R (correlation coefficient) = -0.9238
R-sq = 0.85337335
Estimate of error standard deviation: 12.576572

Parameter estimates:

Parameter	Estimate	Std. Err.	DF	T-Stat	P-Value
Intercept	195.46848	15.240338	9	12.82573	<0.0001
Slope	-20.261261	2.7995107	9	-7.237429	<0.0001

Analysis of variance table for regression model:

Source	DF	SS	MS	F-stat	P-value
Model	1	8285.014	8285.014	52.380383	<0.0001
Error	9	1423.5315	158.17017		
Total	10	9708.546			

a. The fourth line of Printout 14.6 gives the regression equation; thus, $\hat{y} = 195.46848 - 20.261261x$.
b. The seventh line of Printout 14.6 provides the coefficient of determination; thus, $r^2 = 0.85337335$.

c. The **SS** column of the **Analysis of variance table for regression model** contains the regression sum of squares (Model), the error sum of squares (Error), and the total sum of squares (Total); consequently, $SSR = 8285.014$, $SSE = 1423.5315$, and $SST = 9708.546$.

As you can see, the StatExplore regression output contains considerably more information than just the regression equation, coefficient of determination, and three sums of squares. In particular, it provides the correlation coefficient which, in this case, is $r = -0.9238$. *Note:* There is a separate StatExplore program available for obtaining just the correlation coefficient (Choose **Stat ➤ Summary Stats ➤ Correlation**).

STATEXPLORE EXERCISES For the plant-emissions data from Exercise 14.45 on page 713, use StatExplore to solve the following problems:

a. Obtain a scatter diagram for the data and use that diagram to decide whether finding a regression line is reasonable.

b. Determine the regression equation.

c. Find the coefficient of determination and interpret the result.

d. Determine the three sums of squares and show that they satisfy the regression identity.

e. Obtain the correlation coefficient without doing a regression analysis.

> To access StatExplore, go to the student content area of your Weiss MyMathLab course.

Internet Project

Assisted Reproductive Technology

Since 1981, assisted reproductive technology (ART) has been used in the United States to help women achieve pregnancy. Many U.S. women have received some type of fertility service, such as in vitro fertilization (IVF) or egg transfer.

Several factors can influence the chance of having a child by using ART, one of the most important being the age of the prospective mother. In this Internet project, you are to examine how a woman's age can influence the chance for success in assisted reproductive technology. Specifically, you are to consider the relationship between the variables reproductive success rate and age.

URL for access to Internet Projects Page: www.aw-bc.com/weiss

**Focusing on
Data Analysis**

The Focus Database

Recall from Chapter 1 (see page 35) that the Focus database contains information on the undergraduate students at the University of Wisconsin - Eau Claire (UWEC). Statistical analyses for this database should be done with the technology of your choice.

a. Obtain the cumulative GPAs (GPA) and high school percentiles (HSP) of a simple random sample of 50 UWEC undergraduate students.
b. Determine the linear correlation coefficient of the data obtained in part (a).
c. Repeat parts (a) and (b) for cumulative GPA and each of ACT English score (EN-GLISH), ACT math score (MATH), and ACT composite score (COMP). Be sure to retain all four pairs of samples.
d. Among the variables high school percentile, ACT English score, ACT math score, and ACT composite score, identify the one that appears to be the best predictor of cumulative GPA. Explain your reasoning.

Now perform a regression analysis on cumulative GPA, using the predictor variable identified in part (d), as follows:

e. Obtain the regression equation for cumulative GPA.
f. Find the coefficient of determination and interpret your answer.
g. Determine and interpret the three sums of squares SSR, SSE, and SST.

Case Study Discussion

Shoe Size and Height

At the beginning of this chapter, we presented data on shoe size and height for a sample of students at Arizona State University. Now that you have studied regression and correlation, you can analyze the relationship between those two variables. We recommend that you use statistical software or a graphing calculator to solve the following problems, but they can also be done by hand:

a. Separate the data in the table on page 693 into two tables, one for males and the other for females. Parts (b)–(k) are for the male data.

b. Draw a scatter diagram for the data on shoe size and height for males.

c. Does obtaining a regression equation for the data appear reasonable? Explain your answer.

d. Find the regression equation for the data, using shoe size as the predictor variable.

e. Interpret the slope of the regression line.

f. Use the regression equation to predict the height of a male student who wears a size $10\frac{1}{2}$ shoe.

g. Obtain and interpret the coefficient of determination.

h. Compute the correlation coefficient of the data and interpret your result.

i. Identify outliers and potential influential observations, if any.

j. If there are outliers, remove them and repeat parts (b)–(h).

k. Decide whether any potential influential observation is in fact an influential observation. Explain your reasoning.

l. Repeat parts (b)–(k) for the data on shoe size and height for females. For part (f), do the prediction for the height of a female student who wears a size 8 shoe.

Internet Resources: Visit the Weiss Web site www.aw-bc.com/weiss for additional discussion, exercises, and resources related to this case study.

Biography

ADRIEN LEGENDRE: Introducing the Method of Least-Squares

Adrien-Marie Legendre was born in Paris, France, on September 18, 1752, the son of a moderately wealthy family. He studied at the Collège Mazarin and received degrees in mathematics and physics in 1770 at the age of 18.

Although Legendre's financial assets were sufficient to allow him to devote himself to research, he took a position teaching mathematics at the École Militaire in Paris from 1775 to 1780. In March 1783, he was elected to the Academie des Sciences in Paris and, in 1787, he was assigned to a project undertaken jointly by the observatories at Paris and at Greenwich, England. At that time, he became a fellow of the Royal Society.

As a result of the French Revolution, which began in 1789, Legendre lost his "small fortune" and was forced to find work. He held various positions during the early 1790s as, for example, commissioner of astronomical operations for the Academie des Sciences, professor of pure mathematics at the Institut de Marat, and head of the National Executive Commission of Public Instruction. During this same period, Legendre wrote a geometry book that became the major text used in elementary geometry courses for nearly a century.

Legendre's major contribution to statistics was the publication, in 1805, of the first statement and the first application of the most widely used, nontrivial technique of statistics: the method of least squares. In his book, *The History of Statistics: The Measurement of Uncertainty Before 1900* (Cambridge, Mass.: Belknap Press of Harvard University Press, 1986), Stephen M. Stigler writes "[Legendre's] presentation … must be counted as one of the clearest and most elegant introductions of a new statistical method in the history of statistics."

Because Gauss also claimed the method of least squares, there was strife between the two men. Although evidence shows that Gauss was not successful in any communication of the method prior to 1805, his development of the method was crucial to its usefulness.

In 1813, Legendre was appointed Chief of the Bureau des Longitudes. He remained in that position until his death, following a long illness, in Paris on January 10, 1833.

Inferential Methods in Regression and Correlation

GENERAL OBJECTIVES In Chapter 14, you studied descriptive methods in regression and correlation. You discovered how to determine the regression equation for a set of data points and how to use that equation to make predictions. You also learned how to compute and interpret the coefficient of determination and the linear correlation coefficient for a set of data points.

In this chapter, you will study inferential methods in regression and correlation. In Section 15.1, we examine the conditions required for performing such inferences and methods for checking whether those conditions are satisfied. In presenting the first inferential method, in Section 15.2, we show how to decide whether a regression equation is useful for making predictions.

In Section 15.3, we investigate methods for estimating the mean of the response variable corresponding to a particular value of the predictor variable and for predicting the value of the response variable for a particular value of the predictor variable. We also discuss, in Section 15.4, the use of the linear correlation coefficient of a set of data points to decide whether the two variables under consideration are linearly correlated and, if so, the nature of the linear correlation.

Additionally, we present in Section 15.5 an inferential procedure for testing whether a variable is normally distributed.

Shoe Size and Height

As mentioned in the Chapter 14 case study, most of us have heard that tall people generally have larger feet than short people. To examine the relationship between shoe size and height, Professor Dennis Young obtained data on those two variables from a sample of students at Arizona State University. We presented the data obtained by Professor Young in the Chapter 14 case study and repeat them here in the table below. Height is measured in inches.

At the end of Chapter 14, on page 740, you were asked to conduct regression and correlation analyses on these shoe-size and height data. The analyses done there were descriptive. At the end of this chapter, you will be asked to return to the data to make regression and correlation inferences.

Shoe size	Height	Sex	Shoe size	Height	Sex
6.5	66.0	F	13.0	77.0	M
9.0	68.0	F	11.5	72.0	M
8.5	64.5	F	8.5	59.0	F
8.5	65.0	F	5.0	62.0	F
10.5	70.0	M	10.0	72.0	M
7.0	64.0	F	6.5	66.0	F
9.5	70.0	F	7.5	64.0	F
9.0	71.0	F	8.5	67.0	M
13.0	72.0	M	10.5	73.0	M
7.5	64.0	F	8.5	69.0	F
10.5	74.5	M	10.5	72.0	M
8.5	67.0	F	11.0	70.0	M
12.0	71.0	M	9.0	69.0	M
10.5	71.0	M	13.0	70.0	M

15.1 The Regression Model; Analysis of Residuals

Before we can perform statistical inferences in regression and correlation, we must know whether the variables under consideration satisfy certain conditions. In this section, we discuss those conditions and examine methods for deciding whether they hold.

The Regression Model

TABLE 15.1
Age and price data for a sample of 11 Orions

Age (yr) x	Price ($100) y
5	85
4	103
6	70
5	82
5	89
5	98
6	66
6	95
2	169
7	70
7	48

Let's return to the Orion illustration used throughout Chapter 14. In Table 15.1, we reproduce the data on age and price for a sample of 11 Orions.

With age as the predictor variable and price as the response variable, the regression equation for these data is $\hat{y} = 195.47 - 20.26x$, as we found in Chapter 14 on page 702. Recall that the regression equation can be used to predict the price of an Orion from its age. However, we cannot expect such predictions to be completely accurate because prices vary even for Orions of the same age.

For instance, the sample data in Table 15.1 include four 5-year-old Orions. Their prices are $8500, $8200, $8900, and $9800. This variation in price for 5-year-old Orions should be expected because such cars generally have different mileages, interior conditions, paint quality, and so forth.

We use the population of all 5-year-old Orions to introduce some important regression terminology. The distribution of their prices is called the **conditional distribution** of the response variable "price" corresponding to the value 5 of the predictor variable "age." Likewise, their mean price is called the **conditional mean** of the response variable "price" corresponding to the value 5 of the predictor variable "age." Similar terminology applies to the standard deviation and other parameters.

In general, there is a population of Orions for each age. The distribution, mean, and standard deviation of prices for that population are called the *conditional distribution, conditional mean,* and *conditional standard deviation,* respectively, of the response variable "price" corresponding to the value of the predictor variable "age."

With the preceding discussion in mind, we now state the conditions—as Key Fact 15.1—required for using inferential methods in regression analysis.

KEY FACT 15.1

What
Does it Mean?

Assumptions 1–3 require that there are constants β_0, β_1, and σ such that, for each value x of the predictor variable, the conditional distribution of the response variable, y, is a normal distribution having mean $\beta_0 + \beta_1 x$ and standard deviation σ. These assumptions are often referred to as the **regression model**.

Assumptions for Regression Inferences

1. *Population regression line:* There are constants β_0 and β_1 such that, for each value x of the predictor variable, the conditional mean of the response variable is $\beta_0 + \beta_1 x$.

2. *Equal standard deviations:* The conditional standard deviations of the response variable are the same for all values of the predictor variable. We denote this common standard deviation σ.

3. *Normal populations:* For each value of the predictor variable, the conditional distribution of the response variable is a normal distribution.

4. *Independent observations:* The observations of the response variable are independent of one another.

Note the following:

- We refer to the straight line $y = \beta_0 + \beta_1 x$—on which the conditional means of the response variable lie—as the **population regression line** and to its equation as the **population regression equation.**
- The condition of equal standard deviations is called *homoscedasticity.* When that condition fails, we have what is called *heteroscedasticity.*

The inferential procedures in regression are robust to moderate violations of Assumptions 1–3 for regression inferences. In other words, the inferential procedures work reasonably well provided the variables under consideration don't violate any of those assumptions too badly.

Example 15.1 further explains and illustrates graphically the assumptions for regression inferences.

▌▌Example 15.1　Assumptions for Regression Inferences

Age and Price of Orions　Discuss what satisfying the regression-inference Assumptions 1–3 would mean for Orions, with age as the predictor variable and price as the response variable. Display those assumptions graphically.

Solution　Satisfying regression-inference Assumptions 1–3 requires that there are constants β_0, β_1, and σ so that for each age, x, the prices of all Orions of that age are normally distributed with mean $\beta_0 + \beta_1 x$ and standard deviation σ. Thus the prices of all 2-year-old Orions must be normally distributed with mean $\beta_0 + \beta_1 \cdot 2$ and standard deviation σ, the prices of all 3-year-old Orions must be normally distributed with mean $\beta_0 + \beta_1 \cdot 3$ and standard deviation σ, and so on.

To display the assumptions for regression inferences graphically, let's first consider Assumption 1. This assumption requires that for each age, the mean price of all Orions of that age lies on the straight line $y = \beta_0 + \beta_1 x$, as shown in Fig. 15.1.

FIGURE 15.1

Population regression line

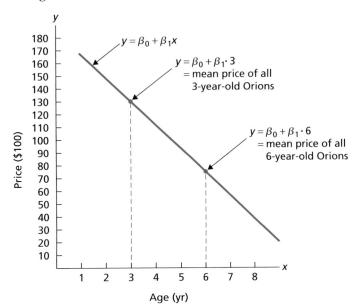

Next, we present graphs that depict Assumptions 2 and 3. These assumptions require that the price distributions for the various ages of Orions are all normally distributed with the same standard deviation, σ, as illustrated in Fig. 15.2 for the price distributions of 2-year-old, 5-year-old, and 7-year-old Orions. The shapes of the three normal curves in Fig. 15.2 are identical because normal distributions that have the same standard deviation have the same shape.

FIGURE 15.2

Price distributions for 2-, 5-, and 7-year-old Orions under Assumptions 2 and 3 (The means shown for the three normal distributions reflect Assumption 1)

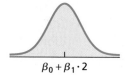 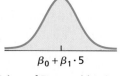 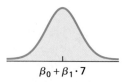

Assumptions 1–3 for regression inferences, as they pertain to the variables age and price of Orions, can be portrayed graphically by combining Figs. 15.1 and 15.2 into a three-dimensional graph, as displayed in Fig. 15.3.

FIGURE 15.3

Graphical portrayal of Assumptions 1–3 for regression inferences pertaining to age and price of Orions

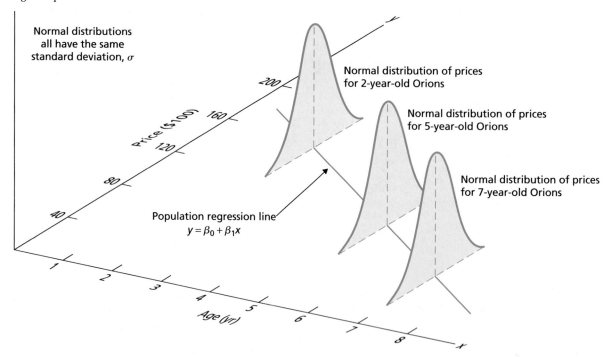

Figure 15.3 depicts the various price distributions for Orions under the condition that Assumptions 1–3 are true for the variables age and price of Orions. Whether this condition is actually the case remains to be seen.

Estimating the Regression Parameters

Suppose that we are considering two variables, x and y, for which the assumptions for regression inferences are met. Then there are constants β_0, β_1, and σ so that, for each value x of the predictor variable, the conditional distribution of the response variable is a normal distribution having mean $\beta_0 + \beta_1 x$ and standard deviation σ.

The parameters β_0, β_1, and σ are usually unknown and must therefore be estimated from sample data. Point estimates for the y-intercept, β_0, and slope, β_1, of the population regression line are provided by the y-intercept, b_0, and slope, b_1, respectively, of a sample regression line.

Another way of looking at this situation is that a sample regression line is used to estimate the population regression line. Of course, a sample regression line ordinarily will not be the same as the population regression line, just as a sample mean, $\bar{x}$, generally will not equal the population mean, μ. We illustrate this situation for the Orion example in Fig. 15.4. (Although the population regression line is unknown, we have drawn it to illustrate the difference between the population regression line and a sample regression line.)

FIGURE 15.4

Population regression line and sample regression line for age and price of Orions

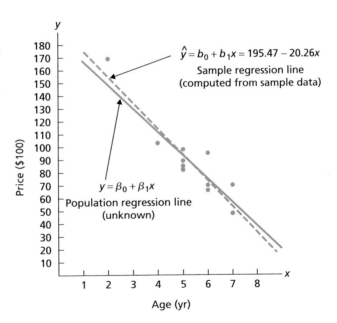

The solid line in Fig. 15.4 is the population regression line; the dashed line is a sample regression line. This sample regression line is the best approximation that can be made to the population regression line by using the sample data in Table 15.1 on page 744. A different sample of Orions would almost certainly yield a different sample regression line.

The statistic used to obtain a point estimate for the common conditional standard deviation σ is called the **standard error of the estimate** or the **residual standard deviation** and is defined as follows.

DEFINITION 15.1

Standard Error of the Estimate

The *standard error of the estimate, s_e,* is defined by

$$s_e = \sqrt{\frac{SSE}{n-2}},$$

where $SSE = \Sigma(y - \hat{y})^2 = S_{yy} - S_{xy}^2/S_{xx}$.

In Example 15.2, we illustrate the computation and interpretation of the standard error of the estimate by returning to the Orion illustration.

Example 15.2 **Standard Error of the Estimate**

Age and Price of Orions Refer to the age and price data for a sample of 11 Orions displayed in Table 15.1 on page 744.

a. Compute and interpret the standard error of the estimate.

b. Presuming that the variables age and price for Orions satisfy the assumptions for regression inferences, interpret the result from part (a).

Solution

a. On page 719, we found that $SSE = 1423.5$. So the standard error of the estimate is

$$s_e = \sqrt{\frac{SSE}{n-2}} = \sqrt{\frac{1423.5}{11-2}} = 12.58.$$

INTERPRETATION Roughly speaking, the predicted price of an Orion in the sample differs, on average, from the observed price by $1258.

b. Presuming that the variables age and price for Orions satisfy the assumptions for regression inferences, the standard error of the estimate, $s_e = 12.58$, or $1258, provides an estimate for the common population standard deviation, σ, of prices for all Orions of any particular age.

Analysis of Residuals

Now that we have examined the assumptions for regression inferences, we need to discuss how the sample data can be used to decide whether we can reasonably presume that those assumptions are met. We concentrate on Assumptions 1–3; checking Assumption 4—the independence assumption—is more involved and is best left for a second course in statistics.

The method for checking Assumptions 1–3 relies on an analysis of the errors made by using the regression equation to predict the observed values of the response variable, that is, on the differences, $y - \hat{y}$, between the observed and predicted values of the response variable. Each such difference is called a **residual**, generically denoted e. Thus

$$\text{Residual} = e = y - \hat{y}.$$

Figure 15.5 provides a graphical representation for the residual of a single data point.

FIGURE 15.5

Residual, e, of a data point

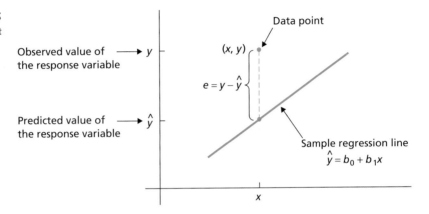

We can express the standard error of the estimate, s_e, in terms of the residuals. Referring to Definition 15.1, we find that the standard error of the estimate can be written as

$$s_e = \sqrt{\frac{SSE}{n-2}} = \sqrt{\frac{\Sigma(y - \hat{y})^2}{n-2}} = \sqrt{\frac{\Sigma e^2}{n-2}}.$$

We can show that the sum of the residuals is always 0, which, in turn, implies that $\bar{e} = 0$. Consequently, the standard error of the estimate is essentially the same as the standard deviation of the residuals.[†] Thus the standard error of the estimate is sometimes called the *residual standard deviation*.

We can analyze the residuals to decide whether Assumptions 1–3 for regression inferences are met because those assumptions can be translated into conditions on the residuals. To show how, let's consider a sample of data points obtained from two variables that satisfy the assumptions for regression inferences.

In light of Assumption 1, the data points should be scattered about the (sample) regression line, which means that the residuals should be scattered about the x-axis. In light of Assumption 2, the variation of the observed values of the response variable should remain approximately constant from one value of the predictor variable to the next, which means the residuals should fall roughly in a horizontal band. In light of Assumption 3, for each value of the predictor variable, the distribution of the corresponding observed values of the response variable should be approximately bell shaped, which implies that the horizontal band should be centered and symmetric about the x-axis.

[†]The exact standard deviation of the residuals is obtained by dividing by $n - 1$ instead of $n - 2$.

Furthermore, considering all four regression assumptions simultaneously, we can regard the residuals as independent observations of a variable having a normal distribution with mean 0 and standard deviation σ. Thus a normal probability plot of the residuals should be roughly linear.

In summary, we have the criteria presented in Key Fact 15.2 for deciding whether Assumptions 1–3 for regression inferences are met by the two variables under consideration.

KEY FACT 15.2

Residual Analysis for the Regression Model

If the assumptions for regression inferences are met, the following two conditions should hold.

- A plot of the residuals against the values of the predictor variable should fall roughly in a horizontal band centered and symmetric about the x-axis.
- A normal probability plot of the residuals should be roughly linear.

Failure of either of these two conditions casts doubt on the validity of one or more of the assumptions for regression inferences for the variables under consideration.

A plot of the residuals against the values of the predictor variable, called a **residual plot,** provides approximately the same information as does a scatter diagram of the data points. However, a residual plot makes spotting patterns such as curvature and nonconstant standard deviation easier.

Because the residual plot in Fig. 15.6(a) is roughly linear and the scatter about the line remains roughly constant, it suggests that the linearity assumption (Assumption 1) and the constant-standard-deviation assumption (Assumption 2) appear to be met for the variables under consideration. The residual plot in Fig. 15.6(b) suggests that the relation between the variables appears to be curved, indicating that the linearity assumption may be violated. The residual plot in Fig. 15.6(c) suggests that the conditional standard deviations appear to increase as x increases, indicating that the constant-standard-deviation assumption may be violated.

FIGURE 15.6

Residual plots suggesting (a) no violation of linearity or constant standard deviation, (b) violation of linearity, and (c) violation of constant standard deviation

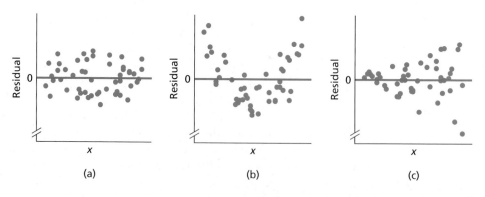

Our previous data analyses show that deciding on the appropriateness of a model is often difficult when we are dealing with small samples—say, of size 20 or less. The same difficulty holds in regression: For small samples, we must be

more liberal in allowing moderate departures from the idealized patterns when analyzing residual plots and normal probability plots to ascertain whether the assumptions for regression inferences are met. There are no definite rules— only judgment based on experience. Example 15.3 illustrates the application of such an analysis.

Example 15.3 Analysis of Residuals

Age and Price of Orions Perform a residual analysis to decide whether we can reasonably consider the assumptions for regression inferences to be met by the variables age and price of Orions.

Solution We apply the criteria presented in Key Fact 15.2. The ages and residuals for the Orion data are displayed in the first and fourth columns of Table 14.8 on page 719, respectively. We repeat that information in Table 15.2.

TABLE 15.2

Age and residual data for Orions

Age x	Residual e	Age x	Residual e
5	−9.16	6	−7.90
4	−11.42	6	21.10
6	−3.90	2	14.05
5	−12.16	7	16.36
5	−5.16	7	−5.64
5	3.84		

Figure 15.7(a) shows a plot of the residuals against age, and Fig. 15.7(b) shows a normal probability plot for the residuals.

FIGURE 15.7

(a) Residual plot; (b) normal probability plot for residuals

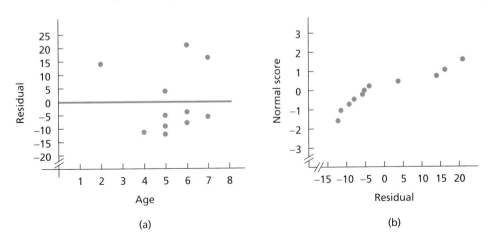

(a)

(b)

Taking into account the small sample size, we can say that the residuals fall roughly in a horizontal band that is centered and symmetric about the x-axis. We can also say that the normal probability plot for the residuals is (very) roughly linear, although the departure from linearity is sufficient for some concern.[†]

INTERPRETATION There are no obvious violations of the assumptions for regression inferences for the variables age and price of Orions (in the age range from 2 to 7 years). ▬

The Technology Center

Most statistical technologies, including Minitab and Excel, provide the standard error of the estimate as part of their regression analysis output. For instance, consider the regression output in Printout 14.2 on page 709 for the age and price data of 11 Orions.

In the Minitab output, the standard error of the estimate is the first entry in the seventh line: S = 12.5766. (Minitab uses S instead of s_e to denote the standard error of the estimate.) In the Excel output, the standard error of the estimate is the first entry in the fourth line: s = 12.58. (Excel uses s instead of s_e to denote the standard error of the estimate.)

Although the TI-83/84 Plus does not display the standard error of the estimate, it can easily be obtained after the regression procedure is run. See the *TI-83/84 Plus Manual* for details.

We can also use statistical technology to obtain a residual plot and a normal probability plot of the residuals. Example 15.4 provides the details for doing so with Minitab, Excel, and the TI-83/84 Plus.

Example 15.4 **Using Technology to Obtain Plots of Residuals**

Age and Price of Orions Use Minitab, Excel, or the TI-83/84 Plus to obtain a residual plot and a normal probability plot of the residuals for the age and price data of Orions given in Table 15.1 on page 744.

Solution Printout 15.1 shows output that results from applying the programs for obtaining a residual plot and a normal probability plot of the residuals for the age and price data.

Note that, in this case, Minitab's normal probability plot uses percents instead of normal scores on the vertical axis. Also note that Excel plots the residuals against the predicted values of the response variable instead of the observed values of the predictor variable. These modifications, however, do not affect the use of the plots as diagnostic tools to help assess the appropriateness of regression inferences. ▬

Obtaining the Output (Optional)

Printout 15.1 provides output from Minitab, Excel, and the TI-83/84 Plus for a residual plot and a normal probability plot of the residuals for the age and

---

[†]Recall, though, that the inferential procedures in regression analysis are robust to moderate violations of Assumptions 1–3 for regression inferences.

PRINTOUT 15.1
Residual plots and normal probability
plots of the residuals for the age and
price data of 11 Orions

EXCEL

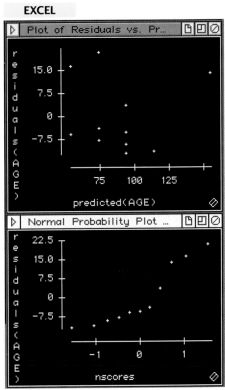

MINITAB

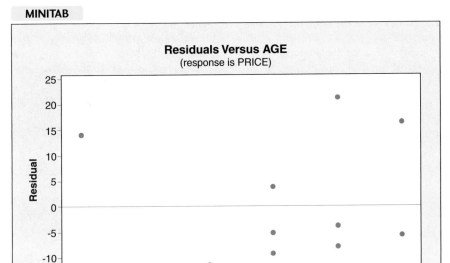

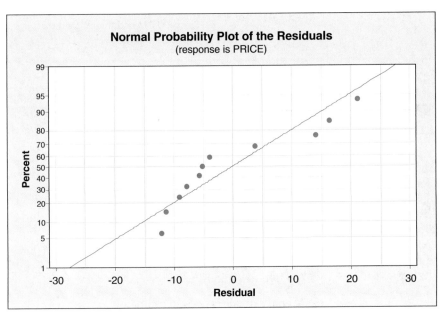

TI-83/84 PLUS

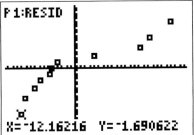

price data of 11 Orions. The following are detailed instructions for obtaining that output. First, we store the age and price data from Table 15.1 in columns (Minitab), ranges (Excel), or lists (TI-83/84 Plus) named AGE and PRICE, respectively. Then, we proceed as follows.

MINITAB	EXCEL	TI-83/84 PLUS
1 Choose **Stat ➤ Regression ➤ Regression...**	1 Choose **DDXL ➤ Regression**	1 Clear the **Y=** screen or turn off any equations located there
2 Specify PRICE in the **Response** text box	2 Select **Simple Regression** from the **Function type** drop-down list box	2 Press **STAT**, arrow over to **CALC**, and press **8**
3 Specify AGE in the **Predictors** text box	3 Specify PRICE in the **Response Variable** text box	3 Press **2nd ➤ LIST**, arrow down to AGE, and press **ENTER**
4 Click the **Graphs...** button	4 Specify AGE in the **Explanatory Variable** text box	4 Press **, ➤ 2nd ➤ LIST**, arrow down to PRICE, and press **ENTER** twice
5 Select the **Regular** option button from the **Residuals for Plots** list	5 Click **OK**	5 Press **2nd ➤ STAT PLOT** and then press **ENTER** twice
6 Select the **Individual plots** option button from the **Residual Plots** list	6 Click the **Check the Residuals** button	6 Arrow to the first graph icon and press **ENTER**
7 Select the **Normal plot of residuals** check box from the **Individual plots** list		7 Press the down-arrow key
8 Click in the **Residuals versus the variables** text box and specify AGE		8 Press **2nd ➤ LIST**, arrow down to AGE, and press **ENTER** twice
9 Click **OK**		9 Press **2nd ➤ LIST**, arrow down to RESID, and press **ENTER** twice
10 Click **OK**		10 Press **ZOOM** and then **9** (and then **TRACE**, if desired)
		11 Press **2nd ➤ STAT PLOT** and then press **ENTER** twice
		12 Arrow to the sixth graph icon and press **ENTER**
		13 Press the down-arrow key
		14 Press **2nd ➤ LIST**, arrow down to RESID, and press **ENTER** twice
		15 Press **ZOOM** and then **9** (and then **TRACE**, if desired)

EXERCISES 15.1

Statistical Concepts and Skills

15.1 Suppose that x and y are predictor and response variables, respectively, of a population. Consider the population that consists of all members of the original population that have a specified value of the predictor variable. The distribution, mean, and standard deviation of the response variable for this population are called the ____, ____, and ____, respectively, corresponding to the specified value of the predictor variable.

15.2 State the four conditions required for making regression inferences.

In Exercises 15.3–15.6, assume that the variables under consideration satisfy the assumptions for regression inferences.

15.3 Fill in the blanks.
a. The straight line $y = \beta_0 + \beta_1 x$ is called the ____.
b. The common conditional standard deviation of the response variable is denoted ____.
c. For $x = 6$, the conditional distribution of the response variable is a ____ distribution having mean ____ and standard deviation ____.

15.4 What statistic is used to estimate
a. the y-intercept of the population regression line?
b. the slope of the population regression line?

c. the common conditional standard deviation, σ, of the response variable?

15.5 Based on a sample of data points, what is the best estimate of the population regression line?

15.6 Regarding the standard error of the estimate,
a. give two interpretations of it.
b. identify another name used for it and explain the rationale for that name.
c. which one of the three sums of squares figures in its computation?

15.7 The difference between an observed value and a predicted value of the response variable is called a _____.

15.8 Identify two graphs used in a residual analysis to check the Assumptions 1–3 for regression inferences and explain the reasoning behind their use.

15.9 Which graph used in a residual analysis provides roughly the same information as a scatter diagram? What advantages does it have over a scatter diagram?

*In Exercises **15.10–15.13**, we repeat the information from Exercises 14.42–14.45. For each exercise here, discuss what satisfying Assumptions 1–3 for regression inferences by the variables under consideration would mean.*

15.10 Tax Efficiency. *Tax efficiency* is a measure—ranging from 0 to 100—of how much tax due to capital gains stock or mutual funds investors pay on their investments each year; the higher the tax efficiency, the lower is the tax. The paper "At the Mercy of the Manager" (*Financial Planning*, Vol. 30(5), pp. 54–56) by Craig Israelsen examined the relationship between investments in mutual fund portfolios and their associated tax efficiencies. The following table shows percentage of investments in energy securities (x) and tax efficiency (y) for 10 mutual fund portfolios.

x	3.1	3.2	3.7	4.3	4.0	5.5	6.7	7.4	7.4	10.6
y	98.1	94.7	92.0	89.8	87.5	85.0	82.0	77.8	72.1	53.5

15.11 Corvette Prices. The *Kelley Blue Book* provides information on wholesale and retail prices of cars. Following are age and price data for 10 randomly selected Corvettes between 1 and 6 years old. Here, x denotes age, in years, and y denotes price, in hundreds of dollars.

x	6	6	6	2	2	5	4	5	1	4
y	270	260	275	405	364	295	335	308	405	305

15.12 Custom Homes. Hanna Properties specializes in custom-home resales in the Equestrian Estates, an exclusive subdivision in Phoenix, Arizona. A random sample of nine custom homes currently listed for sale provided the following information on size and price. Here, x denotes size, in hundreds of square feet, rounded to the nearest hundred, and y denotes price, in thousands of dollars, rounded to the nearest thousand.

x	26	27	33	29	29	34	30	40	22
y	290	305	325	327	356	411	488	554	246

15.13 Plant Emissions. Plants emit gases that trigger the ripening of fruit, attract pollinators, and cue other physiological responses. N. G. Agelopolous, K. Chamberlain, and J. A. Pickett examined factors that affect the emission of volatile compounds by the potato plant *Solanum tuberosom* and published their findings in the *Journal of Chemical Ecology* (Vol. 26(2), pp. 497–511). The volatile compounds analyzed were hydrocarbons used by other plants and animals. Following are data on plant weight (x), in grams, and quantity of volatile compounds emitted (y), in hundreds of nanograms, for 11 potato plants.

x	57	85	57	65	52	67	62	80	77	53	68
y	8.0	22.0	10.5	22.5	12.0	11.5	7.5	13.0	16.5	21.0	12.0

*In Exercises **15.14–15.17**,*
a. compute the standard error of the estimate and interpret your answer.
b. interpret your result from part (a) if the assumptions for regression inferences hold.
c. obtain a residual plot and a normal probability plot of the residuals.
d. decide whether you can reasonably consider Assumptions 1–3 for regression inferences to be met by the variables under consideration. (The answer here is subjective, especially in view of the extremely small sample sizes.)

15.14 Tax Efficiency. Use the data on percentage of investments in energy securities and tax efficiency from Exercise 15.10.

15.15 Corvette Prices. Use the age and price data for Corvettes from Exercise 15.11.

15.16 Custom Homes. Use the size and price data for custom homes from Exercise 15.12.

15.17 Plant Emissions. Use the data on plant weight and quantity of volatile emissions from Exercise 15.13.

15.18 Figure 15.8 shows three residual plots and a normal probability plot of residuals. For each part, decide whether the graph suggests violation of one or more of the assumptions for regression inferences. Explain your answers.

15.19 Figure 15.9 displays three residual plots and one normal probability plot of residuals. For each part, decide whether the graph suggests violation of one or more of the assumptions for regression inferences. Explain your answers.

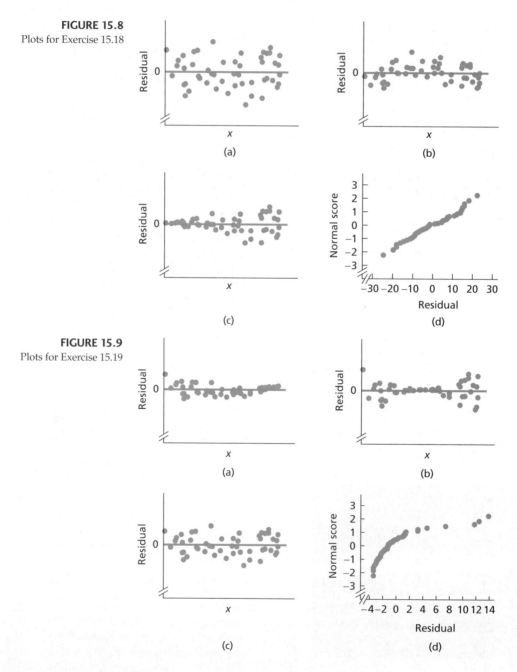

FIGURE 15.8
Plots for Exercise 15.18

FIGURE 15.9
Plots for Exercise 15.19

Using Technology

*For Exercises **15.20–15.25**, use the technology of your choice to*
a. obtain and interpret the standard error of the estimate.
b. obtain a residual plot and a normal probability plot of the residuals.
c. decide whether you can reasonably consider Assumptions 1–3 for regression inferences to be met by the variables under consideration.

15.20 Batting and Scoring. Is the number of runs a baseball team scores in a season related to its team batting average? ESPN compiles end-of-season statistics for Major League Baseball and maintains them on its Web site. The following table provides season team batting averages and total runs scored for a sample of major league baseball teams.

Average	Runs	Average	Runs
.294	968	.267	793
.278	938	.265	792
.278	925	.256	764
.270	887	.254	752
.274	825	.246	740
.271	810	.266	738
.263	807	.262	731
.257	798	.251	708

15.21 Body Fat. In the paper "Total Body Composition by Dual-Photon (^{153}Gd) Absorptiometry" (*American Journal of Clinical Nutrition*, Vol. 40, pp. 834–839), R. B. Mazess et al. studied methods for quantifying body composition. Eighteen randomly selected adults were measured for percentage of body fat, using dual-photon absorptiometry. The following table shows the results of the measurements and the ages of the adults.

Age	% Fat	Age	% Fat	Age	% Fat
23	9.5	45	27.4	56	32.5
23	27.9	49	25.2	57	30.3
27	7.8	50	31.1	58	33.0
27	17.8	53	34.7	58	33.8
39	31.4	53	42.0	60	41.1
41	25.9	54	29.1	61	34.5

15.22 PCBs and Pelicans. Polychlorinated biphenyls (PCBs), industrial pollutants, are a great danger to natural ecosystems. In a study by R. W. Risebrough titled "Effects of Environmental Pollutants Upon Animals Other Than Man" (*Proceedings of the 6th Berkeley Symposium on Mathematics and Statistics, VI*, University of California Press, pp. 443–463), 60 Anacapa pelican eggs were collected and measured for their shell thickness, in millimeters (mm), and concentration of PCBs, in parts per million (ppm). The data are presented on the WeissStats CD.

15.23 Gas Guzzlers. The magazine *Consumer Reports* publishes information on automobile gas mileage and variables that affect gas mileage. In the April 1999 issue, data on gas mileage (in mpg) and engine displacement (in liters, L) were published for 121 vehicles. Those data are stored on the WeissStats CD.

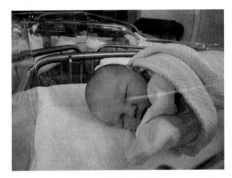

15.24 Estriol Level and Birth Weight. J. Greene and J. Touchstone conducted a study on the relationship between the estriol levels of pregnant women and the birth weights of their children. Their findings, "Urinary Tract Estriol: An Index of Placental Function," were published in the *American Journal of Obstetrics and Gynecology* (Vol. 85(1), pp. 1–9). The data points are provided on the WeissStats CD, where estriol levels are in mg/24 hr and birth weights are in hectograms (hg).

15.25 Shortleaf Pines. The ability to estimate the volume of a tree based on a simple measurement, such as the diameter of the tree, is important to the lumber industry, ecologists, and conservationists. Data on volume, in cubic feet, and diameter at breast height, in inches, for 70 shortleaf pines was reported in C. Bruce and F. X. Schumacher's *Forest Mensuration* (New York: McGraw-Hill, 1935) and analyzed by A. C. Akinson in the article "Transforming Both Sides of a Tree" (*The American Statistician*, Vol. 48, pp. 307–312). The data are provided on the WeissStats CD.

15.2 Inferences for the Slope of the Population Regression Line

In this section and Section 15.3, we examine several inferential procedures used in regression analysis. Strictly speaking, these inferential techniques require that the assumptions for regression inferences given in Key Fact 15.1 on page 744 be satisfied by the variables under consideration. However, as we noted earlier, these techniques are robust to moderate violations of those assumptions.

The first inferential methods we present concern the slope, β_1, of the population regression line. To begin, we consider hypothesis testing.

Hypothesis Tests for the Slope of the Population Regression Line

Suppose that the variables x and y satisfy the assumptions for regression inferences. Then, for each value x of the predictor variable, the conditional distribution of the response variable is a normal distribution with mean $\beta_0 + \beta_1 x$ and standard deviation σ.

Of particular interest is whether the slope, β_1, of the population regression line equals 0. If $\beta_1 = 0$, then, for each value x of the predictor variable, the conditional distribution of the response variable is a normal distribution having mean β_0 $(= \beta_0 + 0 \cdot x)$ and standard deviation σ. Because x does not appear in either of those two parameters, it is useless as a predictor of y.[†]

Hence, we can decide whether x is useful as a (linear) predictor of y—that is, whether the regression equation has utility—by performing the hypothesis test

$$H_0\colon \beta_1 = 0 \ (x \text{ is not useful for predicting } y)$$
$$H_a\colon \beta_1 \neq 0 \ (x \text{ is useful for predicting } y).$$

TABLE 15.3

Age and price data for a sample of 11 Orions

Age (yr) y	Price ($100) x
5	85
4	103
6	70
5	82
5	89
5	98
6	66
6	95
2	169
7	70
7	48

We base hypothesis tests for β_1 (the slope of the population regression line) on the statistic b_1 (the slope of a sample regression line). To explain how this method works, let's return to the Orion illustration. The data on age and price for a sample of 11 Orions are repeated in Table 15.3.

With age as the predictor variable and price as the response variable, the regression equation for these data is $\hat{y} = 195.47 - 20.26x$, as we found in Chapter 14 on page 702. In particular, the slope, b_1, of the sample regression line is -20.26.

We now consider all possible samples of 11 Orions whose ages are the same as those given in the first column of Table 15.3. For such samples, the slope, b_1, of the sample regression line varies from one sample to another and is therefore a variable. Its distribution is called the **sampling distribution of the slope of the regression line.** From the assumptions for regression inferences, we can

---

[†]Although x alone may not be useful for predicting y, it may be useful in conjunction with another variable or variables. Thus, in this section, when we say that x is not useful for predicting y, we really mean that the regression equation with x as the only predictor variable is not useful for predicting y. Conversely, although x alone may be useful for predicting y, it may not be useful in conjunction with another variable or variables. Thus, in this section, when we say that x is useful for predicting y, we really mean that the regression equation with x as the only predictor variable is useful for predicting y.

show that this distribution is a normal distribution whose mean is the slope, β_1, of the population regression line. More generally, we have Key Fact 15.3.

KEY FACT 15.3

The Sampling Distribution of the Slope of the Regression Line

Suppose that the variables x and y satisfy the four assumptions for regression inferences. Then, for samples of size n, each with the same values $x_1, x_2, \ldots, x_n$, for the predictor variable, the following properties hold for the slope, b_1, of the sample regression line.

- The mean of b_1 equals the slope of the population regression line; that is, we have $\mu_{b_1} = \beta_1$.
- The standard deviation of b_1 is $\sigma_{b_1} = \sigma/\sqrt{S_{xx}}$.
- The variable b_1 is normally distributed.

As a consequence of Key Fact 15.3, the standardized variable

$$z = \frac{b_1 - \beta_1}{\sigma/\sqrt{S_{xx}}}$$

has the standard normal distribution. But this variable cannot be used as a basis for the required test statistic because the common conditional standard deviation, σ, is unknown. We therefore replace σ with its sample estimate s_e, the standard error of the estimate. As you might suspect, the resulting variable has a t-distribution, as presented in Key Fact 15.4.

KEY FACT 15.4

t-Distribution for Inferences for β_1

Suppose that the variables x and y satisfy the four assumptions for regression inferences. Then, for samples of size n, each with the same values $x_1, x_2, \ldots, x_n$, for the predictor variable, the variable

$$t = \frac{b_1 - \beta_1}{s_e/\sqrt{S_{xx}}}$$

has the t-distribution with df $= n - 2$.

In light of Key Fact 15.4, for a hypothesis test with the null hypothesis H_0: $\beta_1 = 0$, we can use the variable

$$t = \frac{b_1}{s_e/\sqrt{S_{xx}}}$$

as the test statistic and obtain the critical values or P-value from the t-table, Table IV in Appendix A. We refer to this hypothesis-testing procedure as the **regression t-test.** Procedure 15.1 provides a step-by-step method for performing a regression t-test by using either the critical-value approach or the P-value approach.

| PROCEDURE 15.1 | Regression *t*-Test |

Purpose To perform a hypothesis test to decide whether a predictor variable is useful for making predictions

Assumptions

The four assumptions for regression inferences

STEP 1 **The null and alternative hypotheses are**

> H_0: $\beta_1 = 0$ (predictor variable is not useful for making predictions)
>
> H_a: $\beta_1 \neq 0$ (predictor variable is useful for making predictions)

STEP 2 **Decide on the significance level, α.**

STEP 3 **Compute the value of the test statistic**

$$t = \frac{b_1}{s_e/\sqrt{S_{xx}}}$$

and denote that value t_0.

CRITICAL-VALUE APPROACH

STEP 4 **The critical value(s) are $\pm t_{\alpha/2}$ with df $= n - 2$. Use Table IV to find the critical values.**

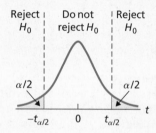

STEP 5 **If the value of the test statistic falls in the rejection region, reject H_0; otherwise, do not reject H_0.**

P-VALUE APPROACH

STEP 4 **The *t*-statistic has df $= n - 2$. Use Table IV to estimate the *P*-value, or obtain it exactly by using technology.**

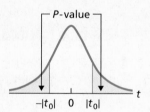

STEP 5 **If $P \leq \alpha$, reject H_0; otherwise, do not reject H_0.**

STEP 6 **Interpret the results of the hypothesis test.**

Example 15.5   The Regression t-Test

Age and Price of Orions   The data on age and price for a sample of 11 Orions are displayed in Table 15.3 on page 758. For ease of reference, we repeat that table here. At the 5% significance level, do the data provide sufficient evidence to conclude that age is useful as a (linear) predictor of price for Orions?

Age (yr) y	Price ($100) x
5	85
4	103
6	70
5	82
5	89
5	98
6	66
6	95
2	169
7	70
7	48

Solution   As we discovered in Example 15.3, we can reasonably consider the assumptions for regression inferences to be satisfied by the variables age and price for Orions, at least for Orions between 2 and 7 years old. So we apply Procedure 15.1 to carry out the required hypothesis test.

STEP 1 State the null and alternative hypotheses.

Let β_1 denote the slope of the population regression line that relates price to age for Orions. Then the null and alternative hypotheses are

H_0: $\beta_1 = 0$ (age is not useful for predicting price)

H_a: $\beta_1 \neq 0$ (age is useful for predicting price).

STEP 2 Decide on the significance level, α.

We are to perform the hypothesis test at the 5% significance level, or $\alpha = 0.05$.

STEP 3 Compute the value of the test statistic

$$t = \frac{b_1}{s_e/\sqrt{S_{xx}}}.$$

In Example 14.4 on page 702, we found that $b_1 = -20.26$, $\Sigma x^2 = 326$, and $\Sigma x = 58$. Also, in Example 15.2 on page 748, we determined that $s_e = 12.58$. Therefore, because $n = 11$, the value of the test statistic is

$$t = \frac{b_1}{s_e/\sqrt{S_{xx}}} = \frac{b_1}{s_e/\sqrt{\Sigma x^2 - (\Sigma x)^2/n}} = \frac{-20.26}{12.58/\sqrt{326 - (58)^2/11}} = -7.235.$$

CRITICAL-VALUE APPROACH	P-VALUE APPROACH

STEP 4 The critical values are $\pm t_{\alpha/2}$ with df $= n - 2$. Use Table IV to find the critical values.

From Step 2, $\alpha = 0.05$. For $n = 11$, df $= n - 2 = 11 - 2 = 9$. Using Table IV, we find that the critical values are $\pm t_{\alpha/2} = \pm t_{0.025} = \pm 2.262$, as depicted in Fig. 15.10A.

FIGURE 15.10A

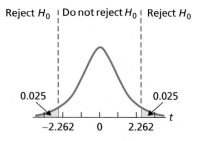

STEP 5 If the value of the test statistic falls in the rejection region, reject H_0; otherwise, do not reject H_0.

The value of the test statistic, found in Step 3, is $t = -7.235$. Because this value falls in the rejection region, we reject H_0. The test results are statistically significant at the 5% level.

STEP 4 The t-statistic has df $= n - 2$. Use Table IV to estimate the P-value or obtain it exactly by using technology.

From Step 3, the value of the test statistic is $t = -7.235$. Because the test is two tailed, the P-value is the probability of observing a value of t of 7.235 or greater in magnitude if the null hypothesis is true. That probability equals the shaded area shown in Fig. 15.10B.

FIGURE 15.10B

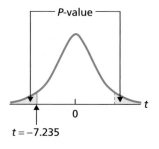

For $n = 11$, df $= 11 - 2 = 9$. Referring to Fig. 15.10B and to Table IV with df $= 9$, we find that $P < 0.01$. (Using technology, we obtain $P = 0.0000488$.)

STEP 5 If $P \leq \alpha$, reject H_0; otherwise, do not reject H_0.

From Step 4, $P < 0.01$. Because the P-value is less than the specified significance level of 0.05, we reject H_0. The test results are statistically significant at the 5% level and (see Table 9.12 on page 440) provide very strong evidence against the null hypothesis.

STEP 6 Interpret the results of the hypothesis test.

INTERPRETATION At the 5% significance level, the data provide sufficient evidence to conclude that the slope of the population regression line is not 0 and hence that age is useful as a (linear) predictor of price for Orions.

Other Procedures for Testing Utility of the Regression

We use Procedure 15.1 on page 760, which is based on the statistic b_1, to perform a hypothesis test to decide whether the slope of the population regression line is not 0 or, equivalently, whether the regression equation is useful for making predictions.

In Section 14.3, we introduced the coefficient of determination, r^2, as a descriptive measure of the utility of the regression equation for making predictions. Thus we should also be able to utilize the statistic r^2 as a basis for performing a hypothesis test to decide whether the regression equation is useful for making predictions—and indeed we can. However, we do not cover the hypothesis test based on r^2 because it is equivalent to the hypothesis test based on b_1.

We can also use the linear correlation coefficient, r, introduced in Section 14.4, as a basis for performing a hypothesis test to decide whether the regression equation is useful for making predictions. That test too is equivalent to the hypothesis test based on b_1 but, because it has other uses, we discuss it in Section 15.4.

Confidence Intervals for the Slope of the Population Regression Line

Recall that the slope of a straight line represents the change in the dependent variable, y, resulting from an increase in the independent variable, x, by 1 unit. Also recall that the population regression line, whose slope is β_1, gives the conditional means of the response variable. Therefore β_1 represents the change in the conditional mean of the response variable for each increase in the value of the predictor variable by 1 unit.

For instance, consider the variables age and price of Orions. In this case, β_1 is the amount that the mean price decreases for every increase in age by 1 year. In other words, β_1 is the mean yearly depreciation of Orions.

Consequently, obtaining an estimate for the slope of the population regression line is worthwhile. We know that a point estimate for β_1 is provided by b_1. To determine a confidence-interval estimate for β_1, we apply Key Fact 15.4 on page 759 to obtain Procedure 15.2, which we refer to as the **regression t-interval procedure.**

PROCEDURE 15.2

Regression t-Interval Procedure

Purpose To find a confidence interval for the slope, β_1, of the population regression line

Assumptions

The four assumptions for regression inferences

STEP 1 For a confidence level of $1 - \alpha$, use Table IV to find $t_{\alpha/2}$ with df $= n - 2$.

STEP 2 The endpoints of the confidence interval for β_1 are

$$b_1 \pm t_{\alpha/2} \cdot \frac{s_e}{\sqrt{S_{xx}}}.$$

STEP 3 Interpret the confidence interval.

Example 15.6   The Regression t-Interval Procedure

Age and Price of Orions   Use the data in Table 15.3 on page 758 to obtain a 95% confidence interval for the slope of the population regression line that relates price to age for Orions.

Solution   We apply Procedure 15.2.

STEP 1 For a confidence level of $1 - \alpha$, use Table IV to find $t_{\alpha/2}$ with df $= n - 2$.

For a 95% confidence interval, $\alpha = 0.05$. Because $n = 11$, df $= 11 - 2 = 9$. From Table IV, $t_{\alpha/2} = t_{0.05/2} = t_{0.025} = 2.262$.

STEP 2 The endpoints of the confidence interval for β_1 are

$$b_1 \pm t_{\alpha/2} \cdot \frac{s_e}{\sqrt{S_{xx}}}.$$

From Example 14.4, $b_1 = -20.26$, $\Sigma x^2 = 326$, and $\Sigma x = 58$. Also, from Example 15.2, $s_e = 12.58$. Hence the endpoints of the confidence interval for β_1 are

$$-20.26 \pm 2.262 \cdot \frac{12.58}{\sqrt{326 - (58)^2/11}},$$

or -20.26 ± 6.33, or -26.59 to -13.93.

STEP 3 Interpret the confidence interval.

INTERPRETATION   We can be 95% confident that the slope of the population regression line is somewhere between -26.59 and -13.93. In other words, we can be 95% confident that the yearly decrease in mean price for Orions is somewhere between $1393 and $2659. ∎

The Technology Center

Procedure 15.1 provides a step-by-step method for performing a hypothesis test to decide whether the slope of a population regression line is not 0 and hence whether the regression equation (i.e., predictor variable) is useful for making predictions. Most statistical technologies, including Minitab and Excel, provide the information needed to perform this test as part of their regression analysis output. However, some statistical technologies, such as the TI-83/84 Plus, require the implementation of a dedicated procedure.

For instance, consider the regression output in Printout 14.2 on page 709 for the age and price data of 11 Orions. In the Minitab output, the value of the *t*-statistic and the *P*-value for the hypothesis test appear as the last two entries in the fifth line. In the Excel output, those two items appear as the last two entries in the last line.

For the TI-83/84 Plus, we implement the hypothesis test to decide whether the slope of a population regression line is not 0 by using the **LinRegTTest** procedure. See the *TI-83/84 Plus Manual* for details.

EXERCISES 15.2

Statistical Concepts and Skills

15.26 Explain why the predictor variable is useless as a predictor of the response variable if the slope of the population regression line is 0.

15.27 For two variables satisfying Assumptions 1–3 for regression inferences, the population regression equation is $y = 20 - 3.5x$. For samples of size 10 and given values of the predictor variable, the distribution of slopes of all possible sample regression lines is a ____ distribution with mean ____.

15.28 Consider the standardized variable

$$z = \frac{b_1 - \beta_1}{\sigma/\sqrt{S_{xx}}}.$$

a. Identify its distribution.
b. Why can't it be used as the test statistic for a hypothesis test concerning β_1?
c. What statistic is used? What is the distribution of that statistic?

15.29 In this section, we used the statistic b_1 as a basis for conducting a hypothesis test to decide whether a regression equation is useful for prediction. Identify two other statistics that can be used as a basis for such a test.

*In Exercises **15.30–15.33**, we repeat information from Exercises 15.10–15.13. Presuming that the assumptions for regression inferences are met, perform the required hypothesis tests, using either the critical-value approach or the P-value approach. (Recall that you previously obtained the sample regression equations in Exercises 14.42–14.45 and the standard errors of the estimate in Exercises 15.14–15.17.)*

15.30 Tax Efficiency. Following are the data on percentage of investments in energy securities and tax efficiency from Exercise 15.10.

x	3.1	3.2	3.7	4.3	4.0	5.5	6.7	7.4	7.4	10.6
y	98.1	94.7	92.0	89.8	87.5	85.0	82.0	77.8	72.1	53.5

At the 5% significance level, do the data provide sufficient evidence to conclude that the slope of the population regression line is not 0 and hence that percentage of investments in energy securities is useful as a predictor of tax efficiency for mutual fund portfolios?

15.31 Corvette Prices. Following are the age and price data for Corvettes from Exercise 15.11.

x	6	6	6	2	2	5	4	5	1	4
y	270	260	275	405	364	295	335	308	405	305

At the 10% significance level, do the data provide sufficient evidence to conclude that the slope of the population regression line is not 0 and hence that age is useful as a predictor of price for Corvettes?

15.32 Custom Homes. Following are the size and price data for custom homes from Exercise 15.12.

x	26	27	33	29	29	34	30	40	22
y	290	305	325	327	356	411	488	554	246

Do the data suggest that size is useful as a predictor of price for custom homes in the Equestrian Estates? Perform the required hypothesis test at the 0.01 level of significance.

15.33 Plant Emissions. Following are the data on plant weight and quantity of volatile emissions from Exercise 15.13.

x	57	85	57	65	52	67	62	80	77	53	68
y	8.0	22.0	10.5	22.5	12.0	11.5	7.5	13.0	16.5	21.0	12.0

Do the data suggest that weight is useful as a predictor of quantity of volatile emissions for the potato plant *Solanum tuberosom*? Use $\alpha = 0.05$.

*In Exercises **15.34–15.37**, apply Procedure 15.2 on page 763 to obtain the required confidence intervals.*

15.34 Tax Efficiency. Refer to Exercise 15.30.
a. Find a 95% confidence interval for the slope, β_1, of the population regression line that relates tax efficiency to percent of investments in energy securities.
b. Interpret your answer to part (a).

15.35 Corvette Prices. Refer to Exercise 15.31.
a. Find a 90% confidence interval for the slope, β_1, of the population regression line that relates price to age for Corvettes.
b. Interpret your answer to part (a).

15.36 Custom Homes. Refer to Exercise 15.32.
a. Find a 99% confidence interval for the slope of the population regression line that relates price to size for custom homes in the Equestrian Estates.
b. Interpret your answer to part (a).

15.37 Plant Emissions. Refer to Exercise 15.33.
a. Obtain a 95% confidence interval for the slope, β_1, of the population regression line that relates quantity of volatile emissions to weight for the potato plant *Solanum tuberosom*.
b. Interpret your answer to part (a).

Using Technology

In Exercises 15.38–15.42, use the technology of your choice to
a. decide at the 5% significance level whether the regression equation is useful for making predictions.
b. repeat part (a), but first remove all influential observations and outliers.
c. Compare your answers in parts (a) and (b) and state your conclusions.

15.38 Batting and Scoring. Following are the data on season team batting average and total runs scored for a sample of major league baseball teams from Exercise 15.20.

Average	Runs	Average	Runs
.294	968	.267	793
.278	938	.265	792
.278	925	.256	764
.270	887	.254	752
.274	825	.246	740
.271	810	.266	738
.263	807	.262	731
.257	798	.251	708

15.39 Body Fat. Following are the age and body fat data for 18 randomly selected adults from Exercise 15.21.

Age	% Fat	Age	% Fat	Age	% Fat
23	9.5	45	27.4	56	32.5
23	27.9	49	25.2	57	30.3
27	7.8	50	31.1	58	33.0
27	17.8	53	34.7	58	33.8
39	31.4	53	42.0	60	41.1
41	25.9	54	29.1	61	34.5

15.40 PCBs and Pelicans. Use the data points given on the WeissStats CD for shell thickness and concentration of PCBs for 60 Anacapa pelican eggs referred to in Exercise 15.22.

15.41 Gas Guzzlers. Use the data on the WeissStats CD for gas mileage and engine displacement for 121 vehicles referred to in Exercise 15.23.

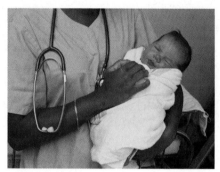

15.42 Estriol Level and Birth Weight. Use the data on the WeissStats CD for estriol levels of pregnant women and birth weights of their children referred to in Exercise 15.24.

15.3 Estimation and Prediction

In this section, we examine how a sample regression equation can be used to make two important inferences:

- estimating the conditional mean of the response variable corresponding to a particular value of the predictor variable; and
- predicting the value of the response variable for a particular value of the predictor variable.

We again use the Orion data, in Example 15.7, to illustrate the pertinent ideas. In doing so, we presume that the assumptions for regression inferences (Key Fact 15.1 on page 744) are satisfied by the variables age and price for Orions. Example 15.3 on page 751 shows that to presume so is not unreasonable.

Example 15.7   Estimating Conditional Means in Regression

TABLE 15.4

Age and price data for a sample of 11 Orions

Age (yr) *x*	Price ($100) *y*
5	85
4	103
6	70
5	82
5	89
5	98
6	66
6	95
2	169
7	70
7	48

Age and Price of Orions   The data on age and price for a sample of 11 Orions are repeated in Table 15.4. Use the data to estimate the mean price of all 3-year-old Orions.

Solution   By Assumption 1 of the assumptions for regression inferences, the population regression line gives the mean prices for the various ages of Orions. In particular, the mean price of all 3-year-old Orions is $\beta_0 + \beta_1 \cdot 3$. Because β_0 and β_1 are unknown, we estimate the mean price of all 3-year-old Orions ($\beta_0 + \beta_1 \cdot 3$) by the corresponding value on the sample regression line, namely, $b_0 + b_1 \cdot 3$.

Recalling that the sample regression equation for the age and price data in Table 15.4 is $\hat{y} = 195.47 - 20.26x$, we estimate that the mean price of all 3-year-old Orions is

$$\hat{y} = 195.47 - 20.26 \cdot 3 = 134.69,$$

or $13,469. Note that the estimate for the mean price of all 3-year-old Orions is the same as the predicted price for a 3-year-old Orion. Both are obtained by substituting $x = 3$ into the sample regression equation.

The estimate of $13,469 for the mean price of all 3-year-old Orions is a point estimate. Having some idea of how accurate that point estimate is would be more informative; in other words, providing a confidence-interval estimate for the mean price of all 3-year-old Orions would be better. We now explain how to obtain such confidence-interval estimates.

Confidence Intervals for Conditional Means in Regression

To develop a confidence-interval procedure for conditional means in regression, we must first identify the distribution of the predicted value of the response variable for a particular value of the predictor variable. Let's return to the Orion illustration and consider the particular value 3 of the predictor variable, that is, 3-year-old Orions.

As shown in Example 15.7, based on the sample data in Table 15.4, the predicted price for a 3-year-old Orion is 134.69 ($13,469). We now consider all possible samples of 11 Orions whose ages are the same as those given in the first column of Table 15.4.

For such samples, the predicted price of a 3-year-old Orion varies from one sample to another and is therefore a variable. Using the assumptions for regression inferences, we can show that its distribution is a normal distribution with a mean equalling the mean price of all 3-year-old Orions. More generally, we have Key Fact 15.5.

KEY FACT 15.5

Distribution of the Predicted Value of a Response Variable

Suppose that the variables x and y satisfy the four assumptions for regression inferences. Let x_p denote a particular value of the predictor variable and let $\hat{y}_p$ be the corresponding value predicted for the response variable by the sample regression equation; that is, $\hat{y}_p = b_0 + b_1 x_p$. Then, for samples of size n, each with the same values, $x_1, x_2, \ldots, x_n$, for the predictor variable, the following properties hold for $\hat{y}_p$.

- The mean of $\hat{y}_p$ equals the conditional mean of the response variable corresponding to the value x_p of the predictor variable: $\mu_{\hat{y}_p} = \beta_0 + \beta_1 x_p$.
- The standard deviation of $\hat{y}_p$ is

$$\sigma_{\hat{y}_p} = \sigma \sqrt{\frac{1}{n} + \frac{(x_p - \Sigma x/n)^2}{S_{xx}}}.$$

- The variable $\hat{y}_p$ is normally distributed.

In particular, for fixed values of the predictor variable, the possible predicted values of the response variable corresponding to x_p have a normal distribution with mean $\beta_0 + \beta_1 x_p$.

In light of Key Fact 15.5, if we standardize the variable $\hat{y}_p$, the resulting variable has the standard normal distribution. However, because the standardized variable contains the unknown parameter σ, it cannot be used as a basis for a confidence-interval formula. So we replace σ by its estimate s_e, the standard error of the estimate. The resulting variable has a t-distribution. More precisely, we have Key Fact 15.6.

KEY FACT 15.6

t-Distribution for Confidence Intervals for Conditional Means in Regression

Suppose that the variables x and y satisfy the four assumptions for regression inferences. Then, for samples of size n, each with the same values, $x_1, x_2, \ldots, x_n$, for the predictor variable, the variable

$$t = \frac{\hat{y}_p - (\beta_0 + \beta_1 x_p)}{s_e \sqrt{\frac{1}{n} + \frac{(x_p - \Sigma x/n)^2}{S_{xx}}}}$$

has the t-distribution with df $= n - 2$.

Recalling that $\beta_0 + \beta_1 x_p$ is the conditional mean of the response variable corresponding to the value x_p of the predictor variable, we can apply Key Fact 15.6 to derive a confidence-interval procedure for means in regression. We refer to that procedure as the **conditional mean *t*-interval procedure** and present it as Procedure 15.3.

PROCEDURE 15.3

Conditional Mean *t*-Interval Procedure

Purpose To find a confidence interval for the conditional mean of the response variable corresponding to a particular value of the predictor variable, x_p

Assumptions

The four assumptions for regression inferences

STEP 1 For a confidence level of $1 - \alpha$, use Table IV to find $t_{\alpha/2}$ with df $= n - 2$.

STEP 2 Compute the point estimate, $\hat{y}_p = b_0 + b_1 x_p$.

STEP 3 The endpoints of the confidence interval for the conditional mean of the response variable are

$$\hat{y}_p \pm t_{\alpha/2} \cdot s_e \sqrt{\frac{1}{n} + \frac{(x_p - \Sigma x/n)^2}{S_{xx}}}.$$

STEP 4 Interpret the confidence interval.

Example 15.8 The Conditional Mean t-Interval Procedure

Age and Price of Orions Use the sample data in Table 15.4 on page 767 to obtain a 95% confidence interval for the mean price of all 3-year-old Orions.

Solution We apply Procedure 15.3.

STEP 1 For a confidence level of $1 - \alpha$, use Table IV to find $t_{\alpha/2}$ with df $= n - 2$.

We want a 95% confidence interval, or $\alpha = 0.05$. Because $n = 11$, we have df $= 9$. From Table IV, $t_{\alpha/2} = t_{0.05/2} = t_{0.025} = 2.262$.

STEP 2 Compute the point estimate, $\hat{y}_p = b_0 + b_1 x_p$.

Here, $x_p = 3$ (3-year-old Orions). From Example 15.7, the point estimate for the mean price of all 3-year-old Orions is

$$\hat{y}_p = 195.47 - 20.26 \cdot 3 = 134.69.$$

STEP 3 **The endpoints of the confidence interval for the conditional mean of the response variable are**

$$\hat{y}_p \pm t_{\alpha/2} \cdot s_e \sqrt{\frac{1}{n} + \frac{(x_p - \Sigma x/n)^2}{S_{xx}}}.$$

In Example 14.4, we found that $\Sigma x = 58$ and $\Sigma x^2 = 326$; in Example 15.2, we determined that $s_e = 12.58$. Also, from Step 1, $t_{\alpha/2} = 2.262$ and, from Step 2, $\hat{y}_p = 134.69$. Consequently, the endpoints of the confidence interval for the conditional mean are

$$134.69 \pm 2.262 \cdot 12.58 \sqrt{\frac{1}{11} + \frac{(3 - 58/11)^2}{326 - (58)^2/11}},$$

or 134.69 ± 16.76, or 117.93 to 151.45.

STEP 4 **Interpret the confidence interval.**

INTERPRETATION We can be 95% confident that the mean price of all 3-year-old Orions is somewhere between \$11,793 and \$15,145. ∎

Prediction Intervals

A primary use of a sample regression equation is to make predictions. As we have seen, for the Orion data in Table 15.4 on page 767, the sample regression equation is $\hat{y} = 195.47 - 20.26x$. Substituting, for example, $x = 3$ into that equation, we obtain the predicted price for a 3-year-old Orion of 134.69, or \$13,469. However, because the prices of such cars vary, finding a **prediction interval** for the price of a 3-year-old Orion makes more sense than giving a single predicted value.

Prediction intervals are similar to confidence intervals. The term *confidence* is usually reserved for interval estimates of parameters, such as the mean price of all 3-year-old Orions. The term *prediction* is used for interval estimates of variables, such as the price of a 3-year-old Orion.

To develop a prediction-interval procedure, we must first identify the distribution of the difference between the observed and predicted values of the response variable for a particular value of the predictor variable. Let's again return to the Orion illustration and consider the particular value 3 of the predictor variable, that is, 3-year-old Orions.

Based on the sample data for 11 Orions displayed in Table 15.4, the predicted price, in hundreds of dollars, for a 3-year-old Orion is 134.69. Suppose that we observe the price of a 3-year-old Orion to be 144.12. Then the difference between the observed price and predicted price is $144.12 - 134.69$, or 9.43.

We now consider all possible samples of 11 Orions whose ages are the same as those given in the first column of Table 15.4. For such samples, the predicted price of a 3-year-old Orion varies from one sample to another and is therefore a variable. The observed price of a 3-year-old Orion is also a variable. Consequently, the difference between the observed price and predicted price is also a variable. Using the assumptions for regression inferences, we can show that its distribution is a normal distribution with a mean of 0. More generally, we have Key Fact 15.7.

KEY FACT 15.7

Distribution of the Difference Between the Observed and Predicted Values of the Response Variable

Suppose that the variables x and y satisfy the four assumptions for regression inferences. Let x_p denote a particular value of the predictor variable and let $\hat{y}_p$ be the corresponding value predicted for the response variable by the sample regression equation. Furthermore, let y_p be an independently observed value of the response variable corresponding to the value x_p of the predictor variable. Then, for samples of size n, each with the same values, $x_1, x_2, \ldots, x_n$, for the predictor variable, the following properties hold for $y_p - \hat{y}_p$, the difference between the observed and predicted values.

- The mean of $y_p - \hat{y}_p$ equals zero: $\mu_{y_p - \hat{y}_p} = 0$.
- The standard deviation of $y_p - \hat{y}_p$ is

$$\sigma_{y_p - \hat{y}_p} = \sigma \sqrt{1 + \frac{1}{n} + \frac{(x_p - \Sigma x/n)^2}{S_{xx}}}.$$

- The variable $y_p - \hat{y}_p$ is normally distributed.

In particular, for fixed values of the predictor variable, the possible differences between the observed and predicted values of the response variable corresponding to x_p have a normal distribution with a mean of 0.

In light of Key Fact 15.7, if we standardize the variable $y_p - \hat{y}_p$, the resulting variable has the standard normal distribution. However, because the standardized variable contains the unknown parameter σ, it cannot be used as a basis for a prediction-interval formula. So we replace σ by its estimate s_e, the standard error of the estimate. The resulting variable has a t-distribution, as presented in Key Fact 15.8.

KEY FACT 15.8

t-Distribution for Prediction Intervals in Regression

Suppose that the variables x and y satisfy the four assumptions for regression inferences. Then, for samples of size n, each with the same values, $x_1, x_2, \ldots, x_n$, for the predictor variable, the variable

$$t = \frac{y_p - \hat{y}_p}{s_e \sqrt{1 + \frac{1}{n} + \frac{(x_p - \Sigma x/n)^2}{S_{xx}}}}$$

has the t-distribution with df $= n - 2$.

Using Key Fact 15.8, we can derive a prediction-interval procedure. We refer to that procedure as the **predicted value t-interval procedure** and present it as Procedure 15.4.

> ## PROCEDURE 15.4
>
> ### Predicted Value t-Interval Procedure
>
> *Purpose* To find a prediction interval for the observed value of the response variable corresponding to a particular value of the predictor variable, x_p
>
> *Assumptions*
>
> The four assumptions for regression inferences
>
> **STEP 1** For a prediction level of $1 - \alpha$, use Table IV to find $t_{\alpha/2}$ with df $= n - 2$.
>
> **STEP 2** Compute the predicted value, $\hat{y}_p = b_0 + b_1 x_p$.
>
> **STEP 3** The endpoints of the prediction interval for the observed value of the response variable are
>
> $$\hat{y}_p \pm t_{\alpha/2} \cdot s_e \sqrt{1 + \frac{1}{n} + \frac{(x_p - \Sigma x/n)^2}{S_{xx}}}.$$
>
> **STEP 4** Interpret the prediction interval.

Example 15.9   The Predicted Value t-Interval Procedure

Age and Price of Orions   Using the sample data in Table 15.4 on page 767, obtain a 95% prediction interval for the price of a 3-year-old Orion.

Solution   We apply Procedure 15.4.

STEP 1 For a prediction level of $1 - \alpha$, use Table IV to find $t_{\alpha/2}$ with df $= n - 2$.

We want a 95% prediction interval, or $\alpha = 0.05$. Also, because $n = 11$, we have df $= 9$. From Table IV, $t_{\alpha/2} = t_{0.05/2} = t_{0.025} = 2.262$.

STEP 2 Compute the predicted value, $\hat{y}_p = b_0 + b_1 x_p$.

As previously shown, the sample regression equation for the data in Table 15.4 is $\hat{y} = 195.47 - 20.26x$. Therefore, the predicted price for a 3-year-old Orion is

$$\hat{y}_p = 195.47 - 20.26 \cdot 3 = 134.69.$$

STEP 3 The endpoints of the prediction interval for the observed value of the response variable are

$$\hat{y}_p \pm t_{\alpha/2} \cdot s_e \sqrt{1 + \frac{1}{n} + \frac{(x_p - \Sigma x/n)^2}{S_{xx}}}.$$

From Example 14.4, $\Sigma x = 58$ and $\Sigma x^2 = 326$; from Example 15.2, we know that $s_e = 12.58$. Also, $n = 11$, $t_{\alpha/2} = 2.262$, $x_p = 3$, and $\hat{y}_p = 134.69$. Consequently, the endpoints of the prediction interval are

$$134.69 \pm 2.262 \cdot 12.58 \sqrt{1 + \frac{1}{11} + \frac{(3 - 58/11)^2}{326 - (58)^2/11}},$$

or 134.69 ± 33.02, or 101.67 to 167.71.

STEP 4 Interpret the prediction interval.

INTERPRETATION We can be 95% certain that the observed price of a 3-year-old Orion will be somewhere between \$10,167 and \$16,771.

We just demonstrated that a 95% prediction interval for the observed price of a 3-year-old Orion is from \$10,167 to \$16,771. In Example 15.8, we found that a 95% confidence interval for the mean price of all 3-year-old Orions is from \$11,793 to \$15,145. We show both intervals in Fig. 15.11.

FIGURE 15.11
Prediction and confidence intervals for 3-year-old Orions

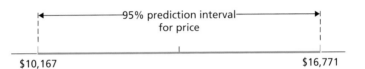

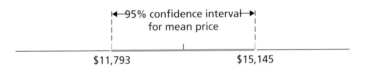

What
Does it Mean?

More error is involved in predicting the price of a single 3-year-old Orion than in estimating the mean price of all 3-year-old Orions.

It is important to note that the prediction interval is wider than the confidence interval, an outcome to be expected, for the following reason: The error in the estimate of the mean price of all 3-year-old Orions is due only to the fact that the population regression line is being estimated by a sample regression line, whereas the error in the prediction of the observed price of a 3-year-old Orion is due to the error in estimating the mean price plus the variation in prices of 3-year-old Orions.

The
Technology
Center

Some statistical technologies (e.g., Minitab) provide dedicated programs to automatically obtain confidence intervals for means and prediction intervals in regression, but others (e.g., Excel, TI-83/84 Plus) do not. For a statistical technology that does not have such programs, you can often use the macro capabilities of the statistical technology to write programs. Refer to the technology manuals for more details.

EXERCISES 15.3

Statistical Concepts and Skills

15.43 Without doing any calculations, fill in the blank and explain your answer. Based on the sample data in Table 15.4, the predicted price for a 4-year-old Orion is $11,443. A point estimate for the mean price of all 4-year-old Orions, based on the same sample data, is _____.

In Exercises **15.44–15.47**, *we repeat information from Exercises 15.10–15.13. Presuming that the assumptions for regression inferences are met, determine the required confidence and prediction intervals. (Recall that you previously obtained the sample regression equations in Exercises 14.42–14.45 and the standard errors of the estimate in Exercises 15.14–15.17.)*

15.44 Tax Efficiency. Following are the data on percentage of investments in energy securities and tax efficiency from Exercise 15.10.

x	3.1	3.2	3.7	4.3	4.0	5.5	6.7	7.4	7.4	10.6
y	98.1	94.7	92.0	89.8	87.5	85.0	82.0	77.8	72.1	53.5

a. Obtain a point estimate for the mean tax efficiency of all mutual fund portfolios with 6% of their investments in energy securities.
b. Determine a 95% confidence interval for the mean tax efficiency of all mutual fund portfolios with 6% of their investments in energy securities.
c. Find the predicted tax efficiency of a mutual fund portfolio with 6% of its investments in energy securities.
d. Determine a 95% prediction interval for the tax efficiency of a mutual fund portfolio with 6% of its investments in energy securities.
e. Draw graphs similar to those in Fig. 15.11 on page 773, showing both the 95% confidence interval from part (b) and the 95% prediction interval from part (d).
f. Why is the prediction interval wider than the confidence interval?

15.45 Corvette Prices. Following are the age and price data for Corvettes from Exercise 15.11.

x	6	6	6	2	2	5	4	5	1	4
y	270	260	275	405	364	295	335	308	405	305

a. Obtain a point estimate for the mean price of all 4-year-old Corvettes.

b. Determine a 90% confidence interval for the mean price of all 4-year-old Corvettes.
c. Find the predicted price of a 4-year-old Corvette.
d. Determine a 90% prediction interval for the price of a 4-year-old Corvette.
e. Draw graphs similar to those in Fig. 15.11 on page 773, showing both the 90% confidence interval from part (b) and the 90% prediction interval from part (d).
f. Why is the prediction interval wider than the confidence interval?

15.46 Custom Homes. Following are the size and price data for custom homes from Exercise 15.12.

x	26	27	33	29	29	34	30	40	22
y	290	305	325	327	356	411	488	554	246

a. Determine a point estimate for the mean price of all 2800-sq-ft Equestrian Estate homes.
b. Find a 99% confidence interval for the mean price of all 2800-sq-ft Equestrian Estate homes.
c. Find the predicted price of a 2800-sq-ft Equestrian Estate home.
d. Determine a 99% prediction interval for the price of a 2800-sq-ft Equestrian Estate home.

15.47 Plant Emissions. Following are the data on plant weight and quantity of volatile emissions from Exercise 15.13.

x	57	85	57	65	52	67	62	80	77	53	68
y	8.0	22.0	10.5	22.5	12.0	11.5	7.5	13.0	16.5	21.0	12.0

a. Obtain a point estimate for the mean quantity of volatile emissions of all (*Solanum tuberosom*) plants that weigh 60 g.
b. Find a 95% confidence interval for the mean quantity of volatile emissions of all plants that weigh 60 g.
c. Find the predicted quantity of volatile emissions for a plant that weighs 60 g.
d. Determine a 95% prediction interval for the quantity of volatile emissions for a plant that weighs 60 g.

Extending the Concepts and Skills

Margin of Error in Regression. In Exercises 15.48 and 15.49, you are asked to examine the magnitude of the margin of error of confidence intervals and prediction intervals in regression as a function of how far the specified value of the predictor variable is from the mean of the observed values of the predictor variable.

15.48 Age and Price of Orions. Refer to the data on age and price of a sample of 11 Orions given in Table 15.4 on page 767.
a. For each age between 2 and 7 years, obtain a 95% confidence interval for the mean price of all Orions of that age. Plot the confidence intervals against age and discuss your results.
b. Determine the margin of error for each confidence interval that you obtained in part (a). Plot the margins of error against age and discuss your results.
c. Repeat parts (a) and (b) for prediction intervals.

15.49 Refer to the confidence interval and prediction interval formulas in Procedures 15.3 and 15.4, respectively.
a. Explain why, for a fixed confidence level, the margin of error for the estimate of the conditional mean of the response variable increases as the value of the predictor variable moves farther from the mean of the observed values of the predictor variable.
b. Explain why, for a fixed prediction level, the margin of error for the estimate of the predicted value of the response variable increases as the value of the predictor variable moves farther from the mean of the observed values of the predictor variable.

Using Technology

Use the technology of your choice in Exercises 15.50–15.54.

15.50 Batting and Scoring. Following are the data on season team batting average and total runs scored for a sample of major league baseball teams from Exercise 15.20.

Average	Runs	Average	Runs
.294	968	.267	793
.278	938	.265	792
.278	925	.256	764
.270	887	.254	752
.274	825	.246	740
.271	810	.266	738
.263	807	.262	731
.257	798	.251	708

a. Obtain a point estimate for the mean total runs scored by all major league teams with a season team batting average of .270.

b. Find a 95% confidence interval for the mean total runs scored by all major league teams with a season team batting average of .270.
c. Find the predicted total runs scored by a major league team with a season team batting average of .270.
d. Determine a 95% prediction interval for the total runs scored by a major league team with a season team batting average of .270.
e. Remove the influential observation and repeat the analyses in parts (a)–(d). Compare your results with and without the removal of the influential observation.

15.51 Body Fat. Following are the age and body fat data for 18 randomly selected adults from Exercise 15.21.

Age	% Fat	Age	% Fat	Age	% Fat
23	9.5	45	27.4	56	32.5
23	27.9	49	25.2	57	30.3
27	7.8	50	31.1	58	33.0
27	17.8	53	34.7	58	33.8
39	31.4	53	42.0	60	41.1
41	25.9	54	29.1	61	34.5

a. Obtain a point estimate for the mean percentage of body fat of all 30-year-old adults.
b. Find a 95% confidence interval for the mean percentage of body fat of all 30-year-old adults.
c. Find the predicted percentage of body fat of a 30-year-old adult.
d. Determine a 95% prediction interval for the percentage of body fat of a 30-year-old adult.

15.52 PCBs and Pelicans. Use the data points given on the WeissStats CD for shell thickness and concentration of PCBs for 60 Anacapa pelican eggs referred to in Exercise 15.22.
a. Obtain a point estimate for the mean shell thickness of all Anacapa pelican eggs with a PCB concentration of 200 ppm.
b. Find a 95% confidence interval for the mean shell thickness of all Anacapa pelican eggs with a PCB concentration of 200 ppm.
c. Find the predicted shell thickness of an Anacapa pelican egg with a PCB concentration of 200 ppm.
d. Determine a 95% prediction interval for the shell thickness of an Anacapa pelican egg with a PCB concentration of 200 ppm.

15.53 Gas Guzzlers. Use the data on the WeissStats CD for gas mileage and engine displacement for 121 vehicles referred to in Exercise 15.23.

a. Find a 95% confidence interval for the mean gas mileage of all vehicles with an engine displacement of 3.0 L.

b. Determine a 95% prediction interval for the gas mileage of a vehicle with an engine displacement of 3.0 L.

c. Compare and discuss the differences between the confidence interval that you obtained in part (a) and the prediction interval that you obtained in part (b).

15.54 Estriol Level and Birth Weight. Use the data on the WeissStats CD for estriol levels of pregnant women and birth weights of their children referred to in Exercise 15.24.

a. Find a 95% confidence interval for the mean birth weight of all children whose mothers had an estriol level of 18 mg/24 hr.

b. Determine a 95% prediction interval for the birth weight of a child whose mother had an estriol level of 18 mg/24 hr.

c. Compare and discuss the differences between the confidence interval that you obtained in part (a) and the prediction interval that you obtained in part (b).

15.4 Inferences in Correlation

Frequently, we want to decide whether two variables are linearly correlated, that is, whether there is a linear relationship between the two variables. In the context of regression, we can make that decision by performing a hypothesis test for the slope of the population regression line, as discussed in Section 15.2.

Alternatively, we can perform a hypothesis test for the **population linear correlation coefficient, ρ** (rho). This parameter measures the linear correlation of all possible pairs of observations of two variables in the same way that a sample linear correlation coefficient, r, measures the linear correlation of a sample of pairs. Thus, ρ actually describes the strength of the linear relationship between two variables; r is only an estimate of ρ obtained from sample data.

The population linear correlation coefficient of two variables, x and y, always lies between -1 and 1. Values of ρ near -1 or 1 indicate a strong linear relationship between the variables, whereas values of ρ near 0 indicate a weak linear relationship between the variables.

If $\rho > 0$, the variables are **positively linearly correlated variables,** meaning that y tends to increase linearly as x increases (and vice versa), with the tendency being greater the closer ρ is to 1. If $\rho < 0$, the variables are **negatively linearly correlated variables,** meaning that y tends to decrease linearly as x increases (and vice versa), with the tendency being greater the closer ρ is to -1. If $\rho = 0$, the variables are **linearly uncorrelated variables,** meaning that there is no linear relationship between the variables. If $\rho \neq 0$—that is, the variables are not linearly uncorrelated or, equivalently, are either positively linearly correlated or negatively linearly correlated—the variables are **linearly correlated variables.**

As we mentioned, a sample linear correlation coefficient, r, is an estimate of the population linear correlation coefficient, ρ. Consequently, as you might expect, we can use r as a basis for performing a hypothesis test for ρ. For a hypothesis test with the null hypothesis H_0: $\rho = 0$ (i.e., the variables are linearly uncorrelated), we employ Key Fact 15.9.

KEY FACT 15.9 | **t-Distribution for a Correlation Test**

Suppose that the variables x and y satisfy the four assumptions for regression inferences. Then, for samples of size n, the variable

$$t = \frac{r}{\sqrt{\dfrac{1 - r^2}{n - 2}}}$$

has the t-distribution with df $= n - 2$ if the null hypothesis $\rho = 0$ is true.

In light of Key Fact 15.9, for a hypothesis test with the null hypothesis $H_0: \rho = 0$, we can use the variable

$$t = \frac{r}{\sqrt{\dfrac{1 - r^2}{n - 2}}}$$

as the test statistic and obtain the critical values or P-value from the t-table, Table IV. We refer to this hypothesis-testing procedure as the **correlation t-test.** Procedure 15.5 on the following page provides a step-by-step method for performing a correlation t-test by using either the critical-value approach or the P-value approach.

Example 15.10 The Correlation t-Test

Age and Price of Orions Refer to the age and price data for a sample of 11 Orions given in Table 15.4 on page 767. At the 5% significance level, do the data provide sufficient evidence to conclude that age and price of Orions are negatively linearly correlated?

Solution As we discovered in Example 15.3 on page 751, considering that the assumptions for regression inferences are met by the variables age and price for Orions is not unreasonable, at least for Orions between 2 and 7 years old. Consequently, we apply Procedure 15.5 to carry out the required hypothesis test.

STEP 1 State the null and alternative hypotheses.

Let ρ denote the population linear correlation coefficient for the variables age and price of Orions. Then the null and alternative hypotheses are

$H_0: \rho = 0$ (age and price are linearly uncorrelated)

$H_a: \rho < 0$ (age and price are negatively linearly correlated).

Note that the hypothesis test is left tailed because a less-than sign ($<$) appears in the alternative hypothesis.

STEP 2 Decide on the significance level, α.

We are to use $\alpha = 0.05$.

PROCEDURE 15.5   Correlation t-Test

Purpose To perform a hypothesis test for a population linear correlation coefficient, ρ

Assumptions

The four assumptions for regression inferences

STEP 1 The null hypothesis is H_0: $\rho = 0$, and the alternative hypothesis is

$$H_a: \rho \neq 0 \quad \text{or} \quad H_a: \rho < 0 \quad \text{or} \quad H_a: \rho > 0$$
$$\text{(Two tailed)} \qquad \text{(Left tailed)} \qquad \text{(Right tailed)}$$

STEP 2 Decide on the significance level, α.

STEP 3 Compute the value of the test statistic

$$t = \frac{r}{\sqrt{\dfrac{1 - r^2}{n - 2}}}$$

and denote that value t_0.

CRITICAL-VALUE APPROACH	P-VALUE APPROACH
STEP 4 The critical value(s) are $\pm t_{\alpha/2}$ or $-t_{\alpha}$ or t_{α} (Two tailed) (Left tailed) (Right tailed) with df $= n - 2$. Use Table IV to find the critical value(s).	**STEP 4** The t-statistic has df $= n - 2$. Use Table IV to estimate the P-value, or obtain it exactly by using technology.

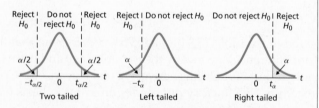

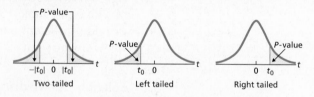

STEP 5 If the value of the test statistic falls in the rejection region, reject H_0; otherwise, do not reject H_0.

STEP 5 If $P \leq \alpha$, reject H_0; otherwise, do not reject H_0.

STEP 6 Interpret the results of the hypothesis test.

STEP 3 Compute the value of the test statistic

$$t = \frac{r}{\sqrt{\dfrac{1 - r^2}{n - 2}}}.$$

In Example 14.10 on page 727, we computed the sample linear correlation coefficient for the age and price data displayed in Table 15.4. We found that $r = -0.924$, so the value of the test statistic is

$$t = \frac{-0.924}{\sqrt{\dfrac{1 - (-0.924)^2}{11 - 2}}} = -7.249.$$

CRITICAL-VALUE APPROACH	P-VALUE APPROACH

CRITICAL-VALUE APPROACH

STEP 4 The critical value for a left-tailed test is $-t_\alpha$, with df $= n - 2$. Use Table IV to find the critical value.

For $n = 11$, df $= 9$. Also, $\alpha = 0.05$. From Table IV, for df $= 9$, $t_{0.05} = 1.833$. Consequently, the critical value is $-t_{0.05} = -1.833$, as shown in Fig. 15.12A.

FIGURE 15.12A

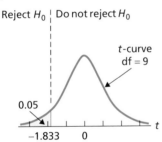

STEP 5 If the value of the test statistic falls in the rejection region, reject H_0; otherwise, do not reject H_0.

The value of the test statistic, found in Step 3, is $t = -7.249$. Figure 15.12A shows that this value falls in the rejection region, so we reject H_0. The test results are statistically significant at the 5% level.

P-VALUE APPROACH

STEP 4 The t-statistic has df $= n - 2$. Use Table IV to estimate the P-value or obtain it exactly by using technology.

From Step 3, the value of the test statistic is $t = -7.249$. Because the test is left tailed, the P-value is the probability of observing a value of t of -7.249 or less if the null hypothesis is true. That probability equals the shaded area shown in Fig. 15.12B.

FIGURE 15.12B

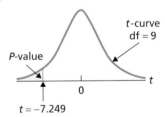

For $n = 11$, df $= 9$. Referring to Fig. 15.12B and Table IV, we find that $P < 0.005$. (Using technology, we obtain $P = 0.0000244$.)

STEP 5 If $P \leq \alpha$, reject H_0; otherwise, do not reject H_0.

From Step 4, $P < 0.005$. Because the P-value is less than the specified significance level of 0.05, we reject H_0. The test results are statistically significant at the 5% level and (see Table 9.12 on page 440) provide very strong evidence against the null hypothesis.

STEP 6 Interpret the results of the hypothesis test.

INTERPRETATION At the 5% significance level, the data provide sufficient evidence to conclude that age and price of Orions are negatively linearly correlated. Prices for Orions tend to decrease linearly with increasing age, at least for Orions between 2 and 7 years old.

The Technology Center

Some statistical technologies have programs that automatically conduct a correlation test. For instance, the TI-83/84 Plus program **LinRegTTest**, which we mentioned at the end of Section 15.2, can be used to perform such a test.

Minitab's **Correlation** program, examined at the end of Section 14.4, provides r and the P-value for a two-tailed correlation test but currently doesn't give the P-value for one-tailed tests. Excel's **Correlation** program, also examined at the end of Section 14.4, provides r but not a P-value.

The time consuming part of conducting a correlation test is computing the sample linear correlation coefficient, r. Most statistical technologies (including Minitab, Excel, and the TI-83/84 Plus) have programs for automatically obtaining r. See the technology manuals for details of performing a correlation test.

EXERCISES 15.4

Statistical Concepts and Skills

15.55 Identify the statistic used to estimate the population linear correlation coefficient.

15.56 Suppose that, for a sample of pairs of observations from two variables, the linear correlation coefficient, r, is positive. Does this result necessarily imply that the variables are positively linearly correlated? Explain.

15.57 Fill in the blanks.
a. If $\rho = 0$, the two variables under consideration are linearly _____.
b. If two variables are positively linearly correlated, one of the variables tends to increase as the other _____.
c. If two variables are _____ linearly correlated, one of the variables tends to decrease as the other increases.

*In Exercises **15.58–15.61**, we repeat information from Exercises 15.10–15.13. Presuming that the assumptions for regression inferences are met, perform the required correlation tests, using either the critical-value approach or the P-value approach. (Recall that you previously obtained the sample linear correlation coefficients in Exercises 14.86–14.89.)*

15.58 Tax Efficiency. Following are the data on percentage of investments in energy securities and tax efficiency from Exercise 15.10.

x	3.1	3.2	3.7	4.3	4.0	5.5	6.7	7.4	7.4	10.6
y	98.1	94.7	92.0	89.8	87.5	85.0	82.0	77.8	72.1	53.5

At the 2.5% significance level, do the data provide sufficient evidence to conclude that percentage of investments in energy securities and tax efficiency are negatively linearly correlated for mutual fund portfolios?

15.59 Corvette Prices. Following are the age and price data for Corvettes from Exercise 15.11.

x	6	6	6	2	2	5	4	5	1	4
y	270	260	275	405	364	295	335	308	405	305

At the 5% level of significance, do the data provide sufficient evidence to conclude that age and price of Corvettes are negatively linearly correlated?

15.60 Custom Homes. Following are the size and price data for custom homes from Exercise 15.12.

x	26	27	33	29	29	34	30	40	22
y	290	305	325	327	356	411	488	554	246

At the 0.5% significance level, do the data provide sufficient evidence to conclude that, for custom homes in the Equestrian Estates, size and price are positively linearly correlated?

15.61 Plant Emissions. Following are the data on plant weight and quantity of volatile emissions from Exercise 15.13.

x	57	85	57	65	52	67	62	80	77	53	68
y	8.0	22.0	10.5	22.5	12.0	11.5	7.5	13.0	16.5	21.0	12.0

Do the data suggest that, for the potato plant *Solanum tubersom*, weight and quantity of volatile emissions are linearly correlated? Use $\alpha = 0.05$.

15.62 Height and Score. A random sample of 10 students was taken from an introductory statistics class. The following data were obtained, where x denotes height, in inches, and y denotes score on the final exam.

x	71	68	71	65	66	68	68	64	62	65
y	87	96	66	71	71	55	83	67	86	60

At the 5% significance level, do the data provide sufficient evidence to conclude that, for students in introductory statistics courses, height and final exam score are linearly correlated?

15.63 Is ρ a parameter or a statistic? What about r? Explain your answers.

Using Technology

Use the technology of your choice in Exercises 15.64–15.68.

15.64 Batting and Scoring. Refer to the data on season team batting average and total runs scored for a sample of 16 major league baseball teams given in Exercise 15.20 on page 757.
a. At the 2.5% significance level, do the data provide sufficient evidence to conclude that season team batting average and total runs scored are positively linearly correlated for major league baseball teams?
b. Remove the influential observation and repeat part (a).
c. Compare your results with and without the removal of the influential observation and state your conclusions.

15.65 Body Fat. Refer to the data on age and body fat for 18 randomly selected adults given in Exercise 15.21 on page 757.
a. Do the data provide sufficient evidence to conclude that, for adults, age and percentage of body fat are positively linearly correlated? Use $\alpha = 0.025$.
b. Remove the potential outlier and repeat part (a).
c. Compare your results with and without the removal of the potential outlier and state your conclusions.

15.66 PCBs and Pelicans. Use the data points given on the WeissStats CD for shell thickness and concentration of PCBs for 60 Anacapa pelican eggs referred to in Exercise 15.22. Do the data provide sufficient evidence to conclude, at the 5% significance level, that concentration of PCBs and shell thickness are linearly correlated for Anacapa pelican eggs?

15.67 Gas Guzzlers. Use the data on the WeissStats CD for gas mileage and engine displacement for 121 vehicles referred to in Exercise 15.23. Do the data provide sufficient evidence to conclude that engine displacement and gas mileage are negatively linearly correlated? Use $\alpha = 0.025$.

15.68 Estriol Level and Birth Weight. Use the data on the WeissStats CD for estriol levels of pregnant women and birth weights of their children referred to in Exercise 15.24.
a. At the 0.05 level of significance, are estriol level and birth weight linearly correlated?
b. At the 0.025 level of significance, are estriol level and birth weight positively linearly correlated?

15.5 Testing for Normality

Several descriptive methods are available for assessing normality of a variable based on sample data. As you learned in Section 6.4, one of the most commonly used methods is the normal probability plot, that is, a plot of the normal scores against the sample data.

If the variable is normally distributed, a normal probability plot of the sample data should be roughly linear. We can thus assess normality as follows.

- If the plot is roughly linear, accept as reasonable that the variable is approximately normally distributed.

- If the plot shows systematic deviations from linearity (e.g., if it displays significant curvature), conclude that the variable probably is not approximately normally distributed.

This visual assessment of normality is subjective because what constitutes "roughly linear" is a matter of opinion. To overcome this difficulty, we can perform a hypothesis test for normality based on the linear correlation coefficient: If the variable under consideration is normally distributed, the correlation between the sample data and their normal scores should be near 1 because the normal probability plot should be roughly linear.[†]

So, to perform a hypothesis test for normality, we compute the linear correlation coefficient between the sample data and their normal scores. If the correlation is too much smaller than 1, we reject the null hypothesis that the variable is normally distributed in favor of the alternative hypothesis that the variable is not normally distributed. Of course, we need a table of critical values, which is provided by Table IX in Appendix A, to decide what is "too much smaller than 1."

We use the letter w to denote the normal score corresponding to an observed value of the variable x. And, for this special context, we use R_p instead of r to denote the linear correlation coefficient. Hence, in view of the computing formula for the linear correlation coefficient given in Definition 14.6 on page 724, the correlation between the sample data and their normal scores is

$$R_p = \frac{S_{xw}}{\sqrt{S_{xx}S_{ww}}},$$

where $S_{xw} = \Sigma xw - (\Sigma x)(\Sigma w)/n$, $S_{xx} = \Sigma x^2 - (\Sigma x)^2/n$, and $S_{ww} = \Sigma w^2 - (\Sigma w)^2/n$ (see Definition 14.3 on page 702).

However, as the sum of the normal scores for a data set always equals 0, we can simplify the preceding displayed equation to

$$R_p = \frac{\Sigma xw}{\sqrt{S_{xx}\Sigma w^2}}$$

and use this variable as our test statistic for a **correlation test for normality.**

Because the null hypothesis of normality is rejected only when the test statistic is too small, the rejection region is always on the left; that is, the hypothesis test is always left tailed. Consequently, we have Procedure 15.6 (next page).

In Example 6.14 on page 301, we considered data on adjusted gross incomes for a sample of 12 federal individual income tax returns. In Table 6.4 on page 301, we obtained the normal scores for the data and, in Fig. 6.23 on page 301, we drew a normal probability plot. Because the normal probability plot shows significant curvature, we concluded that adjusted gross incomes are probably not normally distributed.

That conclusion was a subjective one, based on a graph. We now apply Procedure 15.6 to reach an objective conclusion. Example 15.11, which begins at the top of page 784, provides the details.

[†]Because large normal scores are associated with large observations and vice versa, the correlation between the sample data and their normal scores cannot be negative.

| **PROCEDURE 15.6** | Correlation Test for Normality |

Purpose To perform a hypothesis test to decide whether a variable is not normally distributed

Assumption
Simple random sample

STEP 1 **The null and alternative hypotheses are**

H_0: The variable is normally distributed.
H_a: The variable is not normally distributed.

STEP 2 **Decide on the significance level, α.**

STEP 3 **Compute the value of the test statistic**

$$R_p = \frac{\Sigma xw}{\sqrt{S_{xx}\Sigma w^2}},$$

where x and w denote observations of the variable and the corresponding normal scores, respectively. Denote the value of the test statistic R_p^0.

CRITICAL-VALUE APPROACH	**P-VALUE APPROACH**

STEP 4 **The critical value is R_p^*. Use Table IX to find the critical value.**

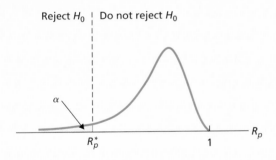

STEP 5 **If the value of the test statistic falls in the rejection region, reject H_0; otherwise, do not reject H_0.**

STEP 4 **Use Table IX to estimate the P-value, or obtain it exactly by using technology.**

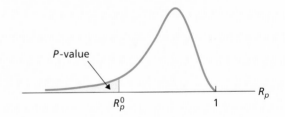

STEP 5 **If $P \leq \alpha$, reject H_0; otherwise, do not reject H_0.**

STEP 6 **Interpret the results of the hypothesis test.**

Note: In addition to the correlation test, there are several other tests for normality. However, the correlation test for normality is one of the most powerful.

Example 15.11   The Correlation Test for Normality

Adjusted Gross Incomes   The Internal Revenue Service publishes data on federal individual income tax returns in *Statistics of Income, Individual Income Tax Returns*. A random sample of 12 returns from last year revealed the adjusted gross incomes, in thousands of dollars, shown in Table 15.5. At the 5% significance level, do the data provide sufficient evidence to conclude that adjusted gross incomes are not normally distributed?

TABLE 15.5
Adjusted gross incomes ($1000s)

9.7	93.1	33.0	21.2
81.4	51.1	43.5	10.6
12.8	7.8	18.1	12.7

Solution   We apply Procedure 15.6.

STEP 1   State the null and alternative hypotheses.

The null and alternative hypotheses are

H_0: Adjusted gross incomes are normally distributed
H_a: Adjusted gross incomes are not normally distributed.

STEP 2   Decide on the significance level, α.

We are to perform the hypothesis test at the 5% significance level, or $\alpha = 0.05$.

STEP 3   Compute the value of the test statistic

$$R_p = \frac{\Sigma xw}{\sqrt{S_{xx} \Sigma w^2}}.$$

To compute the test statistic, we need a table of values for x, w, xw, x^2, and w^2, as displayed in Table 15.6. The normal scores (values of w) are from Table III in Appendix A.

TABLE 15.6
Table for computing R_p

Adjusted gross income x	Normal score w	xw	x^2	w^2
7.8	−1.64	−12.792	60.84	2.6896
9.7	−1.11	−10.767	94.09	1.2321
10.6	−0.79	−8.374	112.36	0.6241
12.7	−0.53	−6.731	161.29	0.2809
12.8	−0.31	−3.968	163.84	0.0961
18.1	−0.10	−1.810	327.61	0.0100
21.2	0.10	2.120	449.44	0.0100
33.0	0.31	10.230	1,089.00	0.0961
43.5	0.53	23.055	1,892.25	0.2809
51.1	0.79	40.369	2,611.21	0.6241
81.4	1.11	90.354	6,625.96	1.2321
93.1	1.64	152.684	8,667.61	2.6896
395.0	0.00	274.370	22,255.50	9.8656

Substituting the sums from Table 15.6 into the equation for R_p yields

$$R_p = \frac{\Sigma xw}{\sqrt{S_{xx}\Sigma w^2}} = \frac{\Sigma xw}{\sqrt{[\Sigma x^2 - (\Sigma x)^2/n][\Sigma w^2]}}$$

$$= \frac{274.370}{\sqrt{[22{,}255.50 - (395.0)^2/12] \cdot 9.8656}} = 0.908.$$

CRITICAL-VALUE APPROACH	*P*-VALUE APPROACH

CRITICAL-VALUE APPROACH

STEP 4 The critical value is R_p^*. Use Table IX to find the critical value.

We have $\alpha = 0.05$ and $n = 12$. From Table IX, the critical value is $R_p^* = 0.927$, as shown in Fig. 15.13A.

FIGURE 15.13A

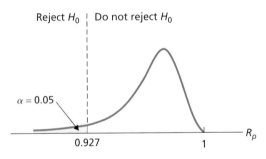

STEP 5 If the value of the test statistic falls in the rejection region, reject H_0; otherwise, do not reject H_0.

From Step 3, the value of the test statistic is $R_p = 0.908$ which, as Fig. 15.13A shows, falls in the rejection region. So we reject H_0. The test results are statistically significant at the 5% level.

***P*-VALUE APPROACH**

STEP 4 Use Table IX to estimate the *P*-value, or obtain it exactly by using technology.

From Step 3, the value of the test statistic is $R_p = 0.908$. Because the test is left tailed, the *P*-value is the probability of observing a value of R_p of 0.908 or less if the null hypothesis is true. That probability equals the shaded area shown in Fig. 15.13B.

FIGURE 15.13B

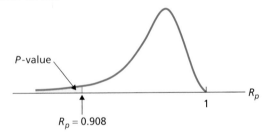

Referring to Fig. 15.13B and to Table IX with $n = 12$, we find that $0.01 < P < 0.05$. (Using technology, we obtain $P = 0.0266$.)

STEP 5 If $P \leq \alpha$, reject H_0; otherwise, do not reject H_0.

From Step 4, $0.01 < P < 0.05$. Because the *P*-value is less than the specified significance level of 0.05, we reject H_0. The test results are statistically significant at the 5% level and (see Table 9.12 on page 440) provide strong evidence against the null hypothesis of normality.

STEP 6 Interpret the results of the hypothesis test.

INTERPRETATION At the 5% significance level, the data provide sufficient evidence to conclude that adjusted gross incomes are not normally distributed.

The Technology Center

In Example 15.11, we presented the details of applying Procedure 15.6 so that you could see exactly how a correlation test for normality works. But, generally, correlation tests for normality are carried out by computer.

Some statistical technologies have a dedicated program for conducting a correlation test for normality, but others don't. Currently, Minitab has such a program, but neither Excel nor the TI-83/84 Plus do. For a statistical technology that doesn't have a dedicated program for conducting a correlation test for normality, you can often use the macro capabilities of the statistical technology to write a program. Refer to the technology manuals for more details.

Here, we briefly explain one method for using Minitab to perform a correlation test for normality. Choose **Stat ➤ Basic Statistics ➤ Normality Test...**, specify the variable in the **Variable** text box, select the **Ryan-Joiner** option button from the **Tests for Normality** field, and then click **OK**. The resulting output provides, among other things, the value of the test statistic R_p (labeled RJ in the output) and the P-value.

EXERCISES 15.5

Statistical Concepts and Skills

15.69 Regarding normal probability plots,
a. what are they?
b. what is an important use for them?
c. how is one used to assess the normality of a variable?
d. why is the method described in part (c) subjective?

15.70 In a correlation test for normality, what correlation is computed?

15.71 If you examine Procedure 15.6, you will note that a correlation test for normality is always left tailed. Explain why this is so.

15.72 Suppose that you perform a correlation test for normality at the 1% significance level. Further suppose that you reject the null hypothesis that the variable under consideration is normally distributed. Can you be confident in stating that the variable is not normally distributed? Explain your answer.

In Exercises 15.73–15.76, perform a correlation test for normality, using either the critical-value approach or the P-value approach. (Recall that you obtained the normal scores for these data sets in Exercises 6.65–6.68.)

15.73 Exam Scores. A sample of the final exam scores in a large introductory statistics course is as follows.

88	67	64	76	86
85	82	39	75	34
90	63	89	90	84
81	96	100	70	96

At the 5% significance level, do the data provide sufficient evidence to conclude that final exam scores in this introductory statistics class are not normally distributed?

15.74 Cell Phone Rates. In the February 2003 issue of *Consumer Reports*, different cell-phone providers and plans were compared. The monthly fees, in dollars, for a sample of the providers and plans are shown in the following table.

40	110	90	30	70
70	30	60	60	50
60	70	35	80	75

Do the data provide sufficient evidence to conclude that monthly fees for cell phones are not normally distributed? Use $\alpha = 0.05$.

15.75 Thoroughbred Racing. The following table displays finishing times, in seconds, for the winners of fourteen

1-mile thoroughbred horse races, as found in two recent issues of *Thoroughbred Times*.

94.15	93.37	103.02	95.57	97.73	101.09	99.38
97.19	96.63	101.05	97.91	98.44	97.47	95.10

Do the data provide sufficient evidence to conclude that the finishing times for the winners of 1-mile thoroughbred horse races are not normally distributed? Use $\alpha = 0.10$.

15.76 Beverage Expenditures. The U.S. Bureau of Labor Statistics publishes information on average annual expenditures by consumers in the *Consumer Expenditure Survey*. In 2000, the mean amount spent by consumers on nonalcoholic beverages was $250. A random sample of 12 consumers yielded the following data, in dollars, on last year's expenditures on nonalcoholic beverages.

361	176	184	265
259	281	240	273
259	249	194	258

At the 10% significance level, do the data provide sufficient evidence to conclude that last year's expenditures by consumers on nonalcoholic beverages are not normally distributed?

Using Technology

15.77 Exam Scores. Use the technology of your choice to perform the hypothesis test in Exercise 15.73.

15.78 Cell Phone Rates. Use the technology of your choice to perform the hypothesis test in Exercise 15.74.

In Exercises 15.79–15.81, use the technology of your choice to conduct a correlation test for normality at the 5% significance level. Interpret your results.

15.79 Fat Consumption in Vegetarians. A paper by Shao-Chun Lu et al., titled "LDL of Taiwanese Vegetarians are Less Oxidizable Than Those of Omnivores" (*Human Nutrition and Metabolism*, Vol. 130(6), pp. 1591–1596) compared the fat consumption of vegetarians and omnivores. The following table displays the amount of fat consumed, in grams per day, for 28 vegetarians in the study.

20.5	31.4	35.7	52.8	27.0	40.3	45.7
19.7	32.5	33.5	58.5	30.1	61.4	33.3
35.3	54.7	54.1	56.7	35.9	58.8	25.7
66.3	35.9	35.7	47.1	38.7	16.4	42.0

15.80 Chips Ahoy! 1,000 Chips Challenge. Students in an introductory statistics course at the U.S. Air Force Academy participated in Nabisco's "Chips Ahoy! 1,000 Chips Challenge" by confirming that there were at least 1000 chips in every 18-ounce bag of cookies they examined. As part of their assignment, they concluded that the number of chips per bag is approximately normally distributed. Their conclusion was based on the following data, which give the number of chips per bag for 42 bags. Do you agree with the conclusion of the students? Explain your answer. [SOURCE: Brad Warner and Jim Rutledge, "Checking the Chips Ahoy! Guarantee," *Chance*, Vol. 12(1), pp. 10–14.]

1200	1219	1103	1213	1258	1325	1295
1247	1098	1185	1087	1377	1363	1121
1279	1269	1199	1244	1294	1356	1137
1545	1135	1143	1215	1402	1419	1166
1132	1514	1270	1345	1214	1154	1307
1293	1546	1228	1239	1440	1219	1191

15.81 Finger Length of Criminals. In 1902, W. R. Macdonell published the article "On Criminal Anthropometry and the Identification of Criminals" (*Biometrika*, 1, pp. 177–227). Among other things, the author presented data on the left middle finger length, in centimeters (cm). The following table provides the midpoints and frequencies of the finger length classes used.

Midpoint (cm)	Frequency	Midpoint (cm)	Frequency
9.5	1	11.6	691
9.8	4	11.9	509
10.1	24	12.2	306
10.4	67	12.5	131
10.7	193	12.8	63
11.0	417	13.1	16
11.3	575	13.4	3

15.82 Gestation Periods of Humans. For humans, gestation periods are normally distributed with a mean of 266 days and a standard deviation of 16 days.
a. Simulate four random samples of 50 human gestation periods each.
b. Perform a correlation test for normality on each sample in part (a). Use $\alpha = 0.05$.
c. Are the conclusions in part (b) what you expected? Explain your answer.

15.83 Emergency Room Traffic. Desert Samaritan Hospital, in Mesa, Arizona, keeps records of emergency room traffic. Those records reveal that the times between arriving patients have a special type of reverse J-shaped distribution called an *exponential distribution*. The records also show that the mean time between arriving patients is 8.7 minutes.

a. Simulate four random samples of 75 interarrival times each.
b. Perform a correlation test for normality on each sample in part (a). Use $\alpha = 0.05$.
c. Are the conclusions in part (b) what you expected? Explain your answer.

CHAPTER REVIEW

You Should Be Able To

1. use and understand the formulas in this chapter.

2. state the assumptions for regression inferences.

3. understand the difference between the population regression line and a sample regression line.

4. estimate the regression parameters β_0, β_1, and σ.

5. determine the standard error of the estimate.

6. perform a residual analysis to check the assumptions for regression inferences.

7. perform a hypothesis test to decide whether the slope, β_1, of the population regression line is not 0 and hence whether x is useful for predicting y.

8. obtain a confidence interval for β_1.

9. determine a point estimate and a confidence interval for the conditional mean of the response variable corresponding to a particular value of the predictor variable.

10. determine a predicted value and a prediction interval for the response variable corresponding to a particular value of the predictor variable.

11. understand the difference between the population correlation coefficient and a sample correlation coefficient.

12. perform a hypothesis test for a population linear correlation coefficient.

*13. perform a correlation test for normality.

Key Terms

REVIEW TEST

Statistical Concepts and Skills

1. Suppose that x and y are two variables of a population with x a predictor variable and y a response variable.
a. The distribution of all possible values of the response variable y corresponding to a particular value of the predictor variable x is called a _____ distribution of the response variable.
b. State the four assumptions for regression inferences.

2. Suppose that x and y are two variables of a population and that the assumptions for regression inferences are met with x as the predictor variable and y as the response variable.
a. What statistic is used to estimate the slope of the population regression line?
b. What statistic is used to estimate the y-intercept of the population regression line?

c. What statistic is used to estimate the common conditional standard deviation of the response variable corresponding to fixed values of the predictor variable?

3. What two plots did we use in this chapter to decide whether we can reasonably presume that the assumptions for regression inferences are met by two variables of a population? What properties should those plots have?

4. Regarding analysis of residuals, decide in each case which assumption for regression inferences may be violated.
a. A residual plot—that is, a plot of the residuals against the observed values of the predictor variable—shows curvature.
b. A residual plot becomes wider with increasing values of the predictor variable.
c. A normal probability plot of the residuals shows extreme curvature.
d. A normal probability plot of the residuals shows outliers but is otherwise roughly linear.

5. Suppose that you perform a hypothesis test for the slope of the population regression line with the null hypothesis $H_0: \beta_1 = 0$ and the alternative hypothesis $H_a: \beta_1 \neq 0$. If you reject the null hypothesis, what can you say about the utility of the regression equation for making predictions?

6. Identify three statistics that can be used as a basis for testing the utility of a regression.

7. For a particular value of a predictor variable, is there a difference between the predicted value of the response variable and the point estimate for the conditional mean of the response variable? Explain your answer.

8. Generally speaking, what is the difference between a confidence interval and a prediction interval?

9. Fill in the blank: $\bar{x}$ is to μ as r is to _____.

10. Identify the relationship between two variables and the terminology used to describe that relationship if
a. $\rho > 0$.   **b.** $\rho = 0$.   **c.** $\rho < 0$.

11. Graduation Rates. Graduation rate—the percentage of entering freshmen, attending full time and graduating within 5 years—and what influences it have become a concern in U.S. colleges and universities. *U.S. News and World Report*'s "College Guide" provides data on graduation rates for colleges and universities as a function of the percentage of freshmen in the top 10% of their high school class, total spending per student, and student-to-faculty ratio. A random sample of 10 universities gave the following data on student-to-faculty ratio (S/F ratio) and graduation rate (grad rate).

S/F ratio x	Grad rate y
16	45
20	55
17	70
19	50
22	47
17	46
17	50
17	66
10	26
18	60

Discuss what satisfying the assumptions for regression inferences would mean with student-to-faculty ratio as the predictor variable and graduation rate as the response variable.

12. Graduation Rates. Refer to Problem 11.
a. Determine the regression equation for the data.
b. Compute and interpret the standard error of the estimate.
c. Presuming that the assumptions for regression inferences are met, interpret your answer to part (b).

13. Graduation Rates. Refer to Problems 11 and 12. Perform a residual analysis to decide whether considering the assumptions for regression inferences to be met by the variables student-to-faculty ratio and graduation rate is reasonable.

In the remainder of this review test, presume that the variables student-to-faculty ratio and graduation rate satisfy the assumptions for regression inferences.

14. Graduation Rates. Refer to Problems 11 and 12.
a. At the 5% significance level, do the data provide sufficient evidence to conclude that student-to-faculty ratio is useful as a predictor of graduation rate?
b. Determine a 95% confidence interval for the slope, β_1, of the population regression line that relates graduation rate to student-to-faculty ratio. Interpret your answer.

15. Graduation Rates. Refer to Problems 11 and 12.
a. Determine a point estimate for the mean graduation rate of all universities that have a student-to-faculty ratio of 17.
b. Determine a 95% confidence interval for the mean graduation rate of all universities that have a student-to-faculty ratio of 17.
c. Find the predicted graduation rate for a university that has a student-to-faculty ratio of 17.

d. Obtain a 95% prediction interval for the graduation rate of a university that has a student-to-faculty ratio of 17.

e. Explain why the prediction interval in part (d) is wider than the confidence interval in part (b).

16. Graduation Rates. Refer to Problem 11. At the 2.5% significance level, do the data provide sufficient evidence to conclude that the variables student-to-faculty ratio and graduation rate are positively linearly correlated?

*17. In a correlation test for normality, the linear correlation coefficient is computed for the sample data and _____.

*18. Mileage Tests. Each year, car makers perform mileage tests on their new car models and submit their results to the U.S. Environmental Protection Agency (EPA). The EPA then tests the vehicles to check the manufacturers' results. In 2000, one company reported that a particular model averaged 29 mpg on the highway. Suppose that the EPA tested 15 of the cars and obtained the following gas mileages.

27.3	31.2	29.4	31.6	28.6
30.9	29.7	28.5	27.8	27.3
25.9	28.8	28.9	27.8	27.6

At the 5% significance level, do the data provide sufficient evidence to conclude that gas mileages for this model are not normally distributed? Perform a correlation test for normality.

Using Technology

19. Graduation Rates. Use the technology of your choice to carry out the residual analysis in Problem 13.

20. Graduation Rates. Refer to Problem 11. Use the technology of your choice to
a. obtain the sample regression equation.
b. determine the standard error of the estimate.

21. Graduation Rates. Use the technology of your choice to do the following.
a. Carry out the hypothesis test in Problem 14(a).
b. Obtain the confidence interval in Problem 15(b) and the prediction interval in Problem 15(d).

22. Graduation Rates. Use the technology of your choice to perform the correlation test in Problem 16.

*23. Mileage Tests. Use the technology of your choice to carry out the correlation test for normality in Problem 18.

24. Fat Consumption and Prostate Cancer. Researchers have asked whether there is a relationship between nutrition and cancer, and many studies have shown that there is. In fact, one of the conclusions of a study by B. Reddy et al., "Nutrition and Its Relationship to Cancer" (*Advances in Cancer Research*, Vol. 32, pp. 237–345), was that "…none of the risk factors for cancer is probably more significant than diet and nutrition." One dietary factor that has been studied for its relationship with prostate cancer is fat consumption. On the WeissStats CD, you will find data on fat consumption and prostate cancer death rate for nations of the world. The data were obtained from a graph—adapted from information in the article mentioned—in John Robbins's classic book *Diet for a New America* (Walpole, NH: Stillpoint, 1987). Use the technology of your choice to solve the following problems:
a. Obtain the sample regression equation with fat consumption as the predictor variable for prostate cancer death rate.
b. Perform a residual analysis to decide whether considering Assumptions 1–3 for regression inferences to be satisfied by the variables fat consumption and prostate cancer death rate appears reasonable.
c. Obtain and interpret the standard error of the estimate.
d. At the 5% significance level, do the data provide sufficient evidence to conclude that fat consumption is useful for predicting prostate cancer death rate for nations of the world?
e. Find a point estimate for the mean prostate cancer death rate for nations with a fat consumption of 140 g per day.
f. Obtain a 95% confidence interval for the mean prostate cancer death rate for nations with a fat consumption of 140 g per day. Interpret your answer.
g. Determine the predicted prostate cancer death rate of a nation with a fat consumption of 140 g per day.

h. Find a 95% prediction interval for the prostate cancer death rate of a nation with a fat consumption of 140 g per day. Interpret your answer.

i. At the 5% significance level, do the data provide sufficient evidence to conclude that fat consumption and prostate cancer death rate are positively linearly correlated?

25. Payroll and Success. Steve Galbraith, Morgan Stanley's Chief Investment Officer for U.S. Equity Research, examined several financial aspects of baseball in the article "Finding the Financial Equivalent of a Walk" (*Perspectives*, October 2003, pp. 1–3). The article includes a table containing the 2003 payrolls (in millions of dollars) and winning percentages (through August 3) of major league baseball teams. We have reproduced the data on the WeissStats CD. Use the technology of your choice to do the following:

a. Find the regression equation of the data with payroll as the predictor variable and winning percentage as the response variable. Interpret your result.

b. Perform a residual analysis to decide whether considering Assumptions 1–3 for regression inferences to be satisfied by the variables payroll and winning percentage appears reasonable.

c. Obtain and interpret the standard error of the estimate.

d. At the 5% significance level, do the data provide sufficient evidence to conclude that payroll is useful for predicting winning percentage?

e. Determine a point estimate for the mean winning percentage of major league baseball teams with a payroll of $50 million.

f. Obtain a 95% confidence interval for the mean winning percentage of major league baseball teams with a payroll of $50 million. Interpret your answer.

g. Determine the predicted winning percentage of a major league baseball team with a payroll of $50 million.

h. Find a 95% prediction interval for the winning percentage of a major league baseball team with a payroll of $50 million. Interpret your answer.

i. Repeat parts (e)–(h) for a payroll of $100 million.

j. At the 5% significance level, do the data provide sufficient evidence to conclude that payroll and winning percentage are positively linearly correlated?

StatExplore in MyMathLab
Analyzing Data Online

You can use StatExplore to carry out most of the statistical analyses discussed in this chapter. To illustrate, we show how to perform a residual analysis and conduct several regression inferences.

EXAMPLE Residual Analysis

Age and Price of Orions Table 15.1 on page 744 displays data on age and price for a sample of 11 Orions. Use Stat-Explore to perform a residual analysis by obtaining a residual plot and a normal probability plot of the residuals. Interpret the results.

SOLUTION To perform a residual analysis, we first store the age and price data in columns named AGE and PRICE. Then we proceed as follows.

> 1 Choose **Stat ➤ Regression ➤ Simple Linear**
> 2 Select AGE for the **X-Variable**
> 3 Select PRICE for the **Y-Variable**
> 4 Click **Next →** twice
> 5 Select the **QQ plot of residuals** and **Residuals vs. X-values** check boxes
> 6 Click **Calculate**
> 7 Choose **Results ➤ Display in ➤ right half**

The resulting **Simple Linear Regression** output shows the regression output obtained in Printout 14.6 on page 737 plus the required plots. The first plot shown, labeled QQ plot of residuals, is a normal probability plot of the residuals, as shown in Printout 15.2.

PRINTOUT 15.2
StatExplore normal probability plot of residuals

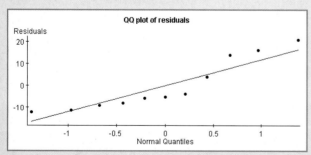

To obtain the residual plot (a plot of the residuals against the values of the predictor variable), we click the **Next →** button at the bottom of the results panel. The resulting plot is shown in Printout 15.3.

PRINTOUT 15.3
StatExplore residual plot

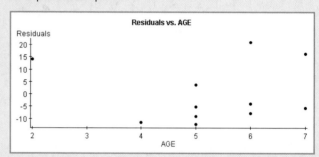

The plots in Printouts 15.2 and 15.3 show that assuming the validity of the assumptions for regression inferences is reasonable and therefore that conducting regression inferences is appropriate.

EXAMPLE Regression Inferences

Age and Price of Orions Using the age and price data for the sample of 11 Orions, apply StatExplore to do the following:
a. Decide at the 5% significance level whether age is useful for predicting price of Orions.
b. Obtain a point estimate and a 95% confidence interval for the mean price of all 3-year-old Orions.
c. Determine the predicted price and a 95% prediction interval for the price of a 3-year-old Orion.

SOLUTION We first store the age and price data in columns named AGE and PRICE. Then we proceed as follows.

> 1 Choose **Stat ➤ Regression ➤ Simple Linear**
> 2 Select AGE for the **X-Variable**
> 3 Select PRICE for the **Y-Variable**
> 4 Click **Next →**
> 5 Select the **Predict Y for X=** check box and then click in the **Predict Y for X=** text box and type <u>3</u>
> 6 Click **Calculate**

The resulting output is shown in Printout 15.4.

PRINTOUT 15.4

StatExplore output for regression analysis

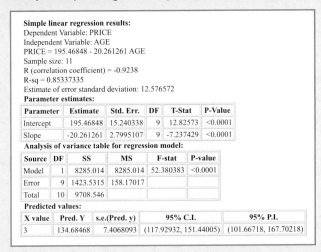

Simple linear regression results:
Dependent Variable: PRICE
Independent Variable: AGE
PRICE = 195.46848 − 20.261261 AGE
Sample size: 11
R (correlation coefficient) = -0.9238
R-sq = 0.85337335
Estimate of error standard deviation: 12.576572

Parameter estimates:

Parameter	Estimate	Std. Err.	DF	T-Stat	P-Value
Intercept	195.46848	15.240338	9	12.82573	<0.0001
Slope	-20.261261	2.7995107	9	-7.237429	<0.0001

Analysis of variance table for regression model:

Source	DF	SS	MS	F-stat	P-value
Model	1	8285.014	8285.014	52.380383	<0.0001
Error	9	1423.5315	158.17017		
Total	10	9708.546			

Predicted values:

X value	Pred. Y	s.e.(Pred. y)	95% C.I.	95% P.I.
3	134.68468	7.4068093	(117.92932, 151.44005)	(101.66718, 167.70218)

The portion of Printout 15.4 up to but not including the **Predicted values** table is identical to that obtained in Printout 14.6 on page 737. In any case, by using the output here, we can now answer the questions posed at the beginning of this example.

a. The information required for deciding, at the 5% significance level, whether age is useful for predicting price of Orions is contained in the row labeled Slope in the **Parameter estimates** table. In particular, we see from the **T-Stat** column that the value of the test statistic for a regression *t*-test is −7.237429, and the corresponding *P*-value is given as <0.0001. Because the *P*-value is less than the specified significance level of 0.05, we reject the null hypothesis of no utility and conclude that age is useful for predicting price of Orions.

b. A point estimate for the mean price of all 3-year-old Orions is provided as **Pred. Y** in the **Predicted values** table, which is 134.68468, or $13,468.47. A 95% confidence interval for the mean price of all 3-year-old Orions is presented as **95% C.I.**, which is (117.92932, 151.44005), or from $11,792.93 to $15,144.01.

c. The predicted price of a 3-year-old Orion is the same as the point estimate for the mean price of all 3-year-old Orions, which we found in part (b) to be $13,468.47. A 95% prediction interval for the price of a 3-year-old Orion is presented as **95% P.I.** in the **Predicted values** table, which is (101.66718, 167.70218), or from $10,166.72 to $16,770.22.

STATEXPLORE EXERCISES For the plant-emissions data from Exercise 15.13 on page 755, use StatExplore to solve the following problems:

a. Perform a residual analysis by obtaining a residual plot and a normal probability plot of the residuals. Interpret the results.

b. Obtain and interpret the standard error of the estimate.

c. Decide at the 5% significance level whether weight is useful for predicting quantity of volatile emissions.

d. Obtain a point estimate and a 95% confidence interval for the mean quantity of volatile emissions for all (*Solanum tuberosom*) plants that weigh 60 g.

e. Determine the predicted quantity of volatile emissions and a 95% prediction interval for the quantity of volatile emissions of a plant that weighs 60 g.

> To access StatExplore, go to the student content area of your Weiss MyMathLab course.

Internet Project

Assisted Reproductive Technology (Revisited)

In this Internet project, you are to take a deeper look into the data on Assisted Reproductive Technology (ART). In your previous exploration of the data in Chapter 14, you found a linear relationship between the age of the prospective mother who uses ART and her chances for reproductive success. Now you will further analyze the results from that regression.

Another interesting relationship appears in the ART data. When eggs from a donor are implanted (i.e., when the implanted eggs are not the prospective mother's), the relationship between success rate and age changes in a surprising way. This part of the study will encourage you to think about the real situation that the analysis is attempting to explain.

URL for access to Internet Projects Page: www.aw-bc.com/weiss

**Focusing on
Data Analysis**

The Focus Database

Recall from Chapter 1 (see page 35) that the Focus database contains information on the undergraduate students at the University of Wisconsin - Eau Claire (UWEC). Statistical analyses for this database should be done with the technology of your choice.

a. Obtain the cumulative GPAs (GPA) and high school percentiles (HSP) of a simple random sample of 50 UWEC undergraduate students.

In parts (b)–(h), use high school percentile as the predictor variable and cumulative GPA as the response variable.

b. Using the sample data from part (a), perform a residual analysis to decide whether considering the assumptions for regression inferences to be satisfied appears reasonable.

Presuming that the assumptions for regression inferences hold for the variables high school percentile and cumulative GPA, use the data from part (a) to solve parts (c)–(h).

c. Obtain and interpret the standard error of the estimate.
d. At the 5% significance level, do the data provide sufficient evidence to conclude that high school percentile is useful for predicting cumulative GPA of UWEC undergraduates?
e. Determine a point estimate for the mean cumulative GPA of all UWEC undergraduates who had high school percentiles of 74.
f. Find a 95% confidence interval for the mean cumulative GPA of all UWEC undergraduates who had high school percentiles of 74.
g. Determine the predicted cumulative GPA of a UWEC undergraduate who had a high school percentile of 74.
h. Obtain a 95% prediction interval for the cumulative GPA of a UWEC undergraduate who had a high school percentile of 74.

Case Study Discussion

Shoe Size and Height

At the beginning of this chapter, we repeated data from Chapter 14 on shoe size and height for a sample of students at Arizona State University. In Chapter 14, you used those data to perform some descriptive regression and correlation analyses. Now you are to employ those same data to carry out several inferential procedures in regression and correlation. We recommend that you use statistical software or a graphing calculator to solve the following problems, but they can also be done by hand:

a. Separate the data in the table on page 743 into two tables, one for males and the other for females. Parts (b)–(j) are for the male data.

b. Obtain the sample regression equation with shoe size as the predictor variable for height.

c. Perform a residual analysis to decide whether considering Assumptions 1–3 for regression inferences to be satisfied by the variables shoe size and height appears reasonable.

d. Obtain and interpret the standard error of the estimate.

e. Determine the P-value for a test of whether shoe size is useful for predicting height. Then refer to Table 9.12 on page 440 to assess the evidence in favor of utility.

f. Find a point estimate for the mean height of all males who wear a size $10\frac{1}{2}$ shoe.

g. Obtain a 95% confidence interval for the mean height of all males who wear a size $10\frac{1}{2}$ shoe. Interpret your answer.

h. Determine the predicted height of a male who wears a size $10\frac{1}{2}$ shoe.

i. Find a 95% prediction interval for the height of a male who wears a size $10\frac{1}{2}$ shoe. Interpret your answer.

j. At the 5% significance level, do the data provide sufficient evidence to conclude that shoe size and height are positively linearly correlated?

k. Repeat parts (b)–(j) for the unabridged data on shoe size and height for females. Do the estimation and prediction problems for a size 8 shoe.

l. Repeat part (k) for the data on shoe size and height for females with the outlier removed. Compare your results with those obtained in part (k).

Internet Resources: Visit the Weiss Web site www.aw-bc.com/weiss for additional discussion, exercises, and resources related to this case study.

Biography   *SIR FRANCIS GALTON: Discoverer of Regression and Correlation*

Francis Galton was born on February 16, 1822, into a wealthy Quaker family of bankers and gunsmiths on his father's side and as a cousin of Charles Darwin's on his mother's side. Although his IQ was estimated to be about 200, his formal education was unfinished.

He began training in medicine in Birmingham and London, but quit when, in his words, "A passion for travel seized me as if I had been a migratory bird." After a tour through Germany and southeastern Europe, he went to Trinity College in Cambridge to study mathematics. He left Cambridge in his third year, broken from overwork. He recovered quickly and resumed his medical studies in London. However, his father died before he had finished medical school and left to him, at 22, "a sufficient fortune to make me independent of the medical profession."

Galton held no professional or academic positions; nearly all his experiments were conducted at his home or performed by friends. He was curious about almost everything, and carried out research in fields that included meteorology, biology, psychology, statistics, and genetics.

The origination of the concepts of regression and correlation, developed by Galton as tools for measuring the influence of heredity, are summed up in his work *Natural Inheritance*. He discovered regression during experiments with sweet-pea seeds to determine the law of inheritance of size. He made his other great discovery, correlation, while applying his techniques to the problem of measuring the degree of association between the sizes of two different body organs of an individual.

In his later years, Galton was associated with Karl Pearson, who became his champion and an extender of his ideas. Pearson was the first holder of the chair of eugenics at University College in London, which Galton had endowed in his will. Galton was knighted in 1909. He died in Haslemere, Surrey, England, in 1911.

chapter 16

Analysis of Variance (ANOVA)

GENERAL OBJECTIVES In Chapter 10, you studied inferential methods for comparing the means of two populations. Now you will study **analysis of variance,** or **ANOVA,** which provides methods for comparing the means of more than two populations. For instance, you could use ANOVA to compare the mean energy consumption by households among the four U.S. regions. Just as there are several different procedures for comparing two population means, there are several different ANOVA procedures.

In Section 16.1, to prepare for the study of ANOVA, we consider the *F*-distribution. Then we introduce one-way analysis of variance, the simplest type of ANOVA, and examine the logic behind it in Section 16.2. We discuss the one-way ANOVA procedure in Section 16.3.

If you conduct a one-way ANOVA and decide that the population means are not all equal, you may then want to know which means are different, which mean is largest, and, in general, the relation among all the means. *Multiple comparison* methods, which we discuss in Section 16.4, are used to tackle these types of questions.

In Section 16.5, we examine the Kruskal–Wallis test. This hypothesis-testing procedure is a generalization of the Mann–Whitney test to more than two populations and provides a nonparametric alternative to one-way ANOVA.

Heavy Drinking Among College Students

Professor Kate Carey of Syracuse University surveyed 78 college students, all of whom were regular drinkers of alcohol. Her purpose was twofold—to identify interpersonal and intrapersonal situations associated with excessive drinking among college students and to detect situations that differentiate heavy drinkers from light and moderate drinkers. She published her findings in the paper "Situational Determinants of Heavy Drinking Among College Students" (*Journal of Counseling Psychology*, Vol. 40, pp. 217–220).

To assess the frequency of excessive drinking in interpersonal and intrapersonal situations, Carey utilized the short form of the Inventory of Drinking Situations (IDS). The following table gives the sample size, sample mean, and sample standard deviation of IDS scores for each drinking category and situational context.

IDS subscale	Light drinkers ($n_1 = 16$)		Moderate drinkers ($n_2 = 47$)		Heavy drinkers ($n_3 = 15$)	
	$\bar{x}_1$	s_1	$\bar{x}_2$	s_2	$\bar{x}_3$	s_3
Interpersonal situations						
Conflict with others	1.23	0.27	1.53	0.49	1.79	0.49
Social pressure to drink	2.64	0.80	2.91	0.55	3.51	0.51
Pleasant times with others	2.21	0.67	2.53	0.51	3.03	0.38
Intrapersonal situations						
Unpleasant emotions	1.22	0.35	1.61	0.69	1.68	0.46
Physical discomfort	1.03	0.08	1.19	0.29	1.40	0.32
Pleasant emotions	2.09	0.73	2.61	0.58	3.03	0.30
Testing personal control	1.52	0.74	1.56	0.56	1.53	0.48
Urges and temptations	1.80	0.56	1.96	0.51	2.33	0.58

At the end of this chapter, you will analyze these data to decide, for each IDS category, whether a difference exists in mean IDS scores among the three drinker categories.

16.1   The *F*-Distribution

Analysis-of-variance procedures rely on a distribution called the *F-distribution,* named in honor of Sir Ronald Fisher. See the Biography at the end of this chapter for more information about Fisher.

A variable is said to have an **F-distribution** if its distribution has the shape of a special type of right-skewed curve, called an **F-curve.** There are infinitely many *F*-distributions, and we identify an *F*-distribution (and *F*-curve) by stating its number of degrees of freedom, just as we did for *t*-distributions and chi-square distributions.

But an *F*-distribution has two numbers of degrees of freedom instead of one. Figure 16.1 depicts two different *F*-curves; one has df = (10, 2), and the other has df = (9, 50).

FIGURE 16.1

Two different *F*-curves

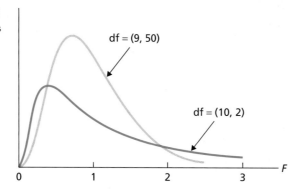

The first number of degrees of freedom for an *F*-curve is called the **degrees of freedom for the numerator,** and the second is called the **degrees of freedom for the denominator.** (The reason for this terminology will become clear in Section 16.2.) Thus, for the *F*-curve in Fig. 16.1 with df = (10, 2), we have

$$df \; = \; (10, 2)$$

Degrees of freedom
for the numerator

Degrees of freedom
for the denominator

Some basic properties of *F*-curves are presented in Key Fact 16.1.

KEY FACT 16.1

Basic Properties of F-Curves

Property 1: The total area under an *F*-curve equals 1.

Property 2: An *F*-curve starts at 0 on the horizontal axis and extends indefinitely to the right, approaching, but never touching, the horizontal axis as it does so.

Property 3: An *F*-curve is right skewed.

Using the *F*-Table

Percentages (and probabilities) for a variable having an *F*-distribution equal areas under its associated *F*-curve. To perform an ANOVA test, we need to know how to find the *F*-value having a specified area to its right. The symbol F_α is used to denote the *F*-value having area α to its right.

Table VIII in Appendix A provides *F*-values corresponding to several areas for various degrees of freedom. The degrees of freedom for the denominator (dfd) are displayed in the outside columns of the table, the values of α in the next columns, and the degrees of freedom for the numerator (dfn) along the top. In Example 16.1, we show how to use Table VIII.

▐▌ Example 16.1   Finding the *F*-Value Having a Specified Area to Its Right

For an *F*-curve with df = (4, 12), find $F_{0.05}$; that is, find the *F*-value having area 0.05 to its right, as shown in Fig. 16.2(a).

FIGURE 16.2
Finding the *F*-value having area 0.05 to its right

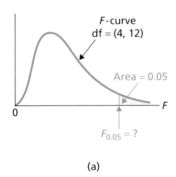

(a)

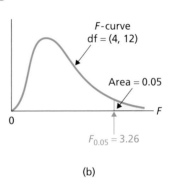

(b)

Solution   To obtain the *F*-value, we use Table VIII. In this case, $\alpha = 0.05$, the degrees of freedom for the numerator is 4, and the degrees of freedom for the denominator is 12. We first go down the dfd column to "12." Next, we go across the row for α labeled 0.05 to the column headed "4." The number in the body of the table there, 3.26, is the required *F*-value; that is, for an *F*-curve with df = (4, 12), the *F*-value having area 0.05 to its right is 3.26: $F_{0.05} = 3.26$, as shown in Fig. 16.2(b).

▬▬■

▌ EXERCISES 16.1

Statistical Concepts and Skills

16.1 How do we identify an *F*-distribution and its corresponding *F*-curve?

16.2 How many degrees of freedom does an *F*-curve have? What are those degrees of freedom called?

16.3 What symbol is used to denote the *F*-value having area 0.05 to its right? 0.025 to its right? α to its right?

16.4 Using the F_α-notation, identify the *F*-value having area 0.975 to its left.

16.5 An *F*-curve has df = (12, 7). What is the number of degrees of freedom for the
a. numerator?   **b.** denominator?

16.6 An *F*-curve has df = (8, 19). What is the number of degrees of freedom for the
a. denominator?   **b.** numerator?

*In Exercises **16.7–16.10**, use Table VIII in Appendix A to find the required F-values. Illustrate your work with graphs similar to that shown in Fig. 16.2 on page 801.*

16.7 An F-curve has df $= (24, 30)$. In each case, find the F-value having the specified area to its right.
a. 0.05   **b.** 0.01   **c.** 0.025

16.8 An F-curve has df $= (12, 5)$. In each case, find the F-value having the specified area to its right.
a. 0.01   **b.** 0.05   **c.** 0.005

16.9 For an F-curve with df $= (20, 21)$, find
a. $F_{0.01}$.   **b.** $F_{0.05}$.   **c.** $F_{0.10}$.

16.10 For an F-curve with df $= (6, 10)$, find
a. $F_{0.05}$.   **b.** $F_{0.01}$.   **c.** $F_{0.025}$.

Extending the Concepts and Skills

16.11 Refer to Table VIII in Appendix A. Because of space restrictions, the numbers of degrees of freedom are not consecutive. For instance, the degrees of freedom for the numerator skips from 24 to 30. If you had only Table VIII and you needed to find $F_{0.05}$ for df $= (25, 20)$, how would you do it?

16.2   One-Way ANOVA: The Logic

In Chapter 10, you learned how to compare two population means, that is, the means of a single variable for two different populations. You studied various methods for making such comparisons, one being the pooled t-procedure.

Analysis of variance (ANOVA) provides methods for comparing several population means, that is, the means of a single variable for several populations. In this section and Section 16.3, we present the simplest kind of ANOVA, **one-way analysis of variance.** This type of ANOVA is called *one-way* analysis of variance because it compares the means of a variable for populations that result from a classification by *one* other variable, called the **factor.** The possible values of the factor are referred to as the **levels** of the factor.

For example, suppose that you want to compare the mean energy consumption by households among the four regions of the United States. The variable under consideration is "energy consumption" and there are four populations—households in the Northeast, Midwest, South, and West. The four populations result from classifying households in the United States by the factor "region," whose levels are Northeast, Midwest, South, and West.

One-way analysis of variance is the generalization to more than two populations of the pooled t-procedure. As in the pooled t-procedure, we make the assumptions listed in Key Fact 16.2.

KEY FACT 16.2

Assumptions for One-Way ANOVA

1. *Simple random samples:* The samples taken from the populations under consideration are simple random samples.

2. *Independent samples:* The samples taken from the populations under consideration are independent of one another.

3. *Normal populations:* For each population, the variable under consideration is normally distributed.

4. *Equal standard deviations:* The standard deviations of the variable under consideration are the same for all the populations.

One-way ANOVA has the same robustness properties as those of the pooled *t*-procedure. It is robust to moderate violations of the normality assumption (Assumption 3). It is also reasonably robust to moderate violations of the equal standard deviations assumption (Assumption 4) provided that the sample sizes are roughly equal.

Generally, normal probability plots are effective in detecting gross violations of the normality assumption. The equal standard deviations assumption is usually more difficult to check. As a rule of thumb, we consider that assumption satisfied if *the ratio of the largest to the smallest sample standard deviation is less than 2.* For convenience, we call this rule of thumb the **rule of 2.**

Additionally, we can assess the normality and equal standard deviations assumptions by performing a residual analysis, in a way similar to what we did in regression. (See Section 15.1 for a discussion of the analysis of residuals.)

In ANOVA, the **residual** of an observation is the difference between the observation and the mean of the sample containing it. If the normality and equal standard deviations assumptions are met, a normal probability plot of (all) the residuals should be roughly linear. Moreover, a plot of the residuals against the sample means should fall roughly in a horizontal band centered and symmetric about the horizontal axis.

The Logic Behind One-Way ANOVA

The reason for the word *variance* in *analysis of variance* is that the procedure for comparing the means analyzes the variation in the sample data. To examine how this procedure works, let's suppose that independent random samples are taken from two populations—say, Populations 1 and 2—having means μ_1 and μ_2. Further, let's suppose that the means of the two samples are $\bar{x}_1 = 20$ and $\bar{x}_2 = 25$. Can we reasonably conclude from these statistics that $\mu_1 \neq \mu_2$, that is, that the population means are different? To answer this question, we must consider the variation within the samples.

Suppose, for instance, that the sample data are as displayed in Table 16.1 and depicted in Fig. 16.3.

TABLE 16.1
Sample data from Populations 1 and 2

Sample from Population 1	21	37	11	20	8	23
Sample from Population 2	24	31	29	40	9	17

FIGURE 16.3
Dotplots for sample data in Table 16.1

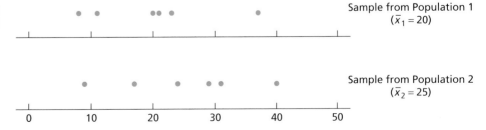

For these two samples, $\bar{x}_1 = 20$ and $\bar{x}_2 = 25$. But we cannot infer that $\mu_1 \neq \mu_2$ because it is not clear whether the difference between the sample means is due to a difference between the population means or to the variation within the populations.

However, suppose that the sample data are as displayed in Table 16.2 and depicted in Fig. 16.4.

TABLE 16.2
Sample data from Populations 1 and 2

Sample from Population 1	21	21	20	18	20	20
Sample from Population 2	25	28	25	24	24	24

FIGURE 16.4
Dotplots for sample data in Table 16.2

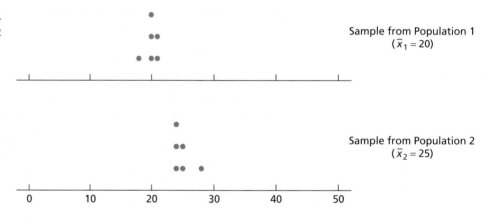

Again, for these two samples, $\bar{x}_1 = 20$ and $\bar{x}_2 = 25$. But this time, we *can* infer that $\mu_1 \neq \mu_2$ because it seems clear that the difference between the sample means is due to a difference between the population means, not to the variation within the populations.

The preceding two illustrations reveal the basic idea for performing a one-way analysis of variance to compare the means of several populations:

1. Take independent simple random samples from the populations.
2. Compute the sample means.
3. If the variation among the sample means is large relative to the variation within the samples, conclude that the means of the populations are not all equal.

To make this process precise, we need quantitative measures of the variation among the sample means and the variation within the samples. We also need an objective method for deciding whether the variation among the sample means is large relative to the variation within the samples. We address these issues now and then apply them in Example 16.2.

Mean Squares and *F*-Statistic in One-Way ANOVA

We first consider the measure of variation among the sample means. In hypothesis tests for two population means, we measure the variation between the two sample means by calculating their difference, $\bar{x}_1 - \bar{x}_2$. When more than two populations are involved, we cannot measure the variation among the sample means simply by taking a difference. However, we can measure that variation by computing the standard deviation or variance of the sample means or, for that matter, by computing any descriptive statistic that measures variation.

In one-way ANOVA, we measure the variation among the sample means by a weighted average of their squared deviations about the mean, $\bar{x}$, of all the sample data. That measure of variation is called the **treatment mean square, *MSTR*,** and is defined as

$$MSTR = \frac{SSTR}{k-1},$$

where k denotes the number of populations being sampled and

$$SSTR = n_1(\bar{x}_1 - \bar{x})^2 + n_2(\bar{x}_2 - \bar{x})^2 + \cdots + n_k(\bar{x}_k - \bar{x})^2.$$

The quantity **SSTR** is called the **treatment sum of squares.**

We note that *MSTR* is similar to the sample variance of the sample means. In fact, if all the sample sizes are identical, then *MSTR* equals that common sample size times the sample variance of the sample means.

Next we consider the measure of variation within the samples. This measure is the pooled estimate of the common population variance, σ^2. It is called the **error mean square, *MSE*,** and is defined as

$$MSE = \frac{SSE}{n-k},$$

where n denotes the total number of observations and

$$SSE = (n_1 - 1)s_1^2 + (n_2 - 1)s_2^2 + \cdots + (n_k - 1)s_k^2.$$

The quantity **SSE** is called the **error sum of squares.**[†][‡]

Finally, we consider how to compare the variation among the sample means, *MSTR*, to the variation within the samples, *MSE*. To do so, we use the statistic $F = MSTR/MSE$, which we refer to as the **F-statistic.** Large values of F indicate that the variation among the sample means is large relative to the variation within the samples and hence that the null hypothesis of equal population means should be rejected.

We summarize the previous discussion in Definition 16.1.

What
Does it Mean?

MSTR measures the variation among the sample means.

What
Does it Mean?

MSE measures the variation within the samples.

What
Does it Mean?

The *F*-statistic compares the variation among the sample means to the variation within the samples.

[†]The terms **treatment** and **error** arose from the fact that many ANOVA techniques were first developed to analyze agricultural experiments. In any case, the treatments refer to the different populations and the errors pertain to the variation within the populations.

[‡]For two populations (i.e., $k = 2$), *MSE* is the pooled variance, s_p^2, defined in Section 10.2 on page 492.

DEFINITION 16.1

Mean Squares and F-Statistic in One-Way ANOVA

Treatment mean square, MSTR: The variation among the sample means: $MSTR = SSTR/(k-1)$, where $SSTR$ is the treatment sum of squares and k is the number of populations under consideration.

Error mean square, MSE: The variation within the samples: $MSE = SSE/(n-k)$, where SSE is the error sum of squares and n is the total number of observations.

F-statistic, F: The ratio of the variation among the sample means to the variation within the samples: $F = MSTR/MSE$.

Example 16.2 Introducing One-Way ANOVA

Energy Consumption　The U.S. Energy Information Administration gathers data on residential energy consumption and expenditures and publishes its findings in *Residential Energy Consumption Survey: Consumption and Expenditures*. Suppose that we want to decide whether a difference exists in mean annual energy consumption by households among the four U.S. regions.

Let μ_1, μ_2, μ_3, and μ_4 denote last year's mean energy consumptions by households in the Northeast, Midwest, South, and West, respectively. Then the hypotheses to be tested are

H_0: $\mu_1 = \mu_2 = \mu_3 = \mu_4$ (mean energy consumptions are all equal)
H_a: Not all the means are equal.

The basic strategy for carrying out this hypothesis test follows the three steps mentioned earlier on page 804.

Step 1. Independently and randomly take samples of households in the four U.S. regions.

Step 2. Compute last year's mean energy consumptions, $\bar{x}_1$, $\bar{x}_2$, $\bar{x}_3$, and $\bar{x}_4$, of the four samples.

Step 3. Reject the null hypothesis if the variation among the sample means is large relative to the variation within the samples; otherwise, do not reject the null hypothesis.

This process is illustrated in Fig. 16.5.

In Steps 1 and 2, we obtain the sample data and compute the sample means. Suppose that the results of those steps are as shown in Table 16.3 on the next page, where the data are displayed to the nearest 10 million BTU.

In Step 3, we compare the variation among the four sample means, shown at the bottom of Table 16.3, to the variation within the samples. To accomplish that, we need to compute the treatment mean square (*MSTR*), the error mean square (*MSE*), and the *F*-statistic. We refer to Table 16.3 and to Definition 16.1 (top of this page).

FIGURE 16.5

Process for comparing four population means

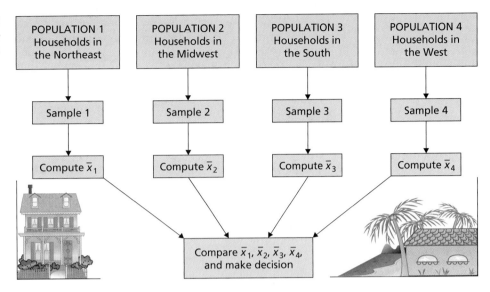

TABLE 16.3

Samples and their means of last year's energy consumptions for households in the four U.S. regions

Northeast	Midwest	South	West
15	17	11	10
10	12	7	12
13	18	9	8
14	13	13	7
13	15		9
	12		
13.0	14.5	10.0	9.2	← *Means*

First, we determine *MSTR*. We have $k = 4$, $n_1 = 5$, $n_2 = 6$, $n_3 = 4$, $n_4 = 5$, $\bar{x}_1 = 13.0$, $\bar{x}_2 = 14.5$, $\bar{x}_3 = 10.0$, and $\bar{x}_4 = 9.2$. To obtain the overall mean, $\bar{x}$, we need to divide the sum of all the observations in Table 16.3 by the total number of observations:

$$\bar{x} = \frac{\Sigma x}{n} = \frac{15 + 10 + 13 + \cdots + 7 + 9}{20} = \frac{238}{20} = 11.9.$$

Therefore,

$$SSTR = n_1(\bar{x}_1 - \bar{x})^2 + n_2(\bar{x}_2 - \bar{x})^2 + n_3(\bar{x}_3 - \bar{x})^2 + n_4(\bar{x}_4 - \bar{x})^2$$
$$= 5(13.0 - 11.9)^2 + 6(14.5 - 11.9)^2 + 4(10.0 - 11.9)^2 + 5(9.2 - 11.9)^2$$
$$= 97.5.$$

Hence,

$$MSTR = \frac{SSTR}{k - 1} = \frac{97.5}{4 - 1} = 32.5,$$

which is the measure of variation among the four sample means shown at the bottom of Table 16.3.

Next, we determine MSE. We have $k = 4$, $n_1 = 5$, $n_2 = 6$, $n_3 = 4$, $n_4 = 5$, and $n = 20$. Computing the variance of each sample gives $s_1^2 = 3.5$, $s_2^2 = 6.7$, $s_3^2 = 6.\overline{6}$, and $s_4^2 = 3.7$. Consequently,

$$SSE = (n_1 - 1)s_1^2 + (n_2 - 1)s_2^2 + (n_3 - 1)s_3^2 + (n_4 - 1)s_4^2$$
$$= (5 - 1) \cdot 3.5 + (6 - 1) \cdot 6.7 + (4 - 1) \cdot 6.\overline{6} + (5 - 1) \cdot 3.7 = 82.3.$$

Hence,

$$MSE = \frac{SSE}{n - k} = \frac{82.3}{20 - 4} = 5.144,$$

which is the measure of variation within the samples in Table 16.3.

Finally, we determine F. As $MSTR = 32.5$ and $MSE = 5.144$, the value of the F-statistic is

$$F = \frac{MSTR}{MSE} = \frac{32.5}{5.144} = 6.32.$$

Is this value of F large enough to conclude that the null hypothesis of equal population means is false? To answer that question, we need to know the distribution of the F-statistic. We discuss that distribution in Section 16.3 and then return to complete the hypothesis test. ▬

EXERCISES 16.2

Statistical Concepts and Skills

16.12 State the four assumptions required for one-way ANOVA. How crucial are these assumptions?

16.13 One-way ANOVA is a procedure for comparing the means of several populations. It is the generalization of what procedure for comparing the means of two populations?

16.14 If we define $s = \sqrt{MSE}$, of which parameter is s an estimate?

16.15 Explain the reason for the word *variance* in the phrase *analysis of variance*.

16.16 The null and alternative hypotheses for a one-way ANOVA test are

$$H_0: \mu_1 = \mu_2 = \cdots = \mu_k$$
$$H_a: \text{Not all means are equal.}$$

Suppose that, in reality, the null hypothesis is false. Does that mean that no two of the populations have the same mean? If not, what does it mean?

16.17 In one-way ANOVA, identify the statistic used
a. as a measure of variation among the sample means.
b. as a measure of variation within the samples.

c. to compare the variation among the sample means to the variation within the samples.

16.18 Explain the logic behind one-way ANOVA.

16.19 What does the term *one-way* signify in the phrase *one-way ANOVA*?

16.20 Figure 16.6 shows side-by-side boxplots of independent samples from three normally distributed populations having equal standard deviations. Based on these boxplots, would you be inclined to reject the null hypothesis of equal population means? Explain your answer.

16.21 Figure 16.7 shows side-by-side boxplots of independent samples from three normally distributed populations having equal standard deviations. Based on these boxplots, would you be inclined to reject the null hypothesis of equal population means? Explain your answer.

Extending the Concepts and Skills

16.22 Show that, for two populations, $MSE = s_p^2$, where s_p^2 is the pooled variance defined in Section 10.2 on page 492. Conclude that $\sqrt{MSE}$ is the pooled sample standard deviation, s_p.

16.23 Suppose that the variable under consideration is normally distributed on each of two populations and that the population standard deviations are equal. Further suppose that you want to perform a hypothesis test to decide whether the populations have different means, that is, whether $\mu_1 \neq \mu_2$. If independent simple random samples are used, identify two hypothesis-testing procedures that you can use to carry out the hypothesis test.

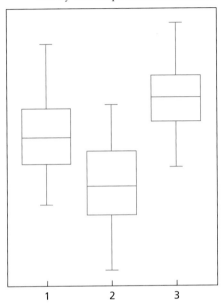

FIGURE 16.6
Side-by-side boxplots for Exercise 16.20

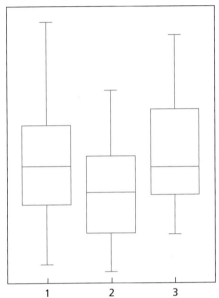

FIGURE 16.7
Side-by-side boxplots for Exercise 16.21

16.3 One-Way ANOVA: The Procedure

In this section, we present a step-by-step procedure for performing a one-way ANOVA to compare the means of several populations. To begin, we need to identify the distribution of the variable $F = MSTR/MSE$, introduced in Section 16.2. We do that in Key Fact 16.3.

KEY FACT 16.3

Distribution of the F-Statistic for One-Way ANOVA

Suppose that the variable under consideration is normally distributed on each of k populations and that the population standard deviations are equal. Then, for independent samples from the k populations, the variable

$$F = \frac{MSTR}{MSE}$$

has the F-distribution with df $= (k - 1, n - k)$ if the null hypothesis of equal population means is true. Here, n denotes the total number of observations.

We have now covered all the elements required to formulate a procedure for performing a one-way analysis of variance. Before presenting the procedure, however, we need to consider two additional concepts.

One-Way ANOVA Identity

First, we define another sum of squares—one that provides a measure of total variation among all the sample data. It is called the **total sum of squares, *SST***, and is defined by

$$SST = \Sigma(x - \bar{x})^2,$$

where the sum extends over all n observations. If we divide SST by $n - 1$, we get the sample variance of all the observations. So SST really is a measure of total variation.

For the energy consumption data in Table 16.3 on page 807, $\bar{x} = 11.9$, and therefore

$$SST = \Sigma(x - \bar{x})^2 = (15 - 11.9)^2 + (10 - 11.9)^2 + \cdots + (9 - 11.9)^2$$
$$= 9.61 + 3.61 + \cdots + 8.41 = 179.8.$$

In Section 16.2, we found that, for the energy consumption data, $SSTR = 97.5$ and $SSE = 82.3$. Because $179.8 = 97.5 + 82.3$, we have $SST = SSTR + SSE$. This equation is always true and is called the **one-way ANOVA identity**, which we emphasize as Key Fact 16.4.

KEY FACT 16.4

One-Way ANOVA Identity

The total sum of squares equals the treatment sum of squares plus the error sum of squares: $SST = SSTR + SSE$.

What
Does it Mean?

The total variation among all the sample data can be partitioned into a component representing variation among the sample means and a component representing variation within the samples.

We provide a graphical representation of the one-way ANOVA identity in Fig. 16.8.

FIGURE 16.8
Partitioning of the total sum of squares into the treatment sum of squares and the error sum of squares

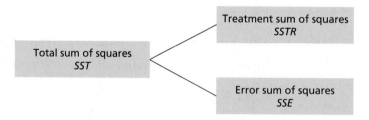

One-Way ANOVA Tables

One-way ANOVA tables are useful for organizing and summarizing the quantities required for performing a one-way analysis of variance. The general format of a one-way ANOVA table is as shown in Table 16.4.

For the energy consumption data in Table 16.3, we have already computed all quantities appearing in the one-way ANOVA table. Table 16.5 displays the one-way ANOVA table for those data.

TABLE 16.4

ANOVA table format for a one-way analysis of variance

Source	df	SS	MS = SS/df	F-statistic
Treatment	$k-1$	SSTR	$MSTR = \dfrac{SSTR}{k-1}$	$F = \dfrac{MSTR}{MSE}$
Error	$n-k$	SSE	$MSE = \dfrac{SSE}{n-k}$	
Total	$n-1$	SST		

TABLE 16.5

One-way ANOVA table for the energy consumption data

Source	df	SS	MS = SS/df	F-statistic
Treatment	3	97.5	32.500	6.32
Error	16	82.3	5.144	
Total	19	179.8		

Performing a One-Way ANOVA

To perform a one-way ANOVA, we need to obtain the three sums of squares, *SST*, *SSTR*, and *SSE*. We can do so by using the defining formulas introduced earlier. Generally, however, when calculating by hand from the raw data, computing formulas are more accurate and easier to use. Both the defining formulas and their computing equivalents are presented in Formula 16.1.

FORMULA 16.1

Sums of Squares in One-Way ANOVA

For a one-way ANOVA of k population means, the defining and computing formulas for the three sums of squares are as follows.

Sum of square	Defining formula	Computing formula
Total, SST	$\Sigma(x-\bar{x})^2$	$\Sigma x^2 - (\Sigma x)^2/n$
Treatment, SSTR	$\Sigma n_j(\bar{x}_j - \bar{x})^2$	$\Sigma(T_j^2/n_j) - (\Sigma x)^2/n$
Error, SSE	$\Sigma(n_j - 1)s_j^2$	$SST - SSTR$

In this table, we used the notation

$$n = \text{total number of observations}$$
$$\bar{x} = \text{mean of all } n \text{ observations;}$$

and, for $j = 1, 2, \ldots, k,$

$$n_j = \text{size of sample from Population } j$$
$$\bar{x}_j = \text{mean of sample from Population } j$$
$$s_j^2 = \text{variance of sample from Population } j$$
$$T_j = \text{sum of sample data from Population } j.$$

Keep the following facts in mind when you use Formula 16.1.

- Only two of the three sums of squares need ever be calculated; the remaining one can always be determined from the other two by using the one-way ANOVA identity, Key Fact 16.4 on page 810.
- Summations involving no subscripted variables are over all n observations; those involving subscripts are over the k populations.
- When using the computing formulas, the most efficient formula for calculating the sum of all n observations is $\Sigma x = T_1 + T_2 + \cdots + T_k$.

We now present, in Procedure 16.1 (next page), a step-by-step method for conducting a **one-way ANOVA test** by using either the critical-value approach or the P-value approach. Because the null hypothesis is rejected only when the test statistic, F, is too large, a one-way ANOVA test is always right tailed.

▌▌▌Example 16.3 **The One-Way ANOVA Test**

Energy Consumption   Recall that independent simple random samples of households in the four U.S. regions yielded the data on last year's energy consumptions shown in Table 16.6.

Northeast	Midwest	South	West
15	17	11	10
10	12	7	12
13	18	9	8
14	13	13	7
13	15		9
	12		

At the 5% significance level, do the data provide sufficient evidence to conclude that a difference exists in last year's mean energy consumption by households among the four U.S. regions?

Solution   First we check the four assumptions required for performing a one-way ANOVA test. Because the samples are independent simple random samples, Assumptions 1 and 2 are satisfied.

Normal probability plots (not shown) of the four samples in Table 16.6 reveal no outliers and are roughly linear, indicating no gross violations of the normality assumption; thus we can consider Assumption 3 satisfied.

The sample standard deviations of the four samples in Table 16.6 are 1.87, 2.59, 2.58, and 1.92, respectively. Because the ratio of the largest to the smallest standard deviation is $2.59/1.87 = 1.39$, which is less than 2, we can, in light of the rule of 2, consider Assumption 4 satisfied. A residual analysis (not shown) further indicates that we can reasonably consider Assumptions 3 and 4 satisfied.

| PROCEDURE 16.1 | One-Way ANOVA Test |

Purpose To perform a hypothesis test to compare k population means, $\mu_1, \mu_2, \ldots, \mu_k$

Assumptions

1. Simple random samples
2. Independent samples
3. Normal populations
4. Equal population standard deviations

STEP 1 The null and alternative hypotheses are

$$H_0: \mu_1 = \mu_2 = \cdots = \mu_k$$
$$H_a: \text{Not all the means are equal.}$$

STEP 2 Decide on the significance level, α.

STEP 3 Obtain the three sums of squares, SST, $SSTR$, and SSE.

STEP 4 Construct a one-way ANOVA table to obtain the value of the F-statistic. Denote that value F_0.

Source	df	SS	MS = SS/df	F-statistic
Treatment	$k-1$	$SSTR$	$MSTR = \dfrac{SSTR}{k-1}$	$F = \dfrac{MSTR}{MSE}$
Error	$n-k$	SSE	$MSE = \dfrac{SSE}{n-k}$	
Total	$n-1$	SST		

| CRITICAL-VALUE APPROACH | P-VALUE APPROACH |

STEP 5 The critical value is F_α with df = $(k-1, n-k)$. Use Table VIII to find the critical value.

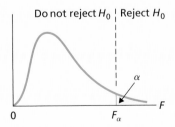

STEP 5 The F-statistic has df = $(k-1, n-k)$. Use Table VIII to estimate the P-value or obtain it exactly by using technology.

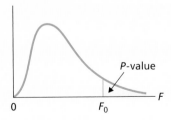

STEP 6 If the value of the F-statistic falls in the rejection region, reject H_0; otherwise, do not reject H_0.

STEP 6 If $P \leq \alpha$, reject H_0; otherwise, do not reject H_0.

STEP 7 Interpret the results of the hypothesis test.

As it is reasonable to presume that the four assumptions for performing a one-way ANOVA test are satisfied, we now apply Procedure 16.1 to carry out the required hypothesis test.

STEP 1 State the null and alternative hypotheses.

Let μ_1, μ_2, μ_3, and μ_4 denote last year's mean energy consumptions for households in the Northeast, Midwest, South, and West, respectively. Then the null and alternative hypotheses are

H_0: $\mu_1 = \mu_2 = \mu_3 = \mu_4$ (mean energy consumptions are equal)
H_a: Not all the means are equal.

STEP 2 Decide on the significance level, α.

We are to perform the hypothesis test at the 5% significance level; consequently, $\alpha = 0.05$.

STEP 3 Obtain the three sums of squares, *SST*, *SSTR*, and *SSE*.

Although we obtained these sums earlier by using the defining formulas, we determine them again to illustrate use of the computing formulas. Referring to Formula 16.1 on page 811 and Table 16.6, we find that

$$k = 4$$

$$n_1 = 5 \qquad n_2 = 6 \qquad n_3 = 4 \qquad n_4 = 5$$
$$T_1 = 65 \qquad T_2 = 87 \qquad T_3 = 40 \qquad T_4 = 46$$

and

$$n = 5 + 6 + 4 + 5 = 20$$
$$\Sigma x = 65 + 87 + 40 + 46 = 238.$$

Summing the squares of all the data in Table 16.6 yields

$$\Sigma x^2 = (15)^2 + (10)^2 + (13)^2 + \cdots + 7^2 + 9^2 = 3012.$$

Consequently,

$$SST = \Sigma x^2 - (\Sigma x)^2/n = 3012 - (238)^2/20 = 3012 - 2832.2 = 179.8,$$
$$SSTR = \Sigma(T_j^2/n_j) - (\Sigma x)^2/n$$
$$= (65)^2/5 + (87)^2/6 + (40)^2/4 + (46)^2/5 - (238)^2/20$$
$$= 2929.7 - 2832.2 = 97.5,$$

and

$$SSE = SST - SSTR = 179.8 - 97.5 = 82.3.$$

STEP 4 Construct a one-way ANOVA table to obtain the value of the *F*-statistic.

Table 16.5 on page 811 is the one-way ANOVA table for the energy consumption data. It reveals that $F = 6.32$.

| CRITICAL-VALUE APPROACH | P-VALUE APPROACH |

CRITICAL-VALUE APPROACH

STEP 5 The critical value is F_α with df = $(k-1, n-k)$. Use Table VIII to find the critical value.

From Step 2, $\alpha = 0.05$. Also, Table 16.6 shows that four populations are under consideration, or $k = 4$, and that the number of observations total 20, or $n = 20$. Therefore, we have df = $(k-1, n-k) = (4-1, 20-4) = (3, 16)$. From Table VIII, the critical value is $F_{0.05} = 3.24$, as shown in Fig. 16.9A.

FIGURE 16.9A

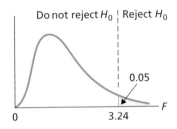

STEP 6 If the value of the F-statistic falls in the rejection region, reject H_0; otherwise, do not reject H_0.

From Step 4, the value of the F-statistic is $F = 6.32$, which, as Fig. 16.9A shows, falls in the rejection region. Thus we reject H_0. The test results are statistically significant at the 5% level.

P-VALUE APPROACH

STEP 5 The F-statistic has df = $(k-1, n-k)$. Use Table VIII to estimate the P-value or obtain it exactly by using technology.

From Step 4, the value of the F-statistic is $F = 6.32$. Because the test is right tailed, the P-value is the probability of observing a value of F of 6.32 or greater if the null hypothesis is true. That probability equals the shaded area in Fig. 16.9B.

FIGURE 16.9B

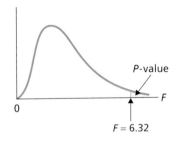

From Table 16.6, four populations are under consideration, or $k = 4$, and the number of observations total 20, or $n = 20$. Thus, we have df = $(k-1, n-k) = (4-1, 20-4) = (3, 16)$. Referring to Fig. 16.9B and to Table VIII with df = $(3, 16)$, we find $P < 0.005$. (Using technology, we obtain $P = 0.00495$.)

STEP 6 If $P \le \alpha$, reject H_0; otherwise, do not reject H_0.

From Step 5, $P < 0.005$. Because the P-value is less than the specified significance level of 0.05, we reject H_0. The test results are statistically significant at the 5% level and (see Table 9.12 on page 440) provide very strong evidence against the null hypothesis.

STEP 7 Interpret the results of the hypothesis test.

INTERPRETATION At the 5% significance level, the data provide sufficient evidence to conclude that a difference exists in last year's mean energy consumption by households among the four U.S. regions. Evidently, at least two of the regions have different mean energy consumptions.

Other Types of ANOVA

We can consider one-way ANOVA to be a method for comparing the means of populations classified according to one factor. Put another way, it is a method for analyzing the effect of one factor on the mean of the variable under consideration, called the **response variable.**

For instance, in Example 16.3, we compared last year's mean energy consumption by households among the four U.S. regions (Northeast, Midwest, South, and West). Here, the factor is "region" and the response variable is "energy consumption." One-way ANOVA permits us to analyze the effect of region on mean energy consumption.

Other ANOVA procedures provide methods for comparing the means of populations classified according to two or more factors. Put another way, these are methods for simultaneously analyzing the effect of two or more factors on the mean of a response variable.

For example, suppose that you want to consider the effect of "region" and "home type" (the two factors) on energy consumption (the response variable). Two-way ANOVA permits you to determine simultaneously whether region affects mean energy consumption, whether home type affects mean energy consumption, and whether region and home type interact in their effect on mean energy consumption (e.g., whether the effect of home type on mean energy consumption depends on region).

Two-way ANOVA and other ANOVA procedures, such as randomized block ANOVA, are treated in detail in the chapter *Design of Experiments and Analysis of Variance,* on the WeissStats CD accompanying this book and on the Weiss Web site, www.aw-bc.com/weiss.

The Technology Center

Most statistical technologies have programs that automatically perform a one-way analysis of variance. In this subsection, we present output and step-by-step instructions to implement such programs.

Example 16.4   Using Technology to Conduct a One-Way ANOVA Test

Energy Consumption   Table 16.6 on page 812 displays last year's energy consumptions for independent samples of households in the four U.S. regions. Use Minitab, Excel, or the TI-83/84 Plus to perform the hypothesis test in Example 16.3.

Solution   Let μ_1, μ_2, μ_3, and μ_4 denote last year's mean energy consumptions for households in the Northeast, Midwest, South, and West, respectively. We want to perform the hypothesis test

H_0: $\mu_1 = \mu_2 = \mu_3 = \mu_4$ (mean energy consumptions are equal)

H_a: Not all the means are equal

at the 5% significance level.

Printout 16.1 shows the output obtained by applying the one-way ANOVA programs to the energy-consumption data presented in Table 16.6.

The outputs in Printout 16.1 show that the *P*-value for the hypothesis test is 0.005 (to three decimal places). Because the *P*-value is less than the specified significance level of 0.05, we reject H_0. At the 5% significance level, the

PRINTOUT 16.1
One-way ANOVA procedure output for
the energy consumption data

MINITAB

One-way ANOVA: ENERGY versus REGION

```
Source   DF      SS      MS      F       P
REGION    3   97.50   32.50   6.32   0.005
Error    16   82.30    5.14
Total    19  179.80


S = 2.268   R-Sq = 54.23%   R-Sq(adj) = 45.64%
```

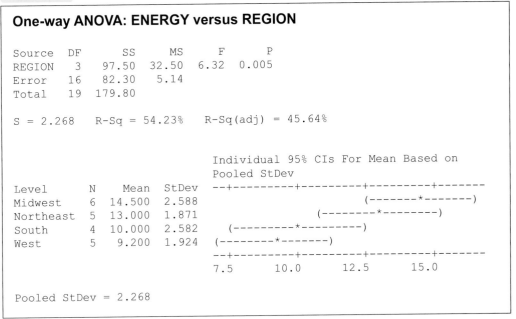

```
                             Individual 95% CIs For Mean Based on
                             Pooled StDev
Level       N    Mean  StDev  --+---------+---------+---------+-------
Midwest     6  14.500  2.588                      (-------*-------)
Northeast   5  13.000  1.871               (--------*--------)
South       4  10.000  2.582       (---------*---------)
West        5   9.200  1.924  (--------*-------)
                             --+---------+---------+---------+-------
                             7.5      10.0      12.5      15.0
```

```
Pooled StDev = 2.268
```

EXCEL

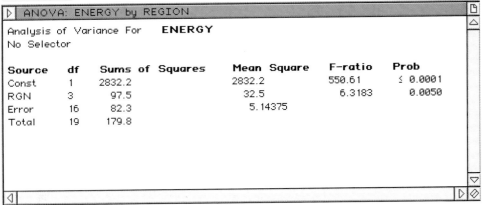

ANOVA: ENERGY by REGION

Analysis of Variance For   **ENERGY**
No Selector

Source	df	Sums of Squares	Mean Square	F-ratio	Prob
Const	1	2832.2	2832.2	550.61	≤ 0.0001
RGN	3	97.5	32.5	6.3183	0.0050
Error	16	82.3	5.14375		
Total	19	179.8			

TI-83/84 PLUS

```
One-way ANOVA
 F=6.318347509
 p=.0049510617
 Factor
  df=3
  SS=97.5
↓ MS=32.5
■
```

```
One-way ANOVA
↑ MS=32.5
 Error
  df=16
  SS=82.3
  MS=5.14375
 Sxp=2.26798369
```

data provide sufficient evidence to conclude that last year's mean energy consumptions for households in the four U.S. regions are not all the same. See the technology manuals for further interpretation of Printout 16.1.

Obtaining the Output (Optional)

Printout 16.1 provides output from Minitab, Excel, and the TI-83/84 Plus for a one-way ANOVA based on the samples of energy consumptions in Table 16.6. The following are detailed instructions for obtaining that output. For the TI-83/84 Plus, store the four samples in Table 16.6 in lists named NE, MW, SO, and WE, respectively.

For Minitab and Excel, first store all 20 energy consumptions from Table 16.6 in a column (Minitab) or range (Excel) named ENERGY. Then, in a column or range named REGION, store the regions corresponding to the energy consumptions in ENERGY. For instance, suppose that you store the sample data for the Northeast in the first five rows of ENERGY, that for the Midwest in the next six rows of ENERGY, and so on. Then you would store the word Northeast in the first five rows of REGION, the word Midwest in the next six rows of REGION, and so on.

Now, we proceed as follows.

MINITAB	**EXCEL**	**TI-83/84 PLUS**
1 Choose **Stat ➤ ANOVA ➤ One-Way...** 2 Specify ENERGY in the **Response** text box 3 Specify REGION in the **Factor** text box 4 Click **OK**	1 Choose **DDXL ➤ Regression** 2 Select **1 Way ANOVA** from the **Function type** drop-down box 3 Specify ENERGY in the **Response Variable** text box 4 Specify REGION in the **Factor Variable** text box 5 Click **OK**	1 Press **STAT**, arrow over to **TESTS**, and press **ALPHA ➤ F** 2 Press **2nd ➤ LIST**, arrow down to NE and press **ENTER** 3 Press **, ➤ 2nd ➤ LIST**, arrow down to MW, and press **ENTER** 4 Press **, ➤ 2nd ➤ LIST**, arrow down to SO, and press **ENTER** 5 Press **, ➤ 2nd ➤ LIST**, arrow down to WE, and press **ENTER** 6 Press **)** and then **ENTER**

EXERCISES 16.3

Statistical Concepts and Skills

16.24 Suppose that a one-way ANOVA is being performed to compare the means of three populations and that the sample sizes are 10, 12, and 15. Determine the degrees of freedom for the F-statistic.

16.25 We stated earlier that a one-way ANOVA test is always right tailed because the null hypothesis is rejected only when the test statistic, F, is too large. Why is the null hypothesis rejected only when F is too large?

16.26 Following are the notations for the three sums of squares. State the name of each sum of squares and the source of variation each sum of squares represents.
a. SSE
b. $SSTR$
c. SST

16.27 State the one-way ANOVA identity and interpret its meaning with regard to partitioning the total variation in the data.

16.28 True or false: If you know any two of the three sums of squares, *SST*, *SSTR*, and *SSE*, you can determine the remaining one. Explain your answer.

16.29 Fill in the missing entries in the following partially completed one-way ANOVA table.

Source	df	SS	MS = SS/df	F-statistic
Treatment	2		21.652	
Error		84.400		
Total	14			

16.30 Fill in the missing entries in the following partially completed one-way ANOVA table.

Source	df	SS	MS = SS/df	F-statistic
Treatment		2.124	0.708	0.75
Error	20			
Total				

In Exercises 16.31 and 16.32,
a. *compute SST, SSTR, and SSE, using the defining formulas.*
b. *verify that the one-way ANOVA identity holds.*
c. *compute SST, SSTR, and SSE, using the computing formulas, and compare your answers with those obtained in part (a).*

16.31 Use the following three samples.

A	B	C
1	5	2
9	2	8
	4	5
	3	
	1	

16.32 Use the following four samples.

A	B	C	D
6	9	4	8
3	5	4	4
3	7	2	6
	8	2	
	6	3	

In Exercises 16.33 and 16.34, construct the one-way ANOVA table for the data. Compute SSTR and SSE by using the defining formulas given in Formula 16.1 on page 811.

16.33 The Assembly Line. The times required by three workers to perform an assembly-line task were recorded on five randomly selected occasions. Here are the times, to the nearest minute.

Hank	Susan	Joseph
8	8	10
10	9	9
9	9	10
11	8	11
10	10	9

(*Note:* $\bar{x}_1 = 9.6$, $\bar{x}_2 = 8.8$, $\bar{x}_3 = 9.8$, $s_1^2 = 1.3$, $s_2^2 = 0.7$, $s_3^2 = 0.7$, and $\bar{x} = 9.4$.)

16.34 Monthly Rents. The U.S. Bureau of the Census collects data on monthly rents of newly completed apartments and publishes the results in *Current Housing Reports*. Independent random samples of monthly rents for newly completed apartments in the four U.S. regions yielded the following data, in dollars.

Northeast	Midwest	South	West
905	870	911	865
798	748	650	852
848	699	881	930
1081	814	1056	766
1144	721		1109
	606		

(*Note:* $\bar{x}_1 = 955.2$, $\bar{x}_2 = 743.0$, $\bar{x}_3 = 874.5$, $\bar{x}_4 = 904.4$, $s_1^2 = 22{,}548.7$, $s_2^2 = 8{,}476.8$, $s_3^2 = 28{,}239.0$, $s_4^2 = 16{,}492.3$, and $\bar{x} = 862.7$.)

Preliminary data analyses indicate you can reasonably consider the assumptions for one-way ANOVA to be satisfied in Exercises 16.35–16.38. In each exercise, perform the required hypothesis test, using either the critical-value approach or the P-value approach.

16.35 Copepod Cuisine. Copepods are tiny crustaceans that are an essential link in the estuarine food web. Marine scientists G. Weiss, G. McManus, and H. Harvey at the Chesapeake Biological Laboratory in Maryland designed an experiment to determine whether dietary lipid (fat) content is important in the population growth of a Chesapeake Bay copepod. Their findings were published as the paper "Development and Lipid Composition of the Harpacticoid Copepod Nitocra Spinipes Reared on Different Diets" (*Marine Ecology Progress Series*, Vol. 132, pp. 57–61). Independent random samples of copepods were placed in containers containing lipid-rich diatoms, bacteria, or leafy macroalgae. There were 12 containers total with four replicates per diet. Five gravid (egg-bearing) females were placed in each container. After 14 days, the number of copepods in each container were as follows.

Diatoms	Bacteria	Macroalgae
426	303	277
467	301	324
438	293	302
497	328	272

At the 5% significance level, do the data provide sufficient evidence to conclude that a difference exists in mean number of copepods among the three different diets? (*Note:* $T_1 = 1828$, $T_2 = 1225$, $T_3 = 1175$, and $\Sigma x^2 = 1,561,154$.)

16.36 In Section 16.2, we considered two hypothetical examples to explain the logic behind one-way ANOVA. Now, you are to further examine those examples.
a. Refer to Table 16.1 on page 803. Perform a one-way ANOVA on the data and compare your conclusion to that stated in the corresponding "What Does it Mean?" box. Use $\alpha = 0.05$.
b. Repeat part (a) for the data displayed in Table 16.2 on page 804.

16.37 Weekly Earnings. The U.S. Bureau of Labor Statistics publishes data on weekly earnings of nonsupervisory workers in *Employment and Earnings*. The following data, in dollars, were obtained from independent random samples of full- and part-time nonsupervisory workers in five service-producing industries.

Transp. and Pub. util.	Wholesale trade	Retail trade	Finance, Insurance, Real estate	Services
632	613	349	571	497
672	558	277	525	509
633	538	259	607	516
677	607	387	493	432
724	591	274		469
655		368		406

At the 5% significance level, do the data provide sufficient evidence to conclude that a difference exists in mean weekly earnings of nonsupervisory workers among the five industries? (*Note:* $T_1 = 3993$, $T_2 = 2907$, $T_3 = 1914$, $T_4 = 2196$, $T_5 = 2829$, and $\Sigma x^2 = 7,540,345$.)

16.38 Permeation Sampling. Permeation sampling is a method of sampling air in buildings for pollutants. It can be used over a long period of time and is not affected by humidity, air currents, or temperature. In the paper "Calibration of Permeation Passive Samplers With Silicone Membranes Based on Physicochemical Properties of the Analytes" (*Analytical Chemistry*, Vol. 75, No. 13, pp. 3182–3192), B. Zabiegata, T. Gorecki, and J. Namiesnik obtained calibration constants experimentally for samples of compounds in each of four compound groups. The following data summarize their results.

Esters	Alcohols	Aliphatic hydrocarbons	Aromatic hydrocarbons
0.185	0.185	0.230	0.166
0.155	0.160	0.184	0.144
0.131	0.142	0.160	0.117
0.103	0.122	0.132	0.072
0.064	0.117	0.100	
	0.115	0.064	
	0.110		
	0.095		
	0.085		
	0.075		

At the 10% significance level, do the data provide sufficient evidence to conclude that a difference exists in mean calibration constant among the four compound groups? (*Note:* $T_1 = 0.638$, $T_2 = 1.206$, $T_3 = 0.870$, $T_4 = 0.499$, and $\Sigma x^2 = 0.456919$.)

Extending the Concepts and Skills

16.39 Political Prisoners. Journal articles and other sources frequently provide only summary statistics of data. This exercise gives you practice in working with such data in the context of ANOVA. According to the American Psychiatric Association, posttraumatic stress disorder (PTSD) is a common psychological consequence of traumatic events that involve threat to life or physical integrity. During the Cold War, some 200,000 people in East Germany were imprisoned for political reasons. Many of these prisoners were subjected to physical and psychological torture during their imprisonment, resulting in PTSD. Ehlers, Maercker, and Boos studied various characteristics of political prisoners from the former East Germany and presented their findings in the paper "Posttraumatic Stress Disorder (PTSD) Following Political Imprisonment: The Role of Mental Defeat, Alienation, and Perceived Permanent Change" (*Journal of Abnormal Psychology*, Vol. 109, pp. 45–55). The researchers randomly and independently selected 32 former prisoners diagnosed with chronic PTSD, 20 former prisoners diagnosed with PTSD after release from prison but subsequently recovered (remitted), and 29 diagnosed with no signs of PTSD. The ages, in years, of these people at time of arrest were as follows.

PTSD	n_j	$\bar{x}_j$	s_j
Chronic	32	25.8	9.2
Remitted	20	22.1	5.7
None	29	26.6	9.6

At the 1% significance level, do the data provide sufficient evidence to conclude that a difference exists in mean age at time of arrest among the three types of former prisoners? (*Note:* Use the formula

$$\bar{x} = \frac{n_1\bar{x}_1 + n_2\bar{x}_2 + n_3\bar{x}_3}{n_1 + n_2 + n_3}$$

to obtain the mean of all the observations.)

Confidence Intervals in One-Way ANOVA. Assume that the conditions for one-way ANOVA are satisfied and let $s = \sqrt{MSE}$. Then we have the following confidence-interval formulas.

- A $(1 - \alpha)$-level confidence interval for any particular population mean, say, μ_i, has endpoints

$$\bar{x}_i \pm t_{\alpha/2} \cdot \frac{s}{\sqrt{n_i}}.$$

- A $(1 - \alpha)$-level confidence interval for the difference between any two particular population means, say, μ_i and μ_j, has endpoints

$$(\bar{x}_i - \bar{x}_j) \pm t_{\alpha/2} \cdot s\sqrt{(1/n_i) + (1/n_j)}.$$

In both formulas, df $= n - k$, where, as usual, k denotes the number of populations and n denotes the total number of observations. Apply these formulas in Exercise 16.40.

16.40 Monthly Rents. Refer to Exercise 16.34.
a. Find and interpret a 99% confidence interval for the mean monthly rent of newly completed apartments in the Midwest.
b. Find and interpret a 99% confidence interval for the difference between the mean monthly rents of newly completed apartments in the Northeast and South.
c. What assumptions are you making in solving parts (a) and (b)?

16.41 Monthly Rents. Refer to Exercise 16.40. Suppose that you have obtained a 99% confidence interval for each of the two differences, $\mu_1 - \mu_2$ and $\mu_1 - \mu_3$. Can you be 99% confident of both results simultaneously, that is, that both differences are contained in their corresponding confidence intervals? Explain your answer.

Using Technology

In Exercises **16.42–16.45**, *use the technology of your choice to:*
a. *Obtain individual normal probability plots and the standard deviations of the samples.*
b. *Perform a residual analysis.*
c. *Decide whether conducting a one-way ANOVA on the data is reasonable. If so, also do parts (d) and (e).*
d. *Perform the specified hypothesis test.*
e. *Interpret the results of your hypothesis test.*

16.42 Cuckoo Care. Many species of cuckoos are brood parasites. The females lay their eggs in the nests of smaller bird species who then raise the young cuckoos at the expense of their own young. Data on the lengths, in millimeters, of cuckoo eggs found in the nests of three bird species—the Tree Pipet, Hedge Sparrow, and Pied Wagtail—are provided on the WeissStats CD. These data were collected by the late O. M. Latter in 1902 and used by L. H. C. Tippett in his text *The Methods of Statistics* (New York: Wiley, 1952, p. 176). At the 10% significance level, do the data provide sufficient evidence to conclude that a difference exists in the mean lengths of cuckoo eggs among the three bird species?

16.43 Magazine Ads. Advertising researchers Shuptrine and McVicker wanted to determine whether there were significant differences in the readability of magazine advertisements. Thirty magazines were classified based on their educational level—high, mid, or low—and then three magazines were randomly selected from each level. From each magazine, six advertisements were randomly chosen and examined for readability. In this particular case, readability was characterized by the numbers of words, sentences, and words of three syllables or more in each ad. The researchers published their findings in the *Journal of Advertising Research* (Vol. 21, p. 47). The number of words of three syllables or more in each ad are provided on the WeissStats CD. At the 5% significance level, do the data provide evidence of a difference in the mean number of words of three syllables or more among the three magazine levels?

16.44 Sickle Cell Disease. A study published by E. Anionwu et al. in the *British Medical Journal* (Vol. 282, pp. 283–286) measured the steady state hemoglobin levels of patients with three different types of sickle cell disease: HB SS, HB ST, and HB SC. The data are presented on the WeissStats CD. At the 5% significance level, do the data suggest a difference in mean steady state hemoglobin levels among the three types of sickle cell disease?

16.45 Prolonging Life. Vitamin C (ascorbate) boosts the human immune system and is effective in preventing a variety of illnesses. In a study by E. Cameron and L. Pauling, published in the *Proceedings of the National Academy of Science USA* (Vol. 75, pp. 4538–4542), patients in advanced stages of cancer were given a Vitamin C supplement. Patients were grouped according to the organ affected by cancer: stomach, bronchus, colon, ovary, or breast. The study yielded the survival times, in days, given on the WeissStats CD. At the 5% significance level, do the data provide sufficient evidence to conclude that, for patients in advanced stages of cancer who are given a Vitamin C treatment, a difference exists in mean survival time among the five different types of cancer?

16.4   Multiple Comparisons[†]

Suppose that you perform a one-way ANOVA and reject the null hypothesis. Then you can conclude that the means of the populations under consideration are not all the same. Once you make that decision, you may also want to know which means are different, which mean is largest, or, more generally, the relation among all the means. Methods for dealing with these problems are called **multiple comparisons.**

Several multiple-comparison methods are available. In this book, we discuss the **Tukey multiple-comparison method.** Other commonly used multiple-comparison methods are the Bonferroni method, the Fisher method, and the Scheffé method.

One approach for implementing multiple comparisons is to obtain confidence intervals for the differences between all possible pairs of population means. Two means are declared different if the confidence interval for their difference does not contain 0. (If a confidence interval for the difference between two population means does not contain 0, we can reject the null hypothesis that the two means are equal in favor of the alternative hypothesis that the two means are different and vice versa. See Exercise 10.13 on page 490.)

[†]If you plan to study the chapter *Design of Experiments and Analysis of Variance* on the WeissStats CD, you should cover this section.

In multiple comparisons, we must distinguish between the *individual confidence level* and the *family confidence level.* The **individual confidence level** is the confidence we have that any particular confidence interval contains the difference between the corresponding population means. The **family confidence level** is the confidence we have that all the confidence intervals contain the differences between the corresponding population means.

The Studentized Range Distribution

The Tukey multiple-comparison method is based on the **studentized range distribution,** or simply the **q-distribution.** A variable has a q-distribution if its distribution has the shape of a special type of right-skewed curve, called a **q-curve.** Actually, there are infinitely many q-distributions (and q-curves); a particular one is identified by two parameters, which we denote κ (kappa) and ν (nu).

Percentages and probabilities for a variable having a q-distribution equal areas under its associated q-curve. To perform a Tukey multiple comparison, we need to know how to find the q-value having a specified area to its right. We use the symbol q_α to denote the q-value having area α to its right. Values of $q_{0.01}$ and $q_{0.05}$ are presented in Tables X and XI, respectively, in Appendix A. In Example 16.5, we explain how to use Table XI.

Example 16.5 **Finding the q-Value Having a Specified Area to Its Right**

For the q-curve with parameters $\kappa = 4$ and $\nu = 16$, find $q_{0.05}$; that is, find the q-value having area 0.05 to its right, as shown in Fig. 16.10(a).

FIGURE 16.10
Finding the q-value having area 0.05 to its right

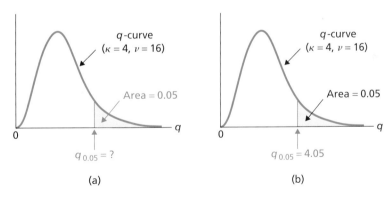

(a)   (b)

Solution To obtain the q-value in question, we use Table XI with $\kappa = 4$ and $\nu = 16$. We first go down the outside columns to the row labeled "16." Then we go across that row to the column headed "4." The number in the body of the table there, 4.05, is the required q-value. That is, for the q-curve with parameters $\kappa = 4$ and $\nu = 16$, the q-value having area 0.05 to its right is 4.05, as shown in Fig. 16.10(b).

The Tukey Multiple-Comparison Method

The formulas used in the Tukey multiple-comparison method for obtaining confidence intervals for the differences between means are similar to the pooled *t*-interval formula (Procedure 10.2 on page 497). The essential difference is that, in the Tukey multiple-comparison method, we consult a *q*-table instead of a *t*-table.

We now present, in Procedure 16.2, a step-by-step method for performing a Tukey multiple comparison. As you can see, the assumptions for its use are the same as those for a one-way ANOVA test.

PROCEDURE 16.2

Tukey Multiple-Comparison Method

Purpose To determine the relationship among k population means, $\mu_1, \mu_2, \ldots, \mu_k$

Assumptions

1. Simple random samples
2. Independent samples
3. Normal populations
4. Equal population standard deviations

STEP 1 Decide on the family confidence level, $1 - \alpha$.

STEP 2 Find q_α for the *q*-curve with parameters $\kappa = k$ and $\nu = n - k$, where n is the total number of observations.

STEP 3 Obtain the endpoints of the confidence interval for $\mu_i - \mu_j$:

$$(\bar{x}_i - \bar{x}_j) \pm \frac{q_\alpha}{\sqrt{2}} \cdot s \sqrt{(1/n_i) + (1/n_j)},$$

where $s = \sqrt{MSE}$. Do so for all possible pairs of means with $i < j$.

STEP 4 Declare two population means different if the confidence interval for their difference does not contain 0; otherwise, do not declare the two population means different.

STEP 5 Summarize the results in Step 4 by ranking the sample means from smallest to largest and by connecting with lines those whose population means were not declared different.

STEP 6 Interpret the results of the multiple comparison.

Performing a Tukey Multiple Comparison

In Example 16.3 on page 812, we conducted a one-way ANOVA to decide whether a difference exists in last year's mean energy consumption by households among the four U.S. regions. Specifically, we performed the hypothesis test

H_0: $\mu_1 = \mu_2 = \mu_3 = \mu_4$ (mean energy consumptions are equal)

H_a: Not all the means are equal,

at the 5% significance level, where μ_1, μ_2, μ_3, and μ_4 denote last year's mean energy consumptions for households in the Northeast, Midwest, South, and West, respectively.

The test results were statistically significant; that is, we rejected H_0. Thus we can conclude at the 5% significance level that, for last year, at least two of the regions have different mean energy consumptions. The Tukey multiple-comparison method allows us to elaborate on this conclusion, as we show in Example 16.6.

Example 16.6   The Tukey Multiple-Comparison Method

Energy Consumption   Apply the Tukey multiple-comparison method to the energy consumption data, repeated in Table 16.7.

TABLE 16.7

Last year's energy consumptions for samples of households in the four U.S. regions

Northeast	Midwest	South	West
15	17	11	10
10	12	7	12
13	18	9	8
14	13	13	7
13	15		9
	12		

Solution   We apply Procedure 16.2.

STEP 1   Decide on the family confidence level, $1 - \alpha$.

As we have done previously in this application, we use $\alpha = 0.05$, so the family confidence level is 0.95 (95%).

STEP 2   Find q_α for the q-curve with parameters $\kappa = k$ and $\nu = n - k$, where n is the total number of observations.

From Table 16.7, $\kappa = k = 4$ and $\nu = n - k = 20 - 4 = 16$. In Table XI, we find that $q_\alpha = q_{0.05} = 4.05$.

STEP 3   Obtain the endpoints of the confidence interval for $\mu_i - \mu_j$:

$$(\bar{x}_i - \bar{x}_j) \pm \frac{q_\alpha}{\sqrt{2}} \cdot s \sqrt{(1/n_i) + (1/n_j)},$$

where $s = \sqrt{MSE}$. Do so for all possible pairs of means with $i < j$.

To begin, we construct a table that gives the sample means and sample sizes. From the data in Table 16.7, we obtain the entries for Table 16.8.

Region	Northeast	Midwest	South	West
j	1	2	3	4
$\bar{x}_j$	13.0	14.5	10.0	9.2
n_j	5	6	4	5

From Step 2, $q_\alpha = 4.05$. Also, on page 808, we found that $MSE = 5.144$ for the energy consumption data. Now, we are ready to obtain the required confidence intervals. The endpoints of the confidence interval for $\mu_1 - \mu_2$ are

$$(13.0 - 14.5) \pm \frac{4.05}{\sqrt{2}} \cdot \sqrt{5.144}\sqrt{(1/5) + (1/6)},$$

or -5.43 to 2.43.

Likewise, the endpoints of the confidence interval for $\mu_1 - \mu_3$ are

$$(13.0 - 10.0) \pm \frac{4.05}{\sqrt{2}} \cdot \sqrt{5.144}\sqrt{(1/5) + (1/4)},$$

or -1.36 to 7.36.

Proceeding in the same way, we obtain the remaining confidence intervals. All six confidence intervals are displayed in Table 16.9 where, for ease of reference, we have placed parenthetically the numbers used to represent the regions.

TABLE 16.9

Simultaneous 95% confidence intervals
for the differences between the energy
consumption means

	Northeast (1)	Midwest (2)	South (3)
Midwest (2)	$(-5.43, 2.43)$		
South (3)	$(-1.36, 7.36)$	$(0.31, 8.69)$	
West (4)	$(-0.31, 7.91)$	$(1.37, 9.23)$	$(-3.56, 5.16)$

Each entry in Table 16.9 is the confidence interval for the difference between the mean labeled by the column and the mean labeled by the row. For instance, the entry in the column labeled "Midwest (2)" and the row labeled "West (4)" is $(1.37, 9.23)$. So the confidence interval for the difference, $\mu_2 - \mu_4$, between last year's mean energy consumptions for households in the Midwest and West is from 1.37 to 9.23.

STEP 4 Declare two population means different if the confidence interval for their difference does not contain 0; otherwise, do not declare the two population means different.

Referring to Table 16.9, we see that we can declare the means μ_2 and μ_3 different and the means μ_2 and μ_4 different; all other pairs of means are not declared different.

STEP 5 **Summarize the results in Step 4 by ranking the sample means from smallest to largest and by connecting with lines those whose population means were not declared different.**

In light of Table 16.8, Step 4, and the numbering used to represent the U.S. regions (shown parenthetically), we obtain the following diagram.

West (4)　　South (3)　　Northeast (1)　　Midwest (2)
　9.2　　　　10.0　　　　　13.0　　　　　14.5

STEP 6 **Interpret the results of the multiple comparison.**

INTERPRETATION Referring to the diagram in Step 5, we conclude that last year's mean energy consumption in the Midwest exceeds that in the West and South, and that no other means can be declared different. All of this can be said with 95% confidence, the family confidence level.

Example 16.6 demonstrates that multiple comparisons require extensive computations, even when the number of populations and the sample sizes are quite small. This difficulty explains why multiple comparisons are almost always done by computer.

The
Technology
Center

Some statistical technologies (e.g., Minitab and SPSS) have dedicated programs that automatically perform a Tukey multiple comparison, but others (e.g., Excel and the TI-83/84 Plus) do not. For a statistical technology that does not have a dedicated program for conducting a Tukey multiple comparison, you can often use the macro capabilities of the statistical technology to write such a program. Refer to the technology manuals for more details.

EXERCISES 16.4

Statistical Concepts and Skills

16.46 What is the purpose of doing a multiple comparison?

16.47 Fill in the blank: If a confidence interval for the difference between two population means does not contain _____, we can reject the null hypothesis that the two means are equal in favor of the alternative hypothesis that the two means are different; and vice versa.

16.48 Explain the difference between the family confidence level and the individual confidence level.

16.49 Regarding family and individual confidence levels, answer the following questions and explain your answers.
a. Which is smaller for multiple comparisons involving three or more means, the family confidence level or the individual confidence level?
b. For multiple comparisons involving two means, what is the relationship between the family confidence level and the individual confidence level?

16.50 What is the name of the distribution on which the Tukey multiple-comparison method is based? What is its abbreviation?

16.51 The parameter ν for the q-curve in a Tukey multiple comparison equals one of the degrees of freedom for the F-curve in a one-way ANOVA. Which one?

16.52 Explain the essential difference between obtaining a confidence interval by using the pooled t-interval procedure and obtaining a confidence interval by using the Tukey multiple-comparison procedure.

16.53 Determine the following for a q-curve with parameters $\kappa = 6$ and $\nu = 13$.
a. The q-value having area 0.05 to its right
b. $q_{0.01}$

16.54 Determine the following for a q-curve with parameters $\kappa = 8$ and $\nu = 20$.
a. The q-value having area 0.01 to its right
b. $q_{0.05}$

*In Exercises **16.55–16.57**, we repeat information from Exercises 16.35–16.37. For each exercise here, use Procedure 16.2 on page 824 to perform a Tukey multiple comparison at the 95% family confidence level.*

16.55 Copepod Cuisine. Following are the data on the number of copepods in each of 12 containers after 14 days for three different diets from Exercise 16.35.

Diatoms	Bacteria	Macroalgae
426	303	277
467	301	324
438	293	302
497	328	272

16.56 From Exercise 16.36: In Section 16.2, we considered two hypothetical examples to explain the logic for one-way ANOVA.

a. Refer to Table 16.1 on page 803.
b. Refer to Table 16.2 on page 804.

16.57 Weekly Earnings. Following are the weekly earnings, in dollars, of full- and part-time nonsupervisory workers in five service-producing industries from Exercise 16.37.

Transp. and Pub. util.	Wholesale trade	Retail trade	Finance, Insurance, Real estate	Services
632	613	349	571	497
672	558	277	525	509
633	538	259	607	516
677	607	387	493	432
724	591	274		469
655		368		406

16.58 Driving Golf Balls. Each manufacturer of golf balls always seems to be claiming that its balls go the farthest. A writer for a sports magazine decided to conduct an impartial test. She randomly selected 20 golf professionals and then randomly assigned four golfers to each of five brands. Each golfer drove the assigned brand of ball. The driving distances, in yards, are displayed in the following table.

Brand 1	Brand 2	Brand 3	Brand 4	Brand 5
293	286	277	291	288
283	284	269	278	300
288	291	284	276	283
281	295	287	282	299

a. Perform a Tukey multiple comparison on these data at the 90% family confidence level. (*Note:* For a q-curve with $\kappa = 5$ and $\nu = 15$, we have $q_{0.10} = 3.83$.)

b. Without doing any further work, decide at the 10% significance level whether the data provide sufficient evidence to conclude that a difference exists in mean driving distances among the five brands of golf balls. Explain your reasoning.

Extending the Concepts and Skills

16.59 Political Prisoners. Refer to Exercise 16.39. Following are summary statistics on the ages, in years, at time of arrest for independent random samples of former East German political prisoners classified according to their current status relative to posttraumatic stress disorder (PTSD).

PTSD	n_j	$\bar{x}_j$	s_j
Chronic	32	25.8	9.2
Remitted	20	22.1	5.7
None	29	26.6	9.6

Conduct a Tukey multiple comparison at a family confidence level of 0.99.

16.60 Explain why the family confidence level, not the individual confidence level, is the appropriate level for comparing all population means simultaneously.

16.61 In Step 3 of Procedure 16.2, we obtain confidence intervals only when $i < j$.
a. Explain how to determine the remaining confidence intervals from those obtained.

b. Apply your answer from part (a) and Table 16.9 on page 826 to determine the remaining six confidence intervals for the differences between the energy consumption means.

Using Technology

*In Exercises **16.62–16.65**, we repeat information from Exercises 16.42–16.45. In each of those exercises, you were asked to decide whether conducting a one-way ANOVA on the data is reasonable. For those exercises where it is, use the technology of your choice to perform a Tukey multiple comparison at the designated family confidence level.*

16.62 Cuckoo Care. Use the data on the WeissStats CD for the lengths, in millimeters, of cuckoo eggs found in the nests of three different bird species referred to in Exercise 16.42. Use a family confidence level of 0.90.

16.63 Magazine Ads. Use the data on the WeissStats CD for the number of words of three syllables or more in advertisements from magazines of three different educational levels referred to in Exercise 16.43. Use a family confidence level of 0.95.

16.64 Sickle Cell Disease. Use the data on the WeissStats CD for the steady state hemoglobin levels of patients with three different types of sickle cell disease referred to in Exercise 16.44. Use a family confidence level of 0.95.

16.65 Prolonging Life. Use the data on the WeissStats CD for the survival times, in days, of patients given a Vitamin C supplement who were in advanced stages of cancer of five different types referred to in Exercise 16.45. Use a family confidence level of 0.95.

16.5 The Kruskal–Wallis Test[†]

In this section, we examine the **Kruskal–Wallis test,** a nonparametric alternative to the one-way ANOVA procedure discussed in Section 16.3. The Kruskal–Wallis test applies when the distributions (one for each population) of the variable under consideration have the same shape; it does not require that the distributions be normal or have any other specific shape.

Like the Mann–Whitney test, the Kruskal–Wallis test is based on ranks. When ties occur, ranks are assigned in the same way as in the Mann–Whitney test: *If two or more observations are tied, each is assigned the mean of the ranks they would have had if there were no ties.* We introduce the Kruskal–Wallis test in Example 16.7.

[†]This section continues the coverage of nonparametric statistics.

Example 16.7 Introducing the Kruskal–Wallis Test

Vehicle Miles   The U.S. Federal Highway Administration conducts annual surveys on motor vehicle travel by type of vehicle and publishes its findings in *Highway Statistics*. Independent simple random samples of cars, buses, and trucks provided the data on number of miles driven last year, in thousands, shown in Table 16.10.

TABLE 16.10

Number of miles driven (1000s) last year for independent samples of cars, buses, and trucks

Cars	Buses	Trucks
19.9	1.8	24.6
15.3	7.2	37.0
2.2	7.2	21.2
6.8	6.5	23.6
34.2	13.3	23.0
8.3	25.4	15.3
12.0		57.1
7.0		14.5
9.5		26.0
1.1		

Suppose that we want to use the sample data in Table 16.10 to decide whether a difference exists in last year's mean number of miles driven among cars, buses, and trucks.

a. Formulate the problem statistically by posing it as a hypothesis test.

b. Is it appropriate to apply the one-way ANOVA test here? What about the Kruskal–Wallis test?

c. Explain the basic idea for carrying out a Kruskal–Wallis test.

d. Discuss the use of the sample data in Table 16.10 to make a decision concerning the hypothesis test.

Solution

a. Let μ_1, μ_2, and μ_3 denote last year's mean number of miles driven for cars, buses, and trucks, respectively. Then the null and alternative hypotheses are

$$H_0\text{: } \mu_1 = \mu_2 = \mu_3 \text{ (mean miles driven are equal)}$$
$$H_a\text{: Not all the means are equal.}$$

b. Preliminary data analyses (not shown) suggest that the distributions of miles driven have roughly the same shape for cars, buses, and trucks but that those distributions are far from normal. Thus, although the one-way ANOVA test of Section 16.3 is probably inappropriate, the Kruskal–Wallis procedure appears suitable.

c. To apply the Kruskal–Wallis test, we first rank the data from all three samples combined. The results of this ranking appear in Table 16.11.

TABLE 16.11

Results of ranking the combined data from Table 16.10

Cars	Rank	Buses	Rank	Trucks	Rank
19.9	16	1.8	2	24.6	20
15.3	14.5	7.2	7.5	37.0	24
2.2	3	7.2	7.5	21.2	17
6.8	5	6.5	4	23.6	19
34.2	23	13.3	12	23.0	18
8.3	9	25.4	21	15.3	14.5
12.0	11			57.1	25
7.0	6			14.5	13
9.5	10			26.0	22
1.1	1				
	9.850		9.000		19.167	← *Mean ranks*

The idea behind the Kruskal–Wallis test is simple: If the null hypothesis of equal population means is true, the means of the ranks for the three samples should be roughly equal. Put another way, if the variation among the mean ranks for the three samples is too large, we have evidence against the null hypothesis.

To measure the variation among the mean ranks, we use the treatment sum of squares, *SSTR*, computed for the ranks. To decide whether that quantity is too large, we compare it to the variance of all the ranks, which can be expressed as $SST/(n-1)$, where *SST* is the total sum of squares for the ranks and *n* is the total number of observations.[†] More precisely, the test statistic for a Kruskal–Wallis test, denoted *H*, is

$$H = \frac{SSTR}{SST/(n-1)}.$$

Large values of *H* indicate that the variation among the mean ranks is large (relative to the variance of all the ranks) and hence that the null hypothesis of equal population means should be rejected.

d. For the ranks in Table 16.11, we find that $SSTR = 537.475$, $SST = 1299$, and $n = 25$. Thus the value of the test statistic is

$$H = \frac{SSTR}{SST/(n-1)} = \frac{537.475}{1299/24} = 9.930.$$

Is this value of *H* large enough to conclude that the null hypothesis of equal population means is false? To answer this question, we need to know the distribution of the variable *H*. We present that distribution in Key Fact 16.5 and then complete the hypothesis test considered in this example. ▪

What
Does it Mean?

The *H*-statistic is the ratio of the variation among the mean ranks to the variation of all the ranks.

KEY FACT 16.5

Distribution of the H-Statistic for a Kruskal–Wallis Test

Suppose that the *k* distributions (one for each population) of the variable under consideration have the same shape. Then, for independent samples from the *k* populations, the variable

$$H = \frac{SSTR}{SST/(n-1)}$$

has approximately a chi-square distribution with df $= k - 1$ if the null hypothesis of equal population means is true. Here, *n* denotes the total number of observations.

Note: A rule of thumb for using the chi-square distribution as an approximation to the true distribution of *H* is that all sample sizes should be 5 or greater.

[†]Recall from Sections 16.2 and 16.3 that the treatment sum of squares, *SSTR*, is a measure of variation among means and that the total sum of squares, *SST*, is a measure of variation among all the data. The defining and computing formulas for *SSTR* and *SST* are given in Formula 16.1 on page 811. For the Kruskal–Wallis test, we apply those formulas to the ranks of the sample data, not to the sample data themselves.

Although we adopt that rule of thumb, some statisticians consider it too restrictive. Instead, they regard the chi-square approximation to be adequate unless $k = 3$ and none of the sample sizes exceed 5.

Computing Formula for H

> **What** *Does it Mean?*
>
> This is the computing formula for H, used for hand calculations.

Usually, an easier way to compute the test statistic H by hand from the raw data is to apply the computing formula

$$H = \frac{12}{n(n+1)} \sum_{j=1}^{k} \frac{R_j^2}{n_j} - 3(n+1),$$

where R_1 denotes the sum of the ranks for the sample data from Population 1, R_2 denotes the sum of the ranks for the sample data from Population 2, and so on.

Strictly speaking, the computing formula for H is equivalent to the defining formula for H only if no ties occur. In practice, however, the computing formula provides a sufficiently accurate approximation unless the number of ties is relatively large.

Performing the Kruskal–Wallis Test

Procedure 16.3 on the next page provides a step-by-step method for conducting a Kruskal–Wallis test by using either the critical-value approach or the P-value approach. Because the null hypothesis is rejected only when the test statistic, H, is too large, a Kruskal–Wallis test is always right tailed.

Although the Kruskal–Wallis test can be used to compare several population medians as well as several population means, we state Procedure 16.3 in terms of population means. To utilize the procedure for population medians, simply replace μ_1 by η_1, μ_2 by η_2, and so on.

Regarding the third and fourth assumptions required for employing Procedure 16.3, note the following:

- Assumption 3: For brevity, we use the phrase "same-shape populations" to indicate that the k distributions of the variable under consideration have the same shape.
- Assumption 4: This assumption is necessary only when we are using the chi-square distribution as an approximation to the distribution of H. Tables of critical values for H are available in cases where Assumption 4 fails.

Example 16.8 illustrates application of Procedure 16.3.

Example 16.8 The Kruskal–Wallis Test

Vehicle Miles We now complete the hypothesis test introduced in Example 16.7. Independent simple random samples of cars, buses, and trucks provided the data on number of miles driven last year, in thousands, shown in

PROCEDURE 16.3	Kruskal–Wallis Test

Purpose To perform a hypothesis test to compare k population means, $\mu_1, \mu_2, \ldots, \mu_k$

Assumptions

1. Simple random samples
2. Independent samples
3. Same-shape populations
4. All sample sizes are 5 or greater

STEP 1 The null and alternative hypotheses are

$$H_0: \mu_1 = \mu_2 = \cdots = \mu_k$$
$$H_a: \text{Not all the means are equal.}$$

STEP 2 Decide on the significance level, α.

STEP 3 Construct a work table to rank the data from all the samples combined.

STEP 4 Compute the value of the test statistic

$$H = \frac{12}{n(n+1)} \sum_{j=1}^{k} \frac{R_j^2}{n_j} - 3(n+1)$$

and denote that value H_0. Here, n is the total number of observations and $R_1, R_2, \ldots, R_k$ denote the sums of the ranks for the sample data from Populations 1, 2, ..., k, respectively.

CRITICAL-VALUE APPROACH

STEP 5 The critical value is χ_α^2 with df $= k - 1$. Use Table VII to find the critical value.

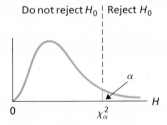

STEP 6 If the value of the test statistic falls in the rejection region, reject H_0; otherwise, do not reject H_0.

P-VALUE APPROACH

STEP 5 The H-statistic has df $= k - 1$. Use Table VII to estimate the P-value or obtain it exactly by using technology.

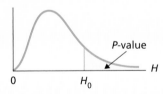

STEP 6 If $P \leq \alpha$, reject H_0; otherwise, do not reject H_0.

STEP 7 Interpret the results of the hypothesis test.

Table 16.10 on page 830. At the 5% significance level, do the data provide sufficient evidence to conclude that a difference exists in last year's mean number of miles driven among cars, buses, and trucks?

Solution We apply Procedure 16.3.

STEP 1 State the null and alternative hypotheses.

Let μ_1, μ_2, and μ_3 denote last year's mean number of miles driven for cars, buses, and trucks, respectively. Then the null and alternative hypotheses are

$$H_0: \mu_1 = \mu_2 = \mu_3 \text{ (mean miles driven are equal)}$$
$$H_a: \text{Not all the means are equal.}$$

STEP 2 Decide on the significance level, α.

We are to perform the hypothesis test at the 5% significance level, or $\alpha = 0.05$.

STEP 3 Construct a work table to rank the data from all the samples combined.

We have already done so—as Table 16.11 on page 830. For ease of reference, we repeat the relevant portion of that table here.

Cars	Rank	Buses	Rank	Trucks	Rank
19.9	16	1.8	2	24.6	20
15.3	14.5	7.2	7.5	37.0	24
2.2	3	7.2	7.5	21.2	17
6.8	5	6.5	4	23.6	19
34.2	23	13.3	12	23.0	18
8.3	9	25.4	21	15.3	14.5
12.0	11			57.1	25
7.0	6			14.5	13
9.5	10			26.0	22
1.1	1				

STEP 4 Compute the value of the test statistic

$$H = \frac{12}{n(n+1)} \sum_{j=1}^{k} \frac{R_j^2}{n_j} - 3(n+1).$$

We have $n = 10 + 6 + 9 = 25$. Summing the second, fourth, and sixth columns of the work table yields $R_1 = 98.5$, $R_2 = 54.0$, and $R_3 = 172.5$. Thus the value of the test statistic is

$$H = \frac{12}{25(25+1)} \left(\frac{98.5^2}{10} + \frac{54.0^2}{6} + \frac{172.5^2}{9} \right) - 3(25+1) = 9.923.$$

CRITICAL-VALUE APPROACH	P-VALUE APPROACH

CRITICAL-VALUE APPROACH

STEP 5 The critical value is χ_{α}^2 with df = k − 1. Use Table VII to find the critical value.

We have $k = 3$—the three types of vehicles—so df $= 3 - 1 = 2$. From Table VII, the critical value is $\chi_{0.05}^2 = 5.991$, as shown in Fig. 16.11A.

FIGURE 16.11A

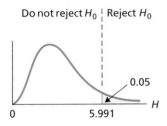

STEP 6 If the value of the test statistic falls in the rejection region, reject H_0; otherwise, do not reject H_0.

From Step 4, we see that the value of the test statistic is $H = 9.923$. Figure 16.11A shows that this value falls in the rejection region. Thus we reject H_0. The test results are statistically significant at the 5% level.

P-VALUE APPROACH

STEP 5 The H-statistic has df = k − 1. Use Table VII to estimate the P-value or obtain it exactly by using technology.

From Step 4, we see that the value of the test statistic is $H = 9.923$. Because the test is right tailed, the P-value is the probability of observing a value of H of 9.923 or greater if the null hypothesis is true. That probability equals the shaded area shown in Fig. 16.11B.

FIGURE 16.11B

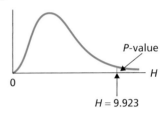

We have $k = 3$—the three types of vehicles—so df $= 3 - 1 = 2$. Referring to Fig. 16.11B and Table VII, we find that $0.005 < P < 0.01$. (Using technology, we obtain $P = 0.007$.)

STEP 6 If $P \leq \alpha$, reject H_0; otherwise, do not reject H_0.

From Step 5, $0.005 < P < 0.01$. Because the P-value is less than the specified significance level of 0.05, we reject H_0. The test results are statistically significant at the 5% level and (see Table 9.12 on page 440) provide very strong evidence against the null hypothesis.

STEP 7 Interpret the results of the hypothesis test.

INTERPRETATION At the 5% significance level, the data provide sufficient evidence to conclude that a difference exists in last year's mean number of miles driven among cars, buses, and trucks.

Comparison of the Kruskal–Wallis Test and the One-Way ANOVA Test

In Section 16.3, you learned how to perform a one-way ANOVA to compare k population means, using independent samples, when the variable under consideration is normally distributed on each of the k populations and the population standard deviations are equal. Because normal distributions that have equal standard deviations have the same shape, you can also use the Kruskal–Wallis test to perform such a hypothesis test.

Which test is better to use under these circumstances? As you might expect, it is the one-way ANOVA test because that test is designed expressly for normal populations—under conditions of normality, the one-way ANOVA test is more powerful than the Kruskal–Wallis test. What is somewhat surprising is that the one-way ANOVA test is not much more powerful than the Kruskal–Wallis test.

However, if the distributions of the variable under consideration have the same shape but are not normal, the Kruskal–Wallis test is usually more powerful than the one-way ANOVA test—often considerably more so. In summary, we have Key Fact 16.6.

KEY FACT 16.6

The Kruskal–Wallis Test Versus the One-Way ANOVA Test

Suppose that the distributions of a variable of several populations have the same shape and that you want to compare the population means, using independent samples. To decide between the one-way ANOVA test and the Kruskal–Wallis test, follow these guidelines: If you are reasonably sure that the distributions are normal, use the one-way ANOVA test; otherwise, use the Kruskal–Wallis test.

The Technology Center

Some statistical technologies have a dedicated program that automatically performs a Kruskal–Wallis test; others do not. For instance, Minitab has such a program but, currently, neither Excel nor the TI-83/84 Plus do. In this subsection, we present output and step-by-step instructions to implement Minitab's Kruskal–Wallis test program.

For a statistical technology that does not have a dedicated program for conducting a Kruskal–Wallis test, you can often use the macro capabilities of the statistical technology to write a program. Refer to the technology manuals for more details.

Example 16.9 **Using Technology to Conduct a Kruskal–Wallis Test**

Vehicle Miles Use Minitab to perform the hypothesis test considered in Example 16.8 on page 832.

Solution Let μ_1, μ_2, and μ_3 denote last year's mean number of miles driven for cars, buses, and trucks, respectively. The task is to use the Kruskal–Wallis procedure to perform the hypothesis test

$$H_0: \mu_1 = \mu_2 = \mu_3 \text{ (mean miles driven are equal)}$$
$$H_a: \text{Not all the means are equal,}$$

at the 5% significance level.

Printout 16.2 shows the output obtained by applying Minitab's Kruskal–Wallis test program to the samples of vehicle miles presented in Table 16.10 on page 830.

PRINTOUT 16.2

Kruskal–Wallis test output for the samples of vehicle miles

MINITAB

Kruskal-Wallis Test: MILES versus VEHICLE

```
Kruskal-Wallis Test on MILES

VEHICLE    N  Median  Ave Rank      Z
Buses      6   7.200       9.0  -1.53
Cars      10   8.900       9.9  -1.75
Trucks     9  23.600      19.2   3.14
Overall   25             13.0

H = 9.92  DF = 2  P = 0.007
H = 9.93  DF = 2  P = 0.007  (adjusted for ties)
```

From the Minitab output in Printout 16.2, the *P*-value for the hypothesis test is 0.007. Because the *P*-value is less than the specified significance level of 0.05, we reject H_0. At the 5% significance level, the data provide sufficient evidence to conclude that a difference exists in last year's mean number of miles driven among cars, buses, and trucks.

Obtaining the Output (Optional)

Printout 16.2 provides output from Minitab for a Kruskal–Wallis test based on the samples of vehicle miles displayed in Table 16.10. The following are detailed instructions for obtaining that output. First, we store all 25 mileages from Table 16.10 in a column named MILES. Next, in a column named VEHICLE, we store the vehicle types corresponding to the mileages in MILES. Then, we proceed as follows.

MINITAB

1 Choose **Stat ➤ Nonparametrics ➤ Kruskal-Wallis...**
2 Specify MILES in the **Response** text box
3 Specify VEHICLE in the **Factor** text box
4 Click **OK**

EXCEL

SEE THE EXCEL MANUAL

TI-83/84 PLUS

SEE THE TI-83/84 PLUS MANUAL

EXERCISES 16.5

Statistical Concepts and Skills

16.66 Of what test is the Kruskal–Wallis test a nonparametric version?

16.67 State the conditions required for performing a Kruskal–Wallis test.

16.68 In the Kruskal–Wallis test, how should you deal with tied ranks?

16.69 Fill in the blank: If the null hypothesis of equal population means is true, the sample mean ranks should be roughly _____.

16.70 For a Kruskal–Wallis test, how do you
a. measure variation among sample mean ranks?
b. measure total variation of all the ranks?
c. decide whether the variation among sample mean ranks is large enough to warrant rejection of the null hypothesis of equal population means?

16.71 For a Kruskal–Wallis test to compare the means of five populations, what is the approximate distribution of the test statistic H?

In Exercises **16.72–16.75**, *perform a Kruskal–Wallis test by using either the critical-value approach or the P-value approach.*

16.72 Entertainment Expenditures. The U.S. Bureau of Labor Statistics conducts surveys on consumer expenditures for various types of entertainment and publishes its findings in *Consumer Expenditure Survey*. Independent random samples yielded the following data, in dollars, on last year's expenditures for three entertainment categories.

Fees and admissions	TV, radio, and sound equipment	Other equipment and services
303	230	130
242	1878	381
152	526	1423
625	130	161
241	600	205
1333	130	1154
739	692	1759
430		232
		1368

At the 5% significance level, do the data provide sufficient evidence to conclude that a difference exists in last year's mean expenditures among the three entertainment categories?

16.73 Lowfat Milk Consumption. Indications are that Americans have become more aware of the dangers of excessive fat intake in their diets, although some reversal of this awareness appears to have developed in recent years. The U.S. Department of Agriculture publishes data on annual consumption of selected beverages in *Food Consumption, Prices, and Expenditures*. Independent random samples of lowfat milk consumption for 1980, 1990, and 2000 yielded the following data, in gallons.

1980	1990	2000
11.1	15.4	11.6
10.7	15.9	13.1
8.6	16.0	17.8
9.4	14.6	17.5
9.2	11.4	13.8
15.1	17.0	11.8
11.6	16.1	14.3
8.3		15.0
		15.6

At the 1% significance level, do the data provide sufficient evidence to conclude that there is a difference in mean (per capita) consumption of lowfat milk for the years 1980, 1990, and 2000?

16.74 Ages of Car Buyers. Information on characteristics of new-car buyers appears in *Buyers of New Cars*, a publication of Newsweek, Inc. Independent random samples of new-car buyers yielded the following data on age of purchaser, in years, by origin of car purchased.

Domestic	Asian	European
41	78	72
42	42	42
51	51	58
47	45	39
33	21	67
83	24	39
35	21	45
69	39	27
50	45	33
60	30	55

Do the data provide sufficient evidence to conclude that a difference exists in the median ages of buyers of new domestic, Asian, and European cars? Use $\alpha = 0.05$.

16.75 Home Size. The U.S. Bureau of the Census publishes information on the sizes of housing units in *Current Housing Reports*. Independent random samples of single-family detached homes (including mobile homes) in the four U.S. regions yielded the following data on square footage.

Northeast	Midwest	South	West
3182	2115	1591	1345
2130	2413	1354	694
1781	1639	722	2789
2989	1691	2135	1649
1581	1655	1982	2203
2149	1605	1639	2068
2286	3361	642	1565
1293	2058	1513	1655

At the 5% significance level, do the data provide sufficient evidence to conclude that a difference exists in median square footage of single-family detached homes among the four U.S. regions?

16.76 Weekly Earnings. Refer to the data in Exercise 16.37 on page 820.
a. Do the data provide sufficient evidence to conclude that a difference exists in mean weekly earnings among nonsupervisory workers in the five industries? Perform a Kruskal–Wallis test at the 5% significance level. (*Note*: Although $n_4 < 5$, most statisticians would consider it reasonable to conduct a Kruskal–Wallis test on these data.)
b. The hypothesis test in part (a) was done in Exercise 16.37 by using the one-way ANOVA test. The assumption in that exercise is that weekly earnings in the five industries are normally distributed and have equal standard deviations. Presuming that to be true, why is performing a Kruskal–Wallis test to compare the means permissible?

In this case, is use of the one-way ANOVA test or the Kruskal–Wallis test better? Explain your answers.

16.77 Suppose that you want to perform a hypothesis test to compare four population means, using independent samples. In each case, decide whether you would use the one-way ANOVA test, the Kruskal–Wallis test, or neither of these tests. Preliminary data analyses of the samples suggest that the four distributions of the variable
a. are not normal but have the same shape.
b. are normal and have the same shape.

16.78 Suppose that you want to perform a hypothesis test to compare six population means, using independent samples. In each case, decide whether you would use the one-way ANOVA test, the Kruskal–Wallis test, or neither of these tests. Preliminary data analyses of the samples suggest that the six distributions of the variable
a. are not normal and have quite different shapes.
b. are normal but have quite different shapes.

Extending the Concepts and Skills

16.79 Vehicle Miles. In this section, we illustrated the Kruskal–Wallis test with data on miles driven by samples of cars, buses, and trucks. The value of the test statistic H was computed on page 831 to be 9.930, whereas on page 834, we found its value to be 9.923.
a. Explain the discrepancy between the two values of H.
b. Does the difference in the values affect our conclusion in the hypothesis test we conducted? Explain your answer.
c. Does the difference in the values affect our estimate of the P-value of the hypothesis test? Explain your answer.

Using Technology

*In Exercises **16.80–16.83**, we repeat the information from Exercises 16.42–16.45. For each exercise, use the technology of your choice to perform the following tasks:*
a. *Decide whether conducting a Kruskal–Wallis test on the data is reasonable. If so, also do parts (b)–(d).*
b. *Perform a Kruskal–Wallis test at the designated significance level.*
c. *Interpret the results of your hypothesis test.*
d. *If a one-way ANOVA was performed on the data in Section 16.3, compare your results there to those obtained here.*

16.80 Cuckoo Care. Use the data on the WeissStats CD for the lengths, in millimeters, of cuckoo eggs found in the nests of three different bird species referred to in Exercise 16.42.

At the 10% significance level, do the data provide sufficient evidence to conclude that a difference exists in the mean lengths of cuckoo eggs among the three bird species?

16.81 Magazine Ads. Use the data on the WeissStats CD for the number of words of three syllables or more in advertisements from magazines of three different educational levels referred to in Exercise 16.43. At the 5% significance level, do the data provide evidence of a difference in the mean number of words of three syllables or more among the three magazine levels?

16.82 Sickle Cell Disease. Use the data on the WeissStats CD for the steady state hemoglobin levels of patients with three different types of sickle cell disease referred to in Exercise 16.44. At the 5% significance level, do the data suggest a difference in mean steady state hemoglobin levels among the three types of sickle cell disease?

16.83 Prolonging Life. Use the data on the WeissStats CD for the survival times, in days, of patients given a Vitamin C supplement who were in advanced stages of cancer of five different types referred to in Exercise 16.45. At the 5% significance level, do the data provide sufficient evidence to conclude that, for patients in advanced stages of cancer who are given Vitamin C treatment, a difference exists in mean survival time among the five different types of cancer?

CHAPTER REVIEW

You Should Be Able To

1. use and understand the formulas in this chapter.

2. use the F-table, Table VIII in Appendix A.

3. explain the essential ideas behind a one-way analysis of variance.

4. state and check the assumptions required for a one-way ANOVA.

5. obtain the sums of squares for a one-way ANOVA by using the defining formulas.

6. obtain the sums of squares for a one-way ANOVA by using the computing formulas.

7. compute the mean squares and the F-statistic for a one-way ANOVA.

8. construct a one-way ANOVA table.

9. perform a one-way ANOVA test.

*10. use the q-tables, Tables X and XI in Appendix A.

*11. perform a multiple comparison by using the Tukey method.

*12. perform a Kruskal–Wallis test.

Key Terms

analysis of variance (ANOVA), *798*
degrees of freedom for the
   denominator, *800*
degrees of freedom for the
   numerator, *800*
error, *805*
error mean square (*MSE*), *806*
error sum of squares (*SSE*), *805*
F_α, *801*
F-curve, *800*
F-distribution, *800*
F-statistic, *806*

factor, *802*
family confidence level,* *823*
individual confidence level,* *823*
Kruskal–Wallis test,* *829, 833*
levels, *802*
multiple comparisons,* *822*
one-way analysis of variance, *802*
one-way ANOVA identity, *810*
one-way ANOVA tables, *810*
one-way ANOVA test, *812, 813*
q_α,* *823*
q-curve,* *823*

q-distribution,* *823*
residual, *803*
response variable, *816*
rule of 2, *803*
studentized range distribution,* *823*
total sum of squares (*SST*), *810*
treatment, *805*
treatment mean square (*MSTR*), *806*
treatment sum of squares (*SSTR*), *805*
Tukey multiple-comparison
   method,* *822, 824*

REVIEW TEST

Statistical Concepts and Skills

1. For what is one-way ANOVA used?

2. State the four assumptions for one-way ANOVA and explain how those assumptions can be checked.

3. On what distribution does one-way ANOVA rely?

4. Suppose that you want to compare the means of three populations by using one-way ANOVA. If the sample sizes are 5, 6, and 6, determine the degrees of freedom for the appropriate F-curve.

5. In one-way ANOVA, identify a statistic that measures
a. the variation among the sample means.
b. the variation within the samples.

6. In one-way ANOVA,
a. list and interpret the three sums of squares.
b. state the one-way ANOVA identity and interpret its meaning with regard to partitioning the total variation among all the data.

7. For a one-way ANOVA,
a. identify one purpose of one-way ANOVA tables.
b. construct a generic one-way ANOVA table.

*8. What is the purpose of conducting a multiple comparison?

*9. Explain the difference between the individual confidence level and the family confidence level. Which confidence level is appropriate for multiple comparisons? Explain your answer.

*10. Identify the distribution on which the Tukey multiple-comparison procedure is based.

*11. Consider a Tukey multiple comparison of four population means with a family confidence level of 0.95. Is the individual confidence level smaller or larger than 0.95? Explain your answer.

*12. Suppose that you want to compare the means of three populations by using the Tukey multiple-comparison procedure. If the sample sizes are 5, 6, and 6, determine the parameters for the appropriate q-curve.

*13. Identify a nonparametric alternative to the one-way ANOVA procedure.

*14. Identify the distribution used as an approximation to the true distribution of the H statistic for a Kruskal–Wallis test.

*15. Explain the logic of a Kruskal–Wallis test.

*16. Suppose that you want to compare the means of several populations, using independent samples. If given the choice between using the one-way ANOVA test and the Kruskal–Wallis test, which would you choose if outliers occur in the sample data? Explain your answer.

17. Consider an F-curve with df = (2, 14).
a. Identify the degrees of freedom for the numerator.
b. Identify the degrees of freedom for the denominator.
c. Determine $F_{0.05}$.
d. Find the F-value having area 0.01 to its right.
e. Find the F-value having area 0.05 to its right.

18. Consider the following hypothetical samples.

A	B	C
1	0	3
3	6	12
5	2	6
	5	3
	2	

a. Obtain the sample mean and sample standard deviation of each of the three samples.
b. Obtain SST, $SSTR$, and SSE by using the defining formulas and verify that the one-way ANOVA identity holds.
c. Obtain SST, $SSTR$, and SSE by using the computing formulas.
d. Construct the one-way ANOVA table.

19. Losses to Robbery. The U.S. Federal Bureau of Investigation conducts surveys to obtain information on the value of losses from various types of robberies. Results of the surveys are published in *Population-at-Risk Rates and Selected Crime Indicators*. Independent random samples of reports for three types of robberies—highway, gas station, and convenience store—gave the following data, in dollars, on value of losses.

Highway	Gas station	Convenience store
797	825	665
841	722	742
684	701	701
933	640	527
869	480	423
	807	435

a. What does *MSTR* measure?

b. What does *MSE* measure?

c. Suppose that you want to perform a one-way ANOVA to compare the mean losses among the three types of robberies. What conditions are necessary? How crucial are those conditions?

20. Losses to Robbery. Refer to Problem 19. At the 5% significance level, do the data provide sufficient evidence to conclude that a difference in mean losses exists among the three types of robberies? Use one-way ANOVA to perform the required hypothesis test. (*Note:* $T_1 = 4124$, $T_2 = 4175$, $T_3 = 3493$, and $\Sigma x^2 = 8,550,628$.)

*21. Consider a q-curve with parameters 3 and 14.

a. Determine $q_{0.05}$.

b. Find the q-value having area 0.01 to its right.

*22. Losses to Robbery. Refer to Problem 19.

a. Apply the Tukey multiple-comparison method to the data. Use a family confidence level of 0.95.

b. Interpret your results from part (a).

*23. Losses to Robbery. Refer to Problem 19.

a. At the 5% significance level, do the data provide sufficient evidence to conclude that a difference in mean losses exists among the three types of robberies? Use the Kruskal–Wallis procedure to perform the required hypothesis test.

b. The hypothesis test in part (a) was done in Problem 20 by using the one-way ANOVA procedure. The assumptions in that exercise are that, for the three types of robberies, losses are normally distributed and have equal standard deviations. Presuming that to be true, why is performing a Kruskal–Wallis test to compare the means permissible? In this case, is use of the one-way ANOVA test or the Kruskal–Wallis test better? Explain your answers.

Using Technology

24. Losses to Robbery. Refer to Problem 19. Use the technology of your choice to

a. obtain normal probability plots for each of the three samples.

b. perform a residual analysis.

c. Does it seem reasonable to consider the assumptions for one-way ANOVA met? Explain your answer.

25. Losses to Robbery. Use the technology of your choice to carry out the one-way ANOVA test required in Problem 20.

*26. Losses to Robbery. Use the technology of your choice to carry out the multiple comparison required in Problem 22.

*27. Losses to Robbery. Use the technology of your choice to carry out the Kruskal–Wallis test required in Problem 23.

28. Rock Sparrows. Rock sparrows breeding in northern Italy are the subject of a long-term ecology and conservation study due to their wide variety of breeding patterns. Both males and females have a yellow patch on their breasts that is thought to play a significant role in their sexual behavior. A. Pilastro, M. Griggio, and G. Matessi conducted an experiment in which they increased or reduced the size of a female's breast patch by dying feathers at the edge of a patch and then observed several characteristics of the behavior of the male. Their results were published in the paper "Male Rock Sparrows Adjust Their Breeding Strategy According to Female Ornamentation: Parental or Mating Investment?" (*Animal Behaviour*, Vol. 66, Issue 2, pp. 265–271). Eight mating pairs were observed in each of three groups: a reduced patch size group, a control group, and an enlarged patch size group. The following data, based on the results reported by the researchers, give the number of minutes per hour that males sang in the vicinity of the nest after the patch size manipulation was done on the females.

Reduced patch	Control	Enlarged patch
1.0	5.0	8.2
5.7	17.4	26.6
0.8	24.0	38.0
7.1	20.6	17.0
0.2	16.4	12.8
0.1	8.5	7.4
8.8	9.3	38.8
13.0	17.5	9.4

a. At the 1% significance level, do the data provide sufficient evidence to conclude that a difference exists in the mean singing rates among male rock sparrows exposed to the three types of breast treatments? Use the one-way ANOVA procedure.

*b. Conduct a Tukey multiple comparison to obtain the relationship among the mean singing rates of males for the three types of breast treatments. Use a 99% family confidence level.

c. Does the rule of 2 hold for the sample data? If not, why is it probably still acceptable to conduct the one-way ANOVA test?

*29. Rock Sparrows. Refer to Problem 28.

a. Perform the hypothesis test required in part (a) of Problem 28 by applying the Kruskal–Wallis procedure.

b. Compare the result obtained in part (a) of this problem to that found in part (a) of Problem 28.

StatExplore in MyMathLab
Analyzing Data Online

You can use StatExplore to carry out most of the statistical analyses discussed in this chapter. To illustrate, we show how to perform a one-way ANOVA test.

EXAMPLE   One-Way ANOVA Test

Energy Consumption Table 16.6 on page 812 displays last year's energy consumptions for independent samples of households in the four U.S. regions. Use StatExplore to decide, at the 5% significance level, whether the data provide sufficient evidence to conclude that a difference exists in last year's mean energy consumption among households in the four U.S. regions. Apply the one-way ANOVA procedure.

SOLUTION   Let μ_1, μ_2, μ_3, and μ_4 denote last year's mean energy consumptions for households in the Northeast, Midwest, South, and West, respectively. The problem is to perform the hypothesis test

$$H_0: \mu_1 = \mu_2 = \mu_3 = \mu_4$$
$$H_a: \text{Not all the means are equal}$$

at the 5% significance level by applying the one-way ANOVA procedure.

To carry out the hypothesis test by using StatExplore, we first store the sample data from Table 16.6 in columns named Northeast, Midwest, South, and West. Then, we proceed as follows.

> 1 Choose **Stat ➤ ANOVA ➤ One Way**
> 2 Select the **Compare selected columns** option button
> 3 Select the columns Northeast, Midwest, South, and West
> 4 Click **Calculate**

The resulting output is shown in Printout 16.3. From Printout 16.3, the *P*-value for the hypothesis test is 0.005. As the *P*-value is less than the specified significance level of 0.05, we reject H_0. At the 5% significance level, the data provide sufficient evidence to conclude that a difference exists in last year's mean energy consumption among households in the four U.S. regions.

PRINTOUT 16.3
StatExplore output for one-way ANOVA

Analysis of Variance results:
Data stored in separate columns.

Column	n	Mean	Std. Error
Northeast	5	13	0.83666
Midwest	6	14.5	1.0567245
South	4	10	1.2909944
West	5	9.2	0.86023253

Source	df	SS	MS	F-Stat	P-value
Treatments	3	97.5	32.5	6.3183475	0.005
Error	16	82.3	5.14375		
Total	19	179.8			

STATEXPLORE EXERCISES Solve the following problems by using StatExplore:

a. Conduct the one-way ANOVA hypothesis test required in Exercise 16.37 on page 820.

*b. Carry out the Kruskal–Wallis hypothesis test required in Example 16.8 on page 832.

> To access StatExplore, go to the student content area of your Weiss MyMathLab course.

Internet Project

Brain Damage and the Courts

In this project, you are to examine data collected within the Australian court system. The subjects in the study are plaintiffs: people who suffered brain damage in automobile accidents and sued for monetary damages based on their injuries.

You will be able to decide whether the (mean) amount of money that plaintiffs request differs among age groups—in other words, whether there is a difference in requested compensation based on age. In a second analysis, you will determine whether there is a difference owing to the plaintiffs' gender.

URL for access to Internet Projects Page: www.aw-bc.com/weiss

Focusing on Data Analysis

The Focus Database

Recall from Chapter 1 (see page 35) that the Focus database contains information on the undergraduate students at the University of Wisconsin - Eau Claire (UWEC). Statistical analyses for this database should be done with the technology of your choice.

  a. Obtain independent simple random samples of size 30 each of the cumulative GPAs (GPA) of freshmen, sophomores, juniors, and seniors at UWEC.

  b. At the 5% significance level, do the data from part (a) provide sufficient evidence to conclude that a difference exists among mean cumulative GPAs for freshmen, sophomores, juniors, and seniors at UWEC? Use the one-way ANOVA procedure.

  *c. Conduct a Tukey multiple comparison on the data from part (a) at the 95% family confidence level. Interpret your results.

  d. Obtain normal probability plots, boxplots, and the sample standard deviations of the GPAs of the sampled students in each class.

  e. Based on your graphs from part (d), do you think that applying the one-way ANOVA procedure in part (b) was reasonable? Explain your reasoning.

  f. Obtain independent simple random samples of size 30 each of the cumulative GPAs of UWEC undergraduates in the five colleges.

  g. At the 10% significance level, do the data from part (f) provide sufficient evidence to conclude that mean cumulative GPAs differ among UWEC undergraduates in the five colleges? Use the one-way ANOVA procedure.

  *h. Conduct a Tukey multiple comparison on the data from part (f) at the 90% family confidence level. Interpret your results.

  i. Obtain normal probability plots, boxplots, and the sample standard deviations of the GPAs of the sampled students in each college.

  j. Based on your graphs from part (i), do you think that applying the one-way ANOVA procedure in part (g) was reasonable? Explain your reasoning.

Heavy Drinking Among College Students

As you learned at the beginning of this chapter, Professor Kate Carey of Syracuse University surveyed 78 college students from an introductory psychology class, all of whom were regular drinkers of alcohol. Her purpose was to identify interpersonal and intrapersonal situations associated with excessive drinking among college students and to detect those situations that differentiate heavy drinkers from light and moderate drinkers.

To quantify drinking patterns, Carey used the time-line follow-back procedure (TLFB). The TLFB is a structured interview that provides reliable estimates of daily drinking. In this case, each student filled in each day of a blank calendar covering the previous month with the number of standard drink equivalents (SDEs) consumed on that day. One SDE is defined to be 1 fluid ounce of hard liquor, 12 fluid ounces of beer, or 4 fluid ounces of wine. Based on the results of the TLFB, the students were divided into three categories according to average quantity of alcohol consumed per drinking day: light drinkers (≤ 3 SDEs), moderate drinkers (4–6 SDEs), and heavy drinkers (> 6 SDEs).

To assess the frequency of excessive drinking in interpersonal and intrapersonal situations, Carey utilized the short form of the Inventory of Drinking Situations (IDS). This form consists of 42 items, each of which is rated on a 4-point scale ranging from (1) never drink heavily in that type of situation to (4) almost always drink heavily in that type of situation. The 42 items are divided into eight subscales, three interpersonal and five intrapersonal, as displayed in the first column of the table on page 799. A subscale score represents the average rating for the items constituting the subscale.

a. For each of the eight IDS subscales, perform a (separate) one-way ANOVA to decide whether a difference exists in mean IDS scores among the three drinker categories. Use $\alpha = 0.05$. (*Note:* For each ANOVA, use the formula

$$\bar{x} = \frac{n_1\bar{x}_1 + n_2\bar{x}_2 + n_3\bar{x}_3}{n_1 + n_2 + n_3}$$

to obtain the mean of all the observations.)

*b. For those one-way ANOVAs in part (a) that are statistically significant, carry out a Tukey multiple comparison with a family confidence level of 0.95.

c. Based on the data in the table on page 799, should any of the eight ANOVAs perhaps not been carried out? Explain your answer. (*Hint:* Rule of 2.)

Internet Resources: Visit the Weiss Web site www.aw-bc.com/weiss for additional discussion, exercises, and resources related to this case study.

Biography

SIR RONALD FISHER: Mr. ANOVA

Ronald Fisher was born on February 17, 1890, in London, England; he was a surviving twin in a family of eight children; his father was a prominent auctioneer. Fisher graduated from Cambridge in 1912 with degrees in mathematics and physics.

From 1912 to 1919, Fisher worked at an investment house, did farm chores in Canada, and taught high school. In 1919, he took a position as a statistician at Rothamsted Experimental Station in Harpenden, West Hertford, England. His charge was to sort and reassess a 66-year accumulation of data on manurial field trials and weather records.

Fisher's work at Rothamsted during the next 15 years earned him the reputation as the leading statistician of his day and as a top-ranking geneticist. It was there, in 1925, that he published *Statistics for Research Workers,* a book that remained in print for 50 years. Fisher made important contributions to analysis of variance (ANOVA), exact tests of significance for small samples, and maximum-likelihood solutions. He developed experimental designs to address issues in biological research, such as small samples, variable materials, and fluctuating environments.

Fisher has been described as "slight, bearded, eloquent, reactionary, and quirkish; genial to his disciples and hostile to his dissenters." He was also a prolific writer—over a span of 50 years, he wrote an average of one paper every 2 months!

In 1933, Fisher became Galton professor of Eugenics at University College in London and, in 1943, Balfour professor of genetics at Cambridge. In 1952, he was knighted. Fisher "retired" in 1959, moved to Australia, and spent the last 3 years of his life working at the Division of Mathematical Statistics of the Commonwealth Scientific and Industrial Research Organization. He died in 1962 in Adelaide, Australia.

Appendixes

Statistical Tables

TABLE I

Random numbers

Line number	Column number									
	00–09		*10–19*		*20–29*		*30–39*		*40–49*	
00	15544	80712	97742	21500	97081	42451	50623	56071	28882	28739
01	01011	21285	04729	39986	73150	31548	30168	76189	56996	19210
02	47435	53308	40718	29050	74858	64517	93573	51058	68501	42723
03	91312	75137	86274	59834	69844	19853	06917	17413	44474	86530
04	12775	08768	80791	16298	22934	09630	98862	39746	64623	32768
05	31466	43761	94872	92230	52367	13205	38634	55882	77518	36252
06	09300	43847	40881	51243	97810	18903	53914	31688	06220	40422
07	73582	13810	57784	72454	68997	72229	30340	08844	53924	89630
08	11092	81392	58189	22697	41063	09451	09789	00637	06450	85990
09	93322	98567	00116	35605	66790	52965	62877	21740	56476	49296
10	80134	12484	67089	08674	70753	90959	45842	59844	45214	36505
11	97888	31797	95037	84400	76041	96668	75920	68482	56855	97417
12	92612	27082	59459	69380	98654	20407	88151	56263	27126	63797
13	72744	45586	43279	44218	83638	05422	00995	70217	78925	39097
14	96256	70653	45285	26293	78305	80252	03625	40159	68760	84716
15	07851	47452	66742	83331	54701	06573	98169	37499	67756	68301
16	25594	41552	96475	56151	02089	33748	65289	89956	89559	33687
17	65358	15155	59374	80940	03411	94656	69440	47156	77115	99463
18	09402	31008	53424	21928	02198	61201	02457	87214	59750	51330
19	97424	90765	01634	37328	41243	33564	17884	94747	93650	77668

TABLE II
Areas under the
standard normal curve

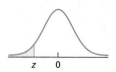

| Second decimal place in z | | | | | | | | | | |
0.09	0.08	0.07	0.06	0.05	0.04	0.03	0.02	0.01	0.00	z
									0.0000†	−3.9
0.0001	0.0001	0.0001	0.0001	0.0001	0.0001	0.0001	0.0001	0.0001	0.0001	−3.8
0.0001	0.0001	0.0001	0.0001	0.0001	0.0001	0.0001	0.0001	0.0001	0.0001	−3.7
0.0001	0.0001	0.0001	0.0001	0.0001	0.0001	0.0001	0.0001	0.0002	0.0002	−3.6
0.0002	0.0002	0.0002	0.0002	0.0002	0.0002	0.0002	0.0002	0.0002	0.0002	−3.5
0.0002	0.0003	0.0003	0.0003	0.0003	0.0003	0.0003	0.0003	0.0003	0.0003	−3.4
0.0003	0.0004	0.0004	0.0004	0.0004	0.0004	0.0004	0.0005	0.0005	0.0005	−3.3
0.0005	0.0005	0.0005	0.0006	0.0006	0.0006	0.0006	0.0006	0.0007	0.0007	−3.2
0.0007	0.0007	0.0008	0.0008	0.0008	0.0008	0.0009	0.0009	0.0009	0.0010	−3.1
0.0010	0.0010	0.0011	0.0011	0.0011	0.0012	0.0012	0.0013	0.0013	0.0013	−3.0
0.0014	0.0014	0.0015	0.0015	0.0016	0.0016	0.0017	0.0018	0.0018	0.0019	−2.9
0.0019	0.0020	0.0021	0.0021	0.0022	0.0023	0.0023	0.0024	0.0025	0.0026	−2.8
0.0026	0.0027	0.0028	0.0029	0.0030	0.0031	0.0032	0.0033	0.0034	0.0035	−2.7
0.0036	0.0037	0.0038	0.0039	0.0040	0.0041	0.0043	0.0044	0.0045	0.0047	−2.6
0.0048	0.0049	0.0051	0.0052	0.0054	0.0055	0.0057	0.0059	0.0060	0.0062	−2.5
0.0064	0.0066	0.0068	0.0069	0.0071	0.0073	0.0075	0.0078	0.0080	0.0082	−2.4
0.0084	0.0087	0.0089	0.0091	0.0094	0.0096	0.0099	0.0102	0.0104	0.0107	−2.3
0.0110	0.0113	0.0116	0.0119	0.0122	0.0125	0.0129	0.0132	0.0136	0.0139	−2.2
0.0143	0.0146	0.0150	0.0154	0.0158	0.0162	0.0166	0.0170	0.0174	0.0179	−2.1
0.0183	0.0188	0.0192	0.0197	0.0202	0.0207	0.0212	0.0217	0.0222	0.0228	−2.0
0.0233	0.0239	0.0244	0.0250	0.0256	0.0262	0.0268	0.0274	0.0281	0.0287	−1.9
0.0294	0.0301	0.0307	0.0314	0.0322	0.0329	0.0336	0.0344	0.0351	0.0359	−1.8
0.0367	0.0375	0.0384	0.0392	0.0401	0.0409	0.0418	0.0427	0.0436	0.0446	−1.7
0.0455	0.0465	0.0475	0.0485	0.0495	0.0505	0.0516	0.0526	0.0537	0.0548	−1.6
0.0559	0.0571	0.0582	0.0594	0.0606	0.0618	0.0630	0.0643	0.0655	0.0668	−1.5
0.0681	0.0694	0.0708	0.0721	0.0735	0.0749	0.0764	0.0778	0.0793	0.0808	−1.4
0.0823	0.0838	0.0853	0.0869	0.0885	0.0901	0.0918	0.0934	0.0951	0.0968	−1.3
0.0985	0.1003	0.1020	0.1038	0.1056	0.1075	0.1093	0.1112	0.1131	0.1151	−1.2
0.1170	0.1190	0.1210	0.1230	0.1251	0.1271	0.1292	0.1314	0.1335	0.1357	−1.1
0.1379	0.1401	0.1423	0.1446	0.1469	0.1492	0.1515	0.1539	0.1562	0.1587	−1.0
0.1611	0.1635	0.1660	0.1685	0.1711	0.1736	0.1762	0.1788	0.1814	0.1841	−0.9
0.1867	0.1894	0.1922	0.1949	0.1977	0.2005	0.2033	0.2061	0.2090	0.2119	−0.8
0.2148	0.2177	0.2206	0.2236	0.2266	0.2296	0.2327	0.2358	0.2389	0.2420	−0.7
0.2451	0.2483	0.2514	0.2546	0.2578	0.2611	0.2643	0.2676	0.2709	0.2743	−0.6
0.2776	0.2810	0.2843	0.2877	0.2912	0.2946	0.2981	0.3015	0.3050	0.3085	−0.5
0.3121	0.3156	0.3192	0.3228	0.3264	0.3300	0.3336	0.3372	0.3409	0.3446	−0.4
0.3483	0.3520	0.3557	0.3594	0.3632	0.3669	0.3707	0.3745	0.3783	0.3821	−0.3
0.3859	0.3897	0.3936	0.3974	0.4013	0.4052	0.4090	0.4129	0.4168	0.4207	−0.2
0.4247	0.4286	0.4325	0.4364	0.4404	0.4443	0.4483	0.4522	0.4562	0.4602	−0.1
0.4641	0.4681	0.4721	0.4761	0.4801	0.4840	0.4880	0.4920	0.4960	0.5000	−0.0

† For $z \leq -3.90$, the areas are 0.0000 to four decimal places.

TABLE II (cont.)
Areas under the
standard normal curve

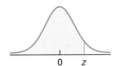

	Second decimal place in z									
z	0.00	0.01	0.02	0.03	0.04	0.05	0.06	0.07	0.08	0.09
0.0	0.5000	0.5040	0.5080	0.5120	0.5160	0.5199	0.5239	0.5279	0.5319	0.5359
0.1	0.5398	0.5438	0.5478	0.5517	0.5557	0.5596	0.5636	0.5675	0.5714	0.5753
0.2	0.5793	0.5832	0.5871	0.5910	0.5948	0.5987	0.6026	0.6064	0.6103	0.6141
0.3	0.6179	0.6217	0.6255	0.6293	0.6331	0.6368	0.6406	0.6443	0.6480	0.6517
0.4	0.6554	0.6591	0.6628	0.6664	0.6700	0.6736	0.6772	0.6808	0.6844	0.6879
0.5	0.6915	0.6950	0.6985	0.7019	0.7054	0.7088	0.7123	0.7157	0.7190	0.7224
0.6	0.7257	0.7291	0.7324	0.7357	0.7389	0.7422	0.7454	0.7486	0.7517	0.7549
0.7	0.7580	0.7611	0.7642	0.7673	0.7704	0.7734	0.7764	0.7794	0.7823	0.7852
0.8	0.7881	0.7910	0.7939	0.7967	0.7995	0.8023	0.8051	0.8078	0.8106	0.8133
0.9	0.8159	0.8186	0.8212	0.8238	0.8264	0.8289	0.8315	0.8340	0.8365	0.8389
1.0	0.8413	0.8438	0.8461	0.8485	0.8508	0.8531	0.8554	0.8577	0.8599	0.8621
1.1	0.8643	0.8665	0.8686	0.8708	0.8729	0.8749	0.8770	0.8790	0.8810	0.8830
1.2	0.8849	0.8869	0.8888	0.8907	0.8925	0.8944	0.8962	0.8980	0.8997	0.9015
1.3	0.9032	0.9049	0.9066	0.9082	0.9099	0.9115	0.9131	0.9147	0.9162	0.9177
1.4	0.9192	0.9207	0.9222	0.9236	0.9251	0.9265	0.9279	0.9292	0.9306	0.9319
1.5	0.9332	0.9345	0.9357	0.9370	0.9382	0.9394	0.9406	0.9418	0.9429	0.9441
1.6	0.9452	0.9463	0.9474	0.9484	0.9495	0.9505	0.9515	0.9525	0.9535	0.9545
1.7	0.9554	0.9564	0.9573	0.9582	0.9591	0.9599	0.9608	0.9616	0.9625	0.9633
1.8	0.9641	0.9649	0.9656	0.9664	0.9671	0.9678	0.9686	0.9693	0.9699	0.9706
1.9	0.9713	0.9719	0.9726	0.9732	0.9738	0.9744	0.9750	0.9756	0.9761	0.9767
2.0	0.9772	0.9778	0.9783	0.9788	0.9793	0.9798	0.9803	0.9808	0.9812	0.9817
2.1	0.9821	0.9826	0.9830	0.9834	0.9838	0.9842	0.9846	0.9850	0.9854	0.9857
2.2	0.9861	0.9864	0.9868	0.9871	0.9875	0.9878	0.9881	0.9884	0.9887	0.9890
2.3	0.9893	0.9896	0.9898	0.9901	0.9904	0.9906	0.9909	0.9911	0.9913	0.9916
2.4	0.9918	0.9920	0.9922	0.9925	0.9927	0.9929	0.9931	0.9932	0.9934	0.9936
2.5	0.9938	0.9940	0.9941	0.9943	0.9945	0.9946	0.9948	0.9949	0.9951	0.9952
2.6	0.9953	0.9955	0.9956	0.9957	0.9959	0.9960	0.9961	0.9962	0.9963	0.9964
2.7	0.9965	0.9966	0.9967	0.9968	0.9969	0.9970	0.9971	0.9972	0.9973	0.9974
2.8	0.9974	0.9975	0.9976	0.9977	0.9977	0.9978	0.9979	0.9979	0.9980	0.9981
2.9	0.9981	0.9982	0.9982	0.9983	0.9984	0.9984	0.9985	0.9985	0.9986	0.9986
3.0	0.9987	0.9987	0.9987	0.9988	0.9988	0.9989	0.9989	0.9989	0.9990	0.9990
3.1	0.9990	0.9991	0.9991	0.9991	0.9992	0.9992	0.9992	0.9992	0.9993	0.9993
3.2	0.9993	0.9993	0.9994	0.9994	0.9994	0.9994	0.9994	0.9995	0.9995	0.9995
3.3	0.9995	0.9995	0.9995	0.9996	0.9996	0.9996	0.9996	0.9996	0.9996	0.9997
3.4	0.9997	0.9997	0.9997	0.9997	0.9997	0.9997	0.9997	0.9997	0.9997	0.9998
3.5	0.9998	0.9998	0.9998	0.9998	0.9998	0.9998	0.9998	0.9998	0.9998	0.9998
3.6	0.9998	0.9998	0.9999	0.9999	0.9999	0.9999	0.9999	0.9999	0.9999	0.9999
3.7	0.9999	0.9999	0.9999	0.9999	0.9999	0.9999	0.9999	0.9999	0.9999	0.9999
3.8	0.9999	0.9999	0.9999	0.9999	0.9999	0.9999	0.9999	0.9999	0.9999	0.9999
3.9	1.0000[†]									

[†] For $z \geq 3.90$, the areas are 1.0000 to four decimal places.

TABLE III
Normal scores

Ordered position	n								
	5	6	7	8	9	10	11	12	13
1	−1.18	−1.28	−1.36	−1.43	−1.50	−1.55	−1.59	−1.64	−1.68
2	−0.50	−0.64	−0.76	−0.85	−0.93	−1.00	−1.06	−1.11	−1.16
3	0.00	−0.20	−0.35	−0.47	−0.57	−0.65	−0.73	−0.79	−0.85
4	0.50	0.20	0.00	−0.15	−0.27	−0.37	−0.46	−0.53	−0.60
5	1.18	0.64	0.35	0.15	0.00	−0.12	−0.22	−0.31	−0.39
6		1.28	0.76	0.47	0.27	0.12	0.00	−0.10	−0.19
7			1.36	0.85	0.57	0.37	0.22	0.10	0.00
8				1.43	0.93	0.65	0.46	0.31	0.19
9					1.50	1.00	0.73	0.53	0.39
10						1.55	1.06	0.79	0.60
11							1.59	1.11	0.85
12								1.64	1.16
13									1.68

TABLE III   (cont.)
Normal scores

Ordered position	n								
	14	15	16	17	18	19	20	21	22
1	−1.71	−1.74	−1.77	−1.80	−1.82	−1.85	−1.87	−1.89	−1.91
2	−1.20	−1.24	−1.28	−1.32	−1.35	−1.38	−1.40	−1.43	−1.45
3	−0.90	−0.94	−0.99	−1.03	−1.06	−1.10	−1.13	−1.16	−1.18
4	−0.66	−0.71	−0.76	−0.80	−0.84	−0.88	−0.92	−0.95	−0.98
5	−0.45	−0.51	−0.57	−0.62	−0.66	−0.70	−0.74	−0.78	−0.81
6	−0.27	−0.33	−0.39	−0.45	−0.50	−0.54	−0.59	−0.63	−0.66
7	−0.09	−0.16	−0.23	−0.29	−0.35	−0.40	−0.45	−0.49	−0.53
8	0.09	0.00	−0.08	−0.15	−0.21	−0.26	−0.31	−0.36	−0.40
9	0.27	0.16	0.08	0.00	−0.07	−0.13	−0.19	−0.24	−0.28
10	0.45	0.33	0.23	0.15	0.07	0.00	−0.06	−0.12	−0.17
11	0.66	0.51	0.39	0.29	0.21	0.13	0.06	0.00	−0.06
12	0.90	0.71	0.57	0.45	0.35	0.26	0.19	0.12	0.06
13	1.20	0.94	0.76	0.62	0.50	0.40	0.31	0.24	0.17
14	1.71	1.24	0.99	0.80	0.66	0.54	0.45	0.36	0.28
15		1.74	1.28	1.03	0.84	0.70	0.59	0.49	0.40
16			1.77	1.32	1.06	0.88	0.74	0.63	0.53
17				1.80	1.35	1.10	0.92	0.78	0.66
18					1.82	1.38	1.13	0.95	0.81
19						1.85	1.40	1.16	0.98
20							1.87	1.43	1.18
21								1.89	1.45
22									1.91

TABLE III (cont.)
Normal scores

Ordered position	n							
	23	24	25	26	27	28	29	30
1	−1.93	−1.95	−1.97	−1.98	−2.00	−2.01	−2.03	−2.04
2	−1.48	−1.50	−1.52	−1.54	−1.56	−1.58	−1.59	−1.61
3	−1.21	−1.24	−1.26	−1.28	−1.30	−1.32	−1.34	−1.36
4	−1.01	−1.04	−1.06	−1.09	−1.11	−1.13	−1.15	−1.17
5	−0.84	−0.87	−0.90	−0.93	−0.95	−0.98	−1.00	−1.02
6	−0.70	−0.73	−0.76	−0.79	−0.82	−0.84	−0.87	−0.89
7	−0.57	−0.60	−0.63	−0.66	−0.69	−0.72	−0.75	−0.77
8	−0.44	−0.48	−0.52	−0.55	−0.58	−0.61	−0.64	−0.67
9	−0.33	−0.37	−0.41	−0.44	−0.48	−0.51	−0.54	−0.57
10	−0.22	−0.26	−0.30	−0.34	−0.38	−0.41	−0.44	−0.47
11	−0.11	−0.15	−0.20	−0.24	−0.28	−0.31	−0.35	−0.38
12	0.00	−0.05	−0.10	−0.14	−0.18	−0.22	−0.26	−0.29
13	0.11	0.05	0.00	−0.05	−0.09	−0.13	−0.17	−0.21
14	0.22	0.15	0.10	0.05	0.00	−0.04	−0.09	−0.12
15	0.33	0.26	0.20	0.14	0.09	0.04	0.00	−0.04
16	0.44	0.37	0.30	0.24	0.18	0.13	0.09	0.04
17	0.57	0.48	0.41	0.34	0.28	0.22	0.17	0.12
18	0.70	0.60	0.52	0.44	0.38	0.31	0.26	0.21
19	0.84	0.73	0.63	0.55	0.48	0.41	0.35	0.29
20	1.01	0.87	0.76	0.66	0.58	0.51	0.44	0.38
21	1.21	1.04	0.90	0.79	0.69	0.61	0.54	0.47
22	1.48	1.24	1.06	0.93	0.82	0.72	0.64	0.57
23	1.93	1.50	1.26	1.09	0.95	0.84	0.75	0.67
24		1.95	1.52	1.28	1.11	0.98	0.87	0.77
25			1.97	1.54	1.30	1.13	1.00	0.89
26				1.98	1.56	1.32	1.15	1.02
27					2.00	1.58	1.34	1.17
28						2.01	1.59	1.36
29							2.03	1.61
30								2.04

TABLE IV
Values of t_α

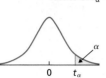

df	$t_{0.10}$	$t_{0.05}$	$t_{0.025}$	$t_{0.01}$	$t_{0.005}$	df
1	3.078	6.314	12.706	31.821	63.657	1
2	1.886	2.920	4.303	6.965	9.925	2
3	1.638	2.353	3.182	4.541	5.841	3
4	1.533	2.132	2.776	3.747	4.604	4
5	1.476	2.015	2.571	3.365	4.032	5
6	1.440	1.943	2.447	3.143	3.707	6
7	1.415	1.895	2.365	2.998	3.499	7
8	1.397	1.860	2.306	2.896	3.355	8
9	1.383	1.833	2.262	2.821	3.250	9
10	1.372	1.812	2.228	2.764	3.169	10
11	1.363	1.796	2.201	2.718	3.106	11
12	1.356	1.782	2.179	2.681	3.055	12
13	1.350	1.771	2.160	2.650	3.012	13
14	1.345	1.761	2.145	2.624	2.977	14
15	1.341	1.753	2.131	2.602	2.947	15
16	1.337	1.746	2.120	2.583	2.921	16
17	1.333	1.740	2.110	2.567	2.898	17
18	1.330	1.734	2.101	2.552	2.878	18
19	1.328	1.729	2.093	2.539	2.861	19
20	1.325	1.725	2.086	2.528	2.845	20
21	1.323	1.721	2.080	2.518	2.831	21
22	1.321	1.717	2.074	2.508	2.819	22
23	1.319	1.714	2.069	2.500	2.807	23
24	1.318	1.711	2.064	2.492	2.797	24
25	1.316	1.708	2.060	2.485	2.787	25
26	1.315	1.706	2.056	2.479	2.779	26
27	1.314	1.703	2.052	2.473	2.771	27
28	1.313	1.701	2.048	2.467	2.763	28
29	1.311	1.699	2.045	2.462	2.756	29
30	1.310	1.697	2.042	2.457	2.750	30
31	1.309	1.696	2.040	2.453	2.744	31
32	1.309	1.694	2.037	2.449	2.738	32
33	1.308	1.692	2.035	2.445	2.733	33
34	1.307	1.691	2.032	2.441	2.728	34
35	1.306	1.690	2.030	2.438	2.724	35
36	1.306	1.688	2.028	2.434	2.719	36
37	1.305	1.687	2.026	2.431	2.715	37
38	1.304	1.686	2.024	2.429	2.712	38
39	1.304	1.685	2.023	2.426	2.708	39
40	1.303	1.684	2.021	2.423	2.704	40
41	1.303	1.683	2.020	2.421	2.701	41
42	1.302	1.682	2.018	2.418	2.698	42
43	1.302	1.681	2.017	2.416	2.695	43
44	1.301	1.680	2.015	2.414	2.692	44
45	1.301	1.679	2.014	2.412	2.690	45
46	1.300	1.679	2.013	2.410	2.687	46
47	1.300	1.678	2.012	2.408	2.685	47
48	1.299	1.677	2.011	2.407	2.682	48
49	1.299	1.677	2.010	2.405	2.680	49

TABLE IV (cont.)
Values of t_α

df	$t_{0.10}$	$t_{0.05}$	$t_{0.025}$	$t_{0.01}$	$t_{0.005}$	df
50	1.299	1.676	2.009	2.403	2.678	50
51	1.298	1.675	2.008	2.402	2.676	51
52	1.298	1.675	2.007	2.400	2.674	52
53	1.298	1.674	2.006	2.399	2.672	53
54	1.297	1.674	2.005	2.397	2.670	54
55	1.297	1.673	2.004	2.396	2.668	55
56	1.297	1.673	2.003	2.395	2.667	56
57	1.297	1.672	2.002	2.394	2.665	57
58	1.296	1.672	2.002	2.392	2.663	58
59	1.296	1.671	2.001	2.391	2.662	59
60	1.296	1.671	2.000	2.390	2.660	60
61	1.296	1.670	2.000	2.389	2.659	61
62	1.295	1.670	1.999	2.388	2.657	62
63	1.295	1.669	1.998	2.387	2.656	63
64	1.295	1.669	1.998	2.386	2.655	64
65	1.295	1.669	1.997	2.385	2.654	65
66	1.295	1.668	1.997	2.384	2.652	66
67	1.294	1.668	1.996	2.383	2.651	67
68	1.294	1.668	1.995	2.382	2.650	68
69	1.294	1.667	1.995	2.382	2.649	69
70	1.294	1.667	1.994	2.381	2.648	70
71	1.294	1.667	1.994	2.380	2.647	71
72	1.293	1.666	1.993	2.379	2.646	72
73	1.293	1.666	1.993	2.379	2.645	73
74	1.293	1.666	1.993	2.378	2.644	74
75	1.293	1.665	1.992	2.377	2.643	75
80	1.292	1.664	1.990	2.374	2.639	80
85	1.292	1.663	1.988	2.371	2.635	85
90	1.291	1.662	1.987	2.368	2.632	90
95	1.291	1.661	1.985	2.366	2.629	95
100	1.290	1.660	1.984	2.364	2.626	100
200	1.286	1.653	1.972	2.345	2.601	200
300	1.284	1.650	1.968	2.339	2.592	300
400	1.284	1.649	1.966	2.336	2.588	400
500	1.283	1.648	1.965	2.334	2.586	500
600	1.283	1.647	1.964	2.333	2.584	600
700	1.283	1.647	1.963	2.332	2.583	700
800	1.283	1.647	1.963	2.331	2.582	800
900	1.282	1.647	1.963	2.330	2.581	900
1000	1.282	1.646	1.962	2.330	2.581	1000
2000	1.282	1.646	1.961	2.328	2.578	2000

	1.282	1.645	1.960	2.326	2.576	
	$z_{0.10}$	$z_{0.05}$	$z_{0.025}$	$z_{0.01}$	$z_{0.005}$	

	n	$W_{0.10}$	$W_{0.05}$	$W_{0.025}$	$W_{0.01}$	$W_{0.005}$	n
TABLE V Values of W_α	7	22	24	26	28	—	7
	8	28	30	32	34	36	8
	9	34	37	39	42	43	9
	10	41	44	47	50	52	10
	11	48	52	55	59	61	11
	12	56	61	64	68	71	12
	13	65	70	74	78	81	13
	14	74	79	84	89	92	14
	15	83	90	95	100	104	15
	16	94	100	106	112	117	16
	17	104	112	118	125	130	17
	18	116	124	131	138	143	18
	19	128	136	144	152	158	19
	20	140	150	158	167	173	20

TABLE VI
Values of M_α

		n_1							
n_2	α	3	4	5	6	7	8	9	10
	0.10	14	20	27	36	45	55	66	78
	0.05	15	21	29	37	46	57	68	80
3	0.025	—	22	30	38	48	58	70	82
	0.01	—	—	—	39	49	59	71	83
	0.005	—	—	—	—	—	60	72	85
	0.10	16	23	31	40	49	60	72	85
	0.05	17	24	32	41	51	62	74	87
4	0.025	18	25	33	43	53	64	76	89
	0.01	—	26	35	44	54	65	78	91
	0.005	—	—	—	45	55	66	79	93
	0.10	18	26	34	44	54	65	78	91
	0.05	20	27	36	46	56	68	80	94
5	0.025	21	28	37	47	58	70	83	96
	0.01	—	30	39	49	60	72	85	99
	0.005	—	—	40	50	61	73	86	101
	0.10	21	29	38	48	59	71	84	98
	0.05	22	30	40	50	61	73	87	101
6	0.025	23	32	41	52	63	76	89	103
	0.01	24	33	43	54	65	78	92	106
	0.005	—	34	44	55	67	80	94	108
	0.10	23	31	41	52	63	76	89	104
	0.05	24	33	43	54	66	79	93	107
7	0.025	26	35	45	56	68	81	95	110
	0.01	27	36	47	58	71	84	98	114
	0.005	—	37	48	60	72	86	101	116
	0.10	25	34	44	56	68	81	95	110
	0.05	27	36	47	58	71	84	99	114
8	0.025	28	38	49	61	73	87	102	117
	0.01	29	39	51	63	76	90	105	121
	0.005	30	40	52	65	78	92	108	124
	0.10	27	37	48	60	72	86	101	116
	0.05	29	39	50	63	76	90	105	121
9	0.025	31	41	53	65	78	93	108	124
	0.01	32	43	55	68	81	96	112	129
	0.005	33	44	56	70	84	99	114	131
	0.10	29	40	51	64	77	91	106	123
	0.05	31	42	54	67	80	95	111	127
10	0.025	33	44	56	69	83	98	114	131
	0.01	34	46	59	72	87	102	119	136
	0.005	36	48	61	74	89	105	121	139

TABLE VII
Values of χ_α^2

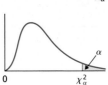

df	$\chi_{0.995}^2$	$\chi_{0.99}^2$	$\chi_{0.975}^2$	$\chi_{0.95}^2$	$\chi_{0.90}^2$
1	0.000	0.000	0.001	0.004	0.016
2	0.010	0.020	0.051	0.103	0.211
3	0.072	0.115	0.216	0.352	0.584
4	0.207	0.297	0.484	0.711	1.064
5	0.412	0.554	0.831	1.145	1.610
6	0.676	0.872	1.237	1.635	2.204
7	0.989	1.239	1.690	2.167	2.833
8	1.344	1.646	2.180	2.733	3.490
9	1.735	2.088	2.700	3.325	4.168
10	2.156	2.558	3.247	3.940	4.865
11	2.603	3.053	3.816	4.575	5.578
12	3.074	3.571	4.404	5.226	6.304
13	3.565	4.107	5.009	5.892	7.042
14	4.075	4.660	5.629	6.571	7.790
15	4.601	5.229	6.262	7.261	8.547
16	5.142	5.812	6.908	7.962	9.312
17	5.697	6.408	7.564	8.672	10.085
18	6.265	7.015	8.231	9.390	10.865
19	6.844	7.633	8.907	10.117	11.651
20	7.434	8.260	9.591	10.851	12.443
21	8.034	8.897	10.283	11.591	13.240
22	8.643	9.542	10.982	12.338	14.041
23	9.260	10.196	11.689	13.091	14.848
24	9.886	10.856	12.401	13.848	15.659
25	10.520	11.524	13.120	14.611	16.473
26	11.160	12.198	13.844	15.379	17.292
27	11.808	12.879	14.573	16.151	18.114
28	12.461	13.565	15.308	16.928	18.939
29	13.121	14.256	16.047	17.708	19.768
30	13.787	14.953	16.791	18.493	20.599
40	20.707	22.164	24.433	26.509	29.051
50	27.991	29.707	32.357	34.764	37.689
60	35.534	37.485	40.482	43.188	46.459
70	43.275	45.442	48.758	51.739	55.329
80	51.172	53.540	57.153	60.391	64.278
90	59.196	61.754	65.647	69.126	73.291
100	67.328	70.065	74.222	77.930	82.358

TABLE VII (cont.)
Values of χ_α^2

$\chi_{0.10}^2$	$\chi_{0.05}^2$	$\chi_{0.025}^2$	$\chi_{0.01}^2$	$\chi_{0.005}^2$	df
2.706	3.841	5.024	6.635	7.879	*1*
4.605	5.991	7.378	9.210	10.597	*2*
6.251	7.815	9.348	11.345	12.838	*3*
7.779	9.488	11.143	13.277	14.860	*4*
9.236	11.070	12.833	15.086	16.750	*5*
10.645	12.592	14.449	16.812	18.548	*6*
12.017	14.067	16.013	18.475	20.278	*7*
13.362	15.507	17.535	20.090	21.955	*8*
14.684	16.919	19.023	21.666	23.589	*9*
15.987	18.307	20.483	23.209	25.188	*10*
17.275	19.675	21.920	24.725	26.757	*11*
18.549	21.026	23.337	26.217	28.300	*12*
19.812	22.362	24.736	27.688	29.819	*13*
21.064	23.685	26.119	29.141	31.319	*14*
22.307	24.996	27.488	30.578	32.801	*15*
23.542	26.296	28.845	32.000	34.267	*16*
24.769	27.587	30.191	33.409	35.718	*17*
25.989	28.869	31.526	34.805	37.156	*18*
27.204	30.143	32.852	36.191	38.582	*19*
28.412	31.410	34.170	37.566	39.997	*20*
29.615	32.671	35.479	38.932	41.401	*21*
30.813	33.924	36.781	40.290	42.796	*22*
32.007	35.172	38.076	41.638	44.181	*23*
33.196	36.415	39.364	42.980	45.559	*24*
34.382	37.653	40.647	44.314	46.928	*25*
35.563	38.885	41.923	45.642	48.290	*26*
36.741	40.113	43.195	46.963	49.645	*27*
37.916	41.337	44.461	48.278	50.994	*28*
39.087	42.557	45.722	49.588	52.336	*29*
40.256	43.773	46.979	50.892	53.672	*30*
51.805	55.759	59.342	63.691	66.767	*40*
63.167	67.505	71.420	76.154	79.490	*50*
74.397	79.082	83.298	88.381	91.955	*60*
85.527	90.531	95.023	100.424	104.213	*70*
96.578	101.879	106.628	112.328	116.320	*80*
107.565	113.145	118.135	124.115	128.296	*90*
118.499	124.343	129.563	135.811	140.177	*100*

TABLE VIII
Values of F_α

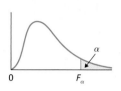

dfd	α	dfn								
		1	2	3	4	5	6	7	8	9
	0.10	39.86	49.50	53.59	55.83	57.24	58.20	58.91	59.44	59.86
	0.05	161.45	199.50	215.71	224.58	230.16	233.99	236.77	238.88	240.54
1	0.025	647.79	799.50	864.16	899.58	921.85	937.11	948.22	956.66	963.28
	0.01	4052.2	4999.5	5403.4	5624.6	5763.6	5859.0	5928.4	5981.1	6022.5
	0.005	16211	20000	21615	22500	23056	23437	23715	23925	24091
	0.10	8.53	9.00	9.16	9.24	9.29	9.33	9.35	9.37	9.38
	0.05	18.51	19.00	19.16	19.25	19.30	19.33	19.35	19.37	19.38
2	0.025	38.51	39.00	39.17	39.25	39.30	39.33	39.36	39.37	39.39
	0.01	98.50	99.00	99.17	99.25	99.30	99.33	99.36	99.37	99.39
	0.005	198.50	199.00	199.17	199.25	199.30	199.33	199.36	199.37	199.39
	0.10	5.54	5.46	5.39	5.34	5.31	5.28	5.27	5.25	5.24
	0.05	10.13	9.55	9.28	9.12	9.01	8.94	8.89	8.85	8.81
3	0.025	17.44	16.04	15.44	15.10	14.88	14.73	14.62	14.54	14.47
	0.01	34.12	30.82	29.46	28.71	28.24	27.91	27.67	27.49	27.35
	0.005	55.55	49.80	47.47	46.19	45.39	44.84	44.43	44.13	43.88
	0.10	4.54	4.32	4.19	4.11	4.05	4.01	3.98	3.95	3.94
	0.05	7.71	6.94	6.59	6.39	6.26	6.16	6.09	6.04	6.00
4	0.025	12.22	10.65	9.98	9.60	9.36	9.20	9.07	8.98	8.90
	0.01	21.20	18.00	16.69	15.98	15.52	15.21	14.98	14.80	14.66
	0.005	31.33	26.28	24.26	23.15	22.46	21.97	21.62	21.35	21.14
	0.10	4.06	3.78	3.62	3.52	3.45	3.40	3.37	3.34	3.32
	0.05	6.61	5.79	5.41	5.19	5.05	4.95	4.88	4.82	4.77
5	0.025	10.01	8.43	7.76	7.39	7.15	6.98	6.85	6.76	6.68
	0.01	16.26	13.27	12.06	11.39	10.97	10.67	10.46	10.29	10.16
	0.005	22.78	18.31	16.53	15.56	14.94	14.51	14.20	13.96	13.77
	0.10	3.78	3.46	3.29	3.18	3.11	3.05	3.01	2.98	2.96
	0.05	5.99	5.14	4.76	4.53	4.39	4.28	4.21	4.15	4.10
6	0.025	8.81	7.26	6.60	6.23	5.99	5.82	5.70	5.60	5.52
	0.01	13.75	10.92	9.78	9.15	8.75	8.47	8.26	8.10	7.98
	0.005	18.63	14.54	12.92	12.03	11.46	11.07	10.79	10.57	10.39
	0.10	3.59	3.26	3.07	2.96	2.88	2.83	2.78	2.75	2.72
	0.05	5.59	4.74	4.35	4.12	3.97	3.87	3.79	3.73	3.68
7	0.025	8.07	6.54	5.89	5.52	5.29	5.12	4.99	4.90	4.82
	0.01	12.25	9.55	8.45	7.85	7.46	7.19	6.99	6.84	6.72
	0.005	16.24	12.40	10.88	10.05	9.52	9.16	8.89	8.68	8.51
	0.10	3.46	3.11	2.92	2.81	2.73	2.67	2.62	2.59	2.56
	0.05	5.32	4.46	4.07	3.84	3.69	3.58	3.50	3.44	3.39
8	0.025	7.57	6.06	5.42	5.05	4.82	4.65	4.53	4.43	4.36
	0.01	11.26	8.65	7.59	7.01	6.63	6.37	6.18	6.03	5.91
	0.005	14.69	11.04	9.60	8.81	8.30	7.95	7.69	7.50	7.34

TABLE VIII **(cont.)**
Values of F_α

				dfn							
10	*12*	*15*	*20*	*24*	*30*	*40*	*60*	*120*	*α*	*dfd*
60.19	60.71	61.22	61.74	62.00	62.26	62.53	62.79	63.06	*0.10*	
241.88	243.91	245.95	248.01	249.05	250.10	251.14	252.20	253.25	*0.05*	
968.63	976.71	984.87	993.10	997.25	1001.41	1005.60	1009.80	1014.02	*0.025*	*1*
6055.8	6106.3	6157.3	6208.7	6234.6	6260.6	6286.7	631.9	6339.4	*0.01*	
24224	24426	24630	24836	24940	25044	25148	25253	25359	*0.005*	
9.39	9.41	9.42	9.44	9.45	9.46	9.47	9.47	9.48	*0.10*	
19.40	19.41	19.43	19.45	19.45	19.46	19.47	19.48	19.49	*0.05*	
39.40	39.41	39.43	39.45	39.46	39.46	39.47	39.48	39.49	*0.025*	*2*
99.40	99.42	99.43	99.45	99.46	99.47	99.47	99.48	99.49	*0.01*	
199.40	199.42	199.43	199.45	199.46	199.47	199.47	199.48	199.49	*0.005*	
5.23	5.22	5.20	5.18	5.18	5.17	5.16	5.15	5.14	*0.10*	
8.79	8.74	8.70	8.66	8.64	8.62	8.59	8.57	8.55	*0.05*	
14.42	14.34	14.25	14.17	14.12	14.08	14.04	13.99	13.95	*0.025*	*3*
27.23	27.05	26.87	26.69	26.60	26.50	26.41	26.32	26.22	*0.01*	
43.69	43.39	43.08	42.78	42.62	42.47	42.31	42.15	41.99	*0.005*	
3.92	3.90	3.87	3.84	3.83	3.82	3.80	3.79	3.78	*0.10*	
5.96	5.91	5.86	5.80	5.77	5.75	5.72	5.69	5.66	*0.05*	
8.84	8.75	8.66	8.56	8.51	8.46	8.41	8.36	8.31	*0.025*	*4*
14.55	14.37	14.20	14.02	13.93	13.84	13.75	13.65	13.56	*0.01*	
20.97	20.70	20.44	20.17	20.03	19.89	19.75	19.61	19.47	*0.005*	
3.30	3.27	3.24	3.21	3.19	3.17	3.16	3.14	3.12	*0.10*	
4.74	4.68	4.62	4.56	4.53	4.50	4.46	4.43	4.40	*0.05*	
6.62	6.52	6.43	6.33	6.28	6.23	6.18	6.12	6.07	*0.025*	*5*
10.05	9.89	9.72	9.55	9.47	9.38	9.29	9.20	9.11	*0.01*	
13.62	13.38	13.15	12.90	12.78	12.66	12.53	12.40	12.27	*0.005*	
2.94	2.90	2.87	2.84	2.82	2.80	2.78	2.76	2.74	*0.10*	
4.06	4.00	3.94	3.87	3.84	3.81	3.77	3.74	3.70	*0.05*	
5.46	5.37	5.27	5.17	5.12	5.07	5.01	4.96	4.90	*0.025*	*6*
7.87	7.72	7.56	7.40	7.31	7.23	7.14	7.06	6.97	*0.01*	
10.25	10.03	9.81	9.59	9.47	9.36	9.24	9.12	9.00	*0.005*	
2.70	2.67	2.63	2.59	2.58	2.56	2.54	2.51	2.49	*0.10*	
3.64	3.57	3.51	3.44	3.41	3.38	3.34	3.30	3.27	*0.05*	
4.76	4.67	4.57	4.47	4.41	4.36	4.31	4.25	4.20	*0.025*	*7*
6.62	6.47	6.31	6.16	6.07	5.99	5.91	5.82	5.74	*0.01*	
8.38	8.18	7.97	7.75	7.64	7.53	7.42	7.31	7.19	*0.005*	
2.54	2.50	2.46	2.42	2.40	2.38	2.36	2.34	2.32	*0.10*	
3.35	3.28	3.22	3.15	3.12	3.08	3.04	3.01	2.97	*0.05*	
4.30	4.20	4.10	4.00	3.95	3.89	3.84	3.78	3.73	*0.025*	*8*
5.81	5.67	5.52	5.36	5.28	5.20	5.12	5.03	4.95	*0.01*	
7.21	7.01	6.81	6.61	6.50	6.40	6.29	6.18	6.06	*0.005*	

TABLE VIII   (cont.)
Values of F_α

		dfn								
dfd	α	1	2	3	4	5	6	7	8	9
	0.10	3.36	3.01	2.81	2.69	2.61	2.55	2.51	2.47	2.44
	0.05	5.12	4.26	3.86	3.63	3.48	3.37	3.29	3.23	3.18
9	0.025	7.21	5.71	5.08	4.72	4.48	4.32	4.20	4.10	4.03
	0.01	10.56	8.02	6.99	6.42	6.06	5.80	5.61	5.47	5.35
	0.005	13.61	10.11	8.72	7.96	7.47	7.13	6.88	6.69	6.54
	0.10	3.29	2.92	2.73	2.61	2.52	2.46	2.41	2.38	2.35
	0.05	4.96	4.10	3.71	3.48	3.33	3.22	3.14	3.07	3.02
10	0.025	6.94	5.46	4.83	4.47	4.24	4.07	3.95	3.85	3.78
	0.01	10.04	7.56	6.55	5.99	5.64	5.39	5.20	5.06	4.94
	0.005	12.83	9.43	8.08	7.34	6.87	6.54	6.30	6.12	5.97
	0.10	3.23	2.86	2.66	2.54	2.45	2.39	2.34	2.30	2.27
	0.05	4.84	3.98	3.59	3.36	3.20	3.09	3.01	2.95	2.90
11	0.025	6.72	5.26	4.63	4.28	4.04	3.88	3.76	3.66	3.59
	0.01	9.65	7.21	6.22	5.67	5.32	5.07	4.89	4.74	4.63
	0.005	12.23	8.91	7.60	6.88	6.42	6.10	5.86	5.68	5.54
	0.10	3.18	2.81	2.61	2.48	2.39	2.33	2.28	2.24	2.21
	0.05	4.75	3.89	3.49	3.26	3.11	3.00	2.91	2.85	2.80
12	0.025	6.55	5.10	4.47	4.12	3.89	3.73	3.61	3.51	3.44
	0.01	9.33	6.93	5.95	5.41	5.06	4.82	4.64	4.50	4.39
	0.005	11.75	8.51	7.23	6.52	6.07	5.76	5.52	5.35	5.20
	0.10	3.14	2.76	2.56	2.43	2.35	2.28	2.23	2.20	2.16
	0.05	4.67	3.81	3.41	3.18	3.03	2.92	2.83	2.77	2.71
13	0.025	6.41	4.97	4.35	4.00	3.77	3.60	3.48	3.39	3.31
	0.01	9.07	6.70	5.74	5.21	4.86	4.62	4.44	4.30	4.19
	0.005	11.37	8.19	6.93	6.23	5.79	5.48	5.25	5.08	4.94
	0.10	3.10	2.73	2.52	2.39	2.31	2.24	2.19	2.15	2.12
	0.05	4.60	3.74	3.34	3.11	2.96	2.85	2.76	2.70	2.65
14	0.025	6.30	4.86	4.24	3.89	3.66	3.50	3.38	3.29	3.21
	0.01	8.86	6.51	5.56	5.04	4.69	4.46	4.28	4.14	4.03
	0.005	11.06	7.92	6.68	6.00	5.56	5.26	5.03	4.86	4.72
	0.10	3.07	2.70	2.49	2.36	2.27	2.21	2.16	2.12	2.09
	0.05	4.54	3.68	3.29	3.06	2.90	2.79	2.71	2.64	2.59
15	0.025	6.20	4.77	4.15	3.80	3.58	3.41	3.29	3.20	3.12
	0.01	8.68	6.36	5.42	4.89	4.56	4.32	4.14	4.00	3.89
	0.005	10.80	7.70	6.48	5.80	5.37	5.07	4.85	4.67	4.54
	0.10	3.05	2.67	2.46	2.33	2.24	2.18	2.13	2.09	2.06
	0.05	4.49	3.63	3.24	3.01	2.85	2.74	2.66	2.59	2.54
16	0.025	6.12	4.69	4.08	3.73	3.50	3.34	3.22	3.12	3.05
	0.01	8.53	6.23	5.29	4.77	4.44	4.20	4.03	3.89	3.78
	0.005	10.58	7.51	6.30	5.64	5.21	4.91	4.69	4.52	4.38

TABLE VIII (cont.)
Values of F_α

10	12	15	20	24	30	40	60	120	α	dfd
				dfn						
2.42	2.38	2.34	2.30	2.28	2.25	2.23	2.21	2.18	0.10	
3.14	3.07	3.01	2.94	2.90	2.86	2.83	2.79	2.75	0.05	
3.96	3.87	3.77	3.67	3.61	3.56	3.51	3.45	3.39	0.025	9
5.26	5.11	4.96	4.81	4.73	4.65	4.57	4.48	4.40	0.01	
6.42	6.23	6.03	5.83	5.73	5.62	5.52	5.41	5.30	0.005	
2.32	2.28	2.24	2.20	2.18	2.16	2.13	2.11	2.08	0.10	
2.98	2.91	2.85	2.77	2.74	2.70	2.66	2.62	2.58	0.05	
3.72	3.62	3.52	3.42	3.37	3.31	3.26	3.20	3.14	0.025	10
4.85	4.71	4.56	4.41	4.33	4.25	4.17	4.08	4.00	0.01	
5.85	5.66	5.47	5.27	5.17	5.07	4.97	4.86	4.75	0.005	
2.25	2.21	2.17	2.12	2.10	2.08	2.05	2.03	2.00	0.10	
2.85	2.79	2.72	2.65	2.61	2.57	2.53	2.49	2.45	0.05	
3.53	3.43	3.33	3.23	3.17	3.12	3.06	3.00	2.94	0.025	11
4.54	4.40	4.25	4.10	4.02	3.94	3.86	3.78	3.69	0.01	
5.42	5.24	5.05	4.86	4.76	4.65	4.55	4.45	4.34	0.005	
2.19	2.15	2.10	2.06	2.04	2.01	1.99	1.96	1.93	0.10	
2.75	2.69	2.62	2.54	2.51	2.47	2.43	2.38	2.34	0.05	
3.37	3.28	3.18	3.07	3.02	2.96	2.91	2.85	2.79	0.025	12
4.30	4.16	4.01	3.86	3.78	3.70	3.62	3.54	3.45	0.01	
5.09	4.91	4.72	4.53	4.43	4.33	4.23	4.12	4.01	0.005	
2.14	2.10	2.05	2.01	1.98	1.96	1.93	1.90	1.88	0.10	
2.67	2.60	2.53	2.46	2.42	2.38	2.34	2.30	2.25	0.05	
3.25	3.15	3.05	2.95	2.89	2.84	2.78	2.72	2.66	0.025	13
4.10	3.96	3.82	3.66	3.59	3.51	3.43	3.34	3.25	0.01	
4.82	4.64	4.46	4.27	4.17	4.07	3.97	3.87	3.76	0.005	
2.10	2.05	2.01	1.96	1.94	1.91	1.89	1.86	1.83	0.10	
2.60	2.53	2.46	2.39	2.35	2.31	2.27	2.22	2.18	0.05	
3.15	3.05	2.95	2.84	2.79	2.73	2.67	2.61	2.55	0.025	14
3.94	3.80	3.66	3.51	3.43	3.35	3.27	3.18	3.09	0.01	
4.60	4.43	4.25	4.06	3.96	3.86	3.76	3.66	3.55	0.005	
2.06	2.02	1.97	1.92	1.90	1.87	1.85	1.82	1.79	0.10	
2.54	2.48	2.40	2.33	2.29	2.25	2.20	2.16	2.11	0.05	
3.06	2.96	2.86	2.76	2.70	2.64	2.59	2.52	2.46	0.025	15
3.80	3.67	3.52	3.37	3.29	3.21	3.13	3.05	2.96	0.01	
4.42	4.25	4.07	3.88	3.79	3.69	3.58	3.48	3.37	0.005	
2.03	1.99	1.94	1.89	1.87	1.84	1.81	1.78	1.75	0.10	
2.49	2.42	2.35	2.28	2.24	2.19	2.15	2.11	2.06	0.05	
2.99	2.89	2.79	2.68	2.63	2.57	2.51	2.45	2.38	0.025	16
3.69	3.55	3.41	3.26	3.18	3.10	3.02	2.93	2.84	0.01	
4.27	4.10	3.92	3.73	3.64	3.54	3.44	3.33	3.22	0.005	

TABLE VIII   (cont.)
Values of F_α

dfd	α	dfn 1	2	3	4	5	6	7	8	9
	0.10	3.03	2.64	2.44	2.31	2.22	2.15	2.10	2.06	2.03
	0.05	4.45	3.59	3.20	2.96	2.81	2.70	2.61	2.55	2.49
17	0.025	6.04	4.62	4.01	3.66	3.44	3.28	3.16	3.06	2.98
	0.01	8.40	6.11	5.18	4.67	4.34	4.10	3.93	3.79	3.68
	0.005	10.38	7.35	6.16	5.50	5.07	4.78	4.56	4.39	4.25
	0.10	3.01	2.62	2.42	2.29	2.20	2.13	2.08	2.04	2.00
	0.05	4.41	3.55	3.16	2.93	2.77	2.66	2.58	2.51	2.46
18	0.025	5.98	4.56	3.95	3.61	3.38	3.22	3.10	3.01	2.93
	0.01	8.29	6.01	5.09	4.58	4.25	4.01	3.84	3.71	3.60
	0.005	10.22	7.21	6.03	5.37	4.96	4.66	4.44	4.28	4.14
	0.10	2.99	2.61	2.40	2.27	2.18	2.11	2.06	2.02	1.98
	0.05	4.38	3.52	3.13	2.90	2.74	2.63	2.54	2.48	2.42
19	0.025	5.92	4.51	3.90	3.56	3.33	3.17	3.05	2.96	2.88
	0.01	8.18	5.93	5.01	4.50	4.17	3.94	3.77	3.63	3.52
	0.005	10.07	7.09	5.92	5.27	4.85	4.56	4.34	4.18	4.04
	0.10	2.97	2.59	2.38	2.25	2.16	2.09	2.04	2.00	1.96
	0.05	4.35	3.49	3.10	2.87	2.71	2.60	2.51	2.45	2.39
20	0.025	5.87	4.46	3.86	3.51	3.29	3.13	3.01	2.91	2.84
	0.01	8.10	5.85	4.94	4.43	4.10	3.87	3.70	3.56	3.46
	0.005	9.94	6.99	5.82	5.17	4.76	4.47	4.26	4.09	3.96
	0.10	2.96	2.57	2.36	2.23	2.14	2.08	2.02	1.98	1.95
	0.05	4.32	3.47	3.07	2.84	2.68	2.57	2.49	2.42	2.37
21	0.025	5.83	4.42	3.82	3.48	3.25	3.09	2.97	2.87	2.80
	0.01	8.02	5.78	4.87	4.37	4.04	3.81	3.64	3.51	3.40
	0.005	9.83	6.89	5.73	5.09	4.68	4.39	4.18	4.01	3.88
	0.10	2.95	2.56	2.35	2.22	2.13	2.06	2.01	1.97	1.93
	0.05	4.30	3.44	3.05	2.82	2.66	2.55	2.46	2.40	2.34
22	0.025	5.79	4.38	3.78	3.44	3.22	3.05	2.93	2.84	2.76
	0.01	7.95	5.72	4.82	4.31	3.99	3.76	3.59	3.45	3.35
	0.005	9.73	6.81	5.65	5.02	4.61	4.32	4.11	3.94	3.81
	0.10	2.94	2.55	2.34	2.21	2.11	2.05	1.99	1.95	1.92
	0.05	4.28	3.42	3.03	2.80	2.64	2.53	2.44	2.37	2.32
23	0.025	5.75	4.35	3.75	3.41	3.18	3.02	2.90	2.81	2.73
	0.01	7.88	5.66	4.76	4.26	3.94	3.71	3.54	3.41	3.30
	0.005	9.63	6.73	5.58	4.95	4.54	4.26	4.05	3.88	3.75
	0.10	2.93	2.54	2.33	2.19	2.10	2.04	1.98	1.94	1.91
	0.05	4.26	3.40	3.01	2.78	2.62	2.51	2.42	2.36	2.30
24	0.025	5.72	4.32	3.72	3.38	3.15	2.99	2.87	2.78	2.70
	0.01	7.82	5.61	4.72	4.22	3.90	3.67	3.50	3.36	3.26
	0.005	9.55	6.66	5.52	4.89	4.49	4.20	3.99	3.83	3.69

TABLE VIII (cont.)
Values of F_α

10	12	15	20	24	30	40	60	120	α	dfd
				dfn						
2.00	1.96	1.91	1.86	1.84	1.81	1.78	1.75	1.72	0.10	
2.45	2.38	2.31	2.23	2.19	2.15	2.10	2.06	2.01	0.05	
2.92	2.82	2.72	2.62	2.56	2.50	2.44	2.38	2.32	0.025	17
3.59	3.46	3.31	3.16	3.08	3.00	2.92	2.83	2.75	0.01	
4.14	3.97	3.79	3.61	3.51	3.41	3.31	3.21	3.10	0.005	
1.98	1.93	1.89	1.84	1.81	1.78	1.75	1.72	1.69	0.10	
2.41	2.34	2.27	2.19	2.15	2.11	2.06	2.02	1.97	0.05	
2.87	2.77	2.67	2.56	2.50	2.44	2.38	2.32	2.26	0.025	18
3.51	3.37	3.23	3.08	3.00	2.92	2.84	2.75	2.66	0.01	
4.03	3.86	3.68	3.50	3.40	3.30	3.20	3.10	2.99	0.005	
1.96	1.91	1.86	1.81	1.79	1.76	1.73	1.70	1.67	0.10	
2.38	2.31	2.23	2.16	2.11	2.07	2.03	1.98	1.93	0.05	
2.82	2.72	2.62	2.51	2.45	2.39	2.33	2.27	2.20	0.025	19
3.43	3.30	3.15	3.00	2.92	2.84	2.76	2.67	2.58	0.01	
3.93	3.76	3.59	3.40	3.31	3.21	3.11	3.00	2.89	0.005	
1.94	1.89	1.84	1.79	1.77	1.74	1.71	1.68	1.64	0.10	
2.35	2.28	2.20	2.12	2.08	2.04	1.99	1.95	1.90	0.05	
2.77	2.68	2.57	2.46	2.41	2.35	2.29	2.22	2.16	0.025	20
3.37	3.23	3.09	2.94	2.86	2.78	2.69	2.61	2.52	0.01	
3.85	3.68	3.50	3.32	3.22	3.12	3.02	2.92	2.81	0.005	
1.92	1.87	1.83	1.78	1.75	1.72	1.69	1.66	1.62	0.10	
2.32	2.25	2.18	2.10	2.05	2.01	1.96	1.92	1.87	0.05	
2.73	2.64	2.53	2.42	2.37	2.31	2.25	2.18	2.11	0.025	21
3.31	3.17	3.03	2.88	2.80	2.72	2.64	2.55	2.46	0.01	
3.77	3.60	3.43	3.24	3.15	3.05	2.95	2.84	2.73	0.005	
1.90	1.86	1.81	1.76	1.73	1.70	1.67	1.64	1.60	0.10	
2.30	2.23	2.15	2.07	2.03	1.98	1.94	1.89	1.84	0.05	
2.70	2.60	2.50	2.39	2.33	2.27	2.21	2.14	2.08	0.025	22
3.26	3.12	2.98	2.83	2.75	2.67	2.58	2.50	2.40	0.01	
3.70	3.54	3.36	3.18	3.08	2.98	2.88	2.77	2.66	0.005	
1.89	1.84	1.80	1.74	1.72	1.69	1.66	1.62	1.59	0.10	
2.27	2.20	2.13	2.05	2.01	1.96	1.91	1.86	1.81	0.05	
2.67	2.57	2.47	2.36	2.30	2.24	2.18	2.11	2.04	0.025	23
3.21	3.07	2.93	2.78	2.70	2.62	2.54	2.45	2.35	0.01	
3.64	3.47	3.30	3.12	3.02	2.92	2.82	2.71	2.60	0.005	
1.88	1.83	1.78	1.73	1.70	1.67	1.64	1.61	1.57	0.10	
2.25	2.18	2.11	2.03	1.98	1.94	1.89	1.84	1.79	0.05	
2.64	2.54	2.44	2.33	2.27	2.21	2.15	2.08	2.01	0.025	24
3.17	3.03	2.89	2.74	2.66	2.58	2.49	2.40	2.31	0.01	
3.59	3.42	3.25	3.06	2.97	2.87	2.77	2.66	2.55	0.005	

TABLE VIII   (cont.)
Values of F_α

dfd	α	1	2	3	4	5	6	7	8	9
	0.10	2.92	2.53	2.32	2.18	2.09	2.02	1.97	1.93	1.89
	0.05	4.24	3.39	2.99	2.76	2.60	2.49	2.40	2.34	2.28
25	*0.025*	5.69	4.29	3.69	3.35	3.13	2.97	2.85	2.75	2.68
	0.01	7.77	5.57	4.68	4.18	3.85	3.63	3.46	3.32	3.22
	0.005	9.48	6.60	5.46	4.84	4.43	4.15	3.94	3.78	3.64
	0.10	2.91	2.52	2.31	2.17	2.08	2.01	1.96	1.92	1.88
	0.05	4.23	3.37	2.98	2.74	2.59	2.47	2.39	2.32	2.27
26	*0.025*	5.66	4.27	3.67	3.33	3.10	2.94	2.82	2.73	2.65
	0.01	7.72	5.53	4.64	4.14	3.82	3.59	3.42	3.29	3.18
	0.005	9.41	6.54	5.41	4.79	4.38	4.10	3.89	3.73	3.60
	0.10	2.90	2.51	2.30	2.17	2.07	2.00	1.95	1.91	1.87
	0.05	4.21	3.35	2.96	2.73	2.57	2.46	2.37	2.31	2.25
27	*0.025*	5.63	4.24	3.65	3.31	3.08	2.92	2.80	2.71	2.63
	0.01	7.68	5.49	4.60	4.11	3.78	3.56	3.39	3.26	3.15
	0.005	9.34	6.49	5.36	4.74	4.34	4.06	3.85	3.69	3.56
	0.10	2.89	2.50	2.29	2.16	2.06	2.00	1.94	1.90	1.87
	0.05	4.20	3.34	2.95	2.71	2.56	2.45	2.36	2.29	2.24
28	*0.025*	5.61	4.22	3.63	3.29	3.06	2.90	2.78	2.69	2.61
	0.01	7.64	5.45	4.57	4.07	3.75	3.53	3.36	3.23	3.12
	0.005	9.28	6.44	5.32	4.70	4.30	4.02	3.81	3.65	3.52
	0.10	2.89	2.50	2.28	2.15	2.06	1.99	1.93	1.89	1.86
	0.05	4.18	3.33	2.93	2.70	2.55	2.43	2.35	2.28	2.22
29	*0.025*	5.59	4.20	3.61	3.27	3.04	2.88	2.76	2.67	2.59
	0.01	7.60	5.42	4.54	4.04	3.73	3.50	3.33	3.20	3.09
	0.005	9.23	6.40	5.28	4.66	4.26	3.98	3.77	3.61	3.48
	0.10	2.88	2.49	2.28	2.14	2.05	1.98	1.93	1.88	1.85
	0.05	4.17	3.32	2.92	2.69	2.53	2.42	2.33	2.27	2.21
30	*0.025*	5.57	4.18	3.59	3.25	3.03	2.87	2.75	2.65	2.57
	0.01	7.56	5.39	4.51	4.02	3.70	3.47	3.30	3.17	3.07
	0.005	9.18	6.35	5.24	4.62	4.23	3.95	3.74	3.58	3.45
	0.10	2.79	2.39	2.18	2.04	1.95	1.87	1.82	1.77	1.74
	0.05	4.00	3.15	2.76	2.53	2.37	2.25	2.17	2.10	2.04
60	*0.025*	5.29	3.93	3.34	3.01	2.79	2.63	2.51	2.41	2.33
	0.01	7.08	4.98	4.13	3.65	3.34	3.12	2.95	2.82	2.72
	0.005	8.49	5.79	4.73	4.14	3.76	3.49	3.29	3.13	3.01
	0.10	2.75	2.35	2.13	1.99	1.90	1.82	1.77	1.72	1.68
	0.05	3.92	3.07	2.68	2.45	2.29	2.18	2.09	2.02	1.96
120	*0.025*	5.15	3.80	3.23	2.89	2.67	2.52	2.39	2.30	2.22
	0.01	6.85	4.79	3.95	3.48	3.17	2.96	2.79	2.66	2.56
	0.005	8.18	5.54	4.50	3.92	3.55	3.28	3.09	2.93	2.81

TABLE VIII (cont.)
Values of F_α

10	12	15	20	24	30	40	60	120	α	dfd
				dfn						
1.87	1.82	1.77	1.72	1.69	1.66	1.63	1.59	1.56	*0.10*	
2.24	2.16	2.09	2.01	1.96	1.92	1.87	1.82	1.77	*0.05*	
2.61	2.51	2.41	2.30	2.24	2.18	2.12	2.05	1.98	*0.025*	*25*
3.13	2.99	2.85	2.70	2.62	2.54	2.45	2.36	2.27	*0.01*	
3.54	3.37	3.20	3.01	2.92	2.82	2.72	2.61	2.50	*0.005*	
1.86	1.81	1.76	1.71	1.68	1.65	1.61	1.58	1.54	*0.10*	
2.22	2.15	2.07	1.99	1.95	1.90	1.85	1.80	1.75	*0.05*	
2.59	2.49	2.39	2.28	2.22	2.16	2.09	2.03	1.95	*0.025*	*26*
3.09	2.96	2.81	2.66	2.58	2.50	2.42	2.33	2.23	*0.01*	
3.49	3.33	3.15	2.97	2.87	2.77	2.67	2.56	2.45	*0.005*	
1.85	1.80	1.75	1.70	1.67	1.64	1.60	1.57	1.53	*0.10*	
2.20	2.13	2.06	1.97	1.93	1.88	1.84	1.79	1.73	*0.05*	
2.57	2.47	2.36	2.25	2.19	2.13	2.07	2.00	1.93	*0.025*	*27*
3.06	2.93	2.78	2.63	2.55	2.47	2.38	2.29	2.20	*0.01*	
3.45	3.28	3.11	2.93	2.83	2.73	2.63	2.52	2.41	*0.005*	
1.84	1.79	1.74	1.69	1.66	1.63	1.59	1.56	1.52	*0.10*	
2.19	2.12	2.04	1.96	1.91	1.87	1.82	1.77	1.71	*0.05*	
2.55	2.45	2.34	2.23	2.17	2.11	2.05	1.98	1.91	*0.025*	*28*
3.03	2.90	2.75	2.60	2.52	2.44	2.35	2.26	2.17	*0.01*	
3.41	3.25	3.07	2.89	2.79	2.69	2.59	2.48	2.37	*0.005*	
1.83	1.78	1.73	1.68	1.65	1.62	1.58	1.55	1.51	*0.10*	
2.18	2.10	2.03	1.94	1.90	1.85	1.81	1.75	1.70	*0.05*	
2.53	2.43	2.32	2.21	2.15	2.09	2.03	1.96	1.89	*0.025*	*29*
3.00	2.87	2.73	2.57	2.49	2.41	2.33	2.23	2.14	*0.01*	
3.38	3.21	3.04	2.86	2.76	2.66	2.56	2.45	2.33	*0.005*	
1.82	1.77	1.72	1.67	1.64	1.61	1.57	1.54	1.50	*0.10*	
2.16	2.09	2.01	1.93	1.89	1.84	1.79	1.74	1.68	*0.05*	
2.51	2.41	2.31	2.20	2.14	2.07	2.01	1.94	1.87	*0.025*	*30*
2.98	2.84	2.70	2.55	2.47	2.39	2.30	2.21	2.11	*0.01*	
3.34	3.18	3.01	2.82	2.73	2.63	2.52	2.42	2.30	*0.005*	
1.71	1.66	1.60	1.54	1.51	1.48	1.44	1.40	1.35	*0.10*	
1.99	1.92	1.84	1.75	1.70	1.65	1.59	1.53	1.47	*0.05*	
2.27	2.17	2.06	1.94	1.88	1.82	1.74	1.67	1.58	*0.025*	*60*
2.63	2.50	2.35	2.20	2.12	2.03	1.94	1.84	1.73	*0.01*	
2.90	2.74	2.57	2.39	2.29	2.19	2.08	1.96	1.83	*0.005*	
1.65	1.60	1.55	1.48	1.45	1.41	1.37	1.32	1.26	*0.10*	
1.91	1.83	1.75	1.66	1.61	1.55	1.50	1.43	1.35	*0.05*	
2.16	2.05	1.94	1.82	1.76	1.69	1.61	1.53	1.43	*0.025*	*120*
2.47	2.34	2.19	2.03	1.95	1.86	1.76	1.66	1.53	*0.01*	
2.71	2.54	2.37	2.19	2.09	1.98	1.87	1.75	1.61	*0.005*	

TABLE IX
Critical values
for a correlation
test for normality

n	0.10	0.05	0.01
		α	
5	0.903	0.880	0.832
6	0.911	0.889	0.841
7	0.918	0.897	0.852
8	0.924	0.905	0.862
9	0.930	0.911	0.871
10	0.935	0.917	0.879
11	0.938	0.923	0.887
12	0.942	0.927	0.894
13	0.945	0.931	0.900
14	0.948	0.935	0.905
15	0.951	0.938	0.910
16	0.953	0.941	0.914
17	0.955	0.944	0.918
18	0.957	0.946	0.922
19	0.959	0.949	0.925
20	0.960	0.951	0.928
21	0.962	0.952	0.931
22	0.963	0.954	0.933
23	0.964	0.956	0.936
24	0.966	0.957	0.938
25	0.967	0.958	0.940
26	0.968	0.960	0.942
27	0.969	0.961	0.944
28	0.969	0.962	0.945
29	0.970	0.963	0.947
30	0.971	0.964	0.949
40	0.977	0.972	0.958
50	0.981	0.976	0.966
60	0.983	0.980	0.971
70	0.985	0.982	0.975
80	0.987	0.984	0.978
90	0.988	0.986	0.980
100	0.989	0.987	0.982
200	0.994	0.993	0.990
300	0.996	0.995	0.993
400	0.997	0.996	0.995
500	0.998	0.997	0.996
1000	0.999	0.998	0.998

TABLE X

Values of $q_{0.01}$

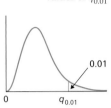

ν	κ									ν
	2	3	4	5	6	7	8	9	10	
1	90.0	135	164	186	202	216	227	237	246	1
2	14.0	19.0	22.3	24.7	26.6	28.2	29.5	30.7	31.7	2
3	8.26	10.6	12.2	13.3	14.2	15.0	15.6	16.2	16.7	3
4	6.51	8.12	9.17	9.96	10.6	11.1	11.5	11.9	12.3	4
5	5.70	6.97	7.80	8.42	8.91	9.32	9.67	9.97	10.2	5
6	5.24	6.33	7.03	7.56	7.97	8.32	8.61	8.87	9.10	6
7	4.95	5.92	6.54	7.01	7.37	7.68	7.94	8.17	8.37	7
8	4.74	5.63	6.20	6.63	6.96	7.24	7.47	7.68	7.87	8
9	4.60	5.43	5.96	6.35	6.66	6.91	7.13	7.32	7.49	9
10	4.48	5.27	5.77	6.14	6.43	6.67	6.87	7.05	7.21	10
11	4.39	5.14	5.62	5.97	6.25	6.48	6.67	6.84	6.99	11
12	4.32	5.04	5.50	5.84	6.10	6.32	6.51	6.67	6.81	12
13	4.26	4.96	5.40	5.73	5.98	6.19	6.37	6.53	6.67	13
14	4.21	4.89	5.32	5.63	5.88	6.08	6.26	6.41	6.54	14
15	4.17	4.83	5.25	5.56	5.80	5.99	6.16	6.31	6.44	15
16	4.13	4.78	5.19	5.49	5.72	5.92	6.08	6.22	6.35	16
17	4.10	4.74	5.14	5.43	5.66	5.85	6.01	6.15	6.27	17
18	4.07	4.70	5.09	5.38	5.60	5.79	5.94	6.08	6.20	18
19	4.05	4.67	5.05	5.33	5.55	5.73	5.89	6.02	6.14	19
20	4.02	4.64	5.02	5.29	5.51	5.69	5.84	5.97	6.09	20
24	3.96	4.54	4.91	5.17	5.37	5.54	5.69	5.81	5.92	24
30	3.89	4.45	4.80	5.05	5.24	5.40	5.54	5.65	5.76	30
40	3.82	4.37	4.70	4.93	5.11	5.27	5.39	5.50	5.60	40
60	3.76	4.28	4.60	4.82	4.99	5.13	5.25	5.36	5.45	60
120	3.70	4.20	4.50	4.71	4.87	5.01	5.12	5.21	5.30	120
∞	3.64	4.12	4.40	4.60	4.76	4.88	4.99	5.08	5.16	∞

TABLE XI
Values of $q_{0.05}$

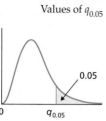

ν	κ									ν
	2	3	4	5	6	7	8	9	10	
1	18.0	27.0	32.8	37.1	40.4	43.1	45.4	47.4	49.1	1
2	6.08	8.33	9.80	10.9	11.7	12.4	13.0	13.5	14.0	2
3	4.50	5.91	6.82	7.50	8.04	8.48	8.85	9.18	9.46	3
4	3.93	5.04	5.76	6.29	6.71	7.05	7.35	7.60	7.83	4
5	3.64	4.60	5.22	5.67	6.03	6.33	6.58	6.80	6.99	5
6	3.46	4.34	4.90	5.30	5.63	5.90	6.12	6.32	6.49	6
7	3.34	4.16	4.68	5.06	5.36	5.61	5.82	6.00	6.16	7
8	3.26	4.04	4.53	4.89	5.17	5.40	5.60	5.77	5.92	8
9	3.20	3.95	4.41	4.76	5.02	5.24	5.43	5.59	5.74	9
10	3.15	3.88	4.33	4.65	4.91	5.12	5.30	5.46	5.60	10
11	3.11	3.82	4.26	4.57	4.82	5.03	5.20	5.35	5.49	11
12	3.08	3.77	4.20	4.51	4.75	4.95	5.12	5.27	5.39	12
13	3.06	3.73	4.15	4.45	4.69	4.88	5.05	5.19	5.32	13
14	3.03	3.70	4.11	4.41	4.64	4.83	4.99	5.13	5.25	14
15	3.01	3.67	4.08	4.37	4.59	4.78	4.94	5.08	5.20	15
16	3.00	3.65	4.05	4.33	4.56	4.74	4.90	5.03	5.15	16
17	2.98	3.63	4.02	4.30	4.52	4.70	4.86	4.99	5.11	17
18	2.97	3.61	4.00	4.28	4.49	4.67	4.82	4.96	5.07	18
19	2.96	3.59	3.98	4.25	4.47	4.65	4.79	4.92	5.04	19
20	2.95	3.58	3.96	4.23	4.45	4.62	4.77	4.90	5.01	20
24	2.92	3.53	3.90	4.17	4.37	4.54	4.68	4.81	4.92	24
30	2.89	3.49	3.85	4.10	4.30	4.46	4.60	4.72	4.82	30
40	2.86	3.44	3.79	4.04	4.23	4.39	4.52	4.63	4.73	40
60	2.83	3.40	3.74	3.98	4.16	4.31	4.44	4.55	4.65	60
120	2.80	3.36	3.68	3.92	4.10	4.24	4.36	4.47	4.56	120
∞	2.77	3.31	3.63	3.86	4.03	4.17	4.29	4.39	4.47	∞

TABLE XII

Binomial probabilities:

$$\binom{n}{x} p^x (1-p)^{n-x}$$

								p					
n	x	0.1	0.2	0.25	0.3	0.4	0.5	0.6	0.7	0.75	0.8	0.9
1	0	0.900	0.800	0.750	0.700	0.600	0.500	0.400	0.300	0.250	0.200	0.100
	1	0.100	0.200	0.250	0.300	0.400	0.500	0.600	0.700	0.750	0.800	0.900
2	0	0.810	0.640	0.563	0.490	0.360	0.250	0.160	0.090	0.063	0.040	0.010
	1	0.180	0.320	0.375	0.420	0.480	0.500	0.480	0.420	0.375	0.320	0.180
	2	0.010	0.040	0.063	0.090	0.160	0.250	0.360	0.490	0.563	0.640	0.810
3	0	0.729	0.512	0.422	0.343	0.216	0.125	0.064	0.027	0.016	0.008	0.001
	1	0.243	0.384	0.422	0.441	0.432	0.375	0.288	0.189	0.141	0.096	0.027
	2	0.027	0.096	0.141	0.189	0.288	0.375	0.432	0.441	0.422	0.384	0.243
	3	0.001	0.008	0.016	0.027	0.064	0.125	0.216	0.343	0.422	0.512	0.729
4	0	0.656	0.410	0.316	0.240	0.130	0.063	0.026	0.008	0.004	0.002	0.000
	1	0.292	0.410	0.422	0.412	0.346	0.250	0.154	0.076	0.047	0.026	0.004
	2	0.049	0.154	0.211	0.265	0.346	0.375	0.346	0.265	0.211	0.154	0.049
	3	0.004	0.026	0.047	0.076	0.154	0.250	0.346	0.412	0.422	0.410	0.292
	4	0.000	0.002	0.004	0.008	0.026	0.063	0.130	0.240	0.316	0.410	0.656
5	0	0.590	0.328	0.237	0.168	0.078	0.031	0.010	0.002	0.001	0.000	0.000
	1	0.328	0.410	0.396	0.360	0.259	0.156	0.077	0.028	0.015	0.006	0.000
	2	0.073	0.205	0.264	0.309	0.346	0.312	0.230	0.132	0.088	0.051	0.008
	3	0.008	0.051	0.088	0.132	0.230	0.312	0.346	0.309	0.264	0.205	0.073
	4	0.000	0.006	0.015	0.028	0.077	0.156	0.259	0.360	0.396	0.410	0.328
	5	0.000	0.000	0.001	0.002	0.010	0.031	0.078	0.168	0.237	0.328	0.590
6	0	0.531	0.262	0.178	0.118	0.047	0.016	0.004	0.001	0.000	0.000	0.000
	1	0.354	0.393	0.356	0.303	0.187	0.094	0.037	0.010	0.004	0.002	0.000
	2	0.098	0.246	0.297	0.324	0.311	0.234	0.138	0.060	0.033	0.015	0.001
	3	0.015	0.082	0.132	0.185	0.276	0.313	0.276	0.185	0.132	0.082	0.015
	4	0.001	0.015	0.033	0.060	0.138	0.234	0.311	0.324	0.297	0.246	0.098
	5	0.000	0.002	0.004	0.010	0.037	0.094	0.187	0.303	0.356	0.393	0.354
	6	0.000	0.000	0.000	0.001	0.004	0.016	0.047	0.118	0.178	0.262	0.531
7	0	0.478	0.210	0.133	0.082	0.028	0.008	0.002	0.000	0.000	0.000	0.000
	1	0.372	0.367	0.311	0.247	0.131	0.055	0.017	0.004	0.001	0.000	0.000
	2	0.124	0.275	0.311	0.318	0.261	0.164	0.077	0.025	0.012	0.004	0.000
	3	0.023	0.115	0.173	0.227	0.290	0.273	0.194	0.097	0.058	0.029	0.003
	4	0.003	0.029	0.058	0.097	0.194	0.273	0.290	0.227	0.173	0.115	0.023
	5	0.000	0.004	0.012	0.025	0.077	0.164	0.261	0.318	0.311	0.275	0.124
	6	0.000	0.000	0.001	0.004	0.017	0.055	0.131	0.247	0.311	0.367	0.372
	7	0.000	0.000	0.000	0.000	0.002	0.008	0.028	0.082	0.133	0.210	0.478

TABLE XII   (cont.)
Binomial probabilities:

$$\binom{n}{x}p^x(1-p)^{n-x}$$

n	x	0.1	0.2	0.25	0.3	0.4	0.5	0.6	0.7	0.75	0.8	0.9
8	0	0.430	0.168	0.100	0.058	0.017	0.004	0.001	0.000	0.000	0.000	0.000
	1	0.383	0.336	0.267	0.198	0.090	0.031	0.008	0.001	0.000	0.000	0.000
	2	0.149	0.294	0.311	0.296	0.209	0.109	0.041	0.010	0.004	0.001	0.000
	3	0.033	0.147	0.208	0.254	0.279	0.219	0.124	0.047	0.023	0.009	0.000
	4	0.005	0.046	0.087	0.136	0.232	0.273	0.232	0.136	0.087	0.046	0.005
	5	0.000	0.009	0.023	0.047	0.124	0.219	0.279	0.254	0.208	0.147	0.033
	6	0.000	0.001	0.004	0.010	0.041	0.109	0.209	0.296	0.311	0.294	0.149
	7	0.000	0.000	0.000	0.001	0.008	0.031	0.090	0.198	0.267	0.336	0.383
	8	0.000	0.000	0.000	0.000	0.001	0.004	0.017	0.058	0.100	0.168	0.430
9	0	0.387	0.134	0.075	0.040	0.010	0.002	0.000	0.000	0.000	0.000	0.000
	1	0.387	0.302	0.225	0.156	0.060	0.018	0.004	0.000	0.000	0.000	0.000
	2	0.172	0.302	0.300	0.267	0.161	0.070	0.021	0.004	0.001	0.000	0.000
	3	0.045	0.176	0.234	0.267	0.251	0.164	0.074	0.021	0.009	0.003	0.000
	4	0.007	0.066	0.117	0.172	0.251	0.246	0.167	0.074	0.039	0.017	0.001
	5	0.001	0.017	0.039	0.074	0.167	0.246	0.251	0.172	0.117	0.066	0.007
	6	0.000	0.003	0.009	0.021	0.074	0.164	0.251	0.267	0.234	0.176	0.045
	7	0.000	0.000	0.001	0.004	0.021	0.070	0.161	0.267	0.300	0.302	0.172
	8	0.000	0.000	0.000	0.000	0.004	0.018	0.060	0.156	0.225	0.302	0.387
	9	0.000	0.000	0.000	0.000	0.000	0.002	0.010	0.040	0.075	0.134	0.387
10	0	0.349	0.107	0.056	0.028	0.006	0.001	0.000	0.000	0.000	0.000	0.000
	1	0.387	0.268	0.188	0.121	0.040	0.010	0.002	0.000	0.000	0.000	0.000
	2	0.194	0.302	0.282	0.233	0.121	0.044	0.011	0.001	0.000	0.000	0.000
	3	0.057	0.201	0.250	0.267	0.215	0.117	0.042	0.009	0.003	0.001	0.000
	4	0.011	0.088	0.146	0.200	0.251	0.205	0.111	0.037	0.016	0.006	0.000
	5	0.001	0.026	0.058	0.103	0.201	0.246	0.201	0.103	0.058	0.026	0.001
	6	0.000	0.006	0.016	0.037	0.111	0.205	0.251	0.200	0.146	0.088	0.011
	7	0.000	0.001	0.003	0.009	0.042	0.117	0.215	0.267	0.250	0.201	0.057
	8	0.000	0.000	0.000	0.001	0.011	0.044	0.121	0.233	0.282	0.302	0.194
	9	0.000	0.000	0.000	0.000	0.002	0.010	0.040	0.121	0.188	0.268	0.387
	10	0.000	0.000	0.000	0.000	0.000	0.001	0.006	0.028	0.056	0.107	0.349
11	0	0.314	0.086	0.042	0.020	0.004	0.000	0.000	0.000	0.000	0.000	0.000
	1	0.384	0.236	0.155	0.093	0.027	0.005	0.001	0.000	0.000	0.000	0.000
	2	0.213	0.295	0.258	0.200	0.089	0.027	0.005	0.001	0.000	0.000	0.000
	3	0.071	0.221	0.258	0.257	0.177	0.081	0.023	0.004	0.001	0.000	0.000
	4	0.016	0.111	0.172	0.220	0.236	0.161	0.070	0.017	0.006	0.002	0.000
	5	0.002	0.039	0.080	0.132	0.221	0.226	0.147	0.057	0.027	0.010	0.000
	6	0.000	0.010	0.027	0.057	0.147	0.226	0.221	0.132	0.080	0.039	0.002
	7	0.000	0.002	0.006	0.017	0.070	0.161	0.236	0.220	0.172	0.111	0.016
	8	0.000	0.000	0.001	0.004	0.023	0.081	0.177	0.257	0.258	0.221	0.071
	9	0.000	0.000	0.000	0.001	0.005	0.027	0.089	0.200	0.258	0.295	0.213
	10	0.000	0.000	0.000	0.000	0.001	0.005	0.027	0.093	0.155	0.236	0.384
	11	0.000	0.000	0.000	0.000	0.000	0.000	0.004	0.020	0.042	0.086	0.314

TABLE XII (cont.)
Binomial probabilities:
$$\binom{n}{x}p^{x}(1-p)^{n-x}$$

							p					
n	*x*	*0.1*	*0.2*	*0.25*	*0.3*	*0.4*	*0.5*	*0.6*	*0.7*	*0.75*	*0.8*	*0.9*
12	0	0.282	0.069	0.032	0.014	0.002	0.000	0.000	0.000	0.000	0.000	0.000
	1	0.377	0.206	0.127	0.071	0.017	0.003	0.000	0.000	0.000	0.000	0.000
	2	0.230	0.283	0.232	0.168	0.064	0.016	0.002	0.000	0.000	0.000	0.000
	3	0.085	0.236	0.258	0.240	0.142	0.054	0.012	0.001	0.000	0.000	0.000
	4	0.021	0.133	0.194	0.231	0.213	0.121	0.042	0.008	0.002	0.001	0.000
	5	0.004	0.053	0.103	0.158	0.227	0.193	0.101	0.029	0.011	0.003	0.000
	6	0.000	0.016	0.040	0.079	0.177	0.226	0.177	0.079	0.040	0.016	0.000
	7	0.000	0.003	0.011	0.029	0.101	0.193	0.227	0.158	0.103	0.053	0.004
	8	0.000	0.001	0.002	0.008	0.042	0.121	0.213	0.231	0.194	0.133	0.021
	9	0.000	0.000	0.000	0.001	0.012	0.054	0.142	0.240	0.258	0.236	0.085
	10	0.000	0.000	0.000	0.000	0.002	0.016	0.064	0.168	0.232	0.283	0.230
	11	0.000	0.000	0.000	0.000	0.000	0.003	0.017	0.071	0.127	0.206	0.377
	12	0.000	0.000	0.000	0.000	0.000	0.000	0.002	0.014	0.032	0.069	0.282
13	0	0.254	0.055	0.024	0.010	0.001	0.000	0.000	0.000	0.000	0.000	0.000
	1	0.367	0.179	0.103	0.054	0.011	0.002	0.000	0.000	0.000	0.000	0.000
	2	0.245	0.268	0.206	0.139	0.045	0.010	0.001	0.000	0.000	0.000	0.000
	3	0.100	0.246	0.252	0.218	0.111	0.035	0.006	0.001	0.000	0.000	0.000
	4	0.028	0.154	0.210	0.234	0.184	0.087	0.024	0.003	0.001	0.000	0.000
	5	0.006	0.069	0.126	0.180	0.221	0.157	0.066	0.014	0.005	0.001	0.000
	6	0.001	0.023	0.056	0.103	0.197	0.209	0.131	0.044	0.019	0.006	0.000
	7	0.000	0.006	0.019	0.044	0.131	0.209	0.197	0.103	0.056	0.023	0.001
	8	0.000	0.001	0.005	0.014	0.066	0.157	0.221	0.180	0.126	0.069	0.006
	9	0.000	0.000	0.001	0.003	0.024	0.087	0.184	0.234	0.210	0.154	0.028
	10	0.000	0.000	0.000	0.001	0.006	0.035	0.111	0.218	0.252	0.246	0.100
	11	0.000	0.000	0.000	0.000	0.001	0.010	0.045	0.139	0.206	0.268	0.245
	12	0.000	0.000	0.000	0.000	0.000	0.002	0.011	0.054	0.103	0.179	0.367
	13	0.000	0.000	0.000	0.000	0.000	0.000	0.001	0.010	0.024	0.055	0.254
14	0	0.229	0.044	0.018	0.007	0.001	0.000	0.000	0.000	0.000	0.000	0.000
	1	0.356	0.154	0.083	0.041	0.007	0.001	0.000	0.000	0.000	0.000	0.000
	2	0.257	0.250	0.180	0.113	0.032	0.006	0.001	0.000	0.000	0.000	0.000
	3	0.114	0.250	0.240	0.194	0.085	0.022	0.003	0.000	0.000	0.000	0.000
	4	0.035	0.172	0.220	0.229	0.155	0.061	0.014	0.001	0.000	0.000	0.000
	5	0.008	0.086	0.147	0.196	0.207	0.122	0.041	0.007	0.002	0.000	0.000
	6	0.001	0.032	0.073	0.126	0.207	0.183	0.092	0.023	0.008	0.002	0.000
	7	0.000	0.009	0.028	0.062	0.157	0.209	0.157	0.062	0.028	0.009	0.000
	8	0.000	0.002	0.008	0.023	0.092	0.183	0.207	0.126	0.073	0.032	0.001
	9	0.000	0.000	0.002	0.007	0.041	0.122	0.207	0.196	0.147	0.086	0.008
	10	0.000	0.000	0.000	0.001	0.014	0.061	0.155	0.229	0.220	0.172	0.035
	11	0.000	0.000	0.000	0.000	0.003	0.022	0.085	0.194	0.240	0.250	0.114
	12	0.000	0.000	0.000	0.000	0.001	0.006	0.032	0.113	0.180	0.250	0.257
	13	0.000	0.000	0.000	0.000	0.000	0.001	0.007	0.041	0.083	0.154	0.356
	14	0.000	0.000	0.000	0.000	0.000	0.000	0.001	0.007	0.018	0.044	0.229

TABLE XII (cont.)

Binomial probabilities:

$$\binom{n}{x}p^x(1-p)^{n-x}$$

n	x	\multicolumn{11}{c}{p}
		0.1	0.2	0.25	0.3	0.4	0.5	0.6	0.7	0.75	0.8	0.9
15	0	0.206	0.035	0.013	0.005	0.000	0.000	0.000	0.000	0.000	0.000	0.000
	1	0.343	0.132	0.067	0.031	0.005	0.000	0.000	0.000	0.000	0.000	0.000
	2	0.267	0.231	0.156	0.092	0.022	0.003	0.000	0.000	0.000	0.000	0.000
	3	0.129	0.250	0.225	0.170	0.063	0.014	0.002	0.000	0.000	0.000	0.000
	4	0.043	0.188	0.225	0.219	0.127	0.042	0.007	0.001	0.000	0.000	0.000
	5	0.010	0.103	0.165	0.206	0.186	0.092	0.024	0.003	0.001	0.000	0.000
	6	0.002	0.043	0.092	0.147	0.207	0.153	0.061	0.012	0.003	0.001	0.000
	7	0.000	0.014	0.039	0.081	0.177	0.196	0.118	0.035	0.013	0.003	0.000
	8	0.000	0.003	0.013	0.035	0.118	0.196	0.177	0.081	0.039	0.014	0.000
	9	0.000	0.001	0.003	0.012	0.061	0.153	0.207	0.147	0.092	0.043	0.002
	10	0.000	0.000	0.001	0.003	0.024	0.092	0.186	0.206	0.165	0.103	0.010
	11	0.000	0.000	0.000	0.001	0.007	0.042	0.127	0.219	0.225	0.188	0.043
	12	0.000	0.000	0.000	0.000	0.002	0.014	0.063	0.170	0.225	0.250	0.129
	13	0.000	0.000	0.000	0.000	0.000	0.003	0.022	0.092	0.156	0.231	0.267
	14	0.000	0.000	0.000	0.000	0.000	0.000	0.005	0.031	0.067	0.132	0.343
	15	0.000	0.000	0.000	0.000	0.000	0.000	0.000	0.005	0.013	0.035	0.206
20	0	0.122	0.012	0.003	0.001	0.000	0.000	0.000	0.000	0.000	0.000	0.000
	1	0.270	0.058	0.021	0.007	0.000	0.000	0.000	0.000	0.000	0.000	0.000
	2	0.285	0.137	0.067	0.028	0.003	0.000	0.000	0.000	0.000	0.000	0.000
	3	0.190	0.205	0.134	0.072	0.012	0.001	0.000	0.000	0.000	0.000	0.000
	4	0.090	0.218	0.190	0.130	0.035	0.005	0.000	0.000	0.000	0.000	0.000
	5	0.032	0.175	0.202	0.179	0.075	0.015	0.001	0.000	0.000	0.000	0.000
	6	0.009	0.109	0.169	0.192	0.124	0.037	0.005	0.000	0.000	0.000	0.000
	7	0.002	0.055	0.112	0.164	0.166	0.074	0.015	0.001	0.000	0.000	0.000
	8	0.000	0.022	0.061	0.114	0.180	0.120	0.035	0.004	0.001	0.000	0.000
	9	0.000	0.007	0.027	0.065	0.160	0.160	0.071	0.012	0.003	0.000	0.000
	10	0.000	0.002	0.010	0.031	0.117	0.176	0.117	0.031	0.010	0.002	0.000
	11	0.000	0.000	0.003	0.012	0.071	0.160	0.160	0.065	0.027	0.007	0.000
	12	0.000	0.000	0.001	0.004	0.035	0.120	0.180	0.114	0.061	0.022	0.000
	13	0.000	0.000	0.000	0.001	0.015	0.074	0.166	0.164	0.112	0.055	0.002
	14	0.000	0.000	0.000	0.000	0.005	0.037	0.124	0.192	0.169	0.109	0.009
	15	0.000	0.000	0.000	0.000	0.001	0.015	0.075	0.179	0.202	0.175	0.032
	16	0.000	0.000	0.000	0.000	0.000	0.005	0.035	0.130	0.190	0.218	0.090
	17	0.000	0.000	0.000	0.000	0.000	0.001	0.012	0.072	0.134	0.205	0.190
	18	0.000	0.000	0.000	0.000	0.000	0.000	0.003	0.028	0.067	0.137	0.285
	19	0.000	0.000	0.000	0.000	0.000	0.000	0.000	0.007	0.021	0.058	0.270
	20	0.000	0.000	0.000	0.000	0.000	0.000	0.000	0.001	0.003	0.012	0.122

Answers to Selected Exercises

Note: Most of the numerical answers presented here were obtained by using a computer. If you solve a problem by hand and do some intermediate rounding or use provided summary statistics, your answer may differ slightly from the one given in this appendix.

Chapter 1

EXERCISES 1.1

1.1 See Definition 1.2 on page 5.

1.3 Descriptive statistics includes the construction of graphs, charts, and tables, and the calculation of various descriptive measures such as averages, measures of variation, and percentiles.

1.5 Inferential

1.7 Inferential

1.9 Descriptive

EXERCISES 1.3

1.13 Conducting a census is generally time consuming and costly, frequently impractical, and sometimes impossible.

1.15 Obtaining a representative sample is important because the sample will be used to draw conclusions about the entire population.

1.17 Dentists form a high-income group whose incomes are not representative of the incomes of Seattle residents in general.

1.19
a. In probability sampling, a random device, such as tossing a coin or consulting a random-number table, is used to decide which members of the population will constitute the sample instead of leaving such decisions to human judgement.
b. No. Because probability sampling uses a random device, it is possible to obtain a nonrepresentative sample.
c. Probability sampling eliminates unintentional selection bias and permits the researcher to control the chance of obtaining a nonrepresentative sample. Also, use of

probability sampling guarantees that the techniques of inferential statistics can be applied.

1.21 Simple random sampling

1.23
a.

G, L, S	G, L, A	G, L, T	G, S, A	G, S, T
G, A, T	L, S, A	L, S, T	L, A, T	S, A, T

b. $\frac{1}{10}, \frac{1}{10}, \frac{1}{10}$

1.25
a.

E, M, P, L	E, M, P, B	E, M, P, F	E, M, L, B	E, M, L, F
E, M, B, F	E, P, L, B	E, P, L, F	E, P, B, F	E, L, B, F
M, P, L, B	M, P, L, F	M, P, B, F	M, L, B, F	P, L, B, F

b. Write the initials of the six artists on separate pieces of paper, place the six slips of paper in a box, and then, while blindfolded, pick four of the slips of paper.
c. $\frac{1}{15}, \frac{1}{15}$

1.27 Answers will vary.

EXERCISES 1.4
1.33
a. The population under consideration consists of the 500 firms from *Fortune Magazine*'s list of "The International 500." The sample consists of 10 of those firms. From Procedure 1.1 on page 17, the method is as follows: *Step 1.* Divide the population size, 500, by the sample size, 10, which gives 50 (rounding down is unnecessary in this case). *Step 2.* Use a random number table (or a similar device) to obtain a number, k,

between 1 and 50. *Step 3.* Select for the sample those firms on the "The International 500" that are numbered k, $k + 50$, $k + 100$, and so on. For example, if $k = 6$, the firms in the sample will be those numbered 6, 56, 106, ...

b. Systematic random sampling

c. Answers will vary.

1.35

a. Number the suites from 1 to 48, use a table of random numbers to randomly select 3 of the 48 suites, and take as the sample the 24 dormitory residents living in the 3 suites obtained.

b. Probably not, because friends often have similar opinions.

c. Proportional allocation dictates that the number of freshmen, sophomores, juniors, and seniors selected be 8, 7, 6, and 3, respectively. Thus a stratified sample of 24 dormitory residents can be obtained as follows: Number the freshman dormitory residents from 1 through 128 and use a table of random numbers to randomly select 8 of the 128 freshman dormitory residents; number the sophomore dormitory residents from 1 through 112 and use a table of random numbers to randomly select 7 of the 112 sophomore dormitory residents; and so on.

EXERCISES 1.5

1.39 Causation (cause and effect)

1.41 Here is one of several methods that could be used: Number the women from 1 to 4753; use a table of random numbers or a random-number generator to obtain 2376 different numbers between 1 and 4753; the 2376 women with those numbers are in one group; the remaining 2377 women are in the other group.

1.43 Designed experiment

1.45

a. The individuals or items on which the experiment is performed

b. Subject

1.47

a. The 20 flashlights

b. Lifetime of a battery in a flashlight

c. Battery brand

d. Four brands of batteries

e. Four brands of batteries

1.49

a. Batches of the product being sold (Some might say that the stores are the experimental units.)

b. Unit sales of the product

c. Display type and pricing scheme

d. Display type has three levels: normal display space interior to an aisle, normal display space at the end of an aisle, and enlarged display space. Pricing scheme has three levels: regular price, reduced price, and cost.

e. Each treatment is a combination of a level of display type and a level of pricing scheme.

1.51 Completely randomized design

REVIEW TEST FOR CHAPTER 1

1. Answers will vary.

2. It is almost always necessary to invoke techniques of descriptive statistics to organize and summarize the information obtained from a sample before carrying out an inferential analysis.

3. Descriptive

4. Descriptive

5. A literature search

6. **a.** A representative sample is a sample that reflects as closely as possible the relevant characteristics of the population under consideration.

   b. In probability sampling, a random device, such as tossing a coin or consulting a table of random numbers, is used to decide which members of the population will constitute the sample instead of leaving such decisions to human judgement.

   c. Simple random sampling is a sampling procedure for which each possible sample of a given size is equally likely to be the one obtained from the population.

7. No, because parents of students at Yale tend to have higher incomes than parents of college students in general.

8. Only (b)

9. **a.**

AA, UA, US	AA, UA, HP	AA, UA, CO
AA, US, HP	AA, US, CO	AA, HP, CO
UA, US, HP	UA, US, CO	UA, HP, CO
US, HP, CO		

   b. $\frac{1}{10}$, $\frac{1}{10}$, $\frac{1}{10}$

10. **a.** Number the athletes from 1 to 100, use Table I to obtain 15 different numbers between 1 and 100, and take as the sample the 15 athletes who are numbered with the numbers obtained.

b. 082, 008, 016, 001, 047, 094, 097, 074, 052, 076, 098, 003, 089, 041, 063

11. See Section 1.4 and, in particular, (a) Procedure 1.1 on page 17, (b) Procedure 1.2 on page 19, and (c) Procedure 1.3 on page 21.

12. **a.** The population under consideration consists of the top 100 North American athletes of the twentieth century. The sample consists of 15 of those athletes. From Procedure 1.1 on page 17, the method is as follows: *Step 1*. Divide the population size, 100, by the sample size, 15, and round the result down, which gives 6. *Step 2*. Use a random number table (or a similar device) to obtain a number between 1 and 6; call it k. *Step 3*. Select for the sample those athletes who are numbered $k, k + 6, k + 12$, and so on. For example, if $k = 4$, then the athletes in the sample will be those numbered 4, 10, 16, ...
 b. Yes, unless for some reason there is a cyclical pattern in the listing of the athletes.

13. **a.** Proportional allocation dictates that 10 full professors, 16 associate professors, 12 assistant professors, and 2 instructors be selected.
 b. The procedure is as follows: Number the full professors from 1 to 205, and use Table I to randomly select 10 of the 205 full professors; number the associate professors from 1 to 328, and use Table I to randomly select 16 of the 328 associate professors; and so on.

14. The statement under the vote tally is a disclaimer as to the validity of the survey. Because the results reflect only responses of Internet users, they cannot be regarded as representative of the public in general. Moreover, because the sample was not chosen at random from Internet users—but rather was obtained only from volunteers—the results cannot even be considered representative of Internet users.

15. **a.** In an observational study, researchers simply observe characteristics and take measurements. In a designed experiment, researchers impose treatments and controls and then observe characteristics and take measurements.
 b. Observational studies can reveal only association, whereas designed experiments can help establish causation (cause and effect).

16. Observational study

17. **a.** Designed experiment
 b. The treatment group consists of the 158 patients who took AVONEX. The control group consists of the 143 patients who were given placebo. The treatments are AVONEX and placebo.

18. See Key Fact 1.1 on page 24.

19. **a.** The batches of doughnuts
 b. Amount of fat absorbed
 c. Fat type
 d. Four types of fat
 e. Four types of fat

20. **a.** The tomato plants in the study (Some might say the plots of land are the experimental units.)
 b. Yield of tomato plants
 c. Tomato variety and planting density
 d. Different tomato varieties and different planting densities
 e. Each treatment is a combination of a level of tomato variety and a level of planting density.

21. **a.** Completely randomized design
 b. Randomized block design; the six different car models
 c. The randomized block design in part (b)

22. Answers will vary.

23. Answers will vary.

Chapter 2

EXERCISES 2.1

2.1 Answers will vary.

2.3 See Definition 2.2 on page 43.

2.5 Qualitative variable

2.7
a. Quantitative, discrete; rank of city by highest temperature
b. Quantitative, continuous; highest temperature, in degrees Fahrenheit
c. Qualitative; state in which a U.S. city is located

2.9
a. Quantitative, discrete; rank of city by percentage of families using the Internet
b. Qualitative; city name
c. Quantitative, discrete; percentage of families in a city that use the Internet

EXERCISES 2.2

2.11 Grouping can help to make a large and complicated set of data more compact and easier to understand.

2.13 See 1–3 on pages 47–48.

2.15 Relative-frequency distributions are better than frequency distributions when two data sets are being compared because relative frequencies are always between 0 and 1 and hence provide a standard for comparison.

2.19 Because each class is based on a single value, the midpoint of each class is the same as the class.

2.21

Speed (mph)	Frequency	Relative frequency	Midpoint
52 ≤ 54	2	0.057	53
54 ≤ 56	5	0.143	55
56 ≤ 58	6	0.171	57
58 ≤ 60	8	0.229	59
60 ≤ 62	7	0.200	61
62 ≤ 64	3	0.086	63
64 ≤ 66	2	0.057	65
66 ≤ 68	1	0.029	67
68 ≤ 70	0	0.000	69
70 ≤ 72	0	0.000	71
72 ≤ 74	0	0.000	73
74 ≤ 76	1	0.029	75
	35	1.001	

2.23

Speed (mph)	Frequency	Relative frequency	Mark
52–53.9	2	0.057	52.95
54–55.9	5	0.143	54.95
56–57.9	6	0.171	56.95
58–59.9	8	0.229	58.95
60–61.9	7	0.200	60.95
62–63.9	3	0.086	62.95
64–65.9	2	0.057	64.95
66–67.9	1	0.029	66.95
68–69.9	0	0.000	68.95
70–71.9	0	0.000	70.95
72–73.9	0	0.000	72.95
74–75.9	1	0.029	74.95
	35	1.001	

2.25

Number of pups	Frequency	Relative frequency
3	2	0.025
4	5	0.063
5	10	0.125
6	11	0.138
7	17	0.213
8	17	0.213
9	11	0.138
10	4	0.050
11	2	0.025
12	1	0.013
	80	1.003

2.27

Network	Frequency	Relative frequency
ABC	5	0.25
CBS	8	0.40
NBC	7	0.35
	20	1.00

EXERCISES 2.3

2.37 A frequency histogram displays the class frequencies on the vertical axis, whereas a relative-frequency histogram displays the class relative frequencies on the vertical axis.

2.39 To avoid confusing bar graphs and histograms, the bars in a bar graph are positioned so that they do not touch each other.

2.41
a. The heights of the bars in the histogram would be comparable to the heights of the columns of dots in the dotplot.
b. No. Each bar in the histogram would represent a range of possible values, whereas, in the dotplot, each column of dots represents one possible value.

2.43
a.

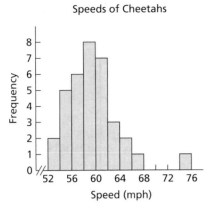

b.

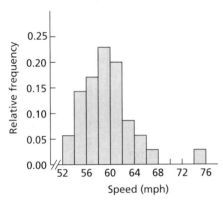

2.45
a.

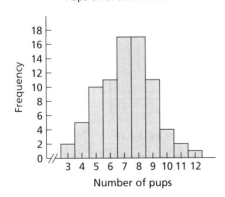

b.

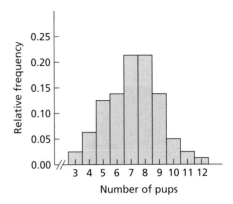

2.47

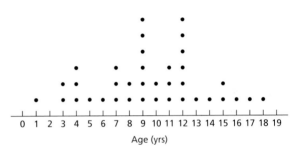

2.49
a.

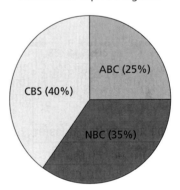

b.

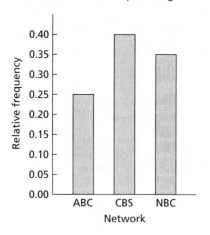

Networks of Top TV Programs

2.51

a. 20%   **b.** 25%   **c.** 7

EXERCISES 2.4

2.59 Histogram. Stem-and-leaf diagrams are generally not useful with large data sets.

2.61 Reconstruct the stem-and-leaf diagram, using more lines per stem.

2.63

a.
5	3 2 4 9 4 7 5
6	0 8 3 7 4 3 6 8 0 6 4 3
7	4 6 3 3 7 1 7 7 1
8	4 0 2

b.
5	2 3 4 4 5 7 9
6	0 0 3 3 3 4 4 6 6 7 8 8
7	1 1 3 3 4 6 7 7 7
8	0 2 4

c.
5	3 2 4 4		5	2 3 4 4
5	9 7 5		5	5 7 9
6	0 3 4 3 0 4 3		6	0 0 3 3 3 4 4
6	8 7 6 8 6		6	6 6 7 8 8
7	4 3 3 1 1		7	1 1 3 3 4
7	6 7 7 7		7	6 7 7 7
8	4 0 2		8	0 2 4

2.65

a.
2	6 4 3 3 6
3	7 2 2 8 2 0 0 5 5 2 1 0 5 0 0 2 3
4	5 2 1 0 5 8 3 4 8 1 0 5 1 3 9 0 6 8 9 5
5	8 7 2 4 5 2 0 1
6	
7	3

b.
2	4 3 3
2	6 6
3	2 2 2 0 0 2 1 0 0 0 2 3
3	7 8 5 5 5
4	2 1 0 3 4 1 0 1 3 0
4	5 5 8 8 5 9 6 8 9 5
5	2 4 2 0 1
5	8 7 5
6	
6	
7	3

c.
2	3 3
2	4
2	6 6
2	
3	0 0 1 0 0 0
3	2 2 2 2 2 3
3	5 5 5
3	7
3	8
4	1 0 1 0 1 0
4	2 3 3
4	5 5 4 5 5
4	6
4	8 8 9 8 9
5	0 1
5	2 2
5	4 5
5	7
5	8
6	
6	
6	
6	
6	
7	
7	3

d. The stem-and-leaf diagram of part (b) is the most useful. That in part (a) is also useful.

EXERCISES 2.5

2.71

a. The *distribution of a data set* is a table, graph, or formula that provides the values of the observations and how often they occur.

b. *Sample data* is a data set obtained by observing the values of a variable for a sample of the population.

c. *Population data* is the data set obtained by observing the values of a variable for an entire population.

d. *Census data* is another name for population data.

e. A *sample distribution* is the distribution of sample data.

f. The *population distribution* is the distribution of population data.

g. *Distribution of a variable* is another name for population distribution.

2.73 Roughly a bell shape

2.75 Answers will vary.

2.77

a. Bell shaped   **b.** Symmetric

2.79

a. Left skewed   **b.** Left skewed

2.81

a. Right skewed   **b.** Right skewed

EXERCISES 2.6

2.87

a. Part of the vertical axis of the graph has been cut off, or truncated.

b. It may be done to present a clearer picture of the ups and downs in a data pattern rather than to mislead the reader.

c. You should start the axis at 0 and put slashes in the axis to indicate that part of the axis is missing.

2.89

c. They give the misleading impression that the district average is much greater relative to the national average than it actually is.

2.91

a. It is a truncated graph.

b.

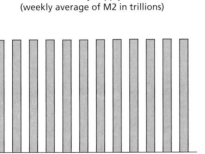

Money Supply
(weekly average of M2 in trillions)

c.

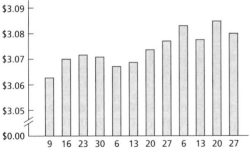

Money Supply
(weekly average of M2 in trillions)

REVIEW TEST FOR CHAPTER 2

1. **a.** A variable is a characteristic that varies from one person or thing to another.
   b. Quantitative variables and qualitative (or categorical) variables
   c. Discrete variables and continuous variables

   d. The information obtained by observing the values of a variable
   e. By the type of variable being observed

2. It helps organize the data, making them much simpler to comprehend.

3. Qualitative data. It makes no sense to look for cutpoints or midpoints for nonnumerical data (or for numerical data obtained by coding nonnumerical data).

4. **a.** 6 and 14 **b.** 18
   c. 22 and 30 **d.** The third class

5. **a.** 10 **b.** 20 **c.** 25 and 35

6. When grouping discrete data in which there are only relatively few distinct observations

7. **a.** The bar for each class extends from the lower cutpoint of the class to the upper cutpoint of the class.
   b. The bar for each class is centered over the midpoint of the class.

8. Pie charts and bar graphs

9. Histogram. Stem-and-leaf diagrams are generally not useful with large data sets.

10. See Figure 2.9 on page 73.

12. **a.** Left skewed. The distribution of a random sample taken from a population approximates the population distribution. The larger the sample, the better the approximation tends to be.
    b. No. Sample distributions vary from sample to sample.
    c. Yes. Left skewed. The overall shapes of the two sample distributions should be similar to that of the population distribution and hence to each other.

13. **a.** Discrete quantitative
    b. Continuous quantitative
    c. Qualitative

14. **a.**

Age at inauguration	Frequency	Relative frequency	Mark
40–44	2	0.047	42
45–49	6	0.140	47
50–54	13	0.302	52
55–59	12	0.279	57
60–64	7	0.163	62
65–69	3	0.070	67
	43	1.001	

b. 40 and 45   **c.** 5

d.

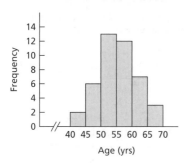

Ages at Inauguration
for First 43 U.S. Presidents

15.

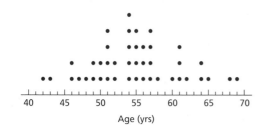

Ages at Inauguration
for First 43 U.S. Presidents

16. a. 4 | 2 3 6 6 7 8 9 9
         5 | 0 0 1 1 1 1 2 2 4 4 4 4 4 5 5 5 5 6 6 6 7 7 7 7 8
         6 | 0 1 1 1 2 4 4 5 8 9

b. 4 | 2 3
     4 | 6 6 7 8 9 9
     5 | 0 0 1 1 1 1 2 2 4 4 4 4 4
     5 | 5 5 5 5 6 6 6 7 7 7 7 8
     6 | 0 1 1 1 2 4 4
     6 | 5 8 9

c. The one in part (b)

17. a.

Number busy	Frequency	Relative frequency
0	1	0.04
1	2	0.08
2	2	0.08
3	4	0.16
4	5	0.20
5	7	0.28
6	4	0.16
	25	1.00

b.

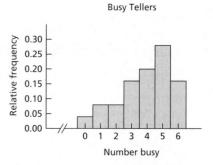

Busy Tellers

18. a.

Class level	Frequency	Relative frequency
Freshman	6	0.150
Sophomore	15	0.375
Junior	12	0.300
Senior	7	0.175
	40	1.000

b.

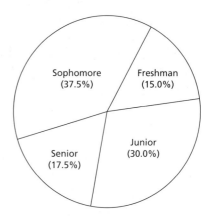

Class Levels for Statistics Students

c.

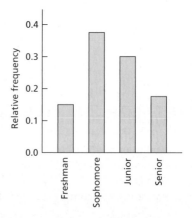

Class Levels for Statistics Students

19. a.

High	Freq.	Relative frequency	Midpoint
0 ≺ 1000	10	0.278	500
1000 ≺ 2000	10	0.278	1500
2000 ≺ 3000	4	0.111	2500
3000 ≺ 4000	4	0.111	3500
4000 ≺ 5000	0	0.000	4500
5000 ≺ 6000	1	0.028	5500
6000 ≺ 7000	1	0.028	6500
7000 ≺ 8000	0	0.000	7500
8000 ≺ 9000	1	0.028	8500
9000 ≺ 10000	1	0.028	9500
10000 ≺ 11000	1	0.028	10500
11000 ≺ 12000	3	0.083	11500
	36	1.001	

b.

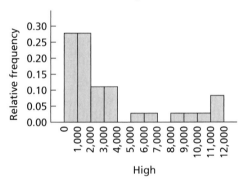

Dow Jones Highs, 1967 – 2002*

* Data from the *World Almanac*

20. a. Bell shaped   **b.** Left skewed

21. Answers will vary, but here is one possibility:

22. b. Having followed the directions in part (a), you might conclude that the percentage of women in the labor force for 2000 is about 3.5 times that for 1960.
   c. Not covering up the vertical axis, you would find that the percentage of women in the labor force for 2000 is about 1.8 times that for 1960.
   d. The graph is potentially misleading because it is truncated. Note that the vertical axis begins at 30 rather than at 0.
   e. To make the graph less potentially misleading, start it at 0 instead of at 30.

Chapter 3

EXERCISES 3.1

3.1 To indicate where the center or most typical value of a data set lies

3.3 The mode

3.5
a. Mean = 5; median = 5.
b. Mean = 15; median = 5. The median is a better measure of center because it is not influenced by the one unusually large value, 99.
c. Resistance

3.7 Median. Unlike the mean, the median is not affected strongly by the relatively few homes that have extremely large or small floor spaces.

3.9 Mean = 7.3 days; median = 6.0 days; modes = 5, 6, 11 days.

3.11 Mean = 78.4 tornadoes; median = 77.0 tornadoes; no mode.

3.13
a. CBS
b. No; neither the mean nor the median can be used as a measure of center for qualitative data.

EXERCISES 3.2

3.21 Mathematical notation allows you to express mathematical definitions and other mathematical relationships much more concisely.

3.23 No; the population mean is a constant. Yes; the sample mean is a variable because it varies from sample to sample.

3.25
a. 46   **b.** 4   **c.** 11.5

3.27
a. 23.3 hours   **b.** 10   **c.** 2.33 hours

3.29 Answers will vary.

EXERCISES 3.3

3.33 To indicate the amount of variation in a data set

3.35 The mean

3.37
a. 2.7   **b.** 31.6   **c.** Resistance

3.39
a. 16.1   **b.** 16.1

3.41
a. 6 days   **b.** 2.6 days   **c.** 2.6 days

3.43

a. 202 tornadoes   **b.** 53.9 tornadoes

c. 53.9 tornadoes

EXERCISES 3.4

3.53 The median and interquartile range are resistant measures, whereas the mean and standard deviation are not.

3.55 No. It may, for example, be an indication of skewness.

3.57

a. A measure of variation

b. Roughly, the range of the middle 50% of the observations

3.59 If the data set has no potential outliers, its boxplot and modified boxplot are identical. This is the case when both the minimum and maximum observations lie within the lower and upper limits; in other words, when the minimum and maximum values are also the adjacent values.

3.61 $Q_1 = 73.5$ games, $Q_2 = 79$ games, $Q_3 = 80$ games

3.63 $Q_1 = 4$ days, $Q_2 = 7$ days, $Q_3 = 12$ days

Note: If you use technology to obtain your results for the remaining answers in this section, they may differ from those presented here because different technologies often use different rules for computing quartiles.

3.65

a. 6.5 games

b. 45, 73.5, 79, 80, 82 games

c. 45 games and 48 games are potential outliers

d. Figure A.1(a) shows a boxplot. Figure A.1(b) shows a modified boxplot.

3.67

a. 8 days

b. 1, 4, 7, 12, 55 days

c. 55 days is a potential outlier.

d. Figure A.2(a) shows a boxplot. Figure A.2(b) shows a modified boxplot.

3.69

a. $Q_1 = 8$ cigs/day, $Q_2 = 9$ cigs/day, $Q_3 = 10$ cigs/day

b. The quartiles for this data set are not particularly useful because of its small range and the relatively large number of identical values. Note, for instance, that Q_3 and Max are equal.

FIGURE A.1

Boxplot and modified boxplot for Exercise 3.65(d)

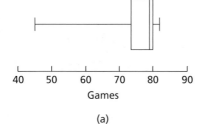

(a)

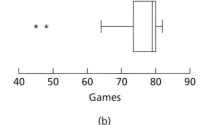

(b)

FIGURE A.2

Boxplot and modified boxplot for Exercise 3.67(d)

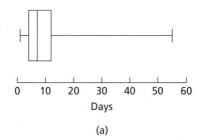

(a)

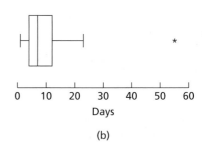

(b)

EXERCISES 3.5

3.79 To describe the entire population

3.81

a. 0, 1

b. the number of standard deviations that the observation is from the mean, that is, how far the observation is from the mean in units of standard deviation

c. above (greater than); below (less than)

3.83 Parameter. A parameter is a descriptive measure of a population.

3.85

a. $\mu = 65.7$ knots   **b.** $\sigma = 26.2$ knots

c. $\eta = 50.0$ knots   **d.** mode $= 50$ knots

e. IQR $= 30$ mph

3.87
a. $z = (x - 16.3)/17.9$ **b.** 0; 1
c. 2.70; −0.68
d. The time served of 64.7 months is 2.70 standard
deviations above the mean time served of 16.3 months;
the time served of 4.2 months is 0.68 standard
deviations below the mean time served of 16.3 months.

3.89 Yes, quite well. Your z-score is 3.50, meaning that
your score of 350 points is 3.50 standard deviations above
the mean score of 280 points. Applying the
three-standard-deviations rule (Key Fact 3.2), almost all
the exam scores lie within three standard deviations to
either side of the mean. So, your score of 350 points is
greater than almost all the other scores on the exam.

REVIEW TEST FOR CHAPTER 3

1. **a.** Numbers that are used to describe data sets are
   called descriptive measures.
   b. Descriptive measures that indicate where the center
   or most typical value of a data set lies are called
   measures of center.
   c. Descriptive measures that indicate the amount of
   variation, or spread, in a data set are called
   measures of variation.

2. Mean and median. The median is a resistant measure,
   whereas the mean is not. The mean takes into account
   the actual numerical value of all observations,
   whereas the median does not.

3. The mode

4. **a.** Standard deviation   **b.** Interquartile range

5. **a.** $\bar{x}$ **b.** s **c.** μ **d.** σ

6. **a.** Not necessarily true   **b.** Necessarily true

7. Three

8. **a.** Minimum, quartiles, and maximum; that is, Min,
   Q_1, Q_2, Q_3, Max
   b. Q_2 can be used to describe center. Max − Min,
   Q_1 − Min, Max − Q_3, Q_2 − Q_1, Q_3 − Q_2,
   and Q_3 − Q_1 are all measures of variation for
   different portions of the data.
   c. Boxplot

9. **a.** An outlier is an observation that falls well outside
   the overall pattern of the data.
   b. First, determine the lower and upper limits—the
   numbers 1.5 IQRs below the first quartile and
   1.5 IQRs above the third quartile, respectively.
   Observations that lie outside the lower and upper
   limits—either below the lower limit or above the
   upper limit—are potential outliers.

10. **a.** Subtract from x its mean and then divide by its
    standard deviation.
    b. The z-score of an observation gives the number of
    standard deviations that the observation is from the
    mean, that is, how far the observation is from the
    mean in units of standard deviation.
    c. The observation is 2.9 standard deviations above
    the mean. It is larger than most of the other
    observations.

11. **a.** 2.35 drinks; 2.0 drinks; 1, 2 drinks
    b. Answers will vary.

12. The median, because it is resistant to outliers and
    other extreme values.

13. The mode; neither the mean nor the median can be
    used as a measure of center for qualitative data.

14. **a.** $\bar{x} = 45.7$ kg   **b.** Range = 17 kg
    c. $s = 5.0$ kg

15. **a.**

$\bar{x}-3s$	$\bar{x}-2s$	$\bar{x}-s$	$\bar{x}$	$\bar{x}+s$	$\bar{x}+2s$	$\bar{x}+3s$
18.3	31.7	45.1	58.5	71.9	85.3	98.7

    b. 18.3 yr, 98.7 yr

16. **a.** $Q_1 = 48.0$ yr, $Q_2 = 59.5$ yr, $Q_3 = 68.5$ yr
    b. 20.5 yr; roughly speaking, the middle 50% of the
    ages has a range of 20.5 yr.
    c. 31, 48.0, 59.5, 68.5, 79 yr
    d. Lower limit: 17.25 yr. Upper limit: 99.25 yr.
    e. No potential outliers
    f.

17. **a.** $\mu = 19.34$ thousand students
    b. $\sigma = 4.07$ thousand students
    c. $z = (x - 19.34)/4.07$ **d.** 0; 1
    f. 1.37; −1.26. The enrollment at Los Angeles is
    1.37 standard deviations above the UC campuses'
    mean enrollment of 19.34 thousand students; the
    enrollment at Riverside is 1.26 standard deviations
    below the UC campuses' mean enrollment of
    19.34 thousand students.

18. **a.** A sample mean. It is the mean price per gallon for
    the sample of 10,000 service stations.
    b. $\bar{x}$
    c. A statistic. It is a descriptive measure of a sample.

Chapter 4

EXERCISES 4.1

4.1 An experiment is an action whose outcome cannot be predicted with certainty. An event is some specified result that may or may not occur when an experiment is performed.

4.3 There is no difference.

4.5 The probability of an event is the proportion of times it occurs in a large number of repetitions of the experiment.

4.7 (b) and (e) because the probability of an event must always be between 0 and 1, inclusive.

4.9
a. 0.209   **b.** 0.670   **c.** 0.017
d. 0   **e.** 1

4.11
a. 0.189   **b.** 0.176   **c.** 0.239   **d.** 0.761

4.13
a. 0.139   **b.** 0.500   **c.** 0.222   **d.** 0.111

4.15 The event in part (e) is certain; the event in part (d) is impossible.

EXERCISES 4.2

4.21 Venn diagrams

4.23 Two or more events are mutually exclusive if at most one of them can occur when the experiment is performed. Thus two events are mutually exclusive if they do not have outcomes in common. Three events are mutually exclusive if no two of them have outcomes in common.

4.25 $A = $ ⚃ ⚄ ⚅

$B = $ ⚃ ⚄ ⚅

$C = $ ⚀ ⚁

$D = $ ⚄

4.27
a. (not A) = ⚀ ⚂ ⚄

   The event that the die comes up odd

b. (A & B) = ⚃ ⚅

   The event that the die comes up 4 or 6

c. (B or C) = ⚀ ⚁ ⚃ ⚄ ⚅

   The event that the die does not come up 3

4.29
a. (not A) is the event the unit has at least five rooms. There are 75,232 thousand units that have at least five rooms.
b. (A & B) is the event the unit has two, three, or four rooms. There are 36,653 thousand units that have two, three, or four rooms.
c. (C or D) is the event the unit has at least five rooms. There are 75,232 thousand units that have at least five rooms. (*Note:* From part (a), (not A) = (C or D).)

4.31
a. No   **b.** Yes   **c.** No
d. Yes, events B, C, and D. No.

4.33 A and C; A and D; C and D; A, C, and D

EXERCISES 4.3

4.39
a. 0.56   **b.** $S = (A$ or B or $C)$
c. 0.01, 0.14, 0.41   **d.** 0.56

4.41
a. 0.070   **b.** 0.120   **c.** 0.698

4.43
a. 0.99   **b.** 0.56

4.45
a. 0.167, 0.056, 0.028, 0.056, 0.028, 0.139, 0.167
b. 0.223   **c.** 0.112   **d.** 0.278   **e.** 0.278

4.47
a. No, because $P(A$ or $B) \neq P(A) + P(B)$.
b. 0.083

EXERCISES 4.4

4.51 Summing the row totals, summing the column totals, or summing the frequencies in the cells

4.53
a. Univariate   **b.** Bivariate

4.55
a. 12   **b.** 91   **c.** 25   **d.** 55   **e.** 15

4.57
a. Second row: 12,838 and 5,864; third row: 2,202 and 11,513; fifth row: 24,754
b. 24,754   **c.** 85   **d.** 22,573
e. 35,386   **f.** 42,492

4.59
a. The player has between 6 and 10 years of experience; the player weighs between 200 and 300 pounds; the player weighs less than 200 lb and has between 1 and 5 years of experience.
b. 0.253; 0.604; 0.066

c.

	Years of experience				
	Rookie Y_1	1–5 Y_2	6–10 Y_3	10+ Y_4	$P(W_i)$
Under 200 W_1	0.044	0.066	0.055	0.022	0.187
200–300 W_2	0.165	0.220	0.187	0.033	0.604
Over 300 W_3	0.066	0.121	0.011	0.011	0.209
$P(Y_j)$	0.275	0.407	0.253	0.066	1.000

(left axis label: Weight (lb))

4.61
a. (i) S_2; (ii) A_3; (iii) (S_1 & A_1)
b. 0.526; 0.190; 0.101
c.

	Age			
	Under 35 A_1	35–44 A_2	45 or over A_3	Total
Family practice S_1	10.1	11.7	4.9	26.7
Internal medicine S_2	20.4	22.2	10.1	52.6
Obstetrics/ gynecology S_3	7.7	8.4	3.8	19.9
Plastic surgery S_4	0.1	0.5	0.2	0.8
Total	38.3	42.7	19.0	100.0

(left axis label: Specialty)

EXERCISES 4.5

4.65
a. 0.077 **b.** 0.333 **c.** 0.077 **d.** 0
e. 0.231 **f.** 1 **g.** 0.231 **h.** 0.167

4.67
a. 0.275 **b.** 0.209 **c.** 0.316 **d.** 0.240
e. 27.5% of the players are rookies; 20.9% of the players weigh more than 300 pounds; 31.6% of the players who weigh more than 300 pounds are rookies; 24.0% of the rookies weigh more than 300 pounds.

4.69
a. 0.529 **b.** 0.153 **c.** 0.084 **d.** 0.549
e. 0.159
f. 52.9% of the residents live with spouse; 15.3% of the residents are over 64; 8.4% of the residents live with spouse and are over 64; of those residents who are over 64, 54.9% live with spouse; of those residents who live with spouse, 15.9% are over 64.

4.71 Answers will vary.

EXERCISES 4.6

4.75 0.090. Of Americans with Internet access, 9.0% are regular Internet users who feel that the Web has reduced their social contact.

4.77
a. 0.255 **b.** 0.265 **c.** 0.491 **d.** 0.509

4.79
a. 0.5, 0.5, 0.083, 0.5
b. 0.5
c. Yes, because $P(B \mid A) = P(B)$.
d. 0.111
e. No, because $P(C \mid A) \neq P(C)$.
f. 0.5
g. Yes, because $P(D \mid A) = P(D)$.

4.81
a. 0.006 **b.** 0.005

4.83
a. 0.928 **b.** 0.072
c. There was a 7.2% chance that at least one "criticality 1" item would fail; in the long run, at least one "criticality 1" item will fail in 7.2 out of every 100 such missions.

4.85 No. If gender and activity limitation were independent, the percentage of males with an activity limitation would equal the percentage of females with an activity limitation, and both would equal the percentage of people with an activity limitation.

EXERCISES 4.7

4.91
a. $P(R_3)$ **b.** $P(S \mid R_3)$ **c.** $P(R_3 \mid S)$

4.93
a. 43.1% **b.** 33% **c.** 39.8%

4.95
a. 34.0% **b.** 35.3%

4.97 0.803

EXERCISES 4.8

4.103 Because the number of possibilities for each action are multiplied to obtain the total number of possibilities

4.105
a. See the tree diagram shown in Fig. A.3.
b. 12
c. $3 \cdot 4 = 12$

FIGURE A.3

Tree diagram for Exercise 4.105(a)

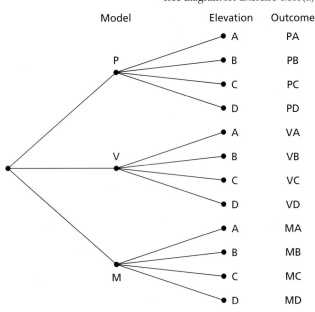

Model	Elevation	Outcome

4.107 1,021,440

4.109 657,720

4.111
a. 35   b. 10   c. 70
d. 1   e. 1

4.113
a. 75,287,520   b. 67,800,320   c. 0.901

4.115
a. 0.125   b. 0.125   c. 0.625

4.117 0.864

4.119
a. 0.99997   b. 0.304

REVIEW TEST FOR CHAPTER 4

1. It enables you to evaluate and control the likelihood that a statistical inference is correct. More generally, probability theory provides the mathematical basis for inferential statistics.

2. a. The experiment has a finite number of possible outcomes, all equally likely.
   b. The probability of an event equals the ratio of the number of ways that the event can occur to the total number of possible outcomes.

3. It is the proportion of times the event occurs in a large number of repetitions of the experiment.

4. (b) and (c), because the probability of an event must always be between 0 and 1, inclusive.

5. Venn diagrams

6. Two or more events are said to be mutually exclusive if at most one of them can occur when the experiment is performed, that is, if no two of them have outcomes in common.

7. a. $P(E)$ b. $P(E) = 0.436$

8. a. False   b. True

9. It is sometimes easier to compute the probability that an event does not occur than the probability that it does occur.

10. a. Univariate   b. Bivariate
    c. Contingency table, or two-way table

11. Marginal

12. a. $P(B \mid A)$ b. A

13. Directly or using the conditional probability rule

14. The joint probability equals the product of the marginal probabilities.

15. Exhaustive

16. See Key Fact 4.2.

17. a.

abc	*abd*	*acd*	*bcd*
acb	*adb*	*adc*	*bdc*
bac	*bad*	*cad*	*cbd*
bca	*bda*	*cda*	*cdb*
cba	*dab*	*dac*	*dbc*
cab	*dba*	*dca*	*dcb*

   b. $\{a, b, c\}, \{a, b, d\}, \{a, c, d\}, \{b, c, d\}$
   c. 24; 4   d. 24; 4

18. a. 0.231   b. 0.363
    c. 0.231, 0.201, 0.147, 0.106, 0.080, 0.177, 0.059

19. a. (not *J*) is the event that the return shows an AGI of at least $100K. There are 7186 thousand such returns.

b. (*H* & *I*) is the event that the return shows an AGI of between $20K and $50K. There are 40,765 thousand such returns.

c. (*H* or *K*) is the event that the return shows an AGI of at least $20K. There are 69,586 thousand such returns.

d. (*H* & *K*) is the event that the return shows an AGI of between $50K and $100K. There are 21,635 thousand such returns.

20. a. Not mutually exclusive
   b. Mutually exclusive
   c. Mutually exclusive
   d. Not mutually exclusive

21. a. 0.510, 0.765, 0.941, 0.235
   b. *H* = (*C* or *D* or *E* or *F*)
      I = (*A* or *B* or *C* or *D* or *E*)
      J = (*A* or *B* or *C* or *D* or *E* or *F*)
      K = (*F* or *G*)
   c. 0.510, 0.765, 0.942, 0.236

22. a. 0.059, 0.333, 0.568, 0.177
   b. 0.941   **c.** 0.568
   d. They are the same.

23. a. 6   **b.** 14,903 thousand
   c. 59,066 thousand   **d.** 3263 thousand

24. a. L_3 is the event that the student selected is in college; T_1 is the event that the student selected attends a public school; (T_1 & L_3) is the event that the student selected attends a public college.
   b. 0.218; 0.864; 0.170. 21.8% of students attend college, 86.4% attend public schools; 17.0% attend public colleges.
   c.

		Type		
		Public T_1	Private T_2	$P(L_i)$
Level	Elementary L_1	0.496	0.068	0.564
	High school L_2	0.198	0.020	0.218
	College L_3	0.170	0.048	0.218
	$P(T_j)$	0.864	0.136	1.000

   d. 0.912   **e.** 0.912

25. a. 0.197; 19.7% of students attending public schools are in college.
   b. 0.197

26. a. 0.136, 0.092
   b. No, because $P(T_2 \mid L_2) \neq P(T_2)$; 9.2% of high school students attend private schools, whereas 13.6% of all students attend private schools.
   c. No, because both events can occur if the student selected is any one of the 1366 thousand students who attend a private high school.
   d. $P(L_1) = 0.564$, $P(L_1 \mid T_1) = 0.574$. Because $P(L_1 \mid T_1) \neq P(L_1)$, the event that a student is in elementary school is not independent of the event that a student attends public school.

27. a. 0.023   **b.** 0.309   **d.** 0.451

28. a. 47.4%   **b.** 20.4%   **c.** 32.3%

29. a. 0.686   **b.** 0.068   **c.** 0.271

30. a. No, because $P(A \& B) \neq 0$ and therefore *A* and *B* have outcomes in common.
   b. Yes, because $P(A \& B) = P(A) \cdot P(B)$.

31. a. 0.278   **b.** 0.192   **c.** 0.165
   d. 27.8% of drivers aged 21–24 at fault in fatal crashes had a BAC of 0.10% or greater; 19.2% of all drivers at fault in fatal crashes had a BAC of 0.10% or greater; of those drivers at fault in fatal crashes with a BAC of 0.10% or greater, 16.5% were in the 21–24 age group.
   e. (b) is prior, (a) and (c) are posterior

32. a. 66   **b.** 1320   **c.** 28; 336

33. a. 635,013,559,600   **b.** 0.213
   c. 0.00045   **d.** 0.032   **e.** 0.013

34. 4,426,165,368

35. a. All households that own a VCR also own a TV.
   b. 83.6%
   c. The percentage of non-TV households that own a VCR.

Chapter 5

EXERCISES 5.1

5.1
a. Probability   **b.** Probability

5.3 {*X* = 3} is the event that the student has three siblings; $P(X = 3)$ is the probability of the event that the student has three siblings.

5.5 The probability distribution of the random variable

5.7
a. 2, 3, 4, 5, 6, 7, 8   **b.** {*X* = 7}
c. 0.021. 2.1% of the shuttle missions between April 1981 and July 2000 had a crew size of 4.

d.

x	2	3	4	5	6	7	8
$P(X = x)$	0.042	0.010	0.021	0.375	0.188	0.344	0.021

e.

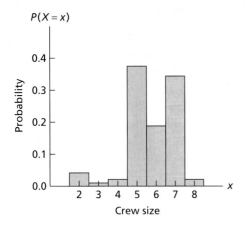

5.9

a. $\{X = 4\}$ **b.** $\{X \geq 2\}$ **c.** $\{X < 5\}$ **d.** $\{2 \leq X < 5\}$
e. 0.212   **f.** 0.922   **g.** 0.523   **h.** 0.445

EXERCISES 5.2

5.15 The mean of a variable of a finite population (population mean)

5.17
a. 5.8 crew members   **b.** 1.3 crew members

5.19
b. -0.052 **c.** $5.2¢$ **d.** $5.20, $52

5.21
a. $\mu_W = 0.25$, $\sigma_W = 0.536$ **b.** 0.25
c. 62.5

EXERCISES 5.3

5.25 Answers will vary.

5.27 6; 5040; 40,320; 362,880

5.29
a. 10   **b.** 1   **c.** 1   **d.** 126

5.31
a. $p = 0.2$

b.

Outcome	Probability
ssss	$(0.2)(0.2)(0.2)(0.2) = 0.0016$
sssf	$(0.2)(0.2)(0.2)(0.8) = 0.0064$
ssfs	$(0.2)(0.2)(0.8)(0.2) = 0.0064$
ssff	$(0.2)(0.2)(0.8)(0.8) = 0.0256$
sfss	$(0.2)(0.8)(0.2)(0.2) = 0.0064$
sfsf	$(0.2)(0.8)(0.2)(0.8) = 0.0256$
sffs	$(0.2)(0.8)(0.8)(0.2) = 0.0256$
sfff	$(0.2)(0.8)(0.8)(0.8) = 0.1024$
fsss	$(0.8)(0.2)(0.2)(0.2) = 0.0064$
fssf	$(0.8)(0.2)(0.2)(0.8) = 0.0256$
fsfs	$(0.8)(0.2)(0.8)(0.2) = 0.0256$
fsff	$(0.8)(0.2)(0.8)(0.8) = 0.1024$
ffss	$(0.8)(0.8)(0.2)(0.2) = 0.0256$
ffsf	$(0.8)(0.8)(0.2)(0.8) = 0.1024$
fffs	$(0.8)(0.8)(0.8)(0.2) = 0.1024$
ffff	$(0.8)(0.8)(0.8)(0.8) = 0.4096$

d. *sssf*, *ssfs*, *sfss*, *fsss*

e. 0.0064. Because each probability is obtained by multiplying three success probabilities of 0.2 and one failure probability of 0.8.

f. 0.0256

g.

y	$P(Y = y)$
0	0.4096
1	0.4096
2	0.1536
3	0.0256
4	0.0016

5.33 The appropriate binomial probability formula is

$$P(Y = y) = \binom{4}{y}(0.2)^y(0.8)^{4-y}.$$

Applying this formula for $y = 0, 1, 2, 3,$ and 4, gives the same result as in part (g) of Exercise 5.31.

5.35
a. 0.161   **b.** 0.332   **c.** 0.468   **d.** 0.821
e.

x	$P(X = x)$
0	0.004
1	0.040
2	0.161
3	0.328
4	0.332
5	0.135

f. Left skewed

g.

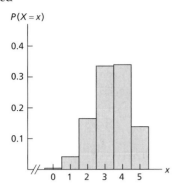

h. $\mu = 3.35$ times; $\sigma = 1.05$ times.
i. $\mu = 3.35$ times; $\sigma = 1.05$ times.
j. On average, the favorite will finish in the money 3.35 times for every 5 races.

5.37
a. 0.201   **b.** 0.666   **c.** 0.046   **d.** 0.954

EXERCISES 5.4

5.47 (1) To model the frequency with which a specified event occurs during a particular period of time; (2) to approximate binomial probabilities

5.49
a. 0.497   **b.** 0.966   **c.** 0.498

5.51
a. $\mu = 0.7$ war. On average, 0.7 war begins during each calendar year.
b. $\sigma = 0.837$ war

5.53
a. 6.667   **b.** 0.352; 0.923

5.55
a. 0.526   **b.** 69,077,553

REVIEW TEST FOR CHAPTER 5

1. **a.** Random variable
   b. Finite; countably infinite

2. The possible values and corresponding probabilities of the discrete random variable

3. Probability histogram

4. 1

5. **a.** $P(X = 2) = 0.386$ **b.** 38.6%
   c. 19.3; 193

6. 3.6

7. X, because it has a smaller standard deviation, therefore less variation.

8. Each trial has the same two possible outcomes; the trials are independent; the probability of a success remains the same from trial to trial.

9. The binomial distribution is the probability distribution for the number of successes in a finite sequence of Bernoulli trials.

10. 120

11. Substitute the binomial (or Poisson) probability formula into the formulas for the mean and standard deviation of a discrete random variable and then simplify mathematically.

12. **a.** Binomial distribution
    b. Hypergeometric distribution
    c. When the sample size does not exceed 5% of the population size because, under this condition, there is little difference between sampling with and without replacement.

13. **a.** 1, 2, 3, 4   **b.** $\{X = 3\}$
    c. 0.2521; 25.2% of undergraduates at ASU are juniors.
    d.

x	1	2	3	4
$P(X = x)$	0.191	0.210	0.252	0.347

   e.

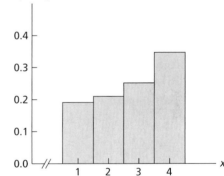

14. **a.** $\{Y = 4\}$ **b.** $\{Y \geq 4\}$ **c.** $\{2 \leq Y \leq 4\}$
    d. $\{Y \geq 1\}$ **e.** 0.174   **f.** 0.322
    g. 0.646   **h.** 0.948

15. **a.** 2.817 lines   **b.** 2.817 lines   **c.** 1.504 lines

16. 1, 6, 24, 5040

17. **a.** 56   **b.** 56   **c.** 1
    d. 45   **e.** 91,390   **f.** 1

18. a. $p = 0.4$

b.

Outcome	Probability
sss	$(0.4)(0.4)(0.4) = 0.064$
ssf	$(0.4)(0.4)(0.6) = 0.096$
sfs	$(0.4)(0.6)(0.4) = 0.096$
sff	$(0.4)(0.6)(0.6) = 0.144$
fss	$(0.6)(0.4)(0.4) = 0.096$
fsf	$(0.6)(0.4)(0.6) = 0.144$
ffs	$(0.6)(0.6)(0.4) = 0.144$
fff	$(0.6)(0.6)(0.6) = 0.216$

d. *ssf*, *sfs*, *fss*

e. 0.096. Each probability is obtained by multiplying two success probabilities of 0.4 and one failure probability of 0.6.

f. 0.288

g.

y	0	1	2	3
$P(Y = y)$	0.216	0.432	0.288	0.064

h. Binomial with parameters $n = 3$ and $p = 0.4$

19. a. 0.3456   **b.** 0.4752   **c.** 0.8704

d.

x	$P(X = x)$
0	0.0256
1	0.1536
2	0.3456
3	0.3456
4	0.1296

e. Left skewed

f.

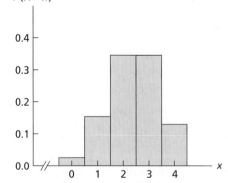

g. The probability distribution is only approximately correct because the sampling is without replacement; a hypergeometric distribution.

h. 2.4 households; on average, 2.4 of every 4 U.S. households live with one or more pets.

i. 0.98 households

20. a. $p > 0.5$ **b.** $p = 0.5$

21. a. 0.266   **b.** 0.099   **c.** 0.826

d.

x	$P(X = x)$
0	0.174
1	0.304
2	0.266
3	0.155
4	0.068
5	0.024
6	0.007
7	0.002
8	0.000

e.

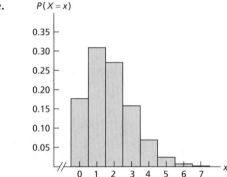

f. Right skewed. Yes, all Poisson distributions are right skewed.

22. a. $\mu = 1.75$ calls; on average, there are 1.75 calls per minute to a wrong number.

b. $\sigma = 1.32$ calls

23. a. 2.4 times   **b.** 0.2613   **c.** 0.6916

Chapter 6

EXERCISES 6.1

6.1 Roughly bell shaped

6.3 They are the same. A normal distribution is completely determined by the mean and standard deviation.

6.5

a. True. They have the same shape because their standard deviations are equal.

b. False. A normal distribution is centered at its mean, which is different for these two distributions.

6.7 The mean μ and standard deviation σ

6.9 They are equal. They are approximately equal.

6.11
a. 55.70%
b. 0.5570; This is only an estimate because the distribution of heights is only approximately normally distributed.

6.13
a.

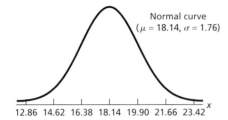

Normal curve
($\mu = 18.14$, $\sigma = 1.76$)

12.86 14.62 16.38 18.14 19.90 21.66 23.42 *x*

b. $z = (x - 18.14)/1.76$
c. Standard normal distribution

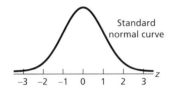

Standard normal curve

−3 −2 −1 0 1 2 3 *z*

d. −1.22; −0.65   **e.** right; 0.49

EXERCISES 6.2

6.19 For a normally distributed variable, you can obtain the percentage of all possible observations that lie within any specified range by first converting to *z*-scores and then determining the corresponding area under the standard normal curve.

6.21 The total area under the standard normal curve equals 1, and the standard normal curve is symmetric about 0. So the area to the right of 0 is one-half of 1, or 0.5.

6.23 0.3336. The total area under the curve is 1, so the area to the right of 0.43 equals 1 minus the area to its left, which is $1 - 0.6664 = 0.3336$.

6.25 99.74%

6.27
a. Read the area directly from the table.
b. Subtract the table area from 1.
c. Subtract the smaller table area from the larger.

6.29
a. 0.9875   **b.** 0.0594   **c.** 0.5
d. 0.0000 (to four decimal places)

6.31
a. 0.9105   **b.** 0.0440   **c.** 0.2121   **d.** 0.1357

6.33
a. 0.7994   **b.** 0.8990   **c.** 0.0500   **d.** 0.0198

6.35
a. 0.0013; 0.0215; 0.1359; 0.3413; 0.3413; 0.1359; 0.0215; 0.0013
b. The entries in the second column of the table are the same as those given in the answer to part (a). The entries in the third column of the table are obtained from those in the second column by multiplying by 100.

6.37 0.67

6.39 0.44

6.41 ±1.645

EXERCISES 6.3

6.45 First express the range in terms of *z*-scores and then determine the corresponding area under the standard normal curve.

6.47
a. 76.47%   **b.** 0.03%

6.49
a. 0.0594   **b.** 0.2699

6.51
a. 68.26%   **b.** 95.44%   **c.** 99.74%

6.53
a. 266.76 yd, 272.20 yd, 277.64 yd
b. 285.56 yd   **c.** 267.98 yd
d. For tee shots on the 1999 men's PGA tour: 25% were less than 266.76 yd, 25% were between 266.76 and 272.20 yd, 25% were between 272.20 and 277.64 yd, 25% were greater than 277.64 yd, 95% were less than 285.56 yd, and 30% were less than 267.98 yd.

EXERCISES 6.4

6.63 When sample sizes are relatively small

6.65
a.

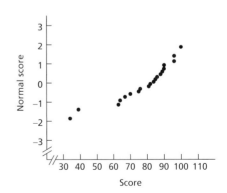

b. 34 and 39 are outliers.
c. Final-exam scores in this introductory statistics class appear not to be normally distributed.

6.67

a.

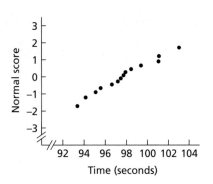

b. No outliers

c. It appears plausible that finishing times for the winners of 1-mile thoroughbred horse races are (approximately) normally distributed.

EXERCISES 6.5

6.77 To ensure that the binomial distribution is sufficiently bell shaped to permit a normal approximation

6.79 The one with parameters $\mu = 15$ and $\sigma = 2.74$

6.81

a. 0.0220　　　　**b.** 0.5780　　　　**c.** 0.5319

6.83

a. 0.0558　　　　**b.** 0.5279　　　　**c.** 0.5279

REVIEW TEST FOR CHAPTER 6

1. It appears again and again in both theory and practice.

2. **a.** A variable is said to be normally distributed if its distribution has the shape of a normal curve.
   b. If a variable of a population is normally distributed and is the only variable under consideration, common practice is to say that the population is a normally distributed population.
   c. The parameters for a normal curve are the corresponding mean and standard deviation of the variable.

3. **a.** False
   b. True. A normal distribution is completely determined by its mean and standard deviation.

4. They are the same when areas are expressed as percentages.

5. Standard normal distribution

6. **a.** True　　　　　　**b.** True

7. **a.** The second curve
   b. The first and second curves
   c. The first and third curves
   d. The third curve　　　**e.** The fourth curve

8. Key Fact 6.2, which states that the standardized version of a normally distributed variable has the standard normal distribution

9. **a.** Read the area directly from the table.
   b. Subtract the table area from 1.
   c. Subtract the smaller table area from the larger.

10. **a.** Locate the table entry closest to the specified area and read the corresponding z-score.
    b. Locate the table entry closest to 1 minus the specified area and read the corresponding z-score.

11. The z-score having area α to its right under the standard normal curve

12. See Key Fact 6.4.

13. The observations expected for a sample of the same size from a variable that has the standard normal distribution

14. Linear

15. **a.**

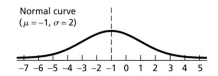

    b.

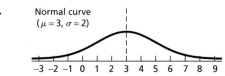

    c.

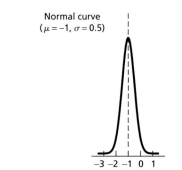

16. **a.**

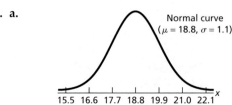

    b. $z = (x - 18.8)/1.1$
    c. Standard normal distribution
    d. 0.8115　　　　　　**e.** Left; -2.55

17. **a.** 0.1469　　**b.** 0.1469　　**c.** 0.7062

18. a. 0.0013   **b.** 0.2709   **c.** 0.1305
    d. 0.9803   **e.** 0.0668   **f.** 0.8426

19. a. −0.52   **b.** 1.28
    c. 1.96; 1.645; 2.33; 2.575   **d.** ±2.575

20. a. 82.76%   **b.** 89.44%   **c.** 99.38%

21. a. 400, 600   **b.** 300, 700   **c.** 200, 800

22. a. 433, 500, 567. Thus, 25% of GRE scores are
    below 433, 25% are between 433 and 500, 25% are
    between 500 and 567, and 25% are above 567.
    b. 733. Thus, 99% of GRE scores are below 733 and
    1% are above 733.

23. a.

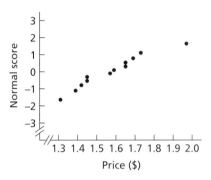

   b. No outliers
   c. It appears plausible that unleaded gas prices were
       (approximately) normally distributed.

24. a. 0.0076   **b.** 0.9505   **c.** 0.9988

Chapter 7

EXERCISES 7.1

7.1 Generally, sampling is less costly and can be done
more quickly than a census.

7.3
a. $\mu = 80$ inches
b.

Sample	Heights (in.)	$\bar{x}$
B, D	79, 84	81.5
B, R	79, 85	82.0
B, G	79, 78	78.5
B, P	79, 74	76.5
D, R	84, 85	84.5
D, G	84, 78	81.0
D, P	84, 74	79.0
R, G	85, 78	81.5
R, P	85, 74	79.5
G, P	78, 74	76.0

c. See Fig. A.4.

FIGURE A.4
Dotplot for Exercise 7.3(c)

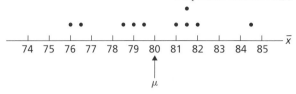

d. 0
e. 0.3. If a random sample of two players is taken, there is
    a 30% chance that the mean height of the two players
    selected will be within 1 inch of the population mean
    height.

7.5
b.

Sample	Heights (in.)	$\bar{x}$
B, D, R	79, 84, 85	82.7
B, D, G	79, 84, 78	80.3
B, D, P	79, 84, 74	79.0
B, R, G	79, 85, 78	80.7
B, R, P	79, 85, 74	79.3
B, G, P	79, 78, 74	77.0
D, R, G	84, 85, 78	82.3
D, R, P	84, 85, 74	81.0
D, G, P	84, 78, 74	78.7
R, G, P	85, 78, 74	79.0

c. See Fig. A.5.

FIGURE A.5
Dotplot for Exercise 7.5(c)

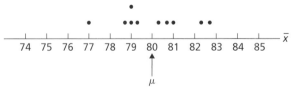

d. 0
e. 0.6. If a random sample of three players is taken, there
    is a 60% chance that the mean height of the three
    players selected will be within 1 inch of the population
    mean height.

7.7
b.

Sample	Heights (in.)	$\bar{x}$
B, D, R, G, P	79, 84, 85, 78, 74	80

c. See Fig. A.6.

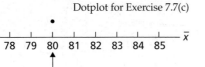

FIGURE A.6

Dotplot for Exercise 7.7(c)

d. 1

e. 1. If a random sample of five players is taken, it is certain (a 100% chance) that the mean height of the five players selected will be within 1 inch of the population mean height.

7.9

a. $\mu = \$25.5$ billion

b.

Sample	Wealth ($billions)	$\bar{x}$
G, B	41, 31	36.0
G, A	41, 26	33.5
G, P	41, 20	30.5
G, T	41, 18	29.5
G, E	41, 17	29.0
B, A	31, 26	28.5
B, P	31, 20	25.5
B, T	31, 18	24.5
B, E	31, 17	24.0
A, P	26, 20	23.0
A, T	26, 18	22.0
A, E	26, 17	21.5
P, T	20, 18	19.0
P, E	20, 17	18.5
T, E	18, 17	17.5

d. $\frac{1}{15}$

e. 0.2. If a random sample of two of the six richest people is taken, there is a 20% chance that the mean wealth of the two people selected will be within 2 (i.e., $2 billion) of the population mean wealth.

7.11

b.

Sample	Wealth ($billions)	$\bar{x}$
G, B, A	41, 31, 26	32.7
G, B, P	41, 31, 20	30.7
G, B, T	41, 31, 18	30.0
G, B, E	41, 31, 17	29.7
G, A, P	41, 26, 20	29.0
G, A, T	41, 26, 18	28.3
G, A, E	41, 26, 17	28.0
G, P, T	41, 20, 18	26.3
G, P, E	41, 20, 17	26.0
G, T, E	41, 18, 17	25.3
B, A, P	31, 26, 20	25.7
B, A, T	31, 26, 18	25.0
B, A, E	31, 26, 17	24.7
B, P, T	31, 20, 18	23.0
B, P, E	31, 20, 17	22.7
B, T, E	31, 18, 17	22.0
A, P, T	26, 20, 18	21.3
A, P, E	26, 20, 17	21.0
A, T, E	26, 18, 17	20.3
P, T, E	20, 18, 17	18.3

d. 0

e. 0.3. If a random sample of three of the six richest people is taken, there is a 30% chance that the mean wealth of the three people selected will be within 2 (i.e., $2 billion) of the population mean wealth.

7.13

b.

Sample	Wealth ($billions)	$\bar{x}$
G, B, A, P, T	41, 31, 26, 20, 18	27.2
G, B, A, P, E	41, 31, 26, 20, 17	27.0
G, B, A, T, E	41, 31, 26, 18, 17	26.6
G, B, P, T, E	41, 31, 20, 18, 17	25.4
G, A, P, T, E	41, 26, 20, 18, 17	24.4
B, A, P, T, E	31, 26, 20, 18, 17	22.4

d. 0

e. $\frac{5}{6}$. If a random sample of five of the six richest people is taken, there is a 83.3% chance that the mean wealth of the five people selected will be within 2 (i.e., $2 billion) of the population mean wealth.

7.15 Sampling error tends to be smaller for large samples than for small samples.

EXERCISES 7.2

7.19 A normal distribution is determined by the mean and standard deviation. Hence a first step in learning how to approximate the sampling distribution of the mean by a normal distribution is to obtain the mean and standard deviation of the variable $\bar{x}$.

7.21 Yes. The standard deviation of all possible sample means (i.e., of the variable $\bar{x}$) gets smaller as the sample size gets larger.

7.23 Standard error (SE) of the mean. Because the standard deviation of $\bar{x}$ determines the amount of sampling error to be expected when a population mean is estimated by a sample mean.

7.25
a. $\mu = 80$ inches   **b.** $\mu_{\bar{x}} = 80$ inches
c. $\mu_{\bar{x}} = \mu = 80$ inches

7.27
b. $\mu_{\bar{x}} = 80$ inches   **c.** $\mu_{\bar{x}} = \mu = 80$ inches

7.29
b. $\mu_{\bar{x}} = 80$ inches   **c.** $\mu_{\bar{x}} = \mu = 80$ inches

7.31
a. The population consists of all babies born in 1991. The variable is birth weight.
b. 3369 g; 41.1 g   **c.** 3369 g; 29.1 g

7.33
a. $\mu_{\bar{x}} = \$51,300$, $\sigma_{\bar{x}} = \$1018.2$. For samples of 50 new mobile homes, the mean and standard deviation of all possible sample mean prices are $51,300 and $1018.2, respectively.
b. $\mu_{\bar{x}} = \$51,300$, $\sigma_{\bar{x}} = \$720.0$. For samples of 100 new mobile homes, the mean and standard deviation of all possible sample mean prices are $51,300 and $720.0, respectively.

EXERCISES 7.3

7.39
a. Approximately normally distributed with a mean of 100 and a standard deviation of 4
b. None
c. No. Because the distribution of the variable under consideration is not specified, a sample size of at least 30 is needed to apply Key Fact 7.4.

7.41
a. Normal with mean μ and standard deviation $\sigma/\sqrt{n}$
b. No. Because the variable under consideration is normally distributed.
c. μ and $\sigma/\sqrt{n}$
d. Essentially, no. For any variable, the mean of $\bar{x}$ equals the population mean, and the standard deviation of $\bar{x}$

equals (at least approximately) the population standard deviation divided by the square root of the sample size.

7.43
a. All four graphs are centered at the same place because $\mu_{\bar{x}} = \mu$ and normal distributions are centered at their means.
b. Because $\sigma_{\bar{x}} = \sigma/\sqrt{n}$, $\sigma_{\bar{x}}$ decreases as n increases. This fact results in a diminishing of the spread because the spread of a distribution is determined by its standard deviation. As a consequence, the larger the sample size, the greater is the likelihood for small sampling error.
c. If the variable under consideration is normally distributed, so is the sampling distribution of the mean, regardless of sample size.
d. The central limit theorem indicates that, if the sample size is relatively large, the sampling distribution of the mean is approximately a normal distribution, regardless of the distribution of the variable under consideration.

7.45
a. A normal distribution with a mean of 1.40 and a standard deviation of 0.064. Thus, for samples of three Swedish men, the possible sample mean brain weights have a normal distribution with a mean of 1.40 kg and a standard deviation of 0.064 kg.
b. A normal distribution with a mean of 1.40 and a standard deviation of 0.032. Thus, for samples of 12 Swedish men, the possible sample mean brain weights have a normal distribution with a mean of 1.40 kg and a standard deviation of 0.032 kg.
c.

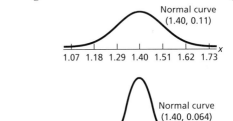

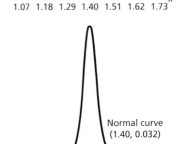

7.47

a. Approximately a normal distribution with mean 6.74 and standard deviation 1.09. Thus, for samples of 200 loans, the possible sample mean loan amounts are approximately normally distributed with a mean of $6.74 million and a standard deviation of $1.09 million.

b. Approximately a normal distribution with mean 6.74 and standard deviation 0.63. Thus, for samples of 600 loans, the possible sample mean loan amounts are approximately normally distributed with a mean of $6.74 million and a standard deviation of $0.63 million.

c. Because, in each case, the sample size is well in excess of 30.

7.49

a. 88.12%. Chances are 88.12% that the sampling error made in estimating the mean brain weight of all Swedish men by that of a sample of three Swedish men will be at most 0.1 kg.

b. 99.82%. Chances are 99.82% that the sampling error made in estimating the mean brain weight of all Swedish men by that of a sample of 12 Swedish men will be at most 0.1 kg.

7.51

a. 0.6424   **b.** 0.8882

7.53 0.9522

REVIEW TEST FOR CHAPTER 7

1. Sampling error is the error resulting from using a sample to estimate a population characteristic.

2. The distribution of a statistic (i.e., of all possible observations of the statistic for samples of a given size) is called the sampling distribution of the statistic.

3. Sampling distribution of the sample mean; distribution of the variable $\bar{x}$

4. The possible sample means cluster closer around the population mean as the sample size increases. Thus, the larger the sample size, the smaller the sampling error tends to be in estimating a population mean, μ, by a sample mean, $\bar{x}$.

5. **a.** The error resulting from using the mean income tax, $\bar{x}$, of the 177,000 tax returns selected as an estimate of the mean income tax, μ, of all 2001 tax returns.
   b. $88
   c. No, not necessarily. However, increasing the sample size from 177,000 to 250,000 would increase the likelihood for small sampling error.
   d. Increase the sample size.

6. **a.** $\mu = \$18$ thousand
   b. The completed table is as follows.

Sample	Salaries	$\bar{x}$
A, B, C, D	8, 12, 16, 20	14
A, B, C, E	8, 12, 16, 24	15
A, B, C, F	8, 12, 16, 28	16
A, B, D, E	8, 12, 20, 24	16
A, B, D, F	8, 12, 20, 28	17
A, B, E, F	8, 12, 24, 28	18
A, C, D, E	8, 16, 20, 24	17
A, C, D, F	8, 16, 20, 28	18
A, C, E, F	8, 16, 24, 28	19
A, D, E, F	8, 20, 24, 28	20
B, C, D, E	12, 16, 20, 24	18
B, C, D, F	12, 16, 20, 28	19
B, C, E, F	12, 16, 24, 28	20
B, D, E, F	12, 20, 24, 28	21
C, D, E, F	16, 20, 24, 28	22

   c.

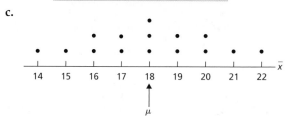

   d. $\frac{7}{15}$
   e. $18 thousand. For samples of four officers from the six, the mean of all possible sample mean monthly salaries equals $18 thousand.
   f. Yes. Because $\mu_{\bar{x}} = \mu$ and, from part (a), $\mu = \$18$ thousand.

7. **a.** The population consists of all new cars sold in the United States in 2001. The variable is the amount spent on a new car.
   b. $21,605; $1442.5
   c. $21,605; $1020.0
   d. Smaller, because $\sigma_{\bar{x}} = \sigma/\sqrt{n}$ and hence $\sigma_{\bar{x}}$ decreases with increasing sample size.

8. **a.** False   **b.** Not possible to tell
   c. True

9. **a.** False   **b.** True   **c.** True

10. **a.** See the first graph that follows.
    b. Normal distribution with a mean of 40 mm and a standard deviation of 6.0 mm, as shown in the second graph that follows.

c. Normal distribution with a mean of 40 mm and a standard deviation of 4.0 mm, as shown in the third graph that follows.

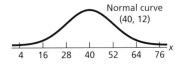

Normal curve
(40, 12)

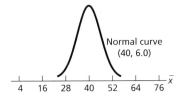

Normal curve
(40, 6.0)

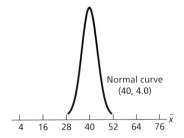

Normal curve
(40, 4.0)

11. **a.** 86.64%   **b.** 0.8664
   c. The probability that the sampling error will be at most 9 mm in estimating the population mean length of all krill by the mean length of a random sample of four krill is 0.8664.
   d. 97.56%. 0.9756. The probability that the sampling error will be at most 9 mm in estimating the population mean length of all krill by the mean length of a random sample of nine krill is 0.9756.

12. **a.** For a normally distributed variable, the sampling distribution of the mean is a normal distribution, regardless of the sample size. Also, we know that $\mu_{\bar{x}} = \mu$. Consequently, because the normal curve for a normally distributed variable is centered at the mean, all three curves are centered at the same place.
   b. Curve B. Because $\sigma_{\bar{x}} = \sigma/\sqrt{n}$, the larger the sample size, the smaller is the value of $\sigma_{\bar{x}}$ and hence the smaller is the spread of the normal curve for $\bar{x}$. Thus, Curve B, which has the smaller spread, corresponds to the larger sample size.
   c. Because $\sigma_{\bar{x}} = \sigma/\sqrt{n}$ and the spread of a normal curve is determined by the standard deviation, different sample sizes result in normal curves with different spreads.
   d. Curve B. The smaller the value of $\sigma_{\bar{x}}$, the smaller the sampling error tends to be.

e. Because the variable under consideration is normally distributed and, hence, so is the sampling distribution of the mean, regardless of sample size.

13. **a.** Approximately normally distributed with mean 4.60 and standard deviation 0.021.
   b. Approximately normally distributed with mean 4.60 and standard deviation 0.015.
   c. No, because, in each case, the sample size exceeds 30.

14. **a.** 0.6212
   b. No. Because the sample size is large and therefore $\bar{x}$ is approximately normally distributed, regardless of the distribution of life insurance amounts. Yes.
   c. 0.9946

15. **a.** No. If the manufacturer's claim is correct, the probability that the paint life for a randomly selected house painted with this paint will be 4.5 years or less is 0.1587; that is, such an event would occur roughly 16% of the time.
   b. Yes. If the manufacturer's claim is correct, the probability that the mean paint life for 10 randomly selected houses painted with this paint will be 4.5 years or less is 0.0008; that is, such an event would occur less than 0.1% of the time.
   c. No. If the manufacturer's claim is correct, the probability that the mean paint life for 10 randomly selected houses painted with this paint will be 4.9 years or less is 0.2643; that is, such an event would occur roughly 26% of the time.

Chapter 8

EXERCISES 8.1

8.1 Point estimate

8.3
a. $18,943.9
b. No. It is unlikely that a sample mean, $\bar{x}$, will exactly equal the population mean, μ; some sampling error is to be anticipated.

8.5
a. $15,321.5 to $22,566.3
b. We can be 95.44% confident that the mean cost, μ, of all recent U.S. weddings is somewhere between $15,321.5 and $22,566.3.
c. It may or may not, but we can be 95.44% confident that it does.

8.7
a. 19.00 gallons. Based on the sample data, the mean fuel tank capacity of all 2003 automobile models is estimated to be 19.00 gallons.
b. 17.82 to 20.18. We can be 95.44% confident that the mean fuel tank capacity of all 2003 automobile models is somewhere between 17.82 gallons and 20.18 gallons.
c. Obtain a normal probability plot of the data.
d. No. Because the sample size is large.

EXERCISES 8.2

8.11
a. Confidence level = 0.90; $\alpha = 0.10$.
b. Confidence level = 0.99; $\alpha = 0.01$.

8.13
a. To ensure that the variable $\bar{x}$ is normally distributed.
b. Because, for large samples, the variable $\bar{x}$ is approximately normally distributed, regardless of the distribution of the variable under consideration.

8.15
a. Assumption 1 is that the sample taken from the population is a simple random sample.

Assumption 2 is that the variable under consideration is normally distributed or that the sample size is large.

Assumption 3 is that the standard deviation of the variable under consideration is known.
b. Actually, the z-interval procedure works reasonably well even when the variable is not normally distributed and the sample size is small or moderate, provided the variable is not too far from being normally distributed.

8.17 Key Fact 8.1 yields the following answers.
a. Reasonable   b. Reasonable
c. Not reasonable   d. Reasonable
e. Not reasonable   f. Reasonable

8.19 a 95% confidence level

8.21
a. 44.16 kg to 46.44 kg
b. We can be 95% confident that the mean weight of all male Ethiopian-born school children, ages 12–15 years old, is somewhere between 44.16 kg and 46.44 kg.

8.23 $2.03 million to $2.51 million. We can be 99% confident that the mean gross earnings of all Rolling Stones concerts is somewhere between $2.03 million and $2.51 million.

8.25
a. 44.55 kg to 46.04 kg
b. It is shorter because the confidence level is smaller.

c.
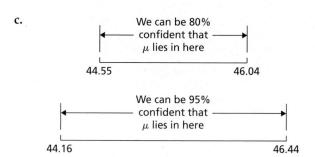

d. The 80% CI is a more precise estimate of μ because it is narrower than the 95% CI.

EXERCISES 8.3

8.33 The margin of error equals the standard error of the mean multiplied by $z_{\alpha/2}$.

8.35
a. Increases the margin of error and hence decreases the precision
b. Decreases the margin of error and hence increases the precision

8.37
a. 10   b. 50 to 70

8.39
a. The sample size (number of observations) cannot be fractional; it must be a whole number.
b. The number resulting from Formula 8.1 is the smallest value that will provide the required margin of error. If that value were rounded down, the sample size thus obtained would be insufficient to ensure the required margin of error.

8.41 $0.49

8.43
a. 1.14 kg   b. 1.14 kg

8.45
a. $0.24 million
b. We can be 99% confident that the error made in estimating μ by $\bar{x}$ is at most $0.24 million.
c. 97 concerts
d. $2.25 million to $2.45 million

EXERCISES 8.4

8.51 When the population standard deviation is unknown, the confidence-interval procedure for a population mean cannot be based on the standardized version of $\bar{x}$. The best that can be done is to replace the population standard deviation, σ, by the sample standard deviation, s, in the formula for the standardized version of $\bar{x}$. The result is the studentized version of $\bar{x}$.

8.53
a. The one-sample *z*-interval procedure
b. The one-sample *t*-interval procedure

8.55 The variation in the possible values of the standardized version is due solely to the variation of sample means, whereas that of the studentized version is due to the variation of both sample means and sample standard deviations.

8.57
a. 1.440   b. 2.447   c. 3.143

8.59
a. 1.323   b. 2.518   c. −2.080   d. ±1.721

8.61 Yes. Because the sample size exceeds 30 and there are no outliers.

8.63
a. $141.95 to $160.13
b. We can be 95% confident that the mean cost for a family of four to spend the day at an American amusement park is somewhere between $141.95 and $160.13.

8.65
a. 14.46 to 15.64 cm
b. We can be 90% confident that the mean depth of all subterranean coruro burrows is somewhere between 14.46 and 15.64 cm.

REVIEW TEST FOR CHAPTER 8

1. A point estimate of a parameter is the value of a statistic that is used to estimate the parameter; it consists of a single number, or point. A confidence-interval estimate of a parameter consists of an interval of numbers obtained from a point estimate of the parameter and a percentage that specifies how confident we are that the parameter lies in the interval.

2. False. The mean of the population may or may not lie somewhere between 33.8 and 39.0, but we can be 95% confident that it does.

3. No. See the guidelines in Key Fact 8.1 on page 364.

4. Roughly 950 intervals would actually contain μ.

5. Look at graphical displays of the data to ascertain whether the conditions required for using the procedure appear to be satisfied.

6. a. The precision of the estimate would decrease because the CI would be wider for a sample of size 50.
   b. The precision of the estimate would increase because the CI would be narrower for a 90% confidence level.

7. a. Because the length of a CI is twice the margin of error, the length of the CI is 21.4.
   b. 64.5 to 85.9

8. a. 6.58
   b. The sample mean, $\bar{x}$

9. a. $z = -0.77$ b. $t = -0.605$

10. a. Standard normal distribution
    b. *t*-distribution with 14 degrees of freedom

11. From Property 4 of Key Fact 8.6 (page 379), as the number of degrees of freedom becomes larger, *t*-curves look increasingly like the standard normal curve. So the curve that is closer to the standard normal curve has the larger degrees of freedom.

12. a. *t*-interval procedure   b. *z*-interval procedure
    c. *z*-interval procedure   d. Neither procedure
    e. *z*-interval procedure   f. Neither procedure

13. 54.3 to 62.8 yr

14. Part (c) provides the correct interpretation of the statement in quotes.

15. a. 11.7 to 12.1 mm
    b. We can be 90% confident that the mean length, μ, of all *N. trivittata* is somewhere between 11.7 and 12.1 mm.
    c. A normal probability plot of the data should fall roughly in a straight line.

16. a. 0.2 mm
    b. We can be 90% confident that the error made in estimating μ by $\bar{x}$ is at most 0.2 mm.
    c. $n = 1692$ d. 11.9 to 12.1 mm

17. a. 2.101   b. 1.734   c. −1.330   d. ±2.878

18. a. 81.7 to 90.7 mm Hg
    b. We can be 95% confident that the mean arterial blood pressure for all children of diabetic mothers is somewhere between 81.7 and 90.7 mm Hg.

Chapter 9

EXERCISES 9.1

9.1 A hypothesis is a statement that something is true.

9.3
a. The population mean, μ, equals some specified number, μ_0; H_0: $\mu = \mu_0$.
b. Two tailed: The population mean, μ, differs from μ_0; H_a: $\mu \neq \mu_0$.
   Left tailed: The population mean, μ, is less than μ_0; H_a: $\mu < \mu_0$.
   Right tailed: The population mean, μ, is greater than μ_0; H_a: $\mu > \mu_0$.

9.5 Let μ denote the mean cadmium level in *Boletus pinicola* mushrooms.
a. $H_0: \mu = 0.5$ ppm   **b.** $H_a: \mu > 0.5$ ppm
c. Right-tailed test

9.7 Let μ denote last year's mean local monthly bill for cell-phone users.
a. $H_0: \mu = \$47.37$ **b.** $H_a: \mu < \$47.37$
c. Left-tailed test

9.9 Let μ denote the mean body temperature of all healthy humans.
a. $H_0: \mu = 98.6°F$ **b.** $H_a: \mu \neq 98.6°F$
c. Two-tailed test

EXERCISES 9.2

9.15 The two types of incorrect decisions are a Type I error (rejection of a true null hypothesis) and a Type II error (nonrejection of a false null hypothesis). The probabilities of these two errors are denoted α and β, respectively.

9.17
a. $z \leq -1.96$ or $z \geq 1.96$ **b.** $-1.96 < z < 1.96$
c. $z = \pm 1.96$ **d.** $\alpha = 0.05$
e.

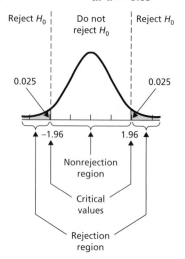

f. Two-tailed test

9.19
a. A Type I error would occur if in fact $\mu = 0.5$ ppm, but the results of the sampling lead to the conclusion that $\mu > 0.5$ ppm.
b. A Type II error would occur if in fact $\mu > 0.5$ ppm, but the results of the sampling fail to lead to that conclusion.
c. A correct decision would occur if in fact $\mu = 0.5$ ppm and the results of the sampling do not lead to the rejection of that fact; or if in fact $\mu > 0.5$ ppm and the results of the sampling lead to that conclusion.
d. Correct decision   **e.** Type II error

9.21
a. A Type I error would occur if in fact $\mu = \$47.37$, but the results of the sampling lead to the conclusion that $\mu < \$47.37$.
b. A Type II error would occur if in fact $\mu < \$47.37$, but the results of the sampling fail to lead to that conclusion.
c. A correct decision would occur if in fact $\mu = \$47.37$ and the results of the sampling do not lead to the rejection of that fact; or if in fact $\mu < \$47.37$ and the results of the sampling lead to that conclusion.
d. Correct decision   **e.** Type II error

9.23
a. A Type I error would occur if in fact $\mu = 98.6°F$, but the results of the sampling lead to the conclusion that $\mu \neq 98.6°F$.
b. A Type II error would occur if in fact $\mu \neq 98.6°F$, but the results of the sampling fail to lead to that conclusion.
c. A correct decision would occur if in fact $\mu = 98.6°F$ and the results of the sampling do not lead to the rejection of that fact; or if in fact $\mu \neq 98.6°F$ and the results of the sampling lead to that conclusion.
d. Type I error   **e.** Correct decision

EXERCISES 9.3

9.31 Critical value: $-z_{0.05} = -1.645$

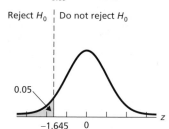

9.33 Critical values: $\pm z_{0.025} = \pm 1.96$

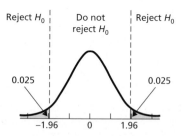

9.35 $H_0: \mu = 0.5$ ppm, $H_a: \mu > 0.5$ ppm; $\alpha = 0.05$; $z = 0.24$; critical value $= 1.645$; do not reject H_0; at the 5% significance level, the data do not provide sufficient evidence to conclude that the mean cadmium level in *Boletus pinicola* mushrooms is greater than the government's recommended limit of 0.5 ppm.

9.37 H_0: $\mu = \$47.37$, H_a: $\mu < \$47.37$; $\alpha = 0.01$; $z = -1.78$; critical value $= -2.33$; do not reject H_0; at the 1% significance level, the data do not provide sufficient evidence to conclude that last year's mean local monthly bill for cell-phone users has decreased from the 2001 mean of $47.37.

9.39 H_0: $\mu = 98.6°F$, H_a: $\mu \neq 98.6°F$; $\alpha = 0.01$; $z = -7.29$; critical values $= \pm 2.575$; reject H_0; at the 1% significance level, the data provide sufficient evidence to conclude that the mean body temperature of healthy human beings differs from 98.6°F.

EXERCISES 9.4

9.45
a. Rejecting a true null hypothesis
b. Not rejecting a false null hypothesis
c. The probability of rejecting a true null hypothesis, that is, of making a Type I error

9.47 The power curve provides a visual display of the overall effectiveness of the hypothesis test.

9.49 Decreasing the significance level of a hypothesis test without changing the sample size increases the probability of a Type II error or, equivalently, decreases the power.

Note: The answers obtained to many of the parts in the remaining exercises of Section 9.4 may vary depending on when and how much intermediate rounding is done. We used statistical software to get the answers to most parts of each of these exercises.

9.51
a. If $\bar{x} \geq 0.6757$ ppm, reject H_0; otherwise, do not reject H_0.
b. 0.05
c.

μ	β	Power	μ	β	Power
0.55	0.8803	0.1197	0.75	0.2433	0.7567
0.60	0.7607	0.2393	0.80	0.1222	0.8778
0.65	0.5950	0.4050	0.85	0.0513	0.9487
0.70	0.4100	0.5900			

d.

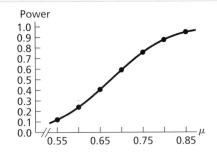

9.53
a. If $\bar{x} \leq \$39.1451$, reject H_0; otherwise, do not reject H_0.
b. 0.01
c.

μ	β	Power	μ	β	Power
29	0.0021	0.9979	41	0.7001	0.2999
32	0.0216	0.9784	44	0.9151	0.0849
35	0.1205	0.8795	47	0.9868	0.0132
38	0.3730	0.6270			

d.

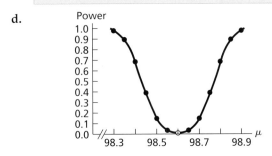

9.55
a. If $\bar{x} \leq 98.4317°F$ or $\bar{x} \geq 98.7683°F$, reject H_0; otherwise, do not reject H_0.
b. 0.01
c.

μ	β	Power	μ	β	Power
98.30	0.0219	0.9781	98.65	0.9645	0.0355
98.35	0.1055	0.8945	98.70	0.8520	0.1480
98.40	0.3136	0.6864	98.75	0.6102	0.3898
98.45	0.6102	0.3898	98.80	0.3136	0.6864
98.50	0.8520	0.1480	98.85	0.1055	0.8945
98.55	0.9645	0.0355	98.90	0.0219	0.9781

d.

9.57
a. If $\bar{x} \geq 0.6361$ ppm, reject H_0; otherwise, do not reject H_0.
b. 0.05

c.

μ	β	Power	μ	β	Power
0.55	0.8509	0.1491	0.75	0.0843	0.9157
0.60	0.6686	0.3314	0.80	0.0238	0.9762
0.65	0.4332	0.5668	0.85	0.0049	0.9951
0.70	0.2199	0.7801			

d.

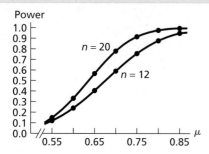

For a fixed significance level, increasing the sample size increases the power.

EXERCISES 9.5

9.63 (1) It allows you to assess significance at any desired level. (2) It permits you to evaluate the strength of the evidence against the null hypothesis.

9.65 The two different approaches to hypothesis testing are the critical-value approach and the *P*-value approach. For a comparison of the two approaches, see Table 9.11 on page 440.

9.67 True

9.69 A *P*-value of 0.02 provides stronger evidence against the null hypothesis because it reflects an observed value of the test statistic that is more inconsistent with the null hypothesis.

9.71
a. 0.0212   **b.** 0.6217

9.73
a. 0.0020   **b.** 0.0156

9.75 H_0: $\mu = 0.5$ ppm, H_a: $\mu > 0.5$ ppm; $\alpha = 0.05$; $z = 0.24$; $P = 0.4052$; do not reject H_0; at the 5% significance level, the data do not provide sufficient evidence to conclude that the mean cadmium level in *Boletus pinicola* mushrooms is greater than the government's recommended limit of 0.5 ppm. The evidence against the null hypothesis is weak or none.

9.77 H_0: $\mu = \$47.37$, H_a: $\mu < \$47.37$; $\alpha = 0.01$; $z = -1.78$; $P = 0.0375$; do not reject H_0; at the 1% significance level, the data do not provide sufficient evidence to conclude that last year's mean local monthly bill for cell-phone users has decreased from the 2001 mean of $47.37. The evidence against the null hypothesis is strong.

9.79 H_0: $\mu = 98.6°F$, H_a: $\mu \neq 98.6°F$; $\alpha = 0.01$; $z = -7.29$; $P = 0.0000$ (to four decimal places); reject H_0; at the 1% significance level, the data provide sufficient evidence to conclude that the mean body temperature of healthy humans differs from 98.6°F. The evidence against the null hypothesis is very strong.

EXERCISES 9.6

9.89 H_0: $\mu = \$1856$, H_a: $\mu \neq \$1856$; $\alpha = 0.05$; $t = -1.056$; critical values $= \pm 2.030$; $P > 0.20$; do not reject H_0; at the 5% significance level, the data do not provide sufficient evidence to conclude that the 2000 mean annual expenditure on apparel and services for consumer units in the Midwest differed from the national mean of $1856.

9.91 H_0: $\mu = 2.30\%$, H_a: $\mu > 2.30\%$; $\alpha = 0.01$; $t = 4.251$; critical value $= 2.821$; $P < 0.005$; reject H_0; at the 1% significance level, the data provide sufficient evidence to conclude that the mean available limestone in soil treated with 100% MMBL effluent exceeds 2.30%.

9.93 H_0: $\mu = 0.9$, H_a: $\mu < 0.9$; $\alpha = 0.05$; $t = -23.703$; critical value $= -1.653$; $P < 0.005$; reject H_0; at the 5% significance level, the data provide sufficient evidence to conclude that women with peripheral arterial disease have an unhealthy ABI.

EXERCISES 9.7

9.101 On the one hand, nonparametric methods do not require normality; they also usually entail fewer and simpler computations than parametric methods and are resistant to outliers and other extreme values. On the other hand, parametric methods tend to give more accurate results when the requirements for their use are met.

9.103 Because the *D*-value for such an observation equals 0, a sign cannot be attached to the rank of |*D*|.

9.105
a. Wilcoxon signed-rank test
b. Wilcoxon signed-rank test
c. Neither

9.107 H_0: $\eta = 35.7$ yr, H_a: $\eta > 35.7$ yr; $\alpha = 0.01$; $W = 33$; critical value $= 50$; $P = 0.305$; do not reject H_0; at the 1% significance level, the data do not provide sufficient evidence to conclude that the median age of today's U.S. residents has increased from the 2000 median age of 35.7 yr.

9.109 H_0: $\mu = \$6735$, H_a: $\mu < \$6735$; $\alpha = 0.10$; $W = 13$; critical value $= 14$; $P = 0.077$; reject H_0; at the 10% significance level, the data provide sufficient evidence to conclude that the mean asking price for 1996 XE King Cab pickups in Phoenix is less than the 2003 *Kelley Blue Book* value.

9.111

a. H_0: $\mu = 2.30\%$, H_a: $\mu > 2.30\%$; $\alpha = 0.01$; $W = 53$; critical value $= 50$; $P = 0.005$; reject H_0; at the 1% significance level, the data provide sufficient evidence to conclude that the mean available limestone in soil treated with 100% MMBL effluent exceeds 2.30%.

b. Because a normal distribution is symmetric

9.113

a. H_0: $\mu = 310$ mL, H_a: $\mu < 310$ mL; $\alpha = 0.05$; $t = -1.845$; critical value $= -1.753$; $0.025 < P < 0.05$; reject H_0; at the 5% significance level, the data provide sufficient evidence to conclude that the mean content is less than advertised.

b. H_0: $\mu = 310$ mL, H_a: $\mu < 310$ mL; $\alpha = 0.05$; $W = 36.5$; critical value $= 36$; $P = 0.054$; do not reject H_0; at the 5% significance level, the data do not provide sufficient evidence to conclude that the mean content is less than advertised.

c. Assuming that the contents are normally distributed, the *t*-test is more powerful than the Wilcoxon signed-rank test; that is, the *t*-test is more likely to detect a false null hypothesis.

EXERCISES 9.8

9.125 See Table 9.16 on page 468.

9.127

a. Yes. Because the sample size is large and the population standard deviation is unknown.

b. Yes. Because the variable under consideration has a symmetric distribution.

c. Because the variable under consideration has a nonnormal symmetric distribution, the preferred procedure is the Wilcoxon signed-rank test.

9.129 *z*-test

9.131 *t*-test

9.133 Wilcoxon signed-rank test

9.135 Consult a statistician.

REVIEW TEST FOR CHAPTER 9

1. **a.** The null hypothesis is a hypothesis to be tested.
   b. The alternative hypothesis is a hypothesis to be considered as an alternate to the null hypothesis.
   c. The test statistic is the statistic used as a basis for deciding whether the null hypothesis should be rejected.
   d. The rejection region is the set of values for the test statistic that leads to rejection of the null hypothesis.
   e. The nonrejection region is the set of values for the test statistic that leads to nonrejection of the null hypothesis.
   f. The critical values are the values of the test statistic that separate the rejection and nonrejection regions. The critical values are considered part of the rejection region.

2. **a.** The weight of a package of Tide is a variable. A particular package may weigh slightly more or less than the marked weight. The mean weight of all packages produced on any specified day (the population mean weight for that day) exceeds the marked weight.
   b. The null hypothesis would be that the population mean weight for a specified day equals the marked weight; the alternative hypothesis would be that the population mean weight for the specified day exceeds the marked weight.
   c. The null hypothesis would be that the population mean weight for a specified day equals the marked weight of 76 oz; the alternative hypothesis would be that the population mean weight for the specified day exceeds the marked weight of 76 oz. In statistical terminology, the hypothesis test would be H_0: $\mu = 76$ oz and H_a: $\mu > 76$ oz, where μ is the mean weight of all packages produced on the specified day.

3. **a.** Obtain the data from a random sample of the population or from a designed experiment. If the data are consistent with the null hypothesis, do not reject the null hypothesis; if the data are inconsistent with the null hypothesis, reject the null hypothesis and conclude that the alternative hypothesis is true.
   b. We establish a precise criterion for deciding whether to reject the null hypothesis prior to obtaining the data.

4. Two-tailed test, H_a: $\mu \neq \mu_0$. Used when the primary concern is deciding whether a population mean, μ, is different from a specified value μ_0.

   Left-tailed test, H_a: $\mu < \mu_0$. Used when the primary concern is deciding whether a population mean, μ, is less than a specified value μ_0.

   Right-tailed test, H_a: $\mu > \mu_0$. Used when the primary concern is deciding whether a population mean, μ, is greater than a specified value μ_0.

5. **a.** A Type I error is the incorrect decision of rejecting a true null hypothesis. A Type II error is the incorrect decision of not rejecting a false null hypothesis.
   b. α and β, respectively
   c. A Type I error   **d.** A Type II error

6. It must be chosen so that, if the null hypothesis is true, the probability equals 0.05 that the test statistic will fall in the rejection region, in this case, to the right of the critical value.

7. **a.** Assumptions: simple random sample; normal population or large sample; σ unknown. Test statistic: $t = (\bar{x} - \mu_0)/(s/\sqrt{n})$.
   b. Assumptions: simple random sample; normal population or large sample; σ known. Test statistic: $z = (\bar{x} - \mu_0)/(\sigma/\sqrt{n})$.
   c. Assumptions: simple random sample; symmetric population. Test statistic: $W = $ sum of the positive ranks.

8. **a.** The true significance level equals α.
   b. The true significance level only approximately equals α.

9. The results of a hypothesis test are statistically significant if the null hypothesis is rejected at the specified significance level. Statistical significance means that the data provide sufficient evidence to conclude that the truth is different from the stated null hypothesis. It does not necessarily mean that the difference is important in any practical sense.

10. It increases.

11. **a.** The probability of rejecting a false null hypothesis
    b. It increases.

12. **a.** The P-value is the probability, calculated under the assumption that the null hypothesis is true, of observing a value of the test statistic as extreme as or more extreme than that observed. By *extreme* we mean "far from what we would expect to observe if the null hypothesis is true."
    b. True   **c.** True
    d. Because it is the smallest significance level for which the observed sample data result in rejection of the null hypothesis.

15. Let μ denote last year's mean cheese consumption by Americans.
    a. H_0: $\mu = 30.0$ lb   **b.** H_a: $\mu > 30.0$ lb
    c. Right tailed

16. **a.** $z \geq 1.28$ **b.** $z < 1.28$
    c. $z = 1.28$ **d.** $\alpha = 0.10$

e.

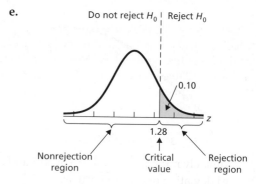

f. Right tailed

17. **a.** A Type I error would occur if in fact $\mu = 30.0$ lb, but the results of the sampling lead to the conclusion that $\mu > 30.0$ lb.
    b. A Type II error would occur if in fact $\mu > 30.0$ lb, but the results of the sampling fail to lead to that conclusion.
    c. A correct decision would occur if in fact $\mu = 30.0$ lb and the results of the sampling do not lead to the rejection of that fact; or if in fact $\mu > 30.0$ lb and the results of the sampling lead to that conclusion.
    d. Correct decision   **e.** Type II error

18. *Note:* The answers obtained to many of the parts of this problem may vary depending on when and how much intermediate rounding is done. We used statistical software to get the answers to most parts of this problem.
    a. 0.10
    b. Approximately normal with a mean of 30.5 and a standard deviation of $6.9/\sqrt{35} \approx 1.17$.
    c. 0.8031
    d. Approximately normal with the specified mean and a standard deviation of $6.9/\sqrt{35} \approx 1.17$. The Type II error probabilities, β, are shown in the table in part (e).
    e.

μ	β	Power	μ	β	Power
30.5	0.8031	0.1969	32.5	0.1944	0.8056
31.0	0.6643	0.3357	33.0	0.0984	0.9016
31.5	0.4982	0.5018	33.5	0.0428	0.9572
32.0	0.3324	0.6676	34.0	0.0159	0.9841

f.

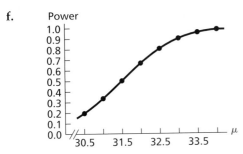

g. Approximately normal with a mean of 30.5 and a standard deviation of $6.9/\sqrt{60} \approx 0.89$.

h. 0.7643

i. Approximately normal with the specified mean and a standard deviation of $6.9/\sqrt{60} \approx 0.89$. The Type II error probabilities, β, are shown in the table in part (j).

j.

μ	β	Power	μ	β	Power
30.5	0.7643	0.2357	32.5	0.0636	0.9364
31.0	0.5631	0.4369	33.0	0.0185	0.9815
31.5	0.3437	0.6563	33.5	0.0041	0.9959
32.0	0.1676	0.8324	34.0	0.0007	0.9993

k.

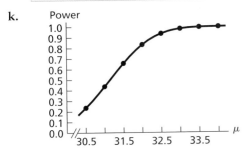

l. For a fixed significance level, increasing the sample size increases the power.

19. a. H_0: $\mu = 30.0$ lb, H_a: $\mu > 30.0$ lb; $\alpha = 0.10$; $z = 0.69$; critical value $= 1.28$; do not reject H_0; at the 10% significance level, the data do not provide sufficient evidence to conclude that last year's mean cheese consumption for all Americans has increased over the 2000 mean of 30.0 lb.

b. A Type II error because, given that the null hypothesis was not rejected, the only error that could be made is the error of not rejecting a false null hypothesis.

20. a. H_0: $\mu = 30.0$ lb, H_a: $\mu > 30.0$ lb; $\alpha = 0.10$; $z = 0.69$; $P = 0.2451$; do not reject H_0; at the 10% significance level, the data do not provide sufficient evidence to conclude that last year's mean cheese consumption for all Americans has increased over the 2000 mean of 30.0 lb.

b. The data provide at most weak evidence against the null hypothesis.

21. H_0: $\mu = \$356$, H_a: $\mu < \$356$; $\alpha = 0.05$; $t = -1.909$; critical value $= -1.796$; $0.025 < P < 0.05$; reject H_0; at the 5% significance level, the data provide sufficient evidence to conclude that last year's mean value lost to purse snatching has decreased from the 2000 mean of $356.

22. a. H_0: $\mu = \$356$, H_a: $\mu < \$356$; $\alpha = 0.05$; $W = 17$; critical value $= 17$; $P = 0.046$; reject H_0; at the 5% significance level, the data provide sufficient evidence to conclude that last year's mean value lost to purse snatching has decreased from the 2000 mean of $356.

b. It is symmetric.

c. Because a normal distribution is symmetric.

23. t-test

24. a. 0 points

b. H_0: $\mu = 0$ points, H_a: $\mu \neq 0$ points; $\alpha = 0.05$; $t = -0.843$; critical values $= \pm 1.96$; $P > 0.20$; do not reject H_0.

c. At the 5% significance level, the data do not provide sufficient evidence to conclude that the population mean point-spread error differs from 0. In fact, because $P > 0.20$, there is virtually no evidence against the null hypothesis that the population mean point-spread error equals 0.

25. It is probably okay to use the z-test because the sample size is large and σ is known. However, it does appear from the normal probability plot that there may be outliers, so one should proceed cautiously in using the z-test.

26. It appears that the variable under consideration is far from being normally distributed and, in fact, has a left-skewed distribution. However, the sample size is large and the plots reveal no outliers. Keeping in mind that σ is unknown, it is probably reasonable to use the t-test.

27. a. In view of the graphs, it appears reasonable to assume that, in Problem 25, the variable under consideration has (approximately) a symmetric distribution, but not so in Problem 26. Consequently, it would be reasonable to use the Wilcoxon signed-rank test in the first case, but not the second.

b. In Problem 25, it is a tough call between the Wilcoxon signed-rank test and the z-test but, considering the possible outliers, the Wilcoxon signed-rank test is probably the better one to use.

Chapter 10

EXERCISES 10.1

10.1
a. Age
b. Population 1 consists of buyers of new domestic cars; Population 2 consists of buyers of new imported cars.
c. Let μ_1 and μ_2 denote the mean ages of buyers of new domestic cars and new imported cars, respectively. The null and alternative hypotheses are H_0: $\mu_1 = \mu_2$ and H_a: $\mu_1 > \mu_2$, respectively.

10.3 Answers will vary.

10.5
a. μ_1, σ_1, μ_2, and σ_2 are parameters; $\bar{x}_1, s_1, \bar{x}_2$, and s_2 are statistics.
b. μ_1, σ_1, μ_2, and σ_2 are fixed numbers; $\bar{x}_1, s_1, \bar{x}_2$, and s_2 are variables.

10.7 So that you can determine whether the observed difference between the two sample means can be reasonably attributed to sampling error or whether that difference suggests that the null hypothesis of equal population means is false and the alternative hypothesis is true.

10.9
a. 0 and 5   b. No.   c. No.

EXERCISES 10.2

10.17 Because it is obtained by combining (pooling) information on variation from the individual samples into one estimate of the common population standard deviation.

10.19 H_0: $\mu_1 = \mu_2$, H_a: $\mu_1 < \mu_2$; $\alpha = 0.05$; $t = -4.058$; critical value $= -1.734$; $P < 0.005$; reject H_0; at the 5% significance level, the data provide sufficient evidence to conclude that the mean time served for fraud is less than that for firearms offenses.

10.21 H_0: $\mu_1 = \mu_2$, H_a: $\mu_1 > \mu_2$; $\alpha = 0.05$; $t = 0.520$; critical value $= 1.711$; $P > 0.10$; do not reject H_0; at the 5% significance level, the data do not provide sufficient evidence to conclude that drinking fortified orange juice reduces PTH level more than drinking unfortified orange juice.

10.23 H_0: $\mu_1 = \mu_2$, H_a: $\mu_1 \neq \mu_2$; $\alpha = 0.01$; $t = -2.935$; critical values $= \pm 2.625$; $P < 0.01$; reject H_0; at the 1% significance level, the data provide sufficient evidence to conclude that the mean daily protein intakes of female vegetarians and female omnivores differ.

10.25
a. −16.92 to 31.72 pg/mL
b. We can be 90% confident that the difference between the mean reductions in PTH levels for fortified and unfortified orange juice is somewhere between −16.92 and 31.72 pg/mL.

10.27
a. −20.61 to −1.15 grams
b. We can be 99% confident that the difference between the mean daily protein intakes of female vegetarians and female omnivores is somewhere between −20.61 and −1.15 g.

EXERCISES 10.3

10.35 The sample standard deviations are 84.7 minutes and 38.2 minutes. These two sample standard deviations provide a fairly clear indication that the two population standard deviations are not equal and thus that the nonpooled *t*-test is more appropriate than the pooled *t*-test.

10.37 The pooled *t*-procedures require equal population standard deviations, whereas the nonpooled *t*-procedures do not.

10.39 H_0: $\mu_1 = \mu_2$, H_a: $\mu_1 \neq \mu_2$; $\alpha = 0.10$; $t = 1.791$; critical values $= \pm 1.677$; $0.05 < P < 0.10$; reject H_0; at the 10% significance level, the data provide sufficient evidence to conclude that a difference exists in the mean age at arrest of East German prisoners with chronic PTSD and remitted PTSD.

10.41 H_0: $\mu_1 = \mu_2$, H_a: $\mu_1 < \mu_2$; $\alpha = 0.05$; $t = -1.651$; critical value $= -2.015$; $0.05 < P < 0.10$; do not reject H_0; at the 5% significance level, the data do not provide sufficient evidence to conclude that the mean number of acute postoperative days in the hospital is smaller with the dynamic system than with the static system.

10.43 H_0: $\mu_1 = \mu_2$, H_a: $\mu_1 > \mu_2$; $\alpha = 0.01$; $t = 3.863$; critical value $= 2.552$; $P < 0.005$; reject H_0; at the 1% significance level, the data provide sufficient evidence to conclude that dopamine activity is higher, on average, in psychotic patients.

10.45
a. 0.2 to 7.2 years
b. We can be 90% confident that the difference between the mean ages at arrest of East German prisoners with chronic PTSD and remitted PTSD is somewhere between 0.2 and 7.2 yr.

10.47
a. −6.97 to 0.69 days
b. We can be 90% confident that the difference between the mean number of acute postoperative days in the hospital with the dynamic and static systems is somewhere between −6.97 and 0.69 days.

EXERCISES 10.4

10.55 Because the shape of a normal distribution is determined by its standard deviation

10.57
a. Pooled t-test   b. Mann–Whitney test

10.59 $H_0: \mu_1 = \mu_2$, $H_a: \mu_1 < \mu_2$; $\alpha = 0.05$; $M = 33$; critical value = 33; $P = 0.0437$; reject H_0; at the 5% significance level, the data provide sufficient evidence to conclude that, in this teacher's chemistry courses, students with fewer than 2 years of high school algebra have a lower mean semester average than those with two or more years.

10.61 $H_0: \eta_1 = \eta_2$, $H_a: \eta_1 < \eta_2$; $\alpha = 0.05$; $M = 44$; critical value = 37; $P = 0.2602$; do not reject H_0; at the 5% significance level, the data do not provide sufficient evidence to conclude that the median number of volumes held by public colleges is less than that held by private colleges.

10.63
a. $H_0: \mu_1 = \mu_2$, $H_a: \mu_1 < \mu_2$; $\alpha = 0.05$; $M = 65.5$; critical value = 83; $P = 0.0016$; reject H_0; at the 5% significance level, the data provide sufficient evidence to conclude that the mean time served for fraud is less than that for firearms offenses.
b. Because two normal distributions with equal standard deviations have the same shape. The pooled t-test is better because, in the normal case, it is more powerful than the Mann–Whitney test.

EXERCISES 10.5

10.75
a. Age
b. Married men and married women
c. Married couples
d. The difference between the ages of a married couple

10.77 Simple random paired sample, and normal differences or large sample. The simple-random-paired-sample assumption is essential. Moderate violations of the normal-differences assumption are permissible even for small or moderate size samples.

10.79
a. Height (of Zea mays)
b. Cross-fertilized Zea mays and self-fertilized Zea mays
c. The difference between the heights of a cross-fertilized Zea may and a self-fertilized Zea may grown in the same pot
d. Yes. Because each number is the difference between the heights of a cross-fertilized Zea may and a self-fertilized Zea may grown in the same pot
e. $H_0: \mu_1 = \mu_2$, $H_a: \mu_1 \neq \mu_2$; $\alpha = 0.05$; $t = 2.148$; critical values = ± 2.145; $0.02 < P < 0.05$; reject H_0; at the 5% significance level, the data provide sufficient evidence to conclude that the mean heights of cross-fertilized and self-fertilized Zea mays differ.
f. $H_0: \mu_1 = \mu_2$, $H_a: \mu_1 \neq \mu_2$; $\alpha = 0.01$; $t = 2.148$; critical values = ± 2.977; $0.02 < P < 0.05$; do not reject H_0; at the 1% significance level, the data do not provide sufficient evidence to conclude that the mean heights of cross-fertilized and self-fertilized Zea mays differ.

10.81 $H_0: \mu_1 = \mu_2$, $H_a: \mu_1 < \mu_2$; $\alpha = 0.05$; $t = -4.185$; critical value = -1.746; $P < 0.005$; reject H_0; at the 5% significance level, the data provide sufficient evidence to conclude that family therapy is effective in helping anorexic young women gain weight.

10.83 $H_0: \mu_1 = \mu_2$, $H_a: \mu_1 > \mu_2$; $\alpha = 0.10$; $t = 1.053$; critical value = 1.415; $P > 0.10$; do not reject H_0; at the 10% significance level, the data do not provide sufficient evidence to conclude that mean corneal thickness is greater in normal eyes than in eyes with glaucoma.

10.85
a. 0.03 to 41.84 eighths of an inch
b. -8.08 to 49.94 eighths of an inch

10.87 -10.30 to -4.23 lb. We can be 90% confident that the weight gain that would be obtained, on average, by using the family therapy treatment is somewhere between 4.23 and 10.30 lb.

EXERCISES 10.6

10.97
a. Yes. Because both assumptions required for a paired t-test are satisfied
b. Yes. Because the paired-difference variable has a normal distribution, its distribution is symmetric.
c. In this case (of normal differences), the paired t-test is preferable because it is more powerful than the paired Wilcoxon signed-rank test.

10.99
a. $H_0: \mu_1 = \mu_2$, $H_a: \mu_1 \neq \mu_2$; $\alpha = 0.05$; $W = 96$; critical values = 25 and 95; $P = 0.044$; reject H_0; at the 5% significance level, the data provide sufficient evidence to conclude that the mean heights of cross-fertilized and self-fertilized Zea mays differ.
b. $H_0: \mu_1 = \mu_2$, $H_a: \mu_1 \neq \mu_2$; $\alpha = 0.01$; $W = 96$; critical values = 16 and 104; $P = 0.044$; do not reject H_0; at the 1% significance level, the data do not provide sufficient evidence to conclude that the mean heights of cross-fertilized and self-fertilized Zea mays differ.

10.101 $H_0: \mu_1 = \mu_2$, $H_a: \mu_1 < \mu_2$; $\alpha = 0.05$; $W = 11$; critical value = 41; $P = 0.001$; reject H_0; at the 5% significance level, the data provide sufficient evidence to conclude that family therapy is effective in helping anorexic young women gain weight.

10.103 H_0: $\mu_1 = \mu_2$, H_a: $\mu_1 > \mu_2$; $\alpha = 0.10$; $W = 20.5$; critical value $= 22$; $P = 0.155$; do not reject H_0; at the 10% significance level, the data do not provide sufficient evidence to conclude that mean corneal thickness is greater in normal eyes than in eyes with glaucoma.

EXERCISES 10.7

10.113 See the first three entries of Table 10.14 (page 554).

10.115
a. Pooled *t*-test, nonpooled *t*-test, and Mann–Whitney test
b. Pooled *t*-test

10.117
a. Pooled *t*-test, nonpooled *t*-test, and Mann–Whitney test
b. Mann–Whitney test

10.119
a. Paired *t*-test and paired Wilcoxon signed-rank test
b. Paired Wilcoxon signed-rank test

10.121 Nonpooled *t*-test

10.123 Nonpooled *t*-test

10.125 Consult a statistician.

REVIEW TEST FOR CHAPTER 10

1. Independently and randomly take samples from the two populations; compute the two sample means; compare the two sample means; and make the decision.

2. Randomly take a paired sample from the two populations; calculate the paired differences of the sample pairs; compute the mean of the sample of paired differences; compare that sample mean to 0; and make the decision.

3. a. The pooled *t*-procedures require equal population standard deviations, whereas the nonpooled *t*-procedures do not.
   b. It is essential that the assumption of independence be satisfied.
   c. For very small sample sizes, the normality assumption is essential for both *t*-procedures. However, for larger samples, the normality assumption is less important.
   d. Population standard deviations

4. a. No. If the two distributions are normal and have the same shape, they have equal population standard deviations; in this case, the pooled *t*-test is preferred. If the two distributions are nonnormal, but have the same shape, the Mann–Whitney test is preferred.
   b. The two distributions are normal. Because, in this case, the pooled *t*-test is more powerful than the Mann–Whitney test

5. By using a paired sample, extraneous sources of variation can be removed. As a consequence, the sampling error made in estimating the difference between the population means will generally be smaller. This fact, in turn, makes it more likely that differences between the population means will be detected when such differences exist.

6. If the paired-difference variable is normally distributed, it would be preferable to use the paired *t*-test because, in that case, it is more powerful than the paired Wilcoxon signed-rank test.

7. a. H_0: $\mu_1 = \mu_2$, H_a: $\mu_1 > \mu_2$; $\alpha = 0.05$; $t = 1.538$; critical value $= 1.708$; $0.05 < P < 0.10$; do not reject H_0; at the 5% significance level, the data do not provide sufficient evidence to conclude that the mean right-leg strength of males exceeds that of females.
   b. $0.05 < P < 0.10$; the evidence against the null hypothesis is moderate.

8. -31.3 to 599.3 newtons (N). We can be 90% confident that the difference between the mean right-leg strengths of males and females is somewhere between -31.3 and 599.3 N.

9. H_0: $\mu_1 = \mu_2$, H_a: $\mu_1 < \mu_2$; $\alpha = 0.01$; $t = -4.118$; critical value $= -2.385$; $P < 0.005$; reject H_0; at the 1% significance level, the data provide sufficient evidence to conclude that, on average, the number of young per litter of cottonmouths in Florida is less than that in Virginia.

10. -3.4 to -0.9 young per litter. We can be 98% confident that the difference between the mean litter sizes of cottonmouths in Florida and Virginia is somewhere between -3.4 and -0.9. With 98% confidence, we can say that, on average, cottonmouths in Virginia have somewhere between 0.9 and 3.4 more young per litter than those in Florida.

11. H_0: $\mu_1 = \mu_2$, H_a: $\mu_1 \neq \mu_2$; $\alpha = 0.05$; $M = 111$; critical values $= 79$ and 131; $P = 0.6776$; do not reject H_0; at the 5% significance level, the data do not provide sufficient evidence to conclude that the mean costs for existing single-family homes differ in New York City and Los Angeles.

12. H_0: $\mu_1 = \mu_2$, H_a: $\mu_1 \neq \mu_2$; $\alpha = 0.10$; $t = -1.766$; critical values $= \pm 1.833$; $0.10 < P < 0.20$; do not reject H_0; at the 10% significance level, the data do not provide sufficient evidence to conclude that there is a difference in effectiveness of the two speed-reading programs.

13. -81.5 to 1.5 words per minute

14. $H_0: \mu_1 = \mu_2$, $H_a: \mu_1 > \mu_2$; $\alpha = 0.05$; $W = 46$; critical value $= 44$; $P = 0.033$; reject H_0; at the 5% significance level, the data provide sufficient evidence to conclude that, on average, the eyepiece method gives a greater fiber-density reading than the TV-screen method.

Chapter 11

EXERCISES 11.1

11.1 A variable is said to have a chi-square distribution if its distribution has the shape of a special type of right-skewed curve, called a chi-square curve.

11.3 The χ^2-curve with 20 degrees of freedom more closely resembles a normal curve. As the number of degrees of freedom becomes larger, χ^2-curves look increasingly like normal curves.

11.5
a. 32.852   **b.** 10.117

11.7
a. 18.307   **b.** 3.247

11.9
a. 1.646   **b.** 15.507

11.11
a. 0.831, 12.833   **b.** 13.844, 41.923

11.13 Because the procedures are based on the assumption that the variable under consideration is normally distributed and are nonrobust to violations of that assumption

11.15 $H_0: \sigma = 0.63°F$, $H_a: \sigma \neq 0.63°F$; $\alpha = 0.05$; $\chi^2 = 96.971$; critical values $= 67.356$ and 120.427; $P = 0.683$; do not reject H_0; at the 5% significance level, the data do not provide sufficient evidence to conclude that the population standard deviation of body temperatures for healthy humans differs from $0.63°F$.

11.17 $H_0: \sigma = 0.27$, $H_a: \sigma > 0.27$; $\alpha = 0.01$; $\chi^2 = 70.631$; critical value $= 21.666$; $P = 1.15 \times 10^{-11}$; reject H_0; at the 1% significance level, the data provide sufficient evidence to conclude that the process variation for this piece of equipment exceeds the analytical capability of 0.27.

11.19 $H_0: \sigma = 0.2$ fl oz, $H_a: \sigma < 0.2$ fl oz; $\alpha = 0.05$; $\chi^2 = 8.317$; critical value $= 6.571$; $P = 0.128$; do not reject H_0; at the 5% significance level, the data do not provide sufficient evidence to conclude that the standard deviation of the amounts being dispensed is less than 0.2 fl oz.

11.21 $0.565°F$ to $0.756°F$. We can be 95% confident that the population standard deviation of body temperatures for healthy humans is somewhere between $0.565°F$ and $0.756°F$.

11.23 0.49 to 1.57. We can be 98% confident that the process variation for this piece of equipment is somewhere between 0.49 and 1.57.

EXERCISES 11.2

11.33 By stating its two numbers of degrees of freedom

11.35 $F_{0.05}$; $F_{0.025}$; F_α

11.37
a. 12   **b.** 7

11.39
a. 1.89   **b.** 2.47   **c.** 2.14

11.41
a. 2.88   **b.** 2.10   **c.** 1.78

11.43
a. 0.12   **b.** 3.58

11.45
a. 0.18, 9.07   **b.** 0.33, 2.68

11.47 Because the procedures are based in part on the assumption that the variable under consideration is normally distributed on each population and are nonrobust to violations of that assumption

11.49 $H_0: \sigma_1 = \sigma_2$, $H_a: \sigma_1 > \sigma_2$; $F = 2.19$; $\alpha = 0.05$; critical value $= 2.03$; $P = 0.035$; reject H_0; at the 5% significance level, the data provide sufficient evidence to conclude that there is less variation among final-exam scores using the new teaching method.

11.51 $H_0: \sigma_1 = \sigma_2$, $H_a: \sigma_1 \neq \sigma_2$; $F = 1.22$; $\alpha = 0.10$; critical values $= 0.53$ and 1.94; $P = 0.624$; do not reject H_0; at the 10% significance level, the data do not provide sufficient evidence to conclude that the variation in anxiety-test scores differs between patients seeing videotapes showing progressive relaxation exercises and those seeing neutral videotapes.

11.53 0.79 to 1.52. We can be 90% confident that the ratio of the population standard deviations of scores for patients seeing videotapes showing progressive relaxation exercises and those seeing neutral videotapes is somewhere between 0.79 and 1.52.

REVIEW TEST FOR CHAPTER 11

1. Chi-square distribution

2. **a.** Right   **b.** Normal

3. The variable under consideration must be normally distributed or nearly so. It is very important because the procedures are nonrobust to violations of that assumption.

4. a. 6.408   **b.** 33.409   **c.** 27.587
   d. 8.672   **e.** 7.564, 30.191

5. The F-distribution

6. a. Right   **b.** Reciprocal; 5, 14
   c. 0

7. The distributions (one for each population) of the variable under consideration must be normally distributed or nearly so. It is very important because the procedures are nonrobust to violations of that assumption.

8. a. 7.01   **b.** 0.07   **c.** 3.84
   d. 0.17   **e.** 0.11, 5.05

9. a. $H_0: \sigma = 16$ points, $H_a: \sigma \neq 16$ points; $\alpha = 0.10$; $\chi^2 = 21.110$; critical values $= 13.848$ and 36.415; $P = 0.736$; do not reject H_0; at the 10% significance level, the data do not provide sufficient evidence to conclude that IQs measured on this scale have a standard deviation different from 16 points.
   b. It is essential because the χ^2-test is nonrobust to violations of that assumption.

10. 12.2 to 19.8 points. We can be 90% confident that the standard deviation of IQs measured on the Stanford Revision of the Binet–Simon Intelligence Scale is somewhere between 12.2 and 19.8 points.

11. a. F-distribution with df $= (14, 19)$
   b. $H_0: \sigma_1 = \sigma_2$, $H_a: \sigma_1 < \sigma_2$; $\alpha = 0.01$; $F = 0.07$; critical value $= 0.28$; $P = 5.79 \times 10^{-6}$; reject H_0; at the 1% significance level, the data provide sufficient evidence to conclude that runners have less variability in skinfold thickness than others.
   c. Skinfold thickness is normally distributed for runners and for others. Construct normal probability plots of the two samples.
   d. The samples from the two populations must be independent simple random samples.

12. a. 0.15 to 0.51
   b. We can be 98% confident that the ratio of the population standard deviations of skinfold thickness for runners and for others is somewhere between 0.15 and 0.51.

Chapter 12

EXERCISES 12.1

12.1 Answers will vary.

12.3 A population proportion is a parameter because it is a descriptive measure for a population. A sample proportion is a statistic because it is a descriptive measure for a sample.

12.5
a. $p = 0.4$
b.

Sample	No. of females x	Sample proportion $\hat{p}$
J, G	1	0.5
J, P	0	0.0
J, C	0	0.0
J, F	1	0.5
G, P	1	0.5
G, C	1	0.5
G, F	2	1.0
P, C	0	0.0
P, F	1	0.5
C, F	1	0.5

c.

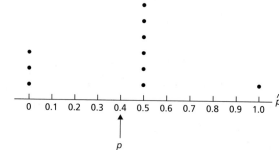

d. 0.4
e. They are the same because the mean of the variable $\hat{p}$ equals the population proportion; in symbols, $\mu_{\hat{p}} = p$.

12.7
b.

Sample	No. of females x	Sample proportion $\hat{p}$
J, P, C	0	0.00
J, P, G	1	0.33
J, P, F	1	0.33
J, C, G	1	0.33
J, C, F	1	0.33
J, G, F	2	0.67
P, C, G	1	0.33
P, C, F	1	0.33
P, G, F	2	0.67
C, G, F	2	0.67

c.

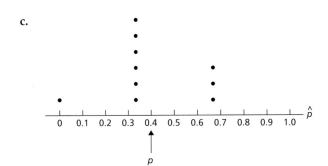

d. 0.4

e. They are the same because the mean of the variable $\hat{p}$ equals the population proportion; in symbols, $\mu_{\hat{p}} = p$.

12.9

b.

Sample	No. of females x	Sample proportion $\hat{p}$
J, P, C, G, F	2	0.4

c.

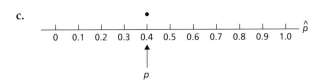

d. 0.4

e. They are the same because the mean of the variable $\hat{p}$ equals the population proportion; in symbols, $\mu_{\hat{p}} = p$.

12.11

a. The No. 1 draft picks in the NBA since 1966

b. Being a center

c. Population proportion. It is the proportion of the population of No. 1 draft picks in the NBA since 1996 who are centers.

12.13

a. 0.00718 **b.** Smaller

12.15

a. 0.4 **b.** 0.5 **c.** 0.7

d. 0.2 **e.** 0.5 **f.** 0.5

g. (a) $0.4 < \hat{p} < 0.6$ (b) None (c) $0.3 < \hat{p} < 0.7$
(d) $0.2 < \hat{p} < 0.8$ (e) None (f) None

12.17

a. 0.626 to 0.674

b. We can be 95% confident that the percentage of Americans who drink beer, wine, or hard liquor, at least occasionally, is somewhere between 62.6% and 67.4%.

12.19

a. 24.0% to 39.8%

b. We can be 90% confident that the percentage of Malaysians infected with the Nipah virus who will die from encephalitis is somewhere between 24.0% and 39.8%.

12.21 Not very well! Procedure 12.1 was applied without checking one of the assumptions for its use; namely, that the number of successes, x, and the number of failures, $n - x$, are both 5 or greater. Because the number of successes here is $0.00361 \cdot 1109 = 4$, Procedure 12.1 should not have been used.

12.23 54.0% to 60.0%

12.25

a. 0.024 **b.** 2401 **c.** 0.611 to 0.649

d. 0.019, which is less than 0.02

e. 2305; 0.610 to 0.650; 0.02

f. By using the guess for $\hat{p}$ in part (e), the required sample size is reduced by 96. Moreover, only 0.1% of precision is lost—the margin of error rises from 0.019 to 0.020.

12.27

a. 7.9% (i.e., 7.9 percentage points) **b.** 271

c. 0.243 to 0.333

d. 0.045, which is less than 0.05 (5%)

e. 260; 0.242 to 0.334; 0.046, which is less than 0.05 (5%)

f. By using the guess for $\hat{p}$ in part (e), the required sample size is reduced by 11. Moreover, only 0.1% of precision is lost—the margin of error rises from 0.045 to 0.046.

12.29 Yes. Because the two confidence intervals, 53.1% to 62.9% and 49.6% to 58.4%, overlap

EXERCISES 12.2

12.37

a. 0.54

b. H_0: $p = 0.5$, H_a: $p > 0.5$; $\alpha = 0.05$; $z = 2.33$; critical value $= 1.645$; $P = 0.0099$; reject H_0; at the 5% significance level, the data provide sufficient evidence to conclude that a majority of Generation Y Web users use the Internet to download music.

12.39 H_0: $p = 0.136$, H_a: $p \neq 0.136$; $\alpha = 0.10$; $z = 2.49$; critical values $= \pm1.645$; $P = 0.0128$; reject H_0; at the 10% significance level, the data provide sufficient evidence to conclude that the percentage of 18–25-year-olds who currently use marijuana or hashish has changed from the 2000 percentage of 13.6%.

12.41

a. H_0: $p = 0.72$, H_a: $p < 0.72$; $\alpha = 0.05$; $z = -4.93$; critical value $= -1.645$; $P = 0.0000$ (to four decimal places); reject H_0; at the 5% significance level, the data provide sufficient evidence to conclude that the percentage of Americans who approve of labor unions now has decreased since 1936.

b. $H_0: p = 0.67$, $H_a: p < 0.67$; $\alpha = 0.05$; $z = -1.34$; critical value $= -1.645$; $P = 0.0901$; do not reject H_0; at the 5% significance level, the data do not provide sufficient evidence to conclude that the percentage of Americans who approve of labor unions now has decreased since 1963.

EXERCISES 12.3

12.45 For a two-tailed test, the basic strategy is as follows: (1) independently and randomly take samples from the two populations under consideration; (2) compute the sample proportions, $\hat{p}_1$ and $\hat{p}_2$; and (3) reject the null hypothesis if the sample proportions differ by too much—otherwise, do not reject the null hypothesis. The process is the same for a one-tailed test except that, for a left-tailed test, the null hypothesis is rejected only when $\hat{p}_1$ is too much smaller than $\hat{p}_2$, and, for a right-tailed test, the null hypothesis is rejected only when $\hat{p}_1$ is too much larger than $\hat{p}_2$.

12.47
a. Uses sunscreen before going out in the sun
b. Teenage girls and teenage boys
c. Sample proportions. Industry Research acquired those proportions by polling samples of the populations of all teenage girls and all teenage boys.

12.49
a. p_1 and p_2 are parameters, and the other quantities are statistics.
b. p_1 and p_2 are fixed numbers, and the other quantities are variables.

12.51
a. $H_0: p_1 = p_2$, $H_a: p_1 < p_2$; $\alpha = 0.01$; $z = -2.61$; critical value $= -2.33$; $P = 0.0045$; reject H_0; at the 1% significance level, the data provide sufficient evidence to conclude that women who take folic acid are at lesser risk of having children with major birth defects.
b. Designed experiment
c. Yes. Because for a designed experiment, it is reasonable to interpret statistical significance as a causal relationship.

12.53 $H_0: p_1 = p_2$, $H_a: p_1 \neq p_2$; $\alpha = 0.10$; $z = -1.52$; critical values $= \pm 1.645$; $P = 0.1286$; do not reject H_0; at the 10% significance level, the data do not provide sufficient evidence to conclude that there is a difference in seat-belt use between drivers who are 25–34 years old and drivers who are 45–64 years old.

12.55
a. The samples must be independent simple random samples; of those sampled whose highest degree is a bachelors, at least five must be overweight and at least

five must not be overweight; and of those sampled with a graduate degree, at least five must be overweight and at least five must not be overweight.
b. $H_0: p_1 = p_2$, $H_a: p_1 > p_2$; $\alpha = 0.05$; $z = 1.41$; critical value $= 1.645$; $P = 0.0793$; do not reject H_0; at the 5% significance level, the data do not provide sufficient evidence to conclude that the percentage who are overweight is greater for those whose highest degree is a bachelors than for those with a graduate degree.
c. $H_0: p_1 = p_2$, $H_a: p_1 > p_2$; $\alpha = 0.10$; $z = 1.41$; critical value $= 1.28$; $P = 0.0793$; reject H_0; at the 10% significance level, the data provide sufficient evidence to conclude that the percentage who are overweight is greater for those whose highest degree is a bachelors than for those with a graduate degree.

12.57
a. -0.0191 to -0.000746, or about -0.019 to -0.001
b. Roughly, we can be 98% confident that the rate of major birth defects for babies born to women who have not taken folic acid is somewhere between 1 per 1000 and 19 per 1000 higher than for babies born to women who have taken folic acid.

12.59 -0.0624 to 0.00240. We can be 90% confident that the difference between the proportions of seat-belt users for drivers in the age groups 25–34 years and 45–64 years is somewhere between -0.0624 and 0.00240.

REVIEW TEST FOR CHAPTER 12

1. **a.** Feeling that marijuana should be legalized for medicinal use in patients with cancer and other painful and terminal diseases
   b. Americans
   c. Proportion of all Americans who feel that marijuana should be legalized for medicinal use in patients with cancer and other painful and terminal diseases
   d. Sample proportion. It is the proportion of Americans sampled who feel that marijuana should be legalized for medicinal use in patients with cancer and other painful and terminal diseases.

2. Generally, obtaining a sample proportion can be done more quickly and is less costly than obtaining the population proportion. Sampling is often the only practical way to proceed.

3. **a.** The number of members in the sample that have the specified attribute
   b. The number of members in the sample that do not have the specified attribute

4. **a.** Population proportion
   b. Normal
   c. np, $n(1 - p)$, 5

5. The precision with which a sample proportion, $\hat{p}$, estimates the population proportion, p, at the specified confidence level

6. **a.** Getting the "holiday blues"
   b. All men, all women
   c. The proportion of all men who get the "holiday blues" and the proportion of all women who get the "holiday blues"
   d. The proportion of all sampled men who get the "holiday blues" and the proportion of all sampled women who get the "holiday blues"
   e. Sample proportions. The poll used samples of men and women to obtain the proportions.

7. **a.** Difference between the population proportions
   b. Normal

8. 37.0% to 43.0%

9. **a.** 19,208   **b.** 14,406

10. 0.574 to 0.629. We can be 95% confident that the percentage of students who expect difficulty finding a job is somewhere between 57.4% and 62.9%.

11. **a.** 0.028   **b.** 2401   **c.** 0.567 to 0.607
    d. 0.020, which is the same as that specified in part (b)
    e. 2367; 0.567 to 0.607; 0.020
    f. By using the guess for $\hat{p}$ in part (e), the required sample size is reduced by 34 with (virtually) no sacrifice in precision.

12. **a.** H_0: $p = 0.25$, H_a: $p < 0.25$; $\alpha = 0.05$; $z = -2.30$; critical value $= -1.645$; $P = 0.0107$; reject H_0; at the 5% significance level, the data provide sufficient evidence to conclude that less than one in four Americans believe that juries "almost always" convict the guilty and free the innocent.
    b. The data provide strong evidence against the null hypothesis and hence in favor of the alternative hypothesis that less than one in four Americans believe that juries "almost always" convict the guilty and free the innocent.

13. **a.** Observational study
    b. Being observational, the study established only an association between height and breast cancer; no causal relationship can be inferred, although there may be one.

14. **a.** H_0: $p_1 = p_2$, H_a: $p_1 < p_2$; $\alpha = 0.01$; $z = -4.17$; critical value $= -2.33$; $P = 0.0000$ (to four decimal places); reject H_0; at the 1% significance level, the data provide sufficient evidence to conclude that the percentage of Maricopa County residents who thought Arizona's economy would improve over the next 2 years was less during the time of the first poll than during the time of the second poll.

    b. The data provide very strong evidence against the null hypothesis and hence in favor of the alternative hypothesis that the percentage of Maricopa County residents who thought Arizona's economy would improve over the next 2 years was less during the time of the first poll than during the time of the second poll.

15. **a.** −0.186 to −0.054
    b. We can be 98% confident that the difference between the percentages of Maricopa County residents who thought Arizona's economy would improve over the next 2 years during the time of the first poll and during the time of the second poll is somewhere between −18.6% and −5.4%.

16. **a.** 0.066; we can be 98% confident that the error in estimating the difference between the two population proportions, $p_1 - p_2$, by the difference between the two sample proportions, −0.12, is at most 0.066.
    b. 0.066   **c.** 3006   **d.** −0.158 to −0.098
    e. 0.03, which is the same as that specified in part (c)

Chapter 13

EXERCISES 13.1

13.1 A variable is said to have a chi-square distribution if its distribution has the shape of a special type of right-skewed curve, called a chi-square curve.

13.3 The χ^2-curve with 20 degrees of freedom more closely resembles a normal curve. As the number of degrees of freedom becomes larger, χ^2-curves look increasingly like normal curves.

13.5
a. 32.852   **b.** 10.117

13.7
a. 18.307   **b.** 3.247

EXERCISES 13.2

13.11 Because the hypothesis test is carried out by determining how well the observed frequencies fit the expected frequencies

13.13
a. Both assumptions are satisfied.
b. Assumption 1 is satisfied, but Assumption 2 fails because 33.3% of the expected frequencies are less than 5.
c. Both assumptions are satisfied. Note that 20% of the expected frequencies are less than 5.

d. Both assumptions are satisfied. Note that 20% of the expected frequencies are less than 5.

e. Assumption 2 is satisfied because only 20% of the expected frequencies are less than 5, but Assumption 1 fails because there is an expected frequency of 0.5 which is less than 1.

f. Assumption 1 is satisfied, but Assumption 2 fails because 25% of the expected frequencies are less than 5.

13.15

a. The population consists of all this year's incoming college freshmen in the United States; the variable is political view.

b. H_0: This year's distribution of political views for incoming college freshmen is the same as the 2000 distribution. H_a: This year's distribution of political views for incoming college freshmen has changed from the 2000 distribution. $\alpha = 0.05$; $\chi^2 = 4.667$; critical value $= 5.991$; $0.05 < P < 0.10$; do not reject H_0; at the 5% significance level, the data do not provide sufficient evidence to conclude that this year's distribution of political views for incoming college freshmen has changed from the 2000 distribution.

c. H_0: This year's distribution of political views for incoming college freshmen is the same as the 2000 distribution. H_a: This year's distribution of political views for incoming college freshmen has changed from the 2000 distribution. $\alpha = 0.10$; $\chi^2 = 4.667$; critical value $= 4.605$; $0.05 < P < 0.10$; reject H_0; at the 10% significance level, the data provide sufficient evidence to conclude that this year's distribution of political views for incoming college freshmen has changed from the 2000 distribution.

13.17 H_0: The color distribution of M&Ms is that reported by M&M/MARS consumer affairs. H_a: The color distribution of M&Ms differs from that reported by M&M/MARS consumer affairs. $\alpha = 0.05$; $\chi^2 = 4.091$; critical value $= 11.070$; $P > 0.10$; do not reject H_0; at the 5% significance level, the data do not provide sufficient evidence to conclude that the color distribution of M&Ms differs from that reported by M&M/MARS consumer affairs.

EXERCISES 13.3

13.25 Answers will vary.

13.27 Cells

13.29 Summing the row totals, summing the column totals, or summing the frequencies in the cells

13.31 Yes. If there were no association between "gender" and "specialty," the percentage of male physicians in residency who specialized in family practice would be

identical to the percentage of female physicians in residency who specialized in family practice. As that is not the case, there is an association between the two variables.

13.33

a.

		College			
		Bus.	Engr.	Lib. Arts	Total
Sex	Male	2	10	3	15
	Female	7	2	1	10
	Total	9	12	4	25

b.

		College			
		Bus.	Engr.	Lib. Arts	Total
Sex	Male	0.222	0.833	0.750	0.600
	Female	0.778	0.167	0.250	0.400
	Total	1.000	1.000	1.000	1.000

c.

		College			
		Bus.	Engr.	Lib. Arts	Total
Sex	Male	0.133	0.667	0.200	1.000
	Female	0.700	0.200	0.100	1.000
	Total	0.360	0.480	0.160	1.000

d. Yes. The tables in parts (b) and (c) show that the conditional distributions of one variable given the other are not identical.

13.35

a.

		Class				
		Fresh.	Soph.	Junior	Senior	Total
Party	Republican	3	9	12	6	30
	Democrat	2	6	8	4	20
	Other	1	3	4	2	10
	Total	6	18	24	12	60

b.

		Class			
		Fresh.	Soph.	Junior	Senior
Party	Republican	0.500	0.500	0.500	0.500
	Democrat	0.333	0.333	0.333	0.333
	Other	0.167	0.167	0.167	0.167
	Total	1.000	1.000	1.000	1.000

c. No. The table in part (b) shows that the conditional distributions of political party affiliation within class levels are identical.

d. Republican 0.500, Democrat 0.333, Other 0.167, Total 1.000

e. True. From part (c), political party affiliation and class level are not associated. Therefore the conditional distributions of class level within political party affiliations are identical to each other and to the marginal distribution of class level.

13.37

a. 8

b. The missing entries, from top to bottom and left to right, are: 30; 145; 654; and 793.

c. 793 thousand   **d.** 145 thousand

e. 654 thousand   **f.** 30 thousand

13.39

a. 1056.5 thousand   **b.** 48.6 thousand

c. 10.6 thousand   **d.** 304.7 thousand

e. 51.9 thousand   **f.** 51.9 thousand

g. 1560.1 thousand

13.41

a.

		Gender		
		Male	Female	**Total**
Race/Ethnicity	White	0.911	0.089	1.000
	Black	0.732	0.268	1.000
	Hispanic	0.814	0.186	1.000
	Other	0.889	0.111	1.000

b. Male 0.825; Female 0.175; Total 1.000

c. Yes. Because the conditional distributions of gender within races are not identical

d. 17.5%   **e.** 8.9%

f. True. Because by part (c), there is an association between the variables "gender" and "race."

g.

		Gender		
		Male	Female	**Total**
Race/Ethnicity	White	0.469	0.216	0.425
	Black	0.338	0.583	0.381
	Hispanic	0.180	0.194	0.183
	Other	0.012	0.007	0.011
	Total	1.000	1.000	1.000

13.43

a.

		Prison facility		
		State	Federal	Local
Educational attainment	8th grade or less	0.142	0.119	0.131
	Some high school	0.255	0.145	0.334
	GED	0.285	0.226	0.141
	High school diploma	0.205	0.270	0.259
	Postsecondary	0.090	0.158	0.103
	College grad or more	0.024	0.081	0.032
	Total	1.000	1.000	1.000

b. Yes. Because the conditional distributions of educational attainment within type of prison facility categories are not identical.

c. 8th grade or less: 0.137; Some high school: 0.273; GED: 0.238; High school diploma: 0.225; Postsecondary: 0.098; College grad or more: 0.029

d.

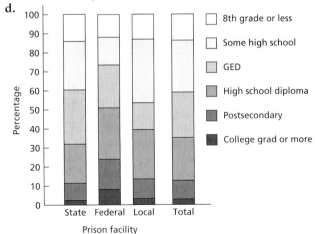

That the bars are not identical reflects the fact that there is an association between educational attainment and type of prison facility.

e. False. Because by part (b), there is an association between educational attainment and type of prison facility.

f.

	Prison facility			
Educational attainment	State	Federal	Local	Total
8th grade or less	0.662	0.047	0.291	1.000
Some high school	0.598	0.029	0.374	1.000
GED	0.768	0.051	0.181	1.000
High school diploma	0.584	0.065	0.352	1.000
Postsecondary	0.590	0.087	0.323	1.000
College grad or more	0.521	0.148	0.331	1.000
Total	0.641	0.054	0.305	1.000

g. 5.4%   **h.** 4.7%   **i.** 11.9%

EXERCISES 13.4

13.49 In most cases, data for an entire population are not available. Therefore inferential methods must usually be applied to decide whether an association exists between two variables.

13.51 If no association exists between marital status and alcohol consumption, the (conditional) distribution of alcohol consumption for married adults is the same as the (marginal) distribution of alcohol consumption for all adults.

13.53 15

13.55 If there is a causal relationship between two variables, they are necessarily associated. In other words, if there is no association between two variables, they could not possibly be causally related.

13.57 H_0: An association does not exist between the ratings of Siskel and Ebert. H_a: An association exists between the ratings of Siskel and Ebert. Assumptions 1 and 2 are satisfied because all expected frequencies are 5 or greater. $\alpha = 0.01$; $\chi^2 = 45.357$; critical value = 13.277; $P < 0.005$; reject H_0; at the 1% significance level, the data provide sufficient evidence to conclude that an association exists between the ratings of Siskel and Ebert.

13.59
a. No. Assumption 2 fails because 33.3% of the expected frequencies are less than 5.
b. Yes. H_0: Social class and frequency of games are not associated. H_a: Social class and frequency of games are associated. Assumptions 1 and 2 are satisfied because all expected frequencies are 5 or greater. $\alpha = 0.01$;

$\chi^2 = 8.715$; critical value = 9.210; $0.01 < P < 0.025$; do not reject H_0; at the 1% significance level, the data do not provide sufficient evidence to conclude that an association exists between social class and frequency of games.

13.61 H_0: Size of city and status in practice are statistically independent for U.S. lawyers. H_a: Size of city and status in practice are statistically dependent for U.S. lawyers. Assumption 1 is satisfied, but Assumption 2 is not because 25% (3 of 12) of the expected frequencies are less than 5. Consequently, the chi-square independence test should not be applied here.

REVIEW TEST FOR CHAPTER 13

1. By their degrees of freedom

2. **a.** 0   **b.** Right skewed   **c.** Normal curve

3. **a.** No. The degrees of freedom for the chi-square goodness-of-fit test depends on the number of possible values for the variable under consideration, not on the sample size.
   b. No. The degrees of freedom for the chi-square independence test depends on the number of possible values for the two variables under consideration, not on the sample size.

4. For both tests, the null hypothesis is rejected only when the observed and expected frequencies match up poorly, which corresponds to large values of the chi-square test statistic. Thus both tests are always right tailed.

5. 0

6. **a.** (1) All expected frequencies are 1 or greater. (2) At most 20% of the expected frequencies are less than 5.
   b. They are very important. If the assumptions are not met, the results could be invalid.

7. **a.** 7.8%   **b.** Roughly 1.9 million
   c. There is an association between race and region of residence.

8. **a.** Obtain the conditional distribution of one of the variables for each possible value of the other variable. If all these conditional distributions are identical, there is no association between the two variables; otherwise, there is an association between the two variables.
   b. No. Because the data are for an entire population, no inference is being made from a sample to the population. The conclusion is a fact.

9. **a.** Perform a chi-square independence test.
   b. Yes. As in any inference, it is always possible that the conclusion is in error.

10. a. 6.408   **b.** 33.409   **c.** 27.587
d. 8.672   **e.** 7.564, 30.191

11. a. H_0: This year's distribution of educational
attainment is the same as the 2000 distribution.
H_a: This year's distribution of educational
attainment differs from the 2000 distribution.
Assumptions 1 and 2 are satisfied because all
expected frequencies are 5 or greater. $\alpha = 0.05$;
$\chi^2 = 2.674$; critical value $= 11.070$; $P > 0.10$; do not
reject H_0; at the 5% significance level, the data do
not provide sufficient evidence to conclude that
this year's distribution of educational attainment
differs from the 2000 distribution.
b. $P > 0.10$. The evidence against the null hypothesis
is weak or none.

12. a.

Party of governor

Region		Democratic	Republican	**Total**
	Northeast	3	6	9
	Midwest	7	5	12
	South	8	8	16
	West	6	7	13
	Total	24	26	50

b.

Party of governor

Region		Democratic	Republican	**Total**
	Northeast	0.125	0.231	0.180
	Midwest	0.292	0.192	0.240
	South	0.333	0.308	0.320
	West	0.250	0.269	0.260
	Total	1.000	1.000	1.000

c.

Party of governor

Region		Democratic	Republican	**Total**
	Northeast	0.333	0.667	1.000
	Midwest	0.583	0.417	1.000
	South	0.500	0.500	1.000
	West	0.462	0.538	1.000
	Total	0.480	0.520	1.000

d. Yes. Because the conditional distributions of party
of governor within regions are not identical

e. 52.0%   **f.** 52.0%   **g.** 50.0%
h. 32.0%   **i.** 32.0%   **j.** 30.8%

13. a. 2046   **b.** 737   **c.** 266
d. 3046   **e.** 5413   **f.** 5910

14. a.

		Control			
		Gov	Prop	NP	**Total**
Facility	General	0.314	0.122	0.564	1.000
	Psychiatric	0.361	0.486	0.153	1.000
	Chronic	0.808	0.038	0.154	1.000
	Tuberculosis	0.750	0.000	0.250	1.000
	Other	0.144	0.361	0.495	1.000

b. Yes. Because the conditional distributions of
control type within facility types are not identical
c. Gov 0.311, Prop 0.177, NP 0.512, Total 1.000
d. See Fig. A.7 on the next page. That the bars are not
identical reflects the fact that there is an association
between facility type and control type.
e. False. By part (b) there is an association between
facility type and control type.
f.

		Control			
		Gov	Prop	NP	**Total**
Facility	General	0.829	0.566	0.905	0.821
	Psychiatric	0.130	0.307	0.034	0.112
	Chronic	0.010	0.001	0.001	0.004
	Tuberculosis	0.001	0.000	0.000	0.001
	Other	0.029	0.127	0.060	0.062
	Total	1.000	1.000	1.000	1.000

g. 17.7%   **h.** 48.6%   **i.** 30.7%

15. H_0: Histological type and treatment response are
statistically independent. H_a: Histological type and
treatment response are statistically dependent.
Assumptions 1 and 2 are satisfied because all expected
frequencies are 5 or greater. $\alpha = 0.01$; $\chi^2 = 75.890$;
critical value $= 16.812$; $P < 0.005$; reject H_0; at the
1% significance level, the data provide sufficient
evidence to conclude that histological type and
treatment response are statistically dependent.

FIGURE A.7
Segmented bar graph for Problem 14(d)

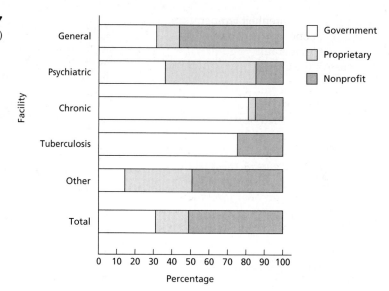

Chapter 14

EXERCISES 14.1

14.1
a. $y = b_0 + b_1 x$
b. b_0 and b_1 represent constants; x and y represent variables.
c. x is the independent variable; y is the dependent variable.

14.3
a. The number b_0 is the y-intercept. It is the y-value at which the straight-line graph of the linear equation intersects the y-axis.
b. The number b_1 is the slope. It measures the steepness of the straight line; more precisely, b_1 indicates how much the y-value on the straight line changes (increases or decreases) when the x-value increases by 1 unit.

14.5
a. $y = 68.22 + 0.25x$ **b.** $b_0 = 68.22$, $b_1 = 0.25$
c.

x	50	100	250
y	80.72	93.22	130.72

d.

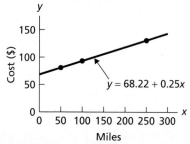

e. About $105; exact cost is $105.72

14.7
a. $b_0 = 32$, $b_1 = 1.8$ **b.** $-40, 32, 68, 212$
c.

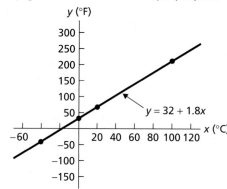

d. About 80° F; exact temperature is 82.4° F

14.9
a. $b_0 = 68.22$, $b_1 = 0.25$
b. The y-intercept, $b_0 = 68.22$, gives the y-value at which the straight line, $y = 68.22 + 0.25x$, intersects the y-axis. The slope, $b_1 = 0.25$, indicates that the y-value increases by 0.25 unit for every increase in x of 1 unit.
c. The y-intercept, $b_0 = 68.22$, is the cost (in dollars) for driving the car 0 miles. The slope, $b_1 = 0.25$, represents the fact that the cost per mile is $0.25; it is the amount the total cost increases for each additional mile driven.

14.11
a. $b_0 = 32$, $b_1 = 1.8$
b. The y-intercept, $b_0 = 32$, gives the y-value at which the straight line, $y = 32 + 1.8x$, intersects the y-axis. The slope, $b_1 = 1.8$, indicates that the y-value increases by 1.8 units for every increase in x of 1 unit.

c. The y-intercept, $b_0 = 32$, is the Fahrenheit temperature corresponding to $0°$ C. The slope, $b_1 = 1.8$, represents the fact that the Fahrenheit temperature increases by $1.8°$ for every increase of the Celsius temperature of $1°$.

14.13
a. $b_0 = 3, b_1 = 4$ **b.** Slopes upward
c.

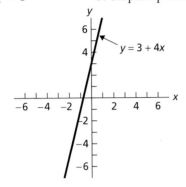

14.15
a. $b_0 = 6, b_1 = -7$ **b.** Slopes downward
c.

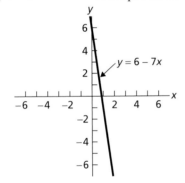

14.17
a. $b_0 = -2, b_1 = 0.5$ **b.** Slopes upward

14.19
a. $b_0 = 2, b_1 = 0$ **b.** Horizontal

14.21
a. $b_0 = 0, b_1 = 1.5$ **b.** Slopes upward

14.23
a. Slopes upward   **b.** $y = 5 + 2x$

c.

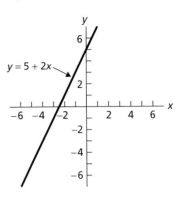

14.25
a. Slopes downward   **b.** $y = -2 - 3x$
c.

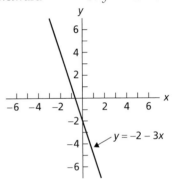

14.27
a. Slopes downward   **b.** $y = -0.5x$

14.29
a. Horizontal   **b.** $y = 3$

EXERCISES 14.2

14.33
a. Least-squares criterion
b. The straight line that best fits a set of data points is the one having the smallest possible sum of squared errors.

14.35
a. Response variable
b. Predictor variable, or explanatory variable

14.37
a. Outlier
b. Influential observation

14.39

a.

Line A: $y = 3 - 0.6x$

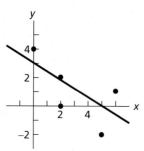

Line B: $y = 4 - x$

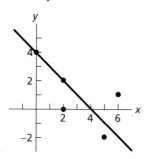

b. Line A: $y = 3 - 0.6x$

x	y	$\hat{y}$	e	e^2
0	4	3.0	1.0	1.00
2	2	1.8	0.2	0.04
2	0	1.8	−1.8	3.24
5	−2	0.0	−2.0	4.00
6	1	−0.6	1.6	2.56
				10.84

Line B: $y = 4 - x$

x	y	$\hat{y}$	e	e^2
0	4	4	0	0
2	2	2	0	0
2	0	2	−2	4
5	−2	−1	−1	1
6	1	−2	3	9
				14

c. Line A

14.41

a. $\hat{y} = 2.875 - 0.625x$

b.

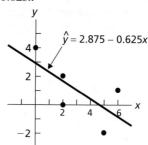

14.43

a. $\hat{y} = 436.6 - 27.9x$

b.

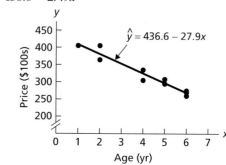

c. Price tends to decrease as age increases.

d. Corvettes depreciate an estimated \$2790 per year, at least in the 1- to 6-year-old range.

e. \$38,080; \$35,289 (see the note at the top of page A-31)

f. The predictor variable is age (in years); the response variable is price (in hundreds of dollars).

g. None

14.45

a. $\hat{y} = 3.52 + 0.16x$

b.

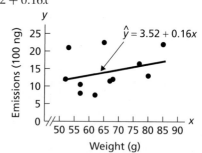

c. Quantity of volatile compounds emitted tends to increase as potato plant weight increases.

d. The quantity of volatile compounds emitted increases an estimated 16 nanograms for each increase in potato plant weight of 1 g.

e. 1574 nanograms (see the note at the top of page A-31)

f. The predictor variable is potato plant weight (in grams); the response variable is quantity of volatile compounds emitted (in hundreds of nanograms).

g. None

14.47 Only the second one

14.49

a. It is acceptable to use the regression equation to predict the price of a 4-year-old Corvette because that age lies within the range of ages in the sample data. It is not acceptable (and would be extrapolation) to use the regression equation to predict the price of a 10-year-old Corvette because that age lies outside the range of the ages in the sample data.

b. Ages between 1 and 6 years, inclusive

14.51

a.

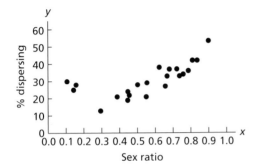

b. No, because the data points are scattered about a curve, not a straight line.

EXERCISES 14.3

14.59

a. The coefficient of determination, r^2

b. The proportion of variation in the observed values of the response variable explained by the regression

14.61

a. $r^2 = 0.920$; 92.0% of the variation in the observed values of the response variable is explained by the regression. The fact that r^2 is near 1 indicates that the regression equation is extremely useful for making predictions.

b. 664.4

14.63

a. $SST = 20$, $SSR = 9.375$, $SSE = 10.625$

b. $20 = 9.375 + 10.625$ **c.** $r^2 = 0.469$

d. 46.9%   **e.** Moderately useful

14.65

a. $SST = 25{,}681.6$, $SSR = 24{,}057.9$, $SSE = 1623.7$

b. 0.937

c. 93.7%; 93.7% of the variation in the price data is explained by age.

d. Extremely useful

14.67

a. $SST = 296.68$, $SSR = 32.52$, $SSE = 264.16$

b. 0.110

c. 11.0%; 11.0% of the variation in the quantity of volatile emissions is explained by potato plant weight.

d. Not very useful

EXERCISES 14.4

14.77 Pearson product moment correlation coefficient

14.79

a. ± 1 **b.** Not very useful

14.81 False. Correlation does not imply causation.

14.83 $r = -0.842$

14.85 $r = -0.685$

14.87

a. $r = -0.968$

b. Suggests an extremely strong negative linear relationship between age and price of Corvettes.

c. Data points are clustered closely about the regression line.

d. $r^2 = 0.937$. This value of r^2 is the same as the one obtained in Exercise 14.65(b).

14.89

a. $r = 0.331$

b. Suggests a weak positive linear relationship between potato plant weight and quantity of volatile emissions.

c. Data points are scattered widely about the regression line.

d. $r^2 = 0.110$. This value of r^2 is the same as the one obtained in Exercise 14.67(b).

14.91

a. $r = 0$

b. No. Only that there is no *linear* relationship between the variables

d. No. Because the data points are not scattered about a straight line

e. For each data point (x, y), the relation $y = x^2$ holds.

f. $r = 0.930$

g. No. A high correlation does not imply a linear relationship between two variables.

i. No. Because the data points are not scattered about a straight line

j. For each data point (x, y), the relation $y = x^3$ holds.

REVIEW TEST FOR CHAPTER 14

1. a. x **b.** y **c.** b_1 **d.** b_0

2. a. $y = 4$ **b.** $x = 0$ **c.** -3

d. -3 units   **e.** 6 units

3. a. True. The *y*-intercept indicates only where the line crosses the *y*-axis; that is, it is the *y*-value when $x = 0$.
   b. False. Its slope is 0.
   c. True. This is equivalent to saying: If a line has a positive slope, then *y*-values on the line increase as the *x*-values increase.

4. Scatter diagram (or scatterplot)

5. A regression equation can be used to predict the response variable for values of the predictor variable within the range of the observed values of the predictor variable.

6. a. Predictor variable, or explanatory variable
   b. Response variable

7. a. Smallest   **b.** Regression   **c.** Extrapolation

8. a. An outlier is a data point that lies far from the regression line, relative to the other data points.
   b. An influential observation is a data point whose removal causes the regression equation (and line) to change considerably.

9. It is a descriptive measure of the utility of the regression equation for making predictions.

10. a. *SST* is the total sum of squares. It measures the variation in the observed values of the response variable.
   b. *SSR* is the regression sum of squares. It measures the variation in the observed values of the response variable explained by the regression.
   c. *SSE* is the error sum of squares. It measures the variation in the observed values of the response variable not explained by the regression.

11. a. Linear   **b.** Increases
   c. Negative   **d.** 0

12. True

13. a. $y = 72 - 12x$ **b.** $b_0 = 72, b_1 = -12$
   c. The line slopes downward because $b_1 < 0$.
   d. $4800; $1200
   e.

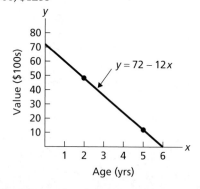

   f. About $2500; exact value is $2400.

14. a.

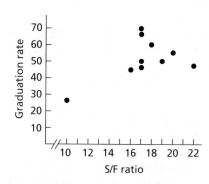

   b. It is reasonable to find a regression line for the data because the data points appear to be scattered about a straight line.
   c. $\hat{y} = 16.4 + 2.03x$

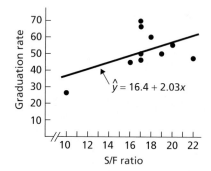

   d. Graduation rate tends to increase as student-to-faculty ratio increases.
   e. Graduation rate increases by an estimated 2.03 percentage points for each increase of 1 in the student-to-faculty ratio.
   f. 50.9%
   g. There are no outliers. The data point $(10, 26)$ is a potential influential observation.

15. a. $SST = 1384.50$; $SSR = 361.66$; $SSE = 1022.84$
   b. $r^2 = 0.261$ **c.** 26.1%
   d. Not very useful

16. a. $r = 0.511$
   b. Suggests a moderately weak positive linear relationship between student-to-faculty ratio and graduation rate.
   c. Data points are rather widely scattered about the regression line.
   d. $r^2 = (0.511)^2 = 0.261$

Chapter 15

EXERCISES 15.1

15.1 Conditional distribution, conditional mean, conditional standard deviation

15.3

a. Population regression line

b. σ

c. Normal; $\beta_0 + 6\beta_1$; σ

15.5 The sample regression line, $\hat{y} = b_0 + b_1 x$

15.7 Residual

15.9 A residual plot, that is, a plot of the residuals against the values of the predictor variable. A residual plot makes it easier to spot patterns such as curvature and nonconstant standard deviation than does a scatter diagram.

15.11 There are constants, β_0, β_1, and σ, such that, for each age, x, the prices of all Corvettes of that age are normally distributed with mean $\beta_0 + \beta_1 x$ and standard deviation σ.

15.13 There are constants, β_0, β_1, and σ, such that, for each weight, x, the quantities of volatile compounds emitted by all potato plants of that weight are normally distributed with mean $\beta_0 + \beta_1 x$ and standard deviation σ.

15.15

a. $s_e = 14.25$; very roughly speaking, on average, the predicted price of a Corvette in the sample differs from the observed price by about $1425.

b. Presuming that, for Corvettes, the variables age (x) and price (y) satisfy the assumptions for regression inferences, the standard error of the estimate, $s_e = 14.25$, provides an estimate for the common population standard deviation, σ, of prices (in hundreds of dollars) for all Corvettes of any particular age.

c. See Fig. A.8.

d. It appears reasonable.

15.17

a. $s_e = 5.42$; very roughly speaking, on average, the predicted quantity of volatile compounds emitted by a potato plant in the sample differs from the observed quantity by about 542 nanograms.

b. Presuming that, for potato plants, the variables weight (x) and quantity of volatile compounds emitted (y) satisfy the assumptions for regression inferences, the standard error of the estimate, $s_e = 5.42$, provides an estimate for the common population standard deviation, σ, of quantities of volatile compounds emitted (in hundreds of nanograms) for all potato plants of any particular weight.

c. See Fig. A.9.

d. Although Fig. A.9(b) shows some curvature, it is probably not sufficiently curved to call into question the validity of the normality assumption (Assumption 3).

15.19 Part (a) is a tough call, but the assumption of linearity (Assumption 1) may be violated, as may be the assumption of equal standard deviations (Assumption 2). In part (b), it appears that the assumption of equal standard deviations (Assumption 2) is violated. In part (d), it appears that the normality assumption (Assumption 3) is violated.

FIGURE A.8

(a) Residual plot and (b) normal probability plot of residuals for Exercise 15.15(c)

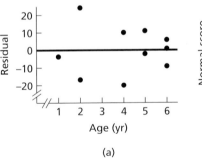

(a)

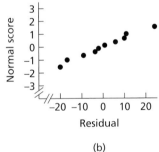

(b)

FIGURE A.9

(a) Residual plot and (b) normal probability plot of residuals for Exercise 15.17(c)

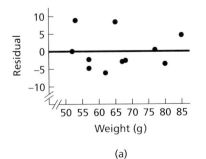

(a)

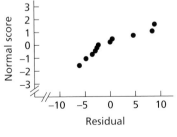

(b)

EXERCISES 15.2

15.27 normal, -3.5

15.29 r^2, r

15.31 $H_0: \beta_1 = 0, H_a: \beta_1 \neq 0; \alpha = 0.10; t = -10.887$; critical values $= \pm 1.860; P < 0.01$; reject H_0; at the 10% significance level, the data provide sufficient evidence to conclude that the slope of the population regression line is not 0 and hence that age is useful as a predictor of price for Corvettes.

15.33 $H_0: \beta_1 = 0, H_a: \beta_1 \neq 0; \alpha = 0.05; t = 1.053$; critical values $= \pm 2.262; P > 0.20$; do not reject H_0; at the 5% significance level, the data do not provide sufficient evidence to conclude that weight is useful as a predictor of quantity of volatile emissions for the potato plant *Solanum tuberosom*.

15.35
a. -32.7 to -23.1
b. We can be 90% confident that, for Corvettes, the decrease in mean price per 1-year increase in age (i.e., the mean annual depreciation) is somewhere between $2310 and $3270.

15.37
a. -0.19 to 0.51
b. We can be 95% confident that, for the potato plant *Solanum tuberosom*, the change in the mean quantity of volatile emissions per 1-gram increase of weight is somewhere between -19 and 51 nanograms.

EXERCISES 15.3

15.43 $11,443. A point estimate for the mean price is the same as the predicted price.

15.45
a. 324.99 ($32,499)
b. 316.60 to 333.38. We can be 90% confident that the mean price of all 4-year-old Corvettes is somewhere between $31,660 and $33,338.
c. 324.99 ($32,499)
d. 297.20 to 352.78. We can be 90% certain that the observed price of a 4-year-old Corvette will be somewhere between $29,720 and $35,278.
e. See Fig. A.10.
f. The error in the estimate of the mean price of all 4-year-old Corvettes is due only to the fact that the population regression line is being estimated by a sample regression line. In contrast, the error in the prediction of the observed price of a 4-year-old Corvette is due to the error in estimating the mean price plus the variation in prices of 4-year-old Corvettes.

15.47 •
a. 13.29 (1329 ng)
b. 9.09 to 17.50. We can be 95% confident that the mean quantity of volatile emissions of all plants that weigh 60 g is somewhere between 909 and 1750 ng.
c. 13.29 (1329 ng)
d. 0.34 to 26.25. We can be 95% certain that the observed quantity of volatile emissions of a plant that weighs 60 g will be somewhere between 34 and 2625 ng.

FIGURE A.10
90% confidence and prediction intervals for Exercise 15.45(e)

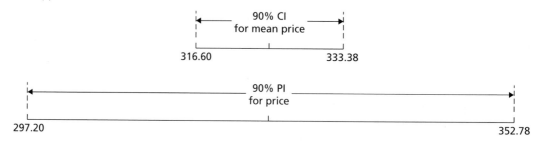

EXERCISES 15.4

15.55 The (sample) linear correlation coefficient, r

15.57
a. Uncorrelated   b. Increases   c. Negatively

15.59 $H_0: \rho = 0, H_a: \rho < 0; \alpha = 0.05; t = -10.887$; critical value $= -1.860; P < 0.005$; reject H_0; at the 5% significance level, the data provide sufficient evidence to conclude that, for Corvettes, age and price are negatively linearly correlated.

15.61 $H_0: \rho = 0, H_a: \rho \neq 0; \alpha = 0.05; t = 1.053$; critical values $= \pm 2.262; P > 0.20$; do not reject H_0; at the 5% significance level, the data do not provide sufficient evidence to conclude that, for the potato plant *Solanum tuberosom*, weight and quantity of volatile emissions are linearly correlated.

15.63 ρ is a parameter; r is a statistic

EXERCISES 15.5

15.69
a. A plot of the normal scores against the sample data
b. Assessing normality of a variable from sample data
c. If the plot is roughly linear, accept as reasonable that the variable is approximately normally distributed. If the plot shows systematic deviations from linearity (e.g., if it displays significant curvature), conclude that the variable probably is not approximately normally distributed.
d. What constitutes *roughly linear* is a matter of opinion.

15.71 If the variable under consideration is normally distributed, the normal probability plot should be roughly linear, which means that the correlation between the sample data and its normal scores should be close to 1. Because the correlation can be at most 1, evidence against the null hypothesis of normality is provided when the correlation is "too much smaller than 1." Thus a correlation test for normality is always left tailed.

15.73 H_0: Final-exam scores in the introductory statistics class are normally distributed. H_a: Final-exam scores in the introductory statistics class are not normally distributed. $\alpha = 0.05$; $R_p = 0.941$; critical value $= 0.951$; $0.01 < P < 0.05$; reject H_0; at the 5% significance level, the data provide sufficient evidence to conclude that final-exam scores in the introductory statistics class are not normally distributed.

15.75 H_0: Finishing times for the winners of 1-mile thoroughbred horse races are normally distributed. H_a: Finishing times for the winners of 1-mile thoroughbred horse races are not normally distributed. $\alpha = 0.01$; $R_p = 0.990$; critical value $= 0.905$; $P > 0.10$; do not reject H_0; at the 1% significance level, the data do not provide sufficient evidence to conclude that finishing times for the winners of 1-mile thoroughbred horse races are not normally distributed.

REVIEW TEST FOR CHAPTER 15

1. **a.** Conditional
   b. See Key Fact 15.1 on page 744.

2. **a.** b_1 **b.** b_0 **c.** s_e

3. A residual plot (i.e., a plot of the residuals against the observed values of the predictor variable) and a normal probability plot of the residuals. A plot of the residuals against the observed values of the predictor variable should fall roughly in a horizontal band, centered and symmetric about the x-axis. A normal probability plot of the residuals should be roughly linear.

4. **a.** Assumption 1   **b.** Assumption 2
   c. Assumption 3   **d.** Assumption 3

5. The regression equation is useful for making predictions.

6. b_1, r, r^2

7. No. Both equal the number obtained by substituting the specified value of the predictor variable into the sample regression equation.

8. The term *confidence* is usually reserved for interval estimates of parameters, whereas the term *prediction* is used for interval estimates of variables.

9. ρ

10. **a.** The variables are positively linearly correlated, meaning that y tends to increase linearly as x increases (and vice versa), with the tendency being greater the closer that ρ is to 1.
    b. The variables are linearly uncorrelated, meaning that there is no linear relationship between the variables.
    c. The variables are negatively linearly correlated, meaning that y tends to decrease linearly as x increases (and vice versa), with the tendency being greater the closer that ρ is to -1.

11. There are constants, β_0, β_1, and σ, such that, for each student-to-faculty ratio, x, the graduation rates for all universities with that student-to-faculty ratio are normally distributed with mean $\beta_0 + \beta_1 x$ and standard deviation σ.

12. **a.** $\hat{y} = 16.4 + 2.03x$
    b. $s_e = 11.31\%$; very roughly speaking, on average, the predicted graduation rate for a university in the sample differs from the observed graduation rate by about 11.31 percentage points.
    c. Presuming that, for universities, the variables student-to-faculty ratio (x) and graduation rate (y) satisfy the assumptions for regression inferences, the standard error of the estimate, $s_e = 11.31\%$, provides an estimate for the common population standard deviation, σ, of graduation rates for all universities with any particular student-to-faculty ratio.

13.

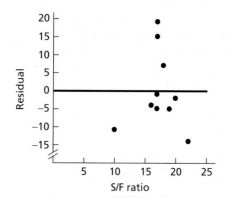

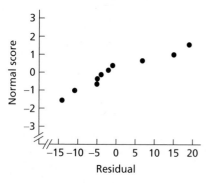

It appears reasonable.

14. a. $H_0: \beta_1 = 0$, $H_a: \beta_1 \neq 0$; $\alpha = 0.05$; $t = 1.682$; critical values $= \pm 2.306$; $0.10 < P < 0.20$; do not reject H_0; at the 5% significance level, the data do not provide sufficient evidence to conclude that, for universities, student-to-faculty ratio is useful as a predictor of graduation rate.

b. -0.75% to 4.80%. We can be 95% confident that, for universities, the change in mean graduation rate per increase by 1 in the student-to-faculty ratio is somewhere between -0.75 and 4.80 percentage points.

15. a. 50.9%

b. 42.6% to 59.2%. We can be 95% confident that the mean graduation rate of all universities that have a student-to-faculty ratio of 17 is somewhere between 42.6% and 59.2%.

c. 50.9%

d. 23.5% to 78.3%. We can be 95% certain that the observed graduation rate of a university that has a student-to-faculty ratio of 17 will be somewhere between 23.5% and 78.3%.

e. The error in the estimate of the mean graduation rate of all universities that have a student-to-faculty ratio of 17 is due only to the fact that the population regression line is being estimated by a sample regression line, whereas the error in the prediction of the observed graduation rate of a university that has a student-to-faculty ratio of 17 is due to the error in estimating the mean graduation rate plus the variation in graduation rates of universities that have a student-to-faculty ratio of 17.

16. $H_0: \rho = 0$, $H_a: \rho > 0$; $\alpha = 0.025$; $t = 1.682$; critical value $= 2.306$; $0.05 < P < 0.10$; do not reject H_0; at the 2.5% significance level, the data do not provide sufficient evidence to conclude that, for universities, the variables student-to-faculty ratio and graduation rate are positively linearly correlated.

17. Their normal scores

18. H_0: Gas mileages for this model are normally distributed. H_a: Gas mileages for this model are not normally distributed. $\alpha = 0.05$; $R_p = 0.983$; critical value $= 0.938$; $P > 0.10$; do not reject H_0; at the 5% significance level, the data do not provide sufficient evidence to conclude that gas mileages for this model are not normally distributed.

Chapter 16

EXERCISES 16.1

16.1 By stating its two numbers of degrees of freedom

16.3 $F_{0.05}$, $F_{0.025}$, F_α

16.5
a. 12   **b.** 7

16.7
a. 1.89   **b.** 2.47   **c.** 2.14

16.9
a. 2.88   **b.** 2.10   **c.** 1.78

EXERCISES 16.2

16.13 The pooled t-procedure of Section 10.2

16.15 The procedure for comparing the means analyzes the variation in the sample data.

16.17
a. The treatment mean square, $MSTR$
b. The error mean square, MSE
c. The F-statistic, $F = MSTR/MSE$

16.19 It signifies that the ANOVA compares the means of a variable for populations that result from a classification by *one* other variable (called the *factor*).

16.21 No. Because the variation among the sample means is not large relative to the variation within the samples

EXERCISES 16.3

16.25 A small value of F results when $SSTR$ is small relative to SSE, that is, when the variation among sample means is small relative to the variation within samples. This result describes what is expected when the null hypothesis is true; thus it doesn't constitute evidence against the null hypothesis. Only when the variation among sample means is large relative to the variation within samples (i.e., only when F is large), is there evidence that the null hypothesis is false.

16.27 $SST = SSTR + SSE$. The total variation among all the sample data can be partitioned into a component representing variation among the sample means and a component representing variation within the samples.

16.29 The missing entries are as follows: In the first row, they are 43.304 and 3.08; in the second row, they are 12 and 7.033; and in the third row, it is 127.704.

16.31
a. $SST = 70, SSTR = 10, SSE = 60$
b. $70 = 10 + 60$
c. $SST = 70, SSTR = 10, SSE = 60$

16.33

Source	df	SS	MS	F
Treatment	2	2.8	1.4	1.56
Error	12	10.8	0.9	
Total	14	13.6		

16.35 $H_0: \mu_1 = \mu_2 = \mu_3$, H_a: Not all the means are equal. $\alpha = 0.05$; $SST = 71,488.667$, $SSTR = 66,043.167$, $SSE = 5,445.50$; $F = 54.58$; critical value $= 4.26$; $P < 0.005$; reject H_0; at the 5% significance level, the data provide sufficient evidence to conclude that a difference exists in mean number of copepods among the three different diets.

16.37 $H_0: \mu_1 = \mu_2 = \mu_3 = \mu_4 = \mu_5$, H_a: Not all the means are equal. $\alpha = 0.05$; $SST = 447,088.667$, $SSTR = 404,258.467$, $SSE = 42,830.200$; $F = 51.91$; critical value $= 2.82$; $P < 0.005$; reject H_0; at the 5% significance level, the data provide sufficient evidence to conclude that a difference exists in mean weekly earnings among nonsupervisory workers in the five industries.

EXERCISES 16.4

16.47 0

16.49
a. The family confidence level. Because the family confidence level is the confidence we have that all the confidence intervals contain the differences between the corresponding population means, whereas the individual confidence level is the confidence we have that any particular confidence interval contains the difference between the corresponding population means.
b. They are identical.

16.51 Degrees of freedom for the denominator

16.53
a. 4.69 **b.** 5.98

16.55 Family confidence level $= 0.95$; $q_{0.05} = 3.95$; simultaneous 95% CIs are as follows.

Means difference	Confidence interval
$\mu_1 - \mu_2$	102.2 to 199.3
$\mu_1 - \mu_3$	114.7 to 211.8
$\mu_2 - \mu_3$	−36.1 to 61.1

The preceding table shows that the following pairs of means can be declared different: μ_1 and μ_2, μ_1 and μ_3. This result is summarized in the following diagram.

Macroalgae (3)	Bacteria (2)	Diatoms (1)
293.75	306.25	457.00

Interpreting this diagram, we conclude with 95% confidence that the mean number of copepods is greater with the diatoms diet than with the other two diets; no other means can be declared different.

16.57 Family confidence level $= 0.95$; $q_{0.05} = 4.20$; simultaneous 95% CIs are as follows.

Means difference	Confidence interval
$\mu_1 - \mu_2$	4.8 to 163.4
$\mu_1 - \mu_3$	270.8 to 422.2
$\mu_1 - \mu_4$	31.9 to 201.1
$\mu_1 - \mu_5$	118.3 to 269.7
$\mu_2 - \mu_3$	183.1 to 341.7
$\mu_2 - \mu_4$	−55.5 to 120.3
$\mu_2 - \mu_5$	30.6 to 189.2
$\mu_3 - \mu_4$	−314.6 to −145.4
$\mu_3 - \mu_5$	−228.2 to −76.8
$\mu_4 - \mu_5$	−7.1 to 162.1

The preceding table shows that the following pairs of means can be declared different: μ_1 and μ_2, μ_1 and μ_3, μ_1 and μ_4, μ_1 and μ_5, μ_2 and μ_3, μ_2 and μ_5, μ_3 and μ_4, μ_3 and μ_5; all other pairs of means are not declared different. This result is summarized in the following diagram.

Retail	Services	Finance	Wholesale	Transport
(3)	(5)	(4)	(2)	(1)
319.0	471.5	549.0	581.4	665.5

Interpreting this diagram, we conclude with 95% confidence that the mean weekly earnings of transportation/public-utility workers exceeds those of the other four industries; the mean weekly earnings of retail-trade workers is less than those of the other four industries; the mean weekly earnings of service workers is less than those of wholesale trade workers; no other means can be declared different.

EXERCISES 16.5

16.67 Simple random samples, independent samples, same-shape populations, and all sample sizes are 5 or greater

16.69 Equal

16.71 Chi-square with df $= 4$

16.73 $H_0: \mu_1 = \mu_2 = \mu_3$, H_a: Not all the means are equal. $\alpha = 0.01$; $H = 11.52$; critical value $= 9.210$; $P < 0.005$; reject H_0; at the 1% significance level, the data provide sufficient evidence to conclude that there is a difference in mean consumption of low-fat milk for the years 1980, 1990, and 2000.

16.75 $H_0: \eta_1 = \eta_2 = \eta_3 = \eta_4$, H_a: Not all the medians are equal. $\alpha = 0.05$; $H = 6.37$; critical value $= 7.815$; $0.05 < P < 0.10$; do not reject H_0; at the 5% significance level, the data do not provide sufficient evidence to conclude that a difference exists in median square footage of single-family detached homes among the four U.S. regions.

16.77
a. Kruskal–Wallis test   b. One-way ANOVA test

REVIEW TEST FOR CHAPTER 16

1. To compare the means of a variable for populations that result from a classification by one other variable (called the *factor*)

2. *Simple random samples:* Check by carefully studying the way the sampling was done. *Independent samples:* Check by carefully studying the way the sampling was done. *Normal populations:* Check by constructing normal probability plots. *Equal standard deviations:* As a rule of thumb, this assumption is considered to be satisfied if the ratio of the largest sample standard

deviation to the smallest sample standard deviation is less than 2.

Also, the normality and equal-standard-deviations assumptions can be assessed by performing a residual analysis.

3. The *F*-distribution

4. df $= (2, 14)$

5. a. *MSTR* (or *SSTR*)   b. *MSE* (or *SSE*)

6. a. The total sum of squares, *SST*, represents the total variation among all the sample data; the treatment sum of squares, *SSTR*, represents the variation among the sample means; and the error sum of squares, *SSE*, represents the variation within the samples.
   b. $SST = SSTR + SSE$; the one-way ANOVA identity shows that the total variation among all the sample data can be partitioned into a component representing variation among the sample means and a component representing variation within the samples.

7. a. For organizing and summarizing the quantities required for performing a one-way analysis of variance
   b.

Source	df	SS	MS = SS/df	F
Treatment	$k-1$	SSTR	$MSTR = \dfrac{SSTR}{k-1}$	$F = \dfrac{MSTR}{MSE}$
Error	$n-k$	SSE	$MSE = \dfrac{SSE}{n-k}$	
Total	$n-1$	SST		

8. Suppose that, in a one-way ANOVA, the null hypothesis of equal population means is rejected. The purpose of a multiple comparison is to then decide which means are different, which mean is largest, or, more generally, the relation among all the means.

9. The individual confidence level is the confidence we have that any particular confidence interval contains the difference between the corresponding population means; the family confidence level is the confidence we have that all the confidence intervals contain the differences between the corresponding population means. It is at the family confidence level that we can be confident in the truth of our conclusions when comparing all the population means simultaneously; thus the family confidence level is the appropriate one for multiple comparisons.

10. Studentized range distribution (or q-distribution)

11. Larger. We can be more confident about the truth of one of several statements than about the truth of all statements simultaneously.

12. $\kappa = 3$, $\nu = 14$

13. Kruskal–Wallis test

14. Chi-square distribution with $df = k - 1$, where k is the number of populations under consideration

15. The Kruskal–Wallis test is based on ranks. If the null hypothesis of equal population means is true, the means of the ranks for the samples should be roughly equal. Put another way, an unduly large variation among the mean ranks provides evidence against the null hypothesis.

16. The Kruskal–Wallis test because, unlike the one-way ANOVA test, it is resistant to outliers and other extreme values.

17. **a.** 2   **b.** 14   **c.** 3.74
     d. 6.51   **e.** 3.74

18. **a.** The sample means are 3, 3, and 6, respectively; the sample standard deviations are 2, 2.449, and 4.243, respectively.
     b. $SST = 110$, $SSTR = 24$, $SSE = 86$; $110 = 24 + 86$
     c. $SST = 110$, $SSTR = 24$, $SSE = 86$
     d.

Source	df	SS	MS = SS/df	F
Treatment	2	24	12.000	1.26
Error	9	86	9.556	
Total	11	110		

19. **a.** The variation among the sample means
     b. The variation within the samples
     c. Simple random samples, independent samples, normal populations, and equal (population) standard deviations. One-way ANOVA is robust to moderate violations of the normality assumption. It is also reasonably robust to moderate violations of the equal-standard-deviations assumption if the sample sizes are roughly equal.

20. H_0: $\mu_1 = \mu_2 = \mu_3$, H_a: Not all the means are equal. $\alpha = 0.05$; $SST = 371{,}141.882$, $SSTR = 160{,}601.416$, $SSE = 210{,}540.467$; $F = 5.34$; critical value $= 3.74$; $0.01 < P < 0.025$; reject H_0; at the 5% significance level, the data provide sufficient evidence to conclude that a difference in mean losses exists among the three types of robberies.

21. **a.** 3.70   **b.** 4.89

22. **a.** Family confidence level $= 0.95$; $q_{0.05} = 3.70$; simultaneous 95% CIs are as follows.

Means difference	Confidence interval		
$\mu_1 - \mu_2$	-65.3	to	323.2
$\mu_1 - \mu_3$	48.4	to	436.9
$\mu_2 - \mu_3$	-71.6	to	298.9

     b. The table in part (a) shows that only μ_1 and μ_3 can be declared different. This result is summarized in the following diagram.

Conven (3)	Gas (2)	Highway (1)
582.2	695.8	824.8

Interpreting this diagram, we conclude with 95% confidence that the mean loss due to convenience-store robberies is less than that due to highway robberies; no other means can be declared different.

23. **a.** H_0: $\mu_1 = \mu_2 = \mu_3$, H_a: Not all the means are equal. $\alpha = 0.05$; $H = 6.82$; critical value $= 5.991$; $0.025 < P < 0.05$; reject H_0; at the 5% significance level, the data provide sufficient evidence to conclude that a difference in mean losses exists among the three types of robberies.
     b. Because normal distributions with equal standard deviations have the same shape. It is better to use the one-way ANOVA test because, when the assumptions for that test are met, it is more powerful than the Kruskal–Wallis test.

Index

Photo Credits

Pages 3 and 36, photo from "Citizen Kane" © Bettman/CORBIS; page 5, photo of Harry Truman © Bettman/CORBIS; page 29, photo of iron lung © Hulton-Deutsch Collection/CORBIS; page 36, image of Florence Nightingale from www.spartacus.schoolnet.co.uk/REnightingale.htm; pages 41 and 89, photo of infants © 2001 PhotoDisc, Inc.; page 43, Svetlana Zakharova, Boston Marathon © Reuters, New Media Inc./CORBIS; pages 55 and 420, photo of cheetahs © 2001 PhotoDisc, Inc.; page 55, photo of great white shark, Corbis Royalty-free; page 57, photo of stockbrokers © 2001 PhotoDisc, Inc.; page 89, image of Adolphe Quetelet courtesy St. Andrews University; pages 91 and 147, Secretariat, 1973 Triple Crown © Jerry Cooke/CORBIS; page 100, photo of driver courtesy AAA Foundation for Traffic Safety; pages 105 and 543, photo of person sleeping, Corbis Royalty-free; page 116, photo of hurricane damage © 2001 PhotoDisc, Inc.; page 131, photo of U.S. Women's World Cup Soccer Team © Thomas E. Witte/NewSport/CORBIS; page 140, photo of hurricane eye © 2001 PhotoDisc, Inc.; page 143, photo of party © 2001 PhotoDisc, Inc.; page 148, image of John Tukey courtesy St. Andrews University; pages 151 and 222, Powerball logo ® the Multi-State Lottery Association; page 159, photo of Funny Cide © Jason Szenes/CORBIS; page 173, photo of oil spill courtesy of NOAA; page 186, photo of property damage © Eye Ubiquitous/CORBIS; page 201, golf photo from psu.edu/sports/golf/women; page 207, photo of horse race, Corbis Royalty-free; page 216, photo of meeting © 2001 PhotoDisc, Inc.; page 217, photo of elementary school children © 2001 PhotoDisc, Inc.; page 217, photo of traffic courtesy of College of Health and Development, The Pennsylvania State University; page 223, image of Andrei Kolmogorov courtesy of St. Andrews University; pages 225 and 270, golf photo © 2001 PhotoDisc, Inc.; page 229, photo of elementary school students © 2001 PhotoDisc, Inc.; page 232, photo of space shuttle courtesy of NASA; page 233, photo of eclipse © 2001 PhotoDisc, Inc.; page 239, photo of factory © 2001 PhotoDisc, Inc.; page 247, photo of couple © 2001 PhotoDisc, Inc.; page 262, photo of ambulance crew and patient © 2001 PhotoDisc, Inc.; page 264, photo of yellow lobster © Joyce Morrill for the Lobster Conservatory; page 271, image of James Bernoulli courtesy St. Andrews University; pages 273 and 320, photo of women's shoes © PhotoDisc, Inc.; page 281, giant tarantula © CORBIS RF; page 298, photo of Jingdong black gibbon © Rod Williams/Bruce Coleman, PictureQuest; page 314, lightning strike © PhotoDisc, Inc.; page 321, image of Carl Friedrich Gauss courtesy of St. Andrews University; pages 323 and 351, photo of Chesapeake & Ohio railroad car © Karen Huntt Mason/CORBIS; page 330, photo of San Antonio Spurs © Joe Mitchell/Stringer/Reuters New Media Inc./CORBIS; page 330, photo of Bill Gates © Reuters NewMedia Inc./CORBIS; page 337, photo of earthquake damage © 2001 CARE photo; page 344, image of brain scans © 2001 PhotoDisc, Inc.; page 347, photo of tax preparation courtesy of Beth Anderson; page 351, image of Pierre-Simon Laplace courtesy St. Andrews University; pages 355 and 395, photo of cookies courtesy of Beth Anderson; page 360, wedding photo courtesy Aaron and Carla Weiss; page 370, photo of the Rolling Stones © Lynn Goldsmith/CORBIS; page 376, photo of common lizard © CORBIS RF; page 387, photo of amusement park ride © Neil Rabinowitz/CORBIS; page 391, photo of seashell © 2001 PhotoDisc, Inc.; page 395, image of William Gosset courtesy of St. Andrews University; pages 397 and 480, photo of Houghton Garden Park © PhotoDisc, Inc.; page 404, photo of cell-phone user © 2001 PhotoDisc, Inc.; page 412, photo of child © 2001 PhotoDisc, Inc.; page 413, photo of trial, Corbis Royalty-free; page 449, photo of water © 2001 PhotoDisc, Inc.; page 454, photo of mother and newborn, Corbis Royalty-free; page 464, photo of family © 2001 PhotoDisc, Inc.; page 475, photo of an arrest © Jonathan Blair/CORBIS; page 480, image of Jerzy Neyman courtesy St. Andrews University; pages 483 and 566, photo of reader, Corbis Royalty-free; page 515, photo of patient © 2001 PhotoDisc, Inc.; page 527, photo of library © 2001 PhotoDisc, Inc.; page 552, photo of tires, Corbis Royalty-free; page 561, photo of snake, Corbis Royalty-free; page 567, image of Gertrude Cox courtesy of Research Triangle Institute; pages 569 and 602, photo of woofer © 2001 PhotoDisc, Inc.; page 594, photo of blood-pressure check © 2001 PhotoDisc, Inc.; page 597, photo of skinfold test © 2001 PhotoDisc, Inc.; page 603, image of W. Edwards Deming © MIT Press; pages 605 and 646, photo of women © PhotoDisc, Inc.; page 617, photo of women, Corbis Royalty-free; page 625, photo of a family opening presents © 2001 PhotoDisc, Inc.; page 638, photo of "Buckle Up" highway sign © 2001 PhotoDisc, Inc.; page 641, photo of jurors, Corbis Royalty-free; page 647, image of Abraham de Moivre courtesy of St. Andrews University; pages 649 and 688, photo of angry driver © 2001 PhotoDisc, Inc.; page 659, photo of M&M's courtesy of Beth Anderson; page 679, photo of Siskel and Ebert © AP Photo/Buena Vista Television; page 683, photo of hospital © David H. Wells/CORBIS; page 689, image of Karl Pearson © Brown Brothers; pages 693 and 740, photo of men © PhotoDisc, Inc.; page 698, photo of thermometer © 2001 PhotoDisc, Inc.; page 715, photo of baseball player © Reuters New Media Inc./CORBIS; page 723, photo of Corvette courtesy of CorvetteMagazine.com; page 724, photo of pelicans © 2001 PhotoDisc, Inc.; page 736, photo of fatty food © 2001 PhotoDisc, Inc.; page 741, image of Adrien Legendre courtesy St. Andrews University; pages 743 and 796, photo of women © PhotoDisc Red; page 755, photo of Corvette courtesy of CorvetteMagazine.com; page 757, photo of newborn, Corbis Royalty-free; page 765, photo of potato plants © Patrick Johns/CORBIS; page 766, photo of newborn © 2001 PhotoDisc, Inc.; page 790, photo of graduate courtesy of Greg Weiss; page 790, photo of hamburger and potato chips © 2001 PhotoDisc, Inc.; page 797, image of Sir Francis Galton courtesy of St. Andrews University; pages 799 and 845, photo of college students © 2001 PhotoDisc, Inc.; page 819, photo of couple in new apartment, Corbis Royalty-free; page 822, photo of Linus Pauling with Vitamin C molecule © Roger Ressmeyer/CORBIS; page 828, photo of copepod © Frank Lane Picture Agency/CORBIS; page 828, photo of driving range, Corbis Royalty-free; page 839, photo of factory workers © 2001 PhotoDisc, Inc.; page 842, photo of rock sparrows © Roger Tidman/CORBIS; page 846, image of Sir Ronald Fisher courtesy St. Andrews University.

Procedure Index

Following is an index that provides page-number references for the various statistical procedures discussed in the book.

Indexes for
Internet Projects, Biographical Sketches, and Case Studies

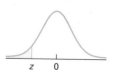

TABLE II
Areas under the standard normal curve

			Second decimal place in z							
0.09	0.08	0.07	0.06	0.05	0.04	0.03	0.02	0.01	0.00	z
									0.0000†	−3.9
0.0001	0.0001	0.0001	0.0001	0.0001	0.0001	0.0001	0.0001	0.0001	0.0001	−3.8
0.0001	0.0001	0.0001	0.0001	0.0001	0.0001	0.0001	0.0001	0.0001	0.0001	−3.7
0.0001	0.0001	0.0001	0.0001	0.0001	0.0001	0.0001	0.0001	0.0002	0.0002	−3.6
0.0002	0.0002	0.0002	0.0002	0.0002	0.0002	0.0002	0.0002	0.0002	0.0002	−3.5
0.0002	0.0003	0.0003	0.0003	0.0003	0.0003	0.0003	0.0003	0.0003	0.0003	−3.4
0.0003	0.0004	0.0004	0.0004	0.0004	0.0004	0.0004	0.0005	0.0005	0.0005	−3.3
0.0005	0.0005	0.0005	0.0006	0.0006	0.0006	0.0006	0.0006	0.0007	0.0007	−3.2
0.0007	0.0007	0.0008	0.0008	0.0008	0.0008	0.0009	0.0009	0.0009	0.0010	−3.1
0.0010	0.0010	0.0011	0.0011	0.0011	0.0012	0.0012	0.0013	0.0013	0.0013	−3.0
0.0014	0.0014	0.0015	0.0015	0.0016	0.0016	0.0017	0.0018	0.0018	0.0019	−2.9
0.0019	0.0020	0.0021	0.0021	0.0022	0.0023	0.0023	0.0024	0.0025	0.0026	−2.8
0.0026	0.0027	0.0028	0.0029	0.0030	0.0031	0.0032	0.0033	0.0034	0.0035	−2.7
0.0036	0.0037	0.0038	0.0039	0.0040	0.0041	0.0043	0.0044	0.0045	0.0047	−2.6
0.0048	0.0049	0.0051	0.0052	0.0054	0.0055	0.0057	0.0059	0.0060	0.0062	−2.5
0.0064	0.0066	0.0068	0.0069	0.0071	0.0073	0.0075	0.0078	0.0080	0.0082	−2.4
0.0084	0.0087	0.0089	0.0091	0.0094	0.0096	0.0099	0.0102	0.0104	0.0107	−2.3
0.0110	0.0113	0.0116	0.0119	0.0122	0.0125	0.0129	0.0132	0.0136	0.0139	−2.2
0.0143	0.0146	0.0150	0.0154	0.0158	0.0162	0.0166	0.0170	0.0174	0.0179	−2.1
0.0183	0.0188	0.0192	0.0197	0.0202	0.0207	0.0212	0.0217	0.0222	0.0228	−2.0
0.0233	0.0239	0.0244	0.0250	0.0256	0.0262	0.0268	0.0274	0.0281	0.0287	−1.9
0.0294	0.0301	0.0307	0.0314	0.0322	0.0329	0.0336	0.0344	0.0351	0.0359	−1.8
0.0367	0.0375	0.0384	0.0392	0.0401	0.0409	0.0418	0.0427	0.0436	0.0446	−1.7
0.0455	0.0465	0.0475	0.0485	0.0495	0.0505	0.0516	0.0526	0.0537	0.0548	−1.6
0.0559	0.0571	0.0582	0.0594	0.0606	0.0618	0.0630	0.0643	0.0655	0.0668	−1.5
0.0681	0.0694	0.0708	0.0721	0.0735	0.0749	0.0764	0.0778	0.0793	0.0808	−1.4
0.0823	0.0838	0.0853	0.0869	0.0885	0.0901	0.0918	0.0934	0.0951	0.0968	−1.3
0.0985	0.1003	0.1020	0.1038	0.1056	0.1075	0.1093	0.1112	0.1131	0.1151	−1.2
0.1170	0.1190	0.1210	0.1230	0.1251	0.1271	0.1292	0.1314	0.1335	0.1357	−1.1
0.1379	0.1401	0.1423	0.1446	0.1469	0.1492	0.1515	0.1539	0.1562	0.1587	−1.0
0.1611	0.1635	0.1660	0.1685	0.1711	0.1736	0.1762	0.1788	0.1814	0.1841	−0.9
0.1867	0.1894	0.1922	0.1949	0.1977	0.2005	0.2033	0.2061	0.2090	0.2119	−0.8
0.2148	0.2177	0.2206	0.2236	0.2266	0.2296	0.2327	0.2358	0.2389	0.2420	−0.7
0.2451	0.2483	0.2514	0.2546	0.2578	0.2611	0.2643	0.2676	0.2709	0.2743	−0.6
0.2776	0.2810	0.2843	0.2877	0.2912	0.2946	0.2981	0.3015	0.3050	0.3085	−0.5
0.3121	0.3156	0.3192	0.3228	0.3264	0.3300	0.3336	0.3372	0.3409	0.3446	−0.4
0.3483	0.3520	0.3557	0.3594	0.3632	0.3669	0.3707	0.3745	0.3783	0.3821	−0.3
0.3859	0.3897	0.3936	0.3974	0.4013	0.4052	0.4090	0.4129	0.4168	0.4207	−0.2
0.4247	0.4286	0.4325	0.4364	0.4404	0.4443	0.4483	0.4522	0.4562	0.4602	−0.1
0.4641	0.4681	0.4721	0.4761	0.4801	0.4840	0.4880	0.4920	0.4960	0.5000	−0.0

† For $z \leq -3.90$, the areas are 0.0000 to four decimal places.